KIRK-OTHMER

ENCYCLOPEDIA OF CHEMICAL TECHNOLOGY

Third Edition

VOLUME 9

**Enamels, Porcelain or Vitreous
to
Ferrites**

EDITORIAL BOARD

HERMAN F. MARK
Polytechnic Institute of New York

DONALD F. OTHMER
Polytechnic Institute of New York

CHARLES G. OVERBERGER
University of Michigan

GLENN T. SEABORG
University of California, Berkeley

EXECUTIVE EDITOR

MARTIN GRAYSON

ASSOCIATE EDITOR

DAVID ECKROTH

KIRK-OTHMER

ENCYCLOPEDIA OF CHEMICAL TECHNOLOGY

THIRD EDITION

VOLUME 9

ENAMELS, PORCELAIN OR VITREOUS
TO
FERRITES

A WILEY-INTERSCIENCE PUBLICATION

John Wiley & Sons

NEW YORK · CHICHESTER · BRISBANE · TORONTO

Copyright © 1980 by John Wiley & Sons, Inc.

All rights reserved. Published simultaneously in Canada.

Reproduction or translation of any part of this work beyond that permitted by Sections 107 or 108 of the 1976 United States Copyright Act without the permission of the copyright owner is unlawful. Requests for permission or further information should be addressed to the Permissions Department, John Wiley & Sons, Inc.

Library of Congress Cataloging in Publication Data:

Main entry under title:
   Encyclopedia of chemical technology.

   At head of title: Kirk-Othmer.
   "A Wiley-Interscience publication."
   Includes bibliographies.
   1. Chemistry, Technical—Dictionaries. I. Kirk, Raymond Eller, 1890–1957. II. Othmer, Donald Frederick, 1904—
      III, Grayson, Martin. IV. Eckroth, David.
V. Title: Kirk-Othmer encyclopedia of chemical technology.

TP9.E685   1978   660'.03   77-15820
ISBN 0-471-02062-1

Printed in the United States of America

CONTENTS

Enamels, porcelain or vitreous, 1
Energy management, 21
Engineering and chemical data correlation, 45
Engineering plastics, 118
Enzyme detergents, 138
Enzymes, immobilized, 148
Enzymes, industrial, 173
Enzymes, therapeutic, 225
Epinephrine and norepinephrine, 241
Epoxidation, 251
Epoxy resins, 267
Esterification, 291
Esters, organic, 311
Ethanol, 338
Ethers, 381

Ethylene, 393
Ethylene oxide, 432
Evaporation, 472
Exhaust control, automotive, 494
Exhaust control, industrial, 511
Expectorants, antitussives, and related agents, 542
Explosives and propellants, 561
Extraction, liquid-liquid, 672
Extraction, liquid-solid, 721
Extractive metallurgy, 739
Fans and blowers, 768
Fats and fatty oils, 795
Feedstocks, 831
Felts, 846
Fermentation, 861
Ferrites, 881

EDITORIAL STAFF FOR VOLUME 9

Executive Editor: **Martin Grayson**
Associate Editor: **David Eckroth**
Production Supervisor: **Michalina Bickford**
Editors: **Galen J. Bushey   Caroline I. Eastman   Anna Klingsberg
Leonard Spiro**

CONTRIBUTORS TO VOLUME 9

Thomas H. Applewhite, *Kraft, Inc., Glenview, Illinois,* Fats and fatty oils
Malcom H. I. Baird, *McMaster University, Hamilton, Ontario, Canada,* Extraction, liquid–liquid
Marilyn Bakker, *Business Communications Co., Inc., Stamford, Connecticut,* Engineering plastics
Hans C. Barfoed, *Nova Industri A/S, Bagszaerd, Denmark,* Enzyme detergents
Gordon Buchi, *Ciba-Geigy Corporation, Ardsley, New York,* Epoxy resins
F. A. M. Buck, *King, Buck & Associates, San Diego, California,* Feedstocks
James N. Cawse, *Union Carbide Corporation, New York, New York,* Ethylene oxide
James H. Conklin, *E. I. du Pont de Nemours & Co., Inc., Wilmington, Delaware,* Exhaust control, industrial
David A. Cooney, *Division of Cancer Treatment, National Cancer Institute, National Institutes of Health, Bethesda, Maryland,* Enzymes, therapeutic
Burton B. Crocker, *Monsanto Company, St. Louis, Missouri,* Fans and blowers
Paul Duby, *Columbia University, New York, New York,* Extractive metallurgy
Edward U. Elam, *Tennessee Eastman Company, Kingsport, Tennessee,* Esters, organic

CONTRIBUTORS TO VOLUME 9

A. L. Friedberg, *University of Illinois, Urbana, Illinois,* Enamels, porcelain or vitreous
John Gannon, *Ciba-Geigy Corporation, Ardsley, New York,* Epoxy resins
Wolf Hamm, *Unilever Limited, Shainbrook, Bedford, England,* Extraction, liquid–solid
Charles M. Heinen, *Chrysler Corporation, Detroit, Michigan,* Exhaust control, automotive
Joseph P. Henry, *Union Carbide Corporation, South Charleston, West Virginia,* Ethylene oxide
W. R. Howell, *Dow Chemical U.S.A., Freeport, Texas,* Epoxy resins
H. N. Jayaram, *Division of Cancer Treatment, National Cancer Institute, National Institutes of Health, Bethesda, Maryland,* Enzymes, therapeutic
Percy P. Kavasmaneck, *Union Carbide Corporation, South Charleston, West Virginia,* Ethanol
Donald E. Keeley, *General Electric Co., Schenectady, New York,* Ethers
Melvin H. Keyes, *Owen-Illinois, Inc., Toledo, Ohio,* Enzymes, immobilized
Ludwig Kniel, *Combustion Engineering, Inc., Bloomfield, New Jersey,* Ethylene
John H. Kroehling, *E. I. du Pont de Nemours & Co., Inc., Wilmington, Delaware,* Exhaust control, industrial
Victor Lindner, *ARRADCOM, Dover, New Jersey,* Explosives and propellants
Teh C. Lo, *Hoffmann LaRoche Inc., Nutley, New Jersey,* Extraction, liquid–liquid
John T. Lutz, *Rohm and Haas Co., Bristol, Pennsylvania,* Epoxidation
Merritt G. Marbach, *Shell Chemical Company, Houston, Texas,* Feedstocks
John R. McLean, *Warner-Lambert/Parke-Davis, Ann Arbor, Michigan,* Epinephrine and norepinephrine
Donald F. Othmer, *Polytechnic Institute of New York, Brooklyn, New York,* Engineering and chemical data correlation
David Perlman, *University of Wisconsin, Madison, Wisconsin,* Fermentation
Mary K. Porter, *Consultant, Troy, New York,* Felts
Thomas G. Reynolds, III, *Ferroxcube Corporation, Saugerties, New York,* Ferrites
Don Scott, *Fermco Biochemics Inc., Elk Grove Village, Illinois,* Enzymes, industrial
Paul Dwight Sherman, Jr., *Union Carbide Corporation, South Charleston, West Virginia,* Ethanol
Stanley Sherman, *Ciba-Geigy Corporation, Ardsley, New York,* Epoxy resins
Donald M. Sowards, *E. I. du Pont de Nemours & Co., Inc., Wilmington, Delaware,* Exhaust control, industrial
Ferris C. Standiford, *W. L. Badger Associates, Inc., Greenbank, Washington,* Evaporation
Chuck Starace, *Novo Laboratories, Inc., Wilton, Connecticut,* Enzyme detergents
George Stergis, *Division of Cancer Treatment, National Cancer Institute, National Institutes of Health, Bethesda, Maryland,* Enzymes, therapeutic
Michael W. Swartzlander, *Union Carbide Corporation, New York, New York,* Ethylene oxide
Chung-Hu Tsai, *Combustion Engineering, Inc., Bloomfield, New Jersey,* Ethylene
Parvez H. Wadia, *Union Carbide Corporation, New York, New York,* Ethylene oxide
A. F. Waterland, *Waterland, Viar & Associates, Inc., Wilmington, Delaware,* Energy management
William J. Welstead, Jr., *A. H. Robins Company, Richmond, Virginia,* Expectorants, antitussives, and related agents
Olaf Winter, *Combustion Engineering, Inc., Bloomfield, New Jersey,* Ethylene
E. G. Zey, *Celanese Corporation, Corpus Christi, Texas,* Esterification

NOTE ON CHEMICAL ABSTRACTS SERVICE REGISTRY NUMBERS AND NOMENCLATURE

Chemical Abstracts Service (CAS) Registry Numbers are unique numerical identifiers assigned to substances recorded in the CAS Registry System. They appear in brackets in the *Chemical Abstracts* (CA) substance and formula indexes following the names of compounds. A single compound may have many synonyms in the chemical literature. A simple compound like phenethylamine can be named β-phenylethylamine or, as in *Chemical Abstracts,* benzeneethanamine. The usefulness of the Encyclopedia depends on accessibility through the most common correct name of a substance. Because of this diversity in nomenclature careful attention has been given the problem in order to assist the reader as much as possible, especially in locating the systematic CA index name by means of the Registry Number. For this purpose, the reader may refer to the CAS Registry Handbook-Number Section which lists in numerical order the Registry Number with the Chemical Abstracts index name and the molecular formula; eg, **458-8-8,** Piperidine, 2-propyl-, (S)-, $C_8H_{17}N$; in the Encyclopedia this compound would be found under its common name, coniine [*458-88-8*]. The Registry Number is a valuable link for the reader in retrieving additional published information on substances and also as a point of access for such on-line data bases as Chemline, Medline and Toxline.

In all cases, the CAS Registry Numbers have been given for title compounds in articles and for all compounds in the index. All specific substances indexed in *Chemical Abstracts* since 1965 are included in the CAS Registry System as are a large number of substances derived from a variety of reference works. The CAS Registry System identifies a substance on the basis of an unambiguous computer-language description of its molecular structure including stereochemical detail. The Registry Number is a machine-checkable number (like a Social Security number) assigned in sequential order to each substance as it enters the registry system. The value of the number lies in the fact that it is a concise and unique means of substance identification, which is independent of, and therefore bridges, many systems of chemical nomenclature. For polymers, one Registry Number is used for the entire family; eg, polyoxyethylene (20)sorbitan monolaurate has the same number as all of its polyoxyethylene homologues.

Registry numbers for each substance will be provided in the third edition index (eg, Alkaloids shows the Registry Number of all alkaloids (title compounds) in a table in the article as well, but the intermediates will have their Registry Numbers shown only in the index). Articles such as Absorption, Adsorptive separation, Air conditioning, Air pollution, Air pollution control methods have no Registry Numbers in the text.

Cross-references have been inserted in the index for many common names and for some systematic names. Trademark names appear in the index. Names that are incorrect, misleading or ambiguous are avoided. Formulas are given very frequently in the text to help in identifying compounds. The spelling and form used, even for industrial names, follow American chemical usage, but not always the usage of *Chemical Abstracts* (eg, *coniine* is used instead of *(S)-2-propylpiperidine, aniline* instead of *benzenamine,* and *acrylic acid* instead of *2-propenoic acid*).

There are variations in representation of rings in different disciplines. The dye industry does not designate aromaticity or double bonds in rings. All double bonds and aromaticity will be shown in the *Encyclopedia* as a matter of course. For example, tetralin has an aromatic ring and a saturated ring and its structure will appear in the

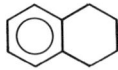

Encyclopedia with its common name, Registry Number enclosed in brackets, and parenthetical CA index name, ie, tetralin [*119-64-2*] (1,2,3,4-tetrahydronaphthalene). With names and structural formulas, and especially with CAS Registry Numbers the aim is to help the reader have a concise means of substance identification.

CONVERSION FACTORS, ABBREVIATIONS, AND UNIT SYMBOLS

SI Units (Adopted 1960)

A new system of measurement, the International System of Units (abbreviated SI), is being implemented throughout the world. This system is a modernized version of the MKSA (meter, kilogram, second, ampere) system, and its details are published and controlled by an international treaty organization (The International Bureau of Weights and Measures) (1).

SI units are divided into three classes:

BASE UNITS

length	meter† (m)
mass‡	kilogram (kg)
time	second (s)
electric current	ampere (A)
thermodynamic temperature§	kelvin (K)
amount of substance	mole (mol)
luminous intensity	candela (cd)

† The spellings "metre" and "litre" are preferred by ASTM; however "-er" are used in the Encyclopedia.
‡ "Weight" is the commonly used term for "mass."
§ Wide use is made of "Celsius temperature" (t) defined by

$$t = T - T_0$$

where T is the thermodynamic temperature, expressed in kelvins, and T_0 = 273.15 K by definition. A temperature interval may be expressed in degrees Celsius as well as in kelvins.

FACTORS, ABBREVIATIONS, AND SYMBOLS

SUPPLEMENTARY UNITS

plane angle — radian (rad)
solid angle — steradian (sr)

DERIVED UNITS AND OTHER ACCEPTABLE UNITS

These units are formed by combining base units, supplementary units, and other derived units (2–4). Those derived units having special names and symbols are marked with an asterisk in the list below:

Quantity	Unit	Symbol	Acceptable equivalent
* absorbed dose	gray	Gy	J/kg
acceleration	meter per second squared	m/s^2	
* activity (of ionizing radiation source)	becquerel	Bq	1/s
area	square kilometer	km^2	
	square hectometer	hm^2	ha (hectare)
	square meter	m^2	
* capacitance	farad	F	C/V
concentration (of amount of substance)	mole per cubic meter	mol/m^3	
* conductance	siemens	S	A/V
current density	ampere per square meter	A/m^2	
density, mass density	kilogram per cubic meter	kg/m^3	g/L; mg/cm^3
dipole moment (quantity)	coulomb meter	C·m	
* electric charge, quantity of electricity	coulomb	C	A·s
electric charge density	coulomb per cubic meter	C/m^3	
electric field strength	volt per meter	V/m	
electric flux density	coulomb per square meter	C/m^2	
* electric potential, potential difference, electromotive force	volt	V	W/A
* electric resistance	ohm	Ω	V/A
* energy, work, quantity of heat	megajoule	MJ	
	kilojoule	kJ	
	joule	J	N·m
	electron volt†	eV†	
	kilowatt-hour†	kW·h†	

† This non-SI unit is recognized by the CIPM as having to be retained because of practical importance or use in specialized fields (1).

Quantity	Unit	Symbol	Acceptable equivalent
energy density	joule per cubic meter	J/m^3	
* force	kilonewton	kN	
	newton	N	$kg \cdot m/s^2$
* frequency	megahertz	MHz	
	hertz	Hz	$1/s$
heat capacity, entropy	joule per kelvin	J/K	
heat capacity (specific), specific entropy	joule per kilogram kelvin	$J/(kg \cdot K)$	
heat transfer coefficient	watt per square meter kelvin	$W/(m^2 \cdot K)$	
* illuminance	lux	lx	lm/m^2
* inductance	henry	H	Wb/A
linear density	kilogram per meter	kg/m	
luminance	candela per square meter	cd/m^2	
* luminous flux	lumen	lm	$cd \cdot sr$
magnetic field strength	ampere per meter	A/m	
* magnetic flux	weber	Wb	$V \cdot s$
* magnetic flux density	tesla	T	Wb/m^2
molar energy	joule per mole	J/mol	
molar entropy, molar heat capacity	joule per mole kelvin	$J/(mol \cdot K)$	
moment of force, torque	newton meter	$N \cdot m$	
momentum	kilogram meter per second	$kg \cdot m/s$	
permeability	henry per meter	H/m	
permittivity	farad per meter	F/m	
* power, heat flow rate, radiant flux	kilowatt	kW	
	watt	W	J/s
power density, heat flux density, irradiance	watt per square meter	W/m^2	
* pressure, stress	megapascal	MPa	
	kilopascal	kPa	
	pascal	Pa	Nm^2
sound level	decibel	dB	
specific energy	joule per kilogram	J/kg	
specific volume	cubic meter per kilogram	m^3/kg	
surface tension	newton per meter	N/m	
thermal conductivity	watt per meter kelvin	$W/(m \cdot K)$	
velocity	meter per second	m/s	
	kilometer per hour	km/h	
viscosity, dynamic	pascal second	$Pa \cdot s$	
	millipascal second	$mPa \cdot s$	
viscosity, kinematic	square meter per second	m^2/s	

xiv FACTORS, ABBREVIATIONS, AND SYMBOLS

Quantity	Unit	Symbol	Acceptable equivalent
	square millimeter per second	mm^2/s	
volume	cubic meter	m^3	
	cubic decimeter	dm^3	L(liter) (5)
	cubic centimeter	cm^3	mL
wave number	1 per meter	m^{-1}	
	1 per centimeter	cm^{-1}	

In addition, there are 16 prefixes used to indicate order of magnitude, as follows:

Multiplication factor	Prefix	Symbol	Note
10^{18}	exa	E	
10^{15}	peta	P	
10^{12}	tera	T	
10^{9}	giga	G	
10^{6}	mega	M	
10^{3}	kilo	k	
10^{2}	hecto	h^a	^{a} Although hecto, deka, deci, and centi are SI prefixes, their use should be avoided except for SI unit-multiples for area and volume and nontechnical use of centimeter, as for body and clothing measurement.
10	deka	da^a	
10^{-1}	deci	d^a	
10^{-2}	centi	c^a	
10^{-3}	milli	m	
10^{-6}	micro	μ	
10^{-9}	nano	n	
10^{-12}	pico	p	
10^{-15}	femto	f	
10^{-18}	atto	a	

For a complete description of SI and its use the reader is referred to ASTM E380 (4) and the article Units and Conversion Factors which will appear in a later volume of the *Encyclopedia*.

A representative list of conversion factors from non-SI to SI units is presented herewith. Factors are given to four significant figures. Exact relationships are followed by a dagger. A more complete list is given in ASTM E 380-76(4) and ANSI Z210.1-1976 (6).

Conversion Factors to SI Units

To convert from	To	Multiply by
acre	square meter (m^2)	4.047 × 10^3
angstrom	meter (m)	1.0 × 10^{-10}†
are	square meter (m^2)	1.0 × 10^{2}†
astronomical unit	meter (m)	1.496 × 10^{11}
atmosphere	pascal (Pa)	1.013 × 10^5
bar	pascal (Pa)	1.0 × 10^{5}†

† Exact.

To convert from	To	Multiply by
barrel (42 U.S. liquid gallons)	cubic meter (m³)	0.1590
Bohr magneton μ_β	J/T	9.274×10^{-24}
Btu (International Table)	joule (J)	1.055×10^3
Btu (mean)	joule (J)	1.056×10^3
Btu (thermochemical)	joule (J)	1.054×10^3
bushel	cubic meter (m³)	3.524×10^{-2}
calorie (International Table)	joule (J)	4.187
calorie (mean)	joule (J)	4.190
calorie (thermochemical)	joule (J)	4.184†
centipoise	pascal second (Pa·s)	1.0×10^{-3}†
centistoke	square millimeter per second (mm²/s)	1.0†
cfm (cubic foot per minute)	cubic meter per second (m³/s)	4.72×10^{-4}
cubic inch	cubic meter (m³)	1.639×10^{-5}
cubic foot	cubic meter (m³)	2.832×10^{-2}
cubic yard	cubic meter (m³)	0.7646
curie	becquerel (Bq)	3.70×10^{10}†
debye	coulomb·meter (C·m)	3.336×10^{-30}
degree (angle)	radian (rad)	1.745×10^{-2}
denier (international)	kilogram per meter (kg/m)	1.111×10^{-7}
	tex‡	0.1111
dram (apothecaries')	kilogram (kg)	3.888×10^{-3}
dram (avoirdupois)	kilogram (kg)	1.772×10^{-3}
dram (U.S. fluid)	cubic meter (m³)	3.697×10^{-6}
dyne	newton (N)	1.0×10^{-5}†
dyne/cm	newton per meter (N/m)	1.00×10^{-3}†
electron volt	joule (J)	1.602×10^{-19}
erg	joule (J)	1.0×10^{-7}†
fathom	meter (m)	1.829
fluid ounce (U.S.)	cubic meter (m³)	2.957×10^{-5}
foot	meter (m)	0.3048†
footcandle	lux (lx)	10.76
furlong	meter (m)	2.012×10^{-2}
gal	meter per second squared (m/s²)	1.0×10^{-2}†
gallon (U.S. dry)	cubic meter (m³)	4.405×10^{-3}
gallon (U.S. liquid)	cubic meter (m³)	3.785×10^{-3}
gallon per minute (gpm)	cubic meter per second (m³/s)	6.308×10^{-5}
	cubic meter per hour (m³/h)	0.2271
gauss	tesla (T)	1.0×10^{-4}
gilbert	ampere (A)	0.7958
gill (U.S.)	cubic meter (m³)	1.183×10^{-4}
grad	radian	1.571×10^{-2}
grain	kilogram (kg)	6.480×10^{-5}
gram force per denier	newton per tex (N/tex)	8.826×10^{-2}
hectare	square meter (m²)	1.0×10^4†
horsepower (550 ft·lbf/s)	watt (W)	7.457×10^2

† Exact.
‡ See footnote on p. xii.

To convert from	To	Multiply by
horsepower (boiler)	watt (W)	9.810×10^3
horsepower (electric)	watt (W)	$7.46 \times 10^{2\dagger}$
hundredweight (long)	kilogram (kg)	50.80
hundredweight (short)	kilogram (kg)	45.36
inch	meter (m)	$2.54 \times 10^{-2\dagger}$
inch of mercury (32°F)	pascal (Pa)	3.386×10^3
inch of water (39.2°F)	pascal (Pa)	2.491×10^2
kilogram force	newton (N)	9.807
kilowatt hour	megajoule (MJ)	$3.6^\dagger$
kip	newton (N)	4.48×10^3
knot (international)	meter per second (m/s)	0.5144
lambert	candela per square meter (cd/m²)	3.183×10^3
league (British nautical)	meter (m)	5.559×10^3
league (statute)	meter (m)	4.828×10^3
light year	meter (m)	9.461×10^{15}
liter (for fluids only)	cubic meter (m³)	$1.0 \times 10^{-3\dagger}$
maxwell	weber (Wb)	$1.0 \times 10^{-8\dagger}$
micron	meter (m)	$1.0 \times 10^{-6\dagger}$
mil	meter (m)	$2.54 \times 10^{-5\dagger}$
mile (U.S. nautical)	meter (m)	$1.852 \times 10^{3\dagger}$
mile (statute)	meter (m)	1.609×10^3
mile per hour	meter per second (m/s)	0.4470
millibar	pascal (Pa)	1.0×10^2
millimeter of mercury (0°C)	pascal (Pa)	$1.333 \times 10^{2\dagger}$
minute (angular)	radian	2.909×10^{-4}
myriagram	kilogram (kg)	10
myriameter	kilometer (km)	10
oersted	ampere per meter (A/m)	79.58
ounce (avoirdupois)	kilogram (kg)	2.835×10^{-2}
ounce (troy)	kilogram (kg)	3.110×10^{-2}
ounce (U.S. fluid)	cubic meter (m³)	2.957×10^{-5}
ounce-force	newton (N)	0.2780
peck (U.S.)	cubic meter (m³)	8.810×10^{-3}
pennyweight	kilogram (kg)	1.555×10^{-3}
pint (U.S. dry)	cubic meter (m³)	5.506×10^{-4}
pint (U.S. liquid)	cubic meter (m³)	4.732×10^{-4}
poise (absolute viscosity)	pascal second (Pa·s)	$0.10^\dagger$
pound (avoirdupois)	kilogram (kg)	0.4536
pound (troy)	kilogram (kg)	0.3732
poundal	newton (N)	0.1383
pound-force	newton (N)	4.448
pound per square inch (psi)	pascal (Pa)	6.895×10^3
quart (U.S. dry)	cubic meter (m³)	1.101×10^{-3}
quart (U.S. liquid)	cubic meter (m³)	9.464×10^{-4}
quintal	kilogram (kg)	$1.0 \times 10^{2\dagger}$

†Exact.

To convert from	To	Multiply by
rad	gray (Gy)	1.0×10^{-2}†
rod	meter (m)	5.029
roentgen	coulomb per kilogram (C/kg)	2.58×10^{-4}
second (angle)	radian (rad)	4.848×10^{-6}
section	square meter (m²)	2.590×10^{6}
slug	kilogram (kg)	14.59
spherical candle power	lumen (lm)	12.57
square inch	square meter (m²)	6.452×10^{-4}
square foot	square meter (m²)	9.290×10^{-2}
square mile	square meter (m²)	2.590×10^{6}
square yard	square meter (m²)	0.8361
stere	cubic meter (m³)	1.0†
stokes (kinematic viscosity)	square meter per second (m²/s)	1.0×10^{-4}†
tex	kilogram per meter (kg/m)	1.0×10^{-6}†
ton (long, 2240 pounds)	kilogram (kg)	1.016×10^{3}
ton (metric)	kilogram (kg)	1.0×10^{3}†
ton (short, 2000 pounds)	kilogram (kg)	9.072×10^{2}
torr	pascal (Pa)	1.333×10^{2}
unit pole	weber (Wb)	1.257×10^{-7}
yard	meter (m)	0.9144†

† Exact.

Abbreviations and Unit Symbols

Following is a list of commonly used abbreviations and unit symbols appropriate for use in the *Encyclopedia*. In general they agree with those listed in *American National Standard Abbreviations for Use on Drawings and in Text (ANSI Y1.1)* (6) and *American National Standard Letter Symbols for Units in Science and Technology (ANSI Y10)* (6). Also included is a list of acronyms for a number of private and government organizations as well as common industrial solvents, polymers and other chemicals.

Rules for Writing Unit Symbols (4):

1. Unit symbols should be printed in upright letters (roman) regardless of the type style used in the surrounding text.

2. Unit symbols are unaltered in the plural.

3. Unit symbols are not followed by a period except when used as the end of a sentence.

4. Letter unit symbols are generally written in lower-case (eg, cd for candela) unless the unit name has been derived from a proper name, in which case the first letter of the symbol is capitalized (W,Pa). Prefix and unit symbols retain their prescribed form regardless of the surrounding typography.

5. In the complete expression for a quantity, a space should be left between the numerical value and the unit symbol. For example, write 2.37 lm, *not* 2.37lm, and 35 mm, *not* 35mm. When the quantity is used in an adjectival sense, a hyphen is often used, for example, 35-mm film. *Exception:* No space is left between the numerical value and the symbols for degree, minute, and second of plane angle, and degree Celsius.

xviii FACTORS, ABBREVIATIONS, AND SYMBOLS

6. No space is used between the prefix and unit symbols (eg, kg).
7. Symbols, not abbreviations, should be used for units. For example, use "A," not "amp," for ampere.
8. When multiplying unit symbols, use a raised dot:

$$N \cdot m \text{ for newton meter}$$

In the case of W·h, the dot may be omitted, thus:

$$Wh$$

An exception to this practice is made for computer printouts, automatic typewriter work, etc, where the raised dot is not possible, and a dot on the line may be used.

9. When dividing unit symbols use one of the following forms:

$$m/s \text{ or } m \cdot s^{-1} \text{ or } \frac{m}{s}$$

In no case should more than one slash be used in the same expression unless parentheses are inserted to avoid ambiguity. For example, write:

$$J/(mol \cdot K) \text{ or } J \cdot mol^{-1} \cdot K^{-1} \text{ or } (J/mol)/K$$

but *not*

$$J/mol/K$$

10. Do not mix symbols and unit names in the same expression. Write:

$$\text{joules per kilogram } or \text{ J/kg } or \text{ J} \cdot \text{kg}^{-1}$$

but *not*

$$\text{joules/kilogram } nor \text{ joules/kg } nor \text{ joules} \cdot \text{kg}^{-1}$$

ABBREVIATIONS AND UNITS

A	ampere	AIChE	American Institute of Chemical Engineers
A	anion (eg, HA)		
a	atto (prefix for 10^{-18})	AIP	American Institute of Physics
AATCC	American Association of Textile Chemists and Colorists	alc	alcohol(ic)
		Alk	alkyl
		alk	alkaline (not alkali)
ABS	acrylonitrile–butadiene–styrene	amt	amount
		amu	atomic mass unit
abs	absolute	ANSI	American National Standards Institute
ac	alternating current, *n.*		
a-c	alternating current, *adj.*	AO	atomic orbital
ac-	alicyclic	APHA	American Public Health Association
ACGIH	American Conference of Governmental Industrial Hygienists		
		API	American Petroleum Institute
ACS	American Chemical Society	aq	aqueous
AGA	American Gas Association	Ar	aryl
Ah	ampere hour	*ar-*	aromatic

as-	asymmetric(al)	D-	denoting configurational relationship
ASHRAE	American Society of Heating, Refrigerating, and Air Conditioning Engineers	d	differential operator
		d-	dextro-, dextrorotatory
		da	deka (prefix for 10^1)
ASM	American Society for Metals	dB	decibel
ASME	American Society of Mechanical Engineers	dc	direct current, *n.*
		d-c	direct current, *adj.*
ASTM	American Society for Testing and Materials	dec	decompose
		detd	determined
at no.	atomic number	detn	determination
at wt	atomic weight	dia	diameter
av(g)	average	dil	dilute
bbl	barrel	*dl-*; DL-	racemic
bcc	body-centered cubic	DMF	dimethylformamide
Bé	Baumé	DMG	dimethyl glyoxime
bid	twice daily	DOE	Department of Energy
Boc	*t*-butyloxycarbonyl	DOT	Department of Transportation
BOD	biochemical (biological) oxygen demand	dp	dew point; degree of polymerization
bp	boiling point		
Bq	becquerel	dstl(d)	distill(ed)
C	coulomb	dta	differential thermal analysis
°C	degree Celsius		
C-	denoting attachment to carbon	(*E*)-	entgegen; opposed
		e	dielectric constant (unitless number)
c	centi (prefix for 10^{-2})		
ca	circa (approximately)	e	electron
cd	candela; current density; circular dichroism	ECU	electrochemical unit
		ed.	edited, edition, editor
CFR	Code of Federal Regulations	ED	effective dose
cgs	centimeter-gram-second	EDTA	ethylenediaminetetraacetic acid
CI	Color Index		
cis-	isomer in which substituted groups are on same side of double bond between C atoms	emf	electromotive force
		emu	electromagnetic unit
		eng	engineering
		EPA	Environmental Protection Agency
cl	carload		
cm	centimeter	epr	electron paramagnetic resonance
CMA	Chemical Manufacturer's Association		
		eq.	equation
cmil	circular mil	esp	especially
cmpd	compound	esr	electron-spin resonance
COA	coenzyme A	est(d)	estimate(d)
COD	chemical oxygen demand	estn	estimation
coml	commercial(ly)	esu	electrostatic unit
cp	chemically pure	exp	experiment, experimental
cph	close-packed hexagonal	ext(d)	extract(ed)
CPSC	Consumer Product Safety Commission	F	farad (capacitance)
		f	femto (prefix for 10^{-15})

FAO	Food and Agriculture Organization (United Nations)	ISO	International Organization for Standardization
fcc	face-centered cubic	IUPAC	International Union of Pure and Applied Chemistry
FDA	Food and Drug Administration	IV	iodine value
FEA	Federal Energy Administration	J	joule
		K	kelvin
fob	free on board	k	kilo (prefix for 10^3)
fp	freezing point	kg	kilogram
FPC	Federal Power Commission	L	denoting configurational relationship
frz	freezing		
G	giga (prefix for 10^9)	L	liter (for fluids only)(5)
g	gram	l-	levo-, levorotatory
(g)	gas, only as in $H_2O(g)$	(l)	liquid, only as in $NH_3(l)$
g	gravitational acceleration	LC_{50}	conc lethal to 50% of the animals tested
gem-	geminal		
glc	gas-liquid chromatography	LCAO	linear combination of atomic orbitals
g-mol wt; gmw	gram-molecular weight	lcl	less than carload lots
		LD_{50}	dose lethal to 50% of the animals tested
grd	ground		
Gy	gray	LED	light emitting diode
H	henry	liq	liquid
h	hour; hecto (prefix for 10^2)	lm	lumen
ha	hectare	ln	logarithm (natural)
HB	Brinell hardness number	LNG	liquefied natural gas
Hb	hemoglobin	log	logarithm (common)
HK	Knoop hardness number	LPG	liquefied petroleum gas
HRC	Rockwell hardness (C scale)	ltl	less than truckload lots
HV	Vickers hardness number	lx	lux
hyd	hydrated, hydrous	M	mega (prefix for 10^6); metal (as in Ma)
hyg	hygroscopic		
Hz	hertz	M	molar
^{i}(eg, Pr^i)	iso (eg, isopropyl)	m	meter; milli (prefix for 10^{-3})
i-	inactive (eg, i-methionine)	m	molal
IACS	International Annealed Copper Standard	m-	meta
		max	maximum
ibp	initial boiling point	MEK	methyl ethyl ketone
ICC	Interstate Commerce Commission	meq	milliequivalent
		mfd	manufactured
ICT	International Critical Table	mfg	manufacturing
ID	inside diameter; infective dose	mfr	manufacturer
IPS	iron pipe size	MIBC	methylisobutyl carbinol
IPT	Institute of Petroleum Technologists	MIBK	methyl isobutyl ketone
		MIC	minimum inhibiting concentration
ir	infrared		
IRLG	Interagency Regulatory Liaison Group	min	minute; minimum
		mL	milliliter

MLD	minimum lethal dose	NSF	National Science Foundation
MO	molecular orbital	NTA	nitrilotriacetic acid
mo	month	NTSB	National Transportation Safety Board
mol	mole		
mol wt	molecular weight	O	denoting attachment to oxygen
mom	momentum		
mp	melting point	o-	ortho
MR	molar refraction	OD	outside diameter
ms	mass spectrum	OPEC	Organization of Petroleum Exporting Countries
mxt	mixture		
μ	micro (prefix for 10^{-6})	OSHA	Occupational Safety and Health Administration
N	newton (force)		
N	normal (concentration)	owf	on weight of fiber
N-	denoting attachment to nitrogen	Ω	ohm
		P	peta (prefix for 10^{15})
n (as n_D^{20}	index of refraction (for 20°C and sodium light)	p	pico (prefix for 10^{-12})
		p-	para
^{n} (as Bu^n),		p.	page
		Pa	pascal (pressure)
n-	normal (straight-chain structure)	pd	potential difference
		pH	negative logarithm of the effective hydrogen ion concentration
n	nano (prefix for 10^{-9}		
na	not available		
NAS	National Academy of Sciences	pmr	proton magnetic resonance
		POP	polyoxypropylene
NASA	National Aeronautics and Space Administration	pos	positive
		pp.	pages
nat	natural	ppb	parts per billion
NBS	National Bureau of Standards	ppm	parts per million
		PPO	poly(phenyl oxide)
		ppt(d)	precipitate(d)
neg	negative	pptn	precipitation
NF	*National Formulary*	Pr (no.)	foreign prototype (number)
NIH	National Institutes of Health	pt	point; part
NIOSH	National Institute of Occupational Safety and Health	PVC	poly(vinyl chloride)
		pwd	powder
		qv	quod vide (which see)
nmr	nuclear magnetic resonance	R	univalent hydrocarbon radical
NND	New and Nonofficial Drugs (AMA)		
		(R)-	rectus (clockwise configuration)
no.	number		
NOI-(BN)	not otherwise indexed (by name)	rad	radian; radius
		rds	rate determining step
NOS	not otherwise specified	ref.	reference
nqr	nuclear quadrople resonance	rf	radio frequency, *n.*
NRC	Nuclear Regulatory Commission; National Research Council	r-f	radio frequency, *adj.*
		rh	relative humidity
		RI	Ring Index
NRI	New Ring Index	RT	room temperature

s (eg, Buˢ);		T	tera (prefix for 10^{12}); tesla (magnetic flux density)
sec-	secondary (eg, secondary butyl)	t	metric ton (tonne) temperature
S	siemens	TAPPI	Technical Association of the Pulp and Paper Industry
(*S*)-	sinister (counterclockwise configuration)	tex	tex (linear density)
S-	denoting attachment to sulfur	tga	thermogravimetric analysis
		THF	tetrahydrofuran
s-	symmetric(al)	tlc	thin layer chromatography
s	second	TLV	threshold limit value
(s)	solid, only as in H_2O(s)	*trans-*	isomer in which substituted groups are on opposite sides of double bond between C atoms
SAE	Society of Automotive Engineers		
SAN	styrene–acrylonitrile		
sat(d)	saturate(d)	TSCA	Toxic Substance Control Act
satn	saturation		
SCF	self-consistent field	Twad	Twaddell
Sch	Schultz number	UNICEF	United Nations (International) Children's (Emergency) Fund
SFs	Saybolt Furol seconds		
SI	Le Système International d'Unités (International System of Units)		
		UL	Underwriters' Laboratory
sl sol	slightly soluble	USDA	United States Department of Agriculture
sol	soluble		
soln	solution	USP	*United States Pharmacopeia*
soly	solubility	uv	ultraviolet
sp	specific; species	V	volt (emf)
sp gr	specific gravity	var	variable
sr	steradian	*vic-*	vicinal
std	standard	vol	volume (not volatile)
STP	standard temperature and pressure (0°C and 101.3 kPa)	vs	versus
		v sol	very soluble
		W	watt
SUs	Saybolt Universal seconds	Wb	Weber
		Wh	watt hour
syn	synthetic	WHO	World Health Organization (United Nations)
ᵗ (eg, Buᵗ),			
t-,		wk	week
		yr	year
tert-	tertiary (eg, tertiary butyl)	(*Z*)-	zusammen; together

Non-SI (Unacceptable and Obsolete) Units — *Use*

Å	angstrom	nm
at	atmosphere, technical	Pa
atm	atmosphere, standard	Pa
b	barn	cm^2
bar†	bar	Pa

† Do not use bar (10^5Pa) or millibar (10^2Pa) because they are not SI units, and are accepted internationally only for a limited time in special fields because of existing usage.

Non-SI (Unacceptable and Obsolete) Units		Use
bhp	brake horsepower	W
Btu	British thermal unit	J
bu	bushel	m^3; L
cal	calorie	J
cfm	cubic foot per minute	m^3/s
Ci	curie	Bq
cSt	centistokes	mm^2/s
c/s	cycle per second	Hz
cu	cubic	exponential form
D	debye	C·m
den	denier	tex
dr	dram	kg
dyn	dyne	N
erg	erg	J
eu	entropy unit	J/K
°F	degree Fahrenheit	°C; K
fc	footcandle	lx
fl	footlambert	lx
fl oz	fluid ounce	m^3; L
ft	foot	m
ft·lbf	foot pound-force	J
gf den	gram-force per denier	N/tex
G	gauss	T
Gal	gal	m/s^2
gal	gallon	m^3; L
Gb	gilbert	A
gpm	gallon per minute	(m^3/s); (m^3/h)
gr	grain	kg
hp	horsepower	W
ihp	indicated horsepower	W
in.	inch	m
in. Hg	inch of mercury	Pa
in. H_2O	inch of water	Pa
in.-lbf	inch pound-force	J
kcal	kilogram-calorie	J
kgf	kilogram-force	N
kilo	for kilogram	kg
L	lambert	lx
lb	pound	kg
lbf	pound-force	N
mho	mho	S
mi	mile	m
MM	million	M
mm Hg	millimeter of mercury	Pa
mμ	millimicron	nm
mph	miles per hour	km/h
μ	micron	μm
Oe	oersted	A/m
oz	ounce	kg
ozf	ounce-force	N
η	poise	Pa·s
P	poise	Pa·s
ph	phot	lx
psi	pounds-force per square inch	Pa
psia	pounds-force per square inch absolute	Pa
psig	pounds-force per square inch gage	Pa

qt	quart	m^3; L
°R	degree Rankine	K
rd	rad	Gy
sb	stilb	lx
SCF	standard cubic foot	m^3
sq	square	exponential form
thm	therm	J
yd	yard	m

BIBLIOGRAPHY

1. The International Bureau of Weights and Measures, BIPM, (Parc de Saint-Cloud, France) is described on page 22 of Ref. 4. This bureau operates under the exclusive supervision of the International Committee of Weights and Measures (CIPM).
2. *Metric Editorial Guide (ANMC-75-1)*, American National Metric Council, 1625 Massachusetts Ave. N.W.; Washington, D.C. 20036, 1975.
3. *SI Units and Recommendations for the Use of Their Multiples and of Certain Other Units (ISO 1000-1973)*, American National Standards Institute, 1430 Broadway, New York, N. Y. 10018, 1973.
4. Based on *ASTM E 380-76 (Standard for Metric Practice)*, American Society for Testing and Materials, 1916 Race Street, Philadelphia, Pa. 19103, 1976.
5. *Fed. Regist.*, Dec. 10, 1976 (41 FR 36414).
6. For ANSI address, see Ref. 3.

R. P. LUKENS
American Society for Testing and Materials

ENAMELS, PORCELAIN OR VITREOUS

Historically, enamel has described decorative and protective glassy coatings on metal as well as glassy, decorative coatings on glass. Enamel has also implied certain organic coatings such as paints or lacquers; these are not discussed here (see Coatings, industrial; Paint). Glaze has most commonly referred to glassy coatings on ceramic bodies (see Ceramics).

In the United States, the term porcelain enamel designates the glassy coating on metal; however, in some other countries, vitreous enamel is the more common term for the same glassy coating. The ASTM defines porcelain enamel as a substantially vitreous or glassy inorganic coating bonded to metal by fusion at a temperature above 425°C (1).

Ceramic coatings, another term used for coatings on metal, connotes emphasis on the protective feature of the coating for the metal (see Refractory coatings). Ceramic coatings are often formulated and designed to contain mainly crystalline rather than glassy material (see also Colorants for ceramics; Glass).

Production processes for glass began about 2500–3000 BC; however, it is not clear when porcelain enamel originated. It is likely that metalsmiths explored the decorative technique of enameling through their use of colored glass inlays or by fabricating patterns of glass pieces on gold, silver, or copper articles. Porcelain-enameled metal art objects have been associated with early civilizations in the Middle East and in the Orient. It is believed that the techniques for the creation of decorative porcelain-enameled objects were passed from one Mediterranean group to another and these groups included Egyptians, Greeks, Romans, Spaniards, and Arabs. The Romans are thought to have introduced enameling to Great Britain. The art of porcelain enameling was traced from the fourth to the eleventh centuries in Byzantium (Istanbul) and was later introduced into Italy and parts of Western Europe (2).

Ancient and more recent enameled art objects evidence four enameling techniques: cloisonné, champlevé, basse-taille, and painting. *Cloisonné* enameling involves the preparation of a design of small wires or partitions (cloisons) which is soldered or fused onto the metal surface. Segments of the design are filled with different colorants of powdered glass and the entire ensemble is heated in order that the enamels are fused to the partitions and the metal substrate. After being ground and polished, this decorated surface displays a colored enameled design outlined by thin wires. *Champlevé* enameling is the technique of carving or gouging a design into the surface of a thick metal-base material such as copper, bronze, gold, or silver. The gouged areas of the design are filled with colored glass and the piece is fired. The polished, finished piece appears to consist of a design of enamel inlays in the metal surface. *Basse-taille* describes the technique of designing the metal surface in low relief and then of covering the entire surface with a transparent porcelain enamel. *Painting* involves enameling in layers. First the entire metal surface is covered with a dark enamel layer, then lighter-colored enamels are applied or painted on the first dark layer.

Ancient metal art objects were made from gold, silver, copper, or bronze. It was not until the early 1800s that porcelain enameling on cast iron was developed and practiced in Central Europe. As the sheet-steel manufacturing process developed about 1850, the industrial development of porcelain enameling of sheet steel naturally followed, first in Germany and Austria. About 1857 porcelain enameling was initiated in the United States by two manufacturers of porcelain-enameled kitchenware: the Grosjean Company of New York and the Vollrath Company of Sheboygan, Wisc.

Porcelain enameling protects against corrosion, decorates, and resists the attack of alkalies, acids, and other chemicals. This material is a nonporous sanitary coating imparting no odors or tastes. Since it is entirely inorganic, it does not serve as a feedstock for microorganisms. Because of its sanitary aspects, its protective and strengthening function, and its decorative character, porcelain enamel has been adopted as the most suitable material for bathtubs, laundry appliances, ranges, sinks, and refrigerator liners. The decorative and corrosion-resistant qualities of this coating have led to its many uses in architectural applications. Because of its extremely low porosity and resistance to chemical attack, porcelain enamel has found widespread use in the dairy, pharmaceutical, brewing, and chemical industries. Porcelain-enameled tanks and vessels made of heavy-gage sheet steel or cast iron are commonly used in the above industrial fields. Porcelain enamel on steel, cast iron, or aluminum is a most desirable composite system of materials. Sinks, stoves, ranges, refrigerators, clothes washers, dishwashers, and dryers represent major uses of these materials in the home-appliance industry. Cooking utensils, architectural panels, signs, silos, bathtubs, lavatories, brewing vessels, chemical storage tanks, gasoline service stations, roofing tile, guard rails, chalkboards, and many other products of commerce indicate the broad spectrum of home and industrial products finished with porcelain enamels.

Several common classifications of porcelain enamels are as follows:

Basis of classification	*Examples*
function	ground coats: single frit; multiple frit cover coat: directly applied to metal; applied to ground-coated metal
service	acid resistant; alkali resistant; hot water resistant; chemical resistant; abrasion resistant; electrically insulating; thermal-shock resistant; catalytic; pyrolytic

composition	alkali borosilicate; titania; lead-bearing; leadless enamels
metal coated	sheet steel, very low carbon (<0.0005% carbon), enameling iron (0.024% carbon or less), cold-rolled steel (0.06–0.10% carbon), hot-rolled steel (1010 or 1020 types; 0.10–0.20% carbon); cast iron; aluminum; copper; gold; silver; stainless steel
decorative character	clear; colored; white; stippled; matte; glossy; semimatte; beading
opacifying material	titania; zirconia; antimony oxide; molybdenum oxide
method of application	wet process: spraying, flow-coating, dipping, electrophoretic, electrostatic dry process: electrostatic; sifting
type of product	appliances; cooking utensils; sanitary ware; chemical equipment; jewelry; architectural panels; signs; hot water tanks; silos
firing temperature	540°C; low temperature, 595–760°C; normal, 790–870°C; high temperature, 870°C

The Enameling Process

The porcelain enameling process involves the re-fusing of powdered glass on the metal surface. The powdered glass is prepared by ball-milling a porcelain enamel glass engineered for specific properties. First, the glass is smelted from raw batch materials such as those listed for enamel glass compositions in Table 1. The enamel smelter is usually a box-shaped tank furnace. Continuous smelters, wherein the thoroughly mixed raw batch is fed in at one end and molten glass is flowing out at the other end, are common in commercial operations. Decomposition, gas evolution, and solution occur during smelting. After the molten glass has been smelted to a homogeneous liquid, it is poured in a thin stream into water or onto cooled metal rollers. This quenched glass, termed frit, is a friable material easily reduced to small particles by a ball-milling operation. Ball-milling the glass frit into small-sized particles can be carried out whether the frit is wet or dry (see Size reduction). Dry powders are used for dry-process cast-iron enameling and for electrostatic application on sheet steel (see Powder coating). Dry powders are also prepared and marketed for the subsequent preparation of slurries and slips used in the wet-process application techniques.

Process flow diagrams for sheet-steel and cast-iron enameling are shown in Figures 1 and 2.

The frit-making process involves all the technology from proper selection of raw materials through thorough mixing and smelting to uniform production of frit. Batch-type smelters, such as crucible furnaces, rotary smelters, or box-shaped smelters, are used for small production requirements or for research and development purposes (see Furnaces).

For conventional wet-process sheet-steel enameling, the porcelain enamel frit is ball-milled with clay, certain electrolytes, and water to form a stable suspension. This clay-supported slurry of small frit particles is called the slip and has a consistency similar to that of a thick coffee cream. The ingredients of the mill batch are carefully controlled. The amount and purity of all materials in the mill, including the clay and

Table 1. Composition of Porcelain Enamel Frits, wt % [a]

	Sheet-steel ground coat	Titania cover enamel	Cast iron, high lead	Cast iron, low lead	Cast-iron ground coat	Cast-iron cover coat
Oxide composition						
KNaO	19.7	(14.0)	(9.4)	(18.9)	(11.4)	(20.8)
K_2O		3.5	2.3	4.6	7.1	4.4
Na_2O		10.5	7.1	14.3	4.3	16.4
B_2O_3	14.6	14.0	8.7	12.0	6.9	2.6
Al_2O_3	7.2		4.6	6.5	11.3	6.0
SiO_2	50.5	45.0	24.6	44.7	51.8	48.9
CaF_2	5.1					3.2
F_2		5.0	6.5	6.3		
CoO	0.6					
MnO	1.9					
NiO	0.2					
TiO_2		20.0				
P_2O_5		2.0				
PbO			33.6		18.4	
CaO			7.1	4.0		
ZnO			5.4			
BaO				4.0		
Sb_2O_5				3.8		8.7
AlF_3						5.0
NaF						7.5
Batch						
feldspar	30.3		17.3	22.4	35.0	22.6
borax (hydrous)	31.6		18.3	26.9	23.0	16.7
borax (anhydrous)		19.1				
quartz	20.0	42.0	10.6	22.3	15.0	26.0
soda ash	6.7			4.3		3.0
$(NaNO_3)$	3.8	7.8	4.8	6.7	4.0	4.5
fluorspar	4.6		8.7	4.6		2.3
cobalt oxide	0.5					
manganese oxide	1.5					
nickel oxide	0.2					
monosodium phosphate		3.2				
zinc oxide			4.8			
red lead			30.7		23.0	
cryolite			2.9	5.5		10.7
boric acid			1.9			
titania		20.2				
Na_2SiF_6		1.2				
K_2SiF_6		7.8				
Sb_2O_5					3.1	
$BaCO_3$					4.2	
$NaSbO_3$						11.4
clay						2.4

[a] Ref. 2.

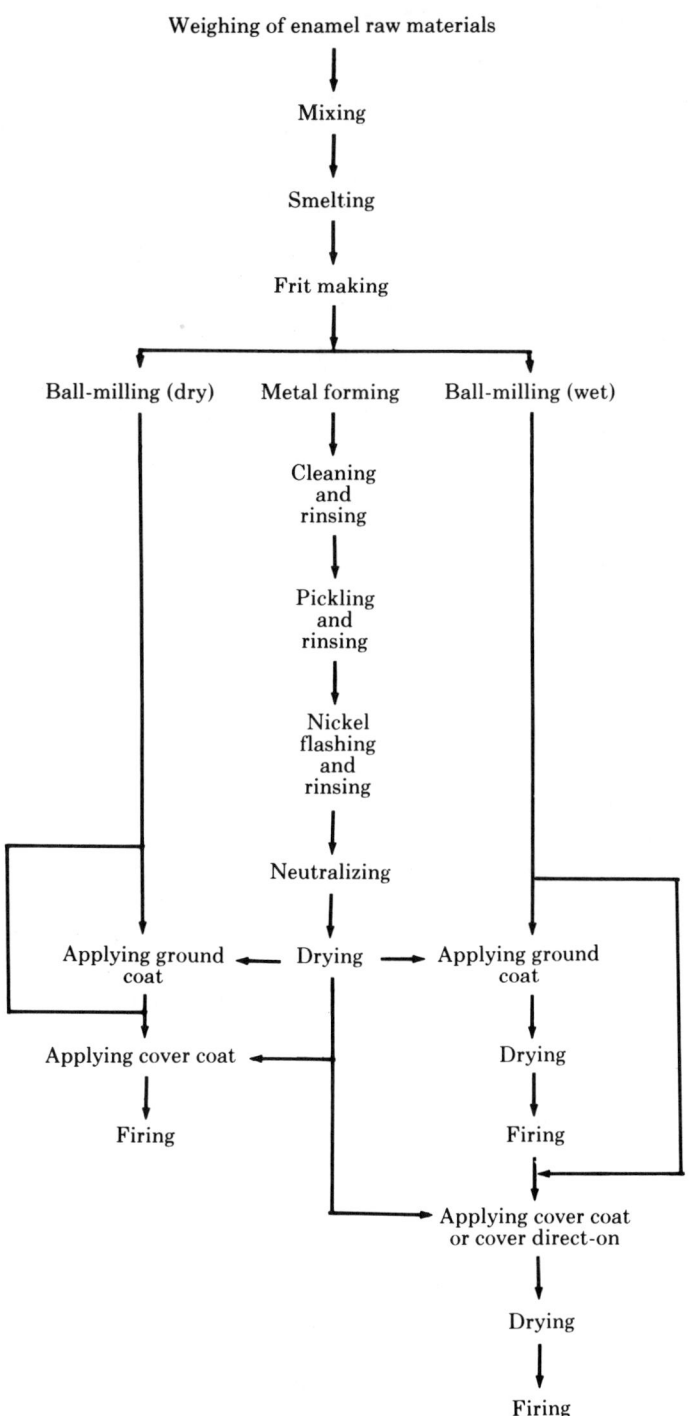

Figure 1. Process flow diagram of sheet-steel enameling.

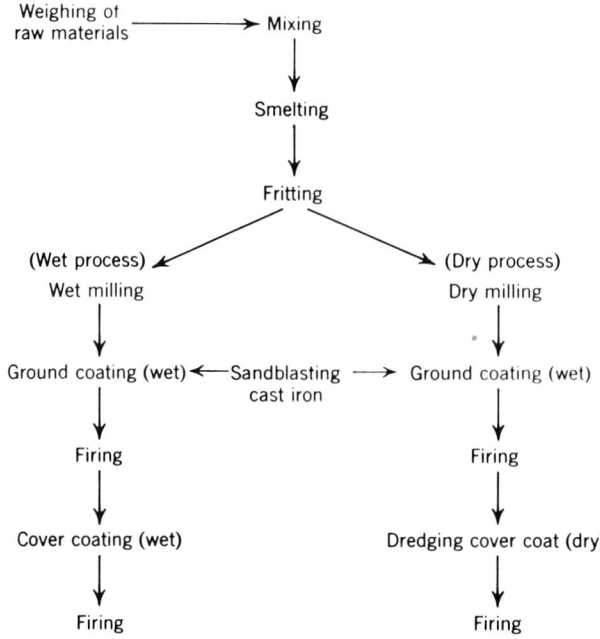

Figure 2. Flow diagram of cast-iron enameling.

water, affect the rheological character of the slip as well as a number of the properties of the fired enamel such as acid resistance (increasing clay content causes lower acid resistance), reflectance, and gloss.

Within the past several years, the development of a one- or two-coat electrostatic application of dry frit particles has proceeded in various parts of the world. Dry ball-milled frit particles coated with a thin organic layer and agitated in a fluidized bed are delivered to spray guns that impart an electric charge to the sprayed particle. The sheet metal article to be coated is at ground potential. The silicone-based coating on the glass particle is applied during the grinding process. The glass particle without the coating has a resistivity of 10^{-6} $\Omega\cdot$cm; with the synthetic coating, the glass particle may achieve a resistivity as great as 10^{-18} $\Omega\cdot$cm. This greater resistivity is necessary so that the charged particle will adhere to the metal surface. An overall resistivity of 10^{-10} $\Omega\cdot$cm permits a coating as thick as 250 μm whereas the maximum thickness achieved with a particle having a 10^{-18} $\Omega\cdot$cm resistivity is about 25 μm. The coating thickness is self-limiting and thickness uniformity is excellent even on curved surfaces (± 6 μm). This dry electrostatic process obviates the need for the drying and intermediate firing for two-coat articles and promises substantial savings in labor, material, and energy costs.

The electrostatic powder processing of porcelain enamel is similar to the electrostatic technique used in preparing organic coatings (see Coating processes). Compared to the conventional wet-process porcelain enameling of sheet steel, the dry system eliminates the need for the clay suspending agents which represent refractory additions. As a consequence, the dry powders permit firing at a lower temperature as compared to that of the wet powders, thereby saving energy. At the lower firing temperatures, approaching the ferrite–austenite transition (ca 725°C), metal sagging and distortion are minimized, and additional materials savings can be achieved as higher carbon, nonpremium steels may then be used and less structural bracing of the steel article is required. The electrostatic dry process also has several additional advantages over the wet process. For example, much less waste of enamel occurs since the dry overspray is airborne and is recycled in a closed system. Productivity of the dry process is higher not only because of the single firing treatment, but also because fewer rejects are experienced. Compared to the wet-process, so-called direct-on cover-coat technique, which requires a decarburized steel and a special pickling treatment that promotes adherence, the dry electrostatic powder process with its two-coat, one-fire feature does not require use of premium steel and requires little or no pickling pretreatment. The problem of disposing of acid pickling wastes, which have a relatively high content of iron sulfates, has become less important in the absence of pickling pretreatment (see Metal surface treatments).

Prior to the recent development of the dry electrostatic process for sheet-steel enameling, the one-coat cover-coat direct-on process had been successfully developed and adapted by a large sector of industrial sheet-steel enamelers. This process requires the use of very low carbon steel (decarburized steel) so that carbon–oxide gas defects are eliminated. Also a special pickling process is required to assure good adherence of the enamel and this involves an accelerated etching of the metal using ferric sulfate additions in the sulfuric acid pickling operation, as well as subsequent heavy nickel flash deposits (see below). Cover enamels are applied by flow-coating, spraying, dip-draining, or electrophoresis.

The electrostatic spray application of enamel slip is a well-developed commercial process. The atomized enamel leaves the spray gun and passes through a high intensity d-c field of 100 kV. The negatively charged enamel particles are attracted to the surface of the metal article which is at ground potential. This method of coating, which is usually automated, produces a uniform coating thickness and good coverage around corners (wraparound).

Electrophoresis also is employed commercially and provides a dense uniform coating. In this process the sheet-metal article is positively charged and the negative electrodes are located in the slip. The metal article, which is immersed in the slip, attracts the charged frit particles which then pack into a tightly adhering coating layer. This process can be automated for uncomplicated shapes.

Ball-mill grinding is accomplished with porcelain or high-alumina balls 1.5–5.0 cm in diameter. Ball size, mill speed, mill charge, and ball charge are important parameters for the determination of the milling time required for optimum size distribution in prepared slip. Ground-coat enamels are ball-milled to a fineness of 95% of solids smaller than 74 μm (200 mesh). Cover coats are ground more finely, to 98% of solids finer than 44 μm (325 mesh).

Ground coats are applied to metal by spraying, dipping, or draining. Large articles may be coated only by spraying or flow-draining the slip, whereas small articles may be dipped into a tank of ground-coat slip. Cover coats may also be applied by similar methods as well as by electrostatic spraying or electrophoresis.

Dry-process cast-iron enameling involves the application of the cover coat by dusting or dredging the dry glass powder (using a long-handled vibrating sifter) onto the heated cast-iron article. The cast iron, which has been previously ground-coated, is removed from the furnace (870°C and higher) for the dusting operation and then is returned to the furnace. Generally, only two dusting and heating cycles are required for the development of a uniform coating.

Preparation of Metals. Sheet-metal parts are formed by the well-known processes of stamping, bending, and shearing. Many parts require welding, and it is important that this be carried out in a uniform, smooth manner so that the welded joint can be enameled without defects (see Welding). Cast-iron parts are formed by the usual cast-iron foundry methods; however, additional care is given to prevent contamination of the surface which causes defects in the enamel, particularly blisters or bubbles.

Enameling cannot be successful unless the metal is thoroughly cleaned and kept clean until the final coat is fired. Simply touching the surface by hand may cause defects. Cast-iron and thick steel parts are sandblasted without danger of excessive loss of metal and excessive warping. Sand, silicon carbide, and steel grit are satisfactory abrasives (qv). Products made from thin sheet materials are most satisfactorily and most economically cleaned by chemical methods which require alkali and soap solutions to remove grease and dirt and acid solutions to remove oxidized metal.

The commercial chemical treatment in the metal preparation process may be carried out with continuous cleaning and pickling equipment. The fabricated sheet-metal articles are supported on a special continuously moving rack which is first passed through the cleaning stage. The ware subsequently is subjected to pickling acids, nickel flashing, and neutralizing treatments. Metal articles also are cleaned and pickled in a batch process whereby baskets of metal articles are immersed in large tanks of various solutions and then water rinses.

The composition of the cleaning solution depends upon the type of oil, grease, and solid material to be removed, including the type and amount of drawing compounds used in the metal-forming operations. Vegetable oils may be saponified and removed by alkalies alone, but mineral oils must be removed with soap. A well-balanced cleaning compound contains an alkali, an alkali salt, which acts as a buffer material to maintain an approximately constant pH, and a soap (qv). Proprietary products that are especially adapted to the cleaning process are available and are usually used at a strength of ca 45 g/L (6 oz/gal) of water at boiling temperatures.

After being cleaned, the ware is immersed successively in one or more tanks of water at 80–95°C and then is transferred to the acid pickling solution. The pickling solution of 6–8% sulfuric acid is contained in lead-lined wooden or stainless steel tanks and is maintained at 60–65°C. The ware remains in the acid solution for as long as it is necessary in order to remove all oxide scale from the metal. Pickling inhibitors are rarely used as they have a tendency to cause enameling defects and prevent etching of the steel which is desirable. Rinse water which removes acid is maintained at a temperature of 80–90°C and rinsing time is usually 3 min. Plating galvanically or flashing a thin film of nickel on the iron after rinsing retards oxidation and enhances enamel bond. The ware is immersed in a solution containing 7.5 g/L (1 oz/gal) of nickel salts, such as nickel sulfate, $NiSO_4.6H_2O$, or the equivalent. The pH is maintained between 3.0 and 3.6 and the temperature at 75°C. The average time of immersion is 4–6 min and the deposit is usually 0.45–1.30 g/m^2 (0.04–0.12 g/ft^2). A wooden or stainless steel tank normally is employed as a container.

For the cover-coat direct-on process, a ferric sulfate etch is included in the metal pretreatment procedure. Hydrogen peroxide is added intermittently to a 1% ferric sulfate solution to reoxidize ferrous sulfate to ferric sulfate. This accelerated pickling procedure usually involves a sulfuric acid–ferric sulfate–sulfuric acid pickling sequence, and is designed to remove ca 20 g/m^2 (2 g/ft^2) of iron from the sheet-metal surface.

After being removed from the nickel bath, the ware is dipped into a hot or cold water rinse, quickly removed, and then transferred to a neutralizing bath where the last traces of acid are removed. Neutralizing with solutions of sodium carbonate and borax is common.

After being removed from the neutralizing solution, the ware is transferred to a dryer where it is maintained at a temperature of about 110–120°C and is provided with good air circulation, which ensures quick and complete drying without rusting of the metal. After being dried, the sheet-steel ware is ready for application of the enamel and after application of the coating, the ware is ready for the firing operation.

Firing. Firing may be carried out in intermittent box-type furnaces or continuous furnaces. The dryer and the furnace may form one continuous unit or separate units in the continuous firing process. In a continuous tunnel-type furnace, each coated item progresses through the furnace supported on firing racks especially designed to withstand long service at the repeated cycles of heating and cooling. Gradual heating and cooling of the ware are more characteristic of continuous firing than of intermittent firing. In the latter process, a batch of ware loaded on a large rack of special support tools is introduced into the box furnace.

Furnace temperature in the hot zone of continuous furnaces and box furnaces matures the coating in a matter of minutes. Ground-coats are fired in a box furnace for 3–6 min at 850–870°C. Cover-coat firing is generally carried out at shorter times and at slightly lower temperatures.

Enamel-firing temperature is related to the coating composition, metal thickness, and the type of metal used. Enamels for aluminum are fired at 510–540°C, whereas coatings for high temperature alloys (qv) may be fired at 930°C.

The industrial porcelain enamel process can be fully automated from the beginning of the metal pretreatment (pickling) procedure through the coating operations and the firing process. Flow-coating, electrostatic spraying, or electrophoretic deposition can be machine-programmed. Robot sprayers controlled by a computerized sensing mechanism can coat nonsimple shapes, and flow-coaters can also be programmed to uniformly coat some rather complicated shapes. Automated coating and firing of cast-iron bathtubs has also been carried out.

Energy Requirements. The relative energy intensiveness of preparing a porcelain enamel compared to two competitive materials, which were designed for a refrigerator liner, was studied by the Porcelain Enamel Institute (3). An acrylic enamel on steel, an acrylonitrile–butadiene–styrene plastic (ABS), and a direct-on white porcelain enamel on steel were compared, considering total energy needs for materials and processing. All three systems were within a 10% range about the arithmetic mean value, eg, 283.6 ± 1.42 MJ/m^2 (ca $25,000 \pm 125$ Btu/ft^2). Because plastic costs are directly related to oil and gas prices, whereas steel and glass costs are related to coal, it is projected that porcelain enamel will be much less energy cost-sensitive than its organic competitors. Furthermore, porcelain enamel manufacturers will probably widen the gap as the dry electrostatic powder process becomes the standard processing procedure for most mass-produced porcelain enameled sheet-steel articles.

Composition

Since porcelain enamels are substantially glassy coatings, the composition of this part of the enamel–metal materials system is based on glass-forming ingredients. The principal glass formers are B_2O_3, SiO_2, and P_2O_5 (see Glass). Other glass formers, such as GeO_2, BeF_2, and As_2O_3, are rarely used as the base for the glass because they are not economical and do not impart especially useful properties. Phosphate glasses, although low melting and commercially economical, are in general not sufficiently resistant to alkali or to hot water attack.

Table 1 lists the compositions in terms of wt % of the oxide components. Raw materials for the glass batch include minerals, such as the feldspars and quartz, since these are inexpensive sources of SiO_2 and Al_2O_3 (see Clays). The batch composition that is especially designed for cover coats is comprised primarily of manufactured chemicals of known, controlled levels of purity.

The composition of porcelain enamel glasses essentially is based on the alkali borosilicate glasses. Both B_2O_3 and SiO_2, the glass formers, are also called network formers, whereas the remainder of the ingredients are called network modifiers. It is considered that BO_3^{3-}, BO_4^{5-}, and SiO_4^{4-} structural units exist in the glass structure, and glass—as a rigid super-cooled liquid—has a short-range order of BO_3^{3-} triangles and BO_4^{5-} and SiO_4^{4-} tetrahedra, but long-range order of the units does not occur. These triangles and tetrahedra are joined at corners, ie, oxygen atoms at corners join two silicon or boron atoms and the continuous three-dimensional network of the glass structure is assured by this arrangement.

In the absence of network-modifier atoms that do not contribute to the continuity of the three-dimensional network of SiO_2 or B_2O_3 glass, these oxide glasses have a very high viscosity at their liquidus (melting temperatures). The viscosity at the liquidus temperature of glasses is in the range of 1–1000 Pa·s ($10–10^4$ P) whereas water nearing its freezing point has a viscosity of about 1 mPa·s (1 cP). This high viscosity attests to the strong interconnecting bonds of the network and to the high degree of association in the liquid glass. Modifier atoms, such as the alkalies, alkaline earths, or halides, cause the number of interconnecting bonds to decrease. Broken bonds resulting from a sodium atom bonding to an oxygen atom (Na—O—Si—O—Na instead of —Si—O—Si—O—Si—) or a fluoride atom bonded to a silicon atom (F—Si—O—Si—F instead of —O—Si—O—Si—O—) are associated with lower viscosity and lower firing temperatures of these modified glasses. Lead oxide, an ingredient in glass, plays the roles of both network former and network modifier. Lead oxide-bearing glasses have been widely used for cast-iron enamels.

Some generalities concerning the effect of specific ingredients of the porcelain enamel can be expressed as follows: an increase in SiO_2 content increases firing temperature and acid resistance, and lowers the expansion coefficient; an increase in alkali content decreases firing temperature and acid resistance, and raises the expansion coefficient; and an increase in the Al_2O_3 and ZrO_2 content increases alkali resistance.

Each porcelain enamel composition is designed to obtain specific performance characteristics of the enamel such as good adherence to the substrate, thermal expansion fit to the metal, desired chemical properties, such as acid resistance, catalytic effectiveness, hot water resistance, or alkali resistance, and desired physical properties, such as abrasion resistance, thermal-shock resistance, high gloss, high reflectance, and desirable color.

Fundamental Considerations in Porcelain Enamels

Adherence of Porcelain Metals. Cobalt-bearing materials incorporated in the frit composition enhance the adherence of the enamel to sheet steel. The mechanism by which glass adheres to metal is still not completely clear and the role of cobalt in promoting adherence is a continuing fundamental research subject. Adherence and wetting are associated: under oxidizing conditions the enamel glass wets the metal, and the oxide on the metal surface tends to dissolve into the enamel layer. Under reducing or neutral conditions in enamel firing, the oxide originally on the metal surface becomes completely dissolved and the glass then fails to wet the metal surface. There are additives or ingredients of the frit other than cobalt which seem to aid adherence. Molybdenum compounds incorporated in the enamel are considered adherence promoters.

The mechanism of adherence can be considered to be of two main types. One is physical adherence, which involves the physical gripping of the glass by a metal surface that has been roughened mechanically prior to enameling or roughened during enameling by the corrosive attack of glass on the metal and by dendritic attachments to the metal formed during the enamel-firing treatment. The other adherence mechanism involves the chemical bonding of the metal, metal oxide, and enamel glass.

Adherence in porcelain enamel terminology generally refers to the amount of glass remaining in an impacted, fractured area of the porcelain enamel system. If after impact the enamel surface is fractured to such an extent that it is stripped clean of the coating, it is said to have poor adherence; excellent adherence requires a large amount of fractured glass remaining on the impacted surface. A standard test has been devised (ASTM C 313-59) (1) using an adherence test apparatus to measure the amount of glass remaining on the base metal.

With respect to sheet-steel enamels, ca 0.25% of cobalt oxide is used to promote adherence of the ground coat to the sheet steel. In the case of cast-iron enameling which assures proper oxidation of the metal, no special adherence additives are needed.

The amount of force required to remove the enamel glass from the metal is not usually measured nor is the strength of the glass–metal bond generally evaluated.

Thermal Fit and Residual Stresses. Thermal expansion measurements, as determined by the interferometer test method (ASTM C 539-66), provide expansion (percent)–temperature data (see Fig. 3). The interferometer consists of optically flat plates of fused silica separated by three fragments of the specimen material. These plates are arranged horizontally and nearly parallel in an electrical furnace so that monochromatic light is reflected from the bottom and top plates, and interference fringes appear in the eyepiece. As temperature is increased, the distance of separation of these fused-silica plates is increased and a movement of interference fringes occurs across the field of view. The softening point refers to the temperature at which the glass specimens can no longer support the load of the top interferometer plate. This is noted by the reversal in the direction of fringe movement.

The expansion coefficient for the metal is constant over the entire temperature range, whereas the coefficient of linear expansion of the enamel glass increases with temperature. However, the glass expansion coefficient reverses above the softening point (Fig. 3**a**).

A porcelain enamel glass becomes less viscous as temperature is increased as in firing. At these elevated temperatures, the enamel is relatively fluid and it conforms to the metal surface.

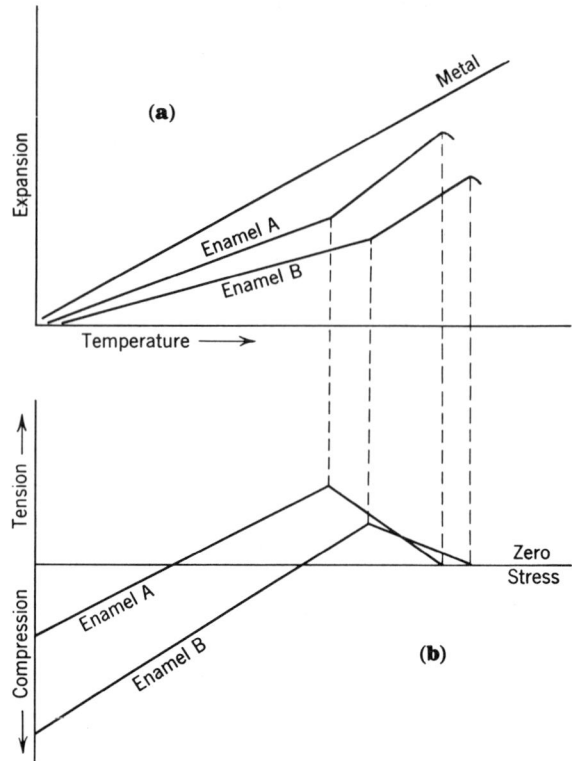

Figure 3. (a) Relative thermal expansion of porcelain enamel and sheet steel. (b) Stress development in enamel layer.

As the porcelain enamel coating is cooled from firing temperatures (750–800°C) to the softening point, the fluid glass does not retain stress. However, as cooling proceeds below the softening point, the coefficient of expansion (or contraction) of the coating exceeds that of the steel, and tensile stresses begin to develop in the coating, as shown in Figure 3b. On further cooling, the stress increases until the temperature at which the expansion coefficient of the glass equals that of the metal. With further cooling the coefficient of expansion of the glass becomes less than that of the metal, coating stresses decrease and compressive stresses develop.

Thermal expansion comparisons of coatings and metals have often been used to determine residual stresses in the coatings. The qualitative residual-stress analysis of Figure 3 shows how enamel A would develop high tensile stress and low residual compressive stress. Enamel B, with its higher softening point and lower expansion coefficient, develops less tensile stress and a much greater residual compressive stress.

Residual-stress analysis must take into account the cooling rates, the viscosity characteristics of the glass, the relative thickness of metal and coating, and the modulus of elasticity of coating and metal throughout the temperature range in which stress is developed. At high temperatures the glass behaves essentially as a viscous material, exhibiting viscous–elastic and then principally elastic properties as the temperature decreases.

Thermal expansion values can be calculated from measurements of thermal deflection of enamel–metal composites. Thermal expansion coefficient in the temperature range of 0–300°C can also be calculated using the additive formula:

$$P = AX_A + BX_B + CX_C + \text{etc}$$

where P is the property, such as linear expansion coefficient; A, B, C are the property factors for each ingredient in the composition; and X_A, X_B, X_C are ingredient compositions in wt %.

Factors for calculating the cubical thermal expansion coefficient of glasses have been determined and are listed in Table 2 (2). Glass compositions high in SiO_2 content have low coefficients of thermal expansion, whereas alkalies and alkaline earth materials raise the expansion coefficient (see Glass).

Residual compression in the coating is desirable since glass, as well as other ceramic materials, is much stronger in compression (about 2070 MPa or 300,000 psi) than in tension (about 69 MPa or 10,000 psi). If residual compression in the enamel layer is too great, the coating may fracture where the radius of curvature of the article is small. Since failure of the coating occurs because the tensile-stress limit has been exceeded, high residual compressive stresses in the coating increase the tensile load-bearing ability by the amount of the residual compressive stress that must be overcome before tension can be induced. If the tensile stress developed during the cooling or reheating of the enamel is too high, the coating may fracture. Crazing results when the damage occurs during cooling; a fine crack pattern called hairlining may be produced during reheating.

The linear expansion coefficient for common materials in the porcelain enamel system is as follows:

steel	11.7×10^{-6} cm/(cm·°C)
ground coat	$10–12.5 \times 10^{-6}$
cover coat	$8–11 \times 10^{-6}$
aluminum	23.5×10^{-6}
cast iron	10.5×10^{-6}

Composite Modulus of Elasticity. The modulus of elasticity of the enamel glass–steel composite system has been shown to lie between the modulus of the glass and that of the metal (5). The composite modulus can be calculated by the following expression:

$$E_c = (E_m - E_e)Q^3 + E_e$$

Table 2. Coefficient of Expansion Factors[a]

Oxide	Factor	Oxide	Factor	Oxide	Factor	Oxide	Factor
SiO_2	0.8	CaO	5.0	SnO_2	2.0	Cr_2O_3	5.1
Al_2O_3	5.0	MgO	0.1	TiO_2	4.1	CoO	4.4
B_2O_3	0.1	BaO	3.0	ZrO_2	2.1	CuO	2.2
Na_2O	10.0	As_2O_5	2.0	Na_3AlF_6	7.4	Fe_2O_3	4.0
K_2O	8.5	P_2O_5	2.0	AlF_3	4.4	NiO	4.0
PbO	4.2	Sb_2O_3	3.6	CaF_2	2.5	MnO	2.2
ZnO	2.1			NaF	7.4		

[a] Summation of (factors × weight percentages) × 10^{-7} = cubical expansion coefficient in vol/(vol·°C) (4). Courtesy of Garrard Press (2).

where E_c = modulus of elasticity of the composite; E_m = modulus of elasticity of the metal; E_e = modulus of elasticity of the enamel; and Q = thickness of the metal per total thickness of the composite.

Residual Compressive Stress. Residual compressive stress in commercial ground-coat enamels varies with enamel thickness as indicated below (6):

Ratio of enamel thickness to metal thickness	Compressive stress, MPa (psi)
0.8	69 (10,000)
0.6	110 (16,000)
0.4	138 (20,000)
0.2	221 (32,000)

Thinner coatings will, other factors remaining equal, yield higher compressive stresses. Higher residual compressive stress in the coating also can be obtained by using enamel glass with a lower expansion coefficient, or a metal with higher expansion coefficient and higher modulus of elasticity.

Maximum Strain. Strain in enamels that leads to failure is on the order of 0.002–0.003 cm/cm. Thinner enamels with their high residual compressive stresses are more flexible and can be strained to a greater degree. Some other physical properties of enamel glass are given in Table 3.

Appearance. Decorative porcelain enamels involve either a one-coat or multiple-coat system; in the latter, the dark cobalt-bearing ground coat is covered with the decorative second coat which affords the desired properties.

The most common cover or direct-on enamel has been white enamel. Whiteness or high diffuse reflectance is called opacity. The white, high-opacity enamel depends on crystalline opacifying agents, such as antimony oxide and zirconium oxide (before 1940) or titanium dioxide, which either remain well-dispersed in the glass during smelting and subsequent firing or are recrystallized from the enamel glass during the firing process. Opacifying pigments (qv) have an index of refraction much different from the 1.50–1.55 range of the glass matrix. The most effective opacifiers are given in Table 4 (7).

Commercial use of titanium dioxide as an opacifier for enamels did not begin until 1946 but recrystallizing white titania enamel is used today in practically all white sheet-steel enameling. Two polymorphic forms of titania, anatase and rutile, may be present in the enamel (see Titanium compounds, inorganic). Anatase is preferred since the anatase crystals are present in correct size range (0.1–0.2 μm) for maximum re-

Table 3. Some Physical Properties of Enamel Glass

Property	Value
density, g/mL	2.5–3.5
hardness (Mohs scale)	5–6
tensile strength, MPa[a]	34–103
compressive strength, MPa[a]	1380–2760
modulus of elasticity, GPa[b]	55–83
dielectric constant	5–10

[a] To convert MPa to psi, multiply by 145.
[b] To convert GPa to psi, multiply by 145,000.

Table 4. Most Effective Opacifiers in Porcelain Enamels

Opacifier	Index of refraction
NaF	1.336
CaF_2	1.434
Sb_2O_3	2.087–2.35
SnO_2	1.997–2.093
ZrO_2	2.13–2.20
TiO_2 (anatase)	2.493–2.554
TiO_2 (rutile)	2.616–2.903

flectance and, therefore, generate the most desirable bluish-white color. Smaller pigment particles provide too much light scattering, thereby giving a more undesirable bluish color; larger particles give a more cream-white color. In contrast to rutile, anatase crystals do not grow or change size with changes in firing temperature.

Spectrophotometer curves for bluish-white cover enamels opacified with oxides of antimony, zirconium, or titanium are shown in Figure 4. The titania enamel shows a characteristic absorption at the violet end of the spectrum.

Color stability of titania enamels can be obtained by adjusting the composition of the glass so that anatase recrystallizes predominantly over a wide temperature range. Figure 5 shows the effect of P_2O_5 content on the relative amounts of anatase and rutile recrystallizing from a titania enamel (7).

The color of cover-coat porcelain enamels can be produced in almost any hue, saturation, and brightness by using clear glass frits milled with colorant oxide pigments (see also Color). A typical mill batch for colored enamels is as follows:

Parts (wt)	*Component*
100	frit (clear alkali borosilicate-type glass)
4	clay (for producing stable suspension)
1/4	bentonite (for supporting suspension)
1/4	sodium nitrite (to disperse flocs)
3	oxide pigment colorant
45	water

Colored enamels may also be prepared by tinting the titania enamels during the smelting operation by the addition of colorant oxides.

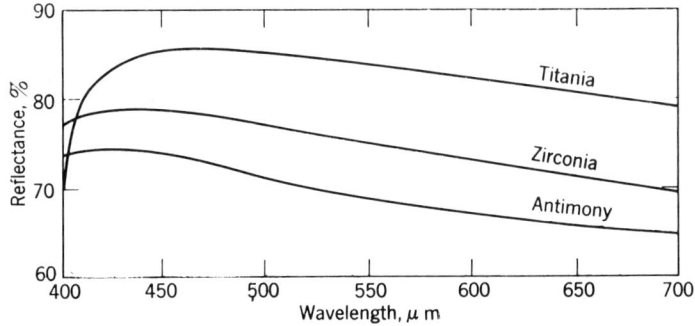

Figure 4. Spectrophotometer curves of titania, zirconia, and antimony enamels. Courtesy of Industrial and Engineering Chemistry (4).

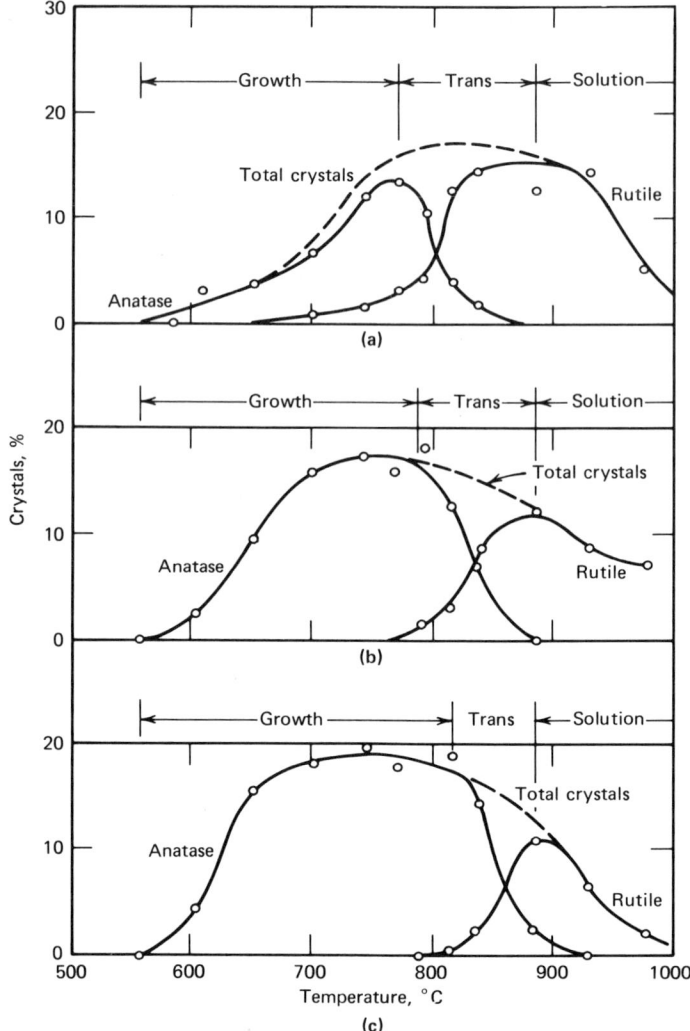

Figure 5. Crystallization of TiO_2 (**a**) in unstabilized enamel, (**b**) in phosphate-stabilized (2% P_2O_5) enamel, and (**c**) in phosphate-stabilized (4% P_2O_5) enamel. Courtesy of the University of Illinois.

Decorative one-coat finishes are also prepared by milling some cover-coat frit with ground-coat frit. This produces a fine speckled coating. Single-coat finishes of ground coat stippled with a white enamel prior to firing are also common.

Titania white enamels for sheet steel and antimony oxide-opacified enamels for cast iron exhibit reflectance values of 75% or higher. Although cast-iron ground coats are not colored, as are the sheet-steel ground coats which have dark cobalt-blue color, the thickness of cast-iron enamels is much greater than that of sheet-steel enamel coats. Dry-process cast iron requires generally two coating cycles to build a smooth and uniform coating and to achieve high reflectance.

Titania enamels have excellent hiding power. One thin coating of 0.05 or 0.08 mm completely masks the dark-colored ground coat and produce a white glossy coating with diffuse reflectance (green filter) of 80% or higher.

Decorating. In the process of commercial porcelain enameling, the decorating and painting steps involve coatings of various colors and textures. Common techniques for decorating include the following: stencil method of spraying an enamel on a stencil-covered surface; brushing the dried, unfired coating using a stencil or similar design; silk screen process (see Reprography); stamping a design using a rubber stamp with ceramic colors as inks, or dusting of a color pigment onto the area stamped with a gum or varnish; decalcomania using ceramic decals; graining, marbleizing, and other transfer processes whereby the designs impressed on an inked rubber cylinder from one flat designed plate are transferred by rolling the cylinder on the enameled surface (cylinders with the designed patterns are also used, after first inking with ceramic colors on a flat plate, then rolling onto the surface to be decorated); stippling or splattering droplets of the slip of a different colored enamel; and mottling to produce granite ware by adding cobalt or nickel sulfates to the ground-coat slip. These additions cause selective rusting and a mottled pattern in the fired finish.

Microstructure and Thickness. Sheet-steel enamels consisting of a ground coat and cover enamels have many bubbles of gas entrapped in the glass layer. Figure 6 is a cross-section photomicrograph of the enamel layer on sheet steel. The grains of ferrite in the cold-rolled steel can be identified in this acid-etched cross section. The ground coat contains more bubbles than the cover coat for three reasons: (*1*) the dissolved moisture in the ground-coat glass reacts with the metal at the firing temperature to produce hydrogen gas (some of this gas dissolves in the steel and some may produce bubbles in the molten glass); (*2*) carbon from the steel forms carbon oxide gases; and (*3*) there are more gas-producing, decomposable materials in the ground-coat mill batch such as carbonates and organic matter in the clay.

Total thickness of sheet-steel enamels consisting of a ground coat and a titania cover enamel is 0.12–0.20 mm. Cast-iron enamel coatings are much thicker, 0.50 mm or greater for dry-process coatings.

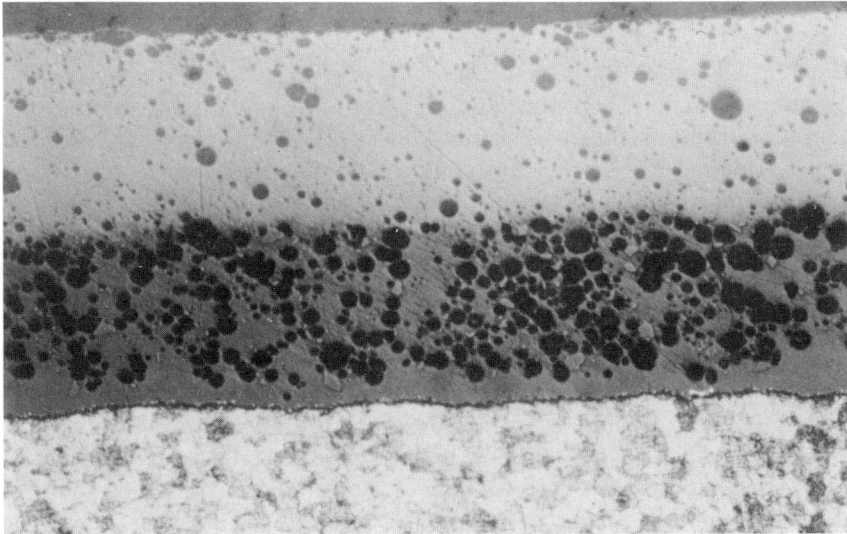

Figure 6. Cross-section photomicrograph of porcelain-enameled sheet steel. Titania cover coat over cobalt ground coat (150×).

One-coat titania enamels directly on steel are successful with certain special steels such as titanium-killed steels and extra-low-carbon steels. The thicknesses of such direct-on white enamels are 0.08–0.13 mm. Carbon or carbides at the surface of the metal form CO or CO_2 gases that produce a gas defect called primary boiling with ground-coat processing. For direct-white enameling, the carbon or carbides in these special steels must be eliminated or inactivated. As noted above, one-coat white enameling using wet or dry processes has been adopted by a substantial number of the industrial producers of sheet-steel porcelain enamels.

Enamel Testing

Abrasion Resistance. Porcelain enamel is the most scratch-resistant and hardest of commercial coatings (see Hardness). This property is used to distinguish between porcelain enamel and organic enamel or painted coatings. The rate of abrasive wear in surface abrasion increases with time, and the subsurface abrasion which follows exhibits a higher but constant rate of wear. Abrasion resistance can be evaluated by the loss of gloss or weight (ASTM C 448-64 Abrasion Resistance of Porcelain Enamels) (1).

Impact Resistance. Tests for impact resistance of porcelain enamels include falling weight tests such as a free-falling ball or a pendulum striking a rigidly held specimen. In such tests successively higher heights of fall are used until a visibly noticeable failure occurs. Factors affecting the impact resistance of porcelain enamels include the following: (*1*) lower modulus of elasticity of the metal, contributes to higher impact resistance; (*2*) the larger the radius of curvature of the article, the greater the impact resistance; (*3*) the thicker the metal and enamel, the greater the resistance to impact (because of the lower modulus of elasticity of the enamel, for equal thickness of the composite, the greater the ratio of enamel to metal, the greater the impact resistance); and (*4*) the physical structure of enamel, eg, excessive bubble content, large crystalline inclusions, and other discontinuities, often contribute to lower impact resistance. Weak bonding of the enamel to the metal also contributes to lower impact resistance.

The size of the fracture after impact failure has occurred is generally independent of the impact resistance. The size of the fracture is larger with a larger radius of curvature, a greater enamel thickness, and a poorer bond of enamel and metal.

Thermal Shock Resistance. Resistance of porcelain enamels to failure by thermal shock was developed for enamels for cooking utensils and other items subjected to high temperatures in service. Thermal shock is experienced by a heated enamel article on quenching with cold water. Thermal-shock tests involve repeated cycles of heating and quenching with water (heating for each successive cycle is carried out at progressively higher temperatures). A visible fracturing, as evidenced by spalling, constitutes a failure.

Water is an especially severe quenchant. Thermal-shock failures with water result from the water vapor entering the enamel layer through small, submicroscopic cracks formed at the instant of shock. The water vapor condenses in the crack and in the bubbles of the enamel near the cracks. On subsequent heating, the vapor from the entrapped water expands to cause spalling of the enamel layer. Other quenchant liquids, such as toluene, oils, and other organic liquids, also cause fine, almost invisible cracks but thermal-shock failures do not result with these quenchants on subsequent heating (ASTM C 385-58).

Thermal-shock resistance is a direct function of enamel thickness. The greater the residual compressive stress in the porcelain enamel, the greater is the resistance to thermal-shock failure. Thin coatings (with their greater residual compressive stress), such as one-coat enamels or the two-coat enamels with a low-expansion titania cover coat, provide excellent thermal-shock resistance.

Resistance to Chemical Attack. The resistance to alkali and acid attack is evaluated on the basis of loss in weight, loss in gloss, or cleanability of the surface.

A spot test with 10% citric acid is used at room temperature for acid resistance of glossy, light-colored enamels. In the spot test, loss of gloss and cleanability are determined in a qualitative manner (ASTM C 282-67). Resistance to boiling 6% citric acid is determined by the loss in weight (ASTM C 283-54). Lower alkali content of the glass yields higher acid resistance. Acid-resistant enamels for chemical service are compositions very high in SiO_2 and TiO_2 content. Alkali resistance is improved with increasing Al_2O_3 and ZrO_2 content. Resistance of enamels to water attack is also introduced in coatings for domestic water heaters (see also Coatings, resistant).

Other Properties and Tests. Physical and chemical properties of porcelain enamels can be evaluated by the following ASTM tests (1): C 282-67 Acid Resistance of Porcelain Enamels (Citric Acid Spot Test); C 313-59 Adherence of Porcelain Enamel and Ceramic Coatings to Sheet Metal; C 614-74 Alkali Resistance of Porcelain Enamels; C 282-67 Boiling Acid Resistance of Porcelain Enameled Utensils; C 538-67 Color Retention of Red, Orange, and Yellow Porcelain Enamels; C 743-73 Continuity of Porcelain Enamel Coatings; C 314-62 Flatness of Porcelain Enameled Panels; C 374-70 Fusion Flow of Porcelain Enamel Frits (Flow-Button Methods); C 346-59 Gloss of Ceramic Materials, 45-Deg Specular; C 540-67 Image Gloss of Porcelain Enamel Surfaces; C 539-66 Linear Thermal Expansion of Porcelain Enamel Frits by the Interferometric Method; C 632-69 Reboiling Tendency of Sheet Steel for Porcelain Enameling; C 347-57 Reflectivity and Coefficient of Scatter of White Porcelain Enamels; C 285-54 Sieve Analysis of Wet Milled and Dry Milled Porcelain Enamel; C 703-72 Spalling Resistance of Porcelain Enameled Aluminum; C 385-58 Thermal Shock Resistance of Porcelain-Enameled Utensils; C 409-60 Torsion Resistance of Laboratory Specimens of Porcelain Enameled Iron and Steel; and C 448-64 Abrasion Resistance of Porcelain Enamels.

Enamel Defects. Deviations from perfect continuity in porcelain enamel coatings and unusual departures from smoothness are described by special terms. For example, roughness is called orange peel; a severe case is called alligator hide. Blisters, pinholes, black specks, dimples, tool marks, and chipping are well-understood terms. Copperheads are defects of iron rust spots in the fired ground coat. Hairlines are defects of a strain pattern originating in the first part of the cover-coat firing and healed in the later stages of firing. Defects may result from accidents occurring at almost every stage of the enameling process. Porcelain enamel terminology is defined in "Standard Definitions of Terms Relating to Porcelain Enamel and Ceramic-Metal Systems," ASTM C 286-73 (1).

BIBLIOGRAPHY

"Enamels, Porcelain or Vitreous" in *ECT* 1st ed., Vol. 5, pp. 718–735, by R. M. King, The Ohio State University; "Enamels, Porcelain or Vitreous" in *ECT* 2nd ed., Vol. 8, pp. 155–173, by A. L. Friedberg, University of Illinois.

1. *1977 Book of ASTM Standards,* American Society for Testing and Materials, Philadelphia, Pa., 1977, Part 17.
2. A. I. Andrews, *Porcelain Enamels,* 2nd ed., Garrard Press, Champaign, Ill., 1961, p. 633.
3. *Porcelain Enamel and Energy,* Porcelain Enamel Institute, Inc., Arlington, Va., 1976.
4. P. Strong and S. Strong, *Ind. Eng. Chem.* **42**(2), 253 (1950).
5. P. S. Wolford and G. E. Selby, *J. Am. Ceram. Soc.* **29**(6), 162 (1946).
6. R. A. Jones and A. I. Andrews, *J. Am. Ceram. Soc.* **31**(10), 274 (1948).
7. R. D. Shannon and A. L. Friedberg, *University of Illinois Bull. Eng. Exp. Sta. Bull. No. 456* **57**(44), (Feb. 1960).

General References

A. E. Farr and G. Carini, "Evaluation of Dry Powder Porcelain Enamel Coatings," *Proc. Porcelain Enamel Institute Technical Forum* **37,** 45 (1975).
A. Lacchia, "Status and Trends of Electrostatic Enamel Powder-Coating Installations in Europe," *Proc. Porcelain Enamel Institute Technical Forum* **38,** 78 (1976).
L. M. Dunning, "A Status Report on Porcelain Enamel Powder Coating Developments," *Proc. Porcelain Enamel Institute Technical Forum* **38,** 37 (1976).
D. R. Dickson and D. R. Larson, "Advances in Porcelain Enamel Powder Frit Materials," *Proc. Porcelain Enamel Institute Technical Forum* **39,** 102 (1977).
W. H. Steenland, "The PEI Energy Usage Survey," *Proc. Porcelain Enamel Institute Technical Forum* **39,** 79 (1977).
D. R. Sauder, "Use of Cold Rolled Steel in Enameling," *Proc. Porcelain Enamel Institute Technical Forum* **39,** 48 (1977).

A. L. FRIEDBERG
University of Illinois

ENCAPSULATION. See Embedding; Microencapsulation.

ENDORPHINS. See Anesthetics; Antibiotics—Peptides: Hormones; Hypnotics, sedatives, and anticonvulsants.

ENERGY MANAGEMENT

By almost any standard the chemical industry is, throughout the world, one of the largest industrial classifications. It is also perhaps the most energy intensive. The chemical industry's dependence on energy is only exceeded by the world economy's dependence on its products.

Table 1 is an estimate of energy use within the United States chemical industry for 1976 (1). The total of 6.3×10^9 GJ/yr (6×10^{15} Btu/yr) represents ca 28% of the energy consumption of the industrial sector in the United States. This energy consumption reflects both the fuels and feedstocks used in the chemical industry. Energy conservation measures discussed in this article refer to fuels only, not feedstocks (see Fuels, survey; Feedstocks).

Even more significant than the gross quantities of energy consumed is the rapidly increasing impact of energy cost on the economics of the chemical industry. In the 1950s and 1960s the chemical industry was undergoing a period of extremely rapid expansion. Energy was universally inexpensive and available. The chemical industry did, however, begin to concentrate in those areas of the United States where energy was most abundant, and the massive amounts of energy involved in the new super-capacity single-train processes mandated energy-oriented designs. Except for these factors energy received very little management attention, for energy costs in many cases were often less than 3% of selling price and other severe management problems inherent in a period of rapid expansion and turbulent technological evolution took precedence. Today energy costs for some of the more energy-intensive products of the chemical industry range as high as 30% of the selling price (2), making energy management a major requirement in the chemical industry.

Energy management should be considered in the broadest possible sense. It is not just a plant energy audit or an energy conservation program. In industry, the term management implies all of the activity involved in the control and coordination of the elements of industry to optimize progress towards a desired goal. Energy must be considered an essential element of industry together with land, labor, capital, and raw materials and must be managed in coordination with the other elements if the industrial economy is to survive.

Energy management is not much different from the management of any of the other elements of industry (see Research management). Many of the concepts im-

Table 1. Estimated Annual Energy Use Within the Chemical Industry, 1976

	10^6 GJ/yr	10^{12} Btu/yr	%
natural gas	1730	1640	27.4
fuel oil	330	310	5.2
coal	370	350	5.9
other fuels	600	570	9.5
purchased steam	190	180	3.0
electricity[a]	3090	2930	49.0
Totals	6310	5980	100.0

[a] Based on average utility heat rate of 10,973 kJ/(kW·h) (10,400 Btu/(kW·h)).

portant to the successful management of materials in the chemical industry apply equally to the management of energy. Energy is not, however, another raw material. It is technically unique both in supply and application, and misunderstanding of the economics of value and cost has caused serious managerial errors.

Available Energy Forms

The dominant energy sources in the chemical industry are fuel and purchased electricity. These are often referred to as basic energy commodities. The use of energy often involves a derived energy commodity. The most significant derived energy commodity for this industry is process steam (qv). Others include compressed air, refrigeration (qv), circulated hot oil, and organic heat transfer vapors (see Heat exchange technology).

Another basic energy source that is often important is the raw material or feedstocks entering the process. If the dominant reactions are exothermal, the energy liberated by the process requires the same energy management consideration as energy liberated by burning purchased fuel. A classic example of this is the ammonia oxidation process for the manufacture of nitric acid (qv). Anhydrous ammonia is burned in air on a platinum catalyst at a pressure of ca 0.81 MPa (8 atm). The reaction is highly exothermic and takes place at temperatures of about 900°C. In modern highly developed nitric acid plant designs, this energy is recovered by gas reheaters and waste heat boilers. The plant not only provides all of its own energy requirements, including a large component of prime energy (shaft work) for the air compressor, but can also export a remarkably large quantity of process steam to other operations.

Another possible energy source derived from the feedstocks or raw material is waste for disposal. If these materials have a positive heating value, they can be a source of energy and must be managed from an energy standpoint (see Fuels from waste).

Energy and Feedstocks Interchangeability. The chemical industry, particularly the dominant petrochemicals portion, is unique in that raw materials and energy resources are, to a large extent, interchangeable. When oil or natural gas is in short supply, this factor exerts considerable economic leverage toward the reduction in use of feedstock alternatives as fuels. In the face of growing alarm about natural gas supplies, many ammonia (qv) reformer furnaces were converted to fuel oil during the mid 1970s to conserve natural gas for use as feedstocks. A second step in this direction, conversion of petrochemical furnaces to coal-firing is being considered for 1978 and beyond (3). The trend toward reserving natural gas and petroleum products for the highest economic end use will continue as long as other energy alternatives are available or can be developed.

Trends and Future Prospects. The highest priority for the chemical industry in the next ten years will be those developments that can extend the real benefits obtained from each unit of energy consumed. These will include conventional energy conservation measures and waste heat recovery. In addition, substantial progress is expected in the application of such technology as the reverse Rankine cycle (vapor recompression, heat pump, etc) and topping and bottoming heat cycles. Existing processes will be modified and refined for maximum energy efficiency. Beyond this, entirely new and more energy efficient processes are being invented for many of the products of the chemical industry, and these can be expected to replace much existing capacity entirely. The demand for highly energy intensive products will be reduced by their higher cost, and the demand for replacement items that can be produced with better energy economy will increase. A strong back-to-nature trend does not appear probable. In fact, the range of synthetic fibers, plastics, and other man-made materials dependent on the chemical industry will probably increase substantially in the years ahead.

Another aspect of the problem is new and expanded sources of energy. Energy source trends in the chemical industry are expected to parallel the rest of the industrial sector. The search for new sources will be concurrent with efforts to improve energy utilization described above. Because of the greater need for new technology in energy source development, actual realization of benefit will, in most cases, be somewhat slower. Some of the expanded, alternative, or new energy sources affecting the chemical industry are discussed below.

Coal. Coal (qv) is not new to the chemical industry as an energy source, but its use in an era of environmental concern does require new and improved technology. A good bit of this is chemical technology derived from the chemical industry. A substantial economic incentive will promote expanding use of coal by the industry, especially at larger plants (4). Coal gasification and coal liquification may increase coal use for clean fuel applications. Fluidized-bed combustion (5–6) of coal is a potentially very important emerging technology that could greatly affect the chemical industry's use of coal both for raising steam and for direct process heating (see Fuels, synthetic; Fluidization).

Nuclear. The greatest impact of nuclear energy on the chemical industry will be as electric power provided by nuclear electric power stations. Nuclear cogeneration may become a significant factor, depending on the overall rate of growth of the nuclear power industry. Such an installation would provide both electric power and by-product process steam (see Nuclear reactors; Power generation).

Solar. Solar energy (qv) is best suited to meeting low level thermal requirements in the chemical industry. Some small quantity but very high level requirements might conceivably be met using heliostats (solar furnaces). The most important applications of solar energy in the chemical industry during the next ten years will probably be in evaporation (qv) and drying (qv) processes for which most of the technology is already available (7).

The significance of solar energy as a component of chemical industry energy supply will depend on cost and availability trends of conventional energy sources and trends in the capital cost of solar devices. Technological advances in the development of low cost photovoltaic cells (qv) could sharply reduce cost for the facilities for direct conversion of solar energy into electric power. This might find some application in the chemical industry though it is doubtful that any such development will have significant impact within the next decade.

Synthetic Fuels. The proposals under consideration for the production of synthetic fuels are virtually too numerous to list (see Fuels, synthetic). Several of these are described in reference 8. The greatest potential for synthetic fuels in the chemical industry, should large-scale use become economically attractive, will probably be their production rather than their consumption. Synthetic fuels are limited by the high capital cost of the process plants; by the energy intensity of some of the processes themselves; by the origin and transportation of base materials; and by the distribution and marketing of the product. Synthetic fuels have greatest potential value for specialized purposes under stringent environmental requirements, or where the base material is plentiful, easy to obtain, and inexpensive. An example of specialized fuel

24   ENERGY MANAGEMENT

requirement would be the potential use of methyl alcohol as a motor fuel for the purpose of reducing objectionable emissions (see Gasoline; Methanol).

Other Sources. These include geothermal energy (qv) (9), wind power, and the harnessing of waves and tides (see Solar energy). The exploitation of these opportunities will be limited by economics and factors of location. Technological advances, nevertheless, can be expected to result in a steady increase in application of these sources. Use by the chemical industry will probably be in proportion to that by other sectors (see also Fusion energy; Hydrogen energy).

Concepts and Techniques of Energy Management

Energy Technology. Successful energy management requires the merging of managerial and administration skills with a very high level of technical competence, particularly with respect to practical energy technology. By itself, energy technology is a very wide field requiring understanding of basic thermodynamics, heat transfer, combustion chemistry, high speed rotating machinery, fluid flow, systems analysis, and materials of construction. True competence in energy technology also requires extensive practical experience, demonstrated sound judgment, as well as the ability to understand and apply the theoretical concepts involved.

Two important concepts that should be understood by industrial energy managers are the first and second laws of thermodynamics (see Thermodynamics). The first law is the law of conservation of energy. This states that energy can neither be created nor destroyed. Or, perhaps more understandably, that the total of energy entering a device, system, or industrial plant must equal the total of all of the energy rejections. For a chemical manufacturing plant, or any other industrial plant, the energy consumed, that is the energy rejected to a heat sink, is the difference between the total energy represented by the fuel, electricity, and raw materials entering the plant and the energy represented by the finished products and waste materials leaving the plant. This relationship can be expressed as:

$$Q_r = (Q_m + Q_e + Q_f) - (Q_p + Q_w)$$

Q_r = energy rejected; Q_m = energy (mass times enthalpy plus mass times heating value) for raw materials or feedstocks; Q_e = energy equivalent of electric power entering; Q_f = energy in fuel entering; Q_p = energy in products leaving; and Q_w = energy in waste materials leaving.

For a factory manufacturing automobiles, furniture, or any other product in which the manufacturing process involves no chemical transformations (or even processes in which there is no change in the chemical energy between the raw materials and the finished products), all of the energy entering as fuel and electricity is wasted to the atmosphere or to a cooling system designed for energy rejection. This concept is illustrated in Figure 1. In petrochemicals, petroleum refining (qv), pulp (qv) and paper (qv), and quite a few other industrial plants, there are energy resources other than purchased fuel and electricity that must be accounted for. The total energy can be determined by direct application of the equation above. This is not just a theoretical exercise but a practical and useful determination that includes not only the commodities purchased as energy but also the energy opportunities or liabilities represented by the raw materials or feedstocks, by the products and by-products, and by the waste materials. For example, in petroleum refining, the total volume of the crude

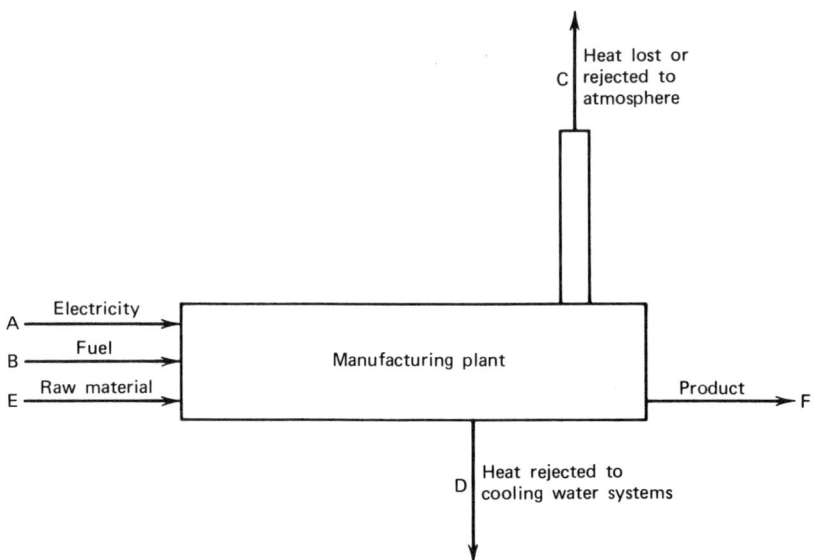

Figure 1. Energy balance in a manufacturing plant, valid only for cases in which there is no change in the enthalpy or chemical energy between the raw material and the product streams. A + B = C + D.

charge minus the total volume of all products is sometimes referred to as refinery shrinkage. In a typical refinery, a large portion of the energy used in the refining process is low market-value gases and residuals derived from the process. The crude charge times its heating value minus the summation of all of the products times their heating value yields an energy shrinkage equaling the amount of such derived energy that must be accounted for in evaluating the effectiveness of energy utilization.

The second law of thermodynamics states that energy will always degrade to a lower level. Or, as it is sometimes stated, energy only runs downhill. This is a very important concept since energy derives its ability to do work or to produce a useful result only by virtue of its level above an available heat sink. Putting energy into an engine, an evaporator, or distillation column will accomplish nothing (other than the possible eventual destruction of the device) unless energy is continuously removed at a lower level. No power can be produced from a steam turbine without an exhaust opening. Nor can power be produced by the turbine unless the steam pressure at the exhaust is lower than that at the throttle. Likewise, a distillation column will not operate unless the temperature of the heating medium in the reboiler is higher than the temperature of the cooling medium in the vapor condenser (see Distillation).

The second law is of vital importance to practical energy management. Energy is a fleeting, intangible thing that has economic value only because it can perform useful and necessary tasks. This ability is solely a function of the energy level above an available heat sink. Prime energy or shaft work (sometimes called mechanical energy) is the highest form of energy because it is the most universally useful. Per joule, it is also usually the most expensive to produce. From a practical standpoint, electricity is usually considered to be prime energy because it can readily be converted to shaft work using an electric motor and because most of the electricity consumed by industry is used to power motors.

The energy management implications of the first and second law of thermodynamics can be stated briefly as follows: all energy used in an industrial plant is ultimately wasted to the atmosphere or to a heat rejection system. The objective of energy management is to ensure that the minimum amount of energy is expended and that maximum benefit is obtained from every unit of energy throughout its unavoidable path of degradation from the highest practical release level to the lowest available rejection level.

Economics of Energy Levels. The value of energy is related to both its cost and its use. This leads to some complex and sometimes confusing economic relationships. The problem is simplified somewhat if energy value is always thought of as being related to energy level. This is reasonable since the amount of work or other benefit that can be derived from energy is a function solely of its level. For instance, steam at 10.34 MPa (1500 psia) and 538°C has an enthalpy of 3462.5 kJ/kg (1490.1 Btu/lb). This energy is far more valuable than an equivalent number of heat units in saturated steam at say 68.9 kPa (10 psia) because the steam at high temperature and pressure can be expanded in a steam turbine to produce high value prime energy and still leave the exhaust of the turbine with enough heat at a usable level for process or space heating.

The most economically defensible method of determining value is based on the cost of the next available energy alternative. In the example cited above, the value of the exhaust steam would be the cost of the high pressure steam less the net cost of providing the power developed by the turbine with an electric motor instead. This concept, of course, applies only if the process chemical plant is expanding essentially all of its process heating steam from the high pressure level through steam turbines. The shaft work developed in this way is often referred to as by-product power. Such processes are referred to in the more recent literature as dual energy use systems and cogeneration (10). Since the heat in the exhaust is required for process use and would be provided directly from a low pressure boiler if it were not obtained from the turbine, the only heat actually charged to the turbine is the thermal (joule) equivalent of the shaft work produced. By-product power obtained in this way represents a considerable energy economy. Typically, its cost is 20–40% of the cost of shaft work from an electric motor using purchased electricity. The exact cost advantage depends on the purchased electric power cost and fuel cost at the specific plant location. A typical comparative example is diagramed in Figures 2 and 3 (see Power generation).

In most cases, the subatmospheric condensing turbine should not be used in an industrial plant. Condensing machines of the size that would normally be found in a chemical plant and operating under usual conditions reject as much as 80% of the throttle steam energy to the condenser. A public utility, operating large efficient units at super-critical throttle conditions with several stages of reheat and five or six stages of regenerative feed preheat, can do a far better job generating power with condensing equipment. One instance where subatmospheric condensing turbines might be efficient in an industrial plant is in a highly exothermic process from which large quantities of waste-heat steam at useful power levels can be extracted and where there is a smaller demand for process heat. In this case, the best use for the surplus steam is often a condensing turbine (11).

Capturing high-value prime energy as energy degrades from a high to a lower level is extremely important in process energy economics. Indeed it is the key to securing maximum benefit from energy as it is degraded in use. There are, however, a family

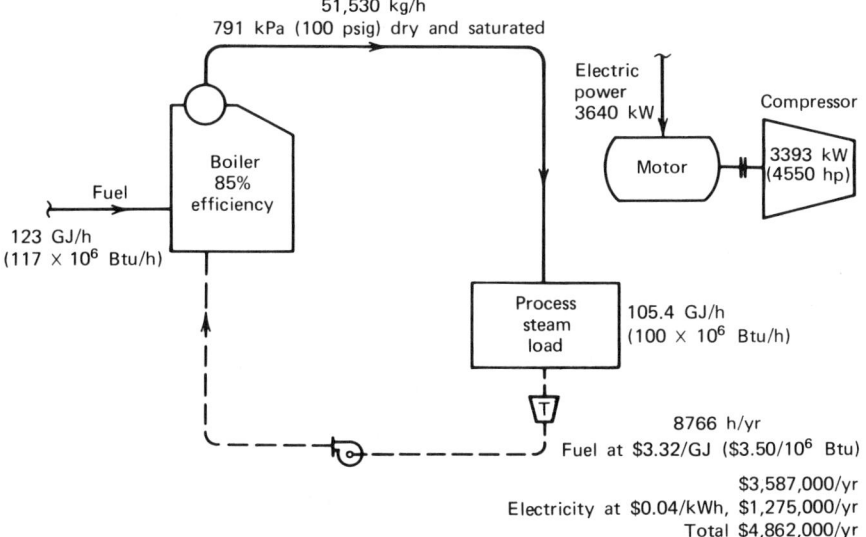

Figure 2. Diagram showing economics of a dual energy use system using purchased electric power (compare with Fig. 3).

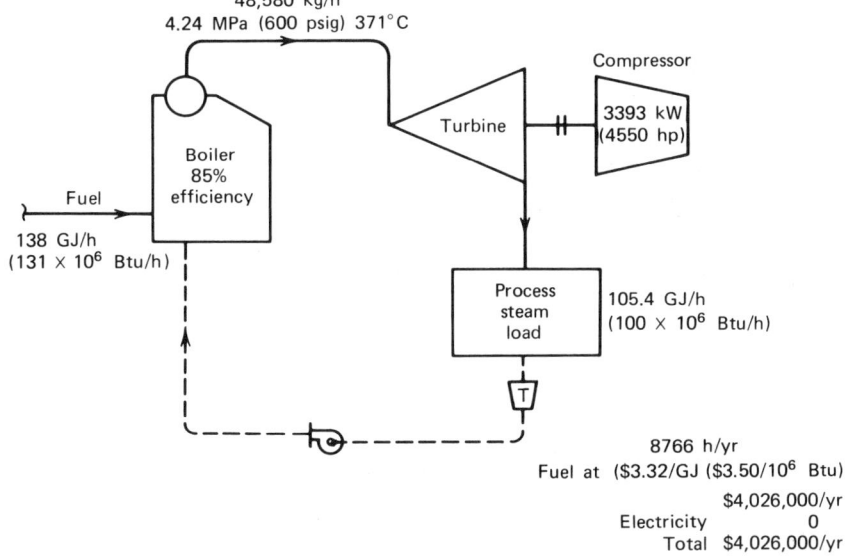

Figure 3. Diagram showing economics of a system using turbine power (compare with Fig. 2).

of largely unexploited energy concepts that could prove even more economically significant in the future. This is the actual recycling of energy by applying relatively small amounts of shaft work to restore the level of the energy—the basis of the heat pump or vapor recompression. In Figures 4 and 5 the economics of vapor compression evaporation are contrasted with those of conventional single-effect steam evaporation. Other similar applications include vapor-compression distillation, vapor-compression

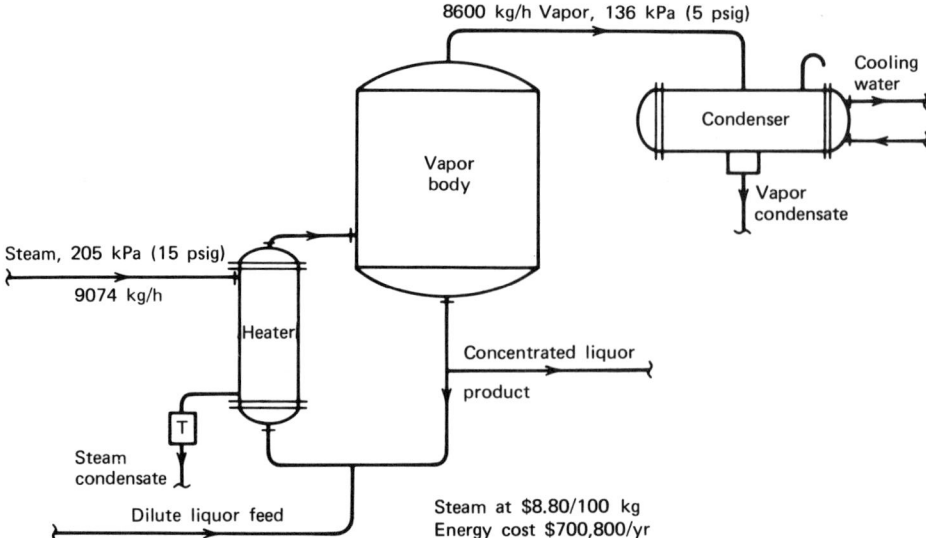

Figure 4. Single effect evaporator on steam.

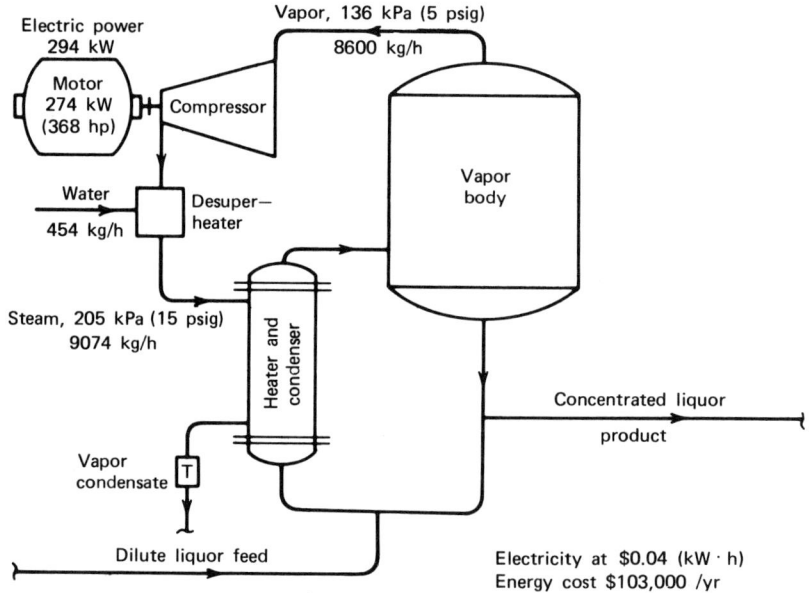

Figure 5. Vapor compression evaporator.

vacuum crystallization, and the so-called organic heat pump. In the latter device, an organic fluid such as a halogenated hydrocarbon refrigerant is vaporized in a heat exchanger by a low-level waste heat source. The vapor is then compressed and the high pressure vapor condensed in another heat exchanger against the higher level heat requirement. An application of the organic heat pump is diagramed in Figure 6. The operating principle is identical to that of the vapor-compression refrigeration cycle

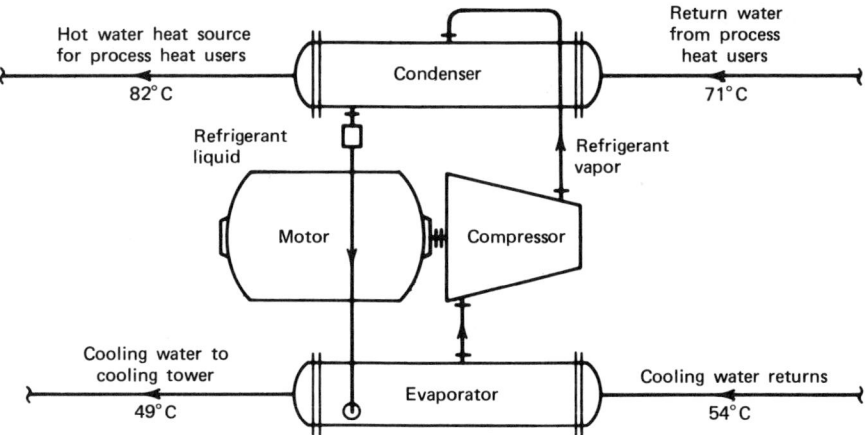

Figure 6. Application of organic heat pump.

except that the heat pump operates with both the evaporator and the condenser at a higher energy level with the intent of applying heat to the condensing medium rather than removing heat from the evaporating medium (see Refrigeration). Vapor compression or heat-pump applications are evaluated by the coefficient of performance (COP). The COP is the ratio of the amount of heat delivered to the heat use divided by the thermal (joule) equivalent of the work required to raise the energy level. The COP is a function of the difference in energy level between the heat source and the heat use; the amount of heat transfer surface involved; its effectiveness; and the efficiency of the compression device. To be useful for an application, the COP must usually be substantially greater than the cost of shaft work (purchased electricity) in energy units divided by the cost of thermal energy (fuel or steam) in the same energy units.

The heat pump has gained some prominence in heating, ventilating, and air conditioning (qv) work in residential, commercial, and institutional applications. In these applications, the same device serves as an air conditioner in the summertime and a heater in the winter. Large-scale applications of vapor recompression and the heat pump in industrial processes are known but not currently widespread. Considering the very high COPs of some industrial applications, the vapor compression concept is thought to hold exceptional promise as a means of saving industrial energy.

In an industrial process plant, the heat source for the reverse Rankine cycle or heat pump is usually a process or service heat rejection. As such, they are heat recovery applications. Alternatively, the low level heat source for the heat pump can be solar collectors. These are called solar-boosted heat pumps and have a much better COP than a space heating application for which ambient air is the heat source.

Some chemical process plants have large quantity low- and intermediate-level heat rejections but insufficient use for this heat at a level that permits direct recovery or heat-pump recovery with a reasonable COP. In such cases, the low temperature Rankine cycle can sometimes be economically employed to gain useful shaft work and lower the final heat rejection level. If the low- or intermediate-level heat rejection is in the form of hot water, low pressure steam, volatile hot process liquid, or process vapor, it may be practical to use the material directly as a vapor-phase working fluid by expanding it through a turbine and condensing it with cooling water at a lower pressure. When it is not feasible to use the heat-containing material directly as a working fluid, an intermediate working fluid such as a halogenated hydrocarbon refrigerant or another organic material is employed.

Waste Heat Recovery. The term waste heat recovery has become important in the lexicon of energy management even though its meaning is imprecise. As noted, some energy must be rejected (ie, wasted) before work can be done or other beneficial results achieved. The real effect of heat recovery is to lower the final energy rejection level so that additional benefit can be obtained before the energy is ultimately wasted.

A broader term, energy recovery, includes power recovery from high pressure liquids and gases that must be decreased in pressure, and the recovery of energy from combustible waste materials (waste fuel utilization) as well as heat recovery. Some examples of heat recovery found in chemical plants include:

Boiler Economizers. Heat exchangers that recover some of the heat in boiler flue gases to preheat the boiler feedwater.

Furnace Convection Sections. Heat exchangers added to industrial process furnaces that use flue gas heat to preheat incoming process material or to provide part of the heat for another process.

Combustion Air Preheaters. Flue-gas-to-air heat exchangers that recover heat from flue gases to heat the combustion air.

Product-to-Feed Heat Exchange. Heat exchangers that cool the product of a thermal process by preheating the raw material feed to that process (see Heat exchange technology).

There are literally hundreds of other examples that could be named. With the recent marked increases in energy costs as a proportion of total chemical production costs, heat recovery has become not a desirable feature but an absolute economic necessity for competitive operation of many processes. It is not enough simply to recover heat, the quantity of heat recovered must be maximized, and the recovered heat must be used in a way that gives maximum benefits.

Individual chemical process units and entire plants consisting of many units must be thoroughly and systematically examined for heat recovery opportunities, and every effort made to exploit these opportunities to the fullest practical extent. There are three essential and interrelated ingredients for successful heat recovery: (*1*) a source of waste heat at a high enough energy level and in a quantity to be economically usable; (*2*) a use for this energy at a low enough level and in a quantity reasonably consistent with the source; and (*3*) a technically and economically practical means of conveying the heat from the source to the use.

All heat rejections are potential heat recovery sources. The practicality of the source is a function of the cost of energy and the cost of the recovery equipment or system. For a specific source, the first criteria are energy level and energy quantity. Table 2 illustrates a way in which heat rejections can be cataloged as a first step toward identifying heat recovery opportunities.

Similar criteria and selection procedures apply in finding appropriate uses for recovered heat. In this case, every heat entry has potential for heat recovery. Frequently, finding appropriate uses for recovered heat is far more difficult than finding sources. Rather extensive modifications of the energy-using equipment are sometimes required to permit use of recovered heat to replace the original energy source. This

Table 2. Process Area No. 4: Heat Rejections with Initial Process Temperatures of 93°C or Higher

Heat exchanger no.	Name	Process flow, kg/h	°C In	°C Out	Approx. heat duty GJ/h[a]
250-21	material A column condenser	6497	152	129	2.52
250-28	heavy residue cooler	454	166	65	0.07
250-31	finishing column bottoms cooler	2446	184	65	0.61
250-33	finishing column condenser	15331	165	146	5.85
250-40	material B stripper condenser (chilled water)	1039	105	17	0.91
260-44	stripper bottoms cooler	3117	105	56	0.34
280-17	1st stage material C condenser	2581	105	43	2.01
280-25	2nd stage material C condenser (brine cooled)	299	105	−4.4	0.18
310-94	material D condenser	3103	133	49	3.35
340-02	material D distillation column condenser	5018	111	56	3.26
340-05	recycle material E condenser	278	151	146	0.16
	Total				19.26

[a] To convert GJ/h to 10^6 Btu/h, divide by 1.054.

might include additions of large amounts of heat transfer surface, revised piping configurations, and operation of the equipment at lower pressures or even under vacuum. Returning recovered heat to the item of equipment or unit from which it came is ideal but should not be considered as a requirement.

Other factors that must be considered in the selection of heat recovery sources and uses are corrosion, abrasion, fouling, and safety. These factors severely limit the practicality of certain heat recovery opportunities. It should not be assumed, however, that because these problems exist they cannot be overcome. In many cases the sheer magnitude of the potential cost savings provides adequate incentive for overcoming special problems and limitations. Another important consideration affecting the economic practicality of heat recovery, despite the absence of technological problems, is the matter of utilization and scheduling. Space heat, for instance, is an excellent low-level energy requirement that is frequently satisfied using recovered heat. The source of the heat, however, can only be used during the winter space-heating season which at times makes it difficult to justify recovery.

The third essential ingredient of heat recovery is the means of conveying the heat from the source to the use. Some of the equipment and systems employed are described briefly below.

Direct Mixing, Contact Condensing, and Recycling. An example of direct mixing is the recovery of heat from steam condensate by returning the condensate to the boiler plant and mixing it with cold make-up water that then enters the boiler feed preheating system. In contact condensing, heat is recovered from a vapor and transferred to a liquid by contacting the vapor with a cooler liquid in a spray chamber or packed tower. In a dryer, it is sometimes practical to return a portion of the vented drying medium to the dryer inlet, thus recovering sensible heat by recycling (see Drying).

Conventional Heat Exchangers. These include shell-and-tube heat exchangers, plate exchangers, and various tubular and plate-coil arrangements (see Heat exchange technology).

Regenerative Heat Exchangers. In this type of exchanger, a thermal mass is first heated by direct contact with the high-temperature heat recovery source stream. The thermal mass is then placed in contact with the use stream where it gives up heat. Two examples of the regenerative heat exchanger are the rotary regenerative gas-to-gas heat exchanger known as the Ljungstrom preheater and the two-chamber brick checker-work regenerators used in metallurgical and glass furnaces for preheating combustion air.

Heat Pipe. The heat pipe (12) is a space-age by-product with substantial potential for heat recovery applications. The device is a pipe sealed on both ends and partly filled with a vapor–liquid phase heat transfer medium. When heat is applied on one end of the pipe, the medium is vaporized and the vapor moves to the other end. There it is condensed by the removal of heat to the heat use. The liquid then returns to the heat intake end by gravity or by capillary wicking (see Heat exchange technology, heat pipe).

Liquid Run-Around Systems. Liquid run-around heat transfer systems (Fig. 7) make it practical to supply recovered heat to a use that is widely separated from the source without the expense of extensive duct work and fans. They are applicable in gas-to-gas or gas-to-air heat recovery.

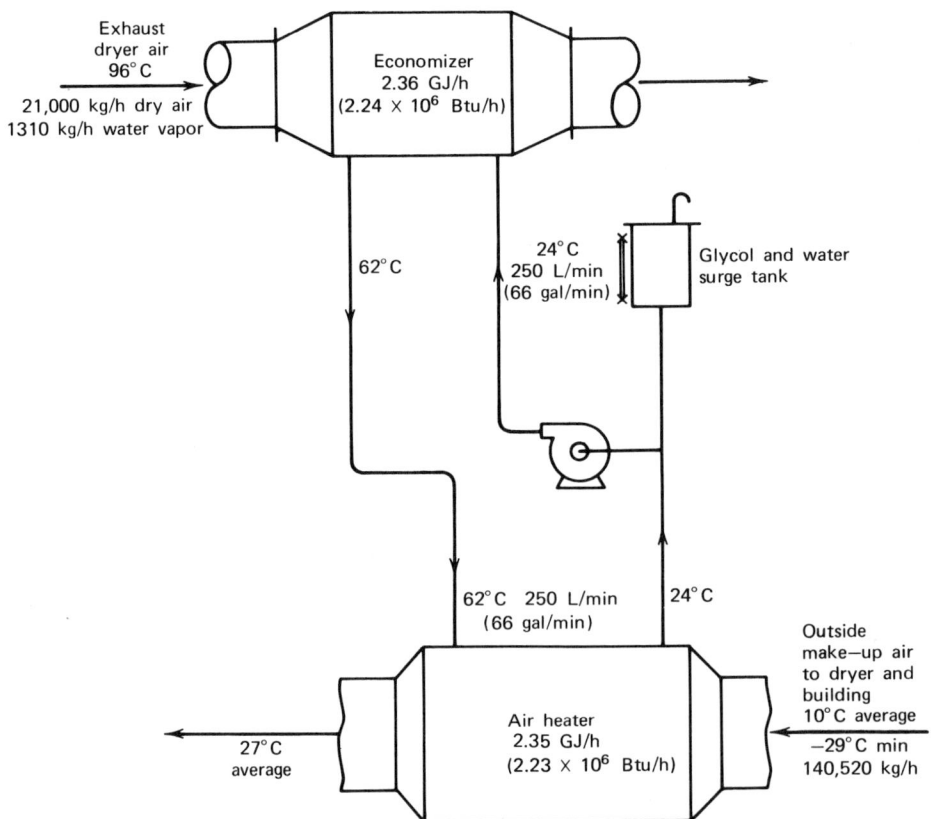

Figure 7. Air-to-air heat recovery in a liquid run-around system.

Waste-Heat Boilers. In most chemical process plants, the steam system is the major and often the only integrating energy system. Recovering waste heat by generating steam makes the heat usable in any part of the plant served by the steam system and results in a real decrease in boiler fuel use. Waste-heat steam can be generated from hot gases, hot liquids, or condensing process vapors. There are many waste-heat boiler designs available to meet various types of applications in the chemical industry (13).

Energy Balances. An energy balance is a tabular or diagrammatic summary of all of the energy sources and all of the energy rejections for a single item of process equipment, for a process unit, or for an entire manufacturing plant. Table 3 is a tabular balance for a propane-fired dryer. The energy balance is a very useful tool for analyzing an industrial operation for energy conservation opportunities. By defining the thermodynamic and heat transfer relationships mathematically, the energy balance can be developed into a mathematical model of the process. Operational changes, system configuration alterations, and equipment changes can then be synthesized and the effects on energy requirements or benefits evaluated. Programming the energy balance model on a digital computer greatly facilitates its use as an analytical tool (see Computers).

As noted, the steam system is the dominant integrating energy system in most chemical process plants, and the steam balance is the most important plantwide energy balance. Other important plantwide balances include the fuel balance, the cooling water balance, and the compressed air balance. Only the steam balance is discussed here but many of the principles and concepts apply equally to the balancing of other utility energy commodities.

Steam derives its value from only two important uses. First, it acts as a heat transfer medium conveying and distributing thermal energy from the heat released

Table 3. Product Dryer Analysis Heat Balance, As Found

	kg/h	Temperature, °C	kJ/kg[a]	Energy, MJ/h[b]
Inputs				
fuel, C_3H_8	130	16	50254	6553
combustion air	6,817	16	15.6	106
secondary air	14,846	16	15.6	232
air in-leakage	4,289	16	15.6	67
water with product	1,731	49	204	354
dry product solids	4,478	49	55.64	249
Totals in	32,291			7561
Outputs				
water vapor from product	1,445	74	2632	3808
water vapor from combustion of H_2	212	74	2632	560
air and combustion products	25,870	74	74.61	1933
Subtotal to bag house	27,527[c]			6301
dry product solids	4,478	122	102.1	458
water with product out	286	122	510.4	146
radiation and convection losses				656
Totals out	32,291			7561

[a] At 0°C; to convert kJ/kg to Btu/lb, divide by 2.32.
[b] To convert MJ/h to Btu/h, divide by 1.054×10^{-3}.
[c] Air to bag house: rh, 24.5%; dew point, −83°C.

by burning fuel in the boiler to the various process and service requirements for heat. Second, it functions as a working fluid. Steam expanding from a higher to a lower pressure through a turbine or engine produces usable shaft work.

The first of these functions is, in most cases, the basic reason for having a steam system in a chemical process plant. The second is a potential additional benefit that is frequently incorrectly or inadequately exploited because similarities between the process steam generation and distribution system and the utility steam power-station heat/power cycle are not properly understood (14).

A steam balance is a rigid accounting of how all steam is generated and ultimately used at a plant. It shows each process or service requirement for steam, including the unavoidable losses incurred in getting the steam to the use. The use of steam as a working fluid to develop prime energy is also identified. A good steam balance is an indispensable energy management tool.

There are certain important limitations that restrict the usefulness of a steam balance if they are not understood. A steam balance depicts the steam flows and their interrelationships as though they actually existed at a given point in time. In fact, the steam balance is based on average or typical data for the various components. The steam flows that exist in each branch of a complex steam system on a cold weekday morning in January are quite different from the patterns existing on a Sunday in July. Neither matches the annual average steam balance flows.

Some large continuous-process industrial plants that spend more than 30 million dollars a year for boiler fuel continuously monitor hundreds of steam flows and feed the data to a computer that displays an instantaneous steam balance. This approach is not practical for many smaller plants, but it does illustrate the importance of steam-balance data.

In general practice, there are two ways of presenting steam balance data. These are schematic and tabular. For both types of presentation, the energy level technique is very helpful in drawing attention to the importance of temperature and pressure. In a schematic balance, horizontal lines are drawn across the page representing nominal pressures in the steam system. The various items of steam-using equipment and steam-system devices are entered between the lines with all individual flows shown on vertical downward lines (Fig. 8). In the tabular energy level balances, each nominal pressure level is shown as a double column for supply and use with the highest pressure at the left of the page. The lower pressures are in descending order to the right. Table 4 contains the same data as Figure 8.

Steam balances are also classified as either mass balances or mass and energy balances. The mass and energy balance is more accurate and more useful in some applications since it recognizes that the energy content in a unit mass of steam changes as it passes through the steam system, particularly when it is expanded through a turbine or engine. The mass and energy balance of complex industrial steam systems can be programmed on a digital computer to become a working model of the system. These balances are extremely useful tools for developing a process plant steam system with minimum energy use and maximum benefit from all energy expended.

Energy Management Programs. An energy management program is a broad coordinated plan for providing, on a continuing basis, all of the managerial information required to achieve the economic minimum energy cost. To be effective the program should provide: thorough and detailed monitoring of energy use and its benefits; accurate detailed knowledge of true energy costs and an understanding of energy-com-

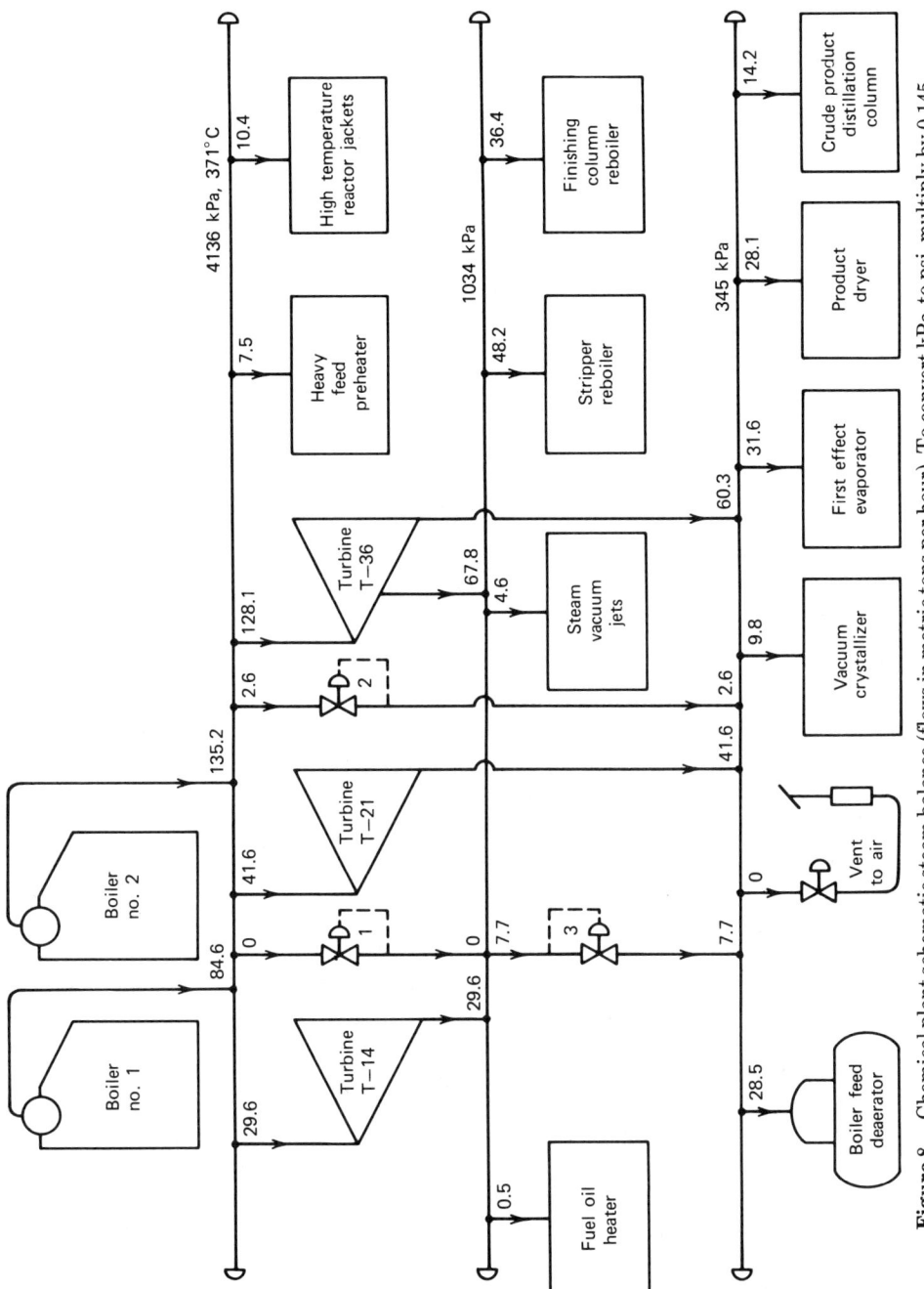

Figure 8. Chemical plant schematic steam balance (flows in metric tons per hour). To convert kPa to psi, multiply by 0.145.

Table 4. Chemical Plant Tabular Steam Balance, Flows in Metric Tons per Hour

	4136 kPa[a], 371°C		1034 kPa[a]		345 kPa[a]	
	Supply	Use	Supply	Use	Supply	Use
boiler no. 1	84.6					
boiler no. 2	135.2					
turbine T-14		29.6	29.6			
turbine T-21		41.6			41.6	
turbine T-36		128.1	67.8		60.3	
pressure reducing valve no. 1		0	0			
pressure reducing valve no. 2		2.6			2.6	
pressure reducing valve no. 3				7.7	7.7	
heavy feed preheater		7.5				
high temperature reactor jackets		10.4				
fuel oil heater				0.5		
steam vacuum jets				4.6		
stripper reboiler				48.2		
finishing column reboiler				36.4		
boiler feed deaerator						28.5
vent to air						0
vacuum crystallizer						9.8
1st effect evaporator						31.6
product dryer						28.1
crude product distillation column						14.2
Totals	*219.8*	*219.8*	*97.4*	*97.4*	*112.2*	*112.2*

[a] To convert kPa to psi, multiply by 0.145.

modity price trends; expeditious identification and evaluation of energy-saving opportunities; coordinated selection of the energy-saving actions that will optimize plantwide results; rapid implementation of selected action items; and continual auditing of results and other measures to ensure that progress is sustained.

Effective monitoring of energy use involves much more than just noting the gross monthly fuel and electricity consumption. As a minimum, all basic and derived energy commodities should be metered to each major functional subdivision of the chemical plant. This provides for an effective system of end-user accountability and also gives essential information on the distribution of energy to various plant operations and the form in which it is distributed. Achieving adequate use monitoring frequently requires additional meters and sometimes also wiring and piping system modifications to enable metering that is compatible with organizational divisions.

Energy cost is not simply a matter of a price times a quantity. The many complex economic and technical concepts involved must be understood by plant management and those responsible for administering energy management programs. The economics of energy levels was discussed above. Another important concept of energy economics is incremental cost. Nearly all economic considerations involving energy should be based on incremental cost rather than on average cost or total allocated cost. The incremental cost of an energy commodity is the true net change in cash flow resulting from an incremental change in the use of that energy commodity. The total allocated cost of steam, eg, includes the cost of fuel plus the cost of maintenance, operations, supervision, and depreciation on steam production facilities investment. The incremental energy cost, however, generally is only the change in fuel cost that results from a given change in steam consumption. The incremental cost of steam is usually higher than the average fuel cost; it is the cost of producing steam in the least efficient boiler that must be operated burning the most expensive fuel that must be used. For example, if a plant has a combustible by-product waste material in sufficient quantity to generate half of the steam required and the rest of the steam is produced by burning purchased fuel oil, then the incremental cost of steam is based on the cost of the fuel oil. The incremental cost of electricity, on the other hand, is usually slightly less than the average cost. This is because most utilities have descending rate scales for high volume industrial customers.

The identification, evaluation, and selection, of energy savings opportunities and their implementation are the central objective of an energy management program. The chemical industry is so varied that there are literally thousands of energy savings opportunities, of which perhaps one to fifty might apply at a specific plant. Some of the opportunities require capital investment for implementation. In others, savings can be achieved by changes in operating and maintenance methods. The savings items that do not require capital investment should not be considered necessarily to have a low implementation cost. Frequently these items are more difficult and costly to implement than capital investment savings. They may involve substantial increases in operator attention, costly operator training, increased supervision, testing and engineering study, and costly remedial or sustaining maintenance effort.

Reducing energy cost is neither easy nor cheap. Problems are encountered and many factors conspire to discourage energy cost-control efforts. Sometimes good energy conservation projects are implemented then by-passed and abandoned at the first sign of corrosion, control problems, or an assumed adverse effect on product quality or production rate. These problems can often be avoided by careful design of the installation. The problems that still occur can nearly always be corrected at a cost well within the economic incentive limits.

In any case, constant attention and effort are required to ensure that energy conservation progress is sustained. A program to test and repair steam traps can involve much effort and cost and produce sizable energy savings. It the traps are then forgotten, however, all of the gains may erode in less than one year. The energy management program must include effective provisions for continual auditing and monitoring of results. An energy budgeting and end-user accountability system that is well designed and conscientiously applied by plant management is among the best ways of ensuring continued progress and sustained results (15) (see Maintenance).

Energy Systems and Equipment

An energy system is a combination of devices and connecting links for the purpose of converting energy from one form to another, transporting the energy from one place to another, and applying the energy to the final use. The steam system, the dominant process chemical plant energy system, typically consists of the steam boilers that convert the chemical energy of the fuel into potential and thermal energy in the high pressure, high temperature steam. Steam turbines convert part of the energy in the steam to usable shaft work. Steam distribution systems consisting of piping mains and laterals with block valves, control valves, and other accessories distribute the steam to the various points of process heat requirement. In these heaters the steam is condensed releasing its latent heat. Condensate flowing from the equipment is controlled by steam traps or other devices and flows through the condensate return system back to the boiler plant. All of these components acting together constitute the plant steam system.

The plant electrical system is another very important energy system. The electrical system consists of the utility company's service entry substation, in-plant generating equipment (when used), primary distribution feeders, secondary substations and transformers, final distribution cables, and various items of switch-gear, protective relays, motor starters, lighting control panels, static capacitors, etc. Other energy systems frequently present in process chemical plants are the compressed air system, the fuel distribution system, the cooling water system, various types of refrigeration systems, and a range of systems that distribute energy via pressurized hot water or specialized heat transfer fluids. Cooling and refrigeration systems are energy systems and must be treated as such in energy management considerations even though their function is to facilitate the rejection of energy rather than its delivery (see Refrigeration).

The overall energy efficiency of a plant depends heavily on the efficiency of the individual energy systems. Some losses are inherent in all energy delivery systems; in most cases these losses cannot be completely eliminated but very substantial and economically attractive savings can be realized by minimizing them. For the steam system, the dominant losses are boiler flue gas losses, steam leaks, steam vents, hot condensate losses, and thermal losses through uninsulated or inadequately insulated piping and equipment surfaces. All of these can be controlled (see Steam).

Electrical distribution system losses are usually less significant than those from the steam system. An electrical system that is poorly maintained or overloaded will, however, cause direct energy losses. Correcting the deficiencies will produce immediate energy savings. The justification for correcting these deficiencies, however, is mainly improved reliability and safety rather than energy economy. This is not to suggest that electric power savings are not a significant part of energy savings; electric power conservation efforts are a very important and very fruitful part of any energy conservation program. Most of the savings, however, result from decreases in the amount of electricity used and improvements in the way it is used (load factor, power factor, etc) rather than from the reduction of distribution-system energy losses.

In the sections that follow, some of the most significant items of energy system equipment found in chemical manufacturing plants are briefly discussed in relation to energy economy.

Boilers and Process Furnaces. Fuel-fired boilers (16) and process furnaces absorb the energy released from burning fuel to generate steam or to provide high level heat directly to a process fluid (see Furnaces, fuel-fired). The most significant energy losses in this process are the latent and sensible heat in the flue gases discharged to the stack. The sensible heat losses can be minimized by the use of fuel-burning equipment that permits low excess air operation and by careful control of the fuel–air ratio to minimize excess air (see Burner technology). Latent heat losses are largely a characteristic of the fuel being burned. The hydrogen component of a hydrocarbon fuel forms water vapor in the combustion products. Because it is impractical to condense this water vapor, the equivalent heat of vaporization is lost energy. Flue gas losses are minimized by the application of sufficient heat transfer surface and heat traps (economizers, air preheaters, etc) to decrease the final flue gas rejection temperature. When fuels containing sulfur are burned, the final exit flue gas temperature is usually not permitted to go below about 149°C because of the severe problems resulting from sulfuric acid energy used in the chemical industry goes to just three process operations: distillation (qv), drying (qv), and evaporation (qv). Not only are these operations energy intensive, they are also susceptible to considerable energy use reduction. Ironically, all three of these processes, though essential to the chemical industry, are not chemical but physical operations. All involve separation through vaporization without any change in the chemical energy of raw materials or products. This means that all the energy applied is degraded and accounted for as losses or intentional heat rejections. Energy for distillation, drying, and evaporation is discussed below.

Distillation. Figure 9 is a diagram of a typical distillation column. Most distillation operations are continuous, although batch distillation is still employed in some segments of the chemical industry. In distillation, the condenser heat rejection load is usually a close approximation to the reboiler heat requirement. The only corrections are material enthalpy changes between feed and outputs, and the radiation and convection losses. Heat application to the reboiler is most often from a plant utility steam system. Direct-fired reboilers are common on large stills in the petroleum industry. Not infrequently these are heated by a waste heat source.

For practical energy conservation, each distillation column must be examined both individually and in context with the rest of the process. In designing a new more energy efficient system, the specific requirements of the process, both in respect to

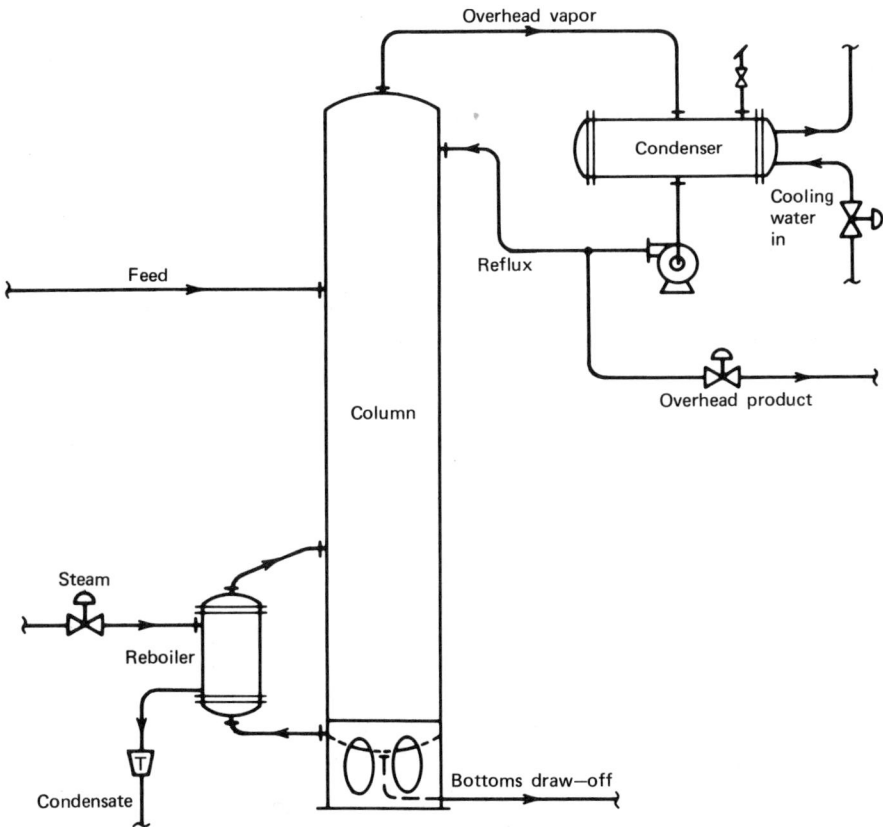

Figure 9. Typical distillation column.

the separations required and the quantity of materials handled, must be considered. Distillation seldom constitutes an entire process, and the upstream and downstream components must be integrated into the energy system. For distillation, the following energy conservation considerations are important:

Reflux Rate Reduction. Unnecessary reflux represents direct energy waste. The desired separations can often be effected with lower reflux if the column design is very carefully fitted to its actual purpose and to the required throughput rate. Column modifications that permit decreased reflux ratios include addition of trays, increased tray efficiency, and different elevations for feed and reflux introduction. Operationally, reflux rates that are higher than necessary are often used to provide a margin of control. Application of an appropriate automatic control system can reduce this requirement (20). In other cases, reflux rates are kept high when column throughput is reduced. Minimum required reflux rates at different feed rates can be determined by test. If the required reflux at part load is still found to be excessive, the column should be campaigned (operated for two or three weeks at full rate and then shut down for an appropriate length of time) during the period of reduced production requirement. If reduced rate is not a transient condition, and the column is simply too large for the job it is doing, replacing the tower with a more appropriately sized unit may be justified on the basis of energy savings.

Heat Exchange. Overhead-to-feed heat exchange, bottoms-to-feed heat exchange, and various schemes for intermediate heat exchange can reduce energy consumption if carefully applied. Product-to-feed heat exchange and other distillation column heat exchange schemes may not save a unit of energy for every unit of energy transferred, but nevertheless they frequently result in savings that more than justify the additional investment.

Heat-Exchange Surface Optimization in Reboiler and Condenser. The temperature differential in a distillation column is a function of the difference in boiling point between the feed components being separated. It is usually thought of as the bottom temperatures minus the overhead temperature. The actual column differential temperature, however, is the difference between the temperature of the heating medium and that of the entering condenser cooling medium. If the amount of heat transfer surface provided in the reboiler or the condenser, or both, is small, then the true column differential can be extremely high. Since energy conservation is really the conservation of energy levels, increased heat transfer surface will significantly decrease the extent to which the energy applied is degraded. For instance, increasing the heat transfer surface in the reboiler makes possible the use of surplus low pressure steam, rather than high pressure boiler steam for heating. Likewise, more condenser heat transfer surface may make it possible to recover heat from this source for other process uses.

Thermal Losses. Distillation columns are large in surface area and most often exposed to the elements. Effective, well-maintained thermal insulation of the column, reboiler, associated piping, and other items of equipment that operate at high temperatures is a requisite of good energy management and is virtually always well justified (see Insulation, thermal).

Multiple-Effect Distillation. Using the overhead vapor from one column as the heat source for another column such that the second column's reboiler becomes the first column's condenser is an intriguing energy conservation concept that can cut energy consumption to one half or one third depending on the number of effects. It also reduces investment cost in heat exchangers and reduces the load on the cooling water system. Where a train of distillation columns is employed or where several distillation operations are conducted in proximity to each other, the potential for multiple-effect distillation should be investigated.

Vapor Compression. Since nearly all the heat added in the reboiler appears in the overhead vapor at a degraded level, restoration of the energy level could permit reuse of the same energy over and over again to operate the distillation column. This is practical when the boiling points of the components being separated are relatively close together. Reuse is accomplished by compressing the overhead vapor and using it as the heat source for the reboiler. The column reboiler then becomes its own condenser.

Drying. Industrial dryers (21) vary greatly in size and type, and handle a wide range of process materials at a correspondingly wide range of drying temperatures. In most cases, water is the material being removed and air is the drying medium.

Many industrial dryers in use today were selected or designed on the basis of lowest first cost with very little consideration given to energy economy and are inefficiently operated. First priority in energy conservation goes to improved operational control of existing equipment; and second priority to improved designs that will improve operating efficiency.

All of the energy used is accounted for as sensible and latent heat in the stack; as added sensible heat in the product; and as radiation or convection losses from the device. Each dryer installation must be thoroughly analyzed to determine the most appropriate energy conservation approaches from the standpoint of practicality and economics. A typical dryer mass and energy balance is shown in Table 3. Some of the energy conservation approaches applicable to dryers are: (1) improving the dewatering of feed material; (2) reducing flow of drying medium to the minimum essential for adequate drying; (3) improving surface contact between the drying medium and the material being dried; (4) recycling stack vapor or exchanging heat in stack vapor with incoming drying medium; (5) recovering sensible heat from product; (6) controlling stack or product temperature (and/or product moisture if practical); and (7) applying waste heat from another operation or process.

Evaporation. In most evaporation systems used in the chemical industry, the objective is concentration of aqueous solutions. Similar equipment is also used extensively for desalination of salt or brackish water. In the latter, the condensed vapors, rather than the concentrated brines, become the product (see Water, supply and desalination).

In evaporation, essentially all of the energy provided is present in the vapor but at an increased entropy. A single-effect evaporator produces slightly less than a kilogram of vapor per kilogram of steam fed to the heater. By using the vapor produced by a first-effect as the heat source for a second-effect evaporator and so on for three, four, or even seven consecutive effects, the factor of performance can be improved almost in proportion to the number of effects employed. Six- and seven-effect evaporators are common in the wood pulp industry for concentration of black liquor prior to firing in chemical and heat recovery units (see Pulp) (22). Since the capital cost per unit of throughput capacity increases almost in direct proportion to the number of effects, most industrial applications are single- or double-effect. Three- and four-effect evaporators are just emerging in the chlor–alkali industry for the concentration of caustic soda (see Alkali and chlorine products). Frequently, the process of crystallization (qv) and evaporation are combined in a single item of equipment. In this case, and in evaporators handling temperature-sensitive materials, boiling must take place at low temperatures, requiring the evaporator to operate at a deep vacuum.

Energy conservation in the evaporation process (23) must be approached on the basis of a specific application. As with nearly all unit-process evaluations for energy economy, generalities are useless. The following factors, however, should be considered: appropriate economic number of effects; optimum steam- and vapor-chest venting; maximum vacuum compatible with available liquid-vapor disengaging area; appropriate feed preheat provisions for each effect; first and subsequent effect condensate heat recovery; minimization of thermal losses from vapor bodies and heaters; practical elimination of air in-leakage; energy efficiency of the vacuum-producing devices (24); best liquor-to-vapor flow arrangement where liquor boiling point increases with concentration; dynamic stability of multiple-effect evaporation systems; control and monitoring of performance; and vapor-compression evaporation (see Fig. 5).

BIBLIOGRAPHY

1. U.S. Department of Commerce and U.S. Federal Energy Administration, *Voluntary Industrial Energy Conservation Program, Progress Report No. 5,* 1977-720-250/8808-31, U.S. Government Printing Office, Washington, D.C., July 1977.
2. U.S. Federal Energy Administration, Office of Industrial Programs, *The Data Base, FEA-180.D,* 1st ed., Vol. 1, Conservation Paper No. 9, U.S. Government Printing Office, Washington, D.C., 1975.
3. T. F. O'Sullivan and co-workers, *Hydrocarbon Process.* 57(7), 95 (1978).
4. C. L. Richards, *Combustion,* (Apr. 1978).
5. P. C. Finlayson and A. J. Grant, "Low Grade Fuels Utilization," *10th Annual Stal-Laval Industrial Power Symposium (SLIPS X), Winnipeg, Manitoba, Canada, Sept. 15–16, 1977* (includes list of references on fluidized bed firing).
6. E. C. McKenzie, *Chem. Eng.* 85(18), 116 (1978).
7. W. A. Beckman, S. A. Klein, and J. A. Duffie, *Solar Heating Design,* Wiley-Interscience, New York, 1978.
8. D. M. Considine, ed., *Energy Technology Handbook,* McGraw Hill Book Company, New York, 1977, pp. 2–120.
9. G. Parkinson, *Chem. Eng.* 85(13), 83 (1978).
10. *A Proven Way to Save Energy—Cogeneration, 4M/5-78,* Manufacturing Chemist's Association, Washington, D.C., 1978.
11. A. F. Waterland, "Energy Conservation in an Industrial Plant," *Trans. ASME, Power Division,* Paper 73-IPWR, 1973.
12. D. A. Reay, *Industrial Energy Conservation,* Pergamon Press, Oxford, Eng., 1977, pp 199.
13. *Ibid.,* pp. 227.
14. W. L. Viar, *Thermodynamic Similarities Between Power Cycles and Industrial Cycles,* Workshop Proceedings, Dual Energy Use Systems, EPRI EM-718-W, Electric Power Research Institute, Palo Alto, Calif., May 1978, pp. 151.
15. A. F. Waterland and W. L. Viar, *Power,* 77 (Aug. 1978).
16. *Steam—Its Generation and Use,* The Babcock and Wilcox Company, New York, 1963.
17. B. G. A. Skrotzki and W. A. Vopat, *Steam and Gas Turbines,* McGraw-Hill, New York, 1950.
18. E. S. Monroe, Jr., *Chem. Eng.* 82(18), 99 (1975); *ibid.* 83(1), 129; (8), 119; (10), 121 (1976).
19. V. K. McElheny, "Technology—A Promising Design for Electric Motors," *N.Y. Times* (May 11, 1977).
20. F. G. Shinskey, *Distillation Control for Productivity and Energy Conservation,* McGraw Hill, New York, 1977.
21. "Industrial Drying Systems" in *1973 Systems Handbook,* American Society of Heating Refrigeration and Air Conditioning Engineers (ASHRAE), New York, 1977 (includes good list of refs).
22. A. Schwartz, *Chem. Eng.* (1978).

23. U.S. Energy Research and Development Administration, *Upgrading Existing Evaporators to Reduce Energy Consumption, Technology Applications Manual COO/2870-2, 1977-740-094/126*, U.S. Government Printing Office, Washington, D.C., 1977.
24. E. S. Monroe, Jr., *Oil Gas J.* **73**(5), 126 (1975).

General Bibliography

O. Lyle, *The Efficient Use of Steam,* 9th impression, Her Majesty's Stationery Office, London, Eng., 1968.
C. L. Wilson, *Energy: Global Prospects 1985–2000, Report of the Workshop on Alternative Energy Strategies (WAES),* McGraw-Hill, New York, 1977.
U.S. Department of Commerce and U.S. Federal Energy Administration, *Waste Heat Management Guidebook, National Bureau of Standards Handbook No. 121 (SD Catalog No. C12.11:121),* U.S. Government Printing Office, Washington, D.C., 1976.
C. B. Smith, *Efficient Electricity Use,* Pergamon Press, New York, 1976.
D. N. Lapedes, *Encyclopedia of Energy,* McGraw Hill, New York, 1976.
G. R. Fryling, *Combustion Engineering,* Combustion Engineering, Inc., New York, 1967.
R. A. Budenholzer, "Power Generation" in A. Standen, ed., *Encyclopedia of Chemical Technology,* 2nd ed., Vol. 16, John Wiley & Sons, Inc., New York, 1968, pp. 436–469.
H. T. Chen and D. F. Othmer, "Thermodynamics" in A. Standen, ed., *Encyclopedia of Chemical Technology,* 2nd ed., Vol. 20, John Wiley & Sons, Inc., New York, 1969, pp. 118–146.
A. Parker and D. M. Himmelblau, "Fuels" in A. Standen, ed., *Encyclopedia of Chemical Technology,* 2nd ed., Vol. 10, John Wiley & Sons, Inc., New York, 1966, pp. 179–220.

A. F. WATERLAND
Waterland, Viar & Associates, Inc.

ENGINEERING AND CHEMICAL DATA CORRELATION

Fitting curves to variables, 46
Electronic calculations, 56
Three variables and three dimensions, 56
Absolute correlations of individual properties, 58
The reference substance concept, 59
Properties of liquids, 62
Vaporization functions of pure liquids, 72
Vapor pressures of solutions of nonvolatile solutes, 84
Binary solutions of volatile liquids, 89
Properties of gases, 95
Examples of use of vapor pressure equations and correlations, 104
Chemical reactions, equilibrium and rate constants, 112
Bibliography, 115

Quantitative measurements are required to establish even the most philosophical theories of chemistry, as well as its practice, and the design of the equipment and plants of the chemical engineer. There never are enough exact experimental data, and therefore estimations and predictions of the most probable values become necessary.

Often there are too many data of quite divergent values, even from the same source, and widely deviating when from several sources.

Methods are necessary for fixing the values to be used from those available, which may vary in amount from paucity to plenitude. Sometimes the scientific laws governing the relation of two or more properties are known as mathematical equations. Experimental values may be conformed to minimize or average their deviations from these equations. The most probable values may be obtained then by interpolation or extrapolation from the equation or a graph. Often the equation is more or less theoretical and requires practical assumptions or empirical corrections to represent the properties of real matter. The model may be entirely empirical and require considerable search for many constants.

Mathematical methods are presented for these curve-fitting operations. However, the interrelation of different properties of matter helps the quantitative understanding of each, as does also the comparison of how the properties of different similar materials vary with changes of conditions. This comparison usually allows the reduction of complicated relations to linear equations. The following presentation shows how to quantify from a minimum of experimental points many chemical and physical properties. Equations are developed from exact thermodynamic relations and are as nearly rigorous as possible with a minimum of constants or coefficients, usually a maximum of two or three. These have been tested with tens of thousands of data points for hundreds of pure gases, liquids, and solids or their mixtures. Table 1 lists some of the representative properties which may be correlated by methods outlined in this article.

With the use of a physical approach rather than one quite empirical and mathematical, absolute correlations are those that study the function of a property dependent on some parameter based on an understanding of the functions of the molecules of each particular substance, ie, their shape, size, motions, mutual attractions, etc. Thus in applying the gas laws, van der Waals estimated molecular attributes in correlating data for real gases.

More simple and readily made are relative correlations based on knowing that two different substances have a particular property that varies with some parameter according to the same laws controlling the several functions of their respective molecules. Exact data may be available for one, thoroughly studied, material which may be used as a reference substance. By comparing the relation of the variation of the property for the substance with that of the reference under the same conditions, a good correlation or prediction of values of data may be made simply. An early example is Dühring's rule. If a general thermodynamic expression for the function is available and is then related to both substances, the correlation often may be thermodynamically rigorous. Many of the correlations demonstrated below are relative and use a reference substance in one of several ways. Because of the interdependence of many physical properties, it has been possible thus to develop a whole system of relative correlations for a large number of physical properties, and to show their interrelation. Absolute correlations usually are confined to a single property.

Fitting Curves to Variables

Ever since man began making measurements in the world around him, he has tried to find mathematical relationships between the measured variables. Usually he has

Table 1. Representative Properties and Variables Correlated and/or Predicted in this Article

Liquids and their solutions of solids

activities and activity coefficients	distribution of liquid gases and solids between two liquids	heats of solution
activities of electrolytes	elevations of boiling points	latent heats
		liquid–gas (humidity) charts
adsorption	emf of cells and half cells	
critical temperatures	enthalpies	osmotic pressures
critical pressures	entropies	solubilities
densities	freezing points	specific volumes (saturated vapors)
diffusion coefficients	heats of fusion	
diffusivities	heats of hydration	surface tensions
		vapor pressures
		viscosities

Solutions of volatile liquids[a]

activity coefficients	equilibrium flash vaporization of petroleum crudes and fractions	partial pressures
azeotropy		total pressures
distribution of solutes (between two liquids)	heats of mixing	vapor compositions
enthalpies	latent heats and partial latent heats	
entropies		

Pure gases and gas mixtures

adsorption heats	entropies	Henry's constant
adsorption pressures	equation of state for pure gases and for mixtures	permeability through membranes
diffusion in gases		pressure–temperature–volume
diffusion in liquids		solubilities of gases
enthalpies	heats of solution of gases	viscosities

Functions for chemical reactions

equilibrium constants	heats of reaction	velocity constants
heats of dissociation	ionization constants	

Miscellaneous

electromotive fores of cells	enthalpy temperature charts	enthalpy–entropy charts

[a] Includes the pertinent properties under liquids and their solutions of solids.

had two distinct objectives: to find the physical law that determines the quantities observed, and to obtain a working formula, with or without a physical basis, that allows at least an approximate value of one variable to be deduced from a given value of some other variable. In the first case, correctness of mathematical form is of primary concern. In the second case, correctness of form is generally subordinate to simplicity and convenience. Most practical correlations are, in fact, a compromise with the form suggested from theoretical considerations and the constants determined by the fitting of experimental data.

Most common are relations of one variable to another, but often three variables or more must be considered. Many meaningful correlations are concerned with and made possible by consideration of apparently quite dissimilar physical properties or functions. These are most interesting and valuable because they show the interrelation of many properties of even quite dissimilar materials.

Today science and technology are developing at a rapid rate and, whereas most developments lead to an increased understanding of fundamental principles, all have an apparently insatiable need for more data. This perpetuates the necessity for extrapolation and interpolation of available data. Representation of experimental data by mathematical expressions or models is extremely desirable since it facilitates this interpolation and extrapolation and minimizes experimental requirements. Prediction of data is thus possible with surety if a model of the variation of a property is known. In this age of computer applications this model allows the development of computer programs, a shorthand method of expression, and use of what would otherwise be vast tables of data.

Developing such a mathematical model is a three-step process: (1) establishing the form of the equation, (2) determining the numerical values of the constants in the equation, and (3) making a quantitative appraisal of the reliability of the resulting equation.

Determining a suitable form for the correlation is generally based on a consideration of underlying theoretical principles as well as considerations of simplicity and ease of use. For example, theoretical considerations embodied in the Clausius-Clapeyron equation:

$$\frac{dP}{P} = \frac{LdT}{RT^2} \tag{1}$$

suggest that a proper form for the correlation of vapor pressure data is $\ln P$ vs $1/T$, which should result in a nearly linear relationship with all the added convenience attendant upon linear relations in general. The simplest forms do not correlate the data adequately which then results in elaboration of the simple forms until satisfactory correlation is achieved. Vapor pressure data have been described over the years by at least a hundred equations, some of which contain more than a score of constants. Nevertheless, the advantages of linear relationships are so numerous as to justify a considerable effort in attempting to achieve one.

There may be no theoretical basis for the relations of two variables or they may be complex or obscure. Thus the search for a suitable linear form may at first be along empirical lines. A simple plot of the data on ordinary rectangular graph paper gives an immediate indication of the essential form of the data (see Fig. 1 and Table 2).

On each of the graphs shown in Figure 1 an equation is suggested by which, having the same general form as the graph, data may be correlated. For most of these types, the method of reduction to a linear form may be immediately apparent, possibly through a transformation of variables. The constants of the resulting linear equation can be found directly from the slope and intercept of the straight line graph, or by means of analytical techniques some of which are discussed below.

The simple plot of one variable against the other may produce a type of curve, allowing an alternative plot which will produce lines which are nearly straight.

Having determined the form of the correlation, the values of the constants in the equation must be determined so that the differences between calculated and observed values are within the range of assumed experimental error in the original data. However, when there is some scatter in a plot of the data, what is the best line that can be drawn representing the data? If it is assumed that all experimental errors are in the y values and that the x values are known exactly, then the constants of the best line are those that minimize the sum of the squares of the residuals (ie, the difference be-

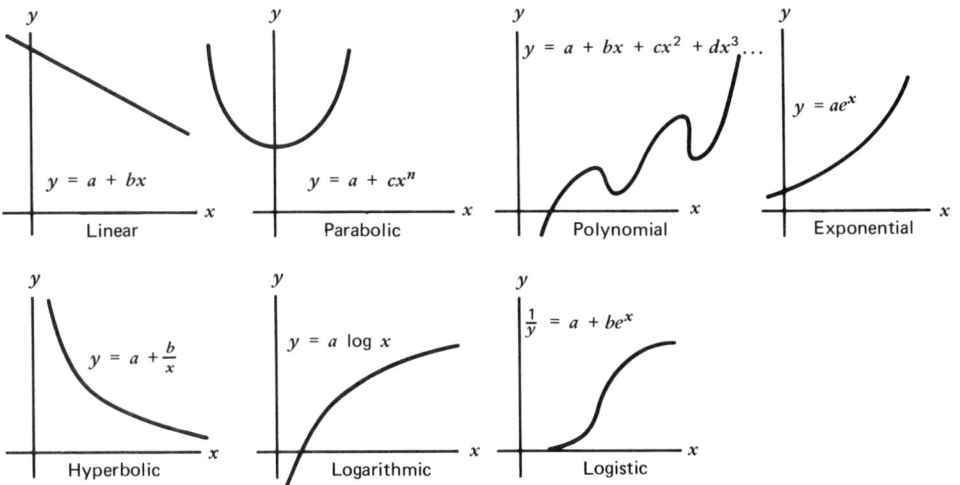

Figure 1. Some common curves and their equations.

tween the observed y values and the calculated y values). For a linear correlation, calling the sums of the squares of the residual R, this would be:

$$R = \sum_{i=1}^{n} (y_i - a - bx_i)^2 \tag{2}$$

where n is the number of data points for which a correlation is required. Values of the constants that minimize R in this equation are found by taking its partial derivative with respect to each of the constants and setting these derivatives equal to zero:

$$\frac{\partial R}{\partial a} = -2 \sum_{i=1}^{n} (y_i - a - bx_i) = 0 \tag{3}$$

$$\frac{\partial R}{\partial b} = -2 \sum_{i=1}^{n} x_i(y_i - a - bx_i) = 0 \tag{4}$$

Rearranging these equations gives the so-called normal equations:

$$na + b \Sigma x_i - \Sigma y_i = 0 \tag{5}$$

$$a\Sigma x_i + b\Sigma x_i^2 - \Sigma x_i y_i = 0 \tag{6}$$

They can be readily solved for a and b as follows:

$$b = \frac{n\Sigma x_i y_i - \Sigma x_i \Sigma y_i}{n\Sigma x_i^2 - (\Sigma x_1)^2} \tag{7}$$

$$a = \frac{\Sigma yi - b\Sigma x_i}{n} \tag{8}$$

Thus finding the best values for the slope and intercept of a straight line simply involves developing the sums indicated and combining them in equations 7–8.

This same procedure can be applied to the transformed variables of any of the cases described above where a simple transformation of one or both of the variables

Table 2. Some Common Curves and Their Equations

Curve	Equation	Method
linear	$y = a + bx$	plot of y vs x; a = intercept on y axis at $x = 0$ and b = slope
parabolic through origin	$y = cx^n$	plot log y vs log x directly or on log (double) coordinate paper; c = antilog of the y intercept at $x = 1$ and n = slope
general	$y = a + cx^n$	first obtain a as the y intercept on plot of y vs x, then plot log $(y - a)$ vs log x and proceed as in parabolic through origin
polynomial quadratic	$y = a + bx + cx^2$	first obtain a as the y intercept on a plot of y vs x, then plot $y - y_n/x - x_n$ vs x where y_n, x_n are the coordinates of any point on a smooth curve through the experimental points. The slope of this new plot gives c, and the intercept is equal to $b + cx_n$
general	$y = z + bc + cx^2 + dx^3 + \ldots$	graphical procedures are almost impossible, but analytical procedures are easy to use and work well
exponential	$y = ab^x$	plot log y vs x directly or on semilogarithmic coordinate paper with y on the log scale; a = antilog of intercept and b = antilog of slope
	$y = ae^{bx}$	plot ln y vs x directly or on semilog paper as above, a = antilog of intercept and b = slope
hyperbolic	$y = a + b/x$	plot y vs $1/x$; a = intercept and b = slope
logarithmic	$y = a \log x$	plot y vs log x directly or on semilog paper with x on the log scale; a = slope
logistic	$1/y = a + be^x$	plot $1/y$ vs e^x; a = intercept and b = slope; alternatively, plot $1/y$ vs x, a = intercept; then plot $1/y - a$ vs x on double log paper; b = antilog of y intercept

resulted in a linearized expression. The sums required would, of course, be formed from the transformed variables rather than from the original data.

Finally, how well the developed equation fits the data is given by the correlation coefficient (0.0 = poor fit, ±1.0 = excellent fit) which is the square root of the coefficient of determination. Before attempting a correlation of the data, all that can be considered is the scatter of all of the y_i values about their mean or:

$$\sum (y_i - \bar{y})^2 \tag{9}$$

After obtaining a correlation, it is more meaningful to speak of the scatter of the y_i values around the correlating function or:

$$\sum (y_i - y)^2 \tag{10}$$

Subtracting $(y_i - y)$ from $(y_i - \bar{y})$ gives $(y - \bar{y})$, which is that portion of the deviation of any data point from the mean of all data points which is explained by the correlation. The coefficient of determination (a statistical concept which also varies from 0.0 to 1.0) is defined as the ratio of this latter quantity to the total deviation of that data point from the mean. The physical significance of this ratio is apparent from the graph shown in Figure 2.

Thus the coefficient of determination γ^2 is given by:

$$\gamma = \frac{\sum (y - \bar{y})^2}{\sum (y_i - \bar{y})^2} \tag{11}$$

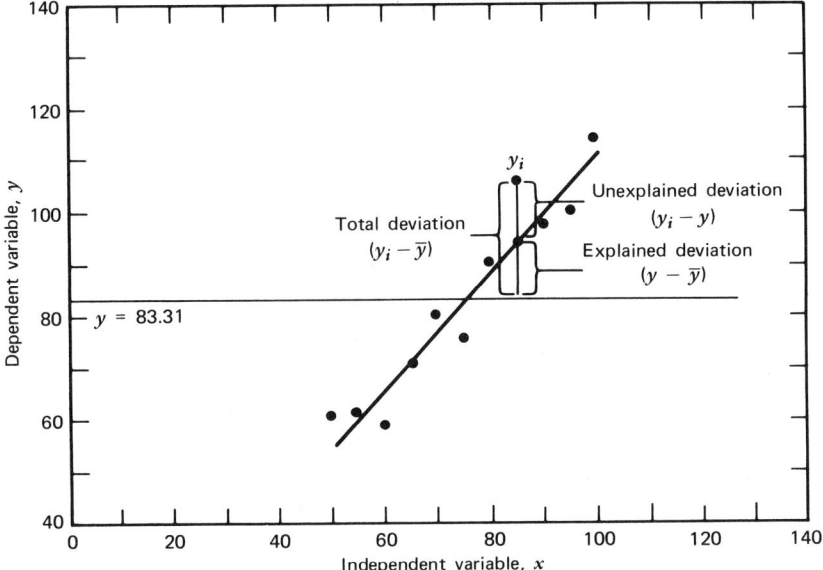

Figure 2. Coefficient of determination.

Simple algebraic transformation gives the following more useful form:

$$\gamma^2 = \frac{\left[\Sigma x_i y_i - \dfrac{\Sigma x_i y_i}{n}\right]}{\left[\Sigma x_i^2 - \dfrac{(\Sigma x_i)^2}{n}\right]\left[\Sigma y_i^2 - \dfrac{(\Sigma y_i)^2}{n}\right]} \quad (12)$$

This it will be noted, involves the same summations of data values which were required in the evaluation of the constants a and b as given above and, therefore, can be determined at the same time as the constants a and b.

With the advent of the computer, these equations can be evaluated readily even for large sets of data, and now also by the modern personal calculator.

The linear correlation of y vs x can be extended to the correlation of y vs multiple independent variables generating an equation of the form:

$$y = a + bx_1 + cx_2 + dx_3 + \ldots \Omega x n \quad (13)$$

where $x_1, x_2 \ldots x_n$ are n different variables. For example, y might be the yield of a reaction product, whereas x_1 might be the reaction temperature, x_2 the reaction pressure, x_3 the concentration of a catalyst, x_4 the space velocity, and so on.

The coefficients of this equation can be determined by a logical extension of the procedure outlined above, namely, to minimize the sum of the squares of the deviations of observed y values from those that would be predicted by the model:

$$R = \sum_{i=1}^{n} (y_i - a - bx_{1i} - cx_{2i} - dx_{3i} - \ldots \Omega x_{ni}) \quad (14)$$

by setting

$$\frac{\partial R}{\partial a} = 0, \quad \frac{\partial R}{\partial b} = 0, \text{ etc} \quad (15)$$

which results in the following set of simultaneous equations. These determine simple sums of terms involving the experimentally determined data values.

$$ma + b\Sigma x_1 + c\Sigma x_2 + d\Sigma x_3 + \ldots \Omega\Sigma x_n = \Sigma y \tag{16}$$

$$a\Sigma x_1 + b\Sigma x_1^2 + c\Sigma x_1 x_2 + d\Sigma x_1 x_3 + \ldots \Omega\Sigma x_1 x_n = \Sigma x_1 y \tag{17}$$

$$a\Sigma x_2 + b\Sigma x_1 x_2 + c\Sigma x_2^2 + d\Sigma x_2 x_3 + \ldots \Omega\Sigma x_2 x_n = \Sigma x_2 y \tag{18}$$

$$d\Sigma x_n + b\Sigma x_1 x_n + c\Sigma x_2 x_n + d\Sigma x_3 x_n + \ldots \Omega\Sigma x_n^2 = \Sigma x_n y \tag{19}$$

In principle, this set of equations can be solved for the constants, a–Ω, just as a and b were obtained previously. In practice, however, the actual numerical evaluation involves considerable computation in all but the simplest examples. It is customary to submit such problems to a rather large computer for solution by any of the usual matrix techniques designed specifically to handle this type of data correlation problem.

Curve fitting has been shown to be achieved best when the form of the equation used is based on a known theoretical relationship between the variables associated with the data points. When no theoretical relationship exists, polynomials are used to describe the curve. Polynomials are easy to evaluate, their unknown coefficients occur linearly, and their degree, ie, the highest power appearing in the equation, affords a convenient measure of smoothness.

Fitting polynomials to data points involves essentially the same technique as indicated above for correlation of multiple variables. The simple expedient of replacing x_1 through x_n in the equation above for multiple independent variables by x^1 through x^n results in the polynomial

$$y = a + bx + cx^2 + dx^3 + \ldots \Omega x^n \tag{20}$$

Continuing this substitution through the resulting set of simultaneous equations and solving them generates the coefficients of this last polynomial equation. This process is termed polynomial regression or curvilinear regression. It should be noted that equation 20 is still linear in the coefficients a through Ω.

A nonlinear model would involve one or more of the coefficients in nonlinear form such as:

$$y = a + be^{-cx} \tag{21}$$

or

$$\log P = A - \frac{B}{T + c} \tag{22}$$

The nonlinear constant c in the above two equations cannot be evaluated directly by the methods described herein. With a little ingenuity, however, even forms such as these can be handled. For example, subtracting a trial value of a from y and taking logarithms transforms the exponential equation into the linear form:

$$\ln(y - a) = \ln b - cx \tag{23}$$

from which the constants b and c as well as the correlation coefficient can be determined. Repetition with different trial values of the constant a quickly establishes the value of a which results in the best fit. Following well-known techniques (1), the familiar Antoine vapor pressure equation (eq. 22) can be evaluated as follows:

$$\text{let} \quad a = A$$
$$b = AC - B$$
$$c = -C$$

this equation then becomes:

$$aT + b - c \log P - t \log P = 0 \qquad (24)$$

which can be handled quite easily by standard techniques and the substitution of sufficient data to yield the constants a, b, and c after which:

$$\text{let} \quad A = a$$
$$B = -(ac + b)$$
$$C = -c$$

A program for personal calculators has been designed (2) to generate the constants a, b, and c of equation 24 from three or more vapor pressure values p at temperatures T.

Returning to the subject of polynomial regression, the coefficients can be obtained from the same program used for multiple regression. The procedure may be simplified considerably however if the values of the independent variable are equidistant, a condition not usually found with experimental data but which is easily achievable if it is necessary to fit some points from a smooth curve to an algebraic expression.

The normal equations for a fourth degree polynomial employing seven data points, equally spaced in X, and with the coordinates chosen so that $X = 0$ for the central point of the seven have been shown (3) to be:

$$7a + 28h^2c + 196h^4e - \Sigma y = 0 \qquad (25)$$
$$28h^2b + 196h^4d - h\Sigma ky = 0 \qquad (26)$$
$$28h^2a + 196h^4c + 1{,}588h^6e - h^2\Sigma k^2y = 0 \qquad (27)$$
$$196h^4b + 1{,}588h^6d - h^3\Sigma k^3y = 0 \qquad (28)$$
$$196h^4a + 1{,}588h^6c + 13{,}636h^8e - h^4\Sigma k^4y = 0 \qquad (29)$$

where h is the size of the X interval and k is the coefficient in the equation $x = kh$ in the transposed coordinate system, ie, $X_1 = -3h$, $X_2 = -2H$, $X_3 = -h$, $X_4 = 0$, $X_5 = h$, $X_6 = 2h$, and $X_7 = 3h$. Recognizing that the constants of all polynomials up through the fourth order can be obtained from subsets of these equations, direct solutions for these constants can be developed and are presented in Table 3. A computer program has been developed (2) to evaluate constants a through d for this special case. The program may be used also on personal calculators.

Although polynomial relaxation is a powerful technique frequently employed in the correlation of data, it has the serious disadvantage of requiring the solution of a set of simultaneous equations that may be ill conditioned, often severely so when the order of the polynominal is large. A second disadvantage of equations of this type is that the number of terms must be predetermined and cannot conveniently be expanded or curtailed to suit the range and accuracy of experimental results. These

Table 3. Summary of Solutions

	Linear	Quadratic	Cubic	4th Order
In the transposed coordinate system				
a'	$\Sigma y/7$	$1/21(7\,\Sigma y - \Sigma k^2 y)$	$1/21(7\,\Sigma y - \Sigma k^2 y)$	$1/924(524\,\Sigma y - 245\,\Sigma k^2 y + 21\,\Sigma k^4 y)$
b'	$1/28\,h(\Sigma ky)$	$1/28\,h(\Sigma ky)$	$1/1512\,h(397\,\Sigma ky - 49\,\Sigma k^3 y)$	$1/1512\,h(397\,\Sigma ky - 49\,\Sigma k^3 y)$
c'		$1/84\,h^2(\Sigma k^2 y - 4\Sigma y)$	$1/84\,h^2(\Sigma k^2 y - 4\Sigma y)$	$1/3168\,h^2(697\,\Sigma k^2 y - 840\,\Sigma y - 67\,\Sigma k^4 y)$
d'			$1/216\,h^3(\Sigma k^3 y - 7\,\Sigma ky)$	$1/216\,h^3(\Sigma k^3 y - 7\,\Sigma ky)$
e'				$1/3168\,h^4(7\,\Sigma k^4 y - 67\,\Sigma k^2 y + 72$
In the original coordinate system[a]				
a	$a' + b'\delta$	$a' + b'\delta + c'\delta^2$	$a' + b'\delta + c'\delta^2 + d'\delta^3$	$a' + b'\delta + c'\delta^2 + d'\delta^3 + e'\delta^4$
b	b'	$b' + 2c'\delta$	$b' + 2c'\delta + 3d'\delta^2$	$b' + 2c'\delta + 3d'\delta^2 + 4e'\delta^3$
c		c'	$c' + 3d'\delta$	$c' + 3d'\delta + 6e'\delta^2$
d			d'	$d' + 4e'\delta$
e				e'

[a] $\delta = -x(4)$.

defects can be removed by using orthogonal polynomials (4). That is to say, the model equation can be expressed in the form:

$$y = C_o T_o(x) - C_1 T_1(x) + \ldots C_i T_i(x) \qquad (30)$$

where $T_i\,x$ is a polynomial of degree i satisfying the orthogonality condition, and the C values are the correlation constants.

By definition, two functions $f_n(x)$ and $f_m(x)$ are said to be orthogonal with respect to a weighting function $w(x)$ over the interval a to b if the integral of the product wf_1f_2 over that interval vanishes or:

$$\int_a^b w(x) f_n(x) f_m(x) = 0, \quad n \neq m \qquad (31)$$

For present purposes the weighting function $w(x)$ can be set equal to 1.0.

Equation 31 can be abbreviated as:

$$y = \sum_{i=0}^{n} c_i T_i(x) \qquad (32)$$

where n is the order of the polynomial. Any coefficient c_j can be represented by:

$$C_j = \frac{\sum_{i=1}^{m} y_i T_j(x_i)}{\sum [T_j(x_i)]^2} \qquad (33)$$

Thus if a correlation is developed for i terms ($1 \leq i \leq m$), the addition of the $(i + 1)$th term would not alter any of the coefficients c_i already determined.

The set of polynomials represented by the general equation:

$$T_j(x) = \cos(j\,\cos^{-1} x) \qquad (34)$$

is known as the Chebyshev polynomials. Over the range $-1.0 \leq x \leq 1.0$ they are well suited to correlation procedures. It should be pointed out that any finite interval a

$\leq Z \leq b$ can be transformed into the interval $-1.0 \leq x \leq 1.0$ by the change of the variable:

$$X = \frac{2Z - b - a}{b - a} \tag{35}$$

Equation 35 can be examined for several values of j. With $j = 0$:

$$\left. \begin{aligned} T_0(x) &= \cos(0 \cdot \cos^{-1} x) \\ &= \cos 0 \\ &= 1.0 \end{aligned} \right\} \tag{36}$$

With $j = 1$:

$$\left. \begin{aligned} T_1(x) &= \cos(\cos^{-1} x) \\ &= X \end{aligned} \right\} \tag{37}$$

With $j = 2$:

$$T_2(x) = \cos(2 \cos^{-1} x) \tag{38}$$

But since $\cos 2\theta = 2\cos^2\theta - 1$ gives:

$$\left. \begin{aligned} T_2(x) &= 2\cos^2(\cos^{-1} x) - 1 \\ &= 2x^2 - 1 \end{aligned} \right\} \tag{39}$$

or in general:

$$T_j(x) = 2x T_{j-1}(x) - T_{j-2}(x) \tag{40}$$

The development of the first nine Chebyshev polynomials may be tabulated:

$$T_0 = 1 \tag{41}$$
$$T_1 = x \tag{42}$$
$$T_2 = 2x^2 - 1 \tag{43}$$
$$T_3 = 4x^3 - 3x \tag{44}$$
$$T_4 = 8x^4 - 8x^2 + 1 \tag{45}$$
$$T_5 = 16x^5 - 20x^3 + 5x \tag{46}$$
$$T_6 = 32x^6 - 48x^4 + 18x^2 - 1 \tag{47}$$
$$T_7 = 64x^7 - 112x^5 + 56x^3 - 7x \tag{48}$$
$$T_8 = 128x^8 - 256x^6 + 160x^4 - 32x^2 + 1 \tag{49}$$
$$T_9 = 256x^9 - 576x^7 + 432x^5 - 120x^3 + 9x \tag{50}$$

The Chebyshev polynomials can be used to determine the coefficients of a correlating equation (4). An algorithm has also been developed (2). In a modification of this technique, the amount of required computer storage space is reduced while the required computation time is increased (5). Where large sets of data are to be correlated, this modification could be very significant. These techniques can be used to represent the vapor pressure of water from the triple point to the critical point (6). The Chebyshev polynomial with 11 terms was found to be the lowest order equation which would correlate the available vapor pressure data for water within the estimated errors of the observed data points (6).

Electronic Calculations

The low-cost pocket calculator permits calculations to be made with a logic and speed comparable to that obtainable on the few, large, and very expensive computers of a generation ago. Particularly the pocket calculators which allow the use of prerecorded programs comprising up to several hundred or more steps allow immediate correlation of large amounts of data. When these can be reduced to linear equations, as for the numerous properties discussed here, the prediction of unknown values can be done promptly, usually with great confidence. By minimizing the effect of deviations of data from such linear equations by least square fits, the calculator makes the best use of all data points in the correlation.

Most of the numerous functions discussed here have had large numbers of data points correlated by programmable calculators to be represented by linear equations. These may use as a model functions which have greater or less thermodynamic rigor.

Often used is the reference substance technique that may employ a temperature scale based on the vapor pressure of a reference substance for which much data are available. These vapor pressure data have been correlated by a suitable empirical and nonlinear equation, sometimes with many constants. That equation, including the constants for the reference substance is prerecorded on a program. This program includes the solution for the vapor pressures of the reference substance from the multiconstant equation as well as all other necessary steps, including the usual taking or using of logarithms and other mathematical operations. Thus in effect, the linear reference-substance-correlating equation is used without necessitating looking up, eg, handbook values of logarithmic functions and vapor pressures of a reference, eg, water. Only the particular condition need be entered on the keys, and the required value is calculated by the program in a matter of seconds (see Programmable pocket calculators).

Three Variables and Three Dimensions

Thermodynamic and other physical–chemical data of one property of a substance are most often functions of (or interrelated with) two or more other properties or conditions such as temperature, pressure, and concentration. The entire expression for three variables represents a three-dimensional surface or solid, defined by the usual axes perpendicular to each other of x = concentration or other variable being studied, T = temperature, and P = pressure. This solid surface is then projected on one of three perpendicular planes, x, T, and P. Obviously the choice of the spacing of the coordinates is important in handling a correlation of the data. An impractical case can be considered wherein data which are to be treated might define a surface shaped like a bowling pin. The intersection of two planes should be on the axis of the pin while the third should be perpendicular to the axis. If the axis was away from the planes and at an angle other than perpendicular or parallel to all of them, the correlation would be difficult.

Another method of plotting a three-dimensional or three-variable system is the common nomograph with two parallel lines as coordinates X and Y. A straight line on any usual Y vs X plot is reduced to a point on this nomograph. A family of straight lines is thus reduced to a series of points which may be connected to give a line labeled, eg, Z. This Z line, in relation to the parallel X and Y lines, defines a solid surface of three dimensions. However, although this method of plotting is more difficult to visualize than the surface of a solid, it has the advantage that it can be expanded readily to include four, five, or more dimensions.

Many physical data are obtained in thermostats at constant temperatures; thus isotherms often are used conveniently to present the experimental data. There may first result a plot of x, the concentration, vs P at constant values of temperature, the third parameter. Geometrically, this is slicing the three dimensional surface by planes parallel to the x vs P plane. The resulting family of lines, the isotherms, may be quite curvilinear, but logically they may be plotted as a first attempt because the raw data are in this form. Equations for these isotherms are often derived in terms of x and P. Curves of power functions are usually obtained.

Lines or equations of constant composition, or of constant pressure, as the third parameter, may be more easily handled geometrically or algebraically. They are obtained by slicing the three dimensional surface to give traces which are projected on another plane, of P–T, or X–T, respectively. These isosteres (constant composition) or isobars (constant pressure), by themselves, may be linear—or in many cases they may be converted to linearity or simplified otherwise much more readily than the isotherms. If this is so, it may be quite advantageous to crossplot the experimental data so as to be able to study them in a form relatively more easy to correlate linearly. (This means choosing values where evenly spaced ordinates intersect the curves and plotting the values of these intersections against those of the corresponding abscissa.) The new lines are those of the constant values of the ordinate. Alternatively values may be chosen as intersections on a constant abscissa and plotted against values on the respective ordinate.

As noted above, a straight line reduces to a point on a parallel coordinate nomograph. Construction of the nomograph becomes much more difficult for nonlinear functions, and is not discussed here. In general, considerable effort in the treatment of data is justified if linear plots are finally obtained which have these geometric advantages and which give the algebraic advantages to the corresponding linear equations.

Special Coordinate Plotting. Many types of logarithmic and other graph papers are available. Either one or both of the rectangular coordinates or of those used in a parallel coordinate nomograph may be specially calibrated with a function so as to make possible a straight line plot. Calibrations on a line are made according to one parameter, then values of that parameter at some condition, eg, temperature, are indicated at uniform values of the temperature. Thus a temperature scale on the X axis of the sheet of logarithmic graph paper may be indicated wherein values are ticked off of the logarithm of the vapor pressure of a substance at uniform divisions of temperature. At these points ordinates are erected, marked with the corresponding temperatures, and used to make the plot (against temperature) in question. In effect this is the calibration of a temperature scale according to the variation with temperature of the logarithm of the vapor pressure of a particular substance, often water, instead of according to the variation with temperature of the more usual thermal expansion of a gas, liquid, or solid.

Absolute Correlations of Individual Properties

Many of the hundreds of reported correlations of chemical and engineering data are for individual properties, based on an absolute method. They represent attempts to build up from molecular functions a model of the physical mechanism controlling the property, into which some numerical values are substituted, usually to evaluate constants in the equations. Since the aspects of molecules are not known exactly, the solution or correlation usually requires numerous assumptions and empirical constants, unique for each substance. Often there are subsidiary equations and tables of values of different parameters to be inserted.

It has been said that the chemical engineer has greatest need for correlations of liquid–vapor equilibrium and of enthalpy. For binary liquid systems, vapor compositions may be correlated and predicted for either isothermal or isobaric conditions as indicated below (7–14) for the ranges of temperature and pressure most used in practice.

More sophisticated approaches and equations have been developed for binary and multicomponent systems. Often these methods require large computers and their programming, along with extensive tables of constants for each individual liquid and correction factors for other parameters, etc. Sometimes assumed values and trial-and-error calculations are necessary.

Calculations and correlation data for vapor–liquid equilibria in the critical range using activity coefficients, tables of solubility parameters, and equations of state have been published (15–28). On the other hand, ref. 29 states: "However, the accuracy of the results is often poor, especially if calculations are performed at conditions remote from those used to establish the hypothetical parameters."

Similarly liquid–liquid solubility data may be correlated by methods that are very much more sophisticated and more nearly absolute than those discussed below (30–34). Using molecular group interaction parameters, activity coefficients, including their residual parts, and the solution of the groups concept of solubility in various tables that have been developed for volume, surface, and interaction parameters, the mutual solubilities of two liquids can be calculated (35–36).

Thermodynamic evaluation and correlation of ternary liquid solubility data may be calculated with elaborate computer programs and iterations, if more nearly absolute solutions are desired than those discussed below (31–34). Great care must be exercised in selecting the 6 to 9 parameters required in such equations and computer programs (25,37–39). If not well selected, quite incorrect results are obtained (29).

For gas solubilities, if the relative or reference substance system (40) which is simple and gives good results is not used, difficulties in finding a more nearly absolute and thermodynamic system have been encountered, particularly at other than room temperatures. This is partly owing to confusion (29) in methods of expressing solubilities which has been surmounted by a logical system (40). Various assumptions, however, allow some success in using correlations that are more nearly absolute. Several attempts to quantify gas solubilities based on corrections for polar systems (41–44) have been made.

An absolute system for correlating enthalpy and entropies of gases has been developed as an alternative to the relative method below (45). This reference method gives a linear equation for enthalpy or entropy, usually with only a single term, when the straight lines pass through the origin. The Lee-Kessler correlation (46) is based on several interlocking equations using the properties established or assumed for the molecules of each particular gas, several constants for the generalized Benedict-Webb-Rubin equation of state (BWR), the critical constants, and compressibility factors as tabulated for each reduced temperature. Adherence to the exact use of these values must be exercised, and they are not intended to cover polar fluids (29).

The Reference Substance Concept

Thermodynamic data usually vary with changes in temperature and pressure. Many other data vary with time. Thus the weight of a growing animal or other biological organism increases with time until maturity. Presumably, substantially the same mechanism of weight increase during the growth period would be followed by all animals of a given species under comparable conditions. Hence, weights of a single specimen compared with those of a standard or reference specimen at the same age are meaningful, whereas the absolute value for the weight of any one growing specimen is meaningless unless the age is known and comparison can be made against others at the same condition of age.

Similarly, the mechanism of the change of a thermodynamic or other physical or chemical property, and hence of its values for different materials under a varying condition, may reasonably be expected to be the same. However, the change of properties of each specific material deviates from the changes of a perfect or ideal gas, liquid, or solid; eg, with change of temperature or of other controlling variables. Thus a meaningful method of expression is the comparison of the value of the property for one real substance against those of another reference substance—well known and studied—always at the same condition of temperature or pressure, or much less seldom of both.

Experimental data are usually obtained as a function of the values of a controlling condition such as temperature. Any correlation is then made either theoretically or empirically, referring to the performance of one particular material. This method compares the absolute values of the property of the substance at different values of the controlling variable. Most often, isotherms are considered, since data may have been determined using a constant temperature bath. Deviations from a smooth curve drawn through the data, or deviations from a theoretical algebraic expression for the mechanism, may be regarded as first differences from true values of a first order.

Instead, the experimental data may be related against the corresponding values of data for another material, which is well known and well studied, always taken at the same condition, eg, of temperature. This comparison of what may be regarded as a ratio of the variation of the property of one in relation to that of the established and true values of its variation correspondingly with values of the reference substance may be made by a table of data, an algebraic equation, or a geometric plot. More nearly correlated results are always obtained for these ratios than for the absolute values, because the variations for both substances follow the same physical mechanisms or laws under the particular condition, eg, of temperature.

Thus any deviations obtained from this reference-substance equation or reference-substance plot which compares the change of the properties of the substance against the change of the properties of the standard material for which values are well known, will actually be much smaller than those in the absolute plot. Errors, if any, are usually quite small compared to the magnitudes of the first-order differences between the true and absolute values of the property and those given by the experimental measurements, and these deviations may be regarded as second-order differences.

The reference substance concept has long been used in various operations. Thus specific gravity is defined as the ratio of the density of the substance to that of water always taken at the same temperature. For gases, specific gravity is the ratio of density referred to that of dry air always taken at the same temperature and the same pressure unless otherwise specified.

Desirably, the reference substance selected should be one which is physically and chemically as similar to the substance of interest as possible, so that the physical mechanism controlling the variance with changes in the property in question may be expected to pertain as closely as possible in both cases. In addition, this reference substance should be one which is well studied with complete and accurate data for the properties being considered throughout wide ranges. Water is often suitable as the standard or reference substance because its properties have been precisely evaluated especially in its liquid state. Thus it is universally used to define specific gravities of liquids. Particular advantages and one disadvantage are noted later.

The reference substance relation, either as an equation or a graph is simple in use, and exact in representation. A minimum of data are required to define the function. Usually it may be presented as a linear algebraic function, a straight line on a graph, or a single point in a nomogram. It also smooths effects of irregularities of experimental data by making obvious the best representation thereof, or allow the simplest consideration of all values, eg, by method of least squares.

Thus a simpler and much more rigorous thermodynamic expression usually may be written to relate the ratio of value of the property with that of a reference substance (both taken at the same values of the specified condition) than can be expressed for the absolute variation of the property of a single substance as a function of the specified condition. The expressions, even for physical or chemical mechanisms which are very complicated, when expressed by an absolute value for a single substance, often are rigorous and quite simple and correspondingly free of errors when the ratio with the values of a reference substance relation is used.

Linear relations, with their simplicity of both algebraic and geometric treatment, usually result. Not more than two terms express most such relations, ie, those of a linear equation: (1) the constant m, expressing the slope of a straight line, and (2) the additive or intercept constant C. Often a family of such lines has a common point and the second constant becomes zero to give an equation with a single constant, that of the slope m. This slope constant is often a function of the activation energy, ie, latent heat or some other related term.

Linear equations are most conveniently used as straight lines when plotted or for hand calculations using the well-known data of the reference substance from tables in handbooks. Equally valuable they are for use in pocket calculators whether these can be fed with prerecorded programs or not. Usually data are plotted. Two points determine the line, or if there are more points, the best line is drawn through them by inspection. (A black thread stretched and moved over the plot aids.) An even better use of all points is by least squares fit with a pocket calculator. In either case, the important constant, the slope m or overall term for heat—or its ratio—is determined without regard to its physical components. Often these valuable functions may then be calculated from m immediately.

Literature data are often smoothed from those of the experimental determinations by an equation arrived at either empirically or semiempirically from reported experimental values. Handbooks do not usually indicate whether experimental or smoothed data are being presented. When literature-smoothed values are correlated by the reference substance technique, straight lines may be obtained, particularly if the data for the reference substance used have also been smoothed by the same system. The reference substance technique converts data, which may be expressed in handbooks by complicated equations, to a linear form. By balancing any discrepancies from ideality of one substance against those of another—the reference—an excellent correlation is obtained.

It may be necessary or desirable to express the absolute variation of a property by some empirical expression. Likewise, it may then be worthwhile to do the same for a reference substance which has been more thoroughly or more accurately evaluated. Thus the algebraic expressions resulting usually express the values within the experimental error, since the relation of the data for the one compound to the data for the reference substance achieves a linearity of expression, and the accompanying advantages may be realized. This treatment usually results in cancelling most of the constants or terms which are common to the same complex equation used for both compounds. Often the expression is converted to a linear form.

When using a pocket calculator which can receive a prerecorded program, it is often possible to store all of the physical property data of the reference substance. This storage may be in the form of an empirical equation with six or more arbitrary constants which have been evaluated from the plenitude of data on that particular substance. The equation is thus called on in the calculation just as, or in addition to, the use of logarithms by the calculator. Such programs are available (2) for many of the properties discussed in this article.

Reference Substance and Corresponding States. The reference substance is a somewhat simpler concept than the so-called principle of corresponding states, also used for correlating data, which grew out of the reference substance concept. It assumes that similar thermodynamic behavior is possessed by different substances when compared at the same reduced conditions. Corresponding states refers to conditions where pressure, temperature, and volume are at an equal fraction of their critical values. All gases are thus considered at equally reduced and generalized properties, less satisfactorily for liquids.

Correlating charts based on corresponding states of all compounds cannot approach in precision and accuracy the presentation of the properties of one substance versus the properties of one other reference substance because of the very generality of attempting to encompass all compounds on one graph. With the reference substance method a single line or equation gives exactly the properties for one substance, most often without working with both reduced temperatures and pressures, or with either.

Charts of corresponding states convert the P, T, V variables for every substance to an involved system of empirical curves. These curves are too difficult to prepare for any single use, but are available in handbooks. The very substantial errors in the generalized correlations of the principle of corresponding states must be tolerated when their use is the only way to estimate data.

Properties of Liquids

The thermodynamic scale of temperatures for gases uses absolute zero, $-273°C$ as the starting or zero point. This point has relatively little meaning in relation to the molecular structure or properties of liquids. A more relevant base or starting point for referring properties of liquids to temperature has been shown (47) to be the critical point of the particular liquid. (No other temperature is as critical to the behavior of a liquid). Thus the temperature difference or interval between the value is measured downwardly, usually expressed as $T_c - T$, where T_c = critical temperature. This difference of temperature below this fixed critical point is more logical, and more useful, as a means of expressing temperatures of liquids than by absolute temperatures or any other temperature value above any other fixed point. Hereinafter it is referred to as the critical difference temperature T_D.

Surface Tension. The surface tension γ has been related (47) to the critical difference temperature T_D:

$$\log \gamma = a \log T_D + b \tag{51}$$

However, if γ' is the surface tension of a reference liquid and if the values of surface tensions of both liquids are always taken at the same value of T_D, then the equation for $\log \gamma'$ may be combined with equation 51 on combining constants to give new constants m and C:

$$\log \gamma = m \log \gamma' + C \tag{52}$$

Here m represents probably the ratio of the surface activation energy of the two liquids. Numerous tests with over a hundred liquids show this linear equation to represent surface tensions within the accuracy of experimental data.

From these relations, and some simplifications which can be utilized in a graphical construction, a thermodynamically correct nomogram was constructed (47) good for all liquids and solutions therein at any temperature. Water requires special treatment because of its abnormality which shows up as a break at about 40°C, where there is evidence that its molecular structure is changing (48). Thus benzene may be a better reference substance.

Density. Similarly, an almost exact linear equation was derived (47) and tested with thousands of points of density data for 100 compounds:

$$\log d = m \log d' + C \tag{53}$$

Here the slope constant m is almost unity for most organic liquids, d is density of the material, d' is density of the reference substance at the same critical difference temperature T_D, and C is a constant. This nearly exact equation predicts densities either graphically or algebraically if densities at two points are known (see Fig. 3). Water is represented by two straight lines intersecting with only slight difference in slopes (47) at about 40°C. Sulfur trioxide has a similar deviation. Again, using graphical techniques, a nomograph was demonstrated (47) to be even more nearly exact, and to eliminate the critical phenomena for scores of compounds, including for liquid solutions.

Most of the temperatures throughout which a substance is liquid are usually above atmospheric pressure. Liquid densities are not greatly affected by pressure but are measured under the corresponding saturated vapor pressures. Below the atmospheric boiling point, the pressures have insignificant effect on the density.

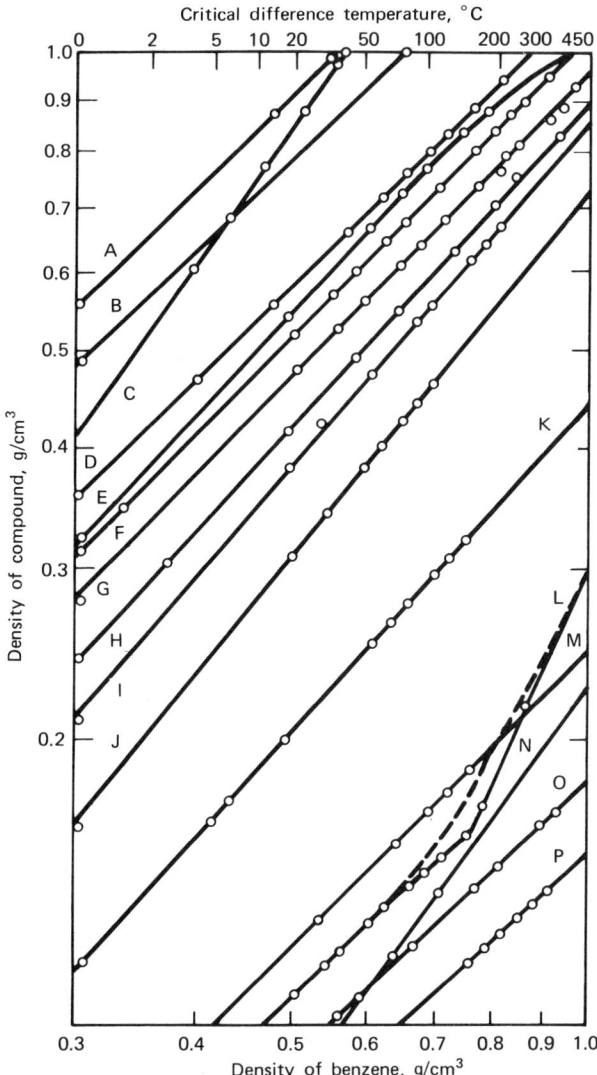

Figure 3. Density of various compounds vs density of benzene at same critical difference temperature. A, carbon tetrachloride; B, bromobenzene; C, oxygen; D, acetic acid; E, water; F, propyl formate; G, propyl alcohol; H, propionitrile; I, ethylene; and J, methane. Plotted densities for the following lines should be multiplied by 10: K, xenon; L, sulfur trioxide; M, stannic chloride; N, oxygen, O, carbon tetrachloride; and P, bromobenzene.

It was found (47,49) that $\log d$ has a direct linear relation with $\log T_D$:

$$\log d = s \log T_D + c \qquad (54)$$

This linear equation, with s and c as constants, is more convenient than the previous one with the reference substance; however, data deviate from it slightly more at higher temperature (49). Studies with many liquids showed that it may be used without substantial error throughout the entire liquid range up to within 25°C of the critical temperature where $T_D = 25°$ or $\log T_D = 1.4$.

A generalized equation, good for all liquids (also good to within 25°C of the critical) was obtained (49) which requires only one value of the density of a particular liquid, d_1, at a temperature T_{D1}. When these values are substituted, then:

$$\log d_1 = s \log T_{D1} + c \tag{55}$$

If this is subtracted from the previous equation, the constant c is eliminated:

$$\log d = \log d_1 + s \log T_D - s \log T_{D1} \tag{56}$$

or

$$d = d_1 (T_D/T_{D1})^s \tag{57}$$

However, since d_1 and T_{D1} are known for the liquid and s is a constant, when these values are inserted, the equation reverts to its original linear logarithmic form.

Over 3500 experimental points were collected in a comprehensive review of density data and correlations (50), and the value of s was averaged to be 0.2437 for all compounds. The average deviation of these values from using those of this general linear logarithmic equation was 1.15% up to $T_D = 25°C$.

Because of the slight curvature of linear relations near the critical temperature, a better fit of the same data was obtained by adding, empirically, another term to the linear logarithmic relation between d and T_D to give a quadratic equation:

$$\log d = c_1 + c_2 \log T_D + c_3 \log^2 T_D \tag{58}$$

Again using the available data (50), and the least squares fit to evaluate the constants for each compound, the average deviation in relation to these thousands of points was 0.37%.

Most other correlations require the density at the critical point which has been measured for only a comparatively few liquids; equations 56–58 do not require this. The generalized correlation requires only one value of density; the quadratic equation three; and the reference equation and the direct expression of $\log d$ in terms of $\log T_D$ each require two points. All have the advantage of being linear.

Viscosity. The viscosity function of liquids, like densities and surface tensions is by itself an important property in many operations. In addition, viscosity also appears in many multivariable equations and dimensionless groups, eg, of heat transfer and mass transfer. Thus a simple correlation of viscosities with temperature is useful for insertion algebraically into more complicated equations, and one is desired that is simple to use and has a theoretical foundation based on the physics and mechanics of fluids so that it can be extrapolated with confidence.

A simple and exact expression (51) is based on a temperature scale depending not on the usual expansion of a gas, liquid, or solid but on the vapor pressure of a liquid, eg, water. Temperatures are thus expressed as a direct function of vapor pressure and on a single line, against the vapor pressure of water.

The ordinates of Figure 4 are so constructed. The viscosities of various liquids are plotted there against the temperatures corresponding to the vapor pressure of water (51). Straight lines are obtained for all liquids plotted; here again viscosities of water itself show up with two straight lines intersecting at a very small angle at about 35–40°C. This anomaly may be owing to the above mentioned possible change in the molecular structure of water at about that temperature (48). This slight deviation does not show up in a usual curvilinear equation or plot where it is simply unnoticed as a

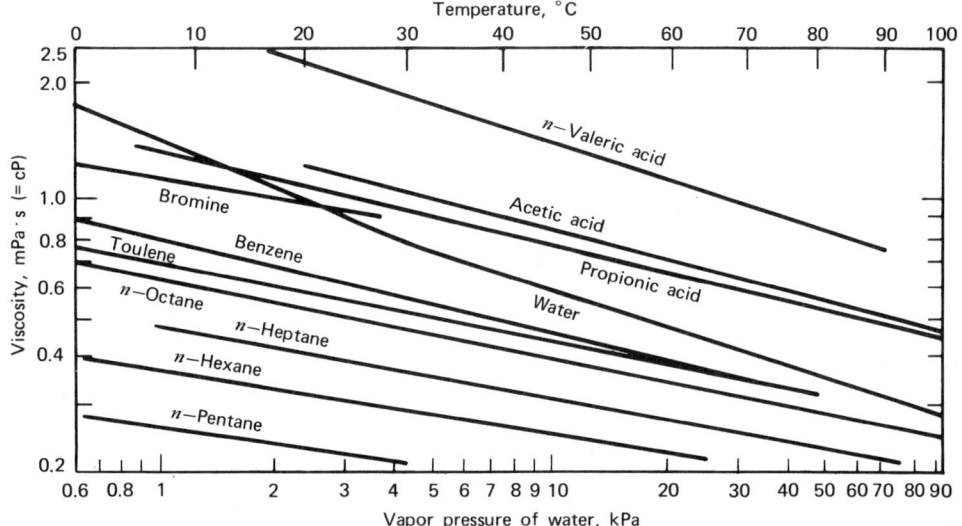

Figure 4. Plot of viscosities of eleven representative liquids against temperatures which are obtained from the corresponding vapor pressures of water. To convert kPa to mm Hg, multiply by 7.5. Courtesy of *Industrial and Engineering Chemistry*.

slight change in the curvature but does become obvious immediately when the correlation is linear.

A nomograph with parallel coordinates may be constructed as for surface tensions and densities. Alternatively, the viscosities of one liquid may be plotted against those of another, a reference substance, always at the same temperatures. Furthermore, the horizontal temperature scale may be constructed for the values of the critical difference temperature T_D or for the reduced temperatures, corresponding to the vapor pressures of water. Straight lines are obtained in all of these plots but the additional complications in plotting give no advantages.

Of more interest is the theoretical consideration. The horizontal calibration, the vapor pressure of water, is defined by the Clausius-Clapeyron equation:

$$\frac{d \ln P}{dT} = \frac{L}{RT^2} \tag{59}$$

where T is temperature, L is latent heat, P is vapor pressure of water, and R is the gas constant.

A similar expression has been developed (51) for viscosities:

$$\frac{d \ln u}{dT} = -\frac{E}{RT^2} \tag{60}$$

where u is viscosity and E is the activation energy of viscosity. If values are always considered at the same temperatures, the equations may be combined to give:

$$\frac{d \ln u}{d \ln P} = -\frac{E}{L} \tag{61}$$

This may be integrated to give approximately:

$$\log u = -\frac{E}{L} \log P + C \quad \text{or} \quad \log u = -m \log P + C \tag{62}$$

66 ENGINEERING AND CHEMICAL DATA CORRELATION

This linear equation is correct within the assumptions of the basic equations, and because $-E/L$ is known to be constant (51) over wide temperature ranges, this equation represents the lines of Figure 4. For large groups of substances $-E/L$ is substantially the same. Slight variations are owing to differences in molecular structure and related properties.

Viscosities of Aqueous Solutions—Caustic Soda. Viscosities of liquid solutions have generally not been correlated. Data are usually presented as isotherms, the viscosity as a curvilinear function of concentration, the independent variable. Such curves are not adapted to interpolation and extrapolation.

Viscosities of pure liquids are linear functions on logarithmic paper of the vapor pressure of a reference material at the same temperature. Water deviates and gives two straight lines intersecting at about 40°C, as shown by the lowest line (zero concentration) in Figure 5.

Caustic soda solutions are often present in both the manufacture and use of NaOH. Their viscosities are thus important. All lines of constant concentration of aqueous solutions of caustic soda show a break as in Figure 5 at the same temperature as does water (52). The same characteristics were shown by a plot of viscosities of aqueous sucrose solutions (53) and of other aqueous solutions of different concentrations of both ionic and nonionic substances.

This break at about 30–40°C in the line for the viscosities of water conforms to the break for densities (47), surface tensions (47), and other properties (48). The log-

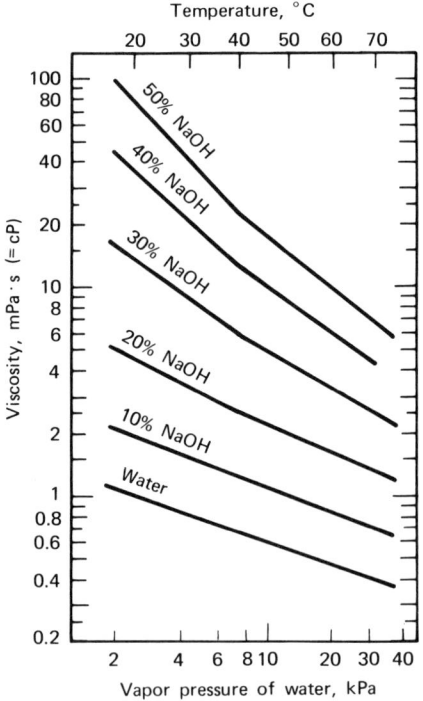

Figure 5. Logarithmic plot of viscosity of caustic soda solutions against temperature scale based on vapor pressures of water. As the viscosity decreases with increasing temperature, the calibrations of viscosity are reversed on the horizontal axis to have the temperatures increase from left to right. To convert kPa to mm Hg, multiply by 7.5.

arithmetic plot gives straight lines and shows up changes of the general function. These deviations are too small to be noted on the usual curvilinear plots or in the constants in the corresponding equation. The straight-line function and plot immediately shows even such very slight changes. Unfortunately, the data are usually correlated by smoothing the data through or around this break point with the result that a single function, curve, or set of constants is in error more or less over the entire temperature range involved.

However, the viscosity of a material is a linear logarithmic function of the viscosity of a reference material, always at the same temperature. This includes viscosities of aqueous solutions using viscosities of water as the reference, as shown by the single straight line, without a break, in Figure 6 throughout the range for each isostere or line of constant composition. This may be demonstrated also algebraically and illustrates the general premise of reference substance plots; ie, plotting data for one substance against those for another tends to eliminate the effect of the irregularities of each. Here the break points in the lines for constant concentration for caustic solutions are balanced exactly by the one for water (52). Linear equations and straight lines are obtained, as shown in Figure 6, for caustic soda solutions, sucrose solutions (53), and waste liquors resulting from pulping operations, and may be expected for other aqueous solutions.

A nomograph has been made (52) using the X and Y coordinates of Figure 6 as parallel lines with concentrations plotted on a curved line between.

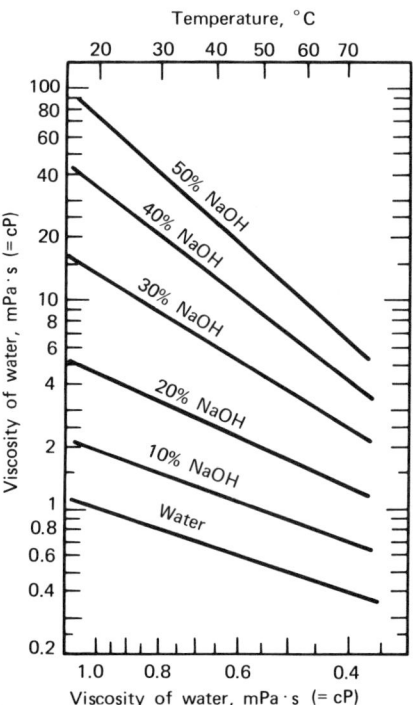

Figure 6. Logarithmic plot of viscosity of caustic soda solutions against temperature scale based on viscosity of water. As the viscosity decreases with increasing temperature, the calibrations of viscosity are reversed on the horizontal axis to have the temperatures increase from left to right.

Solubilities in Liquids. These most important data may be readily correlated, also against a temperature scale based on vapor pressures, eg, of water (30). A special case of the van't Hoff equation is often cited:

$$d \log x = (\overline{h}_l - h_s)/RT^2 dT \tag{63}$$

where x is the solubility in mole fraction of a solid in solution, $\overline{h}_l$ is partial molal heat of the solute in its saturated solution, h_s is the molal heat content of the pure solid, R is the gas constant, and T is the temperature. This may be divided by the Clausius-Clapeyron equation, with values always being taken at the same temperature. If the resulting differential equation is integrated on the assumption of the constancy over the narrow temperature range of $(\overline{h}_l - h_s)$, there results (30) approximately:

$$\log x = \frac{(h_l - h_s)}{L'} \cdot \log P' + C \tag{64}$$

where, at the same temperature as the solubility in question, L' is the molal latent heat of the reference substance (here water) and P' is its vapor pressure.

For many solutions of solids, this equation has been found to be linear within the experimental error and to give better correlation than the sometimes used $\log x$ vs $1/T$ plot. Particularly interesting is the change in values of the constants for integration and for slope when a solid becomes hydrated or otherwise changes its molecular form. Sometimes the slope changes from a plus to a minus value.

Distribution Coefficients Between Two Liquids. Distribution coefficients of a solute between two solvent layers can be correlated similarly (30) as has been shown in many examples (31). These are often used in extraction calculations.

$$\log K_D = \Delta H/L' \log P' + C \tag{65}$$

where K_D = distribution coefficient at the given temperature, ie, concentration of solute in one layer divided by the concentration in the other layer, and ΔH is molal heat of transfer of solute between the two layers.

For both solubilities and distribution coefficients, the linear equations and the straight lines plotted on logarithmic paper against the vapor pressures of water may have slopes either positive or negative, from which heats of solution may be obtained.

Mutual solubility data for a liquid mutually distributed between two, more or less immiscible, solvents are usually plotted in the familiar way on triangular coordinates. In Figure 7 for the system benzene–acetic acid–water, the solid-line curve bounds the area of immiscibility. Straight tie lines between the left or benzene side of the solubility curve and the right or water side indicate the respective compositions of the two coexisting liquid phases. Considerable data and a simple method of determination and correlation are available (31–32).

The correlation and prediction of the mutual solubilities or tie lines from the ternary solubility curve has been based on the triangular plot (33); and tie-line data were also reduced (34) to a linear, logarithmic plot and to the linear equation:

$$\log [(1 - a)/a] = n \log [(1 - b)/b] + C \tag{66}$$

where a is the weight fraction of the solvent in the solvent phase, b is the weight fraction of the water in the water phase, and n and C are constants. Once again, the advantages of a linear equation are important, and data plotted as in Figure 8 allow ready inter-

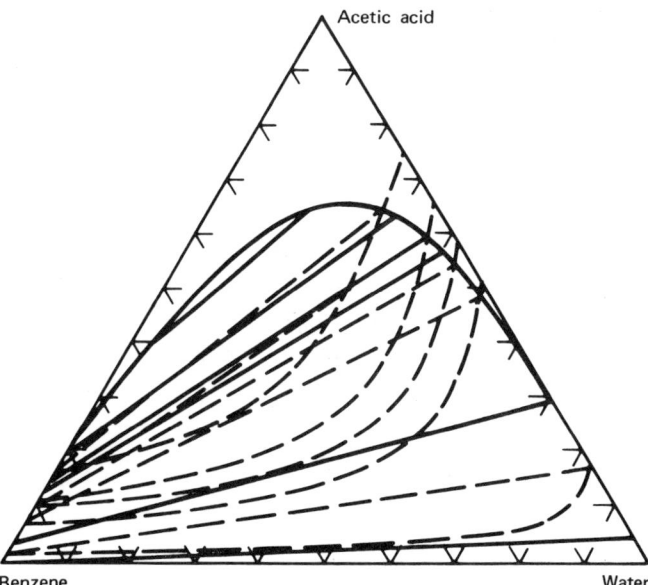

Figure 7. Ternary diagram and solubility curve of benzene-water-acetic acid system. Observed tie lines are solid; lines of constant partial pressures of acetic acid are plotted in broken curved lines, and the predicted tie lines are drawn in broken lines between the two intersections of these isobars with the solubility curve. Courtesy of *Industrial and Engineering Chemistry*.

polation for indication of the actual tie lines for use in calculations for the design of extracting and other equipment (33).

A liquid distributed between two liquids and their immiscible phases presents some thermodynamic similarities to the distribution of one component of a volatile liquid mixture between the liquid and the vapor phase, as noted below. Also, it is obvious that the partial vapor pressures of each of the three liquids (the two solvents and the consolute) must be the same at both ends of a tie line, the intersections with the solubility curve. (Since the two liquid phases with the consolute dissolved between them are in equilibrium, if the total vapor pressures or any partial pressures exerted by the two phases were not the same, there would be a distillation and a condensation taking place between the phases until they were the same.)

Ternary solubilities which express solubility equilibria between the ends of tie lines thus also express equality of partial pressures in the two phases. A method was developed for calculating data of one type from data of the other (34). Broken curved lines of constant partial pressure determined from a correlation of vapor pressure result that are developed from the vapor pressure curve and also the broken straight lines of additional tie lines which are predicted by this method. The agreement is excellent, and again the linear equations are simple in evaluation and use, as shown in Figure 7.

Diffusion Coefficients. Diffusions of gases, liquids, and solids in liquids, often water, and particularly their rates are of great importance in transport calculations and operations. Limited experimental data, if correlated properly, may allow prediction of other values over the entire range of interest. A differential equation has been derived (54):

$$d \log D = (dT/RT_2)E_d \tag{67}$$

70   ENGINEERING AND CHEMICAL DATA CORRELATION

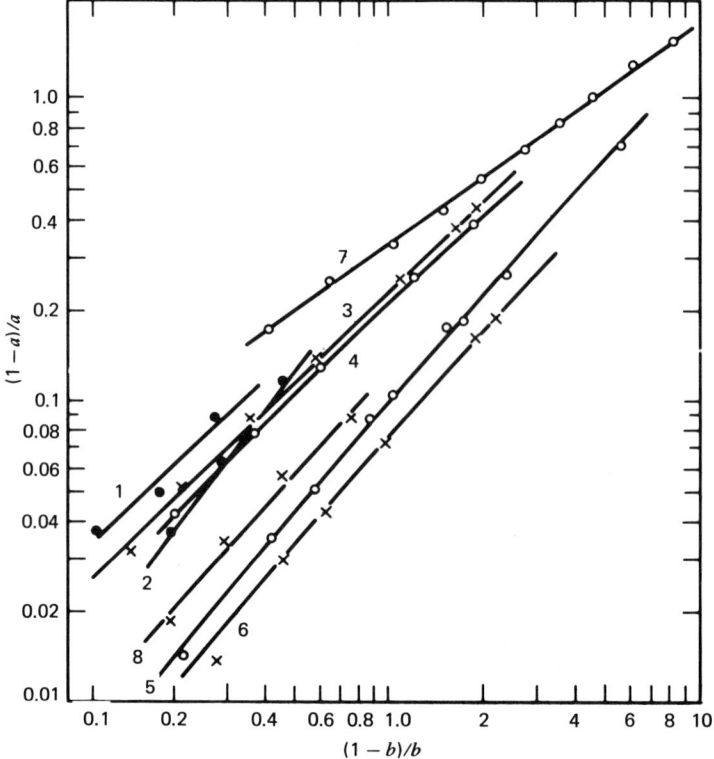

Figure 8. Plot on logarithmic coordinates of $(1 - a)/a$ vs $(1 - b)/b$ for systems of various investigators. Courtesy of *Industrial and Engineering Chemistry.*

Line	System	Temperature, °C
1	methanol, water, 1-butanol	60
2	methanol, water, 1-butanol	15
3	acetic acid, water, chloroform	60
4	acetone, water, chloroform	25
5	ethanol, water, benzene	25
6	acetic acid, water, toluene	25
7	ethanol, water, isoamyl alcohol	15
8	acetone, water, furfural	25

where D is the diffusion coefficient (cm²/s) and E_d is the energy of activation for diffusion. This may be combined with Clausius-Clapeyron equation in differential form, values being taken always at the same temperature. If the result is integrated, assuming the ratio of heats is constant over the narrow temperature range, there is obtained approximately:

$$\log D = (E_d/L) \log P' + C \qquad (68)$$

This linear equation correlates data very well (54). When water is involved, the linear equation or plot has the same break point noted above at about 35°C.

There is a close relation or dependence of diffusion in a liquid to its viscosity. Thus it is logical to combine the two properties in a single function. The differential form of the diffusion equation above was divided by the differential form of the Clausius-

Clapeyron equation so as to define the temperature function as one based on the vapor pressure of water. If instead it is divided by the differential form of the equation developed above for liquid viscosity, always at the same temperatures, the temperature function is based on viscosities, and there results (54):

$$\log D = (-E_d/E_v) \log u_w + C' \qquad (69)$$

where u_w represents viscosity values for the reference substance taken at the same temperatures from tables; and E_v represents the energy of activation for viscosity, as before. This equation represents linearly and accurately all systems with which it has been tried. For aqueous systems using water as the reference, the break point of the linear function is eliminated. A single straight line or linear equation throughout the temperature range results as shown in Figure 9.

However, E_v is closely related to and is approximately equal to E_d for any particular liquid and at any particular temperature. Thus the constant for the ratio of the activation energies of diffusion and viscosity for many systems has been found (54) to vary between -1.07 and -1.15 with an average of -1.1. Accordingly, if the value of the diffusion coefficient is known at only one temperature, substitution of this value in the equation or plot allows prediction of values at other temperatures.

The intercept constant C in the last equation represents the value for water of $\log u = 0$, which is at 1 mPa·s (=1 cP). At 20°C, water has a viscosity of 1.002 mPa·s, and this temperature may be used without significant error as the temperature for determining intercepts ($\log 1 = 0$). This is a fortunate occurrence since most diffusion coefficients are reported at 20°C.

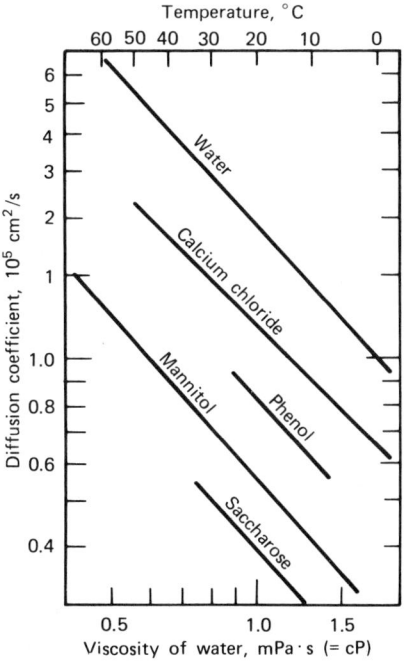

Figure 9. Logarithmic plot of diffusion coefficient ($\times 10^5$) in water of various materials vs temperature scale derived from viscosity of water.

The volume of a molecule v has been estimated as the sum of its atomic volumes; and the rate of diffusion will obviously be a function of the reciprocal of v. By plotting experimental data for diffusion in water of many gases, liquids, and solids, with molecular weights up to almost 600 (54) an equation was developed expressing all available data at 20°C for the coefficient of diffusion D within an average deviation of 5.05%.

$$D = \frac{0.00014}{v^{0.6}} \tag{70}$$

Since this equation also correlates the intercepts of lines of diffusion coefficients, it and the value of m, the average slope of -1.1, may be substituted in the preceding equation to give:

$$\log D = -1.1 \log u_w + 0.00014/v^{0.6} \quad \text{or} \quad D = 0.00014/v^{0.6}\, u_w\, 1.1 \tag{71}$$

Extensive experimental data have been evaluated (29) for the diffusion of gases, liquids, and solids at different concentrations in water correlated by the several equations proposed by various workers in this field. They show that equation 71 gives a low apparent error compared with others. A nomogram has been constructed (54) utilizing other graphical techniques to allow direct prediction of diffusion phenomena.

Vaporization Functions of Pure Liquids

Vapor Pressures, Reference Equation at the Same Temperatures.

Vapor pressures, P for one substance, and P' for a reference substance, are defined in terms of temperature, T, molar latent heats, L and L', and the gas constant R by the Clausius-Clapeyron equation. Values of P and P' and L and L' are taken always at the same temperature:

$$\frac{dP}{LP} = \frac{dT}{RT^2} \tag{72}$$

and

$$\frac{dP'}{L'P'} = \frac{dT}{RT^2} \tag{73}$$

The right sides of equations 72–73 are the same, and since $dP/P = d \ln P$ and $dP'/P = d \ln P$, division gives:

$$\frac{d \log P}{d \log P'} = \frac{L}{L'} = m_T \tag{74}$$

If m_T, which is the ratio of the latent heats per mole is assumed to be constant, when values are always taken at the same temperature, integration gives almost exactly (7):

$$\log P = m_T \log P' + C \tag{75}$$

This is a very useful linear relation and equation. Some of its many aspects are considered from the numerous occasions where it has been used to present various types of pressure and heat functions. Its simple form and some related ones developed below have been shown (55–57) to be more readily used in representing experimental data and predicting unknown values than any previous non-reduced equation, even those with several terms and constants. However, a modification (57) explained below increases its accuracy in representing vapor pressure data.

Since equation 75 is linear, it also represents a straight line on a logarithmic plot of P for one liquid against P' of a reference substance at the same temperature. All that is needed to determine either equation 75 or the straight line of its plot throughout a wide range of vapor pressures is two points of vapor pressure, one of which may be the normal boiling point. The vapor pressures for water, as the reference substance, at the particular temperatures are available in every handbook. These values are substituted in equation 75 to determine m_T and C.

The constants m_T and C of about 500 compounds have been determined, using for each material the known vapor pressures at about 10 values of temperature to define the function. Values of m_T and C have been tabulated (29,55) to allow ready calculation of the logarithm of any vapor pressure desired, and thence from tables, or with a calculator, of the vapor pressure itself. Tabulated values for water of its vapor pressures and latent heats and their logarithms (55) also assist calculations. From these values a nomogram was developed (29,55).

Vapor pressures, latent heats, and various other functions may be determined directly for any of the 500 compounds within a few seconds and from other unlisted compounds, solutions, hydrates, etc, when m and C values or two values of vapor pressures are known. The parallel scale nomograph may be used readily for any group of compounds, solutions, hydrates, etc.

Programs have been developed for pocket calculators (2) to supply the vapor pressure of water at any temperature and the complete calculation for either P in terms of known T or T in terms of known P. The tabulated values of m and C may be used in these programs. Alternatively, if for a new compound two values of vapor pressure are available, or one value (eg, the boiling point) and the latent heat, these programs or the other methods, all give prompt calculation of the vapor pressures at any temperature. If several or more vapor pressure points are known, the calculator program gives the best fit of the linear equation 75 to these points, then the vapor pressure at any desired temperature.

Figure 10 shows the logarithmic plot of vapor pressures of representative hydrocarbons against the vapor pressure of hexane. The correlation of hundreds of experimental points is seen to be excellent, as it has been found to be for many thousands of points for hundreds of other substances. Figure 10 and similar plots of equation 75 are drawn by four simple steps: (1) a standard sheet of logarithmic paper with logarithms of the vapor pressure of a reference substance, eg, water, on the X-axis has values superimposed to indicate the temperatures taken from a handbook corresponding to these calibrations of vapor pressures; (2) at these points, ordinates representing temperatures are erected; (3) two or more points of vapor pressure of the substance in question are plotted on these temperature ordinates, from values (logarithmic) on the Y axis; and (4) the vapor pressure function is drawn as a straight line through these points.

A graph such as Figure 10 has many uses, eg, to check m and C values determined otherwise. This graph, like equation 75 and the resulting nomograms, when obtained from experimental data, automatically includes the assumption of ideal gas laws, while the ratio substantially corrects for those errors inherent in this assumption.

The constant C in equation 75 has been evaluated for several members of a ho-

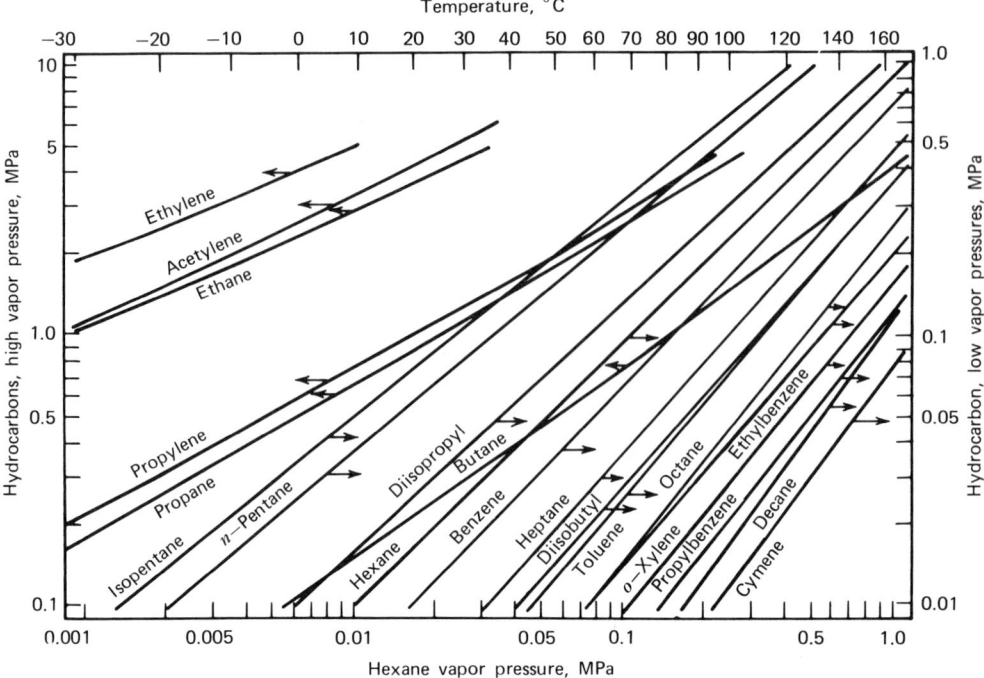

Figure 10. Integral plot of representative hydrocarbons against hexane. Those up to and including C_4 have pressures on the left, above C_4 on the right. The temperature ordinates are erected from points on the horizontal axis of corresponding vapor pressures of hexane; the plot is constructed on these ordinates. To convert MPa to atm, divide by 0.101.

mologous series. It has a linear relation with n, the number of carbon atoms. Moreover, the constant m, the ratio of the molar latent heats, has a linear logarithmic relation with n (56). These functions have been combined for several homologous series to give a single equation for all homologues of a series. An example is that of the straight-chain hydrocarbons of Figure 10 (n between 5 and 17) where:

$$\log P = (0.1852\, n^{0.758} \log P') - 0.352\, n + 2.51 \qquad (76)$$

Latent Heats, Reference Equation at the Same Temperatures. The slope, ie, the tangent of a line in Figure 10 or a similar plot, is the constant m in equation 75. It may be used immediately in determining the ratio of the molar latent heat to the tabulated one in the handbook for the reference (eg, water) at the desired temperature. A simple program for direct determination of the value of latent heat using a hand calculator is available (2). There are other uses for m, some of which are indicated below. Algebraically, m may be determined readily, from values of the coordinates X and Y for two vapor pressure points on the line.

$$m = L/L' = (Y_2 - Y_1)(X_2 - X_1) \qquad (77)$$

Graphically, the angle may be measured in degrees and its tangent determined. It is more convenient and accurate to use a 10 cm rule than to use a small protractor. Measure 10 cm from any point on the line horizontally along an abscissa 10 cm. At that value measure vertically along the ordinate to the line. This distance in centimeters divided by the 10 cm base is the tangent—correct usually to about three decimal places.

Equation 75 allows values of m to be determined therefrom algebraically (55). The tabulated value of latent heat of the reference, water, is taken from the steam tables, or others, at any temperature. This is multiplied by 18, the molecular weight, to give the latent heat of a mole of water. When multiplied by m, the molar latent heat of the substance in question at that temperature is found. If this value is divided by the molecular weight of the substance, its unit latent heat (per gram) is obtained. Latent heats so determined usually check those determined calorimetrically within their experimental errors (56–57).

Values of L and L' as used and obtained from equation 75 are subject to correction since the Clausius-Clapeyron equation assumes the ideal gas laws. However, since both vary in the same way with temperature, their ratio is very close to being constant over a wide range. Thus for most engineering applications, L/L' may be used to give results which are correct within a few percent error (56).

Reference Equations at Same Reduced Temperatures. An equation with only one term expresses a linear logarithmic relation of reduced vapor pressures, P_R to that of the reference substance P'_R always taken at the same reduced temperatures (58). The single constant m_{TR} represents the slope:

$$\log P_R = \frac{LT'_c}{L'T_c} \log P'_R \quad \text{or} \quad \log P_R = m_{TR} \log P'_R \tag{78}$$

This precise linear equation is much simpler and more readily used than other reduced equations for vapor pressures. It represents all vapor pressures of all materials at all temperatures. Vapor pressures of each material are on one of a sheaf of straight lines passing through the critical point: $-\log T_R = 0$, or $T_R = 1$; and $\log P_R = 0$, or $P_R = 1$. The integration constant has disappeared to simplify the algebra and to give this sheaf of lines on a logarithmic graph. Determinations from this equation are made readily with pocket calculators and with nomographs (2,59) which automatically evaluate the critical functions.

Again the value $(Z'_G - Z'_L)/(Z_G - Z_L)$ as later discussed would be a correction factor for the slope term m_{TR} when absolute values of the terms are substituted. However, again this is irrelevant when the equation is based on experimental vapor pressure data to be correlated or predicted using the reference material. Furthermore, because of the operation at the same reduced temperatures, the compressibilities, their differences, and particularly the ratio of their differences, almost balance each other over a very wide range, for all practical purposes, and $m_{TR} = LT'_c/L'T_c$, as indicated. This slope m_{TR} under the same values of reduced temperature is constant over wide ranges, as shown in various reviews (29).

Vapor pressures and latent heats are thus defined very accurately merely by knowing P and T conditions at the boiling point and at the critical point. Predictions with equation 78 have been found (56–57) to be more accurate than with other reduced methods. Some of these have as many as four subsidiary equations with several constants and gave errors of over 10% with from one seventh to over one half of liquids tested. Latent heats when correlated by equation 78 with one constant may be predicted as closely as with any other equation (56–57), when subjected to a correction factor depending on the molecular group.

The linear equation 78 is plotted in Figure 11 for lower hydrocarbons. Each of

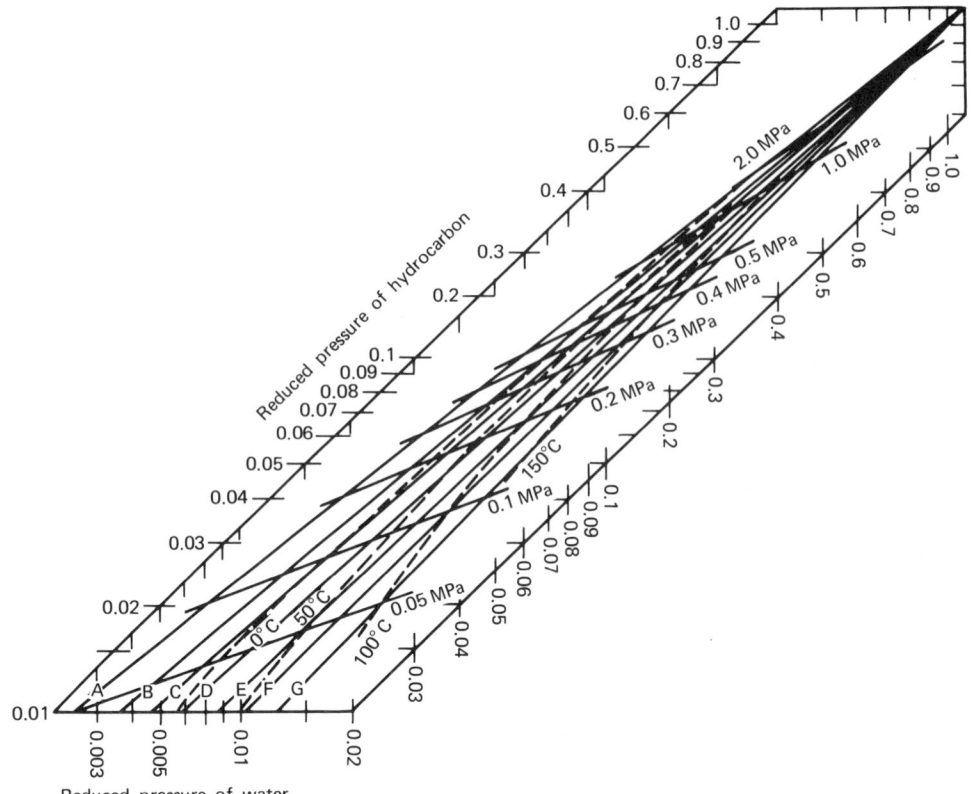

Figure 11. Plot of reduced vapor pressures of lower hydrocarbons (reduced pressure = $T_{observed}/T_{critical}$). Lines represent vapor pressures of: A, ethane; B, propane; C, butane; D, pentane; E, hexane; F, heptane; and G, octane. Note the grid of normal pressures (solid lines) and normal temperatures (broken lines) which is superimposed. To convert Pa to mm Hg, multiply by 7.5×10^{-3}. To convert MPa to atm, divide by 0.101.

the sheaf of straight lines requires only the boiling point and the critical point to allow accurate prediction at all other temperatures of the vapor pressures and latent heats. The grid work of straight lines are indicated for conventional pressures, and the dashed curves for conventional temperatures. (This is a good example of the development of one grid to give linear functions of the data of the desired functions, and another grid superimposed shows conditions as conventionally expressed.)

Values of m_{TR} for the several homologues of a series, as in Figure 11, have been found (56–57) to have a linear logarithmic relation with n, the number of carbon atoms. Thus the vapor pressures can be expressed within about 2% for all normal hydrocarbons always taken at the same reduced temperatures from ethane up to decane or higher, by:

$$\log P_R = 0.7025\, n^{0.1802} \log P'_R \tag{79}$$

Reference Equations, Values of Temperatures at the Same Pressures. In 1940 (7) the early, empirical, and quite inaccurate Dühring relation, was noted as the first reference substance concept. Methods such as were used in obtaining equation 75 were used again but with corresponding data taken always at the same pressure:

$$\log T = m_P \log T' + C \quad \text{where} \quad m_P = \frac{L'\Delta V}{L\Delta V'} \tag{80}$$

Equation 80 depends only on the rigorous Clapeyron equation, without assumption of the Clausius modification, gas laws, compressibilities, etc. Only the boiling point and one other vapor pressure point are needed. More recently (57), the accuracy, ease, and flexibility of use of this equation over the widest possible temperature range were demonstrated.

Equation 80 may be considered in relation to equation 75 and 78. Equation 75 (under conditions of the same temperature, and with water as the reference) may be used only in the temperature range at which water may be a liquid, ie, from 0–375°C. Other reference substances are used for lower or higher temperatures. Equation 78 is less easily used, but is valuable over a complete range of temperatures at which another material is liquid, because ratios of temperatures (reduced) are used instead of absolute values. However since it involves critical properties it cannot be used for many solutions and under conditions where critical properties are questionable, eg, with aqueous salt solutions.

On the other hand, equation 80 may be used without the reduced concept and over the entire range of temperatures for any substance. It is possible to use water at any temperature, because water as one of its many unusual properties, is in the liquid state throughout the greatest range of saturation pressures of any material, ie, has the highest critical pressure. Thus a temperature for water corresponding to any vapor pressure of any substance is always available, since the pressure in question corresponds to a temperature however high or low it may be for that substance.

Water may be used in equation 80 as the single reference substance required at the same pressures for the complete temperature range of any liquid. By comparison equation 75 can use water as the reference only in the temperature range from its freezing point 0° to its critical point at 273°C; the vapor pressures of other materials can be expressed only in that range. Many liquids have vapor pressures outside this temperatures range. But water in equation 80 (at the same pressures) can be used in the pressure range from its freezing point at 60.78 Pa (6×10^{-3} atm) to its critical point, at 22 MPa (218 atm). No liquid has a vapor pressure higher than this. Thus the range of vapor pressures of water between 0–375°C which is greater than that of any other material may serve as a reference at the same pressures, and the corresponding pressures for those of any material from the cryogenic temperatures of the noble gases to the incandescent temperatures of the noble metals.

Equation 80 may be the most inclusive and for some purposes the most convenient nonreduced equation available. It has better accuracy in prediction even than equation 75, one reason being that it is used without the errors inherent to the Clausius assumption which are not involved in equation 80 but are in equations 75 and 78. Even so, equation 88 below (a corrected form of eq. 80, see Table 4) is slightly more accurate. In order to define the two constants, two points of vapor pressures (P_1 at T_1 and P_2 at T_2) are necessary one of which may be the boiling point. At these two points of pressures, find corresponding T'_1 and T'_2 for water from the steam tables; evaluate the slope m_p from an algebraic evaluation of all data points available, possibly by least squares, or from only two points by the relation:

$$m_p = \frac{\log T_1 - \log T_2}{\log T'_1 - \log T'_2} \tag{81}$$

To obtain C, substitute in equation 80 values for m_p also values for T and T' at the same pressure.

Equation 80 may also be used readily to obtain P from T, when m_p and C have been determined from two vapor pressure points: (1) insert log T along with m_p and C, determine log T' and hence T'; (2) from T', tables give P' for water, this is the value sought since $P' = P$.

Note that the slope m_P in equation 80 is substantially the reciprocal of the slope m_T in equation 75.

Another reference equation that considers values of vapor pressures always taken at the same pressure is obtained in a reciprocal (7) instead of a logarithmic form by combining the Clausius-Clapeyron equation for two substances:

$$\frac{RdP}{P} = \frac{LdT}{T^2} = \frac{L'dT'}{(T')^2} \quad \text{or} \quad \frac{dT/T^2}{dT'/(T')^2} = \frac{L'}{L} = m_P \tag{82}$$

The integral form is:

$$\frac{1}{T} = m_P \frac{1}{T'} + C \tag{83}$$

Equation 83 has the same advantages of equation 80 in being applicable at all temperatures where any substance has a vapor pressure, again with water used as a reference substance. Unlike equation 80, equation 83 has the Clausius disadvantage, as does equation 75. Like equation 75 however, this is all but eliminated because of errors balancing themselves out almost to zero as a consequence of the reference substance principle. Equations 75 and 83 have reciprocal values of the slope. In fact, the straight lines in graphs such as Figure 12 (see p. 85), which has both sets of coordinates drawn, are all represented by both these equations (7). Both the method of plotting and of using equation 83 follow the techniques advanced above. The fact that equation 83 is not logarithmic has both advantages and disadvantages.

Reference Equations at Same Reduced Pressures. The expression for reduced temperatures at the same reduced pressures (57) comes, as did equation 83, from the rigorous Clapeyron equation, with no assumption of gas laws or compressibilities:

$$\log T_R = m_{PR} \log T'_R \quad \text{where} \quad m_{PR} = \frac{L'T_c \Delta V}{LT'_c \Delta V'} \tag{84}$$

Equation 84 loses its constant of integration as did equation 78. The single constant m_{PR} represents the slope, also very nearly constant over a wide range. Equation 84 represents, as does also equation 78, a sheaf of lines through the point corresponding to the critical. This is the point 1.0,1.0—which since the scale is logarithmic and log $1 = 0$—corresponds to the point 0,0 which is the origin. To use equation 84, only values of the critical conditions must be known and the boiling point, or some other vapor pressure point. Like equation 80 it can be used with any liquid, regardless of temperature range.

Similarly, if reduced temperatures are taken at the same values of reduced pressures a reduced reciprocal temperature equation 85 may be written:

$$1/T_R = m_{PR} \cdot 1/T'_R + C'$$

or very closely

$$T_R = 1/m_{PR} \cdot T'_R + C \tag{85}$$

$$T_R = 1/m_{PR'} T'_R + (1 - m_{PR})$$

The third form of equation 85 follows since $T_C/T'_C = 1$ then $C = 1 - m_{PR}$. This very simple equation for the vapor pressure phenomenon has a single constant and neither logarithms nor reciprocals of T or P. Particular equations or lines for each substance are in a sheaf, all passing through the point 1,1 since $T_C/T'_C = 1$. Unlike equations 78 and 84 this is not a logarithmic form and the point 1,1 is not the origin. Equation 85 is very nearly exact for every substance throughout its entire liquid range and can always use water as the reference substance, regardless of the temperature range. All that is required for any liquid is its pressure and temperature at the critical, and at one other point, ie, its boiling point.

Equations Corrected for Variations in Slope m. The respective slopes m of these five equations 75, 78, 80, 83, 84 are always very nearly constant. Hence latent heats may be determined therefrom at any temperature, usually within the errors of calorimetric experiments. However, m does vary ever so slightly, over the very wide range of vapor pressures up to the critical point. The lines may have a slight bow upward, particularly in their upper reaches (57). It has been found empirically that when m is evaluated for any of equations 75, 78, 80, or 84 as the first differential throughout the length of the line, these values give a straight line when plotted against $\log P$ or $\log T$ as the case may be. This function combined with equation 75, with P and P' always at the same temperature gives (57) approximately:

$$\log (\log P) = A \log P' + B \tag{86}$$

Where A and B are empirical constants. Equation 86 was found (57) to give a better representation of vapor pressure data than its parent, equation 75, because of this empirical reduction of the slight error in the constancy of the slope. It is a straight-line function, more near to being linear than equation 75.

The slope values for the reduced equation 78 were also found to vary linearly with $\log P_R$; and the corrected result was:

$$\log (\log P) = A (\log P') + B \tag{87}$$

Values of pressure are taken here at the same values of reduced temperatures. Although it is a reduced equation, the double log form allows the constant for the critical pressure to be included in the empirical constants, ie, to drop out of consideration. This allows pressures rather than reduced pressures to be used, with greater simplicity (57).

The corresponding m values for equations 75 and 80 are substantially reciprocal when viewed in relation to values of pressure at the same temperature as compared to the same values when considered at the same pressures. Hence it was expected that the same correction of slopes would pertain for the temperature equations as for the pressure equations. Values of the slopes m_P were calculated by equation 80 at different temperatures for several compounds such as refrigerants whose thermodynamic properties have been tabulated over a wide temperature and pressure range. These calculated values were found to vary linearly with the log of the absolute temperature. From this there may be derived for values of T and T' taken at the same pressures:

$$\log (\log T) = A \log T' + B \tag{88}$$

Equation 88 has been found (57) to represent and to allow the prediction of data with less error than other equations which do not require the calorimetric determination of latent heats. The reduced form is even slightly better, and only reduced pressures are involved because the constants include those for critical pressures. Thus when T and T' are taken at the same reduced temperatures:

$$\log(\log T) = A \log T' + B \tag{89}$$

Nonreduced equations 80 and 88 and reduced equations 84 and 89, usually express vapor pressure data within 1%, as shown by testing with thousands of experimental points (57). This degree of accuracy may be within the usual experimental or correlation errors of the data available in most sources. If desired, the constants can be separated to give the values of the slopes and rather exact values of the latent heats.

Critical Properties. Critical pressures are much more difficult to determine than critical temperatures, and thus are not tabulated for as many compounds. However, they may be calculated precisely using equation 78 from critical temperatures if the boiling point is also known and one additional vapor pressure point (56–57). These two vapor pressures, P_1 and P_2, at temperatures T_1 and T_2 have corresponding pressures P_1' and P_2' for water, and allow the determination of the corresponding reduced pressures P_{R1}' and P_{R2}' for water. These are at the same values of the reduced temperature for water and for the substance. From these the critical pressure of the substance is evaluated using equation 78 in the two point form:

$$\log P_c = \frac{\log P_2 \log P_{r1}' - \log P_1 \log P_{R2}'}{\log P_{R1}' - \log P_{R2}'} \tag{90}$$

Alternatively, the value of P_c can be calculated readily when the following variables are known: (1) the normal boiling point, (2) the corresponding P_R' for water at this value of reduced temperature, (3) the latent heat L, and (4) the critical temperatures for T_c and for water T_c'.

$$\log P_c = -m_{TR} \log P_R' \tag{91}$$

In the same way equation 84 allows calculation (57) of critical temperature, T_c from: (1) the critical pressure of the compound P_c and of water P_c', (2) two temperatures T_1 and T_2 where vapor pressures are P_1 and P_2, and (3) the corresponding reduced temperatures for water at the same reduced pressures T_{R1}' and T_{R2}'. Then:

$$\log T_c = \frac{\log T_2 \log T_{R1}' - \log T_1 \log T_{R2}'}{\log T_{R1}' - \log T_{r2}'} \tag{92}$$

Alternatively, the slope m_{PR} is found when the values known are: (1) the normal boiling point, (2) the corresponding T_R' for water at this value of reduced pressure, (3) the latent heat L, and (4) the critical pressures P_c and that of water, P_c'.

$$\log T_c = \log T_b - m_{PR} \log T_R' \tag{93}$$

From the group contribution theory, it is possible to determine P_c without knowing T_c, using values of m_R for homologues of a series (57).

Saturated Molar Vapor Volumes. The slope m_P in equation 80 includes vapor volumes (57) not considered in using two points of vapor pressures for its evaluation.

The line of this thermodynamic function is determined by two data points of vapor pressures. It is substantially exact because it comes directly from the Clapeyron equation without the Clausius assumptions. Equation 80 allows calculation of m_P, its slope, always operating at the same pressures. In addition, L may be known at one temperature, usually the boiling point, or it may be determined readily from equation 78, or from other suitable relation. Accurate values of latent heat L' and of the specific volumes of both liquid and vaporous water—the reference substance—are listed at every pressure and temperature in the steam tables. (As noted above, since reference is made at the same pressure as water, and water has a higher range of vapor pressures than any other substance, any corresponding temperature may be involved far above or below that of liquid water.)

From these known quantities, ΔV the difference of the molar volumes of vapor and liquid, is readily determined. At low pressures, the liquid volume is insignificantly small; and ΔV approximates V very closely. However, the density of the liquid at one temperature is usually known and may be determined at any desired temperature by the reference substance method (49) discussed above. Thus saturated vapor volumes are simply and precisely determined, and 245 data points for 27 compounds gave average deviations from tabulated data (always previously correlated by other methods) of 2.38–2.92%, depending on input data to the equation. This probably is not much greater than the errors of most determinations of saturated vapor volumes.

From the same data points, vapor volumes were determined also using equation 84 at the same values of reduced pressures. The reduced equation gave corresponding average deviations from 1.74 to 2.35. Other, much more complicated nonlinear equations, when used with the same data on the compounds reported, showed errors up to several times as large (57). Vapor volumes may be calculated from equations 88 and 89. These give somewhat more accurate results but the latent heat and liquid volume must be known at one point (57).

Latent Heats of Vaporization, Correction Using Compressibility Factor. All of the equations 75, 78, 80, 83–85, and 86–89 as well as various other forms of the Clapeyron equation for vapor pressure include latent heats as part of a slope function. Other related equations for pressures of solids, liquids out of solutions with other liquids or solids, gases absorbed in solutions or on solids, and for equilibrium or reaction rate constants, and functions, such as activities and vapor compositions, include the respective heat quantities accompanying the corresponding phase or chemical change.

The derived equations for numerous physical properties are given in Table 4 with the respective activation energies determining the slopes of lines (57–58). The slope, ie, tangent of the line, is the constant m in the equations. A correction may be considered (59) to account for the assumption of the use of the gas laws in the Clausius modification of the Clapeyron equation, although the ratio in m has reduced this error substantially. To make this correction, L should be multiplied by the difference of the compressibility factor for the gas $Z_g = PV/RT$ and Z_L for the liquid, thus by $(Z_g - Z_L)$ or ΔZ. When this is considered for both liquids:

$$m_T = \frac{L}{L'} \cdot \frac{\Delta Z'}{\Delta Z} = \frac{L}{L'} \frac{P' \Delta V'}{P \Delta V} \tag{94}$$

It is important that L and $(Z_G - Z_L)$ are always varying in the same direction and in the same order of magnitude as do L' and $(Z'_G - Z'_L)$, whereas T is always the same for both. The quotient, ie, the value m_T, thus is almost a constant. Moreover, particularly since $(Z_G - Z_L)$ approaches unity with decreasing pressure for both substances, the ratio of this difference always taken at the same temperature for two substances

Table 4. Integral Forms of Clapeyron Equation through Division by Same Equation for Reference Substance (Data Required)[a]

Equation no.	Same values of	Equation	Slope of line, m	To evaluate, besides vapor pressure
75	T	$\log P = m_T \log P' + C$	$m_T = \dfrac{L}{L'} \cdot \dfrac{P'\Delta V'}{P\Delta V} = \dfrac{L}{L'} \cdot \dfrac{\Delta Z'}{\Delta Z}$	latent heats, internal energies, compressibilities
78	T_R	$\log P_R = m_{TR} \log P_{R'}$	$m_{TR} = \dfrac{LT'_c}{L'T_c}$	latent heats, critical pressures
80	P	$\log T = m_P \log T' + C$	$m_P = \dfrac{L'}{L} \cdot \dfrac{P\Delta V}{P'\Delta V'} = \dfrac{L'\Delta V}{L\Delta V'}$	vapor volumes, internal energies, latent heats
83	P	$\dfrac{1}{T} = m_P \dfrac{1}{T'} + C$	$m_P = \dfrac{L'}{L} \cdot \dfrac{P\Delta V}{P'\Delta V'} = \dfrac{L'\Delta V}{L\Delta V'}$	vapor volumes, internal energies, latent heats
84	P_R	$\log T_R = m_{PR} \log T_{R'}$	$m_{PR} = \dfrac{L'T_c \Delta V}{LT'_c \Delta V'} = \dfrac{\Delta S' T_c P'_c}{\Delta S T'_c P_c}$	vapor volumes, internal energies, latent heats, critical temperatures, entropies of vaporization
85	P_R	$T_R = T_R'/m_{PR} + C$ and $C = 1 - m_{PR}$	$m_{PR} = \dfrac{L'T_c \Delta V}{LT'_c \Delta V'} = \dfrac{\Delta S' T_c P'_c}{\Delta S T'_c P_c}$	same as equation 84
86	T	$\log(\log P) = A \log P' + B$		
87	T_R	$\log(\log P) = A \log P' + B$		this is a reduced equation, but usual (nonreduced) values of pressures are used because of double log form
88	P	$\log(\log T) = A \log T' + B$		
89	P_R	$\log(\log T) = A \log T' + B$		this is a reduced equation, but usual (nonreduced) values of temperatures are used because of double log form

for critical pressure: T_c and two vapor pressure points known (P_1 at T_1 and P_2 at T_2)

$$\log P_c = \frac{\log P_2 \log P'_{R1} - \log P_1 \log P'_{R2}}{\log P'_{R1} - \log P'_{R2}}$$

or, if T_c, L, and normal boiling point are known:

$$\log P_c = -m_{TR} \log P'_R$$

for critical temperature: P_c and two vapor pressure points known (P_1 at T_1 and P_2 at T_2)

$$\log T_c = \frac{\log T_2 \log T'_{R1} - \log T_1 \log T'_{R2}}{\log T'_{R1} - \log T'_{R2}}$$

or, if P_c, L, and normal boiling point T_b are known:

$$\log T_c = \log T_b - m_{PR} \log T'_R$$

[a] Boiling point and one other vapor pressure point (reduced forms require T_c and P_c, as one point).

approaches unity even more rapidly. Hence $m_T = L/L'$ quite closely at pressures up to several kPa (atm), and the value of the molar latent heat is thus immediately determinable as the ratio of that for the substance to that for the reference substance, eg, water, taken from a handbook table for the desired temperature.

For any substance (59), $\Delta Z = 1.003 - 0.66\, P_R^{0.74}$ within a close approximation. But L' is the molar latent heat of water which equals Ml, where l is the latent heat per unit of mass in grams, and the molecular weight $M' = 18$. Then $m'_T = 18\, m_T/M$, where M is the molecular weight for the substance in question. Thus:

$$l' = m'_T l' \frac{\Delta Z}{\Delta Z'} = m'_T l' \frac{1.003 - 0.66\, P_R^{0.74}}{1.003 - 0.66\, P'_R{}^{0.74}} \tag{95}$$

Values of m'_T for some 500 compounds (29) have been calculated from m_T and the respective molecular weights (59). They allow the immediate determination of the unit latent heat l at any temperature from the pressure and the critical pressure. The critical pressure may be calculated (57) if the critical temperature is known, as demonstrated above.

The heat equivalent of external work $P\Delta V$ may be evaluated readily since, at the identical temperatures of equation 75, the ratio of the ΔZ terms to its equivalent noted above equals the ratio of the $P\Delta V$ terms.

Somewhat simpler to use and more accurate is equation 78 directly because $m_{TR} = LT'_c/L'T_c$ is almost constant. For the 93 substances evaluated, this relation held over a wide temperature range within 3% (56). If m_{TR} is multiplied by a correction factor K for the members of the different series of polar compounds, the above equation gives an over-all percentage error for the same 93 compounds of about 1.5%. This factor K is 1.14 for phenols and alcohols above methanol, 1.065 for amines and anilines, and 1.08 for esters.

Entropies of Vaporization. The correction by the ΔZ function in equation 75 also allows (29,59) the ready and exact calculation of entropies of vaporization, $\Delta S = L/T$ using the tabulated values (59) of m'_T, of the nomogram (55) or of simple algebraic solution by a calculator. Entropy values $\Delta S'$ for water are obtained with a high degree of accuracy from the steam tables. Since equation 75 is always at the same values of T for both substances, the T terms cancel, and the relation is the same as for latent heats, ie, $m_T = \Delta S/\Delta S'$ very nearly, or exactly $m_T = \Delta S \Delta Z'/\Delta S' \Delta Z$. Similarly, latent heats determined from equation 78 may be divided by the temperature T to determine entropies.

Equation 80 is applicable for correlating and predicting latent heats, as it is for vapor pressures for any liquid throughout its entire temperature range, as temperatures are taken at the same values of vapor pressures. Again, since the slope of equation 80 is the reciprocal of that of equation 75, latent heats may be determined almost exactly. For entropies $m_P = \Delta S' \Delta Z/\Delta S \Delta Z'$ or very nearly $m_P = \Delta S'/\Delta S$. Equation 80 may be more convenient to use in determining entropies. The reduced form, equation 84 has been shown (57) to be even more accurate, and is readily usable in simplified form:

$$m_{PR} = A\, \Delta S'/\Delta S \tag{96}$$

Here A is an easily evaluated and exact constant depending on the ratios of the respective critical constants.

Choice of Equations. Each of the linear equations 75, 78, 80, 83, 84, and 85–89 express very closely the relation of vapor pressures, temperatures, and latent heats. Different uses, advantages, and particularly adaptability or ease in use (see Table 4) govern the choice for a particular application. Programs for pocket computers have been developed (2) for calculations using these equations to determine the relation between temperatures, vapor pressures, latent heats, vapor volumes, and other properties for most materials. It is necessary to know vapor pressures at two or more temperatures or at one temperature (ie, the boiling point) and the latent heat.

Most often, in working with these functions, known values of vapor pressures are used graphically, algebraically, or with a calculator to determine the slope of the line of any of these equations, and thus the numerical value of m. When known data for the reference substance are then substituted in the respective equation for m in Table 4, corresponding properties of the material in question may be evaluated. The form of m in these equations as a ratio of latent heats is almost exact for most substances over wide ranges. It is very much more nearly exact than the assumption that L or L' or that $(L/Z_G - Z_L)$ or $L'/(Z'_G - Z'_L)$ alone is constant. Such assumptions give the quite inaccurate integral form of the Clausius-Clapeyron equation:

$$\log P = L/T + C \tag{97}$$

Plots of $\log P$ versus L/T give curves for this nonlinear expression, which can be used only over a very limited temperature range. Many corrections have been attempted by mathematical curve fitting. Modifications using 4 to 10 or more empirical constants have been accurate within 5–10% over a wider range, as has been shown (56,60).

Many of the correlations of other properties that are developed as examples have been based on equation 75. However, some of these may be performed more readily or more accurately using another one of the equations 78, 80, 83–89.

Furthermore, as has been noted, the scale of temperatures developed from the corresponding logarithms of the vapor pressures of a reference substance may be very useful in developing linear equations to correlate quite different properties. Similarly, there may be used a scale based on reduced temperatures. Other scales of pressure may be based on pressures corresponding to the logarithm of the temperatures or of the reciprocal of the corresponding temperatures. Related correlations and predictions may be also developed using reduced properties.

Vapor Pressures of Solutions of Nonvolatile Solutes

Molecules of a solvent are usually hindered from evaporation by those of the solute. Thus the vapor pressure of the liquid is lowered and the boiling point is elevated. The mechanism and the physical laws expressing vapor pressures are similar for solvents when pure and when containing a solute. For fixed concentrations, solutions have very similar relations, which may be correlated immediately and/or predicted by the nonreduced equations 75, 80, 83, 86, and 88.

A different linear equation, or line on a graph, with different values of the constants (m and C) represents each different fixed concentration of the solution. The reference substance suitably may be the pure solvent or water, especially for aqueous solutions.

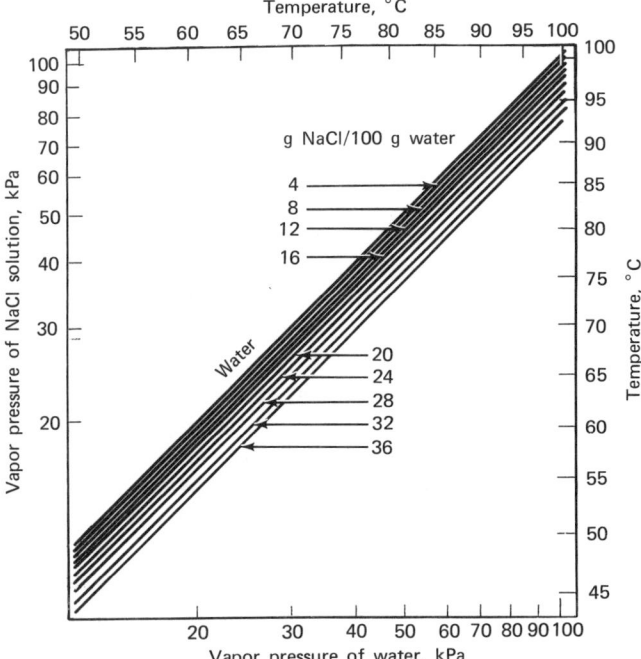

Figure 12. Plot of aqueous sodium chloride solutions of different concentrations against water. To convert kPa to mm Hg, multiply by 7.5.

Thus Figure 12 is a plot of the vapor pressures of aqueous sodium chloride solutions on logarithmic paper against the vapor pressures of pure water at the same temperature (7). The straight lines and the linear equation each representing a particular concentration have different m and C values in the basic equation 75, wherein values of pressure are always taken at the same values of temperature.

Figure 12 also shows the interrelation of this method of plotting to that of equation 82 wherein values of temperature are always taken at the same values of pressures. Equation 82 is an improvement of the Dühring relation. It was shown (7) to be as nearly thermodynamically correct as is equation 75, whereas equation 80, with the same slope, does not have the Clausius assumption to the Clapeyron equation. The horizontal calibrations of temperatures of water also represent the corresponding temperatures for the same vapor pressures of solutions. The plot as developed here is seen to be simultaneously that of equations 75 and 83, with reciprocal values of the slope m term as called for in those equations.

Heats of Solution (Differential) and Dilution. The m values or slopes of the vapor-pressure lines for solutions represent the ratio of the total energy required to disengage a mole of liquid from an infinite amount of solution of a given concentration to the latent heat of the solvent (reference liquid) at the same temperature. For the isosteres considered, this total energy includes the latent heat of the solvent plus the heat of solution under the given conditions. Hence the slopes of the lines may be either greater or less than that of the vapor-pressure line of the pure solvent, depending on whether the heat of solution is positive or negative. In this case for salt in water, the heat of solution is very slightly positive.

The slopes of the linear equations are the lines represented in Figure 12. These may be taken as the values of m in equation 75 and indicate directly the ratio of the heat of evaporation plus the heat of solution, to that of the heat of evaporation of the pure liquid:

$$m = \frac{(L' - \Delta H)}{L'} \quad \text{or} \quad (m - 1) = \frac{-\Delta H}{L'} \quad \text{or} \quad -\Delta H = L'(m - 1) \qquad (98)$$

Differential heats of solution ΔH may be evaluated immediately from these slopes of the lines of data for vapor pressures, which are much more plentiful in the literature and much more readily determined than are calorimetric data of heats of solution. Heats of solution of various systems with water determined in this way have been found to correlate well against calorimetric measurements (7–8) and against values calculated from thermoelectric measurements (8).

Heats of dilution are of importance, eg, on diluting a solution by adding more solvent or on concentrating by the evaporation of a dilute solution. They are of more use than differential heats of solution (at a fixed concentration).

A curve may be drawn representing the slopes of the linear equations of vapor pressure or of the lines minus unity, ie, $m - 1$. This is plotted versus the concentration of the solution expressed as the ratio of the mole fraction of the two components. Integration of this curve between the two concentrations of interest gives immediately the integral heat of solution between the high and low concentrations (7,61). Charts of the enthalpy of the solution versus temperature for all concentrations can be prepared using this relation by techniques discussed below. This is shown in Figure 13 for the ammonium nitrate–water system (62), and in the same manner (61) for solutions of two volatile liquids.

Elevation of Boiling Points. Another useful vapor pressure relation of solutions of solids is the elevation of the boiling point of solutions, which represents the increase in the temperature to which a solution must be heated in order to give the same vapor pressure as that of the pure solvent. This is important in the theory of solutions and is basic in the design of evaporation processes and equipment for solutions of solids. By combining the Clapeyron equation and Raoult's law, an equation has been derived to correlate the elevation of the boiling points at different pressures of solutions, even of high concentration. This relationship (53) for a solution of fixed concentration is:

$$\Delta T_A / \Delta T_P = (T_A / T_P)^2 (L_P / L_A) \qquad (99)$$

Here ΔT_P and ΔT_A are respectively the elevation of boiling points in degrees centigrade from the solution boiling at T_P and T_A under pressures of P and 101.3 kPa (760 mm Hg) (atmospheric pressure) where the latent heats of water are L_P and L_A, respectively. A similar equation applies for every different concentration. However, for each fixed concentration, the boiling temperatures and the latent heats are constant for each of the two pressures:

$$\Delta T_P = a \, \Delta T_A \qquad (100)$$

The constant a is dependent on the concentration. The linear equation shows the relation of the elevation of the boiling point of solutions at different pressures as a function of the elevation of the boiling point at atmospheric pressure in the form $y = ax$. A sheaf of lines through the origin each representing a different constant concentration results. The slope a of each varies with the concentration. This may readily

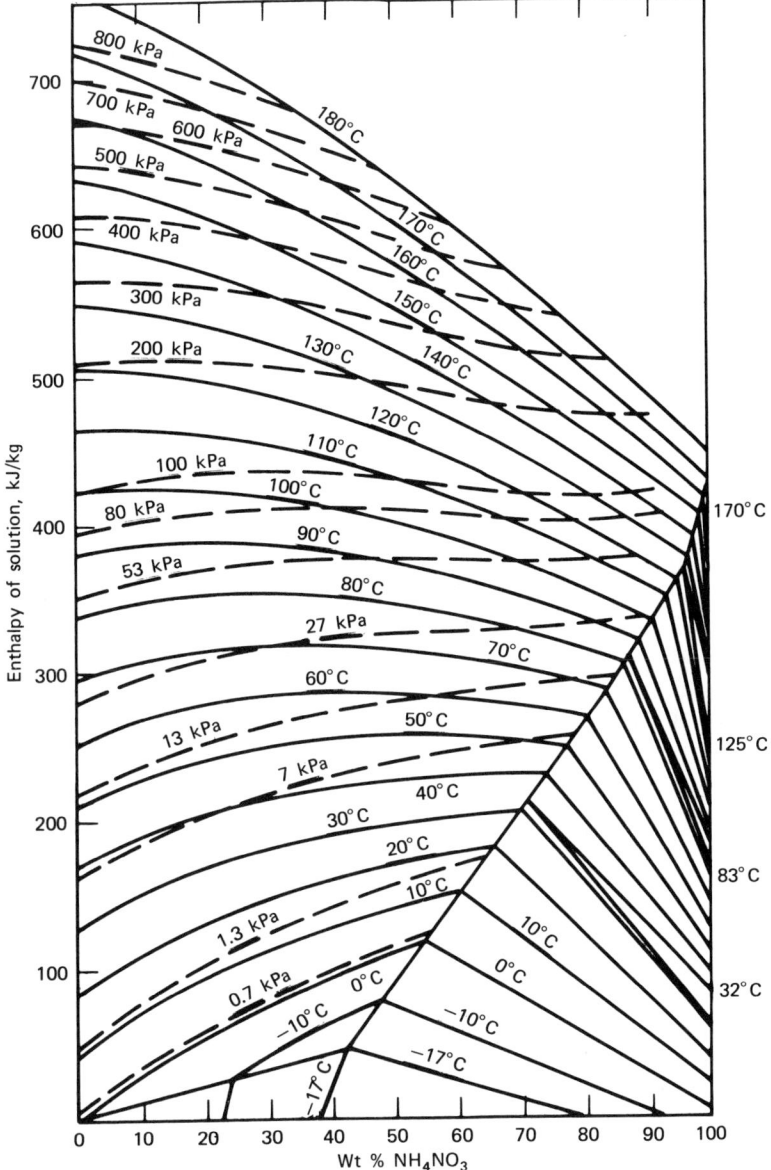

Figure 13. Enthalpy-concentration diagram for the system water–ammonium nitrate. Solid lines are isotherms, dashed lines are isobars. To convert kJ to kcal, divide by 4.184.

be expressed in the form of a nomograph. In the important case of sugar solutions (53) the additional variable of purity of the sugar solution has also been correlated.

Hydrates. The vapor pressure of water from hydrates, ie, crystals containing water of hydration, can be dealt with similarly. Again, the linear equations 75 or 80 represent the equilibrium pressure of the solids with water. The slopes of these equations or those of straight lines on a logarithmic plot of equation 75 represent the ratio of the sum of the heat of vaporization plus the heat of efflorescence (hydration) to the heat of vaporization of pure water. For solids with more than one hydrate, there is a vapor pressure line for each. Usually there are lower pressures and greater slopes of lines for the smaller number of molecules of water of hydration (7).

Similarly, the vapor pressure of solid crystals undergoing sublimation may be correlated. The heat of sublimation would be represented by the slope of the line based on the heat of vaporization plus the latent heat of crystallization or freezing (7). This vapor-pressure line intersects the vapor-pressure line of the liquid at the triple point, which is often very close to the freezing point.

Freezing Points. Figure 14 shows the vapor pressure of water cooled below its freezing point as a direct extension of the normal 45° line (7). However, the line for ice is steeper. This is owing to the fact that the heat of sublimation represented by this slope must include that of fusion in addition to that of evaporation. When the slope is thus determined and used, the heat of fusion is calculated to give a value very close to that which has been precisely determined calorimetrically.

Figure 13 also indicates another interesting correlation (7). The dashed lines are a schematic of vapor-pressure lines for solutions of different salt concentrations. These are extensions of the lines of vapor pressures of salt solutions such as those of Figure 12. Solutions have freezing points progressively lower than that of water as the concentration increases. At these freezing points which are on the ice-equilibrium line, ice is produced as the pure solid freezes. Thus the solution in equilibrium must have the same vapor pressure as the pure ice, ie, the intersections of these dashed lines with the vapor pressure line for ice must be at the freezing points of the solution.

In reverse, if freezing points, which are very easily determined, are known for

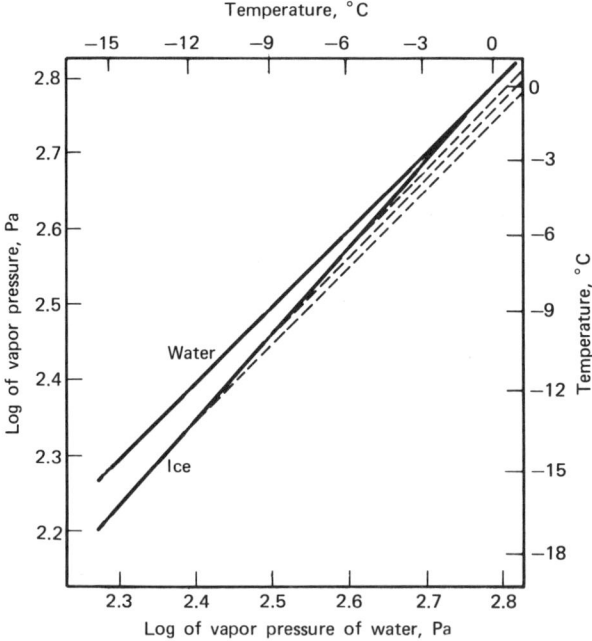

Figure 14. Plot of liquid water and ice against water. The dotted lines are schematic representation of lines for aqueous solutions intersecting the ice line (values of logarithms of pressures are indicated). To convert Pa to mm Hg, multiply by 7.5×10^{-3}.

solutions of different concentrations, these points are at the intersection of the dotted lines for vapor pressures of the solutions with the line for ice. Consequently, if vapor pressures of the same solutions are known at only one other temperature, eg, the boiling point, straight lines may be drawn immediately. They will pass through the freezing temperature on the ice curve, and give the vapor pressures between the freezing point and the boiling point of the solution. The linear equations may be established through these two points and they may also be extrapolated safely. Their slopes or those of the lines may be used to determine heats of solution and thence the enthalpy-concentration diagrams (61–62). Alternatively, the extrapolation of known vapor pressure lines for given solutions intersect the vapor pressure line of ice, to give the freezing temperature for solutions of those concentrations.

Osmotic Pressure. Because of many formal similarities, osmotic pressures of solutions of fixed concentration are found to be governed by the same linear equations as those for vapor pressures. Indeed the mathematical combination of the controlling equations of both phenomena also allows the correlation and prediction of these all-liquid thermodynamic manifestations by the same techniques and linear mathematical models that have been advanced above for equilibria and transport between vapor and liquid phases.

Binary Solutions of Volatile Liquids

Vapor Pressures and Heat Quantities. In systems of two or more volatile components, in addition to the variables of temperature and total pressure, the variables of compositions of both the liquid x and of the equilibrium vapor y phases must be considered along with the related partial pressures of each component. The linear equations and plots using the reference substance technique are applied directly to solutions of two or more volatile liquids. These give the familiar linear functions for total pressures, partial pressures, vapor compositions, activity coefficients, relative volatilities, and equilibrium constants. It is convenient, but not necessary, to use one of the two liquids as the reference material. Usually it will be more convenient to use those reference equations where values are taken at the same temperature. However, in other cases, the advantages of those equations taken at the same pressures may be desired.

The interrelation and indeed the dependence of the vapor composition on the heats of mixing of the components was understood as long ago as 1940 (7). In the linear equations or lines demonstrating vapor pressures for different constant liquid compositions x, the slopes or m values represented the partial latent heat of the evaporation. This includes the partial heat of solution, as also in water out of, say, salt solutions.

An excellent review of the thermodynamics of the relation of heat of mixing to vapor composition and industrial distillation and its equipment is available (63).

Considering the case of an ideal solution, ie, one obeying Raoults law, there is no heat of solution of one component dissolved in another, (eg, of benzene–toluene).

Different values of m and C and different linear equations or straight lines represent each composition (8). These vary in an exactly regular manner for ideal solutions as predicted by Raoult's law. The values of m vary between those of the vapor pressure lines of benzene and toluene. They are directly dependent upon the molar ratio of the two constituents, and the latent heat of each composition depends directly on the sum of the latent heats for the amounts of benzene and toluene present, since there is no heat of solution for these liquids completely soluble in each other.

The vapor–pressure relation of one liquid, entirely immiscible with another, has also been represented in equation 75. Each of the two liquids exerts its vapor pressure quite independently of the other. The additive vapor pressure of the two components also is represented fairly well by equation 75.

Nonideal solutions have heats of mixing; and the partial heats of solution may be calculated by this method from the slopes of the vapor pressure lines. These heats correlate well with thermal data (7–8). The system of ammonia and water has been correlated completely with nomographs (64), also the system of hydrochloric acid and water (8–9).

Nonideal miscible liquid mixtures are by far the most common type met with in practice (see Azeotropic and extractive distillation). Again the partial pressures of each component, and the total pressures as well, may be represented by equation 75. The total pressures of nonideal solutions of benzene and ethanol are illustrated in Figure 15. Here the confusion of crossing lines is prevented by drawing only one half of each line. It has been shown (7–8) that, regardless of the extent of deviations of the systems from ideality, the vapor pressure relations may be represented by equation 75. Even more interesting is the fact that the m values, which are based in part on the heat of mixing, were shown (7) to be directly related and in fact to predict and to determine the vapor composition curve. This is because the heat of mixing adds with the latent heats to determine the m values of the linear relations. They also predict

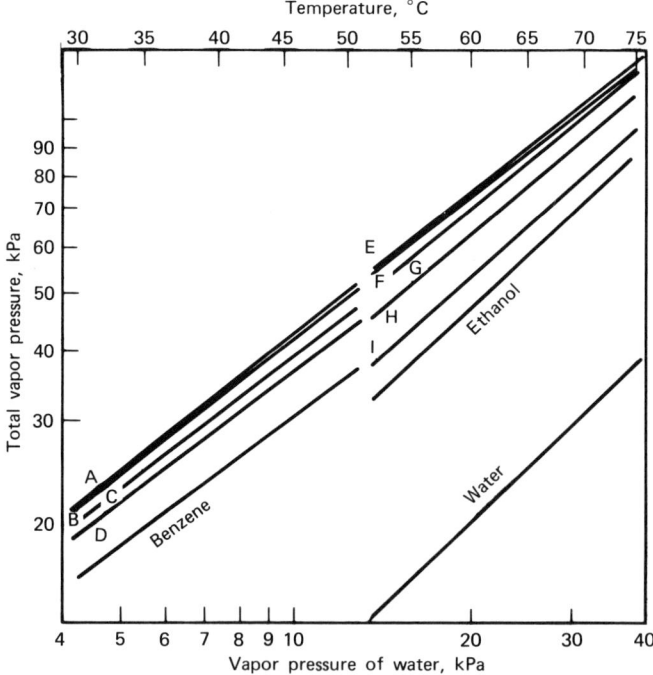

Figure 15. Plot of benzene, ethanol, and their solutions against water illustrating two volatile liquids with a minimum constant-boiling mixture. Ethanol, wt %: A, 17.3; B, 9.47; C, 3.85; D, 2.07; E, 32.1; F, 50.1; G, 64.9; H, 79.8; and I, 93.2. To convert kPa to mm Hg, multiply by 7.5.

the azeotropic point of a system (7–8,10). These points are shown in Figure 16 which is derived from the data (7) represented in Figure 15.

Isobaric and Isothermal Conditions. Either isothermal or isobaric conditions of vapor pressure determination, correlation, or use may be considered. In fact, both may be interrelated by the same lines on a duplex grid (7) which simultaneously would be the plot of equations 75 and 83 as in Figure 12. Most distillations are made under a constant pressure; however, physical chemists often prefer to obtain background data under constant temperature. The same equations and constants as well as plots can be shown readily to accommodate both, not only for vapor pressure presentations, but also for vapor compositions, activities, relative volatilities, and other functions (8).

Thus the normal boiling points of mixtures for either an ideal or a nonideal system such as Figure 15 would be the temperature points on all of the vapor-pressure lines at the intersections with a horizontal line at atmospheric pressure. Similarly, a vertical line or ordinate of constant temperature, eg, 60°C, would cut the family of lines at each individual value of pressure. Either or both types of experimental data may thus be used to augment each other and simultaneously be correlated together, or predicted from such a duplex grid. This flexibility of interchange of conditions holds for most of the correlations discussed. As indicated above, equation 83 with pressures always taken at the same values of temperature of the reference substance (and the corresponding plots) are exactly the same on a grid as those for temperatures at the same values of pressure with reciprocal values of m as the slope constant. Equation 80 has the same slope as equation 83.

Partial Pressures. As may be expected, the partial pressures of any one component of a mixture of volatile liquids are also presented and correlated by the linear equations 75 and 80. Most often one of the liquids is used as the reference substance.

This has been shown to be true for many systems, at a constant value of x the liquid composition. The same is true for a constant value of y the vapor composition. Slopes of the lines have an m function that includes the heat of mixing. This has been discussed in detail and compared with results from calorimetric and other methods of heat determination (8).

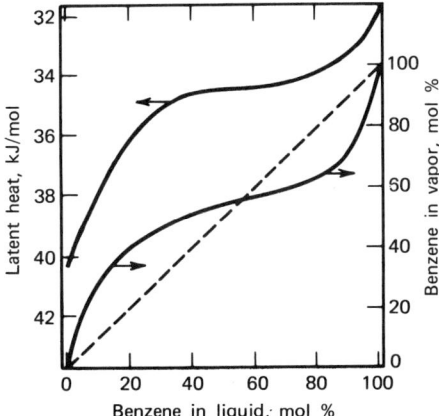

Figure 16. Plot of molal latent heat and of vapor composition for benzene–ethanol solutions against mole percent benzene in liquid. To convert kJ to kcal, divide by 4.184.

Plots of Total Pressures of System. Another simple method of correlation is shown to give (8), without the reference concept, linear equations or plots representing experimental data taken at constant total pressure; eg, the partial pressure of water from acetic acid solutions of fixed concentration is plotted against the total pressure existing on the system. The slopes of lines of logarithmic plots of partial pressures and of total pressure against the reference substance, in this case water, are constant and equal, respectively, to:

$$d \log p_A/d \log P' \quad \text{and} \quad d \log P/d \log P' \tag{101}$$

where p_A is the partial pressure of component A, P' is the normal vapor pressure at the same temperature of component A which is the reference substance, and P is total pressure of the solution. The quotient of these two terms, $d \log p_A/d \log P$, must therefore also be constant. It is the value of the slopes of the linear equation of $\log p_A$ vs $\log P$. Hence, these lines of partial pressure directly plotted against total pressure must be straight, as they are shown to be (8). (This linear equation is not of a reference substance type, although the proof is based on the linearity of reference substance plots.) Heat quantities as represented by the slopes of the lines may be derived and utilized.

Composition of Vapors in Equilibrium with Liquid Composition. Distillation calculations and the design of distillation equipment are based on vapor compositions which may be defined by Dalton's law:

$$y = p/P \quad \text{or} \quad \log y = \log p - \log P \tag{102}$$

where p is the partial pressure of the more volatile component, y is its mole fraction in the vapors in equilibrium with the liquid, and P is the total pressure. If both sides are differentiated with respect to $\log P'$, which is the logarithm of the vapor pressure of the reference substance at the same temperature, there results (8):

$$d \log y/d \log P' = (d \log p/d \log P') - (d \log P/d \log P') \tag{103}$$

The first term on the right side of this equation represents the slopes in the plot of partial pressure versus the vapor pressure of the reference substance at the same temperature, which is known to give straight lines, thus m in equation 75 is known to be constant. The last term is the slope of the lines representing the total pressure of the system versus the vapor pressure of the reference substance at the same temperature, which also is constant due to the constancy of the values of the respective m. Hence, the term on the left also must be a constant. Thus a linear equation or a straight line of constant slope results when the vapor composition y is related to the value of the vapor pressure of the reference substance at the same temperature.

A linear form is also given by combining the corresponding form of equation 75.

$$\log y = (L - L_y)/L' \log P' + C \tag{104}$$

where L is the molar latent heat of the more volatile component from the solution, and L_y is the average molar latent heat of the vapors in equilibrium with the liquid of composition x.

If the logarithmic form of Dalton's law is differentiated with respect to the logarithm of the total pressure P (at the same temperature), the term $(d \log y/d \log P)$ is also a constant:

$$\log y = [(L/L_y) - 1] \log P + C \tag{105}$$

This is a linear logarithmic relation of the vapor composition of any liquid mixture in terms of the total pressure of the system itself. This is not a reference-substance relationship, and temperature is not involved. Thus the logarithmic relation of vapor compositions to total pressure P is a linear equation, or the plot gives straight lines. This has been shown to be true in plots for many systems (8). This is a different plot than those against the vapor pressure of water, which really present a temperature scale. It is particularly useful in correlating experimental data which are usually taken at constant total pressure, and thus may be plotted directly on a standard sheet of logarithmic paper to give straight lines.

The data point dots in such a plot are taken at several constant total pressures. Thus they fall on ordinates. A reference-substance plot of vapor pressures represents the same data which now (8) do not fall on the ordinates which really refer to temperatures. The data points form nearly vertical curves, and these are curves of constant total pressures. If the vapor compositions had been determined at constant temperature, as is sometimes done, these dots would fall exactly on an ordinate on the reference substance plot and a nearly vertical curve on the plot against total pressures.

Activity Coefficients, Equilibrium Constants, and Relative Volatilities. These are used in many theoretical and practical applications; and the usual linear equations and corresponding straight line plots have been shown to pertain when considered in relation to the reference substance taken as one of the component liquids. A useful linear equation or logarithmic plot is that of activity coefficient on the total pressure scale just indicated as a nonreference-substance plot. The linear equation or plot of the reference substance is also readily derived from either the vapor pressure of water or that of one of the components of the mixture. In representing activity coefficients of water from acetic acid solutions, the vertical scale is expanded considerably (because of the small range of values) by plotting logarithms on the ordinary vertical ruling of semilog paper.

This expansion can always be carried as far as might be desirable to accentuate the spread of the lines or their slopes (8).

From the definition of the activity coefficient: $\gamma_A = p_A/P' x_A$, if logarithms are taken, there results:

$$\log \gamma_A = \log p_A - \log P' - \log x_A \tag{106}$$

with P' the vapor pressure of pure A, here also the reference substance. This may be differentiated with respect to $\log P'$; since x_A is a constant its term disappears:

$$\frac{d \log \gamma_A}{d \log P'} = \frac{d \log p_A}{d \log P'} - \frac{d \log P'}{d \log P'} \tag{107}$$

However, the first term on the right of the equality sign represents the value of m, in the reference substance equation, or the slope of the partial pressure lines on a log plot. The second term to the right of the equality sign is equal to unity. Thus both are constant, and hence the left side of the equality sign is also a constant. This term represents the slope of the lines of the logarithmic plot of activity coefficients vs the vapor pressure of the reference substance at the same temperature, and is equal to the slope of the respective partial pressure line.

The nonreference linear equation of the logarithm of the activity coefficient versus the logarithm of the total pressure on the system has also been found to correlate activities very well.

The advantage of these two methods of plotting the various properties of binary solutions is that they give straight lines, even for the most irregular solutions, including those that form maximum and minimum boiling points, and those (9) wherein ionization and chemical combinations are important in making solutions highly nonideal (eg, hydrochloric acid and water).

Other important relations, theoretically and also for the design of distillation processes and equipment, are equilibrium constants and relative volatilities (10). These have been shown to give linear relations either with respect to the reference substance vapor pressure or temperature scale or as a simple function of the logarithm of the total pressures. These functions can be fed into computer programs, or plotted as straight lines.

Equations comparable to equation 75, ie, values of data taken at the same temperatures, have been indicated for vaporization relations of systems of two volatile liquids. However, other vapor pressure equations may also be used, where advantageous, particularly equations 80 or 83, utilizing values of data at the same pressures. Furthermore, although the use of isosteres of constant liquid composition x is usually preferred, isosteres of constant vapor composition y give equally good results.

Reduced values of temperature and/or pressure have been noted above to have advantages in correlating and predicting data for liquids containing only a single volatile component. Similarly they have been found to have advantages in correlating and predicting two-component vaporization. Thus a linear equation correlates vapor–liquid equilibria when, at constant composition, the logarithms of relative volatilities are taken as a function of the ratios for a reference substance of the specific volume of the vapor to the specific volume of the liquid, all at the same value of the reduced temperature. Other uses of reduced conditions give excellent linear correlations (11) except in the critical region.

Azeotropic Mixtures. Azeotropic mixtures are those nonideal liquid mixtures wherein the liquid boiling at one composition has the same composition as the vapor. This boiling point is usually lower than that of either liquid and at the temperature of the azeotrope, the total pressure is greater than that of either pure component. Such a mixture is called a minimum boiling azeotrope. More rarely, maximum boiling azeotropes are formed where the situation is reversed. The azeotropic composition changes with temperature or pressure, and azeotropic behavior may disappear entirely in certain ranges (see Azeotropic and extractive distillation).

When equation 75 is applied to the total pressures of different liquid mixtures of constant composition, it follows that, instead of forming a sheaf of lines with a regular gradation of slopes, as for ideal solutions, lines near those of a minimum boiling azeotropic composition are found above the line for the pure, more volatile liquid, as shown in Figure 15. Usually, this is accompanied by a marked variation in the slopes of the lines, indicative of a marked change in the heat of solution of one component in the other (13).

The corresponding effect of an azeotropic mixture on the lines of vapor composition may result in a change of the sign of m, the slope of the lines, from plus to minus or vice versa. If this occurs the value of m and of the slope at some value of the liquid composition must be zero. Hence, there is no change in the composition of the vapors with a change of the boiling-point temperature (or pressure). This interprets a seemingly unrelated phenomenon for some liquids which form azeotropes, namely, that all of the plots of their vapor composition versus liquid composition pass through a common point of vapor and liquid composition (12). This point does not coincide with the azeotropic composition.

Of greater use in correlating azeotropic data is the fact (13) that the total and partial pressures at the azeotrope, vapor compositions, and activity coefficients, are represented by equations identical to equations 75, 80, 83, 86 and 88. The equations derived therefrom may be used immediately as functions of a single material, and data may be correlated or predicted by the same techniques developed above. Thus there may be inserted in the equations 75, 80, 83, 86, and 88 a superscript A representing P^A as the total pressure of the azeotrope, p^A as the partial pressure of either component desired, y^A as the vapor composition, or γ^A as an activity coefficient. Furthermore, these may be considered in relation to P', the vapor pressure of the reference substance to gain a temperature scale effectively, or P^A, the total pressure on the azeotropic system. All of these correlations may be used with assurance for correlating and hence predicting data. The corresponding heat terms, that define m the slopes of the lines, and that may often be developed from other considerations, allow the interrelation of vapor pressure, vapor compositions, and calorimetric data to predict those which are unavailable.

Properties of Gases

Reference Equation of State for Gases. *Pure Gases.*

Equations of state express the relation between the pressure, volume, and temperature of a gas. Van der Waals in 1879 correlated data for real gases and corrected the simple equation for a perfect gas, $PV = RT$, by assuming the size and interaction of molecules. Empirical corrections made since have added numerous constants in fitting data to curves, without assessment of the physical meaning, if any, of these constants. However, the compressibility factor $Z = PV/RT$, is the precise criterion of the individual gas and its equation of state. It is expressed in the virial form for a pressure series:

$$Z = \frac{PV}{RT} = 1 + \frac{B}{RT}P + \frac{(C - B^2)}{R^2T^2}P^2 + \frac{(D - 3BC + 2B^3)}{R^3T^3}P^3 + \ldots \quad (108)$$

where B is the second, C the third, and D the fourth of the series of virial coefficients. The coefficients B, C, and D, as well as Z might possibly be calculated if sufficient information were available on the fundamental characteristics of molecules. The important coefficient, B, depends only on the temperature and the attraction between molecules, ie, molecular force constants. These, and thus B, may be derived directly from critical constants T_c and P_c. But C and D are difficult or impossible to evaluate either theoretically or empirically using present understanding of molecular physics.

The reference-substance technique is a valuable aid, inasmuch as the same laws govern the P–V–T relations of any two gases at comparable conditions. By considering this relation of properties at the same reduced temperatures and reduced pressures, Z may be determined with a tolerable accuracy without the use of the third and higher virial terms. The equation first used (65), has been developed (66) in a different form:

$$Z = \phi(RT + BP) \quad \text{where} \quad \phi = Z'/(RT + B'P_R P'_c T_c/T'_c) \quad (109)$$

A more useful form may be $V = \phi RT(RT/P + B)$, where ϕ includes known values and is evaluated from the given temperature T and pressure P; T_c and P_c; T'_c and P'_c; R, the gas constant; B', the second virial coefficient of the reference substance; and Z', the compressibility of the reference substance. (The values with primes represent properties of the reference substance.) The compressibility factors Z and Z' are at the same reduced temperature and the same reduced pressure, and the second virial coefficients B and B' are at the same reduced temperature.

This relation was tested with those gases for which sufficient P, V, T data were available (65). In all, 21 gases were tested with over 2000 experimental points. Nitrogen was used as the reference substance. The average deviation in predicting values comparable to each of the experimental points was less than 1%. Thus the reference technique correlates all available data to an equation of state using only one readily calculable virial coefficient B. The only data required for any gas are its critical temperature and critical pressure.

Values of Z so calculated (65) agree with literature values, within their experimental errors, and very much more closely than are given either by the virial equation or the generalized compressibility chart. From Z can be calculated exactly the volumes V using $Z = PV/RT$. Thus this calculation predicts V directly and relatively accurately at the desired T and P, whereas empirical equations of state usually give P as a function of T and V, a much less useful form for most engineering calculations.

Alternatively, the equation expresses V directly:

$$V = \phi RT(RT/P + B) \tag{110}$$

The Martin-Hou equation of state has been modified using 21 constants (67). These empirical constants, which must be evaluated for a particular gas, were determined by fitting to experimental data, in the range available, data for one pure gas, chlorodifluoromethane. The P, V, T data were available throughout some of the rather narrow range $T_R = 0.791$–1.146 and $P_R = 0.131$–2.63. Evaluation and adjustment of the 21 empirical constants allowed calculation of P from T and V so as to obtain a good fit to the 181 data points.

The values of the chlorodifluoromethane system were then predicted (68) using the simple equation of state, $Z = \phi(RT + BP)$, wherein the single constant B is determined knowing only the critical conditions, T_c, V_c, and P_c. Also needed are P, V, T data for a reference substance, always considered at the same values of reduced temperature and pressure.

Nitrogen was used as the reference substance. The compressibility factors for nitrogen at various temperatures and pressures were predicted as before (66). In the entire temperature range of the gas, and at pressures up to 2.4 MPa (350 psia), only the critical values for the material are needed. The average deviation of the prediction from the experimental value is less than 0.7% as calculated by this reference substance equation with only one constant, B to be evaluated. Most engineering work with nitrogen would probably be within this range of accurate prediction, ie, below 35.5 MPa (350 psia).

Another reference-substance technique (69) starts with the Benedict-Webb-Rubin equation of state (BWR). The use of the theorem of corresponding states greatly reduced the effort necessary to evaluate the 8 BWR constants. A large amount of experimental data must be available for a closely related compound, and these are available for only a few gases. Many calculated values were discarded, others checked closely.

Equation 110 and only the critical constants were used (68) to correlate all 50 points of the experimental data. The temperatures were within 3°C of the critical, a very bad range for prediction. The average deviation of all calculated values from experimental values was 3.16%, maximum 5.82%.

Mixtures of Gases. Equation 110 correlates and predicts data for gas mixtures (66,68). Only the pseudo values of critical temperature and critical pressure must be determined which represent the mixture effectively.

Critical constants depend upon the force of interaction between molecules. The effective critical constants for a mixture may be calculated if these forces are known. The mixture of gases is assumed to have the same thermodynamic properties as those of some hypothetical or pseudo-pure gas. The force constants of this pseudo gas are assumed as equal to the force constants of the gas mixture in calculating the effective critical properties. These calculated, ie, effective, critical constants for the mixture may be used in equation 110 and possibly elsewhere.

Again this method requires as input data, besides the known data for a reference substance, only the critical constants of the individual components. In many cases, as noted above, these may be estimated from the vapor-pressure equations. Critical values are used to calculate: (*1*) the pseudo or effective force constants that are assumed to represent those of the mixture, (*2*) the pseudo or effective critical constants of the mixture, (*3*) the compressibility factors of the gas mixture, and, from these (*4*) the complete P, V, T relations. This method has been found to hold for gas mixtures where the components are: (*1*) nonpolar; (*2*) polar, or (*3*) nonpolar and polar.

The literature contains more than 1000 values of P, V, T data for 34 binary and four ternary systems. For each of these points, values were calculated, ie, predicted, using only the critical constants of the individual components, also using other methods. The predictions based on this reference substance equation showed an average deviation from the experimental data of all published work of 2.34%. Some of these results are illustrated in Table 5 (66).

Thus in the absence of experimental data, this simple equation of state using a reference substance, may be used to predict P, V, T data for gas mixtures with a reasonable expectation of a close approximation to actual values.

By this method, the volume of the substance V is calculated directly for each indicated pressure and temperature. This volume would be most commonly needed in engineering practice, rather than P which results as a function of V and T, as in the Martin-Hou equation, the BWR equation, and most others.

Conventional equations of state with 8 to 21 empirical constants require extensive experimental data. Sophisticated methods are available for drawing the best curve through many points. This is entirely empirical and does not concern the physical mechanism or trends which might allow prediction in other ranges. Contrariwise, only the critical constants T_c, V_c, and P_c of the gas in question are needed for the reference equation of state. Even for gas mixtures only these critical data for the individual components are necessary for a reasonably close prediction.

As usual with the reference-substance technique, data may be predicted in any range of conditions for which data are known for the reference substance. Possible errors in experimental data may be demonstrated by deviations larger than expected.

Table 5. Comparison of P, V, T Data Determined by Different Methods

System	Temperature, K	Pressure, MPa[a]	Composition, mol fraction	Ref. 66 Max	Ref. 66 Av	Generalized table Max	Generalized table Av	Virial equation Max	Virial equation Av
methane-carbon dioxide	511.11	551–5513	methane = 0.8469	3.619	1.353	6.640	3.449	20.957	6.187
methane-isopentane	477.7	273–896	methane = 0.4976	3.838	1.247	6.465	5.597	30.122	12.144
ethylene-argon	298	304–1013	ethylene = 0.5986	4.987	2.632	6.990	3.273	25.732	14.210
propane-benzene	411.1	138–552	propane = 0.8827	2.428	1.051	16.955	5.046	39.211	8.548
n-butane-nitrogen	427.7	414	n-butane = 0.1–0.9	2.273	1.415	11.667	4.600	10.000	6.758
toluene-n-hexane	saturated conditions		toluene = 0.6344	5.348	3.549	16.461	10.030	55.282	28.061
hydrogen-nitrogen-methane	373	1013–7091	hydrogen = 0.5266 nitrogen = 0.2611	4.033	3.103	11.973	8.949	22.082	9.466

[a] To convert MPa to atm, divide by 0.101.

Solubilities. So many empirical methods have been used for expressing solubility data for gases in liquids, that correlation of experimental data has been hampered. Furthermore, the usual practice of working with isotherms has been confusing in dealing with two of the three parameters, pressure, temperature, and concentration. The use of isosteres, ie, lines of constant concentration, allows a simple correlation; ie, the pressure of the solution of a given concentration of gas at a given temperature can be treated like vapor pressure (40).

If x is the concentration of the gas dissolved at the temperature T under its partial pressure P, the same correlations as have been used for vapor pressures give partial pressures of a fixed amount of gas dissolved (constant x) as a linear function of the vapor pressure of water at the same temperatures (40). A straight line (isostere) expresses the pressure–temperature relation for each fixed concentration. The slope of the line is the ratio of the partial heat of solution Q of the gas to the latent heat of the reference substance (see eq. 75, where $m = Q/L'$).

Since data are often taken at constant temperatures, the experimental results may be plotted first as isotherms of concentration on the Y axis vs pressure on the X axis. Horizontal lines at fixed concentrations are drawn and values at the intersections of these lines of constant concentration with the isotherms are chosen. The pressure value of each intersection is plotted logarithmically against the temperature value based on the logarithm of the vapor pressure of the reference substance at the same temperature measured along the X axis.

Henry's law constant H in the expression $P = Hx$, holds for one temperature. It varies with the temperature as a linear logarithmic function, and correlates on a logarithmic plot as in Figure 17 to be the familiar straight-line function of the vapor pressure of water at the same temperature (eq. 75, where $m = Q/L'$).

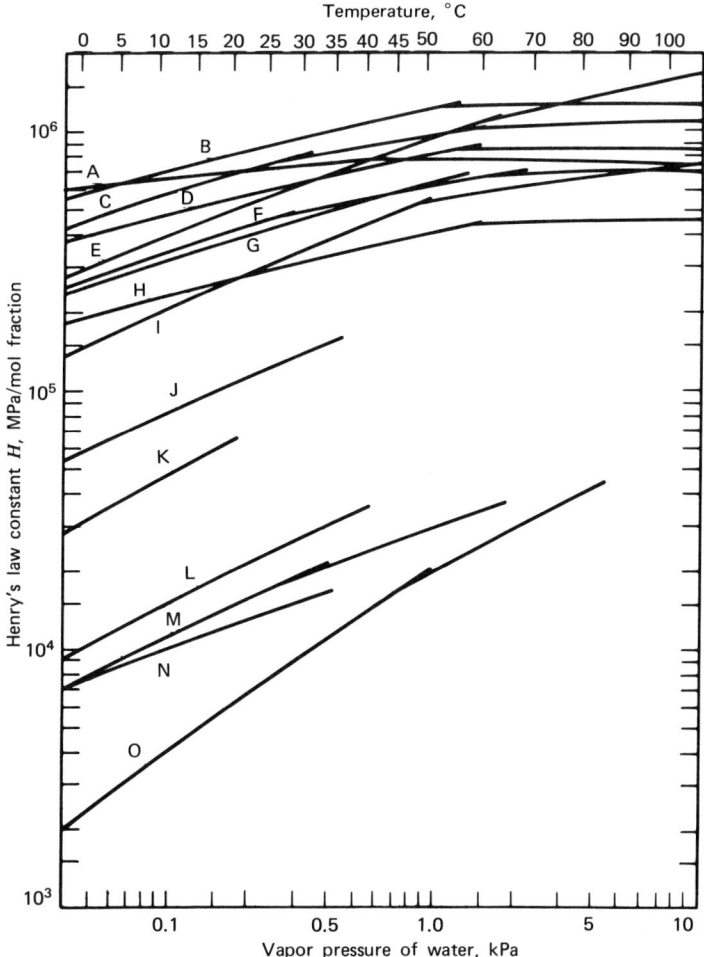

Figure 17. Plot of Henry's law constant for different cases vs vapor pressures of water at the same temperatures. A, H_2; B, N_2; C, air; D, CO; E, H_2S; F, O_2; G, CH_4; H, NO; I, C_2H_6; J, C_2H_4; K, C_3H_6; L, N_2O; M, CO_2; N, C_2H_2; and O, Br_2. To convert kPa to mm Hg, multiply by 7.5. To convert MPa to atm, divide by 10.13.

For a few compounds the straight line relations for gas partial pressures and Henry's law constants have a slight break at some temperature. This may indicate at that temperature a change in the relation of the gas molecules to the solvent molecules, eg, of hydration. In any case, the heat of solution Q changes abruptly, and so does the slope of the line. The amount of change of the slope may indicate the heat of hydration. A major advantage of reducing data to a linear relation is that even a slight break in the function can be clearly seen. If the correlation is curvilinear, the break would pass unnoticed in smoothing the constants to an unnoticed irregularity.

These linear relations have been expressed, eg, in a general equation for SO_2 in water throughout the whole gas–liquid system (56). The m and C constants were evaluated as a function of the concentration x and assembled in an overall equation.

100 ENGINEERING AND CHEMICAL DATA CORRELATION

This overall equilibrium equation for gas solubilities and partial pressures at different temperatures, when used in conventional or computerized gas absorption calculations, aids the design of gas-absorption equipment. The molal heat of solution can be determined immediately from the slopes of the isosteres. It may be expressed separately as an additional quantitative input into the calculations.

In some uses, equation 80, where data values of pressure are taken at the same temperatures as those of the reference, may be more useful for expressing solubilities and Henry's constants for gases. The representation is equally good.

Adsorption. Adsorption pressures of gases from a solid adsorbent, eg, activated carbon, give linear logarithmic relations with a reference substance for the isosteres, the lines of constant composition (70–71). Straight-line isosteres are obtained by cross plotting of isotherms as explained above. In Figure 18 the isotherms representing the experimental data taken at constant temperatures are first plotted on logarithmic paper to show the classic Freundlich isotherms curves and the relation thereto. A single sheet of logarithmic paper is used as shown with concentrations indicated on the left of the X axis. These values are relatively easy to correlate and use when plotted as isosteres, then by horizontal projection to the right, where values of the temperature are indicated from vapor pressures of the reference substance. From the slopes of the isosteres the heats of adsorption can be calculated readily.

Here again, the equations for the linear isosteres may be combined into a general equation for a particular adsorbate–adsorbent system to allow ready evaluation of adsorption design calculations, particularly if computerized. Heats from the slopes of the lines are important. Again, equation 80 may be used if pressures at the same temperatures seem to be the desirable way to use the reference substance. The method has been expanded (71) to allow also use at any temperature when data at only one point are available.

The same relation holds for adsorption of materials (usually liquids or solids) from liquids to solids. Again, isosteres are used to express the pressure or tendency of the adsorbate to leave the adsorbent at different temperatures.

Viscosities. When gases are maintained at constant pressure, their compressibility is not involved. Hence, the same analysis may be used as for liquids, and the same linear equations were found to represent the physical functions of viscosity; eg, the familiar equation form of log viscosity u vs log vapor pressure of a reference substance P' at the same temperature:

$$\log u = -m \log P' + C \quad \text{(liquids and gases, isobaric)} \tag{111}$$

Another example is the equation form of log viscosity vs log viscosity of a reference substance at the same temperature:

$$\log u = m \log u' + C \quad \text{(liquids and gases, isobaric)} \tag{112}$$

Both equations express all available data for liquids and also for gases when at constant pressure. To correlate data on gas viscosities at constant temperatures, it was necessary to consider kinematic pressure P_k and its ratio to gas density in order to develop a linear equation. Isotherms of the logarithm of the gas viscosity $\log u$ are represented by a linear function of the logarithm of the kinematic viscosity divided by the density d (72):

$$\log u = a \log (P_k/d) + C' \quad \text{where} \quad P_k = P + (a/V^2) \tag{113}$$

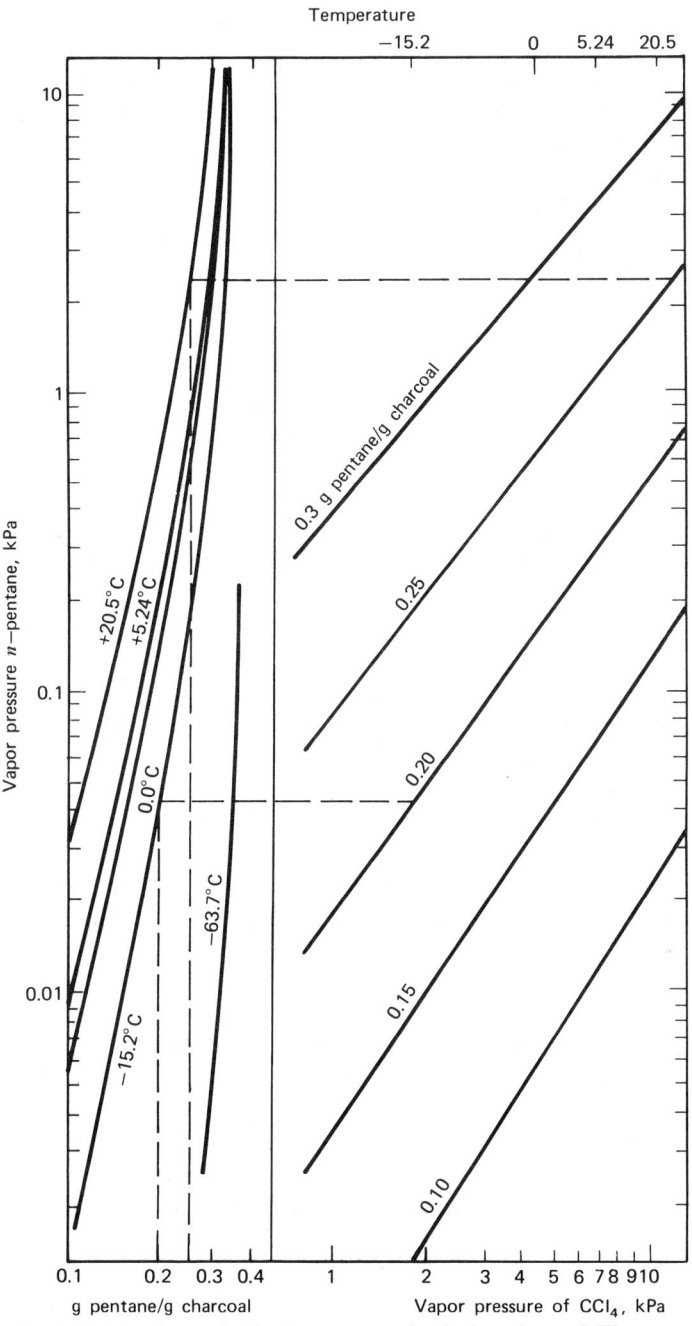

Figure 18. Freundlich isotherms (left) of n-pentane adsorbed on charcoal. The corresponding straight lines of constant composition (right) may be interrelated by projecting horizontally the intersections of the isotherms with the ordinates of concentration, as shown. Courtesy of *Industrial and Engineering Chemistry*. To convert kPa to mm Hg, multiply by 7.5.

Here a is a constant. By considering the various molecular properties of gases (molecular weight, number of molecules per cubic centimeter, mean-free path, velocity, density, etc), the constants a and C' could be evaluated (72) for each considered value of temperature expressed as reduced temperature. Experimental data were found to be well correlated, and a nomograph was prepared which is a general representation over a wide range of temperatures and pressures. It predicts values closely for eleven gases for which data were available at different pressures and temperatures. The linearity of the basic equation has also allowed a simple program to be prepared (2) for pocket calculators.

Diffusion. Gaseous diffusion, ie, mass transfer, controls many industrial processes (see Diffusion separation methods). The mathematical expression is thus of considerable theoretical interest. The mechanisms involved are obviously related to those of viscosity, and thus also should be their mathematical expressions. It has been shown (73) that, at a constant pressure, the diffusion coefficient D of one gas in another is a linear function of temperatures developed from viscosity u also at constant pressure of any gas as a reference (eg, air). The straight line slope is 2.74. This straight-line function has been used to correlate data for many gases. Deviations are within the experimental error.

This simple linear relation developed from kinetic relations of the molecules, is very useful in predicting diffusivities at, say, atmospheric pressure from viscosities because of the relative difficulty of determinations of diffusivities compared to those of viscosities:

$$\log (DP) = 2.74 \log u + C \tag{114}$$

The pressure P is maintained constant. The constant C has been evaluated from the molecular and physical properties considered. A resulting general equation has been derived that includes the temperature variation of diffusion coefficients of <200 to >1400 K. This theoretical representation was tested with 59 binary systems, and compared favorably with the best of six other methods proposed, usually more complicated in use. The linearity of the basic relation has simplified the program development for pocket calculators (2) and the construction of a nomograph (73) that allows the immediate evaluation of diffusion coefficients. Required for the equation, calculator program, or nomograph methods are merely the molecular weights and the critical molal volumes of the two gases.

Permeability of Membranes. A large-volume product of chemical industry is membranes which compare with, and often replace paper products, animal skins, and ultra-thin metal foils. The membranes range in thickness up to that of synthetic leather.

The largest field of industrial use is in films for packaging or wrapping. Here, a plastic film may be desired which is as nearly impermeable as possible, particularly to moisture, as well as to other gases which might cause a deterioration of the product. Alternatively, the membrane desired may be permeable or selectively permeable to one gas, eg, oxygen may be highly desirable to kill anaerobic bacteria.

Permeability relates the two functions of absorption or solution and diffusion; both vary with temperature. (Dissolution or desorption may be included in solution.) An increase in temperature increases the vibrations of the polymeric molecules of the membrane. Thus the diffusion rate increases with increasing temperature. It has been shown above that diffusion as well as solubility and absorption are correlated by a temperature scale based on the vapor pressure of water. Hence, permeability might also be so correlated.

Well recognized equations based on these theories have been combined (74). In these, all values are considered at the same temperature. The equation so obtained was in a differential form so that it could be divided by the Clausius-Clapeyron equation to give, on integration:

$$\log \pi = [(E_D - \Delta H)/L'] \log P' + C \quad \text{or} \quad \log \pi = m \log P' + C \quad (115)$$

where π is permeability constant, E_D is the activation energy of diffusion, ΔH is the heat of solution of the gas in the plastic material of the membrane, P' and L' are the vapor pressure and molar latent heat of water at the same temperature, and m and C are constants. The difference of the heats of diffusion and of solution appears to be very logical in this relation. The correlation has been shown to hold within the accuracy of the experimental data for thousands of points for over 150 systems with temperatures ranging from -50 to $130°C$ and permeability constants varying over 10,000-fold. The equation may be used with confidence directly or in calculator or computer programs. A nomogram is presented in ref. 74, giving values within the probable experimental errors and useful with any other system.

Enthalpies and Entropies. Vast amounts of thermal data in tabular form are available for many compounds, eg, the steam tables, but relatively little for others. Exact equations and computer programs are available for calculating entropies and enthalpies for many gases. Design calculations for many others may have to be made from only scattered data. Utilizing the reference-substance technique, linear equations may be formulated readily, and complete data may be easily calculated directly or through programs for pocket calculators or computers. Mollier diagrams and entropy and enthalpy charts may be prepared quickly on ordinary (not logarithmic) graph paper using only straight lines throughout wide ranges of temperatures and pressures.

The differential entropy S of a gas or gas mixture, and of a reference gas S' of similar structural characteristics—both taken above the same temperature, at constant pressure, and having values always taken at the same temperature—are defined (45) in relation to their specific heats c_p and c'_p at constant pressure:

$$dS = c_p(dT/T) \quad \text{and} \quad dS' = c'_p(dT/T) \quad (116)$$

since T is the same:

$$dS = (c_p/c'_p)dS' \quad \text{or} \quad \Delta S = m\Delta S' \quad (117)$$

The integrated form is exact. For the members of homologous series, while individual specific heats vary markedly their ratios have been found to be constant even over a temperature range of thousands of degrees. This is a very simple family of linear equations, giving a sheaf of straight lines through the origin, one for each homologous gas. See ref. 45 for entropy above $25°C$ for paraffin hydrocarbon gases at atmospheric pressures.

Other constant pressures give other straight lines that are parallel to the line at atmospheric pressure.

Similarly entropy changes with pressure changes may be evaluated at a constant temperature for a change in pressure between P_1 and P_2:

$$\Delta S = R \ln (P_1/P_2) \tag{118}$$

and at different constant temperatures:

$$\Delta S_{T_1} = \Delta S_{T_2} \tag{119}$$

This assumes some properties of a perfect gas. Corrections can be made that are less than 1% at usual temperatures and pressures. The vertical distances (differences of entropy) between lines of constant pressure are equal, thus the linear isobars are parallel. Hence, each is established by only a single experimental point, as shown in ref. 45, for the entropy of superheated methane. It may also be shown that the reference relation of entropy at constant gas volume is linear; and dashed lines of constant volume may be drawn to serve as an additional grid. Lines of constant enthalpy are slightly curved, and another grid for them also may be indicated.

Linear enthalpy equations, computer programs, and graphs using a reference material may be prepared similarly and with the same slight errors and possible corrections. At constant pressure above the same base temperature, enthalpies ΔH and $\Delta H'$ are related:

$$dH = (c_p/c'_p)dH' \tag{120}$$

integration gives:

$$\Delta H = m \Delta H' \tag{121}$$

Again, a family of linear equations or straight lines through the origin represent all substances, and each line represents all pressures for that substance. With water as the reference, its values come from the steam tables.

Examples of Use of Vapor Pressure Equations and Correlations

Many immediate and direct methods have been suggested above of correlating known and predicting unknown data from the interrelation of vapor pressures and latent heats in several linear equations and graphical methods. This includes heats or activation energies of hydration, absorption, adsorption, freezing, solution, dilution, molecular change, as well as freezing and other transition points, etc. Many other representations, correlations, and predictions of data are also possible. A few are given below.

Heats of Solution and Enthalpy Charts. A solution with respective concentrations of two components x_A and x_B has an enthalpy h_S above a fixed temperature, which is the sum of the enthalpies of the individual components h_A and h_B and of that of mixing or dissolution. Thus if q is the integral heat of solution for one mole of solute:

$$h_S = x_A h_A + x_B h_B - x_A q \tag{122}$$

A solution in a solvent has for each of many different concentrations x of the solutes an example of equation 75. Each represents the vapor pressure relations and has a different value of m for each one of the family of lines of different concentrations between x_A and x_B. The reference substance is the solvent itself. The integral heat of solution between these two concentrations can be shown to be equal to an integral

Φ times the latent heat of the solvent (61–62). For convenience, let r equal the ratio x_B/x_A. Then:

$$q = L' \int_0^r (m-1) dr = L'\Phi \tag{123}$$

The value of m and hence of $(m-1)$ is readily obtained for any concentration by measuring the slope of the partial pressure lines on the logarithmic reference-substance plot, or by deriving it algebraically from the corresponding linear equations. Successive values of Φ may be evaluated through stepwise graphical integration of a plot of $(m-1)$ against the ratio $(r = x_B/x_A)$.

The relationship between the heat of solution and the temperature at any concentration is proportional to the latent heat of the pure solvent L'. Latent heats of pure compounds have been shown to vary according to that of water. Thus Φ in any system varies with temperature as does the latent heat of water, decreasing slightly with increasing temperature.

The reference state for a solid solute is usually the state of infinite dilution; thus:

$$h_S = x_A \overline{h}_A + x_B h_B + x_A q_\infty \tag{124}$$

Here, instead of enthalpy of a pure component, the partial enthalpy of the solute at infinite dilution $\overline{h}_A$ has been used and the finite heat of solution is substituted by the heat of infinite dilution. Again $r = x_B/x_A$.

$$q_\infty = L' \int_r^\infty (m-1) dr \tag{125}$$

Graphical evaluation of Φ for the ethanol–chloroform system has given good agreement with previous methods (61) as shown by considering the heat involved per unit weight, using the respective molecular weights M_A and M_B:

$$Q = x_A q/(x_A M_A + x_B M_B) \tag{126}$$

Figure 19 shows a comparison (75). The entire enthalpy-concentration diagram is readily constructed (61). Usually called a Ponchon diagram as in Figure 20 it allows the much more accurate calculations for distillation columns than is possible with the McCabe-Thiele system, which assumes no sensible heat or heat of solution effects.

From vapor pressure data, with calorimetric data when available, the above equations also allow the construction of complete enthalpy plots for solutions of a nonvolatile solid in a solvent, eg, water. Figure 14 was constructed in this way for the water–ammonium nitrate system (62).

Prediction of Vapor–Liquid Equilibria. Because of industrial importance in all distillation calculations and designs, vapor–liquid equilibria (often called simply x, y) have been carefully studied thermodynamically. Noted above has been the correlation of partial pressures, vapor compositions, and heats of solution of mixtures of volatile liquids, and the plotting of enthalpy concentration charts (Ponchon diagrams) from these data.

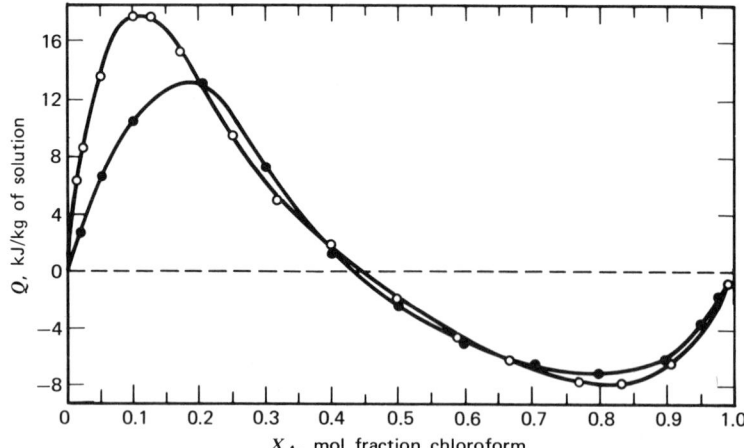

Figure 19. Agreement of data obtained by this method with those calculated by other methods is good except in the region below $X_A = 0.15$. ●, Calculated by method of Scatchard and Raymond (75); ○, calculated by present method. To convert kJ/kg to Btu/lb, divide by 2.324. To convert kJ/kg to Btu/lb, divide by 2.324.

Just as vapor pressure evaluations always are based on an attempt to integrate the Clausius-Clapeyron equation, x, y correlations always attempt an integration of the Gibbs-Duhem equation. The latter is derived for constant temperature, although distillations are conducted at constant pressure. However, here again the heats of mixing are important.

Solutions of these linear equations of vapor pressures and vapor–liquid compositions at the same pressures allow the slope constants, m, or intersections, C, of the corresponding equations or lines of a graph, through thermodynamic derivations, to represent vapor–liquid equilibrium data at constant pressure. Also used is a means of representing data for equilibria in two phases (14) earlier used (33) for liquid–liquid equilibria.

For binary systems at constant pressure, relations were derived (14) for x, molar composition of the liquid; y, molar composition of the vapor; and T, absolute temperature.

$$\left\{\frac{x}{y} - \frac{1-x}{1-y}\right\}\left(\frac{dy}{dx}\right)_P = \frac{L_x}{RT^2}\left(\frac{dT}{dx}\right)_P \tag{127}$$

This equation is correct, thermodynamically, within the limitations of Dalton's law of partial pressures and the Clausius-Clapeyron equation.

Equation 127 does not assume that activity coefficients for a given composition are constant with varying temperature. This is a major assumption in the Redlich and Kister equation, which also is based on the Gibbs-Duhem equation for x, y, and T at constant pressure (76). In the above more nearly rigorous equation, L_x represents the actual molar latent heat of vaporization (including the heat of solution).

Equation 127 may be evaluated with either vapor pressure data or boiling point-composition data. Neither requires the experimental determination of vapor compositions, which requires the analysis of the vapor compositions, usually also of liquid compositions. Adequate vapor pressure data with fewer chances of error may

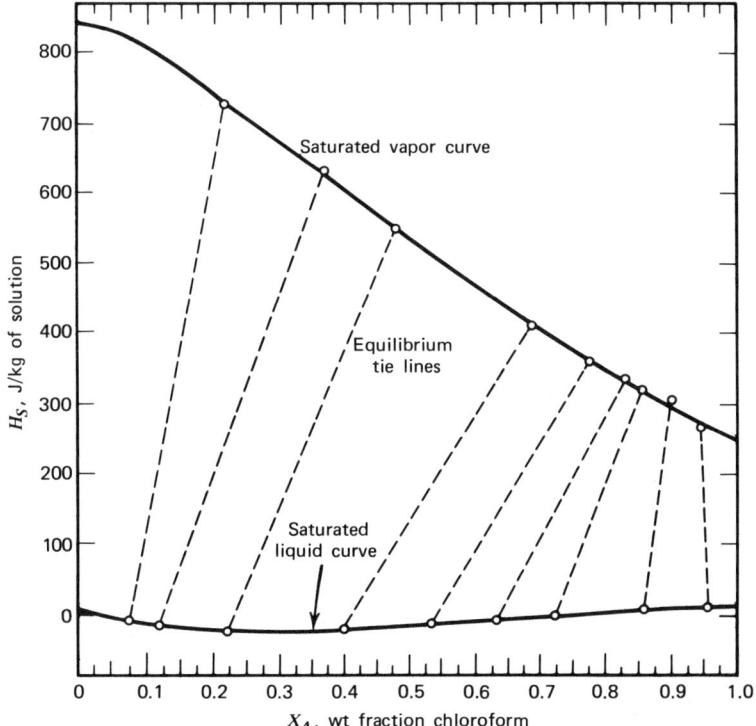

Figure 20. Ponchon diagram for ethanol–chloroform includes tie lines connecting equilibrium compositions on saturated liquid and saturated vapor curves.

be obtained readily and accurately from the family of straight lines of reference logarithmic plots of constant composition (or their corresponding linear algebraic equations).

These require only a relatively few vapor pressure points to define or determine P, T, X data for the entire system. The slopes of the vapor pressure lines (values of m) allow the obtaining of the sum of the individual molal latent heats of the components in any particular composition plus the heat of mixing (8). The heat of mixing thus can be determined by appropriate subtraction if desired. These vapor pressure lines of different compositions thus demonstrate the vapor compositions and all heat quantities (7).

With the reference-substance method of evaluating vapor pressures and latent heats, the above differential equation may be used by evaluation of its right side through substitution of the data, and then proceeding by step-wise integration. From the boiling point composition (T, x) diagram, and starting at $x = 0, y = 0$, it is possible, by taking small increments of Δx and ΔT to determine successive values of y so as to obtain the entire x, y curve. The curves resulting from such integration for the system ethanol–water (with an azeotropic point) and acetone–water (nonideal and nonazeotropic) by taking increments of Δx of 5%, are shown in Figure 21 as representative of many (14) so determined as a check against the experimental determinations of curves of vapor–liquid equilibria.

An even more rigorous test of experimental data is the evaluation of both sides

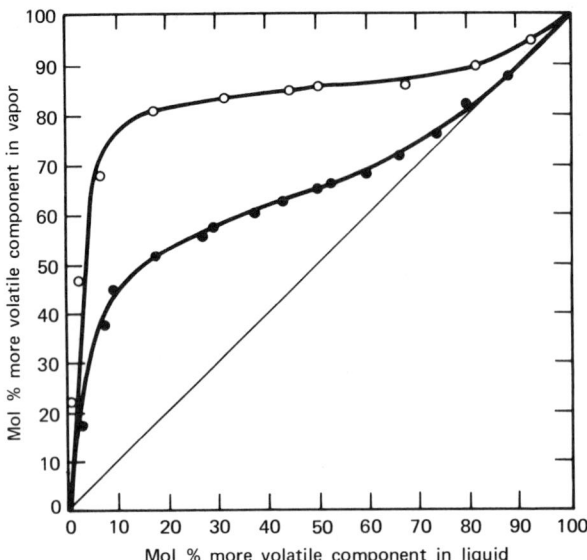

Figure 21. Plot of mole percent more volatile component in vapor vs mole percent more volatile component in liquid. —, Calculated; ○, acetone–water experimental; ●, ethanol–water experimental.

of equation 127 (14). These should always be equal if correct values are used. Algebraic relations were also developed from this technique for expressing the constant pressure or constant temperature vapor–liquid equilibrium relations for both ideal and nonideal binary or ternary systems. Linear equations represent a straight line or a broken line of three sections on a logarithmic plot. These permit the ready evaluation, correlation, and prediction of x, y, T, P data.

Evaluations and correlations of vapor–liquid equilibria for some 38 representative binary systems were thus reduced (14) to linear functions as straight lines and correlated against each other. A plot at constant pressure of $X = x/(x - 1)$ vs $Y = y/(y - 1)$ gives a line made up of three straight sections for any system. The two outer sections are always straight and have a slope very close to unity; the center straight section may have a slope greater or less than unity. For ideal systems, all three sections have a slope of unity and a single straight line results. Data (14) for several systems both ideal and nonideal are plotted in Figure 22.

Extensions of the algebraic treatment of these functions Y and X have made possible the ready and accurate prediction of vapor–liquid equilibria for both binary and ternary systems (14).

Equilibrium Flash Vaporization (EFV). Petroleum fractionators separate many components and are dependent on data of the vapor–liquid equilibria of these components. Often obtained from true boiling point distillations, the EFV data may be obtained from pilot-plant operations. More conveniently, a special equilibrium recycling still made of glass (77) determines EFV data in the laboratory for the equilibrium temperatures at one or several constant pressures corresponding to the volume percentages of the material vaporized. The data may be used directly in the design of distillation and related equipment for petroleum refining.

Such data are reduced to linear equations for ready use in computer or hand

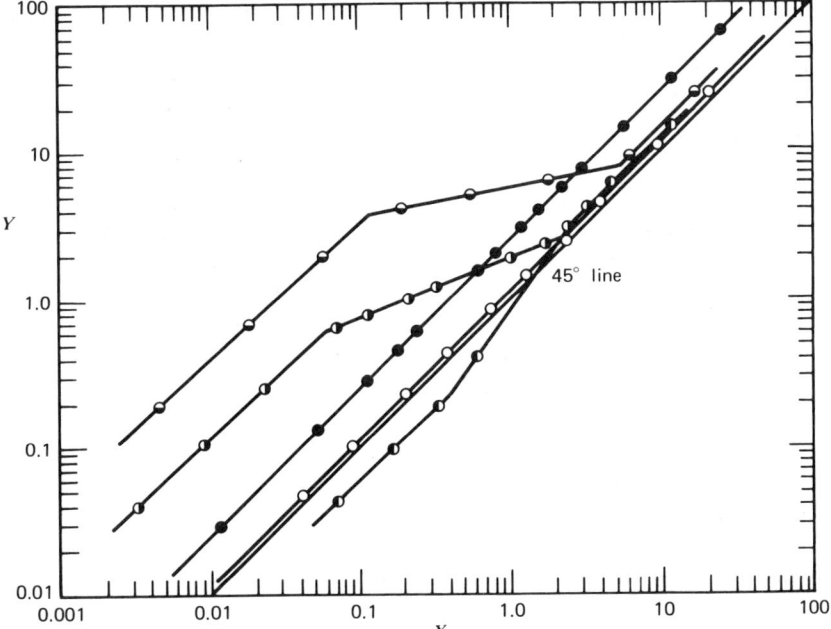

Figure 22. Plot of Y vs X on logarithmic paper for ideal systems (●, benzene–toluene; ○, benzene–ethylene dichloride) and nonideal systems (◐, ethanol–water; ◑, chloroform–acetone; ◒, acetone–water).

calculations. In addition, a logarithmic plot of the total pressures of these determinations vs. the vapor pressure of the reference substance gives a sheaf of straight lines through a single point (78). Each of these lines represents the vapor pressure for a different percentage vaporized of the original liquid crude or fraction. The effective heats of vaporization were readily determinable. From these data and the linear equations obtained from worldwide representative crudes, a nomogram has been developed. Either the equations or the nomograph allows calculations within a few percent for plant design at any pressure or vacuum. In fact the optimum pressure for operation may be so determined. The input requirements are merely the boiling point data that are determined at any pressure point, not necessarily that of the operation being designed.

Activities of Electrolytes. Closely related to vapor pressure is another thermodynamic property, involved in considering the theory and operation of electromotive cells. This is the activity coefficient γ for the solvent in a solution which is defined as:

$$\gamma = P/P'x \qquad (128)$$

where P is the vapor pressure, P' is the vapor pressure of pure solvent at the same temperature, and x is the concentration of solvent in the solution. At the same temperature, equation 128 combined (8) with equation 75 gives:

$$d \log \gamma / d \log P' = \Delta H/L' \quad \text{or} \quad \log \gamma = (\Delta H/L') \log P' + C \qquad (129)$$

where the respective variables are taken at the same temperature and same concentration, and ΔH is the partial heat of solution of component 1, and L' is the latent heat of one mole of pure solvent. Their ratio is constant over a wide range of temperatures. Activity coefficients may thus be directly correlated by a linear function of the temperature, expressed by the vapor pressure of a reference substance such as water.

Although activity coefficients are used with and obtained from P, T, x, y data as noted above, the activity coefficients for the system hydrochloric acid and water were determined precisely electrochemically (79). A correlation for the isotherms was developed which expressed $\log \gamma$ as a function of (1) the concentration raised to the one half, first, second, third, and fourth powers, (2) the Debye-Hückel constant, and (3) four additional empirical constants. For each value of the temperature, new values of the five constants were determined. The multiterm equation, giving waves on plotting, was used to plot isotherms against concentration as a family of curves. This empirical correlation was good but tedious to use.

Alternatively, correlation of $\log \gamma$ on lines of constant concentration by the use of the equation 129 which is linear gives equally good results with only one constant of integration C. Straight lines are obtained on logarithmic paper, the slopes of which are the ratios of the partial heats of solution to the latent heat of water. This is shown for the activities of water from aqueous hydrochloric acid (8). Here again, the choice of concentration rather than temperature as the fixed variable allows a linear relation to be used.

The electrochemical potentials of the cell may be measured readily and accurately. These data give linear functions of the activity coefficients from which may be predicted immediately, in a reverse direction from that above, the heats of solution and the vapor pressures (8–9,80). Thus the data from many types of experimental measurements are interrelated, including voltages of an electrical system, heats from a calorimeter, vapor pressures, and vapor compositions. Data chosen, as taken with least errors, may be selected from any one of these (probably emf), and thus used to predict accurately data of other types, either unknown or difficult to measure as accurately.

Electrochemical Potentials. The electromotive force E of a cell is related to the absolute temperature T, the gas constant R, the Faraday constant F, the valence N, and the activity coefficient γ of water in the cell as:

$$E = (RT/NF) \ln \gamma x \quad \text{or} \quad E/T = (R/NF) \ln \gamma x \qquad (130)$$

Differentiation (8) with respect to $\log P'$ at constant x, and combination with equation 129 for the log of the activity coefficient in the cell, gives:

$$E/T = (R\Delta H/NFL') \log P' + C' \qquad (131)$$

The quotient E/T may thus be correlated as a linear function of the temperature expressed, as in the case of vapor pressures above, by the logarithm of the vapor pressure of water since R/NF is a thermodynamic constant and $\Delta H/L$ has been shown to be constant over the temperature range involved. Many experimental data for the electromotive forces of different cells at different temperatures have been found (8) to be represented by this linear equation within the experimental error. Values of E/T as a function of the temperature represented by the logarithm of the vapor pressure of water at the same temperature give a semilog linear relation or plot.

The Clausius-Clapeyron equation for water as a single component is:

$$\log P' = (L/RT) + C' \qquad (132)$$

Equations 131 and 132 may be combined to give:

$$E/T = (\Delta H/NFT) + C'' \quad \text{or} \quad E = (\Delta H/NF) + C''T \text{ (approximately)} \quad (133)$$

However $\Delta H/NF$ is substantially constant for some range of temperatures. This is obviously the linear form $E = b + aT$. It explains the long-known empirical equation in this linear form for the correction of voltage of standard cells for changes in temperature. It is not nearly as accurate as equation 133 for E/T, but suffices for the small normal variation of cells throughout the range of ambient temperatures (8).

Psychrometric or Humidity Charts. Psychrometric or humidity charts represent the vapor pressure, mass, and heat relations of a vapor, usually water, in a mixture with air or other gas. Conventional humidity charts for meteorological use or for drying calculations and dryer design are of one of several standard forms readily available. Usually they read as the percentages of the normal saturated vapor pressure of water or the percentages of the mass of water per unit mass of air which the air can carry at saturation.

Improved charts may be prepared, but the standard ones are always available and quite satisfactory. Special ones may have to be prepared from fewer data than are available for air and water. For example, a gas other than air may be required for drying a material sensitive to the oxygen of the air, or a liquid other than water, eg, a solvent may be evaporating from a solid or a coating. Water may be evaporating to the air from, or condensing from the air to, or be in equilibrium with, an aqueous solution of a hydroscopic salt.

A minimum of physical data allows the preparation of a chart (68) of considerable flexibility and utility on a sheet of log paper:

(1) The left or Y-axis (logarithmic) is calibrated directly for values of vapor pressure of the liquid.

(2) The temperature scale on the X-axis is calibrated as usual, at even values of temperature, from the tabular values of the corresponding vapor pressures of the liquid. The vertical lines representing dry bulb temperatures are drawn, as ordinates.

(3) The vapor pressure line of the liquid is drawn (pressure liquid vs pressure liquid) as a 45° straight line. This is the saturation line and also represents the maximum vapor which air can carry at any temperature.

(4) Parallel lines are drawn (45°) to the saturation lines at 1, 5, 10, 20%, etc of the vapor pressures at each temperature.

(5) The right side, thus also the Y-axis (logarithmic) is calibrated for humidity as units of mass of vapor per unit of mass of dry air, corresponding to given vapor pressures of the liquid.

Evenly calibrated values in units of mass of the vapor are desirable. A plot of the mass of vapor vs its vapor pressure at the same temperature is almost a straight line and gives values of vapor pressure corresponding to even values of mass of vapor. Horizontal lines are drawn against these evenly spaced values of mass of vapor. Thus the 45° lines first drawn indicate at any temperature, read from the horizontal scale, the percent vapor pressure in MPa (psi) from the left scale, or the units of mass of vapor per unit of mass of air from the right scale.

(6) The adiabatic-saturation or cooling lines are for the liquid very nearly identical with the wet-bulb lines. They are lines of constant enthalpy, with varying temperatures and varying percentage relative humidities. Points on these curves may be calculated from the known P, V, T and thermal data.

112 ENGINEERING AND CHEMICAL DATA CORRELATION

Thus charts may be prepared (68,81) for use with aqueous solutions of different concentrations of hydrophilic agents, such as calcium chloride, lithium bromide, sulfuric acid, or glycerin. They may be prepared readily by the same means for drying of solvents instead of water, by air or noncombusting gas, or even for mixtures of water and solvent as they occur in the water scrubbing of a gas laden with vapors of a water-soluble solvent, such as acetone, ethanol, or methanol.

The physical data needed for preparation of such a chart are: (*1*) the vapor pressures of the liquid in contact with the gas at two or more points, from which latent heats (including the heat of dilution of water from a hydrophilic liquid) also may be evaluated; (*2*) the specific heat of the carrier gas, usually air; and (*3*) the specific heat of the vapors.

Because of the linearity of the percent vapor pressure and relative humidity functions, general programs for pocket calculators are readily prepared and available (2). These values for any desired gas and liquid, or solution, may be inserted to allow immediate calculation of drying, gas absorption, refrigeration, and related problems.

The advantages of such a humidity chart are: (*1*) easy preparation from a minimum of data for systems of vapor in a carrier gas; (*2*) coordinates are from standard logarithmic sheets, parallel, and at right angles; (*3*) percentage relative humidity and percentage vapor pressures are drawn as the same, parallel, 45° lines; thus values of either may be used or obtained by reading from right or left vertical scales, respectively; and (*4*) accuracy is good at low humidities; the same accuracy is obtained in use throughout the entire range because of the logarithmic grid.

Chemical Reactions, Equilibrium and Rate Constants

Reaction kinetics phenomena and data may be correlated by the reference-substance system, with the same advantages of linear functions and straight-line plots. This allows ready checking of experimental data, extrapolation and interpolation from a few experimental points (82). The similarity to pressure–temperature relations may be expected, since the phase change of vaporization may be regarded as comparable to a chemical reaction.

The Van Hoff isochore expresses the equilibrium constant K_P of a chemical reaction as a function of T and ΔH, the heat of reaction.

$$d \ln K_P = \frac{\Delta H}{R} \frac{dT}{T^2} \qquad (134)$$

The Clausius-Clapeyron equation for a reference substance at the same temperature is:

$$d \ln P = \frac{L'}{R} \frac{dT}{T^2} \qquad (135)$$

If these are combined to eliminate the temperature terms and then integrated:

$$\log K_P = \Delta H/L' \log P + C \quad \text{or} \quad \log K_P = m \log P' + C \qquad (136)$$

Similarly, if the equilibrium constant is expressed in terms of concentrations:

$$K_P = K_c (RT)^{\Delta n} \quad \text{or} \quad \log K_c = \Delta E/L' \ln P + C \qquad (137)$$

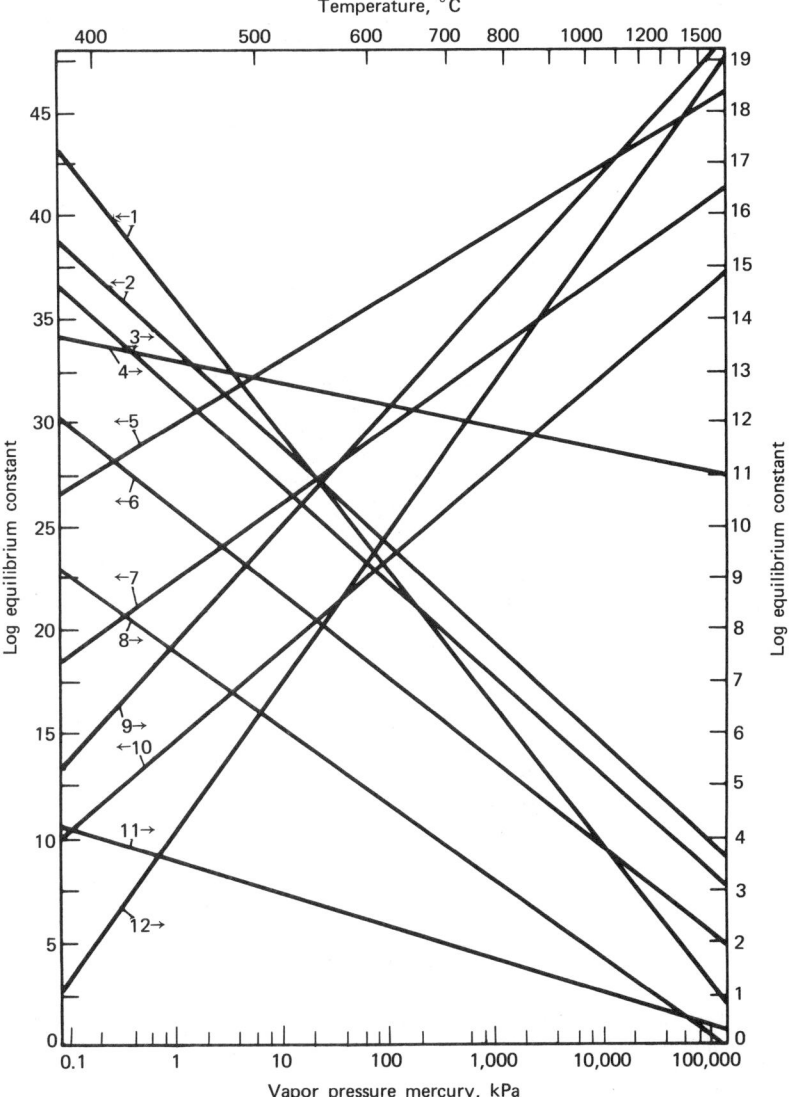

Figure 23. Plot of logarithm of equilibrium constant vs vapor pressure of mercury at same temperature. Arrows indicate whether left- or right-hand scale is to be used for log of equilibrium constant. Because of the wide range of values of K encountered (over 10^{40} times) it is necessary to add the indicated number to the respective scale reading to give log K. Courtesy of *Industrial and Engineering Chemistry*. To convert kPa to mm Hg, multiply by 7.5.

Line no.	Reaction	Add to vertical scale	Line no.	Reaction	Add to vertical scale
1	$C + O_2 \rightarrow CO_2$	+10, left	7	$CH_4 + H_2O \rightarrow CO + 3 H_2$	−35, left
2	$CO + \frac{1}{2} O_2 \rightarrow CO_2$	−5, left	8	$C + 2 H_2 \rightarrow CH_4$	−3, right
3	$C + \frac{1}{2} O_2 \rightarrow CO$	+5, right	9	$C + H_2O \rightarrow CO + H_2$	−16, right
4	$CH_4 + \frac{1}{2} O_2 \rightarrow CO + 2 H_2$		10	$CH_4 + CO_2 \rightarrow 2 CO + 2 H_2$	−30, left
5	$CH_4 + 2 H_2O \rightarrow CO_2 + 4 H_2$	−40, left	11	$CO + H_2O \rightarrow CO_2 + H_2$	−1, right
6	$H_2 + \frac{1}{2} O_2 \rightarrow H_2O$		12	$C + CO_2 \rightarrow 2 CO$	−15, right

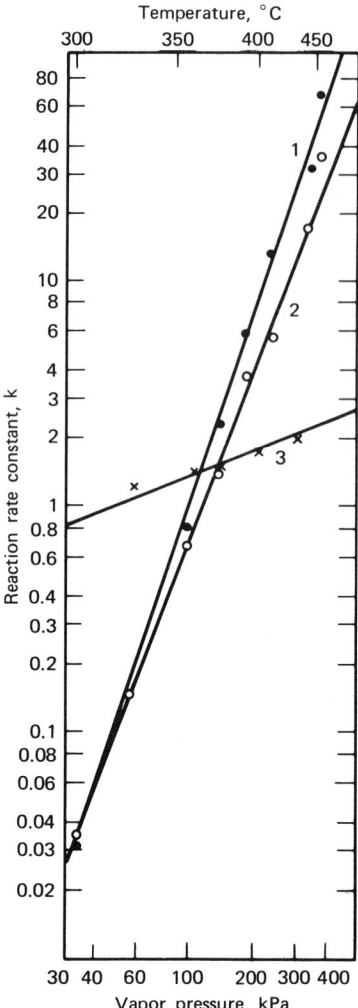

Figure 24. Log plot of reaction rate constant vs vapor pressure of mercury at same temperature. To convert kPa to mm Hg, multiply by 7.5.

Line no.	Reaction	Constant plotted	k, multiply vertical scale by
1	$2\,HI \rightarrow H_2 + I_2$	k_1	10^{-4}
2	$I_2 + H_2 \rightarrow 2\,HI$	k_2	10^{-2}
3	$2\,HI \rightleftharpoons H_2 + I_2$	$K_c = k_1/k_2$	10^{-2}

where Δn is the change in number of moles in the reaction and K_P is equal to K_c when Δn is zero and ΔE is the increase in total internal energy.

These same equations can be developed from those developed above for activities since the log of the equilibrium constant is equal to the difference of the sums of the logs of the activities of the reactants and of the products (82). Figure 23 is a plot of K_P

against vapor pressure of mercury for several reactions with respective values taken always at the same temperatures. Slopes m of lines are ratios of heats of reaction to latent heat of mercury at the same temperature. Positive slopes indicate endothermic reactions; negative slopes, exothermic reactions. All of the lines are straight, many over a range up to 1500 centigrade degrees.

The wide range of K values encountered makes necessary the severe compression of the logarithmic scale, to indicate only values of 5, ie, 10^5. However, the heats of reaction, depending on the slopes, may be calculated readily, and have been found to agree well with calorimetric data.

Equilibrium constants are a ratio of two reaction rate constants for the forward and the reverse reactions, $K = k_1/k_2$. Thus by the Van Hoff equation:

$$-\frac{d \ln k}{dT} = \frac{\overline{E}}{RT^2} \qquad (138)$$

where $\overline{E}$ is activation energy. Combined with the Clausius-Clapeyron equation, this gives (always at same temperature) for the reaction in either direction:

$$\log k = m \log P' + C \qquad (139)$$

Figure 24 plots the reaction rate constant k_1 for the decomposition of hydrogen iodide, reaction rate constant k_2 for the formation of hydrogen iodide from hydrogen and iodine, and the equilibrium constant calculated from the ratio of k_1 and k_2.

There are four useful types of equilibrium constants, and all correlate well using as the temperature scale that of the vapor pressure of water or that defined by some standard equilibrium constant or van't Hoff isochore (82). These four types are: K_c, expressed in concentrations; K_p, in partial pressures; K_i, the ionization constant; and K_{sp}, the solubility product constant.

Interestingly, ionization constants in water all show a break of the linear function at 25–35°C, comparable to the breaks with other properties noted above (48).

Acknowledgment

Sincere appreciation is expressed to R. F. Benenati for his suggestions and contributions in this article.

BIBLIOGRAPHY

"Data Correlation" in *ECT* 1st ed., Vol. 4, pp. 846–873, by D. F. Othmer and E. S. Roszkowski, Polytechnic Institute of Brooklyn; "Data—Interpretation and Correlation" in *ECT* 2nd ed., Vol. 6, pp. 705–755, by D. F. Othmer, Polytechnic Institute of Brooklyn.

1. D. Rossini and co-workers, *Natl. Bur. Stand. Res. Publ.* **1670,** 219 (1945).
2. R. F. Benenati, unpublished research, Polytechnic Institute of New York, Brooklyn, N.Y., 1979.
3. T. K. Sherwood and C. E. Reed, *Applied Mathematics in Chemical Engineering*, McGraw-Hill Co., Inc., 1939.
4. G. E. Forsythe, *J. Soc. Ind. App. Math.* **5**(2), 74 (1957).
5. C. W. Clenshaw, *Comput. J.* **2,** 170 (1960).
6. D. Ambrose and I. J. Lawrence, *J. Chem. Thermodyn.* **4,** 755 (1972).
7. D. F. Othmer, *Ind. Eng. Chem.* **32,** 841 (1940).
8. D. F. Othmer and R. Gilmont, *Ind. Eng. Chem.* **36,** 858 (1944).
9. D. F. Othmer and L. M. Naphtali, *Chem. Eng. Data* **1,** 6 (1956).

10. R. Gilmont and co-workers, *Ind. Eng. Chem.* **42**, 120 (1950).
11. R. Gilmont, D. Zudkevitch, and D. F. Othmer, *Ind. Eng. Chem.* **53**, 223 (Mar. 1961).
12. D. F. Othmer and S. A. Savitt, *Ind. Eng. Chem.* **40**, 168 (1948).
13. D. F. Othmer and E. H. Ten Eyck, *Ind. Eng. Chem.* **41**, 2897 (1949).
14. D. F. Othmer, L. G. Ricciardi, and M. S. Thaker, *Ind. Eng. Chem.* **45**, 1815 (1953).
15. K. C. Chao and G. D. Seader, *AIChE J.* **7**, 598 (1961).
16. K. C. Chao and R. A. Greenkorn, *Thermodynamics of Fluids,* Marcel Dekker, Inc., New York, 1975.
17. B. I. Lee, J. H. Erbar, and W. C. Edmister, *AIChE J.* **19**, 349 (1973).
18. J. M. Prausnitz and P. L. Chueh, *Computer Calculations for High Pressure Vapor–Liquid Equilibria,* Prentice-Hall, Englewood Cliffs, N.J., 1968.
19. J. M. Prausnitz and co-workers, *Computer Calculations for Multi-Component Vapor–Liquid Equilibria,* Prentice-Hall, Englewood Cliffs, N.J., 1967.
20. T. J. J. Edwards and co-workers, *AIChE J.* **21**, 248 (1975).
21. G. M. Wilson, *J. Am. Chem. Soc.* **86**, 127, 133 (1964).
22. D. Tassios, *AIChE J.* **17**, 367 (1971).
23. K. F. Wong and C. A. Eckert, *Ind. Eng. Chem. Fundam.* **10**, 20 (1971).
24. M. Hiranuma and K. Honma, *Ind. Eng. Chem. Proc. Des. Dev.* **14**, 221 (1975).
25. D. S. Abrams and J. M. Prausnitz, *AIChE J.* **21**, 116 (1975).
26. M. M. Abbott and co-workers, *AIChE J.* **21**, 72 (1975).
27. M. M. Abbott and H. C. Van Ness, *AIChE J.* **21**, 62 (1975).
28. J. G. Hayden and J. P. O'Connell, *Ind. Eng. Chem. Proc. Des. Dev.* **14**, 3 (1975).
29. R. C. Reid and T. K. Sherwood, *Properties of Gases and Liquids,* 1st ed., 1958, 2nd ed., 1966, R. C. Reid, J. M. Prausnitz, and T. K. Sherwood, 3rd ed., 1977, McGraw-Hill Book Co., New York.
30. D. F. Othmer and M. S. Thakar, *Ind. Eng. Chem.* **44**, 1654 (1952).
31. D. F. Othmer, R. E. White, and E. Trueger, *Ind. Eng. Chem.* **33**, 1240 (1941).
32. D. F. Othmer and P. E. Tobias, *Ind. Eng. Chem.* **34**, 690 (1942).
33. *Ibid.,* p. 693.
34. *Ibid.,* p. 696.
35. E. L. Derr and C. H. Deal, *Inst. Chem. Eng. Symp. Ser. London* 3(32), 40 (1969).
36. A. Fredenslund and co-workers, *AIChE J.* **21**, 1086 (1975).
37. E. Bender and U. Block, *Verfahrenstechnik (Mainz)* **9**, 106 (1975).
38. J. F. Fabries and H. Rennon, *AIChE J.* **21**, 735 (1975).
39. C. J. King in *Separation Processes,* McGraw-Hill Book Co., New York, 1971, Chapt. 11.
40. D. F. Othmer and R. E. White, *Ind. Eng. Chem.* **34**, 952 (1942).
41. L. R. Field and co-workers, *J. Chem. Thermodyn.* **6**, 237 (1974).
42. W. Hayduk and H. Laudie, *AIChE J.* **19**, 1233 (1973).
43. J. P. O'Connell, *AIChE J.* **12**, 658 (1971).
44. E. W. Tiepel and K. E. Gubbins, *Ind. Eng. Chem. Fundam.* **12**, 18 (1973).
45. D. F. Othmer and R. D. Beattie, *Pet. Refiner,* (Apr. 1952).
46. B. I. Lee and M. G. Kessler, *AIChE J.* **21**, 510 (1975).
47. D. F. Othmer, S. Josefowitz, and A. Schmutzler, *Ind. Eng. Chem.* **40**, 883, 886 (1948).
48. N. E. Dorsey, *Properties of Ordinary Water Substance,* Reinhold, New York, 1940.
49. D. F. Othmer, A. Sze, and J. Isaac, unpublished research.
50. C. F. Spencer and R. P. Danner, *J. Chem. Eng. Data* **13**, 307 (1968); private communication from R. P. Danner.
51. D. F. Othmer and J. W. Conwell, *Ind. Eng. Chem.* **37**, 1112 (1945).
52. D. F. Othmer and S. J. Silvis, *Ind. Eng. Chem.* **42**, 527 (1950).
53. D. F. Othmer and S. J. Silvis, *Sugar* **43**(5), 32 and (7) (1948).
54. D. F. Othmer and M. S. Thakar, *Ind. Eng. Chem.* **45**, 589 (1953).
55. D. F. Othmer and co-workers, *Ind. Eng. Chem.* **49**(1), 125 (1957).
56. D. F. Othmer and H. N. Huang, *Ind. Eng. Chem.* **57**(10), 42 (1965).
57. D. F. Othmer and E. S. Yu, *Ind. Eng. Chem.* **60**, 22 (1968).
58. D. F. Othmer, *Ind. Eng. Chem.* **34**, 1074 (1942).
59. D. F. Othmer and D. Zudkevitch, *Ind. Eng. Chem.* **51**, 791 (1959).
60. D. G. Miller, *Ind. Eng. Chem.* **56**, 46 (1964).
61. D. F. Othmer, R. C. Kowalski, and L. M. Naphtali, *Ind. Eng. Chem.* **51**, 89 (1959).
62. D. F. Othmer and G. J. Frohlich, *AIChE J.* **6**, 210 (1960).

63. A. K. S. Murthy and D. Zudkevitch, private communication, Jan. 1979; *Symposium on Distillation,* London, Eng., Apr. 1979.
64. D. F. Othmer, *Chem. Met. Eng.* **47,** 551, 631 (1940).
65. A. H. Boas, Ph.D. dissertation, unpublished, Polytechnic Institute, Brooklyn, N.Y., 1962.
66. H. T. Chen and D. F. Othmer, *AIChE J.* **12,** 489 (1966).
67. J. J. Martin, *Ind. Eng. Chem.* **59,** 34 (1967).
68. D. F. Othmer and H. T. Chen, *Ind. Eng. Chem.* **60,** 39 (1968); *Applied Thermodynamics,* American Chemical Society, Washington, D.C., 1968, p. 115.
69. T. G. Kaufman, *Ind. Eng. Chem. Fundam.* **7,** 115 (1968).
70. D. F. Othmer and F. G. Sawyer, *Ind. Eng. Chem.* **35,** 1269 (1943).
71. D. F. Othmer and S. Josefowitz, *Ind. Eng. Chem.* **40,** 723 (1948).
72. D. F. Othmer and S. Josefowitz, *Ind. Eng. Chem.* **38,** 111 (1946).
73. D. F. Othmer and H. T. Chen, *Ind. Eng. Chem. Process Des. Dev.* **1,** 249 (1962).
74. D. F. Othmer and G. J. Frohlich, *Ind. Eng. Chem.* **47,** 1034 (1955).
75. G. Scatchard and C. L. Raymond, *J. Am. Chem. Soc.* **60,** 1278 (1938).
76. O. Redlich and A. T. Kister, *Ind. Eng. Chem.* **40,** 341, 345 (1948).
77. D. F. Othmer, E. H. Ten Eyck, and S. Tolin, *Ind. Eng. Chem.* **43,** 1607 (1951).
78. E. H. Ten Eyck and D. F. Othmer, *Pet. Refiner* (Sept.–Oct. 1953).
79. G. Akerlof and J. W. Teare, *J. Am. Chem. Soc.* **59,** 1855 (1937).
80. J. J. Fritz and C. R. Fuget, *Chem. Eng. Data* **1,** 10 (1956).
81. D. F. Othmer, *Chem. Met. Eng.* **47,** 296 (1940).
82. D. F. Othmer and A. H. Luley, *Ind. Eng. Chem.* **38,** 408 (1946).

General References

H. T. Chen and D. F. Othmer, "Thermodynamics" in A. Standen, ed., *Encyclopedia of Chemical Technology,* 2nd ed., Vol. 20, John Wiley & Sons, Inc., New York, 1969, pp. 118–146.
D. F. Othmer and H. T. Chen in *Applied Thermodynamics,* American Chemical Society, Washington, D.C., 1968, Chapt. 7.
R. C. Reid, J. M. Prausnitz, and T. K. Sherwood, *Properties of Gases and Liquids,* 3rd ed., McGraw-Hill Book Co., New York, 1977; also 1st ed., 1958 and 2nd ed., 1966 by R. C. Reid and T. K. Sherwood.
American Petroleum Institute Technical Data Book—Petroleum Refining, 3rd ed., API, New York, 1977.
J. H. Perry, *Chemical Business Handbook* (Section 20), McGraw-Hill Book Co., New York, 1954.

DONALD F. OTHMER
Polytechnic Institute of New York

ENGINEERING PLASTICS

Any discussion of engineering plastics falters at the outset on the matter of definition. The implication that there is a well-defined group of engineering plastics implies that there must also be a well-defined group of nonengineering plastics. Here the discussion breaks down, because there is virtually no plastic material that cannot, in some form, be termed an engineering plastic.

A practical definition must include not only property/performance criteria, but also market/pricing criteria that place certain resins in the engineering category to the exclusion of all others. Those resins that meet both sets of criteria are: nylon [32131-17-2], acetal [9002-81-7], thermoplastic polyester molding compounds, poly(phenylene oxide) [25667-40-9]-based resin (PPO), and polycarbonate [25971-63-5]. These five resin families have been the engineering plastics of the 1970s. The 1980s may see new members entering the group.

A definition based on property/performance criteria alone would include special grades and compounds of the commodity thermoplastics and a variety of polymer alloys and copolymers (qv). It would include all of the specialty thermoplastics that offer high strength along with high temperature performance, and it would include the thermoset resins that were the forerunners of the engineering thermoplastics.

The following discussion focuses on the five engineering resin families mentioned above, emphasizing their comparative advantages and how these have led to current usage patterns. To the extent that other resins compete against them, those resins are mentioned as well. Particular reference is made to acrylonitrile–butadiene–styrene (ABS), poly(phenylene sulfide), and polysulfone, which are recognized borderline engineering thermoplastics, as well as to phenolic molding compounds, and other thermosets, and to the fluoropolymers which offer outstanding but essentially different property advantages (see Acetal resins; Acrylonitrile polymers; Fluorine compounds, organic; Polyamides; Phenolic resins; Polyesters; Polycarbonates; Polyethers; Polymers containing sulfur).

In terms of properties, engineering plastics have a good balance of high tensile properties, stiffness, compressive and shear strength, as well as impact resistance, and they are easily moldable. Their high physical strength properties are reproducible and predictable, and they retain their physical and electrical properties over a wide range of environmental conditions (heat, cold, chemicals). They can resist mechanical stress for long periods of time. Flame retardance is not an essential requirement, but it has become an important added asset (see Flame retardants).

All of the individual physical properties of a resin can be quantified and compared through established testing procedures, but the balance of properties essential to a true engineering resin requires a broader view. A balance of properties exists when the achievement of one property does not demand a trade-off in another, ie, stiffness for low-temperature impact strength. Certain properties of commodity thermoplastics can be improved through the use of stabilizers, fibrous reinforcements, and particulate fillers (qv) to produce grades that compete directly with engineering plastics, but always with a corresponding reduction in other properties (1–2) (see Composite materials; Heat stabilizers; Uv absorbers).

The introduction of the term commodity leads to the market/pricing criteria that place the engineering plastics in a class of their own. Engineering plastics form a distinct group as compared to the high-volume/low-price commodity plastics and the low-volume/high-price specialty plastics (3).

The characteristics of the engineering plastics in these terms stand out in Table 1, which lists a sampling of thermoplastic resins along with average price, number of suppliers, and total consumption in 1978. The absolute figures have all changed since then, but the essential differences still hold.

The commodity plastics are characterized by average prices under $1.10/kg, sales in the 500–3300 thousand metric ton range, and large numbers of suppliers. The engineering plastics are priced close together in the $2.20–2.60/kg range on average, are sold in the 20–120 thousand metric ton range, and have few suppliers. The specialty plastics are priced at over $4.40/kg, some of them well over, and annual consumption of all of them is roughly 10–15 thousand metric tons. Each of the specialty molding compounds has just one supplier.

These and other general characteristics of engineering plastics are cited in reference 4 as follows: predictable properties over a wide range of loadbearing conditions, patent protection on composition or process, or polymer and/or monomer, specialty chemical raw material bases, minimum number of competitors, high development costs and high cost of supplying markets, low volume as compared to commodities, high selling price, and high capital cost/output.

Nylon is an exception to some of these rules: its patents have expired, and its

Table 1. Thermoplastic Resins: Comparative Market Data[a]

Resin	Mid-1978 average list price, $/kg	U.S. 1978 consumption, thousands of metric tons	Number of producers
Commodity			
low-density polyethylene	0.69	3248	13
high-density polyethylene	0.67	1893	13
polypropylene	0.66	1388	12
polystyrene	0.65	1759	15
poly(vinyl chloride)	0.60	2641	20
acrylonitrile–butadiene–styrene (ABS)	1.06	508	7
Engineering			
nylon	2.56	120	9[b]
acetal	2.20	45	2
thermoplastic polyester (TP) molding compounds	2.31	23	5
PPO-based resin[c]	2.51	64	1
polycarbonate	2.49	100	2
Specialty			
poly(phenylene sulfide), 40% glass-reinforced	4.52	2	1
polysulfone	6.50	5	1
poly(ether sulfone)	15.43	0.03	1
poly(amide–imide) molding compounds	22.92	0.2	1
aromatic polyester molding compounds	66.12	0.01	1

[a] Consumption figures for the commodity resins and nylon are from the SPI Committee on Resin Statistics as compiled by Ernst & Ernst, New York. All other consumption figures are Business Communications Co., Inc. estimates.
[b] Polymerizers only.
[c] PPO = poly(phenyl oxide).

economics are fiber-based. This is evidenced in Table 1, which shows that nylon has more suppliers than ABS, and still more if reprocessors of fiber scrap are counted in (see also Recycling). There are just five major suppliers of the virgin nylon-6,6 and -6 resins and compounds that account for most of the special grades of nylon used in engineering applications.

Properties

Since the 1950s when DuPont introduced nylon as the first engineering thermoplastic, it has been joined by four other resin families that have broadened the capabilities of plastics in replacement of metals, glass, and thermosetting resins. The relative capabilities of each of the major engineering thermoplastics are revealed by examination of key properties.

General-purpose nylon-6,6 has a heat deflection temperature (HDT) at 1.82 MPa (264 psi) of 104°C dry, as molded, but at equilibrium the HDT drops to 75°C (5). Its Thermal Index, as determined by Underwriters Laboratories (UL), is 75°C (6). Its flexural modulus is 2.8 GPa (410,000 psi) dry, but at equilibrium only 1.2 GPa (175,000 psi) for nylon-6,6, 965 MPa (140,000 psi) for nylon 6 (7). Notched Izod impact strength for nylon-6,6 is 1.12 J/cm (2.1 ft-lb/in.) of notch at equilibrium, but only 0.53 J/cm (1.0 ft-lb/in.) as molded (8). The Izod test, with its severe notch, is not necessarily the best method for comparing impact resistance, but it is frequently used in comparisons of engineering plastics.

Nylon's performance in a given application is dependent, therefore, on moisture content, and although impact resistance improves as equilibrium is reached, modulus drops sharply. Through modification, all of these properties have been improved, but these are the key properties on which engineering thermoplastics are based (9).

Acetals were introduced in 1960 (the DuPont homopolymer) and 1962 (the Celanese copolymer). Acetals brought somewhat higher heat deflection temperatures and higher U.L. Thermal Index values, but most important, their resistance to moisture pick-up made physical properties independent of moisture content. Flexural modulus of acetals as molded is the same (homopolymer) or slightly lower (copolymer) than nylon-6,6 as molded, but it is relatively constant after that. However, impact resistance does not improve with moisture pick-up as it does with nylon, but it is slightly higher to begin with.

Nylon and acetal, both crystalline resins, surpassed amorphous ABS in such properties as tensile strength, long-term heat resistance, and chemical resistance, but neither of them could approach ABS in notched impact resistance. In the 1960s, two amorphous resins were introduced that offered impact resistance equal to or better than ABS along with other important advantages.

With the introduction of polycarbonate in the early 1960s, a substantial gain was made in the U.L. Thermal Index rating. Polycarbonate can be used continuously at 115°C, 25 degrees higher than acetal, and 40 degrees higher than nylon. Polycarbonates also represented a major improvement in impact resistance, with a notched Izod value of 7.47 J/cm (14.0 ft-lb/in.) in 3.2 mm thickness. In addition, this resin was the first to offer transparency.

General Electric introduced its Noryl, a styrene-modified poly(phenylene oxide) resin, in 1966. Its heat deflection temperature, U.L. Thermal Index, and flexural modulus were not too different from acetal's, but with a notched Izod value of 2.67

J/cm (5.0 ft-lb/in.), impact resistance had been significantly improved. Noryl was between ABS and polycarbonate in price.

With the introduction of poly(butylene terephthalate) [26062-94-2] (PBT) thermoplastic polyester in 1970, new potential opened for thermoplastics in high-heat applications, but not among the unreinforced resins. Unreinforced PBT, as a crystalline resin, offers chemical resistance similar to nylon and acetal and certain common properties like lubricity, but without glass reinforcement its tensile strength and impact strength are relatively low, and its heat deflection temperature under 1.8 MPa (264 psi) load is very low (see Polyesters, thermoplastic).

An outstanding feature of PBT is the extent to which its heat resistance was improved by the addition of glass fibers (see Glass; Composite materials). With 30% glass reinforcement, PBT has a heat deflection temperature of 213°C, and a U.L. Thermal Index of 140°C. Glass-reinforced nylon has a higher HDT, but its U.L. rating is lower, and PBT has the important added advantages of lower moisture absorption and availability in flame-retardant grades without loss of properties.

All of these resins are now available in many different grades, and new grades that offer special price or property advantages are regularly introduced by their suppliers. Table 2 lists the key properties of the unfilled and glass-reinforced general purpose grades of these resins.

Table 2. Basic Property Comparison, Filled and Unfilled Engineering Resins[a]

	HDT, °C	U.L. Thermal Index, °C	Flex. modulus GPa[b]	Notched Izod, J/cm
Unfilled resins				
nylon-6,6 dry	104	75	2.8	0.53
nylon-6,6 (50% rh)	175	75	1.2	1.12
acetal copolymer	110	90	2.6	1.3
PPO-based resin[c]	129	90	2.5	2.67
polycarbonate	132	115	2.4	7.47
PBT[d]	54	120	2.3	0.53
Glass-reinforced (GR) resins				
nylon-6,6 dry 33% GR	249	105	8.9	1.07
nylon-6,6 (50% rh) 33% GR			6.2	
acetal copolymer 25% GR	163	95	7.6	0.85
PPO-based resin 20% GR	143	90	5.2	1.23
PPO-based resin 30% GR	149	90	7.6	1.23
polycarbonate 20% GR	146	120	5.5	1.34
PBT 30% GR	214	140	7.6	0.96

[a] Values are those for general purpose resins as they exist today, not necessarily as they existed at the time of their introduction. Data are from supplier literature, and are intended for rough comparison only.
[b] To convert GPa to psi, multiply by 145,000.
[c] PPO = poly(phenyl oxide).
[d] PBT = poly(butylene terephthalate).

As the table shows, the various resins are affected by glass reinforcement in some markedly different ways. Glass reinforcement raises flexural modulus for all of them. Heat deflection temperatures go up as well, as can be expected because glass fibers physically prevent deflection, but in the case of nylon and PBT, heat deflection temperatures soar. The 249°C HDT for nylon makes it eminently qualified for short-term high-heat exposure, and its long-term capability increases as well. In the case of PBT, glass reinforcement makes a true engineering resin out of a material that is not outstanding unfilled. Most important, its U.L. Thermal Index rises to 140°C, highest of all the resins in this group, which is particularly important with regard to its retention of electrical properties.

Glass reinforcement improves these properties in acetal compared to the neat resin, but in comparison to other glass-reinforced materials it is not outstanding. Glass reinforcement of PPO-based resin and polycarbonate results in decreased impact resistance, and in the case of polycarbonate, it also does away with transparency (see Laminated and reinforced plastics).

These brief comparisons do not accurately reflect the reasons for use of glass-reinforced resins because they ignore other significant properties (10). For example, in recent years, U.L.'s flammability ratings have become critical in the choice of one resin over another.

Along with the advantages offered by each resin, there are disadvantages. Overcoming the disadvantages is, of course, a major concern of resin suppliers. Ultimate goals for these resins would be: a moisture-resistant nylon without price penalty; a flame-retardant acetal; a PPO-based resin with better surface finish; polycarbonate without stress cracking; and an impact-resistant PBT without loss of modulus.

Processing

The melt-processability of thermoplastic resins is a characteristic that distinguishes them from thermosets. This pertains not only to the advantages of injection molding as compared to compression or transfer molding, but also to the variety of processing alternatives that extend the utility of the thermoplastics. There are thermosets that can be injection molded, but only the thermoplastics offer the options of extrusion into sheet, film and profiles, or blow molding (see Plastics technology).

The degree to which each of the engineering plastics is amenable to alternative processing methods varies, and the relative potential of each of them depends also on their potential in alternative processes, not just on their utility in injection molding.

In addition to its use in injection molding, nylon is extruded into monofilament and brush filament. Nylon-6 [25038-54-4] is used for sewing thread, fishing line, household/industrial brushes, and level-filament paint brushes. Nylon-6,6 [9011-55-6], stiffer than nylon-6, is used for sewing thread and household/industrial brushes. Nylon-6,12 [24936-74-1] dominates in personal-care brushes, and although the future may see some competition from poly(ethylene terephthalate) [25038-59-9] (PET), it is not evident as yet. In tapered-filament paint brushes, nylon-6,12's leading position has been taken over by PBT, a more expensive but more versatile filament.

Nylon is used as a wire coating, primarily as a protective abrasion-resistant coating over PVC-insulated wire (see Insulation, electric). Nylon film can be cast or blown, or extrusion-coated onto various substrates. Most nylon film is cast, and virtually all is sold to converters who add a sealant layer of low density polyethylene (LDPE), ethylene–vinyl acetate copolymer (EVA), or ionomer (see Film and sheeting materials). Its major market is vacuum packages of processed meats and cheese, usually combined with a PET sealant cover web. It is also used for fresh-meat packaging, and a new market has opened in medical-device packaging using techniques similar to those for formed-meat packaging. The most important properties in these applications are formability and heat resistance (see Packaging Materials).

Nylon strapping began replacing steel strapping in the early 1960s, even at higher cost, because of the general advantages of nonmetallic strapping. In recent years, nylon has met increasing competition in this market from polypropylene and PET.

Nylon is also extruded into rods, tubes, and shapes for machining, an important option for low-volume runs. In blow molding, nylon has been held back partly by cost, and partly because of the difficulties inherent in crystalline resins because of their sharp melting point. Nylon blow-molding resins have been developed with high melt strengths for parison forming, and the material is used to some extent for monolayer and coextruded bottles and for gas tanks in small equipment. Nylon-6 is also cast to produce very large bearings (see Bearing materials).

Nylon-11 [25035-04-5] is used for powder coatings and for flexible tubing. Nylon-12 [24937-16-4] is used for the same purposes, but to a greater extent in Europe than in the U.S. These resins have exceptional moisture resistance, but they are considerably less stiff than nylon-6 or -6,6. They are used to some extent in rotational molding.

In contrast to nylon, acetal offers few options outside of the injection molding category. An acetal terpolymer is available for injection blow molding, but apart from some carburetor floats, it has found little usage. Acetal is difficult to extrude, but it is extruded, as is nylon, into shapes for subsequent machining. Almost all acetal consumption is in injection molding, a factor that reduces its ultimate consumption.

As discussed below, the PET thermoplastic polyesters used for film and sheet and blow molding are not the same as those used for injection-molded engineering applications. PBT can be blow-molded, but rarely is. PBT is used almost entirely in injection molding, but there is some usage in tapered brush filaments and in extruded strip for small electrical parts.

Noryl's consumption in extrusion is relatively minor compared to injection molding, but it is used to some extent for stock shapes and to an increasing extent for sheet and profiles. Noryl sheet competes with flame-retardant ABS as it does in injection molding, and it can compete with less expensive resins like ABS and PVC where its properties permit the extrusion of thinner walls (11).

The transparency of polycarbonate, combined with its extrudability and impact resistance, made it a strong competitor for acrylic sheet in replacement of flat glass. Extruded sheet for glazing, lighting, and signs accounted for approximately 20% of polycarbonate's volume in 1978. Use in extruded profiles was minor, but polycarbonate has found a place in blow molding for 19-L (5-gal) water bottles, returnable milk bottles, baby nurser bottles, and miscellaneous packaging.

Economics and Marketing

Compared to the commodity plastics, investment costs per unit weight of engineering plastics are higher, capacity utilization rates are lower, and lead time before actual use is longer. One of the chief concerns for a producer of commodity plastics is whether the product can be made cheaply enough. The concerns of a producer of engineering plastics are whether the product can be made at all, and if so, whether it can be sold (3).

U.S. consumption of the five principal engineering plastics in 1978 is given in Table 3 by type of market.

Table 4 lists companies with their engineering resin products and shows the concentration of engineering resins among the various companies. The oil companies have only peripheral involvement with specialty plastics.

This article emphasizes the comparative aspects of engineering plastics in order to show where each has been successful. With this emphasis, a certain amount of perspective has been lost. There is a limit to the extent to which engineering plastics compete with one another in a given application because, by the end of the selection process, the property requirements usually match up with just one resin. Where more than one resin can do the job, price is the deciding factor.

The crystalline and amorphous resins rarely compete directly because of their basic property differences. Within each group, however, limited competition does exist. This article makes frequent reference, for example, to competition between nylon and acetal. In the context of total nylon consumption, however, there is no competition from acetal in markets that make up about 35% of nylon's volume. It is only in injection molding and some shapes for machining that competition might exist, but not in applications that call for low modulus, glass- and/or mineral reinforcement, special heat-stabilized high-impact grades, or special flammability characteristics.

Table 3. U.S. Consumption of Engineering Plastics, 1978 (Thousand Metric Tons)

	Nylon	Acetal	PBT	PPO-based resin	Polycarbonate
Extrusion					
monofilament, brush filament	9.5		1.8		
wire jacketing	5.0				
film, extrusion coating	10.0				2.3
strapping	2.7				
tubing, shapes	10.0	minor		minor	
sheet					22.7
Total extrusion	37.2		1.8		25.0
Injection molding					
automotive	39.0	13.6	9.0	10.5	10.0
electric/electronic	12.7	0.9	6.8	9.0	11.4
business machines				13.6	11.4
industrial/machinery	10.0	9.0	0.9		5.9
plumbing/hardware	6.4	9.5	1.4	2.7	1.4
appliances	5.0	5.5	1.8	22.7	18.2
consumer goods	5.0	6.4	0.9		3.6
miscellaneous	3.6	0.5	0.5	5.0	9.0
Total injection molding	81.7	45.4	21.3	63.5	70.9
Miscellaneous					
blow molding	<0.5	minor			4.1
stamped sheet	<0.5				
coatings	<0.5				
Total miscellaneous	<1.5				4.1
Total consumption	120.0	45.4	23.1	63.5	100.0

Table 4. Suppliers of Engineering Plastics

Company	Nylon	Acetal	TP^a	PPO^b	PC^c	Other, related
DuPont	Zytel	Delrin	Rynite			fluoropolymers, polyimides, aramid
Celanese		d	Celcon	Celanex		
Monsanto	Vydyne					ABS^e
Allied	Capron					fluoropolymers
American Hoechst	Fostalon					
General Electric				Valox	Noryl	Lexan
Mobay Chemical				(new)		Merlon PC^c–ABS^e alloys
GAF				Gafite		
Eastman				d		
Union Carbide						polysulfones, polyarylate amorphous nylon
Chevron	d					
Phillips						PPS^f
Amoco						poly(amide–imide)
Gulf						polyimides
Arco						Dylark
Exxon						hi-temp film

^{a} TP = thermoplastic polyester.
^{b} PPO = poly(phenylene oxide)-based resin.
^{c} PC = polycarbonate.
^{d} There is no product with a trademark, but a resin is produced.
^{e} ABS = acrylonitrile–butadiene–styrene terpolymer.
^{f} PPS = poly(phenylene sulfide).

The applications that require only those properties that both nylon and acetal can offer represent only 15–20% of total nylon volume, and as long as acetal is more costly per unit volume, nylon will continue to be the resin of choice. In addition, as long as the only suppliers of acetal are also nylon suppliers, there is no reason to expect that the price balance will shift.

Lowering the price of acetal might be justified if it would open the door to nylon's (110–140) × 10^3 metric ton annual volume, but it would only increase potential in sub-markets totaling about 20,000 t.

These considerations among other resins as well have determined the relative positions of the engineering plastics compared to the commodity and specialty plastics, and also the relative positions of the engineering plastics with respect to each other (3).

Engineering Plastics

Nylon. Nylons, also known as polyamides, carry numerical designations that refer to the number of carbon atoms in the amide links. Nylon-6,6 is the reaction product of hexamethylenediamine and adipic acid, each of which contains six carbon atoms. Nylon-6 results from the polymerization of caprolactam (see Polyamides). These two types of nylon account for over 90% of nylon resin consumption. Nylon-6,6 is the leader in the United States, nylon-6 leads in Europe. The relative market strength of the two nylons has more to do with historical development than with property differences, but differences do exist that are reflected in usage patterns.

Nylon-6,6 is stiffer, and predominates in all U.S. injection molding applications. Nylon-6 predominates in extrusion, and is somewhat more susceptible to moisture pick-up. However, the differences should not obscure the basic similarity between the two in comparison with other engineering plastics.

About 68% of nylon consumption in 1978 was in injection molding, and it is in these markets that nylon's capabilities as an engineering resin are most crucial.

In the auto industry, most of the nylon used is unfilled, and many of the applications are replacements for die-cast zinc. Glass- and/or mineral-filled nylons (12) play a major role as well. As in other markets, nylon competes with acetal in uses where wear resistance is required. Where strength and modulus are prime considerations, glass-reinforced nylon competes with polycarbonate. Mineral-nylon and mineral-PBT had been competing for exterior parts, but nylon has been most successful thus far. Both materials have exhibited product warpage problems. In some auto applications, polypropylene is a useful lower-cost alternative to nylon (13).

In industrial/machinery parts, nylon is widely used because of its natural lubricity, wear resistance, and chemical resistance. It is the oldest established resin for mechanical drive components such as bearings, gears, sprockets, pulleys, rollers, races, and chains. In these applications, acetal is nylon's closest competitor. Slightly higher in cost, acetal's resistance to moisture pick-up avoids problems with dimensional stability that require special design considerations with nylon.

The potential for plastics in friction-and-wear applications is continually being broadened by new technological developments in the field of tribology. Various fillers like MoS_2 and graphite improve the performance of a wide range of materials, but usually at a cost premium (14) (see Fillers). Nylon and acetal still dominate the field (see Bearing materials).

PBT has a better coefficient of friction against itself, and glass-reinforced PBT grades perform better than nylons and acetals after long hours at high temperatures (15) but, except in applications where flame retardance is required, unfilled PBT poses no major threats to the other crystalline materials in this market.

Also included in the industrial/machinery category are power tools and housings for equipment using gasoline engines. ABS is the favored resin for consumer electric hand tools, but glass- or mineral-filled nylons are favored for heavy-duty professional handles and housings. Because of their hydrocarbon resistance, nylon compounds perform well in lawn mowers, chain saws, and other gasoline-powered tools. Here again, PBT offers some competition, but mineral nylon has a good balance of properties and impact resistance of mineral grades is continually being improved.

In electric/electronic components, nylon is particularly important in wiring devices such as plugs and connector bodies, receptacles, and other parts, sometimes in replacement of phenolic molding compounds or rubbers (see Electrical connectors). The design freedom offered by thermoplastic nylon as opposed to thermosets has been a key factor, but the difficulty of flame-retarding nylon without loss of properties has been a hindrance in electrical applications, which are strictly tied to Underwriters Laboratories standards (see Flame retardants) (16).

Nylon drips when ignited, and because this removes the flame from the part, nylon's usage in a wide range of small-part electrical applications has been permitted. Glass reinforcement inhibits dripping, and GR parts have a greater tendency to burn with a smoky flame (17). A variety of flame retardants has been employed with nylons, but with limited success. Flame retardant compounds have had relatively low impact strength and poor heat stability in processing, which leads to the generation of toxic fumes, mold corrosion, and a variety of other problems. Except for acetal, which is not available in flame-retardant grades, nylon has the lowest percentage of sales (<5%) in flame-retardant grades among the engineering thermoplastics.

This is one reason why PBT is taking the lead in electric/electronic applications in general. In wiring devices, however, where nylon's properties (ie, elongation) are particularly valuable, its position is being improved by relaxation of U.L. requirements in some applications where the physical property advantages outweigh the lack of flame retardance. Volume losses to PBT in other areas, including electronic connectors, are likely to be regained only by suitable flame-retardant grades, and progress is being made in that direction (see also Insulation, electric).

In appliances, nylon plays a minor role compared to the amorphous resins. Moisture absorption and the lack of flame retardance are deterrents. Nylon use is also limited by its relatively low U.L. Thermal Index and low modulus. Most of nylon's usage in appliances is in moving parts and electrical components.

In the plumbing/hardware category, nylon's usage in plumbing is limited by its moisture pick-up, but in hardware it is used for miscellaneous parts including drapery slides and furniture supports. Lubricity is often a factor, and it competes with acetal and sometimes unfilled PBT. In consumer goods, nylon is used in a very diverse group of parts and products including spatulas, coffee urn spigots, garter grips, and brush backs.

DuPont is the major U.S. supplier of nylon by far, with its Zytel nylon-6,6 and a range of other nylons also under the Zytel name. The other major nylon-6,6 suppliers are Monsanto (Vydyne) and Celanese. Allied Chemical (Capron) and American Hoechst (Fostalon) are leading suppliers of nylon-6. Other nylon producers include Belding Chemical, Custom Resins, Inc., and Firestone Synthetic Fibers. Rilsan Corp. produces nylon-11 and 12. Several other companies supply nylons regenerated from fiber scrap, or monofilament, or special compounds based on resins from polymer producers.

Acetals. Applications for acetals are confined almost entirely to injection molding. They are difficult to extrude, but some stock shapes are made by extrusion and subsequent machining. Acetal homopolymers are made by the polymerization of formaldehyde; the high molecular weight polymers are stabilized by conversion of the end groups to esters. Acetal copolymers are made by the copolymerization of trioxane, a cyclic trimer of formaldehyde, with a small amount of comonomer.

As is the case with nylon-6,6 and 6, there are differences between the two types of acetal, but basic similarities exist between the two in terms of market position vs competing resins. The homopolymer is harder, more rigid, and has higher tensile and flexural strength. The copolymer is more stable in long-term high-temperature service, more resistant to hot water, resistant to strong bases, and has higher elongation.

The role of acetal among the engineering resins can best be described by departing from the standard list of markets (Table 3) which reveals little in the way of comparative data. Acetal is usually less expensive than nylon on a weight basis, but because its specific gravity (1.42) is higher than nylon's (1.13), acetal is more expensive on a ¢/cm^3 (¢/in.^{3}) basis. In applications where the two types of resin compete, therefore, nylon is chosen unless there is a need for special acetal properties, usually moisture resistance, as in gears. Dimensional stability is often critical in assuring a gap between the teeth for expansion due to heat build-up (18). Competition between nylon and acetal in bearings is very strong, but nylon has a slight edge. In all moving parts, acetal's outstanding fatigue resistance is a major asset. PBT has lower moisture absorption than either nylon or acetal below 65°C, but it also has lower unfilled strength.

In automotive applications, nylon and acetal compete in the same ways that they do in other markets, except that the property enhancement afforded to nylon through the addition of inorganic fillers and reinforcements gives it broader usage. In appliances, as with nylon, usage is mainly in moving parts. Acetal's use in electric/electronic applications is very minor. Because of its flammability, in particular, it is not found in wiring devices, connectors, and other U.L.-regulated parts.

Formaldehyde polymers pyrolyze to monomer when they burn. They burn without smoke, because there are no carbon-carbon chains in the structure, and they contribute little in the way of fuel to the fire because the carbon in polyformaldehyde is already partially oxidized, but they do burn. Acetals have the lowest oxygen index of all the plastic resins (see Flame retardants), and 15% oxygen in an O_2/N_2 mixture supports burning (17). Efforts to flame-retard acetals have not been successful owing to the nature of their pyrolysis, and lack of flame retardance has limited acetal's growth among the engineering resins.

Acetal is used in a wide variety of consumer products, including lighters, zippers, fishing reels, and writing pens, but its greatest strengths are in those consumer products that require chemical and water resistance, such as garden-chemical sprayers, household water softeners, paint-mixing paddles and canisters, and particularly plumbing applications where the copolymer exhibits superior resistance to continuous exposure to hot water. Usage of plastics in plumbing parts can require, depending on application, low coefficient of friction, abrasion resistance, strength under pressure, high-temperature chemical resistance, long-term property retention, fatigue resistance to retain burst strength, and other short- and long-term properties. Acetal offers a better balance of these properties than other engineering resins (19).

Celanese is the only U.S. supplier of acetal copolymer (Celcon), and DuPont the only supplier of acetal homopolymer (Delrin). The copolymer has about 55% of the acetal market.

Thermoplastic Polyester Molding Compounds. Thermoplastic polyester *molding compounds* are considered here rather than thermoplastic polyesters in general, because if all thermoplastic polyesters were included (see Table 1), total consumption would exceed that of any of the engineering resins and would be more in the neighborhood of the commodity resins. Poly(ethylene terephthalate) (PET) was used to produce over 136,000 metric tons of film in 1978, and consumption for beverage bottles is expected to be at least that high in the early 1980s (see Barrier polymers; Packaging materials).

This does not place the thermoplastic polyester engineering resins in the commodity category, however, because the resins used for the high-volume film and bottle markets are not the same as those used in engineering applications. The distinction was clear until 1978, because only more costly poly(butylene terephthalate) (PBT) was used for injection molding. Earlier attempts to mold PET had failed because of the slow crystallization rate of PET's shorter chain segments, but in 1978 DuPont introduced a new PET molding compound, and in 1979 Mobay entered the market with a family of thermoplastic polyester engineering resins, including two based on PET.

Post-1978 consumption data, therefore, includes both PBT and PET. This does not mean that engineering PETs are classed among the commodity resins, however, because PET resins for fibers, films, blow molding, and injection molding require separate production technologies. The thermoplastic polyester engineering resins used in 1978 were essentially PBT, and the designation PBT is used here because historical usage patterns have been developed with that resin.

Close to 70% of 1978 PBT volume went into automotive and electric/electronic applications, and much of the automotive usage was in electric/electronic parts. As noted earlier, the two characteristics that set PBT apart from the other crystalline resins are its high Thermal Index and its availability in U.L. 94 V-0 grades without a significant loss of properties.

Some of PBT's properties compared to nylon and acetal have been reviewed above. Generally speaking, PBT's relatively low impact resistance is a drawback, and the stiffness of glass-reinforced grades that makes it ideal for some applications is a detriment in others. PBT is not primarily a competitor for nylon and acetal, however. Its high temperature capability and V-0 rating at relatively low cost enable it to compete against phenolic molding compounds and other thermosets (20–21).

PBT has done well in the electric/electronic category because it brought temperature capability close enough to the thermoset range to allow conversion to more cost-effective thermoplastic processing. It has replaced thermosets in a wide range of electric/electronic components, although the heat resistance of PBT has sometimes been overrated. Although the Thermal Index of PBT is fairly close to the 150°C rating of phenolics, and the 215°C heat deflection temperature is much higher than most thermosets can offer, the actual performance of PBT at high temperatures is quite different. In one published test, thermoset bars retained their shape at 215°C but PBT bars sagged (22).

Phenolic molding compounds, sometimes called the oldest of the engineering resins, had become the standard insulating material in appliances, motors, wiring devices, circuit breakers, etc (23). The low-cost phenolics have been processed by compression and transfer molding in large volumes. In 1978 roughly one-third of the 136,000 t volume was used in appliances, and close to half went into industrial controls, circuit breakers, wiring devices, and other electric/electronic parts (see Phenolic resins).

Though inexpensive, phenolic resins are more costly in use because no effective method has yet been found for re-use of scrap (see Recycling). They are also deficient in impact resistance. More expensive thermosets such as alkyds and diallyl phthalate (DAP)-based products are used instead of phenolics for better arc resistance, higher surface resistivity, and better retention of electrical properties at high temperatures and in humid conditions, but the problems of scrap recovery and impact resistance remain (see Alkyd resins; Allyl monomers and polymers).

The impact resistance of PBT although not outstanding is better than that of the thermosets, and its combination of mechanical and electrical properties and long-term retention of these properties, combined with the advantages of thermoplastic processing, have made it a formidable contender for thermoset markets. PBT compounding has been directed at those markets, and the potential for the resin was enhanced with the development of a PBT with 180-s arc resistance and a U.L. 94 V-0 rating.

In addition to PBT various thermoset polyesters are competing as well in injection molding grades (24–26), and poly(phenylene sulfide) (see below) offers higher use temperature and the convenience of thermoplastic processing. Among the major engineering thermoplastics, however, PBT comes closest to the thermosets in balance of properties. Polycarbonate comes closest in dimensional stability, but PBT is the first of the major engineering thermoplastics to be considered for some of the most demanding thermoset applications.

Apart from electric/electronic usage, PBT is used in relatively small volumes in industrial machinery parts, where certain kinds of chemical resistance are required, or where glass-reinforced grades provide necessary heat resistance. It is used along with acetal in pump impellers and housings to an increasing extent. It is used in appliances to provide color where phenolics cannot be used, and for parts that need resistance to chemicals and fats. There is some usage in consumer products like zippers, where its self-coefficient of friction is better than nylon or acetal.

General Electric (Valox) is the leading supplier of thermoplastic polyester molding compounds, with Celanese (Celanex) a close second. GAF (Gafite) is a third supplier, and Eastman Chemical a fourth. Goodyear has offered PBT molding compounds, but its activities in this area are minor now. DuPont has entered the market in 1978 with its Rynite PET-based compounds, and Mobay announced entry in 1979.

Poly(phenylene Oxide)-Based Resin. The only PPO-based engineering thermoplastic commercially available in the United States is General Electric's Noryl, a blend of poly(phenylene oxide) and impact polystyrene (see Styrene plastics). The success of Noryl has been due to the development of a large family of formulations geared to specific markets. G.E. has patent protection on the basic resin and on the formulations introduced in the 1960s and 1970s. The major competitors for Noryl are high-heat grades of ABS and polycarbonate-ABS alloys.

The greater part of Noryl volume is in injection molding, along with some profile and sheet extrusion. An amorphous resin, it was introduced to fill the price/performance gap between ABS and polycarbonate. It competes most effectively against ABS where low-temperature impact strength, high-heat and moisture resistance and/or flame retardance are essential. Noryl has the lowest water absorption rate of the thermoplastic engineering resins, and it is completely resistant to hydrolysis. Also, flame retardance does not depend on the addition of halogens that would decrease impact strength.

The crystalline resins, in general, have relatively poor impact strength and good wear and chemical resistance. The amorphous resins, in general, have relatively good impact strength and relatively poor wear and chemical resistance. Because of these basic differences, the crystalline and amorphous resins seldom compete directly against each other.

Flame-retardant ABS applications have been a particular target for Noryl products because flame-retardant grades of ABS are very close to Noryl in price per unit weight, and when Noryl's lower specific gravity and shorter cycle times are considered, the latter material can often win out on cost. Metal substitution is also a target, as it is with all engineering resins, and this has been important in automotive applications and water-handling markets.

G.E. offers a range of standard Noryl non-flame-retardant grades with heat deflection temperatures from 113 to 150°C, and a range of flame retardant grades with HDTs from 88 to 150°C. Special grades are made for key markets like automotive and TV parts. Special-purpose formulations can be made by varying the relative content of PPO and polystyrene.

Flame retardance is imparted to Noryl by the incorporation of a proprietary flame retardant into the compounding procedure. Among the engineering resins, it has the highest percentage of volume in flame-retardant grades (about 50%). This feature has been important in appliances, business machines, and other electric/electronic applications. An important and growing market for Noryl is in structural-foam cabinetry for business machines, where it is the volume leader (27).

In the automotive market, Noryl is used for some exterior plated parts, but it has been a leading contender for interior parts that have to withstand greenhouse temperatures of 107–110°C. Appliances represent the largest single use category for Noryl, particularly if TV parts are included (Table 3).

Noryl is used for motor housings because of its heat resistance; for electrical enclosures because of its U.L. approval as sole support for current-carrying parts; for mixer housings, TV deflection yokes, and other parts where flame retardance is required; for dishwasher pump covers where hot-temperature moisture resistance is necessary; and for hydromassagers, steam curlers, and other units using moist heat. The greater percentage of Noryl used in appliances is in small kitchen or personal-care electrics, but it is used in some large appliances as well, particularly in control consoles. It is relatively important in environmental control appliances.

In the electric/electronic category, Noryl is used for outlet boxes, switch plates, connectors, compressor terminal covers, terminal blocks and strips, and other parts where competing materials might give problems with flash, warpage, inferior dimensional stability at high temperatures, or higher cost. In outlet and switch boxes, where demands on thermal and mechanical properties are low, commodity thermoplastics can sometimes do the job at lower cost. But for double-gang and ceiling boxes that have to meet the U.L. 514C standard, Noryl has the necessary strength and heat resistance. It is also used in lighting fixtures.

A major issue with respect to Noryl is whether it will be able to retain its markets in competition with newer grades of other styrenic resins. Where heat deflection temperatures of 105–110°C are required, along with flame retardance, almost no other materials compete with Noryl in impact resistance at the same price.

Where only the high heat-deflection temperature is required, there are several competitors including: high-heat ABS grades, a styrene copolymer (Arco's Dylark), an experimental styrene–maleic anhydride, and polycarbonate–ABS alloys.

For applications requiring flame retardance, but with HDTs below 93°C, ABS grades are available with additive flame retardants, or ABS–PVC alloys can be used. The alloys are ductile, tough materials, but they have disadvantages in processing stability because of the PVC content. The additive flame-retardant grades have better processing stability, but they are not as ductile and tend to be weak at low temperatures. Where both high heat-deflection temperatures and flame retardance are required, in a price range at the high end of the ABS scale, the closest competitor is a polycarbonate/ABS alloy with a V-0 rating (Mobay, Bayblend).

Polycarbonate. Polycarbonate is an amorphous polyester of carbonic acid, produced from dihydric or polyhydric phenols through a condensation reaction with a carbonate precursor. Bayer A.G., FRG, held the composition of matter patent that expired in 1979. G.E. produced the resin (Lexan) under a cross-licensing agreement with Bayer, and Mobay (a Bayer subsidiary) produces Merlon. Bayer and G.E. patents continue to protect certain processes and grades.

Because of its transparency and processability, polycarbonate is used in high-volume glass replacement applications not open to any of the other engineering thermoplastics. In window glazing and lighting it competes with acrylic resins and glass, in blow molding it is used for returnable milk bottles. In injection molding polycarbonate competes with the other engineering resins, and its transparency is often an asset.

Apart from transparency, however, polycarbonate has a unique balance of properties. It is relatively expensive compared to ABS, which can do the job in many applications that might otherwise go to polycarbonate. The cost-conscious auto industry used over 9000 t of polycarbonate in 1978, however, often in applications that required its outstanding dimensional stability.

Polycarbonate is very popular for small appliances, generally the first choice where heat or impact resistance, or both, is required. It has broad usage in newer types of hand-held appliances and in transparent sections of stationary appliances. It competes with Noryl and ABS in some uses like motor housings, and is often a replacement for die-cast zinc.

Polycarbonate is used in electric/electronic applications in a variety of uses that include covers for magnetic storage disks, switches, diode blocks, telephone dial rings and push buttons, and enclosures for control equipment. Its usage in connectors has been limited by its susceptibility to stress cracking.

In business machines, polycarbonate is used for structural-foam molded cabinetry, although Noryl can generally be used unless polycarbonate's superior strength or heat resistance is required. A special grade meets the requirements of U.L. Standard 748 for flammability in large complex assemblies like computer-room equipment.

In industrial machinery the greater part of polycarbonate's volume is in power tools, where it competes with ABS. The rest is in a variety of mechanical drive parts, protective equipment, etc. In power tools, polycarbonate has been meeting increasing competition from high-heat ABS grades.

Except in glazing, polycarbonate finds only relatively minor usage in construction. It is used in miscellaneous consumer and recreation products, as well as in institutional food service equipment where its transparency is an asset vs stainless steel. Where higher temperatures are encountered than polycarbonate can tolerate, polysulfone is used (see below). Polysulfone and TPX polymethylpentene (Mitsui Petrochemical, Japan) compete with polycarbonate in medical and laboratory equipment (see Hydrocarbon resins).

Polycarbonate must compete against many other resins in a variety of markets, including some not classified as engineering resins. Polycarbonate is irreplaceable where the full range of its properties are required, but other resins compete where only some of polycarbonate's advantages are called for. These include: acrylic resins for transparency and weather resistance; TPX for transparency, heat resistance, and broader chemical resistance; polysulfone for transparency, heat resistance, improved stress-crack resistance and mechanical properties (see Polymers containing sulfur); heat-resistant ABS for some opaque applications; mineral polypropylene for custom-formulated uses (see Olefin polymers); and DuPont's ST nylon for impact resistance combined with chemical resistance. Polyarylates (see below) began entering the market as new competitors in 1978.

Other Plastics Used in Engineering Applications. The five engineering plastics discussed in this article compete with other plastics in some markets, and to the extent that these other plastics do compete, they can also be called engineering plastics.

Competition from glass and/or mineral-filled polypropylene and flame-retardant ABS has been mentioned. Noryl also must compete against flame-retardant polystyrene in some electronic cabinetry where a relatively low Thermal Index is acceptable. Ultra-high-molecular weight polyethylene (UHMWPE) offers excellent wear resistance, but limited processability (see Olefin polymers).

Thermosets have long been available as insulators in electric/electronic applications, offering a wide range of capabilities in resistance to heat and environmental conditions. Thermoset molding compounds may be formulated to satisfy one or more important uses. Typical distinctive properties of thermosets include dimensional stability, low-to-zero creep, low water absorption, maximum physical strength, good electrical properties, high heat deflection temperatures, high heat resistance, minimal values of coefficient of thermal expansion, low heat transfer, and specific gravities in the 1.35–2.00 range (28–29).

Processing advantages have allowed thermoplastics to replace thermosets in many markets, but competition remains in some of the more demanding uses (30). Thermoplastics in general offer faster molding, lighter weight, possibility for thinner walls and more complex design, and greater impact resistance. Thermosets in general do not exhibit as much creep at elevated temperatures as thermoplastics, including the reinforced grades (see also Amino resins; Polyesters).

The engineering resins do not compete directly with epoxies, but some of the high-temperature resins are being used for that purpose (see Epoxy resins). The engineering resins rarely compete with polyurethanes or silicones (see Urethane polymers; Silicon compounds).

Fluoropolymers are often categorized along with the engineering plastics, but the two groups seldom compete. As a class, fluoropolymers do not offer the loadbearing capability of the engineering plastics, and loadbearing is generally one of the demands placed on plastics in engineering uses. In nonloadbearing uses, however, fluoropolymers have outstanding and unique properties, including resistance to very high and low temperatures, exceptional electrical properties, and low coefficient of friction (see Fluorine Compounds, organic).

Specialty Plastics

Specialty plastics include a mixed group of materials sold at relatively high prices compared to the engineering plastics and in relatively low volumes. The members of this group generally have high-temperature capability but this capability involves complex costly synthesis and usually difficulty in processing. In this group the polyimides (qv) can be used continuously at temperatures in the 260°C range.

There is no question that such materials can be used in engineering applications, but there are only two materials in this group that compete directly with the engineering plastics as defined here. These are poly(phenylene sulfide) (PPS) and polysulfone.

The relative heat resistance of plastic materials is generally measured either by ASTM D648, Deflection Temperature under Load, or U.L. 746B, Polymeric Materials—Long Term Property Evaluation (Relative Thermal Index). These two testing procedures have been referred to in connection with the main engineering plastics, but they become particularly relevant in attempting to define a high-temperature plastic. No comprehensive body of data exists to compare the mechanical, chemical, and electrical stress behavior of all materials under long-term elevated-temperature conditions (31–32).

The U.L. Thermal Index is a preferable indicator, if a choice must be made, because heat deflection temperature merely indicates the temperature at which a bar will first deflect. It provides no information about time to failure or mode of failure. Glass-reinforced nylon and polyester have very high heat-deflection temperatures, but since they are crystalline resins with sharp melting points there is an inherent risk of catastrophic failure.

The U.L. Thermal Index is a better indicator of continuous performance ability, and an acceptable dividing line between the major engineering thermoplastics and the high-temperature thermoplastics can be drawn at the U.L. 150°C Thermal Index. This is the generic rating for phenolic molding compounds, and most thermosets can be used continuously above that temperature. For thermosets, unlike thermoplastics, 150°C is usually taken for granted.

The significance of thermoplastics that can operate over 150°C does not lie in their role as replacements for thermosets in the low price range. With the exception of PPS they are too expensive. The can function in replacement of metals, glass, epoxies, fluoropolymers, and specialty thermosets in areas where thermoplastic processing advantages make the cost worthwhile (33).

Poly(phenylene sulfide) (Phillips Chemical Company, Ryton) is produced by the reaction of *p*-dichlorobenzene and sodium sulfide in a polar solvent. PPS, which is available for molding only in glass- and/or mineral-reinforced compounds, offers U.L. Thermal Index ratings of 240°C at a relatively low price. PPS also has outstanding chemical resistance, with no known solvents below 205°C. An additional asset is that PPS is inherently flame retardant. It is considered a thermosetting thermoplastic because optimal high-temperature properties can be obtained through annealing, in which some cross-linking takes place. PPS is also being used in high-temperature alloys (qv) (34).

Most PPS applications are in structural electric/electronic parts, where PPS competes against phenolic resins on the basis of cost savings through scrap re-use, and against other more expensive thermosets like diallyl phthalate (DAP) and diallyl isophthalate (DAIP) resins. Among the engineering thermoplastics, its closest competitor is PBT.

Polysulfone, Union Carbide's Udel, is a copolymer of 4,4-dichlorodiphenylsulfone and bisphenol A. This is a transparent amorphous resin with a U.L. Thermal Index of 150°C. Polysulfone is selected instead of polycarbonate where the higher use temperature is required, and sometimes for better stress-crack resistance at lower temperatures. Markets are often in glass and stainless steel replacement based on the advantages of transparency, heat resistance, hydrolysis resistance, suitability for food contact, and resistance to acids and alkalis.

Applications are in medical hardware, food processing and handling equipment, automotive, electric/electronic, and industrial parts, often for corrosion resistance. There is some usage of polysulfone as a substrate for circuit boards instead of the customary epoxy-glass composites (35). It is also used for process pipe, and as a matrix for carbon-fiber advanced composites for aircraft parts, in competition with carbon-epoxy composites at 110°C (see Ablative materials; Composites).

Plastics with still higher heat-resistance capabilities include some thermoplastics (ie, other polysulfones), some thermosets (ie, polyimides and aramids), and some that can be melt-processed but need cross-linking for optimal property development (ie, poly(amide–imide) resins). References 36–37 are exceptionally useful for comparisons of high-temperature thermoplastics (see Aramid fibers; Polyimides).

New Developments in Engineering Plastics

The cost of developing, introducing, and providing the marketing and technical back-up for engineering plastics is very high, and there is some opinion in the plastics industry that future engineering plastics will be chemical or physical modifications of current ones.

Desirable properties can be obtained by physically blending or alloying resins, or by incorporating inorganic or organic fillers and reinforcements. This is being done by primary resin suppliers, who use these methods to produce special grades, and by custom compounders, and to an increasing extent by end-users who tailor-make resins to fill their own specific needs (38–39).

The distinction between blends and alloys is not a clear one, but both terms are used for physical mixtures of two or more structurally different polymers (40–43) (see Copolymers). As compared to copolymers, in which the components are linked by strong chemical bonds, the components in alloys adhere primarily through van der Waal forces, dipole interactions, and/or hydrogen bonding (34).

Some blends are marketed as such, but others are proprietary formulations marketed only with a special trade name or grade designation. Mobay's Bayblend polycarbonate/ABS alloys are marketed as alloys; Uniroyal's Arylon-T poly(aryl ether) (sold to U.S.S. Chemical) is a polysulfone–ABS alloy but it is not marketed as such. Some grades of engineering plastics are generally presumed to be blends, but their composition is not revealed. Elastomers can be blended with engineering resins to increase impact strength (see Elastomers, synthetic).

Filled and reinforced compounds are readily identified, and the closing years of the 1970s have seen major advances in the use of glass fiber reinforcement (ie, stampable polypropylene and nylon sheet), carbon-fiber reinforcement (44), and the use of mineral reinforcements (see Carbon; Fillers; Laminated and reinforced plastics).

Copolymerization is another route to developing new resins. Polyarylates, eg, are polyesters produced from diphenyls and dicarboxylic acids. Carborundum's Ekkcel high-temperature (HDT 300°C) aromatic polyester is a polyarylate produced from 4,4-dihydroxydiphenyl and isophthalic acid. The first commercial polyarylate to be introduced into the U.S. engineering plastic market, based on bisphenol A and phthalic acids, was developed by Unitika in Japan and is marketed by Union Carbide as Ardel. At the close of the decade still another was introduced by Hooker Chemical, and more appeared to be in the developmental stage.

Although modification of existing resins may be the major route to new properties, what appear to be entirely new resins are still being introduced. Rohm & Haas announced a new engineering resin in mid-1979 of a polyimide type.

BIBLIOGRAPHY

1. R. H. Heinold, "Broadening the Capabilities of Polypropylene for Appliance Applications," *Proceedings, Soc. Plastics Engineers National Technical Conference,* Nov. 17–19, 1975, pp. 63–67.
2. J. Houston, "Automotive Market—Impetus to the Growth of Polyolefin Polymers," *Proceedings, Chemical Marketing Research Association meeting,* Feb. 7–10, 1978, pp. 142–168.

3. R. C. Wright, "Engineering Thermoplastics," *Proceedings, Fourth Annual Conference on Contingency Planning for Plastics,* PPC/Plastics Publishing Co., 1978, pp. 103–108.
4. B. Nathanson, "Outlook for Engineering Thermoplastics—United States and Western Europe," *Chemical Marketing Research Association meeting,* May 3–6, 1977.
5. *Deflection Temperature Under Load, ASTM Test Method D648,* American Society for Testing and Materials, annual.
6. *Underwriters Laboratories Recognized Components Index,* Underwriters Laboratories Inc., annual.
7. *Flexural Properties, ASTM Test Method D790,* American Society for Testing and Materials, annual.
8. *Impact Resistance of Plastics and Electrical Insulating Materials, Methods A and C, ASTM Test Method D256,* American Society for Testing and Materials, annual.
9. E. R. Rosenberg, "Plastics vs. Metals—to 1980 and beyond," *Proceedings, Second Annual Conference on Contingency Planning for Plastics,* PPC/Plastics Publishing Co., 1976, pp. 128–138.
10. J. E. Theberge, "Reinforced Thermoplastics" in *Modern Plastics Encyclopedia,* Vol. 55, McGraw-Hill Book Co., New York, 1978–1979, p. 140.
11. M. A. Rehm, "Reinforced Thermosets" in *Modern Plastics Encyclopedia,* Vol. 55, McGraw-Hill Book Co., New York, 1978–1979, p. 142.
12. V. Stayner "Non-Fibrous Property Enhancers" in *Modern Plastics Encyclopedia,* Vol. 55, McGraw-Hill Book Co., New York, 1978–1979, p. 186.
13. J. M. Smart, "Thermoplastics—Raw Materials for Replacing Diecasting Alloys," *paper presented at a Joint Conference of the Institute of Purchasing and Supply and the Diecasting Society,* Coventry, England, Oct. 12, 1977.
14. *Internally Lubricated Reinforced Thermoplastics,* Bulletin 254-278, LNP Corporation, 1978.
15. *What You Always Wanted to Know About Engineering Resins,* Celanese Plastics Company publication, 1976.
16. *Tests for Flammability of Plastics Materials for Parts in Devices and Appliances, U.L. Standard 94,* 2nd ed., Underwriters Laboratories, Inc., 1978; 3rd imp., 1979.
17. "Fire Safety Aspects of Polymeric Materials" in *Materials: State of the Art,* Vol. 1, part of a 10 volume series sponsored by the National Materials Advisory Board, Technomic Publishing Co., 1977.
18. W. McKinlay and S. D. Pearson, *Plastics Gearing,* ABA/PGT Publishing Co., Manchester, Conn., 1976.
19. N. C. Baldwin, "The Use of Acetal Copolymer in Plumbing Applications," *Proc. 33rd Annual Technical Conference, Soc. of Plastics Engineers,* 1975, pp. 30–33.
20. A. M. Houston, *Mater. Eng.,* 42 (Feb. 1976).
21. S. Telofski, "Thermoplastic Forces the Obsolescence of Traditional Materials in Automotive Ignition Systems," *Soc. Plastics Engineers Divisional Technical Conference,* Sept. 27–28, 1977, p. 41.
22. *Plast. Technol.* **24,** 83 (Jan. 1978).
23. B. W. Perry, "Phenolics in the Engineering Plastics Arena," *Soc. Plastics Engineers, Connecticut Section, Regional Technical Conference,* Oct. 10–11, 1977, pp. 87–94.
24. *Plast. World* **33,** 57 (July 21, 1975).
25. D. Portman and J. C. Clark, "Glass-Reinforced Polyester for Circuit Breaker Applications," *Proceedings, 31st Annual Technical Conference,* Soc. of the Plastics Industry, Inc., Reinforced Plastics/Composites Inst., 1976, Section 10-F, p. 1.
26. R. D. Lake, J. T. Shreve, and R. L. Lovell, "Pelletized Thermoset Polyester Molding Compounds," *Proceedings, 31st Annual Technical Conference,* Soc. of the Plastics Industry, Inc., Reinforced Plastics/Composites Inst., 1976, Section 13-C, p. 1.
27. J. L. Throne, "Principles of Thermoplastic Structural Foam Processing: A Review," in N. P. Suh and N. H. Sung, eds., *Science and Technology of Polymer Processing,* MIT Press, 1979, pp. 77–131.
28. M. A. Rehm, "Reinforced Thermosets" in *Modern Plastics Encyclopedia,* McGraw-Hill Book Co., New York, 1978–1979.
29. *Plastics Design Forum* **2,** 58 (Jan./Feb. 1977).
30. *Plast. Eng.* **33,** 40 (April 1977).
31. A. M. Houston, *Mater. Eng.* **81,** 28 (June 1975).
32. *Plastics Design Forum* **2,** 18 (Nov./Dec. 1977).
33. *Plast. World* **34,** 28 (June 21, 1976).
34. R. T. Alvarez, S. B. Driscoll, and T. E. Nahill, "High Temperature Performance Polymeric Alloys," *Soc. of Plastics Engineers, Proceedings, Annual Technical Conference,* 1977, pp. 308–310.

35. *Mod. Plast.* **54,** 52 (June 1977).
36. J. E. Theberge, B. Arkles, and P. Cloud, *Mach. Des.* **47,** 73 (Feb. 6, 1975).
37. *Ibid.,* 79 (March 20, 1975).
38. *Plastics Compounding* **1,** 75 (May/June 1975).
39. P. J. Cloud and R. E. Schulz, *Plast. World* **33,** 36 (Sept. 22, 1975).
40. R. J. Jalbert and J. P. Smejkal, "Alloys" in *Modern Plastics Encyclopedia,* Vol. 53, McGraw-Hill Book Co., New York, 1976–1977, p. 108.
41. *Plast. World* **35,** 56 (Nov. 1977).
42. *Mod. Plast.* **54,** 42 (1977).
43. G. R. Forger, *Mater. Eng.* **86,** 44 (July 1977).
44. J. E. Theberge and R. Robinson, *Mach. Des.* **46,** 2 (Feb. 7, 1974).

General References

C. A. Harper, ed., *Handbook of Plastics and Elastomers.* McGraw-Hill Book Co., 1975.
Modern Plastics Encyclopedia, annual, McGraw-Hill Book Co. See in particular the "Design Guide," Vol. 55, Oct. 1978, pp. 463–497.
Mach. Des., Penton/IPC publication, annual materials reference issues.
R. D. Deanin and S. B. Driscoll, "Buying Properties," *Chemtech,* 209 (April 1978).
Plastic Applications in Automobiles and Trucks, annual listing of automotive applications for glass-reinforced plastics, Owens-Corning Fiberglas.
The International Plastics Selector, Inc., San Diego, Calif., *Desk-Top Data Bank Books, Plastics,* 3 volumes.

MARILYN BAKKER
Business Communications Co., Inc.

ENHANCED RECOVERY. See Petroleum.

ENKEPHALINS. See Hormones.

ENZYME DETERGENTS

Enzymes are a vital part of all living processes and are believed to be as old as life itself. The term enzyme is derived from the Greek word *enzymos* and literally translated means in the cell or ferments. This name was first used by W. Kühne in 1876 when he defined enzymes as unformed or unorganized ferments, whose action may take place in the absence of organisms and outside of organisms. It is now known that enzymes are organic proteinaceous catalysts produced by all living cells. These enzymes are of two types: (1) exoenzymes, which are excreted by the manufacturing cell into the cell's environment, once excreted, penetrate and break down organic matter, such as proteins, starches, and fats, into soluble derivatives that can be absorbed or transported through the cell membrane (as a general rule, exoenzymes produce little energy directly for the cell to use); and (2) endoenzymes, which remain within the living cell, are transformed and/or broken down by the action of coenzymes to produce energy and cell components needed for life. Relatively large amounts of energy accompanies these reactions, which is in turn made available to the host cell. The detergent enzymes described here are exoenzymes, characterized by hydrolytic activity under alkaline conditions (1–2) (see Enzymes, industrial).

The development of enzyme supplements in laundry products is mainly the result of work by O. Röhm, founder of Röhm & Haas, Darmstadt, Federal Republic of Germany, and E. Jaag, of the Swiss firm, Gebrueder Schnyder, Biel. Röhm obtained a patent that describes application of pancreatic enzymes for washing purposes (3). About the same time, Röhm & Haas marketed what appears to be the first known enzymatic presoak detergent, Burnus. This product, which consisted primarily of soda and crude pancreatin, remained in the European marketplace for some fifty years. However, the initial proteolytic activity was low, and the poor stability of the enzyme in the presence of soda and resultant alkalinity further reduced the effect of the product (4). Another development of importance in this field occurred during the 1930s when Jaag in Switzerland developed the enzyme detergent concept further with the product known as Bio 38, based on the use of a pancreatic trypsin preparation.

By the 1940s interest in enzymatic pre-soaking agents received increased impetus due to the war-induced shortage of soap in Europe. About this time, Jaag worked on the development of enzymes that would be more compatible with the normal washing components (5–6). His work was carried on for some years and in 1959 an improved enzyme detergent, Bio 40, was produced by Gebrueder Schnyder in cooperation with the Swiss Ferment Company, Ltd. Bio 40 contained a bacterial protease that had considerable advantages over pancreatic trypsin but still was not sufficiently stable and active at pH 9–10. Research into the application of enzymes for the washing of work clothes used in the fish and meat industries began independently about 1948. It took several years before a bacterial protease was developed that had all of the desirable characteristics with regard to high pH stability and activity with the difficult stains involved (7).

Enzyme detergents first appeared in test market during 1966 (8) in the United States. At the same time commercial acceptance in Europe had assumed a 2–3% market share of the European laundry detergent market. The growth in market share in both Europe and the United States increased dramatically, and by 1969 the market share for enzyme detergents had reached 50% in Europe and almost 45% in the United States.

Today, over 50% of the heavy-duty laundry detergents in Europe contain enzymes. However, owing to the unfavorable publicity attributed to the dusting effect of these powdered enzymes (9), the market share for enzyme detergents in the United States has receded to pre-1968 levels. The recent trend toward lower-temperature washing, reduced phosphate formulations, together with the advanced encapsulating technology of the detergent enzyme manufacturers has contributed to a renewed interest in alkaline-active enzymes (see Microencapsulation). In contrast with the laundry detergents, alkaline-active enzymes have been used continuously since their introduction in the United States in the so-called presoak and "booster" laundry pretreatment products (see Drycleaning and laundering; Surfactants and detersive systems).

Properties

Serine-active alkaline proteolytic enzymes constitute >95% (est) of all enzymes sold worldwide for enzyme laundry detergents. Hence, the following discussion deals with this class of enzymes and detergents derived from them.

The use of effective detergent enzymes from bacterial sources is based primarily on the development of alkaline-stable and active proteases. With the exception of unbuilt heavy-duty liquid laundry detergents, which have entered the United States laundry market in recent years, the pH of laundry detergents is generally in the 9–10.5 range. Another important property for a detergent proteolytic enzyme is the thermal stability, or more appropriately, thermal versatility. The so-called European laundry wash utilizes temperatures from ambient to 100°C; thus the more heat-stable the enzyme, the better. Although thermal requirements in the United States are lower, United States hot-water laundering conditions are considered to be in the range of 50–55°C. A third factor in the suitability of a detergent enzyme is the nature of the active sites for the enzyme. Although papain (a vegetable protease from *Carica papaya*) has several chemical characteristics that would appear to be suitable for a laundry detergent environment (thermal stability, broad nonspecific substrate activity), papain contains cysteine and histidine at its active sites. In order for these enzymes to be active, the sulfur group of the cysteine residue at the active sites should be in the normal thiol form. Under European wash conditions where sodium perborate is included in the formulations, the enzyme is inactivated via the oxidation of the thiol group. In the case of trypsin, which contains six disulfide bridges, an oxidation occurs in the presence of perborate to produce a change in the tertiary protein structure (see Biopolymers). At temperatures above 60°C this inactivation is quite rapid. In the United States, sodium perborate is generally not incorporated in detergent formulations. However, in addition to cost and availability considerations, a laundry detergent containing papain or trypsin should be considered incompatible with the so-called powdered sodium perborate-based laundry boosters, bleaches, and other detergent formulations containing sodium perborate (see Bleaching agents). As can be seen in Tables 1 and 2, the alkaline protease from *Bacillus licheniformis* is much more suited than either papain or pancreatic trypsin in this respect.

In addition to stability at relatively high temperatures and pH values, the enzyme should be compatible with other detergent ingredients, such as surfactants, chelating agents (qv) (phosphates, EDTA, nitrilotriacetic acid (NTA), etc), optical brighteners (see Brighteners), perfumes (qv), etc. This is the case with the serine-active alkaline proteolytic enzymes (10).

Table 1. Properties of Alkaline and Neutral Proteases

Characteristic or test	Alkaline protease	Neutral protease
physical properties, pH optima		
activity	pH 8.0–14	pH 5.5–8.0
stability	pH 5.0–10	pH 5.0–8.0
nature of catalytic site	serine	group II metal
stabilization by Ca^{2+}	0	+
inhibition by:		
DFP^a and $PMSF^b$	+	0
$EDTA^c$ and phosphate	0	+
barley extract	+	0
substrate profiles:		
AA esters and amides	+	0
oxidized B-chain insulin	18 bonds split	5 bonds split
enzyme specificity:		
amino acid residue preference	Leu CO—	—NH Leu
	Cys CO—	—NH Phe
	Glu CO—	

^{a} DFP = diisopropyl fluorophosphate.
^{b} PMSF = phenylmethylsulfonyl fluoride.
^{c} EDTA = ethylenediaminetetraacetic acid.

Table 2. Enzyme Stabilities in the Presence of 0.1% Sodium Perborate

Enzyme	Residual activity^a, %
papain, from *Carica papaya*	0
pancreatic trypsin, from bovine and porcine pancreas	48
alkaline protease, from *Bacillus licheniformis*	90

^{a} After 30 min at 50°C and pH 9.5.

Further research has resulted in the development of enzymes that are more stable and active at high pH values. Such enzymes, produced from alkalophilic strains of *Bacillus*, are well suited for use in heavy-duty laundry detergents and in liquid laundry-product formulations (10–12).

The activities of alkaline proteases are characterized by several well known analytical methods. Each method is based upon a measurement of the enzyme's hydrolytic effect upon a carefully standardized protein substrate under particular conditions. The method of Kunitz (13), eg, is based upon the hydrolysis of casein. The Anson method (14) requires using urea-denatured hemoglobin as the substrate. Another method (15) employs trinitrobenzenesulfonic acid (TNBS) using dimethyl casein as the substrate. This latter method is particularly effective in assaying enzyme solutions in the presence of anionic surfactants since it does not require a precipitation with trichloroacetic acid.

Alkaline and neutral proteases differ basically in the nature of their catalytic centers, and can be further characterized by: (*1*) pH range of activity and stability;

(2) activation–inhibition effects; and (3) substrate profiles (4). Table 1 compares the properties of alkaline and neutral (*Bacillus subtilis*) proteases.

Methods of Production

Detergent enzymes are derived from specific microorganisms. Since these properties are inherent in the organisms, they cannot be altered by variations of the growth conditions. Most detergent enzymes used today are excreted by bacteria of the genus *Bacillus*, especially members of the *Bacillus subtilis* group. This group of bacteria are regarded as harmless saprophytes and have been employed in the production of proteases and amylases for the textile and baking industries for many years. The advantages of detergent proteolytic enzymes from bacteria are improved performance under washing conditions and unlimited supply. The vegetable- and animal-derived proteases fall short of both requirements (16) (see also Microbial transformations).

Fermentation requires isolation of the proper strain of bacterium (17–19). The most effective nutrient medium and growth conditions are then selected. Culturing is initiated in glass laboratory flasks, followed by transfer to pilot-plant fermentors. The medium functions as a source of protein and carbohydrate, and also contains the necessary nutrients and minerals such as phosphorus. Feed materials, such as corn, starch, and soybean meal, are especially suitable in this regard. A typical medium for enzyme production is as follows: 3.0% potato starch; 2.5% soybean meal; 8.0% ground barley; 0.4% $CaCO_3$; 0.4% soybean oil; and ca 85.7% tap water.

Fermentation Process. A typical commercial fermentation (qv) process is shown in Figure 1. Equipment requirements include a stainless-steel mixing tank of medium incorporation and blending, a seed fermentor and a main fermentor. The fermentation tanks are totally enclosed and equipped with air lines, agitators, and jackets or coils for maintenance of constant temperature. Usually provision is also made for automatic

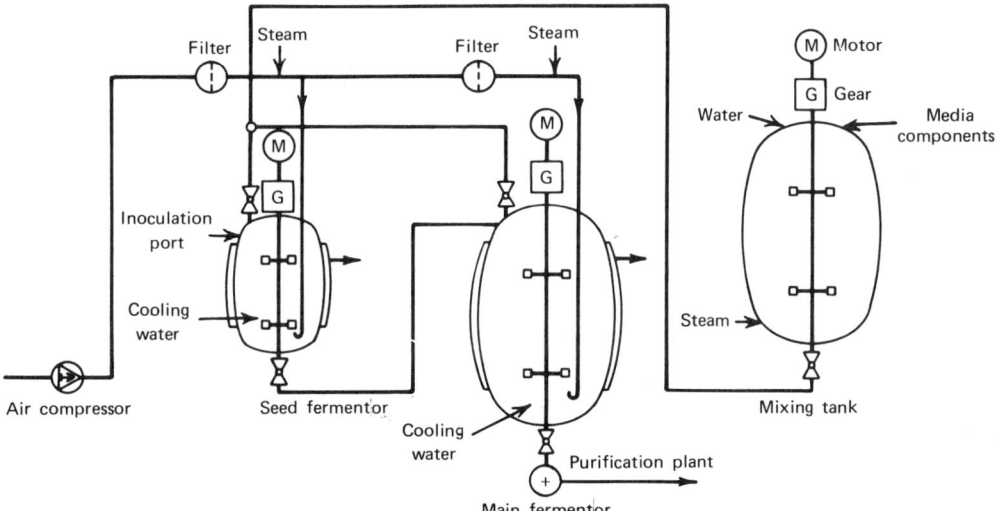

Figure 1. Fermentation process.

foam and pH control (see Defoamers; Hydrogen ion concentration). As the success of the fermentation depends upon the absence of foreign microorganisms, it is important that the air introduced be sterilized by passing through filter pads (see Sterile techniques). All media ingredients are combined with tap water in the mixing tank, transferred to the fermentation vessel and heated to at least 115°C for 1–2 h by steam injection. After cooling the medium to approximately 30°C, a pure culture of the organism is transferred via the inoculation port into the seed fermentor. During the agitation and aeration of the fermentation process, it is important to maintain control over critical parameters such as temperature and dissolved oxygen. Throughout the fermentation samples are withdrawn and their enzyme content determined. As soon as the formation of enzyme has ceased, the contents of the fermentation vessel are moved to the purification stage.

Purification Process. The recovery of enzyme begins as soon as the fermentation has been terminated. The medium is then transported (usually by pumping) through a cooling stage to a separation process utilizing centrifuges where bacteria and gross insoluble substrate components are removed. The remaining particles are then separated by precoated diatomaceous earth filters. The enzyme solution is concentrated by vacuum evaporation or ultrafiltration (qv), if required, and finally the enzyme is removed from the filtrate by the addition of a protein precipitation agent. The enzyme-containing precipitate is isolated further on additional filtering equipment and dried by rotary vacuum dryers, or spray-dried enzyme powder is milled and standardized with substances such as sodium chloride or sodium sulfate. In order to prevent the growth of bacteria during the recovery process, the enzyme precipitate is maintained at the lowest temperature possible, frequently <5°C. Throughout the whole recovery process the pH, temperature, enzyme activity, and bacterial count are monitored on a routine basis.

Finishing of Detergent Enzyme. Until 1971 detergent products were essentially milled and standardized enzymes. In order to eliminate dusting and resulting inhalation difficulties in the detergent plant, processes were developed to encapsulate the enzyme in materials such as nonionic surfactants or a sodium chloride–poly(ethylene glycol) matrix. In this way the dust-forming property of the enzyme is eliminated and the stability of the enzyme in the presence of the laundry detergent powder is enhanced. Several methods of producing stable, virtually dust-free granulates have been reported (20). In the United States stringent guidelines for low dust emission with regard to detergent factory workers have been recommended by the ACGIH (21).

Detergent Applications

Heavy-Duty Laundry Detergents. As a general rule, there are no special formulation requirements when enzymes are to be included in heavy-duty laundry detergents. On the basis of 1.5 Anson unit/g activity, the enzyme concentration of European laundry detergents is normally in the 0.7–1.2% range. During the period between 1968–1971, United States detergent products included enzymes at lower levels, most often between 0.25–0.5%. More recently, the trend has been toward higher amounts. This trend can be traced to two separate factors: lower wash temperatures and reduced levels of sodium tripolyphosphate in the detergents. Enzyme activity is reduced by 50% for approximately each 10°C drop in wash-water temperature.

European wash conditions are especially suited to the action of an enzyme detergent. Because the time–temperature relationship is such that the wash-water temperature increases from 25 to 90°C in ca 35 min, the enzyme is afforded ample reaction time on the protein-based soil before 60°C is reached. Thereafter, the enzyme is gradually inactivated by heat and by the hydrogen peroxide from the sodium perborate which becomes increasingly active. Table 3 describes a basic European detergent composition.

In the United States, standard hot-water wash conditions are 11–15 min at 45–55°C. However, wash temperatures are believed to vary considerably from one region of the country to another, as well as from one wash load to another within the same household. This is owing to the variety of fabrics, ranging from cold water for wool, to warmer 35–40°C water for synthetics, to hot water (50–55°C) for white cotton fabrics.

In Figure 2 the effect of a *Bacillus*-derived highly alkaline-active proteolytic enzyme in the presence of mixtures of sodium tripolyphosphate, sodium carbonate,

Table 3. Enzyme Parameters and Active Sites

Enzyme	Molecular weight	Number of amino acids	Active sites	Number of disulfide bridges
alkaline protease from *B. lichenformis*	ca 27,000	274	serine + histidine	0
pancreatic trypsin	24,000	230	serine + histidine	6
papain	23,000	211	cysteine + histidine	3

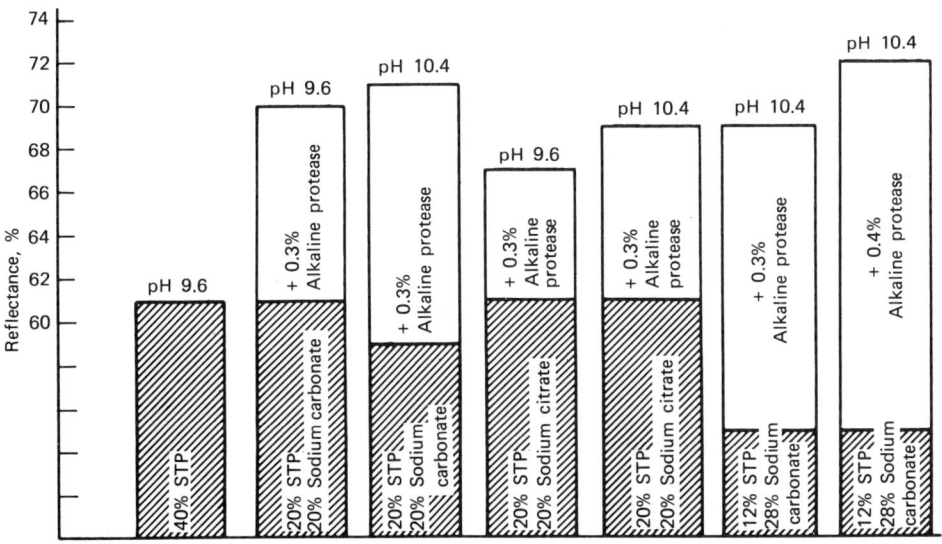

Figure 2. Effect of enzyme in formulations with reduced amounts of phosphate. Reflectance values: ▨, detergent without enzyme; ▢, detergent with enzyme.

test method: CSMA wash test
test cloth: EMPA 116
detergent: standard, less the normal 40% builder content (concentration is 2 g/L)
builder: as indicated (STP = sodium tripolyphosphate)

and sodium citrate is measured using the CSMA (Cosmetic Specialty Manufacturers' Association) wash test on EMPA 116 (22) test fabric (blood, milk, and ink on 100% cotton) (23). The enzyme action on protein-based stains may be utilized in enzyme detergents under cold-water wash conditions provided there is a presoak cycle prior to the actual washing cycle. However, detergent enzymes will not contribute significantly to washing efficacy in a ten minute cold-water (20°C) wash alone. Therefore, it is necessary under cold-water wash conditions to soak the laundry with an enzyme detergent for at least 30 min before the actual wash takes place.

Christensen and co-workers have reported on the use of enzyme detergents under cold-water conditions as a means of energy conservation (17). Tables 4 and 5 illustrate some of the possible enzyme detergent formulations. Figure 2 demonstrates the effect of a highly alkaline-active protease (23).

Specialty Products. In recent years there has been a growth in laundry pretreatment products. Among these, the laundry booster products, usually intended to enhance the alkalinity, water softening, bleaching and stain removal action, are the best known. The laundry presoak is another of the enzyme laundry adjunct products in the United States (19,24). Another area of interest in the specialty field is the so-called pre-spotter agents. These may be further divided into aqueous-based and solvent-

Table 4. European Enzyme Detergent Composition

	Composition, %
sodium LAS[a] (40%)	25.0
alkylphenol ethoxylate(11 EO)[b]	3.0
soap, high-titer (80%)	3.0
CMC[c] (59%)	1.6
sodium tripolyphosphate (STP)	38.0
sodium perborate tetrahydrate	25.0
sodium metasilicate	1.0
alcalase 1.5M	0.9
sodium sulfate	ca 2.5
Total	100.0%

[a] Linear alkanesulfonate surfactant.
[b] Average of 11 ethylene oxide units (see Surfactants).
[c] Carboxymethyl cellulose, sodium salt.

Table 5. United States Enzyme Detergent Formulas

Constituent	Composition, %	
	Phosphate	Nonphosphate
sodium LAS[a]	10–20	10–22
tallow alcohol sulfate	0–10	
tallow ether sulfate	0–30	0–12
sodium tripolyphosphate	up to 34	
sodium carbonate		20
sodium silicate	up to 12	up to 20
sodium carboxymethyl cellulose (CMC)	0–0.5	0–1.0
fluorescent whiteners	0.1–0.5	0.1–0.5
alkaline protease	1.0	1.25
sodium sulfate and perfume	qs	qs

[a] Linear alkanesulfonate surfactant.

based systems. The former generally are designed for protein, and an assortment of vegetable-based pigment stains, and the latter are generally directed at oil- and grease-based soil. Both proteolytic and amylolitic enzymes are used in all of the above specialty products with the exception of solvent-based pre-spot liquids, since enzymes are water-soluble proteins (25–26).

Another area for enzyme-detergent application is chlorine-free heavy-duty dishwashing detergents. An alkaline α-amylase can remove starch-based soils on solid surfaces at higher temperatures (26).

The application of a dual enzyme system for the removal of dental plaque from dentures has been reported (27). This system requires the use of a bacterial protease in combination with a fungal enzyme mutanase under mildly acidic pH conditions (28–29) (see Dentifrices).

Analytical and Test Methods

Generally, the evaluation of the washing efficiency of detergents is done in three steps: (1) screening tests in the laboratory, (2) bundle tests in the laboratory and field, and (3) practical tests in the field. The screening tests in the laboratory are carried out in model-washing apparatuses like the Terg-O-Tometer and Launder-Ometer (30). These models simulate the washing processes as performed in an agitator and drum household washing machine, respectively. The advantage of using the model apparatus is the much higher testing capacity compared with the household machine. Thus the Terg-O-Tometer can handle four subsystems at a time, and the Launder-Ometer, two (31).

In these laboratory tests preliminary evaluation is carried out of different parameters such as enzyme type and concentration, water hardness, type of soil, etc. Later in the development process the findings may be confirmed by the bundle and practical tests. The bundle and practical tests are performed with suitable controls in household washing machines with garments and other items soiled under actual use conditions.

For laboratory evaluations different types of test soilings are used (32). The composition of the stains depends on the type of enzyme under evaluation. For proteolytic enzymes, stains, such as blood plus milk, cocoa, egg, grass, etc, are often used. These substances are applied to various textiles such as cotton, cotton–polyester blends, and polyacrylics. Some of these test materials are commercially available, eg, the well-known test fabric EMPA 116 (22), whereas others are developed and produced in the laboratory.

Health and Safety Factors (Toxicology)

Two properties of detergent enzymes are important regarding their safety: (1) as is the case with other proteinaceous material, repeated inhalation of enzyme dust is associated with a comparatively high risk of respiratory allergy (hay fever, asthma) in susceptible persons; and (2) detergent enzymes are proteinases and will, therefore, irritate moist skin, the eyes, and mucous membranes.

The allergy problem has been largely eliminated by encapsulation of the enzymes in tough granulates and by the introduction of suitable methods of handling and transport of the granulates in well-ventilated working areas. This has reduced the amount of airborne enzyme dust in the workplace. New and more sensitive methods of analysis for airborne dust have been introduced and in many plants the situation is best described as long periods of zero dust levels. However, even in this improved working environment it may not be possible to avoid accidental sensitization of susceptible persons. Such persons are identified by scratch testing and transferred to enzyme-free work areas. The irritation problem has been eliminated by avoiding direct contact with enzymes, which may have an irritating effect even in granular form. The equipment is enclosed and workers wear protective gloves and goggles.

Detergent enzymes must comply with strict product specifications that guarantee very low microbial contamination similar to the requirements for food additives (33). Detergent enzyme manufacturers perform a series of animal (and human) studies to ensure that the products are harmless in use. The most important of such studies are those designed to judge the pathogenicity of the producing microorganism; the acute and subacute toxicity (34); the inhalation toxicity; and the sensitization potential and irritant effect (35).

BIBLIOGRAPHY

"Enzyme Detergents" in *ECT* 2nd ed., Suppl. Vol., pp. 294–309, by Robert P. Langguth and Raymond L. Liss, Monsanto Company.

1. *Chem. Eng. News,* 27 (Aug. 18, 1975).
2. U.S. Pat. 3,451,935 (June 24, 1969), A. Roald, W. Brabant, and N. T. de Oude (to The Procter & Gamble Company).
3. Ger. Pat. 283,923 (1913), O. Röhm.
4. C. Dambmann, P. Holm, and V. Jensen, *Dev. Ind. Microbiol.* **12,** 11 (1971).
5. O. Viertel, *Fette Seifen Anstrichm.* **51,** 145 (1944).
6. E. Jaag, *Chimia* **1**(3), 57 (1949).
7. E. Jaag, *Seifen Ole Fette Wachse* **24,** 789 (1962).
8. Ref. 4, p. 18.
9. I. V. Sollins, *What are the Facts About Enzymes and Enzyme Detergents?* pamphlet, Novo Laboratories, Inc., Wilton, Conn., 1970.
10. U.S. Pat. 3,674,643 (July 4, 1972), K. Aunstrup, O. Andresen, and H. Outtrup (to Novo Terapeutisk Laboratorium A/S).
11. G. Jensen, *paper presented at the meeting in Deutsche Gesellschaft fur Fettwissenschaft, Giessen, FRG, Oct. 8–12, 1972.*
12. C. L. Dornbusch, *paper presented at the Third Biennial Symposium sponsored by the Soap and Detergent Association, St. Louis, Mo., Apr. 13, 1978.*
13. M. Kunitz, *J. Gen. Phys.* **30,** 291 (1947).
14. M. L. Anson, *J. Gen. Phys.* **22,** 79 (1939).
15. S. D. Friedman and S. M. Barkin, *J. Am. Oil Chem. Soc.* **46,** 81 (1969)
16. J. L. A. Briggs, "The Manufacture of Enzymatic Detergents," *paper presented at the II Simposio Jugislavi and Tensioattivi, Bled, Yuguslavia, 1969.*
17. Brit. Pat. 1,253,784 (Aug. 25, 1971), K. Aunstrup, O. Andresen, and H. Outtrup.
18. U.S. Pat. 3,723,250 (Mar. 27, 1973), K. Aunstrup and H. Outtrup (to Novo Terapeutisk Laboratorium A/S).
19. U.S. Pat. 3,827,938 (Aug. 6, 1974), K. Aunstrup and O. Andresen (to Novo Terapeutisk Laboratorium A/S).
20. T. den Ouden, *Tenside Deterg.* **14,** 209 (1977).
21. *Documentation of Threshold Limit Values for Substances in Workroom Air,* ACGIH, Washington, D.C., 1971.
22. *Eidgenossiche Material Prüfungs und Versuchsanstalt für Industrie* (*EMPA*), Bauwesen und Gewerbe, St. Gallen, Switz., or Testfabrics Inc., New York.
23. P. Holm, unpublished data of Novo Industri A/S.

24. P. N. Christensen, P. Holm, and B. Soender, *J. Am. Oil Chem. Soc.* **55,** 101 (1978).
25. U.S. Pat. 3,773,674 (Nov. 20, 1973), W. E. Adams and C. Barrat (to The Procter & Gamble Company).
26. G. Jensen, *The Development of a New Amylase and its Use in Dishwashing Detergents, Novo Information Publication 743139-C/GJE/JNT,* Novo Industri A/S, Bagsvaerd, Denmark, 1973.
27. E. B. Jörgensen, *J. Biol. Biccale* **5,** 239 (1977).
28. *Enzyme Denture Cleansers, Novo Technical Research Booklet 173a-GB500,* Novo Laboratories, Inc., Wilton, Conn., 1977.
29. B. Guggenheim, *Helv. Odontol. Acta* **14**(Suppl. V), 89 (1970).
30. T. Cayle, *J. Am. Oil Chem. Soc.* **46,** 515 (1969).
31. G. Jensen, *Process Biochem.* **7**(8), 23 (1972).
32. B. W. Terry and W. L. Groves, *paper presented at the Annual Spring Meeting of the American Oil Chemists Society, San Francisco, Calif., Apr. 1969.*
33. *First Supplement to the Food Chemicals Codex,* 2nd ed., 1974.
34. *RB 204 118, Report of the Ad Hoc Committee on Enzyme Detergents,* Division of Medical Sciences, National Academy of Sciences–National Research Council, Washington, D.C., 1971, pp. 10–18.
35. E. A. Newman and G. A. Nixon, *Food Cosmet. Toxicol.* **7,** 581 (1969).

General References

A. Zscharn, "Enzyme sparen Seife," *Fette und Seifen* **10,** 463 (1940).
K. Pfleiderer, E. Richter, and W. Wertel, *Wäscherei Plättorei Ztg.* **1,** 1 (1944).
F. M. Schimmel, *Biochemische Waswerking,* Nederlandse Instituute te Huishoudelijke Onderzoek, *Publication No. 97,* 1966.
A. Suter, *"Detersivi Enzimatici," paper presented at the 2nd Symposium about Detergents, Milan, Italy, Oct. 1966.*
H. Braun, *Jahrb. Textil-Reinigungsgewerbe* **9,** 160 (1967).
K. Düsing, "Proteolytische Enzyme in Waschmitteln," *Fette Seifen Anstrichm.* **10,** 738 (1967).
J. C. Hoogerheide and co-workers, *Kemain Teollisuus* **3,** 207 (1967).
"Biologische Waschmittel," *Seifen, Öle, Fette, Wachse* **8,** 203 (1968).
R. Delecourt, "Detergents Enzymatiques," *Proc. Vth Int. Congr. Surface Active Substances, Barcelona, Spain, 1968,* Vol. III, p. 179.
W. Fries *Seifen, Öle, Fette, Wachse* **94,** 467 (1968).
S. R. Green, "Alkali-Resistant Enzyme for Detergents," *Soap Chem. Spec.* **5,** 86 (1968).
J. C. Hoogerheide, "Die Entwicklung von Enzymen zur Verwendung in Waschmitteln," *Fette Seifen Anstrichm.* **10,** 743 (1968).
A. Horowitz and I. Norstedt, "Untersuchung über die Bedeutung des Vorwaschens und des Einweichens bei der Weisswäsche in Trommelwaschmaschinen," *Seifen, Öle, Fette, Wachse* **15,** 467 (1968).
R. L. Liss and R. P. Langguth, "Enzymes in Detergents," *paper presented at the Fall Meeting of the American Oil Chemists' Society, New York, Oct. 1968.*
P. Mannheim, "Enzymatic Laundering," *Soap Perfum. Cosmet.* **41,** 750 (1968).
O. Oldenroth, "Untersuchung über die Wirkung enzymatischer Vorwaschmittel," *Fette Siefen Anstrichm.* **1,** 24 (1968).
S. V. Vaeck, "Entfernung der eiweisshaltigen Flecken beim Einweichen, Waschen und Bleichen," *Jahrbuch für das Textil- und Reinigungsgewerbe* **10,** 49 (1968).
H. W. Worne, "The Role of Enzymes in Detergent Products," *Deterg. Age* **9,** 19 (1968).
"Enzyme-Active Laundry Products," *Consumer Bull. (U.S.A.)* **51,** 4, 39 (1968).
M. Boue-Swinnen and co-workers, "Poudres a Lessiver Enzymatiques" *L'Ingenieur Chimiste,* 50 (1969).
J. L. A. Briggs, "The Manufacture of Enzymatic Detergents," *paper presented at the II Simposio Jugislavi sui Tensioattivi, Bled, Yuguslavia, 1969.*
T. Cayle, "Evaluation of Enzymes for Laundry Products," *J. Am. Oil Chem. Soc.* **10,** 515 (1969).
A. Chwala, "Die Funktion von Enzymen in modernen Waschmitteln, *Seifen, Öle, Fette, Wachse* **15,** 539 (1969).
J. Dobson, "Evaluation of Procedures Utilizing Enzyme Products for Stain Removal from Polyester–Cotton Fabrics with Functional Finishes," thesis, Purdue University, Lafayette, Ind., 1969.
R. P. Langguth and L. W. Mecey, "Detergent Ingredients' Effect on Enzymes," *Soap Chem. Spec.* **9,** 60 (1969).

M. E. Purchase, "Enzyme Products Compared to Other Laundry Agents," *New York's Food and Life Sciences,* **2**(2), (1969).

A. J. Wieg, "Enzymes in Washing Powders," *Process Biochem.* 30 (Feb. 1969).

"An Institute Report on Enzymes in the Laundry," *Good Housekeeping's Laundry Laboratory* (Jan. 1970).

A. Arpino and P. Guastalla, "Analisi e Valutazione dei Detersiyi Enzimatici," *paper presented at the IVth Symposium about Detergents, Sorrento, Italy, Nov. 1970.*

"Enzyme Detergents—Effective and Safe?" *Consumer Bull.* (*U.S.A.*) **53,** 27 (Apr. 1970).

W. A. Farone and A. Cahn, "The Effect of Enzymes on the Performance of Detergent Formulations," *paper presented at the Meeting of the Society for Industrial Microbiology, University of Rhode Island, Kingston, R.I., 1970.*

M. H. Nielsen, "How Enzymes Got Into Detergents," *paper presented at the Meeting of the Society for Industrial Microbiology, University of Rhode Island, Kingston, R.I., 1970.*

M. H. Nielsen, "Enzyme Technology," *Mod. Kemi* **7/8,** 35 (1971).

R. J. Vogels, "Enzymes in Washing Powders," *Symposium on Enzymes in Industry,* **30,** 3 (1971).

CHUCK STARACE
Novo Laboratories Inc.

HANS C. BARFOED
Novo Industri A/S

ENZYMES, IMMOBILIZED

Enzymes, the natural linear polypeptides that catalyze biochemical reactions, have a mol wt of 10^5–10^6. The single polypeptide chains of enzymes can aggregate in solution to yield molecular weights as high as several million.

To date, approximately three thousand enzymes have been extracted and purified from microorganisms, plants, and animals. Regardless of the source, the weight of enzyme present per weight of biomass usually ranges from one tenth to one ten-thousandth part of the total biomass (1). Extraction and purification is often a complex and expensive process owing to the wide spectrum of materials present in the biological starting material.

The properties of purified enzymes can make their use very difficult. Generally, purified enzymes can be stored dry and cool for periods of months or even years, but they lose their catalytic activity, often unpredictably, in a matter of minutes or hours when in solution. It is also well known that the folded structure by which catalytic activity is achieved is destroyed >50–70°C (see Biopolymers). Although many enzymes are sensitive to transition metal ions that act as inhibitors of enzyme activity, the particular transition metal ions that inhibit are characteristic of each enzyme. Likewise an enzyme may be affected by organic inhibitors, activators, or both.

Enzymes have been used traditionally in an impure form because of the characteristics of purified enzymes described above, and the high cost of purification. Historically, enzymes were used in the production of wine, cheese, and bread without the knowledge of the nature of their catalytic activity. During this century, enzymes have found uses in the conversion of food products, the manufacture of textiles, and paper adhesives as well as in sewage disposal, garment cleaning (see Enzyme detergents), animal feeding, in chemical analysis, and in biomedical research (see also Yeast; Antibiotics, peptides).

Methods of Immobilization

Many of the characteristics that make enzymes difficult to use *in vitro* can be overcome by immobilizing the enzyme on a solid support. Generally, immobilized enzymes are more stable to changes in pH and temperature than are their soluble counterparts. When suspended in water, these insoluble enzyme derivatives are usually stable for months, providing the opportunity of reusing the catalyst many times. Separation of products and reactants is also quite easy since these compounds are usually soluble and can simply be rinsed from the insoluble catalyst material. Of course, the reuse of expensive immobilized enzymes can make enzymes economical catalysts where the cost of soluble enzymes had traditionally been prohibitive (see Catalysis).

Enzymes can be immobilized on nearly any material, either organic or inorganic. The method of immobilization is dictated by the support and enzyme required for the particular application. The possible combinations of chemical reagents required to immobilize an enzyme on support materials are far too numerous to list here, but it is possible to describe the types of immobilization procedures typically employed. One of the oldest methods of immobilization consisted of adsorption of the enzyme onto a charged surface, such as an ion-exchange resin. Unfortunately, enzymes insolubilized in this manner slowly leach back into solution. To eliminate loss of enzyme during use, the adsorbed enzyme can be intermolecularly cross-linked to form a sheet of enzyme material on the surface of the support. Alternatively, enzyme molecules can be covalently bonded to the surface of chemically reactive support materials or enzyme molecules may be entrapped in the interstices of polymer structures by copolymerization of enzyme species with the suitable monomer. Figure 1 illustrates some of the general methods of immobilization.

Several review articles are available that describe in detail these methods of immobilization. One of the first thorough reviews is that of Goldman and co-workers in 1971 (2). A few years later, Zaborsky's review updated the methods of immobilization and added considerable information concerning applications (3). Many other texts include review articles detailing the chemical procedures for immobilization (4–7).

Common Methods of Immobilization. Probably more enzymes have been immobilized by activation of polysaccharides with cyanogen bromide than by any other method. In this method, cellulose or a commercial beaded polysaccharide is allowed to react at an alkaline pH with cyanogen bromide. The intermediate is usually not stored but allowed to react immediately with the soluble enzyme. The popularity of this method results in a large part from the ready availability of the support material. This method is easy to perform and usually results in relatively high enzyme loading. Stability of the soluble enzyme in the alkaline region is usually necessary, although some enzymes have been immobilized at neutral pH. The method can seldom be used in large-scale applications because cyanogen bromide is poisonous and expensive.

150   ENZYMES, IMMOBILIZED

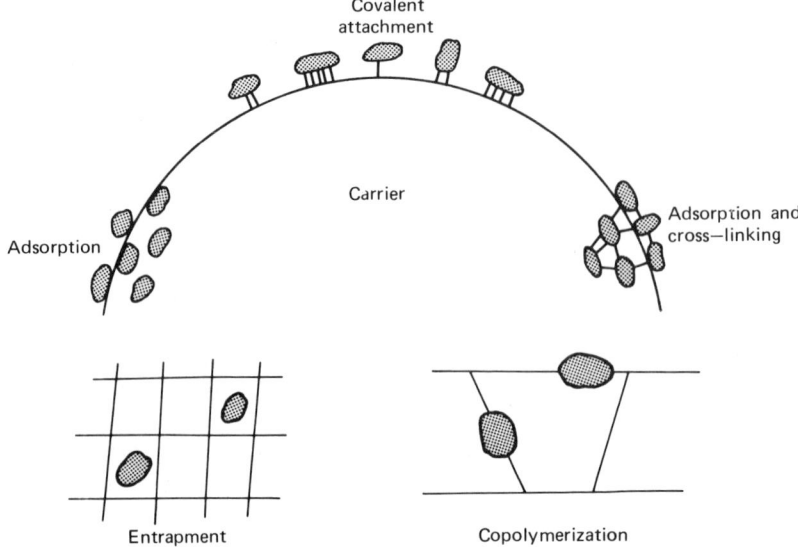

Figure 1.   Some general methods of immobilization.

Another very popular method of immobilization uses glutaraldehyde for enzyme cross-linking. The soluble enzyme can be immobilized without a support material by addition of glutaraldehyde to the enzyme solution which usually results in precipitation of a polymerized enzyme. This method is often used in an application requiring a very high enzyme activity per volume of immobilized-enzyme material which is commonly called enzyme loading. However, if an enzyme is first adsorbed onto a support material, addition of glutaraldehyde leads to the formation of a sheet of polymerized enzyme on the surface of the support material. Typically, the enzyme sheet does not redissolve. Use of glutaraldehyde is an extremely simple procedure although there are some difficulties in achieving reproducibility since a polymeric glutaraldehyde derivative often found as a contaminant reacts with the enzyme molecule (8).

Most studies of enzymes immobilized by entrapment have employed polyacrylamide. Since the immobilization is accomplished without the formation of an enzyme derivative, the method can often be used with enzymes that lose considerable activity during derivatization. Some problems can result from the diffusion of reactants and products in the polyacrylamide gel, particularly when they are of large molecular weight (see Acrylamide polymers).

Physical Properties

The physical properties of immobilized enzymes are determined by the support material. Particulate, rigid, inorganic support materials usually result in low to moderate enzyme loading. These supports have excellent flow properties and, when used in a packed bed, can usually be used at higher flow rates than enzyme derivatives prepared with organic support materials.

Many organic support materials offer a wider choice of functional groups than typical inorganic supports, thus higher enzyme loading is possible which is important in applications where the space available to house the catalyst is limited.

Chemical Properties

The chemical properties of an enzyme often change during immobilization because of changes in the microenvironment or in the tertiary or folded structure of the enzyme molecule. Changes in microenvironment may affect the catalytic activity as a function of pH, temperature, ionic strength, etc. Changes in apparent activity may also result from changes in the solubility of reactant molecules between the bulk solution and the microenvironment. In other cases, apparent changes in specificity of the enzyme and reaction rate are due to the rates of diffusion of reactants and products. These chemical properties of an immobilized enzyme depend on the method of immobilization employed and the support material used (9–14).

Storage Stability. The variation in storage stability for many different methods of immobilization has been reported (15). Since an increase in stability is desirable, preparations that have greater stability are reported most often in the literature. Reported enhanced stability can be the result of diffusional limits on the observed activity rather than actual enhanced stability of the immobilized-enzyme molecule (16). Nevertheless, a constant activity over an extended period often permits the efficient use of the immobilized enzyme.

Stability of Immobilized Enzymes in Use. An enzyme can usually be immobilized with proper support and immobilization reagents so that it loses activity very slowly. The extension of catalytic activity of the enzyme upon immobilization appears to be the most important advantage of immobilized enzymes over their soluble counterparts. Usually, the mechanism that causes loss in activity is not determined, but certainly losses result from the destruction by attrition of some of the support material (17).

Stability at High Temperatures. An immobilized enzyme may have more stability at elevated temperatures than the corresponding soluble enzyme. Recent work indicates that by the proper choice of the method of immobilization an enzyme derivative can be obtained whose folded active structure is much more stable, particularly at elevated temperatures of 60–100°C (18). This increase in thermal stability is accomplished by covalent bonding of the enzyme to a polymer support in a multipoint fashion. Thus the folded structure appears to be "frozen" in (19).

Stability and Activity as a Function of pH. The effect of enzyme immobilization onto charged surfaces has been studied in some detail (2). The pH optimum for enzyme activity is shifted to a lower value by immobilization onto a polycationic polymer, although the use of a polyanionic support shifts the pH optimum to a more alkaline value. Shifts in pH optimum are due primarily to the pH of the microenvironment which is different than the pH of the bulk solution. Counterions that are present to balance the charge on the polymer surface alter the pH of the microenvironment. As would be expected, the change in the pH optimum of enzyme activity is reduced by increasing the ionic strength of the solution in contact with the immobilized enzyme.

Conformational Changes Upon Immobilization. When conformational changes are studied for soluble enzymes, a spectrophotometric method is often the most convenient. Unfortunately, the measurements of physical properties of the immobilized enzyme become difficult by spectrophotometric techniques because of light scattering, and very few studies have been performed to determine the conformation change, if any, during immobilization. Likewise, very little work has been reported on changes in conformation of an immobilized enzyme during storage or use. One report concerns the conformational changes of immobilized trypsin [9002-07-7] and chymotrypsin [9004-07-3] prepared by attachment to CNBr-activated Sephadex (Pharmacia, Inc.); very little change in the fluorescence spectrum occurred upon immobilization of either enzyme (20). The fluorescence spectra of native and immobilized chymotrypsin followed the same pattern as the temperature was raised from 25 to 60°C. Upon cooling, neither the spectrum of the native or immobilized chymotrypsin was identical to unheated samples. Although the native enzyme was not active after heating, the immobilized chymotrypsin with the altered fluorescence spectrum was still active. Clearly, conformations different from that of the native enzyme can sometimes occur with an active immobilized enzyme.

Calorimetry (qv) has also been used to study the conformation of immobilized enzymes (21–22). The thermal stability of liver alcohol dehydrogenase [37205-43-9] and lactate dehydrogenase [9001-60-9] immobilized to Sephadex was only slightly greater than the corresponding native enzyme. Ribonuclease A [9001-99-4] was immobilized onto Sephadex with one to eight points of attachment. The calorimetrically measured transition temperature increased as the number of points of attachment increased. Conversely, the specific catalytic activity decreases with multiple points of attachment. Regardless of the points of attachment, the activity recovered after heating and cooling was proportional to the peak area of the thermogram upon reheating.

Specificity. The enzyme is rare that catalyzes only one reaction. In a few cases, exhaustive studies have shown that an enzyme is absolutely specific, but usually an enzyme catalyzes reactions of several very similar compounds (23). An immobilized enzyme is usually chemically modified, and its microenvironment is changed. Thus the specificity of the immobilized enzyme may be somewhat different from that of its soluble counterpart. Changes in specificity resulting from immobilization of an enzyme are discussed in refs. 3, 24–27. Specificity is particularly important when the immobilized enzyme is used in an analytical application where high specificity is required to reduce interferences. In this section, examples of changes in specificity brought about by immobilization are discussed.

Trypsin bound to agarose has been studied in relationship to catalysis of both large and small substrates and the specificity has been determined to be essentially the same as the native trypsin (28–29). Trypsin derivatives are reported to be formed with water-insoluble poly(maleic acid-co-ethylene) (30). Immobilized-trypsin samples were prepared in weight ratios of bound protein to carrier of 1:20 to 3:1. When the preparation was rich in enzyme, little difference in activity between large and small substrates was detected. In contrast, preparations containing little protein were approximately four to five times lower in activity toward hemoglobin or casein than would be expected from the activity toward ethyl N^2-benzoyl-L-arginate. Steric hinderance caused by the polyelectrolyte network of the carrier plays an important role in determining the rate of interaction between bound trypsin and the high molecular weight substrates. Similar results were obtained for water-insoluble polytyrosyl trypsin (31). An even greater decrease in the rate of reaction is noted at pH 9 when both the carrier and the macromolecular substrates are negatively charged. An immobilized trypsin obtained by incorporation into a polymerizing silicic acid sol exhibits esterase activity to low molecular weight substrates, but no activity to casein (32). Even soluble derivatives of trypsin have been shown to have lower activity to large substrates than would be expected from its activity to low molecular weight substrates (33). Similar results have been reported for derivatives of ficin [9001-33-6] (34–35), bromelain [9001-00-7] (36), papain [9001-73-4] (37), subtilopeptidase A [9014-01-1] (37–38), elastase [9001-08-5] (39), and chymotrypsin (40). Reference 41 supports the conclusion that this change in specificity related to the molecular weight of the substrate can be caused by steric hinderance; the rate of hydrolysis of proteins in the presence of immobilized pronase [9036-06-0] is inversely proportional to the square root of molecular weight. Thus exclusion of the macromolecular substrates from the interior of the support is probably analogous to the mechanism of molecular sieving in liquid chromatography (see Analytical methods; Molecular sieves).

Enzymes immobilized on polyanionic or polycationic supports have differing specificities for uncharged and charged substrates. Specificity changes of this type are almost certainly due to the concentration of substrate in the microenvironment of the enzyme molecule. These effects have been thoroughly investigated and are reviewed in reference 2.

The specificity of an enzyme derivative can be different from the soluble counterpart if the support material creates a hydrophobic environment for the enzyme molecule. This effect was observed when alcohol dehydrogenase was immobilized on acrylic copolymers (42). The apparent Michaelis-Menton constant for the more hydrophobic substrate n-butanol is four times smaller than for the immobilized enzyme, whereas for ethanol there was no affect on the constant.

All the changes in specificity caused by immobilization thus far discussed relate to changes in microenvironment. It is conceivable that the processes of immobilization in which an enzyme is chemically modified might result in changes in specificity because the tertiary structure of the enzyme has been changed. Such specificity changes are noted quite often when soluble enzyme derivatives are formed (43–45). With immobilized enzymes, it is difficult to separate the effects due to changes in enzyme environment from those due to changes in the enzyme molecule itself. Nevertheless, several examples of specificity changes upon immobilization are reported that are not clearly explained by the microenvironment of the enzyme molecule, eg, an immobilized trypsin prepared by polymerization of the enzyme with maleic acid and ethylene copolymers (46). Soluble trypsin will hydrolyze fifteen peptide bonds in pepsinogen whereas the number of bonds broken by the immobilized trypsin never exceeds ten. Small changes in specificity using a variety of low molecular weight substrates were observed when aminoacylase [9012-37-7] was covalently attached to halogenoacetyl cellulose (47). The rates of hydrolysis of more than twenty substrates of D-amino acid oxidase [9000-88-8] were compared for the soluble enzyme, aminoalkyl- and carboxylalkyl-immobilized derivatives (48). Although the differences in rates when the immobilized derivatives were used rather than the soluble enzyme were not large, some substrates had increased rates of hydrolysis and others had decreased rates.

Biochemical Processing

The first step in the development of a commercial process is to select the type of reactor that best fits the process being considered. The advantages and disadvantages of each type of reactor have been reviewed in the literature (3,49–52) and are discussed briefly below. After the type of reactor has been selected, the enzyme or enzymes required for the process can be immobilized and tested. The method of immobilization selected depends not only upon the enzyme or enzymes required, but the reactor design as well. The types of reactors are shown in Figure 2. Generally reactors can be classified into two types: tank reactors and column reactors. The simplest tank reactor is a batch reactor in which a soluble or immobilized enzyme is mixed with reactant solution until the product is formed. Then the tank must be cleaned for the next batch. To accomplish continuous operation, the initial solution can be pumped into the tank while solution containing product exits from the other side of the tank. Such a reactor is designated as a continuously stirred tank reactor (CSTR). Finally, to eliminate losses of immobilized enzyme or insoluble reactant, the reactor can be modified still further by the inclusion of a membrane or ultrafilter at the product stream outlet (see Ultrafiltration). Thus all insoluble or large molecular materials, or both, are retained within the reactor.

The most commonly used column reactor is the packed-bed type in which the immobilized enzyme is packed in a cylinder through which reactant solution is pumped. In some instances when a packed bed tends to plug or flow erratically due to the type

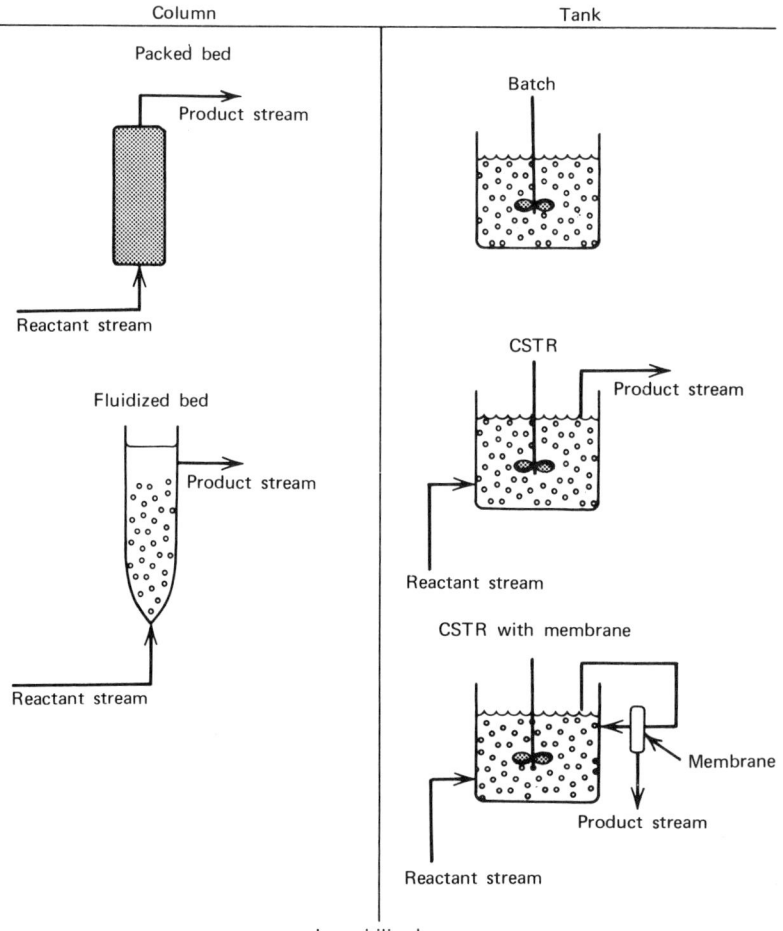

Figure 2. Types of immobilized-enzyme reactors.

of immobilized enzyme or reactant, a fluidized-bed reactor is used. In a fluidized bed, the immobilized enzyme is placed loosely in a column and the upward flow adjusted to cause the particles of immobilized enzyme to entrain or boil (see Fluidization). The different types of reactors are discussed below (see also Reactor technology).

Batch Reactors. A batch reactor is the simplest type of reactor to be used in biochemical processing. Reactant and either soluble or insoluble enzyme are kept in the reactor until the reaction is complete when the entire batch, including the product, is removed and new materials are added. If the enzyme is immobilized, it can be recovered for reuse. Each run requires cleaning of the tank, handling of immobilized enzyme, and replacement of lost enzyme. Immobilized enzyme activity can be lost during handling owing to mechanical breakage, or changes in environment such as pH, or drying which can cause destruction of the active tertiary structure of the enzyme. Thus the choice of immobilized enzyme depends on the handling techniques between batches. It is unlikely, however, that batch reactors will be used extensively with immobilized enzymes in industry. The cleaning steps and handling of enzyme would cause losses of enzyme activity and would be laborious and inefficient. Batch reactors are discussed in refs. 32, 49–51, 53.

Continuous-Flow Stirred-Tank Reactor (CSTR). A batch reactor can be converted to a CSTR by the addition of an inlet port for reactant solution and an exit port for solution containing product. Monitoring equipment can easily be added to this tank design. Additional immobilized enzyme, oxygen, or other cofactors can be added as required.

Since good mixing is easy to accomplish, the reactant concentration is always low and the product concentration is high throughout the reactor. Therefore, this type of reactor is well suited to enzyme-catalyzed reactions that are reactant-inhibited, but inefficient if the enzyme is inhibited by product. The low reactant concentration during continuous operation also requires the presence of high enzyme activity in order to convert most of the reactant to product. The good mixing characteristic also allows the application of a CSTR to colloidal or large reactants.

The immobilized-enzyme support should be tested for mechanical breakage during operation of the CSTR. As is evident from Table 1, organic supports are most often chosen for use with the CSTR. The preference for organic supports probably results from the flexibility of available reactive groups that can be used to obtain high enzyme loading. Another very important consideration is the cost of the support material since large amounts of immobilized enzyme are required in industrial applications. DEAE-cellulose is a support material which is often used because of its relatively low cost.

CSTR With Membrane. A very useful modification of the CSTR is the inclusion of a membrane or ultrafilter at the exit port (55,58–59). This membrane allows the retention of soluble and insoluble macromolecules. If no effective means of immobilization can be found for a particular enzyme, it can be used in soluble form. When the reactant is insoluble, the enzyme must be in soluble form either as the naturally occurring enzyme or as a soluble derivative. Sometimes a soluble derivative is formed by bonding the enzyme molecule to another soluble macromolecule, such as a polysaccharide, which often results in improved enzyme stability.

Packed-Bed Reactors. A packed-bed reactor is the most common type of column reactor. The immobilized enzyme is packed into a cylinder through which the reactant solution is pumped. It is analogous to a chromatographic column and thus requires

Table 1. Immobilized Enzymes for CSTR

Reactant/product	Enzyme(s)	Support	Immobilization method	Reference
maltose/glucose	amylo-glucosidase	DEAE-cellulose	covalent bond (2-amino-4,6-dichloro-s-triazine)	54
glucose/gluconic acid	glucose oxidase (catalase)	polyacrylamide	entrapment	55
lysis of *M. lysodeikticus*	lysozyme	polyacrylamide or cellulose	covalent bond (diazonium intermediate)	56
cornstarch/glucose	glucoamylase	glass	covalent bond (two methods)	57
starch/glucose	glucoamylase	DEAE-cellulose	ionic	58
starch/glucose	glucoamylase	DEAE-cellulose	covalent bond	59
proteins/amino acids	trypsin	alumino-silicates (molecular sieves)	adsorption	60

special supports that give good flow properties for the large columns required in industrial processing. The major characteristics required of the immobilized enzyme are: (*1*) good flow characteristics of the support, (*2*) low to moderate enzyme loading, and (*3*) good stability in use. Table 2 lists some examples which illustrate the use of packed-bed reactors. Note that the supports tend to be rigid beads and are often inorganic.

Analysis of the kinetics of reaction for the packed-bed reactor versus the CSTR shows the packed-bed reactor to be generally more efficient (51,69–70). The reaction rate is usually greater than zero order, resulting in a reaction much faster than would be obtained with the same quantity of enzyme in a CSTR. The lower requirement for

Table 2. Immobilized Enzymes for Packed-Bed Reactors

Reactant/product	Enzyme(s)	Support	Immobilization method	Reference
starch/maltose	β-amylase	cellulose beads	covalent bond	61
sucrose/glucose	invertase	glass/cellulose	covalent bond (azo linkage)	62
N-acetyl-DL-methionine/methionine	aminoacylase	ceramic	covalent bond (glutaraldehyde)	63
lactose/glucose	lactase	ceramic	covalent bond (glutaraldehyde)	64
lactose/glucose	lactase	phenol-formaldehyde resin	cross-linking (glutaraldehyde)	65
dextrin/glucose	glucoamylase	porous silica	covalent bond (glutaraldehyde)	66
glucose/fructose	glucose isomerase	porous glass beads	covalent bond (azo linkage)	67
glucose/fructose	glucose isomerase	chitin	cross-linking (glutaraldehyde)	68

enzyme activity is obviously a very important economic consideration. Packed-bed reactors are particularly advantageous when enzyme catalysis is inhibited by product since the concentration of product is low at the top of the column and only reaches high concentration at the base of the column. Only when the reaction is almost zero order is there no advantage in using the packed-bed reactor over the CSTR (54). The high efficiency of the packed-bed reactor may be the reason that this type is the only one known to be used commercially.

Fluidized-Bed Reactors. The velocity of flow in a fluidized-bed reactor is high enough to lift the particles so that they are suspended in the column, but are not swept away. In other types of chemical processing, fluidized beds are important when good mass and heat transfer characteristics are required, but for use with immobilized enzymes the primary advantage is improved flow characteristics. Fluidized-bed reactors do not plug up when the reactant stream contains traces or even large quantities of insoluble material. The primary disadvantages are the large power requirements to fluidize the bed and the requirement that all of the immobilized-enzyme particles ebulliate at the same velocity. Dense supports such as alumina and stainless steel are useful when using immobilized enzymes in fluidized-bed reactors (71). One application explored was immobilized pepsin [9001-75-6] for the coagulation of milk (72). Cellulase [9012-54-8] was immobilized with collagen on glass beads and used in a fluidized bed to produce glucose (73).

High-Fructose Corn Syrup. For many years, corn and other vegetable starches were converted to glucose by acid hydrolysis (74). Because many side reactions take place in the acid environment, the glucose syrup produced was often of poor quality. The introduction of enzymes to catalyze the breakdown of corn starch to glucose has largely eliminated the problem of side reactions. Unfortunately, the corn syrup is not sufficiently sweet to compete with sucrose in many applications (see Sugar; Sweeteners).

The discovery of microorganisms that produce glucose isomerase [9001-41-6] has now made it possible to prepare sweeter corn syrups (75). Glucose isomerase converts glucose to fructose until an equilibrium mixture is obtained. The equilibrium concentrations vary somewhat depending on conditions, but generally about the same concentration for glucose and fructose is obtained. This corn syrup is referred to as high fructose corn syrup or simply HFCS.

HFCS was first produced commercially in Japan using soluble glucose isomerase (76). Later the process was improved by heat fixation of cells of a *Streptomyces* species (77). The first commercial process in the United States used immobilized glucose isomerase adsorbed to DEAE-cellulose. This process used by Clinton Corn Processing is believed to be described in reference 78. At present, the following companies produce HPFS: A. E. Staley Manufacturing Company; Clinton Corn Processing, a division of Standard Brands, Inc.; Amstar Corporation; Corn Sweeteners Division of Archer Daniels Midland Company; CPC International, Inc.; Hubinger Company, a subsidiary of H. J. Heinz Company; CAR-MI, Inc., a joint venture of Cargill, Inc. and Miles Laboratories, Inc.; and a joint venture of American Maize-Products Company with Amalgamated Sugar Company (79).

The two different grades of HFCS presently available are 42 HFCS and 55 HFCS which contain 42 and 55% fructose, respectively, compared to glucose on a weight basis (80). The 55 HFCS is more expensive than 42 HFCS probably because of the need to have a higher activity per unit volume of immobilized glucose isomerase, a longer retention time, or both, in order to achieve equilibrium conditions. Also currently available is 90 HFCS in which 90% of the glucose-fructose mixture is fructose. This product is presumably prepared in a two-step process since ca 55% fructose is present in the typical equilibrium solution using glucose isomerase. Another process may be used for further conversion beyond 55% fructose (81–82).

The production of HFCS is the largest commercial application of immobilized enzymes. The capacity for production of 42 HFCS and 55 HFCS is estimated to be 1.7×10^6 and 7.5×10^5 metric tons, respectively, in 1979 (79). The amount of HFCS produced in 1979 is 1.4×10^6 t (est). Thus, utilization is ca 55%. Apparently, the market limitation for HFCS at this time is determined by the soft drink industry (79–80). This industry accounts for 36.9% of all industrial sweetener usage in the U.S., but only 4.4% is in the form of HFCS. Both Coca-Cola Company and Pepsico are reluctant to use HFCS in their cola products which represent the largest portion of the total market. 42 HFCS is presumably not used in cola drinks because it is not sweet enough to compete on a weight basis with sucrose. Concern about quality control and availability of 55 HFCS and 90 HFCS appear to limit the use of these products (80) (see Carbonated beverage).

Several commercial processes are available for the production of HFCS. Clinton Corn Processing is thought to be the first company to commercially produce HFCS by use of immobilized glucose isomerase. Initially, the process of using glucose isomerase obtained from the cells of the *Streptomyces* genus was licensed from the Agency of Industrial Science and Technology, Tokyo, Japan (83–84). Patents assigned directly to Standard Brands, Inc., disclose the immobilization of whole cells in beds having a depth-to-width ratio of <2 (85). The system is most likely a series of shallow beds which resemble filter presses and allow operation with very low pressure drop (86). More recent patents assigned to this company relate to the immobilization of purified or partially purified glucose isomerase to anion-exchange cellulose and synthetic anionic-exchange resin (87–88).

Novo Laboratories, Inc., claims to be the biggest manufacturer of immobilized enzymes for the production of HFCS (89). Amstar Corporation, Archer Daniels Midland Company's Corn Sweetened Division, and American Maize-Products Company receive enzymes or technology, or both, required for production of HFCS from Novo Laboratories, Inc. (90). The Novo system uses bacterial sources to obtain glucose isomerase, and the microbial cells are homogenized to form a concentrate that is allowed to react with glutaraldehyde (91–93). The product containing glucose isomerase can be shaped into particulate form by removal of water. This particulate material is used in packed-bed reactors for the continuous production of HFCS (74,94). With a column having a volume of 60 L, the residual activity after 650 h of operation is 40% (74). Particle size is important to prevent excessive back pressure when the process is scaled up. The particle size should be greater than 100 μm; it is generally from 150 to 2800 μm (94).

Several other companies have developed alternative processes and methods for immobilization of glucose isomerase. Corning Glass Works has developed several methods to immobilize enzymes on porous inorganic supports. Two processes are particularly tailored to the immobilization of glucose isomerase on porous ceramics (95–96). The application of these ceramic–glucose isomerase composites to the preparation of HFCS is particularly useful since back pressure is generally minimal with inorganic supports. The technology is licensed to CPC International, Inc. which has also patented several processes in this area (97). Miles Laboratories, Inc. has discovered a novel glucose isomerase and method to stabilize this enzyme (97). This technology is used as a basis for the joint effort called CAR-MI, Inc. (90). Moreover, processes to use either immobilized whole cells or immobilized glucose isomerase developed by R. J. Reynolds Tobacco Company are being used by ICI Americas, Inc. (97). A quite different column apparatus using a regenerated sponge material as a support has been developed by Penick and Ford, Ltd. (97).

Starch Hydrolysis. Starch is generally hydrolyzed to glucose in a two-step process (99). First, the starch is hydrolyzed to low molecular weight dextrins either by acid or the use of soluble α-amylase [9000-92-4]. The α-amylase is stable for several minutes at 105°C and is particularly useful in this initial hydrolysis step. The second step converts the dextrins to glucose by use of soluble glucoamylase [9032-08-0]. The corn syrup thus produced can be converted to HFCS by the use of immobilized glucose isomerase described above.

Although the use of immobilized glucoamylase in the second hydrolysis step is not commercial at this time, much work has been carried out in the laboratory on this application. The immobilization of glucoamylase on cellulose support for hydrolysis of starch in a CSTR is described in reference 58. Other groups have reported on the immobilization of glucoamylase on other organic supports, and their use in packed-bed reactors for starch and dextrin hydrolysis (66,100–101). A study comparing the same immobilized glucoamylase in a packed-bed reactor and a CSTR is reported in reference 54.

The most extensive work reported on this application of immobilized glucoamylase appears in reference 102. Numerous samples of immobilized glucoamylase have been prepared on a variety of inorganic supports (102). It was determined that controlled-pore glass could not be used economically as a support for glucoamylase because of its high cost and the dissolution of glass with time (103). Six other porous ceramic supports show considerably better results. However, the half-life of immobilized glucoamylase is greatly reduced above 50°C. Thus, in operation of the pilot plant, the temperature of the packed bed reactor was maintained at 38–42°C but the conversion of starch hydrolysate to glucose fell short of the required commercial level (103). The report included discussion of some difficulty with microbial contamination which required the use of saturated chloroform solutions to clean the reactors. Apparently, the commercial application of this process awaits the demonstration of cost-effective conversion of starch hydrolysate to glucose. The process cost may be reduced considerably by use of several reactors in parallel (103). Such a multiple-column system would allow the times for column start-up or reloading to be staggered and the production rate could thus be kept within desired limits.

Immobilized Lactase. Whey is a major by-product of cheese making and is a difficult material to dispose of in the environment (104). The composition of whey is ca 5 wt % lactose, ca 1 wt % protein, and 0.5 wt % salts. The balance of the solution is water. Although the protein is often recovered by ultrafiltration (qv), and used in food products, the lactose solution remaining is of little value. The conversion of lactose to glucose and galactose is being explored as means of making this solution useful as a food sweetener (89, 104).

The immobilization of lactase [9031-11-2] on phenol-formaldehyde resin using glutaraldehyde is reported in references 65 and 105. Apparently, lactase can also be stabilized by combination with tannic acid and glutaraldehyde (106). The enzyme immobilized on resin material has been used in pilot plant operation for the hydrolysis of lactose in whey.

A process developed at Corning Glass Works involves the immobilization of lactase to a porous silica support (107). As expected, the half-life of the immobilized lactase decreases with increasing temperature and 38°C is optimum for operating the reactor. A pilot unit has been operated with the Corning process involving 45 kg of immobilized lactase in a packed-bed reactor. This unit has a capacity of 6.8 m^3 (1800 gal) per day of liquid whey. The problems that must be solved before commercialization are the high cost of the immobilized lactase and the high overall product cost, which appears considerably higher than HFCS. To increase the sweetener value and thus the value of the product, a process has been developed at Corning Glass Works to combine the use of immobilized lactase with immobilized glucose isomerase (108).

Another use for immobilized lactase is to hydrolyze lactose in milk. There is a high incidence of lactose intolerance among people in Asia, Africa, Latin America, and the U. S. Midwest. Reduction of the concentration of lactose in milk should allow it to be used more widely throughout the world. A method has been developed using a fiber-entrapped lactase by Snamprogetti S.p.A., Milan, Italy (109). A pilot plant that treated 200 L (53 gal) of milk in a recirculating manner has been tested.

Immobilized Aminocylase. The first industrial application of this immobilized enzyme is considered to be the continuous production of L-amino acids from DL-amino acids (110) (see Amino acids). This process was developed at Tanabe Pharmaceutical Company in 1969. From 1954 to 1969, the L-amino acids were produced using soluble aminoacylase by the following procedure: (1) the desired amino acid is synthesized yielding a racemic mixture; (2) the acyl derivative is formed and L-aminoacylase is added to the solution; (3) after the enzyme-catalyzed reaction, the L-amino acid can be easily separated from the unreacted D-acylamino acid by the difference in solubility of the two compounds; and (4) the D-acylamino acid is racemized and the enzyme treatment repeated. The use of immobilized aminoacylase could improve this process since the enzyme would not have to be removed by pH and/or heat treatments. Moreover, the process could be made continuous instead of batchwise which requires considerable labor.

The immobilization of aminoacylase was studied exhaustively by physical adsorption, ionic binding, covalent binding, cross-linking, and entrapment. The method selected for commercialization was ionic adsorption onto DEAE-Sephadex to yield an inexpensive catalyst with good operational stability. Furthermore, it is possible to regenerate the catalyst by adding fresh, soluble aminoacylase. Considerable savings have resulted from the utilization of immobilized aminoacylase by reducing labor cost and cost of enzyme required. In addition, a higher yield and cleaner product is produced.

Other Commercial Processes. Several other processes have been developed at Tanabe Pharmaceutical Company (110). Aspartic acid is produced by conversion of fumaric acid. In this application the required enzyme is not extracted and purified, but rather entire cells of *Escherichia coli* are immobilized by entrapment in acrylamide gel (see Acrylamide polymers). Whole cells are immobilized when the cost of enzyme extraction and purification is prohibitive or the enzyme molecule, removed from its native environment, is unstable (110). The most common disadvantage of immobilized whole cells is that side reactions often occur which must be eliminated by inhibition or denaturation of the undesired enzyme molecules (111). The techniques of immobilization of whole cells have been reviewed (112). A packed-bed reactor of immobilized cells containing aspartase [9027-30-9] is used with an operational half-life of 120 d at 37°C. A cost savings of 40% over the use of the soluble-enzyme process is claimed (110).

In a similar process, L-citrulline is prepared using immobilized cells of *Pseudomonas putida* and urocanic acid with immobilized cells of *Achromobacter liquidium*. A detailed account is given in reference 113 (see Amino acids).

It has been reported that several corporations produce 6-aminopenicillanic acid by using systems that include immobilized enzymes (86). Snamprogetti of Rome, Italy, uses an enzyme entrapped in spun fibers of cellulose acetate, and Hindustan Antibiotics Limited in India uses a covalently bonded enzyme on a cellulose support.

Potential Commercial Processes. The hydrolysis of protein by immobilized proteases could be used to solubilize, texturize, and increase digestibility. Studies at Corning Glass Works have shown that casein and soybean protein can be hydrolyzed to free amino acid and soluble peptides (114). A combination of pronase and aminopeptidase [9031-94-1] can hydrolyze proteins to nearly 90% free amino acids (115). Furthermore, immobilized pepsin [9001-75-6], papain [9001-73-4] and rennin [9001-98-3] can be used for the hydrolysis of skim milk in manufacture of cheese (116). A very attractive application is in the treatment of milk with immobilized trypsin, a process that enhances shelf life and prevents loss of flavor (117) (see Milk products). Immobilized enzymes have also been investigated in the treatment of beer to prevent the formation of haze (118) (see Beer).

Immobilized enzymes have also been investigated to convert large quantities of cellulose in the biomass to glucose (119–120). Glucose syrup could be produced in very large quantities from this source for use in fermentation (qv) to food and food products.

One of the most promising commercial applications is in the preparation of drugs (89,121–122). Many drugs are prepared by fermentation (qv) where growing microbes supply the enzymes to convert the starting material to useful drugs (see Enzymes, therapeutic). Fermentations are wasteful because of the large quantities of microbes that must be discarded. Nutrients and a ready carbon source must be available for microbial growth. If the proper enzyme or enzymes could be isolated and immobilized, the process could be performed cleanly, eliminating undesirable by-products. Even immobilization of whole cells reduces waste since the cells are usually not viable and thus do not require nutrients and a carbon source. Many steroid transformations can be performed in this manner (105,123) (see Steroids).

Analytical Applications

There are two types of analytical instruments designed to use immobilized enzymes. The type of instrument determines the method and support for immobilization of the required enzyme or enzymes. The first type of instrument is usually referred to as an enzyme electrode in which the immobilized-enzyme derivative is placed on the tip of an electrochemical sensor. An enzyme electrode requires an enzyme derivative that has high enzyme loading (see Ion-selective electrodes). If the immobilized enzyme has low enzyme loading, the enzyme layer must be thick and the response of the enzyme electrode will be very slow. However, an enzyme electrode does not require high physical durability since the immobilized enzyme is permanently placed on the tip of the sensor, and the sample solution need only diffuse through the enzyme layer to the detector.

Usually the support materials are organic polymers containing a high density of reactive side chains so that sufficient enzyme loading can be achieved. In the many examples described in the literature, the enzyme is simply immobilized by reaction with glutaraldehyde or similar reagent in the absence of a support. The resulting insoluble protein mass usually has very high enzyme activity.

The first enzyme electrode described in the literature was designed to determine glucose amperometrically using an O_2 electrode coated with two layers of cuprophane membranes (124). Glucose oxidase [9001-37-0] was used in solution between the two membranes. Another early enzyme electrode measured urea (125). Insoluble urease [9002-13-5] positioned on the surface of an ammonium-ion electrode produced ammonium ions and bicarbonate ions from urea. The initial rate of formation of ammonium ions is proportional to the concentration of urea. Eventually a steady state potential is reached where the rate of diffusion of urea into the membrane and subsequent reaction of ammonium ions is equal to the rate of diffusion of ammonium ions. The steady state potential is proportional to the log of the urea concentration. An ammonia electrode can be used instead of a specific-cation electrode, but the optimum pH for measuring dissolved ammonia gas with this electrode is in an alkaline region in which immobilized urease will rapidly lose activity (126). Thus a compromise pH of 8–9 is used with considerable loss in sensitivity of the detector.

None of the commercial instruments presently available have maintained the simple design of a "dunking" or simple dip probe. Interference from compounds other than those formed by the enzyme-catalyzed reaction, and the requirement for rather large sample sizes necessitate considerable modifications in design of the enzyme electrode. The glucose instrument marketed by Yellow Springs Instrument Company and the instruments to measure organophosphates or carbamates in air or water developed by Midwest Research Institute are examples of commercial enzyme electrodes. Many other enzyme electrodes that measure biochemicals have been reported. Notice in Table 3 the similarity of the methods of immobilization, all of which result in high enzyme loading. Although incomplete, Table 3 indicates the type of immobilization and the variety of materials that can be measured.

The two examples of commercial enzyme electrodes discussed below illustrate how an immobilized enzyme can be combined successfully with electrochemical sensors. The design in both cases enhances operator convenience and speed of detection with a minimum of interferences.

Glucose Analyzer. The semiautomatic glucose analyzer developed by the Yellow Springs Instrument Company (YSI) was the first commercial product to incorporate an immobilized-enzyme reagent (134–135). The probe and membrane system within the YSI glucose analyzer is shown in Figure 3. The H_2O_2 decomposes at the platinum anode set at +0.7 V and the silver cathode completes the electrochemical cell. The immobilized glucose oxidase is sandwiched between the cellulose acetate membrane on the face of the electrode and polycarbonate membrane in contact with the solution. The glucose oxidase affixed to a polycarbonate membrane using glutaraldehyde to cross-link to collagen need not be firmly attached since the membranes on either side do not allow the migration of glucose oxidase out of the compartment. Moreover, the polycarbonate-enzyme composite need not have high physical strength since the so-

Table 3. Enzyme Electrodes

Detected substance	Enzyme	Support	Immobilization method	Commercial instruments	Research instruments, ref.
glucose	glucose oxidase (catalase)	polycarbonate	glutaraldehyde + collagen	Yellow Springs Instrument Company	
organophosphates carbamates	cholinesterase	polyurethane foam	starch entrapped	Midwest Research Institute	
penicillin	penicillinase	polyacrylamide	entrapment		127
uric acid	uricase	none	glutaraldehyde		128
amygdalin	β-glucosidase	membranes	solution entrapment		129
galactose	galactose oxidase	cellulose acetate	glutaraldehyde		130
urea	urease	bovine serum albumin	glutaraldehyde		131
sucrose	invertase mutarotase glucose oxidase	collagen	electrochemical		132
5′-AMP	5′-AMP deaminase	membranes	solution entrapment		133

lution does not flow through the membrane system; instead reactants and products simply diffuse in and out. The outer polycarbonate membrane also serves to keep the catalase [*9001-05-2*], present in the sample, from diffusing through to the immobilized glucose oxidase.

The following reactions take place in the YSI glucose instrument. A sample containing glucose is added to the reaction chamber. Glucose diffuses through the polycarbonate membrane to the polycarbonate-glucose oxidase composite and is converted to H_2O_2 and gluconic acid. The H_2O_2 which diffuses into the reaction chamber is destroyed by catalase added to the sample solution so that no back diffusion can occur. The balance of the H_2O_2 diffuses to the platinum electrode where it is detected by the reaction that converts it to oxygen and water. The cellulose acetate membrane allows the H_2O_2 to diffuse to the electrode surface, but this membrane does not allow uric acid, ascorbic acid, or other larger, electrochemically active molecules to reach the electrode surface.

These membranes were carefully selected to eliminate possible interferences and glucose oxidase was immobilized in such a way as to result in high enzyme loading and with only the required limited durability. Membranes containing glucose oxidase can be used for ca 300 samples (136). Certainly the convenience of an immobilized glucose oxidase is greater because no enzyme solution need be prepared or stored (see also Biomedical automated instrumentation).

164 ENZYMES, IMMOBILIZED

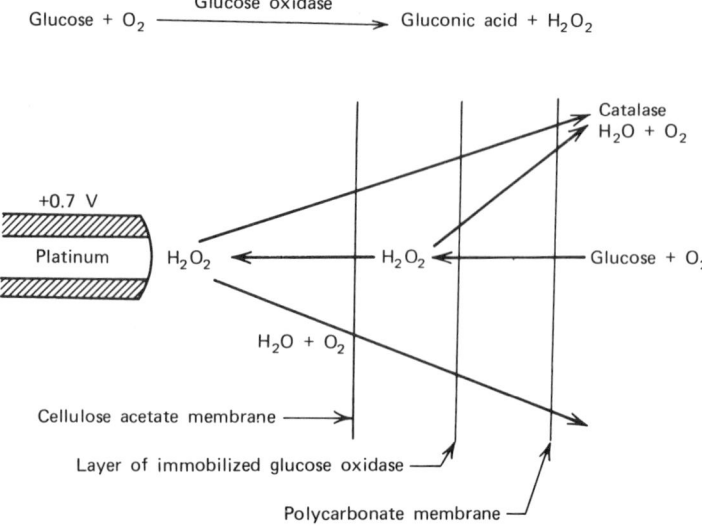

Figure 3. The YSI glucose analyzer.

Pesticide Analyzers. The only other commercial instruments that might be considered enzyme electrodes are those developed by Midwest Research Institute to measure organophosphates and carbamates (137–139) (see Insect control technology). These instruments are unique in that they measure inhibitors of an enzyme reaction rather than the reactant. The immobilized-enzyme monitor (IEM) measures air samples, and the continuous aqueous monitor (CAM) is designed for water samples. Both instruments measure organophosphates or carbamates by the degree of inhibition of catalysis by cholinesterase [9001-08-5]. In their absence, cholinesterase catalyzes the conversion of butyrlthiocholine iodide to butyric acid and thiocholine iodide. The immobilized cholinesterase is sandwiched between two electrodes which measure the concentration of thiocholine as it is formed. A constant 2 μA current is applied resulting in 200 mV in the absence of inhibitors and as much as 500 mV when inhibitors are present. Figure 4 illustrates the IEM's electrochemical cell containing immobilized cholinesterase and the connecting instrumental parts that enable the inhibitors present in air to be extracted at a rate of 60–100 L/min into an aqueous stream with a flow rate of 2 mL/min (see Cholinesterase inhibitors).

Unlike the classical enzyme electrode or the YSI glucose analyzer, these instruments require an immobilized enzyme that allows good flow characteristics with either air or water. The physical properties of immobilized cholinesterase must also permit good contact between the electrochemical cells.

The support for immobilization of cholinesterase is an open-pore polyurethane impregnated with a polysaccharide or a water-insoluble inorganic salt (see Urethane polymers). Most of these variations in procedure result in a usable immobilized cholinesterase. For example, the starch gel-entrapped enzyme pads are adequate for either air or water analysis for periods of ca 1 h but need to be replaced frequently because of loss in enzyme activity. Only one method of immobilization can be used for days with retention of significant enzyme activity. In this method, aluminum hydroxide precipitates from an enzyme solution. The suspension of enzyme on aluminum hy-

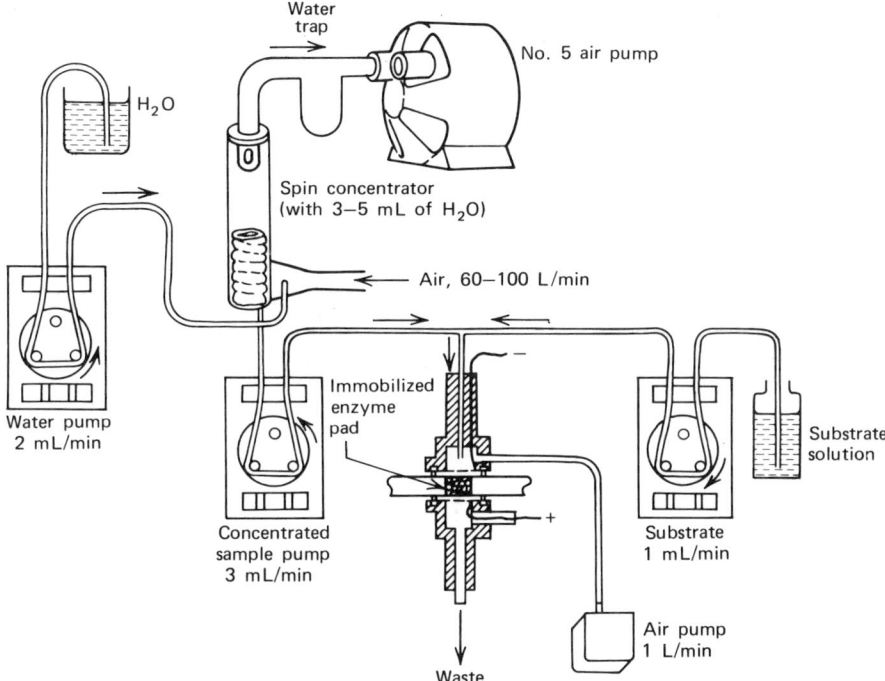

Figure 4. Inhibitor measurement.

droxide is mixed with a starch slurry and the entire material deposited on the open-pore polyurethane pad. This is the only method that gives an immobilized cholinesterase that allows sufficient flow rate, good contact between the pads and the electrode, and retention of sufficient enzyme activity during use over a period of days.

Partitioned-Enzyme Sensors. Most of the other instruments developed commercially are of the partitioned-enzyme-sensor type. These instruments separate the immobilized-enzyme component from the detector and connect them by means of a flowing stream. The advantage of this arrangement is that the immobilized enzyme or enzymes can be separated to allow optimum reaction conditions with regard to temperature, ionic strength, and pH while the detector is in a different environment. An immobilized enzyme designed for this type of instrument must have sufficient enzyme activity and allow a constant, rapid flow rate. The instrument can be designed to operate in a kinetic mode in which only a portion of the substrate is converted to product. When an instrument operates in kinetic mode, it has the advantages of allowing a more rapid measurement and requiring less enzyme loading than the equilibrium mode. When an instrument is operated in equilibrium mode, both sufficient enzyme loading and contact time with the reactant must be present to establish equilibrium between reactants and products. In most cases, where equilibrium mode is employed, it results in the conversion of essentially all the reactant to product, which requires much higher enzyme loading than the kinetic mode. The excess of enzyme activity minimizes the effects of slow loss of enzyme activity or the presence of enzyme inhibitors, or both.

Immobilized enzymes used in partitioned-enzyme-sensor systems are usually prepared on inorganic supports or on the inside surface of organic tubing. These supports allow rapid, constant flow rates and are not easily plugged by particulate waste; Table 4 lists some examples. The porosity of inorganic supports is required to allow sufficient enzyme loading. Immobilization of enzymes in these systems is accomplished by covalent bonding or by cross-linking of enzyme molecules.

Several commercial instruments are available that partition the immobilized enzyme and the detector in a flowing steam (Table 5). Note that the types of supports are the same as those reported in the literature, ie, porous inorganic materials and organic tubing (Table 4).

Glucose analyzer. Technicon offers tubing with the inner surface coated with immobilized enzymes which is used in the SMA and SMAC units (see Biomedical instrumentation). Measurement of glucose is by the hexokinase [$9001\text{-}51\text{-}8$] method.

Table 4. Partitioned-Enzyme-Sensor Instruments

Detected substance	Enzyme	Support	Immobilization method	Reference
glucose	glucose dehydrogenase	nylon tube	glutaraldehyde	140
nitrate ion	nitrate reductase	controlled-pore glass	diazonium salt	141
disaccharides	invertase, glucose oxidase	nylon tube	glutaraldehyde	142
glucose	glucose oxidase	porous glass	diazonium salt	143
penicillin	penicillinase	porous glass	N-hydroxysuccinimide	144
galactose	galactose oxidase	porous silica	p-benzoylazide	145
amino acids	L-amino acid oxidase	controlled-pore glass	glutaraldehyde	146

Table 5. Partitioned-Enzyme-Sensor Commercial Instruments

Detected substance	Enzyme	Support	Immobilization method	Company
glucose	glucose oxidase	porous glass beads	covalent bond (benzoylazo link)	Leeds & Northrup Company
sucrose	glucose oxidase	porous glass beads		
glucose	invertase mutarotase			
lactose	glucose oxidase	porous glass beads		
glucose	lactase			
glucose	hexokinase glucose-6-phosphate dehydrogenase	hollow nylon tubing		Technicon Instruments Corporation
urea	urease	porous alumina particles	cross-linking by disulfide rearrangement	Owens-Illinois, Inc., Kimble Division
glucose	glucose oxidase			Eppendorf Geratebau Netheler
lactose	lactose oxidase			Eppendorf Geratebau Netheler

The product, NADH, is measured spectrophotometrically. The method has the advantage of being compatible with sample introduction and segregation systems traditional to this manufacturer of clinical instruments.

Enzymax systems. Leeds & Northrup Company manufactures instruments to measure the concentration of sugars important in many industrial food processes (147–148). The simplest measures glucose by use of immobilized glucose oxidase. Also available are two instruments, each of which measures a pair of sugars, either glucose/sucrose or glucose/lactose. A diagram of the glucose-sucrose dual instrument is shown in Figure 5. The basic instrument is included within the dotted rectangle but the balance of the equipment is necessary for continuous operation. The single-channel glucose instrument is of the same design except the second channel for sucrose is not included.

In operation, the sample is dialyzed, diluted with buffer, and a single stream split into two flowing streams. In the upper stream, glucose is monitored by using immobilized glucose oxidase followed by an amperometric three-electrode detector that is sensitive to a product of the reaction, hydrogen peroxide. The bottom stream incorporates a column containing immobilized mutarotase [9031-76-9], invertase [9001-57-4], and glucose oxidase, respectively. These enzymes convert sucrose to hydrogen peroxide and gluconic acid. The hydrogen peroxide is detected by the same kind of detector as used in the upper stream and all analyses are done in the equilibrium mode. The lower stream actually measures the sum of the glucose and sucrose concentrations since glucose oxidase is required in the sucrose monitoring stream.

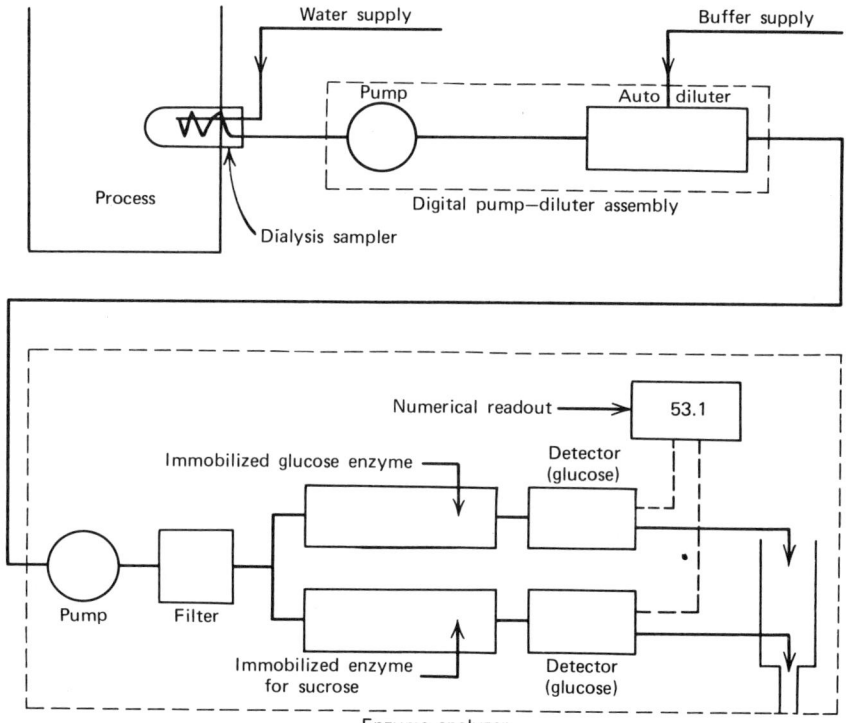

Figure 5. Glucose-sucrose enzymatic analysis.

Fortunately, the response of detectors is linear with the concentration of hydrogen peroxide present so that the signal detected by the upper stream can be subtracted from the lower stream signal to obtain the concentration of sucrose.

All the enzymes immobilized for the Leeds & Northrup Instruments are prepared on porous glass beads which give the excellent flow characteristics required for partitioned-enzyme-sensor instruments and acceptable enzyme loading. The dual instruments also show that suitable combinations of immobilized enzymes can be used in sequence to obtain the desired product.

BUN analyzer. The blood urea–nitrogen analyzer offered by the Kimble Division of Owens-Illinois, Inc., uses a gas-sensing electrode coupled with a cartridge of immobilized urease for measuring ammonia (149–153). The flow design of the Kimble BUN Analyzer is illustrated in Figure 6. This instrument is designed to measure up to 40 samples per hour in operation at the equilibrium mode. To accomplish this, the flow rate must be 1 mL/min through the column filled with immobilized urease whose total volume is only 0.5 mL. As is the case with partitioned-enzyme-sensor instruments, the support for urease must withstand fast flow rates without breakage or plugging. It was also necessary to use a support that has sufficient surface area to result in high enzyme loading so that the analyzer can operate in the equilibrium mode. The organic supports tested often resulted in high enzyme loading but poor flow characteristics. Many nonporous inorganic supports gave excellent flow characteristics when packed

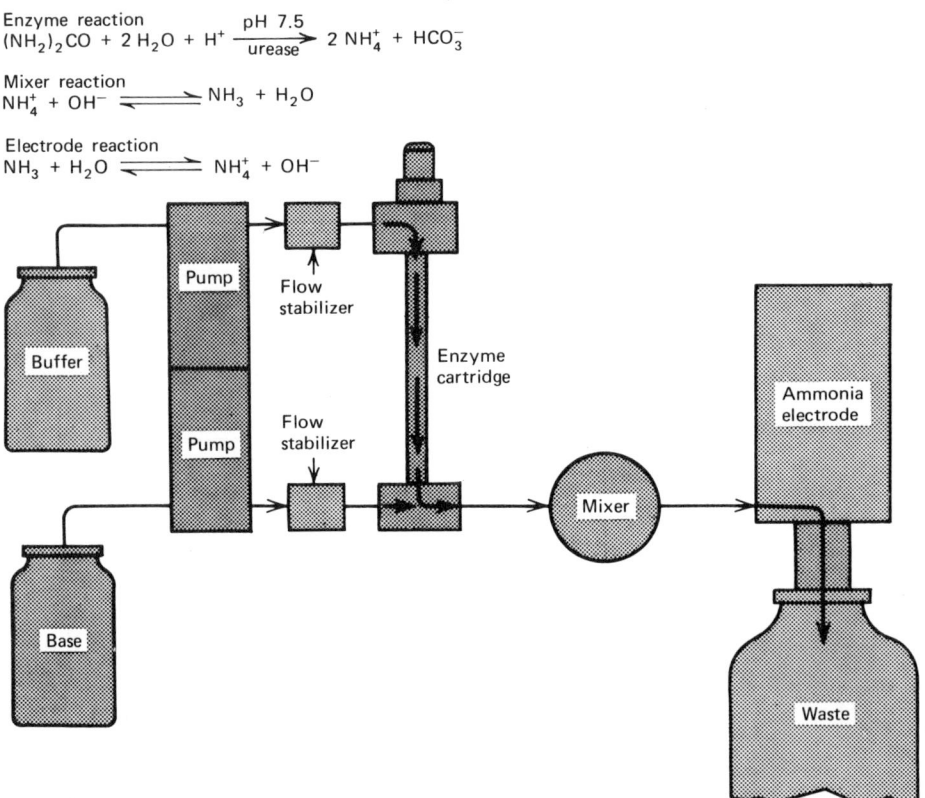

Enzyme reaction
$(NH_2)_2CO + 2 H_2O + H^+ \xrightarrow[\text{urease}]{\text{pH 7.5}} 2 NH_4^+ + HCO_3^-$

Mixer reaction
$NH_4^+ + OH^- \rightleftharpoons NH_3 + H_2O$

Electrode reaction
$NH_3 + H_2O \rightleftharpoons NH_4^+ + OH^-$

Figure 6. The Kimble BUN Analyzer.

in a column but were unable to bind sufficient urease. Porous silica and glass also resulted in low enzyme loading. The best material tested was a porous particulate alumina produced especially for this application. Using this support, not only were the flow characteristics excellent, but sufficient urease was immobilized to allow approximately one thousand tests before replacement.

Other Applications

At present immobilized enzymes are used in analytical testing and in various industrial processes. The selection of the methods of immobilization for each application and the advantages achieved by immobilization are described above. Additional applications of immobilized enzymes have been suggested but have not proved to be practical without further research. For example, in the treatment of many diseases that result in the deficiency of one or more enzymes, the use of an immobilized enzyme would allow the enzyme catalyst to be localized within an area of the body and be able to function over a long period of time (154). Treatment of a patient with an enzyme immobilized on a support rather than using the soluble enzyme may also result in reduced immuno response (see Immunotherapeutic agents).

Research has also begun on the use of immobilized enzymes in fuel cells and in the utilization of solar energy (qv) (155–159). The use of enzymes in fuel cells should make possible the preparation of stable cells based on reactions that would only take place at elevated temperatures in the absence of an enzyme (see Batteries). Research on the conversion of solar energy to chemical energy by use of enzyme catalysts follows the natural process of photosynthesis.

Immobilization of enzymes has also been useful in biochemical research. An immobilized enzyme may be used to synthesize a biochemical compound to be used in research or to study the enzyme in a manner not possible with a soluble enzyme. By attachment of enzyme subunits to a matrix, it is possible to study the function of the subunit under conditions in which it would be dimeric or further aggregated if free in solution (160). There are many examples in nature in which the product of a reaction catalyzed by one enzyme is the reactant for another enzyme-catalyzed reaction. When two such enzymes are immobilized together, the overall reaction rate is faster than when they are immobilized on separate supports (161). It would follow that nature has ordered enzymes in the cell so that reactions can take place rapidly in series. Enzymes immobilized together give the biochemist an opportunity to develop a model cell so that enzyme functions can be studied.

BIBLIOGRAPHY

1. M. Dixon, and E. C. Webb, *Enzymes,* Academic Press, New York, 1964, p. 29.
2. R. Goldman, L. Goldstein, and E. Katchalski in G. R. Stark, ed., *Biochemical Aspects of Reactions on Solid Supports,* Academic Press, New York, 1971.
3. O. R. Zaborsky, *Immobilized Enzymes,* CRC Press, Cleveland, 1973.
4. M. Salmona, C. Saronia, and S. Garattini, eds., *Insolubilized Enzymes,* Raven Press, New York, 1974.
5. R. A. Messing, ed., *Immobilized Enzymes for Industrial Reactors,* Academic Press, New York, 1975.
6. L. Goldstein and G. Manecke in L. B. Wingard, Jr., and E. Katchalski-Katzir, eds., *Applied Biochemistry and Bioengineering,* Vol. 1, 1976, p. 23.
7. *Methods Enzymol.* **44**, (1976).

8. F. M. Richards and J. R. Knowles, *J. Mol. Biol.* **37,** 231 (1968).
9. W. M. Herring, R. L. Lawrence, and J. R. Kittrell, *Biotechnol. Bioeng.* **14,** 975 (1972).
10. H. D. Brown and co-workers, *Biochim. Biophys. Acta* **279,** 356 (1972).
11. A. Svenson and B. Andersson, *Anal. Biochem.* **83,** 739 (1977).
12. F. C. Capet-Antonini, M. Trimard, and S. Tamerasse, *Thromb. Res.* **2,** 479 (1973).
13. A. Flynn and D. B. Johnson, *Int. J. Biochem.* **8,** 507 (1977).
14. G. Manecke, G. Günzel, and H. J. Förster, *J. Polym. Sci., Part C* **30,** 607 (1970).
15. H. H. Weetall, *Biochim. Biophys. Acta* **212,** 1 (1970).
16. J. M. Engasser and P. R. Coulet, *Biochim. Biophys. Acta* **485,** 29 (1977).
17. D. L. Regan, P. Dunnill, and M. D. Lilly, *Biotechnol. Bioeng.* **16,** 333 (1974).
18. K. Martinek and co-workers, *Biochim. Biophys. Acta* **485,** 1 (1977).
19. *Ibid.,* 13 (1977).
20. D. Gabel, I. Z. Steinberg, and E. Katchalski, *Biochem.* **10,** 4661 (1971).
21. A. C. Koch-Schmidt and K. Mosbach, *Biochem.* **16,** 2101 (1977).
22. *Ibid.,* 2105 (1977).
23. T. E. Barman, *Enzyme Handbook,* Vol. 1, Springer-Verlag, New York, 1969, p. 6.
24. L. Goldstein in K. Mosbach, ed., *Method of Enzymology,* Academic Press, Vol. 44, New York, 1976, p. 397.
25. E. Brown and A. Rocois, *Bull Soc. Chim. Fr.,* **3-4,** 743 (1974).
26. W. R. Veith and K. Venkatasubramanian, *Chemtech* **4,** 309 (1974).
27. V. H. Lang and co-workers, *Chem. Ztg.* **11,** 595 (1972).
28. R. J. Knights and A. Light, *Arch. Biochem. Biophys. Acta* **429,** 950 (1976).
29. B. Walter, *Biochim Biophys. Acta* **429,** 950 (1976).
30. Y. Levin and co-workers, *Biochem* **3,** 1905 (1964).
31. A. Bar-Eli and E. Katchalski, *J. Biol. Chem.* **238,** 1690 (1963).
32. P. Johnson and T. L. Whateley, *J. Colloid Interface Sci.* **37,** 557 (1971).
33. B. U. Von Specht and W. Brendel, *Biochim. Phys. Acta* **484,** 109 (1977).
34. W. E. Hornby, M. D. Illy, and E. M. Crook, *Biochem. J.* **98,** 420 (1966).
35. M. D. Lilly, W. E. Hornby, and E. M. Crook, *Biochem. J.* **108,** 845 (1968).
36. C. W. Wharton, E. M. Crook, and K. Brocklehurst, *Eur. J. Biochem.* **6,** 565 (1968).
37. L. Goldstein and co-workers, *Biochem.* **9,** 2322 (1970).
38. L. Goldstein, *Biochim. Biophys. Acta* **315,** 1 (1973).
39. L. Sundberg and T. Kristiansen, *FEBS Lett.* **22,** 175 (1972).
40. T. S. Lai and P. S. Cheng, *Biotechnol. Bioeng.* **20,** 773 (1978).
41. P. Cresswell and A. R. Sanderson, *Biochem. J.* **119,** 447 (1970).
42. A. C. Johansson and K. Mosbach, *Biochim. Biophys. Acta* **370,** 348 (1974).
43. M. Kanazawa, N. Yoshida, and S. I. Ishii, *Biochim. Biophys. Acta* **250,** 372 (1971).
44. A. Light and co-workers, *J. Biol. Chem.* **244,** 6289 (1969).
45. R. Reiner and co-workers, *J. Mol. Catal.* **2,** 335 (1977).
46. E. B. Ong, Y. Tsang, and G. E. Perlman, *J. Biol. Chem.* **241,** 5661 (1966).
47. T. Sato and co-workers, *Arch. Biochem. Biophys.* **147,** 788 (1971).
48. M. Naoi, M. Naoi, and K. Yagi, *Biochim. Biophys. Acta* **523,** 19 (1978).
49. S. P. O'Neill, *Rev. Pure and Appl. Chem.* **22,** 133 (1972).
50. W. R. Vieth and co-workers, *Appl. Biochem. Bioengineering* **1,** 222 (1976).
51. M. D. Lilly and P. Dunnill, *Methods Enzymol.* **44,** 717 (1976).
52. W. J. Pitcher Jr., in R. A. Messing, ed., *Immobilized Enzymes for Industrial Reactions,* Academic Press, New York, 1975, p. 151.
53. M. D. Lilly, D. L. Regan, and P. Dunnill in E. K. Pye and L. B. Wingard, eds., *Enzyme Engineering,* Vol. 2, Plenum Press, New York, 1974, p. 245.
54. S. P. O'Neill, P. Dunnill, and M. D. Lilly, *Biotechnol. Bioeng.* **13,** 337 (1971).
55. K. Buchholz and B. Godelmann, *Biotechnol. Bioeng.* **23,** 1201 (1978).
56. R. Datta, W. Armiger, and D. F. Ollis, *Biotechnol. Bioeng.* **15,** 993 (1973).
57. H. H. Weetall and N. B. Havewala, *Biotechnol. Bioeng. Symp.* **3,** 241 (1972).
58. K. L. Smiley, *Biotechnol. Bioeng.* **13,** 309 (1971).
59. J. Kucera, *Collect. Czech. Chem. Commun.* **41,** 2978 (1976).
60. R. N. Mukherjea and co-workers, *Biotechnol. Bioeng.* **19,** 1259 (1977).
61. H. Maeda, G. T. Tsao, and L. F. Chen, *Biotechnol. Bioeng.* **20,** 383 (1978).
62. R. D. Mason and H. H. Weetall, *Biotechnol. Bioeng.* **14,** 637 (1972).

63. H. H. Weetall and C. C. Detar, *Biotechnol. Bioeng.* **16,** 1537 (1974).
64. H. H. Weetall and co-workers, *Biotechnol. Bioeng.* **16,** 698 (1974).
65. A. C. Olson and W. L. Stanley, *J. Agric. Food Chem.* **21,** 440 (1973).
66. D. D. Lee and co-workers, *Biotechnol. Bioeng.* **18,** 253 (1976).
67. G. W. Strandberg and K. L. Smiley, *Biotechnol. Bioeng.* **14,** 509 (1972).
68. W. L. Stanley and co-workers, *Biotechnol. Bioeng.* **18,** 439 (1976).
69. M. D. Lilly and A. K. Sharp, *Chem. Eng.* **215,** CE12 (1968).
70. H. H. Weetall, *Food Product Development* **7,** 94 (1973).
71. F. X. Hasselberger and co-workers, *Biochem. Biophys. Res. Comm.* **57,** 1054 (1974).
72. M. J. Taylor and co-workers, *Biotechnol. Bioeng.* **19,** 683 (1977).
73. I. Karube and co-workers, *Biotechnol. Bioeng.* **19,** 1183 (1977).
74. P. B. Poulsen, and L. Ziltan, *Methods Enzymol.* **44,** 809 (1976).
75. R. O. Marshall and E. R. Kooi, *Science* **125,** 648 (1957).
76. Y. Takasaki, *Agric. Biol. Chem.* **30,** 1247 (1966).
77. Y. Takasaki, Y. Kosugi, and A. Kanbayashi in D. Perlman, ed., *Fermentation Advances,* Academic Press, New York, 1969, p. 561.
78. B. J. Schnyder, *Staerke* **26,** 409 (1974).
79. B. Metz, *Beverage Industry,* 132 (Nov. 24, 1978).
80. B. Metz, *Beverage Industry,* 5 (July 7, 1978).
81. M. P. J. Kierstan, *Biotechnol. Bioeng.* **20,** 447 (1978).
82. U.S. Pat. 3,044,904 (July 17, 1962), G. R. Serbia and P. R. Aguirre (to Central Aguirre Sugar Company).
83. *Chem. Eng. News* **51,** 36 (Apr. 23, 1973).
84. U.S. Pat. 3,616,221 (Oct. 26, 1971), Y. Takashi and O. Tanabe (to the Agency of Industrial Science and Technology, Tokyo, Japan).
85. U.S. Pat. 3,694,314 (Sept. 26, 1972), N. E. Lloyd and co-workers (to Standard Brands, Inc.).
86. H. H. Weetall in Z. Bohak and N. Sharon, eds., *Biotechnological Applications of Proteins and Enzymes,* Academic Press, New York, 1977, p. 104.
87. U.S. Pat. 3,788,945 (Jan. 29, 1974), K. N. Thompson, R. A. Johnson, and N. E. Lloyd (to Standard Brands, Inc.).
88. U.S. Pat. 3,909,354 (Sept. 30, 1975), K. N. Thompson, R. A. Johnson, and N. E. Lloyd (to Standard Brands, Inc.).
89. R. Greene, *Chem. Eng.* **85,** 78 (Apr. 10, 1978).
90. M. D. Rosenzweig, *Chem. Eng.* **83,** 54 (Sept. 27, 1976).
91. U.S. Pat. 3,979,261 (Sept. 7, 1976), H. Outtrup (to Novo Industri A/S).
92. U.S. Pat. 4,042,460 (Aug. 16, 1977), I. V. Diers (to Novo Industri A/S).
93. U.S. Pat. 3,980,521 (Sept. 14, 1976), S. Amotz, T. K. Nielson, and N. O. Thiesen (to Novo Industri A/S).
94. U.S. Pat. 4,025,389 (May 24, 1977), P. B. R. Poulsen and L. E. Zittan (to Novo Industri A/S).
95. U.S. Pat. 3,982,997 (Sept. 28, 1977), D. L. Eaton and R. A. Messing (to Corning Glass Works).
96. U.S. Pat. 3,992,329 (Nov. 16, 1976), D. L. Eaton and R. A. Messing (to Corning Glass Works).
97. R. D. Sweigart, *Food Eng.* **50,** 80 (1978).
98. U.S. Pat. 4,033,820 (July 5, 1977), R. E. Brouillard (to Penick and Ford, Ltd.).
99. K. J. Skinner, *Chem. Eng. News* **53,** 22 (Aug. 18, 1975).
100. S. J. Swanson, A. Emery, and H. C. Lim., *AIChE J.* **24,** 30 (1978).
101. C. Gruesbeck and H. F. Rose, *Ind. Eng. Chem. Prod. Res. Develop.* **11,** 74 (1972).
102. H. H. Weetall and N. B. Havewalo, *Biotechnol. Bioeng. Symp.* **3,** 241 (1972).
103. H. H. Weetall and co-workers, *Methods Enzymol.* **44,** 776 (1976).
104. *Chem. Week,* 37 (Nov. 14, 1977).
105. W. L. Stanley and R. Palter, *Biotechnol. Bioeng.* **15,** 597 (1973).
106. U.S. Pat. 3,736,231 (May 29, 1973), W. L. Stanely and A. C. Olseon (to the United States).
107. W. H. Pitcher, J. R. Ford, and H. H. Weetal, *Methods Enzymol.* **44,** 792 (1976).
108. U.S. Pat. 3,852,496 (Dec. 3, 1974), H. H. Weetall and S. Yaverbaum (to Corning Glass Works).
109. M. Pastore and F. Morisi, *Methods Enzymol.* **44,** 822 (1976).
110. I. Chibata and T. Tosa in L. B. Wingard, Jr., E. Katchalski-Katzir, and L. Goldstein, eds., *Applied Biochemistry and Bioengineering,* Vol. 1, Academic Press, New York, 1976, p. 329.
111. S. Ohlson, P. O. Larsson, and K. Mosbach, *Biotechnol. Bioeng.* **20,** 1267 (1978).
112. T. R. Jack and J. E. Zajic, *Adv. Biochem. Eng.* **5,** 125 (1977).

113. K. Yamada, *Biotechnol. Bioeng.* **19,** 1563 (1977).
114. R. D. Mason, C. C. Detar, and H. H. Weetall, *Biotechnol. Bioeng.* **17,** 1019 (1975).
115. G. P. Royer and J. P. Andres, *J. Macromol. Sci. Chem.* **7,** 1167 (1973).
116. C. L. Hicks and co-workers, *J. Dairy Sci.* **58,** 177 (1974).
117. E. C. Lee, G. F. Senyk, and W. F. Shipe, *J. Dairy Sci.* **58,** 473 (1975).
118. P. R. Witt, Jr. and co-workers, *Brew. Dig.* **45,** 70 (1970).
119. B. J. F. Hudson, *Chem. Indy* **20,** 1059 (1975).
120. J. C. Davis, *Chem. Eng.* **81,** 52 (Aug. 19, 1974).
121. *Chem. Eng. News* **52,** 19 (Feb. 25, 1974).
122. *Eur. Chem. News* **28,** 32 (March 7, 1975).
123. H. S. Yang and J. F. Studebaker, *Biotechnol. Bioeng.* **20,** 17 (1978).
124. J. J. Updike and G. P. Hicks, *Nature* **214,** 986 (1962).
125. G. G. Guilbault and J. G. Montalvo, *J. Am. Chem. Soc.* **92,** 2533 (1970).
126. M. Mascini and G. G. Guilbault, *Anal. Chem.* **49,** 795 (1977).
127. L. F. Cullen and co-workers, *Anal. Chem.* **46,** 1955 (1974).
128. M. Nanju and G. G. Guilbault, *Anal. Chem.* **46,** 1769 (1974).
129. M. Mascini and G. G. Guilbault, *Anal. Chem.* **49,** 795 (1977).
130. P. J. Taylor, E. Kmetec, and J. M. Johnson, *Anal. Chem.* **49,** 789 (1977).
131. M. Mascini and A. Liberti, *Anal. Chim. Acta* **68,** 177 (1974).
132. I. Satoh, I. Karube, and S. Suzuki, *Biotechnol. Bioeng.* **18,** 269 (1976).
133. D. S. Papastalkopoulos and G. A. Rechnitz, *Anal. Chem.* **48,** 862 (1976).
134. U.S. Pat. 3,539,455 (Nov. 10, 1970), L. C. Clark, Jr.
135. Service Manual, Glucose Analyzer Model 23A Yellow Springs Instrument Company, 1975.
136. K. S. Chua and I. K. Tan, *Clin. Chem.* **24,** 150 (1978).
137. L. H. Goodson and W. B. Jacobs, *Anal. Biochem.* **51,** 362 (1973).
138. E. K. Bauman and co-workers, *Anal. Chem.* **37,** 1378 (1965).
139. E. K. Bauman, L. H. Goodson, and J. R. Thomson, *Anal. Biochem.* **19,** 587 (1967).
140. E. Bisse and D. J. Vonderschmitt, *FEBS Lett.* **81,** 326 (1977).
141. D. R. Senn and P. W. Carr, *Anal. Chem.* **48,** 954 (1976).
142. D. J. Inman and W. E. Hornby, *Biochem. J.* **137,** 25 (1974).
143. M. K. Weibel and co-workers, *Anal. Biochem.* **52,** 402 (1973).
144. J. F. Rusling and co-workers, *Anal. Chem.* **48,** 1211 (1976).
145. S. K. Dahodwala, M. K. Weibel, and A. E. Humphrey, *Biotechnol. Bioeng.* **18,** 1679 (1976).
146. G. Johansson, K. Edström, and L. Ögren, *Anal. Chim. Acta* **85,** 55 (1976).
147. L. Trauberman, *Food Eng.* **47**(2), 58 (1975).
148. H. W. Levin and M. K. Weibel, paper presented at the AIChE 77th National Meeting, Pittsburgh, Pa., June 2–5, 1974.
149. B. Watson and M. H. Keyes, *Anal. Lett.* **9,** 713 (1976).
150. D. J. Hanson and N. S. Bretz, *Clin. Chem.* **23,** 477 (1977).
151. M. H. Keyes and R. Barabino in E. K. Pye and H. Weetall, eds., *Enzyme Engineering,* Vol. 3, Plenum Press, New York, 1977, p. 51.
152. U.S. Pat. 4,008,136 (Feb. 15, 1977), M. H. Keyes (to Owens-Illinois, Inc.).
153. U.S. Pat. 3,933,589 (Jan. 20, 1976), M. H. Keyes (to Owens-Illinois, Inc.).
154. T. M. S. Chang, *Artificial Cells,* Charles C Thomas, Springfield, 1972.
155. L. Karube and co-workers, *Biotechnol. Bioeng.* **19,** 1927 (1977).
156. T. G. Young, I. Hadjipetrow, and M. D. Lilly, *Biotechnol. Bioeng.* **8,** 581 (1966).
157. H. A. Bidela and A. J. Arvia, *Biotechnol. Bioeng.* **17,** 1529 (1975).
158. R. Blumenthal, S. R. Caplan, and O. Keden, *Biophys. J.* **7,** 735 (1967).
159. M. Gibbs and co-workers, *Workshop on Bio-Solar Conversion,* supported under NSF Grant 40253 (RANN) at Indiana University, 1973.
160. W. W. C. Chan and H. M. Mawer, *Arch. Biochem. Biophys.* **149,** 136 (1972).
161. K. Mosbach and co-workers in E. K. Pye and L. B. Wingard, Jr., eds., *Enzyme Engineering,* Vol. 2, Plenum Press, New York, 1974, p. 143.

MELVIN H. KEYES
Owens-Illinois, Inc.

ENZYMES, INDUSTRIAL

All synthetic and degradative reactions carried out by living creatures or extracts or segments of living matter are mediated and, in effect, made possible by enzymes. They digest food, synthesize tissue, make possible the reactions that give the energy needed to move and maintain body temperature and thought, make antibiotics, hormones, prostaglandins, etc, *in vivo* and, in an increasing number of instances, *in vitro*. Enzymes are sometimes very pure, single substances, and at other times, used in very crude form, such as a tissue macerate or immobilized whole cells (see Enzymes, immobilized).

The closeness of fermentations and enzymes is obvious when one considers that enzymes were at one time called ferments. Enzyme reactions are differentiated from fermentation reactions (carried out, of course, via the enzymes in the cells) on the basis of cell proliferation: if a transformation involves cell growth and multiplication it is, properly, a fermentation; if, however, there is no cell proliferation, then it is an enzymic reaction catalyzed by crude enzymes, eg, resting or killed cells.

Industrial enzymes may be defined as enzyme preparations manufactured for use as catalysts either directly or for use in the production of a manufactured commodity (see Antibiotics, β-Lactams; Microbial transformations; Beer; Fermentation; Yeasts).

Nomenclature

Enzymes, particularly industrially important enzymes, are known principally by their trivial (common, historical) names. These were given names based on the source of the enzyme by adding the suffix -in or -ain to a root indicating the source (eg, papain from papaya or pancreatin from pancreas), or named after the substrate or action by adding the suffix -ase to a root indicative of the substrate (eg, lactase acts on lactose, cellulase hydrolyzes cellulose, glucose oxidase oxidizes glucose).

In 1956, the Third International Congress of the International Union of Biochemistry (IUB) established a Commission on Enzymes to develop a systematic approach to naming enzymes. They developed a system applicable where the reaction catalyzed is fully defined by combining a numbering and a naming system. The systematic name is derived from the names of the substrate and the product and the type of reaction. Where two reactants are involved, both are named, separated by a colon. In general, proteolytic enzymes are not sufficiently limited or defined to apply short systematic names.

There are six classes of enzymes in this system (1):

(*1*) *Oxidoreductases* are active in biological oxidation and reduction, and therefore in respiration and fermentation processes. This class includes not only the dehydrogenases and oxidases, but also the peroxidases which use H_2O_2 as the oxidant, the hydroxylases which introduce hydroxyl groups and the oxygenases, which add molecular O_2 to a double bond in the substrate.

(*2*) *Transferases* catalyze the transfer of one-carbon groups (methyl-, formyl-, carboxyl- groups), aldehydic or ketonic residues, acyl, glucosyl, or alkyl groups, nitrogenous groups, and phosphorus- and sulfur-containing groups.

(*3*) *Hydrolases* include the esterases, phosphatases, glycosidases, peptidases, and others.

(*4*) *Lyases* remove groups from their substrates (not by hydrolysis) and may or may not add groups to double bonds. Included are decarboxylases, aldolases, dehydratases, and others.

(*5*) *Isomerases* include racemases, epimerases, cis-trans isomerases, intramolecular oxidoreductases, and intramolecular transferases.

(*6*) *Ligases* catalyze the joining together of two molecules coupled with the breakdown of a pyrophosphate bond in adenosine triphosphate (ATP) or a similar triphosphate (also known as synthetases).

The first element of the code gives the class of enzyme (*1–6* above). Table 1 presents the first three elements of the system. The fourth element is the serial number of the enzyme in its subclass.

The Commission recognized the importance of using trivial names and also set forth some recommendations for consistency and significance in continuing the use of these names, trying to eliminate those that had been applied to net reactions involving more than one enzyme. The use of an IUB designation is not without pitfalls. For example, it does not differentiate between enzymes from different sources, even though they may have very different optimum pH and other properties, if they catalyze the same reaction. Therefore, data from one particular enzyme from a specified source should not be generalized to other enzymes from different sources, even though they all bear the same name and numerical IUB designation.

Characteristics

Unlike other chemical entities, enzymes are characterized primarily on a functional basis. An enzyme is identified by what it does rather than its chemical composition. For example, a protein that speeds the decomposition of hydrogen peroxide to water and molecular oxygen is a catalase. The catalase derived from horse liver can be readily differentiated from the catalase derived from the mold *Aspergillus niger* but they are both catalases; poisoning or heating them eliminates the ability to speed the decomposition of hydrogen peroxide and they are no longer catalases, even though every atom comprising the active material is still there. Catalytic ability and specificity are the hallmarks of enzymes.

The protein nature of enzymes explains many of the problems and properties of enzymes. As proteins they are antigenic. The protein nature, and hence the catalytic properties, are affected by temperature and pH changes, water activity, ionic strength, and similar variables. The same factors that affect the ability of a protein to function as an enzyme also affect the manner in which it acts as an enzyme.

The enzyme molecule consists of a known chain of amino acids which has a particular geometric configuration specific for that arrangement of amino acids. The twisting and turning forms some locations that are catalytically active and these are referred to as the active sites. Sometimes there is one per enzyme molecule and sometimes more. It is this twisting and turning that provides some geometric specificity and, because of the characteristics of amino acids comprising the active site, the specificity.

Catalytic Action. Most reactions in biological systems proceed almost infinitely slowly in the absence of the specific enzyme catalyst, although at tremendous speed in its presence. This applies even to simple reactions, such as the formation of carbonic acid from carbon dioxide and water, a reaction of great importance in rapidly moving carbon dioxide from site of formation in the tissues to the lungs for expiration:

$$CO_2 + H_2O \rightarrow H_2CO_3$$

Table 1. Key to Enzyme Commission Numbering and Classification System[a]

1. Oxidoreductases
- 1.1 Acting on the CH—OH group of donors
  - 1.1.1 with NAD or NADP as acceptor
  - 1.1.2 with a cytochrome as an acceptor
  - 1.1.3 with O_2 as acceptor
  - 1.1.99 with other acceptors

- 1.2 Acting on the aldehyde or keto-group of donors
  - 1.2.1 with NAD or NADP as acceptor
  - 1.2.2 with a cytochrome as an acceptor
  - 1.2.3 with O_2 as acceptor
  - 1.2.4 with lipoate as acceptor
  - 1.2.99 with other acceptors

- 1.3 Acting on the CH—CH group of donors
  - 1.3.1 with NAD or NADP as acceptor
  - 1.3.2 with a cytochrome as an acceptor
  - 1.3.3 with O_2 as acceptor
  - 1.3.99 with other acceptors

- 1.4 Acting on the CH—NH_2 group of donors
  - 1.4.1 with NAD or NADP as acceptor
  - 1.4.3 with O_2 as acceptor

- 1.5 Acting on the C—NH group of donors
  - 1.5.1 with NAD or NADP as acceptor
  - 1.5.3 with O_2 as acceptor

- 1.6 Acting on reduced NAD or NADP as donor
  - 1.6.1 with NAD or NADP as acceptor
  - 1.6.2 with a cytochrome as an acceptor
  - 1.6.4 with a disulfide compound as acceptor
  - 1.6.5 with a quinone or related compound as acceptor
  - 1.6.6 with a nitrogenous group as acceptor
  - 1.6.99 with other acceptors

- 1.7 Acting on other nitrogenous compounds as donors
  - 1.7.3 with O_2 as acceptor
  - 1.7.99 with other acceptors

- 1.8 Acting on sulfur groups of donors
  - 1.8.1 with NAD or NADP as acceptor
  - 1.8.3 with O_2 as acceptor
  - 1.8.4 with a disulfide compound as acceptor
  - 1.8.5 with a quinone or related compound as acceptor
  - 1.8.6 with a nitrogenous group as acceptor

- 1.9 Acting on heme groups of donors
  - 1.9.3 with O_2 as acceptor
  - 1.9.6 with a nitrogenous group as acceptor

- 1.10 Acting on diphenols and related substances as donors
  - 1.10.3 with O_2 as acceptor

- 1.11 Acting on H_2O_2 as acceptor

- 1.12 Acting on hydrogen as donor

Table 1 (*continued*)

1.13 Acting on single donors with incorporation of oxygen (oxygenases)

1.14 Acting on paired donors with incorporation of oxygen into one donor (hydroxylases)
    1.14.1   using reduced NAD or NADP as one donor
    1.14.2   using ascorbate as one donor
    1.14.3   using reduced pteridine as one donor

2. *Transferases*
   2.1 Transferring one-carbon groups
      2.1.1   methyltransferases
      2.1.2   hydroxymethyl-, formyl-, and related transferases
      2.1.3   carboxyl- and carbamoyltransferases
      2.1.4   amidinotransferases

   2.2 Transferring aldehydic or ketonic residues

   2.3 Acyltransferases
      2.3.1   acyltransferases
      2.3.2   aminoacyltransferases

   2.4 Glycosyltransferases
      2.4.1   hexosyltransferases
      2.4.2   pentosyltransferases

   2.5 Transferring alkyl or related groups

   2.6 Transferring nitrogenous groups
      2.6.1   aminotransferases
      2.6.3   oximinotransferases

   2.7 Transferring phosphorus-containing groups
      2.7.1   phosphotransferases with an alcohol group as acceptor
      2.7.2   phosphotransferases with a carboxyl group as acceptor
      2.7.3   phosphotransferases with a nitrogenous group as acceptor
      2.7.4   phosphotransferases with a phospho group as acceptor
      2.7.5   phosphotransferases, apparently intramolecular
      2.7.6   pyrophosphotransferases
      2.7.7   nucleotidyltransferases
      2.7.8   transferases for other substituted phospho groups

   2.8 Transferring sulfur-containing groups
      2.8.1   sulfurtransferases
      2.8.2   sulfotransferases
      2.8.3   CoA-transferases

3. *Hydrolases*
   3.1 Acting on ester bonds
      3.1.1   carboxylic ester hydrolases
      3.1.2   thiolester hydrolases
      3.1.3   phosphoric monoester hydrolases
      3.1.4   phosphoric diester hydrolases
      3.1.5   triphosphoric monoester hydrolases
      3.1.6   sulfuric ester hydrolases

   3.2 Acting on glycosyl compounds
      3.2.1   glycoside hydrolases

Table 1 (*continued*)

      3.2.2   hydrolyzing N-glycosyl compounds
      3.2.3   hydrolyzing S-glycosyl compounds

  3.3 Acting on ether bonds
      3.3.1   thioether hydrolases

  3.4 Acting on peptide bonds (peptide hydrolases)
      3.4.1 α-amino-acyl-peptide hydrolases
      3.4.2   peptidyl-amino-acid hydrolases
      3.4.3   dipeptide hydrolases
      3.4.4   peptidyl-peptide hydrolases

  3.5 Acting on C—N bonds other than peptide bonds
      3.5.1   in linear amides
      3.5.2   in cyclic amides
      3.5.3   in linear amidines
      3.5.4   in cyclic amidines
      3.5.5   in cyanides
      3.5.99 in other compounds

  3.6 Acting on acid anhydride bonds
      3.6.1   in phosphoryl-containing anhydrides

  3.7 Acting on C—C bonds
      3.7.1   in ketonic substances

  3.8 Acting on halide bonds
      3.8.1   in C-halide compounds
      3.8.2   in P-halide compounds

  3.9 Acting on P—N bonds

4. *Lyases*

  4.1 Carbon-carbon lyases
      4.1.1   carboxy lyases
      4.1.2   aldehyde lyases
      4.1.3   ketoacid lyases

  4.2 Carbon-oxygen lyases
      4.2.1   hydro lyases
      4.2.99 other carbon-oxygen lyases

  4.3 Carbon-nitrogen lyases
      4.3.1   ammonia lyases
      4.3.2   amidine lyases

  4.4 Carbon-sulfur lyases

  4.5 Carbon-halide lyases

  4.99 Other lyases

5. *Isomerases*

  5.1 Racemases and epimerases
      5.1.1   acting on amino acids and derivatives
      5.1.2   acting on hydroxy acids and derivatives

Table 1 (*continued*)

	5.1.3	acting on carbohydrates and derivatives
	5.1.99	acting on other compounds

5.2 *Cis-trans* isomerases

5.3 Intramolecular oxidoreductases
    5.3.1   interconverting aldoses and ketoses
    5.3.2   interconverting keto and enol groups
    5.3.3   transposing C=C bonds

5.4 Intramolecular transferases
    5.4.1   transferring acyl groups
    5.4.2   transferring phosphoryl groups
    5.4.99   transferring other groups

5.5 Intramolecular lyases

5.99 Other isomerases

6. *Ligases*
  6.1 Forming C—O bonds
      6.1.1   amino-acid-RNA ligases

  6.2 Forming C—S bonds
      6.2.1   acid-thiol ligases

  6.3 Forming C—N bonds
      6.3.1   acid-ammonia ligases (amide synthetases)
      6.3.2   acid-amino-acid ligases (peptide synthetases)
      6.3.3   cyclo-ligases
      6.3.4   other C—N ligases
      6.3.5   C—N ligases with glutamine as N-donor

  6.4 Forming C—C bonds

[a] Reprinted from Ref. 2, courtesy of Elsevier Publishing Co.

This reaction is catalyzed by the enzyme carbonic anhydrase. It proceeds ten million times faster in the presence of the enzyme than in its absence. Carbonic anhydrase has a catalytic center activity (turnover number) of 100,000 (3). The catalytic center activity is the number of molecules of reactant that can be converted to end product by one catalytic center of an enzyme in one second. In enzyme reactions, the reactant or substance converted by the enzyme is called the substrate.

    The speed during the initial phase of the reaction depends on the amount of enzyme present, as long as the amount of the substrate is in excess and all conditions are constant. Under such conditions, the reaction speed increases in linear proportion with the amount of enzyme, that is, in the presence of a double amount of enzyme the same change requires one-half the reaction time.

    An enzyme can be compared to a hole punched in the bottom of a pail of water: the larger the hole the faster the water runs out; but the total amount of water that runs out is a function of the amount of water in the pail and the level of water in the container it is running into, not on the amount of enzyme used. The amount of enzyme affects only the rate at which equilibrium for the reaction catalyzed is reached.

If the amount of enzyme in the system is kept constant, the reaction time is influenced by a number of other factors including concentrations of substrate and product, presence of activators or inhibitors, temperature, and pH.

An optimum temperature and an optimum pH are characteristic of every enzyme, and so are the minimum and maximum values of pH and temperature at which the enzyme is active (see Figs. 1–2).

At the optimum pH the enzyme's activity is highest, all other conditions being equal. The optimum for practical purposes may cover one or several units of the pH scale. The activity decreases rapidly on both sides of the optimum range until the enzyme is completely inactive. Change of the pH toward the optimum may reactivate the enzyme, and change of pH above or below the levels of temporary inactivation may gradually damage the enzyme and cause permanent inactivity. The optimum of the

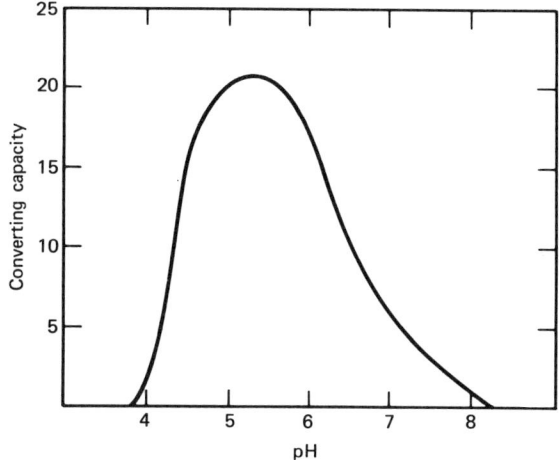

Figure 1. The effect of pH on the activity of malt amylase.

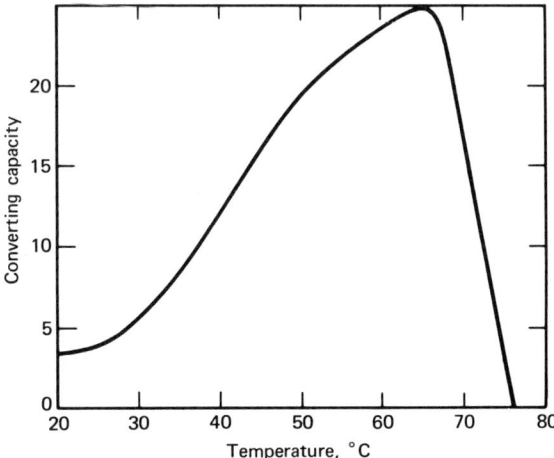

Figure 2. The effect of temperature on the activity of malt amylase.

same enzyme may vary with the substrate on which it acts. If the reaction is reversible, a specific pH may promote the reaction in the opposite direction.

The optimum pH values of different enzymes may cover a wide range. For example, pepsin exhibits highest activity on egg albumin at a pH of 1.5, whereas a bacterial alkaline protease has an optimum pH of 11.

The pH range for optimum enzyme stability (shelf life) is usually within several units, up and down, from the optimum pH.

The effect of temperature for most enzymes roughly follows the rule that every 10°C increase in temperature about doubles the activity as long as the temperature does not damage the enzyme. In other words, 1 g enzyme preparation catalyzes in 1 h at 35°C the transformation of twice as much substrate as in 1 h at 25°C.

Specificity. Most enzymes are highly specific and catalyze only one specific reaction or act upon only one isomer of a particular compound. Some enzymes are less specific and are able to catalyze several, usually related, reactions. The same reaction may be catalyzed by a large number of enzymes, different in their specific characteristics, and produced by different types of cells. The transformation of a compound, expressed by a simple chemical equation, quite frequently requires the cooperation of a number of enzymes. This explains the presence of hundreds of different enzymes in a single cell. The number of known individual enzymes is over 2000, of which ca 10% have been isolated in crystalline form.

According to the purpose of the particular cell, some cells produce more and others produce fewer different individual enzymes. Most enzymes are produced in the quantities needed, but some are produced in extreme excess. The enzyme formation inside the cells is controlled by the dominant genes. Certain types of cells stand out in the excess production of one or a few specific enzymes. Cells of animal glands and mucous membranes, cells in certain plant tissues of seeds, fruits, and leaves, and the cells of specific microorganisms biosynthesize one or a few enzymes in great abundance. Those are the cells utilized in the production of industrial enzymes.

Pure enzymes form the following four major groups with regard to the degree of specificity (4):

(1) *Absolute enzymes* catalyze the reaction of only one substrate. For example, urease breaks down urea into CO_2 and ammonia, but does not attack any other substrate.

(2) *Stereospecific enzymes* catalyze reactions with one type of optical isomer, but may react with a series of related compounds of the same configuration. For example, many proteolytic enzymes hydrolyze only peptide bonds linking L-amino acids (natural amino acids).

(3) *General hydrolyzing enzymes* react with a specific type of chemical linkage. For example, most lipases hydrolyze a wide range of organic esters and many phosphatases break down phosphate esters into phosphoric acid and alcohol.

(4) *Enzymes that attack certain specific points of a molecule.* For example, some proteolytic enzymes act on points where the adjacent amino acid contains a benzene ring. Some hydrolytic enzymes attack at the center of the molecule and others at the ends. For example, α-amylase attacks the center of the starch molecule and of the long glucose chains deriving from starch, whereas β-amylase attacks the ends, splits off two glucose units, and thus forms maltose; amyloglucosidase attacks the nonreducing ends of starch or its hydrolytic products and splits off single glucose units.

Composition and Chemical Nature. All enzymes are proteins, metalloproteins, or conjugated proteins. Metals are often an integral part of the enzyme, where separation can result in irreversible loss of activity (eg, Fe^{3+} in catalase). In other cases, metal ions are required for activity or stability (eg, Ca^{2+} for bacterial α-amylase, Mg^{2+} or Co^{2+} for glucose isomerase). In some cases, the activator for one is a poison for the other; for example, iron poisons urease but is essential for catalase activity, Ca^{2+} stabilizes the starch-liquefying bacterial α-amylase but inactivates the glucose isomerase that may be subsequently used. The nonprotein portion may be other than a metal ion. Flavine adenine dinucleotide (FAD) can be separated from D-amino acid oxidase thereby inactivating the enzyme, and then be recombined to restore the activity to the enzyme. The FAD is more tightly held by glucose oxidase which does not give up FAD on dialysis. The nonprotein portion of an active enzyme separated from the protein portion is called the prosthetic group or coenzyme; the remaining portion is called the apoenzyme and the combined apoenzyme and prosthetic group is called the holoenzyme.

Cofactors. The more general term, cofactor, includes the metallic portions of metalloprotein enzymes, and the coenzyme or prosthetic group portions of holoenzymes. In many enzymes, the coenzyme portion is the reactive part. Most coenzymes contain a nucleotide, a vitamin, a five-carbon sugar, and possibly other compounds tied together. The vitamins in coenzymes are almost exclusively members of the vitamin B group.

Thiamin (vitamin B_1) is part of the thiamin pyrophosphate coenzyme [154-87-0] active in decarboxylation of α-keto acids and in certain reactions of keto sugars; *riboflavin* (vitamin B_2) is part of the flavin mononucleotide coenzyme [130-40-5] and of the flavin adenine dinucleotide coenzyme [146-14-5] which takes part in several oxidation–reduction reactions; *pyridoxin* (vitamin B_6) is part of pyridoxal phosphate coenzyme [54-47-7], active in amino acid decarboxylation, transamination, and other reactions; *niacin*, part of the coenzymes nicotinamide adenine dinucleotide (NAD) [53-84-9] and nicotinamide adenine dinucleotide phosphate (NADP) [53-59-8] mediates many oxidation–reduction reactions; pantothenic acid, part of a coenzyme A, participates in catalyzing reactions of fatty acids, principally the transfer of acetyl groups; *biotin*, part of the biotin coenzyme, catalyzes some carbon dioxide-fixation reactions; *folic acid*, catalyzes reactions of single-carbon compounds; *cyanocobalamin* (vitamin B_{12}) is part of the cobalamide coenzymes (see Vitamins).

Kinetics. The enzyme combines with the substrate and then dissociates into the enzyme and the reaction products (5–8). In the presence of low substrate concentrations, the rate of enzyme action is a function of the amount of substrate and a constant fraction of substrate is converted per unit time. In the presence of excess substrate, the degree of enzyme action is thus a function of enzyme concentration and, for a given enzyme concentration, the amount of substrate converted per unit time is a constant at least as long as the remaining substrate is still in excess.

Figure 3 shows the effect of substrate concentration on reaction rate, assuming constant enzyme concentration (9). In Phase 1, the enzyme molecules are not saturated by substrate (active sites are not filled) and thus the reaction rate varies with the substrate concentration. As the number of substrate molecules increases, the active sites are filled completely and the enzyme approaches full capacity. The relation of the reaction rate (velocity) to substrate concentration is expressed in the Michaelis–Menten equation:

$$v = \frac{V_{\max}[S]}{K_m + [S]}$$

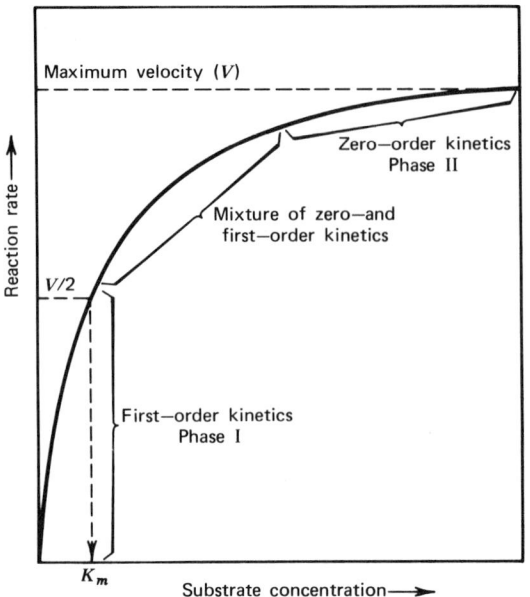

Figure 3. Effect of substrate concentration on reaction rate, assuming that enzyme concentration is constant (9).

where v is the observed velocity at the given substrate concentration S, K_m is the Michaelis constant expressed in units of concentration, and V_{max} is the maximum velocity for a saturated substrate.

A somewhat more useful plot is obtained by taking the reciprocal of the Michaelis–Menten equation to obtain the Lineweaver–Burke equation, a straight-line equation:

$$\frac{1}{v} = \frac{K_m}{K_{max}}\left(\frac{1}{[S]}\right) + \frac{1}{V_{max}}$$

The usefulness of this equation is apparent from Figure 4 if one considers the reciprocal of velocity and the corresponding reciprocal of substrate concentration taken as the variables, wherein K_m/V_{max} is the slope of the straight line and $1/V_{max}$ is the intercept (10).

Inhibition. Various types of inhibitions can be observed in enzyme work, such as (a) irreversible and reversible, and (b) competitive, noncompetitive, and uncompetitive.

Irreversible inhibition is the chemical inactivation of the enzyme itself by a substance that combines with the enzyme proper rendering it inactive. An example would be the nerve gas diisopropyl fluorophosphate, which reacts irreversibly with the amino acid serine, which is an important part of the active site of such enzymes as chymotrypsin (see Cholinesterase inhibitors). In this case, the decrease in activity is a function of how much inhibitor was added and the effect is the same as simply having added less enzyme. Because different inhibitors attach themselves to different amino acids and groupings, a knowledge of the extent to which known inhibitors alter the activity of a particular enzyme gives considerable knowledge of the nature of the

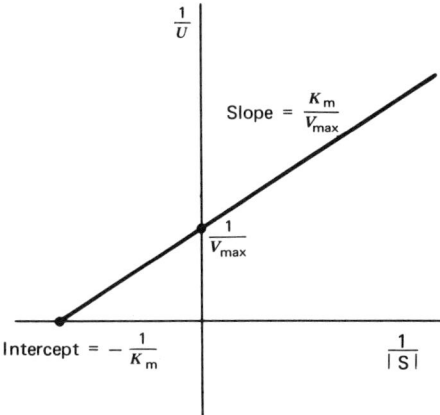

Figure 4. A typical Lineweaver-Burk plot. Lines are extended to $1/v = 0$ to obtain greater accuracy in determining the constants (10).

active site. In irreversible inhibition by an added substance, the complex, once formed, cannot be broken by such simple means as dialysis or dilution.

In reversible inhibition, the complex formed, whether it is with an enzyme, a substrate, or an enzyme–substrate complex, is dissociable. This is often caused by materials that are chemically very similar to the substrate. Reversible inhibition is manifested as follows:

Competitive inhibition is a reversible inhibition in which the inhibitor is in itself a second substrate or reacts like a substrate and competes for the same site as the main substrate. The extent to which this takes place depends on the relative affinity of the enzyme for the two substrates and the relative concentrations of the two substrates. An example is the addition of xylose to a glucose solution containing glucose isomerase. Glucose isomerase acts preferentially on xylose, that is, it has a smaller K_m for xylose than for glucose, so that small amounts of xylose slow the reaction. Eventually the reaction goes to completion, and this is illustrated in the Michaelis–Menten plot (11) in Figure 5.

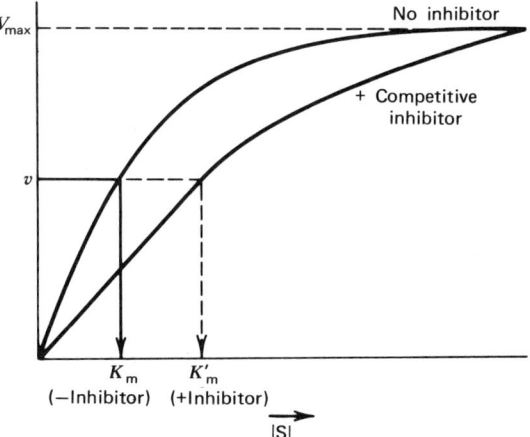

Figure 5. The relation between v, the rate of reaction, and substrate concentration [S] with and without the competitive inhibitor. Note the change in K_m in the absence and presence of inhibitor with no shift in V_{max} (11).

Noncompetitive inhibition would be accomplished by compounds that react with the enzyme substrate complex and also with the enzyme itself or where the inhibitor is or is related to a product of the reaction. For example, the addition of acetaldehyde to ethanol being acted on by alcohol dehydrogenase (which forms acetaldehyde as a product), shown in Figure 6 where the maximum velocity is reduced but the K_m is not altered (12).

Uncompetitive inhibition refers to inhibitors combining only with the enzyme–substrate complex. In this case, both the maximum velocity and the K_m are decreased.

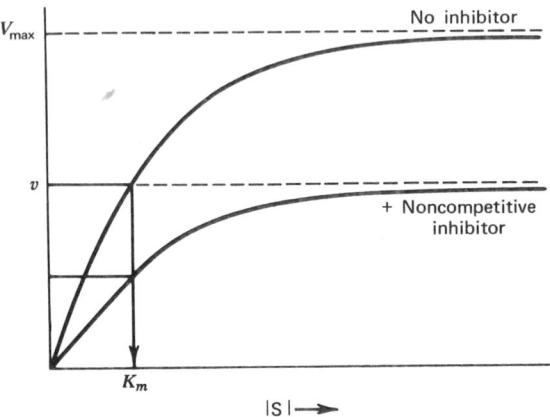

Figure 6. The reaction between v and substrate concentration [S] with and without the noncompetitive inhibitor. Note the shift in V_{max} but the lack of any change in K_m (12).

Production

Industrial enzymes vary from very pure crystalline material (such as glucose isomerase produced by Clinton Corn Processing Co. for its own use) to dried ground tissue (pancreas glands, dried and ground to form pancreatin, or ground germinated barley as malt) and whole microbial cells (immobilized and fixed with gluteraldehyde, for example).

The fundamental process for producing enzymes is similar, regardless of the enzyme source (animal, higher plant, or microbial).

Extracellular enzymes are usually the hydrolytic type. Thus ficin is a protease found in the sap of certain fig trees, papain occurs in papaya latex, pepsin in stomach secretions, and microbial hydrolytic enzymes are secreted by the cells into the growth medium (eg, lactase, pectinase, protease, amylase, etc).

Constitutive enzymes are produced regularly by certain cells (animal, plant, microbial) and tend to be associated with cell mass. Others are produced in response to specific stimuli (eg, glucose isomerase levels are greatly increased by growing cells in the presence of xylose) and are *adaptive*.

Collection or Production of Enzyme-Rich Material. *Animal enzymes* are generally derived from specific organs (eg, lipase from throat glands of young animals, pancreatin from pancreas) although urokinase [9039-53-6], a proteolytic enzyme used to dissolve blood clots *in vivo,* is extracted from human urine.

Plant enzymes from higher plants are obtained from the roots (horseradish, peroxidase), seeds (malt), fruit (papain), sap (ficin), or other plant parts.

Microbial enzymes are obtained by growing the desired organism either on a semisolid medium or in a submerged culture (agitated tank) virtually identical to the equipment and means developed for antibiotic production (see Antibiotics). In general, control of pH, temperature, aeration, medium composition, age, etc, is important in obtaining maximum yield. Enzyme yield may be associated with a particular phase of the growth cycle, or the exhausting or addition of a specific metabolite. Both submerged and surface cultures need temperature controls.

Surface cultures can be stationary, as in trays or piles, or tossed in a slowly rotating drum. It is easier to control temperature, air, and humidity in a rotating drum than in high piles. Often temperature is controlled by controlling the rate of exothermic microbial growth by limiting moisture.

Fermentors used for enzyme production by submerged culture may be quite small 1–10 m^3 (250–2500 gal) but usually have a capacity of 20 to 200 m^3 (5000–50,000 gal). Often equipment too small for economies of scale required for antibiotic production is diverted to enzyme production (see Fermentation).

The strain of organism used for production of microbial enzymes must exhibit high biological stability. Its selection and even development is a highly significant factor in final yield, often a matter of several orders of magnitude. At first, commercially significant strains were selected from nature by isolating and testing many pure cultures. Sometimes they are obtained by a selection process based on nutrients. The selection medium for a dextranase-producing organism might contain dextran as the only significant carbon or energy source. A refinement on selections from nature is the deliberate production of mutants by radiation or chemical means, and testing the mutants. More recently, genetic engineering (qv) has become important in making cells more productive by doubling up on the genes responsible for production of a specific enzyme. Genetic mapping and recombination, ie, hybridization, is a subject unto itself.

Extraction. Not all enzymes are extracted from the enzyme-rich source before use. Malted barley, as a combined source of amylase and carbohydrate (starch) is simply dried, separated from the rootlet formed during sprouting (a process that enriches the enzyme content), and sometimes ground. Other enzymes self-extract by being secreted into the growth medium. The amylase of *Bacillus subtilis,* produced in deep-tank submerged culture, is secreted by the growing cells into the aqueous medium. This medium, freed of cells, is concentrated, preserved, and used directly as a liquid enzyme preparation for the desizing of textiles (see Textiles).

Obviously, intracellular enzymes, if used as a cell-free preparation, must be extracted. Similarly, intra- and extracellular enzymes produced on a semisolid medium must be extracted from the semisolid mass.

Although the concept of extraction is very simple and straightforward, the process is not as simple as the concept because cells must be opened up before the enzymes within can be extracted. Even then, the enzyme might be bound to the cell membrane or contained within an organized organelle, such as a mitochondria. Thus, before or during extraction the cell must be ruptured and the enzyme solubilized. Often the process is spontaneous, following death of the cell, carried out by autolytic enzymes released within the cell. *Penicillium notatum* mycelium killed by toluene undergoes rapid autolysis, for example. On the other hand, *Saccharomyces cerevisiae*, although readily plasmolyzed by acetate, requires high shear to break it up enough to rapidly release enzymes. Cell rupture can be accomplished by enzyme addition, ultrasonics, ball-milling, grinding, freezing and thawing, and homogenization (see Mixing and blending; Ultrasonics). Enzyme solubilization is accomplished spontaneously by indigenous enzymes (autolysis), surfactants, adding enzymes such as lysozyme, pressurization and sudden release, grinding in frozen state, and similar operations.

Once solubilized, the extraction is merely a washing out process. This may be a simple filtration where cell density is minimal in an aqueous form, or a true countercurrent extraction process for enzymes on semisolid medium.

Purification, Concentration. For most industrial enzymes, the aqueous extract is concentrated and stabilized. Purification, if any, is an incidental benefit. Traditionally the crude enzyme was precipitated from the aqueous extract by the addition of water-miscible solvents such as acetone, methanol, ethanol, isopropanol, or mixtures of these. Judicious selection of pH, added solutes, and solvent ratios do effect a fractionation of the proteins and, of course, separation from the bulk of the nonprotein materials in the aqueous extract. With acetone a ratio of solvent to aqueous extract of 0.5–1.5:1 is used, and with alcohols a ratio of 2.5:1–4.0:1. For specific enzymes, ratios outside these ranges may be used. Because the enzymes are generally unstable in these mixtures, they are generally precipitated at a very low temperature, 0–10°C. Solvent contact time is kept to a minimum, often only a few minutes.

The precipitated enzyme is separated by filtration, as in a plate-and-frame filter with the addition of diatomaceous earth filter aid, or separated by centrifugation through a high-force centrifuge such as a Sharples supercentrifuge. Basket centrifuges are also used for some large scale separations (see Centrifugal separation).

Enzymes precipitated by solvents are rapidly dried at low temperature (20–50°C), often under vacuum. The evaporating solvent keeps the temperature of the enzyme preparation low until it is essentially dry and less heat sensitive. The dry enzyme is then pulverized, assayed, standardized, and packaged.

Ecological concerns, regulations on organic vapor in factory environments and exhausts, and the sharp rise in solvent costs since 1973 have caused a major revision in enzyme processing. Ultrafiltration (qv) is now often used to substantially reduce the volume of aqueous phase, and hence the volume of solvent needed to precipitate the enzyme. Indeed, ultrafiltration has, in conjunction with the trend to liquid enzyme preparations, often eliminated the use of solvents (see Filtration).

For some enzymes a high degree of purification is needed. Since enzymes are protein in nature, all techniques usable with proteins are applied here as well. These include other precipitating agents such as ammonium sulfate, magnesium sulfate, as well as freeze-drying, which is the gentlest method of producing a dry preparation. Other techniques used include ion-exchange resin pick up and elution, affinity chromatography, electrodecantation, gel filtration, and nonspecific adsorption, among others.

Purification may be required to eliminate or reduce objectionable side activity. *Mucor miehei*, eg, produces large amounts of lipase/esterase at the same time that it produces acid protease used as a microbial rennet. If not reduced, the lipase could result in hydrolytic rancidity in the final cheese. The lipase can be selectively inactivated by heating and low pH.

Neutral and alkaline protease can be removed from solutions of mixed proteases by adsorption on Duolite C-10 ion-exchange resin, leaving the rennet-like acid protease in solution. In this case the adsorbed enzyme can be eluted and recovered as well (13).

Transglucosidase as a contaminant in glucoamylase preparations is undesirable because it diverts some of the glucose formed to isomaltose; when glucoamylase acts on solubilized starch, it can be separated by coprecipitation of the transglucosidase with maleic anhydride copolymers (14).

Formulations, Stabilization, Standardization. Once manufactured, the enzyme must still be put into a readily usable form, one in which it is sufficiently stable to be physically transferred from point of manufacture to point of use. Commercial enzyme preparations must be formulated to contain the necessary activities and made stable to microbial degradation. The enzyme activity also must be stabilized so that the initial concentration activity is substantially unchanged when used; and the preparation must be standardized to uniform activity and, finally, put into a form that is convenient to use.

Added materials are of various types: (a) pH modifiers: phosphates, citrates, caustic, inorganic and organic acids, etc; (b) preservatives: (1) antimicrobial agents such as benzoate or benzoate ester, sorbate, and (2) hyperosmic agents such as NaCl, glycerol, sorbitol, propylene glycol, sugars, other solutes; (c) sequestrants such as citrate, EDTA; (d) activators such as calcium salts, sulfite, cobalt salts; (e) other enzymes such as pepsin mixed with rennet; and (f) standardizing materials (diluents) such as water, diatomaceous earth, whey, whey solids, lactose, sawdust, flour, NaCl, mannitol.

Traditionally, enzymes have been stabilized by drying and sold in dry powdered form. In the past twenty-five years, stable liquids have become more popular. In some cases, eg, glucose oxidase, the aqueous form is more stable than the powder. In most instances, standardizing materials are added as required to adjust the concentration of active ingredient to a predetermined level, to permit the user always to use the same physical quantity of enzyme material for a batch. Standardized enzyme preparations have the same content of main active enzyme per unit volume or per unit weight; however, the other activities vary from batch to batch. The amount of standardizing material added depends on the difference in activity between the concentration prior to standardization and the final activity selected.

Some enzyme companies offer unstandardized materials on a factor basis. For example, if the standardized version of enzyme A is 10 units per gram, and the concentrate before standardization has 25 units per gram, the concentrate would be sold with a factor of 2.5. Of course, a kilogram of the concentrate with a factor of 2.5, would be priced at 2.5 times the price of a kilogram of standardized material, less a small savings in standardizing and transportation costs. In some industries, especially baking, enzymes have been tableted to facilitate use (see Bakery processes and leavening agents).

Amino Acid Synthesis. L-Alanine and L-aspartic acid are made enzymatically. L-Alanine is made by decarboxylating aspartic acid with L-aspartate-4-carboxylase [9024-57-1] (15). Aspartates are made by the condensation of fumarates and NH_3.

Synthetic racemic mixtures are often resolved by acetylation; the acetylated ra-

Table 2. Amino Acids Production by Enzymatic Method[a]

Amino acid	Reaction	Enzyme	Enzyme source
L-amino acid	N-acetyl-L-amino acid + H_2O → L-amino acid + acetic acid	aminoacylase	*Aspergillus*
L-alanine	L-aspartic acid → L-alanine + CO_2	aspartate decarboxylase	*Pseudomonas dacunhae*
L-aspartic acid	fumaric acid + NH_4^+ → L-aspartic acid	aspartase	*Escherichia coli*
L-citrulline	L-arginine + H_2O → L-citrulline + NH_3	arginine deiminase	*Escherichia coli*
L-cysteine	DL-2-amino-Δ^2-thiazoline-4-carboxylic acid + $2H_2O$ → L-cysteine + NH_3 + CO_2	ATC racemase, L-ATC hydrolase, S-carbamyl-L-cysteine hydrolase	*Pseudomonas thiazolinophilum*
	β-chloro-L-alanine + Na_2S (H_2S) → L-cysteine + NaCl + NaOH	cysteine desulfhydrase	*Aerobacter aerogenes*
L-Dopa	pyrocatechol + pyruvate + NH_4^+ → L-Dopa	β-tyrosinase	*Erwinia herbicola*
L-glutamic acid	DL-hydantoin-5-propionate + $2 H_2O$ → L-glutamic acid + CO_2 + NH_3	L-glutamic acid hydrolase	*Bacillus brevis*
5-hydroxy-L-tryptophan	5-hydroxyindole + pyruvate + NH_4^+ → 5-hydroxy-L-tryptophan + H_2O	tryptophanase	*Proteus rettgeri*
L-lysine	DL-α-amino-ϵ-caprolactam (ACL) + H_2O → L-lysine	ACL racemase, L-ACL hydrolase	*Achromobacter obae, Cryptococcus laurentii*
L-phenylalanine	DL-phenylalanine-hydantoin + $2 H_2O$ → L-tryptophan + CO_2 + NH_3	L-phenylalanine-hydantoin hydrolase; N-carbamyl-L-phenylalanine hydrolase	*Flavobacterium aminogenes*
L-tryptophan	DL-tryptophan-hydantoin + $2 H_2O$ → L-tryptophan + CO_2 + NH_3	L-tryptophanhydantoin hydrolase; N-carbamyl-L-tryptophan hydrolase	*Flavobacterium aminogenes*
	indole + pyruvate + NH_4^+ → L-tryptophan + H_2O	tryptophanase	*Proteus rettgeri*
	indole + L-serine → L-tryptophan + H_2O	tryptophan synthetase	*Escherichia coli*
	indole + L-serine → L-tryptophan + H_2O	tryptophanase	*Proteus rettgeri*
L-tyrosine	phenol + pyruvate + NH_4^+ → L-tyrosine + H_2O	β-tyrosinase	*Erwinia herbicola*

[a] Ref. 17.

cemate is then subjected to the action of acylase [9012-37-7] which only hydrolyzes the acetyl-L-amino acid. The resulting mixture of L-amino acid and acetylated D-amino acid is easily separated.

A list of enzymatically made amino acids is presented in Table 2 (16) (see Amino acids).

Enzymic Modification and Antibiotic Synthesis. Penicillin acylase [9041-06-6] (also called amidase) is derived from *Escherichia coli*. It removes the side chain from penicillin to form 6-aminopenicillanic acid (6-APA) at mildly alkaline pH. At mildly acidic pH, the reaction equilibrium is shifted towards synthesis (18).

Like the penicillins, the cephalosporins are an important group of antibiotics in which various carboxylic acids are attached to the amino group of 7-aminocephalosporanic acid (7-ACA). An acylase prepared from *Bacillus megatherium,* and its use in the synthesis of Cephalexin has been described (17) and may be in use in Japan (see Antibiotics).

Penicillinase [*9001-74-5*] (penicillin beta lactamase) is produced by a number of organisms resistant to penicillin. The enzyme hydrolyzes the beta lactam ring (see Fig. 7).

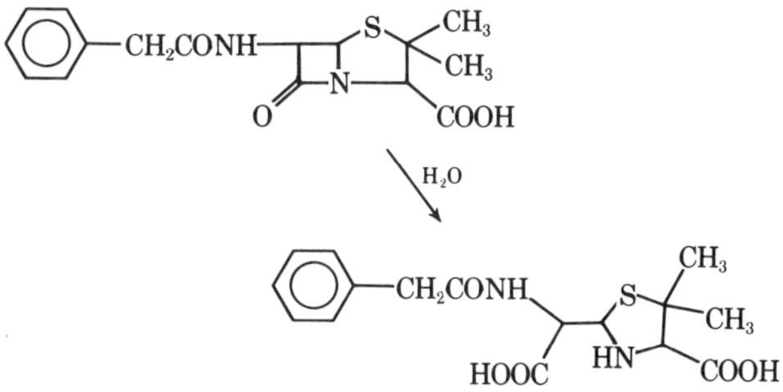

Figure 7. Hydrolysis of benzylpenicillin by penicillinase.

Penicillinase, produced from *Bacillus cereus* or *B. subtilis* is used in patients hypersensitive to penicillin and in penicillin assays. It does not hydrolyze many semisynthetic penicillins.

Important Enzymes

Lipases. Lipases are esterases hydrolyzing esters of glycerol and fatty acids. Since fatty acids are attached to carbons 1, 2, and 3, the hydrolysis of a triglycerides results in the formation of a diglyceride, a monoglyceride, and finally glycerol as one, two, and finally three fatty acid ester linkages are hydrolyzed. Thus, in the course of hydrolysis of a fat, at least three potential substrates exist: triglyceride, diglyceride, and monoglyceride (see ref. 19 for an excellent review).

Lipases show different affinities for these substrates. Thus pancreatic and *Rhizopus arrhizus* lipases substantially hydrolyze all of the triglyceride with accumulation of diglyceride before major attack on the diglyceride. They show least affinity for the monoglyceride. Other lipases preferentially hydrolyze diglycerides or monoglycerides first.

Lipases show different affinities for esters with different chain lengths. Although pancreatic lipase shows greatest activity on long-chain fatty acid (C_{12} or longer) chains, others such as pregastric lipases of young ruminants, act preferentially on short-chain fatty acid esters.

The study of lipases (and, indeed, the action of lipases) is complicated by the insolubility of natural oils and fats in water (some synthetic triglycerides, such as triacetin, the ester of glycerol and acetic acid, have slight water solubility). The enzyme attack on the fat is therefore on the surface, and it is necessary to emulsify the fat in order to approach efficient action. Nature provides bile to emulsify fats in the intestines for efficient digestion by pancreatic lipase. Other emulsifiers commonly used include gum arabic and poly(vinyl alcohol), as well as the bile salts and taurocholic and glycocholic acids.

Animal Lipases. The most important lipase is pancreatic lipase [9001-12-1]. It is almost always used as a crude dried pancreas powder blended with other materials. Proteolytic enzymes (eg, trypsin and chymotrypsin) and amylases are also present. In addition to a high degree of emulsification, a number of other factors promote the hydrolysis of fats by pancreatic lipase. Cations, particularly Na^+ and Ca^{2+} promote lipolysis. Many compatible theories have been proposed regarding Na^+. Calcium promotes the activity by tying up the fatty acids formed that otherwise would cause a back-up of the reaction, that is an inhibition of further lipolysis by the accumulated products of the lipolysis, the fatty acids. Pancreatic lipase functions best at pH 8–9.

Pancreatin [8049-47-6] is used in digestive aids, in removing tissue from hides destined for leather, in septic tank and grease trap cleaners, as enzyme-spotting agent for dry cleaners, for yolk fat removal from egg whites to promote whipping, for the development of flavor in dog food, and similar applications.

The saliva of calves and other young ruminants (kid, lamb) contains a pregastric lipase that hydrolyzes milk triglycerides in the stomach. These lipases act at lower pH than pancreatic lipase, usually at pH 6.5–7.5, and often lower. These enzymes are usually called esterases rather than lipases because they preferentially hydrolyze the short-chain fatty acid esters, releasing butyric, caproic, caprylic acid, and others, but not long-chain fatty acids. They are quite important commercially in the preparation of lipolyzed butter (partially hydrolyzed butter) and for natural flavors characteristic of certain natural cheeses such as Italian cheeses. Commercial sources are the throats of calves, kids, and lambs.

Plant Lipases. Oil seeds (like soy, corn, castor) contain lipases that are important in the synthesis of the oils. However, when these seeds are damaged (as by crushing to press out the oil), the lipases are activated and may cause rancidity in the oil. Certain oil seed lipases have been subjected to close academic scrutiny, but these lipases are not used industrially.

Microbial Lipases. A search is underway for an alkaline lipase active at elevated temperature for use in detergents (see Enzyme detergents). It is likely that by 1985 this will be a commercial reality. Alkaline lipase from *Pseudomonas* is known.

Microbial lipases are important in the analytical procedures for serum triglycerides; these methods are based on the enzymatic determination of glycerol liberated from the serum triglycerides by the action of the lipase. The major microbial enzyme used for this purpose is obtained from *Rhizopus arrhizus*, often extended with lipase from other microbial sources such as *Candida cylindraceae* or *Chromobacterium viscosum*. Other important organisms producing lipases are *Geotrichium candidum, Aspergillus niger, Mucor miehei, Mucor pusillus, Penicillium roqueforti, Corynebacterium acne, Propionibacterium shermanii*, and *Achromobacter lipolyticum*.

Pectic Enzymes. Pectic enzymes take part in the hydrolytic degradation of pectic substances.

The pectic substances universally present in plant tissues, principally in fruits, are carbohydrate derivatives. The term includes a multitude of compounds whose main characteristic is that they contain polygalacturonic acids and their derivatives as the main constituent. Galacturonic acid is formed by the oxidation of the sixth carbon of galactose; dehydration of many galacturonic acid molecules leads to the formation of polygalacturonic acid. The principal structure of pectic substances is composed of straight chains of anhydrogalacturonic acid residues, predominantly connected by (1 → 4)-glycosidic linkages. Most of the carboxylic groups are esterified, mainly with methyl alcohol. The chains vary in size, and the molecular weight of purified pectins may range from 30,000 to 300,000, depending on the source and the method of purification. Crude pectin preparations include a number of related compounds, with properties similar to pectins. On hydrolysis, arabinose, sorbose, rhamnose, and acetyl groups are frequently found besides galactose.

Because of the more than 100 terms used to name the various substances belonging to the group collectively called pectins, a *Revised Nomenclature of the Pectic Substances* was prepared in 1943. The nomenclature was adopted as official by the American Chemical Society in April, 1944 (20–21).

Pectins or pectic substances are not soluble in water but, because of their highly hydrophilic colloidal nature, they disperse in water, readily forming a very viscous liquid. They also form a semisolid jelly with sugars and acids, and a semisolid gel with small amounts of bivalent ions.

The high viscosity of the fruit juices is due to the presence of pectic substances that prevent the quick sedimentation of the dispersed particles. This is highly desirable in some instances, as in apricot nectar, or tomato and orange juices, but is objectionable in many instances as in apple juice, grape juice, and others.

The pectic enzymes acting directly on pectic substances are comprised of three enzymes, of which polygalacturonase [*9032-75-1*] (PG) and pectin methylesterase [*9025-98-3*] esterase (PE), are hydrolytic. The third is nonhydrolytic and is known as pectate lyase [*9015-75-2*] or pectin transeliminase (PTE). Some authors have classified the pectic enzymes into many more divisions, including exo- and endoenzymes in each class, reflecting whether the enzyme attacks the pectic substance chains from the ends or in the middle in a random fashion. (Exo- refers to end attacks, endo- to random attacks along a chain.) There are further differences between the pectic enzymes indigenous to fruits, for example, those responsible for softening during ripening, and the same pectic enzymes from mold (fungal) sources. However, all commercial pectic enzyme preparations are fungal; all those marketed in the United States are from *Aspergillus niger* and virtually all those imported are made by surface culture on wheat bran or other solid media. Production in the United States is shifting to submerged fermentation with very high energy (agitation) and oxygen level requirements.

Because pectic enzymes are produced on complex media with many natural substrates, the commercial enzymes contain, in addition to varying amounts of the three main enzymes, significant activities of other enzymes such as cellulase, hemicellulase, protease, and others.

Pectic substances are essentially chains of anhydrogalacturonate residues, with varying amounts of methyl ester linkages. These chains are attacked at random by PTE and are thus reduced to short pieces of 6 to 8 methoxy anhydrogalacturonate residues which cannot be attacked by PTE. Thus PTE rapidly reduces viscosity caused by pectic substances, but does not break the polymer into its monomeric constituents.

Polygalacturonase breaks the chain into monomers, but only breaks down the polymer after hydrolysis of the methyl ester. Thus PG in combination with PE breaks pectic substances down completely to galacturonic acid and methanol. In the absence of PE and PTE, PG partially hydrolyzes pectic substances. In fact, it acts on the insoluble protopectin to solubilize it and, hence, PG alone can be used to increase viscosity and stabilize the cloudiness in fruit and vegetable juices. Alone, PG is also used to separate plant cells which are held together by protopectin (22–23).

Comparison of different commercial pectic enzyme preparations shows major differences in the content of PG, PE, PTE, cellulase, hemicellulase, and protease. Some are virtually devoid of PTE, others almost devoid of PG. Yet they can function interchangeably because pectic enzymes are evaluated under actual use conditions based on the amount of enzyme for a given price, not on comparison of units. Pectic enzyme preparations vary in price from ca 2 to >200 $/kg. Yet these preparations are competitive on a cost-performance basis.

An example of evaluation of pectic enzymes for apple juice clarification illustrates why a side-by-side comparison on a split batch of juice is essential. Apple juice is pressed from crushed apples by various means that give a product containing different amounts of soluble and suspended pectic substances. The variety of apple is a factor, as is the time of the year since apples contain starch early in the season which must be removed or hydrolyzed. Later in the year, apples are replaced by storage apples which have a much higher content of soluble pectin. The amount of enzyme a particular plant uses may be ten times as high later than early in the season. The comparison test is done by treating juice under factory conditions, then periodically drawing off a sample of juice and gently mixing an equal volume of juice and 95% ethanol and observing the behavior of the precipitate formed. When depectinization is complete, the other suspended material generally settles and a clear juice is obtained. If the juice has starch in it, this starch can be filtered provided the juice has not been heated. If the juice has been heated sufficiently, the starch has been gelatinized, and it clouds the final juice unless it is decomposed by added amylase.

The insolubles in the crushed apple function as a filter aid, allowing a better yield of juice with proper enzyme-controlled breakdown of soluble pectic substances without solubilizing the protopectin through proper enzyme use. Often other colloids are needed to help in the agglomeration and settling; these may include gelatin (100–200 Bloom), bentonite (Western or sodium form) and/or colloidal silica.

The principal uses of pectic enzymes are: (*1*) production of fruit juices and fruit-juice products; (*2*) production of wines; (*3*) fermentation of coffee and cocoa beans; (*4*) rehydration of dehydrated foods; (*5*) production of galacturonic acid and low-methoxyl pectin; and (*6*) recovery and stabilization of citrus oil (see Food processing).

Production of Fruit Juice and Fruit Juice Products (24–25). Freshly pressed fruit juices (qv) contain colloidal material that keeps dispersed solids in suspension; most of it is pectin. In citrus and tomato juices, where the high viscosity is a required advantage, the pectic enzymes of the juice are destroyed by pasteurization in order to maintain the high viscosity. In other fruit juices, such as apple and grape juices, a clear flowing liquid is required where the suspended material settles quickly. For such juices, a commercial mixture of the pectic enzyme is usually added to the fruit during or after crushing; this also increases the yield.

If the hydrolysis of the pectins goes too far, the viscosity drops and many solids may precipitate and settle with the original sediment leaving the juice too flat and watery.

For making low-sugar juices or jellies for diabetics, the pectin is hydrolyzed only by pectic esterase. The resultant low-ester pectin gels in the presence of Ca^{2+} ions. In place of sugars, sorbitol is added as preservative and nonassimilable sugar substitute.

Jelly manufacturers destroy the native pectic substances in order to obtain a clear juice, and then make the jelly by the addition of commercial pectins, acids, and sugars. In the production of jellies, polygalacturonic acid and Ca^{2+} ions are added to the clarified juice. High concentrates with reduced viscosity can be made from prunes.

Wine Production. In making wine (26), the addition of pectic enzyme preparations offers a number of processing advantages at different stages of the operations (see Wine). When the enzyme is added to the crushed grapes, it increases the volume of the free-flowing juice, reduces the pressing time, and increases the final yield of the juice. The addition of the enzymes increases the extraction of color when the grapes are heat-extracted or fermented on the skins. Pectic enzymes may be added to the must before fermentation to settle much of the suspended particles, including part of the unwanted microorganisms. Alternatively, the enzyme preparation may be added during fermentation. In either case, the yeast sediment (lee) will be firmer, more compact (less voluminous) at the time of the first filling, and the wine clearer. Finally, the enzyme may be added to the fermented wine to increase the filtration rate and to produce a clearer wine. A good clarification eliminates the floating microorganisms and much of the precipitated proteins. The reduction of protein improves the stability of the wine.

The processing phase at which the enzyme should be added depends on the quality of the grapes and local conditions. The wine maker can decide on the best treatment, and must make sure that the taste of his product is not changed for the sake of clarity. Depending on the brand of enzyme preparation, the recommended quantities may vary. The commonly used ratios are 50–100 g enzyme preparation per metric ton of grapes or must or wine.

It is claimed that enzyme treatment does not change the characteristic bouquet of wines, but helps to develop a more mature flavor in a shorter time, and that the flavor of treated wines is superior to the flavor of untreated wines. Port wines made from enzyme-treated must fermented on skins contain a greater amount of color. It is further claimed that fruit wines made of enzyme-treated blackberries, currants, loganberries, and other berries, peaches, apples, and other fruits are superior to wines made of untreated fruit juices.

Pectic enzyme preparations are active at 0–65°C. A typical preparation is destroyed in 40 min at 40°C, in 5 min at 82.5°C, and in 2 s at 99°C.

An added effect is the increase in methanol content. The methanol content of wines made of enzyme-treated juices is higher than of wines made from untreated juices; however, it is still well below the permissible level. For instance, an untreated red wine may contain 0.019% methanol, and the enzyme-treated wine 0.023%.

Other Uses. Citrus oils, especially lemon oil, are made with pectic enzymes. The oil is collected from the lemon peel as an emulsion in water, and pectic substances keep the oil in emulsion. Pectic enzymes destroy the emulsifying ability of the pectic substances, releasing the oil and greatly improving its yield. Instead of enzymes, surfactants can be used to break the emulsion.

Coffee beans, as harvested, are the seeds in the coffee cherry. To obtain the seed from the fruit (cherry) the pulpy layer must be removed (27). At least three-fourths of the dry pulpy layer is pectic substance. Traditionally, it is removed by fermentation after the fruit is carefully crushed to break up the pulpy layer without damaging the seed. Pectic enzymes are used to a small extent to hasten the pulping to avoid microbially caused off-flavors that may affect the bean (see Coffee).

Lysozyme. Lysozyme [9001-63-2], a mucopolysaccharidase, is a globulin protein consisting of a single polypeptide chain of 129 amino acid residues cross-linked by four or five disulfide bridges. The enzyme is stable between pH 2.8 and 11.8, and is most soluble at pH 4.5. Its molecular weight is 14,400.

The enzyme as found in egg whites, animal bodies, and plant tissues is present in high concentration in tear fluid, mucus, heart, spleen, and liver. Besides its ability to lyse certain bacterial cells, it is active in precipitating insulin, acting on nucleic acids, mucus, etc. Trypsin, chymotrypsin, and papain have no effect on lysozyme activity.

Lysozyme, produced from chicken egg whites, is commercially available as the isoelectric product and as chloride salt for greater solubility. It is principally applied in scientific research and clinical practice. It is being used, or suggested to be used in the treatment of cancer, virus diseases, eye infections, blood diseases, infectious postoperative complications, hemorrhagic conditions, varicose ulcers, multiple sclerosis, and as a food preservative; in Europe lysozyme is sold as an over the counter "cure-all."

Hyaluronidase. Hyaluronidase [9001-54-1], a transglucosidase which also hydrolyzes the mucopolysaccharide hyaluronic acid is present in snake venom, leeches, in some pathogenic bacteria, and in the testicles of mammals. Commercial hyaluronidase preparations are made from beef seminal vesicles. Hyaluronic acid in tissue acts as a barrier to diffusion. Hyaluronidase reduces the viscosity of hyaluronic acid, and thus facilitates the diffusion and absorption of subcutaneous injections. The optimum pH is 5.3 and the optimum temperature is 37°C. The enzyme is destroyed at a pH below 3.3 and above 7.0, and also at temperatures above 60°C.

Hyaluronidase is used together with local anesthetic agents in surgery and dentistry, in insulin-shock therapy, and with injections of antibiotics, adrenaline, and heparin.

Anthocyanase. An enzyme of fungal origin, anthocyanase [54427-02-0] is capable of destroying (decolorizing) the pigment that gives the color to berries like blackberries and red grapes. The anthocyanin pigment is present in blackberries in such an excess that jams and jellies made from it are dark to the extent of being unattractive. Furthermore, the excess pigment in blackberry wines frequently precipitates on the neck and bottom of the bottles, making an otherwise good product unsalable. The excess anthocyanin may be removed, and attractive jellies, jams, and wine may be produced by treating the blackberries with anthocyanase (0.1 g enzyme preparation per 100 g juice for 8 h or less at 26°C).

In order to produce white wines or light-red wines from red grapes, it is claimed that the red color may be partially or completely removed by this enzyme (the normal commercial practice to make white wines from red grapes is to remove the skins containing the pigment before fermentation.

Naringinase. Naringinase [9068-31-9] hydrolyzes naringin, a glucoside, in two steps: first it splits off rhamnose, resulting in another glucoside, prunin, then it splits off glucose resulting in naringenin.

Naringin, 4',5,7-trihydroxyflavanone 7-rhamnosido-β-glucoside ($C_{27}H_{3.2}O_{14} \cdot 2H_2O$), present mainly in the albedo and membranous tissues of grapefruit, is responsible for the bitter taste of grapefruit juice. Its presence in 0.07% quantities is objectionable, whereas 0.05% is acceptable. The hydrolytic products, prunin and naringenin, are not bitter.

A naringinase preparation in 0.01% quantities debitters grapefruit pulp and juice to acceptable levels in 1–4 h at 50°C and at a pH of 3.1.

Naringinase, present in several plants, is produced commercially from *Aspergillus niger*. Its pH optimum is 3.5–5.0, and the temperature range is 25–50°C.

Carbohydrases. Carbohydrates constitute the largest group of naturally occurring organic materials. More than 50% of the dry substance of all plant material is made up of carbohydrates, which include cellulose, hemicellulose, starches, and complex and simple sugars (see Carbohydrates).

Invertase (28). Also called sucrase, or saccharase, invertase or β-fructofuranosidase, derived from *Saccharomyces cerevisiae* is one of the simplest commercial carbohydrases. It catalyzes the hydrolysis of the β-D-fructofuranosyl linkage in sucrose, raffinose, gentibiose, methyl- and β-fructofuranoside. Invertase produces invert sugar by hydrolyzing sucrose into glucose and fructose. Invert sugar is widely used in confectioneries, candies, syrups, cordials, ice cream, and sweets because it is sweeter than sucrose and begins to crystallize (harden) at much higher concentrations than glucose or sucrose syrups. Liquid- and soft-center candies are made with sucrose and invertase. The sucrose makes a hard enough mass to cover with chocolate; after 3–4 weeks the invertase has converted enough sucrose to the more soluble glucose and fructose to give the liquid or soft center.

Amylases. The most important carbohydrases, produced in larger quantities than any other enzyme, belong to the group of amylase or diastase enzymes, which hydrolyze starch and its hydrolytic degradation products. Although all amylases hydrolyze the D-glycosidic linkage, they differ in many respects. Most of them hydrolyze only linkages between the carbons 1 and 4; others, in addition to the (1 $\rightarrow$ 4) linkage also hydrolyze bonds between carbons 1 and 6.

α-Amylase [9000-90-2] is more active in hydrolyzing larger molecules, cleaving them at random close to the middle of a long glucose chain; β-amylase [900-09-3] splits off maltose molecules at the nonreducing end of a chain; and glucoamylase splits off single glucose units, attacking at nonreducing ends of both long and short chains, and cleaves both (1 $\rightarrow$ 4) and (1 $\rightarrow$ 6) linkages. The various amylases differ also in their thermal deactivation temperature, pH optimum, and other characteristics. The individual amylases alone, or in certain combinations, are used for different specific purposes.

α-Amylase, present in pancreas, saliva, plants, molds, and bacteria, hydrolyzes starch, glycogen, and dextrins. Before attack by amylase, native starch must be gelatinized by heating a starch slurry. Potato and most other starches gelatinize at moderate temperatures, but corn starch requires higher temperatures of about 75°C. On gelatinization the viscosity rises very sharply. α-Amylase is also called the liquefying enzyme because by randomly cleaving the large starch molecules and the long chains of dextrins formed by the primary and secondary cleavages, it rapidly liquefies the thick, starchy paste to a thin, free-flowing solution. It also slowly produces some maltose and possibly glucose. α-Amylases vary in some characteristics, depending on their origin. Some α-amylases of microbial origin are still active above 100°C in the presence of substrate, whereas the malt β-amylase is rapidly destroyed above 63°C.

The importance of α-amylase is twofold. If the purpose of the starch conversion is to only liquefy the starch with the production of dextrins, (eg, for the manufacture of glue, starchy syrups, or other food ingredients) α-amylase alone suffices. Consequently, for such purposes industrial enzymes containing principally α-amylase, with no or little content of other amylases are manufactured.

For manufacturers of glucose from corn starch by the enzyme/enzyme process, the gelatinized starch is first liquefied with α-amylase and then broken down almost completely to glucose by glucoamylase.

α-Amylase facilitates the action of other amylases. If the purpose of the starch conversion is to produce maltose, a sugar fermentable by yeast, β-amylase alone would do the job very slowly and incompletely by splitting one maltose molecule at a time from the nonreducing end of the long chains of the starch molecule. The added β-amylase would not mix with the thick starch paste without the liquefying action of α-amylase.

The complete starch molecule has only a few chain ends, and the action of β-amylase stops at the branching points of the chains, the enzyme being incapable of splitting (1 → 6) linkages occurring at the branchings. α-Amylase, by rapidly liquefying the starch paste, not only accelerates the physical mixing of the saccharifying β-amylase with the substrate, but produces many short chains, with many ends, where the saccharifying enzyme can act quickly. α-Amylase breaks the chains also between the branches, making those portions of the main chains available for the action of the saccharifying enzyme. Therefore, the presence of a small amount of α-amylase accelerates the saccharification of starch by other amylases.

Conversion of grain starch into dextrins and maltose is the purpose of the mashing (brewing) step in the production of beer, ale, and malt champagne (see Beer). The possible complete conversion of starch to maltose is the main purpose in preparing the distillers' grain mash for the production of grain neutral spirits and whiskies (29) (see Beverage spirits, distilled).

In both processes malt provides the needed enzyme mixture. Brewers' malt is produced and used for beer, distillers' malt for whiskey, and gibberellin-treated malt for grain neutral spirits. The principal difference between these malts is in the α-amylase content, which is the lowest in brewers' malt (25–30 units) at 20°C, about twice as high in standard distillers' malt (50–60 units), and the highest in gibberelin-treated malt (80–100 units).

The saccharifying enzyme, β-amylase, usually increases the α-amylase content in the malt. In brewing malts, the saccharifying enzyme is always measured. In distillers' malt, only the α-amylase is determined because usually this is the relative minimum.

In preparing starch glue or brewers' mash, the conversion of the starch is usually terminated by heating to destroy the enzyme.

In distillery practice, care is exercised to protect the continued activity of the enzyme during the fermentation process. With a relatively quick conversion, the enzymes are exposed only for a period of ten minutes or less to 60°C. During this period, a sufficient amount of maltose is formed to promote the development of the yeast and a strong fermentation. As the fermentation progresses, the maltose–dextrin equilibrium changes, and if sufficient enzyme remains, the conversion progresses to completion. The progressive decrease in pH slows down this secondary conversion during fermentation. The conversion is inactivated at a pH of 4 or below (see Fermentation).

The active substance of the malt is water soluble, and the bulk of the enzymes may be obtained in a cold aqueous solution containing about 25% of the barley malt. Concentrates may be prepared from malt by evaporating such extracts. High diastatic malt extracts with considerable maltose content are marketed for many purposes such as brewery and bakery adjuncts.

Commercial α-amylase preparations of microbial origin (*A. niger* and *B. subtilis*), practically free from other amylases, are offered for the production of adhesives from starch, and the liquefaction of grain mashes in distillery practice. They are employed in the production of dextrose from starch together with another enzyme preparation containing principally glucoamylase. α-Amylase splits the large molecules of starch and dextrins, and the glucoamylase splits off single glucose molecules from the nonreducing ends of the chains.

β-Amylase, though widely distributed, is principally found in germinating seeds (malt) and produces maltose from starch, glycogen, and dextrins by successively removing maltose units at nonreducing chain terminations. It occurs in great abundance in brewers', distillers', and gibberellin-treated malts, and is the principal saccharifying enzyme in barley malt. Most commercial amylases and many protease products contain some β-amylase. Pure β-amylase is used as an analytical reagent for measuring α-amylase activity in various enzyme sources, principally in malt.

Glucoamylase (or amyloglucosidase) produces glucose by removing glucose molecules in succession from the nonreducing chain terminations from starch, glycogen, dextrins, and maltose molecules. It also catalyzes transglycosylation. It is present in blood, molds, and bacteria. Barley malt is low in glucoamylase. Commercial glucoamylase preparations are produced from molds.

Glucoamylase is used by several distillers in conjunction with gibberellin-treated malt in the conversion of grain starch to fermentable sugars. Its main application is the production of glucose syrup, glucose paste, and crystalline glucose from starch in the dual-enzyme conversion process.

In the manufacture of dextrose by the dual-enzyme process, a 30–35% starch–water suspension is prepared, the pH is adjusted to 5.5, and α-amylase dissolved in water is added. The mixture is heated under agitation to 85°C, and is kept at that temperature for 40 min. The starchy paste is well liquefied. The mixture is then cooled to 60°C, the pH is adjusted to 4.5 with hydrochloric acid, and glucoamylase is added in form of an aqueous suspension or solution. The starch–enzyme mixture is kept at 60°C for 72–96 h. The dextrin equivalent (DE) is ca 98 (this means that 98% of the total solids is reducing sugar calculated as dextrose).

This process is more economical and results in a better product than either the acid conversion or the acid–enzyme dual conversion.

In the conversion of any starchy product to dextrose, the starch content of grains and by-products derived from processing grains may be converted to glucose by the dual-enzyme process. Such products may be utilized by the feed industry or in various fermentation industries which use glucose as a substrate or as an energy source.

Amylo-1,6-α-glucosidase [9012-47-9] (debranching enzyme) hydrolyzes (1 → 6)-α-linkages, the branching-off linkage in starch, glycogen, and dextrins. It is present in animal tissues, plants, yeast, and other microorganisms. The enzyme is used in two distilleries to convert the starch content of the mashed grains.

Oligo-1,6-α-glucosidase [9032-15-9] or limit dextrinase, present in intestines, molds, and in limited quantities in malt, hydrolyzes (1 → 6) α-linkages in isomaltose, panose (maltotriose), and limit dextrins.

Isomaltase [9032-15-9] and *maltotriase* [9076-78-2] hydrolyze isomaltose and maltotriose, respectively.

Maltase and *α-glucosidase* hydrolyze maltose into two molecules of glucose.

Some authors deny the existence of the last five enzymes listed, as well as the existence of transglucosidase, and attribu' the enzymic activity ascribed to them to the enzyme glucoamylase (amyloglucosidase).

Uses of amylases. It can be assumed that the amylase complex contains even more than the individual enzymes discussed above. The production of glucose may be explained without the existence of the last three enzymes by the action of glucoamylase. Glucoamylase, however, catalyzes transglycosylation reactions with the formation of $(1 \rightarrow 6)$ linkages. On the other hand, there are industrial enzymes that produce glucose almost free from transglycosidase activity. Furthermore, some microorganisms can assimilate and ferment maltose, but cannot ferment maltotriose, dextrins, or starch.

Because of the convenient and relatively cheap production of glucose, the cheapest sugar made from starch, the demand for the production of industrial amylase preparations is in constant increase. The value of amylase preparations exceeds the value of all other enzyme preparations produced in the United States.

Amylases are used in considerable quantities in feed formulations and in limited quantities in pharmaceuticals as dietary aids.

Other amylases. Other enzymes that hydrolyze glucosidic linkages in substrates other than starch or its degradation products are cellulase, lichenase [37288-51-0], inulase [9025-67-6], xylanase [37278-89-0], cyclohexagluconase [9025-68-7], cycloheptagluconase [9025-69-8], chitinase [9001-06-3], polygalacturonase [9032-75-1], lysozyme, α-1,3-glucosidase, α-glucosidase, β-glucosidases, α-galactosidase, trehalase [9025-52-9], and β-galactosidase. Other enzymes may be present in considerable or in trace quantities in commercial amylase preparations.

Lactase a β-galactosidase that hydrolyzes lactose into glucose and galactose, is present in bakers' yeast, in many Aspergilli, and in the digestive secretions of mammals. Lactase is utilized in the manufacture of ice cream to prevent lactose crystallization, in whey to produce more fermentable sugars, and in milk to prepare a product acceptable to those with lactase insufficiency.

Transglucosidase [9031-48-5] is a contaminant in enzyme preparations and is undesirable. Much of the patent literature on glucoamylase deals with removal of transglucosidase.

Pullulanase, isoamylase. In producing maltose from starch the 1,6-bonds of the amylopectin are hydrolyzed by pullulanase or isoamylase [9067-73-6]. Pullulanase is produced by *Klebsielle aerogenes* (30). It hydrolyzes 1,6 linkages on a straight chain (as in pullulan) and at branching points (as in amylopectin or limit dextrin). Isoamylase, produced by *Pseudomonas deramora* hydrolyzes the branch-type 1,6 linkages, but not the straight chain type.

Raffinase [57657-61-1], *Mellibiase*. The α-galactosidase from *Mortierella ovinacea* decomposes raffinose in beet molasses to galactose and sucrose and thus improves the yield of crystalline sucrose since raffinose inhibits the crystallization of sucrose.

β-Glucanase from *Bacillus subtilis,* and *Aspergillus niger,* attacks the 1,3-β and 1,4-β linkages in yeast cell walls and barley. The barley β-glucans are solubilized at 60–65°C, the temperature at which starch is gelatinized in mashing for beer production. At this temperature the β-glucanase present in barley is destroyed; addition of microbial β-glucanase reduces viscosity and facilitates filtration of the mash.

Maltose production by malt β-amylase is likely to be replaced or augmented by β-amylase from *Bacillus circulans*, though this process is not commercialized yet. This β-amylase gives a 60% conversion of starch to maltose; if used with pullulanase to attack the 1,6 branch linkages, yields of 80–90% are achieved. High maltose syrups are in demand for the candy industry.

Other enzymes, such as xylanase, hemicellulase, and mannanase [60748-69-8] are not of commercial importance. However, the trend to making alcohol for use as a motor fuel and as a raw material for single cell protein and other chemicals is moving ahead rapidly because it involves the use of plant materials, a major renewable resource (as compared to fossil fuels). This will increase the importance of many enzymes.

Glucose isomerase, also known as xylose isomerase, rose to prominence in the sucrose shortage of the early 1970s. This enzyme converts glucose to fructose (31). The early processes yield a product containing only about 42% fructose (dry basis) used much like invert sugar, but now 90% conversion is obtained.

Glucose isomerase is produced by many organisms. However, many cannot be used commercially because of the cofactors required, such as arsenate or cobalt. Glucose isomerase is used in immobilized form. Cobalt or magnesium ions are removed by deionization after processing.

The form of the enzyme used varies from entrapped whole cells to virtually crystalline enzyme bound on ion-exchange resin. A number of commercial glucose isomerases from different organisms are available (see under Uses).

Cellulase. A good cellulase preparation has a great economic potential for the hydrolysis of waste and other cellulosic materials to glucose (fermentable). Commercial cellulases are available. They work well on cellulose derivatives, like carboxymethyl cellulose, but do not work or only very slowly on native cellulose. Commercial cellulases are made from *Aspergillus niger* for food use, and from *Trichoderma viride* for non-food uses. At present, the only use is in digestive aids.

Proteases. No single classification has been found suitable for proteases. The following schemes are used:

(*1*) Source (plant, animal, fungal, bacterial).

(*2*) Exo vs endo. Amino exopeptidases attack the end with a free amino group, whereas carboxy exopeptidases attack the end with the free carboxy group. Endo peptidases act on peptide bonds in the interior of the chain, and are further separated according to the nature of the peptide bond broken down.

(*3*) Nature of the active site. This classification is based on the specific inhibitors. Thus, serine proteases are those inhibited by diisopropylfluorophosphate (DFP), and are all endopeptidases, including trypsin, elastase, chymotrypsin, and subtilisin.

Sulfhydryl enzymes are known to have an SH group at the active site. They are affected by metal ions that bind at the SH (thiol) group, and by oxidizing and alkylating agents that also bind at the SH site. This includes the plant proteases (papain, bromelain, ficin) and some microbial proteases.

Metalloenzymes are those whose activity depends on the presence of a metal ion, such as Mg^{2+}, Zn^{2+}, Co^{2+}, Fe^{2+}, Hg^{2+}, Cd^{2+}, Cu^{2+}, Ni^{2+}. In some cases, the metal ion can be removed by sequestering agents such as EDTA; in these instances, addition of the appropriate metal ion restores activity. Cyanide generally inhibits these enzymes. This group includes carboxypeptidase A, some aminopeptidases, and some bacterial proteinases.

Acid proteases such as rennin, pepsin, and some fungal amylases active at pH 2–4 are inhibited by p-bromophenacylbromide or diazo reagents.

(4) pH optima. This relates to the property of the enzyme and the condition of best use. Thus acid proteases are those that should be used for milk clotting, neutral proteases for barley protein hydrolysis in brewing, and alkaline proteases for detergent use.

The assay of proteases presents particular problems. The action of exopeptidases produces amino acids, whereas the endopeptidases produce proteoses and peptones. All proteases cause a generation of free amino and carboxyl groups. The substrates used for protease assay are both natural and synthetic; natural substrates are mainly milk, casein, gelatin, and hemoglobin (one test for papain for use as a meat tenderizer actually uses hamburger); synthetic esters include benzoylarginine ethyl ester and benzoyltyrosine ethyl ester.

All proteases acting on milk first cause it to clot. Milk-clotting units, originally used for papain, use a reconstituted dry milk preparation, with the tube rotated at about a 60° angle to provide a thin film at the surface, which is examined under bright light for the formation of clots. The method is highly reproducible. Enzyme concentration is inversely related to clotting time.

When a complex protein is acted on, it is possible to separate the substrate from the products by precipitating the unattacked substrate with trichloracetic acid. The solubilized portion can then be examined by measuring absorbance at 280 nm to determine solubilized aromatic (eg, tyrosine-containing) groupings, or by reaction with Folin–Ciocalteu phenol reagent and reading at 700 nm. Small peptides can be determined by Lowry's reagent, an alkaline copper solution.

The free carboxyl groups can be determined by stopping the reaction with alkaline (pH 8.5) formalin, and back-titrating the reaction mixture to pH 8.5; the more carboxyl groups liberated, the more alkali is required (formol titration method).

The free amino groups can be determined by 2,4,6-trimethylbenzenesulfonic acid at 420 nm.

Other methods are turbidimetric, based on reduction of turbidity on protease action. In the photographic film digestion method, the protease attacking the gelatin releases some of the silver, this is followed by optically measuring the density of silver remaining. Pepsin and pancreatin proteolytic activity are sometimes given in N.F. units based on digestion of coagulated egg white.

The individual enzymes vary widely in their specific activities, in their optimum pH and pH range, in heat sensitivity, and in the presence of significant other activities. The recommended optimum for both temperature and pH for proteases covers a wide range, depending on the origin of the enzyme and on its use.

The commercial preparations, which usually contain a number of individual proteolytic enzymes, are recommended for specific purposes on the basis of practical experiments carried out by the producers, who select the proper type (origin) of enzyme for every purpose. The practical utilizations include cheese making, meat tenderizing, bread baking, haze elimination from beer (and other beverages), preparation of digestive aids, cleaning of food spots from garments, preparations of pharmaceuticals, surgical applications in wound cleaning, solubilizing protein, and others.

Plant Proteases. *Papain.* Papain is the principal enzyme for beer chillproofing, meat tenderizing and for softening wheat gluten for crackers. The consumption for 1980 is estimated at ca 600 tons papain latex. Papain is derived from the green fruit of *Carica papaya*. It takes ca 9 months for a tree to begin to bear fruit, and ca 3 years for the tree to grow too tall for collecting latex. The latex is collected by making incisions in the unripe fruit during the early morning hours, when the fruit is liquid-laden, allowing the latex to run out and congeal on the surface of the fruit; the congealed latex readily comes off the waxy fruit coating and is collected. The latex at this point is thixotropic, and reliquefies on stirring. Traditionally it has been sun-dried, but a better quality crude product is obtained by tray drying the latex at 55°C in a crude forced-air dryer right at the plantation. The crude latex yield is ca 500 grams per tree per year, corresponding to about 130 grams of dried latex. In one process the collected coagulated latex is promptly liquefied, filtered free of insect fragments, and spray-dried to give a superior product. The crude dried latex must be extracted, filtered, and reprecipitated with ethanol to yield refined papain suitable for use in foods. As a sulfhydryl enzyme, the activity is often protected by the addition of sodium bisulfite. During assay the enzyme is usually activated by cysteine.

Papain tends to cleave the peptide at the arginine or lysine carbonyl group. Commercial papain usually also contains chymopapain, which preferentially cleaves peptide bonds where the carbonyl group is from aspartate or glutamate. Papain is very active over the pH range 3 to 9; it chillproofs beer at pH 4.5, and tenderizes meat at ca pH 7.0. It is the enzyme of choice as a tenderizer because it can be injected antimorten into the animal for wide distribution of the enzyme or applied to the surface of meat and remain relatively inactive until the meat is cooked. It acts well even at 70°C.

In addition to the uses given above, papain is used in digestive aids, wound debridement, protein hydrolysates, liquefying protein waste products, tooth-cleaning powders, and for the recovery of silver from used film.

Papain is imported from Sri Lanka, Zaire, and Central America and refined in the United States.

Bromelain. Bromelain (bromelin) is a plant protease manufactured from the hard stumps left after harvesting pineapples. The stems are collected, cleaned, or peeled (to eliminate fungal oxidases that inactivate bromelain), and crushed in mills used to crush sugar cane (Cuba mills). The juice is then pressed from the crushed stumps and precipitated with solvents such as acetone, ethanol, or isopropanol. Bromelain is also present in the raw fruit and juice; this is why raw pineapple cannot be used in gelatin desserts without breaking down the gel. Bromelain preferentially cleaves peptide bonds where the carbonyl group comes from a basic amino acid like arginine, or an aromatic amino acid like phenylalanine or tyrosine.

Bromelain has properties and uses much like papain, except it is inactivated at a much lower temperature. It is produced domestically in Hawaii; the pineapple variety grown in Puerto Rico is not used for bromelain. Considerable bromelain is imported from Taiwan. Bromelain consumption in the U.S. is probably under 25 tons per year for all uses, compared to over 400 tons of papain.

Ficin is derived from the latex of wild fig trees (*Ficus glabrata*) in Peru. The latex is collected in much the same manner as maple syrup sap. However, in the mid-1970s the native collectors found that much more latex could be collected by cutting down the mature trees. As a result ficin became virtually unavailable in 1975–1977 and began reappearing in only small quantities in 1978–1979. Ficin preferentially cleaves peptide linkages where the carbonyl group is from phenylalanine and tyrosine. It is very similar to bromelain in properties and uses.

Pancreatin is a proteolytic/lipolytic/amylolytic preparation made from hog and beef pancreas. The proteolytic activity is derived from its content of chymotrypsin and trypsin. It is used in spot removers, leather manufacture, digestive aids, and the like.

Chymotrypsin derived from animal pancreas, preferentially hydrolyzes peptides where the carbonyl group is from tyrosine, tryptophan, or phenylalanine. It is most active at pH 8–9. The purified enzyme has medical uses in asthma.

Trypsin, also from animal pancreas, preferentially cleaves peptide linkages where the carbonyl group is from arginine or lysine. Like chymotrypsin, it is most active at pH 8–9.

Bacterial alkaline proteases. Subtilisin, derived from *Bacillus licheniformis* and other *Bacillus* species, is used mainly in detergents, leather tanning, for protein hydrolysates, silver recovery from film, and in brewing (as part of the *B. subtilis* cooker enzyme). Although active over the pH range 6 to 11, the optimum is at pH 8–10; optimum temperature is at 55–65°C. This preparation often contains a neutral protease, amylase, and other enzymes. World consumption has been estimated at 500 t of enzyme protein annually.

Acid Proteases. The acid proteases include the milk-clotting enzymes used in cheese, pepsin from stomachs of adult pigs and cattle, and rennin (rennet) from the stomachs of calves and other suckling animals, and microbial rennets. Also included is a fungal acid protease derived from *Aspergillus oryzae* used in chillproofing applications and protein hydrolysates; this protease does not clot milk. The acid proteases work best at pH of ca 4 to 6, except for pepsin, which has an optimum pH of ca 2.0.

Neutral proteases are produced by both bacterial (*Bacillus subtilis, Bacillus licheniformis*) and fungal (*Aspergillus* sp.) sources in conjunction with acid or alkaline proteases. The *Bacillus subtilis* enzymes are metallo enzymes requiring zinc. Calcium ions act as stabilizers.

Glucose Oxidase. Glucose oxidase or β-D-glucopyranose aerodehydrogenase, a typical aerobic dehydrogenase enzyme produced by *Aspergillus niger, A. oryzae,* and *Penicillium notatum,* oxidizes glucose to gluconic acid in the presence of molecular oxygen.

$$\text{glucose} + O_2 + H_2O \xrightarrow{\text{glucose oxidase}} \text{glucono-}\delta\text{-lactone} + H_2O_2$$
$$\downarrow H_2O$$
$$\text{gluconic acid}$$

The hydrogen peroxide formed is decomposed by the enzyme catalase, which is either produced by the same organism, or is added to the commercial glucose oxidase preparation.

$$2 H_2O_2 \xrightarrow{\text{catalase}} 2 H_2O + O_2$$

The enzyme is almost specific for glucose which it oxidizes 400 times faster than other sugars. Its antibacterial property is attributed to the H_2O_2 formation. In the presence of catalase, owing to decomposition of the H_2O_2, there is no antibacterial action.

At room temperature the optimum pH is 5.5, with an active range of pH 3–8.5. The enzyme is destroyed (90%) within 2 min at 80°C, 3 min at 70°C, and 10 min at 65°C.

The prosthetic group of the enzyme is alloxazine-adenine dinucleotide or flavine-adenine dinucleotide [146-14-5] (FAD). It is a dehydrogenase (uses oxygen as hydrogen acceptor) rather than an oxidase enzyme.

First marketed as a commercial enzyme in 1952, glucose oxidase is obtained from the cells of *Aspergillus niger* cultivated in submerged culture.

The practical application of glucose oxidase preparations is based on the removal of oxygen from beverages or from the air space in a closed food container, or on the removal of glucose from a food ingredient or food product.

The presence of oxygen may change the flavor and color of a product and may hasten the corrosion of cans containing carbonated beverages, etc. The presence of glucose may cause darkening in some foods when drying. Glucose oxidase is used when the removal of either oxygen or glucose is desirable. For example:

(1) In the production of dried egg powder, a small amount of glucose reacts with the egg proteins causing maillard darkening of the product and some loss in flavor. Oxidation of the glucose with glucose oxidase before drying the egg preserves the flavor and prevents darkening of the product. Glucose oxidase/catalase, together with some hydrogen peroxide, is added to the egg. Oxygen is supplied to the reaction as liberated from H_2O_2 by the catalase. The use of 25 mL glucose oxidase preparation per 500 kg egg white (or 750 mL per 500 kg whole eggs) at 30°C, with the frequent addition of calculated amounts of 35% H_2O_2, is recommended.

(2) Glucose oxidase with low cellulase content, added to orange soft drinks, preserves the freshness of flavor. The oxygen is removed by the enzyme from the head space, as well as from the liquid in bottled drinks. About 2 to 12 manometric glucose oxidan units of enzyme per 350-mL (12-oz) bottle is added.

(3) Added to canned beverages, glucose oxidase preparations impede fading of sensitive colors, and retard iron pickup.

(4) Oxidative deterioration of dehydrated foods such as milk powder, cake mixes, encapsulated flavors, etc, is prevented, and the shelf life of the products is extended by the use of glucose oxidase-catalase.

(5) Glucose oxidase preparations preserve the freshness of salad dressings and mayonnaise.

(6) Wrappers for cheese are coated on the inside with glucose oxidase and glucose to prevent growth of aerobic organisms on the surface of the cheese by total removal of oxygen. Process cheese darkening by oxygen is also prevented.

(7) Glucose oxidase is used as an analytical tool in clinical diagnosis in determining the sugar content in blood and in urine.

Lipoxidase. Lipoxidase is added to bread doughs in the form of enzyme-active soy flour at ca 0.7% of the wheat flour used and results in a whiter bread. The wheat pigments are bleached by the H_2O_2 formed by the action of the lipoxidase on wheat lipids.

Peroxidase. Peroxidase catalyzes the oxidation of a substrate by hydrogen peroxide. It is used in analytical chemistry to detect hydrogen peroxide formed by other enzymes, such as cholesterol oxidase [9028-76-6] acting on cholesterol, etc. The commercial source is horseradish root. Peroxidase from milk, so-called lacto peroxidase, is used to introduce radioactive iodine into protein molecules as tags for radioimmunassays, generally in conjunction with glucose oxidase to form H_2O_2 *in situ*.

Catalase. Catalase catalyzes the decomposition of H_2O_2 to water and oxygen. It is a conjugated protein found in most living cells and has been prepared in crystalline form.

Microbial catalase is commercially produced from *Aspergillus niger* at 1000 Baker units/mL. Liver catalase is commercially produced from slaughtered animals in liquid form (100 Keil units/mL).

Besides its use in conjunction with glucose oxidase and other applications, it is employed in the cold sterilization of milk. By the flash method, milk is sterilized by the addition of 0.02% H_2O_2 (135 mL of 35% H_2O_2 per 450 kg milk, and heating by high-temperature short-time, plate pasteurizer (HTST) to 49–54°C for 25–30 s).

The hydrogen peroxide is decomposed 30 min after its addition to the milk by the addition of catalase. The use of 12 mL liquid catalase preparation per kg of 35% H_2O_2 is determined by suitable tests before the milk is used for making various cheeses.

Fungal catalase acts at very low pH, 2–8, whereas liver catalase does not function below ca pH 5.5. It also has higher temperature tolerance. Further applications are in the peroxide treatment of whey, and removing residual peroxide from bleached wood and also human hair.

The commercial unit for liver catalase is the Keil unit, based on initial reaction rate. The Baker unit is used for fungal catalase; it is based on an exhaustion method.

Economic Aspects

Worldwide consumption of industrial enzymes amounts to ca $290 million; about half of this is consumed in the United States. Alkaline protease for detergents and amylases and isomerase for corn syrup production are the major uses. World production of papain is ca 600 t/yr, selling at $25–60/kg based on activity, or ca $30 million for the crude dried latex. Enzymes used in leather making, mainly based on pancreatin with some microbial enzymes added, probably total ca $30–50 million annually. Microbial enzyme consumption is estimated at ca $210 million for 1980. Figures on consumption of some enzymes, such as glucose isomerase and other enzymes used by the giant corn processing industry, are difficult to estimate because most of the production is for internal use (ca 70%). These estimates exclude malt, malt product, and yeast used for leavening. A rough breakdown of worldwide enzyme sales is given in Table 3. Published figures on enzyme sales are difficult to interpret because the level is not clear. Obviously there is a substantial difference between the gross sales volume of enzymes used in baking, eg, at the enzyme component sale level as compared to the formulated tableted product. Double counting is also a problem, where enzymes are resold (see also refs. 32–33).

Table 3. Estimated Worldwide Enzyme Sales, 1980

Enzyme	Millions of dollars
alkaline protease	80
glucose isomerase	45
leather bating	30
papain	30
rennets (animal and microbial)	30
glucoamylase	25
amylases (other than glucoamylase)	21
pectinase	6
bromelain	3
all others	20
Total	290

Commercial Information, Labeling

A few specific data usually listed in the manufacturer's descriptive literature and sales bulletins are needed by the user of industrial enzymes: (1) trade name; (2) name of principal enzyme and of additional enzymes, if present; (3) principal and possible side reactions catalyzed; (4) physical appearance (powder, liquid, syrup) and color; size and types of containers; (5) solubility of the active component in water and occasionally in other liquids, the amount and nature of the inactive soluble and insoluble components; (6) general field and specific purposes of utilization, and the processing step or steps for which the preparation is recommended; (7) the optimum pH, the practical working pH range, the minimum and the maximum pH which inhibits the enzyme, and the minimum and maximum pH which permanently inactivates the enzyme; (8) the optimum temperature and the temperature of thermal destruction of the enzyme (occasionally the time needed to destroy 50 and 100% of the activity at a specified pH and temperature) is also presented; (9) the activity, given in defined units if available, or the recommended enzyme/substrate ratio; (10) nature and amount of activators to be added, if needed; (11) nature of possible inactivators to be avoided; (12) conditions of storage to maintain activity and the anticipated loss with time; (13) the FDA regulations concerning the preparations, if the enzyme is recommended in connection with food, beverages, and pharmaceutical products; and (14) cost per kilogram or per unit of activity.

Labeling. Labeling regulations depend on the product. For enzyme preparations used for septic tank systems or for analytical purposes, the identity and concentration of active enzymes must be stated.

The labeling of enzymes used in processing presents problems. Should an enzyme whose activity is destroyed by heat, but which is not physically removed, be declared as an ingredient? If it is so declared, would this not be more misleading, since the active enzyme is no longer present? This matter may be resolved based on the definition of processing aid adopted by the Codex Committee on Food Additives (34):

"A processing aid is a substance or material, not including apparatus or utensils, and not consumed as a food ingredient by itself, intentionally used in the processing of raw materials, foods or its ingredients, to fulfill a certain technological purpose during treatment or processing and which may result in the nonintentional but unavoidable presence of residues or derivatives in the final product."

The committee concluded that enzyme preparations should be treated as processing aids. The participating governments were asked to report the names of those substances that should be listed under the different categories. The Ad Hoc Enzyme Technical Committee, a U.S. group representing major microbial food enzyme manufacturers, had asked the U.S. FDA to recommend inclusion of all enzyme preparations described in the *Food Chemicals Codex,* 2nd ed. (35) in the Committee's final list as processing aids.

Assay

Only a few enzymes are marketed in pure or nearly pure form, mostly for research or analytical purposes. The pure enzyme content of industrial enzyme preparations varies widely, and its numerical value is not known in most products. The enzyme content of a material is assayed and expressed entirely in terms of activity. The activity is the basis of evaluation, ie, the knowledge of its value is essential to both the producers and users.

Through assay procedures, the manufacturer checks the enzyme activity of raw materials and intermediates during the enrichment, purification, and standardization procedures. The price of the final product is a function of the assayed activity. The user pays for the product on the same basis, defines the amount of preparation to be used in his process according to the activity, which is expressed in units.

Standard units and methods have been developed by researchers, manufacturers, consumers, and governmental groups. Often a number of different assay methods and units are employed for the same enzyme, perhaps for different uses. Some industrial enzyme units have quasi-official recognition, such as the methods and units included in the first supplement of the second edition of the *Food Chemicals Codex.* For example, enzyme units concerning pharmaceutical enzymes are interchangeably published in the USP (36), NF (37), and the United States Dispensatory (38). Concerning several enzymes utilized in the preparation of food and beverages, "enzyme units" are established by the AOAC (Association for Official Agricultural Chemists (39), by AACC (American Association of Cereal Chemists) (40), and by the ASBC (American Society of Brewing Chemists) (41). Units for enzymes used by the paper industry are established by TAPPI.

Units are also established by the individual enzyme manufacturers and by enzyme users. The same manufacturer may use different units to express the activity of a preparation, depending on the purpose for which it will be used.

The most meaningful assay measures an activity directly related to the use. Sometimes an assay procedure is designed to simulate actual use (eg, Swift's hamburger test for papain used for meat tenderization). Other assay methods have no theoretical, but only an empirical relationship (eg, milk clotting tests used for determining activity of proteases for chillproofing beer).

Many assay methods do not specify units as such. The method results in some numerical value which may be converted into units, frequently defined by the individual who wants to use them.

The results of enzyme assays are expressed in many different ways. They define one of the following data: (1) the amount of change that has taken place as a result of the action of a specified amount of enzyme preparation, under specific conditions of pH, temperature, time, substrate concentration, and enzyme/substrate ratio; (2) the amount of time needed for a given amount of enzyme preparation to accomplish a certain amount of chemical change under closely specified conditions; and (3) the initial reaction-velocity constant measured, under closely specified conditions, where the reaction speed is a linear function of the enzyme concentration, usually in low-enzyme high-substrate systems. If the activity (expressed in units) of the enzyme preparation is known and the user knows the enzyme activity in the application, the time required by a given amount of enzyme under specified circumstances to complete the action desired can be estimated. Ultimately, the enzyme unit is the basis for selling, buying, and using industrial enzymes.

All types of units establish a workable relation between the enzyme activity and the practical utilization of the enzyme preparation, regardless of which factor is determined as a variable in the assay.

Where a first-order constant may be determined without much effort, such as the hydrolysis of sucrose by invertase, an enzyme unit can be established to equal the amount which, under specified conditions, causes the first-order constant to have an arbitrarily selected numerical value. The more general practice is to determine the initial rate of the reaction in terms of grams substrate converted by grams of preparation per minute.

Most units are arbitrary and are intended to express some convenient value. For this reason, wherever possible, assay conditions should be selected with regard to temperature, pH, substrate concentration, and enzyme/substrate ratio to be similar to the conditions of the practical utilization. The assay method should also be specified for the process for which the enzyme is to be used.

A good assay method should yield results with high precision ($\pm(1-5\%)$), within a relatively short period of time, and with a limited amount of work. Automated equipment developed for clinical chemistry is often used in enzyme assays (see Biomedical instrumentation).

In all enzyme assays, an exact amount of enzyme preparation (or its aqueous solution) is permitted to act upon an exact amount of a well-defined (often standardized) substrate. The assay is run in an aqueous solution or emulsion at a specified concentration, at a specific temperature and pH (mostly buffered), for a definite time, or until a definite amount of change has been accomplished. Then the action of the enzyme is measured by determining one of the following: (*a*) The amount of substrate unchanged during the period; (*b*) the amount of product obtained during the period; and (*c*) the time necessary to degrade a substrate to an easily measurable fixed point such as the time necessary for a 50% conversion of the substrate; or if the progress of the reaction may be followed by a color reaction, the time until a certain color is reached; or, if acids are produced, the time until a certain amount of neutralizing chemical is used up; or, if gas is produced, the time until a predetermined amount of gas is produced; or, if change in rotation is one of the consequences of the enzyme action, the time to reach a certain point of optical rotation.

Following are examples for the measurement of enzyme activity:

The amount of residual unchanged substrate is measured in catalase assays, where the quantity of unchanged peroxide is determined after a certain reaction time (42). A similar principle is applied in some penicillinase assays.

A physical property of the unchanged substrate and product mixture is measured by viscosity in the assay of α-amylase preparations made for the liquefaction of starch (43). Similarly, in invertase assays, the optical rotation of the sucrose–invert sugar mixture is measured; in this case, the change in rotation is one of the results of the enzyme action.

In the assay of saccharifying activity of various amylases, the amount of product formed is measured by determining the reducing power of the digested substrate. The results are calculated as maltose (ME) or dextrose equivalents (DE), depending on the type of enzyme activity. The reducing sugars are determined by the ferricyanide or the Fehling procedure. In some assays the maltose and glucose content of the digest is determined individually.

In assaying proteolytic activities, the amount of newly formed free amino acids may be measured by the Sorensen–Van Slyke method, the change in soluble nitrogen content of the digested substrate can be determined by the Kjeldahl method, or liberated tyrosine determined colorimetrically.

The progress of lipolysis is followed by determining the free acids formed from glycerides. Pectin methylesterase activity can be measured by determining the increase in methanol content of the digested substrate; methanol may be determined by gas chromatography.

In the α-amylase method, in the Lasche' test, in the test for rennin, and in a number of other assay methods, the time necessary to bring about a definite amount of change in a substrate under specified conditions is measured.

In addition to standard chemical methods, visible and ultraviolet spectrophotometric, polarographic, photometric, microscopic, manometric, and other techniques are frequently applied (see Analytical methods).

Health and Safety; Waste Treatment

Enzymes are consumed in large quantities in raw foods; in general, pure enzymes are inherently safe for industrial use. However, at least three problems have to be considered: (*a*) it is impractical to use only pure crystalline enzymes, but in enzyme preparations used in foods, the impurities must not include deleterious substances, such as mycotoxins; (*b*) enzymes are protein in nature and, like all proteins, can cause allergic reactions; and (*c*) some enzymes, such as proteolytic enzymes like bromelin or papain, can cause skin irritation after frequent or lengthy exposure.

The problems of allergies, or of skin irritation in certain work environments, are handled in much the same way as with nonenzyme moieties under similar circumstances. Dusting as a problem in enzyme detergents is controlled by prilling the enzyme; pineapple workers wear gloves to protect against the bromelain in the raw fruit and unheated juice (44).

For analytical use enzymes are highly purified, often crystalline. For septic tank, leather, and most nonfood, nonanalytical uses impurities accompanying the desired activity are not objectionable and may, in fact, contribute to the desired action.

Waste Treatments. Preparations for the enzymatic degradation of normal household wastes in grease traps and septic tanks, and the acceleration of compost heap conversions have been on the market for some time. Commercial preparations keyed to this application are available. Septic tank preparations are, generally, mixtures of very crude amylase, protease, and lipase with sequestrants, buffers, many other trace enzymes, plus high counts of viable organisms for the *in situ* manufacture of enzymes. Grease trap cleaners are reinforced in lipase and sequestering action. Crude cellulases speed the breakdown of grass clippings and other garden refuse in the compost heap.

Pesticide residues and industrial wastes present special problems. However, the ubiquitous nature of microorganisms affords the opportunity to derive enzymes useful for detoxification even where the organism itself would not survive. Thus immobilized cells (*Candida tropicalis* or *Pseudomonas* sp.) degrade phenol and immobilized *Mi-*

crococcus denitrificans or *Pseudomonas* sp. can reduce the nitrate and nitrite content of waste waters. Even pesticides can be degraded by the use of enzymes (45). The 1980 pesticide worldwide production is estimated at over two million metric tons; 500–600 million pesticide containers are used annually. Enzymes active against all of the major classes of pesticides (phenylcarbamates, acid anilides, phenoxyacetates, phenylureas, and organophosphates) have been produced in the laboratory.

Legal Aspects

The manufacture and use of enzymes is generally government controlled. The USDA Meat Inspection Division controls the marketing of meat tenderizers, whereas the Bureau of Alcohol, Tobacco, and Firearms (BATF) of the U.S. Department of the Treasury has jurisdiction over the use of pectic and other enzymes in wine and beer. The FDA controls all uses of enzymes in foods, pharmaceuticals, as medical diagnostic analytical reagents, and related areas. All products sold come under the Federal Trade Commission (FTC) in addition to governmental agencies relating to the specific products.

Factories where enzymes are produced must use good manufacturing practice, and are inspected by OSHA, in addition to state and local regulatory authorities.

Use of enzymes in animal (nonhuman) feeds is under the jurisdiction of the Association of Feed Control Officials. If the manufacture is not covered by other agencies, the EPA exercises supervision by virtue of the Toxic Substances Control Act (TSCA).

There has been some concern over the use nonenzyme or non-main-enzyme constituents of enzyme preparations for food. U.S. food law provides for the approval of enzyme preparations "generally recognized as safe" (GRAS) by experts qualified by scientific training and experience to evaluate safety, provided it was used in foods before January 1, 1958. An enzyme introduced after January 1, 1958 for use in foods is technically a food additive requiring a food additive petition. Economic reality forced the U.S. enzyme industry, following the 1958 amendment to seek a letter of opinion from the FDA in order to be able to give legal assurance of permitted use to its customers; in 1970 all such letters were revoked and all foods and food ingredients were subject to review and reaffirmation of GRAS status except those for which a specific food additive petition has been filed and approved. Now, a decade later, the GRAS affirmation of enzymes by the FDA has not been finalized. However, indirect approval of the use of many enzymes has been provided by the publication of the *Food Chemicals Codex,* a series of food-grade quality specifications for substances added to foods, published by the NAS under the guidance of the Committee on Specifications, Food Chemical Codex, of the Committee on Food Protection of the NRC. This Codex has been given official recognition by the FDA in the *Federal Register* (46).

In 1974 the NAS issued the first supplement to the second edition of the *Food Chemicals Codex* (47), containing a general description of and specifications for enzymes commonly used in foods in the United States. Assay methods are included. In general, this supplement conforms to the report of the Joint FAO/WHO Experts Committee on Food Additives published in 1972, with subsequent publication of reports in 1975 and 1978. This last publication (48), an unofficial opinion of the Joint Experts Committee portends the future, and likely reflects the position of the FDA:

"The use of enzymes in food is increasing both as processing aids and as additives intended to remain in food products. They are obtained either from animal or plant tissues or from cultures of microorganisms. Only in exceptional cases are these enzymes used as crystallized, pure substances or in highly purified form. In general they are obtained from extracts or fermentation broths and partially purified. From the safety point of view the presence of potentially harmful contaminants and by-products is of concern.... Moreover, when microorganisms are used for the production of enzymes, mutations might occur that could lead to the emergence of new, potentially toxic products ... it [is] now felt that chemical and microbiological specifications and the biological control of strains of microorganisms used to produce these food enzymes are of the utmost importance in assuring the safety of these materials.

The increasing use of immobilized enzymes calls for the evaluation of the immobilizing substrates as well as of the immobilizing techniques."

The Committee proposes five major classes of enzyme preparations:

"(1) Enzymes obtained from edible tissues of animals commonly used as foods. These are regarded as foods and consequently considered acceptable, provided satisfactory chemical and microbiological specifications can be established.

(2) Enzymes obtained from edible portions of plants. These are also regarded as foods and consequently considered acceptable, provided that satisfactory chemical and microbiological specifications can be established.

(3) Enzymes derived from microorganisms that are traditionally accepted as constituents of foods or are normally used in the preparation of foods. These products are regarded as foods and consequently acceptable, provided satisfactory chemical and microbiological specifications can be established.

(4) Enzymes derived from nonpathogenic microorganisms commonly found as contaminants of foods. These materials are not considered as foods. The Committee considers it necessary to establish chemical and microbiological specifications and to conduct short-term toxicity experiments to ensure the absence of toxicity. Each preparation must be evaluated individually and an average daily intake (ADI) must be established.

(5) Enzymes derived from microorganisms that are less well known. These materials also require chemical and microbiological specifications and more extensive toxicological studies, including a long-term study in a rodent species."

The expansion of approved food enzymes in the U.S. is severely restricted by the much higher cost of proving safety of enzymes derived from organisms other than the established source of U.S. enzymes, ie, *Aspergillus niger, Aspergillus oryzae, Bacillus subtilis*. Only an application of the economic magnitude of a glucose isomerase, a cellulase, or a microbial rennet, can justify the expense.

Uses

The key question in the evaluation of an enzyme for industrial use is a cost/benefit analysis: the use of the enzyme to convert A to B is practical only where the increment of cost involved is less than the incremental increase in value of B over A. Information obtained from studies does not answer all questions. For example, the temperature at which an enzyme might be most active may not be used as it might promote microbial decomposition. The optimum substrate concentration for an amylase may not be attainable because the viscosity may be too high for efficient stirring. Cobalt may be the best activator for glucose isomerase but the cost of removing it may make a less effective cofactor a better economic choice. Although the academic enzymologist talks of optima in terms of reaction rate, the industrial enzymologist must think of total material transformed to usable product.

Industrial enzyme preparations include pure crystalline enzymes separated from classified into production of (1) foods and food ingredients, (2) beverages, and (3) miscellaneous goods (49–51). Food and food ingredients include bread and other bakery goods, cereals, confections, sugars, syrups, meat, fish, vegetables, milk, cheese, eggs, chocolate, cocoa, coffee, food flavors, and others (see Table 4). The beverages include fruit juices, fruit drinks, carbonated beverages, beer, wine, and distilled beverages. The miscellaneous group includes pharmaceuticals, tobacco, adhesives, paper, textiles, cleaning compositions, clinical and analytical reagents, and others.

The survey form sent to enzyme users by the Ad Hoc Enzyme Technical Committee in conjunction with the NAS, for the FDA (October 1978 to March 1979) lists 28 enzymes as direct additives and four enzymes used in immobilized form in foods in the United States (see Table 5). (53). These include certain enzyme systems as single entities (eg, "pectinase" for preparations containing polygalacturonase and/or pectin methyl esterase and/or pectin transeliminase).

Table 6 lists the enzymes commercially available in the United States. It does not list the captive production enzymes made on a contract basis or made by the user himself. This latter category includes some very important enzymes such as penicillin amidase for hydrolysis of penicillin G in the production of 6-APA, and *B. subtilis* acetyl esterase [9000-82-2] for the enzymatic deacetylation of cephalosoporin. Not listed are also formulated enzyme products that constitute an economically important aggregate of products. These include: feed enzymes (mixtures of crude *Aspergillus oryzae* and *Bacillus subtilis* enzymes, to aid in digestion and utilization of feeds); grease trap cleaners (based on lipase for use in homes and restaurants); septic tank cleaners (used to speed the breakdown of household refuse entering the septic tank system; in addition to the enzymes, high content of visible microorganisms is an important attribute of these products); and leather-bating enzymes (used to deflesh hides, usually based on pancreatin).

In Medicine. Since the human body produces and uses enzymes for every activity, enzyme preparations lend themselves readily to uses where the body's enzyme supply is not sufficient to meet specific needs (see Pharmaceuticals). Enzymes can be administered either orally, or topically, or in the form of injections. They are also useful tools in medical diagnostic tests (see Medical diagnostic reagents). Medicinal enzymes are used in the following fields:

(1) Digestive aids aim to supplement the digestive enzymes in which the body is deficient; lack of those enzymes can cause gastrointestinal disturbances such as indigestion, upset stomach, hyperacidity, gas, cramps, etc.

The principal enzymes, involved singly or formulated jointly, are amylases, proteases, lipases, and cellulases. The first three may originate from animal, plant, or fungal sources; cellulase is of fungal origin. Fungal amylase, lipases, and cellulase (to digest cabbage and salads) derive from *Aspergillus* and *Penicillium*; bacterial amylase and protease are produced by *Bacillus subtilis*; animal diastase and pancreatin, containing trypsin and pancreatic lipase, are made of hog pancreas; the powerful proteases bromelain, ficin, and papain, are made of plants, the last group is on a prescription only basis.

Table 4. Enzymes Important in Food Processing [a]

Trivial name	CAS Registry No.	Systematic name	EC No.
Oxidoreductases			
glucose oxidase, notatin	[9001-37-0]	β-D-glucose: O_2 oxidoreductase	1.1.3.4
catechol oxidase, polyphenol oxidase, catecholase, phenolase	[9002-10-2]	o-diphenol: O_2 oxidoreductase	1.10.3.1
catalase	[9001-05-2]	H_2O_2: H_2O_2 oxidoreductase	1.11.1.6
peroxidase	[9003-99-0]	donor: H_2O_2 oxidoreductase	1.11.1.7
lipoxygenase, lipoxidase, carotenase	[9029-60-1]		1.13.1.13
Transferases			
catecholmethyltransferase	[9012-25-3]	S-adenosylmethionine: catechol O-methyltransferase	2.1.1.6
sucrose 6-glucosyltransferase, dextran sucrase	[9032-14-8]	α-1,6 glucan: D-fructose 2-glucosyltransferase	2.4.1.5
B. macerans enzyme	[9030-09-5]	α-1,4-glucan 4-glycosyltransferase (cyclizing)	2.4.1.19
Hydrolases			
carboxyl esterase, aliesterase, B esterase	[9016-18-6]	carboxylic ester hydrolase	3.1.1.1
arylesterase, A. esterase	[9032-73-9]	aryl ester hydrolase	3.1.1.2
lipase	[9001-62-1]	glycerol ester hydrolase	3.1.1.3
pectin esterase, pectin methylesterase, pectase	[9025-98-3]	pectin pectyl-hydrolase	3.1.1.11
α-amylase	[9000-90-2]	α-1,4-glucan 4-glucanohydrolase	3.2.1.1
β-amylase	[9000-91-3]	α-1,4-glucan maltohydrolase	3.2.1.2
glucoamylase, α-amyloglucosidase	[9032-08-0]	α-1,4-glucan glucohydrolase	3.2.1.3
cellulase	[9012-54-8]	β-1,4 glucan 4-glucanohydrolase	3.2.1.4
amylopectin-1,6-glucosidase, R enzyme	[9012-47-9]	amylopectin 6-glucanohydrolase	3.2.1.9
oligo-1,6-glucosidase, limit dextrinase	[9032-15-9]	oligodextrin 6-glucanohydrolase	3.2.1.10
polygalacturonase, pectinase	[9032-75-1]	polygalacturonide glycanohydrolase	3.2.1.15
α-glucosidase	[9001-42-7]	α-D-glucoside glucohydrolase	3.2.1.20
β-glucosidase	[9001-22-3]	β-D-glucoside glucohydrolase	3.2.1.21
β-galactosidase, lactase	[9031-11-2]	β-D-galactoside galactohydrolase	3.2.1.23
pepsin	[9001-75-6]		3.4.4.1
rennin	[9001-98-3]		3.4.4.3
trypsin	[9002-07-7]		3.4.4.4
chymotrypsin	[9004-07-3]		3.4.4.5
pancreatopeptidase E, elastase	[9004-06-2]		3.4.4.7
papain	[9001-73-4]		3.4.4.10
chymopapain	[9001-09-6]		3.4.4.11
ficin	[9001-33-6]		3.4.4.12
bromelain	[9001-00-7]		3.4.4.24
subtilipeptidase A, subtilisin bacterial protease	[9014-01-1]		3.4.4.16
aspergillopeptidase A, fungal protease	[9025-49-4]		3.4.4.17
clostridiopeptidase A, collagenase	[9001-12-1]		3.4.4.19
Isomerases			
glucosephosphate isomerase, glucose isomerase	[9001-41-6]	D-glucose-6-phosphate ketol-isomerase	5.3.1.9

[a] Ref. 52.

Table 5. Enzyme Preparations Used in Food Processing[a]

Name of enzyme	CAS Registry Number	E.C. Number	NAS Substance Number[b]
Soluble enzymes			
α-amylase	[9000-90-2]	3.2.1.1	5801
β-amylase	[9000-91-2]	3.2.1.2	5802
bromelain	[9001-00-1]	3.4.4.24	0368
catalase	[9001-05-2]	1.11.1.6	0466
cellulase	[9012-54-8]	3.2.1.4	5803
dextranase	[9025-70-1]	3.2.1.11	5804
ficin	[9001-33-6]	3.4.4.12	0381
α-galactosidase	[9025-35-8]	3.2.1.22	5805
β-glucanase	c	c	5806
glucoamylase (amyloglucosidase)	[9032-08-0]	3.2.1.3	5807
glucose isomerase	[9001-41-6]	5.3.1.9	5808
glucose oxidase	[9001-37-0]	1.1.3.4	5809
hemicellulase	c,d	c,d	5810
invertase	[9001-57-4]	3.2.1.26	0469
isoamylase	[9067-73-6]	3.2.1.68	5811
lactase	[9031-11-2]	3.2.1.23	5812
lipase-esterase	c	3.1.1.3[c], 3.1.1.2	0386
lipoxygenase	[9029-60-1]	1.13.1.13	5813
melibiase	[9025-35-8]	3.2.1.22	5814
pancreatin	e	e	5815
papain	[9001-73-4]	3.4.4.10	0142
pectinase (polygalacturonase + pectin methylesterase + pectin transeliminase)	c	c	0392
pepsin	[9001-75-6]	3.4.4.1	0296
protease	f	f	5816
pullulanase	[9012-47-9]	3.2.1.44	5817
rennet and other milk clotting enzymes	c	3.4.4.3 and others	0168
transglucosidase	g	g	5818
trypsin	[9002-07-7]	3.4.4.4	5819
Immobilized enzymes			
glucoamylase	[9032-08-0]	3.2.1.3	5807
glucose isomerase	[9001-41-6]	5.3.1.9	5808
invertase	[9001-57-4]	3.2.1.26	0469
lactase	[9031-11-2]	3.2.1.23	5812

[a] Ref. 53.
[b] NAS = National Academy of Sciences.
[c] This is a group of enzymes.
[d] Ref. 54.
[e] This is the dried pancreas gland, and it contains many enzymes.
[f] This is a group of enzymes in the 3.4.4. category.
[g] This name refers to a group of glucosyl transferases in the 2.4.1 grouping; the final number identifies the particular transglucosidase.

(2) Properly selected and administered, proteolytic enzymes are applied for debridement, that is, they selectively digest dead skin, necrotic tissues, and debris on wounds, without harming healthy tissues or active blood vessels which feed the wound. They are also useful in treating burns.

(3) A group of enzymes used as anti-inflammatory agents include streptokinase

Table 6. Enzymes Commercially Available in the United States

Enzyme	Vendor[a]	Source	Discussion
α-Amylase			The heat resistance of *Bacillus* α-amylase makes it particularly suitable for starch liquefaction at starch gelatinization (solubilization) temperatures, and unsuitable for baking as it would continue to act during and after baking. Also used for removing starch sizing from fabrics, for production of dextrins and adhesives from starch, to reduce viscosity of starch-containing foods, in waste treatment, as aid in utilization of malt adjuncts (rice, grits, etc) in brewing, wallpaper removal (where starch paste used), detergents, etc. *B. subtilis* enzymes are active at 75–85°C and pH 5.5–9.0 if Ca^{2+} and Na^+ present. Random action on starch produces mixture of glucose, maltose, and other oligosaccharides. Generally contain significant quantities of neutral and/or alkaline protease. *B. licheniformis* preparations are similar to *B. subtilis* preparations, but act at higher temperatures (100°C and above), require less Ca^{2+}, may contain β-glucanase as well as proteases. *Aspergillus* amylases, particularly *A. oryzae* act well at lower pH than *Bacillus* enzymes and are rapidly destroyed above about 60°C. *Aspergillus* preparations may contain some protease, and also substantial amounts of glucoamylase. Fermex, is an *A. oryzae* preparation sold as a combination α-amylase/glucoamylase product. Certain *B. subtilis* preparations are promoted as antistaling agents in breads based on low (for *B. subtilis* amylase) heat stability.
Alphamyl	ED	*Bacillus subtilis*	
Amylase WT	GB	*Bacillus subtilis*	
Amyliq	GB	*Bacillus subtilis*	
Aquazyme	N	*Bacillus subtilis*	
Asperzyme	ED	*Aspergillus oryzae*	
BAN 120 L	N	*Bacillus subtilis*	
Clarase	M	*Aspergillus oryzae*	
Clinvert	CC	*Bacillus subtilis*	
Dex-Lo	GB	*Bacillus subtilis*	
Fermex	GB	see discussion	
Fresh N	GB	*Bacillus subtilis*	
Fungamyl	N	*Aspergillus oryzae*	
Hazyme DS	GB	*Aspergillus niger*	
Milezyme	M	*Bacillus subtilis*	
Mycolase	GB	*Aspergillus oryzae*	
Mycozyme	P	*Aspergillus oryzae*	
Mylase	GB	*Aspergillus oryzae*	
Rapidase	GB	*Bacillus subtilis*	
Rohalase	F	*Aspergillus oryzae*	
Rhozyme, H39, 86L, GC	RH	*Bacillus subtilis*	
Rhozyme S	RH	*Aspergillus oryzae*	
Taka-Therm	M	*Bacillus licheniformis*	
Takamyl	M	*Aspergillus oryzae*	
Tenase	M	*Bacillus subtilis*	
Termamyl	N	*Bacillus licheniformis*	
Veron AC	F	*Aspergillus oryzae*	
Veron AV	F	*Aspergillus oryzae*	
WC-8	GB	*Bacillus subtilis*	
β-Amylase			Used for maltose production, in brewing, and in combination with α-amylase for baking; very active at pH 4–6.5, up to 55–60°C.
Malt	many	germinated barley	
Bromelain	M, D,	*Ananas bracteatus*	Digestive aid; with papain as a meat tenderizer.
	ED	*Ananas comosus* (pineapple)	
Catalase			Used in milk sterilization with H_2O_2, to remove residual H_2O_2; after bleaching hair, textiles, wood, etc; in conjunction with glucose oxidase; to provide oxygen; with H_2O_2 as a leavening agent.
Catalase L	M	liver	
Fermcolase	F	*Aspergillus niger*	
Hyperlase	CH	*Aspergillus niger*	
in glucose oxidase preparations	M, F	*Aspergillus niger*	

Table 6. (continued)

Enzyme	Vendor[a]	Source	Discussion
Cellulase	M, F, RH, GB, ED	Aspergillus niger	A. niger used for digestive aids; T. viride is much more active and restricted to nonfood in U.S.
	ED	Trichoderma viride	A. niger cellulase acts only on modified cellulase whereas T. viride has, in addition, another cellulase active on native cellulose. All crude cellulase preparations contain hemicellulase, other carbohydrases, pectinase, etc. Some have been reported to show some ligninase activity.
Dextranase		Penicilliun funculosum or Penicillium bilacinum	Used in manufacture of sucrose to hydrolyze dextran formed by Leuconostoc mesenteroides contamination of sugar cane juice; and in dental preparations to dissolve the dextran "plaque cement" formed from sucrose in the mouth by Streptococcus mutans.
Ficin	DM, ED	Ficin glabrata (wild fig tree, Peru)	Like papain and bromelain; also used as protein hydrolysates and for recovery of silver from photographic film.
α-Galactosidase	captive	Mortierella vinaceae var. raffinosutilizer	Breaks down raffinose to galactose and sucrose, permitting better crystallization of sucrose from beet sugar molasses.
β-Glucanase			Lowers beer or wort viscosity by degrading barley glucans to facilitate filtration; often found in varying amounts in B. subtilis amylase and protease preparations.
Cereflo 200	N	Bacillus subtilis Candida utilis	
Glucoamylase (amyloglucosidase)			Produces glucose from gelatinized starch thinned with Bacillus sp. α-amylase (contaminating transglucosidase must be reduced to prevent diversion of glucose to isomaltose). Converts residual oligosaccharides to glucose in light beer production; used in immobilized form at pH 3.3–4.8 and 55–60°C, reaction time 2–4 days. Also used for saccharification in distilling industry; very important for ethanol production for gasoline/alcohol blends for motor vehicle fuel.
AMG	N	Aspergillus niger	
Amigase	GB	Aspergillus niger	
Diazyme	M	Aspergillus niger	
Fermex	GB	Aspergillus oryzae (see under amylase)	
Spiritamylase	N	Aspergillus niger	
Glucose isomerase (xylose isomerase)	ICI	Arthrobacter globiformis	In immobilized form converts glucose to fructose pH, temperatures, requirement for metal coenzyme varies with organism. Invert sugar is substituted for by the 42–45% (up to 90%) fructose formed in a glucose solution subjected to the enzyme. HFCS (high fructose corn syrup) is replacing sucrose in many applications.
Immobiliase GI	GB	Actinoplanes missouriensis	
Maxizyme GIMO	N	Bacillus coagulans	
Sweetzyme	M	Streptomyces olivaceus	
captive production, Corn Products Corp.		Streptomyces olivochromogenes	
captive production, Clinton Corn Processing		Streptomyces rubiginosus	

Table 6. (continued)

Enzyme	Vendor[a]	Source	Discussion
Glucose oxidase			In combination with catalase, removes oxygen from mayonnaise, beer, soft drinks, etc, in the presence of excess glucose to prevent oxidative deterioration; in presence of excess oxygen it removes glucose from eggs before drying, thus prevents maillard browning. Preparations differ markedly in trace enzyme present, which may include cellulase, amylase, or invertase. Free of catalase, glucose oxidase is useful in the quantitation of glucose. Fermcozyme 952DM is a stabilized combination of catalase-reduced glucose oxidase with horseradish peroxidase for quantitation of glucose. Dip sticks containing glucose oxidase, peroxidase, and a chromogen are used for urine analysis in diabetes. Glucose oxidase acts well as pH 3.5–8.0, up to about 65°C.
Dee O	M	*Aspergillus niger*	
Fermcozyme CBB	F	*Aspergillus niger*	
Fermcozyme M	F	*Aspergillus niger*	
Fermcozyme 653AM	F	*Aspergillus niger*	
Fermcozyme 1307	F	*Aspergillus niger*	
Fermcozyme 952DM	F	*Aspergillus niger* and horseradish (see discussion)	
GOC	F	*Aspergillus niger*	
Glucose oxidase, purified	F, M	*Aspergillus niger*	
LC 5000	F	*Aspergillus niger*	
Ovazyme	F	*Aspergillus niger*	
Hemicellulase			Often present in cellulase, β-glucanase, and pentosanase preparations. Used as antistaling agent, to improve water-holding capacity of rye flours, clean mucilage from coffee beans to accelerate fermentation; to break down oil-well fracturing gels based on guar gum.
(pentosanases)	ED, M, N	*Aspergillus niger*	
Rhozyme HP 150	RH	*Aspergillus niger*	
Veron HE	F	*Aspergillus niger*	
Invertase			Hydrolyzes sucrose to more soluble and sweeter equimolar mixture of glucose and fructose. Used to make invert sugar, and soft- and liquid-centered candies; optimum at pH 3.5–5.5, up to 70°C.
Maxinvert	GB	*Saccharomyces cerevisiae*	
Fermvertase	F		
Sucrovert	IT		
Lactase			Hydrolyzes lactose to more soluble glucose and galactose to prevent crystallization in ice cream; pretreatment of milk for those with lactose intolerance; treatment of whey to facilitate fermentation and/or protein recovery.
Lactase	M	*Aspergillus oryzae*	
Lactase N	GB	*Aspergillus niger*	
Lactozyme 750L	N	*Kluyveromyces fragilis* (*Saccharomyces*)	
Maxilact	GB	*Saccharomyces lactis* *Lactobacillus helveticus*	
Lipase/Esterase			Removes traces of yolk fat from egg white to improve whipping; develops Italian cheese and other cheese flavors; lipolyzes butter for flavor development; in digestive aids; the *Rhizopus arrhizus* enzyme, highly purified, is used in triglyceride assay.
pancreatin	many	hog and bovine pancreas	
Capalase	DL	calf pregastric glands	
Fermlipase	F	animal pancreas	
Italase	DL	goat pregastric glands	
Lipase	F	*Rhizopus arrhizus*	
Lipase	F	*Aspergillus niger*	
Lipase	GB	*Mucor miehei*	
Lipase di agnello	CH	lamb pregastric glands	
Lipase di capretto	CH	goat pregastric glands	
Lipase di vitello	CH	calf pregastric glands	
Lipase 30	SP	animal pancreas	
Pancrelipase	SP	animal pancreas	

Table 6. (*continued*)

Enzyme	Vendor[a]	Source	Discussion
Lipoxidase (lipoxygenase)	many	soybeans	Bleaching agent for white bread and carotene in presence of oxygen and linoleic acid; a flour improver for flours high in protease; H_2O_2 formed by enzyme action inhibits the native protease.
Papain	many	*Carica papaya*	Meat tenderizer, prevents haze in chilled beer, softens high-gluten flours for cracker production, used in fish stick liquor. Acts well up to 70°C, at pH range 4–9. Commercial preparations generally activated and stabilized with bisulfite; minor uses in wound debridement, blood typing, dentifrices, digestive aids.
Pancreatin	many	ovine and bovine pancreas	Leather bating (breakdown of noncollagen components of hide to facilitate removal), digestive aids.
Pectinase			Pectic enzymes, sometimes blended with gelatin and/or amylases; commercial preparations generally contain cellulase, hemicellulase, sometimes protease. Clarify apple, berry, and citrus juices; increase free run of juice from grapes (makes superior wine to finally pressed juice); prevent gelling of frozen orange juice; reduce viscosity of concentrated juices; prepare citrus pulp wash and natural clouding agent, break citrus oil emulsions to improve oil yield.
Clarex	M	*Aspergillus niger*	
Extractase	F	*Aspergillus niger*	
Irgazyme	N	*Aspergillus niger*	
Klerzyme	GB	*Aspergillus niger*	
Pectinex	N	*Aspergillus niger*	
Pectinol (in U.S.)	RH	*Aspergillus niger*	
Rapizyme	GB	*Aspergillus niger*	
Rohapect	F	*Aspergillus niger*	
Spark L	M	*Aspergillus niger*	
Pectin glycosidase			Prepare stable nonseparating fruit and vegetable juices, purees.
Extractase PC	F		
Rohament P	F	*Aspergillus sp.*	
Pentosanase			Contains glucanase, cellulase, hemicellulase; used in rye flours to improve tenderness, leavening properties, by facilitating water penetration and holding capacity.
Veron HE	F	*Aspergillus niger*	
Rhozyme HP 150	RH		
Pepsin	many	pig stomach mucosa	Used as an extender for rennet for curdling milk for rennin (rennet); digestive aid; predigests cereals; animal feed additive; in synthesis of proteins (plastein formation); digests at pH 1–4; plastein formation pH 4.5–5.0.
Peroxidase	many	horseradish root	Used in conjunction with aerodehydrogenases (oxidases) forming H_2O_2 such as glucose oxidase, cholesterol oxidase, etc, for analytical purposes; attached as "tag" in enzyme immunoassays; lactoperoxidase (from milk) used for biological iodination.
Protease, alkaline			Detergent enzymes; clean raw silk; dehair hides; used in gelatin manufacture.
Alcalase	N	*Bacillus licheniformis*	
Esperase	N	*Bacillus sp.*	
201 P	GB	*Bacillus sp.*	

218   ENZYMES, INDUSTRIAL

Table 6. (continued)

Enzyme	Vendor[a]	Source	Discussion
Pullulanase	ED	*Klebsiella aerogenes*	Improves utilization of available carbohydrates in brewing.
Rennet			Rennin is derived from the stomach mucosa of suckling animals. As the animal ages the enzyme gradually shifts from rennin to pepsin; the age of the animal determines the ratio of rennin to pepsin in a rennet. Microbial rennets are acid proteases, and are highly useful for cheeses that are not aged extensively. Microbial rennets may contain other proteases that can contribute to off-flavors, cellulase that can deteriorate cheese clothes.
American	DL	calf	
Bovin	CH	bovine	
Econozyme	P	porcine rennin/pepsin	
Emporase	DL	*Mucor pusillus*	
Fromase	GB	*Mucor miehei*	
Hannilase	CH	*Mucor miehei*	
Marzyme	M	*Mucor miehei*	
Milezyme	M	*Aspergillus niger*	
Morcurd	P	*Mucor miehei*	
Rennilase	N	*Mucor miehei*	
Sure-Curd	CH	bovine	
Trypsin	SP	ovine and bovine pancreas	see Pancreatin

[a] Enzyme Development Corp., New York, ED; Fermco Biochemics Inc., Elk Grove Village, Ill., F; G. B. Fermentation Industries, Des Plaines, Ill., GB; Miles Laboratories Inc., Elkhart, Ind., M; Novo Laboratories Inc., Wilton, Conn., N; Pfizer Inc., New York, P; Rohm & Haas Company, Philadelphia, Pa., RH; Clinton Corn Processing, Clinton, Iowa, CC; Dairyland Food Labs.; Inc., Waukesha, Wis., DL; Dole, Redwood City, Ca., D; Chr. Hansen's Laboratories Inc., Milwaukee, Wisc., CH; Ingredient Technology Corp., Woodbridge, N.J., IT; Dr. Madis Laboratories Inc., South Hackensack, N.J., DM; Scientific Protein, Madison, Wisc., SP.

all inert material, partially purified enzyme concentrates free from cell residue and most inert material, or cells rich in certain enzymes. The cells may include microorganisms cells, plant tissue (eg, barley malt), or animal tissue, (eg, dried pancreas gland). The microorganisms cells may be used directly as crude enzyme preparations or as a source of self-reproducing enzyme supply. An example for the first type is dried bakers' yeast which reproduce only seldom or not at all during its utilization in household process. Example for the second type is every microbial inoculum used in a microbial process involving cell reproduction (see Fermentation).

Most industrial enzyme preparations are dry, some are liquids, others are syrups. The dry products are usually pulverized (consistency of flour or bran); the malts maintain the original appearance of the barley grain. The activity of most preparations is set by the manufacturer to a predetermined level, by the admixture of inactive substances that do not interfere with the activity of the enzyme concentrate. In many instances the manufacturer markets the same enzyme preparation at different strengths of activity for different purposes. Usually the higher the activity per unit of weight in the product, the higher is the cost per unit of activity.

Most enzyme preparations excel in one principal enzyme and contain other enzymes in considerable quantities or just in traces, usually all derived from a common source. If the use so requires and justifies the additional purification cost, the preparation contains a single enzyme. The usual accompanying enzymes are eliminated, and this is so stated on the label and in the specifications. For special purposes, carefully blended mixtures of enzymes, so-called formula enzymes, are available.

Industrial enzyme preparations are used in almost every field where organic products of plant and animal origin are processed into consumer goods. This can be

Table 7. Enzymes Used in Medicine[a]

Trivial name	CAS Registry Number	Name	E.C. Number	Source	Reaction catalyzed	Uses
alcohol dehydrogenase	[9031-72-5]	alcohol: NAD oxidoreductase	1.1.1.1	yeast, horse liver	$C_2H_5OH + NAD^+ \rightleftharpoons CH_3CHO + NADH + H^+$	to determine alcohol; indicator enzyme for lipase determination
aldehyde dehydrogenase	[9028-86-8]	aldehyde: NAD oxidoreductase	1.2.1.3	*Saccharomyces cerevisiae*	$CH_3CHO + NAD^+ \rightleftharpoons$ acetate $+ NADH + H^+$	indicator enzyme for uricase determination
catalase	[9001-05-2]	hydrogen peroxide: hydrogen peroxide oxidoreductase	1.11.1.6	bovine liver *Aspergillus niger*	$2 H_2O_2 \rightarrow 2 H_2O + O_2$ (as catalase) $C_2H_5 + H_2O_2 \rightleftharpoons CH_3CHO + 2 H_2O$ (as a peroxidase)	intermediate enzyme for uricase determination
cholesterol esterase	[9026-00-0]	sterol-ester hydrolase	3.1.1.13	bovine pancreas	cholesteryl ester + $H_2O \rightarrow$ cholesterol + fatty acid	hydrolyze cholesterol esters for cholesterol determination
cholesterol oxidase	[9028-76-6]	cholesterol: oxygen oxidoreductase	1.1.3.6	*Nocardia etythropolis*, *Brevibacterium* (sp)	cholesterol + $O_2 \rightleftharpoons \Delta^4$ cholestenone + H_2O_2	oxidize cholesterol to form H_2O_2 which is then determined
creatinijnase	[37289-15-9]	creatinine deiminase: creatinine imidohydrolase	3.5.4.21	*Pseudomonas* strain	creatinine + $H_2O \rightleftharpoons N$-methylhydantoin + NH_3	to determine creatinine
diaphorase	[9001-18-7]	NADH: lipoamide oxidoreductase	1.6.4.3	pig heart	tetrazolium dye + NADH + $H^+ \rightarrow NAD^+$ + formazan	indicator enzyme for several reactions
galactose oxidase	[9028-79-9]	D-galactose: oxygen 6-oxidoreductase	1.1.3.9	*Polyporus circinatus*	D-galactose + $O_2 \rightleftharpoons$ D-galactohexodialdose+ H_2O_2	to determine galactose
glucose oxidase	[9001-37-0]	β-D-glucose: oxygen 1-oxidoreductase	1.1.3.4	*Aspergillus niger*	β-D-glucose + $H_2O + O_2 \rightleftharpoons$ D-glucono-δ-lactone + H_2O_2	to determine glucose
glucose-6-phosphate dehydrogenase	[9001-40-5]	D-glucose-6-phosphate: NADP oxidoreductase	1.1.1.49	*Leuconostoc mesenteroides*	D-glucose-6-phosphate + $NADP^+ \rightleftharpoons$ D-gluconate-6-phosphate + $NADPH + H^+$	indicator enzyme for creatine kinase determination indicator enzyme for glucose determination
β-glucuronidase	[9001-45-0]	β-D-glucuronide glucuronosohydrolase	3.2.1.31	*Escherichia coli*, *Helix pomatia*	β-D-glucuronide + $H_2O \rightleftharpoons$ alcohol + D-glucuronate	used to hydrolyze steroid conjugates of glucuronic acid
glucose dehydrogenase	[9028-53-9]	D-glucose: NAD 1-oxidoreductase	1.1.1.47	microbial	β-D-glucose + $NAD^+ \rightleftharpoons$ D-glucono-δ-lactone + NADH	indicator enzyme for amylase determination; to determine glucose
glutamate dehydrogenase	[9029-12-3]	L-glutamate: NAD(P) oxidoreductase (deaminating)	1.4.1.3	bovine liver	L-glutamate + $NAD(P)^+ + H_2O \rightleftharpoons$ 2-oxoglutarate + NADPH + NH_4^+	indicator enzyme for urea determination using urease

Table 7 (continued)

Trivial name	CAS Registry Number	Name	E.C. Number	Source	Reaction catalyzed	Uses
glyceraldehyde-3-phosphate dehydrogenase	[9001-50-7]	O-glyceraldehyde 3-phosphate: NAD oxidoreductase (phosphorylating)	1.2.1.12	rabbit muscle	glyceraldehyde 3-phosphate + NAD$^+$ + Pi^b → glycerate-1,3-diphosphate + NADH + H$^+$	indicator reaction for ATP determination
glycerol dehydrogenase	[9028-32-4]	glycollate: NAD oxidoreductase	1.1.1.6	Enterobacter aerogenes, spinach	glyoxylate + NADH + H$^+$ ⇌ glycollate + NAD$^+$	indicator reaction for triglyceride determination
glycerol-3-phosphate dehydrogenase	[9075-65-4]	glycerol-3-phosphate: NAD 2-oxidoreductase	1.1.1.8	rabbit muscle	L-glycerol-3-phosphate + NAD$^+$ ⇌ dyhydroxyacetone phosphate + NADH + H$^+$	intermediate reaction for triglyceride determinations
L-α-glycerol phosphate oxidase	[9046-23-0]		1.1.3		L-α-glycerolphosphate + O$_2$ ⇌ dehydroxy acetone phosphate + H$_2$O$_2$	intermediate reaction for triglyceride determinations
glycerokinase	[9030-66-4]	ATP: glycerol 3-phosphotransferase	2.7.1.30	Escherichia coli, Candida mycoderma	glycerol + ATP ⇌ glycerol-3-phosphate + ADP	intermediate reaction for triglyceride determinations
hexokinase	[9001-51-8]	ATP: D-hexose 6-phosphotransferase	2.7.1.1	Candida utilis	D-glucose + ATP ⇌ D-glucose-6-phosphate + ADP	intermediate reaction for glucose determination
lactate dehydrogenase	[9001-60-9]	L-lactate: NAD oxidoreductase	1.1.1.27	pig muscle, rabbit muscle, beef heart	L-lactate + NAD$^+$ ⇌ pyruvate + NADH + H$^+$	indicator reaction for triglyceride determination indicator reaction for transaminase determination
lactoperoxidase	[9003-99-0]	donor: hydrogen peroxide: oxidoreductase	1.11.1.7	milk	donor + H$_2$O$_2$ ⇌ oxidized donor + 2 H$_2$O	iodination of proteins ("tags")
lipase	[9001-62-1]	triacylglycerol: acylhydrolase	3 1.1.3	wheat germ, pig pancreas, Candida cylindracea Rhizopus arrhizus	triglycerides → glycerol + fatty acids	to determine triglycerides
luciferase	[9014-00-0]			Photonius pyralis	luciferin + ATP $\xrightarrow{Mg^{2+}}$ adenyl luciferin adenyl luciferin + O$_2$ → adenyl oxyluciferin + light	to determine ATP

Enzyme	CAS	EC	Source	Reaction	Use
malate dehydrogenase	[9001-64-3]	1.1.1.37	pig heart, *Rhizopus arrhizus*	L-malate + NAD$^+$ ⇌ oxaloacetate + NADH + H$^+$	indicator reaction for CO$_2$ determinations
maltase	[9001-42-7]	3.2.1.20	*Saccharomyces italicus*	maltose + H$_2$O → 2 α-D glucose	intermediate reaction for the determination of amylase
peroxidase	[9003-99-0]	1.11.1.7	horseradish root	donor + H$_2$O$_2$ ⇌ oxidized donor + 2 H$_2$O	indicator reaction for many reactions producing H$_2$O$_2$
phenylalanine ammonia lyase	[9024-28-6]	4.3.1.5	*Rhotorula glutinis*	L-phenylalanine → *trans*-cinnamate + NH$_4$ L-tyrosine → *trans-p*-coremarate + NH$_4$	to determine phenylalanine and tyrosine
phosphoenolpyruvate carboxylase	[9067-77-0]	4.1.1.31	plant source	phosphoenolpyruvate + HCO$_3^-$ → oxalacetate + P$_i$[b]	to determine carbon dioxide
β-phosphoglucomutase	[9001-81-4]	2.7.5.1	microbial	β-glucose-1-phosphate ⇌ glucose-6-phosphate	intermediate reaction to determine amylase
6-phosphogluconate dehydrogenase	[9073-95-4]	1.1.1.44	yeast	6-phosphogluconate + NADPH + H$^+$ + CO$_2$	
phosphorylase	[9035-74-9]	2.4.1.1	rabbit muscle	oligosaccharide + P$_i$[a] ⇌ limit dextrin + glucose-1-phosphate	intermediate reaction for amylase determination
pyruvate kinase	[9001-59-6]	2.7.1.40	rabbit muscle	pyruvate + ATP ⇌ phosphoenolpyruvate + ADP	intermediate reaction for triglyceride determination
urease	[9002-13-5]	3.5.1.5	*Canavalia ensiformis* ("jack bean")	urea + 3 H$_2$O = 2 NH$_4$OH + CO$_2$	to determine urea
uricase	[9002-12-4]	1.7.3.3	pig liver, *Candida utilis*	urate + 2 H$_2$O + O$_2$ → allantoin + H$_2$O$_2$ + CO$_2$	to determine uric acid

[a] Ref. 54.
[b] P$_i$ = inorganic phosphate.

[9002-01-1] and streptodornase [37340-82-2] (produced by *Streptococcus*), papain, bromelain, bacterial amylases, and proteases. The efficacy of these has been challenged by the FDA.

(4) Streptokinase and urokinase (from human urine) and associated enzymes are used to dissolve blood clots (thrombosis) and to reduce the clotting tendency of blood.

(5) Purulent exudates in wounds contain much DNA and are highly viscous. Streptodornase reduces the viscosity effectively by depolymerizing DNA, and thus facilitates the cleaning of wounds.

(6) Hyaluronidase accelerates the diffusion and action of locally administered anesthetics used by dentists and surgeons in minor surgery by partial hydrolysis of hyaluronic acid in tissues.

(7) Proteolytic enzymes with mucolytic properties reduce the viscosity of secretions in various mucous membranes and thus facilitate the elimination of such excretions.

(8) Penicillin may cause acute reactions, usually skin rashes, in individuals allergic to this antibiotic. The enzyme penicillinase administered to the patient destroys the penicillin in the body, and removes the cause of the allergy.

(9) Glucose oxidase is extensively used in medical diagnostics for determining the sugar content of blood and urine; urease is used to determine the content of urea in blood.

(10) Lysozyme preparations are recommended for the treatment of nervous ulcers, multiple sclerosis, leg ulcers, measles, varicose ulcers, nose, throat, and blood diseases, eye infections, hemorrhagic conditions, postoperative conditions, virus infections, skin diseases, and for many other purposes. Lysozyme chloride [52219-07-5] is well tolerated. Quantities of 5–10 mg per kg body weight caused no ill effects in rats. The enzyme preparation may be administered orally, topically, or by injection.

(11) Lactase is used to predigest the lactose (milk sugar). In addition to these few medically important enzymes, many others are important in analytical chemistry, particularly clinical chemistry. A listing of these enzymes and their sources is presented in Table 7 (55).

Outlook

The most important enzyme likely to be commerciallized over the next decade is a cellulase/ligninase to utilize the forests of the world and the agricultural wastes for energy and as raw materials for fermentation—derived products. Certainly the effort devoted to this is on a scale sufficient to make the necessary economic breakthrough.

Other studies utilizing enzymes, especially immobilized enzymes, involve continuous production of beer, steroid transformations, hydrogen, and even electricity (see Enzymes, immobilized).

Many traditional fermentation processes are likely to be replaced by enzyme processes, since the enzyme preparation can be better controlled and wastes less substrate by side reactions than the living cell. A major oil company is reported to have funded a $10 million effort to produce enzymatically some of today's most common petrochemicals (56). Hydrolytic enzymes may be used to produce the more complex substrates through conditions that favor the reverse reaction, such as plastein formation from peptides by pepsin.

BIBLIOGRAPHY

"Enzymes and Enzymology" in *ECT* 1st ed., Vol. 5, pp. 735–762, by A. K. Balls, Agricultural Research Administration, U.S. Department of Agriculture; "Enzymes, Industrial" Suppl. 1, pp. 294–312, by Gerald Reed, Red Star Yeast and Products Company; "Enzymes, Industrial" in *ECT* 2nd ed., Vol. 8, pp. 173–230, by George I. deBecze, Schenley Distillers, Inc.

1. E. E. Conn and P. K. Stumpf, *Outlines of Biochemistry,* John Wiley & Sons, Inc., New York, 1976, p. 173.
2. M. Florkin and E. H. Stotz, *Comprehensive Biochemistry,* Vol. 13, 2nd ed., Amsterdam, 1965, pp. 52–55; *Biochem. Biophys. Acta* **429,** 1 (1976).
3. L. Stryer, *Biochemistry,* W. H. Freeman and Co., San Francisco, 1975, p. 115.
4. F. C. Webb, *Biochemical Engineering,* D. Van Nostrand Co., Inc., New York, 1964.
5. P. C. Engel, *Enzyme Kinetics,* John Wiley & Sons, Inc., New York, 1977.
6. C. J. Gray, *Enzyme Catalyzed Reactions,* Van Nostrand Reinhold Co., New York, 1971.
7. E. Zeffran and P. L. Hall, *The Study of Enzyme Mechanisms,* John Wiley & Sons, Inc., New York, 1973.
8. J. M. Reiner, *Behavior of Enzyme Systems,* 2nd ed., Van Nostrand Reinhold Co., New York, 1969.
9. Ref. 1, p. 159.
10. Ref. 1, p. 165.
11. Ref. 1, p. 178.
12. Ref. 1, p. 179.
13. J. Meyrath and G. Volavsek in G. Reed, *Enzymes in Food Processing,* 2nd ed., Academic Press, New York, 1975, p. 272.
14. *Food Chemicals Codex,* First Supplement, National Academy of Sciences, Washington, D.C., 1974, p. 293.
15. U.S. Pat. 3,991,215 (Nov. 9, 1976), E. A. Robbins.
16. Y. Hirose, K. Sano, and H. Shibai in D. Perlman, *Annual Reports on Fermentation Processes,* Vol. 2, 1978, pp. 178–179.
17. T. Fujii, K. Matsumoto, and T. Watanabe, *Process Biochem.,* 21 (Oct. 1976).
18. M. O. Moss "Enzyme Alterations of Penicillins and Cephalosoporins" in A. Wiseman, ed., *Topics in Enzyme and Fermentation Technology,* Vol. 1, Ellis Horwood, Chichester, 1977.
19. H. Brockerhoff and R. G. Jensen, *Lipolytic Enzymes,* Academic Press, New York, 1974.
20. Z. I. Kertesz, *The Pectic Substances,* Interscience Publishers Inc., New York, 1951.
21. Z. I. Kertesz and co-workers, *Chem. Eng. News* **22,** 105 (1944).
22. M. H. Spalding and G. E. Edwards, *Planta* **141,** 59 (1978).
23. G. Enamann-Becker, *Z. Naturforsch.* **28c,** 470 (1973).
24. F. M. Rombouts and W. Pilnik, *Process Biochem.,* 9 (Aug. 1978).
25. E. Grampp, *Flussige Obst.* **10,** 1 (1976).
26. R. Urlaub, *Process Biochem.,* **13,** 14 (Aug. 1978).
27. H. V. Amorim, and V. L. Amorim, "Coffee Enzymes and Coffee Quality" in R. L. Ory and A. J. St. Angelo, eds., *Enzymes in Food and Beverage Processing,* ACS Symposium Series 47, American Chemical Society, Washington, D.C., 1977, p. 39.
28. C. Neuberg and I. S. Roberts, *Invertase Monograph,* Sugar Research Foundation, New York, 1946.
29. J. DeClerck, *Textbook of Brewing,* two volumes, Chapman and Hall, Ltd., London, 1957.
30. *Food Chemical News* **20,** 15 (Jan. 29, 1979).
31. C. Bucke, "Industrial Glucose Isomerase" in A. Wiseman, *Topics in Enzyme and Fermentative Biotechnology,* Vol. 1, John Wiley & Sons, Inc., New York, 1977, pp. 147–171.
32. *Chem. Week,* **123,** 41 (Aug. 9, 1978).
33. *Food Prod. Dev.* **12**(6), 41 (July 1978).
34. *Codex Committee on Food Additives,* 12th Session, the Hague, Netherlands, Oct. 10–16, 1978.
35. *Food Chemicals Codex,* 2nd ed., National Academy of Sciences, Washington, D.C., 1974.
36. *The United States Pharmacopeia, XX-NFXV,* The United States Pharmacopeial Convention, Inc., Rockville, Md., 1980.
37. *The National Formulary XIV,* The American Pharmaceutical Association, Washington, D.C., 1975.
38. *The Dispensatory of the United States,* J. B. Lippincott Co., Philadelphia, Pa.
39. *Official and Tentative Methods of Analysis,* Association of Official Agricultural Chemists, Washington, D.C.

40. *Cereal Laboratory Methods,* American Association of Cereal Chemists.
41. *Methods of Analysis,* American Society of Brewing Chemists
42. D. Scott and F. E. Hammer, *Enzymologia* **22,** 194 (1960).
43. M. J. Mason and R. J. Horst, *Tappi* **35,** 25 (1952).
44. M. Flindt, *Process Biochem.,* 3 (Aug. 1978).
45. D. M. Munnecke, *Process Biochem.,* 14, 31 (Feb. 1978).
46. *Fed. Regist.* **36,** 12093 (June 25, 1971).
47. *Food Chemicals Codex,* First Supplement, National Academy of Sciences, Washington, D.C., 1974; *Food Chemicals Codex,* 3rd ed., 1980.
48. "Evaluation of Certain Food Additives" in *World Health Organization Technical Report Series 617,* World Health Organization, Geneva, Switzerland, 1978, pp. 8–9.
49. G. Reed, *Enzymes in Food Processing,* 2nd ed., Academic Press, New York, 1975.
50. J. R. Whitaker, *Principles of Enzymology for the Food Sciences,* Marcel Dekker Inc., New York, 1972.
51. F. W. Cooler, *Enzyme Use and Control in Foods,* Institute of Food Technologists, Chicago, 1976.
52. *Ibid.,* pp. 4–9.
53. *1978 Enzyme Users Survey,* Appendix I, Ad Hoc Enzyme Technical Committee, 1978.
54. R. F. H. Dekker and G. N. Richards "Hemicellulases" in R. S. Tipson and D. Horton, eds., *Advances in Carbohydrate Chemistry and Biochemistry,* Vol. 32, Academic Press, New York, 1976, pp. 277–352.
55. F. E. Hammer, *Enzymes Used in Clinical Analytical Chemistry,* private communication, 1979.
56. *Economist,* 102 (Dec. 2, 1978).

General References

The expanded fundamentals of applied enzymology can be obtained by studying two books together: G. Reed, *Enzymes in Food Processing,* 2nd ed., Academic Press, New York, 1975, and J. Whitaker, *Principles of Enzymology for the Food Scientist,* Marcel Dekker, New York, 1972.

The state of the art can be substantially evaluated by following 2 annuals: A. Wiseman, *Topics in Enzyme and Fermentation Biotechnology,* Ellis Horwood, Chichester, Vol. 1 was in 1977, and D. Perlman, *Annual Reports on Fermentation Processes,* Academic Press, New York, Vol. 1 was in 1977; and two monthly journals: *Process Biochem.* and *Biotechnol. Bioeng.*

L. Sinnott, *Chemistry in Britain,* 293–297 (1979).

Additional sources that can provide useful and interesting information are the enzyme manufacturers listed in the text, and RANN (Research Applied to Needs Program, Division of Advanced Energy and Resources Research and Technology, National Sciences Foundation, Washington, D.C., 20550).

DON SCOTT
Fermco Biochemics Inc.

ENZYMES, THERAPEUTIC

The manufacture or processing of enzymes for use as drugs is a minor but important facet of today's pharmaceutical industry. This importance derives from several cardinal attributes of enzymes: they are, in the first place, catalysts of remarkable efficiency; because of the unique configuration of amino acid residues at their catalytic centers, they accelerate highly specific chemical and, therefore, pharmacologic reactions; as drugs must, they function effectively under chemically mild conditions, and often optimally under physiologic conditions; and, they are eliminated from the body in a reasonable time. Attempts to capitalize on these advantageous attributes of enzymes as drugs are now being made at virtually every pharmacologic research center in the world. Described below are certain of the pioneering experiments that laid the groundwork for the present intensive interest in enzymes as drugs.

Since the latter years of the 19th century, crude proteolytic enzymes have been used for gastrointestinal disorders, eg, pepsin for dyspepsia (see Gastrointestinal agents). However, controlled trials of their efficacy were seldom attempted. In fact, other than as digestants, enzymes were largely ignored as drugs until Emmerich and his associates observed in 1902 that an extracellular secretion of *Bacillus pyocyaneus* was capable of killing anthrax bacilli and of protecting mice from otherwise lethal innocula of the bacterium (1). Emmerich deduced that the secretion in question was a nuclease, ie, it was acting by enzymatically decomposing nucleic acids (Table 1). This milestone study gradually opened the way for the use of parenteral enzymes in the treatment first of infections, then of cancer, and finally of a diverse spectrum of diseases.

Table 1. Landmarks in the History of Therapeutic Enzymes

Year	Authors	Discovery	Refs.
1902	Emmerich, Low, and Korschun	extracellular secretion (nuclease) of *B. pyocyaneus* capable of protecting mice from lethal effects of anthrax bacilli	1
1931	Avery and Dubos	protective action of a polysaccharide-degrading enzyme against pneumococcal infection in mammals	2
1933	Tillett and Garner	lysis of fibrin by streptokinase from culture filtrates of *Streptococci*	3
1961	Broome	inhibition of lymphoma cells by L-asparaginase from guinea-pig serum	4–5
1965	Fletcher and co-workers	the development of urokinase as a thrombolytic agent	6
1967	Hug and Schubert	infusion of glucosidase from *Aspergillus niger* into patients with Type II glycogenosis (Pompe's disease)	7

Physical and Chemical Properties

Insofar as is known, all enzymes are proteins (see Enzymes, industrial). Attached to the enzyme may be lipid, carbohydrate, nucleotide or metal-containing prosthetic groups (8–13). The pH-activity optima of the therapeutic enzymes range from <2.0 for pepsin to >8.0 in the case of L-asparaginase. Needless to say, sharp optima at pHs removed from neutrality mitigate against activity under physiologic circumstances.

The activity–temperature optima of most therapeutic enzymes are ca 37–47°C. To date the use of enzymes from thermophilic bacteria for preparative pharmacologic purposes (eg, in extracorporeal reactors run at elevated temperatures to maximize the rate of catalysis) has not been attempted.

The types of reactions catalyzed by therapeutic enzymes are, in general, restricted to those able to operate with cofactors or cosubstrates available in the extracellular environment. This restriction is dictated by the relative impermeability of the ordinary cell membrane to molecules having the dimensions of proteins. For this reason, the majority of the therapeutic enzymes used to date are hydrolases [(EC 3.1 through 3.11), where EC stands for the enumeration system of the Enzyme Commission of IUPAC (14)], using water as cosubstrate (Table 2). An additional group of enzymes with pharmacologic uses have built-in cofactors in the form, eg, of pyridoxal phosphate, flavin nucleotides, or zinc (15). On the other hand, the synthetase and analogous multisubstrate enzymes requiring high energy phosphate are not, so far, available for use as drugs because the requisite cosubstrates are absent from the extracellular space or present in concentrations too low to permit adequate catalysis (16).

It is obvious that high maximal velocities are necessary for therapeutic efficacy. In the case of Michaelis constants, only enzymes with constants in the neighborhood of $1 \times 10^{-5} M$ are ordinarily effective as drugs *in vivo* (17). This relationship is a consequence of the fact most substrates are present in body fluids at submillimolar concentrations.

Shipment

Although commercial enzymes can be shipped as suspensions or solutions in stabilizers such as glycerol or as ammonium sulfate pastes, these methods are not suitable for parenteral drugs. Therefore, dry preparations without high concentrations of salts, excipients, or reducing agents have, in large part, been adopted for the preparation of therapeutic enzymes; lyophilization in the presence of mannitol and a physiologic buffer is the most common preparative method for enzymes. Even in the dry state, many proteins are denatured by heat so that refrigeration in transit, often with dry ice, is mandatory.

Economic Aspects

As of 1977 the economic importance of therapeutic enzymes, especially those intended for parenteral use, has been minor: For example, the manufacture of the therapeutic enzymes L-asparaginase, streptokinase, urokinase, and Arvin represents a very minor fraction of the sales of the pharmaceutical houses marketing them.

Table 2. Principal Therapeutic Enzymes

Name of enzyme	CAS Registry No.	Enzyme Commission No.	Catalysis	Use
neuraminidase	[9001-67-6]	3.2.1.18	hydrolysis of terminal acylneuraminyl residues	antineoplastic
ribonuclease	[9001-99-4]	2.7.7.16	RNA → oligoribonucleotides	antineoplastic
proteases		3.4	protein → amino acids and peptides	antineoplastic
L-α-arabinofuranosidase	[9067-74-7]	3.2.1.55	L-α-arabinofuranoside → alcohol + L-arabinose	antineoplastic
brinase	[9074-07-1]	3.4.21.15	fibrinogen → fibrin	fibrinolytic
α-glucosidase	[9001-42-7]	3.2.1.20	D-1,4α-glucoside → D-α-glucose	metachromatic; leukodystrophy
β-glucosidase	[9001-22-3]	3.2.1.21	D-1,4β-glucoside → D-β-glucose	Type A glycogenosis
arylsulfatase	[9016-17-5]	3.1.6.1	phenolsulfate → phenol + sulfate	metachromatic; leukodystrophy
α-galactosidase	[9025-35-8]	3.2.1.22	D-α-galactoside → D-α-galactose	Fabry's disease
β-galactosidase	[9031-11-2]	3.2.1.23	D-β-galactoside → D-β-galactose	Fabry's disease
bromelain	[37189-34-7]	3.4.22.4	protein → amino acids and peptides	digestive insufficiency
collagenase	[9001-12-1]	3.4.24.3	collagen → amino acids and peptides	dermal ulcers
papain	[9001-73-4]	3.4.22.2	proteins → amino acids and peptides	reduction of edema after dental surgery
L-asparaginase	[9015-68-3]	3.5.1.1	L-asparagine → L-aspartic acid + NH_3	antineoplastic
streptokinase	[9025-51-8]	3.4.22.10	plasminogen → plasmin	thrombolytic
Arvin	[9046-56-4]	3.4.21.5	fibrinogen → fibrin	fibrinolytic
lipase	[9001-62-1]	3.1.1.3	carboxylic ester → alcohol + carboxylic acid	pancreatic deficiency
L-glutaminase	[9001-47-2]	3.5.1.2	L-glutamine → L-glutamic acid + NH_3	antineoplastic
L-phenylalanine ammonialyase	[9024-28-6]	4.3.1.5	L-phenylalanine → transcinnamic acid + NH_3 or L-tyrosine → transcoumaric acid	antineoplastic
L-arginase	[9000-96-8]	3.5.3.1	L-arginine → L-ornithine + urea	antineoplastic
L-tyrosinase	[9002-10-2]	1.10.3.1	L-tyrosine + O_2 → dihydroxyphenylalanine + H_2O	antineoplastic
L-serine dehydratase	[9014-27-1]	4.2.1.13	L-serine → pyruvate + NH_3	antineoplastic
L-threonine deaminase	[9024-34-4]	4.2.1.16	L-threonine → 2-ketobutyric acid + NH_3	antineoplastic
L-tryptophanase	[9024-00-4]	4.1.99.1	L-tryptophan → indole + pyruvate + NH_3	antineoplastic
deoxyribonuclease	[9003-98-9]	3.1.4.5	DNA → oligodeoxyribonucleotides	chronic bronchitis
trypsin	[9002-07-7]	3.4.21.4	protein → peptides	athletic injuries
chymotrypsin	[9004-07-3]	3.4.21.1	protein → peptides	athletic injuries

Table 2 (continued)

Name of enzyme	CAS Registry No.	Enzyme Commission No.	Catalysis	Use
hyaluronidase	[37326-33-3]	3.2.1.35	random hydrolysis of 1,4-linkage between 2-acetamido-2-deoxy-D-β-glucose and D-glucuronate residues in hyaluronate	acute myocardial infarction
penicillinase	[9001-74-5]	3.5.2.6	penicillin → penicilloate	penicillin allergy
superoxide dismutase	[9054-89-1]	1.15.1.1	$O_2^{\perp} + O_2^{\perp} + 2H^+ \rightarrow O_2 + H_2O_2$	anti-inflammatory
catechol-1,2-dioxygenase	[9027-16-1]	1.13.11.1	catechol → cis-cis-muconate	poison ivy
lysozyme	[9001-63-2]	3.2.1.17	hydrolysis of 1,4-β-linkages between N-acetylmuramic acid and 2-acetamido-2-deoxy-D-glucose residues in a mucopolysaccharide or mucopeptide	antibiotic

Specification, Standards, and Quality Control

The FDA generally accepts enzymatic assay of activity as an index of correctness of labeling and for purposes of dosage. However, such assays provide no information on the purity of a given preparation. In the case of oral enzymes, such as digestive supplements, total activity, not purity, is the critical attribute inasmuch as the gastrointestinal tract can readily dispose of most inert contaminants. With parenteral formulations, however, the manufacturer must assure that the enzyme is reasonably pure. Tests for separating the components of proteinaceous mixtures are ordinarily required to establish this point. Immunologic tests provide added evidence of homogeneity. Of the impurities that must be excluded, the toxic bacterial lipopolysaccharide endotoxin is the most prominent and the most dangerous, inasmuch as miniscule amounts of it are capable of provoking fever and vascular collapse. For this reason, all preparations of enzymes intended for parenteral use must be nonpyrogenic in the USP rabbit assay (18). Similarly, all such preparations must be sterile. Since many therapeutic enzymes are derived from bacterial sources, these requirements can be formidable for a manufacturer.

Analytical Methods

Historically, colorimetric or spectrophotometric analyses have been used most frequently for therapeutic enzymes. At present, however, given the widespread availability of a variety of advanced analytical instruments, no generalizations are possible. Thus nephelometry is widely used to measure lipase in oral pancreatic supplements (19) and fluorimetry with umbelliferone conjugates of target sugars as substrates (eg, with 4-methylumbelliferone-N-acetylglucosaminate), and to estimate enzymes designed to overcome the in-born errors in metabolism (20). The sensitivity of this technique is derived from the intense fluorescence of the reaction product 4-methylumbelliferone. A still more sensitive general approach to analysis is required for the measurement of the *in vivo* residence times of enzymes in body fluids. Parenteral administration dilutes the dose greatly. Radiometric methods using substrates of high specific activity have been employed, eg, to measure L-asparaginase in body fluids at concentrations as low as 1×10^{-13} M (21). Lastly, as mentioned earlier, techniques useful in separating and measuring proteins as a class have been widely used to assure homogeneity; these techniques include isoelectric focusing, gel and starch electrophoresis, analytical ultracentrifugation, sucrose gradient ultracentrifugation (see Centrifugal separation), ion exchange and affinity chromatography (see Chromatography, affinity), as well as immunological or serological techniques.

Toxicology

In general, the FDA requires that the safety of any enzyme intended for parenteral use be established in test animals (eg, rat and dog) under the conditions anticipated in clinical trials. Model protocols have been issued (22). Toxicological examination of each lot is usually unnecessary when the purification procedures yield highly reproducible preparations.

Uses

Therapeutic enzymes have a broad variety of specific uses: as oncolytics, as anticoagulants or thrombolytics, and as replacements for metabolic deficiencies. Additionally, there is a growing group of miscellaneous enzymes of diverse function.

Oncolytic Enzymes. The oncolytic enzymes fall into two major classes: those that degrade small molecules for which neoplastic tissues have a requirement, and those that degrade macromolecules such as membrane polysaccharides, structural and functional proteins, or nucleic acids (see Chemotherapeutics, antimitotic). At present, tumor-cell specificity is observed only in the former category. An example is the typical oncolytic enzyme L-asparaginase (4). Certain tumor cells are deficient in their ability to synthesize the nonessential amino acid L-asparagine, and are forced to extract it from body fluids; by contrast, most normal cells can produce their own L-asparagine (5). Thus L-asparagine can be viewed as a quasi-essential amino acid for protein synthesis whose removal from select tumors can induce a state of fatal starvation. L-Asparaginase given parenterally acts in this way in many susceptible tumors. Unfortunately, only acute lymphocytic leukemia ordinarily responds to chemotherapy with the enzyme (17). Nevertheless, the response of this one tumor type is promising (60% incidence of complete remissions in almost 6000 cases), and the search is being extended to other enzymes that degrade small molecules as indicated in Table 3. A bifunctional amidohydrolase, L-glutaminase-L-asparaginase (L-glutamine and L-asparagine amidohydrolase), is undergoing clinical trials in the United Kingdom and shows activity in other diseases (23–24).

L-Methioninase (L-methionine-γ-lyase), which effectively dismantles L-methionine to yield methanethiol, ammonia, and α-ketobutyric acid, is effective against several murine tumors, but no clinical trials have been undertaken (25–26). The same is true for L-phenylalanine ammonialyase which deaminates both L-phenylalanine and L-tyrosine yielding *trans*-cinnamic and *trans*-coumaric acids, respectively (27–28). In the case of both of these enzymes, mammalian cells are incapable of reconstructing the substrate from the products, so the reaction is effectively irreversible *in vivo*. Other

Table 3. Oncolytic Enzymes That Degrade Small Molecules

Findings	Refs.
antilymphoma effects of L-asparaginase from Guinea pig serum	4–5
inhibition of tumor growth by L-glutaminase	23
treatment of acute leukemia in humans by L-glutaminase-L-asparaginase [39335-03-0]	24
antitumor activity of L-methioninase [42616-25-1]	25–26
effect of L-phenylalanine ammonialyase on lymphocytic leukemia	27–28
antitumor activity of L-arginase	29
inhibition of tumor growth by L-tyrosinase	30
treatment of experimental tumors with L-serine dehydratase	31
antitumor activity of L-threonine deaminase	32
antitumor activity of L-tryptophanase	33
selective toxicity of diphtheria toxin for malignant cells	34
inhibition of leukemia cells by L-cystine-cysteine degrading enzymes	15

amino-acid-degrading enzymes with oncolytic activity vs experimental tumors include: L-arginase (L-arginine amidinohydrolyase) (29); L-tyrosinase (30); L-serine dehydratase (31); L-threonine deaminase (32); and, indolyl-3-alkane hydroxylase [9024-00-4] which decomposes L-tryptophan (33). This list is expanding at a notable rate since the technique of enrichment culture yields microbial enzymes capable of decomposing amino acids in novel ways.

Diphtheria toxin, a different type of oncolytic enzyme still in the experimental stage, catalyzes transfer of the adenosine diphosphate ribose (ADP-ribose) moiety of nicotinamide adenine dinucleotide (NAD) to elongation factor 2 (33,35). This enzyme process halts protein synthesis. Most important from a chemotherapeutic standpoint is the observation that protein synthesis in tumor cells is one hundred to ten thousand times more sensitive to this toxin than the analogous process in normal cells (34).

Among the oncolytic enzymes that degrade macromolecules neuraminidase (36), ribonuclease (37), and a diverse group of proteases (38) are the most prominent examples (Table 4). Neuraminidase removes sialic acid residues from the surface of (neoplastic) cells, thereby altering their immunogenicity, and rendering them sensitive to immune response (41). To date this effect has been studied mainly in experimental trials. In addition, several ribonucleases have shown modest activity against experimental murine neoplasms, but their use is beset by the problem of forcing these molecules into the cytoplasm where the substrate ribonucleic acid (RNA) occurs (see Derivatives).

Pepsin, given intralesionally, was one of the first enzymes used for the chemotherapy of cancer, but its clinical use was surrounded by controversy and has ceased (39). On the other hand, a mixture of vitamins and proteolytic enzymes, marketed

Table 4. Oncolytic Enzymes That Degrade Macromolecules

Findings	Ref.
antitumor activity of neuraminidase [9001-67-6]	36
antitumor activity of pancreatic ribonuclease [9001-99-4]	37
antitumor activity of a mixture of proteolytic enzymes	38
treatment of inoperable malignant tumors with pepsin [9001-75-6]	39
antitumor activity of carboxypeptidase G_1 [9054-73-3]	40

under the name Wobe Mugos, is widely prescribed for the control of cancer in Europe and appears to be of some use in the palliation of the disease (42). The carboxypeptidases are catalysts that cleave the carboxyl-terminal residue of many peptides; certain of these enzymes also are capable of hydrolyzing the L-glutamyl moiety of folic acid (43). In so doing, they achieve a state of folic acid deficiency deleterious to the tumor cell. Use of this approach has, so far, been restricted to test animals, but human trials are beginning with a preparation designated carboxypeptidase G_1. Because carboxypeptidase G_1 can decompose the drug methotrexate (a folic acid analogue and antagonist), the enzyme is also envisioned as an antidote to overdoses of methotrexate (40).

Enzymes as Activators of Oncolytic Prodrugs. Another innovative use of enzymes as therapeutic agents entails their administration to tumor-bearing subjects along with a prodrug conjugated to a functionality susceptible to attack by the enzyme (see Microbiological transformations). In order to achieve the requisite selectivity, advantage is taken of two features: the acidic intracellular environment of many neoplasms as compared to normal tissues, and an enzyme with an acidic pH-activity optimum.

Using a combination of α-L-arabinofuranosidase [9067-74-1] from *Aspergillus niger* and β-peltatin-α-L-arabinofuranoside, workers at the Central Cancer Research Institute in Berlin have successfully used this technique to depress [^{3}H]thymidine incorporation by mammary adenocarcinomas (44). Other applications of this approach can be anticipated.

Enzymes as Anticoagulants. Historically, the management of thromboembolic vascular disease has relied on the use of anticoagulants, such as heparin and coumarin, to inhibit the formation of fibrin clots. However, the recognition that lysis of preformed fibrin could be accomplished *in vivo* by a process involving the conversion of inactive plasminogen to the fibrinolytic enzyme plasmin (Fig. 1) led to an alternative approach. Subsequently it was found that plasminogen activators, as opposed to preformed plasmin itself, can produce controlled enzymatic fibrinolysis *in vivo*. Two of these plasminogen activators, streptokinase and urokinase, as well as other fibrinolytic enzymes are discussed below (see Blood, coagulants and anticoagulants).

In 1933 it was discovered that filtrates of cultures of hemolytic streptococci were capable of lysing fibrin (3). The active principle, streptokinase, was subsequently purified and its properties investigated (45). The mechanism of interaction between streptokinase and plasminogen remains controversial. Native streptokinase *in vitro* inefficiently catalyzes the conversion of human plasminogen to plasmin *via* the hydrolytic cleavage of a single L-arginyl-L-valyl bond. However, streptokinase *in vivo* is probably activated first by complexing with plasminogen in a one-to-one complex (Fig. 1). This streptokinase–plasminogen complex in turn converts free plasminogen to plasmin with high efficiency (46–47).

Early studies demonstrated that thrombi induced in the marginal ear vein of rabbits could be lysed by the systemic infusion of streptokinase (48). Artificially induced thrombi in the forearm veins of human volunteers were similarly shown to be lysed by intravenously administered streptokinase (49). This enzyme has been shown to be of considerable use in the treatment of deep venous thrombosis (50) and pulmonary embolism (51). It may also be of utility in the treatment of clotted hemodialysis shunts, priapism, and disseminated intravascular coagulation (52–55) (Table 5).

Urokinase is another enzyme that acts as a plasminogen activator. Its presence in human urine was first reported in 1947 (64). Since that time, methods for the limited

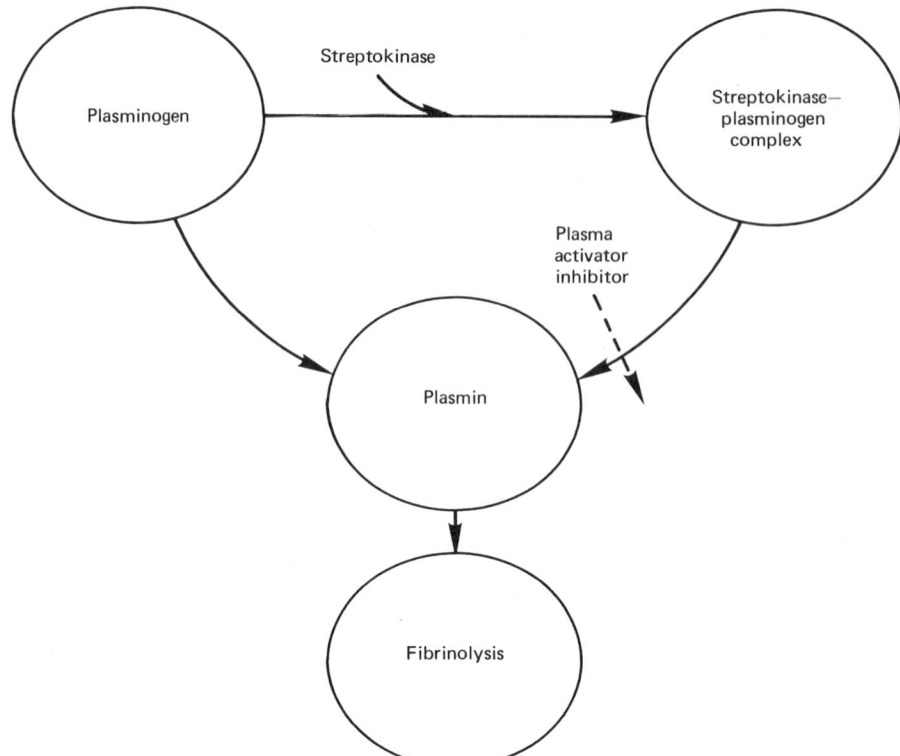

Figure 1. Activation of plasminogen by streptokinase.

production of urokinase from human urine and kidney tissue culture have been reported (65). Urokinase is a trypsin-like enzyme capable of hydrolyzing L-lysine and L-arginine esters. Its activation of plasminogen probably involves a first-order reaction in which a single L-arginyl-L-valyl bond in the plasminogen molecule is cleaved (66).

As compared to streptokinase, clinical experience with urokinase is limited. However, when preliminary studies in man showed that infusions of urokinase could produce a sustained thrombolytic state and could lyse experimental forearm venous thrombi, large-scale controlled studies in the treatment of pulmonary emboli were undertaken. The data showed that a 12-h infusion of urokinase significantly accelerated lysis of pulmonary emboli, improved perfusion, and reversed hemodynamic abnormalities (6,51). Nevertheless, urokinase did not influence mortality or recurrences. This study concluded that urokinase is more effective than heparin alone in the rapid resolution of pulmonary emboli. For this reason, urokinase may assume a place in treatment of acute pulmonary embolism, particularly in candidates who are poor surgical risks for embolectomy. Urokinase has also been of use in the therapy of massive cerebral sinus thrombosis (67).

In 1963 it was observed that the blood of victims of the bite of the Malayan pit viper *Ancistrodon rhodostoma* was incoagulable. It was subsequently established that this effect was due to marked hypofibrinogenemia (68). Paradoxically, little bleeding was seen in the clinic. These findings suggested that the active component responsible for the depletion of plasma fibrinogen ought to be of use in the control of thrombotic states. Subsequently, this component was purified and named Arvin (69). The activity

Table 5. Uses of Anticoagulant Enzymes

Thrombolytic enzyme	CAS Registry No.	Condition				
		Deep vein thrombosis	Peripheral arterial occlusion	Clotted hemodialysis shunt	Pulmonary embolism	Myocardial infarction	Stroke
streptokinase	[9025-51-8]	beneficial[a]	probably not beneficial[b]	beneficial[c]	beneficial[d]	probably not beneficial	probably not beneficial
urokinase	[9039-53-6]	beneficial[e]	probably not beneficial	probably beneficial[i]	beneficial[d]	probably not beneficial[f]	not beneficial[g]
Arvin	[9046-56-4]	beneficial[h]	probably not beneficial		NT[j]	NT[j]	NT[j]
reptilase	[9039-61-6]	beneficial[k]	NT[j]	beneficial[k]	NT[j]	NT[j]	NT[j]
brinase	[9074-07-1]	beneficial[l]	of limited benefit[m]	beneficial[l]	beneficial[m]	NT[j]	NT[j]

[a] Ref. 50.
[b] Ref. 56.
[c] Ref. 55.
[d] Ref. 51.
[e] Ref. 57.
[f] Ref. 54.
[g] Ref. 58.
[h] Ref. 59.
[i] Ref. 60.
[j] NT = definitive clinical trials not published.
[k] Ref. 61.
[l] Ref. 62.
[m] Ref. 63.

of Arvin is due to its action on fibrinogen. Normally, thrombin cleaves two peptides, A and B, from fibrinogen. The resulting fibrin monomer then polymerizes into a stable clot. Arvin, however, splits only peptide A from fibrinogen (70). This partially degraded fibrin monomer also polymerizes, but into a conformation different from that of normal fibrin (71). The resultant clot is comparatively friable and readily dissolved. Arvin makes fibrinogen unavailable for the action of thrombin.

In human trials, the intravenous injection of Arvin produced the expected hypofibrinogenemia. Arvin was found not to alter other coagulation factors, reduce the platelet count, nor fragment erythrocytes (72). In subsequent clinical studies, Arvin has been used successfully to treat deep venous thrombosis with fewer side effects and more controlled thrombolysis as compared to streptokinase. Complications are rare: bleeding is seen in fewer than 5% of patients, a lower incidence than with heparin (59). Recently, subcutaneous administration of Arvin has been shown to be of benefit in the treatment of intermittent claudication (73).

Reptilase, a fraction extracted from the venom of the South American snake *Bothrops atrox*, is very similar to Arvin. It also splits the A peptide fragment from fibrinogen and and presumably exerts its defibrinating effect in the same fashion (74). Likewise, hypofibrinogenemic states can be induced without concomitant bleeding or other side effects. Clinical investigation of reptilase has been limited as compared to Arvin. Uncontrolled studies involving a number of patients treated simultaneously with anticoagulants and reptilase suggest some therapeutic benefits. Moreover, no adverse effects were noted (75). Recently, reptilase was shown to be of use as an alternative to heparin or dicoumarol in the prevention of re-thrombosis in patients undergoing surgical venous embolectomy; bleeding complications were rare (61).

Brinase, a plasmin-like proteolytic enzyme found in extracts of the mold *Aspergillus oryzae*, is capable of hydrolyzing fibrin and fibrinogen as well as casein and gelatin. Studies have shown that patients on chronic hemodialysis with clotted arteriovenous cannulas demonstrated restored vessel patency within several minutes after treatment with brinase (62,76). Disadvantages of brinase therapy include the requirement of prior administration of a neutralizing dose to saturate serum inhibitors. Toxic side effects, moreover, are occasionally severe: marked edema, hematoma, and local pain were reported in a significant number of patients in a recent study (63). However, in most investigations, these side effects are less marked.

Enzyme Replacement Therapy. The treatment of enzyme deficiency states represents an obvious use of enzymes. For example, oral pancreatic extracts have commonly been used in the treatment of cystic fibrosis and pancreatic insufficiency. Unfortunately, the wide variety of nonstandardized preparations currently available and the variation in activity of these agents, tend to obscure their clinical efficacy. The most commonly available preparations are of three types: (*1*) pancreatin [8049-47-6], an alcoholic extract of pig pancreas containing amylase [9000-92-4] and trypsin activity, (2) pancrealipase [53608-75-6] (Cotazym), a lipase-enriched extract, and (*3*) viokase [71060-52-1], also enriched in lipase.

More intriguing is the treatment of the inborn errors of metabolism in which deficiency of a single enzyme leads to accumulation of abnormal amounts of substrate. With the recognition that many of these errors are owing to inadequacies of lysosomal enzymatic catabolism, it was reasoned that exogenously administered enzyme might react with and dispose of such accumulations.

The infusion of crude glucosidase from *Aspergillus niger* into patients with type

II glycogenosis, a condition attributed to a deficiency of this enzyme, was reported in the mid 1960s (7). Although these studies indicated a variable increase in hepatic glucosidase activity, no clinical improvements were noted. However, these early enzyme preparations were both pyrogenic and immunogenic. Similar results were obtained from efforts to treat patients homozygous for metachromatic leucodystrophy with intravenous and intrathecal arylsulfatase, ie, increased enzyme levels in serum and hepatic tissue but with pronounced pyrogenicity or hypersensitivity and lack of clinical improvements (77). Infusion of normal plasma, which contains galactosidase, the enzyme missing in Fabry's disease, reduced the amount of abnormal substrate in the plasma and tissues of two patients heterozygous for this disorder (78). Similar therapy with normal plasma or concentrates of leukocytes has been attempted in the treatment of the mucopolysaccharidoses, with inconsistent reduction in substrate levels. At present it is believed that such therapy is likely to be of little benefit (79). A more radical approach involves the transplantation of organs or organ fragments as sources of the deficient enzyme. Despite immunologic complications, this approach has achieved remission of congenital disorders of bilirubin and copper metabolism (80–81).

Depolymerizing Enzymes. Proteolytic enzymes have been widely used as antiinflammatory agents. Reduction of inflammation and edema is ascribed to the dissolution of soft fibrin and to the clearance of proteinaceous debris found in inflammatory exudates. Nevertheless, the use of these enzymes remains controversial. Most studies have been anecdotal. However, several controlled studies are discussed here.

The bromelains, proteolytic enzymes extracted from the stem of the pineapple plant, were demonstrated to produce detectable serum proteolysis after oral administration (82). Later studies showed that orally administered bromelain significantly reduced bruising following obstetrical manipulations (83). Bromelain–pancreatin combinations have been found to be more effective in digestive insufficiency than either pancreatin alone or placebo (84). Bromelains may also be of use for enhancing the efficacy of antibiotics (qv), notably tetracycline, when the two are given concomitantly (85).

Collagenase, an enzyme unique in that it hydrolyzes native collagen and spares hydrolysis of other proteins, has been used in the debridement of dermal ulcers and burns. Its use has also been suggested in the lysis of diseased intervertebral disks (86–87). Another protease, papain, has been shown to produce marked reduction of obstetrical inflammation and swelling, and of the edema following dental surgery (88–89). Streptokinase–streptodornase contains both protease and nuclease activity. Its success in the treatment of inflammatory exudates is based on the action of streptokinase on plasminogen with the lysis of fibrin clots and the ability of streptodornase to depolymerize deoxyribonucleic acid (DNA) and liquefy exudates. Streptokinase–streptodornase has also shown to be of use in the treatment of empyema, hemothorax, and loculated or clotted pleural effusions (90).

Deoxyribonuclease, an enzyme that degrades nucleic acids, has recently been investigated as a mucolytic agent for use in patients with chronic bronchitis (91). Although effective in producing liquefaction of secretions, it offers no advantage over standard mucolytic agents such as N-acetyl-L-cysteine. The enzyme lysozyme hydrolyzes the chitins and mucopeptides of bacterial cell walls (14). Accordingly, it has been used as an antibacterial agent usually in combination with standard antibiotics.

Controlled studies are few; however, one study showed that lysozyme–antibiotic combinations appear to produce better results in the treatment of post-operative wound infections than antibiotic alone (92). Topical applications were also shown to be beneficial in the debridement of serious burns, cellulitis, and dermal ulceration.

The proteolytic enzymes, trypsin and chymotrypsin (or their combination, chymoral), have been successfully used in the treatment of postoperative hand trauma, athletic injuries, and sciatica (93–95).

Hyaluronidase exerts its action by destroying the intracellular ground substance hyaluronic acid, thus allowing diffusion of vital molecules through this normally impermeable connective tissue barrier. In 1959 improvements of the electrocardiograms of patients with acute myocardial infarction were demonstrated following treatment with hyaluronidase (96). Later work showed that hyaluronidase, by increasing diffusion of nutrients through the liquefied ground substance, limits extension of the infarct (97–98).

The earliest antibiotics discovered were proteolytic enzymes, mostly of bacterial origin. Although pyocyanase and proteolytic extracts of *Bacillus mesentericus* and *Serratia marcescens* are currently not first-line antibiotics, interest has, nevertheless, been focused on another bacterial enzyme, lysostaphin [9011-93-2], whose lytic effects on coagulase-positive *Staphylococcus aureus* are presently under considerable study. The literature on lysostaphin has been extensively reviewed (99): it is a protease that lyses susceptible cells in a highly efficient manner, probably by peptidase-like cleavage of the glycoprotein of the bacterial cell wall (100). At present, lysostaphin has been administered in humans only topically for reduction of staphylococcal carrier rate in the nose and throat where it has been found to be effective and nontoxic. Ultimately, the potential drug applications are twofold: (*1*) since lysostaphin is unique among antistaphylococcal agents in that it destroys bacteria, whether they are active or resting, and is thus capable of killing large numbers of organisms, it may be useful in instances of endocarditis and other conditions where an initial and rapid reduction in bacterial count is necessary; (*2*) more significantly, since the *in vivo* effectiveness of this enzyme against methicillin-resistant strains of *S. aureus* has been demonstrated, lysostaphin might prove useful in the treatment of methicillin-resistant staphylococcal infections, of which many have begun to appear in Europe as well as in the United States.

Enzymes as Antidotes. Another recent development of enzyme chemotherapy is their use as antidotes. The first reported example was the administration of rhodanase [9026-04-4] (thiosulfate–sulfurtransferase) along with the cosubstrate, sodium thiosulfate, to counteract cyanide poisoning; this approach proved to be markedly efficacious (101). More recently, superoxide dismutase [9054-89-1], an enzyme that decomposes the highly toxic oxygen free radical, O_2^-, has been put into veterinary use as an anti-inflammatory agent with efficacy in the treatment of conditions such as traumatic arthritis of horses (102) (see Veterinary drugs). The ability of this enzyme to nullify the toxic effect of putative superoxide-generators *in vivo* is also widely used to implicate the oxygen-free radical as a proximate mediator of such effects. Inasmuch as free-radical toxicity is identified as operative in poisonings with an increasing variety of pharmacologic agents, it can be predicted that the use of superoxide dismutase will increase proportionately (102–103). In this regard, it is useful to point out that this protein can be purified from human erythrocytes, and so is unlikely to be immunogenic.

Other examples include the use of penicillinase to counteract allergic reactions to penicillin (104), and the application of a catechol-degrading enzyme to skin contaminated with poison ivy (105). In this case, the topically applied enzyme degrades the toxin, and is, therefore, properly viewed as a prophylactic.

Derivatives of Enzymes

A tremendous amount of effort is being devoted to the derivatization of enzymes in an attempt to improve their chemical, pharmacologic, or immunologic properties. These efforts fall into three main categories: (*1*) immobilization of unaltered enzyme (see Enzymes, immobilized); (*2*) chemical conjugation of small ligands to the parent enzyme; and (*3*) conjugation with large polymeric molecules such as other proteins, polysaccharides or synthetic homopolymers.

Immobilization of therapeutic enzymes has been attempted principally in an effort to circumvent the immunogenicity of these foreign proteins (106). Plates, coils, and tubes of nylon, Dacron, plastic, glass, or polyacrylamide have been used as the inert supports (12,107–109), and coupling has been achieved by conjugation to amino and carboxyl functions on the protein (8–9). Most of the resultant devices are intended for extracorporeal perfusion to prevent enzyme contact with the patient's immune system. As a result such enzyme reactors resemble the hemodialysis units used for renal failure (see Dialysis). In special cases, however, the enzyme has been immobilized on the corrugated Dacron tubing used for vein grafts and implanted as a therapeutic vascular prosthesis in one of the great vessels (106). To prevent thrombosis, such grafts must also be thoroughly impregnated with the anticoagulant, heparin (110) (see Prosthetic and biomedical devices). Although generalizations are hazardous, at present neither approach is compellingly attractive for use with cancer-chemotherapeutic or metabolic-replacement enzymes because of: the comparatively limited number of catalytically active molecules ordinarily conjugated to the support in question; slow leeching of immunogenic protein from any such reactor (108); and the surgical strain involved in the establishment of the requisite vascular circuitry. Moreover, this approach is only possible when the enzyme in question operates by depleting a target molecule from the circulation, a general condition that is seldom met in the metabolic diseases.

The second major means of modifying therapeutic enzymes to enhance their effect entails the attachment of small molecules (mol wt 1000) to reactive functions on the protein, or conversely the detachment of groups from areas of the molecule not directly involved in the catalytic act. For example, chemical attachment of simple sugars or oligosaccharides to the oncolytic enzyme L-asparaginase greatly modifies and, in many cases, prolongs its residence time in plasma (8–9). Since the rate of clearance of a therapeutic enzyme is often directly related to its half-life in body fluids, this expedient can be used to unmask the pharmacologic activity of an otherwise inert molecule. Cleavage of critical charge-bearing groups can also alter the ionization of an enzyme, leading in some cases to enhanced activity under physiologic conditions. For example, deamination of L-asparaginase with nitrous acid leaves intact amidohydrolytic (and therefore oncolytic) activity but prolongs its residence time in the bloodstream (111).

In the third major approach to derivatization, macromolecules are attached to therapeutic enzymes through single or multiple chemical bridges. These modifications can alter the rate of removal of the protein from the circulation in much the same way as small molecules do. However, although immunogenicity is one of the salient drawbacks to the long-term use of enzymes as drugs, the primary goal of this modification is to mask the antigenic centers on the enzyme, while leaving intact its catalytic function. In one successful case, poly(ethylene glycol) conjugates of both insulin and albumin are markedly hypoimmunogenic as compared to the unaltered antigens (112).

By happenstance or design, macromolecules that affix to cellular targets have also been linked to therapeutic enzymes to enhance their selectivity. The plant lectins, eg, when conjugated to L-asparaginase direct the enzyme to the surface of those cells possessing high-affinity sugar receptors for these agglutinins (13,113). Antitumor antibodies can also be conjugated along similar lines.

In addition to modifying the enzyme itself, it is also possible to modify its packaging. As has been noted, enzymes are generally excluded from the intracellular domain. However, it has been observed that incorporation of these proteins in the aqueous phase of man-made liposomes (unilamellar or multilamellar lipid spheroids with an aqueous center) permits entry into the cytoplasm either by ingestion or via fusion of membranes. This approach promises to be of most value in the therapy of the inborn errors of metabolism.

An analogous technique entails the packaging of enzymes in autologous erythrocytes whose pores have been transitorily enlarged by a hypotonic environment. If the extreme lability of the resultant "ghosts" can be overcome, eg, by membrane hardeners (113), then it is possible that enzyme cells whose longevity equals the erythrocyte (120 d in man) could be manufactured easily; such microreactors would theoretically function as destructor sites for any substrate molecules capable of transmembrane diffusion. For impermeant molecules, attempts are being made to conjugate the appropriate enzyme to the exofacial surface of the erythrocyte (114–115).

The use of enzymes in a therapeutic setting is still in its infancy; manipulations and modifications of these remarkable proteins for the purpose of increasing their pharmacologic efficacy are being made at an increasing rate, and are bound to increase their use in the clinic.

BIBLIOGRAPHY

1. R. Emmerich, O. Low, and A. Korschun, *Cent. Bakt.* **31**, 1 (1902).
2. O. T. Avery and R. Dubos, *J. Exp. Med.* **54**, 73 (1931); K. Goodner, R. Dubos, and O. T. Avery, *J. Exp. Med.* **55**, 393 (1931).
3. W. S. Tillett and R. L. Garner, *J. Exp. Med.* **58**, 485 (1933).
4. J. D. Broome, *Nature (London)* **191**, 1114 (1961).
5. J. D. Broome, *Trans. N.Y. Acad. Sci.* **30**, 690 (1968).
6. A. P. Fletcher and co-workers, *J. Lab. Clin. Med.* **65**, 713 (1965).
7. E. Hug and W. K. Schubert, *J. Cell. Biol.* **35**, 1 (1967).
8. J. W. Marsh, J. Denis, and J. C. Wriston, *J. Biol. Chem.* **252**, 7678 (1977).
9. J. W. Baynes and F. Wold, *J. Biol. Chem.* **251**, 6016 (1976).
10. G. Gregoriadis, *Methods Enzymol.* **44**, 698 (1976).
11. Y. Fishman and N. Citri, *FEBS Lett.* **60**, 17 (1975).
12. K. F. Driscoll and co-workers, *J. Pharmacol. Exp. Ther.* **195**, 382 (1975).
13. W. T. Shier, J. T. Trotter, and D. T. Astudillo, *Int. J. Cancer* **18**, 672 (1976).
14. *Enzyme Nomenclature*, American Elsevier Publishing Co., Inc., New York, 1973.
15. J. R. Uren and H. Lazarus, *Proc. Am. Assoc. Cancer Res.* **16**, 144 (1975).

16. C. G. Meadows and co-workers, *Cancer Res.* **36,** 167 (1976).
17. J. C. Wriston and T. O. Yellin, *Adv. Enzymol.* **39,** 185 (1973).
18. *USP XIX,* Mack Publishing Co., Easton, Pa., 1975.
19. B. L. Osler, ed., *Hawk's Physiological Chemistry,* McGraw-Hill Book Co., New York, 1964.
20. G. L. Dale and E. Beutler, *Proc. Natl. Acad. Sci. U.S.A.* **73,** 4672 (1976).
21. D. A. Cooney, H. A. Milman, and B. Taylor, Jr., *Biochem. Med.* **15,** 190 (1976).
22. D. J. Prieur and co-workers, *Cancer Chemother. Rep.* **4,** 1 (1973).
23. F. A. El-Asmar and D. M. Greenberg, *Cancer Res.* **26,** 116 (1966).
24. A. S. D. Spiers and H. E. Wade, *Br. Med. J.* **29,** 1317 (1976).
25. W. Kreis and C. Hession, *Cancer Res.* **33,** 1862 (1973).
26. *Ibid.,* p. 1866.
27. C. W. Abell, W. J. Stith, and D. S. Hodgins, *Cancer Res.* **32,** 285 (1972).
28. C. W. Abell, D. S. Hodgins, and W. J. Stith, *Cancer Res.* **33,** 2529 (1973).
29. M. Wolf and K. Ransberger, *Enzyme Therapy,* Vantage Press, New York, 1972.
30. R. Isaacs, *Fed. Proc.* **6,** 394 (1947).
31. M. D. Prager, "Tumor Inhibitory Enzymes" in R. L. Clark and co-eds., *Oncology 1970,* Vol. 2, Yearbook Medical Publishers, Inc., Chicago, Ill., 1971, p. 237.
32. R. S. Greenfield and D. Wellner, *Cancer Res.* **37,** 2523 (1977).
33. J. Roberts, *Proc. Am. Assoc. Cancer Res.* **18,** 31 (1977).
34. B. H. Iglewski and M. B. Rittenberg, *Proc. Natl. Acad. Sci. U.S.A.* **71,** 2707 (1974).
35. J. Everse and co-workers, *Proc. Natl. Acad. Sci. U.S.A.* **74,** 472 (1977).
36. L. Weiss, B. Fisher and E. R. Fisher, *Cancer* **34,** 680 (1974).
37. G. S. Tarnowski and co-workers, *Cancer Res.* **35,** 4074 (1976).
38. G. Stojanow, O. Weigelt, and K. Maehder, *Wien. Tierarztl. Monatsschr.* **59,** 193 (1971).
39. E. Payr, *Muench. Med. Wochenschr.* **69,** 1330 (1922).
40. B. A. Chabner, P. L. Chello, and J. R. Bertino, *Cancer Res.* **32,** 2114 (1972).
41. R. L. Simmons and co-workers, *Surgery* **70,** 38 (1971).
42. O. Weigelt, *Arzneim. Forsch.* **24,** 549 (1974).
43. C. A. Spillburg, J. L. Bethune, and B. L. Vallee, *Biochemistry* **16,** 1142 (1977).
44. B. Tschiersch and co-workers, *Cancer Treat. Rep.* **61,** 1489 (1977).
45. M. C. Davies, M. E. Englert, and E. C. De Renzo, *J. Biol. Chem.* **239,** 2651 (1964).
46. K. N. N. Reddy and G. Markus, *J. Biol. Chem.* **247,** 1683 (1972).
47. C. N. Chesterman, M. J. Allington, and A. A. Sharp, *Nature (New Biology)* **238,** 15 (1972).
48. A. J. Johnson and W. S. Tillett, *J. Exp. Med.* **95,** 449 (1952).
49. A. J. Johnson and W. R. McCarty, *J. Clin. Invest.* **38,** 1627 (1959).
50. V. V. Kakkar and co-workers, *Br. Med. J.* **1,** 806 (1969).
51. W. R. Bell, *Thrombos. Haemostasis. (Stuttg.)* **35,** 57 (1976).
52. E. M. Kohner and co-workers, *Br. Med. J.* **1,** 550 (1976).
53. J. H. N. Bett and co-workers, *Lancet i,* 57 (1973).
54. C. P. Aber and co-workers, *Br. Med. J.* **2,** 1100 (1976).
55. R. N. Brogden, T. M. Speight, and G. S. Avery, *Drugs* **5,** 357 (1173).
56. M. Verstraete, J. Vermylen, and M. B. Donati, *Ann. Int. Med.* **74,** 377 (1971).
57. V. Tilsner and W. Gruel, *Munch. Med. Wschr.* **117,** 865 (1975).
58. A. P. Fletcher and co-workers, *Stroke* **7,** 135 (1976).
59. D. A. Tibbutt and co-workers, *Br. J. Haematol.* **27,** 407 (1947).
60. G. H. Hall, H. M. Holman, and A. D. B. Webster, *Br. Med. J.* **4,** 591 (1970).
61. P. Olsson and co-workers, *Thromb. Res.* **9,** 277 (1976).
62. W. H. E. Roschlau, *Thromb. Diath. Haemorrh. Suppl.* **47,** 315 (1971).
63. D. Nyman and co-workers, *Thromb. Diath. Haemorrh.* **34,** 498 (1975).
64. R. G. MacFarlane and J. Pilling, *Nature (London)* **159,** 779 (1974).
65. A. Lesuk, L. Terminiello, and J. H. Traver, *Science,* 147, 580 (1965).
66. K. C. Robbins and co-workers, *J. Biol. Chem.* **242,** 2333 (1967).
67. A. P. Fletcher and N. Alkjaersig, *Proc. Serono Symp.* **9,** 203 (1977).
68. H. A. Reid and co-workers, *Lancet i,* 617 (1963).
69. M. P. Esnouf and G. W. Tunnah, *Br. J. Haematol.* **13,** 581 (1967).
70. W. H. Hollerman and L. J. Coen, *Biochem. Biophys. Acta* **200,** 587 (1970).
71. P. J. Gaffney and M. Brasher, *Nature (London)* **251,** 53 (1974).
72. W. R. Bell, G. Bolton, and W. R. Pitney, *Br. J. Haematol.* **15,** 589 (1968).

73. J. A. Dormandy, K. B. Goyle, and H. L. Reid, *Lancet i,* 625 (1977).
74. P. Mattock and M. P. Esnouf, *Nature (London)* **233,** 277 (1971).
75. N. Egberg and co-workers, *Thromb. Diath. Haemorrh. Suppl.* **47,** 379 (1977).
76. E. P. Frisch, *J. Clin. Pathol.* **25,** 653 (1972).
77. H. L. Greene, G. Hug, and W. K. Schubert, *Arch. Neurol. (Chicago)* **20,** 147 (1969).
78. C. A. Mapes and co-workers, *Science* **169,** 987 (1970).
79. A. J. Desnick, S. R. Thorpe, and M. B. Fiddler, *Physiol. Rev.* **56,** 57 (1976).
80. M. Phillippart, S. S. Franklin, and A. Gordon, *Ann. Int. Med.* **77,** 195 (1972).
81. R. S. DuBois and co-workers, *Lancet i,* 505 (1971).
82. R. D. Smyth, R. M. Brennan, and G. J. Martin *Am. J. Pharm.* **133,** 294 (1961).
83. G. I. Zatuchni and D. J. Colombi, *Obstet. Gynecol.* **29,** 275 (1967).
84. A. Capria and M. Marchioro, *Minerva Gastroenterol.* **17,** 84 (1971).
85. N. Vespa, *Minerva Med.* **63,** 3219 (1972).
86. B. J. Sussman, *J. Neurosurg.* **42,** 389 (1975).
87. C. Watts, R. Knighton, and G. Roulhac, *J. Neurosurg.* **42,** 374 (1975).
88. P. J. Pollack, *Corr. Ther. Res.* **4,** 229 (1962).
89. G. D. Magnes, *J. Am. Dent. Assoc.* **72,** 1920 (1966).
90. S. Sherry and A. P. Fletcher, *Clin. Pharmacol. Ther.* **1,** 202 (1960).
91. J. Lieberman, *Am. J. Med.* **49,** 1 (1970).
92. R. Pelligrini and P. Vartova, *Arzneim. Forsch.* **19,** 149, 375 (1975).
93. W. F. Rathgeber, *S. Afr. Med. J.* **45,** 181 (1971).
94. G. Gaspardy and co-workers, *Rheumatol. Phys. Med.* **11,** 14 (1971).
95. P. C. Shaw, *Br. J. Clin. Pract.* **23,** 25 (1969).
96. J. M. Oliveira, R. Carballo, and H. A. Zimmerman, *Am. Heart J.* **57,** 712 (1959).
97. P. R. Maroko and co-workers, *Circulation* **46,** 430 (1972).
98. P. R. Maroko and co-workers, *N. Engl. J. Med.* **296,** 898 (1977).
99. W. A. Zygmunt and P. A. Tavormina, *Prog. Drug Res.* **16,** 309 (1972).
100. C. A. Schindler and V. T. Schuhardt, *Proc. Natl. Acad. Sci. U.S.A.* **51,** 414 (1964).
101. D. M. Greenberg, "Thiosulfate Sulfurtransferase and Mercaptopyruvate Sulfurtransferase" in B. Sorbo, ed., *Metabolism of Sulfur Compounds,* Vol. 7, Academic Press, Inc., New York, 1975, p. 433.
102. S. Carson and co-workers, *Toxicol. Appl. Pharmacol.* **26,** 184 (1977).
103. H. Witschi, *Fed. Proc.* **35,** 1631 (1977).
104. W. C. Cutting, "The Action and Uses of Drugs" in *Handbook of Pharmacology,* Meredith Corp., New York, 1969.
105. U.S. Pat. 4,002,737 (Jan. 11, 1977), D. P. Borris (Research Corporation).
106. T. M. S. Chang, *Methods Enzymol.* **44,** 676 (1977).
107. S. Uptike, C. Prieve, and J. Magnuson, *Birth Defects* **9,** 77 (1973).
108. D. A. Cooney, H. H. Weetall, and E. Long, *Biochem. Pharmacol.* **24,** 503 (1975).
109. P. V. Sundaram and D. K. Apps, *Biochem. J.* **161,** 441 (1977).
110. L. S. Hersh, V. L. Gott, and F. Najjar, *J. Biomed. Mater. Res. Symp.* (3), 85 (1972).
111. O. Wagner and co-workers, *Biochem. Biophys. Res. Commun.* **37,** 383 (1969).
112. A. Abuchowski and co-workers, *J. Biol. Chem.* **252,** 3578 (1977).
113. S. J. Updike, R. T. Wakamiya, and E. N. Lightfoot, *Science* **193,** 681 (1976).
114. M. B. Fiddler, L. D. S. Hudson, and R. J. Desnick, *Biochem. J.* **168,** 141 (1977).
115. E. Butler and co-workers, *Proc. Natl. Acad. Sci. U.S.A.* **74,** 4620 (1977).

DAVID A. COONEY
GEORGE STERGIS
H. N. JAYARAM
Division of Cancer Treatment
National Cancer Institute
National Institutes of Health

EPINEPHRINE AND NOREPINEPHRINE

Epinephrine [51-43-4] and norepinephrine [51-41-2], the hormones of the adrenal medulla, are the final products of the biosynthetic pathway, phenylalanine (1) → tyrosine (2) → dihydroxyphenylalanine or DOPA (3) → dopamine (4) → norepinephrine (5) → epinephrine (6). This pathway, and the enzymes catalyzing the various reactions, are shown in Figure 1.

Epinephrine and norepinephrine released from the adrenal medulla are involved in the control of heart rate, blood pressure, and lipid and carbohydrate metabolism. In the sympathetic nervous system, norepinephrine released from postganglionic nerve terminals acts as a transmitter of the nerve impulse across the gap, or synaptic cleft, between the nerve terminal and a postsynaptic site of action. In certain neurones in the central nervous system, a neurotransmitter function is served by dopamine, norepinephrine, and epinephrine.

Epinephrine was one of the first chemical substances to be isolated from tissues and found to have profound effects on body function. In 1895, Oliver and Schaefer (1) reported that extracts of adrenal glands increased blood pressure and altered the contraction of smooth muscle. In 1901, Takamine (2) isolated a pressor material containing epinephrine along with a smaller amount of norepinephrine from bovine adrenal gland extracts. In 1904, Stolz (3) synthesized epinephrine. The similarity between the effects seen on stimulation of sympathetic nerves and on administration of epinephrine led to the use of the term sympathomimetic for epinephrine, and subsequently for related compounds, also frequently classified as catecholamines. For a number of years after the discovery of epinephrine, it was considered probable that the various effects caused by stimulation of sympathetic nerves were due to the release of epinephrine or a closely related compound from the nerve. In the 1930s, however, norepinephrine was proposed as the compound released (4–5). In 1948, Von Euler (6) identified norepinephrine in nerve extracts; and in 1957 Brown and Gillespie

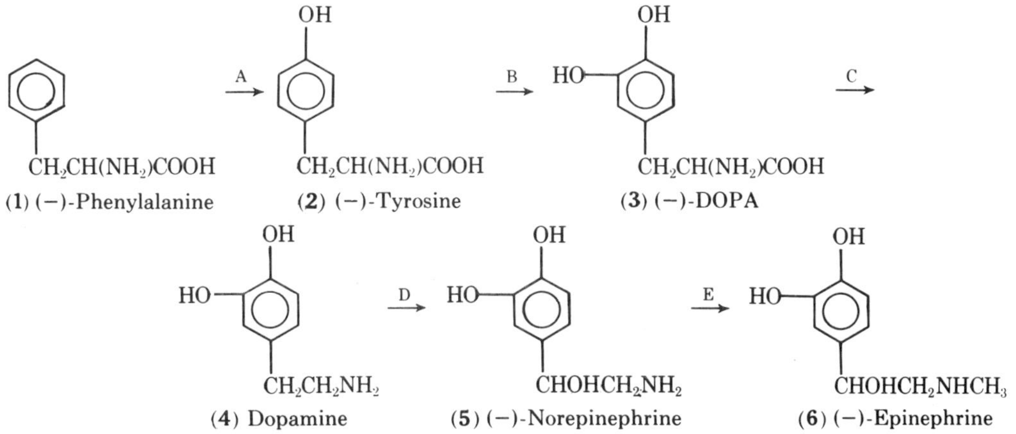

Figure 1. Biosynthesis of epinephrine and norepinephrine. The enzymes catalyzing the reactions are: in the liver, A, phenylalanine hydroxylase; in the adrenal medulla, brain, and peripheral nerves, B, tyrosine hydroxylase, C, DOPA decarboxylase, and D, dopamine-β-oxidase; and in the adrenal medulla with a small amount in the brain, E, phenylethanolamine-N-methyltransferase.

Table 1. Physical Properties of Epinephrine and Norepinephrine[a]

Property	D-(−)-Epinephrine	D-(−)-Norepinephrine
IUPAC name	(−)-3,4-dihydroxy-α-[(methylamino)methyl]benzyl alcohol	(−)-α-(aminomethyl)-3,4-dihydroxybenzyl alcohol
Chem. Abstracts name	(R)-(−)-4-[1-hydroxy-2-(methylamino) ethyl]-1,2-benzenediol	(R)-(−)-4-(2-amino-1-hydroxyethyl)-1,2-benzenediol
other names	(epinephrine[b], adrenaline)	levarterenol, D-arterenol, D-noradrenaline
formula	$C_9H_{13}NO_3$	$C_8H_{11}NO_3$
formula wt	183.2	169.2
description		
appearance	colorless microcrystals; acceptable preparations may be slightly off-white	colorless microcrystals
melting point	ca 209–210°C (dec)	ca 212°C (dec)
solubility	in water, ca 0.1 g/100 mL; insoluble in alcohol and most other organic solvents	slightly soluble in water
stability	Sensitive to air, light, heat, and alkalies. Metals, notably copper, iron, and zinc, destroy its activity. In solution with sulfite or bisulfite, it slowly forms an inactive sulfonate[c]. The red color which forms when neutral or alkaline solutions are exposed to air is caused by adrenochrome.	The compound is unstable in light and air, especially at neutral and alkaline pH. Oxidation to noradrenochrome occurs in the presence of oxygen and such divalent metal ions as copper, manganese, and nickel.
absorption spectrum	0.1 M HCl λ_{max} 221 nm, ϵ ca 6100; λ_{max} 280 nm, ϵ ca 2700	in 0.1 M HCl, λ_{max} 221 nm, ϵ ca 5860; λ_{max} 279 nm, ϵ ca 2560
fluorescence spectrum	in 0.05 M sodium acetate buffer, pH 4, λ_{ex} 283 nm, λ_{em} 337 nm	in 0.05 M sodium acetate buffer, pH 4, λ_{ex} 283 nm, λ_{em} 337 nm
reaction product fluorescence		iodine oxidation in alkaline ascorbate[d], λ_{ex} 412 nm, λ_{em} 505 nm
other tests		color reactions with various reagents[e]
homogeneity	thin-layer chromatography	thin-layer chromatography
system 1	Spot 4 μL of a 0.5% solution in water containing a few drops of formic acid on silica gel F254 Merck (precoated plate). Develop with butanol:acetic acid:water (7:1:2) to a height of 10–11 cm. Detect with uv light or with a spray of Folin's reagent followed by a spray of 10% aqueous sodium carbonate. R_f ca 0.23.	Spot 4 μL of a 0.5% solution in water with a few drops of formic acid on silica gel F254 Merck (precoated plate). Develop with butanol:acetic acid:water (7:1:2) to a height of 10–11 cm. Detect with uv light or Folin's reagent followed by a spray of 10% sodium carbonate. R_f ca 0.27.
system 2	Spot 4 μL of a 0.5% solution in water containing a few drops of formic acid on cellulose F Merck (precoated plate). Develop with butanol:acetic acid:water (7:1:2) to a height of 10–11 cm. Detect	Spot 4 μL of a 0.5% solution in water with a few drops of formic acid on cellulose F Merck (precoated plate). Develop with butanol:acetic acid:water (7:1:2) to a height of 10–11 cm. Detect

Table 1 (continued)

Property	D-(−)-Epinephrine	D-(−)-Norepinephrine
	with uv light or with a spray of Folin's reagent followed by a spray of 10% aqueous sodium carbonate. R_f ca 0.37.	with uv or Folin's reagent. R_f ca 0.32.
specific rotation	$[\alpha]_D^{25}$ −50 to −53.5°, $c = 2$ g/100 mL, 0.5 M HCl[b]. Optical purity determined on the N-acetyl-3,4-di-O-acetyl derivative, $[\alpha]_D^{21}$ −94.7°, $c = 1.01$ g/100 mL, CHCl$_3$[g].	$[\alpha]_D^{25}$ −37.3°, $c = 5$ g/100 mL water containing 1 equiv hydrochloric acid[f]. Optical purity determined on the N-acetyl-3,4-di-O-acetyl derivative, $[\alpha]_D^{25}$ −81.3°, $c = 1$ g/100 mL, CHCl$_3$[f].
source	resolution of (±)-epinephrine with (+)-tartaric acid	resolution of (±)-norepinephrine[h]
likely impurities	(+)-epinephrine	(+)-norepinephrine
storage	protect from light and air	protect from light and air

[a] Data given in: *Specifications and Criteria for Biochemical Compounds, Supplement: Biogenic Amines and Related Compounds*. Courtesy of the National Academy of Sciences.
[b] Ref. 8.
[c] Ref. 9.
[d] Ref. 10.
[e] Ref. 11.
[f] Ref. 12.
[g] Ref. 13; report $[\alpha]_D^{20}$ −87.4°, $c = 1$ g/100 mL, CHCl$_3$ for the triacetyl derivative, and the configuration of natural (−)-epinephrine is the same as D-(−)-mandelic acid.
[h] Ref. 14.

(7) showed that norepinephrine was released when the splenic nerve was stimulated (see Neuroregulators; Psychopharmacological agents).

Physical Properties

Some physical properties are shown in Table 1.

Analytical Methods

Shortly after the isolation of epinephrine, it was noted that oxidation products in alkaline solution were fluorescent. Further study of this reaction led to the development of fluorimetric assay procedures that could be used to determine epinephrine and related compounds in tissues, blood, and urine in nanomolar (10^{-9} mol) quantities. In addition, fluorescence histochemical techniques have been developed that permit the identification of the catecholamine *in situ* in tissue sections. These fluorescence assay methods have been reviewed (15–16).

In recent years, methods have been developed capable of measuring the levels of the catecholamines and their metabolites in the picomole (10^{-12} mol), or femtomole (10^{-15} mol) range. These methods include: gas liquid chromatography (17), gas chromatography combined with mass spectrometry (18), high performace re-

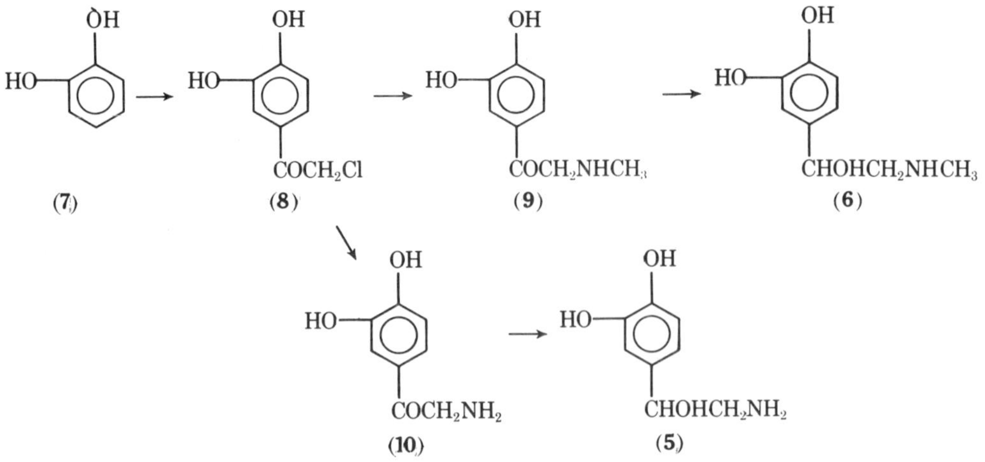

Figure 2. Synthesis of epinephrine and norepinephrine.

versed-phase chromatography with amperometric detection (19), and radioenzymatic analysis (20). Methods of analysis of the catecholamines and their metabolites have been the subject of recent reviews (16,21) (see Analytical methods; Radiochemical technology).

Isolation and Synthesis

The original source of epinephrine for commercial use was bovine adrenal glands from which epinephrine was extracted by Takamine's procedure (2).

Synthetic epinephrine has replaced the natural product as a commercial source. The procedure used is basically that developed by Stolz (3). As shown in Figure 2, Friedel-Crafts acylation of pyrocatechol (7) with chloroacetyl chloride gives the α-chloroacetophenone derivative (8). Displacement of the chlorine by methylamine gives the methylamino derivative (9), which on catalytic reduction, yields ($\pm$)-epinephrine (6). Substitution of ammonia for methylamine in the sequence gives the amino derivative (10), which on reduction yields norepinephrine (5). The isomers are resolved with (+)-tartaric acid to give the physiologically active (−)-isomers. The synthesis of epinephrine and other phenethyl- and phenylpropylamines has been reviewed (22). The synthesis of L-DOPA (23) is shown in Figure 3. Vanillin (qv) (11) and hippuric acid (12) are allowed to react in the presence of sodium acetate, acetic anhydride and N,N-dimethylformamide to give compound (13), catalytic hydrogenation in alkali gives the racemic benzoyl derivative (14). This is resolved with dehydroabietylamine and then hydrolyzed with hydrobromic acid to give L-DOPA (3) (see also Pharmaceuticals, optically active).

Biosynthesis, Distribution, and Metabolic Products

Shortly after the identification of epinephrine in the adrenal gland, tyrosine was suggested as a likely precursor (24). By 1957, the sequence of reactions shown in Figure 1 had been established (25). Phenylalanine is converted to tyrosine in the liver. The adrenal medulla contains the enzymes necessary for the conversion of tyrosine derived

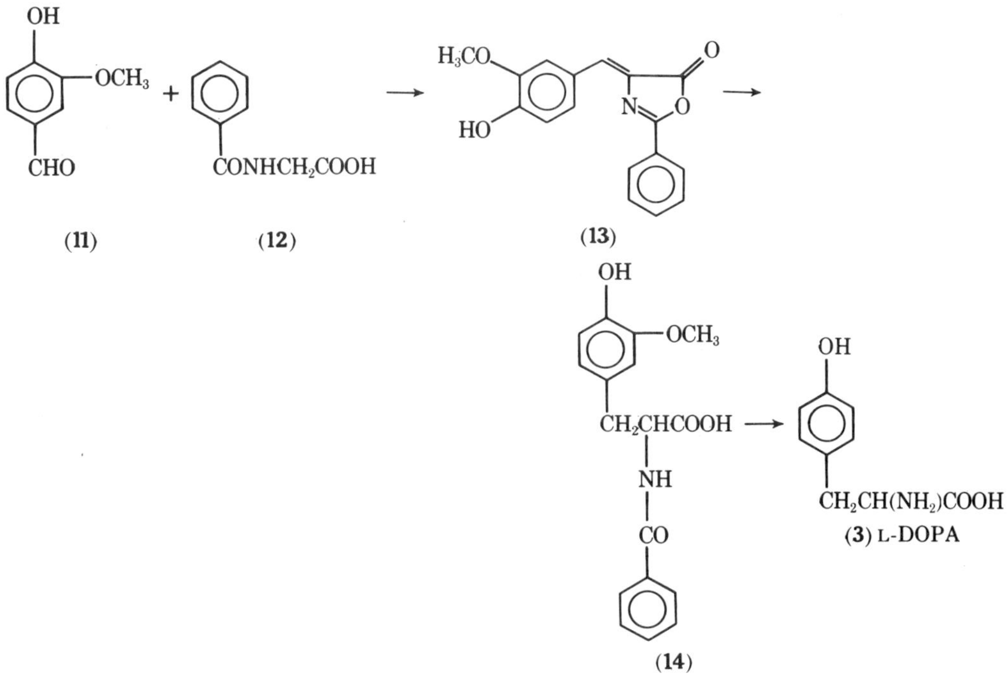

Figure 3. Synthesis of L-DOPA.

from phenylalanine or dietary tyrosine to epinephrine. Other than the adrenal medulla, the catecholamines are synthesized in neurones. The brain has neurones that contain all these enzymes. Accordingly, these neurones contain epinephrine.

The neurones of the sympathetic nervous system and other neurones in the brain lack phenylethanolamine N-methyltransferase (PNMT) and contain predominantly norepinephrine. A third type of neurone in the central nervous system lacks both PNMT and dopamine β-oxidase and, thus, produces dopamine as the end product. The purification and properties of the enzymes involved in the synthesis of dopamine, norepinephrine, and epinephrine have been reviewed (16) (Fig. 1). Per gram of tissue, the adrenal glands of various species contain in excess of 100 μg of epinephrine, with norepinephrine levels of 5–50% of that of epinephrine. Rat brain contains approximately 1.0 μg of dopamine and 0.5 μg of norepinephrine per gram. The amount of epinephrine in the brain is very low, 1 to 5% of the concentration of norepinephrine in various areas of the brain (26–27). Human plasma contains 0.2 μg/L of norepinephrine and 0.05 μg/L of epinephrine (27).

Enzymes are present in tissues that catalyze the conversion of epinephrine, norepinephrine and dopamine to inactive products. The principal reactions shown in Figure 4 are 3-O-methylation by catechol O-methyltransferase (28) and oxidative deamination by monoamine oxidase (29).

The catecholamines may be acted upon by either or both of these enzymes. Accordingly, the possible products are catechol or 3-O-methylcatechols having either amino, carboxy, or alcohol side chains (Table 2). The proportions of these metabolites formed depends on the species and tissue. In the brain, dopamine is converted primarily to the carboxylic metabolites, and norepinephrine to the metabolites having alcohol or glycol side chains (30).

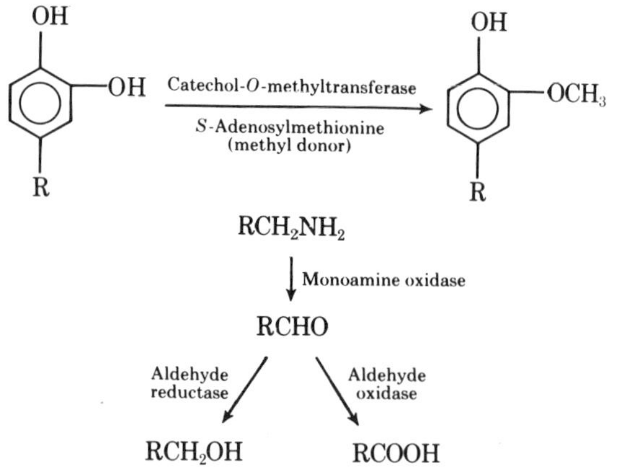

246 EPINEPHRINE AND NOREPINEPHRINE

[Figure: Catechol-O-methyltransferase reaction with S-Adenosylmethionine (methyl donor) converting catechol-R to methoxy-hydroxy-R; followed by RCH₂NH₂ → (Monoamine oxidase) → RCHO → (Aldehyde reductase) RCH₂OH or (Aldehyde oxidase) RCOOH]

Figure 4. Formation of metabolic products.

Table 2. Metabolic Products

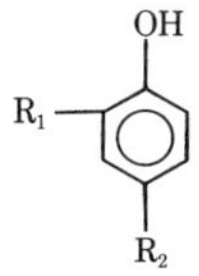

Parent compound	R_1	R_2	Commonly used names and abbreviations	
dopamine	OH	CH_2COOH	3,4-dihydroxyphenylacetic acid	DOPAC
dopamine	OH	CH_2CH_2OH	3,4-dihydroxyphenylethanol	DOPET
norepinephrine and epinephrine	OH	$CH(OH)COOH$	3,4-dihydroxymandelic acid	DHMA
norepinephrine and epinephrine	OH	$CH(OH)CH_2OH$	3,4-dihydroxyphenyl glycol	DHPG, DOPEG
dopamine	OCH_3	$CH_2CH_2NH_2$	3-methoxytyramine	
norepinephrine	OCH_3	$CH(OH)CH_2-NH_2$	normetanephrine	NM
epinephrine	OCH_3	$CH(OH)CH-(NH_2)CH_3$	metanephrine	MN
dopamine	OCH_3	CH_2COOH	3-methoxy-4-hydroxy-phenyl-acetic acid or homovanillic acid	HVA
dopamine	OCH_3	CH_2CH_2OH	3-methoxy-4-hydroxy-phenylethanol	MOPET
norepinephrine and epinephrine	OCH_3	$CH(OH)COOH$	3-methoxy-4-hydroxy-mandelic acid or vanilmandelic acid	VMA
norepinephrine and epinephrine	OCH_3	$CH(OH)CH_2OH$	3-methoxy-4-hydroxyphenyl glycol	MHPG, MOPEG

Physiology

In the chromaffin cells of the adrenal medulla, epinephrine and norepinephrine are present in small vesicles. Stimulation of the planchnic nerve leading to the adrenal medulla causes the secretion of the sympathomimetic amines into the blood for action at receptor sites throughout the body with the exception of the central nervous system. At physiological concentrations, epinephrine and norepinephrine do not pass the blood-brain barrier to any marked extent. Catecholamines located in nerve terminals are released when an electrical impulse or action potential travelling along the nerve reaches the terminal. Although the mechanisms regulating release of catecholamines in the adrenal medulla and nerve terminals are different, in both cases the final step is believed to involve exocytosis. The storage vesicles fuse with the external membrane of the cell, the membrane then opens, releasing the contents of the vesicles into the extracellular space (31).

The rate at which norepinephrine is released from noradrenergic nerve terminals is subject to control by the central nervous system and by local feedback mechanisms. For example, changes in blood pressure activate pressure-sensitive baroreceptors, sending signals to the vasomotor center in the brain. The vasomotor center, in turn, sends signals to the vasoconstrictor or vasodilator nerves, producing the appropriate alteration in activity by way of a decrease or increase in the firing rate of noradrenergic nerves in the periphery. At the local level, the release of norepinephrine can be controlled by receptor sites on presynaptic nerve terminals. When activated by norepinephrine, these presynaptic receptors reduce the release of norepinephrine from the nerve terminal. When the level of norepinephrine in the synaptic cleft is low, the lack of activation of these inhibitory presynaptic receptors will allow increased release of norepinephrine. Accordingly, the presynaptic receptor system will tend to return the rate of release of norepinephrine to a baseline level after an increase or decrease in the firing rate of the nerve. The subject of presynaptic receptor control of neuronal output has recently been reviewed (32).

A third mechanism controlling the level of norepinephrine at postsynaptic effector sites is the reuptake of norepinephrine by the presynaptic neurone terminal. In contrast to the cholinergic nervous system, in which the action of acetylcholine at postsynaptic receptor sites is terminated primarily by enzymatic degradation, the action of the catecholamines is terminated primarily by their being taken up by presynaptic neurones (see Choline; Cholinesterase inhibitors). The uptake through the neuronal membrane is an active transport process capable of taking up the catecholamines against a concentration gradient. The process is stereoselective with a higher affinity for the natural $(-)$-isomers than for $(+)$-isomers. Within the neurone, the catecholamine taken up can either be deaminated by monoamine oxidase (Fig. 4), or taken up into the storage vesicles. The neuronal uptake process has been reviewed (33).

In a stressful situation, epinephrine and norepinephrine are released from the adrenal medulla into the blood stream for action at distant sites, and markedly enhanced activity of the sympathetic nervous system will cause increased local concentration of norepinephrine at sympathetic nerve endings. The increased levels of epinephrine and norepinephrine at receptor sites cause a variety of physiological responses: increased blood pressure, heart rate and cardiac output, glycogenolysis, release of free fatty acids from adipose tissue, and dilatation of the bronchioles (see Antiasthmatic agents). Vasoconstriction in the skin and splanchnic area directs a greater portion of the total blood volume to skeletal muscles. The sum of these effects is an increased ability of the organism for sustained effort, the fight or flight response. In normal or nonstressful situations, small fluctuations in sympathetic discharge act to provide the smaller alterations in function required by small changes in the demands placed on the body.

A basis for classifying and understanding the relationship between the multiple changes produced by epinephrine and norepinephrine is provided in ref. 34 which proposed that there are two types of adrenergic receptors. The α receptor, in general, is responsible for excitatory responses, and the β receptor is primarily involved with inhibitory responses. Subsequently, β receptors were subdivided into β_1 and β_2 receptors. The β_1 receptors are involved in fatty acid metabolism and cardiac stimulation, and the β_2 receptors in glycogenolysis and bronchodilatation (35). A particular sympathomimetic amine may act on both α and β receptors, or act predominantly on one type. Norepinephrine acts primarily on α receptors, epinephrine on both α and β receptors. Isoproterenol (N-isopropylnorepinephrine) acts almost exclusively on β receptors. There are also compounds that selectively antagonize the action of sympathomimetics at α, β_1, or β_2 receptors (see Cardiovascular agents). Compounds that act at these receptors to cause a physiological response are referred to as agonists, and compounds inhibiting the action of the agonists are referred to as antagonists.

The concept of α and β receptors implies the existence of a specific locus at which adrenergic agonists and antagonists act. In recent years, biochemical methods using radioactive adrenergic agonists and antagonists have demonstrated that there are binding sites in tissues that appear to be closely involved with the actions of these agonists and antagonists. Briefly, the method used in such studies consists of determining the amount of radioactive binding agent or ligand bound by the tissue preparation and readily displaced by an excess of nonradioactive ligand. To be considered specific, the binding should be saturable and, if the ligand can exist in isomeric forms, should show the same stereospecificity as the physiological response. Procedures have been developed in which the relative affinity for the binding site of a series of adrenergic agonists or antagonists has been shown to parallel the pharmacological potency of the compounds. These procedures have been reviewed (36–37).

The nature of the reactions initiated by epinephrine and norepinephrine in effector cells is still a subject of study. Epinephrine, norepinephrine, and dopamine, activate adenyl cyclase, an enzyme which converts adenosine triphosphate (ATP) to cyclic adenosine monophosphate (cyclic AMP) and inorganic pyrophosphate. The cyclic AMP, in turn, is known to activate protein kinase enzymes which can phosphorylate other enzymes, thereby altering their activity and thus changing events in the postsynaptic or effector cell. The actions of cyclic AMP and other cyclic nucleotides have recently been reviewed (38).

Therapeutic Use and Toxicity

The multiplicity of physiological actions of epinephrine, norepinephrine, and related compounds gives these agents a variety of clinical uses. Epinephrine is used for its vasoconstrictor property in topical application to reduce bleeding from superficial wounds. Epinephrine acts, in conjunction with a local anesthetic, by constricting blood vessels in the area and reducing local blood flow, to reduce absorption of the anesthetic and prolong its action (see Anesthetics). The vasoconstrictor property of the sympathomimetic amines is also useful in relieving nasal congestion in hay fever and other allergic conditions. In asthma, the bronchodilator property of epinephrine is useful in counteracting bronchospasm. In emergency situations of cardiac arrest, epinephrine can be injected into the heart to restore cardiac rhythm. Sympathomimetic agents are used in the treatment of hypotension and shock. The choice of agent depends on the circumstances responsible for the decreased blood pressure and low rate of tissue perfusion that occurs in shock. In anaphylactic shock, epinephrine is used for its bronchodilator and cardiac stimulant properties. In shock due to blood loss, norepinephrine can be used on an emergency basis to elevate the blood pressure until blood volume can be restored. Isoproterenol is used in conditions in which low cardiac output is the primary problem. Dopamine is also used in shock for its cardiac stimulant properties. The use of dopamine and other sympathomimetics in the treatment of shock has been reviewed (39).

The immediate precursor of dopamine, L-DOPA (Fig. 3), is used in the treatment of Parkinsonism (40), a condition characterized by motor dysfunctions, such as tremor, rigidity, and restricted movement. In this condition, the level of dopamine is low in the corpus striatum of the brain. Dopamine cannot be used to alleviate Parkinsonism since it does not penetrate the blood-brain barrier. However, L-DOPA, the amino acid precursor of dopamine does penetrate the brain where it is converted to dopamine. There are other disturbances of brain function in which the catecholamines are believed to play a role, and in which drugs affecting the actions of the catecholamines have a therapeutic action. In schizophrenia, drugs blocking the action of dopamine receptors have a therapeutic action (41). In the treatment of depression, a number of drugs have a beneficial action by inhibiting the uptake of norepinephrine by neurones and thereby increasing its level at postsynaptic sites in the brain (42). Monoamine oxidase inhibitors, compounds which increase brain levels of norepinephrine by inhibiting its inactivation by deamination (Fig. 4), have also been used in the treatment of depression (43).

Both epinephrine and norepinephrine have an LD_{50} by the oral route of less than 50 mg/kg in rodents and are, therefore, listed as highly toxic substances in the *Registry of Toxic Effects of Chemical Substances* (ROTECS, a data-base system), published by NIOSH. The most serious effects of an overdose of epinephrine are a rapid large increase of blood pressure, cerebral hemorrhage, and cardiac arrhythmias. The effects of norepinephrine are similar, but less severe.

BIBLIOGRAPHY

"Epinephrine" in *ECT* 1st ed., Vol. 5, pp. 763–768 by Oliver Kamm, Parke, Davis & Company; "Epinephrine" in *ECT* 2nd ed., Vol. 8, pp. 231–238, by H. M. Crooks, Jr., Parke, Davis & Company.

1. G. Oliver and E. Schaefer, *J. Physiol.* **18**, 230 (1895).
2. U.S. Pats. 730,175; 730,176; 730,196; 730,197 (June 22, 1903), J. Takamine.
3. F. Stolz, *Chem. Ber.* **37**, 4149 (1904).
4. Z. M. Bacq, *Ann. Physiochem. Biol.* **10**, 467 (1934).
5. C. M. Greer and co-workers, *J. Pharmacol.* **62**, 189 (1938).
6. U. S. Von Euler, *Acta. Physiol. Scand.* **16**, 63 (1948).
7. G. L. Brown and J. S. Gillespie, *J. Physiol.* **138**, 81 (1957).
8. *The U.S. Pharmacopeia, XIX Revision*, Mack Publishing Co., Easton, Pa., 1974, p. 169.
9. L. C. Schroeter, T. Higuchi, and E. E. Schuler, *J. Pharm. Sci.* **47**, 723 (1958).
10. P. A. Shore and J. S. Olin, *J. Pharmacol. Exp. Ther.* **122**, 295 (1958).
11. C. F. Schwender in K. Florey, ed., *Analytical Profiles of Drug Substances*, Vol. 1, Academic Press, New York, 1972, p. 149.
12. L. H. Welsch, *J. Am. Pharm. Assoc. Sci. Ed.* **44**, 507 (1955).
13. L. H. Welsh, *J. Am. Chem. Soc.* **74**, 4967 (1952); P. Pratesi and co-workers, *J. Chem. Soc.* 2069 (1958).
14. B. F. Tullar, *J. Am. Chem. Soc.* **76**, 2067 (1948); U.S. Pat. 2,774,789 (Dec. 18, 1956), B. F. Tullar (to Sterling Drug Inc.).

15. S. Undenfriend, *Fluorescence Assay in Biology and Medicine,* Vol. 2, Academic Press, New York, 1969.
16. T. Nagatsu, *Biochemistry of Catecholamines,* University Park Press, Baltimore, 1973.
17. S. W. Stout and co-workers, *Anal. Biochem.* **76,** 330 (1976).
18. E. P. Abramson, M. W. McCaman, and R. E. McCaman, *Anal. Biochem.* **57,** 482 (1974).
19. P. T. Kissinger and co-workers, *Clin. Chem.* **23,** 1449 (1977).
20. M. Da Prada and G. Zurcher, *Life Sci.* **19,** 1161 (1976).
21. L. L. Iverson, S. D. Iversen, and S. H. Snyder, *Handbook of Psychopharmacology,* Vol. 1, *Biochemical Principles and Techniques in Neuropharmacology,* Plenum Press, New York, 1975.
22. D. Lednicer and L. A. Metscher, *The Organic Chemistry of Drug Synthesis,* John Wiley & Sons, Inc., New York, 1977, pp. 62–82.
23. U.S. Pat. 3,714,242 (Jan. 30, 1973), G. M. Jaffe and W. R. Rehl (to Hoffmann-La Roche Inc.).
24. W. L. Halle, *Beitr. Chem. Physiol. Pathol.* **8,** 276 (1906).
25. H. Blaschko, *Br. Med. Bull.* **13,** 162 (1957).
26. A. M. Sauter and co-workers, *Life Sci.* **21,** 261 (1977).
27. K. Engelman and B. Portnoy, *Circ. Res.* **26,** 53 (1970).
28. J. Axelrod, *Pharmacol. Rev.* **18,** 95 (1966).
29. K. F. Tipton, *Br. Med. Bull.* **29,** 116 (1973).
30. G. R. Breese, T. N. Chase, and I. J. Kopin, *J. Pharmacol. Exp. Therap.* **165,** 9 (1969).
31. R. J. Baldessarini, "Release of Catecholamines" in L. L. Iverson, S. D. Iversen, and S. H. Snyder, eds., *Handbook of Psychopharmacology,* Vol. 3, Plenum Press, New York, 1975, pp. 37–113.
32. K. Starke, *Rev. Physiol. Biochem. Pharmacol.* **77,** 2 (1977).
33. L. L. Iverson in ref. 31, pp. 381–429.
34. R. P. Ahlquist, *Am. J. Physiol.* **154,** 586 (1948).
35. A. M. Land and co-workers, *Nature* **214,** 597 (1967).
36. M. D. Hollenberg and P. Cuatrecasas, "Biochemical Identification of Membrane Receptors: Principles and Techniques" in L. L. Iverson, S. D. Iversen, and S. H. Snyder, eds., *Handbook of Psychopharmacology,* Vol. 2, Plenum Press, New York, 1975, pp. 129–170.
37. S. H. Snyder, *Biochem. Pharmacol.* **24,** 1371 (1975).
38. H. Cramer and J. Schultz, *Cyclic 3′,5′-Nucleolides Mechanism of Action,* John Wiley & Sons Inc., New York, 1977.
39. L. I. Goldberg, *N. Eng. J. Med.* **291,** 707 (1974).
40. A. Barbeau, *Annu. Rev. Pharmacol.* **14,** 91 (1974).
41. A. Carlsson, *Am. J. Psychiatry* **135,** 2 (1978).
42. H. Weil-Malherbe, "The Biochemistry of Affective Disorders," in R. G. Grenell and S. Gabay, *Biological Foundations of Psychiatry,* Raven Press, New York, 1976, pp. 684–713.
43. L. E. Hollister, *Drugs* **4,** 361 (1972).

JOHN R. MCLEAN
Warner-Lambert/Parke-Davis

EPOXIDATION

Epoxidation is the formation of cyclic three-membered ethers (oxiranes) by the reaction of peracids (see Peroxides) and hydrogen peroxide (qv) with olefinic and aromatic double bonds. Oxiranes may also be formed by an internal S_N2 reaction of a chlorohydrin (see Chlorohydrins). The three-membered ethers formed are also designated as 1,2-epoxides.

Epoxides are extremely valuable commercially because of the many reactions they undergo. Most noteworthy of these is the addition of active hydrogen compounds such as ammonia, organic acids, alcohols, and water. Their ability to donate electrons to hydrogen atoms by hydrogen bonding is important in such diverse applications as ulcer therapy and the stabilization of poly(vinyl chloride) resins (see Plasticizers). The fact that epoxides polymerize under thermal ionic and free-radical catalysis has encouraged considerable research on epoxy homopolymers and copolymers for industrial applications. The epoxides of long-chain α-olefins are potentially useful as detergent precursors (see Surfactants and detersive systems). These few examples show the commercial usefulness of epoxides, particularly as chemical intermediates.

Epoxidized materials are manufactured mainly from peracetic acid (peroxyacetic acid) and performic acid (peroxyformic acid). Peracetic acid can be prepared by the oxidation of acetaldehyde by hydrogen peroxide. Performic acid can be prepared by a similar oxidation of formaldehyde. Although there are several routes to these peracids, preformed peracetic acid (via the air oxidation of acetaldehyde) and *in situ* performic acid are the chief reagents. Epoxy plasticizer production in the United States is roughly equally divided between the two processes. On a world-wide basis, performic-process epoxy plasticizers comprise 60–70% of the market.

Epoxidation Processes

Epoxy compounds are characterized by an oxirane group formed by the oxidation of either an olefinic or aromatic double bond (1).

$$\mathrm{C{=}C} \xrightarrow[\text{cat.}]{RO_2H} \mathrm{C\overset{O}{-}C} \tag{1}$$

Findley, Swern, and Scanlan were the first to report the synthesis of epoxidized soybean oil via a preformed peracetic acid technique (2). Terry and Wheeler at General Mills (3) developed epoxy esters based on a preformed-peracetic acid process, and Niederhauser at Rohm and Haas (4) developed epoxidized unsaturated oils based on an *in situ* performic acid process. Niederhauser was the first to report the value of epoxidized soybean oil as a plasticizer for PVC, and his process was used to produce the first commercial epoxy soybean oil plasticizer. Peracetic acid derived from acetic acid and hydrogen peroxide on a commercial scale was used for the epoxidation of unsaturated vegetable oils (5). This preformed peracetic acid can be handled, with proper precautions, in an inert solvent such as ethyl acetate or acetone (6–7). Acetaldehyde monoperacetate, the oxidation intermediate to peracetic acid, is also a useful epoxidizing agent (8). Oxidation of olefins by hydroperoxides has been described (9), but the oxidation processes tend to be less efficient than the peracid processes; therefore, it is highly unlikely that epoxy plasticizers of commercial quality can be produced by hydroperoxide technology in its present state of development.

The epoxidation processes can be divided into two basic types: either the peracid is preformed or it is formed *in situ,* ie, in the primary reaction vessel. Representative schematics are shown in Figure 1. Each has its advantages and disadvantages. Composition and performance of the product can be affected by proprietary processes involving peracid formed *in situ* or preformed peracid with cosolvents, specially selected olefinic substrates and catalysts, methods of addition of components, and posttreatment of the epoxide.

Epoxidation with Peracetic Acid. A catalyst is not necessary for epoxidation at 20–80°C if preformed peracetic acid is used:

$$CH_3COOOH + RCH{=}CHR \rightleftharpoons RCH\underset{\diagdown O \diagup}{-}CHR + CH_3COH \tag{2}$$

The peroxidation of acetic acid with hydrogen peroxide is not efficient except at high molar ratios of acetic acid/hydrogen peroxide; however, large amounts of acetic acid must be removed if high ratios are involved. In addition, concentrations of peracetic acid above 40–45% wt % in acetic acid are explosive at epoxidation temperatures (8), and therefore, related epoxidation processes require large-volume production on an essentially continuous basis since the preformed peracid cannot be safely stored (see Safety).

Epoxidation with Peracid Formed *In Situ,* Acid-Catalyzed. Although no reaction that involves peroxides is without hazard, 30 years of experience has shown the *in situ* processes to be safer than the preformed-peracid processes. Many techniques of *in situ* epoxidation have been developed. In general, a peroxide solution (35–70% H_2O_2 in H_2O) containing small quantities of a mineral acid catalyst (such as sulfuric acid or phosphoric acid) is added to a mixture of an epoxidizable substrate and acetic acid or formic acid. As the reactants mix, the hydrogen peroxide and the organic acid react in the presence of the mineral acid catalyst to form the peracid (eq. 2).

$$RCOH + H_2O_2 \overset{H^+}{\rightleftharpoons} RCOOH + H_2O \tag{3}$$

To prevent uncontrolled exothermic reaction and to optimize epoxidation, the peroxide solution is added incrementally with agitation, and the reaction temperature is maintained at 50–65°C for 10–40 min per addition of peroxide. As a result, only small quantities of peracid are formed (eq. 2) in the presence of the unsaturated substrate. The peracid reacts with the unsaturated portion of the molecule (eq. 2) and is thus quickly depleted which prevents a build-up of detonatable quantities of peroxide compounds.

When the iodine number of the substrate has been reduced to the desired level, the reaction is stopped and the epoxidized substrate is separated from the aqueous layer. The aqueous layer contains a mixture of organic acid as well as some peroxide which can be recycled into the next batch as part of the charge. In the epoxy layer, the acid catalyst is neutralized by a mild base, and residual peroxide is decomposed (see

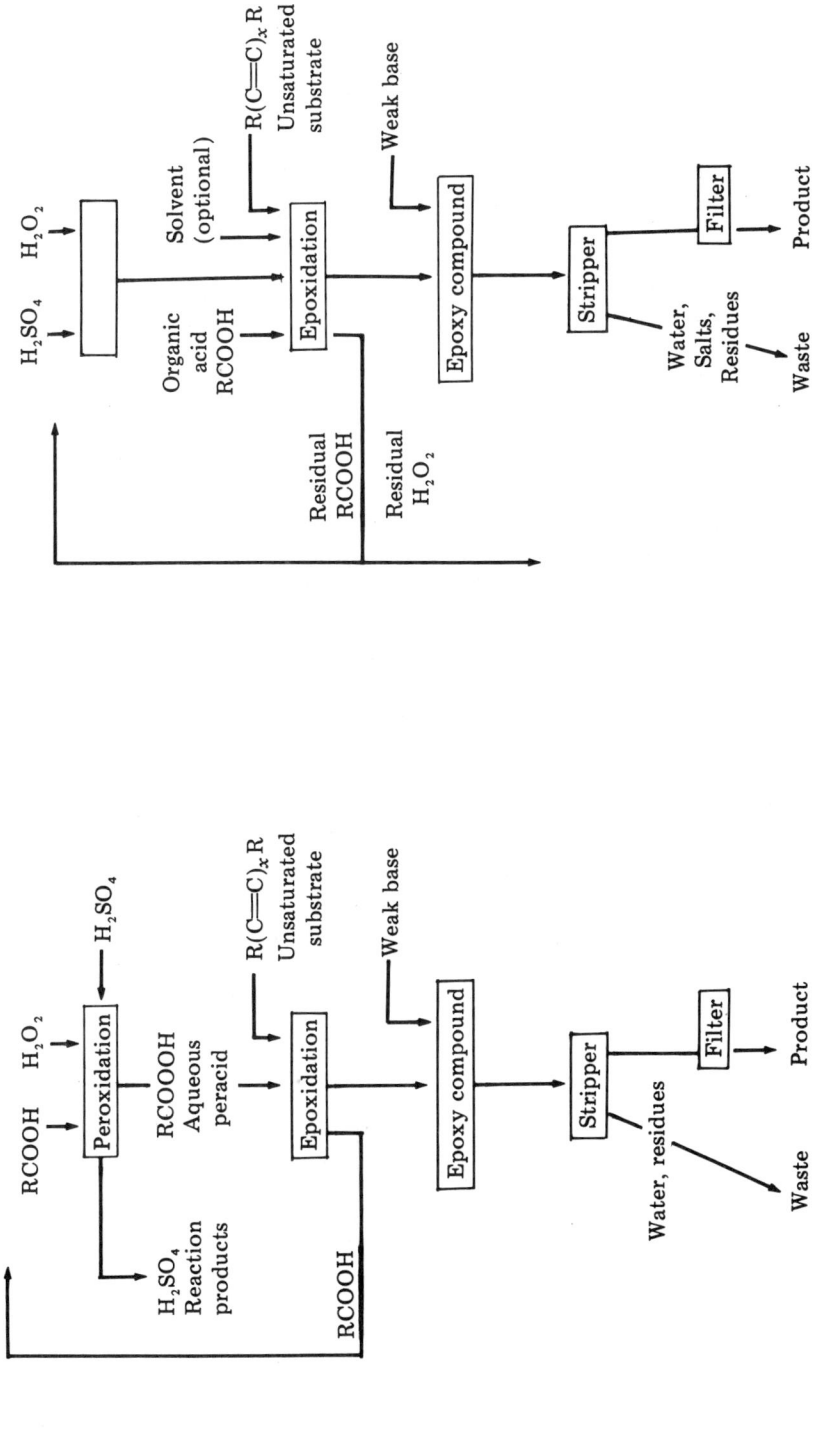

Figure 1. Schematic representation of the two basic epoxidation processes: (a) epoxidation with preformed peracid; and (b) epoxidation with peracid formed *in situ*.

below). The epoxy compound is then washed and transferred to a stripper in which water and nonproduct residues are removed.

Properly selected solvents facilitate epoxidation and the choice of solvent varies with the substrate. Heptane or octane work well with soybean oil and aromatic solvents such as toluene give the best results with linseed oil. A typical *in situ* epoxidation process is as follows (10): (*1*) A mixture is prepared in the reaction vessel containing 400 parts of refined soybean oil (iodine value of 135, or 2.13 mol of ethylenic unsaturation), 100 parts of commercial grade heptane (solvent) and 42 parts of glacial acetic acid (99.7% pure), which corresponds to 0.33 mol of acetic acid per mole of unsaturation; (*2*) the mixture is heated to 56–57°C and agitation is maintained; (*3*) a mixture containing 2.25 parts of conc sulfuric acid (96.5%) and 113.9 parts of 70 wt % aqueous hydrogen peroxide is prepared–1.1 mol of peroxide is used per mole of unsaturation in the soybean oil; (*4*) the peroxide/sulfuric acid mixture is added dropwise to the oil-containing mixture at a uniform rate over a period of one hour with agitation. The temperature is maintained at 56–57°C with cooling; (*5*) after 11.5 h have elapsed, in which time the iodine value has dropped to <6, the reaction is stopped; (*6*) after the agitation ceases, the aqueous phase separates and is discarded; (*7*) Ca(OH)$_2$ (0.4 part) is added to the oil phase in order to neutralize the sulfuric acid (acid residues are detrimental to the color and performance of epoxy plasticizers); (*8*) the oil phase is then stripped under 0.3–0.4 kPa pressure (2–3 mm Hg) up to 107°C; (*9*) the epoxidized oil product is then cooled and filtered; and (*10*) the product has the following constants: oxirane oxygen, 6.9 wt %; iodine value, 3.7; viscosity, 3.8 cm^2/s (=St); and color (Gardner scale), 1–.

Since the epoxidation is reversible and there is considerable potential for side reactions (eq. 4), it is important that the epoxidation be carried out at the lowest temperature and shortest time that are consistent with the desired degree of epoxidation. The rate of ring opening of 9,10-epoxystearic acid (epoxidized oleic acid) has been claimed to be 1.0%/h at 25°C, and 100% in 1–4 h at 65–100°C (2).

The following reactions destroy the oxirane oxygen ring and can take place in the process, although the type and amount of the by-products are difficult to identify.

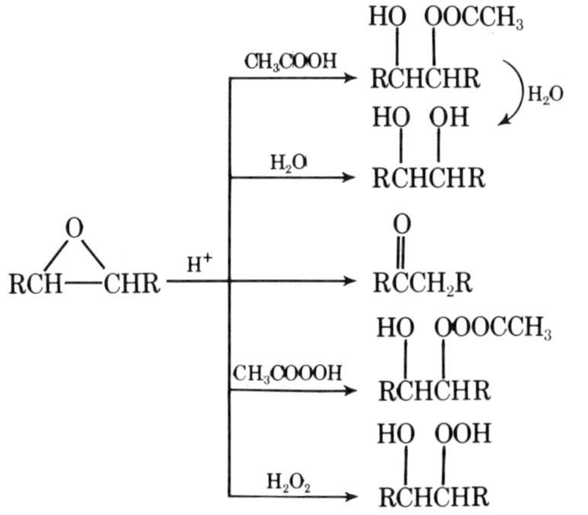

$$\underset{\text{RCH---CHR}}{\overset{\text{O}}{\triangle}} \xrightarrow{\text{H}^+} \begin{cases} \xrightarrow{\text{CH}_3\text{COOH}} \underset{\text{RCHCHR}}{\overset{\text{HO OOCCH}_3}{| \ |}} \\ \xrightarrow{\text{H}_2\text{O}} \underset{\text{RCHCHR}}{\overset{\text{HO OH}}{| \ |}} \\ \longrightarrow \underset{\text{RCCH}_2\text{R}}{\overset{\text{O}}{\|}} \\ \xrightarrow{\text{CH}_3\text{COOOH}} \underset{\text{RCHCHR}}{\overset{\text{HO OOOCCH}_3}{| \ |}} \\ \xrightarrow{\text{H}_2\text{O}_2} \underset{\text{RCHCHR}}{\overset{\text{HO OOH}}{| \ |}} \end{cases} \Bigg\} \text{H}_2\text{O} \quad (4)$$

Ring opening and by-product formation can be minimized by using less than one mole of organic acid (eg, formic acid) for each double bond in the unsaturated substrate. Ratios of 0.25 and 0.5 mol of formic acid per mole of unsaturation are recommended

(4). The use of an inert solvent and the presence of alkali amounting to ≤5 wt % of the organic acid also help to minimize by-product formation (4).

An *in situ* process that included the use of acetic anhydride and a basic salt, and that yielded epoxidized vegetable oil and ester plasticizers with low by-product content has been reported (11).

Oxidation by Hydroperoxides. Hydroperoxides have been observed to convert olefins to epoxides in the presence of transition metal ions such as Mo, W, Cr, or V, as well as Ag for ethylene oxide. Recently, it has been observed that hydrogen peroxide in the presence of catalytic quantities of arsenic compounds reacts with olefins to yield epoxides (12).

Peroxide-derived epoxide processes may lose importance if the disclosure reported in ref. 13 is valid. According to this patent, an olefin is treated directly with oxygen in the presence of a polycyano compound such as tetracyanoethylene (TCNE) (see Cyanocarbons) and arsenic compounds as shown:

$$2\,CH_3CH=CH_2 + O_2 \xrightarrow[\text{As cat., } CH_3COOC_2H_5]{150\,°C/5.9\,\text{MPa}(58\,\text{atm}), \text{TCNE}, 2h} 2\,CH_3CH\!-\!\!\!-\!\!\!-CH_2 \quad (5)$$

48% conversion
52% efficiency

Manufacture of Epoxides

The use of the epoxidation reaction in manufacture has involved hydrogen peroxide as the principal source of epoxidizing agents. Peracetic acid derived from acetaldehyde has become increasingly important in epoxidation. However, this material has not appeared as a merchant commodity, and no plans for shipment to prospective users have been announced. Chances are, therefore, that hydrogen peroxide will remain the principal agent commercially available for epoxidation. However, the manufacturer of epoxides with facilities available for oxidation of acetaldehyde may wish to make use of peracetic acid in an inert solvent in cases in which epoxides are sensitive to ring opening or are difficult to form.

For the purchaser of active oxygen, the essential details of manufacture to be considered involve the conversion of hydrogen peroxide to an effective agent for epoxidation. Other pertinent details are those of arranging proper storage for hydrogen peroxide near the site of operations and installing equipment constructed of materials suitable for epoxidation reactions.

Peracetic Acid. *Sulfuric Acid Catalyst.* Merchant peracetic acid is produced from hydrogen peroxide via sulfuric acid catalysis (14). Peracetic acid (40%) may be prepared at the site of epoxidation, however, by mixing 1.6 mol of glacial acetic acid with 1 mol of 90% hydrogen peroxide in the presence of ca 2–3% sulfuric acid. The mixture attains equilibrium if allowed to stand overnight.

Acetic anhydride may also be added to displace equilibrium by removing water (15). The use of acetic anhydride adds to the hazards of the reaction, however, by making it possible to obtain explosive diacetyl peroxide as a by-product. Peracetic acid solutions prepared by mixing acetic anhydride and 35% hydrogen peroxide at 40°C have been found to contain about 8% diacetyl peroxide. Such mixtures have exploded.

256 EPOXIDATION

Cation-Exchange Resin Catalyst. Information on the use of cation-exchange resins to prepare peroxyacetic acid solutions free of traces of mineral acid has been disclosed by DuPont. This method can be operated continuously or batchwise and is a convenient procedure for the manufacture of equilibrium peroxyacetic acid at the site of operations (16).

The method consists of passing a mixture of glacial acetic acid and hydrogen peroxide through a cation-exchange resin bed confined in a column (see Fig. 2). The column contains poly(styrenesulfonic acid) resin in the acid form which has been treated with acetic acid to remove excess water. The flow of peracetic acid solution from the column as effluent is adjusted to give the peroxide–acetic mixture the proper contact time with the resin bed. For the preparation of 40% peracetic acid solutions, 90% hydrogen peroxide is required (see Figs. 3–4). A 15% peracetic acid solution may be prepared, however, at the rate of ca 120 kg/h by feeding a mixture containing a ratio of 6.33 L of glacial acetic acid to 1 L of 50% hydrogen peroxide to the resin tower. Proper contact times are maintained by charging 2–2.2 kg of such cation-exchange resins as Amberlite IR-120 (Rohm & Haas Company) or Dowex 50-X-8 (Dow Chemical Company) to the tower for each liter per hour of 50% hydrogen peroxide that is introduced.

Generator or Azeotropic Peracetic Acid. Concentrated peracetic acid solutions in water may be prepared by vacuum distillation of a mixture of hydrogen peroxide, acetic acid, and sulfuric acid. The process (see Fig. 5) is operated by continuously feeding 0.5:1 to 2:1 molar ratios of acetic acid to hydrogen peroxide (as a 50% aqueous solution) to a reactor–reboiler (generator) containing 10–20% sulfuric acid catalyst. The reaction products are distilled concomitantly at pressures of 5.3–8 kPa (40–60 mm Hg).

The reactor boiler is maintained at ca 45–50°C during reaction and distillation.

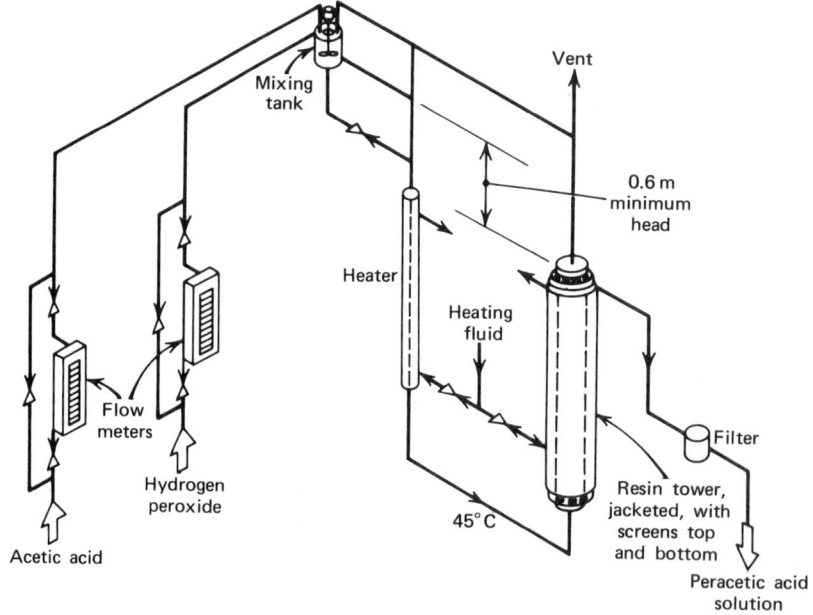

Figure 2. Schematic flow diagram for the continuous preparation of peracetic acid.

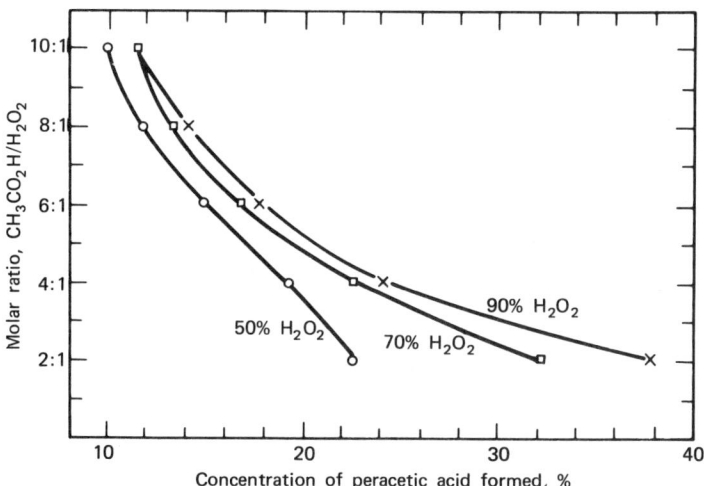

Figure 3. Concentration of peracetic acid formed vs molar ratio of acetic acid/peroxide in the preparation of mineral-free peroxyacetic acid by use of cation-exchange resin.

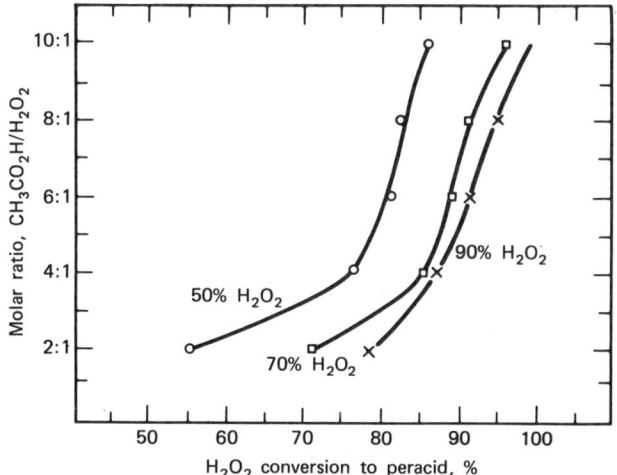

Figure 4. Peroxide conversion to peracid vs molar ratio of acetic acid/peroxide in the preparation of mineral-free peroxyacetic acid by use of cation exchange resin.

Aqueous peracetic acid solutions drawn off overhead vary in concentration between 50 and 60% depending upon the ratios of hydrogen peroxide to acetic acid employed. Higher ratios of peroxide to acetic acid lead to higher concentrations of peracetic acid as acetic acid content is decreased. Based on hydrogen peroxide, yields of peracetic acid are practically quantitative at 97–98%. The aqueous peracetic acid solutions obtained approximate an azeotropic mixture reported to constitute 56.5% peracetic acid and 43.5% water (17).

Certain safety precautions are required in the manufacture of peracids and in their use in epoxidation. Safety factors are discussed at the end of this article.

258   EPOXIDATION

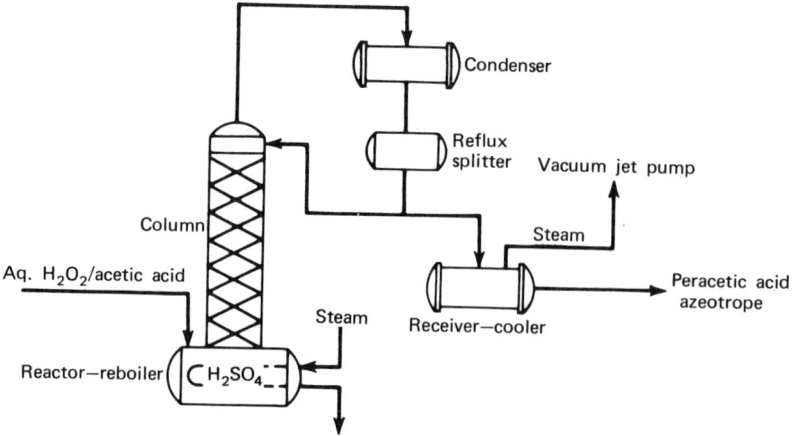

Figure 5. Vacuum-distillation process for preparation of concentrated peracetic acid solutions.

Acetaldehyde Oxidation. The Union Carbide plant, the first of its kind for production of 30% peracetic acid in ethyl acetate, is shown schematically in Figure 6 (18).

In the process, acetaldehyde and solvent are fed in one stream to an oxidizer maintained at 0°C by a refrigeration unit, probably backed by a standby unit for safety. Oxygen, containing perhaps 2% ozone, is obtained as effluent from a small ozone plant and enters the reactor in a separate stream. Acetaldehyde is converted to acetaldehyde monoperoxyacetate, $CH_3CH(OH)OOCOCH_3$, in the oxidizer under ozone catalysis and then decomposed in the converter into peracetic acid and acetaldehyde. These products are separated rapidly as formed to avoid interaction producing two moles of acetic acid. However, if acetaldehyde is taken off immediately and recycled as released in the decomposition of acetaldehyde monoperoxyacetate, a 30% peracetic acid solution in acetone or ethyl acetate results.

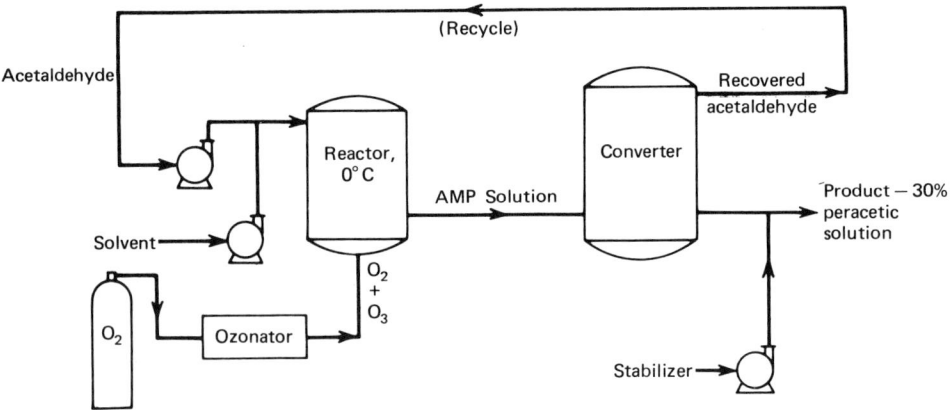

Figure 6. Acetaldehyde oxidation process for manufacture of 30% peracetic acid; AMP = acetaldehyde monoperoxyacetate.

Epoxy Oils and Esters. The epoxidation of fatty oils and esters for commercial manufacture involves making a choice of proper raw materials, choosing a process to meet established product standards, and installing equipment that will ensure safe and efficient operations. To the manufacturer depending upon commercial active oxygen to generate the epoxidizing agent, these tasks are made relatively simple by the use of hydrogen peroxide. In *in situ* systems, such as the hydrogen peroxide–resin procedure for epoxidation, the peracetic acid as formed is utilized by the oxidizable matter. Thus, the removal of peracetic acid from the system causes a shift of the equilibrium to the right resulting in the complete conversion of hydrogen peroxide to peracid. Increased conversion of hydrogen peroxide to peracid results in increased efficiency for such systems.

Sulfuric Acid Processes. *In situ* processes for the epoxidation of soybean oil based on the use of acetic acid with sulfuric acid as a catalyst have been developed by such firms as the Archer-Daniels-Midland Company and the FMC Corporation. In the Archer-Daniels-Midland process, the sulfuric acid catalyst is added last admixed with 0.5 mol of glacial acetic acid (19). Epoxy ring opening by sulfuric acid is minimized because the system is heterogeneous and involves an interaction of an oil phase and an aqueous phase containing hydrogen peroxide and acetic acid. The process is operated by first adding 20% of the total requirement of hydrogen peroxide (50%) to soybean oil and warming the mixture to ca 50°C. A solution of acetic acid containing catalytic amounts of sulfuric acid is then added to the charge over a period of 4 h with separate and simultaneous addition of the remainder of the hydrogen peroxide. The batch is held between 50 and 60°C for ca 13 h when the agitation is halted and the aqueous and oil layers are separated. The oil layer, which contains the epoxy soybean oil product, is refined and dried. The aqueous layer, containing ca 5–6% unused active oxygen, 25–30% acetic acid, and 1–2% sulfuric acid, is not discarded, however, but used to partially epoxidize fresh soybean oil. Utilization of the spent aqueous layer in this fashion is a useful innovation contributing to reaction efficiency and good economics. The aqueous layer is then separated from the partially epoxidized oil, treated with a peroxide decomposition agent, and stripped of acetic acid. The partially epoxidized oil, containing about 18% of its eventual epoxy oxygen content, is treated with fresh hydrogen peroxide and acetic acid, in the manner outlined above, to complete its epoxidation.

A second process based on the use of sulfuric acid for the epoxidation of soybean oil has been disclosed by the FMC Corporation (20). In its usual modification, the process employs an inert solvent, like benzene or hexane, to reduce the effect of sulfuric acid in catalyzing epoxy ring opening. In a typical run (21), a metric ton of soybean oil is charged to a 2.3 m^3 (600-gal) stainless-steel reactor (jacketed) fitted with agitator, cooling coils, reflux condenser, vent, rupture disk, sample line, direct and recording thermometers, feed lines, and manhole. In addition, the reactor is equipped with an automatic system for flooding the reaction with water in emergencies. The oil is heated gently by applying steam (to the jacket) and 180 kg of hexane solvent is added. To the solution of soybean oil in hexane 145 kg of glacial acetic acid is then added followed by 361 kg of 50% sulfuric acid. When the temperature reaches 50°C, the steam is shut off, cooling water is circulated through the coils, and the controlled addition of hydrogen peroxide is begun. About 347 kg of 50% hydrogen peroxide is added over a period of 2 h at 50–60°C. When peroxide addition has been completed, the temperature is allowed to rise to 60–65°C and is maintained in this range until the hydrogen peroxide has been consumed. When the reaction is complete, the organic layer is washed with water and pumped to a vacuum stripping column where water and solvent are removed.

Repeated-Resin Process. The first successful trials of the resin-catalyzed epoxidation of soybean oil and methyl oleate were conducted in 1953 (22). These trials led to the development of the repeated-resin process for epoxidation. Of the processes now available for the epoxidation of fatty oils and esters, the repeated-resin process probably eliminates unsaturation most efficiently and produces the highest epoxy-oxygen values. The process is so named because a relatively large amount of poly(styrenesulfonic acid) resin is used as the catalyst, requiring its reuse in succeeding epoxidation batches for good economics. The general practice is to use 10–15% (dry wt) of the poly(styrenesulfonic acid) based on the weight of the fatty oil or ester to be epoxidized. Advantages which may be listed for this process include high epoxy yields, little by-product formation, nearly complete elimination of unsaturation, low reaction temperatures (60°C), and short reaction periods.

DuPont (23) is credited with being the first to develop processes using styrenesulfonic acids in place of mineral acids as catalysts for epoxidation. It has been proposed (24) that the porous structure of the ion-exchange beads allows the acetic acid and hydrogen peroxide to permeate the bead and react under the catalytic influence of the reactive acid sites in the resin. The exclusion of the larger soybean oil molecules prevents epoxidation from taking place within the resin structure. The degree of cross-linking in the styrenesulfonic acid resin and the absence of metallic (eg, iron) impurities both are important to the efficiency of peracid conversion. Although the ion-exchange resins are expensive, they can be recycled, and activity can be maintained by periodically adding small quantities of resin to the bed. Optimized ion-exchange processes tend to be more efficient and to produce epoxides with less by-product (see Ion exchange).

Operating the resin process simply involves mixing the fatty oil or ester containing 1.0 mol of unsaturation, 0.55 mol of glacial acetic acid, and 12% dry resin based on the weight of epoxidizable material. Hydrogen peroxide (1.1 mol) is added slowly so that a reaction temperature of 60°C is not exceeded. The reaction medium is maintained at 60°C for ca 4 h and then separated from the resin catalyst by decantation or filtration. The resin catalyst remains in the reactor for succeeding runs.

The average poly(styrenesulfonic acid) catalyst available commercially can be reused in approximately 6–8 runs at a 10–12% catalyst level. Factors that contribute to degradation of these catalysts and militate against their indefinite use include: (*1*) the presence of heavy-metal contaminants that induce rapid attack on the resin by hydrogen peroxide, (*2*) the slow degradation of resin cross-linkage by peracid, and (*3*) the physical breakdown of the catalyst beads by mechanical attrition. The life of the resin catalyst is tripled by use of commercial resins with low metals contents.

Although use of the repeated-resin technique produces excellent epoxy products, the procedure is characterized by one difficulty. Degradation of the catalyst produces fine particles which introduce problems in filtration. Therefore, if products with maximum epoxy oxygen values are not required, it is common practice to use much less resin. A smaller amount of resin can be economically discarded after each run. This procedure is termed the *minimal-resin technique.*

Minimal-Resin Process. The minimal-resin process for epoxidation requires the use of ca 2% resin (dry wt) based on the weight of the material to be epoxidized (25). Since the quantity of catalyst in the minimal technique is much less than that used in the repeated-resin process, adjustments in the reaction variables of epoxidation must be made to account for the loss in catalyst surface area. This is partially accomplished by increasing reaction temperature (to 75–80°C) and time (to 7–8 h). Despite these changes, however, the minimal-resin process is 12–15% less effective in producing epoxy oxygen. In this respect, it is approximately equivalent to the sulfuric acid process.

Figure 7 represents a flow sheet for the epoxidation of fatty oils and esters by the minimal resin process. Although the flow sheet is based on extensive laboratory data, it may not be an accurate duplication of a large-scale process (see Fats and fatty oils).

The operations of the minimal technique are essentially those indicated for the repeated-resin and sulfuric acid processes with adjustment of the reaction variables indicated above. After reaction is complete, the resin catalyst is filtered from the reaction mixture and discarded. The reaction mixture is then pumped to a steam-jacketed wash tank equipped with a vacuum jet pump. The oil and aqueous layers are allowed to settle, and a separation of phases is effected. The lower aqueous layer is dropped to the acetic acid still feed tanks and later recovered by azeotropic dehydration using the Othmer method (26). The crude epoxidized product in the wash tank is washed successively with water or treated with alkaline agents and then washed with

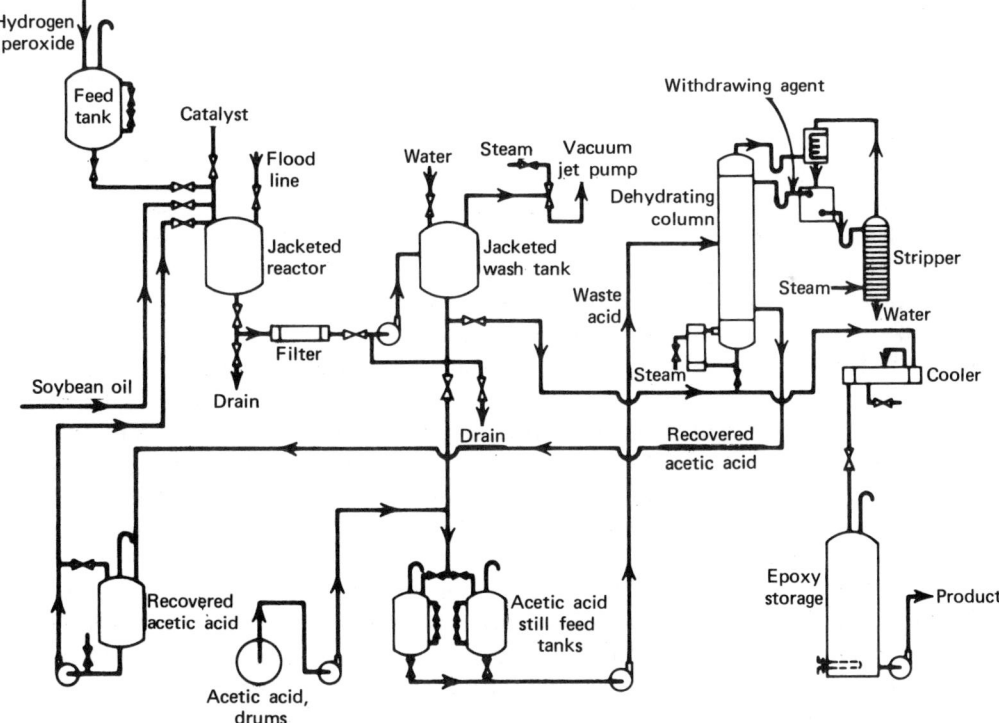

Figure 7. Schematic flow diagram for epoxidation of fatty oils and esters.

water to remove residual acid. Vacuum is applied to the tank containing the washed crude and traces of water are removed at 60–75°C (1.6–2 kPa or 12–15 mm Hg). If a wash tank equipped with vacuum is not available, the washed crude may be pumped to a flash tank or to a vacuum stripping column for drying. The finished product is filtered, if required, and piped to storage.

Continuous Processes. A continuous process for the epoxidation of fatty oils and esters has yet to be installed commercially. Considerable work on continuous processing has been completed, however, and offers promise for the future if the annual consumption of epoxidized oils and monoesters increases at the rate expected. Epoxidation processes dependent upon acetaldehyde-derived peracetic acid, aqueous peracetic acid azeotrope, or *in situ* hydrogen peroxide systems are all amenable to continuous operation.

Economic Aspects

In the United States in 1977 sales included ca 45,000 metric tons of epoxidized oils (primarily epoxidized soybean oil), and ca 14,000 t of epoxidized alkyl fatty acid esters (primarily epoxidized octyl esters of tall oil (qv) fatty acids). In comparison, the estimated sales for phthalate ester plasticizers were ca 545,000 t, and sales for polyester plasticizers were an estimated 23,000 t.

Safety

Special attention must be given to the construction of storage containers and reaction vessels that hold hydrogen peroxide or peracid compositions. Glass-lined kettles and pipes are ideal since they provide an inert contact medium; however, owing to the cost of construction and maintenance of glass-lined production equipment, it is seldom used. High purity aluminum (99.6% pure) does not decompose hydrogen peroxide and is therefore suitable material for storage tanks intended to contain aqueous hydrogen peroxide. Mild steel corrodes in the presence of acids and can catalyze an explosion if used in storage or reaction vessels that are exposed to hydrogen peroxide or peracids. Stainless steel (eg, types 304 and 316) are commonly used for mixing and reaction vessels for epoxidation. Subtle contamination arising from the use of other than the specified stainless steel must be prevented in agitators, piping, and valves. Care also must be taken to avoid contamination during fabrication and repair of the various stainless steel components. Low-carbon steels should be used for welds in order to minimize the corrosion of the adjacent stainless steel. New and repaired equipment materials that come in contact with the reaction medium must be passivated with dilute nitric acid or mixtures of glacial acetic acid (or formic acid if formic acid is used in the reaction) and small amounts of hydrogen peroxide prior to carrying out an epoxidation reaction. Improvement in the passivation of metal occurs with subsequent batch runs.

Hydrogen peroxide is considered to be the most pure and consistent product used in epoxidation. It is supplied in various concentrations (35–70%) in deionized water. Traces of stabilizers are used to prevent decomposition during storage and handling. Care must be taken to prevent contamination of the peroxide with organic or metallic (especially iron) compounds and to follow the manufacturer's instructions for storage and use (see Hydrogen peroxide).

Formic and acetic acids are usually of high purity and require no special care in handling beyond that specified by the manufacturer and consistent with good employee protection.

Uses

Uses for many of the important epoxides are described in the Ethylene oxide, Propylene oxide, Polyethers, Epoxy resins, and Chlorohydrins articles. Types of epoxy plasticizers are described below (qv).

Epoxidized Vegetable Oils. This class of epoxy plasticizers possesses the best combination of economy and stabilizing performance. The vegetable oils (qv) are naturally occurring triglycerides (triesters of glycerol and mixed unsaturated fatty acids). Table 1 contains a list of vegetable oils that have been epoxidized to yield commercially acceptable plasticizers. Because of availability and favorable economics, soybean oil is the predominant base for epoxidized vegetable oil plasticizers. Virtually all animal and vegetable unsaturated oils have been studied as potential substrates for epoxy plasticizers. Many do not contain enough unsaturation to provide the oxirane oxygen content necessary to impart adequate compatibility with PVC and to effectively function as stabilizers against degradation of the PVC by heat and by ultraviolet light. Others, such as castor oil and tung oil, are expensive and/or contain large quantities of eleostearic or ricinoleic acids which are difficult to epoxidize to high yields due to the structural placement of the unsaturation or to the presence of interfering groups. In order to predict the economic success of an epoxidized vegetable oil, one must look beyond the expense of the oil. Low iodine-number oils are less expensive, but they are also less useful; very high iodine-number oils are costly to epoxidize since the peroxide required is one of the more costly ingredients in the epoxidation process.

Epoxidized Alkyl Esters. Alkyl esters of oleic or tall oil fatty acids comprise the second most important group of substrates for epoxy plasticizers.

The alkyl 9,10-epoxystearates have the lowest oxirane oxygen content of the commercial epoxy plasticizers. Consequently, they are less efficient stabilizers than epoxidized vegetable oils and must be considered as secondary plasticizers. However, they are quite efficient for imparting flexibility to PVC at room temperature and at low temperatures.

Epoxidized alkyl esters of tall oil fatty acids (epoxytallates or epoxytofates) are derived from the tall oil fatty acids that are by products of the manufacture of Kraft

Table 1. Vegetable Oil Substrates for Epoxy Plasticizers

Vegetable oil	Iodine number[a]	Double bonds, wt %	Theoretical oxirane oxygen, wt %
linseed	170–204	0.670–0.800	9.67–11.9
safflower	140–150	0.552–0.591	8.11–8.63
soybean	120–141	0.474–0.556	7.03–8.16
corn	102–128	0.406–0.504	6.09–7.44
cottonseed	99–113	0.390–0.445	5.87–6.65
rapeseed	97–108	0.382–0.426	5.76–6.37
peanut	84–100	0.331–0.394	5.03–5.93

[a] g I_2/100 g oil.

paper from pinewood pulp. The tall oil fatty acids are mixtures of oleic and linoleic acids that have iodine values of 80 to 125. The epoxytallates have higher oxirane oxygen contents than their epoxystearate counterparts, and, therefore, they are more compatible with plastics and have more stabilizing functionality. Epoxytallates are more bulky molecules than the epoxystearates, owing to their additional oxirane rings, and they are less efficient as low temperature plasticizers.

Epoxidized Bisalkyl Esters. Preparation of bisalkyl esters by reaction of a 1:2 molar ratio of propylene or ethylene glycol with tall oil fatty acids has been reported (27). The epoxidized bisesters have oxirane oxygen contents of 5.5–5.8%, which is higher than the 4.2–4.6% oxirane oxygen content that is typical of octyl epoxytallates. By virtue of their higher molecular weight and greater epoxy contents, the bisepoxides have low volatility and good compatibility retention, as well as good stabilizing activity.

Miscellaneous Epoxy Plasticizers. The following plasticizers are not of commercial importance. However, they illustrate some of the many alternative applications for the epoxidation reaction.

Aromatic Ester Epoxides. Epoxyhexahydrophthalates, endoalkylene epoxyhexahydrophthalates (28), and alkyl epoxytetrahydrophthalates (29–30) have been patented. These materials behave much like mixtures of diisodecyl phthalates and epoxidized soybean-oil plasticizers. A highly compatible epoxide plasticizer (31) was prepared from a substrate made from naphthenic acid (qv), epoxidized fatty acids, and saturated dihydric or trihydric alcohols (see Alcohols, polyhydric). Epoxy compositions also have been made from polycyclopentadiene (see Cyclopentadiene) with saturated carboxylic acids or epoxidized fatty acids (32), both of which have the following formula:

R = aliphatic or epoxidized fatty acid radical

$n = 1$ or 2

Monoesters of xylenol or cresol with long-chain diepoxy monocarboxylic acids have been described (33) having the formula:

R = H or CH_3 $\mu = 2$–20 $x = 0$–18 $y = 0$–1

This class of plasticizers, eg, o-cresol 9,10-epoxystearate, is claimed (33) to be highly compatible with poly(vinyl chloride) in ratios of up to 1.5 epoxystearate:1 PVC and, at comparable levels, is less volatile than di(2-ethylhexyl) phthalate and imparts greater tensile and tear strength to the PVC compound than the phthalate. Epoxidized copolymers of butadiene and xylene are reported to have 250–1000 mol wt and up to 7% oxirane oxygen content (34).

Alicyclic Epoxy Esters. Alicyclic monoesters similar to the monoesters of xylenol or cresol mentioned above have been developed (35–36) and are typified by the product from reaction of 3,4-epoxycyclohexylmethanol with epoxidized fatty acid. The general formula is:

$$\text{[structure: cyclohexane with } R_1\text{–}R_6 \text{ substituents]} -CH_2OCCH_2CH_2CH-CH(CH_2)_nCH_3$$

R_1 to R_6 = H or a lower alkyl group $\qquad n = 4-12$

Poly(oxyethylene tetrahydrofuran) di(9,10-epoxystearate) has been prepared (37) from the reaction of 4 mol of ethylene oxide with one mole of tetrahydrofuran, esterification of the polymer with 2 mol of oleic acid, and finally, epoxidization of the diester.

Stabilizing Functionality. The oxirane group is a well known acid-accepting functional group.

$$\sim\!\!\sim\!\!HC\overset{O}{-\!-}CH\!\sim\!\!\sim + HX \longrightarrow \sim\!\!\sim\!\!\overset{OH}{CH}\overset{X}{-}CH\!\sim\!\!\sim \qquad (6)$$

X = a halogen or an acid group such as NO_3. The earliest use of vegetable oil epoxides was as heat stabilizers (qv) for PVC. One stabilizing reaction for vegetable oil epoxide/PVC compounds containing no metallic stabilizer is:

$$RCH\overset{O}{-\!-}CHR + \left(\overset{Cl}{\underset{|}{CHCH_2}}\overset{Cl}{\underset{|}{CHCH_2}}\right)\!\!- \overset{\Delta}{\longrightarrow} R\overset{Cl}{\underset{|}{CH}}\overset{OH}{\underset{|}{CHR}} + \!\!-\!\!\left(CH\!\!=\!\!\overset{Cl}{\underset{|}{CH}}CHCH_2\right)\!\!- \qquad (7)$$

Ether formation between the epoxide and the free radical site of the labile hydrogen in the PVC has been suggested (38):

$$RCH\overset{O}{-\!-}CHR + \left(\overset{Cl}{\underset{|}{CHCH}}\overset{Cl}{\underset{|}{CHCH_2}}\right)\!\!- \overset{\Delta}{\rightleftharpoons} \!\!-\!\!\left(\overset{Cl}{\underset{|}{CHCHCHCH_2}}\right)\!\!- \qquad (8)$$
$$\underset{\underset{RCHCHR}{\overset{|}{O}\ \overset{|}{Cl}}}{}$$

BIBLIOGRAPHY

"Epoxidation" in *ECT* 1st ed., Suppl. 2, pp. 325–346, by J. G. Wallace, E. I. du Pont de Nemours & Co., Inc.; "Epoxidation" in *ECT* 2nd ed., Vol. 8, pp. 238–261, by J. G. Wallace, E. I. du Pont de Nemours & Co., Inc.

1. R. J. Gall and co-workers, *Ind. Eng. Chem.* **47,** 147 (1955).
2. T. W. Findley and co-workers, *J. Am. Chem. Soc.* **67,** 412 (1945).
3. U.S. Pat. 2,458,484 (Jan. 4, 1949), D. E. Terry and co-workers (to General Mills).
4. U.S. Pat. 2,458,160 (Oct. 18, 1949), W. D. Niederhauser and co-workers (to Rohm & Haas Co.).
5. U.S. Pat. 2,490,800 (Dec. 13, 1949), F. Greenspan (to Buffalo Electrochemical).
6. *Peracetic Acid and Derivatives, F-40108A,* Union Carbide, 1957.
7. J. A. John and co-workers, *Chem. Ind.,* 62 (Jan. 13, 1962).
8. Brit. Pat. 820,461 (Sept. 23, 1959), J. B. Williamson (to Distillers).
9. R. L. Augustine and D. J. Trecker, *Oxidation,* Vol. 2, Marcel Dekker, Inc., New York, 1977, Chapt. 2.
10. U.S. Pat. 3,360,531 (Dec. 26, 1971), W. H. French (to Ashland Oil & Refining Co.).
11. U.S. Pat. 2,836,605 (June 9, 1954), S. P. Rowland and co-workers (to Rohm & Haas Co.).
12. U.S. Pat. 3,993,673, C. H. McMullen (to Union Carbide Corp.).
13. Jpn. Kokai 76 40,050 (to Mitsubishi Chem Ltd.).
14. F. P. Greenspan, *Ind. Eng. Chem.* **39,** 847 (1947).
15. U.S. Pat. 2,490,800 (Dec. 13, 1949), F. P. Greenspan (to FMC Corp.).
16. Brit. Pat. 776,758 (June 12, 1957), A. T. Hawkinson and W. R. Schmitz (to E. I. du Pont de Nemours & Co., Inc.).
17. B. Phillips, P. S. Starcher, and B. D. Ash, *J. Org. Chem.* **23,** 1823 (1959).
18. U.S. Pat. 2,804,473 (Aug. 27, 1957), B. Phillips, F. C. Frostick, and P. S. Starcher (to Union Carbide Corp.).
19. U.S. Pat. 2,813,878 (Nov. 19, 1957), A. W. Wahlroos (to Archer-Daniels-Midland Co.).
20. U.S. Pat. 2,801,253 (July 30, 1957), F. P. Greenspan and R. J. Gall (to FMC Corp.).
21. *Chem. Week* **75,** 32 (Dec. 25, 1954).
22. Can. Pat. 531,112 (Oct. 2, 1956), A. A. D'Addieco (to E. I. du Pont de Nemours & Co., Inc.).
23. W. R. Schmitz and co-workers, *J. Am. Oil Chem. Soc.* **31,** 363 (1954).
24. F. Chadwick and co-workers, *J. Am. Oil Chem. Soc.* **35,** 355 (1958).
25. A. F. Chadwick and co-workers, *J. Am. Oil. Chem. Soc.* **35,** 355 (1958).
26. D. F. Othmer, *Chem. Met. Eng.* **40,** 631 (1933); **47,** 349 (1940).
27. Brit. Pat. 1,020,866 (Feb. 23, 1960), R. O. Brookings and co-workers (to British Celanese).
28. U.S. Pat. 2,963,490 (Dec. 6, 1960), S. P. Rowland (to Rohm & Haas Co.).
29. U.S. Pat. 3,223,715 (Dec. 14, 1965), W. M. Kraft and co-workers (to Tenneco Chemicals).
30. U.S. Pat. 2,956,975 (Oct. 18, 1960), F. P. Greenspan (to FMC).
31. U.S. Pat. 3,182,034 (May 4, 1965), J. O. VanHook (to Rohm & Haas Co.).
32. Brit. Pat. 873,868 (July 26, 1961), (to Esso Research and Engineering).
33. U.S. Pat. 3,184,425 (May 18, 1965), J. J. Jaruzelski and co-workers (to United States Steel).
34. U.S. Pat. 3,182,035 (May 4, 1965), J. A. Garman (to FMC).
35. Brit. Pat. 827,986 (Feb. 10, 1960), (to Union Carbide).
36. Jpn. Kokai 43, 24,411 (Oct. 22, 1968), (to Dainippon Ink Kagaku Kogyo).
37. U.S. Pat. 3,225,068 (Dec. 21, 1965), J. D. Zech (to Atlas Chemical).
38. D. F. Anderson and co-workers, *J. Polym. Sci. Part A-1* **8,** 2905 (1970).

JOHN T. LUTZ, JR.
Rohm and Haas Co.

EPOXIDE POLYMERS. See Polyethers.

EPOXIDES. See Ethylene oxide; Propylene oxide; Chlorohydrins; Butylenes; Styrenes; Olefins.

EPOXY RESINS

Epoxies are monomers or prepolymers that further react with curing agents to yield high-performance thermosetting plastics. They have gained wide acceptance in protective coatings and electrical and structural applications because of their exceptional combination of properties such as toughness, adhesion, chemical resistance, and superior electrical properties.

Epoxy resins are characterized by the presence of a three-membered cyclic ether group commonly referred to as an epoxy group, 1,2-epoxide, or oxirane.

The most widely used epoxy resins are diglycidyl ethers of bisphenol A [25068-38-6] (1) derived from bisphenol A and epichlorohydrin.

$$CH_2\!-\!\!\!\underset{\underset{\displaystyle O}{\diagup\!\!\!\diagdown}}{CHCH_2}\!\!-\!\!\left[\!O\!-\!\!\!\underset{}{\bigcirc}\!\!-\!\!\underset{\underset{\displaystyle CH_3}{|}}{\overset{\overset{\displaystyle CH_3}{|}}{C}}\!\!-\!\!\underset{}{\bigcirc}\!\!-\!OCH_2\underset{\underset{}{}}{\overset{\overset{\displaystyle OH}{|}}{C}H}CH_2\!\right]_{\!n}\!\!-\!O\!-\!\!\!\underset{}{\bigcirc}\!\!-\!\!\underset{\underset{\displaystyle CH_3}{|}}{\overset{\overset{\displaystyle CH_3}{|}}{C}}\!\!-\!\!\underset{}{\bigcirc}\!\!-\!OCH_2\underset{\underset{\displaystyle O}{\diagdown\!\!\!\diagup}}{CH}\!\!-\!\!CH_2$$

(1)

The outstanding performance characteristics of the resins are conveyed by the bisphenol A moiety (toughness, rigidity, and elevated temperature performance), the ether linkages (chemical resistance), and the hydroxyl and epoxy groups (adhesive properties and formulation latitude, or reactivity with a wide variety of chemical curing agents) (see also Phenolic resins). In addition to bisphenol, other polyols such as aliphatic glycols and novolacs are used to produce specialty resins. Epoxy resins may also include compounds based on aliphatic and cycloaliphatic backbones. Recent developments have utilized a heterocyclic hydantoin (qv) nucleus as a building block to produce a family of epoxies with superior electrical properties and outdoor weatherability. Conventional epoxy resins range from low viscosity liquids to solid resins. Curing agents commonly used to convert epoxies to thermosets include anhydrides, amines, polyamides, Lewis acids, and others.

The commercial significance of bisphenol A–epichlorohydrin-based epoxy resins was first identified by Castan of de Trey Freres, Ltd., Switzerland in 1938 and Greenlee of Devoe and Raynolds (technology now retained by Celanese Corporation) in 1939 (1–2). In 1934, Schlack in Germany first synthesized a polyglycerol ether with an intended use as an intermediate for textile treatments (3). In 1938 Castan allowed an epoxy to react with phthalic anhydride to produce useful materials for casting. In 1943 the basic materials were patented as curing agents, but the compositions were limited to dentistry (1). Ciba Ltd. (now Ciba-Geigy), under license, conducted further development and in 1945 filed the first patent application for epoxy adhesives (qv). These products were offered commercially in 1946. A solid epoxy for coating applications was first produced by Greenlee in 1939. Further significant developments were contributed by Shell Chemical, (as an epichlorohydrin supplier), and Union Carbide (producer of bisphenol and phenolics).

268   EPOXY RESINS

In the late 1950s and early 1960s commercialization of epoxy resins was actively pursued by Shell, Devoe and Raynolds (now Celanese), Union Carbide, Ciba, Dow, and Reichhold. In 1956 Union Carbide developed cycloaliphatic epoxies via peracetic acid synthesis. Commercialization was jointly pursued by Union Carbide and Ciba, who had also developed cycloaliphatic epoxides via an olefin route.

In order to improve high temperature performance and other select properties, various multifunctional resins were synthesized and introduced by Ciba, Celanese, Shell, and Union Carbide in the mid 1960s. Development of epoxy cresol novolacs was achieved by Koppers and subsequently commercialized by Ciba. In the 1970s considerable research and development by Ciba-Geigy on hydantoins resulted in commercialization of resins based on this heterocyclic moiety. In 1976 Shell Chemical introduced hydrogenated bisphenol A materials as uv-weatherable systems to compete with aliphatic polyurethanes (see Urethane polymers).

Resin Properties and Characteristics

Epichlorohydrin and Bisphenol A-Derived Resins. These epoxy resins are most frequently cured with anhydrides, aliphatic amines, or polyamides, depending upon desired end properties. Some of the outstanding properties are superior electrical properties, chemical resistance, heat resistance, and adhesion.

Comparisons of the properties of solvent-free liquid coatings based on unmodified bisphenol A epoxies cured with varying hardeners are shown in Table 1. For solvent-based ambient cure systems, polyamides are often the hardeners of choice. For heat-cured systems, anhydrides are used to provide higher heat resistance systems but at the expense of flexibility. Diluents are commonly used to reduce the viscosity of epoxy systems to aid handling, improve ease of application, and to facilitate higher filler loading to reduce formulation cost. This, however, is achieved at the expense of other properties. To achieve a balance of properties, careful selection of diluent is needed. Table 2 qualitatively shows which diluent should be considered for minimal deterioration of properties.

Specialty Epoxy Resins. In addition to the family of resins derived from epichlorohydrin and bisphenol A, a variety of other important resins have been commercialized containing aliphatic, cycloaliphatic, aromatic, and heterocyclic backbones.

Glycidylation of active-hydrogen-containing structures with epichlorohydrin and epoxidation of olefins with peracetic acid remain the important commercial procedures for introducing the oxirane group into various precursors of epoxy resins (see Epoxidation).

Epoxy Cresol Novolac Resins (ECN). The cresol-novolac epoxy resins (2) are multifunctional, solid polymers characterized by low ionic and hydrolyzable chlorine impurities, high chemical resistance, and thermal performance. ECN resins are widely used as base components in high performance electronic and structural molding compounds, high temperature adhesives, structural molding powders, castings and laminating systems, tooling applications, and powder coatings.

(2) [37382-79-9]

Table 1. Comparative Properties of Bisphenol A-Epoxy Resins Cured with Different Hardeners[a]

	Short-chain aliphatic polyamine	Oxyalkylated short-chain polyamine	Long-chain polyamine adduct	Aromatic polyamine adduct	Polyaminoamide
chemical structure	$H_2NC_2H_4NHC_2H_4NH_2$ diethylenetriamine [111-40-0] $H_2NC_2H_4NHC_2H_4NHC_2H_4NH_2$ triethylenetetramine [112-24-3]	$H_2NC_2H_4NHC_2H_4NHCH_2CH_2OH$ [1965-29-3] $HOCH_2CH_2NHC_2H_4NH$-$C_2H_4NHCH_2CH_2OH$ [4484-60-0]	trimethyl-1,6-hexanediamine [25620-58-0]		reaction product of polyamines with dimer acids
advantages	low viscosity, generally good chemical resistance	low viscosity, generally good chemical resistance, low toxicity	color stability, generally good chemical resistance, good flexibility	long pot-life, good acid resistance, cures in presence of moisture	long pot-life, generally good solvent resistance, good flexibility and adhesion
limitations	dermatitis potential, poor organic acid resistance, mixing ratio critical, film appearance	short pot-life, poor organic acid resistance	short pot-life, low organic acid resistance	high viscosity, color, low resistance to aromatic solvents	poor organic acid resistance, blushing
application–handling					
mixing ratio, wt	100:10 or 13	100:25	100:35	100:60	100:43
viscosity, mPa·s (= cP)	4,500	3,000	7,800	8,000	>10,000
pot-life, min	30–60	<30	23	135	75
color, Gardner	1–4	5–8	1–2	>12	6
dry time (paper-free), h	1.5–3	<10	ca 2	ca 16	
film quality					
flow	poor	poor	good	good	good
surface	blushing	blushing	blush resistant	blush resistant	blushing
color stability	good	good	good	poor	fair

[a] Basis: resin = diglycidyl ether of bisphenol A (wt per epoxide, WPE = 182–196). Courtesy of the Federation of Societies for Coatings Technology (4).

Table 2. Selectivity Chart—Reactive Diluents and Properties of System

Properties	Butyl glycidyl ether [2426-08-6]	C$_8$–C$_{10}$- aliphatic monoglycidyl ether	C$_{12}$–C$_{14}$- aliphatic monoglycidyl ether	Cresyl glycidyl ether [26447-14-3]	Neopentyl glycol diglycidyl ether [17557-23-2]
flexibility		X	X		
efficiency	X				
solvent resistance					X
acid resistance				X	X
pot life extension			X		
SPI rating		3	3		2

The multifunctionality contributes higher reactivity and cross-link density. These factors are especially critical when formulating systems that require improved thermal performance over conventional epichlorohydrin–bisphenol A systems. The melt viscosity of these resins (solids at room temperature) decreases sharply with increasing temperature (see Table 3). This affords the formulator an excellent tool for controlling flow of molding compounds, and facilitating the incorporation of ECN resins into other epoxies, eg, for powder coatings (qv).

The o-cresol novolacs of commercial significance possess degrees of polymerization (n) of 1.7–4.4 and the epoxide functionality of the resultant glycidylated resins varies from 2.7 to 5.4. Softening points (Durran's) of the products are 35–99°C. The glycidylated phenol and o-cresol-novolac resins are soluble in ketones, 2-ethoxyethyl acetate, and toluene solvents. The commercial epoxy novolacs products possess a residual hydrolyzable chlorine content of <0.15 wt % and a total chlorine content of ca 3 wt %.

Epoxy Phenol Novolac Resins. The multiepoxy functionality of the epoxy novolacs (2.2 to >5 epoxy groups per molecule) (3) produce more tightly cross-linked cured systems having improved elevated temperature performance and chemical resistance than the difunctional bisphenol A-based resins.

(3) [9003-35-5]

Table 3. Typical Properties of Epoxy Cresol-Novolac Resins[a]

Property	ECN 1235	ECN 1273	ECN 1280	ECN 1299
mol wt	540	1080	1170	1270
epoxy value, eq/100 g	0.500	0.445	0.435	0.425
softening point, °C	35	73	80	99
density, kg/L	1.11	1.16	1.17	1.19
functionality (see structural formula)				
ether type	0.3	1.2		1.6
epoxide type	2.7	4.8	5.1	5.4

[a] Ref. 5.

The thermal stability of epoxyphenol-novolac resins is useful in adhesives, structural and electrical laminates, coatings, castings, and encapsulations for elevated temperature service (see Table 4). Filament-wound pipe and storage tanks, liners for pumps and other chemical process equipment, and corrosion-resistant coatings are typical applications using the chemically resistant properties of epoxy novolac resins.

Curing agents that give the optimum in elevated temperature properties for epoxy novolacs are those with good high temperature performance such as aromatic amines, catalytic curing agents, phenolics, and some anhydrides. When cured with polyamide or aliphatic polyamides and their adducts, epoxy novolac resins show improvement over bisphenol A-based epoxies, but the critical performance of each cure is limited by the performance of the curing agent.

Bisphenol F Resins. Owing to relatively low viscosity, these resins offer advantages for 100% solids systems. Higher filler levels are possible because of the low viscosity. Faster bubble release is also achieved. Higher epoxy content and functionality of bisphenol F epoxy resins can provide improved chemical resistance compared to conventional epoxies.

Bisphenol F-epoxy resins are used in high-solids–high-build systems such as tank and pipe linings, industrial floors, road and bridge deck toppings, structural adhesives, grouts, coatings, and electrical varnishes. Bisphenol F-epoxy resins are manufactured in Europe and Japan.

Polynuclear Phenol-Glycidyl Ether-Derived Resins. This (4) is one of the first commercially available polyfunctional products. Its polyfunctionality permits upgrading of thermal stability, chemical resistance, and electrical and mechanical properties of bisphenol A-epoxy systems. Used in molding compounds and adhesives,

Table 4. Typical Properties of Epoxy Phenol Novolacs

n value	0.2[a]	1.6[b]	1.8[c]	3.5[d]
epoxide equiv wt	175	178	200	185
epoxy functionality	2.2	3.6	3.8	5.5
viscosity, mPa·s (= cP)	1,400 (52°C)	35,000 (52°C)	3,000 (100°C)	800 (150°C)
softening point, Durran, °C			53	73
color, Gardner	1	2	2	2
heat distortion temperature[e], °C	156(165)	180(207)	182(202)	240

[a] D.E.N. 431, Dow Chemical; EPN 1139, Ciba-Geigy.
[b] D.E.N. 438, Dow Chemical; EPN 1138, Ciba-Geigy.
[c] D.E.N. 439, Dow Chemical.
[d] XD-7855 Experimental Solid Epoxy Novolac, Dow Chemical.
[e] Cured with methylenedianiline: gelled 16 h at 55°C + 2 h at 125°C + 2 h at 175°C (numbers in parenthesis indicate additional 4 h at 200°C).

(4) [27043-37-4]

and as an upgrader in laminating systems, it is available in solid and solution forms (see Laminated and reinforced plastics).

Cycloaliphatic Epoxy Resins. This family of aliphatic, low viscosity epoxy resins consists of two principal varieties, cycloolefins epoxidized with peracetic acid and diglycidyl esters of cyclic dicarboxylic acids. The nonaromatic nature of these materials provides for improved uv resistance and arc-track resistance compared to conventional epoxies. The best properties are generally achieved with anhydride and phenolic curing agents. These cycloaliphatic epoxy resins respond sluggishly to amine curing agents compared to standard bisphenol A-derived products. The reactivity of cycloaliphatic resins in curing reactions is discussed in ref. 6. The cyclohexane ring structure and short side chains give rise to high deflection temperatures but rather brittle products.

Recommended applications include transformers, insulators, bushings, wire and cable, generators, motors and switchgear, additives for adhesives, vinyl stabilization, and as viscosity depressants.

Aromatic and Heterocyclic Glycidyl Amine Resins. Among the specialty epoxy resins containing an aromatic amine backbone, the following are commercially significant.

Tetraglycidylmethylenedianiline-Derived Resins. Resins from aromatic glycidyl amines can be formulated into hot-melt or solution–binder systems with various reinforcements, eg, glass, graphite, boron, aramide. These systems provide for simple cure cycles with excellent mechanical and adhesive properties at room and elevated temperatures. They are utilized for graphite-reinforced composites in aerospace and leisure products (see Ablative materials), structural adhesives, laminates, tooling and casting applications, and structures such as wings and fuselages (see Composite materials; Adhesives). An important new use for the automobile industry may also develop if graphite fiber prices drop as expected from $88/kg to $22/kg.

$$\left[\left(CH_2\!-\!\!\underset{O}{\diagdown\!\diagup}\!\!-\!CHCH_2\!-\! \right)_{\!2}\!\!N\!-\!\!\!\bigcirc\!\!\!-\!CH_2 \right]_2$$

(**5**) [28768-32-3]

Triglycidyl–p-Aminophenol-Derived Resins. This low viscosity resin [5026-74-7] permits cure at low temperatures (70°C) and rapidly develops excellent elevated temperature properties. Used to increase heat resistance and cure speed of bisphenol A-epoxy resins, it has utility in such diverse applications as adhesives, tooling compounds and laminating systems. A molecularly distilled version is used as a binder for solid propellants (see Explosives) and for military flares (see Pyrotechnics). Its chief uses depend on properties of low viscosity, and low temperature reactivity, particularly with carboxy-terminated rubbers.

Triazine-Based Resin. Triglycidyl isocyanurate [2451-62-9] (**6**) is a solid resin that provides superior thermal, electrical, and mechanical properties and is recommended for laminates, insulating varnishes, coatings, and adhesives (see Table 5). Widely used as a curing agent for special polyester-based weatherable powder coatings, it is also used in electronic applications owing to its retention of optical transparency after aging at temperatures up to 150°C and minimal smoke evolution on thermal decomposition (see Embedding).

(6)

Table 5. Typical Properties of Multifunctional Epoxy Resins

Property	Amine based		Other	
	0500[a]	MY 720[b]	0163[c]	PT 810[d]
physical form	liquid	semisolid	solid	solid
viscosity at 25°C, Pa·s[e]	1.5–5.0	10–35[f]		
epoxy value, equiv/100 g, minimum	0.86	0.75	0.38	1.05
functionality	3	4	4	3

[a] Triglycidyl derivative of p-aminophenol (7).
[b] Tetraglycidyl derivative of methylenedianiline (5) (8).
[c] Polynuclear phenol glycidyl ether (9).
[d] Triglycidylisocyanurate (6) (10).
[e] To convert Pa·s to poise, multiply by 10.
[f] Viscosity measured at 50°C.

Hydantoin Epoxy Resins. A series of resins based on the heterocycle hydantoin (7) has been developed by Ciba-Geigy. Extensive reviews of hydantoin-based epoxy resin chemistry and technology are in ref. 11.

The nitrogen-containing heterocycle permits extensive variation in polarity, melting point, viscosity, and hydrophobicity by the choice of alkyl groups. A variety of epoxy resins can be formed through the reaction of hydantoin with epichlorohydrin (analogous to the reaction with bisphenol A). The high polarity of the hydantoin ring provides resins with good adhesion to filler particles, fiberous reinforcements, and a wide variety of substrates. By changing the substituents, resins ranging from low viscosity liquids to relatively high melting solids and from water soluble to water insoluble resins can be obtained.

The main commercial product is based on 5,5-dimethylhydantoin. Curing may be achieved by anhydrides or aromatic amines. These resins offer low viscosities coupled with high epoxy content. Because of the compact structure of the hydantoin nucleus, the resins tend to give high heat deflection temperatures when cured. However, because of their high polarity, resins cured with polar cross-linking agents, such as aliphatic amines, tend to show high moisture sensitivity. Thus the preferred curing

(7); R = R' = CH$_3$ [15336-81-9]

agents are the anhydrides of dicarboxylic acids and aromatic amines. Because of the low viscosity and high polarity of hydantoin-based resins, they generally wet-out fillers and reinforcing fibers extremely well; this can be utilized to provide commercially inexpensive formulations (see Table 6). The excellent wetting out on special fibers such as Kevlar (DuPont) is utilized in tire-cord adhesive applications. Hydantoin epoxies cured with nonaromatic anhydrides, also provide good wet arc-track resistance.

Table 6. Properties of Hydantoin Epoxy Resins

Property	Dimethylhydantoin resin (7)[a]	Blend of dimethylhydantoin resin–diglycidyl ether of bisphenol A (7 + 1)[b]
form	liquid	liquid
viscosity at 25°C, mPa·s (= cP),	2500	5000
epoxy value, equiv/100 g	0.70	0.62

[a] Ref. 12.
[b] Ref. 13.

Resin Synthesis and Manufacture

Epichlorohydrin and Bisphenol A-Derived Resins. Liquid epoxy resins may be synthesized by a two-step reaction of an excess of epichlorohydrin to bisphenol A in the presence of an alkaline catalyst. The reaction consists initially in the formation of the dichlorohydrin of bisphenol A [4809-35-2] and further reaction via dehydrohalogenation of the intermediate product with a stoichiometric quantity of alkali.

The use of a large excess of epichlorohydrin minimizes the formation of higher molecular weight species, ie, further reaction of the diglycidyl ether of bisphenol A with bisphenol A results in the formation of polymeric species (1).

Typical liquid epoxy resins are mixtures of species with n values as follows: $n = 0$, 87.2%; $n = 1$, 11.1%; $n = 2$, 1.5% (14). Moreover, side reactions can reduce the theoretical epoxide functionality:

(1) Hydrolysis:

$$—CH_2CHCH_2 \xrightarrow{H_2O} —CH_2CHCH_2$$
$$\underset{O}{\vee} \underset{OH\ OH}{|\ \ |}$$

α-glycol formation

(2) Formation of bound chlorine by reaction of epichlorohydrin with alcohol OH groups:

$$CH_2\!\!-\!\!CHCH_2Cl + ClCH_2CHCH_2\!\!\sim\!\!\sim \longrightarrow ClCH_2CHCH_2\!\!\sim\!\!\sim \xrightarrow{NaOH} ClCH_2CHCH_2\!\!\sim\!\!\sim$$

(3) Incomplete dehydrohalogenation resulting in a residual amount of saponifiable chlorine (hydrolyzable chlorine).

(4) Abnormal addition of epichlorohydrin, ie, addition to the more substituted carbon atom:

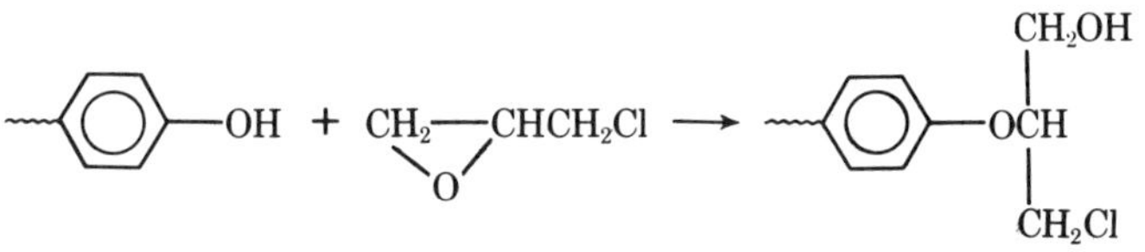

$$\text{---}\underset{}{\bigcirc}\text{---OH} + \text{CH}_2\text{---CHCH}_2\text{Cl} \longrightarrow \text{---}\underset{}{\bigcirc}\text{---OCH}\underset{\text{CH}_2\text{Cl}}{\overset{\text{CH}_2\text{OH}}{|}}$$

The pure diglycidyl ether of bisphenol A [1675-54-3] is a crystalline solid (mp 43°C) with a weight per epoxide (WPE) = 170. The typical commercial unmodified liquid resins are viscous liquids with a viscosity of 11–16 Pa·s (110–160 P) at 25°C, and an epoxy equivalent weight of ca 188. A typical synthesis of a liquid epoxy resin is described in ref. 15. Resins of higher molecular weight can be prepared by two routes:

Taffy Process. Bisphenol A reacts directly with epichlorohydrin in the presence of a stoichiometric amount of caustic. The molecular weight of the product is governed by the ratio of epichlorohydrin–bisphenol A.

In practice, the taffy process is generally employed for only medium molecular weight resins (1), ($n = 1$–4). The polymerization reaction results in a highly viscous product (emulsion of water and resin) and the condensation reaction becomes very dependent on agitation. At the completion of the reaction, the heterogeneous mixture consists of an alkaline brine solution and a water–resin emulsion and recovery of the product is accomplished by separation of phases, washing of the taffy resin with water and removal of water under vacuum.

The relationship between mole ratio of epichlorohydrin–bisphenol A and the resultant epoxy equivalent weight of the resin produced is shown below (16):

Mole ratio epichlorhydrin–bisphenol A	*Epoxy equiv wt*
2.6	249
2.15	345
1.57	516
1.4	582
1.33	730
1.25	862
1.2	1180

Advancement Process. In the advancement process, sometimes referred to as the fusion method, liquid epoxy resin (crude diglycidyl ether of bisphenol A) is chain-extended with bisphenol A in the presence of a catalyst to yield higher polymerized products. The molecular weight of the resin is a function of the ratio of excess liquid epoxy resin to bisphenol A. The terminal groups are preponderantly epoxy groups.

Since no by-products are generated, the advancement process is more convenient than the taffy process and the former can be used to prepare high molecular weight resins directly. Side reactions, which are more likely in the taffy process, lead to branching of the polymer chain to give an increase in melting point and viscosity.

The relationship between n value, epoxy equivalent weight, and melting point is shown in Table 7.

As the molecular weight of the resin increases (greater n value), a corresponding

Table 7. Properties of Commercial Epoxy Resins

Average n value	Epoxy equiv wt	mp (Durran's), °C	Approx mol wt
0.1	185	liquid	360–380
0.3	210	liquid	390–450
0.6	255	liquid	460–560
2.2	487	65–75	850–1100
5.5	950	95–105	1750–2050
14.4	2250	125–135	4000–5000
16.0	3250	145–155	5000–8000

reduction in the terminal epoxy content of the molecule occurs with an increase in the hydroxyl content. This is accompanied by an increase in viscosity and melting point and eventually leads to resins with sufficiently high molecular weight such that the epoxy content is negligible. The resins are thus linear polyethers with pendant hydroxyl groups. Such resins are usually supplied in organic solvents and curing of the polyols with hydroxyl-reactive agents is not necessarily required to develop their properties. A family of higher molecular weight thermoplastic solid and solution resins (Union Carbide) are a related class of polymers or engineering plastics (17).

Gel permeation chromatography studies of epoxy resins prepared by the taffy process show n values = 0, 1, 2, 3, etc, whereas only even-numbered repeat units are observed for resins prepared by the advancement process. This is a consequence of adding a difunctional phenol to a diglycidyl ether derivative of a difunctional phenol in the polymer-forming step.

In recent years, proprietary catalysts for advancement have been incorporated in precatalyzed liquid resins. Thus only the addition of bisphenol A is needed to produce solid epoxy resins. Use of the catalysts is claimed to provide resins free from branching which can occur in conventional fusion processes (18). Additionally, use of the catalysts results in rapid chain-extension reactions due to the high amount of heat generated in the processing.

The preparation of flame-retardant epoxy resins is accompanied by inclusion of tetrabromobisphenol A in the advancement process (see Flame retardants). Products containing ca 20 wt % Br are extensively employed in the printed circuit board industry.

Liquid resins containing bromine (ca 49 wt %) can also be prepared directly from tetrabromobisphenol A and epichlorohydrin and are used for critical applications where a high degree of flame retardancy is required.

SPECIALTY RESINS

Epoxy Cresol-Novolac Resins. The epoxy cresol-novolac resins (2) are prepared by glycidylation of o-cresol–formaldehyde condensates in the same manner as the phenol-novolac resins. The o-cresol–formaldehyde condensates are also prepared under acidic conditions with formaldehyde–o-cresol ratios of less than unity.

Epoxy Phenol-Novolac Resins. Epoxy phenol-novolac resins, represented by the general idealized structure (3), differ from the bisphenol A-epoxy resins in that multiepoxy functionality occurs with increasing mol wt and phenyl ring substitutions become random rather than strongly para-para' directed. This is owing to their preparation from phenol–formaldehyde condensation products rather than phenol–acetone. In the preparation of bisphenol A, the steric hindrance imposed by the isopropylidene linkage strongly limits the product to para-para'-substituted difunctional material. On the other hand, phenol–formaldehyde condensation reactions give methylene bridges with no strong steric limitations. Thus multifunctional products are formed containing a phenolic hydroxyl group per phenyl ring in random para-para', ortho-para', and ortho-ortho' combinations.

Subsequent epoxidation with epichlorohydrin yields the highly functional epoxy novolac. The product can range from a high viscosity liquid of $n = 0.2$ to a solid of n value greater than 3.

Bisphenol F Resin. Bisphenol F epoxy resin [2467-02-9] is of the same general structure as the epoxy novolacs. Whereas the epoxy novolacs vary from viscous liquids to solid materials, the bisphenol F resin has a low viscosity (ca 4 Pa·s (40 P)) and 165 epoxy equivalent weight. Its n value is about 0.15 and crystallization, often a problem with low viscosity conventional bisphenol A resins, is essentially nonexistent with the bisphenol F resin.

Polynuclear Phenol-Glycidyl Ether-Derived Resins. The tetraglycidyl ether of tetrakis(4-hydroxyphenyl)ethane [27043-37-4] (**4**) is prepared by reaction of glyoxal with phenol in the presence of HCl (17). The glycidylated polyphenol (**4**), mp ca 80°C, possesses a theoretical epoxide functionality of four with an epoxy equivalent weight of 185–208.

Cycloaliphatic Epoxy Resins. The epoxidation of cycloolefinic compounds with organic peracids constitutes a class of mono and difunctional epoxy resins with noteworthy features. The cycloaliphatic products are generally liquids of lower viscosity than the standard glycidyl ether resins. The peroxidized resins contain no chlorine and low ash content and their ring-contained oxirane group (cyclohexene oxide type) reacts more readily with acidic curing agents than the bisphenol A-derived epoxy resins.

In addition, glycidyl esters are produced by the reaction of cycloaliphatic carboxylic acids with epichlorohydrin, followed by dehydrohalogenation with caustic. Such products are characterized by low viscosities (ca 500 mPa·s or cP). Reactivity of the glycidyl esters more closely resembles the standard bisphenol–epichlorohydrin resins.

The diglycidyl ether of hydrogenated bisphenol A has recently been commercialized by Shell Chemical (Eponox resins) (20).

Aromatic Glycidyl Amine Resins. Triglycidyl p-aminophenol-derived resin originally developed by Union Carbide Corporation (21) is currently marketed by Ciba-Geigy. Synthesis is conducted by reaction of epichlorohydrin with the phenolic and amino groups followed by dehydrohalogenation. The product is a viscous liquid (1.5–5 Pa·s (15–50 P) at 25°C) which is considerably more reactive toward amines than standard bisphenol A-derived resins. The epoxy equiv wt is 105–114.

N,N,N',N'-Tetraglycidyl 4,4'-diaminodiphenylmethane (**5**) possesses an epoxy equivalent weight of 117–133, and a viscosity of 25 Pa·s (250 P), at 50°C.

The triazine ring-containing product 1,3,5-triglycidyl isocyanurate (**6**) is synthesized by glycidylation of cyanuric acid with epichlorohydrin. The commercial product is a crystalline powder that exhibits an epoxy equivalent weight of ca 108 and softens in the 85–110°C range (see Cyanuric and isocyanuric acids).

Hydantoin is synthesized from a ketone, hydrogen cyanide, ammonia and carbon dioxide. The subsequent glycidylation of the hydantoin is straightforward but may be affected by the solubility behavior of the substituted hydantoin (see Hydantoin).

The most important influence on the properties of hydantoin epoxy resins is the alkyl substitutions in the 5-position of the hydantoin. They are governed by the carbonyl compound chosen in the synthesis of the initial hydantoin.

Aliphatic Glycidyl Ethers. Aliphatic epoxy resins have been synthesized by glycidylation of difunctional or polyfunctional polyols such a 1,4 butanediol, 2,2'-dimethyl-1,3-propanediol (neopentyl glycol), polypropylene glycols, glycerol, trimethylolpropane, and pentaerythritol.

The epoxidation is generally conducted in two steps: (1) the polyol is added to epichlorohydrin in the presence of a Lewis acid catalyst (stannic chloride, boron trifluoride) to produce the chlorohydrin intermediate, and (2) the intermediate is then dehydrohalogenated with sodium hydroxide to yield the aliphatic glycidyl ether. A prominent side-reaction is the conversion of aliphatic hydroxyl groups (formed by the initial reaction) into chloromethyl groups by epichlorohydrin (22). The aliphatic glycidyl ether resins are used as flexibilizers for aromatic resins and as reactive diluents to reduce viscosities in resin systems.

Monofunctional aliphatic glycidyl ethers (eg, based on n-butanol, or mixed C_8–C_{14} alcohols) are used exclusively as reactive diluents to reduce viscosities of epoxy resin systems. Some loss of desirable cured properties results from the lowered functionality of the systems.

Polyfunctional aliphatic resins have exhibited high reactivity and degrees of cure with amines but problems of toxicity have diminished their usefulness and commercial interest.

Curing Reactions

A variety of reagents has been described for converting the liquid and solid epoxy resins to the cured state, which is necessary for the development of the inherent properties of the resins. Liquid epoxy resins contain mainly epoxy groups and solid resins are composed of both epoxy and hydroxyl curing sites.

The curing agents or hardeners are categorized as either catalytic or coreactive and the functional groups of the resins are terminal epoxy together with a pendant hydroxyl per repeat unit of the polymer chain.

Catalytic curing agents initiate resin homopolymerization, either cationic or anionic, as a consequence of using a Lewis acid or base in the curing process. The Lewis acid catalysts frequently employed are complexes of boron trifluoride with amines or ethers and boron trichloride complexes.

$$RCH\!\!-\!\!\!\overset{\displaystyle O}{\diagdown\!\!\diagup}\!\!\!-CH_2 \xrightarrow{BF_3OH^- + H^+} HOCH_2CH\!\!-\!\!\!\underset{R}{\overset{}{|}}\!\!\!\left[\!-OCH_2CH\!\!-\!\!\!\underset{R}{\overset{}{|}}\!\!\!\right]_n\!\!-OCH_2\overset{+}{C}H\!\!-\!\!\!\underset{R}{\overset{}{|}} + \bar{B}F_3OH$$

Mechanisms for polymerization of epoxides by Lewis acids are proposed in refs. 23–26.

The most important Lewis bases are tertiary amines or polyamines converted into tertiary amines upon reaction with epoxide groups.

$$CH_2\!\!-\!\!\!\overset{\displaystyle O}{\diagdown\!\!\diagup}\!\!\!-CHR' + R_3N \longrightarrow R_3\overset{+}{N}CH_2CHR'\!\!-\!\!\!\underset{O^-}{\overset{}{|}} + CH_2\!\!-\!\!\!\overset{\displaystyle O}{\diagdown\!\!\diagup}\!\!\!-CHR' \longrightarrow R_3\overset{+}{N}CH_2CH\!\!-\!\!\!\underset{R'}{\overset{}{|}}\!\!\!\left[\!-OCH_2CH\!\!-\!\!\!\underset{R'}{\overset{}{|}}\!\!\!\right]_n\!\!-O^-$$

Mechanisms for reaction of tertiary amines with epoxides are discussed in refs. 27–28.

Coreactive curing agents are polyfunctional reagents that are employed in stoichiometric quantities with epoxy resins and possess active hydrogen atoms. The important classes include polyamines, polyaminoamides (formed from polyamines and dimerized fatty acids), polyphenols, polymeric thiols, polycarboxylic acids, and anhydrides. Polyamines constitute a large class of hardeners with aliphatic, aromatic, cycloaliphatic, and heterocyclic groups. Cure of liquid epoxy resins is readily accomplished via the epoxy groups at room temperature with nonaromatic amines and at slightly elevated temperatures with aromatic amines, although reaction of the latter can be accelerated to function at RT. The mechanism of the reaction between epoxy resins and primary aliphatic amines is discussed in detail in refs. 29–32. The over-all curing reaction may be depicted as shown:

$$RNH_2 + CH_2CHCH_2OR' \longrightarrow RNHCH_2CHCH_2OR'$$
$$\underset{O}{\diagdown\diagup}\underset{OH}{|}$$

$$RNHCH_2CHCH_2OR' + CH_2\text{—}CHCH_2OR' \longrightarrow RN(CH_2CHCH_2OR')_2$$
$$\underset{OH}{|}\underset{O}{\diagdown\diagup}\underset{OH}{|}$$

Polyaminoamides provide RT cure of epoxy-terminated resins as well as flexibilization; they are derived by reaction of dimerized vegetable oil fatty acids (dimer acids (qv)) with polyamines.

Polyphenols or phenol-terminated resins are utilized to effect chemical cross-linking of epoxy resins with added catalysts or accelerators for the reaction (33).

The reactions of carboxylic acids and anhydrides with epoxy resins has been extensively studied in a variety of investigations, particularly refs. 34–36. The general reaction of epoxide resins and anhydrides is:

Typical epoxy–anhydride systems are described in Table 8.

Table 8. Summary of Epoxy–Anhydride System Handling and Performance[a]

Anhydride	Melting point or viscosity	Mol wt	Recommended concentration, phr[b]	Curing cycles	ASTM deflection temperature, °C
phthalic anhydride	131°C	148	30–75	4 to 24 h at 150°C	110–147
methyl-4-endo-methylene-tetrahydro-phthalic anhydride	175–275 mPa·s[c] at 25°C	178	80–90	2 h at 140°C + 2–192 h at 200°C	150–175
hexahydro-phthalic anhydride	35–37°C	154	50–100	2 h at 100°C, or 2 h at 100°C + 2 h at 149°C	110–130
tetrahydro-phthalic anhydride	99–101°C	152	70–80	2 h at 100°C + 4 h at 150°C	122
dodecenyl-succinic anhydride	290 mPa·s[c] at 25°C	266	130–150	2 h at 121°C or 1 h at 100°C + 2 h at 204°C	60–70

[a] Ref. 29.
[b] phr = parts per hundred parts of resin.
[c] mPa·s = cP.

Epoxy resins contain both epoxy terminal groups and pendant hydroxyl groups and are cured in accordance with the available sites present in a given resin. Low and medium molecular weight resins respond with both epoxy and hydroxyl groups during curing reactions, whereas in high molecular weight resins hydroxyl groups constitute the main source of reactive sites. Low and intermediate weight resins respond to curing by chemicals such as dicyanodiamide (dicy) (see Cyanamides) and imidazoles at elevated temperatures. The curing reactions of these reagents are discussed in refs. 38–39.

Cross-linking with aminoplasts and phenoplasts constitutes an important class of hardeners for high molecular weight epoxy resins that require elevated-temperature cures (see Amino resins). An extensive discussion of aminoplast and phenoplast curing reactions is given in ref. 40.

Economic Aspects

Epoxy sales grew rapidly in the 1960s as new applications developed for the resins. Figure 1 provides sales history and growth rates for epoxy resins in the United States during 1964–1978. As can be seen, some product maturation has developed:

Time period	Compounded growth rate, %/yr
1964–1974	9.5
1968–1978	8
1974–1978	5.7

However, sales of epoxies are exhibiting a resurgence of growth with rapid recovery from the 1975 recession, eg, growth rates of 33, 12, and 14%, respectively, in the years 1976 to 1978 have been observed. Continued strong growth is projected based on new

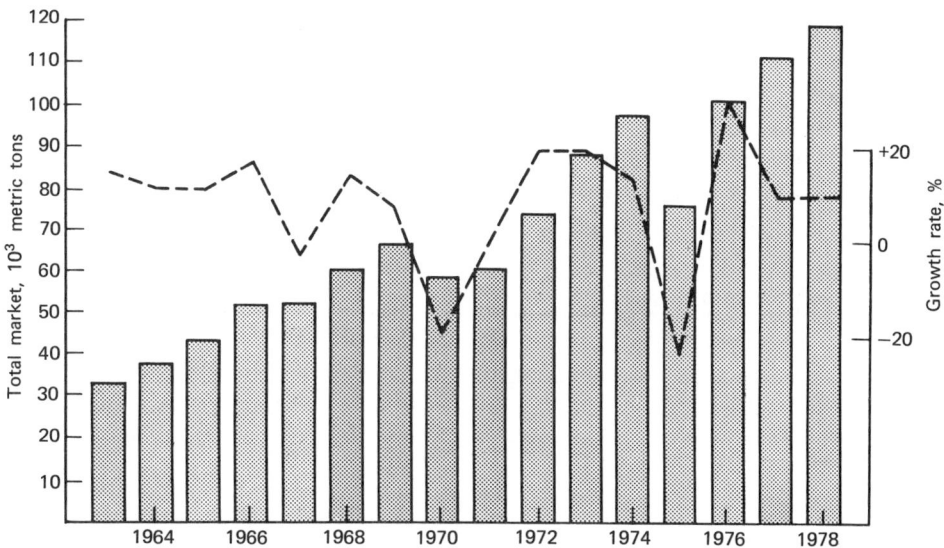

Figure 1. Domestic epoxy market growth from 1963 to 1978. The dashed line represents the growth rate. (Based on data from SPI, Committee on Resins Statistics, as compiled by Ernst and Ernst.)

applications, particularly in the automobile, electronics, and composites industries. Additionally, increasing emphasis on higher performance-longer life materials will provide significant new and increasing uses for epoxies. Although initial costs of epoxy resins often are higher than competitive products, their longer service life provides long-term economy.

Manufacturers. Worldwide capacity for epoxy resins is estimated at over 400,000 metric tons. Over 70% of this capacity is produced by Ciba-Geigy, Dow, and Shell. Epoxies are produced in the following countries: Argentina, Australia, Canada, German Democratic Republic (DDR), England, France, The Netherlands, India, Japan, Mexico, Spain, Switzerland, and the United States.

Trademarks for major resin producers are:

Company	Trade name
Celanese Polymer Specialties	Epi-Rez
Ciba-Geigy	Araldite, Aracast
The Dow Chemical Co.	D.E.R.
Reichhold Chemicals	Epotuf
Shell Chemical Co.	Epon
Union Carbide Corp.	Bakelite

Prices. Selling prices in the United States of various important epoxy resin products are listed in Table 9. As for other petrochemical-based products, prices of epoxies have risen rapidly during the last several years as oil prices have escalated:

Year	Liquid resin selling price (bulk price delivered basis), $/kg
1973	0.97
1974	1.06–1.39
1975	1.46
1976	1.57

1977	1.68
1978	1.79
1979	1.92

As shown in Table 9, prices of specialty products, range from ca $3.10/kg to ca $12.25/kg. These products are often based on more expensive raw materials (than bisphenol A and epichlorohydrin) involving more complex technology and a more costly manufacturing process.

Table 9. United States Epoxy Resin Prices

Epoxy resins type			Selling prices, $/kg (June 1, 1979)	Major applications
liquid, unmodified[a] (WPE = 185–192)			1.88	coatings, casting, tooling, flooring, reinforced pipe
liquid, modified[a]				
Modifier	WPE			
aliphatic glycidyl ether	195–215		1.92	coatings, impregnation, flooring
butyl glycidyl ether	175–195		2.01	
solids[a]				
	450–550		1.85	fiberglass sizing, powder coatings, esters for automotive primers, can and drum coatings, maintenance coatings
	850–975		1.78	
	2000–2500		1.80	
solutions[a]				
Solid (WPE)	Solvent	Solids, %		
450–550	xylene	75	1.55	coatings, printed circuit boards
2000–2500	methyl isobutyl ketone–		1.42	
	toluene	55		
	brominated acetone	80	1.68	
specialties[b]				
epoxy-phenol novolacs			2.93	laboratory bench tops, adhesives, coatings, molding compounds
epoxy-cresol novolacs			4.64	
polyfunctional			6.75–12.21	prepregs for aerospace, leisure goods
cycloaliphatics			3.07	electrical castings, stabilizers, coatings

[a] Truckload prices, delivered basis; WPE = weight per epoxide.
[b] Truckload prices, FOB producing point.

Applications

The major applications and uses for epoxies are shown in Table 10 and Figures 2 and 3.

The largest single use is in the protective coatings market where high chemical resistance and adhesion is important (see Coatings). The overall good growth is based

Table 10. United States Epoxy Resin Sales for 1976–1978[a]

Market	Sales, 10^3 metric tons		
	1976	1977	1978
coatings	50.3	57.2	64.0
laminates–composites	16.3	20.0	24.9
moldings–casting	9.5	10.4	14.1
construction	4.5	6.8	7.7
adhesives	8.6	8.6	8.2
miscellaneous	12.2	10.0	10.0
domestic	101.4	113.0	128.9
export	11.8	13.2	14.1
Total	*113.2*	*126.2*	*143.0*
		+11%	+13%

[a] Based on data from SPI, Committee on Resins Statistics, as compiled by Ernst and Ernst.

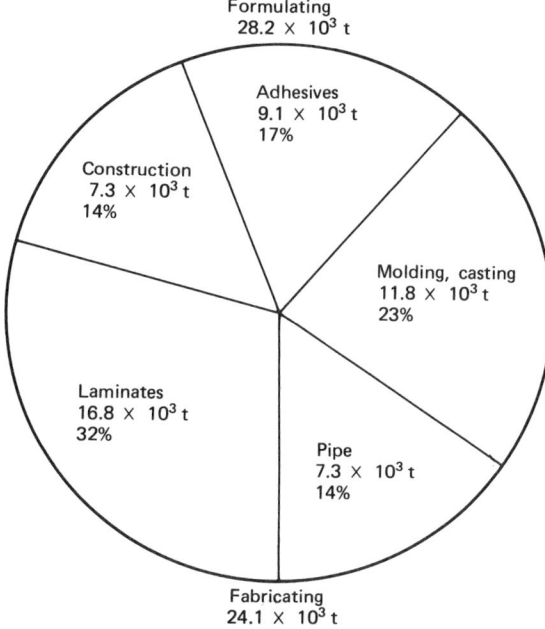

Figure 2. Epoxy resins electrical and structural market for 1978 (total = 52.3×10^3 metric tons).

on increased use in the automotive area where cathodic electrodeposition (see Electroless plating; Electroplating) systems based on epoxies as co-resins continue to gain acceptance (see Electrochemical processing). In container coatings, the continued growth of two-piece cans and increasing use of steel for beer and beverage containers provide further growth of epoxies. In the switch from solvent-based to waterborne systems, epoxies are successfully bridging the gap largely by adaptation of conventional resins by coatings companies. In powder coatings (qv), epoxies are continuing to grow at rates exceeding other technology segments mainly because of the 100% solids feature, improved coverage, and recyclability of material. Pipeline projects, important because

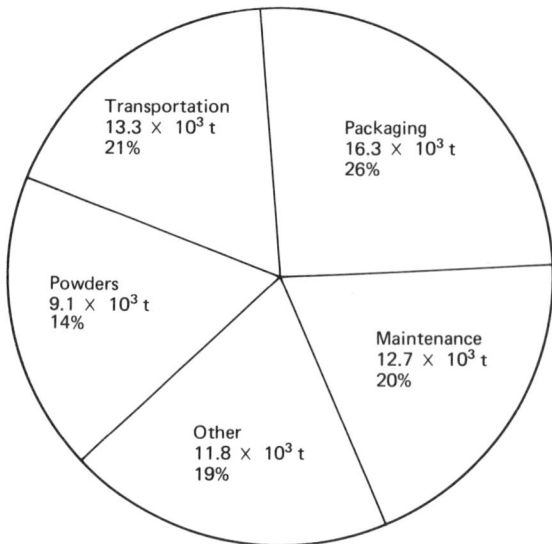

Figure 3. Epoxy resins coatings market for 1978 (total = 63.5×10^3 metric tons).

of worldwide energy problems, are major users of epoxy resins. The maintenance coating segment, although not spectacular in its overall growth, still provides a steady and solid base for epoxies. The value of improved service life is being increasingly accepted even at the somewhat higher material cost of epoxy systems.

The greatest progress in the last several years has been in the laminates-composites segment (Fig. 2) (see Laminated and reinforced plastics). With increasing applications for electronics, (particularly computers (qv)), communications equipment, military and automobile uses, growth for printed circuit boards and molding-casting systems will continue. Because of the new era of commercial aircraft and the military aircraft procurement programs where high strength–weight ratio is exceedingly important, the use of epoxies for composites will undergo significant expansion.

In the adhesives (qv) segment, growth has been slower than in other segments in Figure 2. Increasing competition from other materials, most notably polyurethanes and acrylics is affecting the market (see Urethane polymers; Acrylic ester polymers).

Protective Coatings. The adhesion, toughness and chemical resistance of epoxy resin-based coatings account for their widespread use. Largest uses involve automobile primers and finishes, maintenance and marine coatings (see Coatings, industrial; Coatings, resistant), can coatings, and various other product finishes and industrial specialty finishes (see Fig. 3). Most systems used today are solvent-based but, as a result of new ecology and energy requirements, increasing amounts of waterborne high solids, and solventless systems (powder and liquid) are used. Cure mechanisms can be either one-package, heat-curing (baking) finishes, or room temperature-curing systems (including air-dried esters).

Ambient-cure systems are often based on lower molecular weight solid-epoxy resins cured with aliphatic polyamines or polyamides. Curing normally occurs at ambient temperatures with a working life (pot life) of 8–24 h, depending upon the formulation. Epoxy–polyamine systems are typically used for maintenance coatings in oil refineries, petrochemical plants, and in many marine applications. Such multicoated coverings are applied by spray or brush. These are used widely where water immersion is encountered, particularly in marine applications (see Coatings, marine).

Ester type air-dried coating systems provide better performance than alkyd coatings but lesser performance than straight epoxies (see Alkyd resins). These coatings are simple esterification products of the epoxy resin with a fatty acid, resin acid or a tall oil acid (see Carboxylic acids; Tall oil). They are available in short, medium and long oil-lengths similar to alkyds. Long oil-length esters have limited chemical resistance, whereas the short oil-length esters provide improved chemical resistance. The short oil-length esters promote surface wetting similar to drying oils; therefore, they are used often when optimum surface preparation is not possible. The esters are used as automobile primers in which a good balance of performance and economics is needed.

Heat-curing finishes generally are based on the higher molecular weight, solid epoxy resins cross-linked with aldehyde condensation products. One of the principal applications is for can coatings where the combination of the superior chemical resistance, adhesion, and inertness of the epoxy resin is utilized to provide FDA-acceptable coatings for interior linings.

Normal epoxy–amine ratios in resins are: epoxy–urea, 70:30; epoxy–benzoguanamine, 70:30; epoxy–melamine, 80:20, 90:10. Increasing the epoxy/amine ratio provides: (*1*) improved flexibility, toughness, adhesion and gloss; (*2*) reduced gloss retention, solvent resistance and hardness; and (*3*) higher-temperature baking schedules.

In conjunction with the need to reduce energy consumptions, accelerators are used and faster-curing hardeners are being developed.

Epoxy–phenolic systems offer chemical resistance along with excellent mechanical properties. Resins of 4000–5000 mol wt are used in conjunction with a phenolic that has a low degree of condensation. Part of the methylol groups may be etherified with butanol. Generally, the ratio of epoxy–phenolic ranges from 70:30 to 80:20, depending on the properties desired. Acid catalysts such as phosphoric acid or the morpholine salt of *p*-toluenesulfonic acid are often used to accelerate the cure. Such systems are used for chemically resistant coatings on process equipment, tank and drum linings, and in pipe lining.

Solventless Coatings. Ecological concerns have led to increasing uses of these materials. The advantages include: high build-up in a single application, minimization of surface defects owing to the absence of solvents, excellent heat and chemical resistance, and lower overall application costs. Disadvantages include: poor impact resistance in flexibility, short pot life, and increased sensitivity to humidity.

Attempts to overcome the short pot life have focused on development of special equipment in which the components can be mixed in the proper proportion. Another approach has been to develop curing agents that give long pot life compared to conventional systems.

Powder Coatings. Epoxy-based powder coatings (qv) exhibit useful properties such as excellent adhesion, abrasion resistance, hardness, and chemical resistance. The application possibilities are diverse, including refrigerator liners, oil filters, hospital equipment, primers, shelving, automobile springs, and fire extinguishers.

Interest in powders is primarily related to meeting increasingly stringent antipollution regulations. The development of highly reactive powder systems which cure at low energy (150°C) and the possibility of economical thin films (30–40 μm) have made powder coatings competitive with waterborne and high solids systems. Powder coatings can be applied by fluidized bed (thick films, 50–150 μm) or electrostatic spray (thin films, 30–40 μm).

The higher molecular weight, solid epoxy resins are used in formulations that usually consist of a resin, hardener, reinforcing filler, pigments, flow control agents, and other modifiers. In addition to using conventional hardeners in these formulations, epoxy resins can also be hardened with other resins (acrylics, polyesters).

To improve the weatherability of epoxies (which normally chalk and yellow) epoxy–polyester alloys or hybrids are used. These powders with improved overbake resistance cure at temperatures as low as 130°C. They have film flexibility similar to epoxy resins, but their hardness is slightly decreased. Corrosion resistance is equivalent to epoxy powders in most cases, although solvent and alkali resistance is inferior.

Another approach to weatherable systems is the use of triglycidyl isocyanurate (6) as a curing agent for carboxyl-containing oil-free polyesters. The trifunctionality is responsible for its high reactivity with carboxyl-containing oil-free polyesters. These types of powder coatings are used in the protection of metal window frames, as well as for exterior siding.

Aqueous Powder Suspension (APS). A recent development has been the use of aqueous powder suspension technology to apply a powder coating from water using conventional equipment (41). This enables the manufacturer to obtain cured properties of a powder coating utilizing available solvent-based spray equipment.

An APS system is a dispersion of powdered resins, hardeners, and other additives in water. This suspension contains no solvent or emulsifier. The initial stages of manufacturing APS closely resembles those used to make a powder coating. Bisphenol A-based epoxy resin of 4000–5000 mol wt, a carboxyl-functional polyester, pigments, and other additives, are extruded, cooled, milled, and sieved to produce a powder comprising 100-μm particles. Unlike conventional powder-coating manufacture, this powder is then fine-ground in an aqueous slurry to ca 10-μm particles.

Waterborne Coatings. Waterborne coatings utilize either liquid or solid epoxy resins that have been modified to allow their use with water. They are usually in the form of emulsions, suspensions, dispersions, or water-dilutable resins which can be heat- or RT-cured. They are applied by convenient methods such as roller-coating, dipping, spray, or electrodeposition.

Owing to governmental regulations, considerable research has been expended to develop systems suitable for substitution of solvent-based systems, particularly for automobile and container applications (42–44). Progress has been made in developing emulsions of higher molecular weight epoxy resins. These are important because of the higher chemical and corrosion protection properties of such resins. One recently developed stable, 40% solids emulsion has properties useful for food-can, drum-coating, and coil-primer applications (45).

Significant advances for waterborne coatings have been made by PPG Industries utilizing epoxies as co-resins. These coatings are used in cathodic electrodeposited systems, recently widely accepted for automobile primers. Many patents have been issued for this important technology (46–48).

Electrical, Electronics, and Structural Applications. *Molding Components.* Thermoset epoxy molding compounds are multicomponent mixtures usually based on solid epoxy resins, hardeners, and various fillers and reinforcements. The exact nature of the compounds is usually dictated by the application and the molding procedure to be used—transfer, compression, or injection molding. Because of the use of narrow gates in transfer- and injection-molding processes, the use of long-fiber reinforcement (>3 mm) is usually avoided because of the danger of excessive orientation of the fibers. In compression-molding procedures, this limitation does not apply and long-fiber-reinforced molding compounds are frequently used. Because of the good adhesive properties and low shrinkage on cure of epoxy resins, internal release agents are frequently incorporated into the molding compound formulation. Care must be taken in designing the mold to allow sufficient draft to facilitate easy ejection of the part on demolding.

The main use of epoxy molding compounds is for the encapsulation via transfer molding of solid state devices such as diodes, transistors, and integrated circuits (see Embedding). The molding compounds are formulated to meet stringent requirements regarding flow, reactivity, electrical properties, humidity, and thermal resistance. Extensive testing and experience have shown that the best characteristics are achieved using a solid, multifunctional, epoxidized cresol-novolac resin, and a phenol or cresol novolac as a hardener.

Low ionic impurity levels are imperative. In order to reduce the coefficient of thermal expansion of the final molding, and hence minimize stresses on the encapsulated silicon chip, the highest possible filler loading would be desired. This has to be balanced against the need to maintain as low a melt viscosity as possible to minimize the possibility of damage to the device during the encapsulation process.

Epoxy molding compounds based on solid bisphenol A resins, frequently upgraded with solid epoxy phenol or cresol novolacs, and cured with novolacs or occasionally with solid anhydrides or aromatic amines, are used for encapsulating passive electronic components, ie, coils, capacitors, transformers, ignition coils, etc. In such applications, small amounts of short glass fiber in addition to particulate filler may be used to increase impact strength (see Fillers; Glass). Similar compounds are frequently used for molding high voltage electrical bushings and connectors used in pad-mounted distribution transformers (see Electrical connectors).

Compression molding is mainly used in the manufacture of large, long-glass-fiber (≤25 mm) reinforced parts. Chopped prepreg molding compounds, in which glass roving has been impregnated with a solid resin–hardener formulation and chopped to give strands of the desired length (usually 12–25 mm), and epoxy sheet-molding compounds are mainly used. These are characterized by excellent mechanical strength, rigidity, and good thermal or chemical resistance properties, or both. Components such as pump housings, impellers, pipe fittings, and valves are compression-molded.

Casting and Encapsulation. Cast components are made by filling a mold cavity with a liquid epoxy system and then cross-linking the resin to give an infusible solid part. The resin system used may be formulated to cure either at RT or at elevated temperatures. A major problem with casting large parts is that during cure, the excess heat of reaction must be dissipated to avoid large locked-in thermal stresses in the casting. This is achieved by skillful formulation but results in relatively long mold cycles and poor productivity.

Nevertheless, casting processes have found extensive uses where the good electrical and mechanical properties of epoxy resins offer many design advantages. The most commonly used systems are based on bisphenol A-epoxy resins cured with anhydride hardeners and usually heavily (60–65 wt %) filled with silica fillers. Components using from 50–100 kg of mix are cast on a routine basis. Typical components manufactured by casting techniques are post insulators, bus-bar supports, switchgear components, instrument transformers, distribution transformers, high and low voltage bushings, encapsulated coils, etc. Poor wet arc-track resistance and uv resistance limit bisphenol A-based resins to indoor applications. For outdoor applications, the use of cycloaliphatic resins cured with a nonaromatic anhydride hardener (eg, hexahydrophthalic anhydride) is mandatory. Such systems have increasing use for outdoor insulators, switchgear components, and instrument transformers.

Recently, a new casting process, pressure gelation (49–51), has been developed that overcomes problems associated with the high heat of reaction of an epoxy resin. It allows large parts to be gelled rapidly, reducing the mold cycle from hours to minutes. The increased productivity is responsible for growing worldwide use in the electrical industry (see Embedding).

In addition to electrical uses, epoxy casting resins are utilized in the manufacture of tools (contact and match molds, stretch blocks, vacuum-forming tools, and foundry patterns). Systems consist of a gel-coat formulation designed to form a thin coating over the pattern which provides a perfect reproduction of the pattern detail. This is backed by a heavily filled epoxy system which also incorporates fiber reinforcements to give the tool its strength. For moderate temperature service, a liquid bisphenol A-epoxy resin with an aliphatic amine is used. For higher temperature service, a modified system based on an epoxy phenol novolac and an aromatic diamine hardener may be used.

Electrical Laminates. A major use for epoxy resins is in the manufacture of copper-clad epoxy–glass printed circuit boards. Systems are available that meet the National Electrical Manufacturers Association (N.E.M.A.), G10, G11, FR3, FR4, FR5 specifications. Currently, the majority of boards are manufactured to the flame-retardant FR4 specification. The flame-retardance is achieved by the use of a solid epoxy resin based on tetrabromobisphenol A (see Flame retardants). In combination with dicyandiamide as the hardener, stable impregnation varnishes can be formulated. The impregnated glass fabric can be dried and "B-staged" in a controlled manner by treating in a heated tunnel or tower. The B-staged fabric (prepreg) is cut into sheets and laid up into laminates, with a copper foil top and bottom, and pressed at high pressure and elevated temperature to form a well-consolidated fully cured board.

Identical procedures are used for making structural and electrical laminates usually based on nonflame retardant resins.

Adhesives. Because of excellent adhesion to many substrates, epoxy resins are extensively used for high performance adhesives (qv). These can be categorized into high temperature curing systems (solids and liquids) and room temperature curing systems (liquids).

For the high temperature systems, dicyandiamide, and aromatic-amine hardeners are frequently used. For room temperature systems, polyamide, aminoamide, and aliphatic-amine hardeners are preferred.

Depending on the characteristics and performance requirements, adhesive systems are frequently modified with diluents (reactive and nonreactive) and polyfunctional high performance resins, as well as with fillers of various types.

Structural Composites. Because of their excellent adhesion, good mechanical, humidity, and chemical resistance properties, epoxy resins are used extensively in combination with glass, graphite, boron, and Kevlar fibers in the manufacture of high performance composites (see Composite materials).

An important application is for filament-wound glass reinforced pipe used in oil fields, chemical plant, water distribution, and as electrical conduits. Low viscosity liquid systems having good mechanical properties (elongation at break) when cured are preferred. These are usually cured with liquid anhydride or aromatic-amine hardeners. Similar systems are used for filament-winding pressure bottles and rocket motor casings.

In the aerospace industry, the use of graphite fiber-reinforced composites is growing rapidly because of high strength to weight ratios (see Ablative materials). High performance polyfunctional resins, such as the tetraglycidyl derivative of methylenedianiline in combination with diaminodiphenyl sulfone, are extensively used to provide good elevated temperature properties and humidity resistance. Handling characteristics are well suited to the autoclave molding technique primarily used in the manufacture of such components.

BIBLIOGRAPHY

"Epoxy Resins" in *ECT* 1st ed., Suppl. Vol. 1, pp. 312–329, by R. A. Coderre, Shell Chemical Corporation; "Epoxy Resins" in *ECT* 2nd ed., Vol. 8, pp. 294–312, by Joseph R. Weschler, Ciba Products Company.

1. U.S. Pat. 2,324,483 (July 20, 1943), P. Castan (to Ciba).
2. U.S. Pat. 2,456,408 (Dec. 14, 1948), S. O. Greenlee (to Devoe & Raynolds).
3. U.S. Pat. 2,136,928 (Nov. 15, 1938), P. Schlack (to I. G. Farben).
4. M. Gaschke and B. Dreher, *J. Coatings Technol.* **48**(617), 46 (1976).
5. *Epoxy Cresol Novolac Resins,* technical data bulletin, Ciba-Geigy, Ardsley, N.Y., 1978.
6. W. Hofmann and W. Fisch, *FATIPEC Congr. Rev.,* 243 (1962).
7. *Ciba-Geigy Epoxy Resin 0500, CR544,* technical data bulletin, Ciba-Geigy, Ardsley, N.Y., 1978.
8. *Araldite MY 720, CR6191,* technical data bulletin, Ciba-Geigy, Ardsley, N.Y., 1978.
9. *Ciba-Geigy Epoxy Resin 0163, CR532,* technical data bulletin, Ciba-Geigy, Ardsley, N.Y., 1978.
10. *Araldite PT 810, CR6755,* technical data bulletin, Ciba-Geigy, Ardsley, N.Y., 1979.
11. J. Habermeier, *Angew. Makromol. Chem.* **35,** 9 (1974).
12. *Developmental Resin XB2793, CR455A,* technical data bulletin, Ciba-Geigy, Ardsley, N.Y., 1978.
13. *Developmental Resin XB2826, CR541A,* technical data bulletin, Ciba-Geigy, Ardsley, N.Y., 1978.
14. B. H. Miles, *Am. Chem. Soc. Div. Org. Coat. Plast. Chem. Prepr.* **24**(2), 73 (1964).
15. W. G. Potter, *Epoxide Resins,* Springer-Verlag, New York, 1970, p. 12.
16. U.S. Pats. 2,548, 447 (Apr. 10, 1951), E. C. Shokal (to Shell); 2,500,449 (Mar. 14, 1950), T. F. Bradley (to Shell).
17. H. Lee, D. Stoffey, and K. Neville, *New Linear Polymers,* McGraw-Hill Book Co., New York, 1967, pp. 17–60.
18. U.S. Pats. 3,477,990 (Nov. 11, 1969), M. F. Dante and H. L. Parry (to Shell); 3,978,027 (Nov. 11, 1969), C. D. Marshall (to Shell); 4,048,141 (Sept. 13, 1977) G. A. Doorakian and co-workers (to The Dow Chemical Co.).
19. U.S. Pat. 2,806,016 (Sept. 10, 1957), C. G. Schwarzer (to Shell).
20. *Eponex Resins,* technical bulletin, Shell Oil Company, Houston, Tex.
21. U.S. Pat. 2,951,825 (Sept. 6, 1960), N. H. Reinking and co-workers (to Union Carbide).
22. U.S. Pat. 2,898,349 (Aug. 4, 1959), P. Zuppinger and co-workers (to Ciba-Geigy).
23. J. J. Harris and S. C. Temin, *J. Appl. Polym. Sci.* **10,** 523 (1966).
24. A. J. Landua, *Am. Chem. Soc. Div. Org. Coat. Plast. Chem. Pap.* (42), (1964).
25. L. H. Lee and co-workers, *J. Polym. Sci. Part A* **3,** 2955 (1965).
26. R. J. Arnold, *Mod. Plast.* **41,** 149 (1964).

27. L. Schecter and J. Wynstra, *Ind. Eng. Chem.* **48,** 86 (1956).
28. E. S. Narracott, *Br. Plast.* **26,** 120 (1953).
29. L. Schecter, J. Wynstra, and R. P. Kurjy, *Ind. Eng. Chem.* **48,** 94 (1956).
30. L. J. Gaugh and I. T. Smith, *J. Oil Colour Chem. Assoc.* **43,** 409 (1960).
31. R. E. Parker and N. S. Isaacs, *Chem. Rev.* **59,** 737 (1959).
32. N. B. Chapman, R. E. Parker, and N. S. Isaacs, *J. Chem. Soc.* **2,** 1925 (1959).
33. L. Schecter and J. Wynstra, *Ind. and Eng. Chem.* **48,** 86 (1956).
34. W. Fisch and W. Hofmann, *J. Polym. Sci.* **12,** 497 (1954).
35. W. Fisch, W. Hofmann, and J. Koskikallio, *J. Appl. Chem. London* **6,** 429 (1956).
36. R. F. Fischer, *J. Polym. Sci.* **44,** 155 (1960).
37. *Anhydride Hardeners, 80508,* technical data bulletin, Ciba-Geigy, Ardsley, N.Y., 1969.
38. H. H. Levine, *Am. Chem. Soc. Symp. 148th Meeting,* (1964).
39. A. Farkas and P. F. Strohm *J. Appl. Polym. Sci.* **12,** 159 (1968).
40. D. H. Solomon, *The Chemistry of Organic Film Formers,* John Wiley & Sons, Inc., New York, 1967.
41. E. G. Bozzi, K. Brugger, and H. Lauterbach, "Aqueous Powder Suspension—A New Type of Coating System," *paper presented at the FSCT Meeting, Chicago, Ill., Nov. 1, 1978.*
42. E. G. Bozzi and R. C. Nelson, *J. Coatings Technol.* **49,** 61 (1977).
43. E. G. Bozzi, *Mod. Paint Coatings* **67** (Nov. 1977).
44. M. Gaschke, *Paint Varn. Prod.* **64** (Dec. 1974).
45. E. G. Bozzi, *Am. Chem. Soc. Div. Org. Coat. Plast. Prepr. 39, 176th* ACS Meeting, Miami, Fl., Sept. 1978.
46. U.S. Pat. 3,947,339 (Mar. 30, 1976), R. D. Jerabek and co-workers (to PPG Industries, Inc.).
47. U.S. Pat. 4,009,133 (Aug. 1, 1975), J. E. Jones (to PPG Industries, Inc.).
48. M. W. Ranney, *Epoxy Resins and Products—Recent Advances,* Noyes Data Corp., Park Ridge, N.J., 1977, pp. 47–91.
49. U.S. Pat. 3,777,000 (Dec. 4, 1973), E. Kusenberg and co-workers (to Ciba-Geigy Corp.).
50. G. Buchi and R. Flynn, "Pressure Gelation Process—Recent Developments in Epoxy Resin Casting Processes," *29th Annual Technical Conference,* Reinforced Plastics–Composites Institute, Society of Plastics Industry, New York, 1979.
51. S. E. Ellis, *34th Annual Technical Conference,* Society of Plastics Engineers, Atlantic City, N.J., 1976.

General References

Toxicology

D. J. Birmingham, *Arch. Ind. Health* **19,** 365 (1959).
C. H. Hine and co-workers, *Arch. Ind. Health* **17,** 129 (1958).
C. H. Hine and co-workers, *Arch. Ind. Health* **14,** 250 (1956).
K. F. Malhn, and R. L. Zullmis, *Industrial Toxicology of Plastics,* Elsevier Publishing Company, Amsterdam, The Netherlands, 1964.

STANLEY SHERMAN
JOHN GANNON
GORDON BUCHI (Electrical, Electronics, and Structural Applications)
Ciba-Geigy Corporation

W. R. HOWELL (Epoxy Phenol Novolacs and Bisphenol F Resins)
Dow Chemical U.S.A.

ESTERIFICATION

This article describes methods for the production of carboxylic esters

For the properties of these compounds, see Esters, organic. For esters of inorganic acids, see the articles on Nitric acid, Phosphoric acids, Sulfuric acid, etc.

The most usual method for the preparation of esters is the reaction of a carboxylic acid and an alcohol with the elimination of water. Esters are also formed by a number of other reactions, including the use of acid anhydrides, acid chlorides, amides, nitriles, unsaturated hydrocarbons, ethers, aldehydes, ketones, dehydrogenation of alcohols, and ester interchange. Detailed reviews of esterification are given in refs. 1–7.

A compilation in 1978 indicates that more than 600 esters are currently sold in the United States (8). More than 100 of these esters are available in medium and bulk lots (9).

On the basis of bulk production (10), poly(ethylene terephthalate) manufacture is the most important esterification process today. This polymer is produced by the direct esterification of terephthalic acid and ethylene glycol or by the transesterification of dimethyl terephthalate and ethylene glycol. In 1977 2.3×10^6 metric tons of terephthalic acid and its dimethyl ester were consumed in this process to manufacture polyester fibers ($>1.4 \times 10^6$ t) and thermosetting plastics ($>4.5 \times 10^5$ t) (see Polyesters). As a result of polyester manufacture, the production of dimethyl terephthalate (1.5×10^6 t) is the second most important esterification process (see Phthalic acids).

The phthalates are also important from a volume standpoint. The 1973 tariff production figures indicate that ca 5×10^5 t of phthalate plasticizers (qv) were produced. Of this total, dioctyl phthalates accounted for about 40% of the phthalate plasticizers produced. More than 40 different phthalates are commercially available, including mixed esters such as amyl–decyl, butyl–benzyl, ethyl–phthalyl, ethyl–β-hydroxyethyl, etc.

The acetates of most alcohols are also commercially available and have diverse uses. Because of their high solvent powers, ethyl, isopropyl, butyl, isobutyl, amyl, and isoamyl acetates are used in cellulose nitrate and other lacquer-type coatings (see Cellulose derivatives, esters). Butyl and hexyl acetates are excellent solvents for polyurethane coating systems (see Coatings, industrial; Urethane polymers). Ethyl, isobutyl, amyl, and isoamyl acetates are frequently used as components in flavoring (see Flavors and spices), and isopropyl, benzyl, octyl, geranyl, linalyl, and methyl acetates are important additives in perfumes (qv).

Production of the acrylates represents another important application of commercial esterification (see Acrylic ester polymers). In 1974 2.4×10^5 t of acrylic acid esters were consumed. About 77% of this consumption was used to prepare emulsion polymers (surface coatings, textiles (qv), papers (qv), etc). The remainder was used in other polymer applications such as acrylic fibers. Methyl methacrylate (2.8×10^5 t consumed in 1974) is the monomer used to produce Plexiglass and Lucite clear acrylic plastics (see Methacrylic polymers).

Reactions Between Organic Acids and Alcohols

In the esterification of organic acids with alcohols, it has been shown that in most cases under acid catalysis, the union is between acyl and alkoxy groups, ie, between

$$\mathrm{RC\underset{\parallel}{\overset{O}{}}\!-}$$

and —OR′ rather than between

$$\mathrm{RCO\underset{\parallel}{\overset{O}{}}\!-}$$

and —R′. Acid hydrolysis of acetoxysuccinic acid gave malic acid with retention of configuration at the asymmetric carbon atom (11):

$$\mathrm{HO_2CCH_2CHCOH \atop \underset{OCOCH_3}{|}} \overset{H_2O}{\underset{H^+,\, CH_3COOH}{\rightleftharpoons}} \mathrm{HO_2CCH_2CHCOH \atop \underset{OH}{|}} \tag{1}$$

n-Amyl alcohol produced by basic hydrolysis of n-amyl acetate with ^{18}O-enriched water does not contain ^{18}O (12).

Effect of Structure. The rate at which different alcohols and acids are esterified as well as the extent of the reaction are dependent upon the structure of the molecule and types of radicals present. Specific data on rates of reaction, mechanisms and extent of reactions are discussed below. More details concerning structural effects are given in refs. 6 and 13.

With acetic acid at 155°C, the primary alcohols are esterified most rapidly and completely (methanol gives the highest yield and the most rapid reaction). Ethyl, n-propyl, and n-butyl alcohols react with about equal velocities and limits. Under the same conditions, the secondary alcohols react much slower and have lower limits of esterification; however, wide variations are observed among the different members of this series. The tertiary alcohols react slowly and the limits are generally low (1–10% conversion at equilibrium). Tests with isobutyl alcohol at 155°C and various acids show that those acids containing a straight chain (acetic, propionic, and butyric) and phenylacetic and β-phenylpropionic acids are esterified readily. Formic acid has the highest initial rate of reaction, but the esterification limits of the acid increase with increasing molecular weight of the acid. The introduction of a branched chain in the acid decreases the rate of esterification, and two branches cause a still greater retarding effect. Double bonds also have a retarding influence. However, the limits of esterification of these substituted acids are higher than for the normal straight chain acids. Aromatic acids, benzoic and p-toluic, react slowly but have high esterification limits.

The nitrile group has a pronounced inhibiting effect on the rate of esterification of aliphatic acids. With the chloroacetic acids, the velocity decreases with increased substitution. From tests on substituted acrylic acids, it is shown that an α,β-unsaturated acid is esterified much less easily than its saturated analogue. A triple bond in the α,β position has about the same effect as a double bond. A β,γ double bond has less of a retarding action. If the double bond is sufficiently removed (as in erucic and brassidic acids (see Carboxylic acids)), no effect is noted. Conjugated double bonds, when one is in the α,β position, give a great retarding effect. Cis substituted acids esterify more slowly than the trans isomers.

With anhydrous ethyl alcohol and hydrogen chloride, the rate of esterification of the straight chain fatty acids from propionic through stearic is substantially constant; branching of the chain causes retardation. In the saturated dibasic acids, the rate of esterification is a maximum at glutaric acid. The ease of esterification of the cycloparaffin monocarboxylic acids increases in the order C_3-, C_7-, C_6-, C_5-, and C_4-rings; with the exception of cyclopropanecarboxylic acid, these are esterified more rapidly than the corresponding open-chain acids.

Substitutions that displace electrons towards the carboxyl group of aromatic acids diminish the rate of the reaction (14). The substitution of fluoromethoxy or -ethoxy groups in the ortho position has an accelerating action, whereas iodo, bromo, nitro, or methyl groups produce retardation. The influence of groups in the ortho and para positions is not nearly so marked. Disubstituted benzoic acids are readily esterified except those with 2,6-substitutions (15).

Kinetic Considerations. Extensive kinetic and mechanistic studies have been made on the esterification of carboxylic acids since Berthelot and St. Gilles first studied the esterification of acetic acid (16). Although ester hydrolysis is catalyzed by both hydrogen and hydroxide ions (17–18), a base-catalyzed esterification is not known.

One possible mechanism (6) for the bimolecular acid-catalyzed ester hydrolysis and esterification is shown in equation 2:

$$\text{RCOR}' + \text{H}^+ \underset{\text{fast}}{\overset{K_1}{\rightleftharpoons}} \text{RC}\begin{pmatrix}\text{OH}\\+\\\text{OR}'\end{pmatrix} \underset{\text{fast}}{\overset{k_1, +\text{H}_2\text{O}}{\rightleftharpoons}} \begin{matrix}\text{OH}\\|\\\text{RCOR}'\\|\\\overset{+}{\text{OH}}_2\end{matrix} \underset{\text{fast}}{\overset{\text{fast}}{\rightleftharpoons}} \begin{matrix}\text{OH}\\|\\\text{RCOR}'\\|\\\text{OH}\end{matrix} + \text{H}^+ \underset{\text{fast}}{\overset{\text{fast}}{\rightleftharpoons}}$$

$E \qquad\qquad EH^+$

$$\begin{matrix}\text{OH}\\|\\\text{RC---}\overset{+}{\text{OH}}\\||\\\text{OH}\text{R}'\end{matrix} \underset{k_2}{\overset{\text{fast}, +\text{R}'\text{OH}}{\rightleftharpoons}} \text{RC}\begin{pmatrix}\text{OH}\\+\\\text{OH}\end{pmatrix} \underset{K_2}{\overset{\text{fast}}{\rightleftharpoons}} \text{RCOH} + \text{H}^+ \qquad (2)$$

$\qquad\qquad\qquad\qquad AH^+ \qquad\qquad A$

This mechanism leads to the rate equation (eq. 3) for hydrolysis (and to an analogous expression for the esterification) (13):

$$-\frac{d[E]}{dt} = \frac{k_1 K_1 [E][\text{H}_2\text{O}][\text{H}^+]}{1+\alpha} - \frac{k_2 K_2 [A][\text{ROH}][\text{H}^+]}{1+1/\alpha} \qquad (3)$$

In this expression, α depends on those rate coefficients in the above mechanism whose values are assumed to be high. Other mechanisms for the acid hydrolysis and esterification differ mainly with respect to the number of participating water molecules and possible intermediates (19–21).

Applications of kinetic principles to industrial reactions are often very useful.

Generally, the rate of esterification with acid catalyst is proportional to the acid or hydrogen ion concentrations as well as the concentration of the alcohols and organic acid. The effect of temperature on the reaction rate is given by the well-known Arrhenius equation. These factors are interrelated and may be used to predict optimum operational conditions for the production of a given ester if the necessary data are available, ie, (1) the order of the reaction under the conditions to be used, (2) a mathematical relation describing the yield with time, and (3) an empirical equation relating the reaction rate constant with temperature, catalyst concentration, and proportions of reactants.

However, all esterification reactions do not necessarily permit mathematical treatment. In a study of the esterification of 2,3-butanediol and acetic acid using sulfuric acid catalyst, it was found that the reaction occurs through two pairs of consecutive reversible reactions of approximately equal speeds. These reactions do not conform to any simple first-, second-, or third-order equation, even in the early stages (22).

In a study of the kinetics of the reaction of 1-butanol with acetic acid at 0–120°C using sulfuric acid as catalyst with a mole ratio of 1-butanol to acetic acid of 3:19.6 and a catalyst concentration of 0–0.14 wt %, an empirical equation was obtained that permits estimation of the value of the rate constant with an average deviation of 15.3% from the molar ratio of reactants, catalyst concentration, and temperature (23).

Similar studies have been performed on the formation of n-butyl phthalate at 80–150°C with sulfuric acid catalyst (24). The reaction of phthalic anhydride with monobutyl phthalate is complete in 10 min at 100°C with 1 wt % of catalyst.

Equilibrium Constants. The reaction between an organic acid and an alcohol to produce an ester and water according to equation 4 is a classical example of a reversible equilibrium:

$$\underset{\text{O}}{\overset{\text{O}}{\text{RCOH}}} + \text{R'OH} \rightleftharpoons \underset{\text{O}}{\overset{\text{O}}{\text{RCOR'}}} + \text{H}_2\text{O} \qquad (4)$$

This was first demonstrated in 1862 by Berthelot and St. Gilles (25), who found that when equivalent quantities of ethyl alcohol and acetic acid were heated together the esterification stopped when about two thirds of the acid had reacted. Similarly, when equal molar proportions of ethyl acetate and water were heated together, hydrolysis of the ester stopped when about one third of the ester was hydrolyzed. By varying the molar ratios of alcohol to acid, yields of ester >66% were obtained by displacement of the equilibrium. The results of these tests were in accordance with the mass action law shown in equation 5:

$$K = [\text{ester}][\text{water}]/[\text{acid}][\text{alcohol}] \qquad (5)$$

However, in many cases the equilibrium constant is affected by the proportion of reactants (7,26–27). The catalyst as well as temperature and the presence of salts may also affect the value of the equilibrium constant (28–29).

The effect of water on the equilibrium constant for the reaction of 1 mol of ethanol, 1 mol of acetic acid, and 23 moles of water has been investigated. This mixture has an equilibrium constant of 3.56, compared with 3.79 for the reaction with anhydrous materials (7,30).

Theoretical yields of ester obtainable with varying proportions of reactants are shown in Figure 1 for four values of the equilibrium constant. Thus when K equals 10 (esters of p-toluic acid with primary alcohols), with equivalent amounts of acid and alcohol, a yield of about 76% may be expected.

In general, esters having equilibrium constants below unity are not prepared by direct interaction of alcohol and acid; in these cases, the acid anhydrides or acid chlorides are used. Reactions with the latter agents do not involve reversible equilibria and hence give high yields of esters.

Completion of Esterification. Because the esterification of an alcohol and an organic acid involves a reversible equilibrium, these reactions do not go to completion. Conversions approaching 100% are desirable and can often be achieved by the simple method of upsetting the equilibrium by removing one of the products formed, either the ester or the water. In general, esterifications are divided into three broad classes, depending upon the volatility of the esters.

(1) Esters of high volatility, such as methyl formate, methyl acetate, and ethyl formate, have lower boiling points than those of the corresponding alcohols, and therefore, can be readily removed from the reaction mixture.

(2) Esters of medium volatility are capable of removing the water formed by distillation. Examples of this type are propyl, butyl, and amyl formates, ethyl, propyl, butyl, and amyl acetates, and the methyl and ethyl esters of propionic, butyric, and valeric acids. In some cases, ternary mixtures of alcohol, ester, and water are formed. This group is capable of further subdivision: with ethyl acetate, all of the ester is removed as a vapor mixture with alcohol and part of the water, while the balance of the water accumulates in the system; with butyl acetate, on the other hand, all of the water formed is removed overhead with part of the ester and alcohol, and the balance of the ester accumulates in the system.

(3) Esters of low volatility have several types of esterification. In the case of esters of butyl and amyl alcohols, water is removed as a binary mixture with the alcohol. To produce esters of the lower alcohols (methyl, ethyl, and propyl), it may be necessary to add a hydrocarbon such as benzene or toluene to increase the amount of distilled

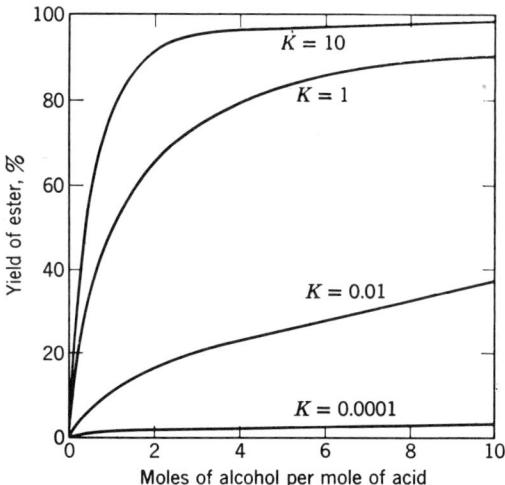

Figure 1. Theoretical yields of ester obtainable with varying proportions of reactants for different values of equilibrium constant.

water. With high boiling alcohols (eg, benzyl, furfuryl, and β-phenylethyl) an accessory liquid is required to eliminate the water by distillation.

Use of Azeotropes. With the aliphatic alcohols and esters of medium volatility, a variety of azeotropes are encountered on distillation (31) (see Azeotropic distillation). Binary azeotropes may be formed between the alcohol and water, the alcohol and ester, and the ester and water. Ternary azeotropes involving the alcohol, ester, and water are also possible. In general, the ternary azeotropes have the lowest boiling points, but the differences between the boiling points of the various combinations in some instances are very small. The ester–water binaries have boiling points very close to those of the ternary mixtures. An extremely efficient fractionating column is required to obtain a pure ternary azeotrope.

Binary azeotropes of the alcohol and water may be utilized with the higher boiling, nonvolatile esters for completion of the reaction (31). Almost all of the alcohols (up to C_{20}-alcohols) except methanol form binary azeotropes with water. The azeotropes formed by water with ethyl, n-propyl, isopropyl, allyl, and *tert*-butyl alcohols are single phase (ie, on condensation of the vapor, the components are completely miscible). Other means to eliminate the water are often necessary, eg, extraction with a water-insoluble solvent, such as benzene or carbon tetrachloride, drying with potassium carbonate, or salting out. The higher alcohols form azeotropes that on condensation separate into two liquid phases; in such a case, the alcohol-rich phase can be separated by further distillation into azeotrope and pure alcohol and the water-rich phase into azeotrope and water. Under certain conditions, entraining gases are used to facilitate the removal of water (32).

Removal of Water by Desiccants and Chemical Means. Another means to remove the water of esterification is calcium carbide supported in a thimble of a continuous extractor through which the condensed vapor from the esterification mixture is percolated (33) (see Carbides).

A column of activated bauxite (Florite) mounted over the reaction vessel has been used to remove the water of reaction from the vapor by adsorption (34).

Catalysts. The choice of the proper catalyst for an esterification reaction is dependent upon several factors (35–37). The most common catalysts used are strong mineral acids but other agents, such as tin salts, organo-titanates, silica gel, and cation-exchange resins, are often employed.

In laboratory preparations, sulfuric acid and hydrochloric acid have classically been used as esterification catalysts. However, formation of alkyl chlorides or dehydration, isomerization, or polymerization side reactions may result. Sulfonic acids, such as benzenesulfonic acid, p-toluenesulfonic acid, or methanesulfonic acid, are used in plant operations because of their less corrosive nature. Phosphoric acid is sometimes employed but it leads to rather slow reactions. Metal salts minimize side reactions but usually require higher temperatures than strong acids.

Acid-Regenerated Cation Exchangers. The use of acid-regenerated cation exchangers (see Ion exchange) as catalysts for effecting esterification offers distinct advantages over conventional methods. The nature of the cation exchanger is of relatively minor importance as long as it contains strongly acidic groups.

Despite the higher cost compared with the ordinary catalysts, such as sulfuric or hydrochloric acid, the cation exchangers present several features that make their use economical. The ability to use these agents in fixed-bed operations makes them attractive for continuous processes.

The esterification of n-butyl alcohol and oleic acid with a phenol–formaldehydesulfonic acid resin (similar to Amberlite IR-100) is essentially second order after an initial slow period (38). The velocity constant is directly proportional to the surface area of the catalyst per unit weight of reactants. A series of tests using Amberlite IR-120 (sulfonated polystyrene resin) to esterify diethylene glycol using toluene as the entrainer for removal of water gave the results in Table 1 (39).

Batch Esterification. The production of esters may be carried out by batch or continuous processing. In general, large-volume production favors continuous esterification methods. However, the older batch processing, based on the use of a still-pot reactor and an ordinary fractionating column (bubblecap or packed type), is still used.

Ethyl Acetate. The production of ethyl acetate is shown in Figure 2 (2). The esterification chamber is a cylindrical tank, or still pot, heated by a closed-coil steam pipe. The reaction charge consists of 30 parts of 8% acetic acid (obtained from fer-

Table 1. Tests Using Amberlite IR-120 to Esterify Diethylene Glycol

Acid	Diethylene glycol, mol/mol acid	Amberlite IR-120, g/100 g acid	Temperature, °C	Reaction time, h	Monoester, %	Diester, %
lauric	1	7.5	140	18	24	71
lauric	4	15.5	130	10	71	21
lauric	6	7.5	140	18	86	11
lauric	12	15.0	132	18	100	
oleic	12	10.6	140	18	100	
stearic	12	8.9	150	18	100	
benzoic	2	24.6	140	4	75	

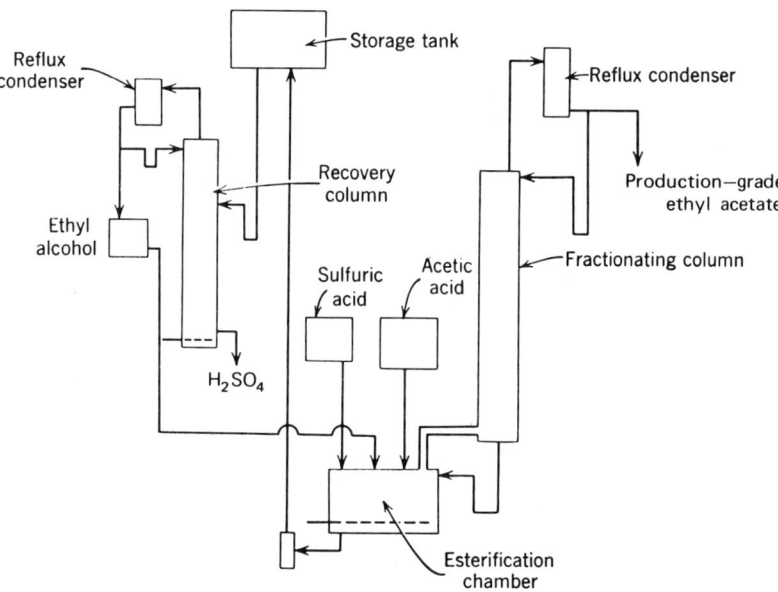

Figure 2. Batch ethyl acetate process (2).

mentation), 30 parts of 95% alcohol, and 1 part 1.53–1.84 sp gr (50–66 Bé) sulfuric acid. The temperature at the top of the fractionating column is maintained at ca 70°C to give a ternary azeotropic mixture of ca 83% ethyl acetate, 9% alcohol, and 8% water. The vapor is condensed, part of it is returned to the top plate of the column as reflux, and the remainder is drawn off to storage. The ternary azeotrope (production-grade ethyl acetate) is satisfactory for many commercial purposes, but for an alcohol-free and water-free ester, further purification is needed.

n-Butyl Acetate. Equipment used for the batch esterification to give butyl acetate is shown in Figure 3. Glacial acetic acid is mixed with an excess of butyl alcohol and a small amount of 1.84 sp gr (66 Bé) H_2SO_4 in the still pot. The mixture is heated for several hours by means of a steam jacket to give esterification equilibrium. After the preliminary heating, slow rectification is permitted to remove the water already formed and thus increase the yield. The esterification is continued until no more water separates. At this point, the temperature at the top of the column rises and the percentage of acetic acid in the distillate increases. It is necessary to neutralize the small amount of acid remaining in the still pot before further distillation. A solution of sodium hydroxide is added to the still pot and the mixture is allowed to stand to form a water layer that is removed. The upper layer (ester) is then washed with water and distilled to obtain a butyl acetate of 75–85% purity, the remainder is butyl alcohol.

Continuous Esterification. The law of mass action, the laws of kinetics, and the laws of distillation all operate simultaneously in a process of this type. Esterification can occur only when the concentrations of the acid and alcohol are in excess of equilibrium values; otherwise, hydrolysis must occur. The equations governing the rate of the reaction and the variation of the rate constant (as a function of such variables

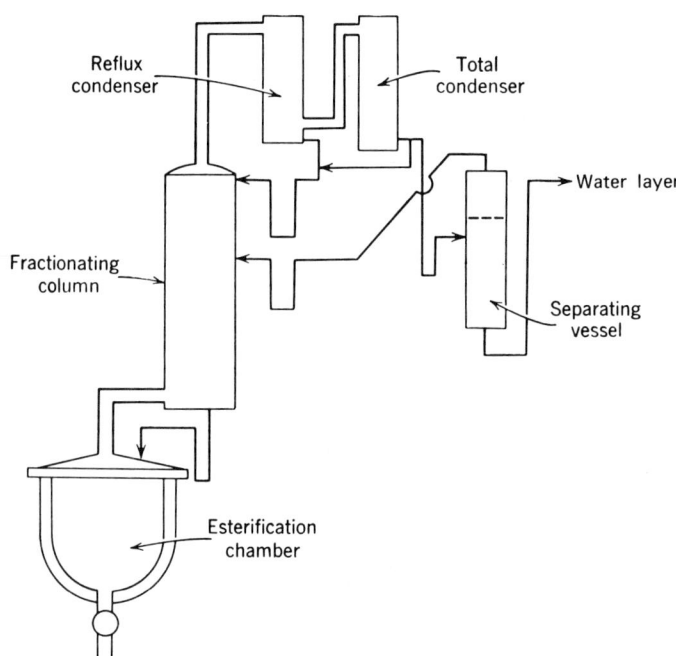

Figure 3. Batch *n*-butyl acetate process (2).

as temperature, catalyst strength, and proportions of reactants) describe the kinetics of the liquid-phase reaction. The usual distillation laws must be modified, since most esterifications are exothermic and moles of reactants disappear while moles of products appear on each plate. Since these kinetic considerations are superimposed on distillation operations, each plate must be treated separately by successive calculations after the extent of conversion has been determined (see Distillation).

Continuous esterification of acetic acid in an excess of n-butyl alcohol with sulfuric acid catalyst using a four-plate single bubblecap column with reboiler has been studied (40). The rate constant and the theoretical extent of reaction were calculated for each plate, based on plate composition and on the total incoming material to the plate, both of which gave good agreement with the analytical data.

A continuous distillation process has been studied for the production of high-boiling esters from intermediate-boiling polyhydric alcohols and low-boiling monocarboxylic aliphatic or aromatic acids (41). The water of reaction and some of the organic acid were continuously removed from the base of the column.

Ethyl Acetate. The production of ethyl acetate by continuous esterification is an excellent example of the use of azeotropic principles to obtain a high yield of ester. The acetic acid, 1.53–1.84 sp gr (50–66 Bé) H_2SO_4, and an excess of 95% ethyl alcohol are mixed in reaction tanks provided with agitators. After esterification equilibrium is reached in the mixture, it is pumped into a receiving tank and through a preheater into the upper section of a bubblecap plate column (see Fig. 4). The temperature at the top of this column is maintained at ca 80°C and its vapor (alcohol with the ester formed and ca 10% water) is passed to a condenser. The first recovery column is op-

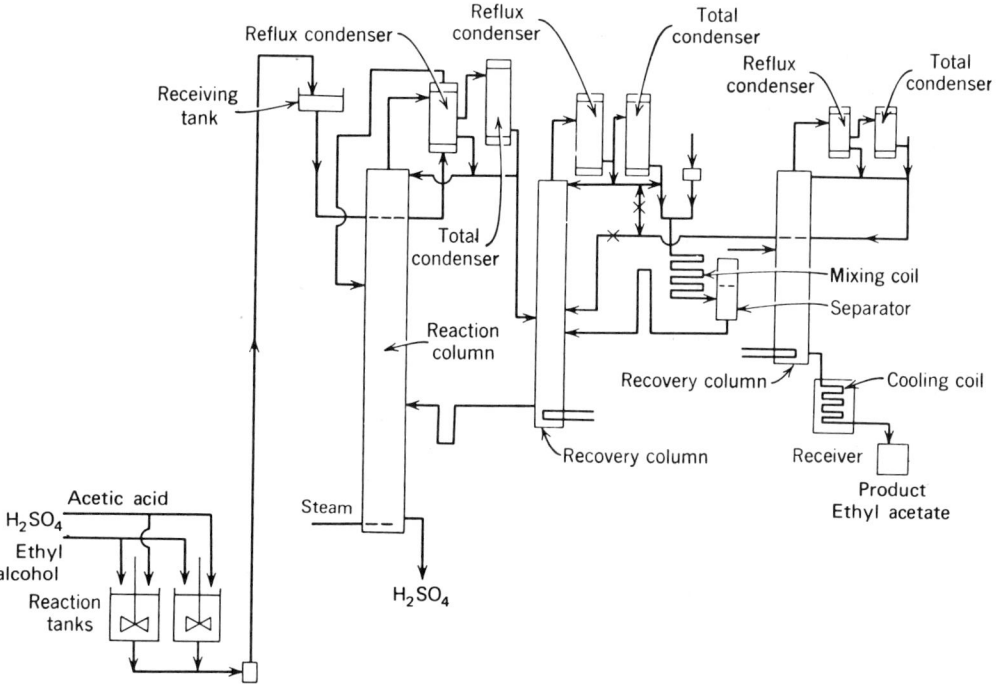

Figure 4. Continuous ethyl acetate process (2).

erated with a top temperature of 70°C, producing a ternary azeotrope of 83% ester, 9% alcohol, and 8% water. The ternary mixture is fed to a mixing coil where water is added in order to form two layers and allowed to separate in a decanter. The upper layer contains ca 93% ethyl acetate, 5% water, and 2% alcohol and is sent to a second recovery or ester-drying column. The overhead from this column is the ternary mixture that is sent back to the mixing coil. The residue from this column is 95–100% ethyl acetate which is sent to a cooler and then to a storage tank. This process also applies to methyl acetate and methyl butyrate.

Vapor-Phase Esterification. Catalytic esterification of alcohols and acids in the vapor phase has received attention because the conversions obtained are generally higher than in the corresponding liquid-phase reactions (7).

Physicochemical Considerations. The determination of the equilibrium constant K_G for the reaction $C_2H_5OH + CH_3COOH \rightleftharpoons C_2H_5OOCCH_3 + H_2O$ has been the subject of a number of investigations over the temperature range of 40–300°C (42). The values of the equilibrium constant range from 6–559 (43) with 71–95% ester as the equilibrium concentration from an equimolar mixture of ethyl alcohol and acetic acid, depending upon the technique used. A study of the reaction mechanism indicates that adsorption of acetic acid is the rate-controlling step; the molecularly adsorbed acetic acid then reacts with alcohol in the vapor phase.

The rate of esterification of acetic acid and ethyl alcohol in equimolar quantities has been studied in a dynamic system using silica gel catalyst at 150–270°C (44).

Ethyl Acetate. Catalysts proposed for the vapor-phase production of ethyl acetate include silica gel, zirconium dioxide, activated charcoal, and potassium hydrogen sulfate. More recently, phosphoric-acid-treated coal (45) and calcium phosphate (46) catalysts have been described.

Other Esters. The esterification of acetic acid with various alcohols in the vapor phase has been studied using several catalysts precipitated on pumice (47).

Esterification of Other Compounds

Acid Anhydrides. The acid produced from an acid anhydride cannot hydrolyze the ester, and hence this reaction goes to completion:

$$(RCO)_2O + R'OH \longrightarrow RCOOR' + RCOOH \tag{6}$$

However, this method is applied only when esterification cannot be effected by the usual means because of the higher cost of the anhydrides. The production of cellulose acetate (see Cellulose acetate and triacetate fibers) and aspirin (acetylsalicylic acid) (see Salicylic acid) are two examples of the large-scale use of acetic anhydride. The speed of acylation is greatly increased by the use of catalysts (48) such as sulfuric acid, perchloric acid, trifluoroacetic acid, trifluoropropionic acid, phosphorus pentoxide, zinc chloride, ferric chloride, sodium acetate, and tertiary amines.

Ref. 49 describes a procedure wherein anhydride incorporated into a polystyrene backbone is treated with ethanol:

$$\text{resin}\begin{array}{c}O\\\diagup\\O\\\diagdown\\C_6H_5\end{array}\hspace{-1em}\begin{array}{c}\\\\O\end{array} + C_2H_5OH \longrightarrow C_6H_5COOC_2H_5 + \text{resin}-COOH \tag{7}$$

The excess reagent can be removed by filtration.

Dibasic acid anhydrides such as phthalic anhydride and maleic anhydride readily react with alcohols to form the monoalkyl ester:

$$R(CO)_2O + R'OH \longrightarrow R(COOR')(COOH) \tag{8}$$

This reaction can be used for identification of individual alcohols because of the wide variations noted in the melting points of monoalkyl esters up to the dodecyl derivatives. The reaction can be used to separate alcohols of various classes (3). Monoesters are converted into the normal diesters by heating with an excess of alcohol and a catalyst; however, diesters are generally formed directly.

Ketene (qv), like acid anhydrides, unites with alcohols to form (acetate) esters:

$$CH_2{=}C{=}O + ROH \longrightarrow CH_3COR \tag{9}$$

Ketene is an efficient acetylating agent with some alcohols but in the absence of catalysts may be either nonreactive or sluggish with others, especially phenols and tertiary alcohols (50).

Acid Chlorides. Acid chlorides react with alcohols to form esters:

$$RCOCl + R'OH \longrightarrow RCOOR' + HCl \tag{10}$$

The acid chlorides are generally more reactive than the corresponding acid anhydrides; the hydrogen chloride liberated acts as a catalyst. However, because of this, the acid chlorides are not used with alcohols susceptible to rearrangement. Acid chlorides are used for quantitative determination of hydroxyl groups and for acylation of sugars. Industrial applications include the formation of the alkyl or aryl carbonates from phosgene (see Carbonic and chloroformic esters) and phosphate esters such as triethyl, triphenyl, tricresyl, and tritolyl phosphates from phosphorus oxychloride (see Phosphoric acids).

Esterification of tertiary alcohols by acid chlorides is described in ref. 51.

The reaction of alcohols and acid chlorides in the presence of magnesium has been described (48). With primary and secondary alcohols the reaction is very smooth, with high, sometimes quantitative, yields. Difficultly esterifiable hydroxy compounds such as tertiary alcohols and phenols can be esterified by this method. The reaction, which is carried out in ether or benzene, is usually very vigorous, with evolution of hydrogen.

An esterification process of acid chlorides that proceeds through the corresponding *tert*-butyl thioate group is described in ref. 52:

$$RCOCl \xrightarrow{TlSC(CH_3)_3} RCOSC(CH_3)_3 \xrightarrow[Hg(OCOCF_3)_2]{HOC(CH_3)_3} RCOOC(CH_3)_3 \tag{11}$$

The relatively stable *S-tert*-butyl thioate group can be used as a carboxylic acid protective group. Esters and lactones obtained in this manner have been produced in high yields.

Amides. Amides may be converted to esters by the following reversible reaction:

$$\underset{\text{O}}{\text{R}\overset{\parallel}{\text{C}}\text{NH}_2} + \text{R'OH} \rightleftharpoons \underset{\text{O}}{\text{R}\overset{\parallel}{\text{C}}\text{OR'}} + \text{NH}_3 \qquad (12)$$

In order to produce high yields of ester it is necessary to remove the ammonia produced, either by heating or by combining with a mineral acid, eg, H_2SO_4 or HCl.

The structural relationships involved in esterification of amides are shown in Table 2 (53). Other methods of converting amides to esters have been described (35). Alkyl halides are treated with amides to give esters (54). Esters can be synthesized from *N*-alkyl-*N*-nitrosoamides, which are derived from the corresponding amides (55).

Nitriles. Alcoholysis of nitriles (qv) offers a convenient way to produce esters without isolating the acid:

$$\text{RCN} + \text{H}_2\text{O} + \text{R'OH} \longrightarrow \underset{\text{O}}{\text{R}\overset{\parallel}{\text{C}}\text{OR'}} + \text{NH}_3 \qquad (13)$$

Acids are used to combine with the ammonia formed. A large excess of alcohol is used but the amount of water is generally kept small. Catalysts such as hydrogen chloride, hydrogen bromide, and sulfuric acid have been employed (48).

One of the most important applications of this process is that of methyl methacrylate manufacture. In this process (56), acetone cyanohydrin is treated with concentrated sulfuric acid at 100°C, affording the corresponding methacrylamide sulfate which is esterified with methanol at 90°C. After purification, methyl methacrylate (99.8% purity) is obtained in a yield of ca 85%.

With aromatic nitriles, a single ortho substituent inhibits the conversion of the nitrile with methanol or ethanol as shown in Table 3 (57).

Unsaturated Hydrocarbons. Olefins (qv) from ethylene through octene have been converted into esters. With ethylene and propylene, only a single ester is produced using acetic acid (ethyl acetate and isopropyl acetate, respectively). With the butylenes, two products are possible: *sec*-butyl esters result from 1- and 2-butenes, whereas *tert*-butyl esters are obtained from isobutylene. The C_5 olefins give rise to three *sec*-

Table 2. Comparative Yields of Esters from Amides or Acids[a]

Methyl ester	Yield of ester, %	
	From amide	From acid
formate	34	
acetate	70	56
monochloroacetate	64	65
dichloroacetate	57	70
trichloroacetate	53	73
phenylacetate	50	86
propionate	80	44
benzoate	15	37

[a] Ref. 53.

Table 3. Yields of Esters from Nitriles[a]

Ester	Yield, %	
	Methyl ester	Ethyl ester
benzoate	65.4	35.1
o-toluate	0	0
m-toluate	92.1	58.7
p-toluate	83	80

[a] Hydrochloric acid catalyst.

amyl esters and one *tert*-amyl ester. As the carbon chain is lengthened, the reactivity of the olefin with organic acids increases.

In the case of ethylene (qv), it is necessary to use high temperatures and pressures as well as active catalyst to effect esterification (58). Yields of 40–50% based on ethylene were obtained with boron trifluoride–hydrogen fluoride mixtures as catalysts at 150°C. 2-Butene under pressure at 115–120°C with an excess of glacial acetic acid containing 10% H_2SO_4 gave as much as a 60% yield of *sec*-butyl acetate (59).

Tert-butyl acetate was prepared by passing isobutylene and acetic acid (2:1 mol ratio) in the liquid phase over a silica catalyst impregnated with vanadium pentoxide and potassium sulfate at 1.7 MPa (250 psi). Conversion of isobutylene to ester increased with increasing temperature and ranged from 10% at 52°C to 24% at 93°C. Based on the acetic acid charged, yields of 31–43% of *tert*-butyl acetate resulted at 93°C (60).

Most of the vinyl acetate produced in the United States today is made by the vapor-phase ethylene process. In this process, a vapor-phase mixture of ethylene, acetic acid, and oxygen is passed at elevated temperature and pressures over a fixed-bed catalyst consisting of supported palladium (61). Less than 70% oxygen, acetic acid, and ethylene conversion are realized per pass. Therefore, these components have to be recovered and returned to the reaction zone. The vinyl acetate yield using this process is typically in the 91–95% range (62). Vinyl acetate can be manufactured also from acetylene (qv), acetaldehyde, and the liquid-phase ethylene process (63) (see Vinyl polymers).

Acetylenes and olefins have recently been used to produce esters via carbonylation techniques (see Acetylene-derived chemicals). Methyl itaconate is synthesized in 63% yield by treating propargyl alcohol with carbon monoxide in the presence of HCl, methanol, and catalytic amounts of palladium (64):

$$HC\equiv CCH_2OH \xrightarrow[Pd, CH_3OH]{HCl, CO} \underset{COCH_3}{\overset{COCH_3}{CH_2=C}} \tag{14}$$

Esters can be obtained from halogenated olefins using a metal carbonyl (qv) catalyst (65), eg, *trans*-1-bromo-2-phenylethylene is treated with nickel carbonyl in the presence of methanol to afford the corresponding methyl cinnamate (see Cinnamic acid):

$$PhCH=CHBr \xrightarrow[CH_3OH, CH_3O^-]{Ni(CO)_4} PhCH=CHCOCH_3 \tag{15}$$

304 ESTERIFICATION

Ethers. In the presence of anhydrous agents such as ferric chloride (66), hydrogen bromide, and acid chlorides, ethers react to form esters (see Ethers). Esters can also be prepared from ethers via an oxidative process (67).

With mixed sulfonic-carboxylic anhydrides, ethers are converted to a mixture of the corresponding carboxylate and sulfonate esters (68):

$$RCOSO_2R + R'OR' \longrightarrow RCOR' + RSO_2OR' \qquad (16)$$

$$RCOSO_2R + (CH_2)_nO \longrightarrow RCO(CH_2)_nOSO_2R \qquad (17)$$

In a similar manner, cyclic ethers are converted to the corresponding diesters (eq. 17). The highly reactive ethers can be cleaved in ca 1 h at 130°C with yields in most cases of >80% (68). Unsaturated esters can be prepared from the corresponding acetylenic ethers with yields in most cases of >50% (69) as in the example shown in equation 18.

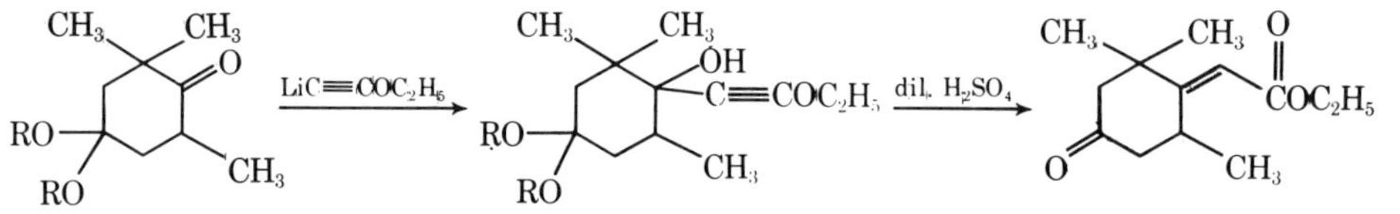

(18)

An esterification process involving ethers of potential interest to industry is that of synthesis of bis-(β-hydroxyethyl) terephthalate from terephthalic acid and ethylene oxide:

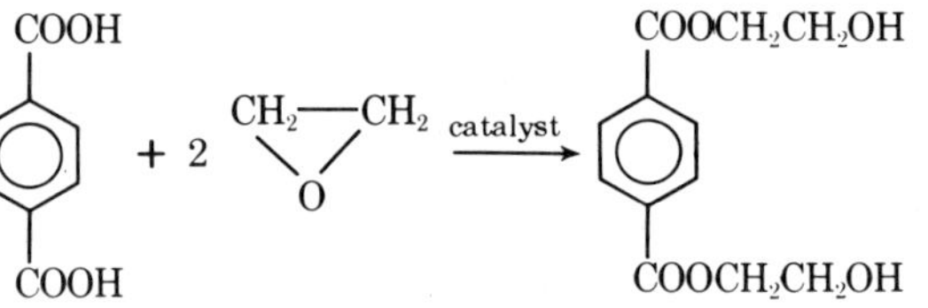

(19)

This ester can be used effectively as a monomer in the production of polyesters. Bis-(β-hydroxyethyl) terephthalate and related compounds can be produced in organic solvents using finely divided carbon catalyst (70). The carbon functions not only as a catalyst but also helps to remove color from the reaction mixture upon removal of the carbon by hot filtration.

Aldehydes and Ketones. Esters are obtained readily by condensation of aldehydes in the presence of alcoholate catalysts such as aluminum ethylate, $Al(OC_2H_5)_3$, by the Tishchenko reaction (48). The alcoholate catalysts may be prepared from commercial aluminum and n-butyl or isobutyl alcohol in the presence of 2–2.5% aluminum chloride (71).

Trihalomethyl ketones react with alcohols in the presence of alkaline catalysts even at room temperature (72):

$$RCCCl_3 + R'OH \longrightarrow RCOR' + CHCl_3 \qquad (20)$$

A variety of esters can be prepared from the corresponding ketones using peracids in a process usually referred to as the Baeyer-Villiger reaction (73); eg, cyclopentanone is converted to δ-valerolactone upon treatment of the ketone with peroxytrifluoroacetic acid:

$$\text{cyclopentanone} + CF_3CO_2H \xrightarrow[10-15\ °C]{CF_3COH} \text{δ-valerolactone} \qquad (21)$$

This conversion can be carried out, in many cases, with >80% yield.

Alcohols. The direct synthesis of esters by dehydrogenation of alcohols offers a simple method for the preparation of certain types of esters, such as ethyl acetate (74):

$$2\ RCH_2OH \xrightarrow{\text{catalyst}} R\overset{O}{\underset{\|}{C}}OCH_2R + 2\ H_2 \qquad (22)$$

The reaction is catalyzed by copper with various promoters or activators and is carried out in the vapor phase at 200–300°C at normal pressure.

Technical Preparation of Esters

The choice of the esterification process to obtain a maximum yield is dependent upon many factors, ie, no single process has universal applicability. Although extensive preparative techniques have been reviewed elsewhere (7,48), the methods given in his section are representative of both laboratory and plant-scale techniques used in batch esterifications.

Methyl Esters. Methyl esters are obtained in good yield using methylene dichloride or ethylene dichloride as solvent (75). The latter is generally preferred but the choice of the solvent depends to some extent upon the boiling point of the desired ester. Also, the toxicity of these solvents should be considered prior to using them (see Industrial hygiene and toxicology, chlorocarbons). The general procedure is as follows: for each mole of aliphatic carboxyl group, 96 g (3 mol) of methanol, 300 mL of ethylene dichloride, and 3 mL of conc H_2SO_4 are used. With aromatic acids, the amount of H_2SO_4 is increased to 15 mL/mol of carboxyl group. The mixture is refluxed for 6–15 h, although in some cases the time limit may be as short as ½ h. Progress of esterification is usually indicated by the development of cloudiness and separation of an upper layer containing water, methanol, and sulfuric acid. After the reaction is completed, the cooled mixture is washed successively with water, sodium bicarbonate solution, and again with water. The ethylene chloride layer is then distilled at atmospheric or reduced pressure and the residual methyl ester is purified by distillation or crystallization.

Medium Boiling Esters. From a technique for esterifying ethyl and propyl alcohols, ethylene glycol, and glycerol with various acids using a third component such as benzene, toluene, hexane, or carbon tetrachloride to remove the water produced, the preparation of ethyl lactate is described as an example of the general procedure (76). A mixture of 1 mol of 80% lactic acid and 2.3 mol of 95% ethyl alcohol is added to a volume of benzene equal to half that of the alcohol (ca 43 mL), and the resulting mixture is refluxed for several hours. Upon cooling the overhead distillate two layers are formed. The lower layer is extracted to recover the benzene and alcohol, and the water is discarded. The upper layer is returned to the column for reflux. After all the water is removed from the reaction mixture, the excess of alcohol and benzene is removed by distillation and the ester is fractionated.

High Boiling Esters. The following procedure can be used for making diethyl phthalate and other high boiling esters (77). Phthalic anhydride (1 equiv) and 2.5 equivalents of ethanol are refluxed for 2 h in the presence of 1% of concentrated H_2SO_4. To produce the monoester, the excess of alcohol is distilled at <100°C. For the diester, a mixture of 67% benzene and 33% alcohol is introduced continuously below the surface of the reaction mixture and the resulting alcohol–water–benzene ternary is distilled and condensed. A yield of diester of >99% is obtained by passing 3.4–7 equivalents of alcohol through the mixture in 4.5–7 h.

Organotitanates, zirconates, or organotin compounds are effective catalysts for the esterification of carboxylic acids or anhydrides with higher boiling monohydroxy alcohols at temperatures which permit the continuous distillation of the water formed (78). Refluxing 1 mol phthalic anhydride with 3 mol 2-ethyl-2-hexanol under stirring with these agents and removing the water by a trap separator gave the corresponding esters in ≥99% yields (see Phthalic acid).

Difficultly Esterifiable Acids. The sterically hindered acids, such as 2,6-disubstituted benzoic acids, cannot usually be esterified by the conventional means. Several esters of sterically hindered acids have been prepared by dissolving 2 g of the acid in 14–20 mL of 100% H_2SO_4 (79). After standing a few minutes at RT, the solution is poured into an excess of cold absolute alcohol. Most of the alcohol is removed under reduced pressure, about 50 mL of water is added, and the distillation is continued under reduced pressure to remove the remainder of the alcohol. The organic matter is extracted with ether and treated with sodium carbonate solution. The ester is then distilled. Yields of esters made in this manner are 57–81%.

Ester Interchange

Ester interchange (transesterification) is a reaction between an ester and another compound, characterized by an exchange of alkoxy groups or of acyl groups, and resulting in the formation of a different ester.

In the best-known types of ester interchange, the compound with which the ester reacts is an alcohol (eq. 23), an acid (eq. 24) or another ester (eq. 25). These ester interchanges may be called, more specifically, ester-alcohol interchange or alcoholysis, ester-acid interchange or acidolysis, and ester-ester interchange, respectively (7).

$$RCOR' + R''OH \rightleftharpoons RCOR'' + R'OH \qquad (23)$$

$$RCOR' + R''COH \rightleftharpoons RCOH + R''COR' \qquad (24)$$

$$RCOR' + R''COR''' \rightleftharpoons RCOR''' + R''COR' \qquad (25)$$

These reactions are reversible and ordinarily do not involve large energy changes.

Ester–Alcohol Interchange (Alcoholysis). *Reaction Conditions.* The reaction commonly takes place in one liquid phase, sometimes with one of the reactants being only partially soluble and going into solution gradually as the reaction proceeds. Unless an excess of one of the reactants is used, or unless one of the products is withdrawn from the reaction phase by vaporization or precipitation, the reaction does not proceed to completion but comes to a standstill with substantial proportions of both alcohols and both esters in equilibrium. The concentrations present at equilibrium depend upon the characteristics of the alcohols and esters involved, but in most practical uses of the reaction, one or both of the devices mentioned are used to force the reaction toward completion.

Temperatures. With alkaline catalysts, the reaction often takes place at RT or even at lower temperatures. With acid catalysts, temperatures near 100°C are commonly used. With no catalyst, temperatures ≥250°C may be required for a practical reaction rate.

Catalysts. Of the alkaline catalysts, alkali metal alkoxides are the most effective; ordinarily, the sodium or potassium alkoxide of the alcohol entering the reaction is preferred. Various other catalysts of milder alkalinity are preferred in special cases. For example, the use of sodium methyl carbonate as catalyst in the methanolysis of poly(vinyl acetate) is said to yield a poly(vinyl alcohol) having improved color. Aluminum alkoxide has been proposed as a catalyst for the alcoholysis of certain unsaturated esters; other sensitive esters have been made with the Grignard reagent as catalyst. Zinc is reported to be an efficient catalyst in the alcoholysis of ethyl esters of α-halogenated aliphatic acids by allyl and methallyl alcohols; conventional catalysts would favor undesirable side reactions. Recently, neutral organic titanates have received much attention (80). Divalent metal salts such as zinc or manganese acetate and organotins such as dibutyltin oxide have been employed.

Among the acid catalysts, sulfuric acid, sulfonic acids, and hydrochloric acid are most used. With polyhydric alcohols, sulfuric acid is preferred to hydrochloric because of the tendency of hydrochloric acid to form chlorohydrins.

Equilibrium. In general, primary alcohols are more active than secondary alcohols (that is, they tend to displace them), and secondary alcohols tend to displace tertiary alcohols, but in addition, there are considerable differences among different members of the same class. Various alcohols have been compared in this way (4,81).

Applications. The applications of alcoholysis in analysis, research, and manufacturing are too numerous to mention in detail (for more details see refs. 7 and 48). A few examples follow.

n-Butyl Oleate. Olive oil, 3 kg, consisting mainly of the glyceryl esters of oleic acid, is refluxed for 20 h with 7 L of n-butyl alcohol containing 150 g of concentrated H_2SO_4. The product contains a small proportion of saturated esters (82).

Poly(vinyl alcohol). Poly(vinyl alcohol) (see Vinyl polymers) is more easily prepared, in a form which can be filtered and washed in a practical way, by alcoholysis of poly(vinyl acetate) than by its saponification in an aqueous system:

$$\left[\!-CH_2CH-\atopOOCCH_3\right]_n + n\,CH_3OH \rightleftharpoons n\,CH_3\overset{\overset{\displaystyle O}{\|}}{C}OCH_3 + \left[\!-CH_2CH-\atopOH\right]_n \qquad (26)$$

The use of a catalytic quantity of alkali equivalent to only a small fraction of the acetate has the advantage that contamination of the poly(vinyl alcohol) with salts, which are difficult to remove, is minimized. A variant of the process is the use of a mixture of alcohol with the acetate ester produced by the alcoholysis as the alcoholyzing agent. This provides a means of controlling the completeness of removal of the acetate groups from the poly(vinyl acetate) (83).

Acrylic Esters. A procedure has been described for preparation of higher esters from methyl acrylate which illustrates the use of an acid catalyst together with the removal of one of the products by azeotropic distillation (84). Aluminum isopropoxide catalyzes the reaction of amino alcohols with methyl acrylate and methyl methacrylate. A review of the synthesis of acrylic esters by transesterification is given in ref. 85 (see Acrylic acid).

Ester–Acid Interchange (Acidolysis). This reaction requires the use of an elevated temperature or the use of an acid catalyst (7), or both. Like alcoholysis, the reaction is reversible and requires the use of an excess of the replacing acid or removal of one of the products from the reaction if a high degree of replacement of the acid radical of an ester by another acid is to be obtained. This can be accomplished by distilling one of the products from the reaction mixture during the acidolysis.

In a series of organic acids of similar type, not much tendency exists for one acid to be more active than another. For example, in the replacement of stearic acid in methyl stearate by acetic acid, the equilibrium constant is 1.0, and even for the replacement of stearic acid by formic acid, it is only 1.3. Branched-chain acids, and some aromatic acids, especially sterically hindered acids such as ortho substituted benzoic acids, would be expected to be less active in replacing other acids. Mixtures of esters are obtained when acidolysis is carried out without forcing the replacement to completion by removing one of the products. The acidolysis equilibrium and mechanism are discussed in detail in ref. 86.

Ester–Ester Interchange. The reaction between two esters to produce two other esters was described by Friedel and Crafts in 1865 but has not been used as much as alcoholysis. The same general principles apply, as to reversibility of the reaction and means of driving the reaction to completion (7). In general, the same catalysts are effective as in alcoholysis. Usually the reaction is slower than alcoholysis of the same esters. Without a catalyst, a reaction time of several hours at >250°C is required to bring two typical esters to equilibrium. Catalysts are almost essential to bring reaction rates into a practical range so that the use of destructive temperatures can be avoided. Tin compounds (qv), especially stannous hydroxide, have been mentioned frequently as catalysts and have the merit that they do not produce much decomposition or discoloration of the esters (87). More effective at lower temperatures are the acid catalysts, such as sulfuric acid and sulfonic acids, and especially the alkaline catalysts such as sodium alkoxides. With an alkaline catalyst, ester–ester interchange can be carried out at temperatures as low as 0°C.

BIBLIOGRAPHY

"Esterification" in *ECT* 1st ed., Vol. 5, pp. 776–817, by C. E. Leyes, Celanese Corporation of America; "Ester Interchange" in *ECT* 1st ed., Vol. 5, pp. 817–823, by E. W. Eckey, E. W. Eckey Research Laboratory; "Esterification" in *ECT* 2nd ed., Vol. 8, pp. 313–356, by Charles E. Leyes, Newark College of Engineering; "Ester Interchange" in *ECT* 2nd ed., Vol. 8, pp. 356–365, by E. W. Eckey, E. W. Eckey Research Laboratory, and E. F. Izard, E. I. du Pont de Nemours & Co., Inc.

1. H. A. Goldsmith, *Chem. Rev.* **33,** 257 (1943).
2. D. B. Keyes, *Ind. Eng. Chem.* **24,** 1096 (1932).
3. E. E. Reid, "Esterification" in P. Groggins, *Unit Processes in Organic Synthesis*, 4th ed., McGraw-Hill Book Co., New York, 1952.
4. S. Patai, *The Chemistry of Carboxylic Acids and Esters,* Wiley-Interscience, New York, 1969.
5. F. A. Lowenheim and M. K. Moran, *Industrial Chemicals,* 4th ed., John Wiley & Sons, New York, 1975.
6. M. L. Bender, *Chem. Rev.* **60,** 53 (1960).
7. K. S. Markley in K. S. Markley, ed., *Fatty Acids,* Part 2, Wiley-Interscience, New York, 1961, p. 757.
8. *1979–1980 Catalog–Handbook of Organic and Biochemicals,* Aldrich Chemical Company, Inc., Milwaukee, Wisc.
9. *Chem. Mark. Rep.,* **212,** 42 (Nov. 7, 1977).
10. *Chem. Eng. News,* **55,** 41 (June 6, 1977).
11. B. Holmberg, *Ber.* **45,** 2997 (1912).
12. M. Polanyi and A. L. Szabo, *Trans. Faraday Soc.* **30,** 508 (1934).
13. E. K. Euranto ref. 4, p. 505.
14. R. J. Hartman, L. B. Storms, and A. G. Gassmann, *J. Am. Chem. Soc.* **61,** 2167 (1939).
15. A. G. Gassmann and R. J. Hartman, *J. Am. Chem. Soc.* **63,** 2393 (1941).
16. M. Berthelot and L. Péan de Saint-Gilles, *Ann. Chim. Phys.* **68,** 225 (1863).
17. E. S. Gould, *Mechanisms and Structure in Organic Chemistry,* Holt, Rinehart, and Winston, New York, 1959, p. 314.
18. J. Hine, *Physical Organic Chemistry,* McGraw-Hill Book Company, New York, 1962, p. 275.
19. C. A. Lane, *J. Am. Chem. Soc.* **86,** 2521 (1964).
20. Y. K. Syrkin and I. I. Moiseev, *Usp. Khim.* **27,** 717 (1958).
21. V. A. Palm and co-workers, *Zh. Fiz. Khim.* **36,** 2499 (1962).
22. N. Schlechter, D. F. Othmer, and S. Marshak, *Ind. Eng. Chem.* **37,** 900 (1945).
23. C. E. Leyes and D. F. Othmer, *Ind. Eng. Chem.* **37,** 968 (1945).
24. S. Berman, A. A. Melnychuk, and D. F. Othmer, *Ind. Eng. Chem.* **40,** 1312 (1948).
25. M. Berthelot and P. St. Gilles, *Ann. Chim. Phys.* **65,** 385 (1862).
26. W. Swietoslawski, *J. Phys. Chem.* **37,** 701 (1933).
27. P. E. Coria, *Rev. Fac. Cienc. Quim. Univ. Nac. La Plata* **10,** 67 (1935); *Chem. Abstr.* **36,** 7427 (1942).
28. C. A. Durruty, *An. Asoc. Quim. Argent.* **19,** 227 (1931); *Chem. Abstr.* **26,** 3721 (1932).
29. H. M. Trimble and E. L. Richardson, *J. Am. Chem. Soc.* **62,** 1018 (1940).
30. W. P. Jencks and M. Gilchrist, *J. Am. Chem. Soc.* **86,** 4651 (1964).
31. L. H. Horsley in R. F. Gould, ed., *Azeotropic Data—III,* American Chemical Society, Washington, D.C., 1973.
32. J. C. Konen, E. T. Clocker, and R. P. Cox, *Oil Soap* **22,** 57 (1945).
33. E. Thielpape and A. Fulde, *Ber.* **B66,** 1454 (1933).
34. P. L. Gordon and R. Aronowitz, *Ind. Eng. Chem.* **37,** 780 (1945).
35. I. T. Harrison and S. Harrison, *Compendium of Organic Synthetic Methods,* Vol. 1, Wiley-Interscience, New York, 1971.
36. I. T. Harrison and S. Harrison, *Compendium of Organic Synthetic Methods,* Vol. 2, Wiley-Interscience, New York, 1974.
37. L. S. Hegedus and L. Wade, *Compendium of Organic Synthetic Methods,* Vol. 3, Wiley-Interscience, New York, 1977.
38. C. L. Levesque and A. M. Craig, *Ind. Eng. Chem.* **40,** 96 (1948).
39. M. J. Astel, B. Schaeffer, and C. O. Obenland, *J. Am. Chem. Soc.* **77,** 3643 (1955).
40. C. E. Leyes and D. F. Othmer, *Trans. Am. Inst. Chem. Eng.* **41,** 157 (1945).
41. U.S. Pat. 2,426,968 (Sept. 2, 1947), H. W. Grubb, L. M. O'Hara, and K. Atwood (to Seagram and Sons).
42. V. I. Goldanskii, *J. Phys. Chem. USSR* **21,** 431 (1947); H. C. Tidwell and E. E. Reid, *J. Am. Chem. Soc.* **53,** 4353 (1931).
43. A. Mailhe and F. deGodon, *Bull. Soc. Chim. Fr.* **29,** 101 (1921); G. Edgar and W. H. Schuyler, *J. Am. Chem. Soc.* **46,** 64 (1924).
44. H. F. Hoerig, D. Hanson, and O. L. Kowalke, *Ind. Eng. Chem.* **35,** 575 (1943).

310 ESTERIFICATION

45. A. M. Chashchin and N. M. Lebedeva, *Gidroliz. Lesokhim. Promst.* **15,** 6 (1962); *Chem. Abstr.* **59,** 2640 (1963).
46. S. Sharipova, A. Arifdzhanov, and A. Sultanov, *Katal. Pererab. Uglevodorodnogo Syrya,* 108 (1967); *Chem. Abstr.* **71,** 12476 (1969).
47. J. F. Spangenberg, *Ind. Quim. Buenos Aires* **7,** 393 (1945); *Chem. Abstr.* **41,** 4028 (1947).
48. C. A. Buehler and D. E. Pearson, *Survey of Organic Synthesis,* Wiley-Interscience, New York, 1970, p. 801.
49. M. B. Shambhu and G. A. Digenis, *Tetrahedron Lett.,* 1627 (1973).
50. C. D. Hurd and A. S. Roe, *J. Am. Chem. Soc.* **61,** 3355 (1939).
51. W. H. Puterbaugh, *J. Org. Chem.* **27,** 4010 (1962).
52. S. Masamune and co-workers, *J. Am. Chem. Soc.* **97,** 3515 (1975).
53. S. G. Toole and F. J. Sowa, *J. Am. Chem. Soc.* **59,** 1971 (1937).
54. J. S. Matthews and J. P. Cookson, *J. Org. Chem.* **34,** 3204 (1969).
55. E. H. White, *J. Am. Chem. Soc.* **77,** 6011 (1955).
56. F. A. Lowenheim and M. K. Moran, *Faith, Keys, and Clark's Industrial Chemicals,* John Wiley & Sons, Inc., New York, 1975, p. 547.
57. P. Pfeiffer, I. Engelhardt, and W. Alfuss, *Ann.* **467,** 158 (1928).
58. J. A. John in S. A. Miller, ed., *Ethylene and Its Industrial Derivatives,* Ernest Benn Ltd., London, Eng., 1969, p. 765.
59. B. T. Brooks, *Ind. Eng. Chem.* **27,** 278 (1935).
60. U.S. Pat. 3,014,066 (Dec. 19, 1961), E. R. Kerr and M. C. Throckmorton (to Texaco).
61. W. Schwerdtel, *Hydrocarbon Process.* **47,** 187 (1968).
62. Brit. Pat. 981,987 (Feb. 3, 1965), (to Farbenfabriken Bayer Aktiengesellschaft).
63. Ref. 58, p. 946.
64. J. Tsuji and T. Nogi, *Tetrahedron Letter.,* 1801 (1966).
65. E. J. Corey and L. S. Hegedus, *J. Am. Chem. Soc.* **91,** 1233 (1969).
66. B. Ganem and V. R. Small, *J. Org. Chem.* **39,** 3728 (1974).
67. E. C. Juenge and D. A. Beal, *Tetrahedron Lett.,* 5819 (1968).
68. M. H. Karger and Y. Mazur, *J. Org. Chem.* **36,** 532 (1971).
69. J. N. Marx and J. Sondheimer, *Tetrahedron* (Suppl. 8, Part 1), 1 (1966).
70. U.S. Pat. 3,652,647 (Mar. 28, 1972), E. G. Zey (to Celanese Chemical Company).
71. M. Y. Kagan and I. A. Sobolev, *J. Chem. Ind.* (*Moscow*) (2), 35 (1933); *Chem. Abstr.* **27,** 4215 (1933).
72. J. Houben and W. Fisher, *Ber.* **B64,** 240, 2636 (1931).
73. H. O. House, *Modern Synthetic Reactions,* W. A. Benjamin, Inc., New York, 1965, pp. 123–129.
74. Ref. 58, p. 764.
75. R. O. Clinton and S. C. Laskowski, *J. Am. Chem. Soc.* **70,** 3135 (1948).
76. U.S. Pat. 1,421,604 (July 4, 1922), J. A. Steffens (to U.S. Industrial Alcohol Co.).
77. U.S. Pat. 2,076,111 (Apr. 6, 1927), W. J. Bannister (to Commercial Solvents Corp.).
78. Brit. Pat. 852,110 (Oct. 26, 1960), (B. F. Goodrich Co.).
79. M. S. Newman, *J. Am. Chem. Soc.* **63,** 2431 (1941).
80. *TYZOR Organic Titanates, Du pont Bulletin D5258,* E. I. Du pont de Nemours & Co., Inc., Wilmington, Del., 1972, 5M, Rev. 377.
81. G. B. Hatch and J. E. Adkins, *J. Am. Chem. Soc.* **59,** 1694 (1937).
82. E. E. Reid and co-workers, in A. H. Blatt, ed., *Organic Synthesis,* Coll. Vol. II, John Wiley & Sons, Inc., New York, 1943, p. 469.
83. U.S. Pat. 2,266,996 (Dec. 23, 1941), N. D. Scott and J. E. Bristol (to DuPont).
84. C. E. Rehberg and C. H. Fisher, *J. Am. Chem. Soc.* **66,** 1203 (1944).
85. J. K. Haken, *Synthesis of Acrylic Esters by Transesterification,* Noyes Development Corporation, Park Ridge, N.J., 1967.
86. J. Koskikallio in ref. 4, pp. 103–136.
87. A. E. Bailey, *Industrial Oil and Fat Products,* Interscience Publishers, Inc., New York, 1945, p. 676.

E. G. Zey
Celanese Corporation

ESTERS, ORGANIC

Esters are compounds that, on hydrolysis, yield alcohols or phenols and acids according to the equation:

$$RA + H_2O \rightleftharpoons ROH + HA$$

where R is a hydrocarbon fragment and A is the anionic portion of an organic or inorganic acid. For carboxylic acid esters, the reaction can be represented as:

$$RCOOR' + H_2O \rightleftharpoons R'OH + RCOOH$$

where R and R' are the same or different hydrocarbon radicals. The reverse reaction constitutes the usual method for preparing esters.

Many molecules contain both carboxy and hydroxy groups, and the specific nature of the esters formed by interaction of these groups depends largely upon the distance between them in the molecule. Aliphatic compounds that contain carboxy and hydroxy groups attached to the same carbon atom usually give a cyclic ester formed from two molecules of the compound; the reaction is illustrated for lactic acid, from which the common name, lactides, for cyclic esters of this type is derived:

[95-96-5]

When the hydroxy and carboxy groups are separated by at least two carbon atoms, internal esters, or lactones, may be formed:

$$HO(CR_2)_n COOH \rightarrow \overline{O(CR_2)_n C}=O + H_2O$$

When $n = 2$ or $n \geq 5$, hydroxycarboxylic acids (qv) have a strong tendency to form polyesters by intermolecular esterification rather than lactones:

$$x\ HO(CR_2)_2COOH \rightarrow H\text{+}O(CR_2)_n CO\text{+}_x OH + x\ H_2O$$

The tendency for lactone or polyester formation depends upon the chain length and upon the amount and location of branching on the carbon chain of the acid (see Polyesters). Lactones containing five or six ring atoms are usually formed to the exclusion of polyesters, whereas those with rings containing more than seven atoms are usually formed to only a small extent. Special techniques are required to prepare large lactones in good yields. Particularly when seven-membered ring formation is possible, both cyclic and linear products are formed and may be interconverted by use of heat.

Mercaptans and carboxylic acids form an analogous series of compounds, the thiol esters:

$$RSH + R'COOH \rightarrow R'\overset{O}{\underset{\|}{C}}SR + H_2O$$

Monocarboxylic acid monoesters of polyhydric alcohols (eg, ethylene glycol and its homologues) can be prepared by the usual methods; they undergo reactions similar to those of esters of monohydric alcohols (see Alcohols, polyhydric; Glycols). In addition, they disproportionate readily into completely esterified and unesterified polyols. With dicarboxylic and polycarboxylic acids, polyhydric alcohols give polymers that are linear if both reactants are difunctional and cross-linked if either or both is polyfunctional. Acylals, $RCH(OCOR')_2$, are esters of 1,1-diols or aldehyde hydrates. Neither their methods of preparation nor their reactions are typical of those of ordinary esters; they have little technical or commercial value. The ortho esters, $RC(OR')_3$, resemble acetals more than they do the simple esters although they give alcohols and carboxylic acids on hydrolysis (1). The methods of preparation and the reactions of vinyl esters, in which the carbon atom carrying the alcohol oxygen atom is unsaturated

$$(RCH{=}CHOCR')$$
$$\text{(with } C{=}O\text{)}$$

are also not typical of those of simple esters (see Vinyl polymers, poly(vinyl acetate)).

Nomenclature

As a rule, esters are named by replacing the endings (-*ic acid* by -*ate*), the acid stem being preceded by the name of an organic group (derived from the alcohol component of the ester) as a separate word (2). For example, $CH_3CH_2CH_2COOCH_3$, the methyl ester of butan*oic* acid, is methyl butan*oate* (or methyl butyr*ate* if the trivial name, buty*ric* acid, is used for the parent acid). Monoesters of dibasic acids can be named either as such (mono*methyl* succinate) or, more explicitly, by inserting the word hydrogen between the names of the alcohol and acid portions to indicate that one acid group retains its acidic hydrogen atom ($CH_3OCOCH_2CH_2COOH$, methyl hydrogen succinate or, more systematically, methyl hydrogen butanedioate) (see Carboxylic acids; Dicarboxylic acids).

Esters of polyhydric alcohols with monobasic acids are usually named analogously to simple esters [ethylene glycol diacetate [111-55-7], ethylene glycol monoacetate [542-59-6]], but the more systematic names, 1,2-ethanediyl diacetate and 2-hydroxyethyl acetate, are preferable. Esters of glycerol (1,2,3-propanetriol) with monobasic acids have been known as glycerides, and trivial names for glycerol esters have been derived by replacing the -*ic* acid ending with -*in* [$CH_3(CH_2)_{16}COO$-$CH_2CH(OH)CH_2OH$), monostearin [31566-31-1]; $CH_3COOCH_2CH(OCOCH_3)$-CH_2OCOCH_3, triacetin [102-76-1]]. However, the use of more systematic names, such as 1,2,3-propanetriyl triacetate, is recommended (see Glycerol).

Lactones have been named using Greek letters ($\alpha, \beta, \gamma, \delta$, etc) to denote the carbon atom to which the hydroxy group of the parent acid is attached, eg, $CH_3\overline{CHCH_2COO}$, β-butyrolactone [3068-88-0]; $\overline{CH_2CH_2CH_2COO}$, γ-butyrolactone [96-48-0]. This nomenclature is still common but the modern trend has been away from the use of Greek letters as locants; thus 3-hydroxybutyric acid lactone is preferable to β-butyrolactone. *Chemical Abstracts* has stopped using the lactone nomenclature entirely and names the compounds as heterocycles (β-butyrolactone is 4-methyl-2-oxetanone [3068-88-0]; γ-butyrolactone, dihydro-2(3*H*)-furanone [96-48-0]); and lactide (as shown previously) is 3,6-dimethyl-1,4-dioxane-2,5-dione [95-96-5].

In naming esters containing one or more substituents, it is necessary to indicate specifically in which portion of the molecule the substituents occur and to distinguish between isomers such as ethyl chloroacetate [105-39-5], $CH_2ClCOOCH_2CH_3$ and 2-chloroethyl acetate [542-58-5], $CH_3COOCH_2CH_2Cl$.

In older literature, esters such as RCOSR′, formed by esterification of a carboxylic acid with a mercaptan, are called thiol esters; and the isomers, RCSOR′, are thione esters. According to recent recommendations, the atom to which the alcohol moiety is attached is specified, eg, $CH_3CSOC_2H_5$, O-ethyl ethanethioate [926-67-0].

When the ester function is named as a substituent of a more complex molecule, it becomes an alkoxycarbonyl group, eg, 2-(methoxycarbonyl)cyclohexanepropanoic acid [71060-54-3]:

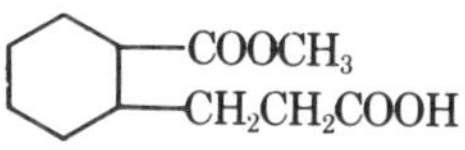

Ortho esters, $RC(OR')_3$, are trivially named on the basis of the ortho acids such as $HC(OC_2H_5)_3$, triethyl orthoformate [122-51-0]. However, more systematic names such as 1,1,1-triethoxymethane are preferred. Imido esters (often called imide ethers or imino ethers), $RC(=NH)OR'$, are preferably named as derivatives of imidic acids $RC(=NH)OH$, eg, $C_6H_5C(=NH)OC_2H_5$ is ethyl benzimidate [825-60-5] or ethyl benzenecarboximidate.

Physical and Chemical Properties

The lower esters are colorless, volatile liquids, and many of them have pleasant odors (see Flavors and spices). Most of the esters of the higher saturated fatty acids are colorless and odorless crystalline solids; those of the very long chain acids and alcohols are hard, brittle, and lustrous crystalline solids and are generally referred to as waxes (qv). Most of these waxes are not known in pure form but only as complex mixtures comprising the naturally occurring waxes.

In general, the melting points of the esters of fatty acids are lower than those of the corresponding acids, and the boiling points of the methyl, ethyl, and propyl esters of the lower acids are less than those of the corresponding acids. With increasing chain length of the alcohol, the boiling points increase and ultimately become much higher than those of the corresponding acids. Since many of the monoesters are relatively stable to heat in the absence of moisture, they are generally distillable without decomposition. Therefore, they are often employed in processes of separation and identification of mixed fatty acids.

The data for the methyl and ethyl esters of the saturated fatty acid series are fairly complete and relatively well known, but data for the esters of the higher alcohols are often lacking. Data on boiling point, density, molar volume, viscosity, solubility, heat of combustion, and other physical constants of esters of aliphatic acids have been tabulated (3).

The esters are generally insoluble in water, but are soluble in various organic liquids. The lower esters are themselves good solvents for many organic compounds, including most liquids. They are especially good solvents for cellulose-based lacquers.

Reactions. The most familiar reactions of esters involve nucleophilic attack at the electron-deficient carbonyl carbon atom and can be generalized by the equation:

$$RCOR' + Nu^- \longrightarrow R\underset{OR'}{\underset{|}{C}}(O^-)Nu \longrightarrow RCNu(=O) + OR'^-$$

where $Nu = OH^-, OR''^-, :NH_3, :NHR''_2$, etc (4).

These reactions are frequently acid catalyzed; the acid polarizes the carbonyl group and increases its susceptibility to nucleophilic attack:

$$RCOR' + H^+ \rightleftharpoons R\underset{:Z}{\overset{OH}{\underset{|}{C}}}{}^{(+)}\!-OR'$$

Hydrolysis. Although hydrolysis of low molecular weight esters to their corresponding acids and alcohols occurs slowly even in the absence of added catalysts, for practical purposes it is usually carried out at elevated temperatures in the presence of either acidic or basic catalysts. The ease of hydrolysis decreases with increasing molecular weight and with the presence of bulky groups which impede attack of the nucleophile. Some esters are hydrolyzed only under quite drastic conditions.

Although esters have been observed to hydrolyze by six different mechanisms, depending upon the ester and the hydrolysis conditions, only two reaction pathways are important for most esters (4–5). The following two equations illustrate the most common paths for base- and acid-catalyzed hydrolysis, respectively.

$$HO^- + \underset{R}{\underset{|}{C}}(=O)OR' \underset{fast}{\overset{slow}{\rightleftharpoons}} \left[\underset{R}{\underset{|}{HO\underset{|}{C}OR'}}(O^-)\right] \underset{slow}{\overset{fast}{\rightleftharpoons}} \underset{R}{\underset{|}{HO\underset{|}{C}}}(=O) + OR'^- \overset{fast}{\longrightarrow} RCOO^- + R'OH$$

(1)

$$RCOR'(=O) + H^+ \rightleftharpoons R\underset{+}{\overset{OH}{\underset{|}{C}}}\!-OR' \underset{-H_2O}{\overset{+H_2O}{\rightleftharpoons}} R\underset{\underset{H\;\;\;H}{\overset{|}{O^+}\!\diagdown}}{\overset{OH}{\underset{|}{C}}}\!-OR' \rightleftharpoons R\underset{OH\;H}{\overset{OH}{\underset{|}{C}}}\!-\overset{+}{O}R' \underset{+R'OH}{\overset{-R'OH}{\rightleftharpoons}}$$

$$\left[R\underset{OH}{\overset{OH}{\underset{|}{C}}}\right]^+ \rightleftharpoons RCOH(=O) + H^+$$

Basic hydrolysis (saponification), which results in the formation of the carboxylate anion in the last step, is irreversible, as indicated in the first equation. Acidic hydrolysis in the second equation is an equilibrium reaction.

The mechanism represented by the first equation correctly predicts observed steric and electronic effects in ester hydrolysis. In the formation of intermediate (**1**), both the density of negative charge at the reaction center and the extent of crowding are increased. Electron-attracting substituents facilitate and bulky substituents retard hydrolysis.

The following saponification rates, relative to that of methyl acetate, illustrate the polar effect. The values shown represent the ratios of the saponification rate constants:

Ester	CAS Registry No.	$k_{ester}/k_{CH_3COOCH_3}$
CH_3COOCH_3	[79-20-9]	1.0
$(COOCH_3)_2$	[553-90-2]	170,000
$CH_2ClCOOCH_3$	[96-34-4]	761
$CH_3COOC_2H_5$	[141-78-6]	0.60
$CHCl_2COOCH_3$	[116-54-1]	16,000
$CH_3COCOOC_2H_5$	[617-35-6]	10,000

The following two series of saponification rates, relative to that of ethyl acetate, illustrate the effect of increasing steric hindrance in the acyl and alkyl substituents, respectively. The values shown represent the ratios of the saponification rate constants:

Ester	CAS Registry No.	$k_{ester}/k_{CH_3COOC_2H_5}$
$CH_3COOC_2H_5$	[141-78-6]	1.0
$C_2H_5COOC_2H_5$	[105-37-3]	0.47
$(CH_3)_2CHCOOC_2H_5$	[97-62-1]	0.10
$(CH_3)_3CCOOC_2H_5$	[3938-95-2]	0.01
$CH_3COOCH_2CH(CH_3)_2$	[110-19-0]	0.70
$CH_3COOCH_2C(CH_3)_3$	[926-41-0]	0.18
$CH_3COOCH_2C(C_2H_5)_3$	[10332-40-8]	0.03

Steric effects in acid-catalyzed hydrolysis are similar to those in base-catalyzed hydrolysis, but polar effects are much less important in acid-catalyzed reactions.

Ester hydrolysis probably has its greatest commercial importance in the manufacture of soap and in related industries, where the raw materials are high molecular weight esters, usually glycerides, of fatty acids. The discussion that follows is in that context (for more detailed information, see Soap).

Basic Hydrolysis. The hydrolysis of esters with alkali is one of the oldest known chemical reactions (6). The use of a soap made from oil and the ash of a plant, now known to be rich in potassium carbonate, for washing wool is referred to in clay tablets dating from about 2500 BC. The early Romans and Gauls treated fats with wood ashes to obtain a solution with detergent properties, but not until the time of Scheele (1779) and Chevreul (1813–1823) was the reaction understood. Today, this reaction is the basis of the soap industry. For complete saponification of esters, slightly more than the stoichiometric amount of alkali is required.

Acidic Hydrolysis. Hydrolysis of esters with dilute acid is slow, owing to the poor emulsifying power of acids (7). In the soap and related industries, this difficulty is circumvented by the use of sulfonated condensation products of oleic acid (9-octadecenoic acid) and aromatic hydrocarbons (Twitchell reagents) that apparently serve chiefly as emulsifying agents. This process is of no apparent importance in the United States at present.

Hydrolysis by Steam. High pressure steam or pressurized water at 185–300°C alone or with catalysts, such as zinc, magnesium, or calcium oxides, hydrolyzes fats directly to high-grade fatty acids. This method offers advantages to the soap industry over basic hydrolysis, and continuous commercial processes have been developed (8).

Hydrolysis by Enzymes. In this process, enzymes prepared from ground, centrifuged, and fermented castor beans are used to effect ester hydrolysis (9) (see Enzyme detergents). The reaction is slow but it has the advantage of taking place at atmospheric pressure and low temperature (ca 35°C). With these enzymes, the reaction is somewhat selective; esters of C_{12}–C_{18}-acids are hydrolyzed more readily than those of either longer or shorter chain length.

Alcoholysis, Acidolysis, and Ester–Ester Interchange (*Transesterification*). See Ester Interchange (see Esterification).

Ammonolysis and Aminolysis. In a reaction that is exactly analogous to their hydrolysis to alcohols and acids, esters react with ammonia to give alcohols and amides (10). The reaction is usually carried out in aqueous or alcoholic ammonia. Simple esters react with ammonia at a satisfactory rate at room temperature; higher esters may require elevated temperatures and pressures. Sometimes ester ammonolysis is carried out at low temperatures to avoid attack at another reactive group:

$$ClCH_2COOC_2H_5 + NH_3 \text{ (aq)} \xrightarrow{0-5°C} \underset{62-87\%}{ClCH_2CONH_2 + C_2H_5OH}$$

Other ammonia derivatives, such as primary and secondary amines, react analogously to give *N*-substituted amides:

$$RCOOR' + R''R'''NH \rightarrow RCONR''R''' + R'OH$$

Hydrazine reacts in the same way to give hydrazides:

$$RCOOR' + H_2NNH_2 \rightarrow RCONHNH_2 + R'OH$$

The mechanism of these reactions has been studied extensively (11).

When esters are passed with ammonia in the vapor phase over contact catalysts at 400–500°C, they are rapidly converted to nitriles. Thus Mailhe (12) in 1920 prepared oleonitrile (9-octadecenitrile) by passing a mixture of ammonia and methyl oleate (methyl 9-octadecenoate) over alumina at 500°C or over thoria at 480–490°C. These reactions can also be carried out in the liquid phase in the presence of suitable catalysts. This conversion of esters to nitriles is actually an ammonolysis followed by catalyzed dehydration of the amide:

$$RCOOR' + NH_3 \rightarrow RCONH_2 + ROH$$

$$RCONH_2 \xrightarrow[\text{catalyst}]{\Delta} RCN + H_2O$$

Fats (qv) [glycerol (1,2,3-propanetriol) esters of fatty acids] are converted to the nitriles of their corresponding acids in 75% yields when heated with ammonia at 285–335°C (13).

Reduction. Hydrogenation of esters over copper chromite catalyst at 200–300°C and 10–30 MPa (100–300 atm) reduces them to alcohols (14):

$$RCOOR' + 2 H_2 \rightarrow RCH_2OH + R'OH$$

When R is saturated and no halogen or sulfur is present, the reaction is smooth and almost quantitative; but when R is an aromatic nucleus, such as benzene or pyrrole, the reaction proceeds beyond the alcohol step, as in the reduction of ethyl benzoate [*93-89-0*] to toluene. This cleavage can be minimized by carrying out the reaction at low temperatures with a high ratio of catalyst to ester. An important commercial use of this reaction is the reduction of coconut oil and other fats to alcohols, chiefly dodecyl and decyl alcohols, which are used to prepare sulfated alcohol-type detergents. The reaction can also be used to prepare diols for use as polymer intermediates, as in the reduction of dimethyl 1,4-cyclohexanedicarboxylate [*94-60-0*] to 1,4-cyclohexanedimethanol.

The use of sodium in alcoholic solutions to reduce esters to alcohols was first reported by Bouveault and Blanc in 1904 (15). Since then it has been used on a large scale for the commercial production of alcohols from fatty esters, but the process has now been replaced by catalytic hydrogenation for the production of the commercially important saturated alcohols. An advantage of this process over the catalytic one is that unsaturated esters can be reduced to the unsaturated alcohols; in catalytic hydrogenation, unsaturated esters are usually reduced to the saturated alcohols.

For laboratory and small-scale commercial use in the production of fine chemicals, the Bouveault-Blanc reaction has been supplanted by reduction with complex metal hydrides (see Hydrides). Practically any ester can be reduced by this method, and alcohols of high purity are obtained in excellent yield. Carbon–carbon double bonds are normally not affected (16). The hazards associated with the use of highly reactive and flammable lithium aluminum hydride can be avoided by use of a complex hydride such as sodium bis(2-methoxyethoxy)aluminum hydride, which is commercially available as a 70% solution in benzene or toluene.

Reaction With Grignard Reagents. The reaction of esters with Grignard reagents is similar to that of ketones: the nucleophilic alkyl or aryl group of the Grignard reagent attacks the carbonyl carbon (see Grignard reaction). Expulsion of ROMgX gives a ketone which usually reacts further to give a tertiary alcohol, except in the special case of formates, which give secondary alcohols:

$$\underset{RCOR'}{\overset{O}{\|}} \xrightarrow{R''MgX} \left[\underset{RCR''}{\overset{O}{\|}} \right] \xrightarrow{R''MgX} \underset{\underset{R''}{|}}{\overset{OMgX}{\underset{|}{RCR''}}} \xrightarrow{H_2O} \underset{\underset{R''}{|}}{\overset{OH}{\underset{|}{RCR''}}}$$
$$+ R'OMgX$$

The overall yield in the preparation of triphenylmethanol from ethyl benzoate and phenylmagnesium bromide is 95%.

Acetoacetic Ester Condensation. In the presence of certain bases (usually sodium alkoxide), an ester having hydrogen on the α-carbon atom reacts with a second molecule of the same ester or with another ester (which may or may not have hydrogen on the α-carbon atom) to form a β-ketoester (17):

$$R_2CHCOOR' + R_3''CCOOR' \xrightarrow{R'O^-} R_3''CCCR_2COOR' + ROH$$

The special case of formic acid ester condensation with esters that contain an α-hydrogen gives aldehyde–esters:

$$R_2CHCOOR' + HCOR' \xrightarrow{R'O^-} HCCR_2COOR' + R'OH$$

These reactions are special cases of the Claisen reaction, which includes ketone–ester condensation to form 1,3-diketones. Mechanistically, they follow the pattern outlined earlier for ester reactions; the base serves to remove a proton from the ester bearing the α-hydrogen, giving an anion that adds to the carbonyl carbon of the second ester. Elimination of RO^- regenerates the catalyst and gives the β-ketoester:

$$R_2CHCOOR' + R'O^- \rightarrow R_2\bar{C}COOR' + R'OH$$

$$R_3''CCOOR' + R_2\bar{C}COOR' \rightarrow R_3''CCCR_2COOR' + R'O^-$$

Although mixed condensations between two esters, each of which bears an α-hydrogen, are of course possible, the result is a mixture of all possible products, and such reactions are usually of no preparative interest. The parent members of this class, the acetoacetic acid esters, are now made entirely by the reaction of diketene [674-82-8] (4-methylene-2-oxetanone) with the appropriate alcohol, rather than by the self-condensation of esters of acetic acid (see Ketene and acetoacetic acid and derivatives).

Preparation of Acyloins. When aliphatic esters are allowed to react with metallic sodium in inert solvents, acyloins (α-hydroxyketones) are formed:

$$2\ RCOOR' + 4\ Na \xrightarrow{-2\ NaOR'} \begin{array}{c} RCONa \\ \| \\ RCONa \end{array} \xrightarrow[-2\ NaOH]{H_2O} \left[\begin{array}{c} RCOH \\ \| \\ RCOH \end{array} \right] \rightarrow \begin{array}{c} RC=O \\ | \\ RCHOH \end{array}$$

Much use has been made of this reaction for the construction of large-ring compounds by intramolecular condensation of esters of long-chain dicarboxylic acids (18).

Pyrolysis. The pyrolysis of simple esters of the formula $RCOOCR'R''CHR_2'''$ to form the free acid and an olefin is a general reaction that is used for producing olefins (19). The pyrolysis is generally carried out at 300–500°C over an inert heat-transfer agent such as Pyrex glass or 96% silica glass chips. Esters of tertiary alcohols are pyrolyzed more readily than esters of secondary alcohols, and esters of primary alcohols are the most difficult to pyrolyze. Some of the higher boiling esters of tertiary alcohols cannot be distilled without decomposition; *tert*-pentyl acetate [625-16-1] (1,1-dimethylpropyl acetate) decomposes to pentenes and acetic acid below 200°C. The reaction has been long believed to be a one-step, gas-phase reaction with a cyclic transition state:

The double bond does not move along the carbon chain, but mixtures of isomeric olefins are produced by pyrolysis of unsymmetrical secondary or tertiary esters that contain more than one carbon atom capable of losing hydrogen by this mechanism. Recently this mechanism has been criticized for its failure to explain the detailed product composition that results from pyrolysis of certain esters, and alternative mechanisms have been proposed involving surface-catalyzed reactions (20).

Methyl esters and esters of the formula $RCOOCH_2CR_3$ that do not contain hydrogen atoms capable of elimination by the mechanism given above must be heated to much higher temperatures before they decompose, and their pyrolysis products are complex. Pyrolysis of these esters is not a synthetically useful reaction.

Miscellaneous Reactions. Ketones can be obtained in substantial yields from fatty acid esters as well as from the fatty acids by heating the esters or acids in the presence of certain metals or their oxides (21):

$$2\ RCOOCH_2CH_2R' \rightarrow RCOR + CO_2 + H_2O + 2\ R'CH{=}CH_2$$

For example, ethyl laurate [106-33-2] (ethyl dodecanoate) in the vapor phase over a thoria-gel catalyst at 300°C gives a 92.5% yield of laurone (12-tricosanone), and ethyl undecenoate [36118-46-4] gives an 86% yield of didecenyl ketone (1,20-heneicosadien-11-one) (22). Other vapor-phase catalysts are manganese chromite and zinc chromite. When the reaction is run in the liquid phase, the yields are lower, and metallic iron is probably the best catalyst. Aluminum, manganese, and the oxides of iron, silicon, copper, zinc, titanium, aluminum, and magnesium have all been reported as catalysts for this reaction to give ketones (23). In most cases, these reactions can be thought of as resulting from decomposition of the ester on the catalyst to give metal carboxylate and alcohol (or olefin) followed by ketonization of the metal carboxylate. The advantage of ester ketonization over ketonization of the corresponding carboxylic acids, which gives the same ketones, is that the esters are more volatile; hence, those of higher acids are more readily brought into contact with the catalyst than are the parent acids.

Esters can be used as acylating agents in the Friedel-Crafts ketone synthesis although, except in special cases, the more reactive acyl halides or anhydrides are usually chosen (24) (see Friedel-Crafts reactions).

Certain esters, particularly those of disubstituted acetic acids, may be dehydrogenated to corresponding α,β-unsaturated esters. Methyl isobutyrate [547-63-7] (methyl 2-methylpropanoate), eg, gives methyl methacrylate [80-62-6] (methyl 2-methylpropenoate) (25):

$$\underset{\substack{|\\CH_3CHCOOCH_3}}{CH_3} \xrightarrow{\text{catalyst}} \underset{\substack{|\\CH_2{=}CCOOCH_3}}{CH_3} + H_2$$

Occurrence and Preparation

Currently, most of the simple esters used commercially are of synthetic origin, although a number of them occur in nature. Some of the naturally occurring esters, other than fats and waxes, and some of their sources are as follows (26): ethyl acetate in many wines, brandy, wine vinegar, and some fruits, such as pineapples; amyl (pentyl) acetate in apples, bananas, and other fruits; geranyl formate (3,7-dimethyl-2,6-octadienyl formate), citronellyl formate, and acetate in geranium oil; terpinyl acetate in cypress oil; bornyl acetate [76-49-3] (1,7,7-trimethylbicyclo[2.2.1]hept-2-yl acetate) in pine-needle oil; geranyl acetate (3,7-dimethyl-2,6-octadienyl acetate) in lemon grass oil; menthyl acetate [89-48-5] (2-methyl-5(1-methylethyl)cyclohexyl acetate) in peppermint oil; benzyl acetate (phenylmethyl acetate) in jasmine, hyacinth, and gardenia; methyl benzoate [93-58-3] in clove oil; methyl salicylate [119-37-8] (methyl 2-hydroxybenzoate) in the oils of wintergreen and sweet birch; ethyl [103-36-6], benzyl (phenylmethyl), and cinnamyl [122-69-0] (3-phenyl-2-propenyl) cinnamates (3-phenyl-2-propenoates) in oil of styrax; and methyl anthranilate (methyl 2-aminobenzoate) in jonquil, tuberoses, ylang-ylang, jasmine, and mandarin-leaf oil (see Oils, essential). Most of these naturally occurring esters have very pleasant odors, and either they or their synthetic counterparts are used in the confectionery, beverage, perfume, cosmetic, and soap industries.

Recovery of naturally occurring esters is accomplished by steam distillation, extraction, or pressing, or by a combination of these processes. Synthetic esters are generally prepared by reaction of an alcohol with an organic acid in the presence of a catalyst such as sulfuric acid or p-toluenesulfonic acid. Ion-exchange resins of the sulfonic acid type can also be used, and an azeotroping agent such as benzene or toluene can be used to remove water and force the reaction to completion (see Esterification).

Analysis, Specifications, and Standards

The analysis of esters usually includes the determination of such physical properties as boiling range, melting point or freezing point, specific gravity, refractive index, amount of nonvolatile residue or ash, and color; and it usually includes such chemical properties as free acidity and actual ester content. The physical properties are determined by well-known methods. Free acidity is determined by titrating a sample in either water or an alcohol-water mixture with standard alkali. The ester content is generally denoted by the saponification number (SN), which is the number of milligrams of potassium hydroxide required to saponify one gram of the ester, or by the saponification equivalent (SE), which is the number of milliequivalents of potassium hydroxide required to saponify one gram of the ester. The saponification number and the saponification equivalent are related by the following formula:

$$SE = \frac{56.1 \times 1000}{SN} \qquad 56.1 = \text{equiv wt of KOH}$$

The RCOOR′ group of simple esters is comparatively easy to identify from characteristic bands in the ir spectrum. The carbonyl group absorbs strongly at 1750–1735 cm^{-1}, and a second strong band, characteristic of the —C—O—C— linkage, is present in the region between 1300–1100 cm^{-1}. The location of these bands gives an indication of the type of acid from which the ester is derived (27).

Although the advent of ir, and nmr spectroscopy, and mass spectrometry has rendered the method obsolete, esters were characterized in the past by the preparation of solid derivatives of the component acids and alcohols. Hydrazides are particularly useful for characterization of the acid moiety, since they can be made by heating the ester with hydrazine as described previously.

An excellent review of the various methods of ester analysis is ref. 28.

The specifications of individual commercial esters vary widely. Typical specifications of a number of esters, as taken from the manufacturers' trade publications, are given in Table 1. These specifications illustrate the tolerances commonly allowed.

Health and Safety Factors

No general statement can be made about the toxicity of esters (for individual toxicities of many esters, see under Esters of Aliphatic Acids). The degree of toxicity covers a wide range. Many are highly volatile and can act as asphyxiants or narcotics. Skin absorption, as well as inhalation, can be a significant hazard with esters that are volatile and have high solvent action. Esters generally hydrolyze upon contact with moisture, so that the toxicity of an ester is generally related to that of its hydrolysis products (29). Because of the generally high solvent power of esters for fats and oils, prolonged or repeated contact of the skin with esters, particularly those of low molecular weight, can produce drying and irritation. Inhalation and exposure of the eyes to high concentrations of vapor can cause irritation of the mucous membranes. Regulations for rail shipment require that containers for all compounds with a flash point below 37.8°C be marked with a DOT shipping name and be labeled and/or placarded with the appropriate "Flammable" label.

Uses

Solvents and Plasticizers. The greatest uses of esters are in the solvent (30) and plasticizer (31–32) fields. The lower esters are used in lacquers, paints (qv), and varnishes, and the higher esters are used chiefly as plasticizers (qv). Table 2 shows the production figures for carboxylic acid esters used in plasticizers. Table 2 also gives data for several acetates (under "Miscellaneous esters") that are indicative of the amounts of esters used in solvents (qv) (33). The United States production of esters for use in surface coatings in 1973 was ca 130,000 metric tons, valued at $40,000,000 (34).

Resins, Plastics, and Coatings. Many polymeric materials in commercial use are based on esters (see Coatings, industrial). These include vinyl polymers made from such unsaturated esters as the acrylates (2-propenoates) (see Acrylic acid), methacrylates (2-methyl-2-propenoates) (see Methacrylic acid), vinyl (ethenyl) acetate (see Vinyl polymers), and their homologues; alkyd resins (qv), which are essentially cross-linked polyesters prepared from polyhydric alcohols and dibasic acids; and polyester resins (qv) and plastics (qv). The last are usually made by transesterification, starting with dimethyl or diphenyl esters of dibasic acids, such as terephthalic or carbonic acids, and dihydric alcohols, such as ethylene glycol (1,2-ethanediol) or 1,4-cyclohexanedimethanol. However, particularly in the case of poly(ethylene terephthalate) the trend is to direct esterification, rather than transesterification. These polymers are thermoplastic materials that are useful as films, fibers, molding plastics, and in some cases as hot-melt adhesives and textile sizes. The properties of these materials can be varied widely by changing either the dibasic acid or the glycol. Incorporation of an unsaturated acid, such as maleic acid ((Z)-butenedioic acid), gives a material that can be cross-linked, and which is, therefore, thermosetting.

Table 1. Typical Specifications of Representative Commercial Esters

Ester	Color, max[a]	n_D^{20}	d_{20}^{20}	Distillation range, °C (101.3 kPa or 1 atm)	Freezing point, °C	Flash point, °C[b]	Autoignition temperature, °C[c]	Flammable limits in air Lower[d]	Flammable limits in air Upper[d]	NFPA Std. No. 30 class[e]
ethyl acetate	10[f]	1.3693[f]	0.884[f]	72.0–78.0[f]	−84[f]	−1[f]	485	2.02 (38)	10.7 (38)	IA
vinyl acetate	10[g]	1.3712[g]	0.901[g]	75.5–78.0[g]	−84[g]	−2[g]				IA
	5		0.9335–0.9345			−8				IA
propyl acetate	15	1.3844	0.885	99.0–103.0	−93	14	457	1.71 (38)	7.95 (93)	IB
isopropyl acetate	10	1.3772	0.872	85.0–90.0	−62	6	479	1.76 (38)	7.2 (41)	IB
butyl acetate	10	1.3497	0.883	122.0–128.0	−74	29	407	1.38 (38)		IC
isobutyl acetate	10	1.3997	0.870	112.0–119.0	−99	24	427	1.27 (93)	7.5 (93)	IB
2-ethylhexyl acetate	15	1.4103	0.872	192.0–205.0	−93	79	268	0.76 (93)	8.14 (149)	IIIA
ethylene glycol diacetate	15	1.4159	1.107	187.0–193.0	−42	99[h]	482	1.6 (135)	8.4 (154)	IIIA
2-ethoxyethyl acetate	15	1.4030	0.973	150.0–160.0	−61	59	382	1.24 (93)	12.7 (135)	II
2-butoxyethyl acetate	15	1.4200	0.94	186.0–194.0	−64	81	340	0.88 (93)	8.54 (135)	IIIA
2-(2-ethoxyethoxy)ethyl acetate	15	1.4230	1.011	214.0–221.0	−25	107[h]	360	0.98 (135)	19.4 (185)	IIIB
2-(2-butoxyethoxy)ethyl acetate	15	1.4265	0.980	235.0–250.0	−32	116[h]	349	0.76 (135)	10.7 (203)	IIIB
glyceryl triacetate	5	1.4296	1.160	258.0[i]	3.2[j]	153[h]	432	1.05 (189)		IIIB
glyceryl tripropionate		1.4314[k]	1.09[l]	285.0[i]	−58	167	421	0.8 (186)		IIIB
methyl acrylate	10		0.9561–0.9576	79.8–80.3		10				IA
ethyl acrylate	10		0.9227–0.9242	98.8–99.8		10				IA
butyl acrylate	10		0.9001–0.9016	145.7–148.0		49				II
2-ethylhexyl acrylate	10		0.8855–0.8870	214.8–218.0		91[h]				IIIA
isobutyl isobutyrate	15	1.3990	0.855	144.0–151.0	−80	38	432	0.96 (93)	7.59 (93)	II
2,2,4-trimethyl-1,3-pentanediol diisobutyrate		1.4300[k]	0.941	280[i]	−70	143[h]	424	0.48 (172)		IIIB

Compound	Color[a,b]	n_D^{25}[k]	Density	bp (°C)[i]	mp/pour (°C)	Flash point (°C)	Fire point (°C)	Vapor pressure[d]	Viscosity	NFPA[e]
methyl stearate	see text for properties									
butyl stearate	see text for properties									
glyceryl monostearate					70–78	230[h]				
bis(2-ethylhexyl) adipate	20	1.4472[k]	0.927	417[i]	<–70	206[h]	377	0.38 (242)		IIIB
dimethyl phthalate	5	1.513[k]	1.192	284[i]	–1	157[h]	490	0.94 (181)		IIIB
diethyl phthalate	10	1.4990[k]	1.120	298[i]	<–50	161[h]	457	0.75 (187)		IIIB
dibutyl phthalate	15	1.4905[k]	1.048	340[i]	–35	190[h]	404	0.47 (236)		IIIB
bis(2-ethylhexyl) phthalate	20–25	1.4836[k]	0.9852	384[i]	–50	216[h]	391	0.28 (246)		IIIB
dimethyl terephthalate	<15[l]			288[i]	140.6	153[m]	519			
bis(2-ethylhexyl) terephthalate	15	1.4867[k]	0.9835	383[i]	–48	238	399			IIIB
bis(2-ethylhexyl) trimellitate	40	1.4832[k]	0.989	600[i]	–38	263	410	0.26	2.5	IIIB

[a] Pt–Co scale.
[b] Tag open cup unless otherwise noted.
[c] ASTM D-2155.
[d] Temperature (°C) at which measured is shown in parentheses.
[e] NFPA (National Fire Prevention Association) Standard No. 30 classifications (flash point and boiling point limits, °C, respectively): IA <23, <38; IB <23, ≥38; IC, ≥23, <38, –; II, ≥38, <60, –; IIIA, ≥60, <93, –; IIIB, ≥93, –.
[f] 85% purity.
[g] 99% purity.
[h] Cleveland open cup.
[i] Boiling point of pure ester.
[j] Supercools to ca –70°C.
[k] n_D^{25}.
[l] Melt color.
[m] Micro-Cleveland open cup.

Table 2. United States Production and Sales of Carboxylic Esters, 1976 [a]

Material	CAS Registry No.	Production, metric tons	Sales Quantity, t	Sales Value, $1000	Unit value, $/kg
Plasticizers					
phthalic anhydride esters, total		473,071	424,824	293,018	0.66
dibutyl phthalate	[84-74-2]	6,215	6,658	5,491	0.82
diethyl phthalate	[84-66-2]	7,319	5,351	4,928	0.93
diisodecyl phthalate	[26761-40-0]	64,923	47,063	30,071	0.62
dimethyl phthalate	[131-11-3]	4,008	3,763	3,053	0.82
dioctyl phthalates, total		142,408	178,470	102,989	0.57
bis(2-ethylhexyl) phthalate	[117-81-7]	134,600	172,500	99,266	0.57
other dioctyl phthalates		7,808	5,970	3,723	0.62
bis(tridecyl) phthalate	[119-06-2]	4,750	6,492	5,075	0.79
hexyl decyl phthalate	[25724-58-7]	8,999	3,960	2,412	0.62
all other phthalic anhydride esters		234,449	193,476	138,999	0.73
trimellitic (1,2,4-benzenetricarboxylic acid) esters, total		10,469	7,758	8,293	1.06
triisooctyl trimellitate	[27251-75-8]	1,134	428	463	1.08
trioctyl trimellitate	[89-04-3]	4,209	3,393	3,558	1.06
all other trimellitic acid esters		5,127	3,938	4,272	1.08
adipic (hexanedioic) acid esters, total		24,511	25,577	27,016	1.04
bis(2-ethylhexyl) adipate	[103-23-1]	17,823	17,100	16,373	0.95
diisodecyl adipate	[1330-86-5]	928	607	904	1.48
octyl decyl adipate	[110-29-2]	1,527			
all other adipic acid esters		4,233	7,870	9,739	1.21
complex linear polyesters and polymeric plasticizers, total		23,985	18,963	29,473	1.57
adipic acid type		15,114	11,344	17,270	1.52
all other		8,871	7,619	12,203	1.61
epoxidized esters, total		53,249	49,477	49,953	1.01
epoxidized linseed oils		2,885	2,608	4,117	1.59
epoxidized soya oils		41,476	38,037	36,604	0.97
all other epoxidized esters		8,888	8,832	9,232	1.04
isopropyl myristate (tetradecanoate)	[110-27-0]	1,527	1,390	1,595	1.15
oleic (9-octadecenoic) acid esters, total		4,507	4,291	4,165	0.97
butyl oleate	[142-77-8]	804	794	805	1.01
methyl oleate	[112-62-9]	1,390	1,310	992	0.75
propyl oleates (including propyl oleate and isopropyl oleate)		259	203	165	0.82
all other oleic acid esters		2,054	1,982	2,203	0.10
sebacic (decanedioic) acid esters		773	337	878	2.60
stearic (octadecanoic) acid esters, total		5,492	5,314	4,632	0.88
butyl stearate	[123-95-5]	3,043	3,039	2,278	0.75
isobutyl stearate	[646-13-9]	694			
all other stearic acid esters		1,755	2,275	2,354	1.04
Surface-active agents					
carboxylic acid esters, total		100,916	82,616	105,397	1.28
anhydrosorbitol esters		11,981	7,058	10,715	1.52
diethylene glycol esters, total		625	589	755	1.28
diethylene glycol distearate	[109-30-8]	215	185	254	1.37
diethylene glycol monostearate	[106-11-6]	117	111	141	1.28
all other		293	293	360	1.23
ethoxylated anhydrosorbitol esters, total		12,210	12,093	15,970	1.37
ethoxylated anhydrosorbitol monostearate	[9005-67-8]	3,827	3,749	4,821	1.28
ethoxylated anhydrosorbitol monooleate	[9005-65-6]	2,281	2,248	3,365	1.50

Table 2 (*continued*)

Material	CAS Registry No.	Production, metric tons	Sales Quantity, t	Sales Value, $1000	Unit value, $/kg
all other		6,102	5,643	7,784	1.37
ethylene glycol esters		1,390	1,343	1,409	1.06
glycerol (1,2,3-propanetriol) esters, total		*38,819*	*33,993*	*38,783*	*1.15*
complex glycerol esters		1,071	1,169	1,824	1.57
glycerol esters of chemically defined acids, total		*11,793*	*11,596*	*12,011*	*1.04*
glycerol monolaurate (monododecanoate)	[27215-38-9]	27	28	53	1.92
glycerol monooleate [mono(9-octadecenoate)]	[25496-72-4]	1,709	1,734	2,489	1.43
glycerol monostearate (monooctadecanoate)	[1319-95-5]	9,719	9,482	8,662	0.90
all other		338	352	807	2.29
glycerol esters of mixed acids, total		25,955	21,228	24,948	1.17
glycerol monoester of hydrogenated cottonseed oil acids	[61789-07-0]	1,289			
glycerol monoester of coconut oil acids	[61789-05-7]	88	88	146	1.63
glycerol monoester of hydrogenated soybean oil acids	[61789-08-0]	3,842	3,045	3,956	1.30
glycerol monoester of lard acids	[61789-10-4]	1,368	912	1,037	1.15
all other		19,368	17,183	19,809	1.15
natural fats and oils, alkoxylated, total		*6,289*	*5,448*	*6,124*	*1.12*
castor oil, ethoxylated	[61791-12-6]	3,689	3,044	3,729	1.23
lanolin ethoxylated	[61790-81-6]	624	501	570	1.15
all other		1,976	1,903	1,825	0.95
poly(ethylene glycol) esters, total		*19,242*	*14,948*	*16,729*	*1.12*
poly(ethylene glycol) esters of chemically defined acids, total		10,650	8,472	12,047	1.43
poly(ethylene glycol) dilaurate	[9005-02-1]	451	440	684	1.56
poly(ethylene glycol) dioleate	[9005-07-6]	1,459	590	811	1.37
poly(ethylene glycol) distearate	[9005-08-7]	1,620	1,556	2,117	1.37
poly(ethylene glycol) monolaurate	[9004-81-3]	1,623	1,579	2,288	1.45
poly(ethylene glycol) monooleate	[9004-96-0]	1,145	917	1,116	1.22
poly(ethylene glycol) monostearate	[9004-99-3]	3,709	2,827	4,212	1.50
all other		643	563	819	1.46
poly(ethylene glycol) esters of mixed acids		8,592	6,476	4,682	0.73
propanediol esters, total		*1,839*	*1,524*	*2,365*	*1.54*
1,2-propanediol monolaurate (monododecanoate)	[27194-74-7]	11	11	30	2.62
1,2-propanediol monostearate (monooctadecanoate)	[1323-39-3]	1,293	1,361	1,940	1.43
all other		535	152	395	2.60
other carboxylic acid esters		8,523	6,074	12,547	2.07
Flavor and perfume materials					
anisyl (4-methoxyphenyl) acetate	[1200-06-2]	6	5	81	14.55
benzyl acetate	[140-11-4]	828	986	2,009	2.05
benzyl cinnamate	[103-41-3]		5	62	13.78
benzyl propionate	[122-63-4]	12	12	40	3.35
benzyl salicylate	[118-58-1]	669	632	2,350	3.73
cinnamyl acetate	[103-54-8]	9	7	66	9.99
cinnamyl anthranilate	[87-29-6]		1	15	20.08

Table 2 (continued)

Material	CAS Registry No.	Production, metric tons	Sales Quantity, t	Sales Value, $1000	Unit value, $/kg
isobutyl phenylacetate	[102-13-6]	14	13	65	4.98
isobutyl salicylate	[87-19-4]		6	20	3.37
isopentyl salicylate	[87-20-7]	427	348	976	2.80
methyl anthranilate	[134-20-3]	128	115	449	3.90
methyl phenylacetate	[101-41-7]	16	11	61	5.73
phenethyl acetate	[103-45-7]		36	219	6.11
phenethyl isobutyrate	[103-48-0]	5	3	32	9.77
2-phenethyl phenylacetate	[102-20-5]	14	9	100	11.13
2-phenoxyethyl isobutyrate	[103-60-6]	27	22	120	5.51
cedryl acetate	[77-54-3]	145	104	973	9.37
guaiac wood acetate	[134-28-1]	18	15	167	10.69
α-terpinyl acetate	[80-26-2]	445	420	921	2.18
vetivenyl acetate	[117-98-6]	11	6	569	97.42
allyl hexanoate	[123-68-2]	17	13	71	5.58
butylbutyryl lactate	[7492-70-8]	27	26	228	8.62
citronellyl acetate	[150-84-5]	20	13	105	7.91
citronellyl formate	[105-85-1]	14	10	96	10.52
citronellyl isobutyrate	[97-89-2]		2	27	13.76
citronellyl propionate	[141-14-0]	2			
3,7-dimethyl-cis-2,6-octadien-1-ol (neryl acetate)	[141-12-8]	9	5	53	11.18
ethyl butyrate	[105-54-4]	253	172	362	2.09
ethyl heptanoate	[106-30-9]	3	4	24	5.71
ethyl hexanoate (ethyl caproate)	[123-66-0]	6	4	18	5.29
ethyl myristate	[124-06-1]	10	9	54	6.02
ethyl nonanoate	[123-29-5]		3	23	7.74
ethyl octanoate	[106-32-1]	2	1	12	8.11
ethyl propionate	[105-37-3]	68	56	124	2.23
geranyl acetate	[105-87-3]	54	44	336	7.56
geranyl formate	[105-86-2]		7	77	11.51
geranyl propionate	[105-90-8]	1			
isopentyl butyrate	[106-27-4]	38	36	97	2.67
isopentyl formate	[110-45-2]	4	3	16	4.85
isopentyl isovalerate	[659-70-1]	7			
Miscellaneous esters					
esters of monohydric alcohols, total		1,659,937	887,679	546,997	0.62
butyl acetate, unmixed	[123-86-4]	51,033	44,638	22,375	0.51
butyl acrylate	[141-32-2]	93,116	52,015	38,262	0.73
dibutyl maleate	[105-76-0]	3,585	3,130	2,627	0.84
bis(2-ethylhexyl) maleate	[142-16-5]	391	171	154	0.90
dilauryl-3,3'-thiodipropionate	[123-28-4]	1,001	885	2,198	2.49
ethyl acetate (85%)	[141-78-6]	97,773	87,845	36,380	0.42
ethyl acrylate	[140-88-5]	133,870	61,446	37,927	0.62
2-ethylhexyl acrylate	[103-11-7]	19,971	19,126	16,190	0.84
propyl acetate	[109-60-4]	19,419	18,546	9,923	0.53
vinyl acetate	[108-05-4]	671,617	322,742	129,054	0.40
all other		568,161	277,135	251,907	0.90
polyhydric alcohol esters, total		47,230	44,302	42,785	0.97
ethylene glycol diacrylate	[2274-11-5]	385			
trimethylolpropane triacrylate	[15625-89-5]	226			
all other		46,619	44,311	42,785	0.97

[a] Ref. 33.

Lubricants. Esters, in the form of natural fats, oils, and waxes have been used as lubricants since ancient times (see Lubrication and lubricants). Animal and vegetable fats and oils have largely been supplanted by petroleum hydrocarbons, but petroleum lubricants are not suitable for turboprop and turbojet aircraft engines that must operate under a wide range of climatic conditions and at very high engine temperatures. Such aliphatic esters as bis(2-ethylhexyl) sebacate [bis(2-ethylhexyl) decanedioate] [122-62-3] are widely used as base oils for lubricating turbojet engines because of their relatively low rate of change of viscosity with temperature and their wide liquid range (35).

Perfumes, Flavors, Cosmetics, and Soap. Compared with the tonnage of esters used in solvents and plasticizers, the tonnage of esters used in improving odors and flavors (qv) is small but nevertheless very important economically and esthetically. The soap (qv) industry, of course, is still a very large consumer of esters in the form of fats and oils from tallow and vegetable oils. Table 2 includes United States production and sales of flavor and perfume (qv) materials (see also Cosmetics).

Surface-Active Agents. Polyol or poly(ethylene oxide) esters of long-chain fatty acids are nonionic surfactants used in foods, pharmaceuticals, cosmetics, textiles, lubricants, cleaning compounds, etc (see Surfactants and detersive systems). Those that are most widely used are included in Table 2.

Medicinals. Although the ester group itself apparently is inert physiologically, esters are used widely in pharmaceuticals (36–37). In general, esterification of a physiologically active alcohol or phenol with an aliphatic acid detoxifies it by decreasing the concentration of active compound present. The active compound is released gradually in the body by hydrolysis of the ester. An example of detoxification is acetylation of salicylic acid to give acetylsalicylic acid [2-(acetyloxy)benzoic acid] [50-78-2], the common analgesic, aspirin. Although sodium salicylate itself is an effective analgesic, it is more prone to cause stomach irritation than is the acetylated compound. Benzocaine [94-09-7], a local anesthetic, is the ethyl ester of 4-aminobenzoic acid. Heroin [561-27-3], an opiate, is the diacetate of morphine; in this case, acetylation of the parent alcohol has increased and somewhat altered the physiological action. Salicylates, some benzoates, and esters of the fatty acids in chaulmoogra oil are also important medicinally (see Pharmaceuticals, controlled release).

Esters of Aliphatic Acids

Table 1 lists the physical properties of some representative esters that are sold in large commercial quantities. These data are taken from manufacturers' literature and represent typical values for the commercial products. An idea of the commercial value of these and other esters can be obtained from Table 2. In Table 3 are listed the physical properties of the pure forms of the esters shown in Table 1, together with those of some of their homologues. Some of the uses and toxicological data for a few of the more important esters of aliphatic acids are listed in the following section. Detailed data on a large number of solvents, including the following esters and many others, have been compiled (37), as well as data on toxicity and handling precautions for these and a large number of other industrial materials (29).

Formic Acid Esters. *Methyl Formate.* Methyl formate, $HCOOCH_3$, has a pleasant, ethereal, nonresidual odor. Its uses are similar to those of ethyl formate. Methyl formate can cause optic neuritis and irritation to the conjunctiva. Exposure of guinea

Table 3. Properties of Pure Esters of Aliphatic Acids

Ester	CAS Registry No.	Mol wt	n_D^{20}	d_{20}^{20}	Boiling point, °C[a]	Freezing point, °C	Flash point, °C[b]
methyl formate	[107-31-3]	60.05	1.344	0.975	32.0	−99.8	−32
ethyl formate	[109-94-4]	74.08	1.3598	0.9236[c]	54.3	−79	−20
methyl acetate	[79-20-9]	74.08	1.3594	0.9330	57	−98.1	−10
ethyl acetate	[141-78-6]	88.10	1.3723	0.9003	77.06	−83.6	4
vinyl acetate	[108-05-4]	86.10	1.3959	0.9317	72.2–72.3	−93.2	−29[d]
propyl acetate	[109-60-4]	102.13	1.3844	0.887	101.6	−92.5	14
isopropyl acetate	[108-21-4]	102.13	1.3773	0.8718	90	−73.4	6
butyl acetate	[123-86-4]	116.16	1.3951	0.882	126	−73.5	33[e]
isobutyl acetate	[110-19-0]	116.16	1.3902	0.8712	117.2	−98.6	18
pentyl acetate	[628-63-7]	130.18	1.4023	0.8756	149.25	−70.8	25
2-ethylhexyl acetate	[103-09-3]	172.26	1.4204	0.8734[c]	199.3	−93	82[f]
ethylene glycol diacetate	[111-55-7]	146.15	1.415	1.128[g]	191	−31	104
2-ethoxyethyl acetate	[111-15-9]	132.16	1.4058	0.9749	156.4	−61.7	49
2-butoxyethyl acetate	[112-07-2]	160.21		0.943	187.8		82
2-(2-ethoxyethoxy)ethyl acetate	[111-90-0]	176.21		1.0114[c]	217.4	−25	107
2-(2-butoxyethoxy)-ethyl acetate	[112-34-5]	204.27		0.981[c]	247	−32.2	115[f]
benzyl acetate	[140-11-4]		1.5232	1.0550	215.5	−51.5	102
glyceryl triacetate	[102-76-1]	218.23		1.161	258	−78	138
methyl propionate	[554-12-1]	88.10	1.3775	0.9150	79.8	−87.5	−2
ethyl propionate	[105-37-3]	102.13	1.3839	0.8917	99.10	−72.6	12
glyceryl tripropionate	[139-45-7]	260.3	1.4318	1.100[h]	175–176 (2.67 kPa, 20 mm Hg)		
methyl acrylate	[96-33-3]	86.09	1.4040	0.9535	80.5	<−75	−3
ethyl acrylate	[140-88-5]	100.11	1.4068	0.9234	99.8	<−72	15
butyl acrylate	[141-32-2]	128.17	1.4185	0.898	69 (6.7 kPa, 50 mm Hg)	−64.6	49
2-ethylhexyl acrylate	[103-11-7]	184.28		0.8869[c]	130 (6.7 kPa, 50 mm Hg)	−90	82
methyl butyrate	[623-42-7]	102.13	1.3878	0.8984	102.3	−84.8	14
ethyl butyrate	[105-54-4]	116.16	1.4000	0.8785	121.6	−100.8	25
butyl butyrate	[109-21-7]	144.22	1.4075	0.8709	166.6	−91.5	53
methyl isobutyrate	[548-63-7]	102.13	1.3840	0.891	92.6	−84.7	
ethyl isobutyrate	[97-62-1]	116.16	1.3903	0.870	110–111	−88	
isobutyl isobutyrate	[97-85-8]	144.22	1.3999	0.875[g]	148.7	−80.7	
methyl stearate	[112-61-8]	298.50			215 (2.0 kPa, 15 mm Hg)	38	153
ethyl stearate	[111-61-5]	312.52		1.057	213–215 (2.0 kPa, 15 mm Hg)	33.7	
butyl stearate	[123-95-5]	340.58		0.855	343	27.5	160
dodecyl stearate	[5303-25-3]	440.80					
hexadecyl stearate	[1190-63-2]	496.91				57	
dimethyl adipate	[627-93-0]	174.20	1.4283	1.0600	115 (1.73 kPa, 13 mm	10.3	

Table 3 (*continued*)

Ester	CAS Registry No.	Mol wt	n_D^{20}	d_{20}^{20}	Boiling point, °C[a]	Freezing point, °C	Flash point, °C[b]
diethyl adipate	[141-28-6]	202.25	1.4372	1.0076	245 (Hg)	−19.8	
bis(2-ethylhexyl) adipate	[103-23-1]	370.58		0.9268[c]	214 (0.67 kPa, 5 mm Hg)	−60	196

[a] At 101.3 kPa (1 atm) unless otherwise noted.
[b] Tag closed cup (ASTM D56-52) unless otherwise noted.
[c] d_{20}^{20}.
[d] Cleveland open cup.
[e] Tag open cup.
[f] Open cup (unspecified).
[g] d_4^0.
[h] d_{18}^{20}.

pigs to 5% concentrations of the vapor in air proved lethal within 20–30 min, whereas 1.5–2.5% concentrations were dangerous in 30–60 min. The maximum concentration tolerated without serious disturbance was 0.5%, and the maximum concentration tolerated for several hours without serious disturbance was 0.15–0.20 vol % in air (29).

Ethyl Formate. Ethyl formate, $HCOOC_2H_5$, has a pleasant, nonresidual odor. It is used like the methyl ester as a fumigant and larvicide, as an intermediate in the synthesis of vitamin B, and in the formulation of synthetic flavors (see Formic acid).

Acetic Acid Esters. Methyl Acetate. Methyl acetate, CH_3COOCH_3, is a colorless liquid with a pleasant odor; it is used as a solvent for cellulose nitrate, cellulose acetate, and many resins and oils, and in the manufacture of artificial leather (see Acetic acid). Methyl acetate is a narcotic but less so than its higher homologues. It has an irritating effect upon the mucous membranes of the eyes and the upper respiratory tract and, in this respect, its action is stronger than that of the higher members of the series. The irritant concentration is about 10,000 ppm. Signs and symptoms include irritation and burning of the eyes, lacrimation, labored respiration (dyspnea), palpitation of the heart, and complaints of depression or dizziness (29). The threshold limit value (TLV) is 200 ppm (610 mg/m^3) (38).

Ethyl Acetate. Ethyl acetate, $CH_3COOC_2H_5$, a stable, colorless, flammable liquid with a pleasant odor, is used widely in formulating printing inks, adhesives, and lacquers; it is an effective solvent for many types of resins. In addition to protective coating applications, ethyl acetate is used extensively as a cellulose nitrate solvent in the manufacture of products such as artificial and patent leathers, inks, cements, photographic films, and linoleum. Solvent systems incorporating ethyl acetate are used in the formulation of such products as candy glaze, cleaning fluids, flavors and perfumes, embalming fluids, and spirit varnishes.

Ethyl acetate is a useful coextractant (usually with ethers) for camphor, fats and oils, antibiotics, and several resins and gums; it can also be used in the recovery of acetic acid from dilute aqueous solutions.

Ethyl acetate has low acute toxicity to laboratory animals. The oral LD_{50} in rats is 11.3 g/kg (39). The LC_{50} for a single 8-h vapor exposure for rats is 1600 ppm (40). Rats exposed to 8000 ppm for 4 h survived but exposure to 16,000 ppm for 4 h was lethal. Deaths were attributed to pulmonary edema, hemorrhage, and hyperemia of the respiratory tract. Repeated exposures of rabbits to 4450 ppm for 1 h daily for 40 d produced secondary anemia, leukocytosis, and cloudy swelling and fatty degeneration of various organs. Guinea pigs exposed to concentrations of 2000 ppm, 4 h/d, 6 d/wk for 65 exposures showed no evidence of harm (41). A concentration of 8,600–20,000 ppm has been considered dangerous to man for short exposures. The vapors can be irritating to mucous membranes but provide good warning since irritation occurs at approximately 400 ppm. Repeated or prolonged direct skin contact will cause drying and cracking of the skin. The threshold limit value (TLV) is 400 ppm (1.4 g/m^3) (38). Appropriate ventilation should be used to maintain the concentration of vapor in the air below the TLV. Repeated or prolonged contact with the skin should be avoided. If contact of the material with the eyes occurs, immediately flush them with water and get medical attention.

Propyl Acetate. Propyl acetate, $CH_3COOCH_2CH_2CH_3$, a colorless liquid with a mild, fruity odor, is a good solvent for cellulose nitrate, chlorinated rubber, and heat-reactive phenolics; its principal use is as a printing ink solvent. Compared with ethyl acetate and isopropyl acetate, propyl acetate has slow evaporation rate and good solvent power which promote improved flow and leveling characteristics in a variety of coating formulations.

This compound is only slightly toxic to laboratory animals. The acute oral LD_{50} is >3.2 g/kg for rats. When held in occluded contact with the skin of guinea pigs for 24 h, it caused slight skin irritation, but gave no evidence of skin penetration (42). Cats exposed 6 h/d for 5 d to vapor concentrations of 5300 ppm exhibited eye irritation and salivation (40). Exposure of cats to vapor concentrations of 24,500 ppm for 30 min resulted in narcosis and death (40). Prolonged or repeated contact with the skin should be avoided. The TLV is 200 ppm (840 mg/m^3) (38). Local exhaust ventilation should be used to maintain the concentration of vapor in the air below the TLV.

Isopropyl Acetate. Isopropyl acetate, $CH_3COOCH(CH_3)_2$, is an active solvent for many synthetic resins, such as ethyl cellulose, cellulose acetate butyrate, and cellulose nitrate, some vinyl copolymers, polystyrene methacrylate resins, and for natural resins such as kauri and manila gums, pontianak, and dammar.

The acute oral LD_{50} in rats is 3.0–6.5 g/kg and the LC_{50} is 32,000 ppm for a single 4-h vapor exposure (40,43). Transient eye irritation and discomfort occurred at approximately 200 ppm in men who were exposed experimentally to the vapor (44). The major effect of exposure to isopropyl acetate vapor is a sensation of irritation to the eyes, nose, and throat. With higher concentrations, these symptoms may be followed by the slow onset of narcosis manifested by drowsiness. If the liquid is splashed into the eye, it will cause moderate irritation. Repeated contact with the skin may cause defatting and dryness, but significant irritation and skin sensitization does not occur (40). The TLV is 250 ppm (950 mg/m^3) (38). Appropriate local exhaust ventilation should be used to keep the concentration in air below the TLV. Repeated or prolonged contact with the skin should be avoided.

Butyl Acetate. Butyl acetate, $CH_3COO(CH_2)_3CH_3$, a colorless, flammable liquid with a pleasant, fruity odor, is widely used in cellulose nitrate lacquers; it is also an active solvent for cellulose acetate butyrate, ethyl cellulose, chlorinated rubber, polystyrene, methacrylate resins, and for many natural gums such as kauri, manila, pontianak, and dammar. This ester is used also as a solvent in the preparation of artificial leather, textiles, and plastics, and as an extraction solvent in processing various oils and pharmaceuticals. It is used as a perfume ingredient and as a component in synthetic flavors such as apricot, banana, butter, pear, quince, pineapple, grenadine, butterscotch, and raspberry.

This compound has low acute oral toxicity to laboratory animals. The acute oral LD_{50} in rats is >6.4 g/kg. Closed contact for 24 h with the skin of guinea pigs did not cause irritation, and there was no evidence of skin penetration at doses as high as 10 mL/kg. Rats exposed to a calculated air concentration of approximately 14,000 ppm for 5 h were narcotized and failed to survive, whereas rats exposed to approximately 1300 ppm for 6 h all survived and did not exhibit adverse symptoms (42). The vapors are irritating to the eyes and respiratory tract. Prolonged or repeated skin contact may result in defatting and cracking of the skin. The TLV is 150 ppm (700 mg/m^3) (38). Local exhaust ventilation should be used to maintain the concentration of vapor in the air below the TLV. Repeated or prolonged contact with the skin should be avoided, and appropriate eye protection should be worn. If contact of the material with the eyes occurs, immediately flush them with water and get medical attention.

Isobutyl Acetate. Isobutyl acetate, $CH_3COOCH_2CH(CH_3)_2$, is a water-white liquid with a mild, fruity ester odor. Isobutyl acetate resembles butyl acetate and methyl isobutyl ketone (4-methyl-2-pentanone) and can be used interchangeably for these solvents in many formulations. It is also used as a component in synthetic flavors of apple, apricot, banana, butter, mirabelle plum, pineapple, rum, and strawberry.

This compound has low toxicity to laboratory animals. The acute oral LD_{50} to rats when administered as a 10% solution in corn oil is 3.2–6.4 g/kg. Rats exposed to a vapor concentration of approximately 3000 ppm for 6 h did not show any adverse symptoms. Rats exposed to a calculated vapor concentration of ca 23,000 ppm for 2.5 h developed prostration and narcosis and failed to survive (42). The TLV for isobutyl acetate is 150 ppm (700 mg/m^3) (38). Local exhaust ventilation should be used to maintain the concentration of vapor in the air at or below the TLV. Repeated or prolonged contact with the skin should be avoided. Appropriate eye protection should be worn. If contact of the material with the eyes occurs, immediately flush them with water and get medical attention.

Amyl Acetates (Pentyl Acetates). Amyl acetate and mixed amyl acetates (a mixture of normal-, secondary-, and isoamyl acetates) are used as lacquer solvents, as extractants in penicillin manufacture, and in the production of photographic film, leather polishes, drycleaning preparations, and flavoring agents. Mixed *sec*-amyl acetates are used as solvents for cellulose compounds and in the production of leather finishes, textile sizes, and printing compounds. Isoamyl acetates are used as solvents and in flavorings and perfumes.

The amyl acetates are moderately toxic. Amyl acetate is irritating to mucous membranes when inhaled in high concentrations; it also has a narcotic effect. A concentration of 1000 ppm, breathed for 0.5 h, has caused headache, fatigue, oppression in the chest, and irritation of the eyes and mucous membranes of the nose and throat, with excessive salivation. A concentration of 5000 ppm produces deep narcosis in cats in 30 min (29). The TLV is 100 ppm (525 mg/m^3) (38).

2-Ethylhexyl Acetate. 2-Ethylhexyl acetate, $CH_3COOCH_2CH(C_2H_5)(CH_2)_3CH_3$, is a high-boiling retarder solvent with limited water solubility used to promote flow of and retard blushing in lacquers, emulsions, and silk-screen inks, and as a flow-control agent in baking enamels. It can also be used as a dispersant for vinyl organosols and as a coalescing aid for latex paints.

This compound is only slightly toxic orally to laboratory animals. The acute oral LD_{50} is >3.2 g/kg for rats and mice (42). Other investigators have reported the oral LD_{50} for rats to be 5.89 g/kg (45). The intraperitoneal LD_{50} is 1.6–3.2 mg/kg for both species. Closed contact for 24 h with the skin of guinea pigs produced moderate skin irritation. Deaths did not occur from skin doses up to 20 mL/kg (42,46). Rats exposed to calculated vapor concentrations of approximately 1100 ppm for 6 h survived and gained weight normally (42). The compound was only slightly irritating to the eyes of rabbits (46). Prolonged or repeated contact with the skin should be avoided.

Ethylene Glycol Diacetate. Ethylene glycol diacetate, $(CH_2OCOCH_3)_2$, is a slowly evaporating solvent that gives good flowout to baking enamels and lacquers. It is used also in formulating cellulose nitrate and cellulose acetate butyrate leather lacquers and printing inks. It is used as a reflow solvent in thermoplastic acrylic coatings, as a solvent for cellulose acetate adhesives, as an extractant for diolefins, and as a perfume fixative. This compound is only slightly toxic to laboratory animals. The acute oral LD_{50} for rats is >1.6 g/kg (42). Other investigators report the acute oral LD_{50} to be 6.86 g/kg for rats and 4.94 g/kg for guinea pigs (40). It was not an irritant when held for 24 h in closed contact with the skin of guinea pigs; and there was no evidence of skin absorption at doses of 20 mL/kg (42).

2-Ethoxyethyl Acetate. 2-Ethoxyethyl acetate, $CH_3COOCH_2CH_2OC_2H_5$, is an excellent solvent for cellulose nitrate, ethyl cellulose, cellulose acetate butyrate, polystyrene, poly(vinyl acetate), acrylics, and many other commonly used coating polymers.

This compound has low acute oral toxicity for laboratory animals. The LD_{50} is 5.10 g/kg for rats, 1.91 g/kg for guinea pigs, and 1.95 g/kg for rabbits. Intraperitoneal injections of 0.5 and 1.0 mL/kg as single doses were well tolerated by guinea pigs, whereas 5.0 mL/kg was fatal for mice. The cutaneous LD_{50} in rabbits is 10.3 g/kg. Dogs survived 120 daily 7-h exposures to vapor concentrations of 600 ppm. Guinea pigs survived a 1-h exposure to air saturated with vapor. Mice, guinea pigs, and a rabbit were unaffected by twelve 8-h exposure to 450 ppm, whereas two cats and another rabbit died (40). The TLV is 100 ppm (540 mg/m^3) (38). Local exhaust ventilation should be used to maintain the concentration of vapor in the air below the TLV. Direct skin contact may result in penetration of the skin if contact is prolonged, and the compound is a slight eye irritant. Contact with the skin should be avoided by the use of protective clothing and gloves, and appropriate eye protection should be worn. If contact of the material with the eyes occurs, immediately flush them with water and get medical attention.

2-Butoxyethyl Acetate. 2-Butoxyethyl acetate, $CH_3COOCH_2CH_2OC_4H_9$, is particularly useful as a coalescing aid for latex paints. It is also useful in multicolor lacquers and lacquer emulsions and as a retarder solvent in the formulation of high–low lacquer thinners, printing inks, and epoxy coatings.

This compound is only slightly toxic orally to laboratory animals. The acute LD_{50} for rats is 1.6–3.2 g/kg, for mice it is 3.2 g/kg (42). Another investigator found the oral LD_{50} for rats to be 7.46 mL/kg (47). Intraperitoneally, the LD_{50} is 1.6 g/kg for mice and 400 mg/kg for rats. When held in occluded contact with the skin of guinea pigs for 24 h, the compound produced only slight irritation of the skin. The compound was absorbed through the skin, causing death in guinea pigs at 5 mL/kg (42). The skin LD_{50} for rabbits is 1.58 mL/kg (47). A drop in the eye of a rabbit produced only slight irritation. Rats exposed to vapor concentrations of 450 ppm for 6 h showed no adverse symptoms. Skin and eye contact should be avoided by the use of good handling techniques and by wearing appropriate protective clothing and eye protection. Contaminated clothing should be removed and washed before reuse. Breathing of vapor should be avoided by use of appropriate ventilation.

2-(2-Ethoxyethoxy)ethyl Acetate. 2-(2-Ethoxyethoxy)ethyl acetate, $CH_3CO(OCH_2CH_2)_2OC_2H_5$, is completely soluble in water. It is used as a coalescing aid in latex paints and in textile and silk-screen printing inks, and in lacquers with slow-drying characteristics.

The acute oral toxicity of this compound is low for laboratory animals. When it is administered as a 50% aqueous solution, the acute oral LD_{50} is 11.0 g/kg for rats and 3.93 g/kg for guinea pigs (40). It was only slightly irritating to the eyes of rabbits. The single skin LD_{50} for rabbits is 15.2 mL/kg (45).

2-(2-Butoxyethoxy)ethyl Acetate. 2-(2-Butoxyethoxy)ethyl acetate, $CH_3CO(OCH_2CH_2)_2OC_4H_9$, has limited water solubility; it is used chiefly as a solvent in printing inks and high-bake enamels, as a coalescing aid in latex paints, in silk-screen inks, and as a component in polystyrene coatings for decals.

This ester has low acute oral toxicity for laboratory animals. The acute oral LD_{50} doses are reported variously as 7 mL/kg and 11.92 g/kg for rats and 2.34 g/kg for guinea pigs. Other acute oral LD_{50} values are 2.8 mL/kg for rabbits, 2.7 mL/kg for guinea pigs, 7.1 mL/kg for rats, 6.6 mL/kg for mice, and 5.0 mL/kg for chickens. The compound is slightly irritating to the skin and can penetrate intact skin. The skin LD_{50} is 5.5 mL/kg for rabbits. There is evidence of cumulative effects of skin contact, since the 90-d LD_{50} for rabbits is 2.0 mL/kg. Kidney damage was produced by repeated contact with the skin. The compound is only slightly irritating to the eyes (40). Exposure to concentrated vapor (concentration not reported) for 8 h caused death in one of six test rats (45). Prolonged or repeated contact with the skin should be avoided. Exposure to high concentrations of vapor should be avoided by the use of appropriate ventilation. Suitable eye protection should be worn to prevent contact with the eyes.

Glyceryl Triacetate. Glyceryl triacetate (triacetin), $CH_3COOCH(CH_2OCOCH_3)_2$, is used mainly as a plasticizer for cellulosic resins and cellulose acetate tow for use in cigarette filters. It is also used as a plasticizer for vinylidene (vinylene) polymers and copolymers, as a solvent and carrier in pharmaceutical preparations, as a solvent and fixative in the compounding of perfumes and flavors, and as an ingredient in inks for printing on plastics and other nonabsorbent surfaces.

Administered undiluted to laboratory animals, triacetin has an oral LD_{50} of 6.4–12.8 g/kg in rats and 3.2–6.4 g/kg in mice. Intraperitoneally, the LD_{50} is 800–1600 mg/kg in both species. When it was held in occluded contact with the skin of guinea pigs for 24 h, practically no irritation resulted, and there was no evidence of skin absorption at dosages as high as 20 mL/kg. Skin sensitization tests in guinea pigs were negative. Inhalation of vapor concentrations averaging 250 ppm, 6 h/d, 5 d/wk for 13 wk, produced no ill effects in rats. Rats survived vapor concentrations of 8200 ppm, 6 h/d for 5 d without adverse effects (42). When placed in the eyes of rabbits, no damage to the eye resulted. Rats gained weight normally when triacetin replaced fat in the diet up to 55% of the total diet (40). No special handling precautions are required.

Benzyl Acetate. Benzyl acetate, $CH_3COOCH_2C_6H_5$, is a colorless, mobile liquid with the distinctive fragrance associated with jasmine. It is almost insoluble in water, but miscible with alcohol, ether, and similar solvents. Benzyl acetate is a component of the extract of gardenia, hyacinth, and ylang-ylang, and is the main component of extract of jasmine. It is prepared by the reaction of benzyl chloride with sodium acetate.

Benzyl acetate is handled in the same manner as benzyl alcohol. The bulk of benzyl acetate is used in soap odors, but it is also popular for other perfumes and is used to a minor extent in flavors.

Propionic Acid Esters. *Glyceryl Tripropionate.* Glyceryl tripropionate (tripropionin), $CH_3CH_2COOCH(CH_2OCOCH_2CH_3)_2$, is used as a specialty plasticizer. It has a very low toxicity for laboratory animals. The acute oral LD_{50} for rats, mice, and guinea pigs is 6.4–12.8 g/kg. Intraperitoneally, the LD_{50} for rats and mice is 1.6–3.2 g/kg and 3.2–6.4 g/kg for guinea pigs. Closed contact for 24 h with the skin of guinea pigs caused very slight irritation and did not produce any systemic toxic effects by skin absorption. The ester did not produce skin sensitization in guinea pigs, and it did not cause any damage when instilled in the eyes of rabbits. Rats exposed to vapor calculated to be 600 ppm in air 6 h/d for 62 d over a 3-mo period did not show any effects (42).

Butyric Acid Esters. *Methyl Butyrate.* Methyl butyrate, $CH_3(CH_2)_2COOCH_3$, is a colorless liquid used in the manufacture of artificial rum and fruit essences. It is a weak local irritant and is moderately toxic when inhaled or ingested.

Ethyl Butyrate. Ethyl butyrate, $CH_3(CH_2)_2COOC_2H_5$, is used in the manufacture of artificial rum, pineapple oil, and in perfumes. The acute oral LD_{50} for rats is 13 g/kg. The acute dermal LD_{50} is >2 g/kg for rabbits. When the undiluted liquid was held in occluded contact with rabbit skin for 24 h, moderate irritation resulted (48).

Isobutyric Acid Esters. *Methyl Isobutyrate.* Methyl isobutyrate, $(CH_3)_2CHCOOCH_3$, is a colorless liquid slightly soluble in water and miscible in all proportions with common organic solvents. Methyl isobutyrate is only slightly toxic having an oral LD_{50} in rats of 8–16 g/kg and in mice of >3.2 g/kg. Intraperitoneally, the LD_{50} in both species is 1.6–3.2 g/kg. When held in occluded contact with guinea pig skin for 24 h it caused slight skin irritation and gave no evidence of being absorbed through the skin. It caused slight irritation of the eye but did not damage the cornea. Atmospheric concentrations of 42,000 ppm caused death in rats exposed for 1 h. Removal of rats from the chamber before that time resulted in recovery. Exposure to 6400 ppm for 6 h caused only mild symptoms and no deaths.

Ethyl Isobutyrate. Ethyl isobutyrate, $(CH_3)_2CHCOOC_2H_5$, is a colorless liquid slightly soluble in water and miscible in all proportions with common organic solvents.

Isobutyl Isobutyrate. Isobutyl isobutyrate, $(CH_3)_2CHCOOCH_2CH(CH_3)_2$, is an active solvent for cellulose nitrate, and can be used as a direct substitute for 4-methyl-2-pentyl acetate in most formulations. The distinct odor and flavor of this ester make it an interesting material for the formulation of perfumes and as a bulk component of flavor essences.

This compound is slightly toxic to laboratory animals. The acute oral LD_{50} for rats and mice is 6.4–12.8 g/kg. The intraperitoneal LD_{50} is 3.2–6.4 g/kg for rats and 800–1600 mg/kg for mice. Closed contact with guinea pig skin (24 h) produced slight skin irritation. The compound has a skin absorption LD_{50} >10 mL/kg. Rats exposed to an atmospheric vapor concentration of ca 650 ppm for 6 h did not show injury. Exposure to a concentration of approximately 5400 ppm for the same period of time caused narcosis and the death of two of three rats. The survivors gained weight normally (42). Inhalation of the vapor should be avoided by the use of appropriate ventilation. Prolonged or repeated contact with the skin should be avoided.

Stearic Acid Esters. Commercial stearic acid esters are normally mixtures of, chiefly, stearic and palmitic acid esters corresponding to the composition of commercial stearic acid (see Carboxylic acids).

Methyl Stearate. Methyl stearate, $CH_3(CH_2)_{16}COOCH_3$, occurs as white crystals, insoluble in water and soluble in alcohol, ether, and similar organic solvents. No red label is required for rail shipment. Its uses are similar to those of butyl stearate. This ester is available commercially as a mixture comprising 95% stearate (octadecanoate), 4% palmitate (hexadecanoate), and 1% oleate (9-octadecenoate): acid value, 0.5 max; saponification value, 188–192; iodine value (Wijs), 1 max; unsaponifiable, 0.5% max; typical mp, 36°C; color (% transmittance) at 440 and 550 nm, 71 and 98%, respectively. A grade comprising 58% stearate, 40% palmitate, and 2% myristate (tetradecanoate) is also available.

Ethyl Stearate. Ethyl stearate, $CH_3(CH_2)_{16}COOC_2H_5$, occurs as white crystals, insoluble in water and soluble in alcohol, ether, and similar organic solvents. The commercial-grade ester has a mp of 20–24°C and a bp of 180°C (0.5 kPa or 4 mm Hg). Its uses are similar to those of butyl stearate.

Butyl Stearate. Butyl stearate, $CH_3(CH_2)_{16}COO(CH_2)_3CH_3$, is a colorless liquid. This ester is available in different grades, with acid no. (max) 1.0–2.0, hydroxyl no. (max) 2.0, iodine values (max) 0.5–2.5; saponification no. 170–177 or 165–180; and color (% transmittance), 90–97% at 440 nm. It is of value in compounding lubricating oils, as a lubricant for the textile and molding trade, in special lacquers, and as a waterproofing agent. In the cosmetic and pharmaceutical fields it is used in vanishing creams, ointments, rouges, lipsticks, and nail polishes. Its oily characteristics have made it of particular value in polishes and in coatings that are to be polished.

Glyceryl Monostearate. The commercial ester, prepared by the glycerolysis of a blend of fully hydrogenated vegetable oils with a monoester content of 90% (min), is a white powder with a congeal point of 70°C and a clear point of 78°C. It is used by macaroni and noodle manufacturers to gain better firmness and nonstick properties. The flash point (Cleveland open cup) is 230°C. The solid moistened with water and held in closed contact with guinea pig skin for 24 h caused no significant irritation and there was no evidence of absorption through the skin at a dose of 10 g/kg. Good industrial hygiene practices, which include the avoidance of dust inhalation, should be followed.

Other Stearates. Dodecyl stearate, $CH_3(CH_2)_{16}COO(CH_2)_{11}CH_3$, is used as a specialty plasticizer. Hexadecyl stearate, $CH_3(CH_2)_{16}COO(CH_2)_{15}CH_3$, is used as a heat-stable textile lubricant. Commercial ethylene glycol distearate, $[CH_2OOC(CH_2)_{16}CH_3]_2$, has the following properties: acid no., 6.0; hydroxyl no., 33–43; iodine no., 1.0; saponification no., 190–200; color (% transmittance), 90% at 550 nm.

Adipic Acid Esters. *Bis(2-ethylhexyl) Adipate.* Bis(2-ethylhexyl) adipate, $[CH_2CH_2COOCH_2CH(C_2H_5)C_4H_9]_2$, is used as a plasticizer to impart low-temperature flexibility to PVC formulations, particularly in vinyl meat-wrapping film (see Adipic acid). Production of this ester was 13.6 t in 1975 (49). It has very low toxicity and is slightly irritating to the skin and eyes; it is not a skin sensitizer (40,42).

Aromatic Acid Esters

Esters of aromatic acids, like those described in the preceding section, have a wide variety of applications. Certain benzoic acid (qv) esters are used in perfumery (Table

336 ESTERS, ORGANIC

3); the phthalates (1,2-benzenedicarboxylates), isophthalates (1,3-benzenedicarboxylates), trimellitates (1,2,4-benzenetricarboxylates), and, to a certain extent, the terephthalates (1,4-benzenedicarboxylates) are used as plasticizers (Table 3) (see Dicarboxylic acids; Phthalic acids and other benzenepolycarboxylic acids). Other esters of dibasic and polybasic aromatic acids, especially isophthalic and terephthalic acid esters, are used as polymer intermediates. Properties of some of the commercially important esters are included in Table 1.

BIBLIOGRAPHY

"Esters, Organic" in *ECT* 1st ed., Vol. 5, pp. 824–950, by T. Earl Jordan, Publicker Industries, Inc.; "Esters, Organic" in *ECT* 2nd ed., Vol. 8, pp. 365–383, by Edward U. Elam, Tennessee Eastman Company.

1. H. W. Post, *The Chemistry of Aliphatic Orthoesters,* Reinhold Publishing Corp., New York, 1943.
2. J. D. Fletcher, O. C. Dermer, and R. B. Fox, eds., *Nomenclature of Organic Compounds,* American Chemical Society, Washington, D.C., 1974, pp. 137–145.
3. K. S. Markley, ed., *Fatty Acids, Their Chemistry, Properties, Production, and Uses,* 2nd ed., Interscience Publishers, Inc., New York, Part 1, 1960; Part 2, 1961.
4. R. T. Morrison and R. N. Boyd, *Organic Chemistry,* 3rd ed., Allyn and Bacon, Inc., Boston, Mass., 1973, pp. 675–684.
5. E. K. Euranto in S. Patai, ed., *The Chemistry of Carboxylic Acids and Esters,* Wiley-Interscience, New York, 1969.
6. M. Levey, *J. Chem. Ed.* **31,** 521 (1954).
7. D. Swern, ed., *Bailey's Industrial Oil and Fat Products,* Wiley-Interscience, New York, 1964, p. 936.
8. *Ibid.,* p. 937.
9. *Ibid.,* p. 940.
10. S. Patai, ed., *The Chemistry of Carboxylic Acids and Esters,* Wiley-Interscience, New York, 1969, pp. 410–431.
11. E. S. Gould, *Mechanism and Structure in Organic Chemistry,* Henry Holt & Co., Inc., New York, 1959, pp. 329–332.
12. A. Mailhe, *Compt. Rend.* **170,** 813 (1920); *Bull. Soc. Chim.* **27,** 226 (1920); *Ann. Chim.* 183 (1920).
13. A. M. Schwartz, J. W. Perry, and J. Berch, *Surface-Active Agents and Detergents,* Vol. II, Interscience Publishers, Inc., New York, 1958, p. 104; G. Reutenauer and M. Lacombe, *Oleagineux* **2,** 500 (1947).
14. H. Adkins in R. Adams and co-eds., *Organic Reactions,* Vol. 8, John Wiley & Sons, Inc., New York, 1954, pp. 1–27.
15. M. W. Formo in E. S. Pattison, ed., *Industrial Fatty Acids and Their Applications,* Reinhold Publishing Corp., New York, 1959, pp. 70–71.
16. N. G. Gaylord, *Reduction With Complex Metal Hydrides,* Interscience Publishers, Inc., New York, 1956, pp. 391–543.
17. C. B. Hauser and B. E. Hudson, Jr., in R. Adams and co-eds., *Organic Reactions,* Vol. 1, John Wiley & Sons, Inc., New York, 1947, pp. 266–302.
18. S. M. McElvain in R. Adams and co-eds., *Organic Reactions,* Vol. 4, John Wiley & Sons, Inc., New York, 1948, pp. 256–269.
19. Ref. 11, pp. 500–504; C. T. Smith and co-workers, *Ind. Eng. Chem.* **34,** 743 (1942).
20. D. W. Wertz and N. L. Allinger, *J. Org. Chem.* **42,** 698 (1977), and references cited therein.
21. P. Sabatier and E. E. Reid, *Catalysis Then and Now,* Part II, Franklin Publishing Co., Englewood, N.J., 1965, pp. 308–314.
22. S. Swann, S. S. Kistler, and E. G. Appel, *Ind. Eng. Chem.* **26,** 1014 (1934).
23. A. W. Ralston, *Fatty Acids and Their Derivatives,* John Wiley & Sons, Inc., New York, 1948, p. 833.
24. Ref. 10, pp. 433–436.
25. E. L. McDaniel and H. S. Young, *Ind. Eng. Chem. Prod. Res. Dev.* **2,** 287 (1963).
26. E. Guenther and D. Althausen, "The Chemistry of Essential Oils" in *The Essential Oils,* Vol. 2, D. Van Nostrand Co., Inc., New York, 1949, pp. 618–656.

27. L. J. Bellamy, *The Infrared Spectra of Complex Molecules,* 3rd ed., John Wiley & Sons, Inc., New York, 1975; L. J. Bellamy, *Advances in Infrared Group Frequencies,* Methuen & Co., Ltd., London, Eng., 1968, pp. 166–168.
28. T. S. Ma in Ref. 10, pp. 871–921.
29. N. I. Sax, *Dangerous Properties of Industrial Materials,* 4th ed., Reinhold Publishing Corp., New York, 1975.
30. T. H. Durrans, *Solvents,* 8th ed., revised by E. H. Davies, Chapman & Hall, Ltd., London, Eng., 1971, pp. 145–165.
31. D. N. Buttrey, *Plasticizers,* 2nd American ed., Franklin Publishing Co., Inc., Palisades, N.J., 1960.
32. I. Mellan, *Industrial Plasticizers,* MacMillan, New York, 1963.
33. "Synthetic Organic Chemicals, U.S. Production and Sales, 1976" in *U.S. International Trade Commission Publication 833,* U.S. Government Printing Office, Washington, D.C., 1977.
34. "Surface-Coating Raw Materials—Estimated Production Value and Production in 1973" in *Chemical Economics Handbook,* Stanford Research Institute, Menlo Park, Calif., Oct. 1975.
35. W. G. Dukek and A. H. Popkin in M. C. Gunderson and A. W. Hart, eds., *Synthetic Lubricants,* Reinhold Publishing Corp., New York, 1962, pp. 151–245.
36. A. Burger, *Medicinal Chemistry,* 3rd ed., Part I, Wiley-Interscience, New York, 1970, pp. 67, 71.
37. C. Marsden and S. Mann, eds., *Solvents Guide,* 2nd ed., Interscience Publishers, a division of John Wiley & Sons, Inc., New York, 1963.
38. *Threshold Limit Values for Chemical Substances and Physical Agents in the Workroom Environment with Intended Changes for 1977,* ACGIH, Washington, D.C., 1976.
39. M. Windholz, ed., *The Merck Index,* 9th ed., Merck & Co., Inc., Rahway, N.J., 1976.
40. F. A. Patty, ed., *Industrial Hygiene and Toxicology,* 2nd revised ed., Vol. II, Interscience Publishers, a division of John Wiley & Sons, Inc., New York, 1963.
41. *Hygienic Guide Series,* American Industrial Hygiene Association, New York.
42. Unpublished data, Health, Safety, and Human Factors Laboratory, Eastman Kodak Company, Rochester, N.Y.
43. E. Browning, *Toxicity and Metabolism of Industrial Solvents,* Elsevier Publishing Co., New York, 1965.
44. L. Silverman and co-workers, *J. Ind. Hyg. Toxicol.* **28,** 262 (1946).
45. *Union Carbide Publication 12/71-10M, Esters,* Union Carbide Corporation, New York.
46. H. E. Smyth, Jr., and C. P. Carpenter, *J. Ind. Hyg. Toxicol.* **26,** 269 (1944).
47. H. E. Smyth and co-workers, *Am. Ind. Hyg. Assoc. J.* **23,** 95 (1962).
48. D. L. J. Opdyke, *Food Cosmet. Toxicol.* **12,** 719 (1974).
49. J. L. Blackford, "Acrylic Acid and Esters" in *Chemical Economics Handbook,* Stanford Research Institute, Menlo Park, Calif., Aug. 1976.

General Reference

S. Patai, ed., *The Chemistry of Carboxylic Acids and Esters,* Wiley-Interscience, New York, 1969; an extensive coverage of the chemistry of esters.

EDWARD U. ELAM
Tennessee Eastman Company

ESTROGENS. See Hormones.

ESTRONE. See Hormones.

ETHANE. See Hydrocarbons C_1–C_6.

ETHANOIC ACID. See Acetic acid.

ETHANOL

Ethanol [64-17-5] or ethyl alcohol, CH_3CH_2OH, has been described as one of the most exotic synthetic oxygen-containing organic chemicals because of its unique combination of properties as a solvent, a germicide, a beverage, an antifreeze, a fuel, a depressant, and especially because of its versatility as a chemical intermediate for other organic chemicals.

Ethanol is the IUPAC name for this chemical; the name ethyl alcohol is also correct. The name alcohol is a generic name derived from two Arabic words, *al* and *kohl,* that described a finely ground powder used by Oriental women to darken their eyebrows. The name, unqualified, gradually became specific for ethyl alcohol, "spirits of wine rectified to the highest degree" (1). Ethyl alcohol is, of course, well known as a constituent of alcoholic beverages (see Beverage spirits, distilled; Beer; Wine).

As a beverage ethanol had been prepared and used long ago by the Egyptian pharaohs (2–3). Some indication of the antiquity of the knowledge of ethyl alcohol is the fact that Noah is believed to have built for himself a vineyard in which he grew grapes that he fermented into a sort of alcoholic beverage (4).

Physical Properties

Ethyl alcohol under ordinary conditions is a volatile, flammable, clear, colorless liquid. Its odor is pleasant, familiar, and characteristic, as is its taste when it is suitably diluted with water.

The physical and chemical properties of ethyl alcohol are primarily dependent upon the hydroxyl group. This group imparts polarity to the molecule and also gives rise to intermolecular hydrogen bonding. These two properties account for the differences between the physical behavior of lower molecular weight alcohols and that of hydrocarbons of equivalent weight. Infrared spectrographic studies (5) have shown that, in the liquid state, hydrogen bonds are formed by the attraction of the hydroxyl hydrogen of one molecule and the hydroxyl oxygen of a second molecule. The effect of this bonding is to make liquid alcohol behave as though it were largely dimerized. This behavior is analogous to that of water, which however, is more strongly bonded and appears to exist in liquid clusters of more than two molecules. The association of ethyl alcohol, it should be noted, is confined to the liquid state; in the vapor state it is monomeric.

A summary of physical properties of ethyl alcohol is presented in Table 1. Detailed information on the vapor pressure, density, and viscosity of ethanol can be obtained from refs. 6–14. A listing of selected binary and ternary azeotropes of ethanol is compiled in ref. 25.

Chemical Properties

The chemistry of ethyl alcohol is largely that of the hydroxyl group, namely, reactions of dehydration, dehydrogenation, oxidation, and esterification. The hydrogen atom of the hydroxyl group can be replaced by an active metal, such as sodium, po-

Table 1. Physical Properties of Ethanol[a]

Property	Value
freezing point, °C	−114.1
normal boiling point, °C	+78.32
critical temperature, °C	243.1
critical pressure, kPa[b]	6383.48
critical volume, L/mol	0.167
critical compressibility factor, z, in $PV = znRT$	0.248
density, d_4^{20}, g/mL	0.7893
refractive index, n_D^{20}	1.36143
$\Delta n_D/\Delta t$, 20–30°C, per °C	0.000404
surface tension, at 25°C, mN/m (= dyn/cm)	231
viscosity, at 20°C, mPa·s (= cP)	1.17
solubility in water, at 20°C	miscible
heat of vaporization, at normal boiling point, J/g[c]	839.31
heat of combustion, at 25°C, J/g[c]	29676.69
heat of fusion, J/g[c]	104.6
flammable limits in air	
lower, vol %	4.3
upper, vol %	19.0
autoignition temperature, °C	793.0
flash point, closed-cup, °C	14
specific heat, at 20°C, J/(g·°C)[c]	2.42
thermal conductivity, at 20°C, W/(m·K)	0.170
dipole moment, liq at 25°C, C·m[d]	5.67×10^{-30}
magnetic susceptibility at 20°C	0.734×10^{-6}
dielectric constant at 20°C	25.7

[a] Refs. 6–14.
[b] To convert kPa to atm, divide by 101.3.
[c] To convert J to cal, divide by 4.184.
[d] To convert C·m to Debye, divide by 3.336×10^{-30} (esu = D $\times 10^{-18}$).

tassium, and calcium, to form a metal ethoxide (ethylate) with the evolution of hydrogen gas (see Alkoxides, metal).

$$2\ CH_2H_5OH + 2\ M \rightarrow 2\ C_2H_5OM + H_2$$

Sodium ethoxide can be prepared by the reaction of absolute ethyl alcohol and sodium, or by refluxing absolute ethyl alcohol with anhydrous sodium hydroxide (16):

$$CH_3CH_2OH + NaOH \rightarrow CH_3CH_2ONa + H_2O$$

Commercially, water is removed by azeotropic distillation (qv) with benzene (17). Sodium ethoxide precipitates upon addition of anhydrous acetone (18). This strong base hydrolyzes readily to give ethyl alcohol and sodium and hydroxyl ions.

$$CH_3CH_2O^-Na^+ + H_2O \rightleftharpoons CH_3CH_2OH + Na^+ + OH^-$$

Sodium ethoxide can also be prepared by the reaction of sodium amalgam with ethyl alcohol.

Sodium ethoxide is used in organic synthesis as a condensing and reducing agent. The reaction between sodium ethoxide and sulfur monochloride yields diethyl thiosulfite (19).

$$2\ CH_3CH_2ONa + S_2Cl_2 \rightarrow (CH_3CH_2O)_2S_2 + 2\ NaCl$$

Barbiturates (see Hypnotics) (Veronal, Barbital, Luminal, Amytal), ethyl orthoformate, and other chemicals are produced commercially from sodium ethoxide.

Aluminum and magnesium also react to form ethoxides, but the reaction must be catalyzed by amalgamating the metal (adding a small amount of mercury).

$$6\ CH_3CH_2OH + 2\ Al \rightarrow 2\ (CH_3CH_2O)_3Al + 3\ H_2$$

$$2\ CH_3CH_2OH + Mg \rightarrow (CH_3CH_2O)_2Mg + H_2$$

Well-cleaned aluminum filings react at room temperature in the presence of mercuric chloride (20–21). In an autoclave, metallic aluminum and ethyl alcohol react without a catalyst at 120°C (22). The reaction can also be promoted by the addition of sodium ethoxide (23).

Other reactions involving the hydrogen atom of the hydroxyl group in ethyl alcohol include the opening of epoxide rings to form hydroxy ethers,

$$CH_3CH_2OH + RCH\underset{O}{-\!\!\!-\!\!\!-}CH_2 \longrightarrow CH_3CH_2OCHRCH_2OH$$

and the addition to acetylene to form ethyl vinyl ether,

$$CH_3CH_2OH + HC\!\equiv\!CH \rightarrow CH_3CH_2OCH\!=\!CH_2$$

These reactions are carried out in the presence of acidic and basic catalysts. The acid-catalyzed addition of ethyl alcohol to acetylene or to a vinyl ether produces acetals (diethers of 1,1-dihydroxyethane). The acid-catalyzed reaction of ethyl alcohol with an aldehyde or ketone also gives acetals.

$$2\ CH_3CH_2OH + RCHO \rightleftharpoons RCH(OCH_2CH_3)_2 + H_2O$$

The hydroxyl group can be replaced by halogens from inorganic acid halides or phosphorus halides to give two different products, ethyl esters of the acid and ethyl halide (1). Phosphorus trihalides, and thionyl chloride, $SOCl_2$, are used to make triethyl phosphite, diethyl sulfite, or ethyl chloride. The ethyl chloride yield is reduced by formation of mixed alkyl esters of phosphites, such as $CH_3CH_2OP(OH)_2$ and $(CH_3CH_2O)_2POH$.

Diethyl sulfite or triethyl phosphite reaction

$$\begin{cases} 3\ CH_3CH_2OH + PCl_3 \xrightarrow[\text{low temp}]{\text{tert-amine}} (CH_3CH_2O)_3P + 3\ HCl \\ 2\ CH_3CH_2OH + SOCl_2 \rightarrow (CH_3CH_2O)_2SO + 2\ HCl \end{cases}$$

Ethyl chloride reaction

$$\begin{cases} 3\ CH_3CH_2OH + PCl_3 \rightarrow 3\ CH_3CH_2Cl + H_3PO_3 \\ CH_3CH_2OH + SOCl_2 \rightarrow CH_3CH_2Cl + SO_2 + HCl \end{cases}$$

Reaction of ethanol with phosphorus trichloride produces mainly triethyl phosphite; ethyl bromide is the principal product of reaction with phosphorus tribromide. However, reaction conditions strongly affect the composition of reaction products.

The halogen acids also produce alkyl halides.

$$CH_3CH_2OH + HX \rightarrow CH_3CH_2X + H_2O$$

The halogen influences the rate of reaction and, in general, the order of reactivity is

HI > HBr > HCl. Important uses of ethyl chloride include the manufacture of tetraethyllead and ethyl cellulose. Ethyl bromide can be used to produce ethyl Grignard reagent and various ethyl amines (see Amines).

Esterification. Esters are formed by the reaction of ethanol with inorganic and organic acids, acid anhydrides, and acid halides. If the inorganic acid is oxygenated (sulfuric acid, nitric acid), the ester has a carbon-oxygen linkage that is easily hydrolyzed (24–26).

$$CH_3CH_2OH + H_2SO_4 \rightarrow CH_3CH_2OSO_3H + H_2O$$

$$2\ CH_3CH_2OH + H_2SO_4 \rightarrow (CH_3CH_2O)_2SO_2 + 2\ H_2O$$

$$CH_3CH_2OH + HONO_2 \rightarrow CH_3CH_2ONO_2 + H_2O$$

Organic esters are formed by the elimination of water between an alcohol and an organic acid.

$$CH_3CH_2OH + RCOOH \rightleftharpoons RCOOCH_2CH_3 + H_2O$$

The reaction is reversible and reaches equilibrium slowly. Generally, acidic catalysts are used, such as strong sulfuric acid, hydrochloric acid, boron trifluoride, and p-toluenesulfonic acid (27). Batchwise and continuous processes are used for the esterification reaction.

Ethyl alcohol also reacts with acid anhydrides or acid halides to give the corresponding esters.

$$CH_3CH_2OH + (RCO)_2O \rightarrow RCOOCH_2CH_3 + RCOOH$$

$$CH_3CH_2OH + RCOCl \rightarrow RCOOCH_2CH_3 + HCl$$

The direct conversion of ethyl alcohol to ethyl acetate is believed to take place via acetaldehyde and its condensation to ethyl acetate (Tishchenko reaction) (28–34).

$$CH_3CH_2OH \rightarrow CH_3CHO + H_2$$

$$2\ CH_3CHO \rightarrow CH_3COOCH_2CH_3$$

An ethyl acetate yield of ca 24% is obtained using a copper oxide catalyst with 0.1–0.2% thoria at 350°C.

Dehydration. Ethyl alcohol can be dehydrated to form ethylene or ethyl ether.

$$CH_3CH_2OH \rightarrow CH_2{=}CH_2 + H_2O$$

$$2\ CH_3CH_2OH \rightarrow CH_3CH_2OCH_2CH_3 + H_2O$$

Generally, both ethylene and ethyl ether are formed to some extent, but the conditions can be altered to favor one reaction or the other.

Dehydrogenation. The dehydrogenation of ethyl alcohol to acetaldehyde can be effected by a vapor-phase reaction over various catalysts.

$$CH_3CH_2OH \rightarrow CH_3CHO + H_2$$

Haloform Reaction. Ethyl alcohol reacts with sodium hypochlorite to give chloroform—the haloform reaction.

$$CH_3CH_2OH + NaOCl \rightarrow CH_3CHO + NaCl + H_2O$$

$$CH_3CHO + 3\ NaOCl \rightarrow CCl_3CHO + 3\ NaOH$$

$$CCl_3CHO + NaOH \rightarrow CHCl_3 + HCOONa$$

Similarly, bromoform, $CHBr_3$, and iodoform, CHI_3, are obtained from sodium hypobromite and hypoiodite, respectively. Ethyl alcohol is the only primary alcohol that undergoes this reaction.

Reactivity Parameters. The reactivity of ethyl alcohol–water mixtures has been correlated with three distinct alcohol concentration ranges (35–36). For example, the chromium trioxide oxidation of ethyl alcohol (37), the catalytic decomposition of hydrogen peroxide (38), and the sensitivity of colloidal particles to coagulation (39) are characteristic for ethyl alcohol concentrations of 25–30%, 40–60%, and above 60% alcohol, respectively. The effect of various catalysts also differs for different alcohol concentrations (35).

Manufacture

Industrial ethyl alcohol can be produced either (1) synthetically from ethylene, (2) as a by-product of certain industrial operations, or (3) by the fermentation (qv) of sugar, starch, or cellulose.

There are two main processes for the synthesis of ethyl alcohol from ethylene. The earliest to be developed (in 1930 by Union Carbide Corporation) was the indirect-hydration process, variously called the strong sulfuric acid–ethylene process, the ethyl sulfate process, the esterification-hydrolysis process, or the sulfation–hydrolysis process. The other synthesis process, designed to eliminate the use of sulfuric acid and which, since the early 1970s, has completely supplanted the old sulfuric acid process, is the direct hydration process. The current synthetic production of industrial alcohol far exceeds its production by fermentation.

Other synthetic methods have been investigated but have not become commercial. These include, for example, the hydration of ethylene in the presence of dilute acids (weak sulfuric acid process); the conversion of acetylene to acetaldehyde, followed by hydrogenation of the aldehyde to ethyl alcohol; and the Fischer-Tropsch hydrocarbon synthesis (see Fuels, synthetic).

Synthetic ethyl alcohol is produced by five U.S. companies via the catalytic vapor phase hydration of ethylene: Union Carbide Corporation (UCC), Publicker Inc., Shell Chemical Company, Tennessee Eastman Company (a Division of Eastman Kodak Company), and U.S. Industrial Chemicals Company (USI, a Division of National Distillers and Chemical Corporation).

Indirect Hydration (Esterification-Hydrolysis) Process. The preparation of ethanol from ethylene by the use of sulfuric acid is a three-step process (Fig. 1).

(1) Absorption of ethylene in concentrated sulfuric acid to form mono- and diethyl sulfates:

$$CH_2{=}CH_2 + H_2SO_4 \rightarrow CH_3CH_2OSO_3H$$
$$\text{ethyl hydrogen sulfate}$$
$$\text{(monoethyl sulfate)}$$
$$2\ CH_2{=}CH_2 + H_2SO_4 \rightarrow (CH_3CH_2O)_2SO_2$$
$$\text{diethyl sulfate}$$

(2) Hydrolysis of ethyl sulfates to ethanol:

$$CH_3CH_2OSO_3H + H_2O \rightarrow CH_3CH_2OH + H_2SO_4$$
$$(CH_3CH_2O)_2SO_2 + 2\ H_2O \rightarrow 2\ CH_3CH_2OH + H_2SO_4$$
$$(CH_3CH_2O)_2SO_2 + CH_3CH_2OH \rightarrow CH_3CH_2OSO_3H + (CH_3CH_2)_2O$$
$$\text{diethyl ether}$$

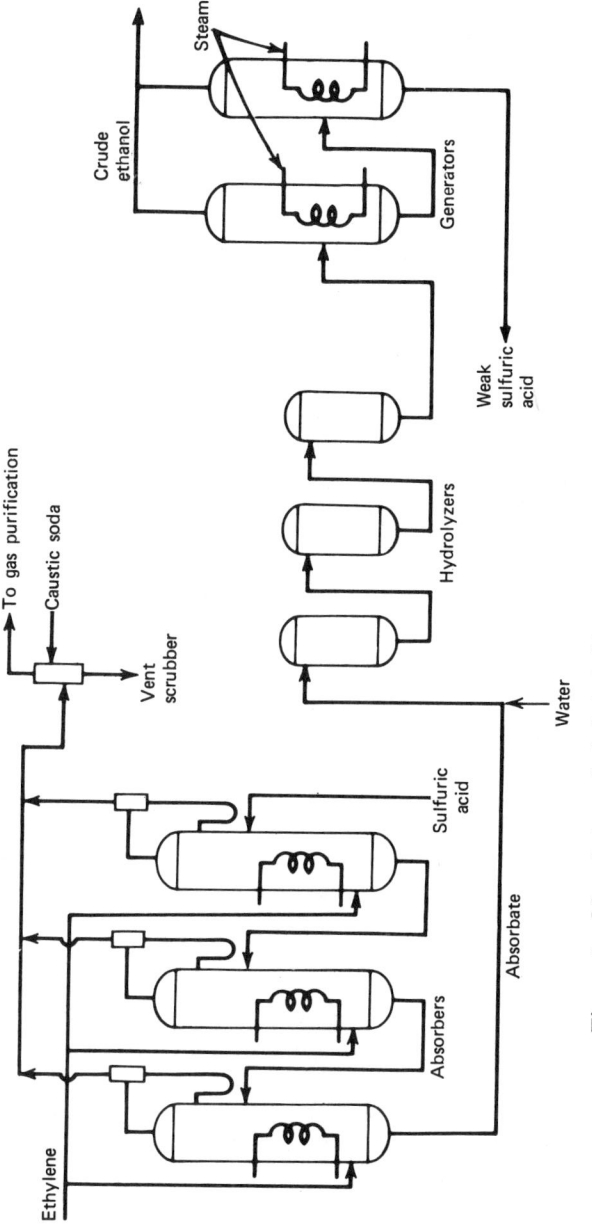

Figure 1. Manufacture of ethyl alcohol by esterification–hydrolysis (indirect hydration).

(3) Reconcentration of the dilute sulfuric acid.

The hydrocarbon feedstock contains 35–95% ethylene; the remaining gases are methane and ethane. Certain unsaturated hydrocarbons are undesirable as their presence leads to the formation of secondary alcohols.

The absorption is carried out by countercurrent passage of ethylene through 95–98% sulfuric acid in a column reactor at 80°C and 1.3–1.5 MPa (180–200 psig) (40). The absorption is exothermic, and cooling is required (41) to keep the temperatures down and thereby limit corrosion problems. The absorption rate increases when ethyl hydrogen sulfate is present in the acid (42–45). This is attributed to the greater solubility of ethylene in ethyl hydrogen sulfate than in sulfuric acid.

The effects of various catalysts (46–47), contaminants (48–49), acid concentration (50), temperature (51), and pressure (52–56) on the rate of absorption have been studied. The patent literature indicates that absorption can be improved by making more efficient the contact between the gaseous ethylene and liquid sulfuric acid (57–60), by suitable design of the absorption tower (61), and by various combinations of absorption and hydrolysis (62–67).

The absorbate containing the mixed ethyl sulfates is hydrolyzed with enough water to give an approximate 50–60% aqueous sulfuric acid solution. The hydrolysis mixture is separated in a stripping column to give dilute sulfuric acid bottoms and a gaseous alcohol–ether–water mixture overhead. The overhead mixture is washed with water or dilute sodium hydroxide and then purified by distillation (62,64–65,67–68).

Diethyl ether is the principal by-product of the reaction of ethyl alcohol with diethyl sulfate. Various methods have been proposed to diminish its formation (69–71), including separation of diethyl sulfate from the reaction product. The diethyl sulfate not only causes an increase in ether formation but is also more difficult to hydrolyze to alcohol than is ethyl hydrogen sulfate. The equilibrium constant for the hydrolysis of ethyl hydrogen sulfate is independent of temperature, and the reaction rate is proportional to the hydrogen ion concentration (72–74).

The reconcentration of dilute (50–60%) sulfuric acid is one of the more costly operations in the manufacture of ethanol by this process. An acid reboiler, followed by a two-stage vacuum evaporation system raises acid concentration to about 90%. The 90% acid is then brought to 96–98% strength by fortification with 103% oleum (fuming sulfuric acid).

The buildup of carbonaceous materials in the sulfuric acid presents one of the most serious problems of acid concentration (75–79). Acid concentration also presents a corrosion problem. The vessels are mild steel lined with lead or brick; the steam-heating elements are composed of silicon, iron, or tantalum, and pipelines are generally constructed of lead (80).

Direct Hydration of Ethylene. Hydration of ethylene to ethanol via a liquid-phase process catalyzed by dilute sulfuric acid was first demonstrated more than a hundred years ago (81). In 1923, the passage of an ethylene–steam mixture over alumina at 300°C was found to give a small yield of acetaldehyde, and it was inferred that this was produced via ethanol (82). Since the late 1920s, several industrial concerns have expressed interest in producing ethanol synthetically from ethylene over solid catalysts. However, not until 1947 was the first commercial plant for the manufacture of ethanol by catalytic hydration started in the U.S. by Shell; the same process was commercialized in the United Kingdom in 1951.

There are two main process categories for the direct hydration of ethylene to ethanol. Vapor-phase processes contact a solid or liquid catalyst with gaseous reactants.

Mixed-phase processes contact a solid or liquid catalyst with liquid and gaseous reactants. Generally, ethanol is produced by a vapor-phase process; mixed-phase processes are used for the analogous hydration of propylene to isopropanol (see Propyl alcohols). Important exceptions to these two generalizations exist, but the discussion that follows emphasizes technology associated with the commercially important vapor-phase direct hydration of ethylene.

Chemistry. The stoichiometric equations pertinent to the vapor-phase hydration of ethylene over a catalyst support impregnated with phosphoric acid have been summarized (83).

$$C_2H_4 + H_2O \rightleftharpoons C_2H_5OH$$

Diethyl ether can be formed from the alcohol, or conversely, ether can be hydrated to ethanol.

$$2\ C_2H_5OH \rightleftharpoons (C_2H_5)_2O + H_2O$$

Acetaldehyde, a deleterious by-product, is most likely formed from trace amounts of acetylene in the feed (84–85).

$$C_2H_2 + H_2O \rightarrow CH_3CHO$$

The acetaldehyde is particularly undesirable as it leads to the formation of crotonaldehyde (86), an impurity that adversely affects ethanol quality even at parts per million levels.

$$2\ CH_3CHO \rightarrow CH_3CH(OH)CH_2CHO \rightarrow CH_3CH{=}CHCHO + H_2O$$

Both aldehydes can be hydrogenated to their saturated normal alcohols.

$$CH_3CHO + H_2 \rightarrow C_2H_5OH$$

$$CH_3CH{=}CHCHO + 2\ H_2 \rightarrow C_4H_9OH$$

Higher hydrocarbons are formed by the polymerization of ethylene. Any higher unsaturated hydrocarbons present are converted to the corresponding alcohols by hydration.

Catalysts. At ambient temperatures, only a relatively small amount of ethanol is present in the vapor-phase equilibrium mixture, and an increase in temperature serves only to decrease the alcohol concentration. An increase in pressure helps to shift the equilibrium toward the production of ethanol because of a decrease in the number of molecules (Le Chatelier's principle). On the other hand, reaction velocity is low at low temperatures. Hence, it is necessary to use catalysts and relatively high temperatures (250–300°C) to approach equilibrium within a reasonably short time.

Many catalysts for the hydration of olefins in general, and of ethylene in particular, are described in the patent literature. Practically all of them are acidic. Ellis presents a patent literature review through 1937 of the types of catalysts used (46–47), and a general review of olefin hydration is given by Tapp (87).

The use of phosphoric acid on a charcoal support was claimed in one of the earliest patents (88) on olefin hydration. Since then many claims have appeared for the use of phosphoric acid on supports and for the use of metal and boron phosphates (89). Phosphoric acid on a porous inert support such as Celite diatomite (calcined diatomaceous earth) has been commercially used by Shell as a catalyst (90) (see Diatomite). The support is impregnated with aqueous phosphoric acid of concentration less than 70% and then dried to give an acid concentration of 75–85%. A catalyst prepared this way appears dry and there is no "drooling" (weeping) of acid from the support. The main factor determining the catalytic activity is the concentration of phosphoric acid (a function of reaction temperature and steam pressure) in the pores of the support. If the concentration falls, the reaction rate declines; if the concentration becomes too high, there is an increasing tendency to polymerize ethylene. It has been pointed out (84) that iron and aluminum oxides in the Celite carrier react with the phosphoric acid. This promotes cracking of the ethylene, loss of physical strength of the support, and an entrainment of fines with resultant plugging of the reactor equipment. Leaching with a mineral acid to reduce iron and aluminum content prior to impregnation with phosphoric acid yields a catalyst that operates at lower temperatures, gives higher conversions of ethylene, and has a longer life (84). Treatment of the calcined support with superheated steam at 200–260°C, prior to leaching with acid, improves the mechanical strength of the catalyst (91–92).

For commercial application, catalyst activity is only one of the factors to be considered. Equally important is catalyst life, but little has been published on this aspect. Partly because of entrainment losses and partly through loss of acid as volatile triethyl phosphate, the catalyst loses activity unless compensating steps are taken. This decline in activity can be counteracted by the periodic or continuous addition of phosphoric acid to the catalyst during use, a fact that seems to have been disclosed as early as 1940 (93). A catalyst subjected periodically to acid addition could remain in service indefinitely according to a report by Shell (90). A later Shell patent (84) states that complete reimpregnation with acid is required every 200 run-days.

Catalyst longevity also requires a support material that does not crumble or disintegrate during preparation and use. Bentonites and montmorillonite (see Clay), extracted with HCl to reduce the alumina content to below 10%, are claimed by Hibernia-Chemie (94) to be better than Celite since they have superior mechanical strength and greater acid absorptivity. Hibernia also claims (95) porous carbon to be a robust and long-lasting support for phosphoric acid. Regular density silica gel as a support is reported to have low mechanical strength (84) and suffer fairly rapid disintegration (96), but a U.S.I. (National Distillers) patent (85) extolled the stability and high activity of their granular silica gel support of low density and high pore volume. A more recent patent (97) assigned to U.S.I. discloses that water vapor treatment of a preformed silica xerogel prior to phosphoric acid impregnation gives a catalyst with superior crush strength and microporosity. Two recent patents claim improved performance for silica gel supports containing 0.3–10 wt % of an alkali hydroxide/carbonate that are subjected to thermal treatment before impregnation.

Many other acids and acidic oxides have been mentioned as catalysts for ethylene hydration (98–102) as have ion-exchange resins (103–104) (see Ion exchange).

Blue tungsten oxide and combinations thereof (105–112) have been the subject of a number of patents as have copper(II) fluoborate (113) and alkali metal or ammonium sulfate/hydrogen sulfate catalysts (114–115).

Reaction Mechanism and Kinetics. The equilibria involved in the hydration/dehydration of ethylene first proposed by Whitmore (116) can be expressed as follows:

$$C_2H_4 \underset{-H^+}{\overset{+H^+}{\rightleftharpoons}} C_2H_5^+ \underset{-OH^-}{\overset{+OH^-}{\rightleftharpoons}} C_2H_5OH$$

$$\underset{-H^+}{C_2H_5OH} \Big\| \underset{+H^+}{C_2H_5OH}$$

$$(C_2H_5)_2O$$

The rate-determining step involves addition/subtraction of the proton (117) which is donated by the catalyst.

With phosphoric acid-based catalysts, in which the active component is liquid acid absorbed in the pores of the support, the reaction probably follows the path proposed by Taft (118) for the hydration of olefins in aqueous solution:

$$>C=C< + H_3O^+ \underset{}{\overset{(fast)}{\rightleftharpoons}} \left[>C\overset{\overset{H}{|}}{=}C< \right]^+ + H_2O$$

π complex

(rate determining step)

$$CH_3CH_2-OH + H_3O^+ \underset{-H_2O}{\overset{+H_2O}{\rightleftharpoons}} CH_3CH_2-\overset{+}{O}H_2 \underset{-H_2O}{\overset{+H_2O}{\rightleftharpoons}} \overset{+}{C}H_3CH_2$$

(fast) oxonium ion (fast) carbenium ion

The kinetics of the ethylene hydration reaction have been investigated for a tungstic oxide/silica gel catalyst, and the energy of activation for the reaction determined to be ~125 kJ/mol (ca 30 kcal/mol) (105,119). Gel'bshtein and co-workers (120) examined the kinetics over a phosphoric acid/silica gel catalyst. Making some simplifying assumptions to Taft's mechanism, they derived a rate equation of the form:

$$\text{rate} = k(P_e - P_a/P_w K_p)$$

where k_1 = rate constant for the forward reaction, P_w = partial pressure of water vapor, P_e = partial pressure of ethylene, P_a = partial pressure of ethanol.

Equilibrium Constant. At the pressures used in commercial production of ethanol (6.1–7.1 MPa or 60–70 atm), alcohol yield per pass is significantly limited by equilibrium considerations. This fact has focused attention on determination of equilibrium constants and equilibrium yields (121–123). The results of these determinations are shown below:

$$\log K_f = 2132/T - 6.241$$
$$\Delta F_f^\circ = 28.6/T - 9.740$$

where f = fugacity, K_f = equilibrium constant based on fugacity of components, F_f = free energy based on fugacity.

On the basis of a calculated equilibrium constant (124), the equilibrium conversions have been calculated at various temperatures, pressures and water/ethylene ratios (Figs. 2–3) (124).

Effect of Process Variables. Several investigators (102,105) have studied the effects of the following reaction parameters:

(1) An increase in pressure causes a corresponding increase in ethanol production rate. Higher pressures also increase polymer formation; hence, there is little practical advantage to be gained above a certain limit.

(2) An optimum temperature exists at which ethanol production rate is maximal. Ethylene conversion is limited by catalyst activity at lower temperatures and by equilibrium considerations at higher temperatures.

(3) An optimum ethylene-to-water ratio exists that gives a maximum ethanol production rate. However, as expected, the highest ethylene conversion is obtained at the lowest ethylene-to-water mole ratio.

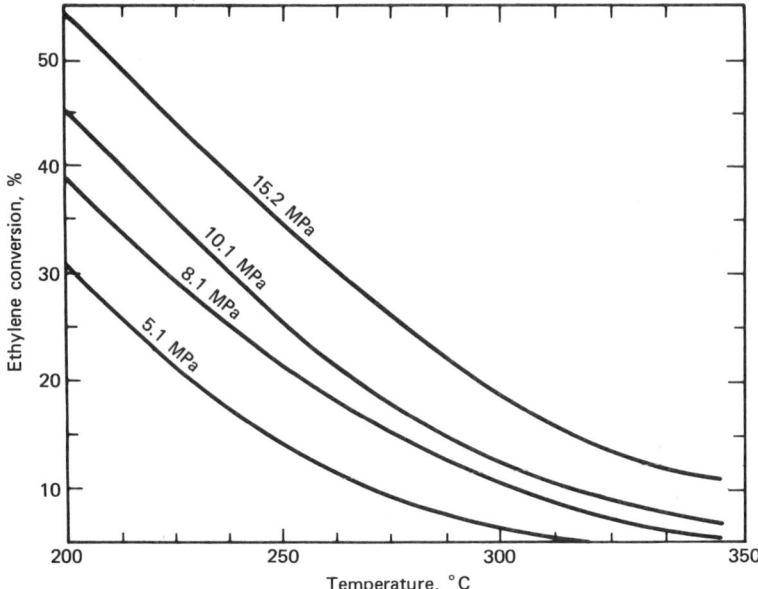

Figure 2. Calculated conversions at equilibrium in the vapor phase hydration with equimolar ethylene and water (124); to convert MPa to atm, divide by 0.101.

(4) An increase in space velocity increases the ethanol production rate but at the expense of higher recycling costs.

Hibernia Chemie has described a vapor-phase process that passes fresh and recycled 85 wt % phosphoric acid over a catalyst of hydrochloric acid-leached bentonite impregnated with phosphoric acid. Catalyst activity was claimed to remain constant over a period of one year at the following conditions (125):

temperature, °C	265
pressure, kPa (atm)	7114.61 (70.23)
space velocity, h^{-1}	1727
mole ratio, ethylene/water at inlet	1.2
conversion/pass	6.18
yield/pass	5.98
selectivity, %	96.8

Process. Figure 4 shows a simplified flow diagram for the process employed by Union Carbide to produce ethanol by the direct hydration of ethylene (126). An ethylene-rich gas is combined with process water, heated to the desired reaction temperature, and passed through a fixed-bed catalytic reactor to form ethanol. The vapor leaving the reactor is slightly hotter than the feed because the reaction is exothermic. The reactor product is cooled by heat exchange with the reactor feed stream and is separated into liquid and vapor streams. The liquid stream goes to the ethanol refining system, and the vapor stream is scrubbed with water to remove the ethanol. The washed gas, mostly unreacted ethylene, is enriched with fresh ethylene feed and recycled to the reactor. A small vent or purge stream is removed from the recycled ethylene to prevent buildup of impurities in the gas cycle.

The liquid product streams are fed to a distillation system to remove the light

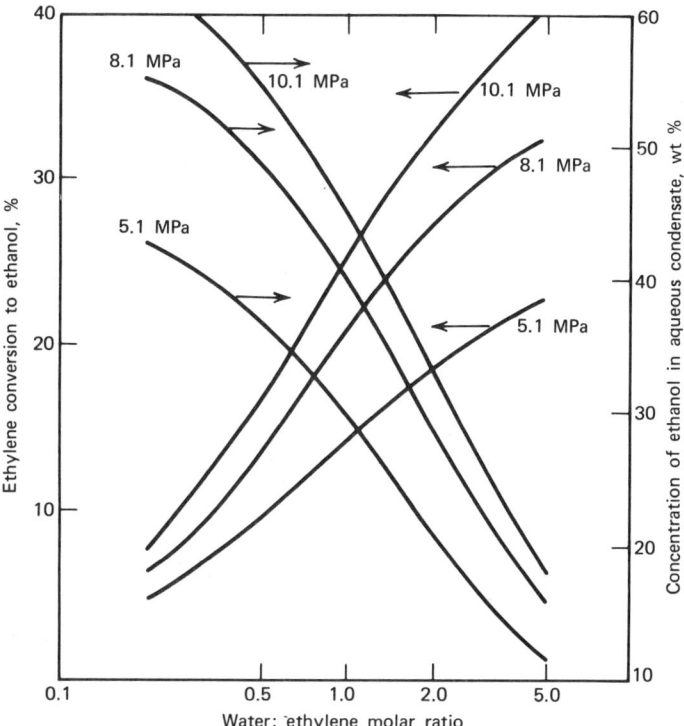

Figure 3. Calculated conversions at equilibrium at 250°C in the vapor phase hydration (124); to convert MPa to atm, divide by 0.101.

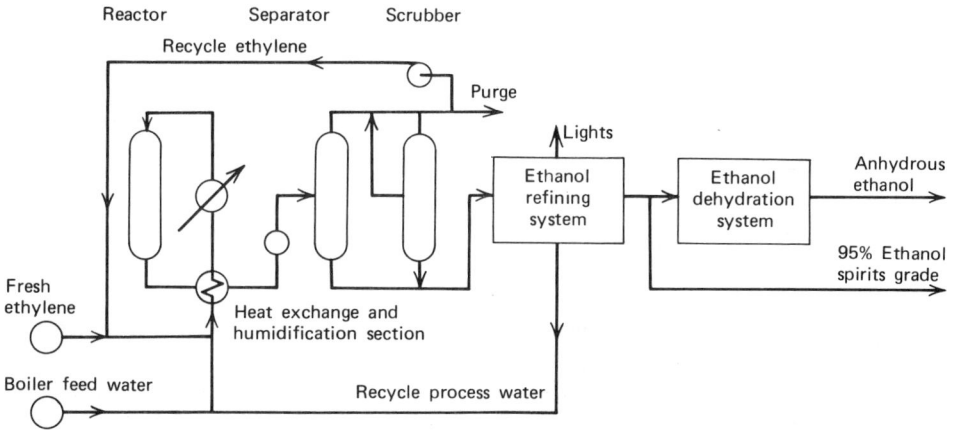

Figure 4. Manufacture of ethanol by direct hydration of ethylene. Courtesy Gulf Publishing Co.

impurities and to recover the ethanol as a 95% volume ethanol–water azeotrope. To produce anhydrous ethanol, the ethanol–water azeotrope is fed to a dehydration system.

An advantage provided by this process is the recycling of process water recovered in the refining stills to the reaction system. This reduces the amount of boiler feed water to less than one-fifth of the total amount of water fed to the reactor. The investment and operating cost required for supporting boiler feed water facilities are correspondingly reduced. Recycling process water also reduces the amount of water discharged to the sewer, thus decreasing the ethanol losses and the load on pollution abatement facilities.

Several patents (127–129) deal with methods for preventing the formation of deposits in heat exchangers, reducing corrosion and avoiding the need for corrosion-resistant materials. Copper is widely used for lining the reactors and for piping, and some heat exchangers are made of phosphor bronze. Eastman Kodak Co. (130) advocates the use of a stainless steel-clad reactor lined with overlapping copper curtains or shingles for corrosion resistance. A Hibernia Chemie patent (95) claims that a copper lining in the reactor has a limited life and can promote the formation of cuprene. They describe a porous carbon brick lining with a Cu-Ag alloy between the brick and the reactor wall (125).

Other Methods of Preparation. In addition to the direct hydration process, the sulfuric acid process, and fermentation routes to manufacture ethanol, several other processes have been suggested. These include the hydration of ethylene by dilute acids, the hydrolysis of ethyl esters other than sulfates, the hydrogenation of acetaldehyde, and the use of synthesis gas. None of the above methods has been successfully implemented on a commercial scale, but the route from synthesis gas is currently receiving a great deal of attention.

Hydration of Ethylene Using Dilute Acids. A review of the early work on the hydration of ethylene using dilute acids, the weak acid process, is given by Ellis (47). The reaction is favored by the use of low temperatures and high pressures. Temperatures in the range of 150–250°C are the most frequently quoted (131–136), although temperatures as low as 80°C have been reported (137).

Equilibrium constants for the hydration of ethylene were measured (138) using a 3 mol % sulfuric acid solution at 176–307°C and 8.1–26.3 MPa (80–260 atm). Above 250°C and 20.3 MPa (200 atm) polymerization of ethylene became important, and below 220°C ethyl ether formation was significant. Both high temperature and high acid concentration favor the reduction of sulfuric acid to sulfur dioxide (139); 65% acid is reduced by ethylene at above 180°C. To control the severe corrosion problems encountered, a tantalum-plated reactor has been recommended (146). Loss of acid catalyst in the off-gas can be compensated by adding ethyl sulfate in the feed (140). Operation at pressures from atmospheric to 2.5 MPa (25 atm) using a battery of reactors have been described (141). Ethylene was absorbed in the acid solution during countercurrent passage through the battery at 130°C.

Hydration of Ethyl Ether. Using the same type of acid catalysts as in the hydration of ethylene to ethanol, ethyl ether can be hydrated to the alcohol. Catalysts that have been used for the hydration of ether include phosphoric acid (142), sulfuric acid (143–144), hydrochloric acid (145), metallic oxides (146–148) and silicates (149). Sulfuric acid concentrations ranging from 5–25% at 200°C (143) to 63–70% at 110–135°C and 1.01–1.42 MPa (10–14 atm) (147) have been claimed.

Aluminum oxide has been the most widely used catalyst (150). At 320°C and at 1.01–1.42 MPa, 50–66% conversion to alcohol based on the ether was obtained. The ethanol produced by the direct hydration of ether generally has a foul odor owing to the presence of polymeric hydrocarbon material, which can be removed by washing the aqueous alcohol with ether (151).

Hydrolysis of Ethyl Esters. The hydrolysis of esters (other than ethyl sulfates) is not a commercial route for producing ethanol. An indirect hydration of ethylene actually takes place during the proposed (152) hydrolysis of ethyl sulfite catalyzed by silver sulfate.

$$2\ C_2H_4 + SO_2 + H_2O \rightarrow (C_2H_5O)_2SO$$

$$(C_2H_5O)_2SO + 2\ H_2O \rightarrow 2\ C_2H_5OH + H_2SO_3$$

The hydrolysis of ethyl acetate, prepared by the reaction of ethylene with acetic acid under pressure (153), and the hydrolysis of the ethyl ester of chlorosulfonic acid (154) have been considered and found to be of little industrial importance.

Hydrogenation of Acetaldehyde. Acetaldehyde made from acetylene can be hydrogenated to ethanol with the aid of a supported nickel catalyst at 150°C (155). A large excess of hydrogen containing 0.3% of oxygen is recommended to reduce the formation of ethyl ether. Anhydrous ethanol has also been made by hydrogenating acetaldehyde over a copper on pumice catalyst (156).

Oxidation of Hydrocarbons. Ethanol is one of a variety of oxygen-containing compounds produced by the oxidation of hydrocarbons (see Hydrocarbon oxidation). Ethanol is reported to be obtained in a yield of 51% by the slow combustion of ethane (157–158). When propane is oxidized at 350°C under a pressure of 17.2 MPa (170 atm) (159–160), 8% of oxygen is converted to ethanol. Lower conversions to ethanol are obtained in oxidizing butane. Other oxidation systems used to produce ethanol and acetaldehyde (161–162) and methods for separating the products have been described in the patent literature.

Synthesis Gas. Since petroleum prices rose abruptly in 1974, the production of ethanol from synthesis gas, a mixture of carbon monoxide and hydrogen, has received considerable attention. The use of synthesis gas as a base raw material has the same drawback as fermentation technology: low yields limited by stoichiometry.

$$2\ CO + 4\ H_2 \rightarrow CH_3CH_2OH + H_2O$$

Its appeal lies in the fact that synthesis gas can be produced from trash, municipal sewage, scrap wood, sawdust, newsprint or other waste (see Fuels, synthetic; Fuels from biomass; Fuels from waste).

The early work of Fischer and Tropsch on the methanol synthesis showed that ethanol could be obtained in the process (164) and that by certain modifications the proportion of ethanol in the product could be increased (165). The literature concerning this method is extensive (166–175). The conditions that favor ethanol formation are 125–175°C and 1.42 MPa (14 atm) in the presence of reduction catalysts such as powdered iron.

Ethanol can also be obtained by the reaction of methanol with synthesis gas at 185°C and under pressure (6.9–20.7 MPa or 68–204 atm) in the presence of a cobalt octacarbonyl catalyst (176). However, although ethanol was the major product, methyl formate and methyl, propyl and butyl acetates and propyl and butyl alcohols and methane were all present in the product (see Oxo process).

A recent Belgian patent (177) claims improved ethanol selectivity of over 62%, starting with methanol and synthesis gas and using a cobalt catalyst with a halide promoter and a tertiary phosphine. At 195°C, and initial carbon monoxide pressure of 7.1 MPa (70 atm) and hydrogen pressure of 7.1 MPa, methanol conversions of 30% were indicated, but the selectivity for acetic acid and methyl acetate was only 7%.

Ruthenium and osmium catalysts (178–179) have also been employed for this reaction. The addition of a bicyclic trialkyl phosphine is claimed to increase methanol conversion from 24% to 89% (180) (see also Methanol).

Fermentation Ethanol. Fermentation (qv), one of the oldest chemical processes known to man, is used to make a variety of products, including foods, flavorings, beverages, pharmaceuticals, and chemicals. At present, however, many of the simpler products such as ethanol, are synthesized from petroleum feedstocks at lower costs. The future of the fermentation industry, therefore, depends on its ability to utilize the high efficiency and specificity of enzyme catalysis to synthesize complex products and on its ability to overcome variations in the quality and availability of raw materials.

Ethanol is made from a variety of agricultural products such as grain, molasses, fruit, whey, and sulfite waste liquor. Ethanol production by fermentation, excluding that for beverages, has been declining in the United States since synthetic ethanol was introduced in the 1930s, because of the low cost and assured availability of ethylene. In 1975, only ca 76×10^6 L of 190 proof industrial ethanol were produced by fermentation, compared to 795×10^6 L by synthesis. Generally, most of the agricultural products mentioned above command higher prices as foods, and others, eg, potatoes, are uneconomical because of their low ethanol yield and high transportation cost. The energy crisis of the early seventies may have generated renewed interest in ethanol fermentation, but its use still depends on the availability and cost of the carbohydrate relative to the availability and cost of ethylene. Sugar and grain prices, like oil prices, have risen dramatically since 1973.

In some European countries (eg, Belgium, France, Italy and the Netherlands) fermentation is still important in producing ethanol. In others (eg, Denmark, England and Germany) synthetic ethanol predominates (181). In Japan more ethanol is fermented than synthesized. The 1973 annual capacity was estimated at ca 250×10^6 L for ethanol fermentation and only 76×10^6 L for synthetic ethanol (182). Current U.S. producers of fermented industrial alcohol include the Chemical Division of Georgia-Pacific, Grain Processing Corporation, and Publicker Industries. Georgia-Pacific is the only producer using sulfite waste liquor as the carbohydrate source. In addition, alcoholic beverage producers cannot be ruled out as potential industrial ethanol manufacturers (see Beverage spirits, distilled).

Ethanol can be derived by fermentation processes from any material that contains sugar. The many and varied raw materials used in the manufacture of ethanol via fermentation are conveniently classified under three types of agricultural raw materials: sugar, starches, and cellulose materials. Sugars (from sugar cane, sugar beets, molasses, fruit) can be converted to ethanol directly (see Sugar). Starches (from grains, potatoes, root crops) must first be hydrolyzed to fermentable sugars by the action of enzymes from malt or molds (see Beer; Yeast). Cellulose (qv) (from wood, agricultural residues, waste sulfite liquor from pulp and paper mills) must likewise be converted to sugars, generally by the action of mineral acids. Once simple sugars are formed, enzymes from yeast can readily ferment them to ethanol (183). Because fermentation ethanol has been thoroughly and repeatedly discussed in the literature, the coverage here is illustrative rather than comprehensive, with special emphasis on the potential raw materials for ethanol production of the future.

Sugars. The most widely used sugar for ethanol fermentation is blackstrap molasses (183–194) which contains about 35–40 wt % sucrose, 15–20 wt % invert sugars such as glucose and fructose, and 28–35 wt % of nonsugar solids. Blackstrap (derived from Java and the Dutch word "stroop," meaning syrup) is collected as a by-product of cane sugar manufacture (see Syrups). The molasses is diluted to a mash containing ca 10–20 wt % sugar. After the pH of the mash is adjusted to about 4–5 with mineral acid, it is inoculated with the yeast, and the fermentation is carried out nonaseptically at 20–32°C for about 1–3 d. The fermented beer, which typically contains ca 6–10 wt % ethanol, is then sent to the product recovery and purification section of the plant.

The direct fermentation of sugar cane juice, sugar beet juice, beet molasses (a by-product in the production of beet sugar), fresh and dried fruits, cane sorghum, whey, and skim milk had been considered as a means of obtaining ethanol, but none of these raw materials could compete economically with molasses. Although the manufacture of ethanol from the sugar-containing waste products of the fruit industry appears to be a highly desirable operation, particularly as a means of reducing stream pollution in the vicinity of canning plants, such production is costly because of the need to remove most of the water (as much as 97%) contained in the waste product.

The quadrupling of the selling price of crude petroleum by the Organization of Petroleum Exporting Countries (OPEC) in 1973 may have a profound impact on fermentation processes for producing ethanol. A primary objective of the U.S. Department of Energy is to develop methods to derive fuels economically from sugar crops and corn, to evaluate the potential feasibility of the various methods, and to suggest means of practical application. The results of the sugar crop research on agronomics and fuels conversion undertaken by Batelle's Columbus Laboratories (185) lists several merits of sugar cane as a candidate energy resource. Sugar cane, a renewable raw material, is renowned for its agricultural productivity, and its juice is directly fermentable to ethanol. On the other hand, sugar cane products are valuable in food and feed applications and their conversion to chemicals and energy can be considered an underutilization of their potential value. By the year 2000 the energy needs of the U.S. are expected to exceed ca 10^{17} kJ (100 quads or 10^{17} Btu) (186–187) and sugar cane is expected to contribute to less than 2×10^{15} kJ (2 quads) of energy. However, as is seen under Manufacturing Costs below, the prospects for sugar cane as a raw material for industrial ethanol are much brighter.

The subject of fermentation alcohol has always been of considerable interest to several tropical countries, but until the oil crisis of 1973, only India appeared to appreciate the importance of fermentation alcohol as a strategic material in its economy. Ethanol prices in India have been maintained at an extremely low level by processing cane molasses, which has been a waste product of negligible value (188).

In 1975, Brazil embarked upon an ambitious program for fermentation ethanol manufacture (189–190) which could provide the necessary blueprint for similar programs in other tropical countries. The five-year project, which is expected to cost the government and private Brazilian investors one billion dollars, involves the planting of "energy plantations" at the rate of an additional 8.1×10^9 m² (2×10^6 acre) of sugar cane and other plants each year. If the national plan achieves its goal, Brazil will be producing 3×10^9 L of ethanol a year by 1980, nearly equal to the current worldwide consumption (192). By 1981, fully one-fifth of the gasoline used in Brazil is planned to be replaced by alcohol which, with government price regulations, could cost ca one-half as much to produce as the retail price of gasoline.

By 1977, annual Brazilian alcohol production had risen to a level of 1.6×10^9 L. That is the equivalent of 12×10^6 bags of sugar, more than 14% of the total Brazilian sugar harvest, itself the largest in the world. In addition to sugar cane, for which the ethanol fermentation technology is the simplest and best understood, the Brazilian government is actively pursuing plans to cultivate the sugar-yielding tuber, manioc (cassava) (192).

Starches. All potable alcohol and most fermentation industrial alcohol is currently made principally from grains. Fermentation of starch from grain is somewhat more complex than fermentation of sugars because starch must first be converted to sugar and then to ethanol. This process was known to the ancient Egyptians and Mesopotamians, who brewed beer almost 5000 years ago (193). The simplified equations for the conversion of starch to ethanol are:

$$C_6H_{10}O_5 + H_2O \xrightarrow{enzyme} C_6H_{12}O_6 \xrightarrow{yeast} 2\ C_2H_5OH + 2\ CO_2$$

Starch is converted enzymatically to glucose either by diastase present in sprouting grain or by fungal amylase. The resulting dextrose is fermented to ethanol with the aid of yeast producing CO_2 as a coproduct. A second coproduct of unfermented starch, fiber, protein and ash known as distillers grain (a high protein cattle feed) is also produced (see Starch; Wheat and other cereal grains).

Among the disadvantages in the use of grain are its fluctuations in price. Each time the price of grain has fallen, there has been renewed interest in the use of grain alcohol as an automotive fuel additive (gasohol). Not until late 1973 could ethanol derived from grain compete with ethanol from ethylene in the industrial marketplace. With higher gasoline prices it has been claimed (194–195) to be an economical additive to unleaded automotive fuel (see Gasoline).

Cellulosic Materials. Approximately 700×10^6 metric tons of carbohydrate-containing cellulosic wastes are generated annually. The technology for converting this material into petroleum feedstock is available, but the stoichiometry of the process is disadvantageous. Even if each step in the process of the conversion of cellulose to ethanol proceeded with 100% yields, almost two-thirds of the mass would disappear during the sequence, most of it as carbon dioxide in the fermentation of glucose to ethanol. This amount of carbon dioxide leads to a disposal problem rather than to a raw material credit (196).

Another problem is that the aqueous acid used to hydrolyze the cellulose in wood to glucose and other simple sugars destroys much of the sugars in the process. Nevertheless, it has been claimed that forests could theoretically provide 50% of the oil and gas used by U.S. utilities in 1978, replacing 20% of annual fossil fuel consumption (197), and a lot sooner than earlier forecasts had anticipated. New ways of reducing the cost of converting cellulosic wastes from wood, newspapers, and municipal garbage into glucose include the use of less corrosive acids and reduced hydrolysis time. One way of making cellulose wastes more susceptible to hydrolysis is by subjecting them to a short burst of high energy electron beam radiation (198). Hydropulping of cellulose feedstocks followed by a 10 μs burst from a 3×10^6-eV electron beam accelerator is claimed to reduce the time of hydrolysis by dilute acid from hours to seconds.

An alternative to acid hydrolysis is the use of enzymes (qv). Although they avoid the corrosion problems and loss of fuel product associated with acid hydrolysis, enzymes have their own drawbacks. Enzymatic hydrolysis slows as the glucose product accumulates in a reaction vessel. This end-product inhibition eventually halts the hydrolysis unless some way is found to draw off the glucose as it is formed. In mid-1978, Gulf Oil researchers described the simultaneous enzymatic hydrolysis of cellulose and fermentation of the resulting glucose to ethanol, removing glucose as it is formed and overcoming the problem of product inhibition of hydrolysis (197). Mutated strains of the common soil mold *Trichoderma viride* can process fifteen times as much cellulose as natural strains. The results have been encouraging; in some cases, cellulose from sawdust, bark, and effluent streams from the pulp and paper industries have produced ethanol in yields approaching 100% of the theoretical value. Design for a commercial-scale plant producing 95×10^6 L of ethanol annually is expected by the end of 1979.

Recovery and Purification. Various distillation and equipment modifications are used to ensure a pure water azeotrope of ethanol (95% by weight ethanol). A review of the patent literature concerning the many methods used for the purification of ethanol obtained by various synthetic processes is given in ref. 199. Some phosphoric acid is vaporized, or entrained as liquid, from the reactor. During recovery the reactor effluent is cooled, causing the condensation of acid on exchanger and piping walls, creating a potential corrosion problem. Corrosion can be prevented by spraying the effluent with caustic solutions so that the resultant condensate has a pH in the range 6.5–7.5. A second benefit is that this treatment aids in achieving an acceptable odor in the purified product (200). Eastman (201) and Hibernia (127) describe the use of a spray directly into the reactor effluent; Shell (200) partially cools the effluent in a heat exchanger before contacting with a spray.

The specifications for an industrial alcohol depend upon the intended use. Probably for this reason noncomparable purification schemes are frequently found in the patent literature (202–206). Premium grade ethanol has low water content, low odor ratings and high permanganate-time test values. Odor is improved by the partial hydrogenation of ethanol using Raney nickel catalyst (207) or by contact with unglazed porcelain and iron (208). The permanganate test is extremely sensitive to the presence of aldehydes. One ppm of crotonaldehyde or 2 ppm of sorbaldehyde is reported to decrease the test time from 60 to 30 min (200). This sensitivity suggests that purification by distillation alone is inadequate for achieving high permanganate test times (200). Carbonyl-containing and unsaturated materials are removed by treatment with sodium borohydride (209–210) and boric acid (211). Other methods used to remove carbonyl impurities include treatment with hydroxylamine hydrochloride, potassium permanganate, or N-hydroxybenzenesulfonamide (211).

Purification schemes have generally emphasized the following techniques: (*1*) extractive distillation using water reflux to distill a large share of the impurities and concentrate the crude alcohol–water mixture (see Azeotropic and extractive distillation), (*2*) efficient fractionation to produce approximately 190 proof alcohol (see Distillation; Beverage spirits), (*3*) hydrogenation to convert aldehyde impurities to alcohols, together with the use of chemicals such as inorganic bases and sodium sulfite (204), and (*4*) ion-exchange resins or azeotropic distillation to dehydrate 190 proof to 200 proof or absolute alcohol.

Manufacturing Costs

The cost of producing ethanol depends upon the location of the manufacturing plant; the design, type, and degree of modernization of equipment; the kind of raw material used; the price paid for the raw material; the relative labor costs represented; the scale of production; and the total investment. Currently, the economics of fermentation ethanol does not compare favorably with that of synthetic ethanol from ethylene because each kilogram of ethanol requires 2 kg sugar, 3.3 kg corn or 4 kg molasses. On the other hand, it takes only 0.6 kg of ethylene to make a kilogram of ethanol. The raw material cost via ethylene is only 15.9¢/kg (or ca 13.2¢/L) at an ethylene cost of 26.4¢/kg in 1977 (212). It is therefore expected that ethylene will continue to be the raw material choice for ethanol, a choice that dictates both availability and price range for the alcohol. Plentiful ethylene means plentiful ethanol relative to the limited demand for the material.

Inasmuch as all fermentation alcohol plants differ in some respects, no accurate, generally applicable, basic investment cost can be formulated, and no standard production technique exists. In general, plants utilizing molasses apply a much simpler process and require less equipment than do grain plants, since the latter entails grain handling and by-product feed recovery operations and has higher steam requirements. With decreasing petroleum and natural gas supplies, increasing attention is being focused on the manufacturing costs and economic feasibility for fermentation processes that use regenerable raw materials.

The efficient production of ethanol from sugar cane is based upon using the fibrous by-products to generate the steam needed to crush and mill the agricultural inputs, with enough steam left over to conduct the ethanol distillation (see Bagasse). Batelle (185) has estimated the manufacturing cost for ethanol on a 265×10^6 L/yr scale, shown in Table 2. This is about 30% of the 1977 U.S. industrial ethanol production, a scale that a 202 km^2 plantation or cooperative might produce. Batelle estimates that cane juice might be obtained for about $0.13/kg of fermentable solids, and fiber might be obtained for about $1.05/GJ ($1.00/10^6 Btu). When an appropriate penalty is applied to the significant quantities of stillage (distiller's dried solubles) produced, because of its relatively high ash and low protein content, the net manufacturing cost of ethanol appears to be approximately 28.5¢/L. A manufacturing cost of 28.5¢ is equivalent to a selling price of about 40¢/L. The 0.5 kg of ethylene contained in a liter of ethanol was worth about 13.2¢ in 1976, and the selling price of denatured ethanol was 30¢/L. An increase in ethylene price to 46¢/kg would lead to a selling price of 40¢/L if the present margin per liter were retained. Because 46¢/kg ethylene could come from $132.1/m^3 ($21/bbl) petroleum, the cross-over point between petrochemical ethanol and fermentation ethanol would be approximately $132.1/m^3 of petroleum. If petrochemical ethanol producers increase their nonraw material charges as ethylene price increases take place, the cross-over could occur at $106.9/m^3 ($17/bbl).

In December, 1976, a three-million-kilometer test program to compare the performance and properties of unleaded gasoline and gasohol (a 10 vol % solution of grain alcohol in gasoline) was begun in Nebraska. Approximately 2.5×10^6 km of driving experience have been accumulated to 1979 with no problems encountered (194–195). The economic feasibility of gasohol in Nebraska is due to several factors, one of which is an 0.85¢/L lower tax on automotive fuel containing at least 10% agriculturally-derived ethanol than on standard gasoline. In addition, the ethanol increases the fuel

Table 2. Estimated Manufacturing Costs of Ethanol From Sugar Cane Juice[a]

	Millions of dollars
Capital investment	
battery limits plant[b]	49.0
utilities, offsites, etc	24.0
interest during construction	9.0
working capital and startup expenses	24.0
Total	106.0
annual capital charges, 20 yr, 9%	11.6
Operating and maintenance	
nutrients and denaturants	2.0
utilities: steam at $6.6 per metric ton	16.8
other	0.6
labor and supervision	1.6
maintenance	1.8
overheads, supplies, indirects	2.4
taxes and insurance	1.6
Total	83.0
Raw materials cost	
cane juice at $0.13/kg.	56.5
Credits	(20.4)
Net manufacturing cost	74.2
Net manufacturing cost (in ¢/L)	28.5

[a] Ref. 185.
[b] Excludes steam plant, includes fermentation at $35 million and distillation at $11 million.

octane number, permitting a less costly petroleum base stock to be used for blending gasohol. It has been found that cars using gasohol consumed ca 6.7% less fuel than the cars operating on unleaded gasoline. Finally, anhydrous ethanol and gasoline display a positive volume change on mixing at alcohol levels below 16 liquid volume percent.

However, the biggest factor in the success of the gasohol program would be the ability to produce anhydrous ethanol from grain at less than $0.32/L. Researchers at the University of Nebraska believe that a conventional fermentation plant, producing 7.6×10^7 L/yr anhydrous ethanol at an estimated cost of 23 million dollars, could lead to a profitable business. The favorable economics shown in Table 3 relies on the partial use of distressed or moldy grain (which could cost as little as one-fifth the 0.074¢/L for first quality grain) and a ready cattle market for the distillers dried grains.

Table 4 indicates that when both the farming operation and the fermentation process are considered, and when proper use is made of field waste, then for every three liters of grain alcohol produced at least one liter is new (fixed-solar) energy entering the economy. In contrast, when a similar analysis is made for the production of synthetic ethanol from ethylene, the energy equivalent of ca two liters of ethanol is consumed for every liter produced (193). Depending on gasoline prices and the cost of the raw material, the production of industrial alcohol by the fermentation of grain could become attractive from both economic and energy considerations (see Gasoline; Fuels from biomass).

358 ETHANOL

Table 3. Economic Analysis For Fermentation Ethanol From Grain, $/yr [a]

Basis		
76 × 10⁶ L ethanol/year		
fixed capital $23 million		
working capital $3.5 million		
total investment $26.5 million		
Expenses		
corn, 204.4 million L/yr, 0.074¢/L	15,125,000	
distressed grain, 67 × 10⁶/yr, 0.037 ¢/L	2,479,000	
conversion cost, 7.9¢/L C_2H_5OH	6,004,000	
Total expenses	*23,609,000*	
Income	C_2H_5OH at 29¢/L	34¢/L
ethanol	22,040,000	25,840,000
distillers dried grains $120/metric ton	8,600,000	8,600,000
carbon dioxide, $2 per metric ton	130,000	130,000
total income	30,770,000	34,570,000
profit: before taxes and depreciation	7,161,000	10,961,000
depreciation, 10 yr straight line	2,300,000	2,300,000
gross profit	4,861,000	8,661,000
taxes at 50%	2,430,500	4,330,500
net profit	2,430,500	4,330,500
return on investment	9.2%	16.3%
cash flow (net and depreciation)	4,730,500	6,630,500

[a] Courtesy of the ACS (193).

Table 4. Overall Energy Balance For Grain Alcohol Production From Corn By Using 75% Of The Field Waste [a]

	kJ[b]/L C_2H_5OH
energy production	
ethanol	21,052
aldehydes, fusel oil	306
75% of stalks, cobs, husks	34,583
Total	*55,941*
energy consumption	
farming operation	12,788
transportation of stalks, etc	334
alcohol plant	30,024
net consumption in by-product production	5,060
Total	*48,206*
total net energy production	7,735

[a] Courtesy of the ACS (193).
[b] To convert kJ to Btu, divide by 1.054.

Economic Aspects

In 1977, worldwide synthetic ethanol capacity amounted to 2.19×10^9 L, with the U.S. having a capacity of 1.173×10^9 L, 55% of the world total (Table 5). Extensive overcapacity exists in the domestic ethanol market; demand for the alcohol has been between $(757-795) \times 10^6$ L annually (Table 6). Growth at a relatively modest rate of 2–3% per year was projected, and production was not expected to reach 85% of capacity

Table 5. Synthetic Ethanol Capacity in 1977 [a]

	$L \times 10^{6a}$
U.S.	1173
Union Carbide	454
National Distillers	246
Publicker	227
Shell Chemical	151
Texas Eastman	95
Canada	64
U.K.	318
France	151
Federal Republic of Germany	159
Eastern Europe	227
Japan	98
Total	*2190*

[a] Courtesy of the ACS (213).

before 1981. The overcapacity has resulted in pressure on prices but they are not expected to decline (212) unless feedstock ethylene prices drop. Ethylene accounts for about half the cost of making synthetic ethanol; in 1977 it was still the least expensive feedstock.

1.514×10^9 L of synthetic ethanol are produced in the U.S., U.K., France, Fed. Rep. of Germany, and Japan; the balance, some 1.893×10^9 L of industrial alcohol, comes via the fermentation route and is tied inextricably to the future of agricultural politics through national monopoly control agencies. Until the early 1960s, the price of industrial alcohol in the United States (Table 7) was closely related to the cost of molasses during normal world conditions and to the cost of grain during emergency times. Now that synthetic ethanol represents most U.S. production, greater price stability has been obtained. Fermentation ethanol became attractive again in late 1973 as hydrocarbon prices rose because the fermentation process requires only solar energy (qv). However, fermentation ethanol is seen as being uneconomic before 1985, largely because ethylene prices did not rise as fast as projected at the end of 1973.

The increase in foreign synthetic ethanol capacity has been dramatic. In 1973, U.S. producers commanded a hefty 74% of the world synthetic ethanol capacity; in 1977 the U.S. had only 54% of the worldwide capacity. By the end of 1978, world capacity is expected to increase another 3.33×10^9 L (224). This will cause the U.S. share to drop to 46%.

Not only did U.S. ethanol exports decline sharply in the late 1970s (Table 8), but ethanol's use as a chemical intermediate suffered considerably from replacement in the production of acetaldehyde, butyraldehyde, acetic acid, and ethylhexanol. The switch from the ethanol route to those products has depressed demand for ethanol by more than 95×10^6 L (25×10^6 gal) since 1971. This decrease reflects the recent process developments for the manufacture of acetaldehyde (qv) and acetic acid, largest use for acetaldehyde, by direct routes using ethylene and butane, respectively (172) (see Hydrocarbon oxidation). For example, U.S. consumption of ethanol for acetaldehyde manufacture (Table 9), declined steadily from 50% in 1962 to 37% in 1964 and 8% in 1976.

Table 6. U.S. Industrial Ethanol Production by Materials Used, 10^6 L (95 vol %)

Fiscal year ending June 30	From fermentation materials[a]					From synthetic materials			Total alcohol
	Molasses	Sulfite liquors	Grain	Other	Total	Ethyl sulfate	Ethylene	Total	
1930	363.7				363.7	3.4		3.4	367.1
1935	307.7		9.8	7.2	324.7	35.2		35.2	359.9
1940	333		28.0	2.6	363.7	121.9		121.9	485.6
1945	391.4	1.9	144.6	131.3	669.2	222.5		222.5	891.7
1950	215.4	6.8	4.9	4.5	231.6	339.5	54.1	393.6	624.5
1955	133.2	11.0	17.0	6.8	168.4	546.9	104.8	652.2	820.6
1960	60.9	15.1	15.9	2.6	94.6	778.2	160.5	938.7	1033.3
1965	8.33	14.0	159.7	47.3	229.4	891.7	234.3	1126.1	1355.4
1970	5.68	14.0	168.8	52.6	241.1	606.8	556.4	1164.3	1404.2
1975	25.7	15.5	222.5	52.2	316.1	60.9	788.8	849.7	1165.78

[a] Production data for fermentation industrial ethanol as reported by the source (214–216) from 1961 to 1975 are overstated due to the inclusion of beverage alcohol and spirits.

Table 7. Ethanol 190 and 200 Proof Price History, $/L

	190 Proof		Tax per proof liter, $	200 Proof	
	Tax free	Tax paid		Tax free	Tax paid
1940	0.06	1.56	0.79	0.083	1.67
1945	0.13	4.65	2.37	0.145	4.90
1950	0.10	4.62	2.37	0.116	4.87
1955	0.10	5.37	2.77	0.119	5.67
1960	0.13	5.41	2.77	0.156	5.70
1965	0.13	5.41	2.77	0.156	5.70
1970	0.14	5.41	2.77	0.16	5.70
1975	0.26	5.54	2.77	0.282	5.83

Table 8. Industrial Ethyl Alcohol Salient Statistics, 10^6 L (95 vol %)

Fiscal year ending June 30	Production	Imports	Exports[a]	Stocks at end[a] of fiscal year
1950	624.5	0.0	3.1	46.2
1955	820.6	0.0	0.9	99.6
1960	1033.3	56.1	11.62	66.2
1965	1385.3	30.3	10.4	380.4
1970	1377.8		30.0	366.4
1971	1135.5		18.96	299.0
1972	1133.6		83.5	199.1
1973	1353.9	2.3	223.8	178.3
1974p	1289.5	52.2	170.1	n/a
1975p	1157.5	43.1	66.6	n/a

[a] In 1961, the method used to report the collected data was changed by the source (217–218). As a result, ethanol data which were reported separately prior to 1961 are now included under the collective heading of "Alcohol and Spirits" and are not comparable to earlier years.

Units, Specifications, Shipping, and Test Methods

Units. The alcohol content of spirits is usually given in terms of proof, an archaic term inherited from early distillers of fermentation alcohol. In England, the proof was to pour some of the spirit over gunpowder, and ignite the spirit; at or above a limiting concentration (eleven parts of alcohol by volume to ten parts of water), the gunpowder would explode. Because volumes were at the time much easier to measure accurately than weights, this measurement of alcohol persisted, even though there is a considerable volume change on mixing ethyl alcohol with water.

In the United States the proof is twice the alcohol content by volume, thus 190° proof alcohol contains 95% ethyl alcohol by volume (Table 10). According to Federal statutes, "... proof spirits shall be held to be that alcoholic liquor which contains one-half of its volume of alcohol of a specific gravity of 0.7939 at 15.5°C." A gallon (3.785 L) of proof spirits can be made by mixing 0.5000 gal (1.8925 L) of absolute alcohol with 0.5373 gal (2.0337 L) water; it contains 42.49% alcohol by weight.

A wine gallon is a measure of quantity, 231 in.^{3} (3.785 L), of any proof. A proof gallon (tax gallon) is a wine gallon of 100° proof spirits, or its equivalent.

Table 9. Specially Denatured Alcohol—Consumption[a], 10^6 L (SDA[b] Formulations)

Fiscal year ending June 30	Acetaldehyde	Acetic acid	Raw material for chemical synthesis					Total	Other	Total
			Vinegar	Acetate	Chloride	Dibromide				
1950	330.0	26.9	22.7	24.2	16.3	1.9		472.7	50.7	676.4
1955	439.1	20.8	31.4	22.7	25.0	18.1		607.5	50.3	824.4
1960	591.6	16.7	41.6	26.9	23.8	0.6		810.3	108.0	1071.9
1965	548.5	19.3	43.5	25.0	24.6	0.4		794.5	133.2	1143.8
1970	305.5	15.9	63.2	22.7	16.3			605.9	182.4	1091.9
1975	87.1	14.8	74.2	14.0	1.1	neg		366.0	174.9	752.1

[a] Ref. 216.
[b] SDA = specially denatured alcohol.

Table 10. Conversion of U.S. Proof to Alcohol by Vol; Alcohol by Wt; and British Proof

U.S. proof at 15.5°C	Alcohol at 15.5°C by vol %	Alcohol, by wt %	British proof[a]
0	0.0	0.00	100.0
2	1.0	0.80	98.4
4	2.0	1.59	96.8
6	3.0	2.39	95.2
8	4.0	3.19	93.6
10	5.0	4.00	91.9
20	10.0	8.05	83.5
30	15.0	12.14	75.0
40	20.0	16.27	66.1
50	25.0	20.44	57.0
60	30.0	24.67	48.0
70	35.0	28.97	39.3
80	40.0	33.36	30.6
90	45.0	37.86	21.7
100	50.0	42.49	12.9
110	55.0	47.24	4.0
120	60.0	52.15	4.8[b]
130	65.0	57.21	13.5[b]
140	70.0	62.44	22.3[b]
150	75.0	67.87	31.3[b]
160	80.0	73.53	39.9[b]
170	85.0	79.44	48.6[b]
180	90.0	85.69	57.3[b]
190	95.0	92.42	66.0[b]
200	100.0	100.0	76.0[b]

[a] Underproof, unless otherwise indicated.
[b] Overproof.

The determination of the proof (the alcohol content) is usually made by measuring specific gravity with hydrometers at the standard temperature of 15.5°C, since only very small amounts of impurities other than water are usually present.

Specifications. The specifications for ethyl alcohol are designed with sufficient latitude to allow for the two principal means of production, synthesis from ethylene and fermentation. The requirements given by the U.S. Pharmacopeia (USP) and the American Chemical Society (ACS) generally form the foundation for the most widely used specifications (219–220). A tabulation of the specifications from the major alcohol producers, with consideration for the USP and ACS requirements, is shown in Table 11 (see also Fine chemicals).

The major producers of ethyl alcohol also market the specially denatured and completely denatured alcohols, as well as various proprietary solvents in which ethyl alcohol is the basic ingredient. These various products can also be described by rigid and descriptive specifications, but the requirements must make allowance for the chemical and physical character of the denaturants.

The use of ethyl alcohol in some medicinal and cosmetic products requires a very meticulous grade, particularly with reference to odor. In some instances, the odor can be correlated with the concentration of certain minor impurities; in most instances it cannot be directly associated with any measurable contaminant, and the quality can be ascertained only by odor comparison with previously accepted material.

Table 11. Typical Ethanol Specifications

Requirement	190° proof	200° proof
specific gravity, 20/20°C max	0.816	0.7905
purity, vol %, min	95	99.9
acidity, wt % as CH_3COOH, max	0.002	0.002
nonvolatile matter, g/100 mL, max	0.001	0.001
miscibility with water	complete	complete
permanganate time test, minutes, min	50	30
odor	no foreign or residual	
color, APHA, max	10	10
water, wt %, max		0.1

Shipping. Commercial ethyl alcohol is shipped in railroad tank cars, tank trucks, 208-L (55-gal) and 19-L (5-gal) drums, and in smaller glass or metal containers having capacities of 0.473 L (one pint), 0.946 L (one quart), 3.785 L (one U.S. gal), or 4.545 L (one Imperial gal). The 208 L drums may be of the unlined iron type. If a guarantee of more meticulous quality is desired, the drums may be lined with phenolic resin. All containers, of course, must comply with the specifications of the U.S. Department of Transportation. Both 190° proof and 200° proof ethyl alcohol are considered as red label (flammable) materials by the DOT, as both have flash points below 37.8°C by the Tag closed-cup method.

Test Methods. The most generally used means of ascertaining the purity of commercial ethyl alcohol is by specific gravity determination, sometimes referred to as alcoholometry. For this reason, the specific gravity should be determined very accurately to the fourth decimal place by means of a calibrated pycnometer or hydrometer (221). The value thereby obtained is referred to standard United States Bureau of Internal Revenue tables relating specific gravity to alcohol content. Of course, this procedure is valid only for undenatured alcohol, since the purity nomograph is based on the two-component alcohol–water mixture (see Beverage spirits, distilled).

The other analytical methods necessary to control the typical specification given in Table 12 are, for the most part, common quality control procedures. When a chemical analysis for purity is desired, acetylation or phthalation procedures are commonly employed. In these cases, the alcohol reacts with a measured volume of either acetic or phthalic anhydride in pyridine solution. The loss in titratable acidity in the anhydride solution is a direct measure of the hydroxyl groups reacting in the sample. These procedures are generally free from interference by other functional groups, but both are affected adversely by the presence of excessive water, as this depletes the anhydride reagent strength to a level below that necessary to ensure complete reaction with the alcohol. Both procedures can be adapted to a semimicro or even micro-scale determination.

Of course, acetylation and phthalation are not selective for ethyl alcohol, but determine any reactive OH group present. A survey of methods applicable to the determination of ethyl alcohol has been published (222).

A method using a solution of acetyl chloride in toluene has been described (223).

Table 12. Respective Volumes of Alcohol and Water and Specific Gravity [a]

U.S. proof	Alcohol, vol	Water, vol	Apparent sp gr, 15.6/15.6°C
101	50.50	53.24	0.93320
102	51.00	52.74	0.93222
103	51.50	52.25	0.93123
104	52.00	51.75	0.93023
105	52.50	51.25	0.92923
106	53.00	50.75	0.92822
107	53.50	50.26	0.92720
108	54.00	49.76	0.92618
109	54.50	49.26	0.92515
110	55.00	48.76	0.92409
120	60.00	43.71	0.91333
130	65.00	38.60	0.90190
140	70.00	33.43	0.88986
150	75.00	28.19	0.87714
160	80.00	22.87	0.86364
170	85.00	17.46	0.84927
180	90.00	11.93	0.83362
188	94.00	7.36	0.81963
189	94.50	6.77	0.81775
190	95.00	6.18	0.81582
191	95.50	5.59	0.81385
192	96.00	4.99	0.81184
193	96.50	4.39	0.80979
194	97.00	3.78	0.80770
195	97.50	3.17	0.80555
196	98.00	2.55	0.80333
197	98.50	1.93	0.80104
198	99.00	1.29	0.79866
199	99.50	0.65	0.79620
200	100.00	0.00	0.79365

[a] Table 12 may also be used to determine the quantity of water needed to reduce the strength of ethanol by a definite amount. Divide the alcohol in the given strength by the alcohol in the required strength, multiply the quotient by the water in the required strength, and subtract the water in the given strength from the product. The remainder is the number of liters (gallons) of water to be added to 378.5 L (100 gal) of spirits of the given strength to produce a spirit of a required strength.

Owing to both the reactivity and volatility of the reagent, special techniques and careful handling are required for accurate analyses, and in general the method is less satisfactory than that using acetic anhydride. To overcome the difficulties of using acetyl chloride solutions, 3,5-dinitrobenzoyl chloride can be used, and after hydrolysis, titration with tetrabutylammonium hydroxide (224) is carried out. Pyromellitic dianhydride has also been used as an acylating agent (225) in the presence of aldehydes.

Colorimetric methods have been successfully used for determining trace amounts of ethanol. Ammonium hexanitratocerate(IV) has been used as a reagent (226) and for continuous automatic analysis. Alcohols form colored complexes with 8-hydroxyquinoline and vanadic compounds. The absorbance of these complexes, measured at 390 μm has been used to provide an analytical procedure (227).

Acetone (228) and ethyl acetate (229) have been used as internal standards for the gas chromatographic determination of ethanol. This technique and enzymatic methods (230) have become widely used for estimation of ethanol in blood samples. Automatic versions of the alcohol dehydrogenase method have been devised (231) and evaluated (232). The content of alcohol in wines has been measured using a differential refractometer (233). A high frequency method has been developed (234) to give a continuous analytical procedure for aqueous ethanol solutions. The determination of ethanol in exhaled air by such equipment as the Breathalyzer (335) and Alcotest (336) depends upon color changes in a solid adsorbent.

Health and Safety Considerations

Ethyl alcohol is a flammable liquid requiring a red label by the DOT and Coast Guard shipping classifications; its flash point is 14°C (Tag closed cup). Vapor concentrations between 3.3 and 19.0% by vol in air are explosive. Liquid ethyl alcohol can react vigorously with oxidizing materials. Ethyl alcohol has found wide application in industry, and experience shows that it is not a serious industrial poison (237–239). If proper ventilation of the work environment is maintained, there is little likelihood that inhalation of the vapor will be hazardous.

The threshold limit value for ethyl alcohol vapor in air has been set at 1000 ppm for an 8-h time-weighted exposure by the ACGIH (1977 listing). The minimum identifiable odor of ethyl alcohol has been reported as 350 ppm. Exposure to concentrations of 5000–10,000 ppm result in irritation of the eyes and mucous membranes of the upper respiratory tract and, if continued for an hour or more, may result in stupor or drowsiness. Concentrations of this latter order of magnitude have an intense odor and are almost intolerable to begin with, but most people can become acclimated to the exposure after a short time. Table 13 gives the effects of exposure to even heavier concentrations.

Ethyl alcohol is oxidized completely to carbon dioxide and water in the body, thus it is not a cumulative poison. An average adult is able to oxidize each hour the equivalent of 19 g of 100 proof whiskey (241). Less than 10% of the absorbed alcohol is excreted, chiefly in urine, measurably in expired air, and detectably in sweat (242). Alcohol poisoning and alcohol intoxication are almost invariably the result of using alcohol as a beverage, rather than inhalation as a vapor. About 75–80 g of ingested alcohol will produce symptoms of intoxication in an average (70 kg) person. About 150–200 g will cause stupor, and 250–500 g may be a fatal dose. Severe hypoglycemia in certain individuals following a heavy drinking bout is now well-established (243).

Table 13. Ethanol Vapor Concentration and Its Effects in Humans[a]

Concentration		
mg/L of air	ppm/vol of air	Effects in humans
10–20	5,300–10,640	some transient coughing and smarting of the eyes and nose, which disappear after 5–10 min; not comfortable but tolerable
30	15,960	continuous lacrimation and marked coughing; could be tolerated but with discomfort
40	21,280	just tolerable for short periods
>40	>21,280	intolerable and suffocating for even short periods

[a] Ref. 240.

Because alcohol intoxication may be simulated by many pathologic conditions, including diabetic acidosis, the postconvulsive depression of epilepsy, uremia, head injuries, and poisonings by any other central nervous depressant and some stimulants (244), a diagnosis of acute alcoholism should not be made casually, chemical testing of blood, urine, or expired air is always desirable.

Some authorities question whether drunkenness can result from the inhalation of ethyl alcohol vapors. Experience has demonstrated that in any event such intoxication is indeed rare (245). There is no concrete evidence that the inhalation of ethyl alcohol vapor will cause cirrhosis. Liver function is definitely impaired during alcohol intoxication (246), making the subject more susceptible to the toxic effects of chlorinated hydrocarbons.

Repeated exposure to ethyl alcohol results in the development of a tolerance as evidenced by decreasing symptomatic reactions. It has been demonstrated that the symptoms of exposure are less clear and the time required to produce them is greater in subjects accustomed to alcohol. There is no proof, however, of physiological adaptation in man in terms of metabolic changes or resistance to cellular injuries. The subject of the interaction of alcohol with other drugs has received much attention (241) and certain general principles are beginning to emerge.

Uses

Industrial ethanol is one of the largest-volume organic chemicals used in industrial and consumer products. The main uses for ethanol are as an intermediate in the production of other chemicals and as a solvent (Table 14). As a solvent, ethanol is second only to water. Ethanol is a key raw material in the manufacture of drugs, plastics, lacquers, polishes, plasticizers, perfumes, and cosmetics. Around 1960, manufacture of ethanol was the top consumer of ethylene in the U.S., but since 1965 it rates below manufacture of ethylene oxide and polyethylene.

In recent years, ethanol's role as a solvent has increased sharply. Its use as an intermediate, on the other hand, has lost ground as other materials have replaced ethanol for making acetaldehyde, butyraldehyde (qv), and ethylhexanol (see Alcohols). Butadiene (qv) was made from ethanol on a large scale during World War II, but this route is no longer competitive with butadiene derived from petroleum operations.

Table 14. Specially Denatured Alcohol Consumption[a], L × 10^6 of SDA[b] Formulations

Fiscal year ending June 30	Solvent					Total
	Resins and lacquers	Proprietary	Toilet preparations and pharmaceuticals	Industrial processing	Cleaning preparations	
1950	33.3	88.4	33.7	37.1	6.8	203.1
1955	32.5	84.8	40.1	40.5	14.0	216.4
1960	28.0	93.1	51.1	34.7	40.1	253.4
1965	23.0	89.3	147.6	36.7	43.9	349.2
1970	20.1	112.4	212.3	59.4	72.3	486.0
1975	12.9	91.2	142.7	47.3	80.2	385.7

[a] Ref. 216.
[b] SDA = specially denatured alcohol.

Ethanol has been used in the United States to a considerable extent as an antifreeze, but it has largely been replaced for such use by ethylene glycol (see Antifreezes).

The fastest growing solvent use of ethanol is in cleaning preparations and liquid detergents. Considerable increase in the use of ethanol for toilet preparations, such as aerosol hair sprays and mouthwashes, has also occurred (see Aerosols). Vinegar based on synthetic ethanol is gaining an increasing share of the market for vinegar used in pickles, ketchup, and mustard. With the high cost of petroleum after 1973, the method of fermentation of corn and sugar cane (widely used before oil was plentiful and cheap) is regaining its former popularity. The use of fermentation alcohol as a automotive fuel additive has been discussed above. Several tropical countries also see ethanol, produced from regenerable resources, as an attractive petrochemical feedstock. With the aid of U.S. companies, the Brazilian government is seeking ways to use ethanol in the production of vinyl acetate, 1,2-dichloroethane and several other products (191) (see Vinyl polymers; Chlorocarbons).

Denatured Ethanol. For hundreds of years alcoholic beverages have been taxed to generate government revenue all over the world. When ethanol emerged as a key industrial raw material, the alcohol tax was recognized as a burden to many essential manufacturing industries. To lift this burden, the Tax-Free Industrial and Denatured Alcohol Act of 1906 was passed in the United States. The United States Treasury, Bureau of Alcohol, Tobacco and Firearms (BATF) now oversees the production, procurement, and use of ethanol in the United States (see Beverage spirits, distilled).

The concern of the government is to prevent tax-free industrial ethanol from finding its way into beverages. To achieve this end, the regulations call for a combination of financial and administrative controls (bonds, permits, and scrupulous record keeping) and chemical controls (denaturants that make the ethanol unpalatable). Regulations establish four distinct classifications of industrial ethanol. The classifications with the most stringent financial and administrative controls call for little or no chemical denaturants. The classifications that call for the most effective chemical denaturants require the least financial and administrative controls. For a list of denaturants currently authorized, see ref. 213.

Completely Denatured Alcohol. Completely denatured alcohol (CDA) escapes the involved financial and administrative controls required of the other classifications of industrial ethanol. No tax is applied, no bond is required, no permit is needed to enable a customer to purchase CDA. Requirements for records by both producer and user are minimal. These simplified regulations are possible because CDA is denatured with substances that render it totally unfit for beverage purposes. It is also unsuitable where odor is objectionable. CDA and products made from it are, however, governed by special labeling requirements of the BATF. Repackaging of completely denatured alcohol is permitted as long as labeling requirements are met.

Proprietary Solvents and Special Industrial Solvents. Proprietary solvents and special industrial solvents can be purchased by customers without payment of tax, without posting a bond for tax, and without securing a permit from the BATF (see also Solvents, industrial). Suppliers are required, however, to notify the BATF of the name, address, type of business, and approximate annual requirements and intended end use for any user buying in bulk (214–215).

Proprietary solvents may be repackaged for retail, wholesale, and industrial sales. Retail sales of special industrial solvents are prohibited. Agents may repackage special industrial solvents for wholesale and industrial sales, but only in drum quantities and with the producer's label. Special labeling requirements of the BATF apply to both proprietary and special industrial solvents.

Specially Denatured Alcohol. Specially denatured alcohols (SDA) are formulations of ethanol containing denaturant substances that generally render them unfit for beverage use but do not limit their use in specified applications. To use a specially denatured alcohol, a manufacturer must apply to the BATF, giving quantitative formulas and processes. Specimen labels and a sample of the finished product are also required. Then, the prospective user must obtain a bond for the total amount of specially denatured alcohol on hand or in transit at any given time.

Pure Ethanol. Undenatured ethanol can be bought on either a tax-free or tax-paid basis. Approved educational, scientific or medical organizations and public agencies can buy tax-free alcohol. Use and withdrawal permits are required, as are a bond and detailed records. Resale of tax-free ethanol is prohibited. Approved industrial uses for pure tax-paid ethanol include pharmaceuticals, cosmetics, flavoring extracts, and foods.

Chemicals Derived From Ethanol

Acetaldehyde. Until the early 1970s, the main use of industrial ethanol was for the production of acetaldehyde (qv). By 1977, the ethanol route to acetaldehyde had largely been phased out as ethylene and ethane became the preferred feedstocks for acetaldehyde production (217–218,247–265). Acetaldehyde usage itself has also changed; two major derivatives of acetaldehyde, acetic acid and butanol, are now produced from feedstocks other than acetaldehyde. It is estimated (213) that by 1981, only 60×10^6 L of ethanol will be used to make acetaldehyde, compared to the peak of over 568×10^6 L in the early 1960s.

There are two ways to produce acetaldehyde from ethanol: oxidation and dehydrogenation

$$C_2H_5OH + \tfrac{1}{2} O_2 \rightarrow CH_3CHO + H_2O + 169.4 \text{ kJ } (40.5 \text{ kcal})$$

$$C_2H_5OH \rightarrow CH_3CHO + H_2 - 72.0 \text{ kJ } (17.2 \text{ kcal})$$

Oxidation of ethanol to acetaldehyde is carried out in the vapor phase over a silver or copper catalyst (266). Conversion is slightly over 80% per pass at reaction temperatures of 450–500°C with air as an oxidant. Chloroplatinic acid selectively catalyzes the liquid-phase oxidation of ethanol to acetaldehyde giving yields exceeding 95%. The reaction takes place in the absence of free oxygen at 80°C and at atmospheric pressure (267). The kinetics of the vapor and liquid-phase oxidation of ethanol have been described in the literature (268–269).

The reaction kinetics for the dehydrogenation of ethanol is also well documented (270–273). The vapor phase dehydrogenation of ethanol in the presence of a chromium-activated copper catalyst at 280–340°C produces acetaldehyde in a yield of 89% and a conversion of 75% per pass (274). Other catalysts used include neodymium oxide and samarium hydroxide (275).

Ethylene. Where ethylene is in short supply and fermentation ethanol is made economically feasible, ethylene is manufactured by the vapor-phase dehydration of ethanol. The production of ethylene from ethanol using naturally renewable resources is an active and useful alternative to the pyrolysis process based on nonrenewable petroleum. This route may make ethanol a major raw material source for producing other chemicals.

Dehydration of ethanol has been effected over a variety of catalysts, among them synthetic and naturally occurring aluminas, silica-aluminas, and activated alumina (276–283), hafnium and zirconium oxides (282), and phosphoric acid on coke (284). Operating space velocity is chosen to ensure that the two consecutive reactions,

$$2\ C_5H_5OH \rightarrow (C_2H_5)_2O + H_2O$$

$$(C_2H_5)_2O \rightarrow 2\ C_2H_4 + H_2O$$

go to completion, avoiding the need to recover and recycle unreacted ethanol. The dehydration is endothermic, and temperature is a critical operating parameter; high temperatures produce aldehydes and low temperatures, ethers. The catalyst is usually regenerated with steam and air every few weeks to remove carbon deposits.

A fluidized-bed catalytic reactor system developed by C. E. Lummus (284) offers several advantages over fixed-bed systems in temperature control, heat and mass transfer, and continuity of operation. Higher catalyst activity levels and higher ethylene yields (99% compared to 94–96% with fixed-bed systems) are accomplished by continuous circulation of catalyst between reactor and regenerator for carbon burn-off and continuous replacement of catalyst through attrition (see Fluidization).

Glycol Ethers. The addition of one mole of ethylene oxide to ethanol gives ethylene glycol monoethyl ether.

$$CH_3CH_2OH + \underset{\underset{O}{\diagdown\diagup}}{CH_2\text{—}CH_2} \rightarrow CH_3CH_2OCH_2CH_2OH$$

Addition of two moles of oxide gives the monoethyl ether of diethylene glycol.

$$CH_3CH_2OH + 2\ \underset{\underset{O}{\diagdown\diagup}}{CH_2\text{—}CH_2} \rightarrow CH_3CH_2OCH_2CH_2OCH_2CH_2OH$$

The oxide–alcohol route is the only commercially important route to glycol ethers now in use. Anhydrous alcohols must be used; otherwise the water present forms contaminating glycols.

The reactions are highly exothermic. Under liquid phase conditions at about 200°C, the overall heat of reaction is −83.7 to −104.6 kJ/mol (−20 to −25 kcal/mol) ethylene oxide reacting (285). The opening of the oxide ring is considered to occur by an ionic mechanism with a nucleophilic attack on one of the epoxide carbon atoms (286). Both acidic and basic catalysts accelerate the reactions, as does elevated temperature. The reaction kinetics and product distribution have been studied by a number of workers (287–288).

Ethylene oxide, propylene oxide, butylene oxide and other epoxides react with ethanol to give a variety of liquid, viscous, semiwax, and solid products. These products are used in the coatings industry as solvents, and as paints, antioxidants, corrosion inhibitors, and special-purpose polymers (see Glycols).

Vinegar. Dilute solutions of alcohol as fermented worts are oxidized by air at 30–40°C in the presence of various organisms such as (289) *Mycoderma aceti*, *B. aceti*, and *B. xylinus*, to produce dilute acetic acid as vinegar (see also Wine). Recently, synthetic ethanol has assumed growing importance as the raw material primarily because of the stability of supply. Vinegar based on synthetic ethanol has a fully acceptable aroma and taste (290) and is gaining an increased share of the market for vinegar used in such products as pickles, ketchup, and mustard. A continued growth of ca 6% per year is predicted (212) leading to a 1981 demand of about 98×10^6 L, roughly 10% of the total ethanol market in that year.

Ethylamines. Mono-, di-, and triethylamines, produced by catalytic reaction of ethanol with ammonia (291), are a significant outlet for ethanol. The vapor-phase continuous process takes place at 1.38 MPa (13.6 atm) and 150–220°C over a nickel catalyst supported on alumina, silica, or silica-alumina. In this reductive amination under a hydrogen atmosphere, the ratio of the mono-, di-, and triethylamine product can be controlled by recycling the unwanted products. Other catalysts used include phosphoric acid and derivatives, copper and iron chlorides, sulfates and oxides, in the presence of acids or alkaline salts (292). Piperidine can be ethylated with ethanol in the presence of Raney nickel catalyst at 200°C and 10.3 MPa (102 atm), to give N-ethylpiperidine (293) (see Amines by reduction; Amines, aliphatic; Amines, cyclic).

Ethyl Acrylate. In the synthesis of ethyl acrylate, the esterification of acrylic acid is a major use for ethanol.

$$CH_2=CHCOOH + C_2H_5OH \rightleftarrows CH_2=CHCOOC_2H_5 + H_2O$$

Acrylic acid can also react with either ethylene or ethyl esters of sulfuric acid.

$$CH_2=CHCOOH + C_2H_4 \rightleftarrows CH_2=CHCOOC_2H_5$$

$$2\ CH_2=CHCOOH + (C_2H_5O)_2SO_2 \rightleftarrows 2\ CH_2=CHCOOC_2H_5 + H_2SO_4$$

These processes have supplanted the condensation reaction of ethanol, carbon monoxide and acetylene as the principal method of generating ethyl acrylate (294). Acidic catalysts, particularly sulfuric acid (295–299), are generally effective in increasing the rates of the esterification reactions. Care is taken to avoid excessive polymerization losses of both acrylic acid and the esters, which is accentuated by the presence of strong acid catalysts. A synthesis for acrylic esters from vinyl chloride (300) has also been examined.

$$CH_2=CHCl + C_2H_5OH + CO \rightarrow CH_2=CHCOOC_2H_5 + HCl$$

Vinyl chloride reacts at 270°C at >6895 kPa (68 atm) with ethanol in the presence of a cobalt and palladium catalyst to give ethyl acrylate in 17% yield (see Acrylic acid and derivatives).

Ethyl Ether. Most ethyl ether is obtained as a by-product of ethanol synthesis via the direct hydration of ethylene. The procedure used for production of ether from ethanol and sulfuric acid is essentially the same as that first described in 1809 (301). The chemical reactions involved in the production of ethyl ether by the indirect ethanol-from-ethylene process are like those for the production of ether from ethanol using sulfuric acid.

$$C_2H_5OH + H_2SO_4 \rightleftarrows C_2H_5HSO_4 + H_2O$$

$$C_2H_5HSO_4 + C_2H_5OH \rightarrow (C_2H_5)_2O + H_2SO_4$$

A more recent concept (302) is that the ether is principally derived from the reaction between diethyl sulfate and ethanol

$$(C_2H_5O)_2SO_2 + C_2H_5OH \rightleftarrows (C_2H_5)_2O + C_2H_5HSO_4$$

An alternative mechanism for the formation of diethyl ether from ethanol and sulfuric acid is by dehydration:

$$2\ C_2H_5OH \rightarrow (C_2H_5)_2O + H_2O$$

the reaction is catalyzed by all but the weakest acids. In the dehydration of ethanol over heterogeneous catalysts, such as alumina (303–307), ether is the main product below 260°C; at higher temperatures both ether and ethylene are produced. Other catalysts used include silica-alumina (308–309), copper sulfate, tin chloride, manganous chloride, aluminum chloride, chrome alum, and chromium sulfate (310–311) (see Ethers).

Ethyl Vinyl Ether. The addition of ethanol to acetylene gives ethyl vinyl ether (312–316).

$$CH_3CH_2OH + HC\equiv CH \rightarrow CH_3CH_2OCH=CH_2$$

The vapor phase reaction is generally run at 1.38–2.07 MPa (13.6–20.4 atm) and temperatures of 160–180°C with alkaline catalysts such as potassium hydroxide and potassium ethylate. High molecular weight polymers of ethyl vinyl ether are used for pressure-sensitive adhesives, viscosity index improvers, coatings and films; lower molecular weight polymers are plasticizers and resin modifiers (see Acetylene-derived chemicals; Vinyl polymers).

Ethyl Acetate. The esterification of ethanol by acetic acid was studied in detail over a century ago (317), and considerable literature exists on determinations of the equilibrium constant for the reaction. The usual catalyst for the production of ethyl acetate is sulfuric acid, but other catalysts have been used, including cation-exchange resins (318), α-fluoronitrites (319), titanium chelates (320) and quinones and their partly reduced products.

Ethyl acetate is made industrially by both batch and continuous processes (321–322). Glacial acetic acid is commonly the starting material, and any water formed during the esterification has to be removed. Sulfuric acid may be added periodically to the reactor to replace the acid lost in side reactions. The vapor-phase esterification of ethanol has also been studied extensively (323–324), but it is not used commercially. The reaction can be catalyzed by silica gel (325–326), thoria on silica or alumina (327), zirconium dioxide (328) and by xerogels and aerogels (329). Above 300°C the dehydration of ethanol becomes appreciable. Ethyl acetate can also be produced from acetaldehyde by the Tischenko reaction (330–332) using an aluminum alkoxide catalyst and, with some difficulty, by the boron trifluoride-catalyzed direct esterification of ethylene with organic acids (333) (see Esterification).

Ethyl Chloride. Previously a major use for industrial ethanol was the synthesis of ethyl chloride for use as an intermediate in producing tetraethyllead, an antiknock gasoline additive. Ethanol is converted to ethyl chloride by reaction with hydrochloric acid in the presence of aluminum or zinc chlorides. However, since about 1960, routes based on the direct addition of hydrochloric acid to ethylene or ethane have become more competitive (334–335) (see Chlorocarbons and chlorohydrocarbons).

Other Derivatives and Reactions. The vapor-phase condensation of ethanol to give acetone has been well documented in the literature (336–345); however, acetone is usually obtained as a by-product from the cumene (qv) process, by the direct oxidation of propylene, or from isopropanol (see Ketones).

The Guerbet reaction (346–349) involving condensation of ethanol in the presence of sodium ethylate, catalyzed by potassium hydroxide and boric anhydride (350–351) or alkaline phosphates (352), gives n-butanol:

$$2\ CH_3CH_2OH + CH_3CH_2ONa \rightarrow CH_3CH_2CH_2CH_2OH + CH_3COONa + H_2$$

However, the oxo reaction starting from propylene and proceeding via the hydrogenation of butyraldehyde, has become the more widely employed commercial route for preparing n-butanol (see Oxo process).

During World War II, production of butadiene (qv) from ethanol was of great importance. About 60% of the butadiene produced in the United States during that time was obtained by a two-step process utilizing a 3:1 mixture of ethanol and acetaldehyde at atmospheric pressure and a catalyst of tantalum oxide and silica gel at 325–350°C (353–357). Extensive catalytic studies were reported (358–361) including a fluidized process (362). However, because of later developments in the manufacture of butadiene by the dehydrogenation of butane and butenes, and by naphtha cracking, the use of ethanol as a raw material for this purpose has all but disappeared.

An extensive listing of 35 other reactions including alkylation, etherification, alcoholysis, and halogenation of ethanol has been compiled by Hatch (1).

BIBLIOGRAPHY

"Alcohol, Industrial" in *ECT* 1st ed., Vol. 1, pp. 252–288, by R. S. Aries, Consulting Chemical Engineer.
"Ethanol" in *ECT* 2nd ed., Vol. 8, pp. 422–470, by C. A. Pentz and G. A. Lescisin, Union Carbide Corp.

1. L. F. Hatch, *Ethyl Alcohol,* Enjay Chemical Co., a Division of Humble Oil and Refining Co., New York, 1962.
2. C. M. Beamer, *Chem. Eng. Prog.* **43**(3), 92 (1947).
3. E. Huber, *Z. Spiritusind.* **50,** 164 (1927).
4. *Ethyl Alcohol Production Technique,* Noyes Development Corp., Pearl River, N.Y., 1964.
5. D. Hodzi, *Hydrogen Bonding,* Pergamon Press, London, 1957.
6. F. K. Beilstein, *Handbuch der organischen Chemie,* E III, Vol. 1, Springer-Verlag, Berlin, 1958, pp. 1229–1236.
7. *Selected Values of Properties of Chemical Compounds,* Manufacturing Chemists' Association Research Project Tables, Chemical Thermodynamic Properties Center, Dept. of Chemistry, A&M College of Texas, College Station, Texas.
8. *Selected Properties of Hydrocarbons and Related Compounds,* American Petroleum Institute Project 44, Chemical Thermodynamic Properties Center, Dept. of Chemistry, A&M College of Texas, College Station, Texas.
9. J. Timmermans, *Physico-Chemical Constants of Pure Organic Compounds,* Elsevier Publishing Co., New York, 1950, pp. 306–312.
10. K. A. Kobe and R. E. Pennington, *Thermochemistry of Petrochemicals,* Reprint No. 44, University of Texas, Bureau of Engineering Research, Austin, Texas, Jan. 1949–Dec. 1951, pp. 56–59.
11. R. C. Reid and J. M. Smith, *Chem. Eng. Prog.* **47**(8), 415 (1951).
12. S. Young, *Sci. Proc. Dublin Soc.* **12,** 374 (1910).
13. A. R. Challoner and R. Powell, *Proc. R. Soc.* (*London*) **238A,** 90 (1956).
14. Unpublished physical property data, Union Carbide Corporation, Chemicals and Plastics Division, 1965.
15. L. H. Horsley, ed., *Azeotropic Data, Advances in Chemistry Series,* Vol. 116, ACS, 1973.
16. D. Williams and R. W. Bost, *J. Chem. Phys.* **4,** 251 (1936).
17. U.S. Pat. 1,712,830 (May 14, 1929), L. P. Kyrides (to National Aniline Division, Allied Chemical & Dye Corp.).
18. U.S. Pat. 1,978,647 (Oct. 30, 1934), E. T. Olson and R. H. Twinning (to Cleveland Cliffs Iron Co.).
19. A. Meuwsen, *Ber.* **68,** 121 (1935).
20. E. W. Zappi and E. Restelli, *An. Assoc. Quim. Argent.* **32,** 89 (1934).
21. F. Henle, *Ber.* **53,** 719 (1920).
22. Brit. Pat. 454,480 (Oct. 1, 1936), (to Consortium fur Elektrochemische Industrie G.m.b.H.).
23. Ger. Pat. 602,376 (Sept. 17, 1934), J. Seib (to Deutsche Gold- und Silber-Scheideanstalt vorm. Roessler).
24. C. M. Suter and E. Oberg, *J. Am. Chem. Soc.* **56,** 677 (1934).
25. M. Gallagher and D. B. Keyes, *J. Am. Chem. Soc.* **56,** 2221 (1934).

26. U.S. Pat. 2,831,882 (April 22, 1958), C. P. Spaeth (to E. I. du Pont de Nemours & Co., Inc.).
27. H. D. Hinton and J. A. Nieuwland, *J. Am. Chem. Soc.* **54,** 2017 (1932).
28. P. Ya. Ivannikov and E. Ya. Gavrilova, *J. Chem. Ind.* (*USSR*) **12,** 1256 (1935).
29. B. N. Dolgov and co-workers, *Zh. Obshch. Khim.* **25,** 693 (1955).
30. B. N. Dolgov and co-workers, *J. Chem. Ind.* (*USSR*) **12,** 1066 (1935).
31. B. N. Dolgov and co-workers, *Org. Chem. Ind.* (*USSR*) **1,** 70 (1936).
32. N. M. Beizel and co-workers, *Org. Chem. Ind.* (*USSR*) **1,** 102 (1936).
33. U.S. Pat. 2,004,350 (June 11, 1935), N. D. Scott (to E. I. du Pont de Nemours & Co., Inc.); Brit. Pat. 424,284 (Feb. 18, 1935), (to E. I. du Pont de Nemours & Co., Inc.).
34. H. Adkins and R. Conner, *J. Am. Chem. Soc.* **53,** 1091 (1931).
35. M. Bobtelsky, *J. Chem. Soc.* Part IV, 3615 (1950).
36. Ref. 1, p. 68.
37. M. Bobtelsky and R. Cohn, *Z. Anorg. Allg. Chem.* **210,** 225 (1933).
38. M. Bobtelsky and coworkers, *J. Am. Chem. Soc.* **67,** 966 (1945).
39. Br. Jirgensons, *Z. Phys. Chem.* **158A,** 56 (1931).
40. Brit. Pat. 273,263 (Sept. 15, 1926), (to Compagnie de Bethune).
41. U.S. Pat. 2,545,161 (March 13, 1951), E. C. Morrell and R. F. Robey (to Standard Oil Development Co.); Brit. Pat. 655,475 (July 25, 1951), (to Standard Oil Development Co.).
42. S. P. G. Plant and N. V. Sidgwick, *J. Soc. Chem. Ind.* (*London*) **40**(2), 14T (1921).
43. F. Vallette, *Chim. Ind.* (*Paris*) **13,** 718 (1951).
44. Brit. Pat. 221,512 (Feb. 19, 1925), (to Compagnie de Bethune).
45. P. Fritsche, *Chem. Ind.* (*London*) **20,** 266 (1897); **21,** 33 (1898).
46. C. Ellis, *Chemistry of Petroleum Derivatives*, Chemical Catalog Co., New York, 1934, p. 301.
47. C. Ellis, *Chemistry of Petroleum Derivatives*, Vol. II, Reinhold Publishing Corp., New York, 1937.
48. W. S. E. Hickson and K. C. Bailey, *Sci. Proc. R. Dublin Soc.* **20,** 267 (1932).
49. K. C. Bailey and W. E. Calcutt, *Sci. Proc. R. Dublin Soc.* **21,** 309 (1936).
50. B. T. Brooks, *Ind. Eng. Chem.* **27,** 278 (1935).
51. B. Neuman, *Gas Wasser* **67,** 1, 14, 53 (1924).
52. Brit. Pat. 308,859 (April 4, 1929), R. E. Slade (to Imperial Chemical Industries Ltd.).
53. U.S. Pat. 1,885,585 (Nov. 1, 1932), B. T. Brooks (to Petroleum Chem. Corp.).
54. U.S. Pat. 1,919,618 (June 25, 1933), B. T. Brooks (to Standard Alcohol Co.).
55. B. G. Simek, *Chem. Prumysl* **7,** 122 (1957); (through) *Erdoel Kohle* **10,** 882 (1957).
56. F. Strahler and F. Hachtel, *Brennst. Chem.* **15,** 166 (1934).
57. C. F. Tidman, *J. Soc. Chem. Ind.* (*London*) **40,** 103R (1921).
58. A. Damiens, *Compt. Rend.* **175,** 585 (1922).
59. A. Damiens, *Bull. Soc. Chim.* **33,** 71 (1923).
60. H. Hennell, *Ann. Chim.* **35**(2), 154 (1827).
61. U.S. Pat. 2,755,297 (July 17, 1956), B. I. Smith and W. H. Rader (to Esso Research and Engineering Co.).
62. U.S. Pat. 2,765,347 (Oct. 2, 1956), B. I. Smith (to Esso Research and Engineering Co.).
63. U.S. Pat. 2,779,803 (Jan. 29, 1957), K. C. Bottenberg (to Phillips Petroleum Co.).
64. U.S. Pat. 2,792,432 (May 14, 1957), W. C. Muller and F. D. Miller (to National Petroleum Chemicals Corp.).
65. U.S. Pat. 2,792,433 (May 14, 1957), W. C. Muller and J. S. Atwood (to National Petro-Chemicals Corp.).
66. U.S. Pat. 2,302,825 (Nov. 24, 1942), C. B. Wilde (to Stauffer Chemical Co.).
67. U.S. Pat. 2,061,810 (Nov. 14, 1937), W. H. Shiffler and M. M. Holm (to Standard Oil of California).
68. U.S. Pat. 2,038,512 (April 21, 1936), R. N. Graham (to Union Carbide Corp.).
69. V. A. Gutuirya and co-workers, *Azerb. Neft. Khoz.* **16**(5), 72 (1936).
70. U.S. Pat. 2,474,568 (June 18, 1949), L. A. Bannon and C. E. Morrell (to Standard Oil Development Co.).
71. U.S. Pat. 2,474,569 (June 18, 1949), L. A. Bannon (to Standard Oil Development Co.).
72. R. Kremann, *Monatsh. Chem.* **31,** 245, 275 (1910); R. Kremann, *J. Soc. Chem. Ind.* (*London*) **29,** 782 (1910).
73. R. Kremann, *Monatsh. Chem.* **28**(3), 13 (1907); **31,** 165 (1910); **38,** 53 (1917).
74. W. A. Drushei and G. A. Linhart, *Am. J. Sci.* **32,** 51 (1911).
75. U.S. Pat. 2,414,759 (Jan. 21, 1947), H. O. Mottern (to Standard Alcohol Co.).

76. U.S. Pat. 2,512,327 (June 20, 1950), T. P. Hawes and A. P. Geraitis (to Standard Oil Development Co.).
77. Can. Pat. 470,593 (Jan. 2, 1951), T. P. Hawes and co-workers (to Standard Oil Development Co.).
78. U.S. Pat. 2,856,265 (Oct. 15, 1958), G. A. Lescisin (to Union Carbide Corp.).
79. Brit. Pat. 638,547 (June 17, 1950), J. Howlett and co-workers (to Distillers Co., Ltd.).
80. T. C. Carle and D. M. Stewart, *Chem. Ind. (London)* (19), 830 (1962).
81. U.S. Pat. 41,685 (1861), E. A. Cotelle.
82. J. P. Wibaut and J. J. Dickman, *Proc. K. Ned. Akad. Wet.* **26,** 321 (1923).
83. C. R. Nelson and co-workers, *Chem. Eng. Progr.* **50,** 562 (1954).
84. U.S. Pat. 2,960,477 (Nov. 15, 1958), W. C. Smith, C. A. MacMurray, and W. L. Holmes (to Shell Oil Co.).
85. Ger. Pat. 2,015,536 (Nov. 5, 1970), O. Frampton and J. Feldman (to National Distillers and Chemical Corp.).
86. T. C. Carle and co-workers, *Chem. Ind.* **19,** 830 (1962).
87. W. J. Tapp, *Ind. Eng. Chem.* **40,** 1619 (1948); **42,** 1698 (1950); **44,** 2020 (1952).
88. Brit. Pat. 308,859 (1929), R. E. Slade (to ICI).
89. Brit. Pat. 378,865 (1932), L. H. Horsley (to ICI).
90. U.S. Pat. 2,579,601 (Dec. 25, 1951), C. R. Nelson and co-workers (to Shell Development Co.).
91. Brit. Pat. 1,144,947 (March 12, 1969), H. G. Hagemeyer and M. Statman (to Eastman Kodak Co.).
92. U.S. Pat. 3,554,926 (Jan. 12, 1971), M. Statman and D. S. Martin (to Eastman Kodak Co.).
93. Brit. Pat. Appln. 15643 (1940), (to Standard Oil Development Co.).
94. U.S. Pat. 3,704,329 (Nov. 28, 1972), E. Rindtorff and W. Ester (to Veba-Chemnie Aktiengesellschaft).
95. U.S. Pat. 3,232,997 (Feb. 1, 1966), W. Ester (to Hibernia-Chemie Gesellschaft nit beschrankter Hafting).
96. M. A. Dalin, *Khem. Nauka i Promy* **1,** 259 (1956).
97. U.S. Pat. 4,012,452 (Mar. 15, 1977), O. D. Frampton (to National Distillers Corp.).
98. Jpn. Kokai 75,146,585 (Nov. 25, 1975), (to Nippon Synthetic Alcohol Co. Ltd.).
99. Jpn. Kokai 76,044,915 (Dec. 1, 1976), (to Nikki Chem. Co. Ltd.).
100. Brit. Pat. 394,376 (1933), H. Dreyfus.
101. U.S. Pat. 2,504,618 (April 18, 1950), R. C. Archibald and R. A. Trimble (to Shell Development Co.).
102. J. Muller and H. I. Waterman, *Brennst. Chem.* **38,** 321 (1957).
103. U.S. Pat. 2,876,266 (March 3, 1959), C. Wegner (to Farbenfabriken Bayer A. G.).
104. U.S. Pat. 2,769,847 (Nov. 6, 1956), R. L. Robinson (to Imperial Chemical Industries, Ltd.).
105. C. V. Mace and C. F. Bonilla, *Chem. Eng. Prog.* **50,** 385 (1954).
106. Brit. Pat. 691,360 (May 13, 1953), R. C. Thomson and R. K. Greenhaig (to Imperial Chemical Industries, Ltd.); Belg. Pat. 499,676 (May 28, 1951).
107. U.S. Pat. 2,725,403 (Nov. 29, 1955), M. A. E. Hodgson (to Imperial Chemical Industries, Ltd.).
108. Brit. Pat. 665,214 (Jan. 16, 1952), J. Thompson and P. W. Reynolds (to Imperial Chemical Industries, Ltd.).
109. Winkler, *Ind. Agr. Aliment (Paris)* **66,** 159 (March, April 1949).
110. U.S. Pat. 2,720,232 (Feb. 15, 1955), H. R. Arnold and J. E. Carnahan (to E. I. du Pont de Nemours & Co., Inc.).
111. U.S. Pat. 2,755,309 (July 17, 1956), P. W. Reynolds (to Imperial Chemical Industries, Ltd.).
112. U.S. Pat. 2,807,655 (Sept. 24, 1957), L. R. Pitwell (to Imperial Chemical Industries, Ltd.).
113. U.S. Pat. 2,763,697 (Sept. 18, 1956), H. J. Hagemeyer, Jr., and W. J. Clegg (to Eastman Kodak Co.).
114. Jpn. Kokai 5,1059,807 (May 25, 1976); 5,1063,105 (June 1, 1976); 5,1075,007 (June 19, 1976); 5,1086,405 (July 29, 1976) and 5,1141,802 (Dec. 7, 1976), (to Mitsui Toatsu Chem. Inc.).
115. Ger. Pat. 2,702,500 (July 28, 1977), (to Mitsui Toatsu Chem. Inc.).
116. F. C. Whitmore, *Ind. Eng. Chem.* **26,** 94 (1934).
117. W. S. Brey and K. A. Krieger, *J. Am. Chem. Soc.* **71,** 3637 (1949).
118. R. W. Taft, *J. Am. Chem. Soc.* **74,** 5372 (1952).
119. G. K. Boreskov and co-workers, *Tr. Nauchno. Issled. Inst. Sint. Spirtov Org. Prod.* 213 (1960).
120. A. I. Gel'bshtein, Yu. M. Bakshi, and M. I. Temkin, *Dokl. Akad. Nauk SSSR* **132,** 384 (1960).
121. M. P. Applebey and co-workers, *J. Soc. Chem. Ind. Trans.* **56,** 279–81 (1937).
122. R. H. Bliss and B. F. Dodge, *Ind. Eng. Chem.* **29,** 19 (1937).

123. Yu. M. Bakshi, *Dokl. Akad. Nauk SSSR* **132,** 157 (1960).
124. J. Muller and H. I. Waterman, *Genie Chim.* **78,** 173 (1957).
125. Belg. Pat. 715,907 (Dec. 2, 1968), (to Hibernia-Chemie).
126. *Hydrocarbon Process.* **52,** 122 (1973).
127. Brit. Pat. 1,106,424 (March 2, 1968), to Hibernia-Chemie.
128. U.S. Pat. 3,156,629 (Nov. 10, 1964), W. Ester (to Bergwerksgesellschaft Hibernia).
129. Brit. Pat. 1,007,096 (Oct. 13, 1965), E. Rindorff and co-workers (to Hibernia-Chemie).
130. Can. Pat. 859,781 (Dec. 29, 1970), D. C. Hull and G. I. LeMaster (to Eastman Kodak Co.).
131. U.S. Pat. 1,951,740 (March 20, 1934), W. H. Shiffler and M. M. Holm (to Standard Oil Development Co.).
132. U.S. Pat. 2,021,564 (Nov. 19, 1935), F. J. Metzger (to Air Reduction).
133. U.S. Pat. 1,607,459 (1927), O. Johannsen and O. Gross.
134. U.S. Pat. 2,179,092 (Nov. 7, 1939), V. N. Ipatieff (to Universal Oil Products).
135. Ger. Pat. 1,035,632 (Aug. 7, 1958), H. G. van Raay (to Farbwerke Hoechst A.G.).
136. *Chem. Week* **84**(6), 58 (1959).
137. Brit. Pat. 446,781 (1936), to Standard Oil Development.
138. E. R. Gilliland and co-workers, *Ind. Eng. Chem.* **38,** 370 (1936).
139. W. H. Shiffler, M. M. Holm, and L. F. Brooke, *Ind. Eng. Chem.* **31,** 1099 (1939).
140. U.S. Pat. 2,044,417 (June 16, 1936), F. R. Balcar (to Air Reduction).
141. Brit. Pat. 511,247 (1960), H. M. Guinot (to Usines de Melle).
142. Brit. Pat. 750,176 (1956), (to Chemische Werke Hüls).
143. A. Vansheidt and M. Lozovskii, *Zh. Obshch. Khim.* **3,** 329 (1933); *Chem. Abs.* **28,** 2323 (1934).
144. U.S. Pat. 3,095,458 (June 25, 1963), C. A. Judice and L. E. Pirkle (to Esso Research and Eng. Co.).
145. U.S. Pat. 2,045,785 (June 30, 1936), W. K. Lewis (to Standard Oil Dev. Co.).
146. N. S. Kozlov and N. Golubovskaya, *J. Gen. Chem. USSR (Eng. Transl.)* **6,** 1506 (1936).
147. U.S. Pat. 2,519,061 (Aug. 15, 1950), R. B. Mason (to Standard Oil Dev. Co.).
148. U.S. Pat. 2,115,874 (May 3, 1938), C. W. Rehm (to Union Carbide and Carbon).
149. T. V. Antipina and O. V. Isaer, *Zh. Fiz. Khim.* **31,** 2245 (1957); *Chem. Abs.* **52,** 9944 (1958).
150. N. Ya. Kagan and co-workers, *J. Gen. Chem. USSR (Eng. Transl.)* **3,** 337 (1933).
151. U.S. Pat. 2,974,175 (March 7, 1961), R. N. Watts and R. B. Mason (to Esso Res. and Eng. Co.).
152. U.S. Pat. 2,472,618 (June 7, 1949), A. S. Ramage (to A. A. F. Maxwell).
153. U.S. Pat. 2,317,949 (April 27, 1963), R. E. Burke (to Standard Oil Co. of Ohio).
154. W. Traube and R. Justh, *Brennst. Chem.* **4,** 150 (1923).
155. P. Sabatier and J. B. Senderens, *Comp. Rend.* **137,** 301 (1903).
156. S. A. Miller, ed., *Ethylene and Its Industrial Derivatives,* Ernest Benn Std., London, 1969.
157. D. M. Newitt and A. M. Block, *Proc. Royal Soc. (London)* **A140,** 426 (1933).
158. D. M. Newitt and D. T. A. Townsend, *J. Inst. Pet. Technol.* **20,** 252A (1934).
159. P. J. Weizevich and P. K. Frolich, *Ind. Eng. Chem.* **26,** 267 (1934).
160. U.S. Pat. 1,858,822 (May 17, 1932), P. K. Frolich (to Standard Oil Development Co.).
161. U.S. Pat. 2,700,677 (Jan. 25, 1955), K. D. Bowen, D. R. Keck, and D. C. Lee, Jr. (to Celanese Corp. of America).
162. U.S. Pat. 2,702,298 (Feb. 15, 1955), N. M. Caruthers (to Stanolind Oil and Gas Co.).
163. U.S. Pat. 3,052,731 (Sept. 4, 1962), C. R. Murphy (to Gulf Research and Development Co.).
164. F. Fischer, *Ind. Eng. Chem.* **17,** 576 (1925).
165. R. Taylor and G. T. Morgan, *Proc. Roy. Soc.* **A131,** 533 (1931).
166. R. S. Aries, *Chem. Eng. News* **25,** 1792 (1947).
167. C. R. Downs and J. H. Rushton, *Chem. Eng. Prog.* **1**(1), 12 (1947).
168. P. C. Keith, *Am. Gas. Assoc. Mon.* **28,** 253 (1946).
169. M. D. Schlesinger and co-workers, *Ind. Eng. Chem.* **46,** 1322 (1954).
170. R. B. Anderson, J. Feldman, and H. H. Storch, *Ind. Eng. Chem.* **44,** 2418 (1952).
171. V. V. Kamzolkin, *Priroda* **45**(11), 93 (1956).
172. I. Wender, R. A. Friedel, and M. Orchin, *Science* **113,** 206 (1951).
173. Fr. Pat. 1,006,012 (April 18, 1953), H. Grasshof.
174. C. E. Morrell and co-workers, *Ind. Eng. Chem.* **44,** 2839 (1952).
175. *Pet. Refiner* **36**(10), 241 (1957).
176. G. T. Morgan and R. Taylor, *Trans. Chem. Eng. Congr. London* **3,** 441 (1936).
177. Belg. Pat. 842,430 (Dec. 1, 1976), L. H. Slaugh (to Shell Intl. Res.).
178. Jpn. Kokai 5,2073,804 (June 21, 1977), (to Mitsubishi Gas Chem. Ind.).

179. U.S. Pat. 4,062,898 (Dec. 13, 1977), M. Dubcek and G. C. Knapp (to Ethyl Corp.).
180. Neth. Pat. 7,606,138 (1977), L. Slaugh (to Shell).
181. *Chem. Mkt. Rep.* **4,** 17 (July 30, 1973).
182. *'73 Annual Survey of Petrochemical Industries in Japan,* Heavy Chemical Industry Press, Tokyo, 1973.
183. *Industrial Alcohol,* Misc. Publ. No. 695, U.S. Dept. Agr., Feb. 1950.
184. *Chem. Eng. News* **29**(47), 4932 (1951).
185. E. S. Lepinsky, *Sugar J.* 27 (Aug. 1976).
186. ERDA, *A National Plan for Energy Research, Development, and Demonstration: Creating Energy Choices for the Future,* Vol. 1 (1976).
187. *National Energy Outlook,* Federal Energy Administration, 1976.
188. *Report of Petrochemical Committee,* Govt. of India; *Journal of Indus. & Trade* **12,** 11 (1962).
189. N. Coutinho, *Bras. Acucareiro,* 19 (1976).
190. E. A. Jackson, *Process Biochem.* 29 (1976).
191. *Bus. Week.,* 841 (Nov. 7, 1977).
192. S. K. Chan, *Tapioca,* Investigations at the Federal Experiment Station, Serdang, Malaysia, 1969.
193. W. A. Scheller and B. J. Mohr, *Chemtech* **7**(10), 616 (Oct. 1977).
194. W. A. Scheller, *Proc. 8th Nat. Conf. on Wheat Util. Res.,* USDA Pub. ARS-W19, Sept., 1974.
195. W. A. Scheller and B. J. Bohr, *Am. Chem. Soc. Div. Fuel. Chem. Prepr.* **20,** 2, 71 (1975).
196. *Chem. Eng. News* **54**(16), 12 (April 12, 1976).
197. *Chem. Week.* **122**(14), 40 (April 5, 1978).
198. *Chem. Week.* **122**(12), 31 (March 22, 1978).
199. Ref. 1, Chapt. 3, pp. 33–38.
200. U.S. Pat. reissue 23,507 (May 27, 1952), R. L. Maycock and co-workers (to Shell Development Co.).
201. U.S. Pat. 2,773,910 (Dec. 11, 1956), R. J. Schrader (to Eastman Kodak Co.).
202. U.S. Pat. 2,981,661 (April 25, 1961), W. Sisco and J. S. Wiederecht (to American Cyanamid Co.).
203. U.S. Pat. 3,014,971 (Dec. 26, 1961), S. W. Wilson (to Esso Research and Engineering Co.).
204. U.S. Pat. 2,892,874 (June 30, 1959), C. J. B. Duculot.
205. U.S. Pat. 2,721,874 (Oct. 25, 1955), A. McIlroy (to Stanolind Oil and Gas Co.).
206. U.S. Pat. 2,892,757 (June 30, 1959), A. E. Markham (to Puget Sound Pulp and Timber Co.).
207. U.S. Pat. 2,944,087 (July 5, 1960), E. W. Mommensen and co-workers (to Esso Research and Engineering Co.).
208. U.S. Pat. 2,857,436 (Oct. 21, 1958), J. R. Mackinder and co-workers (to Shell Development Co.).
209. U.S. Pat. 2,957,023 (Oct. 18, 1960), W. A. Dimler, Jr. (to Esso Research and Engineering Co.).
210. U.S. Pat. 2,867,651 (Jan. 6, 1959), R. H. Wise (to Standard Oil Co.).
211. U.S. Pat. 2,885,446 (May 5, 1959), S. P. Sharp and A. Steitz, Jr. (to Pan American Petroleum Co.).
212. *Chem. Week* **120**(2), 26 (Jan. 12, 1977).
213. *Chem. Eng. News* **55**(2), 12 (Jan. 10, 1977).
214. *Annual Report of the Commissioner of Internal Revenue,* U.S. Treasury Dept., IRS, data for 1950–52.
215. *Alcohol and Tobacco Summary Statistics,* U.S. Treasury Dept., ATF, data for 1953–1973.
216. *Statistical Release, Distilled Spirits,* U.S. Treasury Dept., ATF, preliminary data for 1974–1975.
217. U.S. Pat. 3,076,032 (Jan. 29, 1963), W. Reimenschneider and K. Diaier (to Farbwerke Hoechst A.G.).
218. U.S. Pat. 3,057,915 (Oct. 9, 1962), W. Reimenschneider and co-workers (to Farbwerke Hoechst A.G.).
219. *USP XX-NFXV,* Mack Publishing Co., Easton, Pa., 1980.
220. *Reagent Chemicals, ACS Specifications,* 5th ed., ACS, Washington, D.C., 1974.
221. *1977 Book of ASTM Standards, Part 29, D-891, Method C,* American Society for Testing and Materials, Philadelphia, Pa., 1977.
222. V. C. Mehlenbocher, "Determination of Alcohols" in John Mitchell, Jr., ed., *Organic Analysis,* Vol. 1, Interscience Publishers, Inc., New York, 1953, pp. 1–65.
223. D. M. Smith and W. M. D. Bryant, *J. Am. Chem. Soc.* **57,** 61 (1935).
224. W. T. Robinson and co-workers, *Anal. Chem.* **33,** 1030 (1961).
225. S. Siggia and co-workers, *Anal. Chem.* **33,** 900 (1961).
226. P. R. Ashurst, *J. Inst. Brew.* **69,** 457 (1963).
227. M. Stiller, *Anal. Chim. Acta.* **25,** 85 (1961).

228. G. Machata, *Mikrochim. Acta.* 691 (1962).
229. K. D. Parker and co-workers, *Anal. Chem.* **34**, 1234 (1962).
230. A. R. Alha and V. Tamminen, *Am. Med. Exptl. et Biol. Fenniae (Helsinki)* **38**, 121 (1960); *Chem. Abstr.* **55**, 13528i (1961).
231. H. V. Malmstadt and T. P. Hadjuoannou, *Anal. Chem.* **34**, 455 (1962).
232. H. Croon and W. Croon, *Z. Anal. Chem.* **192**, 378 (1963).
233. P. Jaulmes and J. P. Laval, *Trav. Soc. Pharm. Montpellier* **21**, 21 (1962).
234. S. Musha and M. Takeda, *Kogyo Kajaku Zasshi* **61**, 1143 (1958).
235. B. B. Coldwell and C. L. Grant, *J. Forensic Sci* 8(2), 149 (1963).
236. K. Grosskopf, *Draeger Heft* **248**, 16 (1962); *Chem. Abstr.* **59**, 1023 (1963).
237. C. H. Thienes and T. J. Haley, *Clinical Toxicology,* 4th ed., Lea and Febiger, Philadephia, 1964.
238. F. A. Patty, ed., *Industrial Hygiene and Toxicology,* Vol. 2, 2nd ed., Interscience Publishers, a division of John Wiley & Sons, Inc., New York, 1962.
239. N. I. Sax, *Dangerous Properties of Industrial Materials,* 4th ed., Van Nostrand Reinhold Corp., New York, 1975.
240. L. T. Fairhall, *Industrial Toxicology,* 2nd ed., The Williams and Wilkins Co., Baltimore, 1957.
241. R. B. Forney and F. W. Hughes, *Clin. Pharmcol. Ther.* **4**, 619 (1963).
242. H. W. Haggard and L. A. Greenberg, *J. Pharmacol. Exp. Ther.* **52**, 150 (1934).
243. E. J. Fredericks and M. Z. Lazor, *Am. Intern. Med.* **59**, 90 (1963).
244. R. E. Gosselin and co-workers, *Clinical Toxicology of Common Products, Acute Poisoning,* 4th ed., Williams and Wilkins Co., Baltimore, 1976.
245. A. J. Lanza and J. A. Goldberg, *Industrial Hygiene,* Thomas, Springfield, Illinois, 1939, p. 466.
246. E. Browning, *Toxicity and Metabolism of Industrial Solvents,* Elsevier, Amsterdam, 1965, p. 326.
247. U.S. Pat. 2,870,866 (Jan. 27, 1959), G. Baecklund (to Aktiebolaget Chematur).
248. U.S. Pat. 3,073,752 (Jan. 15, 1963), M. Mention (to Les Usines de Melle S.A.).
249. U.S. Pat. 3,154,586 (Oct. 28, 1964), O. E. Bander and co-workers (to Farbwerke Hoechst A.G.).
250. U.S. Pat. 3,149,167 (Sept. 15, 1964), L. Hornig and co-workers (to Farbwerke Hoechst A.G.).
251. *Chem. Eng. News* **39**, 52 (April 17, 1961).
252. U.S. Pat. 3,172,913 (March 9, 1965), L. Hornig and co-workers (to Farbwerke Hoechst A.G.).
253. U.S. Pat. 3,154,586 (Oct. 17, 1964), O. E. Bander and co-workers (to Farbwerke Hoechst A.G.).
254. U.S. Pat. 3,122,586 (Feb. 25, 1964), W. Berndt (to Consortium fur Elektrochemische Ind. G.m.b.H.).
255. U.S. Pat. 3,106,579 (Oct. 8, 1963), L. Hornig and H. Lenzmann (to Farbwerke Hoechst A.G.).
256. U.S. Pat. 3,119,874 (Jan. 28, 1964), E. Paszthory and W. Reimenschneider (to Farbwerke Hoechst A.G.).
257. U.S. Pat. 3,118,001 (Jan. 14, 1964), W. Reimenschneider (to Farbwerke Hoechst A.G.).
258. U.S. Pat. 3,104,263 (Sept. 17, 1963), W. Reimenschneider (to Farbwerke Hoechst A.G.).
259. U.S. Pat. 3,119,875 (Jan. 28, 1964), A. Steinmetz and co-workers (to Farbwerke Hoechst A.G.).
260. U.S. Pat. 3,087,968 (April 30, 1958), L. Hornig and co-workers (to Farbwerke Hoechst A.G.).
261. U.S. Pat. 3,086,994 (April 23, 1963), J. Smidt and co-workers (to Consortium fur Elektrochemische Ind. G.m.b.H.).
262. U.S. Pat. 3,080,425 (March 5, 1963), J. Smidt and co-workers (to Consortium fur Elektrochemische Ind. G.m.b.H.).
263. U.S. Pat. 3,086,052 (April 16, 1963), J. Smidt and co-workers (to Consortium fur Elektrochemische Ind. G.m.b.H.).
264. U.S. Pat. 3,131,223 (April 23, 1964), J. Smidt and co-workers (to Consortium fur Elektrochemische Ind. G.m.b.H.).
265. U.S. Pat. 2,974,173 (March 7, 1961), R. B. Long and co-workers (to Esso Research and Engineering Co.).
266. U.S. Pat. 3,106,581 (Oct. 8, 1963), S. D. Neely (to Eastman Kodak Co.).
267. U.S. Pat. 3,080,426 (March 5, 1963), I. Kirshenbaum, E. M. Amir, and E. J. Inchalik (to Esso Research and Engineering Co.).
268. C. F. Cullis and E. J. Newitt, *Proc. R. Soc. (London)* **A237**, 530 (1956); **A242**, 516 (1957).
269. J. Klassem and R. S. Kirk, *AIChE J.* **1**, 488 (1955).
270. U.S. Pat. 1,977,750 (Oct. 23, 1935), C. O. Young (to Union Carbide Corp.).
271. A. Bielanski and co-workers, *Bull. Acad. Polon. Sci. Cl. 3* **4**, 533 (1956).
272. A. Bielanski and co-workers, *Bull. Acad. Polon. Sci. Cl. 3* **3**, 497 (1955).
273. A. A. Balandin and P. Titeni, *Dokl. Akad. Nauk SSSR* **113**, 1019 (1957).

274. U.S. Pat. 2,861,106 (Nov. 18, 1958), W. Optiz and W. Urbanski (to Knapsack-Giesheim A.G.).
275. U.S. Pat. 2,884,460 (April 28, 1959), V. I. Komarensky (to Heavy Minerals Co.).
276. A. Kh. Bork and O. A. Markova, *Zh. Fiz. Khim.* **22,** 1381 (1948).
277. S. Abe, *Sci. Papers Inst. Phys. Chem. Res. (Tokyo)* **40,** 331 (1943).
278. Ital. Pat. 509,424 (June 14, 1955), V. Martello and S. Cecotti (to Bombrini Parodi-Defino Societa per Azioni).
279. V. M. Nikitin, *J. Gen. Chem. USSR (Eng. Transl.)* **15,** 273 (1945).
280. V. M. Nikitin and N. V. Razumov, *J. Gen. Chem. USSR (Eng. Transl.)* **11,** 133 (1941).
281. K. V. Topchieva and K. Yun-Pin, *Dokl. Akad. Nauk SSSR* **101,** 305 (1955).
282. S. B. Ansimov and G. I. Khaidarov, *Zh. Obshch. Khim.* **18,** 40 (1948).
283. Z. E. Kosalapov, *J. Gen. Chem. USSR (Eng. Transl.)* **5,** 307 (1935).
284. U. Tsao and J. W. Reilly, *Hydrocarbon Proc.* **57,** 133 (Feb. 1978).
285. A. B. Metzner and co-workers, *AIChE J.* **5,** 496–501 (1959).
286. R. E. Parker and co-workers, *Chem. Rev.* **59,** 737 (1959).
287. B. Weibull and co-workers, *Acta. Chem. Scand.* **8,** 847 (1954).
288. R. M. Laird and co-workers, *J. Chem. Soc. B* **9,** 1062 (1969).
289. T. P. Hilditch, *Catalytic Processes in Applied Chemistry,* Chapman & Hall Ltd., London, 1929, p. 264.
290. *Synthetic Ethyl Alcohol in the Manufacture of Vinegar,* Tech. Bull. No. 19, Enjay Chemical Company.
291. U.S. Pat. 2,085,785 (July 6, 1937), R. R. Bottoms (to the Girdler Corp.).
292. Brit. Pat. 399,201 (Oct. 2, 1933), (to Rohm & Haas Co.).
293. Fr. Pat. 767,771 (July 24, 1934), (to I. G. Farbenindustrial A.G.).
294. U.S. Pat. 2,964,558 (Dec. 13, 1960), J. M. F. Leathers (to Dow Chemical Co.).
295. Neth. Pat. 134,896 (April 17, 1972), (to Distillers Co.).
296. Jap. Pat. 47-37404 (Sept. 20, 1972), (to Nippon Shokubai Kapaku Kogyo).
297. U.S. Pat. 3,458,561 (July 29, 1969), C. T. Kautter and co-workers (to Rohm & Haas Co.).
298. Brit. Pat. 1,257,371 (Dec. 15, 1971) (to Toyo Soda Manufac.).
299. U.S. Pat. 3,442,935 (May 6, 1969), L. A. Pine and co-workers (to Esso Res. and Eng.).
300. U.S. Pat. 3,457,299 (July 22, 1969), R. D. Closson and co-workers (to Ethyl Corp.).
301. Ref. 156, p. 792.
302. B. T. Brooks, *Ind. Eng. Chem.* **27,** 282 (1935).
303. U.S. Pats. 1,873,536 and 1,873,537 (Oct. 23, 1932), R. L. Brown and W. W. Odel.
304. I. E. Adadurov and P. Ya. Krainii, *J. Phys. Chem. USSR* **5,** 136 (1934).
305. J. B. Sendrens, *Ann. Chim. Phys.* **25,** 449 (1912).
306. R. N. Pease and C. C. Yung, *J. Am. Chem. Soc.* **46,** 2397 (1924).
307. *Ibid.,* 390 (1924).
308. T. V. Antipina and O. V. Isaev, *Zh. Fiz. Khim.* **31,** 2078 (1957).
309. T. V. Antipina and M. D. Sinitsyna, *Vestn. Mosk. Univ., Ser. Mat. Mekh. Astron., Fiz. Khim.* **12,** 137 (1957).
310. Brit. Pat. 332,756 (July 31, 1930), (to N. V. de Bataasche Petr. Maatschappij).
311. Brit. Pat. 350,010 (June 5, 1931), (to N. V. de Bataasche Petr. Maatschappij).
312. C. E. Schildknecht, A. O. Zoss, and C. McKinley, *Ind. Eng. Chem.* **39,** 180 (1947).
313. W. G. Hanford and D. L. Fuller, *Ind. Eng. Chem.* **40,** 1171 (1948).
314. U.S. Pat. 1,959,927 (May 22, 1934), W. Reppe (to I. G. Farbenindustrie A.G.).
315. U.S. Pat. 2,021,869 (Nov. 19, 1935), W. Reppe and W. Wolff (to I. G. Farbenindustrie A.G.).
316. U.S. Pat. 2,066,076 (Dec. 19, 1936), W. Reppe and W. Wolff (to I. G. Farbenindustrie A.G.).
317. M. P. E. Berthelot and co-workers, *Am. Chim. Phys.* **65,** 385 (1861).
318. D. I. Saletan and R. R. White, *Chem. Eng. Prog. Symp. Ser.* **48**(4), 59 (1952).
319. USSR Pat. 145,235 (1962), A. A. Skladner and co-workers.
320. Belg. Pat. 613,426 (1962), (to Wacker-Chemie).
321. D. B. Keyes, *2nd Eng. Chem.* **24,** 1096 (1932).
322. E. Chadwick, *Ind. Chem.* **39,** 345 (1963).
323. P. K. Frohlich and co-workers, *J. Am. Chem. Soc.* **52,** 1565 (1930).
324. H. F. Hoering and co-workers, *Ind. Eng. Chem.* **35,** 575 (1943).
325. G. Edgar and W. H. Schuyler, *J. Am. Chem. Soc.* **46,** 64 (1924).
326. H. Essex and J. D. Clark, *J. Am. Chem. Soc.* **54,** 1290 (1932).
327. A. Mailhe and F. de Godon, *Bull. Soc. Chem. Fr.* **29,** 101 (1921).

328. W. J. Knox and T. N. Burbridge, *J. Am. Chem. Soc.* **65,** 999 (1963).
329. K. Kerby and S. Swann, *Ind. Eng. Chem.* **32,** 1607 (1940).
330. V. E. Tischenko, *J. Russ. Phys. Chem. Soc.* **38,** 355 (1906).
331. W. C. Child and H. Adkins, *J. Am. Chem. Soc.* **45,** 3013 (1923).
332. East Ger. Pat. 10,646 (1955), P. Herte and A. Loventz.
333. R. D. Morin and A. E. Bearse, *2nd Eng. Chem.* **43,** 1596 (1951).
334. U.S. Pat. 3,265,748 (1966), D. M. Hurt (to E. I. du Pont de Nemours & Co.).
335. A. W. Fleer and co-workers, *Ind. Eng. Chem.* **47,** 986 (1955).
336. Can. Pat. 321,646 (April 19, 1932), G. Bloomfield and co-workers (to Commercial Solvents Corp.).
337. Brit. Pat. 338,518 (Nov. 19, 1930), H. Dreyfus.
338. Brit. Pat. 347,593 (April 22, 1931), (to Holzverkohlungs-Industrie A.G.).
339. Brit. Pat. 359,430 (Oct. 14, 1931), (to Deutsche Gold- und Silber-Scheideanstalt vorm. Roessler).
340. E. Donath, *Chem. Ztg.* **12,** 1191 (1888); **51,** 924 (1927).
341. U.S. Pat. 1,663,350 (March 20, 1928), K. Roka (to Holzverkohlungs-Industrie A.G.).
342. Brit. Pat. 353,467 (July 16, 1931), H. F. Oxley and co-workers (to British Celanese Co., Ltd.).
343. St. Bakowski and L. Stepniewski, *Przem. Chem.* **20,** 142 (1936).
344. B. A. Bolotov and co-workers, *Zh. Prikl. Khim.* **28,** 299 (1955).
345. B. A. Bolotov and K. P. Katkova, *Zh. Prikl. Khim.* **28,** 414 (1955).
346. M. Guerbet, *Compt. Rend.* **128,** 511 (1899).
347. U.S. Pat. 2,457,866 (Jan. 4, 1949), C. A. Carter (to Union Carbide Corp.).
348. E. F. Pratt and D. G. Kubler, *J. Am. Chem. Soc.* **76,** 52 (1954).
349. J. Bolle and L. Bourgeois, *Compt. Rend.* **233,** 1466 (1951).
350. Brit. Pat. 655,864 (Aug. 1, 1951), M. Sulzbacher.
351. M. Sulzbacher, *J. Appl. Chem. (London)* **5,** 637 (1955).
352. U.S. Pat. 2,762,847 (Sept. 11, 1956), R. E. Miller and G. E. Bennett (to Monsanto Chemical Co.).
353. I. I. Ostromuislenski, *J. Russ. Phys.-Chem. Soc.* **47,** 1472 (1915).
354. W. M. Quattlebaum, W. J. Toussaint, and J. T. Dunn, *J. Am. Chem. Soc.* **69,** 593 (1947).
355. U.S. Pat. 2,421,361 (May 27, 1947), W. J. Toussaint and J. T. Dunn (to Union Carbide).
356. B. B. Corson and co-workers, *Ind. Eng. Chem.* **42,** 359 (1950).
357. I. R. Laszlo and co-workers, *Magy. Kem. Foly.* **60,** 65 (1944).
358. U.S. Pat. 2,800,517 (July 23, 1957), C. Romanovsky and T. E. Jordan (to Publicker Industries, Inc.).
359. G. Natta and R. Rigamonti, *Chim. Ind. (Milan)* **29,** 239 (1947).
360. A. Brenman, *Rev. Chim. (Bucharest)* **8,** 286 (1957).
361. U.S. Pat. 2,850,463 (Sept. 2, 1958), C. Romanovsky and T. E. Jordan (to Publicker Industries, Inc.).
362. M. Essayan and I. Ciolan, *Rev. Chim. (Bucharest)* **7,** 31 (1956).

PAUL DWIGHT SHERMAN, JR.
PERCY R. KAVASMANECK
Union Carbide Corporation

ETHANOLAMINES. See Alkanolamines.

ETHERS

Ethers are compounds of the general formula Ar—O—Ar′, Ar—O—R, and R—O—R′ where Ar is any aryl group and R is any alkyl group. If the two R or Ar groups are identical, the compound is a symmetrical ether. Examples of symmetrical ethers are methyl ether, CH_3OCH_3, and phenyl ether, $C_6H_5OC_6H_5$; examples of unsymmetrical ethers are methyl ethyl ether, $CH_3OCH_2CH_3$, and methyl *tert*-butyl ether, $CH_3OC(CH_3)_3$ (see also Glycols, glycol ethers). Cyclic ethers are oxygen heterocycles such as tetrahydrofuran, $\underline{OCH_2CH_2CH_2CH_2}$, *p*-dioxane, $\underline{OCH_2CH_2OCH_2CH_2}$, and 1,2-propylene oxide, $\underline{OCH_2CHCH_3}$.

Simple ethers derive their name from the two groups attached to the oxygen followed by the word ether, eg, ethyl ether, $CH_3CH_2OCH_2CH_3$. If one group has no simple name, the compound may be named as an alkoxy derivative, eg, 2-ethoxyethanol, $CH_3CH_2OCH_2CH_2OH$.

Physical Properties

In general, ethers are neutral, pleasant-smelling compounds that have little or no solubility in water, but are easily soluble in organic liquids (1). Their boiling points approximate those of hydrocarbons having comparable molecular weights and geometries. Table 1 gives the physical properties of some representative ethers.

Chemical Properties

Ethers are comparatively unreactive compounds because the carbon–oxygen bond is not readily cleaved. For this reason, ethers are frequently employed as inert solvents in organic synthesis. Ethers do react with exceptionally powerful basic reagents, particularly certain alkali–metal alkyls, to give cleavage products (2). Ethers react with less powerful bases to give the same cleavage products but only under the forcing conditions of high temperature and pressure:

$$CH_3Na + CH_3CH_2OCH_2CH_3 \rightarrow CH_4 + CH_2\!\!=\!\!CH_2 + CH_3CH_2ONa$$

The ether linkage can also be cleaved by strong acids, generally at high temperatures (2):

$$(CH_3)_2CHOCH(CH_3)_2 + 2\ HBr \xrightarrow{140°C} 2(CH_3)_2CHBr + H_2O$$

$$C_6H_5OCH_3 + HI \xrightarrow{130°C} C_6H_5OH + CH_3I$$

Other acids that have been used for this cleavage are phosphoric acid, pyridine hydrochloride, boron tribromide, trifluoroacetic acid, and nitric acid. Acid cleavage of aryl alkyl ethers always gives phenol because the aromatic carbon–oxygen bond is much stronger than the aliphatic carbon–oxygen bond. Unsymmetrical aliphatic ethers usually yield a mixture of alkyl halides and alcohols when cleaved by halogen-containing acids.

Ethers are potentially hazardous chemicals since, in the presence of atmospheric oxygen, a radical-chain process can occur, resulting in the formation of peroxides which

Table 1. Physical Properties of Some Representative Ethers

Ether	CAS Registry No.	bp, °C$_{kPa}$[a]	d_4^{20}	n_D^{20}
Aliphatic				
saturated, symmetrical			sp. gr.	
methyl	[115-10-6]	−23.7	1.617 (air)	
2-methoxyethyl (diglyme)	[111-96-6]	162	0.9451_{20}	1.4097
ethyl	[60-29-7]	34.5	0.7146	1.3527
1-chloroethyl	[6986-48-7]	116–117	1.1060^{25}	1.4186^{25}
n-propyl	[111-43-3]	90.5	0.7360	1.3809
isopropyl	[108-20-3]	68.5	0.7257	1.3682
n-butyl	[142-96-1]	142.0	0.7704	1.3981
sec-butyl	[6863-58-7]	122.0	0.7590^{25}	1.3931
isobutyl	[628-55-7]	123.0	0.7612_{15}^{15}	
tert-butyl	[6163-66-2]	108.0	0.7622	1.3946
n-amyl	[693-65-2]	188.0	0.7849	1.4119
isoamyl	[544-01-4]	173.0	0.7777	1.4085
sec-amyl	[56762-00-6]	161.0	0.7830_{20}	1.4058
n-hexyl	[112-58-3]	223_{102}	0.7936	1.4204
n-heptyl	[629-64-1]	258.5_{102}	0.8008	1.4275
n-octyl	[629-82-3]	286–287	0.8063	1.4327
unsymmetrical				
methyl *n*-propyl	[557-17-5]	38.9	0.738	1.3579
methyl isopropyl	[598-53-8]	32.5_{104}	0.7237^{15}	1.3576
methyl *n*-butyl	[628-28-4]	70.5	0.7443	1.3736
methyl isobutyl	[625-44-5]	$105-106_{96}$	0.7549	1.3852^{25}
methyl *tert*-butyl	[1634-04-4]	55.1	0.7406	1.3690
ethyl *n*-butyl	[628-81-9]	91.5	0.7490	1.3818
ethyl *n*-amyl	[17952-11-3]	118.0	0.7622	1.3927
2-ethoxyethanol (cellosolve)	[110-80-5]	135	0.931_{20}	1.406^{25}
2-(2-ethoxy)ethoxyethanol (carbitol)	[111-90-0]	196	0.9855^{25}	1.4273
unsaturated				
vinyl	[109-93-3]	28–31	0.767^{25}	
vinyl methyl	[107-25-5]	5–6	0.7511	
vinyl ethyl	[109-92-2]	35.0	0.7533	1.3739^{25}
vinyl *n*-butyl	[111-34-2]	93.5	0.7735^{25}	1.3997^{25}
allyl	[557-40-4]	95.0	0.8053	1.4163
bis(2-methallyl)	[628-56-8]	105.4	0.8627	1.4206
allyl ethyl	[557-31-3]	67.6	0.765	1.3881
allyl glycidyl	[106-92-3]	$75.0_{6.7}$		1.4310^{30}
ethynyl ethyl	[927-80-0]	49.0	0.8001	1.3796
ethynyl butyl	[3329-56-4]	104.0	0.8078^{25}	1.4033^{25}
Aromatic				
methyl phenyl	[100-66-3]	153.8	0.9954	1.5179
4-methoxytoluene	[104-93-8]	176.5	0.9689	1.5124
ethyl phenyl	[103-73-1]	172	0.9792	1.5076
1-methoxy-4-*trans*-propenylbenzene (*trans*-anethole)	[4180-23-8]	253 mp 21	0.9882	1.5615
1-methoxy-4-allylbenzene (estragole)	[140-67-0]	215	0.9645	1.5230
phenyl	[101-84-8]	258 mp 28	1.0863	1.5780
2-methoxyphenol (guaiacol)	[90-05-1]	205.5	1.1287	1.5385
1,2-dimethoxybenzene (veratrole)	[91-16-7]	206.7 mp 22–23	1.084	
1,4-dimethoxybenzene	[150-78-7]	212.6 mp 58–60	1.0526_{55}^{55}	

Table 1 (continued)

Ether	CAS Registry No.	bp, °C$_{kPa}$^{a}	d_4^{20}	n_D^{20}
2-methoxy-4-allylphenol (4-allyl-guaiacol)	[97-53-0]	255	1.0664	1.5410
1,2-dimethoxy-4-allylbenzene (4-allylveratrole)	[93-15-2]	248	1.055	1.532
1-allyl-3,4-methylenedioxybenzene (safrole)	[94-59-7]	234.5	1.0950	1.5383
1-propenyl-3,4-methylenedioxybenzene (isosafrole)	[93-16-3]	252	1.1224	1.5782
2-methoxy-4-cis-propenylphenol (cis-isoeugenol)	[5912-86-7]	134$_{1.7}$	1.0851	1.5700
2-methoxy-4-trans-propenylphenol (trans-isoeugenol)	[5932-68-3]	141$_{1.6}$	1.0852	1.5782
1-benzyloxy-2-methoxy-4-trans-propenylbenzene (benzyl isoeugenol)	[120-11-6]	mp 58		
butylated hydroxyanisole (BHA), a mixture of:	[25013-16-3, 55949-47-8]	264–270$_{98}$ mp 48–55		
2-tert-butyl-4-methoxyphenol	[121-00-6]			
and 3-tert-butyl-4-methoxyphenol	[88-32-4]			
Cyclic				
ethylene oxide (oxirane)	[75-21-8]	13.5$_{99.4}$	0.8824$^{10}_{10}$	1.3597^7
1,2-propylene oxide (methyloxirane)	[75-56-9]	34.3	0.859^0	1.3670
1,3-propylene oxide (oxetane)	[503-30-0]	47.8	0.8930^{25}	1.3961
tetrahydrofuran (oxolane)	[109-99-9]	64.5	0.8892	1.4050
furan (oxole)	[110-00-9]	31.36	0.9514	1.4214^2
tetrahydropyran (oxane)	[142-68-7]	88	0.8810	1.4200
1,4-dioxane	[123-91-1]	101$_{100}$	1.0337	1.4224

^{a} To convert kPa to mmHg, multiply by 7.5.

are unstable, explosion-prone compounds (3). The reaction may be generalized in terms of the following steps involving initiation, propagation, and termination.

$$initiation: CH_3CH_2OCH_2CH_3 \longrightarrow CH_3\dot{C}HOCH_2CH_3$$

$$propagation: \begin{cases} CH_3\dot{C}HOCH_2CH_3 \longrightarrow CH_3\overset{OO\cdot}{\underset{|}{C}}HOCH_2CH_3 \\ CH_3\overset{OO\cdot}{\underset{|}{C}}HOCH_2CH_3 + CH_3CH_2OCH_2CH_3 \longrightarrow CH_3\overset{OOH}{\underset{|}{C}}HOCH_2CH_3 + CH_3\dot{C}HOCH_2CH_3 \end{cases}$$

$$termination: 2\,CH_3\dot{C}HOCH_2CH_3 \longrightarrow \underset{\underset{CH_3CHOCH_2CH_3}{|}}{CH_3CHOCH_2CH_3}$$

The initiation step, which may occur in a variety of ways, is not known in all cases. Commonly used ethers such as ethyl ether, isopropyl ether, tetrahydrofuran, and p-dioxane are particularly prone to form explosive peroxides on prolonged storage and exposure to air and light (see Peroxides).

Ethers are weakly basic and are converted to unstable oxonium salts by strong acids such as sulfuric acid, perchloric acid, and hydrobromic acid; relatively stable complexes are formed between ethers and Lewis acids such as boron trifluoride, aluminum chloride, and Grignard reagents (qv) (4):

$$CH_3OCH_3 + CH_3BF_4 \rightarrow [(CH_3)_3O]^+BF_4^-$$

Like other aromatic compounds, aromatic ethers can undergo substitution in the aromatic ring with electrophilic reagents, eg, with nitration, halogenation, and sulfonation. They also undergo Friedel-Crafts (qv) alkylation and acylation.

Allyl phenyl ethers rearrange cleanly at high temperatures producing o-allylphenols, or p-allylphenols if both ortho positions are blocked. This reaction is called the Claisen rearrangement (5):

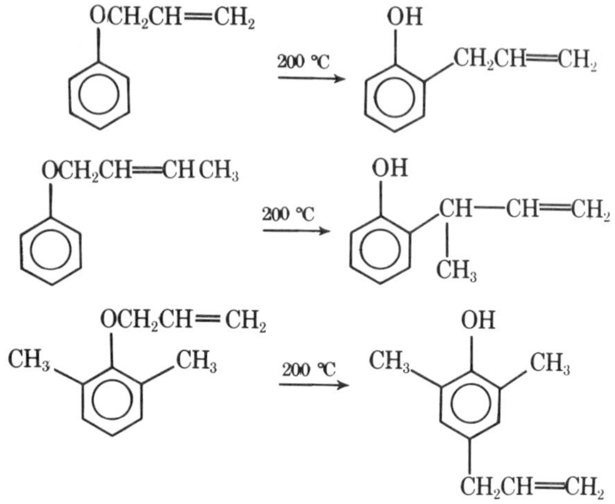

Preparation

The most versatile method of preparing ethers is the Williamson ether synthesis, particularly in the preparation of unsymmetrical alkyl ethers (6–7). The reaction of sodium alcoholates with halogen derivatives of hydrocarbons gives the ethers:

$$RX + NaOR' \rightarrow ROR' + NaX$$

Dialkyl sulfates can replace the halogen derivatives, and this modification is especially useful for the preparation of phenolic ethers:

$$(RO)_2SO_2 + C_6H_5ONa \rightarrow C_6H_5OR + ROSO_3Na$$

Aromatic halides do not react easily with phenoxide ions to produce diaryl ethers unless the aromatic halide is substituted with one or more electron-withdrawing groups, eg, nitro or carboxyl groups. The Ullmann reaction uses finely divided copper or copper salts to catalyze the reaction of phenoxides with aromatic halides to give diaryl ethers.

Alcohols can be dehydrated with strong acid catalysts and high reaction temperatures to produce ethers. This method is particularly useful for the preparation of symmetrical lower alkyl ethers, such as ethyl ether. The reaction gives poor yields of ethers with secondary and tertiary alcohols; dehydration to form the corresponding olefin is a more favorable reaction:

$$2\ CH_3CH_2OH \xrightarrow[H_2SO_4]{140°C} CH_3CH_2OCH_2CH_3 + H_2O$$

The reaction fails for the production of diaryl ethers from phenols.

Ethers can be prepared by the addition of an alcohol or phenol to an olefin under acid catalysis; sulfuric acid, phosphoric acid, hydrochloric acid, and boron trifluoride have all been used as catalysts:

$$CH_3OH + (CH_3)_2C= CH_2 \xrightarrow[100°C]{BF_3} (CH_3)_2CCH_3 \\ \quad\quad\quad\quad\quad\quad\quad\quad\quad\quad\quad\quad\quad\quad\quad | \\ \quad\quad\quad\quad\quad\quad\quad\quad\quad\quad\quad\quad\quad\quad OCH_3$$

Economic Aspects

Table 2 compares the prices and availability of several commercially available ethers as of July, 1978.

Table 2. Price and Availability of Commercial Ethers[a]

Ether	Price, $/kg	Availability
ethyl	0.53	tank car
n-butyl	1.34	tank car
vinyl	1.56	75-mL bottle
	per bottle	
1,4-dioxane	1.61	tank car
tetrahydrofuran	1.65	tank car
isopropyl	0.33	tank car
1-methoxy-4-propenylbenzene	6.40	208-L (55-gal) drum
	per drum	
1,4-dimethoxybenzene	5.51	208-L (55-gal) drum
2-methoxyphenol	5.73	227-kg drum
2-$tert$-butyl-4-methoxyphenol (BHA) 3-$tert$-butyl-4-methoxyphenol	14.00	208-L (55-gal) drum
2-methoxy-4-propenylphenol	15.10	208-L (55-gal) drum
1-benzyloxy-2-methoxy-4-propenylbenzene	34.17	11-kg can

[a] Ref. 8.

Health and Safety Factors

Although ethers are not particularly hazardous, their use involves risks of fire, toxic effects, and several unexpected reactions.

Since almost all ethers burn in air, an assessment of their potential hazards depends upon flash points and ignition temperatures. The flash point of a liquid is the lowest temperature at which vapors are given off in sufficient quantities for the vapor–air mixture above the surface of the liquid to propagate a flame away from the source of ignition. In other words, an explosive vapor–air mixture can form whenever a liquid is used or stored in an open container at a temperature above its flash point. Table 3 lists the flash points of several common ethers.

The ignition temperature is the minimal temperature required to initiate or cause self-sustained combustion. Table 3 lists ignition temperatures of several common ethers. Attention is directed to the particularly low ignition temperatures of ethyl ether and 1,4-dioxane, especially with reference to some common ignition sources such as

Table 3. Flash Points and Ignition Temperatures of Some Common Ethers[a]

Ether	Flash point, °C	Ignition temperature, °C
ethyl	−45	180
isopropyl	−28	443
n-butyl	25	194
vinyl	−30	360
1,4-dioxane	12	180
tetrahydrofuran	−14	321

[a] Ref. 9.

a lighted cigarette (732°C) or a pressurized (0.7 MPa or 100 psi) steam line (180°C).

The effect of ethers owing to ingestion, skin contact, or inhalation may range from drowsiness and lack of coordination to serious injury or death.

Ethers are reported to have a low order of toxicity when ingested, although 30–60 mL may be fatal when swallowed (10), and a case of fatal poisoning owing to ingestion of a large quantity of 2-methoxyethanol has been recorded (11).

Prolonged or repeated contacts of ethers with skin cause tissue defatting and dehydration leading to dermatitis. Some compounds penetrate the skin in harmful amounts. 1,4-Dioxane and 1-allyl-3,4-methylenedioxybenzene (safrole) have recently been listed as Category I carcinogens by OSHA. Category I carcinogens are confirmed cancer-causing agents based on human data, or based on tests in two mammalian species, or in one species if the tests have been replicated.

Inhalation is the most common means by which ethers enter the body. The effects of various ethers may include narcosis, irritation of the nose, throat, and mucous membranes, and chronic or acute poisoning. In general, ethers are central nervous system depressants, eg, ethyl ether and vinyl ether are used as general anesthetics.

Ethers tend to absorb and react with oxygen from the air to form unstable peroxides that may detonate with extreme violence when concentrated by evaporation or distillation, when combined with other compounds that give a detonable mixture, or when disturbed by heat, shock, or friction.

Appreciable quantities of crystalline solids have been observed as gross evidence for the formation of peroxides, and peroxides may form a viscous liquid in the bottom of ether-filled containers. If viscous liquids or crystalline solids are observed in ethers, no further tests for the detection of peroxides are recommended.

Several chemical and physical methods for detecting and estimating peroxide concentrations have been described. Most of the qualitative tests for peroxides are readily performed and strongly recommended when any doubt is present (12).

Applications

Alkyl ethers are used for organic reactions and extractions, as plasticizers, as vehicles for other products, and as anesthetics (qv). Ethers are generally insoluble in water, but dissolve most organic compounds, and therefore, have found wide application in paint and varnish removers (qv), as high boiling solvents for gums, resins, and waxes, and in lubricating oils (see Lubrication). The vapors of certain ethers are toxic to insects and are useful as agricultural insecticides (see Insect control technology) and industrial fumigants (see Poisons, economic). The lower molecular weight technical-grade ethers are the least expensive and most commonly used in industry.

Aryl ethers have distinctive, pleasant odors and flavors which make them valuable to perfume (qv) and flavor industries (see Flavors and spices). Because of their heat stability they are useful as heat-transfer fluids (see Heat exchange technology). Other aryl ethers are useful as food preservatives and antioxidants (see Food additives).

Commerically Important Ethers

ETHYL ETHER

Diethyl ether is probably the most important member of the ether family. It is a colorless, very volatile, highly flammable liquid with a sweet pungent odor and burning taste. As a commercial product it is available in several grades; it is used in chemical manufacture, as a solvent, extractant, or reaction medium, and as a general anesthetic. The physical properties of diethyl ether are given in Table 4.

Manufacture. Almost all of the diethyl ether manufactured today is obtained as a by-product when ethanol (qv) is produced via the vapor-phase hydration of ethylene (qv) over a supported phosphoric acid catalyst. Such a process has the flexibility to adjust to some extent the relative amounts of ethanol and diethyl ether produced in order to meet existing market demands. If additional diethyl ether is required it can be prepared by the vapor phase dehydration of ethanol in a fixed-bed reactor using an alumina catalyst in greater than 95% yield.

The continuous dehydration of ethyl alcohol by sulfuric acid (13–14) was first described by P. Boullay in 1809. This process was later standardized in the United States, and called the Barbet process. In this process (Fig. 1), concentrated sulfuric acid and 95% ethyl alcohol are charged into a lead-lined steel kettle in the ratio of three parts acid to one part alcohol. The reaction is started by heating the mixture to 125–140°C with a steam jacket or internal steam coils. A supply of alcohol vapor is continuously fed into the acid–alcohol mixture at a rate to maintain ca 127°C. The vapor from the still, consisting of ether, alcohol, and water, is passed through a caustic scrubber to remove traces of sulfur dioxide and entrained sulfuric acid. The alkaline solution formed, containing small amounts of ether and alcohol, is passed from the bottom of the scrubber to the lower level of the fractionating column. At this point, the ether and alcohol are removed, and the aqueous solution is discharged as waste. The vapors from the top of the scrubber, consisting of ether, alcohol, and water, are separated in a continuous fractionation. Water passes from the column as waste, and ethyl alcohol (ca 95%) is withdrawn from the center of the column and returned to the vaporizer for recycling. Ether vapors pass from the top of the column through a reflux condenser maintained at 34°C. Fractions boiling above this temperature are returned to the column; ether vapors are condensed and run into a storage tank.

Handling and Shipment. The handling of ethyl ether is hazardous because of its highly flammable properties. Not only is it highly volatile, but it also has a low autoignition temperature and, as a nonconductor, can generate static electrical charges that may result in ignition or vapor explosion. The area in which ethyl ether is handled should be considered a Class I hazardous location as defined by the National Electrical

388 ETHERS

Table 4. Physical Properties of Ethyl Ether

Property	Value			
freezing point, °C	−116.3			
boiling point at 101 kPa (1 atm), °C	34.48			
density at 0.53 kPa (4 mm Hg)				
0°C	0.7364			
20°C	0.7145			
30°C	0.7019			
refractive index, n_D^t				
15°C	1.3555			
20°C	1.3527			
vapor density (air = 1)	2.55			
viscosity at 20°C, mPa.s (=cP)	0.23			
surface tension at 20°C, mN/m (=dyn/cm)	17.30			
specific heat, J/g[a]				
0°C	2.21			
30°C	2.29			
120°C	3.36			
180°C	4.36			
heat of combustion at 20°C, kJ/mol[a]	2726.7			
heat of vaporization at 30°C, kJ/mol[a]	25.59			
heat of formation at 25°C, kJ/mol[a]	−272.8			
critical temperature, °C	194			
critical pressure, MPa[b]	3.60			
critical density, g/mL	0.2625			
flash point (closed-cup), °C	−49			
autoignition temperature, °C	180–190			
explosive range in air, vol %	1.85–48.0			
coefficient of cubical expansion/°C	0.00164			
dielectric constant at 26.9°C, 85.8 kHz	4.197			
vapor pressure	°C	kPa[c]	°C	kPa[c]
	0	184.9	50	1276
	10	290.8	60	1734
	20	439.8	70	2304
	40	921.0		

[a] To convert J to cal, divide by 4.184.
[b] To convert MPa to atm, divide by 0.1013.
[c] To convert kPa to mm Hg, multiply by 7.5.

Code (15). All tools used in making connections or repairs should be of the nonsparking type. All possible care should be taken in loading and unloading tank cars. The tank cars should be properly spotted and the usual caution signs and derails placed in position. Before any connection is made, the tank car should be grounded and bonded. Tank cars should always be unloaded through dome connections rather than through bottom outlets, eg, a pressure-type LPG tank car. A positive suction-type pump or natural gas can be used to remove the ether from the tank car. In no event should air pressure be used.

    Special containers have been developed for anesthetic ether to prevent deterioration before use. Their effectiveness as stabilizers usually depends upon the presence of a lower oxide of a metal having more than one oxidation state. Thus the sides and the bottoms of tin-plate containers are electroplated with copper, which contains a small amount of cuprous oxide. Stannous oxide is also used in the linings for tin con-

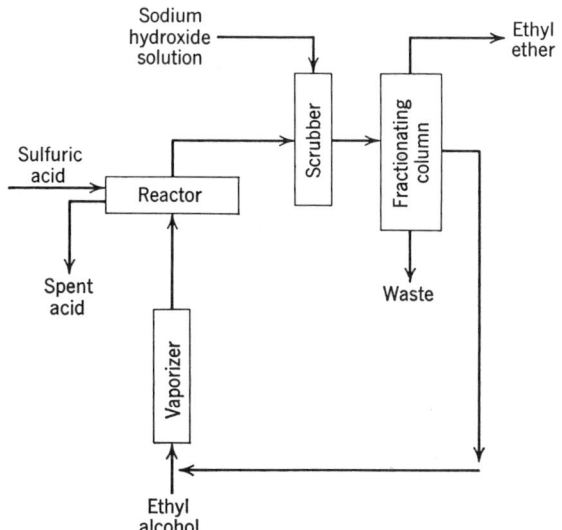

Figure 1. Manufacture of ethyl ether from ethyl alcohol.

tainers. Instead of using special containers, iron wire or certain other metals and alloys or organic compounds have been added to ether to stabilize it.

Economic Aspects. Table 5 gives the total production of all ethyl ether and its unit price. During World War II and the Vietnam War, large quantities of ethyl ether were used as a solvent in the manufacture of smokeless powder which explains the high production in 1945 and the 1960s (see Chemicals in war). The enormous price increase in the 1970s is due largely to the oil-price increases imposed by the OPEC nations.

Uses. Ethyl ether is commercially available in the following grades: USP anesthesia, absolute (ACS), industrial, solvent (conc), and synthetic. Specifications vary, depending on the consumer and use. In many instances the ether has to meet a specific test written into the specification, eg, it may be important that the ether is completely anhydrous or free from alcohol and aldehyde.

Table 5. United States Production and Price of Ethyl Ether, 1940–1977[a]

Year	Production, thousands of metric tons	Unit price, $/kg
1940	7.8	0.24
1950	17.2	0.18
1960	42.5	0.24
1970	47.6	0.26
1971	37.2	0.18
1972	34.0	0.18
1973	32.8	0.18
1974	28.4	0.18
1975	12.5	0.44
1976		0.44
1977		0.44

[a] Ref. 16.

Specifications. The technical concentrated ether (see Fig. 1) contains very small amounts of alcohol, water, aldehydes, peroxides, and other impurities (see Table 6). The more refined grades, such as anesthetic ether, are obtained from technical ether by redistillation and dehydration followed by alkali or charcoal treatment.

Analytical Methods. Most of the analytical and testing methods used for ethyl ether are conventional laboratory methods. Ethyl ether that is to be used for anesthetic purposes or in processes that involve heating or distillation must be peroxide-free and should pass the USP standard test with potassium iodide. This test detects approximately 0.001% peroxide as hydrogen peroxide. Typical specifications and containers for various grades of ethyl ether are given in Table 6.

Ethyl ether is classified by the ICC as a flammable liquid. As such it must be packed in ICC specification containers when shipped by rail, water, or highway; all ICC regulations regarding loading, handling, and labeling must be followed. Each container of ethyl ether must carry an identifying label or stencil. Tank cars and boxcars, either carload or less than carload, must bear the ICC dangerous placard. Each drum or each box with inside containers must bear the ICC red label for flammable liquids.

Toxicology. The toxicity of ethyl ether is low and its greatest hazards in industry are fire and explosion. The vapor is absorbed almost instantly from the lungs and very promptly from the intestinal tract. It undergoes no chemical change in the body. Prevention and control of health hazards associated with the handling of ethyl ether depend primarily upon prevention of exposure to toxic atmospheric concentrations and scrupulous precautions to prevent explosion and fire.

A concentration of 35,000 ppm in the air produces unconsciousness in 30–40 min. This concentration also constitutes a serious fire and explosion hazard and should not be permitted to exist under any circumstance. Any person exposed to ethyl ether

Table 6. Typical Specifications for Various Grades of Ethyl Ether

Specification	Technical refined	ACS, absolute	USP XIX[a]
color, max	water-white	water-white	colorless
d_{25}^{25}	0.710–0.713	0.7079	0.713–0.716
acidity as acetic acid, wt %	0.0025	0.0010	passes test[b]
peroxides, max wt %	passes test[c]	0.0001	passes test[c]
aldehydes, max wt %	passes test[d]	0.0005	passes test[d]
alcohol, max %	0.5 vol %	0.05 wt %	
nonvolatile matter, max wt %	0.002	0.0010	0.003
water, max wt %	0.3	0.0100	
odor	nonresidual	passes test[e]	passes test[e]
net container contents, kg[f]			
19-L (5-gal) drum	13.6	13.6	13.6
208-L (55-gal) drum	147	147	147

[a] For use in anesthesia; the USP (14) also recognizes slightly less pure grades: ethyl oxide (solvent ether), ether abs, and reagent-grade ether.
[b] Free acid, requiring no more than 0.4 mL of 0.02N sodium hydroxide for 25 mL of ethyl ether.
[c] No color with potassium iodide reagent; USP test.
[d] No turbidity with alkaline mercuric chloride–potassium iodide reagent; USP test.
[e] No foreign odor when last traces of ether evaporate from odorless absorbent paper; USP test.
[f] The tech grade can also be shipped in 103W insulated tank cars of 15.1-m³ (4000-gal), 22.7-m³ (6000-gal), 30.3-m³ (8000-gal), and 37.9-m³ (10,000-gal) capacity.

vapor of any appreciable concentration should be promptly removed from the area. Recovery from exposure to sublethal concentrations is rapid and generally complete. Except in emergencies, and then only with appropriate protective equipment, no one should enter an area containing ether vapor until the concentration has been found safe by measurement with a combustile-gas indicator.

If any ethyl ether fire occurs, carbon dioxide, carbon tetrachloride, and dry chemical fire extinguishers meeting National Fire Prevention Association Code 1 and 2 requirements may be used successfully (15). Water may also be effectively applied. (see Plant safety). Hose streams played into open tanks of burning ethyl ether serve only to scatter the liquid and spread the fire. However, ether fires may be extinguished by a high pressure water spray that cools the burning surface and smothers the fire. Automatic sprinklers and deluge systems are also effective.

Uses. Ethyl ether has a wide range of uses in the chemical industry. It is a good solvent or extractant for fats, waxes, oils, perfumes, resins, dyes, gums and alkaloids. When mixed with ethanol, ethyl ether becomes an excellent solvent for cellulose nitrate in the manufacture of guncotton (see Explosives and propellants) collodion solutions (see Membrane technology), and pyroxylin plastics (see Cellulose derivatives, esters). Another important use is as an extractant for acetic acid as well as other organic acids, eg, in the cellulose acetate (qv) and plastic industries to recover acetic acid from dilute aqueous systems. Ethyl ether is also used as a denaturant in several denaturant alcohol formulas. It has been used as a starting fuel for diesel engines and as an entrainer for dehydration of ethanol and isopropyl alcohol. It may be used as an anhydrous, inert reaction medium for the Grignard and Wurtz-Fittig reactions. Ethyl ether is used as a general anesthetic in surgery.

Although ethyl ether is made from ethylene, it is useful as a commercial source of ethylene in plants that do not have access to petroleum refinery gases. The ether, a liquid, is more conveniently transported than ethylene. Ethanol denatured with ethyl ether may also be used. The vapor is passed over alumina at ca 343°C, the yield is very high, and the ethylene produced is very pure.

The estimated use pattern of ethyl ether during 1974 was as follows: solvents and military production of smokeless powder, 65%; chemical synthesis, 25%; general anesthetic and other medicinal uses, 3%; miscellaneous uses, 7% (16).

OTHER ETHERS

***n*-Butyl Ether.** *n*-Butyl ether is prepared by dehydration of *n*-butyl alcohol by sulfuric acid or by catalytic dehydration over ferric chloride, copper sulfate, silica, or alumina at high temperatures. It is an important solvent for the Grignard and other reactions that require an anhydrous, inert medium. *n*-Butyl ether is also an excellent extracting agent for use with aqueous systems owing to its very low water solubility.

Isopropyl Ether. Isopropyl ether is manufactured by the dehydration of isopropyl alcohol with sulfuric acid. It is obtained in large quantities as a by-product in the manufacture of isopropyl alcohol from propylene by the sulfuric acid process, very similar to the production of ethyl ether from ethylene. Isopropyl ether is of moderate importance as an industrial solvent, since its boiling point lies between that of ethyl ether and acetone. Isopropyl ether very readily forms hazardous peroxides and hydroperoxides, much more so than other ethers.

Vinyl Ether. Vinyl ether is manufactured by the pyrolytic dehydrochlorination of 1,1′-dichloroethyl ether. Vinyl ether is used as a general inhalation anesthetic for procedures of short duration. Approximately 4% ethanol is added to the vinyl ether used as an anesthetic to reduce ice formation in the masks used for administration (see also Vinyl polymers).

Methyl *tert*-Butyl Ether (MTBE). Methyl *tert*-butyl ether is finding increasing use as a blending component in high octane gasoline as current gasoline additives based on lead, manganese and bromine are phased out (see Gasoline). Methyl *tert*-butyl ether is easily made from isobutylene and methanol in the presence of an acidic ion-exchange resin catalyst, in the liquid phase and at temperatures below 100°C. Although the market is currently small, it is expected to reach 2.65×10^6 m^3 (700×10^6 gal) by 1985.

2-Methoxyphenol. 2-Methoxyphenol is prepared by methylating 1,2-dihydroxybenzene. It is useful as an antioxidant for fats, oils, and vitamins and as a polymerization inhibitor. 2-Methoxyphenol is effective as an antioxidant (qv) and inhibitor at levels of 10–100 ppm.

Butylated Hydroxyanisole. 2- and 3-*tert*-butyl-4-methoxyphenol (butylated hydroxyanisole (BHA)) is prepared from 4-methoxyphenol and *tert*-butyl alcohol over silica or alumina at 150°C or from hydroquinone and *tert*-butyl alcohol or isobutene, using an acid catalyst and then methylating. It is widely used in all types of foods such as butter, lard and other fats, meat, cereals, baked goods, candies, and beer as an antioxidant (see Food additives). Its antioxidant properties are not lost during cooking so that flour, fats, and other BHA-stabilized ingredients may be used to produce stabilized products.

BIBLIOGRAPHY

"Ethers" in *ECT* 1st ed., Vol. 5, pp. 858–876, by Mary A. Magill, Chemical Abstracts, and J. G. Parks and C. M. Beamer, Enjay Co., Inc; "Ethers" in *ECT* 2nd ed., Vol. 8, pp. 470–498, by Arnold P. Lurie, Eastman Kodak Co.

1. S. Patai, ed., *The Chemistry of the Ether Linkage,* Wiley-Interscience, New York, 1967.
2. E. Staude and F. Tatat in ref. 1, Ch. 2.
3. A. G. Davies, *Organic Peroxides,* Butterworths, London, Eng., 1961, p. 79.
4. S. Searles, Jr., and M. Tamres in ref. 1, Ch. 6.
5. D. L. Dalrymple, T. L. Kruger, and W. N. White in ref. 1, Ch. 14.
6. C. A. Buehler and D. E. Pearson, *Survey of Organic Syntheses,* Vol. 1, Wiley-Interscience, New York, 1970, Ch. 6.
7. C. A. Buehler and D. E. Pearson, *Survey of Organic Syntheses,* Vol. 2, Wiley-Interscience, New York, 1977, Ch. 6.
8. *Chem. Mark. Rep.,* (July 24, 1978).
9. N. V. Steere in ref. 1, pp. 683–685.
10. M. N. Gleason, R. F. Gosselin, and H. F. Hodge, *Clinical Toxicology of Commercial Products,* The Williams and Wilkins Co., Baltimore, Md., 1957.
11. E. G. Young and L. B. Wollner, *J. Ind. Hygiene Toxicol.* **28,** 267 (1947).
12. A. J. Gordon and R. A. Ford, *The Chemist's Companion,* Wiley-Interscience, New York, 1972, p. 427.
13. B. T. Brooks, *Ind. Eng. Chem.* **27,** 278 (1935).
14. *The Pharmacopeia of the United States of America,* 19th ed., (USP XIX), Mack Publishing Co., Easton, Pa., 1975; *USP XX-NFXF,* 1980.

ETHYL ALCOHOL. See Ethanol.

ETHYLBENZENE. See Xylenes and ethylbenzene; Styrene.

ETHYLENE

Ethylene [74-85-1] (ethene $H_2C\!\!=\!\!CH_2$, mol wt 28.0536, is the largest volume organic chemical produced today. It is the most important building block of the petrochemical industry (1) and is converted to a multitude of intermediate and end products on a large scale, mainly polymeric materials such as plastics, resins, fibers, and elastomers. Other important products are solvents, surfactants, coatings, plasticizers, and antifreeze.

Physical Properties

Ethylene, the lightest olefinic hydrocarbon, is a colorless, flammable gas with a slightly sweet odor. Its physical properties are listed in Table 1, the properties of liquid ethylene in Table 2. A Mollier diagram (enthalpy vs pressure) of ethylene for pressures up to ca 1 GPa (10,000 atm) is published in ref. 2. More complete thermodynamic and physical properties of ethylene are given in refs. 2–7.

Chemical Properties

The chemistry of ethylene centers about its double bond, which reacts chiefly by addition to give saturated hydrocarbons, their derivatives, or polymers. The geometry of ethylene is relatively simple, with all six atoms in one plane. The double bond prevents rotation, except at high temperatures.

In molecular orbital theory, the π-bond is represented as an electron cloud above and below the plane of the other three orbitals (bonds). The nucleophilic character of ethylene with its relatively negative π bond is readily understood from this theory. Ethylene reacts with electrophilic reagents such as strong acids (H^+), halogens, and oxidizing agents, but not with nucleophilic reagents such as Grignard reagents and bases.

The chemical reactions of ethylene may be divided into those of commercial

Table 1. Physical Properties of Ethylene

Property	Value
triple point	
temperature, °C	−169.19
pressure, kPa[a]	0.11
latent heat of fusion, kJ/mol[b]	3.350
normal boiling point	
temperature, °C	−103.71
latent heat of vaporization, kJ/mol[b]	13.540
density of the liquid	
mol/L	20.27
d_4^{-104}	0.566
specific heat of the liquid, J/(mol·K)[b]	67.4
viscosity of the liquid, mPa·s (= cP)	0.161
surface tension of the liquid, mN/m (= dyn/cm)	16.4
specific heat of the ideal gas at 25°C, J/(mol·K)[b]	42.84
critical point	
temperature, °C	9.2
pressure, MPa[c]	5.042
density, mol/L	7.635
compressibility factor	0.2813
gross heat of combustion of the gas at 25°C, MJ/mol[b]	1.411
limits of flammability at atmospheric pressure and 25°C	
lower limit in air, mol %	2.7
upper limit in air, mol %	36.0
auto ignition temperature in air at atmospheric pressure, °C	490

[a] To convert kPa to mm Hg, multiply by 7.5.
[b] To convert J to cal, divide by 4.184.
[c] To convert MPa to psi, multiply by 145.

Table 2. Properties of Liquid Ethylene

Temperature, °C	Vapor pressure, kPa[a]	Liquid density, mol/L
−150	2.04	22.5
−125	24.0	21.3
−100	151.0	20.1
−75	424.0	18.7
−50	1150.0	17.2
−25	2200.0	15.2
0	4110.0	12.1

[a] To convert kPa to mm Hg, multiply by 7.5.

significance and those which have mainly academic interest. This division is necessarily arbitrary and reactions of the second category may well be in the first category in the future. The first category comprises in order of importance, polymerization, oxidation, and addition reactions.

Polymerization. Ethylene polymerization represents the largest segment of the petrochemical industry with polyethylene ranking first as an ethylene consumer. Ethylene (99.9+% purity) is polymerized (8) under specific conditions of temperature and pressure and in the presence of an initiator or catalyst:

$$n(CH_2\!\!=\!\!CH_2) \rightarrow +\!CH_2CH_2\!+\!_n$$

The reaction is exothermic and may involve homogenous initiation (radical or cationic) or heterogenous initiation (solid catalyst). The products range in molecular weight from below 1000 up to 5 million or more (ultra-high molecular weight polyethylene) (see Olefin polymers).

There are two commercially important systems for polyethylene production: (1) high pressure polymerization by free-radical initiation, using oxygen, peroxides, or other strong oxidizers as initiators; (2) low pressure polymerization by heterogenous catalysis using transition metal oxides such as molybdenum oxide or chromium oxide supported on inorganic carriers, or Ziegler catalysts such as aluminum alkyls and titanium halides which may also be supported on inorganic carriers.

High pressure processes produce so-called low density polyethylene (LDPE) with a density of 0.910–0.940 g/mL. Pressures of 60–350 MPa (8,700–50,750 psi) at up to 350°C are applied. Low density polyethylene is used primarily in the manufacture of films for injection molding, and as wire, cable, and paper coatings.

The conventional low pressure processes produce so-called high density polyethylene (HDPE) with a density of 0.941–0.970 g/mL, used primarily for blow- and injection-molded products. Novel low pressure processes employ advanced catalyst systems for the production of LDPE (9). Polymerization temperatures range from 50 to 300°C at pressures of 0.1–20 MPa (14.5–2900 psi). The polymerization in low pressure processes may be carried out with a fixed- or fluid-bed catalyst or with the catalyst slurried or suspended in the reactant mixture which may contain a solvent or diluent. Catalysts vary widely but usually fall into one of the two groups listed above.

In order to improve physical properties of HDPE and LDPE, copolymers (10) of ethylene with small amounts of other monomers, such as higher olefins, ethyl acrylate, maleic anhydride, vinyl acetate, and acrylic acid, are produced in significant commercial quantities.

The polymerization of ethylene and propylene for the production of specialty rubbers is carried out with a Ziegler-type catalyst (aluminum alkyl and hydrocarbon soluble vanadium compounds) in the presence of an inert hydrocarbon solvent or liquid propylene which acts as diluent. The two types manufactured are copolymers (11) (EP) from ethylene and propylene and terpolymers (11) (EPT or EPDM) from ethylene and propylene and a small amount of termonomer (see Elastomers, synthetic, ethylene–propylene rubber).

Oxidation. Oxidation reactions of ethylene give ethylene oxide–glycol, ranking second, acetaldehyde ranking sixth, and vinyl acetate ranking eighth as consumers of ethylene in the United States. Direct oxidation processes are preferred today because they are more economical than older multistage processes.

$$CH_2\!\!-\!\!CH_2 / HOCH_2CH_2OH$$
$$\diagdown\!O\!\diagup$$

Ethylene oxide–glycol, above, was at one time exclusively produced by the chlorohydrin process. Ethylene and hypochlorous acid give ethylene chlorohydrin, which is converted either to ethylene glycol by caustic alkaline hydrolysis or to ethylene oxide by treatment with calcium hydroxide.

Direct vapor-phase air oxidation over silver oxide catalysts, at 200–300°C and 1.5–3.0 MPa (217.5–435 psi) pressure, has largely replaced the chlorohydrin process (see Ethylene oxide).

About two thirds of the ethylene oxide produced is converted to ethylene glycol by noncatalytic high pressure, high temperature hydrolysis. Higher glycols such as di- and triethylene glycol are produced as by-products. An improved ethylene glycol process, which avoids the production of ethylene oxide as an intermediate, has been commercialized recently. Ethylene is first oxidized to ethylene glycol mono- and diacetates over a tellurium catalyst in conjunction with 2-bromoethylacetate and then hydrolyzed. The acetic acid is recovered and recycled (12):

$$4\ CH_2=CH_2 + 6\ CH_3COOH + 2\ O_2 \rightarrow 2\ CH_3COOCH_2CH_2OH + 2\ CH_3COOCH_2CH_2OOCCH_3$$

$$2\ CH_3COOCH_2CH_2OH + 2\ CH_3COOCH_2CH_2OOCCH_3 + 6\ H_2O \rightarrow 4\ HOCH_2CH_2OH + 6\ CH_3COOH$$

The direct oxidation of ethylene is the preferred route to acetaldehyde today. Before 1960 acetaldehyde was produced from ethanol, from acetylene, and in smaller quantities from saturated hydrocarbons. In the direct oxidation process palladium chloride is reduced at room temperature to palladium while acetaldehyde is formed. The palladium is then reoxidized by cupric chloride, and oxygen converts the reduced cuprous chloride to cupric chloride:

$$CH_2=CH_2 + PdCl_2 + H_2O \rightarrow CH_3CHO + Pd + 2\ HCl$$

$$Pd + 2\ CuCl_2 \rightarrow PdCl_2 + 2\ CuCl$$

$$2\ CuCl + 2\ HCl + \tfrac{1}{2}\ O_2 \rightarrow 2\ CuCl_2 + H_2O$$

$$CH_2=CH_2 + \tfrac{1}{2}\ O_2 \rightarrow CH_3CHO$$

A vapor-phase ethylene-based process is now the preferred route to vinyl acetate. Palladium on carbon, alumina, or silica–alumina is employed as catalyst at ca 175–200°C and 0.4–1.0 MPa (58–145 psi):

$$CH_2=CH_2 + CH_3COOH + \tfrac{1}{2}\ O_2 \rightarrow CH_2=CHOCOCH_3 + H_2O$$

Addition. Addition reactions of ethylene have considerable importance, in the following order of descending consumption:

Reaction	*For the production of*
halogenation–hydrohalogenation	ethylene dichloride, ethylene chloride, ethylene dibromide, ethyl chloride
alkylation	ethylbenzene, ethyltoluene, aluminum alkyls
oligomerization	alpha olefins, linear primary alcohols
hydration	ethanol
hydroformylation	propionaldehyde

The most important intermediate produced by addition is ethylene dichloride, $ClCH_2CH_2Cl$, which ranks third in United States ethylene consumption (see Chlorocarbons). By far the leading derivative of ethylene dichloride (EDC) is vinyl chloride monomer (VCM) used to produce poly(vinyl chloride) resins and chlorinated hydrocarbons. In the past, VCM was produced from acetylene and hydrogen chloride. Today, it is almost all manufactured from ethylene via EDC by chlorination or oxychlorination processes. Variations are mainly in the catalyst and reaction conditions (see Vinyl polymers).

In the chlorination process, EDC is produced in the liquid or vapor phase from chlorine and ethylene in the presence of metallic chlorides or ethylene dibromide as catalysts at 40–140°C and 0.1–0.5 MPa (14.7–72.5 psi).

In oxychlorination, ethylene, hydrogen chloride, and atmospheric oxygen react in the vapor phase over a supported copper chloride catalyst at 200–350°C and 1 MPa (145 psi) in a packed or fluidized bed. The EDC obtained is then cracked at 450–515°C and 0.6–2.5 MPA (87–362.5 psi) pressure to yield VCM and HCl. The HCl is recycled to the oxychlorination reaction. The overall process is balanced so that only VCM is produced without net production of HCl:

$$CH_2{=}CH_2 + Cl_2 \rightarrow ClCH_2CH_2Cl$$

$$CH_2{=}CH_2 + 2\,HCl + \tfrac{1}{2}\,O_2 \rightarrow ClCH_2CH_2Cl + H_2O$$

$$2\,ClCH_2CH_2Cl \rightarrow 2\,CH_2{=}CHCl + 2\,HCl$$

$$2\,CH_2{=}CH_2 + Cl_2 + \tfrac{1}{2}\,O_2 \rightarrow 2\,CH_2{=}CHCl + H_2O$$

Ethyl chloride and ethylene dibromide, used in gasoline as antiknock additives, are also manufactured by halogenation–hydrohalogenation of ethylene.

$$CH_2{=}CH_2 + HCl \xrightarrow[\text{ACl}_3\text{ catalyst}]{25{-}40°C} CH_3CH_2Cl$$

$$CH_2{=}CH_2 + Br_2 \xrightarrow[\text{liquid phase}]{35{-}85°C} BrCH_2CH_2Br$$

Ethylbenzene, $C_6H_5CH_2CH_3$, the precursor of styrene (qv), ranks fourth in present United States ethylene consumption and is produced from benzene and ethylene in yields over 98% by four different processes (see Xylenes and ethylbenzene).

Most ethylbenzene is produced by a Friedel-Crafts reaction with aluminum chloride catalyst and a small amount of hydrochloric acid or ethyl chloride as initiator at 80–100°C and pressures of 0.1–0.2 MPa (14.5–29 psi). The reaction is complicated by side reactions such as cracking, polymerization, and hydrogen transfer. Polyalkylated reaction products are converted to monoethylbenzene with benzene at 200°C and higher in a reverse alkylation reaction.

An improved process uses aluminum chloride in a homogenous system at 140–200°C, with yields over 99% (13–14).

Another process uses a solid boron trifluoride catalyst supported on modified alumina for the alkylation and transalkylation reactions. The recently developed fourth process, employing a solid, acidic zeolite catalyst and operating in the vapor phase at 400–460°C and pressures of 1.4–2.8 MPa (ca 200–400 psi) provides a yield of over 99% (15–16). Transalkylation is possible at 425–450°C. Phosphoric acid on kieselguhr and silica–alumina have also been used as catalysts.

Ethyltoluene is utilized to make vinyltoluene. Alkylation of toluene with ethylene is carried out with aluminum chloride catalyst at 60–90°C and a pressure of 0.4 MPa (58 psi). Dehydrogenation of the resulting mixture of p- and m-ethyltoluene with zinc oxide catalyst gives a mixture of p- and m-vinyltoluene.

Alkylation products of aluminum with ethylene find application as initiator and starter compounds in the production of α-olefins and linear primary alcohols, as polymerization catalysts, and in the synthesis of some monomers (eg, 1,4-hexadiene). Triethylaluminum, $Al(C_2H_5)_3$, is the most important of the ethylene-derived aluminum alkyls. It may be produced by several methods:

In the *three-for-two process*, two moles of triethylaluminum produce three moles by reaction with aluminum, hydrogen, and ethylene in two steps:

$$2\ Al(C_2H_5)_3 + \tfrac{3}{2}\ H_2 + Al \rightarrow 3\ Al(C_2H_5)_2H$$

$$3\ Al(C_2H_5)_2H + 3\ C_2H_4 \rightarrow 3\ Al(C_2H_5)_3$$

$$Al + \tfrac{3}{2}\ H_2 + 3\ C_2H_4 \rightarrow Al(C_2H_5)_3$$

The *olefin displacement process* is based on the displacement of certain alkyl groups:

$$Al(R'H)(R''H)(R'''H) + 3\ C_2H_4 \rightarrow Al(C_2H_5)_3 + R' + R'' + R'''$$

R', R'', and R''' are oligomers of ethylene having, eg, a branched methyl group on the second carbon from the aluminum or long ligand chains built up by ethylene chain-growth reactions.

In the first method, triethylaluminum may be produced from other commercial aluminum alkyls (eg, triisobutylaluminum). In the second, the displacement reaction converts long-chain aluminum alkyls to α-olefins (see Organometallics).

Short-chain polymers of ethylene (oligomers) produced by Ziegler's ethylene chain-growth technique and recovered as linear α-olefins and primary alcohols in the C_4–C_{20} range are of growing commercial importance.

Linear α-olefins are produced by the cracking of normal paraffin waxes and by the oligomerization of ethylene. These are mainly precursors of plasticizer-range (C_6–C_{10}) and detergent-range (C_{12} up) alcohols. Smaller quantities are used for the production of fatty amine oxides, sulfonates, thermoplastic comonomers, and synthetic lubricants.

The synthesis of α-olefins is based on addition of ethylene to triethylaluminum, which serves as a catalyst system wherein the ligands grow to a certain chain length by oligomerization at 80–120°C and 20 MPa (2900 psi) pressure. The alkyl group is then displaced by ethylene at 245–300°C and 0.7–2.0 MPa:

$$Al(C_2H_5)_3 + n\ C_2H_4 \rightarrow AlR'R''R'''$$

$$AlR''R'''\!-\!CH_2CH_2R + C_2H_4 \rightarrow AlR''R'''C_2H_5 + CH_2\!=\!CHR$$

R, R_1, R_2, and R_3 are ethylene oligomers of the same or different chain lengths. Alternatively, air oxidation of these higher alkyl aluminum compounds to alkoxides, followed by hydrolysis, gives straight-chain alcohols with an even number of carbon atoms:

$$AlR'R''R''' + \tfrac{3}{2}\ O_2 \rightarrow Al(R'O)(R''O)(R'''O)$$

$$Al(R'O)(R''O)(R'''O) + 3\ H_2O \rightarrow R'OH + R''OH + R'''OH + Al(OH)_3$$

The hydration products of ethylene are ethanol and diethyl ether. The latter is produced mostly as a by-product from synthetic ethanol units.

Ethanol is produced from ethylene by way of ethyl sulfate or by direct hydration. In the first case, ethylene is absorbed in 90–98% sulfuric acid at 50–85°C and a pressure of 1.0–1.4 MPa (145–203 psi) to give a mixture of ethyl sulfates, which are hydrolyzed to ethanol and dilute sulfuric acid:

$$CH_2\!=\!CH_2 + H_2SO_4 \rightarrow C_2H_5OSO_2OH$$

$$2\ CH_2{=}CH_2 + H_2SO_4 \rightarrow C_2H_5OSO_2OC_2H_5$$

$$C_2H_5OSO_2OH + C_2H_5OSO_2OC_2H_5 + 3\ H_2O \rightarrow 3\ C_2H_5OH + 2\ H_2SO_4$$

Small amounts of diethyl ether are formed as a by-product:

$$2\ C_2H_5OH \rightarrow C_2H_5OC_2H_5 + H_2O$$

Direct hydration of ethylene, the preferred route today, is accomplished over supported phosphoric acid or tungsten based catalysts at 300°C and 7 MPa (ca 1000 psi) pressure. Acetaldehyde and diethyl ether, formed as by-products, may be recycled.

In hydroformylation (oxo reaction) ethylene reacts with synthesis gas of a hydrogen–carbon monoxide ratio of 1:1 over a cobalt catalyst at 60–200°C and a pressure of 4–35 MPa (580–5075 psi) to form propionaldehyde (see Oxo process):

$$CH_2{=}CH_2 + CO + H_2 \rightarrow C_2H_5CHO$$

Other Reactions. Many other reactions of ethylene are of academic interest only; comprehensive discussions are provided by standard treatises and numerous references (17).

Ethylene may be hydrogenated to ethane under a wide variety of conditions such as passage over finely divided platinum or palladium at RT or over finely divided nickel at 150–300°C and elevated pressure. The reaction is used as a model reaction in chemical engineering research work.

Friedel-Crafts alkylation reactions occur in the presence of ionic catalysts ($AlCl_3$, $SnCl_4$, BF_3 promoted by either hydrogen chloride or ethyl chloride) with paraffins and acyl halides. Alkylation with paraffins give a variety of branched-chain paraffins; eg, 2,3-dimethylbutane is formed from isobutane and ethylene with aluminum chloride catalyst promoted by ethyl chloride at 60°C:

$$CH_2{=}CH_2 + (CH_3)_3CH \rightarrow (CH_3)_2CHCH(CH_3)_2$$

Large-scale alkylation of ethylene with isobutane is not carried out; ethylene is too valuable a chemical to convert to a fuel (see Alkylation; Friedel-Crafts reactions).

Acyl halides may be added to ethylene in the presence of aluminum chloride to form halogenated ketones:

$$CH_2{=}CH_2 + RCOCl \rightarrow RCOCH_2CH_2Cl$$

Ethylene reacts with halogens at low temperature to give the dihaloethanes. The reaction is more rapid with chlorine than with bromine and iodine and is utilized for the production of ethylene dibromide and in conjunction with catalysts on a large industrial scale for the production of ethylene dichloride. The reaction with iodine is incomplete and reversible. Bromine adds irreversibly but the reaction is not as rapid as with chlorine.

$$CH_2{=}CH_2 + Cl_2 \rightarrow CH_2ClCH_2Cl$$

$$CH_2{=}CH_2 + Br_2 \rightarrow CH_2BrCH_2Br$$

$$CH_2{=}CH_2 + I_2 \rightleftarrows CH_2ICH_2I$$

It is possible to chlorinate ethylene directly to trichloroethylene and perchloroethylene at high temperatures:

$$CH_2{=}CH_2 + 3\ Cl_2 \rightarrow ClHC{=}CCl_2 + 3\ HCl$$

$$CH_2{=}CH_2 + 4\ Cl_2 \rightarrow Cl_2C{=}CCl_2 + 4\ HCl$$

The commercial use of these reactions is very limited since a profitable outlet for HCl is needed. Most trichloroethylene and perchloroethylene are produced from EDC by chlorination and dehydrochlorination (see Chlorocarbons).

Hydrogen halides also add to ethylene to give corresponding ethyl halides. The order of reactivity is HCl < HBr < HI. An ionic catalyst such as aluminum chloride or bromide enhances the rate of addition.

The reaction of ethylene with boiling sulfur monochloride gives 2,2'-dichlorodiethyl sulfide (mustard gas). Mustard gas made during World War I was prepared by this reaction:

$$2 \; CH_2\!\!=\!\!CH_2 + S_2Cl_2 \rightarrow (ClCH_2CH_2)_2S + S$$

Manufacture

Single-component feedstocks, such as ethane or propane, or multicomponent hydrocarbon feedstocks, such as natural gas liquids (NGL) and naphthas and gas oils from crude oil, may be used for the production of olefins (see Feedstocks).

In Europe, where there is a lack of natural gas and natural gas condensates, production of olefins is essentially based on light and heavy naphthas (kerosene) obtained from the distillation of crude oil. In the United States, NGL fractions are currently used to the maximum available extent. However, because of the mounting scarcity of such feedstocks there is an increasing trend toward the utilization of heavier naphthas, paraffinic raffinates, and gas oils.

Various pretreatment processes such as hydrodesulfurization and hydrocracking for improving the quality of heavier feedstocks are under development. Full-range crude oils require a large supply of oxygen to process and the economic feasibility of this route to olefins has at present not been fully established. The utilization of coal as the raw material for olefins production is definitely in the more distant future. Olefins, particularly ethylene, used to be recovered from coke oven and refinery off-gases in cryogenic plants on a minor scale. However, coke oven gases, because of their low ethylene content, are no longer considered a usable source of ethylene. The recent cost increase of hydrocarbon feedstocks has revived some interest in the use of ethanol as raw material for ethylene production in countries which produce cane sugar or other crops from which ethanol may be obtained by fermentation.

Today, ethylene is produced almost exclusively via the pyrolysis of hydrocarbons in tubular reactor coils installed in externally fired heaters. So-called autothermic processes in which a part of the feedstock is burned as fuel to furnish the heat of reaction, such as the Hoechst HTP process and others (18–20), have not achieved industrial prominence. Autothermic processes employ very high temperatures, require an oxygen plant and invariably coproduce a substantial amount of acetylene (qv). The separation of the acetylene from the primary ethylene considerably increases the plant cost and thus this process is not a feasible route to cheap ethylene.

The same is true of processes utilizing a heat carrier, such as pebbles (21–25) or a fluidized-bed reactor (26–29) and of regenerative furnaces such as those employed in the Wulff process (18–19,30–32).

Some measure of success has been achieved with the autothermic cracking of full range crude oils for ethylene by several companies, notably by BASF in the late 1960s. Further progress has been reported recently by Union Carbide in conjunction with Kureha Chemical and Chiyoda and Dow Chemical (33–37). By late 1979 Union Carbide is to start up a 2300 metric tons per year experimental plant at Seadrift, Texas, with an advanced prototype reactor. A prototype Dow plant for 11,300 t/yr is under construction at Freeport, Texas.

Ethylene may be produced by dehydration of ethanol in fixed- or fluid-bed reaction systems. Developmental work is also in progress to eventually produce ethylene and other petrochemicals on an industrial scale from coal via synthesis gas and Fischer-Tropsch processes (see Fuels, synthetic). Where cheap propylene is available, the Phillips Triolefin process (38) may be used to convert it to ethylene and n-butenes. Catarole catalysis (39) and the DuPont process (40) have only historical interest today. Processes employing diluents other than steam, eg, the hydropyrolysis process (41–43) using hydrogen under pressure, have thus far had only limited success.

Pyrolysis of Hydrocarbons in Tubular Heaters. The coils installed in externally fired heaters represent a thermal reactor. The properties of the feedstock and the conditions at which the reactor coils are operated determine reactor effluent product distribution. High selectivity toward the production of desired olefins and diolefins (ie, ethylene, propylene, and butadiene) with minimum methane and coking in the coils leading to longer heater runs are achieved by operating the pyrolysis heaters at high temperatures (750–900°C), short residence times (0.1–0.6 s), and low hydrocarbon partial pressure. Steam is added to the feedstock to reduce the hydrocarbon partial pressure and the amount of carbon being deposited on the tube walls. The steam-to-hydrocarbon weight ratios usually vary from 0.3 for ethane to as high as 1.0 for gas oils.

The transformation of paraffinic and naphthenic hydrocarbons into olefins is a highly endothermic process. Primary reactions produce olefins essentially via a free-radical mechanism (44). Secondary reactions involving primarily-produced olefins become important when the feedstock reaches high levels of conversion. Free-radical reactions proceed by a chain mechanism consisting of initiation, propagation, and termination (45). Ethane is a preferred feedstock for ethylene and its reaction mechanism is rather well understood. Reaction schemes involving 6 to 49 reactions have been proposed and product distribution can be predicted over a wide range of operating conditions (46–59). Experimental results and kinetic modeling of propane pyrolysis (60–61) employing tubular flow reactors have been reported (57–58,62–72). Similar studies have also been made for the pyrolysis of other heavier single-component feedstocks like butanes (73–77), pentanes (78–79), and hexanes (78,80–81). Reaction schemes for the cracking of paraffinic mixtures have been proposed (58,82–84).

Straight-run naphthas and gas oils consist of mixtures of paraffins, cycloparaffins (naphthenes), and aromatics. Carbon distribution and ring content can be calculated from the refractive index, density, and molecular weight (ASTM D3238-74) (85). The boiling points of petroleum fractions used for pyrolysis are normally within a selected temperature range. Kinetic models for the pyrolysis of these hydrocarbon mixtures have been developed based on their yields of C_4 and lighter hydrocarbons and on a combination of kinetic relationships applying to the overall decomposition rate and equations representing the curve-fitted yield data (86–89). Correlations of ethylene yields as a function of the n-paraffin–isoparaffin split, the cyclopentane homologues, and the hydrogen content have been reported (90–91) as well as a generalized feed decomposition model for naphtha pyrolysis (92).

The United States Bureau of Mines Correlation Index (BMCI) may be used to correlate successfully the quality of gas oil feedstocks and its effect on yields (93). From published data, the following equation relating ultimate ethylene yields and BMCI for virgin gas oils is obtained:

$$CH_2{=}CH_2 \text{ (ultimate)} = -0.36 \text{ (BMCI)} + 36.75$$

with $15 \leqslant BMCI \leqslant 45$ and $0.75 \leqslant$ steam to oil $\leqslant 1.2$. This relationship holds for maximum severity and includes the yields from acetylene hydrogenation and recycling to extinction of the intermediate ethane. The composite ethylene yield can be calculated from the chemical structure of the gas-oil components, by a procedure (94–95) that gives statistical information on the types of bonding of the components of a gas oil. Linkage elements may be determined by high pressure adsorption chromatography and nmr techniques.

Complete data on a particular feedstock may not always be available and furthermore the coil design may have to accommodate similar but structurally different feedstocks. The so-called average residence time, the equivalent residence time (51,96), the average hydrocarbon partial pressure (96), and the kinetic average partial pressure (KAPP) (97) are used as suitable parameters for correlation purposes.

Conversion for individual components is defined as the fractional disappearance of the reactants. For mixtures of known composition, the conversion of the mixture α is normally defined as the conversion of the individual components X_i according to their weight fraction in the feed mixture Y_i:

$$\alpha = \sum_i Y_i X_i$$

For complex multicomponent feedstocks of unknown composition, a conversion based on the molecular weights of the pyrolysis feed and effluent can be derived from the material balance as shown below:

$$\alpha = \frac{\frac{mf}{me} - 1}{\gamma - 1}$$

where mf = molecular weight of feed, me = molecular weight of effluent, and γ = molal expansion factor, moles of effluent produced per mole of feed converted (constant for any given feedstock).

The conversion of a feedstock of unknown composition can also be expressed by using hydrogen content rather than molecular weight:

$$\alpha = \frac{Y - 1}{CY - 1}$$

where $Y = (H - 6)/(HF - 6)$; H = hydrogen content in C_5 plus product, wt %; HF = hydrogen content of the feed, wt %; and C = a constant for any given feedstock.

Severity is often used to describe the depth of cracking or extent of conversion. Quantitative definitions of severity are given by the kinetic severity function (98) and the so-called overall kinetic severity function (88). More recent severity indicators are the cracking severity index (97) for naphtha cracking and the molecular collision parameter for gas oil cracking (99). Other indicators (90,100) such as coil-outlet temperature and hydrogen content of liquid products, also represent cracking severity. The usefulness of these indexes depends greatly on type of feedstock, operating conditions of the reactor coils, and on the reliability of the effluent analyses. The term maximum severity represents an acceptable optimum when balancing ethylene yield and coil run length. The atomic ratio of hydrogen to carbon of the liquid pyrolysis products is a valuable indicator for determining maximum severity (101).

Selectivity for a given feed may be generally defined as the effect of operating conditions on yield structure for a fixed feed conversion. In the case of ethane pyrolysis, selectivity has been simply defined as the molal or mass yield of ethylene per mole or unit mass of ethane converted for which the methane yield is a good indicator (59). In the pyrolysis of other feedstocks, however, the yield of by-products is significant and the yield of ethylene alone cannot be the single governing parameter. The total yield of ethylene, propylene, and butadiene produced per unit mass of feedstock converted is occasionally used as a selectivity indicator.

Modern ethylene plants are normally designed for near maximum cracking severity because of economic considerations. The production of ethylene from liquid feedstocks is accompanied by the production of other olefins which invariably entails the coproduction of many industrially important petrochemicals. Typical yield distribution patterns for a variety of feedstocks and feedstock quantities required for a modern 450,000 t/yr ethylene plant are illustrated in Table 3. Feedstock properties are shown in Table 4.

Polymer-grade ethylene and propylene are produced from the effluent of an ethylene plant. Other pyrolysis products affect operating economics significantly (see Table 3). With propane as feed, the C_3 fraction of the pyrolysis product may contain 40–60 wt % propylene, and for heavier feedstocks, it may contain 90–95 wt % propylene. The C_4 fraction from naphtha or gas-oil plants may contain up to 50 wt % butadiene. The debutanized gasoline is rich in aromatics and may have various outlets. After a two-stage hydrogenation treatment (102), it is an excellent feedstock for the recovery of aromatics. After a single-stage hydrotreatment, it may also be used as a high-octane blending component in unleaded gasoline. Interest has recently developed in the fraction following EP pyrolysis gasoline in the boiling range of 200–290°C. After suitable pretreatment this fraction may be hydrodealkylated with good yields of naphthalene.

Acetylene must be removed from the raw pyrolysis product to meet the ethylene specification. Most frequently the acetylene is eliminated by hydrogenation to part ethylene and part ethane, although recovery is technically and, under certain circumstances, economically feasible in very large plants.

The residual pyrolysis fuel oil is least valuable. It is thermally unstable so that it cannot be stored for long. It is also nearly always incompatible for blending with fuel oils from other sources. Its sole advantage is its low sulfur content and it is used as fuel under boilers and in pyrolysis heaters.

The Design of Pyrolysis Heaters. The pyrolysis heaters and their adjacent transfer-line exchangers (also called quench coolers) are the heart of an ethylene plant (see Fig. 1). They consist of (1) an upper central part with tubes arranged horizontally in the direction of the heater axis for feedstock preheat, feedwater preheat, and steam superheating coils designed for heat transfer from the rising flue gases essentially by forced convection, and (2) lower heater boxes containing the reactor coils arranged in vertical planes paralleling the heater axis and designed predominantly for radiant heat transfer from heater box walls and radiating flue gases. The feedstock is proportioned to the different coils ahead of the heater entrance and the surface of the preheat coil is dimensioned so that the feedstock, when passing from the preheat to the reactor coil, has about reached its incipient cracking temperature in the range of 560–680°C, depending on the type and properties of the feedstock. Design of the reactor coils is a formidable task (100,103–106), which may be facilitated by a computer.

Table 3. Typical Once-Through Pyrolysis Yields for Various Feedstocks

Pyrolysis yields, wt %	Ethane	Propane	n-Butane	Iso-butane	Light naphtha	Full range naphtha	Extraction raffinate	Kerosene	Atmospheric Light	Atmospheric Heavy	Gas oil Heavy vacuum
hydrogen	3.7	1.31	0.9	1.25	0.98	0.86	0.9	0.65	0.6	0.51	0.43
methane	2.8	25.2	20.9	22.6	17.4	15.3	16.5	12.2	10.6	8.82	7.7
acetylene	0.26	0.65	0.55	0.6	0.95	0.75	0.85	0.35	0.4	0.21	0.16
ethylene	50.5	38.9	37.3	10.7	32.3	29.8	28.4	25.0	24.0	21.36	17.5
ethane	40.0	3.7	4.5	0.6	3.95	3.75	3.9	3.7	3.1	4.54	2.8
methylacetylene and propadiene	0.03	0.6	0.8	3.0	1.25	1.15	1.1	0.75	1.05	0.19	0.45
propylene	0.8	11.5	16.4	21.2	15.0	14.3	14.5	14.5	14.7	13.25	13.4
propane	0.16	7.0	0.15	0.3	0.33	0.27	0.3	0.4	0.45	0.86	0.35
1,3-butadiene	0.85	3.55	3.85	2.15	4.75	4.9	4.8	4.4	4.8	6.15	5.46
butylenes	0.2	0.95	1.8	17.5	4.55	4.15	5.7	4.2	4.4	5.54	6.29
butanes	0.23	0.1	5.0	8.0	0.1	0.22	0.25	0.1	0.1	0.05	0.11
C$_5$s	0.22	1.6	1.6	2.0	3.85	2.35	3.5	2.0	3.3	2.18	5.5
C$_6$–C$_8$ nonaromatics					2.02	2.05	3.8	1.55	1.5	2.51	4.7
benzene	0.2	2.2	2.0	3.06	5.6	6.0	6.1	6.2	5.7	5.43	2.95
toluene	0.05	0.4	0.9	1.4	1.65	4.6	2.35	2.9	3.0	3.27	2.7
xylene and ethylbenzene			0.35	0.4	0.72	1.65	1.8	1.2	1.2	0.74	1.3
styrene						0.85	0.95	0.7	0.7	0.50	0.4
C$_9$ to 200°C		1.0	1.3	3.25	0.65	3.1	0.9	3.1	2.3	3.08	3.0
fuel oil		1.34	1.7	1.99	3.3	3.95	3.4	16.1	18.1	20.81	24.8
Total	100.0	100.0	100.0	100.0	100.0	100.0	100.0	100.0	100.0	100.0	100.0
for production of 450,000 t/yr ethylene											
fresh feed, t/yr	531,978	1,054,776	1,080,592	3,815,603	1,230,538	1,335,121	1,383,365	1,580,514	1,665,862	1,771,268	2,247,211
yield, 200°C+, t/yr		14,134	18,370	75,930	40,608	52,737	47,034	254,463	301,521	368,601	557,308

Table 4. Feedstocks for Olefin Production

	Ethane, wt %	Propane, wt %	Butane, wt %	Light naphtha	Full range naphtha	Extraction raffinate	Kerosene	Gas oil Atmospheric Light	Gas oil Atmospheric Heavy	Gas oil Heavy vacuum
specific gravity				0.68	0.727	0.69	0.77	0.84	0.872	0.919
°API				76.6	63.1	73.6	52.3	37.0	30.7	22.5
ASTM distillation										
initial bp, °C				36	45	C$_5$s	185	184	307	334
10%	95–100			50	77		190	240	348	404
30%		95–100	nC$_4$ 65–70	60	101		200	256	357	465
50%			iC$_4$ 25–30	77	120		210	275	366	495
70%				107	139		220	294	377	520
90%				146	160		234	320	395	559
end point, °C				168	179	C$_9$s	245	363	412	584
PONA^a, wt %										
paraffins				80	69.1	77.5	52	47		
cycloparaffins				15	21.1	16.6	28	30		
aromatics				5	9.8	5.9	20	23	22	42.5
crude source				Middle East	Middle East	reformer	Eastern Europe	Middle East	Libyan	U.S.A.
processing history				straight run	straight run		straight run	straight run	straight run	straight run
remarks				$\frac{\text{np}^b}{\text{ip}} \geq 1$		$\frac{\text{np}^b}{\text{ip}} < 1$				
	C$_1$ 0–3	C$_2$ 0–2	C$_3$ 2–5							
	C$_3$ 0–2	iC$_4$ 0–2	iC$_5$ 0–2							
		nC$_4$ 0–0.5	nC$_5$ 0–0.5							

^{a} Paraffins, olefins, naphthenes (cycloparaffins), aromatics.
^{b} np = n-paraffin; ip = isoparaffin.

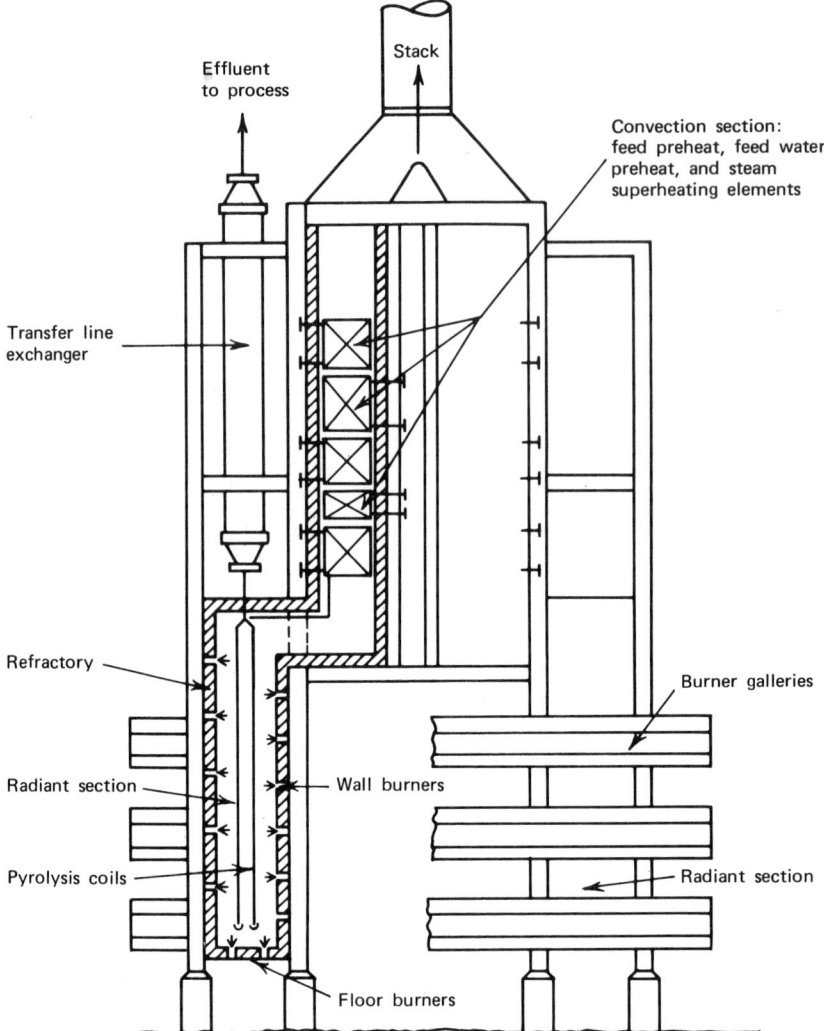

Figure 1. Typical heater configuration.

Figure 2 shows profiles of important variables over coil length which resulted from a coil calculation for a typical naphtha feedstock.

Of particular importance is the choice of metal alloys that can withstand the high temperature and stress–strain relationship and are stable against carburization. For many years, a special 25% chrome–20% nickel alloy (ASTM-HK40) has been a preferred tube material for reactor coils. Lately designers have turned to materials having a still higher nickel content such as ASTM-HP40 which may be alloyed with minor amounts of carbide stabilizers such as tungsten and niobium (columbium). For supports and hangers Cr–Ni steels with a very high chromium content are used.

Several firms design and construct pyrolysis heaters, including Lummus (96,107–110), Kellogg (101,111–112), Kinetic Technology International (113–114), Stone & Webster (45,115), Foster Wheeler (116), Linde A.G., Selas of America

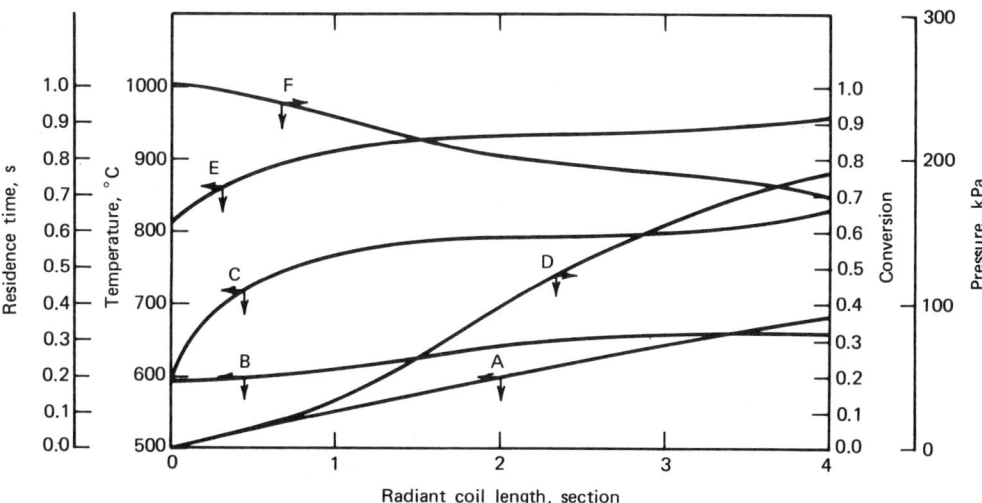

Figure 2. Major profiles for a typical naphtha pyrolysis coil. A, Residence time; B, hydrocarbon partial pressure; C, fluid temperature; D, conversion; E, tube wall temperature; and F, fluid pressure. To convert kPa to atm, divide by 101.3.

(117–120), and Mitsubishi (121–122). Most of these successful designs have followed the outline shown in Figure 1.

Table 5 lists the important characteristics of pyrolysis heaters.

Heater run length and coil maintenance are an important facet of heater operation because of coke deposition (54,72,96,105–106,113,117,123–131). The mechanism of coke formation is not clear and, therefore, it has not been possible to model it in precise mathematical terms. To alleviate or minimize the coking rate and to enhance the heater run length, trace amounts of sulfur compounds such as hydrogen sulfide or ethyl mercaptan (if the feedstock is basically free of sulfur compounds) or antifoulants such as Nalco 262 (132) are added to the charge stocks.

Figure 1 also shows the transfer-line exchanger located adjacent to the heater. In general, upflow is employed through the exchanger and the flow from two reactor coils is combined into one transfer-line exchanger. For example, a heater producing 50,000 metric tons ethylene annually may have eight coils and four transfer-line ex-

Table 5. Pyrolysis Heater Characteristics[a]

Characteristics	Value
number of coils	2–20
coil length, m	50–80
process gas outlet temperature, °C	750–900
tube wall temperature, reactor coil, clean, °C	950–1050
max reactor tube wall temperature, °C	1040–1100
average heat absorption, reactor coil, kW/m^2	50–80
residence time, reactor coil, s	0.15–0.6
inside coil diameter, mm	50–200

[a] Per heater unit; for a yearly ethylene production capacity of 20,000–70,000 t.

changers. The exchangers provide rapid cooling of the heater effluent to prevent further reactions of the effluent from continuing adiabatically which would impair the yield of primary products. The transition piece between heater outlet and exchanger tube sheet is designed to provide minimum residence time. The effluent leaves the heater as a superheated gas and the transfer surface is calculated to make profitable use of its heat content for the generation of steam down to a temperature reasonably above its dew point, to avoid condensation and fouling of the exchanger surface. For heavy feedstocks the steam pressure is in the range of 10.5–12.5 MPa (1522–1812 psi), the steam is used for driving the plant's compressors. Mass velocities in the tubes are purposely held high, above 45 kg/(m^2·s), to promote heat transfer and avoid film condensation near the exchanger outlet. Exchangers are manufactured by Borsig and Schmidt'sche Heissdampf (133–134) in Germany and Mitsubishi Heavy Industries (121–122) and Mitsui (135) in Japan.

Computer control ensures maximum energy conservation and increases ethylene and coproducts yield. The pyrolysis furnace area of an olefins plant is highly suitable for computer control (136–138). It is the most energy-intensive element and is a most important determining factor of the overall profitability of a plant.

Recovery and Purification. The gaseous effluent from the pyrolysis section leaves the transfer-line exchangers at temperatures in the range of 300°C (gaseous feedstocks) to 600°C (liquid feedstocks) to enter the recovery and purification section.

The effluent is resolved into the desired products by compression in conjunction with condensation and fractionation at gradually lower temperatures. Absorption (139) is no longer used for recovery because of its higher energy requirements. Today, separation is carried out by low temperature fractionation. Pyrolysis gases derived from ethane feedstock require the least effort and apparatus for separation, whereas those from heavy naphthas and gas oil producing ethylene, propylene, butadiene, and substantial amounts of liquid products require more complicated treatment.

Figure 3 represents a flow diagram of an ethylene plant for a low-sulfur naphtha feedstock. In a 300,000 t/yr ethylene plant between six to eight heaters are necessary of which one may be a spare. At a point following each exchanger, the effluent is further quenched to 200–220°C by a stream of refractory fuel oil from the primary fractionator to arrest further degradation reactions. The effluent streams are then combined and enter the primary fractionator. Normally, the primary gasoline fractionator is operated to pass overhead all steam and a specified end point pyrolysis gasoline in addition to the gaseous products. The quench tower (140–141) in essence operates as a partial condenser to the primary fractionator, condensing practically all of the steam and most of the pyrolysis gasoline components. A water phase and a gasoline phase separate in the quench water drum, and hot quench water may be used as a process heat source for fractionators. The gasoline phase is recycled to the primary fractionator except for the net feed of gasoline which is routed to the gasoline stripper for stabilization. Care is necessary in selecting the proper materials for the equipment because of the presence of acidic components such as CO_2 and H_2S in the gas.

On leaving the quench tower the pyrolysis gas is compressed to about 3.5 MPa (507.5 psi) in five stages. Polymer formation from highly reactive diolefins is of concern throughout the plant. Compression ratios must be strictly limited to avoid fouling and deposits within the compressors and on the discharge side of each stage.

The acid gas components are removed after the third stage by a scrubbing system employing a dilute caustic soda solution (142). Disposal of the spent caustic solution

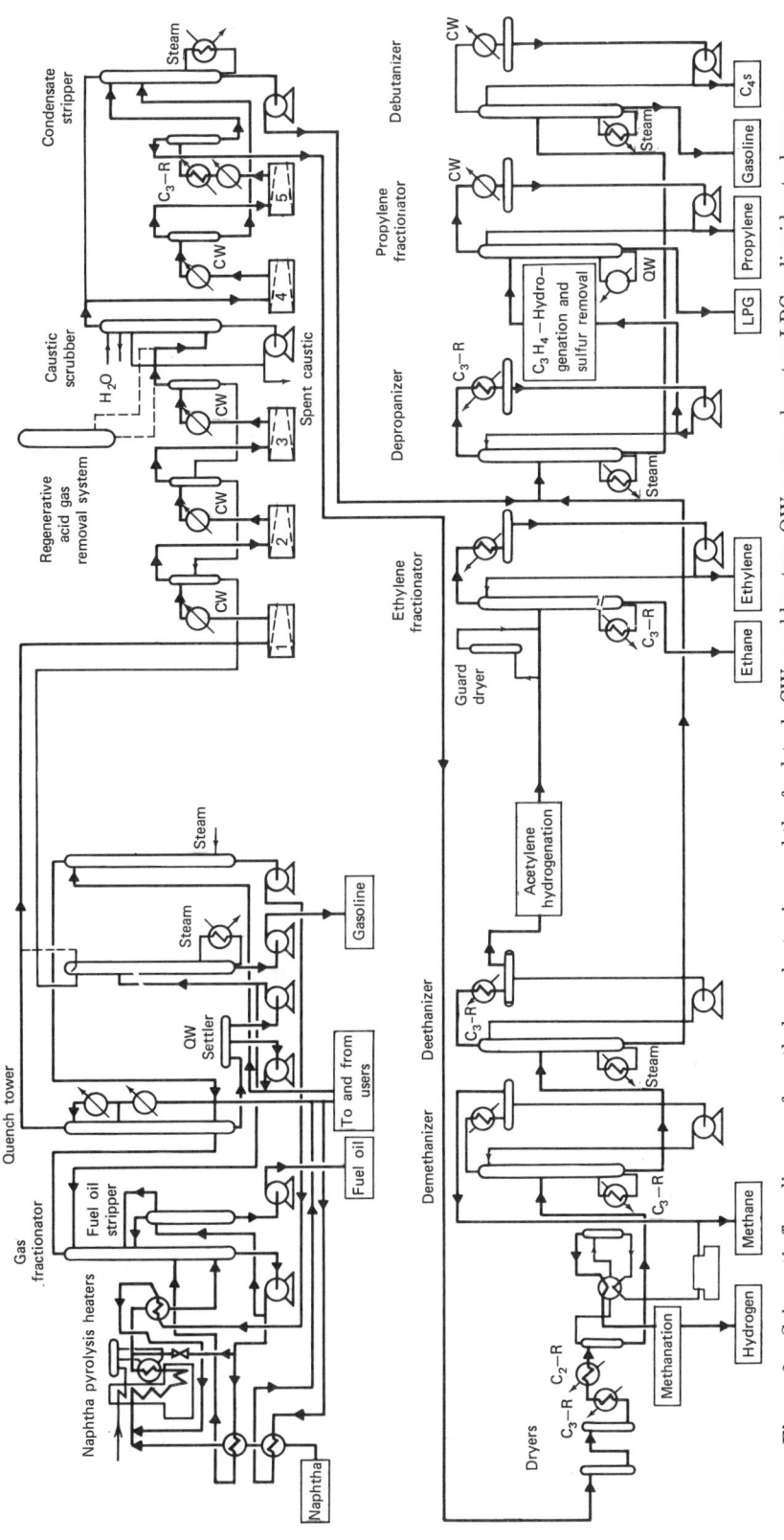

Figure 3. Schematic flow diagram of an ethylene plant using naphtha feedstock. CW = cold water; QW = quench water; LPG = liquid petroleum gas.

can be a troublesome environmental problem. Plants processing feedstocks of higher sulfur content (in excess of 500 ppm) require a regenerative solvent for acid gas removal ahead of the caustic soda polishing stage. This is more economical and avoids an extensive spent caustic disposal problem. The caustic solution is always followed by a water wash to remove any hydroxide carryover.

Following acid gas removal, further compression produces two interstage condensates that are rich in C_2 and C_3 components. To avoid the use of liquid dryers these fractions are stripped of their C_2 components in a second stripper, similar in purpose to the front-end pyrolysis gasoline stripper. The stripper overhead vapors are recycled to a lower compression stage. The stripper base product is thus completely dried by distillation and can be introduced into the tail-end depropanizer. All the ethylene and ethane in the pyrolysis gas is forced to take the route over the dryers to the demethanizer. Molecular sieves are the preferred desiccant because of their improved selectivity over activated alumina.

Up to this point the pyrolysis gas from the transfer-line exchangers has undergone the following treatments: (1) further heat recovery in conjunction with the recovery of the fuel oil and pyrolysis gasoline fractions, (2) compression, (3) removal of acidic components, and (4) drying by solid desiccant of the remaining vapors consisting predominantly of C_2 and lighter hydrocarbons, and coincidental thereto, drying by distillation of a C_3 and heavier liquid stream. The streams obtained are further resolved by low temperature distillation utilizing a refrigeration cascade employing propylene, ethylene, and methane (in the order of decreasing boiling points).

The vapor stream is partially condensed at essentially constant pressure over the stages of the cascade refrigeration system to about $-165°C$ until only the hydrogen remains in the vapor state. The stage condensates (only one is shown in Fig. 3) are introduced at various levels into the demethanizer. Hydrogen, 85–90 vol % pure, is withdrawn from the lowest temperature stage separator. The demethanizer produces an overhead consisting predominantly of methane, with minor amounts of hydrogen, carbon monoxide, and only a very minor amount of ethylene as impurities, and a methane-free base product of ethylene and heavier hydrocarbons. Ethane and propane are removed in a similar fashion. Acetylene is usually not recovered, but hydrogenated to ethylene (143) (see Acetylene).

For the separation of binary mixtures with close boiling points, such as in the ethylene–ethane and propylene–propane fractions, open heat pumps are thermodynamically the most attractive. Both heating and cooling duties have to be incorporated into the cascade refrigeration system for optimum energy utilization while strictly observing constraints on the process side.

Ethane obtained as base product from the C_2 splitter (ethylene–ethane fractionator) is usually recycled to extinction over a pyrolysis heater especially reserved for this purpose (Fig. 3); the depropanizer base product is fractionated in the debutanizer for an overhead product containing the butadiene and a base product of light pyrolysis gasoline that may be combined with the heavier product from the front-end primary gasoline stripper. This product is primarily a source material for the production of aromatics.

Multipass plate-fin type aluminum exchangers are advantageously used for nonfouling services, particularly in the chilling train of the plant. Compressors for pyrolysis gas and refrigeration are the volute type, with split casings and forged steel rotors. Rotors for the ethylene refrigerant compressor are ASTM 387 Grade 22 (2.5%

Cr, 1.0% Mo) steel. Chrome–nickel steels of various specifications are used for vessels and lines in the low temperature section of the plant in conformity with ASME-API codes. Where high hydrogen partial pressures are encountered, molybdenum alloys are preferred.

An ethylene or olefin plant requires extensive support facilities for boiler feed water preparation, treatment of noxious effluents, storage of products and consumables, cooling water plant and, if necessary, for the generation of steam and electric power.

Plants using gaseous feedstocks (ethane and propane) do not produce enough high pressure steam for all the compressors to be steam-turbine driven, hence require additional steam generating capacity. Those charging naphthas and gas oils generate a surplus of steam that may be exported. In all plants an on-site steam boiler is necessary for start-up purposes.

Petrochemical Refineries. Today's trend is in the direction of using heavier, liquid feedstocks for the production of olefins, particularly of ethylene and propylene, for which the market demand seems to grow more rapidly than that for the traditional refinery fuels. An ethylene plant of 450,000 t/yr, based on liquid feedstocks, may require upwards of 1,500,000 t/yr of feedstocks which equals the output of a fairly large refinery. To reduce or eliminate the transportation cost of the feedstocks, large petrochemical complexes have been built in close proximity to oil refineries. The question has arisen to what extent refinery operations producing essentially motor and industrial fuels may economically be integrated with petrochemical operations (144–146) (see also Feedstocks; Petroleum).

Economically the most attractive configuration of a combination plant of this type would produce a relatively constant amount of motor fuel and residual fuel oil. Middle distillates, for which the seasonal demand varies most throughout the year, would be upgraded as required and used as feedstocks for additional petrochemicals. Various processes or process combinations exist by which to perform the upgrading, of which hydrodesulfurization is perhaps the most prominent one. Some forms of hydrotreatment using a nickel–tungsten or cobalt–molybdenum catalyst improve ethylene yields from gas oils (147–149). Gas-oil pretreatment includes the solvent extraction of aromatics to optimize olefin and aromatic yields (see BTX processing).

Ethanol Dehydration. The economic feasibility of producing ethylene from ethanol (qv) depends mainly on the availability and price of fermentation ethanol. High volume production of ethylene from ethanol (derived from fermentable raw materials) cannot normally compete with ethylene produced in large hydrocarbon-based olefins units in the foreseeable future. The process, however, offers several advantages to a country with abundant fermentation materials but limited hydrocarbon resources (150): (*1*) a local market can be served since ethylene can be produced from fermentation ethanol economically in small quantities; (*2*) dependency on foreign supply of crude oil or natural gas liquids for ethylene production is eliminated; (*3*) the total investment is only a fraction of the investment required for a hydrocarbon-based olefins plant with the same ethylene production capacity; (*4*) the relative capital investment per metric ton of ethylene is low since no investments and related annual charges have to be made for the production, recovery, storage, and marketing of salable by-products of typical hydrocarbon-based olefins units; (*5*) a small ethylene plant (eg, 10,000–60,000 t/yr) can be built initially together with a downstream ethylene-derivative unit (ie, polyethylene), additional capacity can be added at a later date without significant economic penalty; (6) abundant agricultural or industrial wastes (eg, waste fruit juices, molasses, sulfite liquors) may be utilized in a productive and profitable way; (7) renewable resources are used as raw material which may provide feedstock for a developing petrochemical industry.

Ethanol dehydration goes back to 1797 when ethylene was first being produced by passing ethanol or ether over heated alumina or silica (151). Later, activated alumina and phosphoric acid on a suitable support became the choices for industrial processes (152).

Despite a great deal of research on the dehydration reactions of ethanol on various catalytic surfaces, the actual reaction mechanism is not fully understood, but the following are possible:

$$\text{consecutive reactions:} \quad \text{ethanol} \xrightarrow{230°C} \text{ether} \longrightarrow \text{ethylene}$$

$$\text{parallel reactions:} \quad \text{ether} \xleftarrow{230°C} \text{ethanol} \xrightarrow{300-400°C} \text{ethylene}$$

$$\text{simultaneous reactions:} \quad \text{ethylene} \longleftarrow \text{ether} \xleftarrow{230°C} \text{ethanol} \xrightarrow{300-400°C} \text{ethylene}$$

In industrial production the ethylene yield is 94 to $\geq 99\%$ of the theoretical value, depending on the processing scheme. Traces of aldehyde, acids, higher hydrocarbons, and carbon dioxide, as well as water, have to be removed.

The fixed-bed ethanol dehydration process was developed in Europe before World War I and has been commercialized in many countries. Ethanol dehydration was widely carried out in Europe, Asia, South America, and Australia on a small industrial scale before low cost ethylene became available from hydrocarbon-based olefins plants. In 1978 small plants were still in operation in Brazil, India, Pakistan, and Peru.

As a result of increasing petroleum prices, ethanol dehydration processes are again being considered in many developing countries. Brazil and India, eg, have developed ambitious plans to utilize the production of ethylene from ethanol using domestic resources as raw materials for their petrochemical industries (see Chemurgy; Fuels from biomass).

Over the years, significant process improvements have been made in ethanol dehydration. A new fluid-bed process has been developed which significantly reduces plant investment and operating costs (153).

Ethylene from Coal. There are several possible routes to ethylene from coal by conventional and by new process technology.

Coal-derived synthesis gas manufactured by long-established coal gasification processes (Koppers-Totzek, Winkler, or Lurgi) may be converted into hydrocarbons by the Fischer-Tropsch (FT) process (see Fuels, synthetic). Conventional FT processes produce higher molecular weight saturated and olefinic hydrocarbons and oxygenated compounds. Some ethylene may be recovered directly or it may be produced by pyrolysis of the hydrocarbons (eg, ethane, propane, butane, FT-naphtha) and by dehydration of recovered ethanol. These processes, however, are not economical today for ethylene production or other petrochemicals and may be considered in special situations only.

At present, the only commercial FT plant (Sasol I) is operating in South Africa where coal is mined at very low cost. It produces a full range of products (including ethylene) which are normally produced from petroleum. A second much larger FT process coal conversion facility (Sasol II) is at present under construction.

Conventional commercial FT processes have been used in the past primarily for fuels manufacture. However, on the development level, numerous efforts have been directed to optimize this technology for ethylene production (154). For example, Ruhrchemie's new FT iron catalyst modified by activators can convert synthesis gas directly to lower olefins, primarily ethylene and propylene, in relatively good yield (155).

In addition, indirect processes for the manufacture of ethylene have been developed that are based on coal-derived methanol. In one process, ethanol is produced by methanol homologation with synthesis gas followed by ethanol dehydration to yield ethylene (154). In the Mobil process, methanol is converted primarily to ethylene and propylene by using a promoted zeolite catalyst (see Molecular sieves). Ethylene and propylene yields of 18.9 and 36.3 wt % are obtained, respectively, at 98% methanol conversion (156).

Propylene Disproportionation. Phillips' Triolefin process utilizes a disproportionation reaction whereby relatively low cost propylene can be upgraded to more valuable ethylene and butenes. For every two moles of propylene converted, approximately one mole of ethylene and one mole of butene are produced (38). It is a means for altering the product mix of refineries, ethylene plants, and other chemical plants. Excess propylene can be converted to ethylene and butenes which may be more readily usable or easier to market. A commercial plant using the process was built by Gulf Oil Canada Ltd. in 1966.

Shipment and Storage

Until the early 1930s ethylene was almost universally recovered from coke-oven gases and shipped in steel cylinders under high pressure. A 5-t truck could carry close to 50 such cylinders. The empty cylinders weighed close to 4 t and the gas content 0.85 t, a most unsatisfactory tare ratio (empty weight of vehicle to payload). To reduce the tare ratio effectively, transport had to be in the liquid phase at greatest density, ie, at about atmospheric pressure.

In 1951 ICI in the UK set up a highway truck transportation system using cryogenic tank trucks, an operation continued successfully for years. In the United States such transports come under the supervision of the ICC.

Simultaneously with highway vehicles, heavily insulated railroad tank cars were developed for the haulage in the liquid state of cryogenic gases, including nitrogen, oxygen, and ethylene (see Cryogenics). The tare ratio for ethylene was thus reduced below 0.8. A standard ethylene railroad tank car, manufactured by Lox Equipment Co. of Livermore, California, can carry a payload of 65 t. Tanks for truck–trailers and railroad cars are designed as double-shell units just as cryogenic storage tanks (see below).

For intercoastal transport (157) the French and Japanese developed the seagoing tanker in 1964–1965. By 1976 the tanker ethylene fleet consisted of 28 vessels, most of them with a capacity of 400–2200 t. There is as yet no regular intercontinental traffic in ethylene although larger tankers, recently designed to carry alternately LNG, ethylene, and LPG, have intercontinental range (158–160). The tankers must have a reliquefaction plant aboard as ethylene is too costly to vent; they are universally designed for atmospheric pressure operation (161).

Much larger volumes are moved by integrated pipeline systems in the United States and Europe than by tanker.

Pipelines. In the United States the largest number of ethylene producer and consumer plants is located in Texas and Louisiana. Other centers of production and processing operate in Southern California and the Midwest. A system of pipelines connecting producer and consumer plants (162) has been developed along the Texan coast as of late 1977, stretching from Lake Charles, La., to Victoria, Tex., including seven large ethylene producer plants, 13 producer–consumer complexes, and 20 consumer plants as well as 11 salt dome storage reservoirs. Each company operates its own system independently with intrasystem transfers possible by contract or agreement. The larger part of the grid is maintained at about the critical pressure of ethylene or slightly above since the average ground temperature at the depth at which the lines are buried rarely approaches the critical temperature. Each system operates automatically, ie, any excess production, on pressure increase, is shunted to the storage compressors and a deficiency, on falling pressure, is made up from the underground storage reservoirs (163) (see Pipelines). A smaller but similar ethylene grid is in operation in Louisiana. The two grids are connected by a proprietary 200 mm Dow Chemical Co. line.

The large European producers and consumers (Germany, Holland, France, and Belgium) are interconnected by an extensive ethylene pipeline system that will eventually reach from the North Sea coast to the Mediterranean coast. In England the ICI road tanker service mentioned earlier was replaced in 1968 by a pipeline from Billingham to Carrington. It might be extended into Northern Scotland if projects for the utilization of North Sea gas mature.

An interconnecting pipeline system relieves producer participants of the need to install expensive above-ground storage of their own and makes available to participants alternative supplies of ethylene under strict quality control from cheaper underground reservoirs during periods of a forced shutdown or turn-around of a producer plant. An ethylene pipeline may be designed to operate at subcritical or supercritical pressure (164).

Proprietary studies have been made of the cost of transporting ethylene by pipeline but very few data have been published. In 1970 Jozon in France (165) calculated supercritical transport costs as between 5.5 and 2.2 ¢/(t·km) for volumes from 100,000 to 600,000 t/yr. These data are in need of updating. Owing to the persistent inflationary trend, today's costs may be about double those calculated for 1970.

Storage. As a liquid, ethylene may be stored at any temperature along its vapor pressure curve, below its critical temperature of 9.2°C (166–167). In ethylene plants a certain amount of operational storage, up to about four hours of production, is normally provided in pressure tanks, held at a temperature where a convenient level of refrigeration is available from inside the plant. Holding or production storage for large volumes of ethylene, because of the high cost of pressure storage, is normally provided as atmospheric pressure storage in tanks designed for cryogenic conditions.

Cryogenic Storage. A great number of cryogenic tanks are in service at LNG peak load and LNG satellite stations as well as in a few LNG base-load storage plants in the United States and abroad. Nitrogen and other cryogenic gases as well as some ethylene have similarly been stored in bulk for many years.

Cryogenic tanks are designed to conform to API specification 620, appendix Q. They provide for storage in a double envelope, an inner tank of stainless or equivalent Ni-steel and an outer envelope of carbon steel. Ethylene tanks differ from LNG tanks only to the extent that a lower-grade steel (5.5% Ni) is allowed for construction of the inner tank. Outer and inner tank are in communication by way of an insulated flat inner deck suspended from the roof. The inner tank is bedded on a load bearing insulating brick foundation and the annulus between tanks, generally wide enough to accommodate scaffolding for tank erection, is filled with Perlite or similar granulated insulation.

When the tank is set flat on the ground, a heating coil must be installed below the foundation to prevent frost heave of the ground. With piled foundations, of course, this provision is not necessary. As ethylene is too costly to vent, a pressure maintenance system coupled with a reliquefaction unit takes care of heat leaks into the tank and of inventory changes (168–169).

Above-ground ethylene tanks have been built in capacities up to 24,000 m^3 (8.5 × 10^5 ft^3) (170). In an ethylene plant producing 300,000 t/yr of ethylene for use in downstream units one such tank will provide for a production reserve of 16 days which is adequate to bridge the annual turnaround period of the ethylene plant.

Salt Domes. Large amounts of ethylene are stored in the dense phase at pressures two to three times the critical pressure in cavities created in underground salt domes. Suitable underground salt domes are found in many parts of the world. By a technique similar to that employed in prospecting for oil and gas, these salt structures may be tapped and leached with fresh or brackish water through a leaching string to create a more or less spindle-sized, brine-filled void that may measure 20–30 m in diameter by 300 m in height. In operation, the lower part of the cavity contains a sizable amount of brine. As ethylene is compressed into the reservoir, it will displace an equal volume of brine into surface ponds. Conversely, ethylene may be withdrawn by returning brine from these ponds. At the subterranean interphase the ethylene pressure balances the weight of the brine column.

As the ethylene becomes wet in storage, withdrawn ethylene must be redried before it is released into a pipeline or processing plant. Salt dome storage in North America and Europe (171) has been very successful, particularly in connection with an ethylene pipeline system that connects ethylene producers and consumers (172).

Caverns. In some areas of the world salt is found in stratified layers at various depths, separated by relatively impervious layers of limestone, dolomite, or other sedimentary rock. The salt may be removed by single or multistring leaching to create underground caverns in a sort of multitiered arrangement for the storage at supercritical pressures of ethylene. The Dow Chemical Co. at Midland, Michigan, has successfully operated a subterranean cavern storage system since the late 1950s.

In-ground storage of ethylene is most economical and effective in combination with ethylene pipeline transportation systems which connect producers and consumers. The largest such distribution grid is located in Texas. Its total storage capacity has been reported to exceed three million cubic meters (1 × 10^8 ft^3) to which must be added the grid's pipeline inventory. If it is assumed that all in-ground storage of the grid is at an average depth of 1000 m, a reasonable figure, where pressure p = 12.1 MPa (1754.5 psi) and temperature t = 50°C and minimum pipeline operating conditions are at the critical pressure, p = 5.04 MPa (730.8 psi) and t = 20°C, three million m^3

storage volume would provide a production reserve (breathing capacity) of 530,000 t ethylene.

Economic Aspects

Worldwide, ethylene is first in total market value and second in metric tons produced among petrochemicals. In 1977 estimated ethylene production worldwide was 34×10^6 t, with a total market value of ca 9.7×10^9 dollars. Table 6 (173–174) shows the growth of United States production which comprises a major part of the world's production.

Table 6. United States Ethylene Production[a]

Year	Production, 10^3 t/yr[b]
1930	9
1940	136
1945	363
1950	680
1955	1,382
1960	2,471
1965	4,340
1970	8,204
1975	9,297
1980[c]	12,000–13,000

[a] Refs. 173–174.
[b] Approximate values.
[c] Estimated.

Table 7. Worldwide Ethylene Production Capacity[a], 10^6 Metric Tons

Country	1977	1982 (projected)	Capacity growth rate[b], % annually
North America			
United States	16.68	19.70	5.5
Canada	1.08	1.57	7.8
Mexico	0.44	1.44	26.8
Central America	0.0	0.0	0.0
South America	0.47	1.79	30.7
Western Europe	13.90	19.39	6.9
Eastern Europe	>2.81	>6.69	19.0
Middle East	0.12	1.13	56.6
Far East	6.49	7.73	3.6
Oceania	0.30	0.30	0.0
Africa	0.15	0.79	39.4
Total	>40.86	>60.53	8.2

[a] Ref. 174.
[b] Predicted in 1977. Note that demand growth rate is likely to be less than this rate of capacity build-up.

Table 8. Approximate Price History of Olefins Plant Products in the United States, $/t^a

Year	Ethylene	Propylene	Butadiene	Benzene	Toluene	Mixed xylenes	Regular gasoline	No. 6 fuel oil[b]
1959	105	40	325	93	76	82	40	11
1963	110	53	258	75	64	76	39	11
1967	88	66	258	82	70	78	41	11
1971	66	62	233	62	67	73	44	28
1974	166	119	320	225	177	146	104	81
1975	222	170	441	225	188	174	110	77
1977	275	205	436	255	168	157	136	81
1978	286	226	460	231	172	175	147	88

[a] Refs. 175–179.
[b] Max 1% sulfur.

The distribution of existing and projected worldwide ethylene production capacity (174) and the average annual growth rates between 1977 and 1982 are shown in Table 7. A strong growth of petrochemical industries in the Middle East, Africa, South America, and Eastern Europe is projected which will reduce expansion in industrialized countries.

The economics of ethylene production is strongly affected by the factors that determine the markets for petroleum and natural gas products. Table 8 shows an approximate price history of olefin plant products in the United States since 1959. The price for ethylene declined from the early 1960s to the early 1970s because of the construction of more efficient plants. These had increasingly larger capacity and steadily declining capital investment and production costs per unit of ethylene produced. However, the large increase in crude oil prices in 1973 combined with the general inflation increased ethylene prices to an all-time high and affected the prices of feedstocks and other products of ethylene production facilities accordingly.

Table 9 shows recent investment and production cost estimates for a 500,000 t/yr ethylene plant using various types of feedstocks. Although the absolute dollar values are quickly outdated, the trend is worthy of examination.

Table 10 gives the assumed basis for operating costs, and Table 11 assumed feedstock and coproduct values.

Production costs (Table 9) should not be interpreted as being indicative of selling prices. Other factors often have a considerable effect on prices. For example, market considerations frequently require operations at less than full capacity. Thus effective unit costs would rise above those shown for the full capacity operation assumed here. The investment costs (Table 9) indicate a significant increase as the feedstocks become heavier. Cracking of light hydrocarbons produces a small quantity of coproducts and requires less feed per ton of ethylene produced. Naphtha and heavier hydrocarbon feeds produce substantially larger quantities of coproducts and, therefore, require more feed and investments in larger and additional equipment. Coproduct credits also increase substantially as the feeds become heavier.

The significant growth in the size of ethylene plants in the past fifteen years illustrates the economic advantages afforded by scale which result basically from the reduction in capital requirements per unit of ethylene produced. Table 12 shows the effect of ethylene plant size on the investment and operating costs for various feedstocks. Reductions in unit production costs are primarily a result of a decrease in capital

Table 9. Ethylene Production Economics[a]

Estimate	Feedstock					
	Ethane	Propane	n-Butane	Full range naphtha	Gas oil Atmospheric	Gas oil Vacuum
plant investment estimate[b], million $	187	215	225	247	260	273
cost of production[c], $/t ethylene						
feed cost[d]	152	307	312	371	432	517
operating costs	101	70	74	78	82	85
coproduct credits[d]	(4)	(155)	(174)	(264)	(359)	(482)
cost of production ex. ROI[e]	249	222	212	185	155	120
ROI[e], 25% of investment before taxes	94	108	112	124	130	136
Total	*343*	*330*	*324*	*309*	*285*	*256*

[a] Basis: 500,000 t/yr ethylene production; plant start-up: March 1978, U.S. gulf coast location, high severity operation; ethane recycled to extinction.
[b] Includes outside battery limits (OSBL) investment at 60% of inside battery limits (ISBL) investment.
[c] See Table 10.
[d] See Table 11.
[e] Return on investment.

Table 10. Assumed Basis for Operating Costs[a]

Expenditures	Value
Annual capital related charges, % of investment	
depreciation	
ISBL[b] facilities	6.7
OSBL[c] facilities	6.7
maintenance	
ISBL[b] facilities	4
OSBL[c] facilities	2
insurance and local taxes	2
Labor costs	
direct labor including benefits, $/man-year	19,100
supervision and plant overhead, % of direct labor	200
Variable costs[d]	
utilities	
fuel, $/GJ[e]	2.46
power, $/kWh	0.024
cooling water, $/m^3	0.025
boiler feed water, $/m^3	0.39
catalyst and chemicals, $/t ethylene	1.7–1.9[f]

[a] March 1978, U.S. gulf coast location, for a plant operating at 100% capacity.
[b] Inside battery limits.
[c] Outside battery limits.
[d] Excluding feedstock.
[e] To convert GJ to Btu, multiply by 9.488×10^5.
[f] Depending on feedstock.

related items (maintenance, insurance, local taxes, depreciation, and return-on-investment) per unit of ethylene produced, as the plant size increases.

The economy of scale, however, diminishes when going to higher plant capacities. In a few instances single-train ethylene plants (no duplication of compressors or other major equipment, except for the cracking heaters) have now reached a capacity of about 700,000 t/yr. As plant sizes increase, more costly field fabrication (as opposed to shop fabrication) of some large equipment items becomes necessary. This limits unit investment savings for large plants. Furthermore, a large initial capital is required, and the risk of possible start-up delays and shutdowns and the effect of changing market conditions have to be considered.

Figure 4 shows the trend of increasing ethylene plant size leveling off, ie, the movement to larger plants is being restrained by diminishing economic benefits and market growth considerations (180).

Specifications and Analysis

Before the large-scale production of polyethylene, ethylene specifications met the requirements of ethanol and ethylene oxide plants, which accepted 97–98 wt % ethylene, with maximum propylene and acetylene contents as high as 0.1 and 1.0 wt %, respectively. Because polyethylene is now the predominant derivative, nearly all

Table 11. Assumed Feedstock Prices and Coproduct Values[a]

	$/t
Feedstocks	
ethane	129
propane	131
n-butane	130
full range naphtha	125
gas oil	
atmospheric	112
vacuum	106
Coproducts	
propylene, chemical-grade	205
mixed C_4 hydrocarbons	219–338[b]
contained butadiene	405
other C_4s at fuel value[c]	112
C_5/200°C pyrolysis gasoline	117–133[b]
contained benzene	157
contained toluene	108
contained xylenes	101
contained C_5s	122
contained C_6–C_8 nonaromatics	101
contained C_9-200°C	86
C_3 LPG	131
methane-rich gas	123
hydrogen-rich gas	366
pyrolysis fuel oil at 90% of fuel value[c]	87

[a] March 1978, U.S. gulf coast location.
[b] Ranges are due to variations in composition resulting from cracking different feedstocks.
[c] Fuel value: $2.46 GJ (to convert GJ to Btu, multiply by 9.488×10^5).

Table 12. Relative Ethylene Plant Investment and Cost of Production Estimate as Function of Feedstock and Plant Capacity[a]

	Feedstock					
				Full range	Gas oil	
Estimate	Ethane	Propane	n-Butane	naphtha	Atmospheric	Vacuum
Relative plant investment[b]						
plant capacity, t/yr						
200,000	1.35	1.55	1.60	1.84	1.95	2.03
300,000	1.17	1.34	1.40	1.59	1.67	1.76
400,000	1.07	1.22	1.28	1.42	1.50	1.57
500,000	1.00[c]	1.15	1.20	1.32	1.39	1.46
600,000	0.94	1.08	1.12	1.23	1.31	1.38
Relative cost of production						
plant capacity, t/yr						
200,000	1.13	1.12	1.10	1.11	1.06	0.98
300,000	1.06	1.04	1.03	1.01	0.94	0.87
400,000	1.03	0.99	0.97	0.94	0.87	0.79
500,000	1.00[c]	0.96	0.94	0.90	0.83	0.75
600,000	0.97	0.93	0.91	0.86	0.80	0.72

[a] Plant startup: March 1978, U.S. gulf coast location.
[b] Includes OSBL (outside battery limit) investment at 60% of ISBL (inside battery limit) investment.
[c] Selected base. For investment and cost of production estimate in dollars, see Table 9.

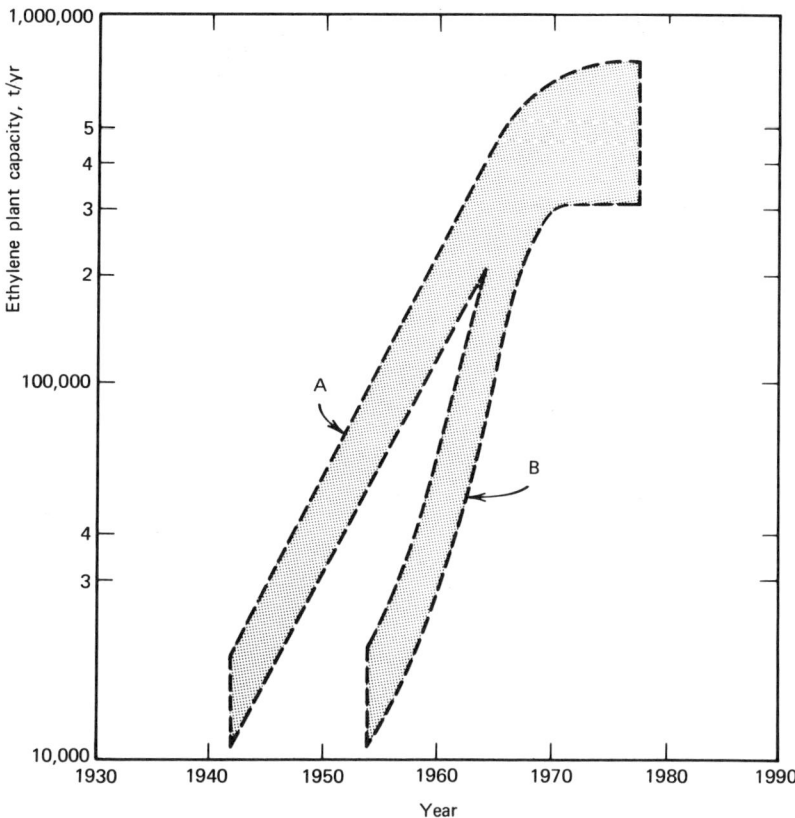

Figure 4. Ethylene plant size increasing with time. A, United States; B, Western Europe and Japan.

new ethylene plants are designed to meet the specifications for polymer-grade ethylene (181). New catalyst developments and requirements for improved polymer product properties have led to a gradual tightening of polymer-grade ethylene specifications in recent years. High density polyethylene is the most demanding end use and sets the standards for a number of maximum impurities concentrations allowable. Generally, polymer-grade ethylene adequate for high density polyethylene production is also suitable for low density polyethylene production. A typical polymerization-grade ethylene specification is given on Table 13.

Ethylene is determined by absorption in bromine solutions, fuming sulfuric acid, or a solution of mercuric sulfate in 25% sulfuric acid, but none of these reagents is selective for ethylene in the presence of other olefins. Modern ethylene plants use gas-chromatographic techniques supplemented by specific test methods for determining trace quantities of sulfur, water, other hydrocarbons, and other elements [ASTM Standard Tests D 2504-67 (1972) (182) and D 2505-67 (1972) (183)].

The Wickbold and Beckman combustion apparatus is suitable for determining trace quantities of total sulfur in ethylene (ASTM 2785-70) (1975) (184). Traces of water in ethylene may be determined by using the Karl Fischer reagent (ASTM E 203-75) (185).

Table 13. Typical Specification for Polymerization-Grade Ethylene[a]

Specification	Value
ethylene, mol %	99.9
impurities, ppm max	
acetylene	2
methane	200
ethane	200
oxygen	2
carbon monoxide	1
carbon dioxide	5
total sulfur	2
propylene	10
$C_{4+}s$	10
water	2
hydrogen	5
methanol	5
total chlorine	2
other compounds	5

[a] Ref. 181.

Safety and Environmental Factors

The MCA has issued a safety data sheet which gives information on safety, toxicology, and the handling of ethylene (186). At RT and atmospheric pressure ethylene is a colorless gas with a mild odor, of a density practically equal to air, nonirritating to eyes and the respiratory system. Released to the atmosphere it quickly diffuses and at the very low concentration of 2.7 vol % forms a flammable mixture with air. As it does not give adequate warning of its presence, any accidental leak must be approached with the greatest caution and in strict observance of safety procedures contained in plant operating manuals. Chronic injuries to humans from the temporary inhalation of ethylene have not been reported.

Care must be used in the design of certain equipment containing ethylene at high pressure and high temperatures. Under extreme conditions (rarely encountered), ethylene can spontaneously decompose into its elements, given a sudden, energy-initiating source (shock wave, hot spots), a phenomenon also encountered in the handling of acetylene (187–189). The reaction is highly exothermic and may produce a decomposing flame front that might rupture piping and equipment.

Within the past several years the United States government has imposed stringent pollution control regulations that particularly affect operations of the refining, chemical and petrochemical industries (190). Plants producing olefins emit aqueous effluents containing spent alkalis and material, such as phenols and polycyclic hydrocarbons requiring much oxygen for degradation, releasing oxides of sulfur and carbon. *The Federal Water Pollution Control Act* as amended in 1972 and *The Clean Air Act* as amended in 1977 are therefore of particular concern to the ethylene industry. These laws specify special effluent aftertreatments (see Wastes, industrial).

Uses

Almost all ethylene is consumed by the chemical industry as feedstock for a variety of petrochemical products. A few direct agricultural uses of ethylene have been developed, such as use as a ripening agent for fruits and vegetables and as an herbicide for witchweed. Table 14 shows the important ethylene derivatives and end uses.

Table 15 shows the 1976 ethylene use pattern for first-generation ethylene derivatives in the three largest markets, the United States, Western Europe, and Japan, with polyethylene being dominant in all three areas. These areas developed separately because ethylene is not easily shipped across oceans (see under Shipment).

Trends

Table 7 shows that, by the end of 1977, available ethylene capacities in the United States, Western Europe, and Japan were 16.7, 13.9, and ca 6.5×10^6 t/yr, respectively. However, these capacities were utilized only to the extent of about 12, 10, and 4.2×10^6 t in the same area. The difference represents surplus capacity.

Historically, ethylene demand has grown very rapidly, reflecting widespread market penetration of ethylene derivative products. From the mid 1960s to 1973, worldwide ethylene production increased more than 15%/yr; 12%/yr in North America and more than 20%/yr in Western Europe and Japan. Although the bulk of consumption will continue to be in the industrialized countries, the share consumed by developing countries will continue to increase. Between 1979 and 1985, demand growth in the developing countries will exceed 15%/yr, whereas growth in the industrialized countries is projected to decline markedly. For example, annual demand growth in the United States between 1979 and 1985 is expected to be in the 5–6% range, and somewhat lower in Western Europe and Japan. Thus worldwide demand growth between 1979 and 1985 is expected to be 7–8% annually. This slowdown is a function of maturing markets and slower economic growth. Nevertheless, ethylene and its derivatives are expected to remain a growth industry, since demand growth is believed to exceed the real growth rate of the gross national product in most countries.

By 1982–1983, an estimated 50% of United States production will be coming from naphtha and gas oils. The corresponding figure for Europe is in excess of 90%.

Unlike United States companies, most European chemical companies show no disposition to become involved in the by-products energy market, and thus it can be assumed that European producers will continue to remain essentially naphtha oriented. In addition, they may obtain additional feedstocks (natural gas liquids and petroleum fractions) from the gradually mounting output of the North Sea fields (see Feedstocks). The situation will change by the mid 1980s as the Persian Gulf nations implement their decision not to permit the flaring of any gas beyond 1985.

This program will make available in the Middle East immense quantities of choice ethylene feedstocks which will put these countries into the petrochemicals business. By 1982 Saudi Arabia alone may have to dispose of nearly 2×10^6 t of ethane (191) and an estimated 4×10^6 t of propane annually and of larger amounts by 1985. Sizable olefin complexes will come onstream in Qatar and Iraq in 1980 and 1982. Plans for six petrochemical complexes in Saudi Arabia are presently being discussed with Western interests and it is likely that some agreements, possibly in the form of joint ventures that would also provide for the distribution and marketing of the products from such

Table 14. Major Ethylene Derivatives and Uses[a]

Process, petrochemical intermediates (uses)	Petrochemical products (uses)
Polymerization (43.0)	
(27.4)	polyethylene, low density (film and sheet; injection molding; paper, wire, and cable coatings)
(14.7)	polyethylene, high density (blow molding; injection molding; pipe, conduit, tubing; film and sheet)
(0.2)	polyethylene copolymers (film; electrical insulations; hot-melt coatings and adhesives)
(0.7)	ethylene–propylene co- and terpolymers (automotive parts; tires; hose and tubing; thermoplastic modifiers; wire and cable)
Oxidation (23.4)	
ethylene oxide (18.6) (surfactants)	
ethylene glycol (antifreeze)	polyesters (filament; fiber; film)
di- and triethylene glycols (dehydration agent; solvents)	
polyols	polyurethane and polyesters (filament; fiber; film; flexible and rigid foams; coatings; elastomers; adhesives; sealants)
morpholine (rubber chemicals; corrosion inhibitors; optical brighteners)	glycol ethers (solvents)
ethanolamines (liquid detergents; acid gas absorbent; textile chemicals)	
acetaldehyde (2.6) and vinyl acetate (2.2)	poly(vinyl acetate) (emulsion paints; adhesives; textile chemicals; paper and surface coatings)
	poly(vinyl alcohol) (adhesives; textile chemicals; coatings; emulsifiers; safety glass)
	poly(vinyl chloride)–acetate copolymers (phonograph records; floor tiles; solution coatings; PVC pipe)
acetic anhydride	cellulose acetate (fibers; cigarette filters; plastics)
	acetyl salicylic acid (aspirin)
acetic acid	acetate esters (solvents)
chloroacetic acid (herbicides)	cellulose ethers (detergent promoters; textile chemicals)
pentaerythritol (alkyd resins; synthetic lubricants)	
peracetic acid (bleaching agent; catalyst)	
Halogenation–hydrohalogenation (16.3)	
ethylene dichloride (14.7)	trichloroethylene (metal degreasing solvent)
	perchloroethylene (dry cleaning and metal cleaning solvent)
	methyl chloroform (metal degreasing and cleaning solvent)
vinyl chloride monomer	poly(vinyl chloride) (pipe and tubing; sheet; construction materials; textile and paper coatings; film; insulation and packaging materials)
vinylidene chloride	poly(vinylidene chloride) and copolymers (film; coatings for food packaging)
ethylenediamine (carbamate fungicides; chelating agent)	
ethyl chloride (1.4) (refrigerant; solvent; ethylation agent)	tetraethyllead (gasoline additive)

Table 14 (*continued*)

Process, petrochemical intermediates (uses)	Petrochemical products (uses)
ethylene dibromide (0.2) (lead scavenging agent)	vinyl bromide (flame retardant for fibers and plastics)
ethyl bromide (<0.1) (ethylation agent; solvent; refrigerant)	
Alkylation (9.3)	
ethylbenzene (9.1), styrene	polystyrene-GP, high, expanded (packaging; houseware; appliances; TV cabinets; furniture; lighting; construction; toys)
	acrylonitrile–butadiene–styrene copolymers (pipe and fittings; appliances; automotive use; business machines; telephones; electrical and electronic application)
	styrene–butadiene latexes (textile and paper coatings; emulsion paint; adhesive foam products)
	styrene–acrylonitrile copolymers (appliances; automotive use; houseware; furniture; packaging)
	styrene–butadiene elastomers (tires; mechanical rubber goods; footwear; adhesives; chewing gum)
	unsaturated polyesters (reinforced boats, bathtubs, tanks, panels, and trays; nonreinforced decorative and electrical applications)
ethyltoluene (0.1), vinyltoluene	unsaturated polyesters (coating resins; electrical insulation)
ethyl anilines (<0.1) (dyes; pharmaceuticals; pesticides)	
1,4-hexadiene (<0.1)	ethylene–propylene terpolymers (automotive parts; tires; hose and tubing; wire and cable)
aluminum alkyls (<0.1) (reaction promotors, initiators, catalysts)	
Oligomerization (4.0)	
α-olefins (1.5) (plasticizers; surfactants)	
linear primary alcohols (2.5) (plasticizers; surfactants)	
Hydration (3.5)	
ethanol (3.5) (toiletries; cosmetics; industrial solvents and thinners; detergents; flavors; disinfectants)	single-cell protein–yeast (animal feed)
	ethyl acetate (solvent)
ethyl amines (corrosion inhibitor; rubber chemicals; insecticides)	
acetaldehyde	
ethyl ether (solvent; medicinals; pharmaceuticals)	
Oxo-reaction (0.5)	
propionaldehyde (0.5)	
propionic acid (grain preservatives; cellulose plastics; herbicides)	
n-propyl alcohol (herbicides; solvent)	
Other (<0.01) (ripening agent for fruit and vegetables; herbicide)	

[a] Numbers indicate percent of total 1976 United States ethylene consumption.

Table 15. Major Ethylene Uses[a], Percent of Total 1976 Consumption

Derivative	United States	Western Europe	Japan
low density polyethylene	26.9	36.5	26.6
ethylene oxide	18.8	13.4	13.8
ethylene dichloride	14.8	17.2	16.0
high density polyethylene	14.8	15.3	13.0
ethylbenzene	9.1	7.7	9.3
others	15.6	9.9	21.3

[a] Ref. 174.

projects, will be reached, as the small populations of the OPEC countries would not be able to absorb them.

If these plans mature, a slate of derivatives requiring in the order of $(2.2–2.5) \times 10^6$ t of ethylene would be produced, mostly from ethane, in Saudi Arabia, Iraq, and Qatar. New markets will have to be found for these products.

The People's Republic of China will probably become self-sufficient in the production of petrochemicals within a decade.

Studies have shown that ethane can profitably be exported in LNG where markets for LNG might be developed (Japan) (192). The ethane would be separated at the LNG receiving terminal and processed for ethylene in a modified plant that would make profitable use of the LNG cold potential.

The Middle East OPEC nations will also have to find outlets for most of their LPG production. LPG can be marketed as propane–air and butane–air fuel in gas-starved areas such as South Africa, India, Japan, South America, and possibly the northeastern section of the North American continent and wherever population density and local climate create a demand for additional, environmentally compliant fuel. On a fuel value basis LPG costs far less to produce and transport than does LNG.

The world demand for ethylene is difficult to forecast beyond 1985. In Europe the feedstock situation seems to be assured beyond that time. In the United States feedstock supplies at favorable prices for ethylene capacity enlargement beyond 1983 appear to be geared to an increase of domestic crude oil production.

Future plants will be designed for greater flexibility to be able to use liquid feedstocks of different type and origin. There is also the possibility of a reversal to LPG as prime feedstock to plants in countries not too distant from sources of LPG as mentioned above. Plant size is not likely to increase beyond present day outputs from single-train plants anywhere in the world.

BIBLIOGRAPHY

"Ethylene" in *ECT* 1st ed., Vol. 5, pp. 880–898, by John Happel, New York University, and E. I. Becker, Polytechnic Institute of Brooklyn; "Ethylene" in *ECT* 2nd ed., Vol. 8, pp. 499–523, by D. L. Caldwell and I. Lichtenstein, The Lummus Company.

1. W. C. Fernelius, H. Wittcoff, and R. E. Varnerin, *J. Chem. Ed.* **56,** 385 (1979).
2. H. Benzler and A. von Koch, *Chem. Ing. Technol.* **27**(2), 71 (1955).
3. S. Angus and co-workers, *Ethylene, 1972 International Thermodynamic Tables of Fluid State*, International Union of Pure and Applied Chemistry, Butterworths, London, Eng., 1972.

4. K. E. Starling, *Fluid Thermodynamic Properties for Light Hydrocarbon Systems,* Gulf Publishing Company, Houston, Tex., 1973.
5. D. R. Douslin and R. H. Harrison, *J. Chem. Thermodyn.* **8,** 301 (1976).
6. R. H. Harrison and D. R. Douslin, *J. Chem. Eng. Data* **22**(1), 24 (1977).
7. F. S. Bonscher, L. M. Shipman, and L. C. Yen, *Hydrocarbon Process.* **53**(1), 145 (1974).
8. S. A. Miller, *Ethylene and Its Industrial Derivatives,* Ernest Benn, Ltd., London, Eng., 1969, p. 335.
9. J. C. Davis, *Chem. Eng.* **85**(1), 25 (1978).
10. Ref. 8, p. 437.
11. Ref. 8, p. 413.
12. A. M. Brownstein, *Chem. Eng. Prog.* **71**(9), 72 (1975).
13. U.S. Pat. 3,848,012 (Nov. 12, 1974), Applegath and co-workers (to Monsanto).
14. A. C. MacFarlane, *Oil Gas J.* **74**(6), 99 (1976).
15. F. G. Dwyer, P. J. Lewis, and F. H. Schneider, *Chem. Eng.* **83**(1), 90, (1976).
16. P. J. Lewis and F. G. Dwyer, *Oil Gas J.* **75**(40), 55 (1977).
17. S. A. Miller, *Ethylene and Its Industrial Derivatives,* Ernest Benn, Ltd., London, Eng., 1969.
18. P. H. Calderbank and co-workers, *J. Appl. Chem. (London)* **7,** 425 (1957).
19. R. M. Deanesly and C. H. Watkins, *Chem. Eng. Prog.* **27,** 134 (1951).
20. H. K. Kamptner, *Hydrocarbon Process.* **45**(4), 187 (1966).
21. S. C. Eastwood and A. E. Potas, *Pet. Eng.* **19**(12), 43 (1958).
22. J. M. Reid and co-workers, *Inst. Gas Technol Res. Bull.* (16), (Nov. 1963).
23. P. W. Sherwood, *Petroleum (London)* **19**(161), 309 (1956).
24. M. O. Kilpatrick and co-workers, *Pet. Refiner* **33**(4), 171 (1954).
25. T. F. Loughry, *Gas (Los Angeles)* **29**(11), 27 (1953).
26. C. H. Chilton, *Chem. Eng.* **66**(13), 37 (1959).
27. P. Schmalfeld, *Hydrocarbon Process. Pet. Refiner* **42**(7), 145 (1963).
28. *Chem. Eng. News* **37**(46), 50 (1959).
29. A. Steinhofer, O. Frey, and H. Nonnenmacher, *Hydrocarbon Process. Pet. Refiner* **42**(7), 119 (1963).
30. J. F. Farnsworth and co-workers, *Ind. Eng. Chem.* **47,** 1517 (1955).
31. G. L. Fleming, *Chem. Eng. Progr.* **52,** 249 (1956).
32. M. Bogart and R. Long, *Chem. Eng. Prog.* **58,** 90 (1962).
33. E. Mosberger, *Chem. Ing. Technol.* **43**(3), 131 (1971).
34. J. D. Kearns, D. Milks, and G. R. Kamm, "Development of Scaling Methods for a Crude Oil Cracking Reactor Using Short Duration Test Techniques," *paper presented at ACS 175th National Meeting, Anaheim, Calif., Mar. 12–17, 1978.*
35. T. Ishikawa and R. G. Keister, "Advanced Cracking Reactor, a Petrochemical Alternative," *paper presented at the NPRA National Meeting, San Antonio, Tex., Apr. 4, 1978.*
36. T. Ishikawa and R. G. Keister, *Hydrocarbon Process.* **57**(12), 109 (1978); *Oil Gas J.* **75**(49), 34 (1977).
37. S. C. Stinson, *Chem. Eng. News* **57**(22), 32 (1979).
38. *Hydrocarbon Process.* **50**(11), 140 (1971).
39. H. J. Andrews and L. W. Pollock, *Pet. Refiner* **31,** 154 (1952).
40. *Chem. Week* **107**(17), 95 (1963).
41. U.S. Pat. 3,842,138 (Oct. 15, 1974), E. Chahvekilian and J. M. Plichon (to Pierrefitte-Auby).
42. C. Barre, E. Chahvekilian, and R. Dumon, *Hydrocarbon Process.* **55**(11), 176 (1976).
43. E. Chahvekilian, "Olefins by Pyrolytic Cracking of Hydrocarbons Under Pressure and in the Presence of Hydrogen," *paper presented at AIChE 83rd National Meeting, Houston, Tex., Mar. 1977.*
44. F. O. Rice and K. F. Herzfeld, *J. Am. Chem. Soc.* **56,** 284 (1934).
45. S. B. Zdonik, E. J. Green, and L. P. Hallee, "Manufacturing Ethylene," The Petroleum Publishing Company, Tulsa, Oklahoma, reprinted from *Oil Gas J.,* (1966–1970).
46. H. G. Davis and K. D. Williamson, *Proceedings of the Fifth World Petroleum Congress, New York, May 31–June 5, 1959,* Vol. IV, 37 (1960).
47. K. J. Laidler and B. W. Wojciechowski, *Proc. Roy. Soc. Ser. A.* **260,** 91 (1961).
48. R. H. Snow, *J. Phys. Chem.* **70,** 2780 (1966).
49. M. C. Lin and M. H. Back, *Can. J. Chem.* **44,** 2357 (1966).
50. *Ibid.,* p. 2369.
51. Ref. 45, Chapt. 5.

52. J. N. Bradley and M. A. Frend, *J. Phys. Chem.* **75,** 1492 (1971).
53. P. D. Pacey and J. H. Purnell, *Ind. Eng. Chem. Fundam.* **11,** 233 (1972).
54. J. J. Dunkleman and L. F. Albright, "Industrial and Laboratory Pyrolysis," *ACS Symposium Series 32,* American Chemical Society, Washington, D.C., 1976, Chapt. 14.
55. J. E. Taylor and D. M. Kulich in ref. 54, Chapt. 5.
56. K. D. Williamson and H. G. Davis in ref. 54, Chapt. 4.
57. L. L. Ross and W. R. Shu, *Oil Gas J.* **75**(43), 58 (1977).
58. K. M. Sundaram and G. F. Froment, *Ind. Eng. Chem. Fundam.* **17**(3), 174 (1978).
59. H. C. Schutt, *Chem. Eng. Prog.* **55**(1), 68 (1959).
60. A. G. Volkan and G. C. April, *Ind. Eng. Chem. Process Des. Dev.* **16**(4), 429 (1977).
61. L. F. Albright, *Ind. Eng. Chem. Process Des. Dev.* **17**(3), 377 (1978).
62. R. J. Laidler, N. H. Sagert, and B. W. Wojciechowski, *Proc. Roy. Soc. Ser. A.* **270,** 242 (1962).
63. L. S. Kershenbaum and J. J. Martin, *AIChE. J.* **13,** 148 (1967).
64. T. Kunugi and co-workers, *Ind. Chem. Eng.* **7,** 550 (1967).
65. A. C. Buekens and G. F. Froment, *Ind. Eng. Chem. Proc. Des. Dev.* **7,** 435 (1968).
66. B. L. Crynes and L. F. Albright, *Ind. Eng. Chem. Process Des. Dev.* **8,** 25 (1969).
67. T. Haraguchi and co-workers, *Kagaku Kogaku* **33**(8), 786 (1969).
68. K. Kubota and N. Morita, *Kagaku Kogaku* **34**(6), 612 (1970).
69. G. E. Herriott and co-workers, *AIChE J.* **18,** 84 (1972).
70. D. Edelson and D. L. Allara, *AIChE J.* **19,** 638 (1973).
71. K. Kubota, *J. Chem. Soc. (Japan) Ind. Chem. Sect.* (8), 1393 (1974).
72. J. J. Dunkleman and L. F. Albright, "Industrial and Laboratory Pyrolysis," *ACS Symposium Series 32,* American Chemical Society, Washington, D.C., 1976, Chapt. 15.
73. J. E. Blakemore, J. R. Barker, and W. H. Corcoran, *Ind. Eng. Chem. Fundam.* **12**(2), 147 (1973).
74. D. R. Powers and W. H. Corcoran, *Ind. Eng. Chem. Fundam.* **13,** 351 (1974).
75. D. L. Allara and D. Edelson, *Int. J. Chem. Kinet.* **7,** 479 (1975).
76. A. G. Buekens and G. F. Froment, *Ind. Eng. Chem. Process Des. Dev.* **10**(3), 309 (1971).
77. G. F. Froment and co-workers, *AIChE J.* **23,** 93 (1977).
78. S. S. Levuch and co-workers, *Neftekhimiya* **10**(5), 656 (1970).
79. D. Kunzru, Y. T. Shah, and E. B. Stuart, *Ind. Eng. Chem. Process Des. Dev.* **12**(3), 339 (1973).
80. P. A. Longwell and B. H. Sage, *J. Chem. Eng. Data* **5,** 322 (1960).
81. V. V. Babash, T. N. Mukhina, and M. E. Aerov, *Int. Chem. Eng.* **15**(2), 368 (1975).
82. K. Kubota and N. Morita, *J. Chem. Soc. (Japan)* **72,** 616 (1969).
83. M. Murata, N. Takeda, and S. Saito, *J. Chem. Eng. (Japan)* **7,** 286 (1975).
84. L. Szepesy, K. Welther, and O. Szalai, *Hung. J. Ind. Chem. Veszprem* **5,** 161 (1977).
85. "Standard Method of Test for Carbon Distribution and Structural Group Analysis of Petroleum Oil by the n-d-M Method," *Annual Book of ASTM Standards,* Part 25, American Society for Testing and Materials, Philadelphia, Pa., 1976, p. 113.
86. M. Hirato, S. Yoshioka, and M. Tanaka, *Hitachi Rev.* **20**(8), 325 (1971).
87. M. Hirato and S. Yoshioka, *Int. Chem. Eng.* **13**(2), 347 (1973).
88. V. Illes, O. Szalai, and Z. Csermely in ref. 72, Chapt. 24.
89. L. Szepesy, K. Welther, and O. Szalai, *Hung. J. Ind. Chem. Veszprem* **5,** 233 (1977).
90. H. G. Davis and R. G. Keister in ref. 72, Chapt. 22.
91. S. B. Zdonik, E. J. Bassler, and L. P. Hallee, *Hydrocarbon Process.* **53**(2), 73 (1974).
92. W. R. Shu and L. L. Ross "A Feed Decomposition Model for Naphtha Pyrolysis," *paper presented at AIChE 71st Annual Meeting, Miami Beach, Florida, Nov. 12–16, 1978.*
93. E. J. Green, S. B. Zdonik, and L. P. Hallee, *Hydrocarbon Process.* **54**(9), 164 (1975).
94. B. Lohr and H. Dittman, *Oil Gas J.* **75**(28), 53 (1977).
95. E. Lassmann and H. J. Wernicke, *Oil Gas J.* **77**(2), 95 (1979).
96. J. M. Fernandez-Baujin and S. M. Solomon in ref. 72, Chapt. 20.
97. W. R. Shu, L. L. Ross, and K. H. Pang, "A Naphtha Pyrolysis Model for Reactor Design," *paper presented at AIChE 85th National Meeting, Philadelphia, Pa., June 4–5, 1978.*
98. Ref. 45, Chapt. 6.
99. R. H. Witt and F. M. Wall, "Theoretical Analysis of Gas Oil Pyrolysis and Product Yield Correlation," *paper presented at AIChE 70th Annual Meeting, New York, Nov. 13–17, 1977.*
100. J. J. Leonard, J. E. Gwyn, and G. R. McCullogh in ref. 72, Chapt. 18.
101. B. P. Ennis, R. Orriss, and H. B. Boyd, "High Temperature, Low Contact Time Pyrolysis," *Symposium Series 43,* Institute of Chemical Engineers, Harrogate, Eng., June 1975.

102. C. T. Adams and C. A. Trevino in ref. 72, Chapt. 23.
103. I. Lichtenstein, *Chem. Eng. Prog.* **60**(12), 64 (1964).
104. H. A. J. Vercammen and G. F. Froment, "Chemical Reaction Engineering-Houston," *ACS Symposium Series 65,* American Chemical Society, Washington, D.C., 1978, Chapt. 22.
105. A. G. Goossens, M. Dente, and E. Ranzi, *Oil Gas J.* **76**(36), 89 (1978).
106. A. G. Goossens, M. Dente, and E. Ranzi, *Hydrocarbon Process.* **57**(9), 227 (1978).
107. J. M. Fernandez-Baujin and A. J. Gambro, "Technology and Economics for Modern Olefins Plants," *paper presented at the Fourth Interamerican Congress of Chemical Engineers, Caracas, Venezuela, July 13–16, 1975.*
108. S. M. Solomon, "Developments in Olefins Production Technology," *paper presented at the Institution of Chemical Engineers, London Congress, London, Eng., May 11–12, 1977.*
109. U.S. Pat. 3,274,978 (Sept. 27, 1966), E. H. Palchik, T. F. O'Sullivan, and W. Tucker (to C-E Lummus).
110. J. Chen and M. J. Maddock, *Hydrocarbon Process.* **52**(5), 147 (1973).
111. J. L. James, M. E. Hopkins, and R. Orriss, "The Optimization of Pyrolysis Furnace Yields by Temperature Profile Control," *paper presented at Journee d'Etude sur le Steam Cracking, Association Francaise des Technicians du Petrole, Paris, Fr., Nov. 6, 1972.*
112. H. P. Leftin, D. S. Newsome, and T. J. Wolff in ref. 72, Chapt. 21.
113. A. Mol, *Hydrocarbon Process.* **53**(7), 115 (1974).
114. J. DeBlieck and A. Mol, *Pet. Int.* **15**(3), 50 (1975).
115. A. J. Gunning and C. M. Robinson, "The Effect of Increased Energy Costs on the Design of High Temperature Furnaces," *paper presented at the University of Salford, the North Western Branch of the Institution of Chemical Engineers, January 7, 1976.*
116. K. D. Demarest, *Chem. Eng. Prog.* **67**(1), 57 (1971).
117. J. L. DeBlieck and A. G. Goossens, *Hydrocarbon Process.* **50**(3), 76 (1971).
118. B. Lohr and H. Dittman, *Oil Gas J.* **75**(43), 65 (1977).
119. T. Bailey and F. M. Wall, *Chem. Eng. Prog.* **74**(7), 45 (1978).
120. T. Bailey and F. M. Wall, *Oil Gas J.* **75**(47), 172 (1977).
121. T. Sato and co-workers, *Chem. Econ. Eng. Rev.* **4**(7), 14 (1972).
122. T. Sato and co-workers, *Technical Review, Mitsubishi Heavy Industries, Ltd.* **10**(3), 35 (1973).
123. C. H. Tsai and L. F. Albright in ref. 72, Chapt. 16.
124. S. M. Brown and L. F. Albright in ref. 72, Chapt. 17.
125. L. F. Albright and C. F. McConnell, "Deposition and Gasification of Coke During Ethane Pyrolysis," *paper presented at ACS 175th National Meeting, Anaheim, Calif., Mar. 12–17, 1978.*
126. L. F. Albright, C. F. McConnell, and K. Welther, "Types of Coke Formed During the Pyrolysis of Light Hydrocarbons," *paper presented at ACS 175th National Meeting, Anaheim, Calif., Mar. 12–17, 1978.*
127. L. F. Albright and C. Y-H. Yu, "Pyrolysis of Acetylene, Butadiene, and Benzene in Different Tubular Reactors," *paper presented at ACS 175th National Meeting, Anaheim, Calif., Mar. 12–17, 1978.*
128. L. F. Albright, C. Y-H. Yu, and K. Welther, "Coke Formation During Pyrolysis Operations," *paper presented at AIChE 85th National Meeting, Philadelphia, Pa., June 4–8, 1978.*
129. Y. T. Shah, E. B. Stuart, and K. D. Sheth, *Ind. Eng. Chem. Process Des. Dev.* **15**(4), 518 (1976).
130. L. E. Chambers and W. S. Potter, *Hydrocarbon Process.* **53**(1), 121 (1974).
131. H. Smolen, "The Carburization and Coking of Reactor Tubes for High Temperature Pyrolysis of Hydrocarbons to Olefins Depending on Quality of Tube Materials, Furnace Heating and Process," *Symposium Series 43,* Institute of Chemical Engineers, Harrogate, Eng., June 1975.
132. *Product Bulletins G-262, G-5263, G-5264, G-5267, G-5268,* Nalco Chemical Company, Petroleum and Process Chemical Division, 1800 Esperson Building, Houston, Tex.
133. U.S. Pat. 3,144,080 (Aug. 11, 1964), F. Vollhardt (to Schmidt'sche Heissdampf G.m.b.H.).
134. U.S. Pat. 3,392,211 (July 9, 1968), K. Buschmann and co-workers (to BASF and Schmidt'sche Heissdampf G.m.b.H.).
135. K. Sato, *Chem. Econ. Eng. Rev.* **9**(6), 16 (1977).
136. B. M. Bergen and M. Asgari, *Oil Gas J.* **76**(36), 86 (1978).
137. R. Saporita and R. McCue in ref. 136, p. 69.
138. T. Perkins and L. Lindsay, *Oil Gas J.* **75**(50), 101 (1977).
139. L. Kniel and W. H. Slager, *Chem. Eng. Prog.* **43**(7), 335 (1947).
140. M. Picciotti, *Hydrocarbon Process.* **56**(5), 183 (1977).

141. S. P. Ho, U. K. Im, and G. E. Tacquart, "Modelling and Optimization of the Oil–Water Quench System in an Olefins Unit," *paper presented at 70th Annual AIChE Meeting, New York, Nov. 13–17, 1977.*
142. M. Raab, *Linde Rep. Sci. Technol.* **23,** 23 (1976).
143. H. Baldus and W. Foerg, *Linde Rep. Sci. Technol.* **9,** 8 (1966).
144. A. J. Gambro, K. Muenz, and M. A. Abrahams, *Hydrocarbon Process.* **51**(3), 73 (1972).
145. K. Stork, M. A. Abrahams, and A. Rhoe, *Hydrocarbon Process.* **53**(11), 157 (1974).
146. S. Sinkar, *Oil Gas J.* **74**(8), 103 (1976).
147. U.S. Pat. 3,720,729 (Mar. 13, 1973), M. C. Sze and N. C. Kafes (to C-E Lummus).
148. U.S. Pat. 3,781,195 (Dec. 25, 1973), D. T. Davis, T. G. Glover, and J. R. Jones (to B.P. Chemicals International, Ltd.).
149. S. B. Zdonik and co-workers, *Hydrocarbon Process.* **55**(1), 149 (1976).
150. O. Winter and M. T. Eng, *Hydrocarbon Process.* **55**(11), 125 (1976).
151. R. R. Hudgins, "Catalytic Dehydration of Ethanol Using a Frequency Response Technique," dissertation, Princeton University, Princeton, N.J., June 1964.
152. M. E. Winfield, *Catalysts,* Vol. 7, Reinhold Publishing Co., New York, 1960, pp. 93–183.
153. O. Winter, "Ethylene from Ethanol," *paper presented at First Brazilian Petrochemical Congress, Rio de Janeiro, Brazil, Nov. 7–12, 1976.*
154. A. M. Brownstein, *Trends in Petrochemical Technology,* Petroleum Publishing Co., Tulsa, Oklahoma, 1976, pp. 69–76, 127.
155. B. Buessemeier, C. D. Frohning, and B. Cornils, *Hydrocarbon Process.* **55**(11), 105 (1976).
156. U.S. Pat. 3,911,041 (Oct. 7, 1975), (to Mobil Oil Corporation).
157. *Eur. Chem. News* **18** (special issue: storage and distribution), (1972).
158. A. L. Waddams and G. Cann, *Hydrocarbon Process.* **53**(11), 135 (1974).
159. K. Ito, *Chem. Econ. Eng. Rev.* **6**(2), 19 (1974).
160. A. Sike, "Maritime Gas Transport in the Development of the World Chemical Industry," *paper presented at World Congress of the Societe de Chimie Industrielle, Valley Forge, Pa., Oct. 18–21, 1976.*
161. R. J. Lackey, "The Imco Code for Gas Tankers: A Review of the Finalized Code," *Proceedings, Gastech 75 LNG & LPG Technology Congress, Paris, France.*
162. *Eur. Chem. News* **25,** 10 (1974).
163. Private communication, Nov. 1978.
164. H. G. B. Nolan, *Oil Gas Int.* **9**(6), 82 (1969).
165. R. Jozon, *Oil Gas Int.* **10**(6), 109 (1970).
166. K. S. Wardale, *Chem. Eng.,* (Nov. 1967).
167. H. C. Schutt, *Oil Gas J.* **56**(31), 74 (1958).
168. A. T. Zeiner, *Eur. Chem. News* **15,** (1969).
169. W. B. Clapp and L. F. Litzinger, *Pipeline Gas J.* **198**(7), 72 (1971); W. H. Litchfield and co-workers, *Chem. Eng. Prog.* **55**(4), 68 (1959).
170. Chicago Bridge & Iron Co., private communication, Apr. 1977.
171. *Rev. Assoc. Fr. Tech. Pet.* **202,** 25 (1970).
172. R. R. Quirio and R. R. Hultin, "Ethylene Storage Near Midland, Michigan," *paper presented at Symposium on Salt, North Ohio Geological Society, Cleveland, Ohio, 1963.*
173. "Ethylene" in *Chemical Economics Handbook,* Stanford Research Institute, Menlo Park, Calif., Jan. 1978.
174. *World Petrochemicals,* SRI International, Menlo Park, Calif., 1978, pp. 3–4, 3–5, 3–9.
175. *Chemical Pricing Pattern, The Chemical Marketing Newspaper,* Schnell Publishing Company, New York, 1971.
176. *Chemical Economics Handbook,* Stanford Research Institute, Menlo Park, Calif., Jan. 1978.
177. *Oil Gas J.,* Petroleum Publishing Company, Tulsa, Oklahoma.
178. *Eur. Chem. News,* IPC Industrial Press, Ltd., London, Eng.
179. C-E Lummus, in-house information.
180. W. Tucker and M. A. Abrahams, "Heavier Feedstock Trends: The Changing Economics of Petrochemical Production," *paper presented at the World Congress of the Societe de Chimie Industrielle, Valley Forge, Pa., Oct. 1976.*
181. A. Halm, A. Chaptal, and T. Sialelli, *Hydrocarbon Process.* **54**(2), 89 (1975).
182. "Petroleum Products and Lubricants II," *Annual Book of ASTM Standards,* Part 24, American Society for Testing and Materials, Philadelphia, Pa., 1976, p. 412.

183. *Ibid.,* p. 415.
184. *Ibid.,* p. 712.
185. "Soap; Engine Coolants; Polishes; Halogenated Organic Solvents; Activated Carbon; Industrial Chemicals," *Annual Book of ASTM Standards,* Part 30, American Society for Testing and Materials, Philadelphia, Pa., 1976, p. 624.
186. *Chemical Safety Data Sheet SD-100, Properties and Essential Information for Safe Handling and Use for Ethylene,* Manufacturing Chemists Assoc., Washington, D.C., 1973.
187. W. W. Lawrence and S. E. Cook, "The Thermal Decomposition of Ethylene," *paper presented at Loss Prevention Symposium, Houston, Tex., 1967.*
188. G. S. Scott and co-workers, *U.S. Bur. Mines Rep. Invest.* **6659**(8), (1965).
189. F. F. McKay and co-workers, *Hydrocarbon Process.* **56**(11), 487 (1977).
190. N. R. Passow, *Chem. Eng.* **85**(26), 173 (1978).
191. *Oil Gas J.* **76**(26), 122 (1978).
192. L. Kniel, "Utilization of the LNG Cold Potential With Reference to the Manufacture of Ethylene," *paper presented at International Conference on Cryogenics, London, Eng., Mar. 28, 1969.*

LUDWIG KNIEL
OLAF WINTER
CHUNG-HU TSAI
Combustion Engineering Inc.

ETHYLENE GLYCOL, CH_2OHCH_2OH.   See Glycols.

ETHYLENEIMINE, $HN(CH_2)_2$.   See Imines.

ETHYLENE OXIDE

Ethylene oxide [75-21-8] (oxirane) was first prepared by Wurtz in 1859, by the reaction of 2-chloroethanol (ethylene chlorohydrin) with aqueous potassium hydroxide (1) (see Epoxidation). He also tried to prepare ethylene oxide by direct oxidation, without success (2). Many other investigators also tried and failed (3–6) before Lefort succeeded in the direct oxidation of ethylene to ethylene oxide over a silver catalyst (7–8). Early manufacture of ethylene oxide was by the chlorohydrin process; since 1940 that process has been almost completely replaced by direct oxidation. Total annual United States sales value of ethylene oxide exceeds 10^9 making it one of the most significant organic chemical products. Because of its high chemical reactivity, most ethylene oxide produced is converted to other compounds, especially ethylene glycol and surfactants (qv).

Physical Properties

Ethylene oxide is a colorless gas condensing at low temperatures to a mobile liquid. It is miscible in all proportions with water, alcohol, ether, and most organic solvents. Its vapors are flammable and explosive. The physical properties of ethylene oxide are summarized in Tables 1–7.

Structure. Ethylene oxide's strained configuration has been a subject for bonding and molecular structure studies. Valence bond and early molecular-orbital studies have been reviewed (25). More recent work uses the INDO (intermediate neglect of differential overlap) method and LMO (localized molecular orbitals) for extensive calculations of electron distribution (26–28). LMO bond density maps show that the

Table 1. Some Physical Constants of Ethylene Oxide, C_2H_4O

Property	Value	Refs.
molecular weight	44.05	
boiling point, °C		
at 101.3 kPa (760 mm Hg)	10.4	9–10
Δbp/pressure at 100 kPa, K/kPa (K/mm Hg)	0.25 (0.033)	9
coefficient of cubical expansion at 20°C, per °C	0.00161	9
critical pressure, MPa (atm)	7.19 (71.0)	10
critical temperature, °C	195.8	10
dielectric constant at 0°C	13.71	11
dipole moment, C·m (Debyes)	6.34×10^{-30} (1.90)	11
explosive limits in air, vol %		
upper	100	9
lower	3	9
flash point, tag open cup, °C	<−18	9
freezing point, °C	−112.5	9
heat of combustion at 25°C, kJ/mol[a]	1306.04	12
heat of fusion, kJ/mol[a]	5.17	13
heat of solution in pure water at 25°C and constant pressure, kJ/mol[a]	6.3	14
ionization potential, J[a]		
experimental	$1.73–1.80 \times 10^{-18}$	15
calculated	1.65×10^{-18}	16
refractive index, n_D^7	1.3597	9, 17

[a] To convert J to cal, divide by 4.184.

Table 2. Physical Properties for Ethylene Oxide Liquid from −40 to +195.8°C

Temperature, °C	Vapor pressure[a], kPa[b]	Enthalpy[a], J/g[c] Liquid	Enthalpy[a], J/g[c] Vaporization	Density[d], kg/L	Heat capacity[d], J/(kg·K)[c]	Surface tension[e], mN/m (= dyn/cm)	Viscosity[f], mPa·s (= cP)	Thermal conductivity[d], W/(m·K)
−40	8.35	0	628.6	0.9488	1878	34.2	0.495	0.20[g]
−30	15.05							
−20	25.73	38.8	605.4	0.9232	1912	30.9	0.400	0.18[g]
−10	42.00							
0	65.82	77.3	581.7	0.8969	1954	27.6	0.325	0.16[g]
10	99.54							
20	145.8	115.3	557.3	0.8697	2008	24.3	0.26[g]	0.15[g]
30	207.7							
40	288.4	153.2	532.1	0.8413	2092	22[g]	0.23[g]	0.14[g]
50	391.7							
60	521.2	191.8	505.7	0.8108	2247	19[g]	0.20[g]	0.14[g]
70	681.0							
80	875.4	232.6	477.4	0.7794	2426	16.5[g]	0.17[g]	0.14[g]
90	1108.7							
100	1385.4	277.8	445.5	0.7443	2782	13.5[g]	0.16[g]	0.13[g]
120	2088	330.4	407.5	0.7052	3293[g]	11[g]		
140	3020	393.5	359.4	0.6609	4225[g]	8[g]		
160	4224	469.2	297.1	0.608		5.5[g]		
180	5741	551.2	222.5	0.533				
195.8	7191							

[a] Ref. 9.
[b] To convert kPa to mm Hg, multiply by 7.50.
[c] To convert J to cal, divide by 4.184.
[d] Refs. 18–19.
[e] Refs. 18, 20.
[f] Refs. 18–20.
[g] Calculated.

bond density is strongly polarized toward the oxygen atom (27). The maximum bond density is outside the CCO triangle, reminiscent of the bent bonds of the valence-bond formulation (29).

The structural parameters of ethylene oxide as determined by microwave spectroscopy and electron diffraction (30) are given in Table 8.

Chemical Properties

Ethylene oxide is a highly reactive molecule; industrially it is primarily a chemical intermediate for a wide variety of compounds (see Derivatives). The three-membered ring is opened in most of its reactions, except in the formation of oxonium salts with strong anhydrous mineral acids.

Reaction of ethylene oxide with many compounds having a labile hydrogen atom introduces the hydroxyethyl group:

$$RH + \underset{O}{CH_2\text{—}CH_2} \longrightarrow RCH_2CH_2OH$$

The terminal hydroxyl thus formed can react further with ethylene oxide:

$$RCH_2CH_2OH + n\ \underset{O}{CH_2\text{—}CH_2} \longrightarrow RCH_2CH_2O\text{−}(CH_2CH_2O)_n\text{−}H$$

Table 3. Physical Properties of Ethylene Oxide Vapor from 298 to 800 K

Temperature, K	Entropy[a,b], J/(mol·K)[c]	Heat of formation[b,d], kJ/mol[c]	Free energy of formation[a,b], kJ/mol[c]	Viscosity[e], μPa·s[f]	Thermal conductivity[g], W/(m·K)	Heat capacity[h], J/(mol·K)[c]
298	242.4	−52.63	−13.10			48.28
300	242.8	−52.72	−12.84	9.0	0.012	48.53
400	258.7	−56.53	1.05	13.5	0.025	61.71
500	274.0	−59.62	15.82	15.4	0.038[b]	75.44
600	288.8	−62.13	31.13	18.2	0.056[b]	86.27
700	302.8	−64.10	46.86	20.9	0.075[b]	95.31
800	316.0	−65.61	62.80		0.090[b]	102.9

[a] Ref. 21.
[b] Calculated.
[c] To convert J to cal, divide by 4.184.
[d] Refs. 12, 21.
[e] Refs. 18–20.
[f] 1 μPa·s = 10^{-5} P.
[g] Refs. 18–19.
[h] Ref. 22.

Table 4. Physical Properties of Aqueous Solutions of Ethylene Oxide

Ethylene oxide Wt %	Mol %	Specific gravity at 15/15°C[a]	Freezing point[b], °C	Boiling point[a], °C
0	0	1.000	0.0	100
2.5	1.0	0.9993	−0.9	70
5	2.1	0.9986	−1.6 (eutectic)	58
10	4.4	0.9973	5.6	42.5
15	6.7	0.9959	8.9	38
20	9.3		10.4	32
30	14.9		11.1 (max)	27
40	21.4		10.4	21
50	29.0		9.3	19
60	38.0		7.8	16
70	48.8		6.0	15
80	62.1		3.7	13
90	78.6		0.0	12
100	100		−112.5	10.4

[a] Ref. 9.
[b] Refs. 9 and 20.

Table 5. Flash Points of Ethylene Oxide–Water Solutions

Ethylene oxide, wt %	Flash point, closed cup, °C
1	31
3	3
5	−2

Table 6. Solubility of Gases in Ethylene Oxide, Henry's Constants, MPa[a,b]

Temperature, °C	Nitrogen	Argon	Methane	Ethane
0	289	171	63	8.6
25	221	144	62	11.0
50	189	131	62	13.3

[a] To convert MPa to atm, divide by 0.101.
[b] Ref. 23.

Table 7. Solubility of Ethylene Oxide in Water, mL Vapor[a]/mL Solvent[b]

Pressure, kPa[c]	Temperature, °C		
	5°C	10°C	20°C
20	45	33	20
27	60	46	29
40	105	76	49
53	162	120	74
67	240	178	101
80		294	134
93			170
101			195

[a] Reduced to 0°C and 101.3 kPa[c].
[b] Ref. 24.
[c] To convert kPa to mm Hg, multiply by 7.5.

Table 8. Structural Parameters of Ethylene Oxide[a]

	Bond distances, nm	Bond angles	
r_m (C—C)	0.1462	<HCH	116.9
r_m (C—H)	0.1086	<COC	61.62
r_m (C—O)	0.1428	θ[b]	21.6
		ϕ[c]	7.97

[a] Ref. 30.
[b] Angle of the C—C bond to the H$_2$C plane.
[c] (<OCC)/(2 − θ).

These reactions give a series of poly(ethylene glycol) [25322-68-3] derivatives of increasing chain length and water solubility. Complete reviews of ethylene oxide reactions are presented in refs. 31 and 32 (see also Polyethers).

Reduction. Catalytic or chemical reduction of ethylene oxide produces ethanol (qv) (33).

Clathrate Formation. Ethylene oxide forms a stable clathrate with water (20) (see Clathration). It is nonstoichiometric, with 6.38 to 6.80 molecules of ethylene oxide to 46 molecules of water in the unit cell (34). The maximum observed melting point is 11.1°C. An x-ray structure of the clathrate found it to be a type I gas hydrate, with six equivalent tetrakaidecahedral (14-sided) cavities fully occupied by ethylene oxide, and two dodecahedral cavities 20–34% occupied (35) (see also Hydrocarbons).

Polymerization. Low molecular weight polymers from ethylene oxide, poly(ethylene glycol), are formed by continuation of the reactions with water or alcohols. The average molecular weight of the final product can be varied from 200 to 14,000 by adjustment of reaction conditions and reactant ratios (36).

High polymers of average molecular weights from 90,000 to 4×10^6 are formed by a process of coordinate anionic polymerization (37). This reaction can be written, for a $FeCl_3$-initiated system (38):

$$FeCl(OR)_2 + CH_2\underset{O}{-}CH_2 \rightarrow CH_2\underset{O}{-}CH_2 \cdot FeCl(OR)_2 \rightarrow ROCH_2CH_2OFeClOR$$

The patent literature describes numerous organometallic compounds, alkaline earth compounds and mixtures as catalysts. This process is also important as the mechanism of formation of nonvolatile residue (nvr) (39) during ethylene oxide storage. The primary catalyst for nvr formation is rust; no inhibitor has been found.

Crown Ethers. Ethylene oxide forms cyclic oligomers (crown ethers) in the presence of fluorinated Lewis acids such as boron trifluoride, phosphorus pentafluoride, or antimony pentafluoride (see Chelating agents). Hydrogen fluoride is the preferred cocatalyst (40). Cyclic polyethers $(\text{-}CH_2CH_2O\text{-})_n$ are formed where $n = 2\text{-}11$. The presence of BF_4^-, PF_6^-, or SbF_6^- salts of alkali, alkaline earth, or transition metal cations directs the oligomerization to the cyclic tetramer, 1,4,7,10-tetraoxacyclododecane [294-93-9] (12-crown-4); pentamer, 1,4,7,10,13-pentaoxacyclopentadecane [33100-27-5] (15-crown-6); and hexamer, 1,4,7,10,13,16-hexaoxacyclooctadecane [17455-13-9] (18-crown-6), by a template effect. Each cation maximizes the formation of the crown ether that best encircles its ionic radius (41).

Other Chemical Reactions. With Water. Wurtz was the first to obtain ethylene glycol when heating ethylene oxide and water in a sealed tube (1). Lourenco first noted the formation of diethylene and triethylene glycols (42). This was the first synthesis of polymeric compounds of well-defined structure. Hydration is slow at ambient temperatures, but is speeded by heat or acid or base catalysts (Table 9). The anion of the catalyzing acid is relatively unimportant (50) (see Glycols).

With Alcohols. These reactions parallel those of ethylene oxide with water. The primary products are monoethers of ethylene glycol; secondary products are monoethers of poly(ethylene glycols) (36). Most are appreciably water soluble (see Glycols, glycol ethers).

Table 9. Rate Constants for the Hydrolysis of Ethylene Oxide [a]

Temperature, °C	Acidic k_a, L/(mol·min) [b]	Neutral $10^4 k_w$, min^{-1}	Basic $10^2 k_b$, L/(mol·min) [b]
20	0.32	0.22	0.34
30	1.00	0.55	1.0
40	2.5	1.9	3.06
60		11.9	17.0
80		60.6	77
113		510	
123		1170	
131		1730	

[a] Refs. 43–49.
[b] Extrapolated to zero salt concentration.

With Organic Acids and Anhydrides. The carboxyl group of an organic acid reacts with ethylene oxide to give initially the corresponding ethylene glycol monoester. This may react further with ethylene oxide to give a poly(ethylene glycol) ester, or with another acid to give the glycol diester. The diacid ester of ethylene glycol may be obtained directly by reaction of ethylene oxide with the acid anhydride. The reaction of ethylene oxide with dibasic acids such as terephthalic acid is one route to polyesters (qv) which are the base of a major part of the fibers industry (51) (see Fibers, chemical).

With Ammonia and Amines. Ethylene oxide reacts with ammonia to form a mixture of mono-, di-, and triethanolamines. The hydrogen atoms on the nitrogen are more reactive than those of the hydroxyl group (52). A small amount of water is essential to the reaction (53) (see Alkanolamines).

Complex nitrogen compounds are formed from alkylamines and ethylene oxide (54). Thus diethylamine and ethylene oxide react to yield diethylaminoethanol. The dialkylaminoethanols react further with ethylene oxide to give amine derivatives of poly(ethylene glycols):

$$R_2NCH_2CH_2OH + n\ CH_2\overset{O}{-\!-\!-}CH_2 \longrightarrow R_2NCH_2CH_2-(OCH_2CH_2)-OH$$

Primary and secondary aromatic amines react in the same way as the corresponding alkylamines.

With Hydrogen Sulfide and Mercaptans. Ethylene oxide reacts with hydrogen sulfide to yield 2-mercaptoethanol and thiodiglycol (bis-2-hydroxyethyl sulfide), a useful solvent (55–56). Reaction conditions determine the proportions of each derivative. Three moles of ethylene oxide react with 1 mole of hydrogen sulfide in water to form the strong base, tris(hydroxyethyl)sulfonium hydroxide [$(HOCH_2CH_2)_3S^1OH^-$]. Addition of ethylene oxide to long-chain alkyl mercaptans yields polyoxyethylene mercaptans, some of which are nonionic surfactants (57).

p-Oxathiane and p-dithiane are formed from ethylene oxide and hydrogen sulfide at 200°C in the presence of aluminum oxide as a catalyst (58) (see Sulfur compounds).

Friedel-Crafts Reactions. In the presence of aluminum chloride, ethylene oxide undergoes a Friedel-Crafts reaction (qv) with aromatic hydrocarbons. Thus ethylene oxide and hydrogen chloride passed into a mixture of benzene and aluminum chloride yields dibenzyl ($C_6H_5CH_2CH_2C_6H_5$) and a small amount of 2-phenylethanol ($C_6H_5CH_2CH_2OH$) (59).

With Acyl Halides, Hydrogen Halides, and Metallic Halides. Ethylene oxide reacts with acetyl chloride at slightly elevated temperatures using hydrogen chloride as a catalyst, to give the acetate of ethylene chlorohydrin (60). Hydrogen halides produce the corresponding halohydrins (61). Ethylene oxide, in aqueous solutions with certain metallic halides, reacts with the acid freed by dissociation and hydrolysis to precipitate the metallic hydroxide (62–63). The halides of aluminum, chromium, iron, thorium, and zinc in dilute solution react with ethylene oxide to form sols or gels of the metal oxide hydrates and ethylene halohydrin (64) (see Chlorohydrins).

Phosphorus oxychloride reacts with ethylene oxide in the presence of aluminum chloride to give tris-2-chloroethyl phosphate, a valuable plasticizer (65).

Phosgene adds to ethylene oxide and other alkylene oxides with the formation of esters of chlorocarbonic acid (66).

With Grignard Reagents (67). Ethylene oxide reacts with Grignard reagents, RMgX, to form primary alcohols containing two more carbon atoms, RCH_2CH_2OH, as the principal product (68–70) (see Grignard reaction).

With Compounds Containing Active Methylene or Methine Groups. Compounds containing active —CH_2— or —CH= groups, such as malonic and monosubstituted malonic esters, ethyl cyanoacetate, and β-keto esters, condense with ethylene oxide and other alkylene oxides to form a wide variety of compounds (61,71). Ethylene oxide and diethyl malonate, in the presence of sodium ethoxide, form principally diethyl (2-hydroxyethyl)malonate, which loses one molecule of water to form α-carbethoxy-γ-butyrolactone.

The sodium salt of ethyl acetoacetate in ethanol reacts at 0°C with ethylene oxide to give 2-acetyl-4-butyrolactone, an intermediate for vitamin B_1 and antimalarials (72).

With Phenols. The 2-hydroxyethyl aryl ethers are prepared from reaction of ethylene oxide with phenols at elevated temperatures and pressures (73–74). 2-Phenoxyethyl alcohol is a perfume fixative. The water-soluble alkylphenol ethers of the higher polyethylene glycols are important surface-active agents. They are made by adding ethylene oxide to the alkylphenol at ca 200°C and 200–250 kPa (2–2.5 atm), using sodium acetate or hydroxide as a catalyst. The properties of these alkylphenol ethers can be varied over a continuous spectrum of solubility and performance characteristics by variation of the alkyl chain(s) and the number of ethylene oxides added (75) (see Alkylphenols).

With Hydrogen Cyanide. Ethylene oxide reacts readily with hydrogen cyanide in the presence of alkaline catalysts, such as diethylamine, to give ethylene cyanohydrin. This is readily dehydrated to give acrylonitrile (qv) in 80–90% yield. Ethylene cyanohydrin can be hydrolyzed or esterified to give acrylic acid (qv) or alkyl acrylates:

$$HCN + \underset{O}{CH_2\text{---}CH_2} \longrightarrow HOCH_2CH_2CN$$

$$HOCH_2CH_2CN \longrightarrow CH_2\text{=}CHCN + H_2O$$

Acrylic acid is also obtained with aqueous sulfuric acid in the absence of alcohol (76).

With Carbonyl Compounds. Ethylene oxide reacts with ketones in the presence of metal halide catalysts, such as boron trifluoride or tin tetrachloride, under anhydrous conditions only, to give cyclic ketals (77). For example, ethylene oxide reacts with acetone to give 2,2-dimethyl-1,3-dioxolane. Considerable ethylene oxide polymer is formed as a by-product. Aldehydes react to give the corresponding acetals. These reactions have some use for formation of protecting groups in organic synthesis (78).

Miscellaneous Reactions. Ethylene oxide is considered a potential environmental pollutant. In the atmosphere it reacts in the normal photochemical cycle to form smog (79) (see Air pollution). Ethylene oxide is thermodynamically unstable, decomposing exothermically at >500°C in the absence of catalyst to methane, carbon monoxide, ethane, hydrogen, and carbon (80–81). The probable course for the pyrolysis is summarized in the key steps (82):

$$\underset{O}{CH_2\text{---}CH_2} \rightleftharpoons \cdot CH_2CH_2O\cdot \longrightarrow (CH_3CHO)^* \longrightarrow \cdot CH_3 + \cdot CHO$$

$$\cdot CH_3 + \cdot CHO \xrightarrow{\text{various radical recombinations}} CO + H_2 + CH_4 + C_2H_6$$

Isomerization of ethylene oxide to acetaldehyde occurs at 200–300°C in the presence of catalysts such as activated alumina, phosphoric acid, and metallic phosphates (83).

Diborane adds rapidly to ethylene oxide at −80°C to form diethoxyborane and a solid polymer containing about eight ethylene oxide units per molecule. Propylene oxide reacts in a similar manner. Both polymers contain boron and are of the type H$(\text{CHRCH}_2\text{O})_n\text{BH}_2$ (84) (see Boron compounds).

Potassium thiocyanate or thiourea reacts in aqueous solution with ethylene oxide to give ethylene sulfide (85).

Ethylene carbonate (1,3-dioxolan-2-one) is commercially prepared from ethylene oxide by addition of carbon dioxide with ammonium or alkali metal salts as catalysts (86):

$$\text{CH}_2\text{—CH}_2 + \text{CO}_2 \xrightarrow{R_4N^+Br^-} \text{ethylene carbonate}$$

Isocyanates similarly yield derivatives of 2-oxazolidinone (86), and carbon disulfide yields ethylene trithiocarbonate (87).

Many other reported reactions of ethylene oxide are only of laboratory significance. These include nucleophilic reactions with amides, alkali metal organic compounds, and pyridinyl alcohols (88); and electrophilic reactions with orthoformates, acetals, titanium tetrachloride, sulfenyl chlorides, halosilanes, and dinitrogen tetroxide (89).

Manufacture

The commercial production of ethylene oxide is by two basic routes: the ethylene chlorohydrin and direct oxidation processes. The first process is the older and involves the reaction of ethylene with hypochlorous acid followed by dehydrochlorination of the resulting chlorohydrin with lime to produce ethylene oxide and calcium chloride. The chlorohydrin process was first introduced during World War I in Germany by Badische Anilin-und Soda-Fabrik (BASF) and others (90). In 1925 Union Carbide Corporation was the first to commercialize this process in the United States. However, Dow Chemical's Freeport, Texas, facility is reportedly the only United States ethylene oxide plant now utilizing this chlorohydrin process (91).

The second process, which completely dominates the field today, is the direct oxidation technology. As the name implies, the fundamental reaction is the catalytic oxidation of ethylene with oxygen over a silver-based catalyst to yield ethylene oxide. Direct oxidation processes are divided into two categories depending on the source of the oxidizing agent: the air-based process and the oxygen-based process. In the first, air or air enriched with oxygen is fed directly to the system. In the second, a high purity oxygen stream (>95 mol %) from an air separation unit is employed as the source of the oxidizing agent. Union Carbide Corporation was the first to commercialize an air-based direct oxidation process in 1937, and the first oxygen-based system was

440 ETHYLENE OXIDE

commercialized by Shell Oil Company in 1958 (92). Extensive information on the early developments of the chlorohydrin and direct oxidation processes are reported in ref. 93.

There are thirteen producers of ethylene oxide in the United States. Table 10 shows the plant locations, estimated capacities, and types of processes employed. The total United States production capacity for 1978 was ca 2.7×10^6 metric tons. The percentages of total domestic production made by the air- and oxygen-based processes are ca 60 and 40%, respectively. The largest producer is Union Carbide Corporation with approximately 42% of the United States ethylene oxide capacity. The top four domestic producers, UCC, Dow, Celanese, and Jefferson, account for 68% of United States capacity. About 93% of domestic ethylene oxide capacity is located on the Gulf Coast or in Puerto Rico near secure and plentiful ethylene supplies. Most ethylene oxide is consumed by its producers in making derivatives. Two additional producers may soon have an impact upon the domestic ethylene oxide supply situation. Imperial Chemical Industries has commissioned design and engineering studies for a 200,000 t/yr ethylene oxide plant to be built in Texas (94–95). Oxirane Corporation began production of ethylene glycol in 1978 via the acetoxylation process at Channelview, Texas. This plant will have an ethylene glycol capacity equivalent to 261,000 t/yr of ethylene oxide representing ca 10% of United States domestic capacity (96).

Technologies for direct oxidation plants for ethylene oxide have been developed by several companies. Union Carbide Corporation and Dow Chemical use their own technologies. Shell Development and Scientific Design license ethylene oxide technology and over 55% of present world capacity is based on their processes. Huels, Snam Progetti, and Japan Catalytic also offer processes for licensing. Shell technology is solely oxygen-based, and Scientific Design offers both air- and oxygen-based processes (92).

Table 10. Domestic Producers, Capacities, Process Types and Technology Used for Ethylene Oxide[a]

Producer	Location	Capacity, 10^3 t/yr	Process oxidant	Technology
BASF Wyandotte	Geismar, La.	144	oxygen	Shell
Calcasieu	Lake Charles, La.	102	oxygen	Shell
Celanese	Clear Lake, Tex.	181	oxygen	Shell
Dow	Freeport, Tex.[b]	91	air	Dow
Dow	Plaquemine, La.	200	air	Dow
Eastman	Longview, Tex.	88	oxygen	Shell
Houston Chemical	Beaumont, Tex.	70	air	Scientific Design
Jefferson	Port Neches, Tex.	227	air	Scientific Design
Northern Petrochemical	Joliet, Ill.	100	oxygen	Scientific Design
Olin	Brandenburg, Ky.	50	oxygen	Shell
PPG	Guayanilla, P.R.	136	oxygen	Scientific Design
Shell	Geismar, La.	122	oxygen	Shell
Sun Olin	Claymont, Del.	45	oxygen	Shell
Union Carbide	Ponce, P.R.	250	air	Union Carbide
Union Carbide	Seadrift, Tex.	384	air	Union Carbide
Union Carbide	Taft, La.	500	air	Union Carbide
Total capacity		2690		

[a] Ref. 91 and miscellaneous sources.
[b] 82×10^3 t/yr additional chlorohydrin capacity available.

Direct Oxidation Processes. The phenomenal growth in United States and world ethylene oxide production capacity since 1940 and the marked trend towards larger single-train plants is chiefly owing to the great commercial success of the direct oxidation process. Compared to the chlorohydrin process, direct oxidation eliminates the need for large volumes of chlorine, there are no chlorinated hydrocarbon byproducts to be sold, and processing facilities can be made simpler and operating costs are lower (97). The main disadvantage of the direct oxidation process is the lower yield or selectivity of ethylene oxide per unit of feed ethylene consumed. The major inefficiency in the process results from the loss of ca 25–30% of the ethylene to carbon dioxide and water. Consequently, operating conditions must be carefully controlled to maximize selectivity.

All ethylene oxide direct-oxidation plants are based on the original process chemistry discovered by Lefort in 1931 (7–8). Although the reactions involved in the direct oxidation process have been the subject of extensive research and several patents, the precise mechanism is not fully understood. According to the best current understanding of the mechanism, ethylene oxide is formed when ethylene reacts with one atom of a diatomic oxygen species adsorbed on a silver surface. Recent work, reviewed in ref. 98, provides the basis for the following discussion.

Oxygen is strongly adsorbed on a silver surface; ethylene is not (99). There are three types of adsorption of oxygen on silver. The first is a dissociative adsorption of oxygen on a clean silver surface:

$$O_2 + 4\ Ag(\text{adjacent}) \rightarrow 2\ O^{2-}_{ads} + 4\ Ag^+$$

The second is nondissociative. It is favored relative to the first when more than 25% of the surface is covered with chloride ions, which can be supplied by organic chlorine inhibitors:

$$O_2 + Ag \rightarrow O^-_{2\,ads} + Ag^+$$

The third, dissociative, occurs when more than 25% of the surface is covered with chloride ions but at higher temperatures than the previous two types:

$$O_2 + 4\ Ag(\text{nonadjacent}) \rightarrow 2\ O^{2-}_{ads} + 4\ Ag^+$$

Ethylene reacts with either the dissociated or the undissociated oxygen species. With undissociated (diatomic) oxygen, ethylene reacts to form ethylene oxide leaving one oxygen atom adsorbed on the surface:

$$C_2H_4 + O^-_{2\,ads} \rightarrow C_2H_4O + O^{2-}_{ads}$$

With dissociated (monatomic) oxygen, ethylene reacts to form carbon dioxide and water:

$$C_2H_4 + 6\ O^{2-}_{ads} \rightarrow 2\ CO_2 + 2\ H_2O$$

Since O^{2-}_{ads} must be removed from the silver surface before more $O^-_{2\,ads}$ can form, the two reactions above must be combined. The stoichiometry for the complete reaction is:

$$7\ C_2H_4 + 6\ O_2 \rightarrow 6\ C_2H_4O + 2\ CO_2 + 2\ H_2O$$

This leads to a limiting selectivity of 6/7 or 85.7% for the reaction forming ethylene oxide from ethylene and adsorbed oxygen. The actual selectivity is further limited by the reaction of ethylene oxide with oxygen to form carbon dioxide and water, and the formation of dissociated oxygen directly from molecular oxygen in two of the three reactions mentioned above:

$$\underset{O}{CH_2\text{---}CH_2} + 2\tfrac{1}{2}\,O_2 \longrightarrow 2\,CO_2 + 2\,H_2O$$

A large amount of heat is released by these reactions. At 600 K, each kg of ethylene converted to ethylene oxide releases 3.756 MJ (3564 Btu); each kg of ethylene converted to carbon dioxide and water releases 50.68 MJ (48,083 Btu). Small quantities of acetaldehyde and traces of formaldehyde are also produced. They generally total less than 0.1% of the ethylene oxide formed. Acetaldehyde is most likely formed by isomerization of ethylene oxide, formaldehyde by direct oxidation of ethylene (99).

Commercial processes operate under recycle conditions in a packed-bed, multitubular reactor. Reaction temperatures of 200–300°C are typical, and operating pressures of 1–3 MPa (10–30 atm) have been reported (92,97,100–102). The reactor is of the shell-and-tube type comprised of several thousand mild steel or stainless steel tubes, 20–50 mm inside diameter (92). Figure 1 is a schematic diagram of an oil-cooled reactor for the production of ethylene oxide. Based on published information regarding catalyst productivities and space velocities, the reactor tube lengths are 6–12 m (92,97,102–108). These tubes are filled with a silver-based catalyst ca 3–10 mm dia and supported on a carrier material with a surface area usually <1 m^2/g (92,97,100–101). The selectivity or yield (moles of product produced per moles of ethylene consumed) is normally 60–77% depending on catalyst type, per pass conversion, reactor design, and a large number of process operating variables.

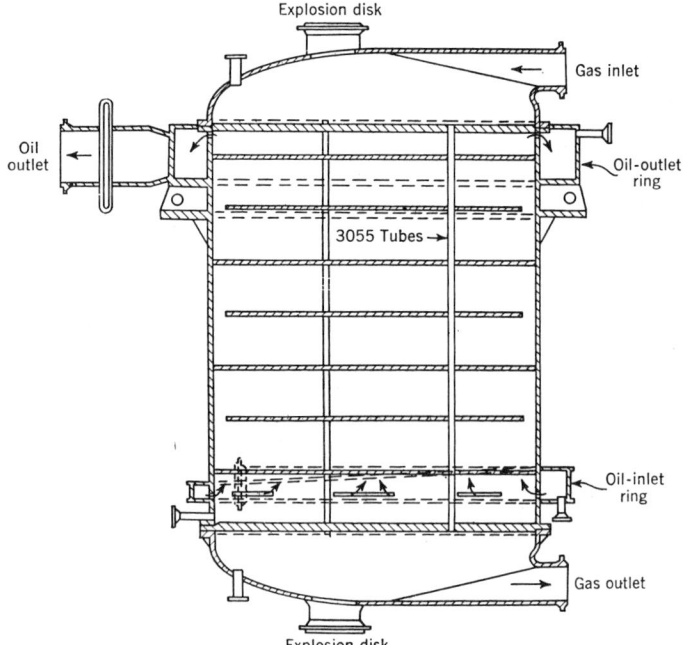

Figure 1. Oil-cooled reactor for the oxidation of ethylene to ethylene oxide.

Technological innovations in catalyst development and process design and engineering have enabled ethylene oxide manufacturers to meet the commercial needs for larger facilities without using a great number of reactors (97,109). For example, in the Shell oxygen-based process, it is claimed that individual reactor capacities have increased from a modest 9,000 t/yr in 1958, to 25,000 t/yr in 1968, to ca 75,000 t/yr in recent designs (109). There is a pronounced trend in both the United States and Europe towards larger single-train plant sizes. In the late 1950s a 30,000 t/yr ethylene oxide unit was considered large, whereas ten years later plant sizes of 100,000–150,000 t/yr were typical (110). Today some producers have plant capacities in excess of 250,000 t/yr.

Air-Based Direct Oxidation Process. A schematic flow diagram of the ethylene oxide process is shown in Figure 2. Published information on the detailed evolution of commercial ethylene oxide processes is very scanty, and Figure 2 does not necessarily correspond to the actual equipment or process employed in any modern ethylene oxide plant. Precise information regarding process technology is proprietary. However, Figure 2 does illustrate all the salient concepts involved in the manufacturing process. The process can be conveniently divided into three major sections: reaction system, oxide recovery, and oxide purification.

In the first section, compressed air is filtered, purified (if necessary), and fed separately with ethylene into a recycle gas stream. This recycle stream feeds a bank of one or more primary multitubular reactors which operate in parallel. The number of primary reactors used depends chiefly on the plant capacity, size of the individual reactors, and the activity of the catalyst used. The fresh air to C_2H_4 ratio is varied within limits in order to ensure, after dilution with recycle gas, a predetermined op-

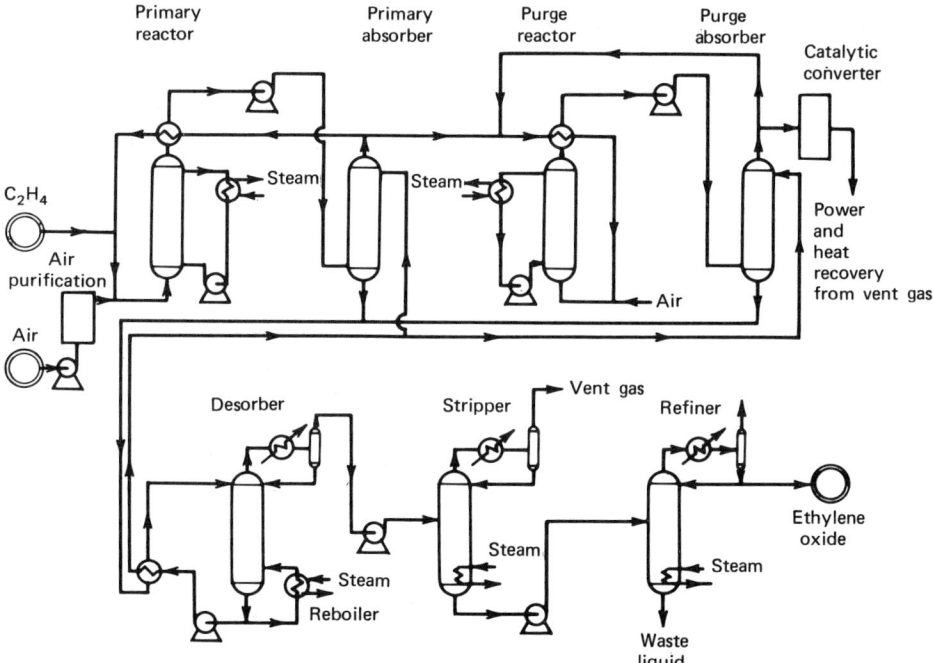

Figure 2. Air-based direct oxidation process for ethylene oxide (92,97,100–101,111–113).

timum $O_2:C_2H_4$ feed ratio to the silver-based catalyst. The ethylene is oxidized to ethylene oxide, carbon dioxide and water in the packed-bed converters, and the heat of reaction is removed by circulating or boiling an organic oil on the shell side (102) (see Heat exchange technology), eg, Dowtherm, tetralin, or other high boiling materials. The hot oil is cooled in a steam generator producing considerable amounts of high pressure steam for the ethylene oxide and other processes at the plant site (92,100).

The per pass ethylene conversion in the primary reactors is maintained at 20–50% in order to ensure catalyst selectivities of 63–75%. Vapor-phase oxidation inhibitors such as ethylene dichloride or vinyl chloride or other halogenated compounds are added to the inlet of the reactors in ppm concentrations to retard carbon dioxide formation (101,114–115). The process stream exiting the reactor may contain 1–2 mol % ethylene oxide. This hot effluent gas is then cooled in a shell-and-tube heat exchanger to around 35–40°C by using the cold recycle reactor feed stream gas from the primary absorber. The cooled crude product gas is then compressed in a centrifugal blower before entering the primary absorber.

The second important step of the process is ethylene oxide recovery from the crude product gas. This is accomplished in the primary absorber by countercurrent scrubbing with cold water in a packed column ca 18–20-m high. The ethylene oxide produced in the reactor is dissolved in the absorber water along with some carbon dioxide and traces of hydrocarbons and aldehydes (97). The aqueous stream is removed from the base of the absorber and sent to a desorber. The unabsorbed gas from the main absorber overhead is split into two portions. The largest portion is recycled to the primary reactor after it cools the hot product gas in the shell-and-tube heat exchanger, and the circulation cycle is repeated. A much smaller fraction of the primary absorber overhead gas stream is heat-exchanged to raise its temperature and it is then fed as the main stream to the secondary or purge reactor system. In the purge reactor, more air may be added to increase the oxygen content of the feed gas. The gases leaving the purge reactor are heat-exchanged against the recycle feed gas to the same reactor and then enter a purge absorber. In the purge absorber ethylene oxide is removed with water in the same manner as in the main absorber.

The chief purpose of the purge reactor system is to allow reaction of a substantial portion of the ethylene content of the purge gas, which must be vented from the main reactor system in order to prevent accumulation of inert gases, primarily nitrogen and carbon dioxide (100). Figure 2 shows a two-stage, air-based plant with a single purge reactor. In larger plants, three or more stages of reaction may be used to improve overall yield of product (100,107). In such cases the flow scheme is virtually the same, except additional purge reactor–absorber systems are added in series to the first purge reactor. The scrubbed gas from the last purge absorber may be partly recycled to the same purge reactor inlet or vented from the system.

In some cases the ethylene content of the vent gas leaving the last purge reactor makes it economical to further process this gas for energy recovery (100,102). Such a scheme not only extracts valuable power from the vent gas but also reduces considerably the hydrocarbon emissions from the process. Several such schemes have been described in refs. 102 and 116–119. The basic scheme involves heating the ethylene lean gas (<1.5 mol %) to ca 200°C and then passing it into a catalytic combustion chamber filled with an active oxidation catalyst containing a noble metal, eg, platinum. Such an active catalyst should burn virtually all of the ethylene and ethane in the vent gas, thereby raising its temperature to 400–600°C. The hot, pressurized gas expands in a turbine that is coupled to the feed gas compressor in the main process. The hot exhaust gases from the turbine are used to generate steam. Using such a scheme, a 10% reduction in the overall cost is claimed for the manufacture of ethylene oxide by the air-based process (119).

The third key section of the process deals with ethylene oxide purification. The ethylene oxide-rich water absorbent streams from both the main and purge absorbers are combined, and after heat exchange are fed to the top section of a desorber where the absorbate is steam-stripped under reduced pressure. The cycle water is thus stripped of its ethylene oxide content. The oxide is distilled at the top and is compressed for further purification. The lean water from the lower section of the desorber is virtually free of oxide and is recirculated to the main and purge absorbers.

The ethylene oxide recovered in the desorber still contains some carbon dioxide, nitrogen, aldehydes, and traces of ethylene and ethane, all of which were removed from the cycle gas in the absorbers. In the stripper the light gases are separated overhead and vented, and the partially purified ethylene oxide is sent from the bottom of the stripper to the mid-section of a final refining column. The ethylene oxide from the refining section should have a purity of >99.5 mol %. The final product is usually stored as a liquid under an inert atmosphere (see Storage).

The overall economics of the process are strongly dictated by the design of the reaction system and the actual operating conditions used. The catalyst properties, as they influence reactor design and operating variables are, therefore, of the greatest significance. Specific information on actual conditions employed in the manufacture is not disclosed. However, the general ranges suggested by literature and patent reviews are summarized in Table 11.

Oxygen-Based Direct Oxidation Process. Even though the fundamental reaction and the ultimate results are the same, there are substantial differences in detail between air- and oxygen-based processes. Virtually all the differences arise from the change in the oxidizing agent from air (ca 20 mol % O_2) to pure oxygen ($\geq$95 mol % O_2).

Table 11. Ranges of Reaction System Variables in the Direct Air-Oxidation Process for Ethylene Oxide[a]

Variable	Range
ethylene, mol %	2–10
oxygen, mol %	4–8
carbon dioxide, mol %	5–10
ethane, mol %	0–1.0
temperature, °C	220–277
pressure, MPa[b]	1–3
space velocity[c], h^{-1}	2000–4500
pressure drop, kPa[d]	41–152
conversion, %	20–65
selectivity or yield (mol basis, %)	63–75

[a] Refs. 97, 105–107.
[b] To convert MPa to psi, multiply by 145.
[c] The space velocity is defined as the standard volume of the reactant stream fed per unit time divided by the volume of reactor space filled with catalyst.
[d] To convert kPa to mm Hg, multiply by 7.5.

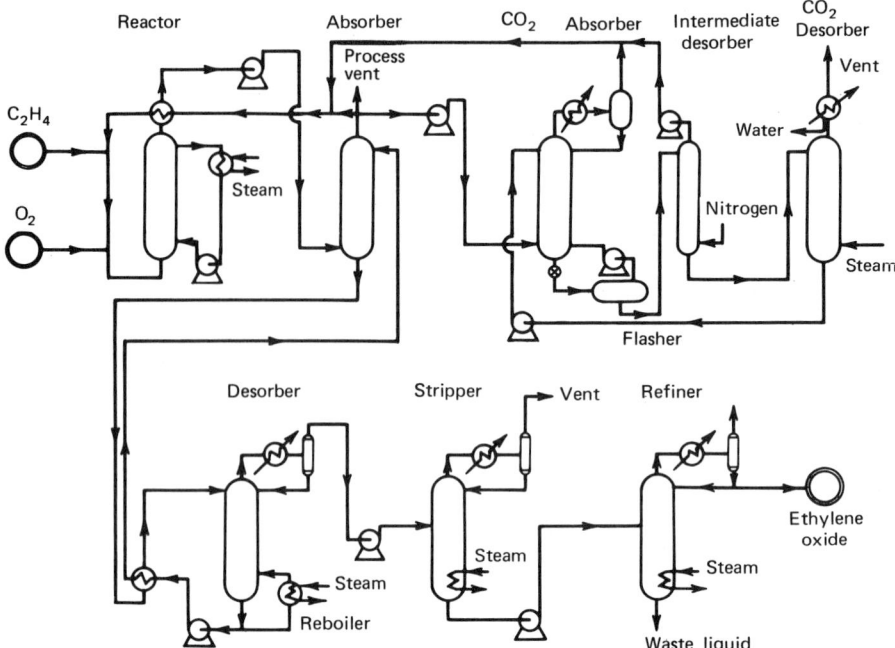

Figure 3. Oxygen-based direct oxidation process for ethylene oxide (92,97,100–101,111–113,121).

Owing to the low per-pass conversion, the need for complete removal of ethylene oxide by absorption, and the accumulation of nitrogen in the cycle, the air process requires a substantial purge stream. As a direct consequence of this purge stream, the air-based process requires the staged reaction–absorption system described earlier. The oxygen-based process uses substantially pure oxygen, reduces the quantities of inert gases introduced into the cycle and thereby results in almost complete recycle of the unconverted ethylene (92,100). This eliminates the need for a purge reactor system in an oxygen-based process. However, as in the air-based process, the volume of carbon dioxide formed is about half the volume of ethylene that reacts at a catalyst selectivity of 70–80%. This CO_2 must be eliminated on a continuous basis in order to control its concentration at a fixed acceptable level in the cycle. Concentrations of CO_2 much in excess of 15 mol % adversely influence catalyst activity (92,97). Therefore, in an oxygen-based system, part of the recycle gas leaving the absorber must be treated in a CO_2 removal unit before it is sent back to the main reaction cycle.

In addition to the CO_2 removal unit purge stream, an additional process vent is required to prevent accumulation of argon in the cycle. Argon is a major impurity in the oxygen supply, and can build up to levels of 30–40 mol % in the cycle gas if no deliberate purge is used (97). When this happens, the selectivity of the process decreases rapidly and the cycle gas, owing to the lower heat capacity of the argon, may enter the flammable region (120). In spite of this additional purge, the total vent stream in an oxygen-based process is much smaller than in an air-based unit. The operation of the main reactor can be at much higher ethylene concentration than that possible in the air-based process. The high ethylene concentration improves the catalyst selectivity because the per pass conversions are lower for a given ethylene oxide production (92,100–101). The small purge rates in an oxygen-based system operated with very high purity oxygen (99.0–99.5 mol %) make it possible to use cycle diluents of improved heat capacity other than nitrogen (110). These diluents facilitate the use of higher oxygen concentrations in the cycle and, therefore, improve selectivity (see Cycle Diluents).

Figure 3 shows a simple schematic diagram of an oxygen-based process. Ethylene, oxygen, and the recycle gas stream are combined before entering the tubular reactors. The basic equipment for the reaction system is identical to that described for the air-based process, with one exception: the purge reactor system is absent and a carbon dioxide removal unit is incorporated. The CO_2 removal scheme illustrated is based on a patent by Shell Oil Company (121), and minimizes the loss of valuable ethylene in the process.

The main process vent stream generally contains a fairly high hydrocarbon concentration, particularly if the diluent is not nitrogen. In such cases the purge stream can be readily used in a boiler or incinerated in a furnace without supplemental fuel (100,122). The ethylene oxide recovery and refining sections for both the air- and oxygen-based processes are almost identical. As with the air-based process, specific operating conditions for the reaction system are proprietary; however, the general ranges reported in the literature and patents are summarized in Table 12.

Process Technology Considerations. Innumerable complex and interacting factors ultimately determine the success or failure of a given ethylene oxide process. Those aspects of process technology that are common to both the air- and oxygen-based systems are reviewed below, along with some of the major differences.

Ethylene Oxide Catalysts. Of all the factors that influence the utility of the direct oxidation process for ethylene oxide, the catalyst used is of the greatest importance. It is for this reason that catalyst preparation and research have been considerable since

Table 12. Ranges of Reaction System Variables in the Oxygen-Based Direct Oxidation Process for Ethylene Oxide [a]

Variable	Range
ethylene, mol %	15–40
oxygen, mol %	5–8.5
carbon dioxide, mol %	5–15
ethane, mol %	0–2
argon, mol %	5–15
nitrogen, mol %	2–60
methane, mol %	1–60
temperature, °C	220–275
pressure, MPa[b]	1–2.2
space velocity[c], h^{-1}	2000–4000
conversion, %	7–15[d]
selectivity or yield (mol basis, %)	70–77

[a] Refs. 102–104, 108, 122.

[b] To convert MPa to psi, multiply by 145.

[c] The space velocity is defined as the standard volume of the reactant stream fed per unit time divided by the volume of reactor space filled with catalyst.

[d] At 30 mol % C_2H_4.

the reaction was discovered. There are four basic components in commercial ethylene oxide catalysts: the active catalyst metal; the bulk support; catalyst promoters which increase selectivity and/or activity, and improve catalyst life; and inhibitors or anti-catalysts which suppress the formation of carbon dioxide and water without appreciably reducing the rate of formation of ethylene oxide (97).

Silver-containing catalysts are used exclusively in all commercial ethylene oxide units, although the catalyst composition may vary considerably (123). Nonsilver-based catalysts such as platinum, palladium, chromium, nickel, cobalt, copper ketenide, gold, thorium, and antimony have been investigated but are only of academic interest (93,124–129). Catalysts using any of the above metals are either too active for the complete oxidation of ethylene at useful operating temperatures, or have very poor selectivities for ethylene oxide production at the conversion levels required for commercial operation.

A variety of different procedures have been reported for silver catalyst preparation on relatively inert support materials. The different methods are: (1) precipitation of silver oxide from aqueous silver nitrate or other salt solutions by alkali or alkaline earth compounds (130–131); (2) thermal decomposition of silver salts, in particular silver oxalate or silver carbonate (131–132); (3) reduction of silver salts by hydrogen, formaldehyde, hydrazine or hydroxylamine (133–134); (4) electrolysis of silver salt solutions (135); and (5) selective removal of secondary metals from silver containing alloys (136–137). The silver is added to the support either as a coating from a suspension or by impregnation with a solution. The coating procedure is claimed to give a catalyst of higher silver content and initial activity. However, the catalyst is susceptible to silver loss by abrasion and tends to lose selectivity to ethylene oxide after use for several months (138). The second approach appears to be more popular in the patent literature, and some investigators have used a combination of coating and impregnation procedures (138–140).

The chemical and physical properties of the support strongly dictate the performance of the finished catalyst. Although nonsupported silver catalysts have been advocated in some patents (141), it is unlikely that they are used commercially since pure silver tends to sinter at reaction temperatures with a resultant activity loss (97). For commercial operation the preferred supports are alundum (α-alumina) and silicon carbide (123,125). Other supports are glass wool, quartz, carborundum, and ion-exchange zeolites (see Molecular sieves). The surface area, porosity, and pore size of the support influence the size of the silver particles on the support and, therefore, affect the performance of the final catalyst (142). High surface area supports (3–100 m^2/g) generally yield poor ethylene oxide catalysts presumably because ethylene reacts in the pores from which the ethylene oxide is released slowly. The combination of slow product release and poor heat conductivity of high surface area supports is claimed to result in the combustion of ethylene oxide (143).

Silver alone on a support does not give rise to a good catalyst (144). However, addition of minor amounts of promoter enhance the catalyst activity and improve its long-term stability. Excess addition lowers the catalyst performance (145–146). The most commonly used promoters are alkaline earth metals, such as calcium or barium, and alkali metals such as cesium, rubidium or potassium. A specific example describes a high performance silver catalyst on an α-alumina support ca 8 mm dia (108,147). The catalyst is of a ring shape and contains 7.8 wt % Ag deposited by an impregnation process which yields silver particle sizes of 50–200 nm. The catalyst also has ca 105 ppm of cesium or other alkali metals like potassium or rubidium.

Many organic compounds, especially the halides, are very effective for suppressing the undesirable oxidation of ethylene to carbon dioxide and water, although not significantly altering the main reaction to ethylene oxide (144). These compounds, referred to as catalyst inhibitors, can be used either in the vapor phase during the process operation or incorporated into the catalyst manufacturing step (97). The inhibitor plays a significant role in the process, although the mechanism of inhibitor action is far from understood. Important gas-phase inhibitors are ethylene dichloride, ethylene dibromide, other alkyl halides, aromatic hydrocarbons, amines, and organometallic compounds (144,148). In a study of the effect of ethylene dichloride on catalyst activity, it was found that small amounts improved catalyst performance; however, excess amounts of ethylene dichloride deactivated the catalyst (123).

In order to assure control of the reaction, the vapor-phase inhibitor concentration must be closely controlled in the ppm range. Although several compounds have been claimed to be useful, it is likely that commercial processes use only ethylene dichloride or some of the simpler chlorinated aromatics (97). In general, the choice between inhibitors is not based on their differences in performance but rather on the designer's preference for dealing with the type of control problems each inhibitor system imposes (97).

Inhibitors added in catalyst manufacture are generally alkali metal halides and cyanides. These compounds are more effective inhibitors than the corresponding silver compounds (97). It is not known whether commercial ethylene oxide catalysts have inhibitors added in the catalyst preparation step. One problem cited in using this approach is catalyst reproducibility (97).

Temperature and Thermal Factors. Proper dissipation of heat, and control of catalyst surface temperatures are important process variables. Unless temperatures are carefully controlled, localized hot spots in the bed of 100–300°C above the desired temperature are possible. Such high temperature excursions can lead to catalyst sintering, loss of activity, equipment damage and costly downtime (114). The actual catalyst temperatures are governed by the balance between the rate at which heat is generated by chemical reaction and the rate at which heat is removed from the reactor. Heat removal is achieved in two ways: loss of sensible heat of the hot product gases from the reactor, and use of the coolant on the shell side of the reactor. The heat generation due to chemical reaction increases almost exponentially with temperature. This is owing to the simultaneous increase in the rate of oxidation to ethylene oxide and the corresponding reduction in catalyst selectivity with increasing temperature; eg, a 200,000 t/yr ethylene oxide unit operating at 70% selectivity generates 381 GJ/h (361×10^6 Btu/h) of heat. If the selectivity dropped to 50% at the same productivity, the heat release would increase by ca 220% (97). Since great quantities of heat are released, acceptable selectivities are usually ensured only within a fairly narrow range of reaction temperatures. Therefore, stable coolant temperature control is required for successful process operation (149). The use of inhibitors to control the heat generation plays an important role in the stable operation of the reactor.

During normal process operation, the rate of reaction is highest when the partial pressures of the reactants are high. This occurs close to the inlet of the catalyst bed, and consequently a hot spot generally develops near the front end of the reactor inlet. Such a hot spot always occurs at commercial productivities even if an efficient cooling system is used (97). In order to ensure stable reactor operation this hot spot temperature should be no more than 30–40°C above the corresponding coolant temperature at that location (150). One solution to this problem involves the use of a gradation in catalyst activity which increases from the inlet to the outlet of the reactor (151). Using this approach, both the selectivity and the productivity were improved by a uniform bed temperature profile (151). Other techniques to improve reactor thermal conductivity include the use of metals such as copper in the form of rods, wires, or lumps embedded in the catalyst tube (97).

Space Velocity. The choice of the optimum space velocity is a crucial design variable and is dependent on several interrelated factors: catalyst type and age, tube diameter, operating temperature and pressure, concentrations of the reactants, and the level of conversion desired. Although several theoretical and experimental studies (123,152–156) have described the relationship of temperature, contact time, and feed concentrations, actual operating data for the space velocities are regarded as valuable know-how based on plant experience. The patent literature claims space velocities of 1000–4500 h^{-1} to be beneficial for the process (103,106,147,157–159). In general, as the space velocity is increased, reactor-bed heat transfer is improved, catalyst selectivity increases, conversion per pass decreases and compression cost increases (123,155–156).

Operating Pressure. Since the total number of moles is decreased as ethylene and oxygen react to yield ethylene oxide, higher pressures favor a higher degree of completion of the reaction. However, only a marginal change in selectivity is observed at higher pressures (97). Higher operating pressures result in greater productivity per unit volume of reaction space, improved heat transfer, better recovery of ethylene oxide from the relatively dilute recycle stream, improved carbon dioxide absorption in the CO_2 removal unit, and lower compression power consumption. However, higher system pressures have a detrimental influence on the explosive limit of the cycle gas, limiting operation to a lower safe oxygen concentration (160). Processes generally operate at 1.0–2 MPa (10–20 atm) (103–106,148,161).

Cycle Diluents. The small vent stream flow in an oxygen-based process makes it economically feasible to use cycle gas diluents other than nitrogen. Shell has patents for the use of methane in its process (122), and Halcon claims similar benefits from ethane (162). Each of these diluents has a higher heat capacity and thermal conductivity than nitrogen, facilitates a higher safe oxygen concentration in the cycle, and assists in moderating the peak reaction temperature (162). The result is a higher yield of ethylene oxide at a fixed productivity, or a higher productivity with the same reactor volume (122,162). The desirable concentrations are 30–60 mol % of the diluent, 15–30 mol % of ethylene, and 4–15 mol % of oxygen, depending on the explosive limit for the gas mixture employed (122,162). The disadvantages of using methane or ethane are: hydrocarbon purification costs to eliminate sulfur and higher paraffinic hydrocarbons; lower steam generation owing to higher sensible heat losses in the cycle gas leaving the reactor, and possible safety concerns associated with vapor clouds as a result of rupture-disk failures or equipment damage.

Raw Material Purity Requirements. Unlike the chlorohydrin process, in the direct oxidation system high purity ethylene is an important requirement for successful commercial operation (114,163). Typical ethylene feed specifications generally require a purity greater than 98 mol % (97). In addition, several specifically undesirable substances must be eliminated from the raw materials supply because they adversely influence catalyst performance and reactor stability (144,148). Acetylene (qv) in trace quantities is a catalyst poison (148). Propylene reacts more rapidly than ethylene and yields undesirable by-products such as acetaldehyde, acetone, methanol, propylene oxide, and propionaldehyde (163–164). Furthermore, the presence of extraneous hydrocarbons may cause the deposition of carbonaceous material on the catalyst surface (97). Oxygenated hydrocarbons should be removed from the feed gas. Hydrogen and carbon monoxide in the reactor feed gas should be minimized, because they are readily oxidized at reaction temperatures (200–300°C). Carbon dioxide, methane, and ethane do not present severe operating problems as long as their concentrations are controllable in the cycle.

If trace amounts of compounds that act as gas-phase inhibitors for the catalyst are present in the ethylene feed, their concentrations must be carefully controlled in order to prevent rapid catalyst deactivation (97,163). Sulfur and its compounds, even in trace amounts, should be removed because of irreversible poisoning of the silver catalyst (114).

For the air-based process, air purification schemes have been used to protect catalyst life and activity (100). For an oxygen-based system the major impurities in the oxygen are nitrogen and argon. Both are inert to the catalyst. However, their concentrations in the oxygen feed should be kept low, otherwise the cost of ethylene losses in the purge stream is prohibitive (92). Most modern oxygen-based units are designed for a feed containing ≥97–99 mol % oxygen (120).

Process Safety Considerations. Results from a kinetic and process optimization study may indicate operating conditions that are unsafe from the viewpoint of fire safety, equipment damage potential and operating sensitivity (149). Over the past several years a number of ethylene oxide units have suffered dangerous fires and deflagrations (97). The design of the plant should be such that the gas mixtures handled are always outside the explosive limit (see Plant safety). The actual safe operating ranges are dependent on operating temperatures, pressures, equipment configurations, gas compositions, ignition sources present, and the dynamics of the catalyst, process, and instrumentation system (97). An oxygen-based system has pipelines containing pure oxygen and requires a fairly complicated and highly instrumented method for safe introduction of pure oxygen into a process stream that is rich in hydrocarbons. Since the air-based processes operate at lower concentrations of reactants, temperature excursions in process piping, in the event of an ignition, should be lower. The selection of safe plant operating conditions and the design of effective process safety systems is a complex job that requires extensive laboratory data on the influence of process variables on explosibility, as well as several years of proven commercial experience. The history of the direct oxidation processes clearly indicates that, given the necessary design expertise and commercial experience, the ethylene oxide process can be operated safely.

Ethylene Oxide Recovery. The product gases from the reactor contain <3 mol % ethylene oxide. An economic process scheme must be designed for essentially complete product recovery and result in very small vent losses of raw materials. Important process constraints also dictate the final design; eg, oxide hydrolysis at high temperatures must be avoided. Vacuum stripping of the ethylene oxide must also be done with great care because the gas mixture may be flammable, depending on the design conditions selected. Refs. 165–169 describe specific techniques for improving the design of the basic recovery scheme (recovery efficiency of 96–98%) shown in Figures 2 and 3. Other references describe additives to water for more efficient absorption (170–171) and other novel ways for product recovery (172–173).

Ethylene Oxide Purification. The main impurities in ethylene oxide are aldehydes. Some of the chief consumers of ethylene oxide including poly(ethylene glycol) and detergent ethoxylates have been requiring more stringent aldehyde specifications. Ten years ago most of the ethylene oxide produced contained <100 ppm aldehydes (109). Recently this level has been reduced to 30 ppm. This requirement results in additional capital costs for larger purification columns to separate the acetaldehyde and formaldehyde from the ethylene oxide at these low levels. At very low total aldehyde concentrations, formaldehyde becomes the principal impurity because of its relative volatility in the purification column (109). Improvements in ethylene oxide catalysts have been claimed to reduce aldehyde generation (109). In addition to conventional distillation, several other approaches have been patented for reducing the aldehyde content of the finished product; eg, the use of molecular sieves (174), treatment with potassium ethylhexanolate followed by distillation (175), and extractive distillation with methanol (176).

Pollution Considerations. A detailed study of the pollution considerations in the manufacture of ethylene oxide by the direct oxidation processes is described in ref. 102 (see Air pollution). The air emissions from all variations of the direct oxidation process are primarily hydrocarbons: ethylene, ethylene oxide, and traces of ethane. In some units, traces of NO_x or SO_x have been reported as a result of combustion operations associated with pollution control or process drive machinery. The biggest air emission in either the air- or oxygen-based processes comes from the main process vent stream. In an air unit, this vent comes from the last purge reactor absorber and consists of spent air (N_2, O_2, and some inert gases), carbon dioxide, traces of ethylene oxide and generally <2 mol % hydrocarbons.

A catalytic converter may be added to the main process vent (see Exhaust control, industrial). The similar vent stream from an oxygen-based system is about one hundred times smaller and much richer in hydrocarbons. Therefore, it can be used as a fuel (100). Table 13 summarizes the approximate composition ranges for the vent streams in both air- and oxygen-based systems.

The other emission occurs in the recovery section of the process where a CO_2-rich purge gas is vented. In an air unit this stream is from the ethylene oxide distillation tower; in an oxygen process it is from the CO_2 removal system vent. Once again these streams are not of comparable composition as can be seen in Table 13. In an oxygen process the CO_2 removal unit can be designed to reduce the ethylene content of this vent to very low levels. Some of the processes use gas-fired turbines to feed air and ethylene to the system (102). Since the combustion in the turbines is not completely efficient, some unburnt hydrocarbons escape to the air. Most ethylene oxide units are known to have flares or rupture disks, or both, for venting the process gas during major upsets. Data on flaring frequency is scanty but a utilization of one to two times a year may be typical (102).

The total air emissions in the air- and oxygen-based processes can be made comparable (102). If the vent gas in an oxygen unit is not burnt, the total hydrocarbon emissions have been estimated as ca 12 g/kg of product (102). If methane is used as a diluent, and the resulting purge gas incinerated, the total emissions can be reduced to ca 4 g/kg of product (see Incinerators). For the air-based process without catalytic conversion of the purge gas, the hydrocarbon emissions are estimated to be >30 g/kg of product (102). However, with the use of a properly designed catalytic converter, the total emissions can be substantially reduced to levels (<15 g/kg) comparable to those of an oxygen process.

Table 13. Typical Vent Gas Composition Ranges for Ethylene Oxide Production by Direct Oxidation[a]

Stream	Reported composition range, mol %	
	Air-based	Oxygen-based
main process vent stream		
nitrogen	85–93	2–35
oxygen	1.0–5	5–7
methane	0–0.9	1–35
ethane	trace–0.2	trace–2.0
ethylene	trace–2.5	13–35
ethylene oxide	0–0.01	0–0.01
carbon dioxide	5–15	5–15
argon		5–15
water	0.1–1.5	0.1–0.5
CO_2-rich purge gas (water-free basis)		
nitrogen	13–25	
oxygen	1–26	0.02
ethylene and hydrocarbons	2.5–8.0	0.3–0.9
ethylene oxide	0–1.0	
carbon dioxide	62–80	99–99.7
inerts		0.005–0.015

[a] Refs. 97, 102–103, 116, 118, 121, 177.

The treatment of liquid effluents is also important. The water produced as a result of the burning reaction of ethylene or ethylene oxide must be continuously purged from the process. Care must be taken to ensure that the purge does not have excessive amounts of ethylene oxide, glycol, or aldehydes (109). The organic content of aqueous wastes from air- and oxygen-based systems are comparable and consist of small quantities of aldehydes, glycols, and heavy glycols. The liquid stream can be adequately treated by conventional in-plant biological systems (102). In the oxygen process, however, the inorganic salts used in the carbon dioxide removal system increase the total water pollution load compared to an air-based system (100). There are no solid wastes resulting from the direct oxidation processes (102).

Air Vs Oxygen Process Differences and Economics. The relative economics of the air versus the oxygen process are reported (92,97,100–101). Two major process characteristics dictate the difference in the capital costs for the two processes. The air process requires additional investment for the purge reactors and their associated absorbers, and for the energy recovery from the vent gas. However, this is offset by the need for an oxygen production facility and a carbon dioxide removal system for an oxygen-based unit. In a comparison of necessary investments for small to medium capacity units (<50,000 t/yr), oxygen-based plants have a lower capital cost even if the air-separation facility is included (100). However, for medium- to large-scale plants (75,000–150,000 t/yr) the air process investment is smaller than that required for the oxygen process and the air-separation unit, unless the oxygen is purchased from a very large air-separation unit serving many customers.

There are also operating cost considerations that differ significantly among the two processes. The costs of silver catalyst, oxygen, and ethylene are critical factors determining the relative economics. For a given catalyst type, the oxygen process operates at a higher selectivity and requires a smaller volume of catalyst. Even though the cost of ethylene comprises ca 60% of the total manufacturing cost in both processes, the incremental product cost between the air and oxygen processes is influenced only slightly by changes in the price of ethylene. For example, in 1976 it was estimated that an ethylene price increase of 2.20 ¢/kg raised the product cost for air oxidation by only 0.088 ¢/kg in excess of that of an oxygen-based unit. On the other hand, the price of oxygen has a much more significant effect on the economics of an oxygen unit. A change in the oxygen price by 0.22 ¢/kg altered the product cost by 0.243 ¢/kg (92).

The oxygen-process plant has no high pressure purge gas stream of sufficient volume to make energy recovery attractive as in an air process. The oxygen process also has a considerable steam requirement for carbon dioxide stripping in the CO_2 removal unit. The total compression costs for an oxygen-based process, including the air separation unit, are slightly higher than for an air-based system (97).

Purity of the feedstocks (oxygen and ethylene) also determine the relative economics of the air- and oxygen-based processes. If the oxygen purity is low, the volume of the ethylene-rich purge gas is increased markedly and the oxygen-based process becomes unattractive (100). In addition to requiring a high oxygen purity, the ratio of argon to argon plus nitrogen in the oxygen feed is critical to the attainment of high yields in the oxygen process (120). Lower values of the ratio improves yields by ca 1–2%. For the large-scale ethylene oxide processes, yield changes of even a fraction of a percent can have an impact on the overall process economics. In general, as the ethylene purity decreases the air process becomes more attractive. However an air-based process may require an air purification system if contaminants such as sulfur, halogens, and heavier hydrocarbons are present (114,163).

Both air- and oxygen-based processes can be designed to be comparable in the following areas: product quality, process flexibility for operation at reduced rates, and on-stream reliability (100,109). For both processes an on-stream value of 8000 h/yr is typical (101). The reliability of the oxygen-based system is closely linked to the reliability of the air-separation plant, and in the air process, operation of the multistage air compressor and power recovery from the vent gas is crucial (100).

For the same production capacity, the oxygen-based process requires fewer reactors, all of which operate in parallel and are exposed to reaction gas of the same composition. However, the use of purge reactors in series for an air-based process in conjunction with the associated energy recovery system increases the overall complexity of the unit. Given the same degree of automation, the oxygen-based unit requires fewer operators if the air-separation plant is outside the battery limits of the ethylene oxide process (100).

From the preceding discussion, it is clear that no meaningful generalizations can be made regarding the overall superiority of either the air- or oxygen-based process.

Chlorohydrin Process. The chlorohydrin process is commercially attractive to a producer only when an adequate supply of captive low cost chlorine and lime or caustic are available. Further, a market for the by-products ($ClCH_2CH_2Cl$; $(ClCH_2CH_2)_2O$; and $CaCl_2$ or $NaCl$) plays an important role in process economics (97,178). The chlorohydrin process accounted for 50–60% of the United States production in 1955, but advantages of the direct oxidation technology have reduced this percentage to ca 3% in 1978. The commercial process is basically conducted in two steps with the production of ethylene chlorohydrin as an intermediate (see Chlorohydrins).

Two alternatives are commercially available for making ethylene chlorohydrin

(179–180). In the first method, ethylene is mixed with a slurry of hydrated lime at a pressure of 20 MPa (2900 psi) and 20°C. The mixture is treated with chlorine to form unstable calcium oxychloride which decomposes in the aqueous media to give hypochlorous acid and calcium chloride. The ethylene chlorohydrin (30–40% soln) is then formed by the reaction of ethylene and hypochlorous acid:

$$CaO + Cl_2 \rightarrow CaCl(OCl)$$

$$CaCl(OCl) + Cl_2 + H_2O \rightarrow CaCl_2 + 2\ HOCl$$

$$HOCl + C_2H_4 \rightarrow HOCH_2CH_2Cl$$

In the second approach, chlorine is dispersed in water at 0.25 MPa (36 psi) and 15–25°C. Ethylene is then sparged into the resulting hypochlorous acid solution to yield ethylene chlorohydrin (4–6% soln).

The ethylene chlorohydrin produced by either method is directly converted to ethylene oxide by dehydrochlorination using either sodium or calcium hydroxide. Commercial processes were designed and operated to minimize the yields of by-products. A metric ton of product requires 800 kg of ethylene, 2 t of chlorine and 1.6 t of calcium oxide (181). The by-products are: 3.2 t calcium chloride, 100–150 kg dichloroethane, 70–90 kg bis(2-chloroethyl) ether, and 5–10 kg acetaldehyde (181). An excellent process description is given in ref. 97.

Although the chlorohydrin process appears simpler, the capital costs are about the same, chiefly owing to the more expensive materials of construction (182). The chlorohydrin process is more ethylene-efficient, but the higher yield is obtained at the expense of purchased chlorine and lime. The product cost was estimated to be about 7–9 ¢/kg higher than for the direct oxidation process in 1969 (97). One of the unique advantages of the chlorohydrin process is that it can utilize a feed gas stream containing relatively dilute ethylene concentrations (ca 6 mol %).

New Routes for Manufacture. If proven successful some of these alternative methods may have a very favorable impact on the overall production costs for making ethylene oxide.

Arsenic-Catalyzed Liquid-Phase Process. An arsenic-catalyzed liquid-phase process for olefin oxides has recently been patented by Union Carbide (183). A selective epoxidation of ethylene by hydrogen peroxide in a 1,4-dioxane solvent in the presence of an arsenic catalyst is claimed. No solvent degradation is observed. Ethylene oxide is the only major product detected:

$$C_2H_4 + H_2O_2 \xrightarrow[60\text{-}130°C]{As} \underset{O}{CH_2\text{---}CH_2} + H_2O$$

The catalyst used may be either elemental arsenic, an arsenic compound, or both.

Thallium-Catalyzed Epoxidation Process. The use of Tl(III) for olefin oxidation to yield glycols, carbonyls or epoxides is well known (184). Since the epoxidation with Tl(III) is stoichiometric to produce Tl(I), reoxidation is needed. Recently Halcon has patented processes based on such epoxidation (185–188) to yield ethylene oxide. The major benefits of such a process are claimed to be: high yields of ethylene oxide, flexibility to produce either propylene oxide or ethylene oxide, and the potential of a useful by-product (acetaldehyde). Recent advances (188) using organic hydroperoxides in place of oxygen for reoxidation offer considerable promise since reaction rates are rapid, and low pressures (100–200 kPa or 1–2 atm) can be used.

Lummus Hypochlorite Process. A Lummus patent claims a process for propylene oxide or ethylene oxide using *tert*-butyl hypochlorite (189). The chemistry for this new process parallels the classical chlorohydrin technology with brine recycle. Advantages claimed are: high ethylene yield (85–90%), reduced reactor size, and lower steam requirements for the saponification step. However, disadvantages include high capital cost for a large chlorine plant, difficult and energy-intensive distillation steps, and losses of *tert*-butyl alcohol in the process.

Electrochemical Process. Several patents (190–192) claim that ethylene oxide is produced in good yields in addition to faradic quantities of substantially pure hydrogen when water and ethylene react in an electrochemical cell to form ethylene oxide and hydrogen. The only raw materials that are utilized in ethylene oxide formation are ethylene, water, and electrical energy. The electrolyte is regenerated *in situ*, ie, within the electrolytic cell. The addition of oxygen to the ethylene is activated by a catalyst such as elemental silver or its compounds at the anode or its vicinity (190). The common electrolytes used are water-soluble alkali metal phosphates, borates, sulfates or chromates (191) at ca 22–55°C. The process can be either batch or continuous (see Electrochemical processing).

Unsteady-State Direct Oxidation Process. Periodic interruption of the feeds can be used to reduce the sharp temperature gradients associated with the conventional oxidation of ethylene over a silver catalyst (193). Steady and periodic operation of a packed-bed reactor has been investigated for the production of ethylene oxide (194). By periodically varying the inlet feed concentration of ethylene or oxygen, or both, considerable improvements in the selectivity to ethylene oxide were claimed.

Fluid-Bed Direct Oxidation Process. The use of fluidized beds for ethylene oxide production was reported to provide good temperature control and inhibition of side reactions using a granular silver catalyst (195). Several additional fluid-bed processes that claim improved product yields were patented later (196–199). However, only one process developed by Vulcan-Atlantic was reported to have been successfully demonstrated in pilot-scale equipment (200). The pilot fluid-bed reactor is described as a multitubed converter that provides uniform heat transfer to a fluid circulating in the shell of the converter, and minimizes back-mixing of the fluidized catalyst (200). A novel circulating fluid-bed process and the associated reaction apparatus have been patented (201–203). By employing high gas velocities, the ethylene oxide productivity per unit catalyst volume is claimed to be three to four times greater than the maximum reported number for conventional fixed-bed tubular reactors (203–204). In spite of the recent rapid advances in fluidization technology, no commercial fluid-bed ethylene oxide processes have been reported (see Fluidization).

Economic Aspects

United States annual production history of ethylene oxide is shown in Figure 4. Annual production exceeded 2×10^6 metric tons in 1975. Approximately 19% of United States ethylene (qv) production is consumed in ethylene oxidation, making ethylene oxide the second largest derivative of ethylene, surpassed only by polyethylene (see Olefin polymers). World ethylene oxide capacity is estimated by country in Table 14. Total world capacity in 1978 was ca 6.5×10^6 t.

More than 98% of the total production of ethylene oxide is converted to derivatives. The consumption pattern of ethylene oxide in the United States is shown in

Table 15. About 60–65% of the ethylene oxide is consumed in manufacture of ethylene glycols (91,96). The largest markets for ethylene glycol are antifreeze (qv) and polyester (qv) intermediates. These markets account for 27.7 and 22.3% of the United States ethylene oxide consumption, respectively. Other major derivatives of ethylene oxide include glycol ethers, ethanolamines, and surfactants. Major uses of these derivatives are shown in Table 15, as well as the expected United States annual growth rates through 1980 of each derivative. Antifreeze, the largest market, is predicted to have the lowest growth rate at 3%/yr. The fastest growing market areas are expected to be ethylene glycol for polyesters (9% annually) and surfactants (10% annually). Overall growth in United States demand for ethylene oxide is ca 6.3%/yr through 1980 (96). This rate of growth is considerably less than the phenomenal 18.7% average annual rate of the 1967–1970 period but is in line with an historical growth rate of 7.0% average annual growth during the 1962–1977 period. Worldwide ethylene oxide demand is expected to grow at a slightly higher rate as characterized by a predicted 7.3% annual growth in Japan and 7% annual growth in the Federal Republic of Germany through 1980 (207–208).

The United States price history for ethylene oxide is shown in Figure 4. List prices were fairly stable from 1956 through 1968 at 34 ¢/kg. The reported list price dropped to 20 ¢/kg in 1969 to more accurately reflect market values at that time. A weak market

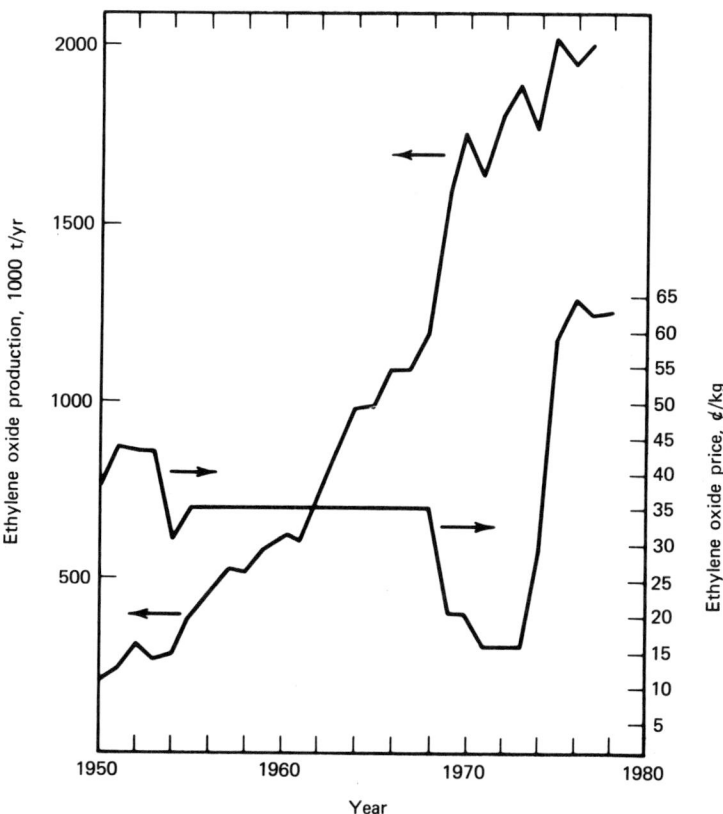

Figure 4. United States ethylene oxide production and sales prices, 1950–1977 (205–206).

Table 14. World Production Capacities for Ethylene Oxide[a]

Country	Total capacity, 1000 t/yr	World capacity, %	No. of companies[b]	Capacity by process, 1000 t/yr				
				UCC	Shell	SD	Dow	Other
United States and Puerto Rico	2690	41	12	1134	832	433	291	
Canada	122	1.9	2	68			54	
Mexico	128	2	1			128		
Brazil	35	0.5	1			35		
Belgium	270	4.1	3	120	120	30		
France	300	4.6	2		200	100		
FRG	626	9.6	5		200	150		276
Italy	310	4.7	3	40	150	45		75
Netherlands	330	5	2		210		120	
United Kingdom	298	4.6	3	18	280			
Sweden	40	0.6	1			40		
Spain	12	0.2	1			12		
Bulgaria	10	0.2	S					10
Czechoslovakia	65	1	S		40			25
DDR	100	1.5	S			100		
Poland	60	0.9	S					60
Romania	40	0.6	S			40		
USSR	85	1.3	S					85
Japan	729	11.1	7		304	110		315
People's Republic of China	85	1.3	S					85
Taiwan	175	2.7	2	125		50		
India	28	0.4	2		12			16
Democratic People's Republic of Korea	10	0.2	S					10
Total	*6548*	*100*		*1505*	*2348*	*1273*	*465*	*957*

[a] Capacities are estimated as of Dec. 1978.
[b] S = all capacity is government owned.

Table 15. United States Consumption Pattern For Ethylene Oxide[a]

Use	Consumption (1976), %	Projected annual growth rate (1976–1980), %	Principal use
ethylene glycol for antifreeze	27.7	3	automotive coolant
ethylene glycol for polyester	22.3	9	polyester fibers and films
glycol ethers	6.5	6	solvents for lacquers, resins, enamels, epoxy coatings, water-based coatings, varnishes, and printing inks; antiicing agent in jet fuel
ethanolamines	6.4	7	soaps and detergents; gas conditioning
surfactants	14.1	10	detergents
all other	23.0	6	

[a] Consumption pattern estimated for 1976 and projected annual growth rates are from ref. 96.

and overcapacity forced prices to 15.5 ¢/kg in 1971. The surge in ethylene oxide price to a current value of about 62 ¢/kg began in 1974. This price increase reflects the tremendous corresponding increase in ethylene prices from 6.8 ¢/kg in the early 1970s to a current level of 26.5–27.5 ¢/kg.

Worldwide capacity for ethylene oxide should be boosted by ca 2×10^6 t/yr to a total of 8.6×10^6 t/yr by 1983. Projects underway are listed in Table 16. Union Carbide Corporation, Imperial Chemical Industries, Jefferson Chemical, and Shell have announced plans for significant expansions and new plants in the United States. In addition, 22 principal ethylene oxide projects are planned or underway in 17 other countries.

Shipment and Storage

Shipment. Small shipments of ethylene oxide are made in the containers described in Table 17. Large shipments of 40–100 m³ (10,000–25,000 gal) are made in insulated, type 105-A-100W or 111-A-100W4 tank cars. A red label is required by the U.S. Bureau of Explosives on all shipments of ethylene oxide. For further detailed information on the shipping and handling of ethylene oxide, see refs. 9, 210, and 211.

Storage. Carbon steel and stainless steel should be used for all equipment in ethylene oxide service. Although ethylene oxide attacks most organic materials (including plastics, coatings, and elastomers), certain fluoroplastics are resistant and can be used in gaskets and seals.

Storage tanks should be designed for at least 618 kPa (75 psig) in accordance with the ASME code for unfired pressure vessels. All-welded construction is recommended. Ethylene oxide storage tanks should be well-insulated, protected by water-spray systems, equipped with cooling coils, and electrically grounded. Safety devices should discharge outdoors to a safe location. New equipment should be cleaned of iron rust and scale and immediately purged with inert gas before charging with ethylene oxide.

The pressure that should be maintained with an inert gas in the storage tank is a function of the partial pressures of the contained gases. The partial pressure exerted by the ethylene oxide is determined by the liquid temperature. A nonexplosive vapor phase is attained by the regulated addition of inert gas at a precalculated total tank

Table 16. Planned Ethylene Oxide Capacity[a]

Company	Planned site	Capacity, 1000 t/yr	Technology[b]
Jefferson Chemical Co.	Port Neches, Tex.	91	Halcon
Shell Chemical Co.	Geismar, La.	182	Shell
Imperial Chemical Ind.	Baytown, Texas	200	Shell
Dow Chemical Canada	Ft. Saskatchewan, Canada	136	Dow
Oxiteno Nordeste SA	Camacari, Brazil	105	SD
Petroleos Mexicanos	La Cangrejera, Mexico	100	SD
Petroleos Mexicanos	Allende, Mexico	200	SD
Petroleos Mexicanos	La Cangrejera, Mexico	200	Halcon
Productora de Alcoholes	El Tablazo, Venezuela	NA	NA
Technoimport	Burgas, Bulgaria	80	SD
SIR Consorzio Ind.	Porto Torres, Italy	90	NA
Liquichimica Sud	Augusta, Italy	150	Shell
Polimex Cekop	Plock, Poland	15	Shell
Ministry Chemical Ind.	Brazi, Romania	35	SD
Union Explosiv Rio Tinto	Huelva, Spain	140	NA
Ind. Quim Asociadas	Tarragona, Spain	70	Shell
Imperial Chemical Ind.	Wilton, England	expansion	
Techmashimport	Nishnekamsk, USSR	200	SD
Techmashimport	Dzerzhinsk, USSR	103	SD
Petkim Petro Kimya AS	Aliaga, Turkey	54	Shell
Indian Petrochemical	Baroda, India	5	Halcon
Mitsubishi Chemical Ind. Ltd.	Mizushima, Japan	50	SD
Mitsubishi Petrochemical Co.	Kashima, Japan	120	Shell
Korea Machinery Import	Pyong Yang, N. Korea	10	
Honan Petrochemical Corp.	Yeo-Cheon, S. Korea	72	Shell

[a] Refs. 94 and 209.
[b] SD = Scientific Design; NA = not available.

Table 17. Container Information[a]

	Type of container			
	Drum	EOX cylinder	FE cylinder	LB bottle
DOT container specification	5P	4BA240	4B240	3E1800
net weight, kg	180	80	7	0.2

[a] Ref. 9.

pressure. Figure 5 shows the operating pressure required for a safe vapor-phase mixture with nitrogen or methane at different liquid temperatures.

Ethylene oxide may undergo slow polymerization during storage. Excessive temperatures or contamination with such impurities as water, alkalies, acids, metal oxides, and iron or aluminum salts could cause rapid polymerization or reaction of ethylene oxide; hence, prolonged storage at elevated temperatures or contact with these impurities should be avoided (9).

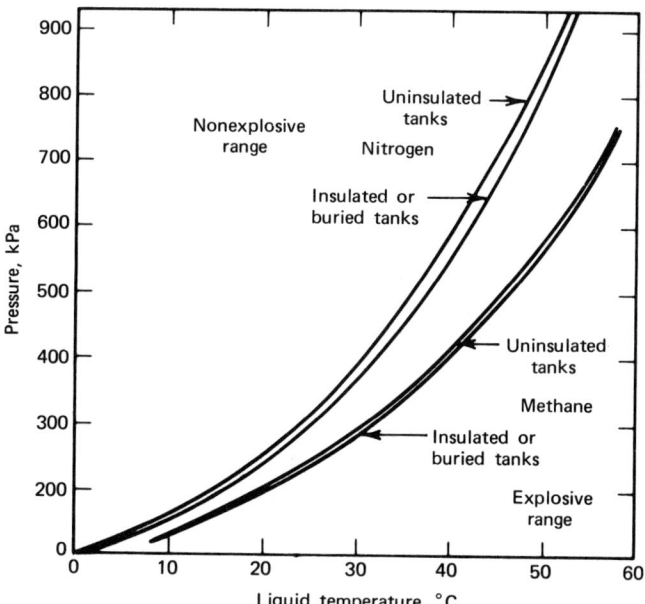

Figure 5. Recommended safe storage pressures for liquid ethylene oxide under nitrogen or methane blanketing gas (9). To convert kPa to psi, multiply by 0.145.

Specifications, Analytical and Test Methods

Pure ethylene oxide is sold only as a high-purity chemical, with typical specifications shown in Table 18. The purity of the commercial product is so high that only impurities are specified. There is normally no assay specification. A complete review and description of analytical and test methods for pure ethylene oxide is given in ref. 212.

An adsorption-gas-liquid chromatographic (glc) method is used for determination of small quantities of ethylene oxide in air (213–214). The method is particularly suited for determination of 8-h time-weighted average exposures to ethylene oxide. Its lower limit of detection is 0.15 ppm (213). Other methods for determination of ethylene oxide in air include a continuous infrared monitor (215), reagents on impregnated papers, titrimetric, and colorimetric methods (216–218) (see Analytical methods).

Table 18. Specifications for Ethylene Oxide[a]

acidity	0.002 wt % max as acetic acid; this is equivalent to 0.019 mg KOH/g
aldehydes	0.0030 wt % max as acetaldehyde
acetylene	none
water	0.03 wt % max
residue	0.005 g/100 mL max
color	10 Pt–Co, max
odor	nonresidual
suspended matter	substantially free

[a] Ref. 9.

Health and Safety Factors

Toxicology. Ethylene oxide is a relatively toxic liquid and gas. The liquid causes severe eye injury, and the gas may cause eye irritation. Vapors are irritating to the nose and throat and their presence is readily detected at first by smell. However, continued exposure results in olfactory fatigue and the warning property of smell is lost (219–220).

A TLV of 50 ppm in air is currently recommended as a safe concentration for daily 8-h exposure in a 40-h wk (221). Significantly exceeding this TLV may result in intractable nausea and vomiting that may, untreated, continue for several hours. The onset of illness is rapid in severe exposures but may be delayed one or two hours after moderate exposures.

Nausea and vomiting must be treated by a physician. Experience has shown that protracted vomiting can be controlled by injections of sodium phenobarbital or prochlorperazine. Such treatment must be given only by a physician (9). If nausea and vomiting are mild in nature, breathing of oxygen for 10–15 min is often sufficient to bring relief. Persons who have experienced nausea and vomiting from overexposure may develop a reduced ethylene oxide tolerance in which systemic illness develops after exposure to increasingly smaller concentrations of the gas. Prolonged breathing of high concentrations of ethylene oxide also causes dizziness, weakness, and chest pain with the possibility of developing pulmonary edema. Ethylene oxide does not accumulate in the body and, therefore, should not produce chronic poisoning.

Liquid or dissolved ethylene oxide on exposed skin does not cause immediate irritation but may cause severe delayed skin burns. A 50% aqueous solution of ethylene oxide is the most irritating combination. Pure ethylene oxide on dry skin will frequently evaporate before causing irritation, but any moist skin areas, eg, under watches or rings, can accumulate irritating concentrations (222). If liquid ethylene oxide is spilled on clothes or shoes, they must be promptly removed, or the absorbed liquid can produce delayed skin burns. Clothing can be freed of ethylene oxide by normal washing and drying, but leather shoes or gloves should be discarded. Liquid or dissolved ethylene oxide may cause severe eye burns.

Acute toxicity and inhalation studies have been conducted on a wide variety of animals and fish (221,223).

Ethylene oxide has been found to be a mutagen in at least 13 test species. The most plausible mechanism for induction of heritable changes by ethylene oxide is the alkylation of DNA. Base-pair substitution and various types of chromosomal aberrations have been observed (221,224–225).

Ethylene oxide has not been found to be a carcinogen in a limited number of studies performed to date. These include a human epidemiology study (5–10 ppm exposure for an average of 10.7 yr); a 1.5 yr average-lifetime mouse skin study; and a rat injection study. In all cases, there was no significant difference between the exposed and control groups (225). In a two-year vapor inhalation study on rats, cytogenic, mutagenic, and teratogenetic evaluations with ethylene oxide have been studied at the 10, 30, and 100 ppm levels for 40 h/wk (226). Other long-term studies are planned by the National Cancer Institute (227) and NIOSH (228).

Regulation of the use of ethylene oxide as a sterilizing agent or a pesticide is being considered by NIOSH (221), the EPA (224), and the FDA (225).

Explosibility and Fire Control. Pure ethylene oxide vapor will explode in the presence of common igniters. The lowest pressure at which an explosion can occur ranges from 68 kPa (500 mm Hg) at 10°C (229) to 35 kPa (290 mm Hg) at 100°C (230). Pure ethylene oxide vapor is difficult to ignite compared to the oxide-air or hydrocarbon-air mixtures, requiring spark energies about ten thousand times larger (231). The commonly accepted value for the autoignition temperature of ethylene oxide vapor is 560°C (9).

Although the explosion pressures are difficult to predict, the surge is immediate. The theoretical explosion pressure ratio is 10.6 for a complete reaction starting at 100°C (230). Lower ratios can be caused by incomplete reaction, higher ratios from the involvement of any liquid ethylene oxide present. In all the tests that have been performed on the explosive decomposition of ethylene oxide vapor, no final pressure higher than sixteen times the initial pressure has been recorded (9).

Among the many sources of ignition for ethylene oxide vapor are acetylides, static electricity, excessive heat, glowing carbon, and open flames. Acetylene may be a trace impurity in some hydrocarbon diluent gases for ethylene oxide; therefore, acetylide-forming metals such as copper, silver, mercury, and their alloys should not be used to handle ethylene oxide. A fused nichrome wire is a convenient igniter for laboratory test deflagrations.

The vapor phase of a tank or reactor that contains ethylene oxide should be diluted with an inert gas to prevent explosive decomposition if an ignition should occur. The amount of diluent required varies with temperature, pressure, and the inerting gas as shown in Figure 5.

Pure liquid ethylene oxide will explode (deflagrate) in the presence of strong igniters, if the temperature and pressure are elevated above a threshold level (Fig. 6) (211). Decompositions occuring in a liquid phase are inherently more dangerous than those in a vapor phase because of the greater concentration of potential energy.

If an open flame must be used in an area where ethylene oxide is handled, all equipment should be first emptied and filled with water, and gas tests for flammable mixtures in the area should be made with an indicator of the type approved for acetylene. As long as the area is not free of the odor of ethylene oxide, no repairs or inspections should be made.

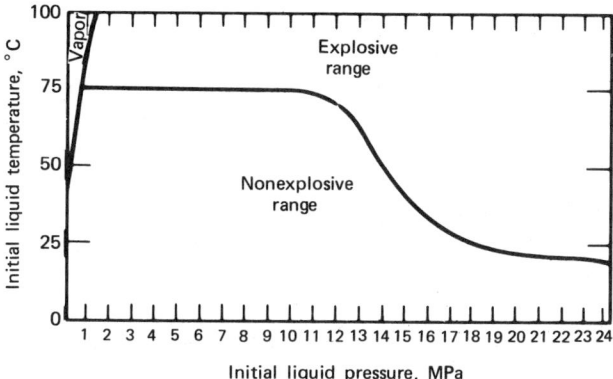

Figure 6. Recommended safe operating pressures and temperatures for pure liquid ethylene oxide (211). Initiator: fused nichrome wire. To convert MPa to psi, multiply by 145.

Ethylene oxide vapor fires should be extinguished in the same manner as any flammable gas fire.

Fires involving large quantities of liquid ethylene oxide are difficult to extinguish, even though ethylene oxide is soluble in water. These fires will continue to burn until the liquid is diluted with at least 22 parts of water by volume. Most small fires can be extinguished with dry chemical agents that are properly applied. There is a serious reflash hazard, as with all liquefied flammable gases.

Uses

Ethylene oxide is an excellent fumigant and sterilizing agent (232). It is generally noncorrosive to metals and leaves no residual odor or taste. Mixtures of 10–30% ethylene oxide in carbon dioxide or 12% ethylene oxide in dichlorodifluoromethane are commonly used. The latter is effective and nonflammable at ordinary temperatures; therefore, it is recommended for use in hospitals. These gas sterilants permit convenient sterilization of delicate instruments and supplies made of almost any material. They readily penetrate deep pores and narrow crevices, and pass through wrappings of polyethylene, paper, and cloth (see Sterile techniques).

A 10% concentration of ethylene oxide in carbon dioxide effectively kills most insect pests at all life stages. Carbon dioxide increases the rate of kill by stimulating insect respiration, and makes the mixture nonflammable. It is used for fumigation of spices, furs, bedding, and transportation equipment (see Poisons, economic).

Ethylene oxide has been studied for use as a rocket fuel (233) and as a component in munitions (234) (see Explosives and propellants).

Derivatives

Ethylene Glycol. Ethylene glycol is mainly used as automotive antifreeze (qv) and as a raw material for poly(ethylene terephthalate) (36) (see Polyesters; Polyester fibers; Film and sheeting materials). Other important uses for ethylene glycol are as a heat transfer fluid, an aircraft and runway deicer, and as a raw material for ethylene glycol dinitrate for use in low freezing dynamites (see Glycols).

Di-, Tri-, and Tetraethylene Glycols. These products are coproducts of ethylene glycol. Diethylene glycol is used as an intermediate for polyester resins, for triethylene glycol (by reaction with ethylene oxide), and for urethane foams (36). It is used for drying of natural gas, as a plasticizer, and as a solvent, emulsifier, humectant, and lubricant. Triethylene glycol is the preferred agent for drying natural gas and in other moisture removing applications. Triethylene glycol's uses overlap those of diethylene glycol, and is frequently preferred when a less volatile compound is required. Tetraethylene glycol has an important value in the extraction of aromatic hydrocarbons from nonaromatic hydrocarbons (see Drying agents; Glycols).

Poly(ethylene Glycols). Low molecular weight poly(ethylene glycols), made by reaction of ethylene oxide with water or ethylene glycol, are a series of water-soluble polymers with molecular weights from 200 to 14,000 (36,235). They have good lubricity, are heat stable, inert to many chemical agents, and do not hydrolyze or deteriorate.

Poly(ethylene glycols) of molecular weight 200–600 are water-white liquids. The higher members of the series range in consistency from a soft white greaselike petrolatum to hard waxy solids. They are used in cosmetics, ointments, and other products where blandness, water solubility, and lubricity are desired. They are also used as water-soluble lubricants for rubber molds, in textile processing and ceramics, and in metal-forming applications. The polyglycols are good plasticizers, and dispersants for casein and gelatin compositions, glues, zein, cork, and special printing inks. They are fine solvents for dyes, resins, proteins, and many medicaments (see Glycols; Polyethers).

Poly(ethylene Oxide). High molecular weight ethylene oxide polymers are made by coordinative anionic polymerization (236) on alkyls or alkoxides of Group IIA or IIIA metals (see Polyethers). Commercially available materials (Polyox, Union Carbide) are available in a molecular weight range of 90,000–3,800,000. These polymers have diverse applications. A very dilute (<100 ppm) solution in water greatly reduces viscous drag. This is used in firefighting to deliver larger quantities of water through a smaller hose. Agriculturally, it is used to encapsulate seeds at regular spacings in a tape (Seed Tape, Union Carbide) leading to regular plant spacing and elimination of thinning. It is also used for warp sizing, coagulation, and as a water-soluble packaging material (235) (see Flocculating agents; Packaging materials).

Crown Ethers. Cyclic oligomers of ethylene oxide are members of the crown ether family. These have been the subject of intense study because of their cation complexing abilities. A cavity in the center of the molecule is lined with oxygen atoms which hold cations by electrostatic attraction. Each ether has a strong preference for cations whose ionic radius best fits the cavity (237) (see Chelating agents).

Applications of crown ethers have been appearing rapidly. Their primary uses in synthetic organic chemistry have been the solubilization of inorganic reagents in organic solvents and the separation of inorganic salts into individual, naked ions (238). In both cases, new or greatly enhanced reactivity appears (239–240). Potential commercial uses of crown ethers include liquid–liquid extraction (qv) based on ion size, liquid-membrane separations, ion-selective electrodes (qv), and catalysis of the synthesis of specialized chemicals (241).

The physiological effects of crown ethers have not yet been studied extensively. Their ability to interact strongly with alkali or alkaline earth cations makes it likely that they would influence vital processes in which those cations participate (242). It has been reported that the inhalation of ppm levels of 12-crown-4 will cause permanent testicular atrophy as well as other effects (243).

Glycol Ethers. These are made by reaction of ethylene oxide with alcohols. They are used for deicing fluids, and as brake fluids, detergents, and solvents (36) (see Glycols).

Ethanolamines. These are made by the reaction of ethylene oxide and ammonia (244). About one third are used in detergents. Other applications include gas purification, cosmetics, textile specialties, and as chemical intermediates (see Alkanolamines).

Nonionic Surface-Active Agents. These are derived by addition of ethylene oxide to fatty alcohols, alkylphenols, tall oil, alkyl mercaptans, and various polyols such as poly(propylene glycol), sorbitol, mannitol, and cellulose. They are used in household detergent formulations and industrial surfactant applications (245–246) (see Emulsions; Surfactants and detersive systems).

Urethane Intermediates. These are essentially random polymers and may contain both ethylene oxide and propylene oxide in their structures (36) (see Urethane polymers).

Other Derivatives. Ethylene carbonate, made from the reaction of ethylene oxide and carbon dioxide, is used as a solvent. Acrylonitrile (qv) can be made from ethylene oxide via ethylene cyanohydrin; however this route has been entirely supplanted by more economic processes.

BIBLIOGRAPHY

"Ethylene Oxide" in *ECT* 1st ed., Vol. 5, pp. 906–925, by R. S. Aries, Consulting Chemical Engineer, and Henry Schneider, R. S. Aries & Associates; Koert Gerzon, Cornell University (Structure and Reactions); "Ethylene Oxide" in *ECT* 2nd ed., Vol. 8, pp. 523–558, by Henry C. Schultze, Union Carbide Corporation.

1. A. Wurtz, *Ann.* **110,** 125 (1859); *Ann. Chim. Phys.* **55,** 433 (1859).
2. A. Wurtz, *Ann. Chim. Phys.* **69,** 355 (1863).
3. W. Bone and R. Wheeler, *J. Chem. Soc.* **85,** 1637 (1904).
4. L. Reyerson and L. J. Swearingen, *J. Am. Chem. Soc.* **50,** 2872 (1928).
5. R. Willstater and M. Bommer, *Ann.* **422,** 136 (1921).
6. Ger. Pat. 168,291; Fr. Pat. 360,785 (1905–1906); Brit. Pat. 21,941 (Oct. 4, 1906), J. Walter.
7. Fr. Pat. 729,952 (Mar. 27, 1931); and additions 41,254 (July 4, 1931); 41,724 (Sept. 10, 1931); 41,484 (Sept. 25, 1931); 41,810, 41,811 (Apr. 21, 1933); Fr. Pat. 739,562 (Oct. 3, 1931), T. E. Lefort (to Societe Francaise de Catalyse Generalisee).
8. Brit. Pat. 402,438, 402,749 (Dec. 4, 1933), T. E. Lefort (to Societe Francaise de Catalyse Generalisee); U.S. Pat. 1,998,878 (Apr. 23, 1935), T. E. Lefort (to Carbide and Carbon Chemicals Corp.); reissued as U.S. Pats. 20,370 (May 18, 1937), 22,241 (Dec. 29, 1942).
9. *Ethylene Oxide, Brochure F-7618E,* Union Carbide Corporation, New York, 1973.
10. C. J. Walters and J. M. Smith, *Chem. Eng. Prog.* **48,** 337 (1952).
11. J. D. Nickerson and R. McIntosh, *Can. J. Chem.* **35,** 1325 (1957).
12. A. S. Pell and G. Pilcher, *Trans. Faraday Soc.* **61,** 71 (1965).
13. W. F. Giauque and J. Gordon, *J. Am. Chem. Soc.* **71,** 2176 (1949).
14. F. S. Bichowsky and F. D. Rossini, *The Thermochemistry of Chemical Substances,* Reinhold Publishing Corp., New York, 1936, p. 46.
15. F. H. Field and J. L. Franklin, *Electron Impact Phenomena,* Academic Press, New York, 1957.
16. R. Rein and co-workers, *J. Chem. Phys.* **45,** 4743 (1966).
17. G. B. Kistiakowsky and W. W. Rice, *J. Chem. Phys.* **8,** 618 (1940).
18. R. W. Gallant, *Hydrocarbon Process.* **46**(3), 143 (1967).
19. C. L. Yaws, *Physical Properties,* McGraw-Hill Publishing Co., New York, 1977, pp. 167–176.
20. O. Maass and E. H. Boomer, *J. Am. Chem. Soc.* **44,** 1709 (1922).
21. D. R. Stull, E. F. Westrum, and G. C. Sinke, *The Chemical Thermodynamics of Organic Compounds,* John Wiley & Sons, Inc., New York, 1969, p. 419.
22. K. A. Kobe and R. E. Pennington, *Pet. Refiner* **29**(9), 135 (1950).
23. J. D. Olson, *J. Chem. Eng. Data* **22,** 326 (1977).
24. A. Seidell, *Solubilities of Organic Compounds,* D. Van Nostrand Co., Inc., New York, 1941.
25. R. E. Parker and N. S. Issacs, *Chem. Rev.* **59,** 737 (1959).
26. R. Bonaccorsi, E. Scrocco, and J. Tomasi, *J. Chem. Phys.* **52,** 5270 (1970).
27. G. Frenking and co-workers, *Bull. Chem. Soc. Jpn.* **48,** 2769 (1975).
28. H. Fujimoto and co-workers, *Bull. Chem. Soc. Jpn.* **49,** 1508 (1976).
29. C. A. Coulson, *Valence,* Oxford University Press, New York, 1976, pp. 215–216.
30. C. Hirose, *Bull. Chem. Soc. Jpn.* **47,** 1311 (1974).
31. A. Rosowsky in A. Weissberger, ed., *Heterocyclic Compounds,* Vol. 19, pt. 1, Wiley-Interscience, New York, 1964.
32. M. S. Malinovskii, *Epoxides and Their Derivatives,* Daniel Davey & Co., Inc., New York, 1965 (especially for references from Eastern European sources).
33. Ref. 31, pp. 181–222.

34. D. N. Glew and N. S. Rath, *J. Chem. Phys.* **44**, 1710 (1965).
35. R. K. McMullan and G. A. Jeffrey, *J. Chem. Phys.* **42**, 2725 (1965).
36. *Glycols, Brochure F-41515B,* Union Carbide, New York, 1978.
37. J. Farakawa and T. Saegusa, *Polymerization of Aldehydes and Oxides,* Wiley-Interscience, New York, 1963.
38. G. Gee, W. C. E. Higginson, and J. B. Jackson, *J. Chem. Soc.,* 231 (1962).
39. T. H. Baize, *Ind. Eng. Chem.* **53**, 903 (1961).
40. J. Dale, G. Borgen, and K. Daasvatn, *Acta Chem. Scand.* **B28**(3), 378 (1974); U.S. Pat. 3,928,386 (Dec. 23, 1975), J. Dale, G. Borgen, and K. Daasvatn.
41. J. Dale and K. Daasvatn, *J. Chem. Soc. Chem. Commun.,* 295 (1976); U.S. Pat. 3,997,563 (Dec. 14, 1976), J. Dale and K. Daasvatn.
42. A. Lourenco, *Ann. Chem. Phys.* **67**, 275 (1863).
43. J. N. Bronsted, M. Kilpatrick, and M. Kilpatrick, *J. Am. Chem. Soc.* **51**, 428 (1929).
44. J. Koskihallio and E. Whalley, *Trans. Faraday Soc.* **55**, 815 (1959).
45. F. A. Long, J. G. Pritchard, and F. E. Stafford, *J. Am. Chem. Soc.* **79**, 2362 (1957).
46. A. M. Eastham and G. A. Latremouille, *Can. J. Chem.* **30**, 169 (1952).
47. P. O. I. Virtanen, *Suomen Kemistilehti* **34B**, 62 (1961).
48. H. J. Lichtenstein and G. H. Twigg, *Trans. Faraday Soc.* **44**, 905 (1948).
49. J. Koskihallio and E. Whalley, *Can. J. Chem.* **37**, 783 (1959).
50. L. Smith, G. Wode, and I. Widhe, *Z. Physik. Chem.* **130**, 154 (1927).
51. R. Landau and R. E. Lidov in S. A. Miller, ed., *Ethylene: Its Industrial Derivatives,* Ernest Benn Ltd., London, Eng., 1969, p. 613.
52. R. F. Goldstein, *The Petroleum Chemicals Industry,* John Wiley & Sons, Inc., New York, 1958, p. 352.
53. L. Knorr, *Chem. Ber.* **32**, 729 (1899).
54. Ref. 52, pp. 354–355.
55. Ref. 31, pp. 327–328.
56. C. C. Culvenor, W. Davies, and N. S. Heath, *J. Chem. Soc.* **I**, 278 (1949).
57. H. Lemaire in M. J. Schick, ed., *Nonionic Surfactants,* Marcel Dekker, Inc., New York, 1966, pp. 177–179.
58. Ref. 31, p. 330.
59. Ref. 31, pp. 432–434.
60. Ref. 31, pp. 350, 432–440.
61. E. L. Gustus and P. G. Stevens, *J. Am. Chem. Soc.* **55**, 378 (1933).
62. W. Deckert, *Angew. Chem.* **45**, 559 (1932).
63. O. F. Lubatti, *J. Soc. Chem. Ind. (London)* **54**, 424T (1935).
64. W. Zeise, *Chem. Ber.* **66B**, 1965 (1933).
65. Ref. 52, p. 358.
66. M. S. Malinovski and N. D. Medjanzema, *J. Gen. Chem. USSR* **23**, 221 (1953).
67. Ref. 31, pp. 386–417.
68. N. G. Gaylord and E. J. Becker, *Chem. Rev.* **49**, 413 (1951).
69. M. S. Kharasch and O. Reinmuth, *Grignard Reactions of Nonmetallic Substances,* Prentice-Hall, Inc., Englewood Cliffs, N.J., 1954.
70. V. Grignard, *Bull. Soc. Chim. Fr.* **29**, 944 (1903); *Compt. Rend.* **136**, 1260 (1903).
71. Ref. 31, pp. 418–428.
72. Ref. 52, p. 358.
73. Ref. 31, pp. 308–315.
74. Ref. 52, pp. 348–352.
75. C. R. Enyeart in ref. 57, pp. 80–81.
76. Ref. 52, p. 357.
77. A. Petrov, *J. Gen. Chem. USSR* **10**, 981 (1940).
78. J. F. W. McOmie, *Protective Groups in Organic Chemistry,* Plenum Press, New York, 1973, p. 326.
79. S. Jaffe in H. M. England and W. T. Beery, eds., *Proceedings, Second International Clean Air Congress,* Academic Press, Inc., New York, 1971, pp. 316–324.
80. S. A. Greene and L. J. Gordon, *Jet Propul.,* 798 (July 1957).
81. F. A. Burden and J. H. Burgoyne, *Nature* **163**, 723 (1949); *Proc. Roy. Soc. London Ser. A* **199**, 328 (1949).

82. G. B. Shah, E. P. Chock, and R. G. Rinker, *Ind. Eng. Chem. Fundam.* **10**(1), 13 (1971).
83. Ref. 52, p. 358.
84. F. G. A. Stone and H. J. Emeleus, *J. Chem. Soc.* **III,** 2755 (1950).
85. Ref. 51, p. 575.
86. Ref. 31, pp. 453–454.
87. Ref. 31, pp. 343–344.
88. Ref. 51, pp. 567–576.
89. Ref. 51, pp. 576–580.
90. J. F. Norris, *J. Ind. Eng. Chem.* **11,** 817 (1919).
91. *Chem. Mark. Rep.* **215**(5), 9 (July 31, 1978).
92. I. Kiguchi, T. Kumazawa, and T. Nakai, *Hydrocarbon Process.* **55**(3), 69 (1976).
93. G. O. Curme and F. Johnston, *Glycols, ACS Monograph No. 114,* Reinhold Publishing Corp., New York, 1952, Chapts. 2 and 5.
94. *Chem. Eng. News* **56,** 10 (May 1, 1978).
95. *Chem. Week* **122,** 17 (May 24, 1978).
96. S. C. Johnson, *Chem. Eng. Prog.* **72**(9), 25 (1976).
97. Ref. 51, pp. 521–563.
98. P. A. Kilty and W. M. H. Sachtler in H. Heinemann and J. J. Carberry, eds., *Catalysis Reviews,* Vol. 10, Marcel Dekker, Inc., New York, 1974, pp. 1–16.
99. G. H. Twigg, *Trans. Faraday Soc.* **42,** 284,657 (1946); *Proc. Roy. Soc. (London) Ser. A* **188,** 92 (1946).
100. M. Gans and B. J. Ozero, *Hydrocarbon Process.* **55**(3), 73 (1976).
101. B. DeMaglie, *Hydrocarbon Process.* **55**(3), 78 (1976).
102. D. E. Field and co-workers, *Engineering and Cost Study of Air Pollution Control for the Petrochemical Industry,* Vol. 6, *EPA Report No. EPA-450/3-73-006-f,* EPA, Washington, D.C., June 1975.
103. W. Ger. Pat. 2,300,512 (July 26, 1973), R. P. Nielsen and co-workers (to Shell International Research).
104. W. Ger. Pat. 2,448,449 (Apr. 30, 1975), P. A. Kilty (to Shell International Research).
105. Belg. Pat. 638,319 (Oct. 8, 1964), R. S. Davis (to Halcon International).
106. Dutch Pat. Appl. 6,502,927 (Sept. 23, 1964), (to The Dow Chemical Company).
107. Brit. Pat. 1,075,454 (July 12, 1967), (to Halcon International).
108. U.S. Pat. 4,012,425 (Mar. 15, 1977), R. P. Nielsen and J. H. LaRochelle (to Shell Oil Company).
109. K. R. Barker and J. C. Zomerdijk, *Eur. Chem. News Chemscope* **24,** 28 (Oct. 12, 1973).
110. R. Landau and co-workers, *Chem. Eng. Prog.* **64**(3), 17 (1968).
111. C. H. Chilton, *Chem. Eng.* **65**(15), 100 (1958).
112. *Hydrocarbon Process.* **54**(11), 145 (1975).
113. *Hydrocarbon Process.* **52**(11), 129 (1973).
114. P. W. Sherwood, *Chim. Ind. (Paris)* **70,** 1078 (1953); *Pet. Process.* **9,** 1592 (1954).
115. W. E. Vaughan and R. M. Goepp, Jr., *U.S. Dept. Comm. Off. Tech. Serv.,* PB Rep. 79607 (1947); *FIAT Final Report 875,* May 2, 1947.
116. U.S. Pat. 3,552,122 (Jan. 5, 1971), M. Parmegiani, S. D. Milanese, and O. Bellofatto (to Snam Progetti).
117. R. J. Ruff, *Chem. Eng. Prog.* **53**(8), 377 (1957).
118. U.S. Pat. 3,603,085 (Sept. 7, 1971), M. Parmegiani and S. D. Milanese (to Snam Progetti).
119. N. Y. Step, O. N. Dyment, and B. B. Chesnokov, *Khim. Prom. (Moscow)* **48,** 654 (1972).
120. U.S. Pat. 3,083,213 (Mar. 26, 1963), M. L. Courter (to Shell Oil Company).
121. U.S. Pat. 3,878,126 (Apr. 15, 1975), E. G. Foster (to Shell Oil Company).
122. U.S. Pat. 3,119,837 (Jan. 28, 1964), H. A. Kingsley and F. A. Cleland (to Shell Oil Co.).
123. E. T. McBee, H. B. Haas, and P. A. Wiseman, *Ind. Eng. Chem.* **37,** 432 (1945).
124. L. Y. Margolis and O. M. Todes, *Isv. Akad. Nauk SSSR Otd. Khim. Nauk,* 52, (1952); *Chem. Abstr.* **46,** 5413 (1952).
125. P. H. Emmett, *Catalysis,* Vol. VII, Reinhold Publishing Corp., New York, 1960, p. 246.
126. L. Y. Margolis, *Usp. Khim.* **28,** 615 (1959).
127. Brit. Pat. 1,329,252 (Sept. 5, 1973), D. Bryce-Smith and co-workers.
128. A. Ayame and co-workers, *Nippon Kagaku Kaishi* **8,** 1189 (1974).
129. O. Svajgl and co-workers, *Chem. Prum.* **22,** 493 (1972).
130. U.S. Pat. 2,605,239 (July 29, 1952), G. W. Sears (to E. I. du Pont de Nemours & Co., Inc.).

131. U.S. Pat. 2,831,870 (Apr. 22, 1958), W. J. McClements and B. E. Elliott (to Allied Chemical & Dye Corp.).
132. Brit. Pat. 754,593 (Aug. 8, 1956), (to N. V. de Bataafasche Petroleum Maatschappij).
133. U.S. Pat. 2,805,207 (Sept. 3, 1957), F. J. Metzger.
134. Brit. Pat. 811,828 (Apr. 15, 1959), (to The Dow Chemical Company).
135. Brit. Pat. 501,278 (Feb. 21, 1939), (to N. V. de Bataafasche Petroleum Maatschappij).
136. U.S. Pat. 2,686,762 (Aug. 17, 1954), E. L. C. Tollefson (to National Research Council of Canada).
137. A. Cambron and W. A. Alexander, *Can. J. Chem.* **34,** 665 (1956).
138. Jpn. Pat. 46-40256 (Nov. 17, 1971), S. Ishii and co-workers (to Nippon Shokubai Kagaku Kogyo).
139. U.S. Pat. 3,793,231 (Feb. 19, 1974), H. Bergmann and co-workers (to Huels).
140. Brit. Pat. 1,300,971 (Dec. 29, 1972), (to Huels).
141. U.S. Pat. 3,843,491 (Oct. 22, 1974), M. Piro and co-workers (to Snam Progetti).
142. P. Harriott, *J. Catal.* **21,** 56 (1971).
143. R. Wolf and co-workers, *Chem. Tech. (Berlin)* **14,** 600 (1962).
144. U.S. Pats. 2,279,469, 2,279,470 (Apr. 14, 1942), G. H. Law and H. C. Chitwood (to Carbide and Carbon Chemicals Corp.).
145. J. Barthory and co-workers, *Proc. Conf. Appl. Phys. Chem.* **2,** 279 (1971).
146. J. J. Carberry and co-workers, *J. Catal.* **26,** 247 (1972).
147. U.S. Pat. 3,962,136 (June 8, 1976), R. P. Nielsen (to Shell Oil Company).
148. P. W. Sherwood, *Oil Gas J.* **39,** 80 (1957); **40,** 150 (1957).
149. S. Carra and P. Forzatti, *Catal. Rev. Sci. Eng.* **15**(1), 1 (1977).
150. G. K. Boreskov and co-workers, *Kinet. Katal.* **3,** 214 (1962).
151. U.S. Pat. 3,121,099 (Feb. 11, 1964), H. H. A. Endler (to Montecatini).
152. S. Wan, *Ind. Eng. Chem.* **45,** 234 (1953).
153. E. J. Mistrik and J. Kostal, *Int. Chem. Eng.* **13,** 622 (1973).
154. V. A. Rastaturin, *Zh. Prikl. Khim.* **43,** 1343 (1970).
155. A. Montalvo, J. L. San Jose, and F. Franssen, *Rev. Inst. Mex. Pet.* **5**(3), 69 (1973).
156. H. Endler and C. Mazzolini, *Chim. Ind.* **38,** 274 (1956).
157. U.S. Pat. 3,758,418 (Sept. 11, 1973), W. J. Leonard (to Shell Oil Company).
158. Brit. Pat. 1,411,315 (Oct. 22, 1975), A. L. Gelbshtein (to the U.S.S.R.).
159. Rom. Pat. 55,800 (Sept. 15, 1973), I. Oprescu and co-workers (to Institutul Petrochim).
160. S. A. Miller in ref. 51, p. 1057.
161. Brit. Pat. 748,957 (May 15, 1956), R. B. Egbert (to Chempatents Inc.).
162. Belg. Pat. 707,567 (June 5, 1968), D. Brown (to Halcon International).
163. F. L. W. McKim and A. Cambron, *Can. J. Res.* **27B,** 813 (1949).
164. L. M. Kaliberdo and co-workers, *Kinet. Katal.* **8,** 463 (1967).
165. U.S. Pat. 3,766,714 (Oct. 23, 1973), J. W. Cunningham (to Shell Oil Company).
166. U.S. Pat. 3,745,092 (July 10, 1973), R. G. Vanderwater (to Shell Oil Company).
167. U.S. Pat. 3,729,899 (May 1, 1973), J. W. Cunningham and co-workers (to Shell Oil Company).
168. U.S. Pat. 3,964,980 (June 22, 1976), B. I. Ozero (to Halcon International).
169. U.S. Pat. 3,856,484 (Dec. 24, 1974), G. Cocuzza (to Societa Italiana Resine).
170. Belg. Pat. 813,653 (July 31, 1974), (to Nippon Shokubai Kagaku Kogyo).
171. Brit. Pat. 1,435,848 (May 19, 1976), (to Nippon Shokubai Kagaku Kogyo).
172. Ger. Pat. 2,364,149 (June 27, 1974), G. Cocuzza and co-workers (to Societa Italiana Resine).
173. U.S. Pat. 3,644,432 (Feb. 22, 1972), R. Hoch and co-workers (to Halcon International).
174. U.S. Pat. 3,335,547 (Aug. 15, 1967), D. Garrett (to Halcon International).
175. Fr. Pat. 1,330,900 (May 20, 1963), Paire and co-workers (to Kuhlmann Co.).
176. U.S. Pat. 3,265,593 (Aug. 9, 1966), D. G. Leis (to Union Carbide Corporation).
177. U.S. Pat. 3,523,957 (Aug. 11, 1970), U. Tsao (to Lummus Co.).
178. G. L. Bata and J. E. Hazell, *Oil Gas J.* **63,** 107 (Oct. 4, 1965).
179. D. G. Weaver and J. L. Smart, "Ethylene Oxide Derivatives, Glycols and Ethanolamines" in W. J. Murphy, ed. *Modern Chemical Processes,* Vol. 6, Reinhold Publishing Corporation, New York, 1961, pp. 92–98.
180. E. T. Borrows and D. A. Caplin, *Chem. Ind. (London),* 532 (1953).
181. W. L. Faith, D. B. Keyes, and R. L. Clark, *Industrial Chemicals,* 2nd ed., John Wiley & Sons, Inc., New York, 1957, pp. 383–388.
182. R. Landau, *Pet. Refiner* **32**(9), 146 (1953).
183. U.S. Pat. 3,993,673 (Nov. 23, 1976), C. H. McMullen (to Union Carbide Corporation).

184. A. G. Lee, *The Chemistry of Thallium,* Elsevier, New York, 1971.
185. U.S. Pat. 4,021,453 (May 3, 1977), W. F. Brill (to Halcon International).
186. Belg. Pat. 853,864 (Oct. 24, 1977), N. Rizkalla and A. N. Naglieri (to Halcon International).
187. U.S. Pat. 4,058,542 (Nov. 15, 1977), N. Rizkalla and A. N. Naglieri (to Halcon International).
188. Belg. Pat. 855,127 (Nov. 28, 1977), (to Halcon International).
189. W. Ger. Pat. 2,541,526 (Apr. 1, 1976), A. P. Gelbein (to Lummus Co.).
190. U.S. Pat. 3,427,235 (Feb. 11, 1969), J. A. M. Le Duc (to Pullman Inc.).
191. U.S. Pat. 3,288,692 (Nov. 29, 1966), J. A. M. Le Duc (to Pullman Inc.).
192. U.S. Pat. 3,723,264 (Mar. 27, 1973), J. A. M. Le Duc and co-workers (to Pullman Inc. and Farbenfabriken Bayer Aktiengesellschaft).
193. A. Renken, M. Muller, and C. Wandrey, *Proc. 4th Intl. Conf. Chem. Reac. Eng. Heidelberg,* 107 (1976).
194. W. Ger. Pat. 2,605,991 (Aug. 25, 1977), A. Renken, C. Wandrey, and M. Muller (to Hoechst A.G.).
195. U.S. Pat. 2,430,443 (Nov. 11, 1947), S. B. Becker (to Standard Oil Company).
196. U.S. Pat. 2,578,841 (Dec. 18, 1951), N. C. Robertson and R. T. Allen (to Celanese Corporation).
197. U.S. Pat. 2,600,444 (June 17, 1952), F. W. Sullivan, Jr. (to GAF Corporation).
198. U.S. Pat. 2,628,965 (Feb. 17, 1953), F. W. Sullivan, Jr. (to GAF Corporation).
199. U.S. Pat. 2,684,967 (July 27, 1954), C. H. O. Berg (to Union Oil Company).
200. T. E. Corrigan, *Pet. Refiner* **32**(2), 87 (1953).
201. U.S. Pat. 3,948,609 (Apr. 6, 1976), N. M. Guseinov and co-workers (to the U.S.S.R.).
202. Brit. Pat. 1,379,797 (July 28, 1972), N. M. Guseinov and co-workers (to the U.S.S.R.).
203. U.S.S.R. Pat. 562,555 (June 25, 1977), N. M. Guseinov and co-workers (to the U.S.S.R.).
204. *Processing* **21**(11), 54 (1973).
205. Selected issues of *Chem. Mark. Rep.,* (1978).
206. U.S. International Trade Commission, Washington, D.C., 1977 data are preliminary.
207. *Chem. Econ. Eng. Rev.* **9,** 45 (Dec. 1977).
208. *Chim. Actual.,* 1, 3 (June 1, 1977).
209. *Hydrocarbon Process.* **57**(6), Section 2 (1978).
210. *U.S. Code of Federal Regulations,* Title 49, paragraph 173.124, U.S. Government Printing Office, Washington, D.C., 1977.
211. Guide to Ethylene Oxide Tank Car Unloading, *Brochure F-47004,* Union Carbide Corporation, New York, 1978.
212. R. L. Anderson, "Ethylene Oxide" in F. D. Snell and L. S. Ettre, eds., *Encyclopedia of Industrial Chemical Analysis,* Vol. 12, John Wiley & Sons, Inc., New York, pp. 317–340, 1971.
213. A.H. Qazi and N. H. Ketcham, *Am. Ind. Hyg. Assoc. J.* (38), 635 (Nov. 1977).
214. *NIOSH Manual of Analytical Methods,* 2nd ed., Vol. 3, U.S. Dept. of Health, Education, and Welfare, Washington, D.C., Apr. 1977, Method No. 5286.
215. *Gas Sterilants, Brochure F-2085E,* Linde Division, Union Carbide Corp., S. Plainfield, N.J., 1978, p. 16.
216. N. E. Bolton and N. H. Ketcham, *Arch. Environ. Health* **8,** 711 (1964).
217. J. C. Gage, *Analyst* **82,** 587 (1957).
218. F. E. Critchfield and J. B. Johnson, *Anal. Chem.* **29,** 797 (1957).
219. *Properties and Essential Information for Safe Handling and Use of Ethylene Oxide, Chemical Safety Data Sheet SD-38,* Manufacturing Chemists Association, Washington, D.C., 1971.
220. *Codes of Practice for Chemicals with Major Hazards: Ethylene Oxide, Brochure CPP4,* Chemical Industry Safety and Health Council of the Chemical Industries Association, London, Eng., 1975.
221. *Special Occupational Hazard Review with Control Recommendations for the Use of Ethylene Oxide as a Sterilant in Medical Facilities, NTIS PB-274795,* National Institute for Occupational Safety and Health, Washington, D.C., 1977.
222. R. J. Sexton and E. V. Hensen, *Arch. Ind. Hyg. Occup. Med.* **2,** 549 (1950).
223. K. Verschueren, *Handbook of Environmental Data on Organic Chemicals,* Van Nostrand Reinhold, New York, 1977, pp. 329–331.
224. *Fed. Reg.* **43,** 3801 (1978).
225. *Fed. Reg.* **43,** 27474 (1978).
226. *Tox-Tips,* Smithsonian Science Information Exchange, Washington, D.C., Feb. 1977, pp. 9–17.
227. *Tox-Tips,* Smithsonian Science Information Exchange, Washington, D.C., Dec. 1976, p. 11.
228. *Tox-Tips,* Smithsonian Science Information Exchange, Washington, D.C., June 1978, pp. 25–31.
229. H. Grosse-Wortmann, *Chem. Ing. Tech.* **42**(2), 85 (1970).

230. J. H. Burgoyne, K. E. Bett, and R. Lee, *Chemical E Symposium Series No. 25,* Institution of Chemical Engineers, London, Eng., 1968, pp. 1–7.
231. W. G. Courtney, W. J. Clark, and C. M. Slough, *ARS J.* **32,** 1530 (1962).
232. Ref. 215, p. 5.
233. B. Kit and D. S. Evered, *Rocket Propellant Handbook,* The MacMillan Co., New York, 1960, pp. 220–225.
234. U.S. Pat. 3,940,443 (Feb. 24, 1976), C. A. Glass (to U.S.A. as represented by the Secretary of the Navy).
235. F. W. Stone and J. J. Stratta, "Ethylene Oxide Polymers" in N. M. Bikales, ed., *Encyclopedia of Polymer Science and Technology,* Vol. 6, Wiley-Interscience, New York, 1967, pp. 103–145.
236. J. Farakawa and T. Saegusa, *Polymerization of Aldehydes and Oxides,* Interscience, New York, 1963.
237. R. M. Izatt and J. J. Christensen, eds., *Synthetic Multitidentate Macrocyclic Compounds,* Academic Press, Inc., New York, 1978.
238. G. Weber and co-workers, *Angew. Chem. Ind. Ed. Eng.* **18,** 226 (1979).
239. G. W. Gokel and H. Dupont Durst, *Synthesis,* 168 (1976).
240. C. L. Liotta in ref. 237, pp. 111–206.
241. R. A. Schwind in ref. 237, pp. 289–307.
242. C. J. Pedersen in ref. 237, pp. 47–48.
243. B. K. J. Leong, T. O. T. Ts'o, and M. B. Chenoweth, *Toxicol. Appl. Pharmacol.* **27,** 342 (1974).
244. *Ethanolamines, Brochure F-41175C,* Union Carbide, New York, 1970.
245. *Chemicals and Plastics Physical Properties, Brochure F-44086,* Union Carbide, New York, 1973.
246. Ref. 57, pp. 44–371.

JAMES N. CAWSE
JOSEPH P. HENRY
MICHAEL W. SWARTZLANDER
PARVEZ H. WADIA
Union Carbide Corporation

ETHYLENIC ALCOHOLS. See Acetylene-derived chemicals.

ETHYLIDENE DIACETATE, $CH_3CH(OOCH_3)_2$. See Acetic acid.

ETHYL MERCAPTAN (ETHANETHIOL), CH_3CH_2S. See Thiols.

ETHYNYLATION. See Acetylene-derived chemicals.

EUTROPHICATION. See Water—Water pollution.

EVAPORATION

Evaporation by its broadest definition is the conversion of a liquid to a vapor and applies to such widely diverse equipment as boilers, cooling towers, dryers, and humidifiers, and losses from fields, storage tanks, and reservoirs. In the narrower chemical engineering sense, evaporation is the removal of volatile solvent from a solution or a relatively dilute slurry by vaporizing the solvent. In nearly all industrial applications, the solvent is water and in most cases the nonvolatile residue is the valuable constituent. Evaporation differs from distillation in that when the volatile stream consists of more than one component no attempt is made to separate these components. Thus production of distilled water from impure feedwater utilizes an evaporator rather than a distillation unit even though small capacity units are usually called stills. Although the product of an evaporator system may be a solid, the heat required for vaporization of the solvent must be transferred to a solution or a slurry of the solid in its saturated solution in order that the device be classified as an evaporator rather than a dryer. It is not unusual for an evaporator to be used to produce a solid as its only product. For instance, table salt is produced by feeding a saturated brine to an evaporator, precipitating the salt as water is removed. A side stream of salt crystals in brine is withdrawn to a filter or centrifuge where the salt is recovered in essentially dry form; the filtrate is returned to the evaporator as a supplementary feed. Thus the heat required for evaporation of the water is transferred to a slurry in the evaporator even though the only material leaving the system is a solid (except for the evaporated water and usually a small bleed of brine that is necessary to purge from the system the impurities entering with the feed brine).

The highly varied purposes for which evaporators are used industrially include: (1) reducing the volume to economize on packaging, shipping, and storage costs (eg, of salt, sugar, caustic soda, orange juice, and milk); (2) obtaining a product in its most useful form (eg, salt from brine or sugar from cane juice); (3) eliminating minor impurities (eg, salt, sugar); (4) removing major contaminants from a product, eg, diaphragm cell caustic soda solutions contain more salt than caustic when produced but practically all the salt can be precipitated by concentrating to a 50% NaOH solution; (5) concentrating a process stream for recovery of resources, eg, pulp mill spent cooking liquor, if concentrated sufficiently in an evaporator, can be burned in a boiler to produce steam, yielding also an ash that can be used to reconstitute fresh cooking liquor; (6) concentrating wastes for easier disposal, such as nuclear reactor wastes, dyestuff plant effluents, and cooling tower blowdown streams; (7) transforming a waste into a valuable product, such as spent distillery slop after alcohol recovery, which can be concentrated to produce an animal feed; and (8) recovering distilled water from impure streams such as sea water and brackish waters.

Salt was produced in prehistoric times from seawater or saline waters by solar evaporation in ponds. This method is still in widespread use and probably accounts for more total evaporation than any other process. The first artificially heated evaporators date back to Roman times, again for salt, in flat pans over wood fires. Originally the pans were of lead, later of wrought iron, and such pans heated by coal were still in use in England in the 1960s (primarily to produce a salt of low bulk density for which there was a substantial export market). Similar open pans, except heated by steam coils immersed in the brine, are still used for production of salt with special grain characteristics in the United States. Evaporators were introduced by the sugar industry, the first steam-heated evaporator about 1800, the first one using vacuum in 1812, the first multiple-effect type in 1843, and the first vapor-compression evaporator about 1880. Today, by far the largest number of evaporators is steam heated and use multiple-effect or vapor compression as the means of reducing the energy required for evaporation.

Steam-Heated Evaporators

The three principal requirements of steam-heated evaporators are: (*1*) transfer to the liquid of large amounts of heat needed to vaporize the solvent, (*2*) efficient separation of the evolved vapor from the residual liquid, and (*3*) accomplishing these aims with the least expenditure of energy justifiable by the capital cost involved. Most steam-heated evaporators use metal tubes as heat transfer surfaces, although some employ flat plates or, for special corrosion problems, impregnated carbon tubes. Final separation of the entrained liquid from evolved vapor may be obtained in external centrifugal or mesh-type separators but preliminary separation is generally required. An evaporator is frequently also used as a crystallizer, and therefore its design may be influenced by the need to produce crystals of the desired size, shape, and purity without adversely affecting the vapor–liquid separation and heat transfer functions (see Crystallization). Many types of evaporators are available to perform these functions. The choice depends on the characteristics of the liquor being handled with regard to its heat transfer properties and its tendencies toward salting, scaling, fouling, foaming, and corrosion. The extent to which it is desired to conserve heat also influences the choice of evaporator type.

Evaporator may refer either to the type of construction utilized or to the entire assemblage of equipment in a single installation. Thus a single multiple-effect evaporator may contain a number of effects of either the same or different evaporator types. An effect is a section of the evaporator heated by steam at one pressure and releasing water (vapor) at a lower pressure to another section. The term steam generally indicates the heat supply, whereas vapor means the material evaporated. Thus vapor from one effect becomes steam at the next effect. The term prime steam identifies the steam supplied from an outside source to operate the evaporator (see also Steam). An effect may consist of several bodies, all operating at the same steam and vapor pressures. The purpose of more than one body in an effect may be to handle liquor at different concentrations, or the result of size limitations or of additions to increase the capacity of an existing evaporator.

Natural Circulation Evaporators. Natural circulation evaporators (see Fig. 1) were the first developed commercially and still represent probably the largest number of units in operation. These evaporators utilize the density difference between the liquid and the generated vapor to circulate the liquid past the heating surface and thereby give good heat-transfer performance. The heat transfer tubes may be either vertical or horizontal, with the liquor either inside or outside the tubes. The horizontal tube type of Figure 1a has the heating steam inside the tubes that are immersed in the boiling liquid. This type was originally built with rectangular cast-iron bodies having a hemicylindrical top that required little floor space or headroom. Currently, the major use is for making distilled water for boiler feed. These evaporators are incorporated in the power plant cycle, usually as single effects heated by turbine bleed steam and

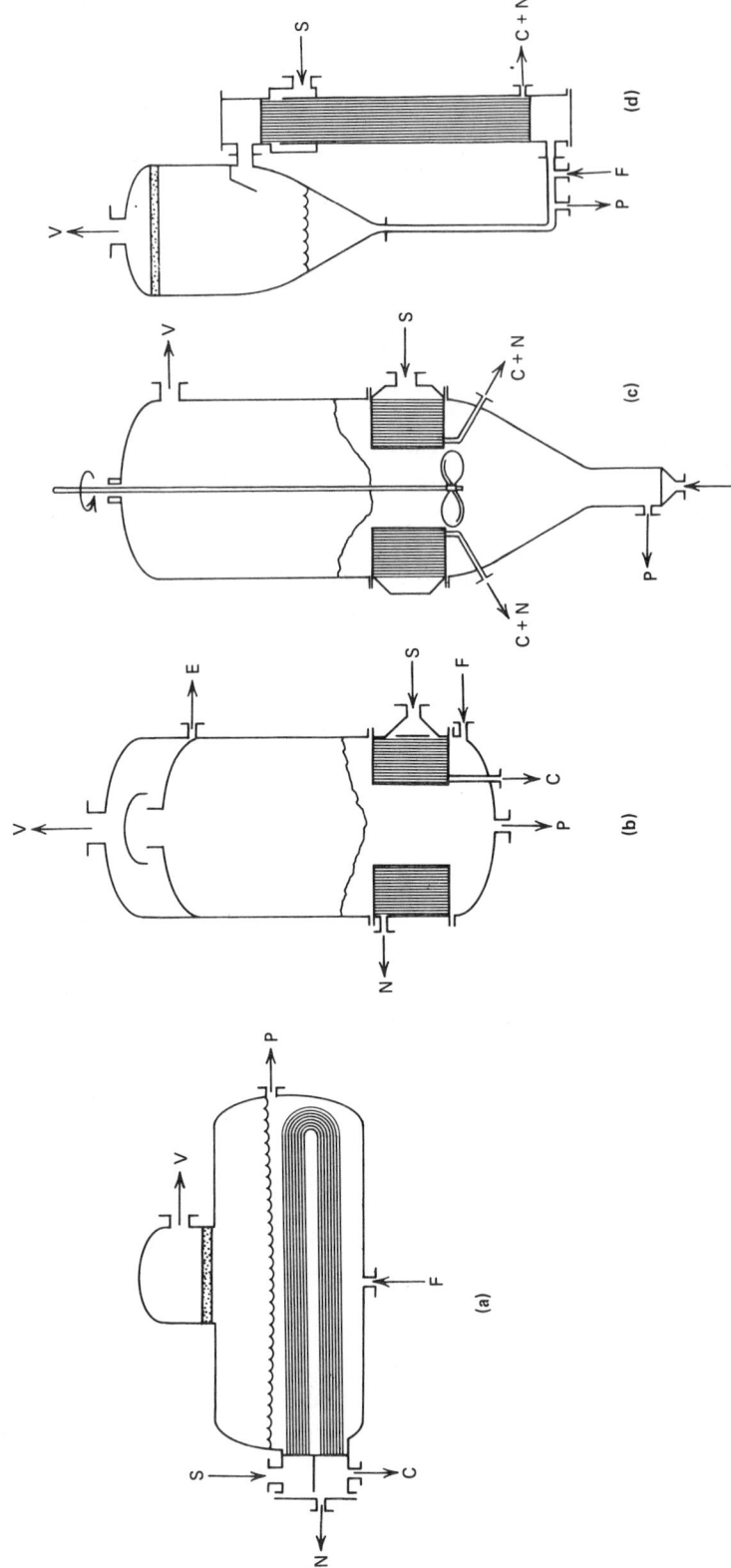

Figure 1. Natural circulation evaporators: (**a**) horizontal-tube, (**b**) short-tube vertical, (**c**) propeller calandria, and (**d**) long-tube recirculating. Symbols: C = condensate, E = entrainment return, F = feed, N = noncondensibles' vent, P = product or concentrate, S = steam, and V = vapor.

exhausting vapor to the feed heater circuit. They frequently operate under considerable pressure, and therefore, employ horizontal cylindrical shells to better withstand the pressures and give a large liquid surface area for efficient disengagement. The type shown in Figure 1**b** was developed long ago, mainly for use in sugar mills, and was called the standard evaporator. Today it is called the calandria or short-tube vertical (STV) evaporator. It employs fairly large diameter (usually 0.05 m) tubes only about 2 m long that are easily cleaned mechanically. Good heat transfer requires downtakes to permit recirculation through the tubes. These must have a flow area on the order of 60% or more of the flow area through the tubes themselves. The use of large diameter short tubes and the need for a large downtake area result in a large body diameter relative to the amount of heating surface provided, usually larger than would be needed for vapor–liquid disengagement alone.

The natural circulation types of Figure 1**a** and 1**b** are not generally suited for handling feeds that deposit appreciable amounts of solids. The reason is that circulation is by boiling action alone and any interruption in boiling, or even a reduction in boiling rate, allows the solids to drop out of suspension, increasing the tendency for subsequent deposition to occur on the heating surface rather than forming new solids in suspension. However, by adding a propeller in the downtake, as in Figure 1**c**, the solids can be kept better in suspension. This type, called the propeller calandria, has been in use for about 70 years for crystallizing table salt from brine. Although it might be thought of as another version of the forced circulation evaporator, its heat transfer performance is about the same as the STV type of Figure 1**b**, indicating that the propeller is not really contributing directly to heat transfer performance but only keeps salts from forming on the tubes, which would then impede heat transfer.

The natural circulation type shown in Figure 1**d** is called the long-tube recirculation (LTR) or recirculating LTV evaporator. It employs longer tubes in an external heating element that is more easily accessible for cleaning. In this case, the vapor-head size is determined only by disengagement requirements and can be appreciably smaller than those shown in Figures 1**a–c.** Smaller vapor-head size with resultant shorter holdup time and the easier cleaning make these evaporators well-suited for degradable products such as milk.

Forced Circulation Evaporators. The forced circulation evaporator, suitable for the largest variety of applications, is the most expensive type. It usually consists of a shell-and-tube heat exchanger, a vapor–liquid separator (variously called vapor head, vapor body, separator, flash chamber, or body), and a pump to circulate the liquor from the body through the heater and back to the body. The system is usually arranged so that there is no boiling in the heater. The heat input is, therefore, absorbed as sensible heat, and vapor liberation does not occur until the liquor enters the flash chamber. Absorption of the heat input as sensible heat results in a temperature rise that reduces the net temperature difference available for heat transfer. To keep this temperature rise to reasonable limits, usually on the order of 3–6 K, requires circulating large volumes of liquor relative to the amount evaporated (ca 0.11 m^3 liquor per kg evaporated for dilute aqueous solutions at a 5 K rise). There is also an upper limit to temperature rise, usually on the order of 10 K, beyond which flashing at the entry to the flash chamber becomes so violent that large masses of liquor are ejected with the vapor. This makes entrainment separation more difficult and may impose structural shock loads on the separator. The head requirements of the circulating pump are generally quite low, consisting primarily of conventional friction and acceleration and deceleration losses at heater and body inlet and outlet, plus vortex losses in the body. The circulating pump is, therefore, usually of the propeller or mixed-flow type.

The vapor body of a forced circulation evaporator is sized primarily for vapor–liquid separation and is arranged so that as much of the liquid as possible entering from the heater "sees" the pressure existing in the vapor space. When that does not occur, some liquid is recycled to the pump without flashing fully to the equilibrium pressure, resulting in another loss in ΔT available for heat transfer, which is variously called the submergence, short-circuiting, nonequilibrium, or Δ' loss. This loss can be minimized by returning liquor above the liquid level in the body (which adds to pump head) or by introducing the liquor tangentially, which sets up secondary circulation patterns in the body, to bring more of the heated liquid to the surface. If the evaporator is used as a crystallizer, the sizing of the vapor body may also be influenced by crystallization requirements. These may dictate the liquor volume in the system, crystal surface area available for growth, temperature rise through the heater, and physical arrangement of the body. If used as a crystallizer, discontinuities should be avoided in the walls near the operating level or in the splash zone, such as peepholes, manholes, and body flanges, on which salts can deposit. Furthermore, body walls should be so smooth that the salt does not adhere, or else so rough that it stays in place until cleaning time. The intermediate situation is to be avoided since it results in salt coming off the walls in thicknesses on the order of 0.01–0.1 m which can be transported by the circulating pump and deposited over the inlet ends of the tubes in the heater.

The heating element of a forced circulation evaporator is usually of the conventional shell-and-tube type, most often single pass and vertical but frequently double pass and horizontal. The heater is usually located far enough below the liquor level in the body so that hydrostatic head prevents boiling in the tubes. In crystallizing service it is desirable, but not always possible, to locate the heater far enough below the liquor level so that boiling does not occur even in a tube which has had its inlet blocked and thus has liquor in temperature equilibrium with the heating medium. This avoids complete filling of the tube with cemented solids which are most difficult to remove. Tube size and length are chosen to give reasonable tube velocities (1.5–3 m/s) for the circulation rate available and the heating surface needed. In crystallizing service, small tube diameters (less than 0.03 m) are avoided to reduce risk of plugging, and the heaters are preferably vertical single pass to afford more uniform distribution of flow to all tubes. Vortices in circulating lines and heater-inlet water boxes may result in such nonuniform velocities that there is little or no flow at all in some tubes and they quickly became plugged with solids.

Several configurations of forced circulation evaporators are shown in Figure 2. The most common arrangement is shown in Figure 2a with an external vertical single-pass heater and a tangential inlet to the body. Figure 2b shows the Oslo type of crystallizing evaporator in which the crystals are retained in a fluidized bed below the flash chamber. Since it is not always necessary to avoid boiling in the heating element, it can project into the vapor head, as shown in Figure 2c. Auxiliaries shown in Figure 2 are not exclusive to each type of unit. Entrainment separators shown are top- and bottom-outlet centrifugal type and knit-mesh, respectively. The evaporator in Figure 2a includes a swirl breaker over the circulating pump inlet to reduce vortex losses in the vapor head and an elutriation leg to size, wash, and cool a crystallized product with part or all of the feed. In all cases, the steam inlet and vent outlet on the heat exchanger are placed so as to provide a positive flow path over the heating surface between the

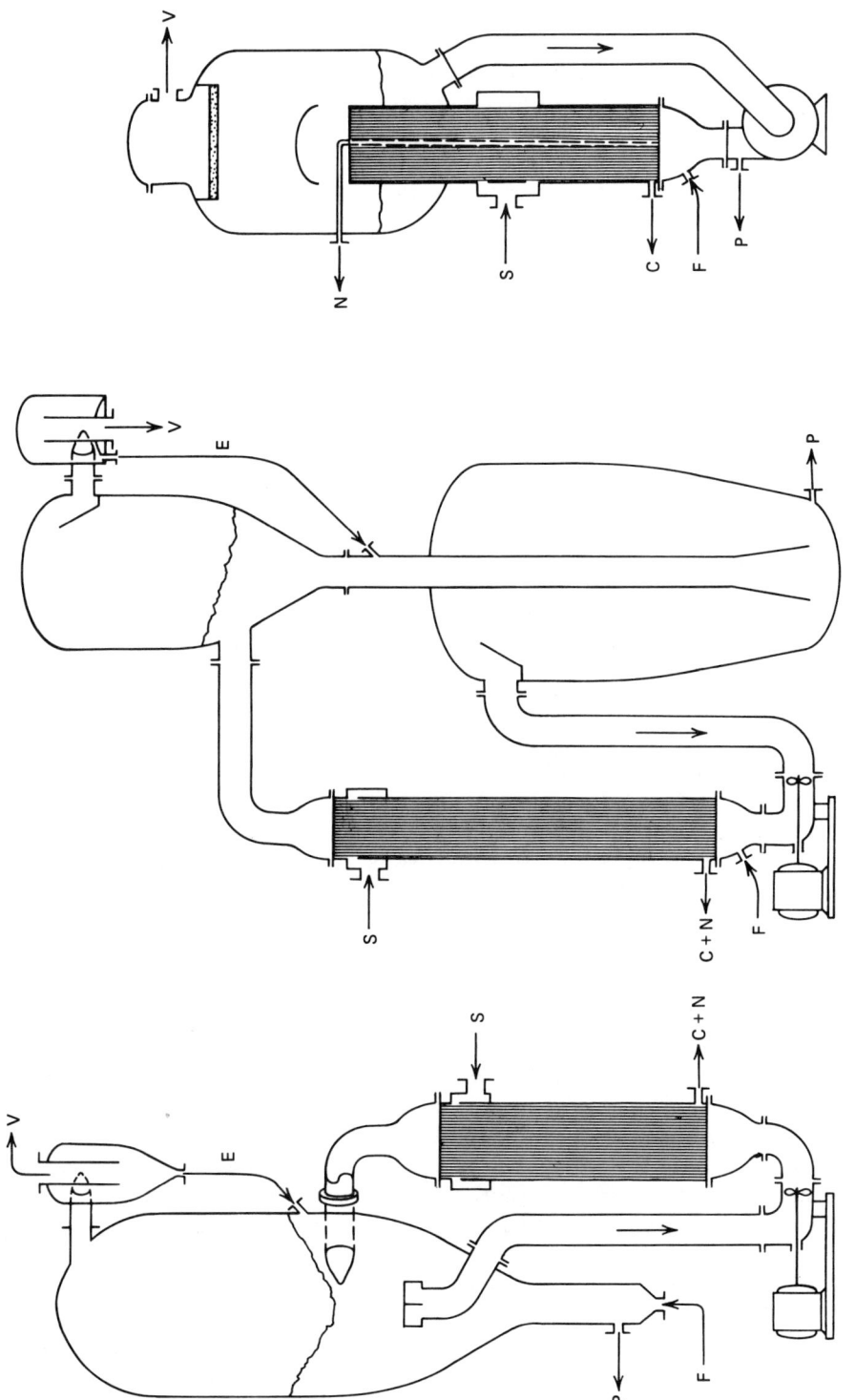

Figure 2. Forced circulation evaporators: (a) submerged-tube, shown as circulating magma crystallizer, (b) submerged-tube, shown as suspension type crystallizer, and (c) boiling type. (See Fig. 1 for symbols.)

two. Forced circulation evaporators can be built for high single-unit capacities, with bodies as large as 15 m in diameter. The circulating pump is usually the limiting factor and it is not unusual to provide a single large body with as many as four or five separate heaters and circulating systems. For extreme fouling or salting conditions, the individual heaters can be arranged so they can be valved from the evaporator and cleaned without interrupting system operation.

Film-Type Evaporators. The film-type evaporators illustrated in Figure 3 were developed 50 to 80 years ago. Figure 3a shows the *rising film* or long-tube vertical (LTV) evaporator most widely used in the United States. It consists of a vertical shell-and-tube heat exchanger surmounted by a vapor–liquid separator. Another version uses an offset vapor head similar to Figure 1d. Tubes are generally 0.05 m or less in diameter and 6–10 m long, which permits packing a large amount of heating surface into a single shippable tube bundle (on the order of 3000 m^2). Because of the simplicity of construction, costs per heating surface area are the lowest of any type, and heat transfer performance is good under most operating conditions. Liquor flow through the LTV is generally once-through (which distinguishes it from the LTR of Figure 1d). However, if feed and discharge liquor properties differ widely, the evaporator may have partitions in the upper vapor head and lower inlet water box so that the effluent from one section can be returned to the inlet of another section of the total number of tubes in the bundle, and only the last group of tubes must handle liquor at the finished density. The LTV-type evaporator cannot handle crystallizing solutions but is excellent for foaming solutions because the deflector above the tube bundle acts as a foam breaker. Its widest use is for kraft-mill black liquor, which has foaming, fouling, and scaling tendencies and becomes quite viscous as it approaches discharge concentration (see Pulp).

The LTV is classified as a film-type evaporator because boiling takes place within the tubes and the vapor–liquid mixture is usually in the annular or film-flow regime for much of the tube length. High vapor velocities are generated and the interfacial shear also causes the liquid to move at high velocity, yielding good heat transfer coefficients. However, frictional pressure drop, acceleration head developed as the vapor is generated, and hydrostatic head of the vapor–liquid mixture all cause the pressure and hence boiling temperature to be higher within the tube than at the tube exit, the temperature and pressure increasing with distance down from the outlet. Thus even if the feed is at its boiling point at the pressure in the vapor head, there is a section in the lower part of the tubes where the liquid cannot boil and is traveling at low velocity, and thus has poor heat transfer characteristics. Proper selection of tube dimensions helps to minimize the adverse effects of reduced available ΔT and poor heat transfer in the nonboiling zone, as does the use of preheaters on the feed. These preheaters may be either external or incorporated as a part of the main tube bundle. Since LTV performance depends on the vapor–liquid velocities developed in the boiling zone, it follows that heat transfer coefficients are a function of the load or of the overall ΔT. Coefficients are low at low temperature differences and low temperatures. This type of evaporator sometimes suffers from instability problems when operated under part-load conditions.

The *falling-film evaporator* shown in Figure 3b is an inverted version of the LTV that greatly reduces the adverse effects of pressure drop on available ΔT exhibited by the rising-film LTV. The hydrostatic head loss is eliminated, acceleration losses are lower because the liquid film is not accelerated substantially by the vapor flow,

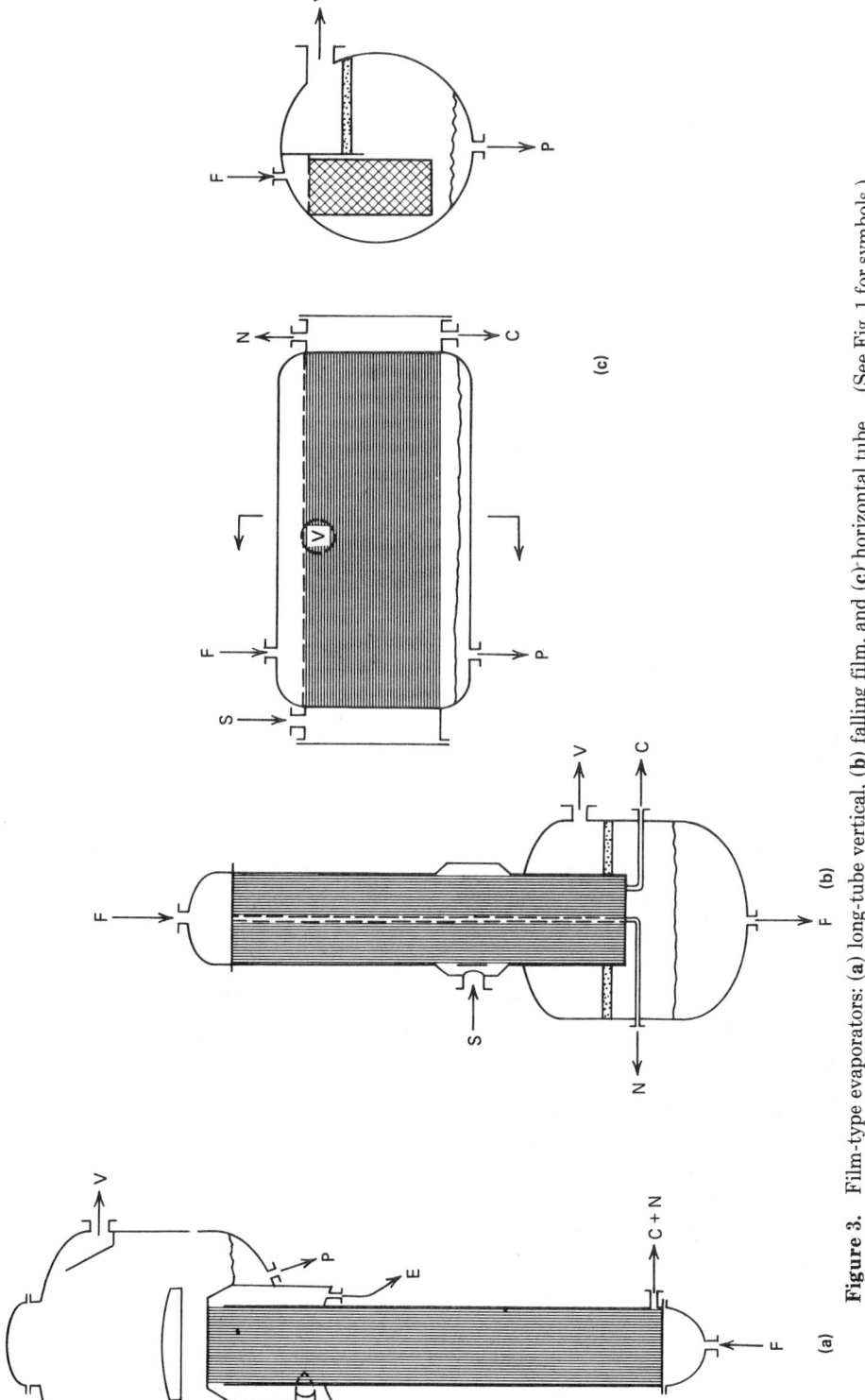

Figure 3. Film-type evaporators: (**a**) long-tube vertical, (**b**) falling film, and (**c**) horizontal tube. (See Fig. 1 for symbols.)

and the frictional pressure drop is generally only a little more than that of vapor flowing alone in a dry tube. In addition, heat transfer performance is practically the same regardless of whether or not the film is boiling. This eliminates a poorly performing nonboiling zone at the inlet end of the tubes, even when the feed enters far below its boiling point. Uniform feed distribution to all tubes is usually the principal difficulty. Current methods include individual distributors in each tube, a full cone spray nozzle in the upper water box, a perforated plate above the top tube sheet with holes allowing feed to impinge on the tube sheet web between tubes or, for high flow rates, no distribution devices at all. Feed rates are generally on the order of $0.4–1.6$ m^3/(s·m) of wetted tube perimeter. Since these are usually higher than the net feed rate, some recirculation is required. Thus the falling-film evaporator also belongs to the forced circulation class. As with the LTV, the inlet water box and the vapor head may be partitioned, permitting evaporation of the feed in stages; only the last stage has to operate at final density. This is advantageous when viscosity or boiling point increase rapidly as discharge concentration is approached but requires separate pumps for each stage.

The principal advantages of the falling-film evaporator are its good heat transfer performance, even at low temperature and low temperature differences, its low initial cost, and its excellent vapor–liquid separation characteristics. Principal applications have been for citrus juices, where performance at low temperature and low holdup is important, and applications requiring operation at low temperature differences, such as vapor compression or multiple-effect evaporators needing a large number of effects to be economical, eg, for producing fresh water from saline waters. The falling-film evaporator is normally not suitable for the usual crystallizing operations but has been very effective in scaling environments, scaling being avoided by maintaining a suspension of the scaling ingredient in the circulating liquid (1).

A special heating surface developed for this type of evaporator is the doubly-fluted tube (2), which gives two to three times the heat transfer coefficient of a smooth wall tube. On the steam side, the condensing coefficient is enhanced by the presence of longitudinal grooves. Surface tension forces draw the condensate into the grooves, leaving the area between the grooves bare so that they give coefficients comparable to those achieved with dropwise condensation. Heat transfer coefficients on the boiling liquor side are also markedly higher than they are on smooth tubes. The principal application for these tubes is for production of distilled water from seawater (see Water).

A combination of rising-film–falling-film evaporator utilizes a two-pass vertical heating element situated above the vapor head. The first pass, fed at the bottom, operates as a rising film and the vapor–liquid mixture then goes through the second pass as a falling film. This type evaporator usually operates with recirculation but the amount of recirculation is less than in a normal forced circulation or falling-film evaporator. It was developed primarily for handling high viscosity fluids and for applications requiring low residence times.

Another version of the film evaporator is that in which liquid is showered over the outside of substantially horizontal tubes as shown in Figure 3c. This was originally called the Lillie evaporator and is now termed the horizontal-tube or *spray-film evaporator*. It has essentially the same advantageous heat transfer characteristics as the falling-film type of Figure 3**b** and in addition requires much less headroom. Its main disadvantages are poorer vapor–liquid separation and greater difficulty of cleaning fouled tubes. Uniform liquor distribution at the top of the tube bundle is usually accomplished by perforated troughs or spray nozzles. Uniform distribution may not prevail further down in the bundle as vapor release from the side or ends of the bundle tends to drive liquid with it instead of the liquid falling vertically from one tube row to the next. In the original Lillie and in some modern versions, the tubes are rolled into a tube sheet at only one end and are sloped uphill so that condensate can drain countercurrent to the steam. The other end is then fitted with a perforated plug to act as a vent. Other versions employ tube sheets at both ends and frequently have several steam side passes, with a smaller number of tubes in each succeeding pass, to provide a tapered flow path from steam inlet to vent outlet.

The *wiped-film* or *agitated-film evaporator,* shown in Figure 4, uses mechanical energy to promote heat transfer. It employs a single large-diameter straight or tapered tube as the heating surface, in which a set of blades is rotated. The blades maintain either a fixed close clearance from the wall or actually ride on the film of liquid and help to carry the liquid as a film around and along the length of the heating surface. The cost of these evaporators per unit of heating surface is very high and the capacity is relatively low, since a maximum of only about 20 m^2 of surface can be provided in a single tube. Thus, these evaporators are used primarily only for materials that cannot be handled in other evaporators, such as highly viscous liquids or liquids requiring very low residence times in contact with the heating surface. Because of the high capital cost and limited heating surface, these units are usually operated as a single effect at high ΔTs. This and structural considerations require that the large diameter tubes be of fairly heavy wall construction. Such evaporators exhibit poor heat transfer performance on low viscosity fluids because of the added resistance of the metal wall.

Energy Conservation

Most of the complexity and cost of an evaporator installation is a result of attempts to reduce energy consumption, which is usually by far the most important element of operating cost. Evaporators are not normally rated directly in efficiency of energy usage since they only have to separate the solvent from the solution, which requires very little theoretical energy in an ideal system. Thus, for separating water from a salt solution having a boiling point rise of 5 K at its atmospheric boiling point, the minimum theoretical energy requirement is only 30.1 kJ/kg of water removed (13.0 Btu of work per lb) whereas it takes 2250 kJ/kg (970 Btu of heat per lb) to vaporize the water. Since heat and work energy are not directly interchangeable in utility or

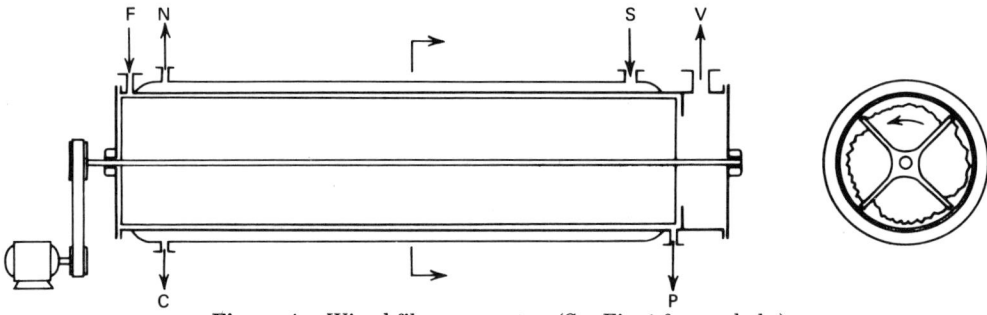

Figure 4. Wiped-film evaporator. (See Fig. 1 for symbols.)

cost, steam-heated evaporators are generally rated in terms of steam economy, ie, kg of water evaporated per kg of steam used, also called gained output ratio or performance ratio, and frequently standardized at pounds of water evaporated per 1000 Btu extracted from the steam (kg evaporated per 2324 kJ). Even evaporators that do use work energy (electrical or mechanical) for their operation are not usually rated in terms of efficiency but instead in such terms as J/m^3 (= 1.05×10^{-6} kW·h/1000 gal) evaporated.

The single-effect evaporator is the simplest arrangement. It uses steam from an outside source and exhausts its vapor to the atmosphere or to an air- or water-cooled condenser. Such an evaporator requires about 1 kg steam per kg of water evaporated and somewhat more if the feed is colder than the product and heat cannot be recovered by preheating the feed with concentrate and condensate. The high steam consumption limits the use of single effects to small capacities and to materials requiring an expensive type of evaporator, such as the wiped-film type, or having a very high boiling point, or to cases where the vapor is contaminated by materials that would cause excessive fouling or corrosion of heating surfaces when condensed, eg, rayon spin-bath liquor. Such evaporators may be operated on a continuous, batch, or semibatch basis with very little difference in heat requirements. In both batch and semibatch operation, final concentration is not reached until the end of the cycle, and these methods are, therefore, used primarily when the heat transfer properties become markedly poorer as final concentration is approached. Semibatch operation is the more common, with feed added continuously during most of the cycle in order to maintain a liquid inventory large enough to permit the evaporator to operate properly.

The single-effect evaporator produces almost as much vapor as the amount of steam used, the only difference being that the vapor is at a lower temperature and hence lower pressure. Compressing the vapor for reuse as the heat source was put into operation in the 19th century. This thermocompression or vapor recompression operation can be accomplished by either mechanical or steam-jet compressors. Mechanical compressors are by far the more efficient and may be driven by electric motors, gas or diesel engines, or steam or gas turbines. Because of the high specific volume of water vapor, positive displacement compressors are suitable for only the smallest capacities. Centrifugal compressors are most frequently used, whereas axial-flow multistage compressors are needed for the highest capacities (ca 50–500 m^3/s or ca $1.3 \times (10^4$–$10^5)$ gal/s). Compressor efficiencies are usually in the range of 70–75% for single-stage centrifugal machines (for compression ratios to about 1.5) and 80–85% for axial-flow machines (for compression ratios of about 1.15 per stage). Steam-jet compressors on the other hand are only about 25–30% efficient. Therefore, if high pressure steam is available and capacity requirements are appreciable, a mechanical compressor driven by a steam turbine is preferred to a steam jet.

The ideal mechanical power requirement of a thermocompression evaporator is given by the Carnot equation:

$$W = Q\Delta T/T$$

where W is the work done, Q is the heat received at absolute suction temperature T, and ΔT is the difference in saturation temperature at compressor discharge and suction pressures. To minimize the work required ΔT is kept low, so that the evaporator must have a large heating surface area. The optimum balance between power consumption and evaporator cost is usually at net ΔT's in the range of 3–10 K. The need to operate at low net ΔTs is a disadvantage of some types of evaporator, such as natural circulation and LTV units, whose coefficients fall off at low ΔT, and the submerged, forced circulation types, which lose ΔT as a result of temperature rise in the heating element and short-circuiting in the vapor head. The most advantageous evaporator type for vapor compression operation, when suitable for the liquor handled, is the falling film, which has very little ΔT loss. Since the ΔT across which the compressor must work also includes the boiling point rise (BPR) and any losses owing to pressure drop in vapor circuits, special care is exercised to minimize these. If the BPR of the product is high, multiple stages on the liquor side are advantageous so that only the last portion of the evaporation occurs at final product concentration. Vapor-side ΔT losses are minimized primarily by adequate duct sizing and the use of an entrainment separator having a low pressure drop, such as knit mesh. Entrainment separation is particularly critical because the vapor becomes superheated on passing through the compressor and any liquid carried with it evaporates, depositing its contained solids on the blades. Entrainment should preferably be held to 5 ppm total solids or less to minimize compressor problems.

A thermocompression evaporator is like a flywheel since the compressor keeps a large amount of heat content circulating. However, heat input must balance heat output in order that the total energy content or temperature of the system remain constant. The work input of the compressor ultimately appears as a heat input to the system, and ideally this should balance the heat losses from the system so that no supplementary source of heat is needed, except at startup. Since the heat input from the compressor is generally quite small, it is necessary to preheat the feed very close to the evaporator temperature by heat exchange with condensate and concentrate. The approach temperature differences required in these feed heaters, in order to achieve a balanced system, decrease with decreasing evaporator ΔT and input power. Thus making a thermocompression evaporator more efficient requires increases in both evaporator and preheater heating surface area. Where possible, supplementary sources of makeup heat are utilized, such as waste-heat boilers on gas turbine drives or diesel engines. If steam-turbine driven or steam-jet compressors are used, a surplus of heat is usually available and it is necessary to bleed vapor at the compressor suction to maintain the evaporator at constant temperature.

The largest number of vapor-compression evaporators have been built for producing potable water from seawater or brackish water (see Water, desalination). These are usually relatively small capacity units for military use, offshore drill rigs, marine vessels, and the like, and use engine-driven compressors and short-tube vertical evaporators. Makeup heat is obtained from engine exhaust and engine cooling water. The principal advantage is the ability to build a complete, compact, highly efficient, easily transportable unit that needs only connection to fuel supply and a feedwater source for its operation. Another use has been for citrus juice concentration, which must be done at low temperatures to avoid degradation (see Fruit juices). The need for low temperatures prevents use of multiple-effect evaporators for heat economy but also causes problems for vapor-compression operation because of the very high specific volume of steam under these conditions. A refrigerant, such as ammonia or Freon, can be employed as a secondary working fluid. The evaporator is then heated by the compressed refrigerant and the evaporator vapor is condensed in the refrigerant reboiler (see Refrigeration). Thus this system has the disadvantage that the heat must be passed through two heating surfaces instead of one. Since the 1973 energy crisis, thermocompression evaporators because of their higher efficiency are now used in fields where they were not previously considered, such as for paper-mill waste cooking liquors and for disposal of cooling-tower blowdown wastes. In the latter application, the cooling-tower water is usually nearly saturated with $CaSO_4$, and thus the evaporator must handle a severely scaling liquor. The falling-film evaporator has been used successfully in this case; seeding with $CaSO_4$ solids can prevent scale formation completely (1). Other uses include applications where lack of need for a heat sink is important and as a preconcentrator to increase the capacity and efficiency of existing multiple-effect installations. Diaphragm-cell caustic soda, for instance, is normally concentrated in forced circulation evaporators because of the large amounts of salt precipitated. However, an appreciable amount of water can be removed from the cell liquor before saturation is first reached with salt and the BPR is still relatively low; therefore, a falling-film thermocompression preconcentrator is advantageous.

If steam instead of power is the source of energy, multiple-effect evaporation is the principal means of energy conservation. In this operation, the vapor from one effect is used to heat another effect boiling at a lower temperature and the vapor from this effect is used to heat yet another effect boiling at still lower temperature. In such evaporators it is desirable to use an initial steam temperature as high as possible in the first effect and a heat sink temperature as low as possible to condense the vapor from the last effect, in order to develop the highest practical total temperature difference for heat transfer. The upper temperature limit is usually set by available steam pressure or scaling, fouling, product degradation, or corrosion characteristics of the liquor. The lower temperature limit is determined either by temperature and availability of cooling water or the need for low temperature steam for some other use. Other factors that may affect the lower temperature limit, which usually corresponds to a high vacuum, are poor heat transfer for most evaporator types at low temperature, high liquor viscosity, the very high vapor volumes from which liquor must be separated, and the cost of removing noncondensibles from the system.

Since each effect of an evaporator produces almost as much vapor as the amount it condenses, the total evaporation accomplished per unit of prime steam, or steam economy, increases in almost direct proportion to the number of effects used. The total heat load is also split up between the effects so that each effect has a much lower heat duty than a single effect for the same total evaporation load. However, the total available ΔT is also split up similarly so that each effect of a multiple effect requires about as much heating surface as a single effect operating over the same total temperature difference. Thus in selecting the number of effects to use in any installation, steam cost savings and capital cost of effects have to be balanced. Even before the energy crisis, evaporators with six or seven effects were common in the pulp and paper industry and as many as seventeen effects had been used in large seawater evaporators.

These relationships between number of effects, steam economy, and heating surface needs are not exact and can only be determined for a particular project by detailed heat and material balances and consideration of the effects of temperature level and temperature difference on heat transfer performance. The steam economy is generally not equal to the number of effects, primarily because of the influence of sensible heating loads. It frequently can be improved by alterations in feed sequence or installation of intereffect heat exchangers. The steam economy is usually close to a fixed fraction of the number of effects. The fraction may be over 1.0 when the feed is hot and may be as high as 0.9 even when the feed is cold, provided efficient means are included for preheating feed and for recovering waste heat at the least practical temperature differences. The most common method of heat recovery is by flashing the condensate produced in each effect to each of the lower temperature effects in turn so that the flash vapor is added to evaporator vapor to accomplish more useful evaporation. Prime steam condensate is usually not flashed since the flash vapor would be working at lower steam economy (over fewer effects) than the prime steam and the condensate would then have to be reheated in the boiler circuit at the same efficiency as generating heat in the prime steam.

Similarly, heating surface area needs are not directly proportional to the number of effects used. For some types of evaporator, heat transfer coefficients decline with temperature difference; as effects are added the surface needed in each effect increases. On the other hand, heat transfer coefficients increase with temperature level. In a single effect, all evaporation takes place at a temperature near that of the heat sink, whereas in a double effect half the evaporation takes place at this temperature and the other half at a higher temperature, thereby improving the mean evaporating temperature. Other factors to be considered are: the BPR, which is additive in a multiple-effect evaporator and therefore reduces the net ΔT available for heat transfer as the number of effects is increased, and the reduced demand for steam and cooling water and hence the capital costs of these auxiliaries as the number of effects is increased.

In a multiple-effect evaporator, the effects are numbered in the direction of steam flow, the first effect being the one heated by prime steam. Liquor feed sequence through the evaporator may be forward, backward, parallel, or mixed. *Backward feed* (to the coolest effect first and then successively through the higher temperature effects) is generally used when the feed is cold, since only a small volume has to be heated to the highest temperature, thereby reducing sensible heating losses and improving steam economy. *Forward feed* is generally used when feed is hot and when the concentrated product is not too viscous at the last effect temperature. Where necessary, forward feed can be used on a cold feed and can give almost the same steam economy as backward feed if the feed is preheated in stages by vapor extracted from each higher temperature effect of the evaporator in turn. Such an arrangement is used for seawater, which can be concentrated only in small amounts at high temperature without scaling but about threefold at the lowest effect temperature. These preheaters add to the complexity of the evaporator but not necessarily to total heating surface needs, since they reduce the heat loads in the effects themselves. *Parallel feed* is frequently used in crystallizing operations and involves feeding to and withdrawing product from each effect. *Mixed feed* operation is common if feed is at some temperature intermediate between first and last effect, the finished product is too viscous to handle at low temperature or liquor at an intermediate concentration and temperature is desired for further processing. All such conditions prevail in a kraft-mill liquor evaporator (see Pulp). The evaporators are usually of the LTV type and the feed heaters are an integral part of the evaporator tube bundles. The highest temperature effect is frequently subdivided so that only a part of the tubes must work on liquor of the highest concentration and viscosity. Other flow sheet variations include evaporators with several bodies in parallel on steam and vapor but in series on feed. Thus the kraft-mill evaporator might be an eight-body seven effect with two bodies in parallel in the last effect position in order to better handle the very large vapor volumes generated at the lowest temperature. Variations may also involve the steam path; eg, a combination double-triple effect, which may have one effect on prime steam with the evolved vapor being split, one part to a single effect and the other to a double effect. This variant is used when the liquor has a high BPR as it approaches final concentration, so high that not all evaporation can be accomplished, in this example, in a triple effect. Other flow sheets have been developed for specific types of evaporator, such as parallel split feed, which is used for desalination evaporators arranged in vertical stacks. In this case, about half the feed goes to the odd-numbered effects in one stack and the other half to the even-numbered effects in a second stack next to the first, with vapor connections crossing from one stack to the other.

Another means of gaining multiple-effect steam economy is by multistage flash operation, developed originally as the Alberger salt process about 50 years ago. In this process, cool liquor is preheated in stages by vapor condensing at successively higher temperatures in a recovery section and finally by prime steam in a brine heater. The heated liquor is then flashed down in stages to generate the vapor used for preheating. The condensate from this vapor is also flashed down in the same manner so that the total flow being flashed is the same as the flow being heated. As a result, the temperature rise and flashing range in all heaters and flashers is about the same. Flashing through a 75 K range evaporates only about 12% of the feed, thus it is necessary to recirculate the liquor through the heaters and flashers when an appreciable degree of concentration is required. This then requires a rejection section in the flash train, containing one or more stages in which liquor is flashed to reject heat to cooling water before being recirculated to the recovery section. Even when recovering fresh water from seawater, recirculation is usually employed to reduce the amount of fresh feed that must be treated for scale prevention and deaerated before introduction into the evaporator. The characteristics of a flash evaporator are entirely different from those of a multiple-effect evaporator. Many more stages than effects are needed to achieve the same steam economy. Whereas in a multiple-effect evaporator the amount of heating surface provided is the most important determinant of capacity, and only the number of effects has a strong influence on steam economy, in a multistage flash evaporator the production rate is proportional only to the product of liquor circulation rate and total flash range. The steam economy is affected by both the number of stages and the amount of heating surface provided. The principal advantages of the multistage flash evaporator are that only one circulating pump is needed and that a number of stages, comprising both flash space and heating elements, can be combined in one vessel. The evaporator is in essence a number of forced circulation evaporator heating elements connected in series on one circulating pump, plus a number of forced circulation flash chambers connected in series. Since all heat is absorbed as sensible heat, the multistage flash evaporator suffers from the same loss in ΔT as the forced circulation evaporator, and also has short-circuiting losses of the same magnitude. The disadvantages of the multistage flash evaporator are therefore primarily this loss in net ΔT available for heat transfer, plus the fact that all liquor in the system is close to discharge concentration.

Although the interrelationships of the variables in a multistage flash evaporator is complex, the sizing of the heating surfaces is relatively simple, using the following factors (3): (1) heat transfer coefficients from conventional correlations for condensing vapor and for liquid heating in tubes; (2) heat load, excluding brine heater: $Q = D\Delta H/(1 - \Delta T/1250)$; (3) weight of brine circulated: $W = Q/C_p \Delta T$; and (4) net Δt for heat transfer = $\Delta T/N \ln (1 - \Delta T/1250 - R \cdot P/\Delta T)/(1 - \Delta T/1250 - R \cdot P/\Delta T - P/N)$, where Q = amount of heat transferred through heating surfaces; D = distillate production rate; ΔH = average latent heat of evaporation; ΔT = total flashing range, K; C_p = specific heat of circulating brine; N = number of flash stages; R = average sum of BPR and short-circuiting losses; and P = steam economy desired based on 2324 kJ/kg (1000 Btu/lb) latent heat. From these, it is observed that heating surface can be traded against number of flash stages but that the number of stages should be on the order of three times the steam economy in order not to seriously decrease the net Δt available for heat transfer. Furthermore, the ratio of heating surface needed to the amount of brine circulated is usually so high that the recirculating liquid must be pumped through long lengths of small diameter tubing in order to achieve reasonable tube velocities and hence reasonable heat transfer coefficients. This need for small diameter tubing and very long flow lengths, which necessitates a number of intermediate water boxes, makes the multistage flash evaporator generally unsuitable for crystallizing service.

It is possible, of course, to combine the different types of evaporator as well as the various methods of energy conservation into a single system. Thus if a crystallizing solution is being handled but saturation is not reached until after an appreciable amount of solvent has been evaporated, the preliminary concentration can be done in several effects of the natural circulation or film type, followed by forced circulation effects for the crystallizing duty. In a thermocompression evaporator of the steam-jet type, one with a steam turbine driven mechanical compressor, or one driven by a gas turbine to which a waste heat boiler can be added, more low pressure steam can be generated than can be used effectively in the thermocompression section. Vapor can, therefore, be bled at the compressor suction and used to operate a multiple-effect or multistage flash-evaporator section. The latter usually can also serve as a preheater for the feed to the thermocompression section. Such evaporators are most useful when the evaporation requirements are the sole or principal reason for installing the facilities for steam or power production. Since it costs only little more to generate steam at a far higher pressure than can be possibly used directly in the evaporator, the addition of a steam jet or steam turbine compressor accomplishes additional evaporation at very little additional operating cost. Similarly, a gas turbine may be relatively inefficient when generating only power but if also used to generate low pressure steam in a waste heat boiler, it becomes as efficient as a more expensive high pressure boiler and steam turbine system.

Temperature Difference and Heat Transfer

The capacity of a steam-heated evaporator is governed by the amount of heat it can transfer, which is determined in turn by the amount of heating surface area and temperature difference available, and the heat transfer coefficients achieved (see Heat transfer technology). There are common conventions in the industry for these terms but they are not universally used; therefore, care must be exercised in interpreting and applying available data. Heating surface area is almost always taken as the total tubing area on the side in contact with the liquor being evaporated, which is different from normal heat exchanger practice, where it is always the outside area of the tubes. The definition of temperature difference is open to the widest interpretation and is the most frequent cause of misapplication of data. Considering only one effect of an evaporator, it is the difference between saturated steam temperature on one side of the heating surface and the liquor temperature on the other side that is effective for heat transfer. However, for the effect as a whole, with its associated liquor circuits, vapor piping, and entrainment separators, it is the difference between saturated steam temperature at the inlet to the heating element and the condensing temperature of the evolved vapor at the exit of the vapor discharge piping that determines how the evaporator effect reacts with its surroundings. Certain losses in temperature difference are inherent, including boiling point rise, vapor circuit losses (those due to friction and acceleration and deceleration heads of steam and vapor flowing into and through the heating element, entrainment separator, and vapor piping) and liquor circuit losses (the difference between the temperature at which heat is absorbed at the heating surface and the temperature at which it is released in the vapor head). Boiling point rise, or boiling point elevation (BPE), is usually the most important of these losses and is the difference between the boiling point of the solution in the evaporator and the boiling point of pure solvent at the same pressure. Since the vapor cannot condense and give up the bulk of its heat content until it has cooled to the boiling point of the pure solvent, BPR represents a loss of available ΔT and is deducted from the overall ΔT before computing heat transfer coefficients. It is usually estimated from known properties of the solution at the discharge concentration of each effect or each body, but care must be exercised here also, since it is sometimes obtained from direct measurements of vapor temperature made at a point where the vapor may still be somewhat superheated. Furthermore, in the falling-film-type evaporator, eg, the discharge is frequently taken from the circulating pump that feeds the heating surface, with the result that the discharge from the heating surface is at a higher concentration than the discharge from the effect.

Any pressure drop losses of steam in the heater or of vapor in the entrainment separator and vapor piping also reduce condensing temperatures, and therefore, represent a loss in ΔT available for heat transfer. These losses become more serious at the lower temperatures, because of the increasing difficulty of handling lower density vapor and the decreasing slope of the vapor pressure curve. In many cases, these vapor circuit losses are tabulated separately and deducted from the overall ΔT before computing heat transfer coefficients. However, at times these losses are ignored or combined with the BPR loss or the heat transfer ΔT, which must be borne in mind when interpreting heat transfer data. Incorporating these losses in the BPR term is especially dangerous because actual chemical BPR changes very little with evaporator load whereas the net ΔT across the heating surface increases in almost direct proportion to load, and vapor circuit losses increase as almost the square of the load.

Losses of ΔT in the liquor circuit may also be substantial but are not always taken into consideration. Only in the forced circulation evaporator are at least part of these losses usually calculated. Since heat is absorbed as sensible heat, it results in a temperature rise that can be either measured or calculated from known heat input and known circulation rate. If this is done, coefficients are reported on the basis of log mean temperature difference. The submergence or short-circuiting losses are frequently included in the coefficient, eg, when the heater inlet temperature is not measured directly but instead is calculated from saturation temperature at the measured vapor head pressure, plus BPR. In natural circulation evaporators the boiling point at the heating surface is higher than that in the vapor space because of the hydrostatic head of liquid above the heating surface. It was once the practice to estimate this loss in ΔT and report heat transfer coefficients corrected for hydrostatic head but today this correction is ignored. Similarly, for LTR or LTV evaporators the course of the liquor temperature as it passes the heating surface cannot readily be measured or calculated and coefficients are based on steam and vapor head temperatures. In general, it is practical to consider these losses only when heat transfer coefficients can be calculated from theory, as in forced circulation and falling-film evaporators. For the latter, the loss caused by friction and acceleration of vapor generated in the tubes must be considered when calculating coefficients from theory, but may or may not have been included in the coefficients reported in the literature.

When designing a new evaporator or analyzing the operation of an existing evaporator, a temperature distribution table first should be prepared. Table 1 gives the data on a quadruple-effect evaporator containing two natural circulation effects followed by two forced circulation effects. Such a table compares the magnitude of the various losses in ΔT and hence indicates the principal obstructions to capacity. Thus if one of the effect's net ΔTs were substantially higher than the others, improvements in performance of that effect would give the greatest gain in capacity. If the above data were taken when the evaporator was operating considerably below design capacity, the last effect vapor-circuit loss would be of special concern. It would increase approximately with the square of capacity and soon would become a much more important obstruction as conditions elsewhere were improved.

When used in design, a temperature distribution table serves as the basis for calculating a heat and material balance around the evaporator to determine heat loads in the individual effects, vapor flows, and liquor concentrations, from which vapor-circuit losses or line sizes and BPR estimates can be reconfirmed. The designer does not simply take the net ΔTs from such a table and the calculated or estimated heat transfer coefficients to arrive at the heating surface needed in each effect. The designer also uses it for optimizing. For instance, the temperature distribution may be altered so that the heating surface areas and heating element designs are identical in two or more effects. In the example of Table 1 the ΔTs might be reduced in the first two effects which are of the natural circulation type and cheaper per unit of heating surface, in order to give greater ΔTs and hence lower surface requirements in the more expensive last two forced circulation effects. The added cost of reducing vapor-circuit

Table 1. Temperature Distribution, °C

Effect	I	II	III	IV	Condenser
steam temperature[a]	125.7	112.7	94.2	72.8	47.0
net ΔT	9.6	13.4	11.9	12.9	3.0[b]
liquor temperature	116.1	99.3	82.3	59.9	
BPR[c]	3.2	4.3	6.9	8.6	
vapor temperature[a]	112.9	95.0	75.4	51.3	
vapor-circuit loss	0.2	0.8	2.6	4.3	
heater inlet			83.1	61.6	30
outlet			86.1	64.5	44
net ΔT_{LM}			9.5	9.7	8.1[d]
submergence loss			0.8	1.7	

[a] Calculated from measured steam chest and vapor head pressures.
[b] Effective ΔT if a direct contact condenser.
[c] From known properties of solution, at measured or calculated concentration.
[d] Effective ΔT if a surface condenser.

losses through use of larger lines and entrainment separators can be compared with the savings resulting therefrom in increased net ΔTs across the heating surfaces.

Actual heat transfer coefficients encountered in evaporators cover a wide range, depending on the physical properties of the solution, its fouling characteristics, the type of evaporator employed, the boiling temperature, and the temperature difference. Only in the submerged-tube forced circulation and the falling-film evaporator can heat transfer coefficients easily be calculated from theory (2). For other types, performance estimates are usually based on earlier experience with the same or a similar liquor. In general, coefficients range from a low of about 175 J/(m²·s·K) (100 Btu/(h·ft²·°F)) to a high of about 3500 (2000). The lowest coefficients are encountered at low temperatures and low ΔTs in film-type or natural circulation evaporators, or at very high viscosities. The highest coefficients are encountered when employing doubly-fluted tubes. These high coefficients are possible only when fouling resistances are very low, eg, when handling seawater that has been properly treated for scale prevention and deaerated (see Dispersants).

Vapor-Liquid Separation

The heating surface usually determines the evaporator cost and the vapor head the space requirements. The vapor-liquid separator must have enough horizontal plan area to allow the bulk of the initial entrainment to settle back against the rising flow of vapor and enough height to smooth out variations in vapor velocity and to prevent splashing directly into the vapor outlet. Separators are usually sized on the basis of the Souders-Brown expression:

$$U = K \sqrt{(\rho_l - \rho_g)/\rho_g}$$

where U = vapor velocity, ρ_l = density of liquid, ρ_g = density of gas. For most types of evaporator, the decontamination factor DF (DF = kg of vapor per kg of entrained liquid) decreases as K increases, approximately as follows (3):

DF	K m/s	ft/s
100	0.051	0.167
200	0.033	0.108
500	0.020	0.067
1000	0.015	0.049
2000	0.011	0.036

In general, it is not economically attractive to provide the full degree of entrainment separation desired in the vapor head alone. Instead, the vapor head is sized for a decontamination factor on the order of 200 and reliance is placed on supplementary separators such as shown in Figures 1–3 for removal of residual entrainment. Somewhat lower separator velocities are frequently used in crystallizing evaporators to help reduce the buildup of salts on the walls of the vessel above the liquor level. In the falling-film evaporator, most of the entrainment separation occurs within the tubes and higher velocities are permissible in the vapor head and through the curtain of liquor falling from the ends of the tubes. Here, decontamination factors of 1000 or more can be achieved at K values of 0.15 m/s under favorable conditions (primarily avoidance of long fall distances from tube ends to the liquid pool below the tubes, which generates finer, more easily entrainable droplets).

Heat Removal and Noncondensible Gases

A single- or multiple-effect evaporator does not consume heat; it merely degrades the heat input and means must be provided for removing the waste heat. Heat is usually rejected to a river, wells, or a cooling tower. The commonest means of heat rejection is by a barometric condenser in which the vapor from the last effect is condensed by direct, countercurrent contact with water cascading over weirs or trays. The condenser is elevated far enough above grade so that water can drain away to a hot well barometrically against the vacuum in the system. Noncondensible gases in the vapor accumulate at the top of the condenser and are cooled close to the temperature of the incoming water, thereby reducing the amount of water vapor associated with the gases. These gases are removed by either a steam-jet ejector (usually of several stages) or a mechanical vacuum pump. Steam-jet ejectors are the most common but are relatively inefficient. At present energy costs, mechanical vacuum pumps have substantially lower operating costs. Another type of direct contact condenser is the cocurrent or jet condenser, in which the water is delivered through nozzles at a velocity high enough to carry the noncondensible gases out the tailpipe. Although this eliminates the need for a separate vacuum pump, the water flow and pressure requirements are sufficiently higher than those of a countercurrent condenser that overall energy consumption is increased. A surface condenser is much more expensive than a direct-contact condenser and is used only when contaminants in the vapor (eg, SO_2) would pollute the condenser water or when the pure condensate has a substantial value. Surface condensers are designed by conventional shell-and-tube methods. Particular care must be taken to minimize shell-side pressure drop, since even a small pressure drop at the high vacuum usually employed represents a substantial loss in available ΔT. Furthermore, in general, more noncondensible gases are present than in the usual steam-heated exchanger and special precautions must be taken to properly channel the vapor flow past the surface between steam inlet and vent outlet, and to allow for the effects of noncondensible gases on heat transfer performance.

Noncondensible gases are more prevalent in an evaporator system than in most other steam-heated equipment and they must be properly handled to avoid serious impairment of heat transfer or reduction in steam economy (5). They may be present as gases dissolved in the feed or liberated by decomposition reactions in the liquor (eg, by bicarbonate breakdown), air in-leakage, and air dissolved in the condenser water. Although venting in practice is almost always empirical and done in excess, the optimum vent rate usually corresponds to a noncondensible gas content of 5–10 vol % in the vents. Vents from heating elements operating under pressure may be released to the atmosphere, whereas vents from vacuum effects are passed through the condenser to remove as much of the associated water vapor as possible ahead of the vacuum pump. Vents are frequently cascaded from one effect to the next on their way to the condenser. Such cascading results in increased concentrations of noncondensible gases in the later effects, and its only potential benefit is the recovery of heat from an upstream vent that has accidentally been opened too wide. In general, noncondensible gases have little effect on heat transfer performance provided that they are properly channeled through the evaporator heating elements. This requires a positive vapor flow path from steam inlet to vent outlet, with no pockets of low velocity where noncondensible gases can be trapped, and no low resistance channels that can bypass steam directly from inlet to outlet.

Other Evaporative Methods

Solar evaporation was the first evaporative method developed and still is in widespread use, primarily for production of salt from seawater (see Chemicals from brine). Evaporation rates vary widely with climate and can be predicted from weather data and the properties of the solution being evaporated (4). Evaporation rates of pure water are measured experimentally and are published by the U.S. Weather Bureau. They are determined in small pans and must be reduced about 30% to yield rates experienced in large ponds or reservoirs. They must be reduced even further if the material has an appreciable BPR. For seawater saturated with salt, the resultant rate is only about half that of fresh water in small pans. This is of critical importance with substantial rainfall, since a net positive rate based on water in small pans may become a negative rate in large solar ponds containing brine. In solar salt plants the yield almost never approaches the yield expected from the evaporation accomplished. The principal cause is seepage of partially concentrated brine through the pond bottoms and dikes. This seepage has been the principal impediment to application of solar ponds to other uses, such as concentration of oil field brines and other waste streams. Present laws require elimination of seepage to avoid groundwater contamination. The cost of providing the seepage barrier exceeds by far all other costs of solar pond development. Solar evaporation has also been used for the production of potable water from seawater. A barrier is required between the ponds and the atmosphere to admit the solar energy and to condense and collect the evaporated water. This barrier is even more expensive than a pond liner, making the method uneconomical even at today's energy prices. Yields are usually less than half the evaporation rates achieved in open ponds.

At the opposite extreme of fossil fuel usage are single-effect evaporative systems. Conventional steam-heated single-effect evaporators have been discussed above. When the BPR is extremely high (as for manufacture of anhydrous NaOH), the evaporator may be heated by molten salt or Dowtherm instead of steam. The LTV-type evaporator is normally used in this service. Wiped-film evaporators also sometimes use these high temperature, low pressure heating media to achieve high ΔTs without need for very heavy heat transfer wall thicknesses. The simplest single-effect evaporative method brings combustion gases into direct contact with the material being concentrated. A spray dryer can be used in this manner to concentrate liquids, and has been so used for high BPR liquids, such as $CaCl_2$. Less expensive in cost and more efficient in fuel consumption is the submerged combustion evaporator. In this case, the burner is immersed in the solution or slurry being concentrated and the combustion gases rise through the liquid to release almost all but the latent heat of the water in the combustion gases. Fuel may be either natural gas or the lighter distillate oils. Such evaporators are inexpensive and well adapted to handling corrosive and severely scaling liquids. They require no heat sink, but since the evolved vapor is mixed with large volumes of combustion gas they make it impractical to achieve much better than single-effect steam economy. At present energy costs, such evaporators are impractical for all except very low capacities or the most refractory scale-forming or corrosive liquids.

BIBLIOGRAPHY

"Evaporation" in *ECT* 1st ed., Vol. 5, pp. 927–947, by W. L. Badger, Consulting Engineer; "Evaporation" in *ECT* 2nd ed., Vol. 8, pp. 559–580, by Ferris C. Standiford, W. L. Badger Associates, Inc.

1. J. H. Anderson, *ASME Paper 76-WA/Pwr-5,* ASME, New York, 1976.
2. J. G. Knudsen in R. H. Perry and C. H. Chilton, eds., *Chemical Engineers' Handbook,* 5th ed., McGraw-Hill Book Co., New York, 1973, Section 10, pp. 32–35.
3. F. L. Rubin in R. H. Perry and C. H. Chilton, eds., *Chemical Engineers' Handbook,* 4th ed., McGraw-Hill Book Co., New York, 1963, Section 11, pp. 35–36.
4. J. Ferguson, *Aust. J. Sci. Res.* **V,** 315 (1952).
5. F. C. Standiford, *Chem. Eng. Prog.* **75**(7), 59 (1979).

General References

References 2 and 3.
W. L. Badger and J. T. Banchero, *Introduction to Chemical Engineering,* McGraw-Hill Book Co., New York, 1955, Ch. 5.
F. C. Standiford, *Chem. Eng.,* 158 (Dec. 9, 1963).
Energy Conservation Tools for the Process Engineer: Upgrading Existing Evaporators to Reduce Energy Consumption, ERDA, Technical Information Center, P.O. Box 62, Oak Ridge, Tenn., 1977.

Evaporator testing and troubleshooting

Equipment Testing Procedures—Evaporators, AIChE, New York, 1978.
F. C. Standiford, *Chem. Eng. Prog.* **58**(11), 80 (1962).
H. H. Newman, *Chem. Eng Prog.* **64**(7), 33 (1968).

Desalination evaporators

The Office of Saline Water (now the Off. of Water Res. Tech.) of the U.S. Dept. of Interior published almost 1000 reports and was responsible for numerous articles on desalination. The Nuclear Desalination Information Center of Oak Ridge National Laboratory maintains a bibliography of these sources of information, which are abstracted and indexed in ORNL-NDIC-11 and -13.
N. Lior, ed., *Measurement and Control in Water Desalination,* Academic Press, Inc., New York, in press.

FERRIS C. STANDIFORD
W. L. Badger Associates, Inc.

EXHAUST CONTROL, AUTOMOTIVE

Although more than 25 years have passed since Haagen-Smit first presented his theory of photochemical oxidant production in which he implicated automobile exhaust as a major source of air pollution, many questions still remain about the role of automobile exhaust in pollution (1) (see Air pollution). Among these are: (1) the atmospheric chemistry and mechanics that produce pollutant levels; (2) the concentration of various pollutants to which people are exposed; (3) the health effects produced by various pollutants at ambient concentrations; (4) the correlation between laboratory measurements of exhaust gases and ambient concentrations; and (5) the effect of vehicle operation on ambient concentrations.

An in-depth study of each of these areas is not required to discuss exhaust controls because the actual limits, according to specific test procedures, that must be observed have been established legislatively (2-3) by a complex political and economic process (4-6).

Atmospheric Chemistry

Automobile exhaust contains carbon monoxide, carbon dioxide, oxides of nitrogen (primarily NO), water, and nitrogen. The exhaust also can contain hundreds of hydrocarbons and particulates including carbon and oxidized carbon compounds, metal oxides, oil additives, fuel additives, and breakdown products of the exhaust system, including the exhaust-control catalysts. These exhaust products can combine in a large variety of ways in the atmosphere, particularly since the amounts of each material change with operating conditions and the mechanical state of the vehicle. The photochemical reaction between oxides of nitrogen and hydrocarbons (HC) that caused the original interest in the automobile as a source of pollution has been investigated extensively (7-9).

$$2\,NO + O_2 \rightarrow 2\,NO_2 \tag{1}$$

$$NO_2 \xrightarrow{sunlight} NO + O \tag{2}$$

$$O + O_2 + HC \rightarrow O_3 + HC \tag{3}$$

$$O_3 + NO \rightarrow NO_2 + O_2 \tag{4}$$

$$NO_2 + O_2 \underset{sunlight}{\rightleftarrows} NO + O_3 \tag{5}$$

Ozone is the principal oxidant produced. However, comparatively low levels of some other ultimate products, such as PAN (peroxyacetyl nitrate), are apparently responsible for two unique effects of Los Angeles smog: plant damage known as silver leaf, and eye irritation. The oxidants produced elsewhere than in the West Coast area only rarely show these effects at comparable levels (10-12).

Formation of aerosols is another effect of photochemical products. This contribution was recognized in 1956 but, since it involves a wide variety of reactions with sulfates from different sources, the precise reactions and their magnitude have not been delineated (13). Substantial studies are in progress in this area (14–16). The contribution made by the original hydrocarbons and other oxidants, particularly ozone, to air pollution has not been quantitatively described. There is substantial evidence that natural sources contribute extensively to the hydrocarbon content of the atmosphere. The recent correlation between years of high rainfall and high levels of oxidant on the West Coast area is the latest contribution to the substantial literature of this hypothesis (17). The question of the importance of the anthropogenic vs natural sources of hydrocarbons was further confused by the findings of high oxidant levels in remote locations (18). Atmospheric hydrocarbon transport has been proposed as the cause for these effects; however, a number of studies have questioned this explanation (19–23). Oxidant increases have been ascribed to exchanges of ozone between the stratosphere and the troposphere (24–26). Finally, each of the hydrocarbons has a unique rate of reaction that varies by the hundred-fold so that a damage figure based on tonnage of hydrocarbons from a particular source is essentially meaningless (27–29).

The need for exhaust control is not at issue since there is little question that the automobile contributes to both of the principal reactants in the photochemical reaction although the degree of control required remains an important issue.

Pollutant Concentrations. Health authorities assess the importance of a particular pollutant not only in regard to total quantity, but also for concentration and the period of exposure. The methodology to be used and the accuracy of measurement with regard to all pollutants has been a subject of considerable study and controversy. In recent years, it has been shown that atmospheric measurement techniques for carbon monoxide, oxides of nitrogen, hydrocarbons, and oxidants have been in error, at times of such magnitude that trends have been difficult to establish (30–33). In addition, there is a wide variation in readings as a result of comparatively small changes in locations of sampling stations (34–36). This has important implications because of the comparative paucity of sampling stations used to categorize large areas. The distinct possibility exists that an area may be incorrectly categorized because readings are available only from a single station. The Federal and State regulatory attitude has been that if even a location with limited sampling shows a reading above the standards, appropriate measures should be taken (see Regulatory agencies). This approach has very serious economic implications. In spite of the many imperfections of measurement technique, there has been a consistent reduction in those pollutants with which the automobile is associated (37–39).

Health Effects. The current ambient air quality standards promulgated by the EPA are shown in Table 1. These have been set after voluminous study of the available literature. The literature has been compiled in Criteria documents and has been reviewed by the National Academy of Sciences as well as by numerous other authorities (40–42). A useful summary has been published (43).

The principal effects of concern are:

Carbon monoxide acts by substituting for oxygen in the blood to form carboxyhemoglobin. The high concentrations of CO in urban atmosphere can threaten the health of individuals with angina or other cardiovascular ailments.

Oxides of nitrogen cause irritation of the respiratory system and may reduce the individual's ability to resist infection.

Oxidants is a term that covers a wide amount of reaction products, and the EPA has proposed that ozone measurement be used as a surrogate (33). Ozone (qv) has

Table 1. National Ambient Air Quality Primary Standards

Pollutant	As promulgated	Concentration	Averaging period
oxidant (ozone)	160 $\mu g/m^3$	0.08 ppm	1 h
carbon monoxide	10 mg/m^3	9 ppm	8 h
	40 mg/m^3	35 ppm	1 h
nitrogen dioxide	100 $\mu g/m^3$	0.05 ppm	annual
sulfur dioxide	80 $\mu g/m^3$	0.03 ppm	annual
	365 $\mu g/m^3$	0.14 ppm	24 h
suspended particulate matter	75 $\mu g/m^3$		annual
	260 $\mu g/m^3$		24 h
hydrocarbons (corrected for methane)	160 $\mu g/m^3$	0.24 ppm	3 h (6–9 AM)

neither the eye-irritating properties nor the phytotoxic effects encountered in the West Coast oxidants (44).

As in the case of oxides of nitrogen, the principal concern is irritation of the respiratory system resulting in stress during exercise and possible infection.

Vehicle Control Requirements in the United States

The first requirements for control of vehicle emissions were set by the Motor Vehicle Pollution Control Board in California in 1961. They specified requirements for crankcase ventilation and levels of carbon monoxide and hydrocarbons in the exhaust (45). Since then, there have been a variety of requirements for both passenger cars and trucks regarding oxides of nitrogen as well as carbon monoxide and hydrocarbons. These have been set by the Congress of the United States and by the California Air Resources Board (the successor to the MVPCB). At the present time, the requirements are as shown in Table 2.

California regulations have generally been more stringent than the Federal regulations because the state's oxidant problem has been especially serious owing to the unusual local meteorology (46). As a result, California has been granted permission by Congress to have more demanding requirements than those of the rest of the country. As can be seen from Table 2, controls on evaporative losses of hydrocarbons have also been added in recent years.

Emission Control Measurements. Studies by California authorities and the automobile industry established that a vehicle operated on a chassis dynamometer could reasonably simulate urban traffic for the purpose of evaluating exhaust emissions (47). However, emission results obtained with this procedure are subject to substantial variation unless a variety of conditions are very carefully maintained and exacting measurements are made at every stage of the operation.

The details of the EPA Exhaust Emission Test procedure have been published (3).

Control Devices

There are three sources of vehicle emissions: (1) hydrocarbons from the crankcase ventilation system; (2) evaporative losses including hydrocarbons from the carburetor

Table 2. Vehicle Control Requirements[a]

Exhaust emission standards		1978		1979		1980		1981	
		g/km	(g/mi)	g/km	(g/mi)	g/km	(g/mi)	g/km	(g/mi)
Passenger cars									
Federal	HC	0.93	(1.5)	0.93	(1.5)	0.25	(0.41)	0.25	(0.41)
	CO	9.32	(15)	9.32	(15)	4.35	(7.0)	2.11	(3.4)
	NO_x	1.24	(2.0)	1.24	(2.0)	1.24	(2.0)	0.62	(1.0)
California	HC	0.25	(0.41)	0.25	(0.41)	0.25	(0.41)	0.25	(0.41)
	CO	5.59	(9.0)	5.59	(9.0)	5.59	(9.0)	2.11	(3.4)
	NO_x	0.93	(1.5)	0.93	(1.5)	0.62	(1.0)	0.62	(1.0)
Light duty trucks (less than 2.7 t or 6,000 lbs GVW^b)									
Federal	HC	1.24	(2.0)	1.06	(1.7)	1.06	(1.7)	1.06	(1.7)
	CO	12.40	(20)	11.20	(18)	11.20	(18)	11.20	(18)
	NO_x	1.93	(3.1)	1.43	(2.3)	1.43	(2.3)	1.43	(2.3)
California	HC	0.56	(0.9)	0.25	(0.41)	0.25	(0.41)	0.25	(0.41)
	CO	10.60	(17)	5.59	(9.0)	5.59	(9.0)	5.59	(9.0)
	NO_x	1.24	(2.0)	0.93	(1.5)	0.93	(1.5)	0.62	(1.0)
evaporative emissions, HC, g/test									
Passenger cars and light duty trucks									
Federal		6.0		6.0		6.0		2.0	
California		6.0		6.0		2.0		2.0	

[a] HC = hydrocarbons; NO_x = nitrogen oxides.
[b] GVW = gross vehicle weight.

and the gas tank as well as the various lines leading from these sources to the engine; and (3) hydrocarbons, carbon monoxide, and oxides of nitrogen from the exhaust system.

Concentrations from these three sources, which are considered representative of an uncontrolled vehicle, are shown in Table 3.

Since both the crankcase and evaporative emissions are ultimately controlled in the engine, controls for the first two are considered before discussing the other engine controls. Figure 1 illustrates a typical crankcase control system.

Crankcase fumes enter the engine during most of the vehicle operation at the intake manifold at a rate that is controlled by a valve of the type shown in Figure 1. Since the lowest air flow occurs during engine idling and deceleration, and since these

Table 3. Representative Uncontrolled Vehicle Emissions (1960 Model Year)

	g/km	(g/mi)
Exhaust emissions		
HC	6.59	(10.6)
CO	52.20	(84)
NO_x	2.55	(4.1)
Crankcase emissions		
HC	2.55	(4.1)
Evaporative emissions		
HC	2.67	(4.3)

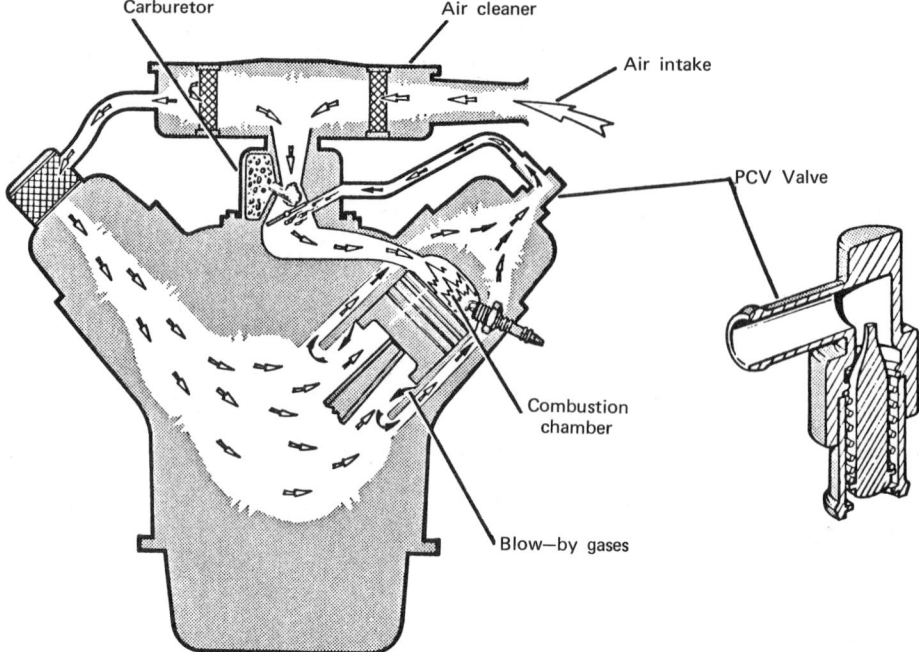

Figure 1. Closed crankcase ventilation system.

conditions also produce the highest vacuum at the intake manifold, the valve moves to a position where the smallest opening is presented. As air flow increases and the vacuum level decreases, the orifice opens allowing a greater amount of fumes to pass. The net result is that disturbances of the fuel intake charge are minimized. The typical evaporative control system as shown in Figure 2 is somewhat more complex.

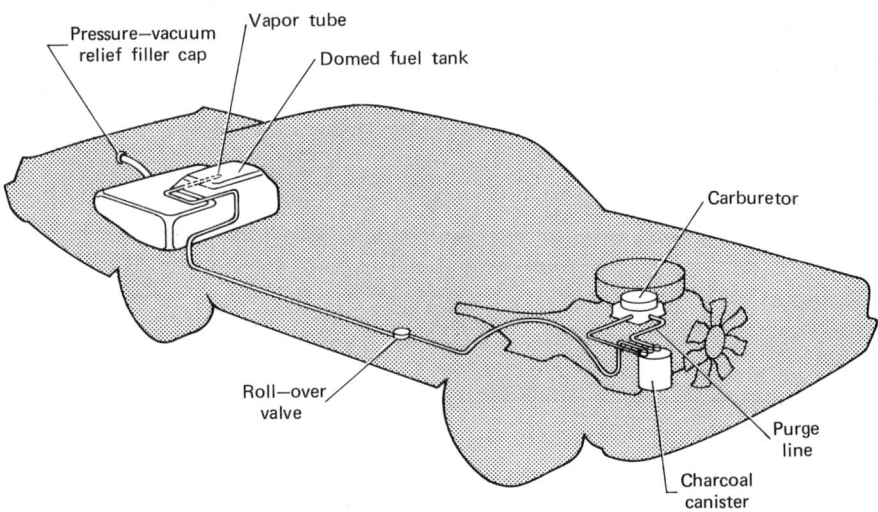

Figure 2. Vapor saver, regular size cars and station wagons.

The gas tank cap must provide a positive seal against exhaust of gasoline vapors. In addition, the cap must be able to pop open if the vapor pressure exceeds the maximum design limit of the tank, ca 13.8 kPa (2 psi). During engine operation, it must be vented to the atmosphere to compensate the resulting vacuum as fuel is pulled out of the tank. While the vehicle is at rest, the charcoal canister collects fumes from the gas tank and from the carburetor; during vehicle operation these fumes are pulled into the intake manifold for further processing by the engine.

Engines. Clearly, the engine is both the principal source of pollution as well as the major point of control. The majority of the engines in automotive service have four cycles, a piston drive, and are gasoline-fueled. Of importance to the study of exhaust emissions is the fact that gasoline consists of a blend of hydrocarbon compounds with carbon contents ranging from C_4 to C_{10} and with a boiling range from 32–210°C with some variations depending on the climate or season (48–49) (see Gasoline and other motor fuels).

During the combustion process the hydrocarbons react partially with the oxygen that is introduced to give hundreds of new compounds. Some of these reach the exhaust. Others are further consumed during the combustion cycle. The most important reactions of combustion are shown below; equations are not balanced but do show reactants and principal products of combustion.

With excess air

$$\underbrace{(HC)_X}_{\text{fuel}} + \underbrace{(O_2 + 3.76\, N_2)_Y}_{\text{air}} \rightarrow \underbrace{CO_2 + H_2O + N_2 + O_2}_{\text{major}} + \underbrace{CO + H_2 + (HC)_Z}_{\text{traces}} \quad (6)$$

$$N_2 + O_2 \rightleftarrows NO \quad (7)$$

With excess fuel

$$\underbrace{(HC)_X}_{\text{fuel}} + \underbrace{(O_2 + 3.76\, N_2)_Y}_{\text{air}} \rightarrow CO_2 + H_2O + CO + H_2 + N_2 + (HC)_Z + O_2 \text{ (trace)} \quad (8)$$

$$N_2 + O_2 \rightleftarrows NO \quad (9)$$

Important hydrocarbon products of combustion are: hydrogen (with excess fuel), methane, ethylene, propylene, toluene, acetylene, butenes, xylenes, pentane, butane, and ethane; and oxygenates such as formaldehyde, acetone, and methanol.

Clearly the most important products are carbon dioxide and water, both of which are harmless with regard to further atmospheric reactions. Since the air contains 79% nitrogen as well as ca 21% oxygen, a secondary reaction occurs during the combustion process. This is shown in equation 7 (see also below). Figure 3 shows the way the exhaust products are affected by various combinations of air and fuel.

As can be seen, the so-called stoichiometric point, which is the point at which there is theoretically enough oxygen to consume all of the hydrocarbons and carbon monoxide, occurs at ca 15 parts of air to one part of fuel by weight. If the reaction were allowed to continue for sufficient time, this theoretical condition would occur. Unfortunately, each combustion cycle occurs several thousand times per minute, and as a result, the reactions do not go to completion. This is shown in the carbon monoxide and hydrocarbon curves, but as would be expected, the carbon monoxide output is lowered by operating with leaner mixtures. The precise curves also vary with the nature of the hydrocarbons in the gasoline mixture. In addition, the carbon monoxide participates in the water gas reaction and the result is the formation of some hydrogen (50–51) (see Fuels, synthetic; Hydrogen).

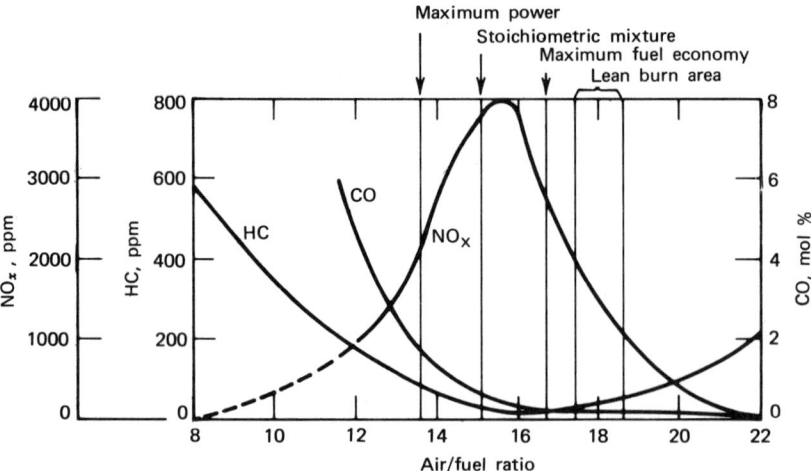

Figure 3. Emissions vs air-fuel ratio.

The hydrocarbon line on the chart also indicates that with greater addition of air the hydrocarbon output is lowered. The hydrocarbons do not go down as rapidly as carbon monoxide with increase of air in the air/fuel mixture. There are a number of explanations that have been given. One is that the hydrocarbons tend to cling to the comparatively cold surfaces of the combustion chamber and consequently are not completely burned (52). Another is that hydrocarbons are trapped in areas such as the gap between the piston ring and the piston (53). A third and extremely important explanation is that the hydrocarbons and air are not completely mixed at the time of combustion so that a portion of them is only partially burned. The literature on this subject is extensive and highly theoretical (54). Factors of importance are preparation of the mixture, method of fuel induction, and combustion chamber design. Each of these is discussed briefly below.

An observation from the chart which is of great importance is that at some point, as air increases in the mixture, there will be a flame-out. In other words, a point at which there is irregular combustion owing to the fact that the flame front was not able to move rapidly enough in the combustion chamber to completely ignite the hydrocarbon. This depends on such factors as compression ratio, location of the spark plug, intensity of the spark, amount of spark advance, etc. Thus the extent is limited to which air can be added to effectively reduce the CO and hydrocarbon concentrations.

As has been indicated above, the formation of the oxides of nitrogen is the result of the direct combination of oxygen and nitrogen at higher temperatures. Figure 4 illustrates the conditions under which this occurs.

As Figure 4 shows, there is very little combination up to the point reached in a normal combustion cycle. After that, there is a very rapid increase with increasing temperatures. The factors that increase this direct combination include high pressures, spark advance, and high combustion temperatures. In other words, almost anything that increases the efficiency of the engine. The control of NO_x emissions becomes one of derating the engine in some fashion. The most common approach is the addition of exhaust gas to the intake charge which has the effect of lowering the combustion temperature (55). There are a number of other reactions which have also been reported (56).

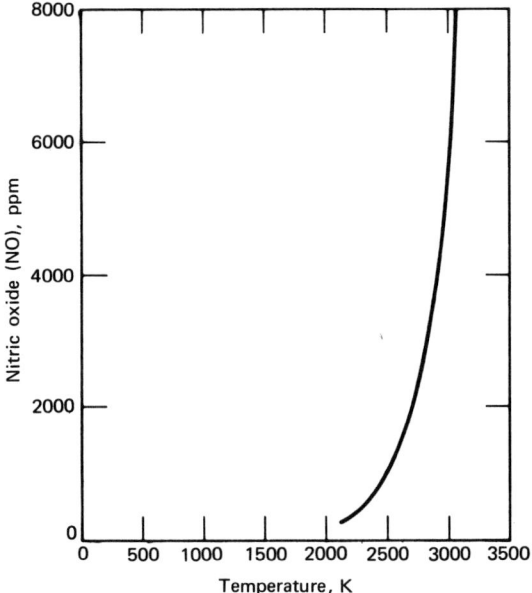

Figure 4. Effect of temperature on nitric oxide formation.

In addition to these components there is also interest in the particulates that are exhausted when the vehicle is operated on the fuel-rich side. If operated to the left of the stoichiometric point, smoke tends to form. This consists mainly of carbon and carbon-rich compounds. An additional source of particulates is the additive that may be used in the gasoline or in the engine lubricant. Thus the regulation of exhaust particulates is being considered (57). A great deal of ingenuity has been used in developing devices that will control the variables that have been discussed above in order to provide as clean an engine exhaust as possible.

Engine Controls. As was indicated above, the mixture of air and fuel that goes to the cylinders is very important. Carburetors are calibrated so that air and fuel combine at a variety of desired air/fuel ratios.

A wide variety of mixtures is required to conform to driving variations. For example, during warmup under cold conditions, a fuel-rich mixture is needed. This is supplied by a choke which essentially restricts the amount of air admitted under the particular operation. The choke action is normally controlled by a bimetallic strip, but in order to keep the enriching effect to a minimum, an electric heating coil is often used so that the action will proceed more rapidly.

Fuel economy is generally promoted by fuel-lean operation and power by somewhat richer operation. The throttle transmits the driver's power needs and the response in terms of air/fuel mixture permits operation at various speeds and under various power demands.

The degree of advancement or retardation of the spark relative to the center of the power stroke also has a substantial effect on the way the engine operates both in regard to fuel economy, exhaust emissions, and power. A graph showing all of the variables becomes extremely complex and computer technology has been used extensively to develop these diagrams (58–59). A typical computer-drawn diagram is

502   EXHAUST CONTROL, AUTOMOTIVE

illustrated in Figure 5 which delineates the relationship between brake-specific nitric oxide emissions (BSNO), brake-specific hydrocarbon emissions (BSHC), and brake-specific fuel economy (BSFC) at a particular engine load and speed.

Adjustment of the spark under a wide variety of conditions is the basis of several electronic controls, one of which is shown in Figure 6.

In an electronic spark controller, signals are fed into a spark-control computer to report the condition of the inlet air temperature, the throttle position, the engine speed, and the intake manifold vacuum. These are then translated by the computer into various commands to the distributor.

Beyond the carburetor, the mixture of air and fuel is distributed to the various cylinders by an intake manifold. Variations in the mixture can easily occur, particularly when the effects of crankcase ventilation and the exhaust gas recirculation (see below) have to be added. Various devices to heat the intake manifold in order to improve distribution have been considered, but it still remains a serious problem. New devices for the control of the carburetor mixture and spark are developed every year.

A typical engine and its many emission control points are shown in Figure 7.

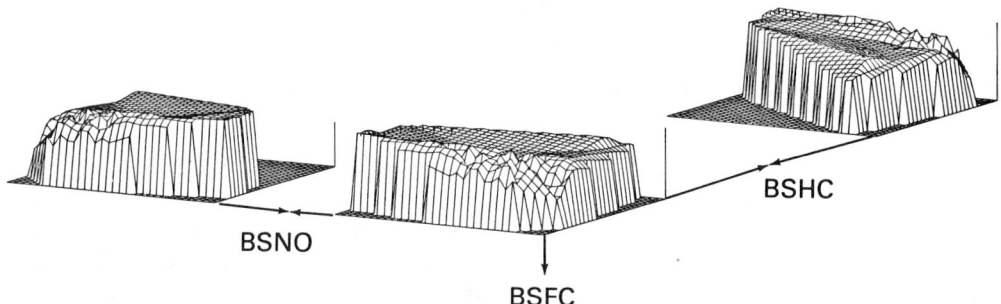

Figure 5. Fuel vs emissions.

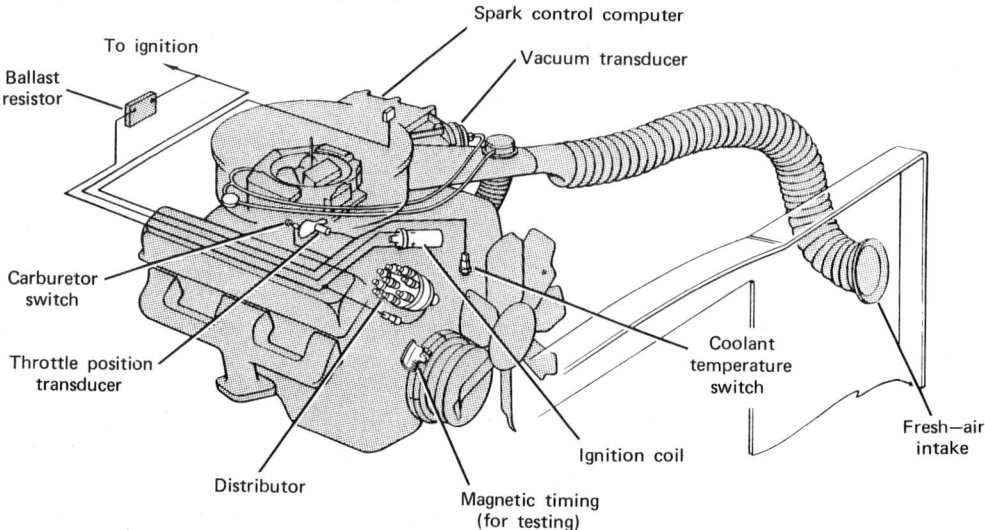

Figure 6. Electronic spark control system.

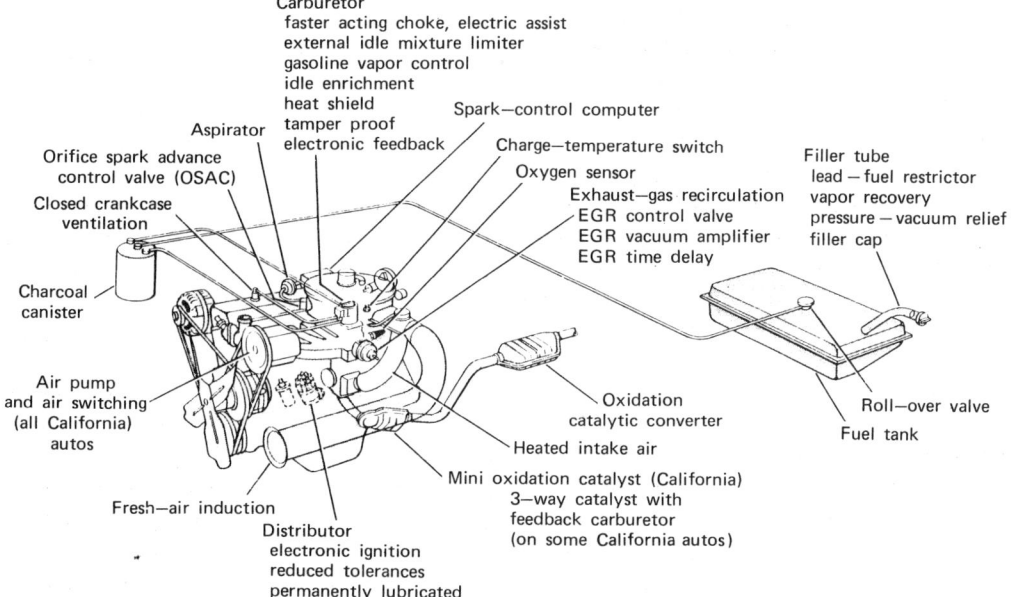

Figure 7. 1979 cleaner air system; comparisons are with precontrolled engines.

A great deal of work is under way to avoid at least some of the complications of the distribution process by injecting the fuel directly into the manifold at the various cylinder points. In most of the systems, the fuel is injected at the cylinder ports (60–62) at relatively low pressure as compared to direct cylinder injection. The fuel to be delivered is determined by signals similar to those discussed under spark timing control.

The exhaust-gas recirculation (EGR) acts to reduce the formation of NO_x during the combustion process. Oxides of nitrogen are formed at the very high temperatures encountered in the combustion flame, and NO_x formation increases as the temperature increases (see also Burner technology). EGR reduces the peak combustion temperature (and also the combustion rate), and thereby reduces the amount of NO_x formed. However, EGR also produces increased hydrocarbon emissions, and these must be controlled by other means.

Recirculated exhaust gas is introduced into the intake manifold by way of a control valve. Care is needed in applying EGR to assure optimum distribution among the cylinders and to minimize any adverse effect on the fuel and air distribution so that uniform combustion is maintained in each of the cylinders for efficient engine operation.

The last important engine control is the combustion chamber itself. Its shape is very important to obtain good mixing and swirling action in the chamber so that complete and rapid combustion can occur. Literally hundreds of approaches have been tried to improve these two factors. Among those that have been put into production recently are chambers in which a fuel-rich mixture is burned in a small prechamber, gases are then inducted into the main chamber where they burn as a leaner mixture (63). Another is the use of a third valve to bring about swirling and, incidentally, lower the temperature of the chamber for NO_x control by using a small amount of air (64).

A third is to use shrouded intake valves to bring about a swirling action (65). Developments in this area are moving very rapidly as a result of the application of advanced computer design technologies to the problem.

Imperfections in fuel preparation, fuel mixture, spark control, and combustion chamber technology have required development of a considerable technology in what can properly be called post-engine controls.

In theory, the control of any carbon monoxide and hydrocarbons that manage to elude the combustion chamber should be comparatively simple. If for some reason the fuel mixture is too rich, additional air would have to be supplied. If time and temperature permitted, all of the carbon monoxide and hydrocarbons should be converted into carbon dioxide and water. In practice, this control is a very difficult task because (1) there is not sufficient time, (2) temperature is too low, and (3) it is difficult to provide the extra air when needed without reducing the temperature. Figure 8 shows the effect of temperature on the reduction in concentration for both hydrocarbons and carbon monoxide at various time intervals.

Figure 9 shows actual engine exhaust temperatures under normal driving conditions. These temperatures are not adequate to bring about any substantial reduction in HC or CO concentration during much of the engine cycle. The first systems that were tried for post-engine control of hydrocarbons and CO featured the introduction of air at the exhaust ports which are the hottest points in the exhaust stream (66). Considerable reduction in hydrocarbons and CO can be brought about by this technique but the amount of air must be very carefully controlled in order to prevent unnecessary cooling of the exhaust stream. In addition, a large number of exhaust-manifold reactors have been tried (67–68) that were designed to retain some of the heat in the exhaust gases so that the reaction can occur for a longer period of time. A typical system is shown in Figure 10.

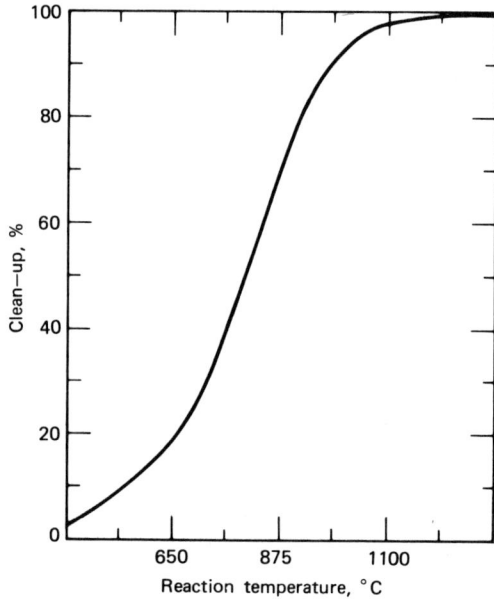

Figure 8. Effect of reaction temperature on reduction in concentration of hydrocarbons and carbon monoxide.

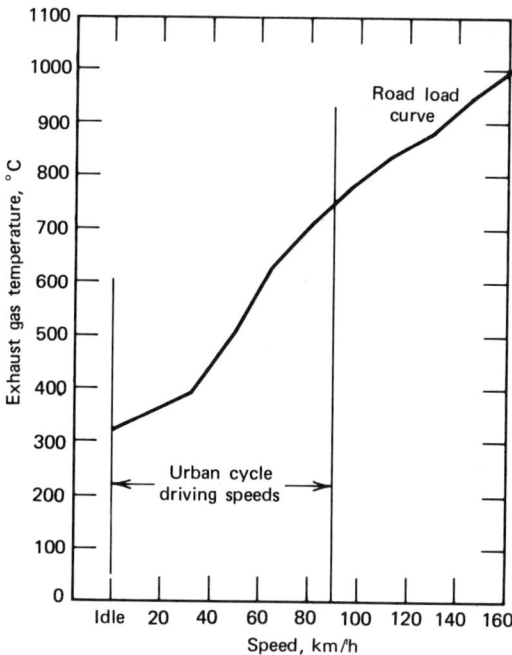

Figure 9. Exhaust gas temperatures at exhaust manifold exit.

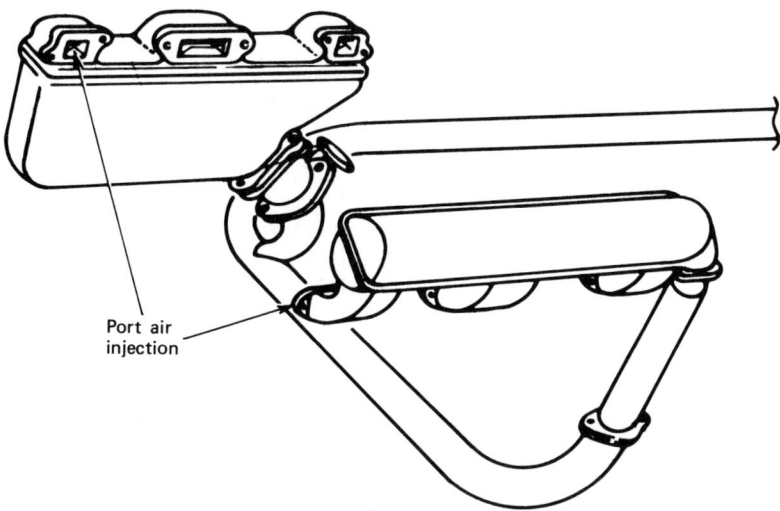

Figure 10. Thermal reactor system.

However, if the gases leaving the exhaust manifold are not sufficiently controlled, it is necessary to apply further treatment. Since these gases are at temperatures that no longer permit a reaction in the available time, catalysts are employed (see Catalysis).

506 EXHAUST CONTROL, AUTOMOTIVE

Catalysts. The first catalysts used were oxidative and required a lean fuel-exhaust mixture. This was accomplished either by the addition of supplementary air by an air pump, as described above, or by an aspirator preceding the catalyst.

Oxidation catalysts promote further burning of hydrocarbons and carbon monoxide in the exhaust gas. Effective operation of this type of catalyst requires temperatures of 315°C or higher as well as an adequate supply of oxygen. Oxidation catalysts in current use normally start oxidizing within two minutes after the start of a cold engine.

Catalytic conversion efficiencies (amount of HC or CO converted by the catalyst as a percentage of the material entering the catalyst), as measured during the official Federal emission test procedure, vary with catalyst location and availability of supplemental air. Catalysts now in use that are located ca 0.7 m downstream in the exhaust from the engine and that do not have an air pump or aspirator have conversion efficiencies of ca 65% for HC and 45% for CO after being driven for 80,500 km. With the addition of an air pump and a start-up catalytic unit designed to operate on cold exhaust gases, mounted at the engine exhaust outlet, catalyst systems have shown efficiencies of ca 80% for HC and ca 75% for CO after 80,500 km.

For most vehicles, oxidation catalysts were instituted when HC and CO limits were reduced by the EPA to 0.93 and 9.32 g/km (1.5 and 15 g/mi), respectively.

Current oxidation catalysts consist of platinum and mixtures of platinum and other noble metals, notably palladium. These metals are deposited on alumina of high surface area. One catalyst is in a form consisting of a mixture of small ceramic (alumina) beads or pellets coated with catalyst material. Another common form consists of a monolith in a honeycomb configuration to provide the necessary surface area and a top layer of deposited catalyst metals (Fig. 11). Both types have advantages and disadvantages. The selection of one or the other is dictated by the kind of vehicle usage (69). In the event that EGR or spark retardation have not provided adequate control of NO_x emissions, catalysts provide a third control option.

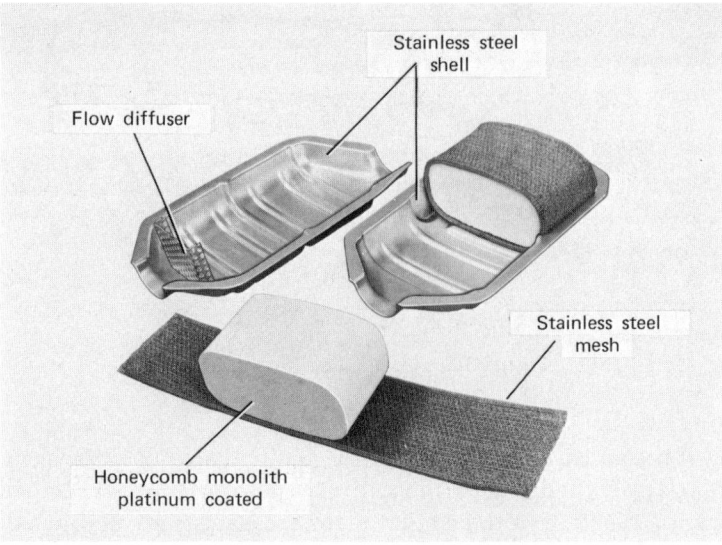

Figure 11. Monolith catalytic converter. Courtesy of the Chrysler Corporation.

Three-Way Catalyst. With emission-standards for HC, CO, and NO_x of 0.25, 2.1, and 0.62 g/km (0.41, 3.4, and 1.0 g/mi), respectively, scheduled for 1981, a three-way catalyst appears to be the most appropriate device available to allow gasoline engines to meet emission control requirements. The most feasible approach to reducing NO_x emissions to levels at or below 0.93 g/km (1.5 g/mi) uses a reducing-type catalyst based on a material such as rhodium. When this catalyst is combined with a conventional oxidation catalyst (using platinum and rhodium as the active materials), it can provide three-way emission control of HC, CO, and NO_x.

Effective catalytic control of all three pollutants is possible only if the exhaust gas contains a very small amount of oxygen. It is necessary, therefore, to maintain precise control of the air-fuel mixture entering the engine, keeping it very close to the stoichiometric level. Figure 12 shows this narrow range in terms of carbon monoxide concentration, along with catalyst efficiencies under steady-state operating conditions with three-way catalysts. A downstream oxidation catalyst with oxygen supplied by an air pump is needed to clean up the remaining HC and CO left by the three-way catalyst.

A critical requirement of the three-way system is the sensing of the exhaust gas composition and the adjustment of the carburetor to provide the desired control. Currently, the sensor consists of sintered zirconium dioxide ceramic material doped with proprietary metal oxides (70–71). A variety of electronic controls are being investigated and no single methodology has been as yet firmly established.

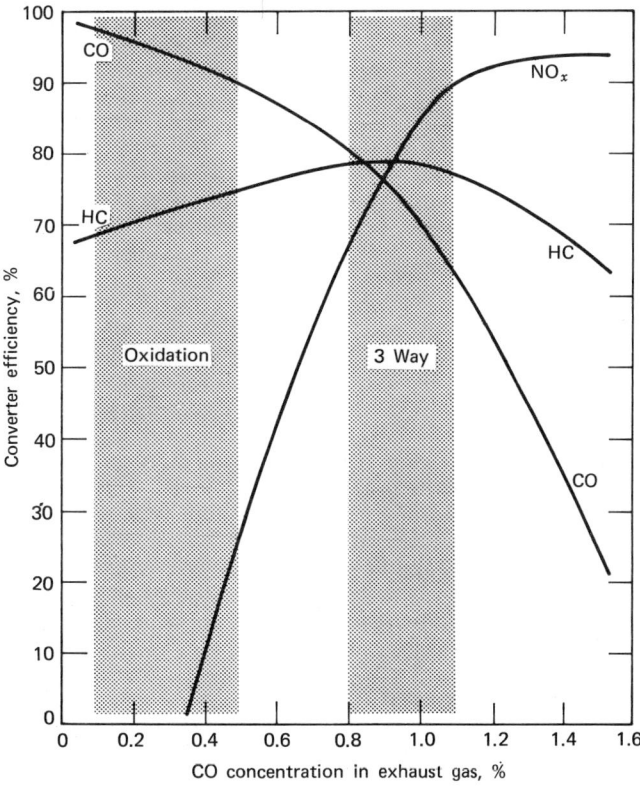

Figure 12. Catalyst operating characteristics. Courtesy of the Chrysler Corporation.

Although considerable work has been done on supports (72), resistance to various poisons (73–76), pore size, pretreatments, and other factors (77–78), it is fairly clear that the important components of catalysts for automobile exhaust involve the precious metals, platinum, palladium, and rhodium, impregnated on an alumina base and possibly pretreated with cerium or heat, or both.

Future Directions

No discussion of emission controls would be complete without some reference to possible alternative power plants that have a more favorable inherent control than the conventional Otto cycle gasoline engine. Among these are the steam engine and the Stirling engine, which rely on external sources of heat. Their continuous burners are easier to control from an emissions point of view. The gas turbine also operates with a continuous feed and at very lean mixtures in order to obtain good efficiency; however, rather high temperatures are necessary and NO_x control may be a problem. In addition, the Diesel engine operates with very lean fuel mixtures and high pressures so that hydrocarbons and CO are not a problem. NO_x is inherently difficult to control, however, and a great deal of study is being expended on methods that would reduce its formation. Electric power plants offer the possibility of complete control of hydrocarbons, CO, and NO_x simply because they do not involve on-board fuel or high temperatures; however, ozone problems may exist and, at the present time, power densities are not adequate to make the electric vehicle truly competitive in anything but very limited applications (see Batteries). Hybrid vehicles consisting of gasoline or diesel engines and electric motors have been produced and have been built on a laboratory scale.

BIBLIOGRAPHY

"Automobile Exhaust Control" in *ECT* 2nd ed., Vol. 2, pp. 814–839, by W. B. Innes and K. Tsu, American Cyanamid Company.

1. A. J. Haagen-Smit, *Ind. Eng. Chem.* **44,** 1342 (1952).
2. *State of California, Administrative Code, Title 13, Motor Vehicles,* Air Resources Board, 1979, Chapt. 3.
3. *Control of Air Pollution from New Motor Vehicles and New Motor Vehicle Engines: Certification and Test Procedures,* Code of Federal Regulations, Title 40, Part 86, 1979.
4. *Report by the Committee on Motor Vehicle Emissions,* National Academy of Sciences, Commission of Socio-Technical Systems, National Research Council, Nov. 1974.
5. F. P. Grad and A. J. Rosenthal, *The Automobile and the Regulation of Its Impact on the Environment,* University of Oklahoma Press, Norman, Okla., Oct. 1975.
6. L. B. Lave and E. P. Seskin, *Air Pollution and Human Health,* Johns Hopkins University Press, Baltimore, Md., 1977.
7. B. Dimitriades, *Development and Utility of Reactivity Scales from Smog Chamber Data, RI 8023,* U.S. Bureau of Mines, Washington, D.C., 1975.
8. K. L. Demerjian, J. A. Kerr, and J. G. Calvert, "The Mechanism of Photochemical Smog Formation" in *Advances in Environmental Sciences and Technology,* Vol. 4, Wiley-Interscience, New York, 1974, pp. 1–262.
9. J. N. Pitts and A. M. Winer, *Mechanisms of Photochemical Reactions in Urban Air,* Vol. II, *Chamber Studies, EPA-600/3-77-014b,* California Univ., Riverside, Feb. 1977.
10. M. D. Thomas, "Effects of Air Pollution on Plants," *U.S. Technical Conference on Air Pollution Proceedings,* World Health Organization, Monograph 46, Geneva, 1961, pp. 233–278.
11. E. R. Stephens and co-workers, *Int. J. Air Water Pollut.* **4,** 79 (June 1961).

12. W. Noble, *J. Agric. Food Chem.* **3,** 330 (April 1955).
13. E. A. Schuck, H. W. Ford, and E. R. Stephens, *Air Pollution Effects of Irradiated Automobile Exhaust as Related to Fuel Consumption,* Report No. 26, Air Pollution Foundation, San Marino, Calif., Oct. 1958.
14. *Determination of the Formation Mechanisms and Composition of Photochemical Aerosols, CAPA-8-71, PB230987/AS,* Calspan Corp., Buffalo, 1973.
15. *Study of Aerosol Formation in Photochemical Air Pollution, CAPA-8-71, PB246060/AS,* Calspan Corp., Buffalo, 1975.
16. *The Formation of Aerosols in a Photochemical Flow Reactor—Final Report, CAPA-8-71, PB275762/AS,* Rockwell Int., Newbury Park, Calif., 1977.
17. J. S. Sandberg, M. J. Basso, and B. A. Okin, *Science* **200,** 1051 (June 2, 1978).
18. *Ozone in Clean Remote Atmospheres: Concentrations and Variabilities, CAPA-15-76, PB272290/AS,* Stanford Research Inst., Menlo Park, Calif., 1977.
19. J. T. Middleton and A. J. Haagen-Smit, *J. Air Pollut. Control Assoc.* **11,** 129 (March 1961).
20. P. Coffey, W. Stassiuk, and V. Mohnen, "Ozone in Urban and Rural Areas of New York State" in *Proc. Int. Conf. on Photochemical Oxidant Pollution and its Control, EPA-600/3-77-0012,* Research Triangle Park, N.C., Jan. 1977.
21. C. W. Spicer and co-workers, *The Transport of Oxidant Beyond Urban Areas, EPA-600/3-76-018,* Battelle Columbus Labs, Ohio, Feb. 1976.
22. W. A. Lonneman, "Ozone and Hydrocarbon Measurements in Recent Oxidant Transport Studies" in *Proc. Int. Conf. on Photochemical Oxidant Pollution and its Control, EPA-60013-001a* and *b,* Raleigh, N.C., Jan. 1977.
23. H. H. Westberg and co-workers, *Measurement of Light Hydrocarbons and Studies of Oxidant Transport Beyond Urban Areas,* Washington State University, Pullman, Nov. 1976.
24. F. W. Gotz, "Ozone in the Atmosphere" in T. F. Malone, ed., *Compendium of Meteorology,* American Meteorological Society, Boston, Mass., 1951, pp. 275–291.
25. H. K. Paetzold, "The Photochemistry of the Atmosphere Ozone Layer" in *Chemical Reactions in the Lower and Upper Atmosphere,* Interscience Publishers, New York, 1961, pp. 181–195.
26. A. W. Bartel and J. W. Temple, *Ind. Eng. Chem.* **44,** 857 (April 1952).
27. *Progress Report on Hydrocarbon Reactivity Study,* California Air Resources Board, June 1975.
28. B. Dimitriades, "The Concept of Reactivity and its Possible Application in Control," *Proceedings of the Solvent Reactivity Conference, EPA-650/3-74-010,* Nov. 1974.
29. K. R. Darnal and co-workers, *Environ. Sci. Technol.* **10,** 690 (1976).
30. G. Tiao, G. E. Box, and W. J. Hamming, "A Statistical Analysis of Los Angeles Ambient Carbon Monoxide Data, 1955–1972," *J. Air Pollut. Control Assoc.* **25,** (1975).
31. T. R. Hauser and C. M. Shy, *Envir. Sci. Tech.* **6,** 890 (Oct. 1974).
32. *Air Quality Criteria for Hydrocarbons,* National Air Pollution Control Administration, March 1970, pp. 41–47.
33. *Fed. Regist.* **43,** 26962 (June 22, 1978).
34. A. J. Hoffman and co-workers, *Science* **190,** 243 (Oct. 17, 1975).
35. F. L. Ludwig and J. H. Kealoha, "Selecting Sites for Carbon Monoxide Monitoring," *EPA 450/3-75-077,* Research Triangle Park, N.C., July 1975.
36. *Guidance for Air Quality Monitoring Network Design and Instrument Siting,* EPA, Jan. 1974.
37. *Air Pollution Control in California—1976,* California Air Resources Board, 1977.
38. *Environmental Quality—1977 Eighth Annual Report,* U.S. Council on Environmental Quality, Washington, D.C., Dec. 1977.
39. *National Air Quality and Emissions Trends Report, 1975, EPA, EPA-450/1-76-002,* Nov. 1976.
40. *Air Quality Criteria for Photochemical Oxidants,* National Air Pollution Control Administration, AP-63, March 1970.
41. *Air Quality Criteria for Carbon Monoxide,* pp. 62, National Air Pollution Control Administration, March 1970.
42. *Air Quality Criteria for Nitrogen Oxides,* EPA, AP-84, Jan. 1974.
43. B. G. Ferris, Jr., *J. Air Pollut. Control Assoc.* **28,** 482 (1978).
44. A. P. Altshuller, *J. Air Pollut. Control Assoc.* **28,** 594 (1978).
45. *Test Procedure for Vehicle Exhaust Emissions,* California Motor Vehicle Pollution Control Board, May 1961.
46. *Air Quality and Meteorology—1975 Annual Report,* Southern California Air Pollution Control District, 1976.

47. *Los Angeles Auto Exhaust Test Station Project, A Joint Agency Report, 1961–1963,* Air Pollution Control District, County of Los Angeles, 1964.
48. E. A. Sheldon, *Motor Gasolines, Winter 1976–77,* ERDA, Berc/PPS-77/3, June 1977.
49. E. A. Sheldon, *Motor Gasolines, Summer 1976,* ERDA, Berc/PPS-77/1, Jan. 1977.
50. B. A. D'Alleva and W. G. Lovell *SAE Trans.* **38,** 90 (1936).
51. E. F. Obert, *Internal Combustion Engines,* 3rd ed., 1968, p. 140, Scranton, Pa., International Textbook Co.
52. W. A. Daniel and J. T. Wentworth, *Exhaust Gas Hydrocarbons—Genesis and Exodus,* SAE paper 468B, March 1962.
53. J. T. Wentworth, *Piston Ring Variables Affect Exhaust Hydrocarbon Emissions,* SAE paper 680109, Jan. 1968.
54. N. C. Blizard and J. C. Keck, *Experimental and Theoretical Investigation of Turbulent Burning Model for Internal Combustion Engines,* SAE paper 740191, Feb. 1974.
55. H. K. Newhall, *Control of Nitrogen Oxides by Exhaust Recirculation—A Preliminary Theoretical Study,* SAE paper 670495, May 1967.
56. E. S. Starkman, H. E. Stewart, and V. A. Zvonow, *An Investigation Into the Formation and Modification of Emission Precursors,* SAE paper 690020, Jan. 1969.
57. *The Clean Air Act,* Section 202 (a) (3) (A) (iii), as amended, Aug. 1977.
58. L. S. Vora, *Computerized Five Parameter Engine Mapping,* SAE paper 770079, Feb. 1977.
59. R. E. Baker and E. E. Daby, *Engine Mapping Methodology,* SAE paper 770077, Feb. 1977.
60. J. Camp and T. Rachel, *Closed-Loop Electronic Fuel and Air Control of Internal Combustion Engines,* SAE paper 750369, Feb. 1975.
61. G. T. Engh and S. Wallman, *Development of the Volvo Lambda-Sond System,* SAE paper 770295, Feb. 1977.
62. I. Gorille, N. Rittmannsberger, and P. Werner, *Bosch Electronic Fuel Injection with Closed-Loop Control,* SAE paper 750368, Feb. 1975.
63. T. Date and S. Yagi, *Research and Development of the Honda CVCC Engine,* SAE paper 740605, Aug. 1974.
64. H. Y. Nakamura, and co-workers, *Development of a New Combustion System (MCA-jet) in Gasoline Engine,* SAE paper 78007, Feb. 1978.
65. E. J. Mitchell, J. M. Cobb, and R. A. Frost, *Design and Evaluation of a Stratified Charge Multifuel Military Engine,* SAE paper 680042, Jan. 1968.
66. D. A. Brownson, R. S. Johnson, and A. Candelise, *A Progress Report on ManAirOx—Manifold Air Oxidation of Exhaust Gas,* SAE paper 486N, March 1962.
67. R. J. Herrin, *Lean Thermal Reactor Performance Characteristics—A Screening Study,* SAE paper 760319, Feb. 1976.
68. A. Jaimee and co-workers, *Thermal Reactor—Design, Development and Performance,* SAE paper 710293, Jan. 1971.
69. M. Teague, *Ceramic Substrate Technology for Automotive Catalysts,* SAE paper 760310, Feb. 1976.
70. W. J. Fleming, *Device Model of the Zirconia Oxygen Sensor,* SAE paper 770400, March 1977.
71. E. Hamann, H. Manger, and L. Steinke, *Lambda-Sensor with Y_2O_3—Stabilized ZrO_2—Ceramic for Application in Automotive Emission Control Systems,* SAE paper 770401.
72. C. A. Duliev and co-workers, *Metal Supported Catalysts for Automotive Applications,* SAE paper 770299, Feb.–March, 1977.
73. L. I. Hegedus and J. C. Summers, *J. Catal.* **48,** 345 (1977).
74. J. C. Schlatter and K. C. Taylor, *J. Catal.* **48,** 42 (1977).
75. J. C. Summers and L. L. Hegedus, *J. Catal.* **51,** 185 (1978).
76. L. L. Hegedus and co-workers, *Poison-Resistant Catalysts for the Simultaneous Control of Hydrocarbon, Carbon Monoxide and Nitrogen Oxide Emissions,* Presented before Div. of Petroleum Chemistry, Inc., at Symposium, Miami Beach, Fla., Sept. 1978.
77. J. J. Mooney, C. E. Thompson, and J. C. Dettling, *Three-way Conversion Catalysts—Part of the New Emission Control System,* SAE paper 770365, Feb.–March 1977.
78. E. Koberstein, *Characterization of Multifunctional Catalysts for Automotive Exhaust Purifications,* SAE paper 770366, Feb. 1977.

CHARLES M. HEINEN
Chrysler Corporation

EXHAUST CONTROL, INDUSTRIAL

Many commercial operations and industrial processes generate gaseous chemical by-products, the disposal of which may cause environmental pollution problems. These pollutants often contribute to changes in the ambient air that have a deleterious effect on living organisms and material objects. The general cost to society of tolerating air pollution has resulted in the regulation of exhaust emissions from both stationary and mobile sources to achieve specific conditions of environmental quality. Currently, the anticipated benefits of improved environmental quality confers a high value on the engineering technology used to meet the regulatory objectives.

A principal element in this technology seeks to minimize the generation of undesirable by-products by modifying specific process materials or operating conditions. Process economics or product quality may restrict the general applicability of this approach. Thus considerable technical activity is directed to developing specialized equipment that would retrofit (modify) or augment existing processing equipment and could capture the objectional by-products or convert them to harmless effluents. However, acceptable technology is now often incorporated into new processing equipment at the conceptual stage in design.

A major developing technology for control of exhaust gas pollutants is their catalyzed conversion into innocuous chemical species, such as water and carbon dioxide. This is typically a thermally activated process commonly called catalytic incineration, or less often, catalytic combustion or oxidation, or catalytic fume abatement; the devices for its application are termed either catalytic or catalyst afterburners, abaters, reactors, or filters. This article is concerned with catalytic incineration of exhaust emissions from stationary industrial sources; vehicular sources are dealt with in the preceding article. Other articles provide more detailed discussions of air pollution control methods (qv) and catalysis (qv) and specific, alternative technologies such as incinerators (qv) (see also Gas cleaning).

Catalytic incineration of industrial exhaust emissions began in the late 1940s (1), primarily for energy recovery rather than improved environmental quality, although odor control was a minor objective at the time. By the mid-1950s there were several dozen catalytic incinerators in California (2), primarily Los Angeles county, the first sizable area within the United States to experience a serious air pollution problem. These early units had not been designed to meet specific emission standards, and their performance was poor relative to thermal incinerators (2). However, it is now generally recognized that both types of incineration systems are capable of a high degree of efficient performance in exhaust emission control.

A comparative view of the technical basis of exhaust emission control systems can distinguish between capture devices, which involve several types of physical interaction, and conversion devices, which primarily involve chemical reaction. Functions of capture devices include condensation, precipitation, filtration, adsorption, and adsorption emulsification, whereas functions of conversion devices include thermal activation, chemical activation (ozone, peroxide, acid–base, etc), electrical activation, and photoactivation.

Condensation precipitation, filtration, adsorption, absorption emulsification, thermal activation, and chemical activation are the basis of conventional commercial systems. Several functions may be incorporated into a single piece of equipment: a chemical scrubber may condense high boiling gases, absorb soluble compounds, trap (filter) insoluble particulates, and react with such materials to produce less objectionable products. However, there are many exhaust emissions that cannot be adequately controlled by chemical scrubbing, or can be controlled more effectively by other types of equipment.

Although installation of emission-control devices requires capital, they may generate useful materials and be net consumers or producers of energy.

Capture of dry cleaning (chlorinated) solvents returns the condensed materials to the process at a fraction of their replacement cost. This approach may not be useful if concentrations are low or flammability is encountered (see Dry cleaning). However, capture devices may be economical when the recovered material has some value, eg, as fuel for combustion equipment. In such cases, consideration should be given to catalytic or thermal incineration with heat recovery by hot gas recirculation or by heat exchange. Whenever the captured products have no value, an additional expense is incurred to provide for their disposal.

Conversion by catalytic oxidation of reduced sulfur compounds (hydrogen sulfide, mercaptans, etc) from paper manufacturing can return appreciable quantities of sulfur trioxide and sulfuric acid to the process. In nitric acid manufacturing the heat generated during catalytic control of nitrogen oxides is used to compress the process feed stream.

Current commercial activity in air pollution control is indicated by the fact that in 1977 the installation of some $470,000,000 worth of control equipment was placed on order in Canada and the United States (3). Future control technology may be anticipated in the recommendations formulated by regulatory agencies (4–9).

CATALYTIC INCINERATION

An incinerator is a chemical reactor in which the reaction is activated by heat and is characterized by a specific rate of consumption of reactants. In this redox reaction there are at least two chemical reactants: an oxidizing and a reducing agent. The rate of reaction is related to the nature and to the concentration of reactants and to the conditions of activation, that is the temperature at which the reactants meet under conditions otherwise suitable for reaction. Implicit in such a simplified description are practical concepts of: temperature (activation), turbulence (proper mixing of reactants), and time (period in which the reaction can accomplish a significant change in the quantities of reactants present).

Exhaust emissions from industrial sources usually contain organic compounds well mixed with oxygen. Imparting the necessary, uniform temperature for reaction within this mixture is of primary importance in the design of incineration equipment.

Thermal incineration (10) relies on a homogenous gas-phase reaction condition. Proper activation requires establishing the minimum required temperature (650–800°C) for an adequate time (0.1 to 0.3 s). General design consideration is given to minimizing heat input and reactor size under the constraints of time, turbulence and temperature.

Catalytic incineration relies on heterogeneous gas-solid interface reaction conditions. Activation necessitates establishing an interaction of one or both of the reactants with the solid catalyst to generate more reactive chemical species, and a required temperature at the catalyst surface for an adequate time, 200–400°C and 0.005 to 0.01 s, respectively, for supported platinum catalysts.

Catalytic incineration performs the same chemical reactions as thermal incineration but at much lower temperatures and, hence, at considerably reduced fuel costs. The time requirement varies inversely with temperature in both systems. Carbon dioxide and water are products of the complete reaction of hydrocarbons with oxygen for both systems.

Typical System

If the proper reaction mixture is available at adequate pressure and temperature, a catalyst bed is simply positioned at a convenient location in the duct work; this is referred to as retrofitting and is discussed in more detail below. However, in most cases the exhaust emissions are at near atmospheric pressure and temperature and the installation of a complete system is required.

A typical system, schematically illustrated in Figure 1, may utilize the following components: (1) blower (primary) to move the gas stream; blower (secondary) to supply air, if required; (2) preheater (burner) or heat exchanger (primary) to heat reactants to operating temperatures; (3) catalyst bed and reactor for the chemical reaction of pollutant; (4) heat exchanger (secondary) or recirculation ducting for the recovery of heat for other uses; (5) instrumentation and controls (not shown) for maintaining operating conditions, and assessing performance variations; and (6) filter/mixer which is the section between preheater and catalyst bed used to assure flow distribution, shield the bed from flame impingment, remove noncombustible particulates, and vaporize entrained liquids and aerosols.

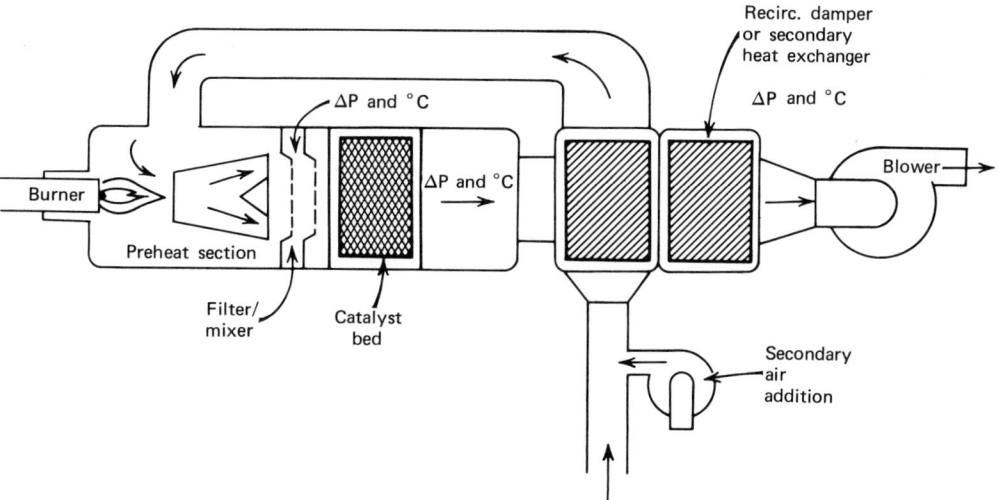

Figure 1. Typical components of a catalytic incinerator. ΔP and °C indicate pressure drop and heated gas, respectively.

Generally, conventional components are used for all of the above items except for the catalyst bed and reactor and filter/mixer.

Rate Processes. The general phenomenon of heterogeneous catalysis involves specific rate processes (types of physical or chemical transitions) that can be visualized in the following simple steps (11): (*1*) gas-phase mass (heat) transfer of reactants to catalyst surface; (*2*) diffusion which establishes distribution of adsorbed reactive species; (*3*) chemical reaction between oxidizing and reducing species; (*4*) diffusion and desorption of reaction products; and (*5*) gas-phase mass (heat) transfer of products away from the catalyst surface. For simplicity, surface diffusion and adsorption (desorption) relative to all reactant (product) species have been combined in steps (*2*) and (*4*) (see Heat exchange technology; Mass transfer).

Thus there is a combination of both physical and chemical rate processes involved in the overall catalyzed reaction. The slowest of these steps controls the apparent rate of reaction per unit surface area of catalyst. The nature of the interaction of these rate processes is illustrated in Figure 2. The combined rates of the physical processes k_p and chemical processes k_c make a varying contribution to the overall (apparent) reaction rate k relative to temperature at the catalyst surface.

The parenthetical heat appears in (*1*) and (*5*) because catalytic incineration requires a specific temperature for reaction to occur and it is an exothermic process. This heat is utilized or generated at the catalyst surface (step *3*, above). Therefore, consideration must be given to heat transfer rates from and to the surface in order to depict local conditions of surface temperature at various points within the catalyst bed.

The most important concept in catalytic incineration is the nature of the con-

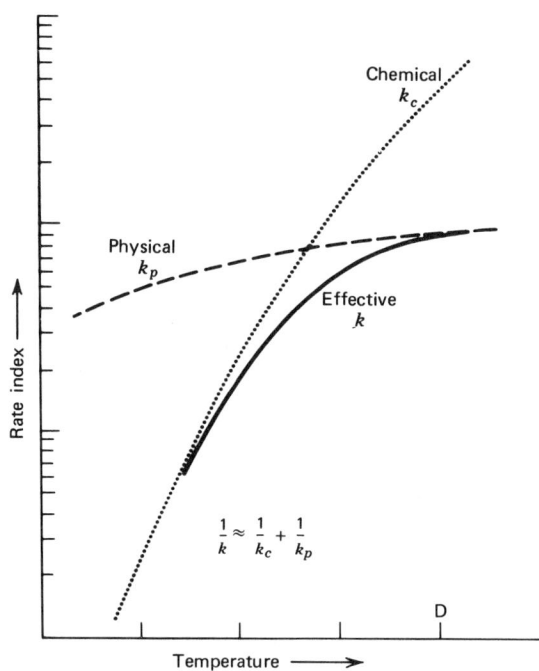

Figure 2. Transfer rate processes as a function of temperature. D is the lowest temperature at which there is adequate mass transfer.

tribution that rate processes make to acceptable performance, with due consideration being given to the variations of operating conditions and catalyst properties that occur during the useful lifetime of the catalyst. The system is static in physical design yet dynamic in operating conditions and catalyst properties. The catalyst typically is designed for maximum activity which dictates that it functions in a relatively activated state which is the condition most conducive to potential deactivation by physical or chemical changes in the catalytically active components. Such changes, although accepted as normal while the system is in use (12), cause changes in the various rate processes that ultimately may be reflected as a variation in performance of the system.

Design Concepts. Optimization of the incineration system is predicated on appraisal of catalyst performance requirements relative to exhaust emissions control and operating conditions. This anticipates the availability of a reasonable description of the characteristics of the exhaust emissions under both normal and anticipated abnormal conditions. Of course, such information is useful in the design of any exhaust emission control system; however, for catalytic incineration it determines the relative order of importance of individual design parameters with respect to the various interacting rate processes discussed in the previous section. An interpretive view of such considerations is represented in Figure 3.

Catalyst Composition

It is recognized that steps (2) and (4) in the rate process are not likely to be rate-controlling (13); this is particularly true if the degree of dispersion and intrinsic activity of the catalyst sites is large. Therefore, the decision is usually made to base

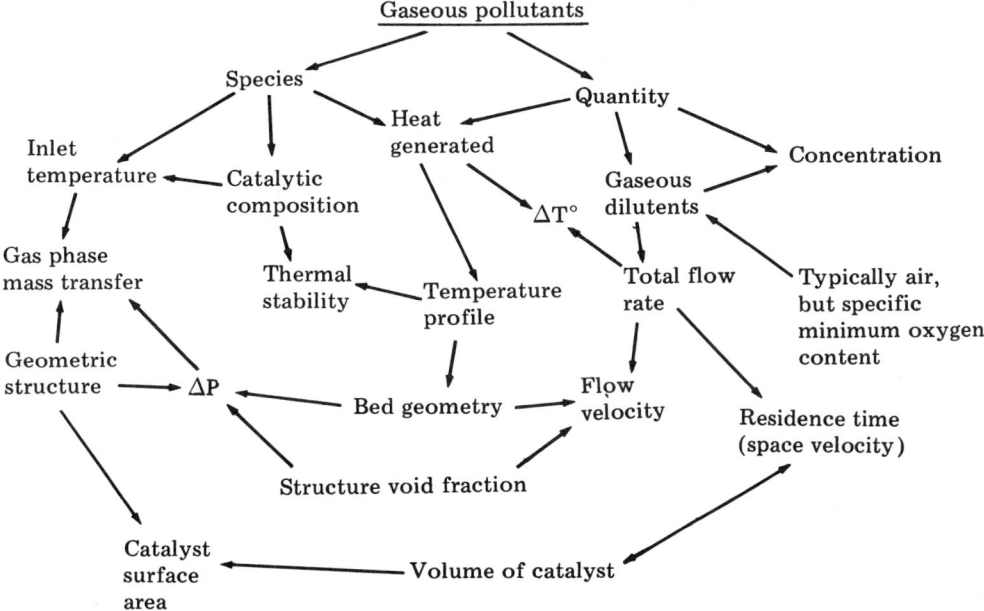

Figure 3. General considerations in catalytic incinerator design.

primary design on steps (1) and (5), rather than the essentially equivalent step (3) because the former has inherently greater stability; ie, changes in hydrodynamic characteristics of the geometric structure, relative to their mass/heat transfer capabilities, do not accrue as rapidly as relative changes in catalytic activity. All of the above dictate that optimization in design is based primarily on catalyst composition and on operating conditions (above temperature D in Fig. 2), and that structural characteristics for mass/heat transfer be given secondary consideration.

Catalytic Species. Precious metal catalysts were found to be superior to base or transition metal oxide catalysts for automobile exhaust emissions control systems (14–15). This is particularly true for catalysts on monolithic supports (15). Of the potentially useful heterogeneous supported catalysts, only the precious metal elements rhodium, palladium, iridium, platinum, and gold (16–17) maintain a stable metal surface under the typical operating conditions. Platinum is the primary choice for catalytic incineration (14,16,18–19). For some systems, palladium shows superior activity and thermal stability relative to platinum (20–22); however, its susceptibility to poisoning does not generally allow its use in an unalloyed composition. The most generally used platinum/palladium alloys are not superior to unalloyed platinum for industrial exhaust emissions control systems (23).

Rhodium is sometimes used in conjunction with platinum or palladium as either an alloying element (24) or as a dispersion stabilizer if it is unalloyed and interacts with the carrier (25). Both approaches are considered to improve catalyst stability. Rhodium is also used as a selective component for reducing nitric oxide to nitrogen (24–25).

Iridium and gold appear not to be used in exhaust-emission catalyst systems.

The precious metal is formulated as a supported catalyst (26) consisting of very small metal particles (3 to 20 nm dia) distributed uniformly upon the surfaces of a support or carrier. The word support has been used both in reference to the catalyst carrier upon which the active catalytic ingredient is distributed, and the geometric structure (pellet, honeycomb, etc) upon which a catalyst coating is applied. The nature of the metal dispersion (eg, the particle size range or the ratio of surface atoms/total atoms of metal present, and the particle distribution upon the carrier) has an important bearing on catalytic activity and stability (27).

The quantity of metal present relative to that of the other components (including the support) is referred to as metal loading. The quantity may vary depending on the degree of dispersion and on the geometric distribution of metal near the surface, which may be different for each catalyst manufacturing process. Geometric distribution of catalyst species on the support surface can influence the performance of the catalyst (28–30). Variations in degree of dispersion for different preparative procedures is found for both platinum (27) and palladium (31) supported on alumina. Primary differences among procedures are related to the mode of localizing the precious metal compound upon the surface of the carrier and to the specific reaction that reduces it to the metallic state. Conventional deposition steps rely on simple solution impregnation (adsorption) or metathetical chemical reaction (surface ion exchange). Conventional reduction reactions involve thermal decomposition or reaction with liquid or gaseous reducing compounds. However, it is possible to prepare the carrier and metal (sol) separately, and to blend them to produce the supported metal catalyst (32). Such variability must be considered with supported catalysts used in catalytic incineration because the reactions can be "demanding, structure-sensitive" ones (33), that is catalyst activity can be highly discriminatory for specific molecular structures.

Catalyst Carriers. The carriers considered in this article are chemically stable metal oxides, such as alumina and silica (26) (see Aluminum compounds; Silicon compounds). When highly active catalytic components are utilized in catalytic incineration, some conventional properties of the catalyst carrier become secondary (34). Because steps (2) and (4) above are not rate-controlling, such factors as pore size and distribution, high specific surface area, and coating thickness are of minor importance except as they may alter the nature and stability of metal dispersion and the thermal stability of the carrier. Several of the anhydrous aluminum oxides are the carriers of choice (35): activated or gamma-alumina is predominant, but eta-, kappa-, and alpha-aluminas are utilized in specific applications (see Aluminum compounds).

Geometric Support. The critical nature of the geometric support structure, relative to economic and operational factors, is reflected by the large amount of technical development work devoted to support structures during the past three decades (10). The early support structures were metal ribbons (1), spherical pellets (36), and ceramic rods that were assembled into bricks resembling small tube banks (37). By the early 1960s, work was being devoted to monolithic supports, primarily ceramic honeycomb. Commercial methods of manufacture (38–39) and design requirements for utilization of supports in catalytic incinerators (40) have been established. Recent developments include metal honeycomb (41), random packings of ceramic-coated metal wires (42) and ordered arrays of ceramic fibers (43) (see Refractory fibers).

The individual matrix elements (eg, nonbonded cylindrical, spherical, random-shaped particles (pellets); fibers; wires; ribbons; rods; monoliths) all form a foraminous substrate when used in catalytic incineration. Selection of a specific type of substrate involves consideration of the following: chemical stability; thermal stability; physical integrity of matrix elements; structural integrity of substrate; catalyzation characteristics; catalyst bed-design requirements; fouling characteristics; maintenance characteristics; and economic factors, including equipment cost and operating cost.

Each of these parameters is considered in view of anticipated operating conditions: temperature profile; flow rate/linear velocity; residence time/space velocity; and pressure drop (gradient). Combinations of parameters result in incremental variation of heat, mass, and momentum transfer (between the gas and solid phases) during movement of gas through the catalyst bed.

Many commercial catalytic incinerators, with the exception of such small units as the automobile exhaust converter, utilize unitary monolithic supports rather than pellets. The reasons for this choice illustrate the practical analytical process in substrate selection.

Pellets have functional deficiencies of nonreproducibility in packing pattern and changes in packing pattern during use. They wear at the surface where the catalyst is located, and therefore, require catalyst coatings diluted with ingredients that impart resistance to spalling and abrasion. The surface near the contact between pellets is poorly exposed and of minimum catalytic effectiveness. Small pellets with high pressure drop characteristics must be used to obtain higher surface area. The high pressure drop requires a thin bed design which leads to sharp temperature gradients within the bed and to low linear flow velocities for the resulting large frontal area. Dense material must be used for abrasion and crush resistance. The increased mass produces high heat capacity and sluggishness in response to heat. Pellets form relatively small flow channels with high macrotortuosity, which results in plugging and nonuniform flow patterns. A principal virtue of pellets is a relatively low selling price.

Metal oxide ceramics such as mullite (aluminum silicate), alpha-alumina, cordierite, and spinel (magnesium aluminate) are the major materials used for monolithic substrates. Nickel alloys are used for metal substrates because they are easy to fabricate, resist oxidation. Regardless of composition, a textured substrate surface is preferred since it enhances coating adhesion by a degree of physical interlocking.

Catalyzation. Application of a coating of supported catalyst to the surface of the geometric substrate often requires special considerations (1,36–37,41,44) as noted earlier. Other considerations include:

Pellets. Preformed pellets with the proper strength are catalyzed by wet impregnation procedures. Inadequate dilution of catalytic material with physically durable diluents results in surface wear which leads to two problems: physical loss of catalytic material and generation of dust that accumulates and plugs gas-flow passages. These problems are intensified whenever smaller particles (ca 3 mm dia) are selected because of the increase in surface area. Abrasion becomes extreme in such cases with fluidized-bed catalysts (45).

Fibers. The very small diameter and interparticle spacing in fiber batts presents catalyst coating problems. The fiber must be dense in order to be sufficiently strong, thus a separate catalyst carrier (washcoat) is required. During application of the conventional carrier material, it tends to filter out onto the fiber batt, thus bridging and plugging the interparticle spacings. Finally, the small fibers can flex which causes abrasion and spalling of the catalyst coating. Most of these problems are reduced with ordered arrays of bonded fibers (43), but this approach sacrifices considerable surface area.

Metal Substrate Surfaces. Direct electrophoretic deposition of catalyst with little or no carrier (1) usually results in poor adhesion and in reduced catalytic activity (10). This deficiency is sometimes exacerbated by oxidation of the metal substrate. Improvements have been achieved by applying dual-function wash coatings (41) that serve as a bonding layer, a barrier to oxidation products, and as carriers for the supported catalyst (42).

Ceramic Honeycomb. Small cell-size (<1.6 mm dia) honeycombs with long cells (>25 mm) require dilute catalyst slurries which form thin, nonuniform coatings. Multiple-coating treatments do not completely eliminate the nonuniformity of coating thickness which is thicker at each cell entrance. Large diameter, short cells can be coated more uniformly with the required thickness of coating and need a minimum of coating binder for adhesion and cohesion. Thus, the deposition of almost pure surfaces of supported catalyst is possible using conventional dispersion procedures (26–27,31–32).

Coating binders may vary since they depend primarily on the specific substrate and supported catalyst used. Soluble compounds that decompose to produce aluminum oxide are effective (36) and are preferred for substrates with very small gas-flow channels. Colloidal dispersions of hydrated aluminum oxide provide improved bonding of thicker coatings and are preferred for substrates with cells of larger size (44). Attention should be given to formulation of coating preparations that minimize binder migration during the drying and curing steps.

The amount of binder remaining in the coating on monoliths is 2–20 wt % after drying and curing. The final step in most curing sequences is the setting of the binder phase by heating at 200–500°C for one or more hours.

Packing Procedures for Monolithic Supports. There are two general types of catalyst packings: fixed (single) bed and modularized bed. The fixed bed is assembled from precut pieces that are packed within a single vessel (shell, spool piece). A resilient, mechanical insulating layer usually is placed between the inside vessel wall and the outer edge of the catalyst bed to permit thermal expansion/contraction and provide a seal to prevent gases from passing around the catalyst. Grates, screens, and metal seal rings are usually placed on each end of large sized beds of, eg, 0.3 m³ (ca 10 ft³).

A modularized bed consists of a number of modules or cartridges of catalyst placed within a single vessel. A typical module consists of a stainless steel can, which is filled with catalyst pieces, with a resilient seal between the can and the catalyst. These modules are gasketed into the openings of an egg crate-type of metal frame and held firmly in position by retaining bars or clips. Precautions are taken to assure that there is no bypassing of the gas. Modules usually have ca 0.1 m² (1 ft²) flow area and are 0.2 to 0.3 m deep. The weight of each module is limited to 14 to 23 kg each for ease of handling. Modular beds are preferred in applications where maintenance treatments are anticipated. They also facilitate partial replacement of a catalyst bed.

Catalytic Reactor Design

The performance of a catalyst bed that is operated adiabatically with gas-phase transport control of an exothermic reaction is determined solely by the rates of changes occurring across the thin stagnant gaseous film adjacent to the gas/solid interface. The transfer of mass in the form of chemical reactants to the solid surface determines the rate at which they react and the rate at which heat is generated at the surface. The transfer of heat from the surface establishes thermal conditions within the film which influence the rate of mass and momentum transport to the surface. The transfer of momentum to the solid surface determines the characteristics of the film and the dominant mode of all transport processes within the film; ie, molecular diffusion or eddy diffusion (turbulent mixing) (see also Reactor technology).

Design Variables. All convective transfer rates are maximized by eddy diffusion under turbulent flow conditions. The interrelationship of these transfer processes is expressed as follows (46):

$$j_D = \frac{k_c}{U}\left(\frac{\mu}{\rho D}\right)^{2/3} = \frac{f}{2} = \frac{h}{C_p U} = \left(\frac{C_p \mu}{K}\right)^{2/3} = j_H \tag{1}$$

The translation from mass transfer j_D to momentum transfer $f/2$ to heat transfer j_H is governed by the Reynolds Number, $Re = 4r_h G/\mu$. The hydraulic radius r_h also appears in the apparent friction factor f and makes possible the expression of the transfer processes in terms of the foraminous substrate characteristics. It is recognized that r_h is a characteristic length, which may not be the most suitable geometric dimension to use for correlation of flow friction among geometrically dissimilar structures (47–48); however, it has been utilized here for conceptual consistency. Free-flow area velocity U is implicitly related to geometric structure void fraction. The fundamental hydrodynamic characteristics of the foraminous substrate are related to the mode of surface distribution, which can be associated with a tortuosity factor and with the following variables: L, substrate length (in direction of flow); S, geometric (peripheral) surface area; and E, void fraction (macroporosity) of the structure. These are generated by the matrix elements which also define hydraulic radius:

$$r_h = \frac{\text{flow channel area}}{\text{flow channel perimeter}}$$

The pressure drop ΔP across the catalyst bed is important in the transport processes and in energy consumption for movement of the gas stream. It appears in the friction factor f under turbulent flow, as:

$$P = f \frac{G^2 L}{2 g_c r_h \rho} \qquad (2)$$

It should be noted that there are two contributions to pressure drop (46–48): viscous shear (skin friction) and pressure force (form drag). The velocity gradient developed by shear stress does not eliminate the contribution of transport rate across the diffusion film; however, the contribution of skin friction to total pressure drop may become undetectable with respect to that from form drag (46).

In addition to selection of operating conditions that enhance transport rates by virtue of eddy diffusion, several other structure-related variables are utilized for turbulence generation in catalytic reactors. These are based on such concepts as:

Surface roughness produces greater shear stress (46), and therefore, catalyst coatings with high roughness factors are preferred (49).

Entrance contraction and exit expansion of the gas-flow path inhibits development of velocity profile (48). Thus multiple thin layers rather than one thick layer of monolith are used (49), and the spacing between layers is adjusted by using alternating layers of different cell sizes.

Irregular flow path geometry (macrotortuosity) enhances form drag. Therefore, unaligned cells are included from layer to layer, or interacting adjacent layers of cells (crossflow monoliths) (50) and particulate and random fiber packings are used.

Theoretical Modeling. The pragmatic commentary in the previous sections should not be construed as a denial of the existence or usefulness of a theoretical basis for catalytic incineration processes. Representative references to detailed discussions of fundamental concepts include transfer phenomena (51–52), catalyzed chemical reactions (53), and general discussions of heterogeneous catalysis (54–55) (see Catalysis).

Representative models deal with specific concepts concerning: differing geometric supports (49); thermal excursions in monolithic supports (34,56–57); transport limitations relative to bed design (58–59); and effects of inaccuracies in data acquisition (60). Ref. 61 presents an instructive discussion of modeling of practical systems and clearly illustrates the degree of complexity of such systems and the restrictions resulting from simplifying assumptions.

Because of limitations in the types of precise measurements that can be made, current catalytic reactor design only relies on the fundamental theoretical models for general guidance in making empirical correlations and using intuitive judgement. Although the data used for some computer modeling programs are similar to variables associated with the theoretical model, the computational results are not considered to be a unique mathematical solution to a formal theoretical expression.

Empirical Correlations. Beginning with Reynolds' work at the turn of the century, the utility of empirical correlations was well established in systems engineering (see Dimensional analysis). An effective approach starts with an analysis of the variables associated with the theoretical model. Then the relative importance of these variables is defined according to experimental data for specific test conditions, the results of which, are used to derive the parameters for the empirical model. This approach has been applied to catalytic incineration to determine how performance varies with changes in operating conditions and in geometric support (40); examples are given below:

Structure-Limited Performance. The extent of reaction is used to determine a performance index. A structure index is defined as (40):

$$E^2 S^2 L \frac{\Delta P}{U^2} \tag{3}$$

A more exact equation would include the gravitational constant and variables associated with gaseous properties, which are omitted here for simplicity.

The correlation for structures described in Table 1 is shown in Figure 4.

The disparity for structure D is obvious, it is less apparent for some of the other structures. Proper adjustment to compensate for variation of hydrodynamic versus geometric parameters is relatively simple for structures D and F. This is illustrated in Figure 5, where the solid lines indicate geometric elements, and the dashed lines indicate their hydrodynamic equivalent (not to scale) as applied to surface area and void fraction in equation 3. When the hydrodynamic values are used, the points shown in Figure 4 (as indicated by the arrow) move toward their proper position on the Structure Limited Performance curve. Similar adjustment for, eg, structures 4C and P are more difficult to compute because of the uncertainties that exist in correction for the tortuosity factor.

Table 1. Substrate Structure Performance

Structure[a]		Geometric		Comments[b]
Code	Point	Void fraction, E	Surface area, S	
4C	X	0.75	188	6-mm cell crossflow honeycomb[c], alternate rows of cells at 45° angle, open connection at cross-over points
2B	○	0.61	268	3-mm cell beehive-type honeycomb
3B	●	0.64	187	5-mm cell beehive-type honeycomb
2B/3B	▲	0.625	228	alternating 25-mm thick layers from above
4C/2B	□	0.68	228	alternating 25-mm thick layers from above
4C/3B	■	0.695	188	alternating 25-mm thick layers from above
D	⊗	0.60	695	590 cells per motor, triangular-cell honeycomb (see Fig. 5)
		(0.30)[d]	(512)[d]	
P	△	0.30	380	pellets, 3 by 3 mm extruded cylinders
F	▽	0.98	1280	ceramic fiber bundles, 50 fibers per bundle (see Fig. 5)
		(0.86)[d]	(206)[d]	
S	▼	0.427	412	sticks, 3 by 32 mm extruded cylinders

[a] Monoliths packed as 25-mm thick layers, data includes bed depth (L) from 1 to 6 layers. Data for pellets covers bed depths (L) from 13 to 100 mm. See Figure 4 for use of data point symbols.
[b] Data includes gas velocity (V) range corresponding to standard gaseous space velocities from 8 to 167/s.
[c] The crossflow structure has a finite tortuosity property relative to the beehive structure.
[d] Hydrodynamic values.

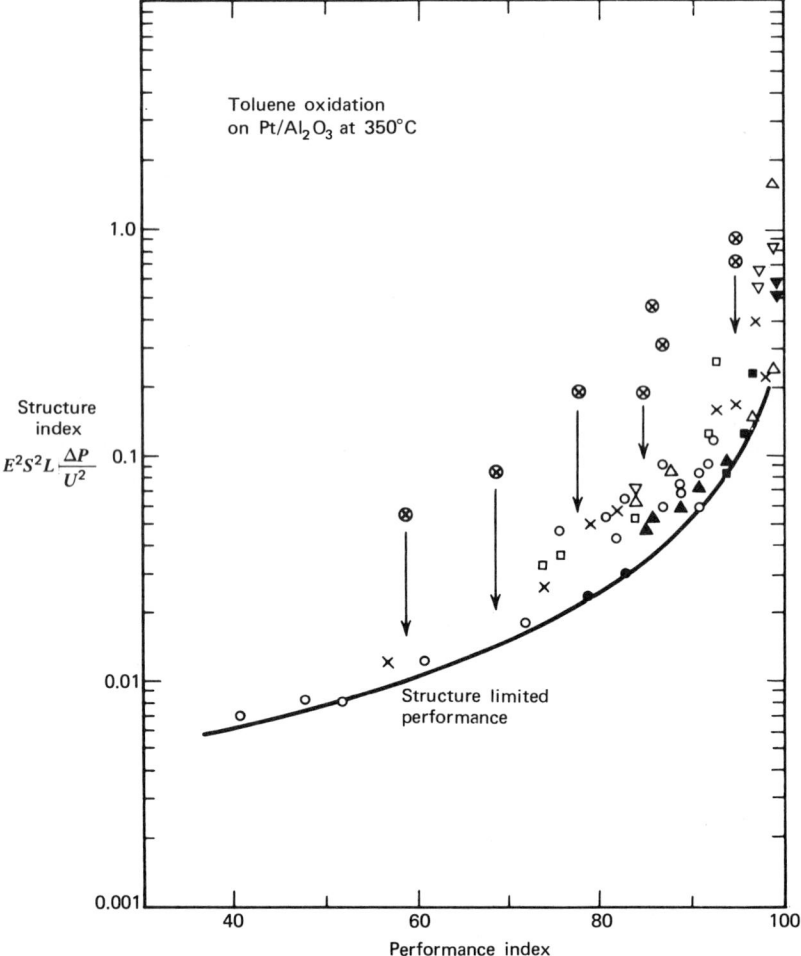

Figure 4. Substrate structure contribution to performance in catalytic incineration.

An important conclusion from Figure 4 is that there is no apparent limitation to the performance capability of a specific geometric structure. It may be inferred that any system operating well above the curve is not of optimum design, or structural variables have not been properly defined.

Operating Variables and Performance. Among the various numerical equivalents, the gas mass flow rate G and space velocity X (standard gas volume flow rate per unit volume of catalyst), are useful for comparative analysis. A correlation of the performance index in terms of these variables and relative to the pressure-drop gradient, $\Delta P/L$, is:

$$\frac{\Delta P}{L^3 G X^2} \tag{4}$$

This correlation is shown in Figures 6–7 where some structural parameters have been omitted from the group of operating variables in order to separate the curves for individual geometric supports.

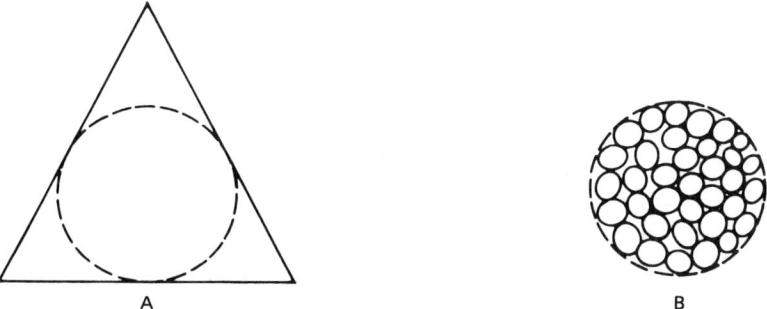

Figure 5. Typical adjustment of substrate geometric parameters: A, triangular cell (internal); B, fiber bundle (external). Hydrodynamic diameter = 4(cell surface area/cell perimeter).

The correlation of Figure 6 is for the single geometric structure 2B (Table 1). The numbers 1–6 associate variation of bed depth L, in cm, with a performance index and the related value of the correlation function. In addition, a grid is formed from the

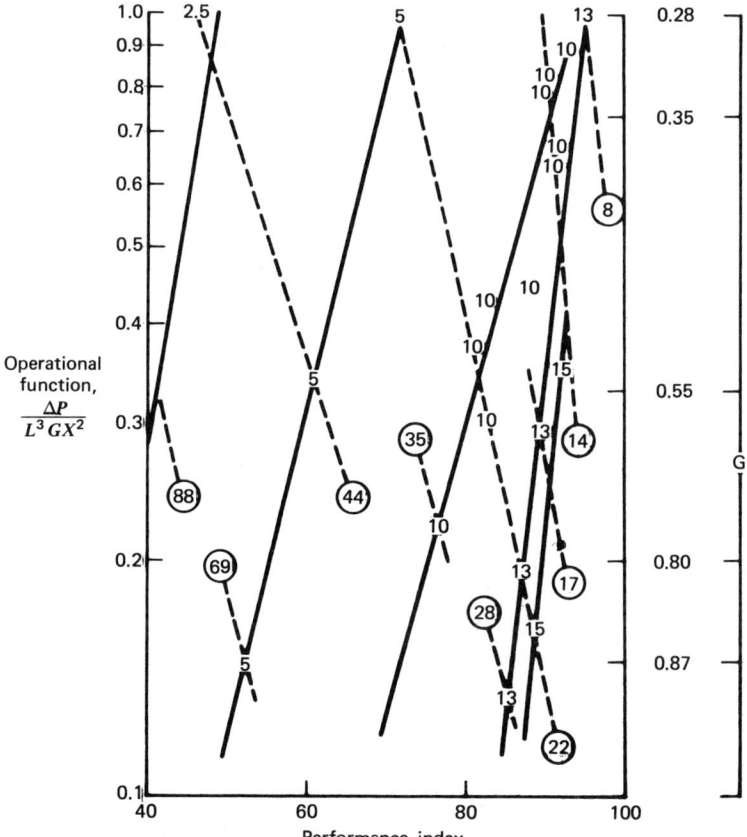

Figure 6. Operating variable contribution to performance for 2B-substrate. Open numbers are in cm. Closed (circled) numbers indicate space velocity (x).

dashed lines describing constant space velocity X (circled number), and the lines describing constant gas mass rate G (shown on right margin). A similar correlation can be generated for each of the structures given in Table 1, but the range of the correlation function varies with each structure (over the same range of G). This is shown by the various points in Figure 7.

Figure 7 also suggests that lines of constant bed depth L, in cm, are collectively common to all structures within the range of the correlation. There is a distinctive range in the function for similar ranges of gas mass rate G for each structure. The concept of a minimum limit of bed depth relative to a specific space velocity and performance index is implied. This minimum limit varies among the structures. It is important that these relationships are explicitly recognized in experimental catalytic incinerator design studies.

Operating Conditions. Catalytic incinerators are expected to function properly under the varying operating conditions that are sometimes required to accomodate variation in the exhaust emissions being treated. This variation rarely includes only the variation encountered in start-up and shut-down procedures. Even for an apparently constant operating condition, nonuniformity of flow and temperature within the catalytic reactor requires design for proper performance within the range of such variation. Thus all commercial equipment is over-designed to some extent both in the

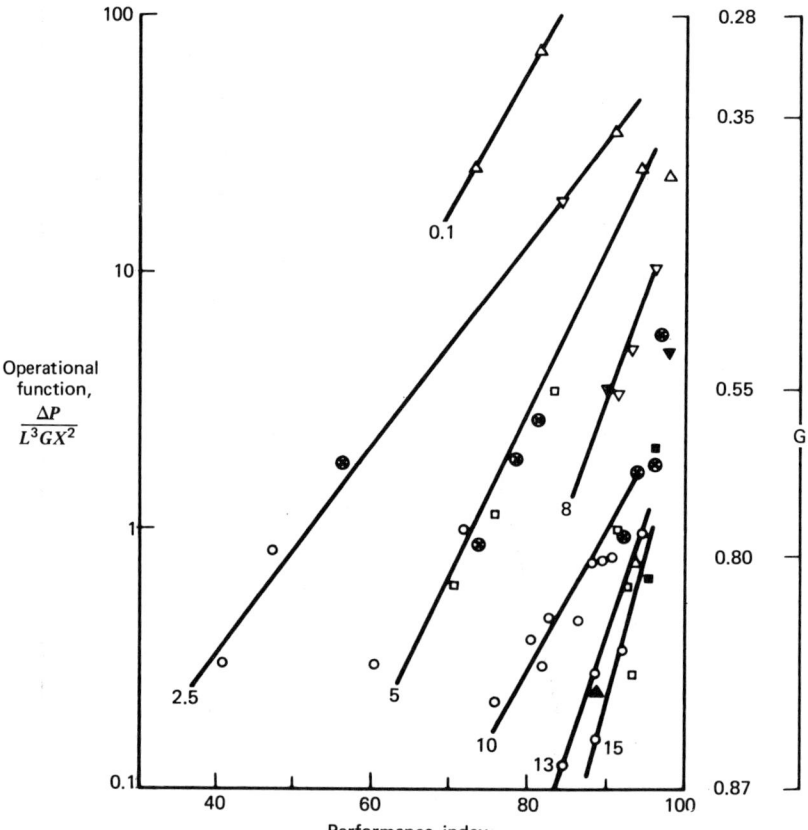

Figure 7. Operating variable contribution to performance for various substrates.

catalytic reactor and in other components such as burners, blowers, and heat exchangers. Although over-design of catalyst and operating conditions for performance exacts a cost for equipment and energy, there is significant compensation in reduced maintenance cost and increased catalyst life.

Variation in flow rate and in gaseous composition of exhaust emissions generally may require changes in operating conditions of a catalytic incinerator. Typical circumstances include varying types of quantities of materials being processed, idling modes of operation during product or process changes (even shift changes, coffee and lunch breaks), and major or minor process upsets. In operations where the catalytic incinerator is supplying energy or materials to other processes, the consequences of varying conditions become more important. This is particularly true for catalytic incinerators with secondary heat recovery which are designed for a high performance index under highly exothermic conditions to maximize energy recovery.

General guidelines are:

Temperature. *Catalyst inlet.* A minimum of 250°C is maintained for complex hydrocarbon pollutants to prevent possible organic char formation. A minimum of 120°C is needed for simple, reactive reducing species to prevent water condensation on the catalyst surfaces.

Catalyst outlet. A maximum of 650°C is used with systems that release nominal heat.

Linear Gas Velocity. 3–6 m/s is required for normal performance index and heat release and higher velocities are used for high heat release systems.

Reactant Concentrations. Concentrations vary with heat of reaction, but should be maximized when concentrations are low, up to 10–15% of lel (lower explosive limit); higher concentration should be reduced toward this level.

Pressure Drop. A minimum of 500 Pa (2 in of water); 750–1250 Pa (3 to 5 in water) is optimum.

These conditions are sometimes compromised for practical reasons such as space or equipment design limitations.

Deactivation and Maintenance

The performance index of a catalyst is highest with initial use, and during use, performance decreases at varying rates. However, there are cases where performance has increased after a deactivation/maintenance cycle. This appears to be a selective activity increase associated with a specific reaction rather than a general increase in activity. This reduction in performance with service is referred to as deactivation and depends upon a number of factors. Some catalyst manufacturing processes yield a product with super activity, ie, there is a detectable loss of activity within a few hours, followed by the slower, normal rate of deactivation (62). Proper reactor design anticipates that operating conditions or specific maintenance procedures will be used to afford acceptable performance during a specific period (63). When this can no longer be achieved, the catalyst is considered to be deactivated and is replaced with fresh material. Much of the precious metal can be recovered from the spent catalyst.

Thermal Aging. Deactivation may be the sole result of physical changes resulting from thermal effects, such as recrystallization of the carrier, the precious metal, or both, and is referred to as thermal aging or sintering. Thermal aging is accompanied by large changes in surface area and by dispersion of the supported metal (20,62–65).

For this reason most supported catalysts are over-designed relative to precious metal composition and loading. Reactors designed for high mass transfer capability have good aging resistance because of the associated high heat transfer rate from the reaction sites which minimizes the nominal temperature of exposure. The typical platinum/alumina supported catalysts show detectable thermal aging during short periods of exposure at 500°C (in air), but they may give satisfactory performance for very extended periods (63).

The nature and physical stability of the carrier is important to the resistance that supported catalysts show to thermal aging. Physical changes in the carrier force movement of the dispersed metal, which enhances the possibility of metal crystal growth and loss of metal distribution and surface area. For physically stable carriers, the chemical nature of the surface can vary the transport rate of precious metal which again affects metal recrystallization rate (10). Thermal aging has an irreversible effect on catalyst activity.

Masking. Deactivation may be the sole result of physical changes associated with the accumulation of contaminates that form a barrier or mask between the gas stream and catalyst surface. This has also been referred to as catalyst blinding. Masking may result from deposition of noncombustible material entrained in the exhaust emissions. Examples include dust in process air, corrosion or oxidation products from process equipment, and material included in process formulations, such as pigments and fillers. Organic phosphates and silicones are substances that require special consideration.

Prevention of noncombustible masking is usually achieved with a properly designed filter/mixer element that is located between the burner and catalyst in the catalytic incinerator (see Fig. 1) and which can be cleaned during normal shutdowns. Such deposits usually can be removed by air lancing or by aqueous washing.

Poisoning. Deactivation also may result from chemical changes caused by the combined effects of thermal condition and contamination and is characterized by chemical reaction of a contaminant with the supported catalyst. The contaminate must be in a specific chemical form in contact with the carrier or precious metal catalyst at a temperature sufficient for reaction; otherwise, the contaminate merely represents a possible source of masking. For example, iron oxide reacts with an aluminum oxide carrier at high temperatures, but to undergo extensive reaction, it must be able to move into the carrier. This may require additional fluxing contaminants or a still higher temperature to facilitate diffusion of the oxide. Formation of the iron-aluminum spinel results in deactivation because it is a different carrier species which can consume the transitory precious metal oxide (66), and thus introduce a process for converting the metal to a noncatalytic form (64–65).

In systems operating at relatively low temperatures under low heat release conditions, iron oxide normally causes deactivation by masking. True catalyst poisoning is encountered infrequently in typical catalytic incineration systems, but when it is encountered, its effect on the catalyst is irreversible (2,10–11,17,21,26,29).

Inhibition. Inhibition has been considered as reversible poisoning, because of its temporary effects. Although its mechanism of deactivation may have features in common with poisoning, the extent of chemical change is quantitatively lower and is reversible. Depending on circumstances, specific contaminates can result in either poisoning or inhibition. Sodium salts, chlorine compounds, and some sulfur and phosphorus compounds have exhibited both types of deactivation. The inhibiting effect of carbon monoxide (67–69), nitrogen oxides (67), some reduced sulfur compounds (70), and some olefins (67) is associated with chemisorption, rather than chemical reaction, and thus they should not be considered as catalyst poisons.

Maintenance. Although modes of deactivation are discussed individually, combinations of deactivation effects are encountered in commercial catalytic incinerators and specific comments appear in subsequent sections below.

In general, contaminates should be minimized both in the emissions and in any general maintenance treatments. Catalyst temperatures should be high enough to prevent deposition of solid combustible material, but low enough to minimize thermal aging (including thermal maintenance treatments). Corrective maintenance procedures should be initiated early in the deactivation process. For example, periodic elevation of catalyst temperature to 400–450°C will effectively remove small quantities of combustible char without deleterious effects on the catalyst, even if some base metal poisons are codeposited with the char. However, if sufficient material accumulates before burn-off, a highly exothermic reaction occurs that may cause either a serious deactivation from irreversible thermal aging, or a poisoning reaction with the base metal contaminates.

Applications

Chemical Processes

Nitric Acid Manufacture. The principal pollutants associated with nitric acid manufacture are the residual unreacted nitrogen dioxide NO_2 and nitric oxide NO from the absorption column (Point A in Fig. 8). These two oxides constitute the NO_x (measured and expressed as NO_2) that must be maintained at or below the regulatory levels. The process of reducing the NO_x to the required emission level is commonly called abatement.

Nonselective Reduction. The predominant abatement method is the catalytic reduction of the NO_x to elemental nitrogen with any available hydrocarbon as the reactant in accordance with the following reactions:

$$C_xH_y + \left(\frac{y}{2} + 2x\right) NO_2 \rightarrow \frac{y}{2} H_2O + x\, CO_2 + \left(\frac{y}{2} + 2x\right) NO \qquad (5)$$

$$C_xH_y + \left(\frac{y}{4} + x\right) O_2 \rightarrow \frac{y}{2} H_2O + x\, CO_2 \qquad (6)$$

$$C_xH_y + \left(\frac{y}{2} + 2x\right) NO \rightarrow \frac{y}{2} H_2O + x\, CO_2 + \left(\frac{y}{4} + x\right) N_2 \qquad (7)$$

Equation 5 often is called the decolorizing reaction because the reddish-brown NO_2 is reduced to colorless NO and the exhaust in the vicinity of the stack is clear. However, the subsequent atmospheric cooling and slow oxidation of the NO to NO_2 often produces detectable color some distance downwind of the emitting location.

Before the promulgation of the 1971 Federal Standards, most nitric acid units were designed and built with catalytic combustors whose sole purpose was the introduction of heat (via eqs. 5–6) into the high pressure (0.7–0.9 MPa or 7–9 atm) gas before expansion through the power-recovery turbine (Fig. 8). The decolorization of the effluent was an incidental benefit. Units commonly were operated with NO_x emissions at 1000–2000 ppm (volumetric).

After 1971, industry focused on equations 6–7 which are known as the abatement reactions.

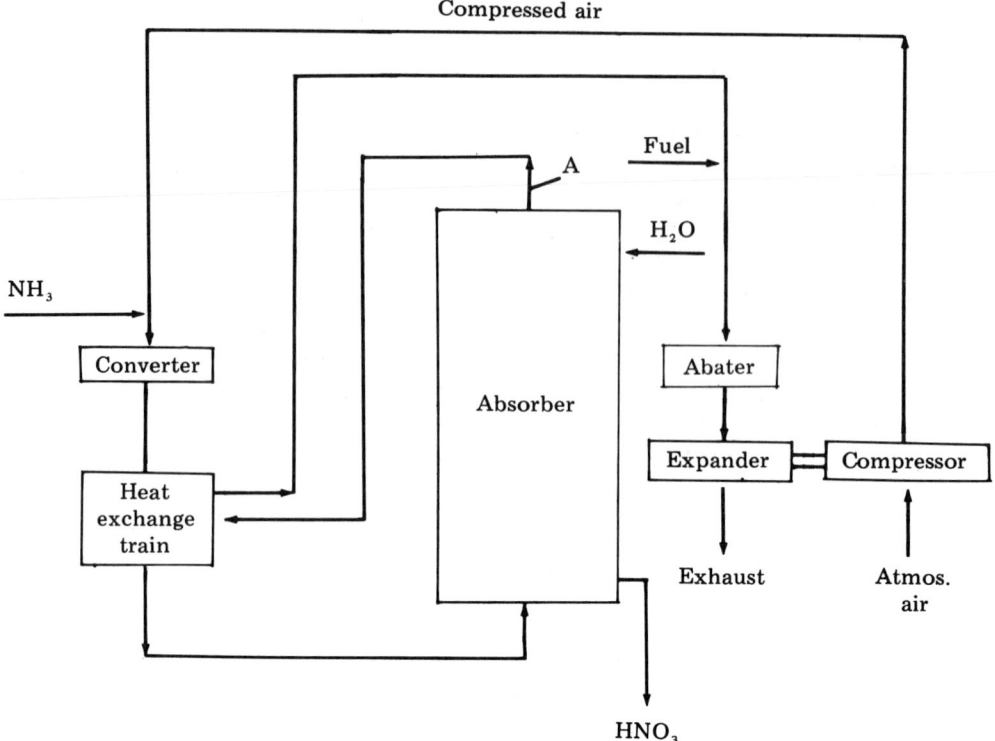

Figure 8. Schematic nitric acid plant. Composition at A, mol %: NO_x, 0.1–0.5; O_2, 1.5–3.5; N_2, balance.

By careful design using the criteria discussed earlier, catalytic abaters have been built and operated to reduce the NO_x levels to less than 200 ppm at fuel efficiencies of 110–125% of stoichiometry with a 2–3 yr catalyst lifetime.

Currently, either platinum or alloys or admixtures of platinum and rhodium are the dominant catalytic metals in use. Palladium exhibits a higher activity and is cheaper than platinum; however, the propensity of palladium to crack hydrocarbon fuels to elemental carbon under upset conditions that produce excessively fuel-rich mixtures (>140% of stoichiometry) have caused a marked decline in its use. Catastrophic melt-downs of the catalyst support have occurred during upset situations in which palladium beds have been operated at overly gas-rich and then gas-lean conditions where the excess oxygen reacts with the deposited carbon and produces a surface temperature sufficiently high to melt the ceramic support. Platinum catalyst beds have been known to operate for extensive periods of time at 150–200% of stoichiometry on natural gas without exhibiting coking.

The preferred catalyst support is the unitary ceramic honeycomb described above. Although beds of pellets and ribbons can be designed to give acceptable abatement levels, it is the relatively low pressure drop and high space velocities of the honeycomb that make it the preferred choice. Expander blade erosion from catalyst fines, which has occurred in several cases with pellet beds, is not experienced with the unitary honeycomb beds.

Because of the low pressure drop that is characteristic of a unitary honeycomb bed, attention must be given to good fuel/tail-gas mixing as well as to adequate flow distribution into the bed. Neglect of mixing and flow parameters results in an apparently under-designed bed that needs to be operated at higher-than-design temperatures and with excessive fuel consumption.

The key to successful abatement (assuming an adequate bed design and active catalyst) is close attention to the operating parameters of the upstream portion of the nitric acid plant: the oxygen content of the tail gas entering the abater must be known and controlled. Thus bleach air, secondary air, and the NH_3/air mixture must be controlled. Over-correction of oxygen or fuel flows in response to abater outlet temperatures is a common error. Even a well-operated nitric acid unit exhibits periodic tail-gas composition changes that are immediately reflected as abater exit temperature fluctuations. Unless the NO_x monitor indicates that the allowable level of emission has been exceeded, no fuel-gas adjustment needs to be made, except perhaps once per shift. These fuel adjustments 2–3 times a day are required in locations at which the diurnal changes in air density result in changes in the mass capacity of the compression train.

The presence of refractory dust, such as is produced by an in-line gas-fired preheater, or from the refractory shield on a heat exchanger/waste-heat boiler tube sheet, are frequent causes of shortened catalyst life by the masking process discussed above.

Upsets in the absorption column that result in 9000–10,000 ppm NO_x have produced an inhibition of catalytic activity by chemisorption as noted earlier. The effects of chemisorption of NO_2 are not permanent and the bed recovers immediately after the upstream abnormality is corrected.

Because of the specific temperature rise associated with the oxygen and NO_x removal, dual-stage abaters with an internal quench section are employed at oxygen contents over 3% (71). The oxygen level limit depends upon the fuel, preheating conditions, bed design, and whether or not some heat-removal device, such as a waste heat boiler, is interposed between the abater exit and the expander inlet. High temperature catalysts have been developed that operate in the 1100°C range. Hence the abater design is limited by the metallurgy of the equipment.

The most commonly used fuel is methane. Because of the possible scarcity of natural gas, other fuels such as ammonia synthesis purge gas, propane, butane, propylene, kerosene, and naphtha are being employed with increasing frequency (see Feedstocks).

Selective Reduction. Spurred by possible natural gas shortages, the selective reduction of NO_x in the presence of oxygen with NH_3 is being used for NO_x abatement. The reactions are as follows:

$$8 \, NH_3 + 6 \, NO_2 \rightarrow 7 \, N_2 + 12 \, H_2O \tag{8}$$

$$4 \, NH_3 + 6 \, NO \rightarrow 5 \, N_2 + 6 \, H_2O \tag{9}$$

The above reactions require close temperature control inasmuch as low temperatures consume unacceptably large quantities of excess ammonia, and high temperatures promote the oxidation of ammonia via the following reaction:

$$2 \, NH_3 + 5/2 \, O_2 \rightarrow 2 \, NO + 3 \, H_2O \tag{10}$$

The optimum operating temperature range varies with the type of catalyst used. In general, precious metal catalysts yield higher conversions of NO_x to N_2 with low excess ammonia usage at lower temperatures than the base metal oxides or zeolites (see Molecular sieves).

Acrylonitrile Manufacture. In the manufacture of acrylonitrile (qv), off-gases containing from 1–3% of CO plus various hydrocarbons are emitted to the atmosphere. Catalytic beds of platinum group metals are used to reduce the regulated compounds to acceptable levels. Close attention to bed design is required to prevent the formation of appreciable quantities of NO_x caused by the fixation of combustion-air nitrogen. Some NO_x also is produced from fuel nitrogen by oxidation. Because of the high thermal energy content of the off-gases, considerable heat recovery is possible in abating acrylonitrile plant emissions.

Phthalic/Maleic Anhydride Manufacture. Partial oxidation of a suitable hydrocarbon feedstock over a fixed-bed catalyst (usually V_2O_5) produces phthalic anhydride or maleic anhydride, or both, which are collected as condensed particulates on oil-filled switch-condenser surfaces (see Maleic anhydride; Phthalic acids). The gaseous effluent of the phthalic anhydride process contains up to 0.5% carbon monoxide and smaller quantities of organic acids (phthalic acid, maleic acid, benzoic acid, etc) derived from the feedstock and its partial oxidation products. Catalytic oxidation of the carbon monoxide and the organic compounds in this exhaust stream has been successfully demonstrated in a year-long large scale (ca 0.94 m^3/s or 2000 SCFM) pilot-plant experiment which utilized the exotherm across the catalyst bed in a high efficiency heat exchanger to preheat the incoming exhaust stream to the temperature required for the catalytic reaction (72). Fuel for preheating the stream is not required after start-up. The first commercial catalytic fume control unit with a recuperative heat exchanger started up in August, 1978.

Potential problems for catalytic oxidation equipment are related to upsets at the front end of the plant that cause condensible particulates to mask the heat exchanger or catalyst surface. Temperature and pressure-drop monitoring controls warn if these conditions develop and actuate safety bypass equipment if necessary.

Kraft Paper Manufacturing. The application of catalytic oxidation to abate air pollution problems in paper making has been limited to pilot-size trial projects. Very encouraging results in field pilot studies with continued laboratory pilot plant work indicate an excess of 95% reduction in concentration of odorous organic sulfides, normally referred to as total reduced sulfur (TRS), and of the accompanying wood-source hydrocarbons. This is accomplished at economically attractive operating temperatures and space velocities (catalyst-bed residence time). Of perhaps greater economic importance is the apparent complete oxidation of the sulfur to SO_3, which when absorbed in a scrubber, becomes a source of weak sulfuric acid for possible return to the process. The present high temperature thermal incineration of this effluent produces SO_2 and requires alkaline scrubbing and subsequent disposal of calcium sulfate (see Sulfur recovery).

The source on which the field pilot catalytic unit experience was gained is typical of the numerous paper mills using the sulfate pulping process. The sparging of black liquor goes to the recovery boilers (where the organic sulfides end up as sulfur dioxide, another air pollution problem) and produces a vent gas containing odorous TRS and aerosol-forming hydrocarbons. These air spargers are known as BLO (black liquor oxidation) units. A BLO unit can be considered an air pollution control device to assist in bringing black liquor recovery boilers into regulatory compliance (see Pulp).

A result of the experience described above has been the application of a small catalytic fume abater to a water-saturated waste-air stream resulting from processing tall oil (qv). It is designed to reduce the small amounts of odorous mercaptans to less than 5 ppm and the other fatty acids, resin, and unsaponifiable organic compounds by 95 to 96% (73).

Vinyl Monomer Manufacturing. Process vent gases containing small quantities of halogenated hydrocarbons and substantial quantities of nonhalogenated hydrocarbons have been successfully reduced to comply with air pollution agency objectives in large-scale laboratory-pilot catalytic fume abaters, with satisfactory long term catalyst performance. The design freedoms offered by precious metals on ceramic honeycomb support catalyst have been demonstrated in equipment that utilizes the heat energy resulting from the substantial exotherm of the nonhalogenated hydrocarbon oxidation to preheat the exhaust gases. Fuel consumption is thereby minimized.

It is not known at this time what effect substantial quantities of halogenated organics in the absence of nonhalogenated organics has on catalyst performance.

Heat Processing and Miscellaneous Operations

Of the many methods used to eliminate or modify air pollution emissions, the usual approach is to control the emission at its source. The sources of emissions requiring control are numerous and of wide variety. The EPA states (4) that 44% of the 1975 estimated 19,000,000 metric tons of emissions from stationary sources were contributed by evaporation of organic solvents. The major sources of these emissions were auto and light truck manufacturing, can manufacturing, coil coating, fabric coating, paper coating, appliance manufacturing, tire manufacturing, printing, pressure-sensitive and magnetic tape manufacturing, dry cleaning, wire coating and textile manufacturing. This list does not include other combustible gaseous emissions such as carbon monoxide, organic sulfides, some organic halides, organic resins, and organic acids that remain in the off-gases from many chemical, paper-making and plywood manufacturing processes.

In the application of catalysis to the low temperature oxidation of these combustible emissions, the technical design parameters can be quite specific if a full description of the reactants is available. The wide variety of solvent vehicles used to apply coating materials to metal, fabrics, wood, paper, etc, have many hydrocarbon solvents in common, but more common is the lack of details available regarding solvent names, types, quantities and/or properties contained in the effluent stream that is to be processed. Consideration of the physical state of the fumes (gaseous or aerosol), as well as the criteria for meeting regulatory objectives (odor or opacity reduction or elimination; a fixed minimum mass flow rate, or a percentage reduction of the pollutants), must be undertaken in the design of any air pollution control device. Because of the often limited information about the potential reactants and noncombustible particulates that can mask the catalyst, successful commercial catalytic air pollution control devices are usually over-designed to compensate for the unexpected. Included in these are the typical process variations in temperature and flow rate.

Surface Coating. Enforcement of air pollution regulations concerning hydrocarbon emissions has been in effect for various surface coating processes since the 1950s.

Surface coating processes produce similar air pollution problems in a number of different industries. A detailed description of each of the numerous types of surface coating processes of each industry discussed below will not be given; however, details have been published by the EPA (4) (see Coating processes; Coatings).

Can Manufacturing. Approximately half of all cans produced are used in the packaging of beer (qv) and other beverages (see Packaging materials). An internal coating is necessary to protect the purity and flavor of the can contents for beverages or any edible product that might react with the container metal. Both the exterior decorative and interior sanitary coatings are applied to the metal surface via rolls or spray guns using a solvent vehicle. In the last three to four years, there has been an effort to utilize water-borne solvent systems which contain up to 20% organics. These coatings require no air pollution control devices unless a visible or odorous effluent results.

Both retrofitted and new catalytic oxidation systems have been used by the principal can manufacturers at an increasing rate over the last six years. It is believed the can manufacturers industry utilizes more catalyst than any of the other surface coating industries. The incentive to convert to catalytic oxidation for control of all solvent hydrocarbons is the increasing cost and scarcity of natural gas and an awareness of the reliability of properly designed and applied catalyst systems based on the recent technology and understanding described earlier.

A large number of diverse solvents are used in exterior and interior coatings in plants for manufacturing both three-piece and two-piece cans.

Most organic vehicle systems are usually found in the cure-oven exhausts at concentrations 2–16% of the lower explosive limit (lel). The oven exhaust volumes are usually 1–35 m^3/s (35–180 ft^3/s). When burned, these concentrations of combustibles provide an exotherm of 30 to 220°C. The heat that is released is used for preheating the incoming effluent and/or heating the cure oven by recycling the hot, cleaned gases to the supply blowers or by heating make-up air via heat exchange. A few plants use the heat of the cleaned exhaust to produce hot water for the two-piece can line washers, hot air for dry-off ovens, or building space heating. For example, one large can company utilizes the heat energy contained in the stream leaving some of their catalytic fume abaters to supply all the heat energy required by the oven's heating zones, which have no burners (Figs. 9–10). The fuel energy supplied to the catalytic fume abater is less than would be needed to heat the oven if the solvent fumes were exhausted directly to the atmosphere without use of the fume abater. The exhaust rate of the oven is adjusted to maintain a solvent concentration of at least 8% of the lel, equivalent to a 110°C temperature differential.

The various reaction rate properties of the different solvents influence the design of a catalytic reactor. For example, for a specific catalyst bed design, an effluent stream containing a preponderance of monohydric alcohols, aromatic hydrocarbons or propylene requires a lower catalyst operating temperature versus that required for solvents such as isophorone and short-chain acetates.

Design considerations and costs. Design considerations and costs of the catalyst, the hardware and the controls of a fume control system are directly proportional to the oven exhaust volume. In terms of volumes to be processed, the size of the catalyst bed often ranges from 1.0 m^3 (35 ft^3 at 0°C and 101 kPa or 1 atm) per 1000 m^3/min (35,000 ft^3/min) of exhaust, to 2 m^3 per 1000 m^3/min of exhaust. The selection of the size is dependent upon the concentration and type solvent, the cell size of the ceramic honeycomb-supported catalyst (surface area, see earlier), and the allowable pressure drop in the system.

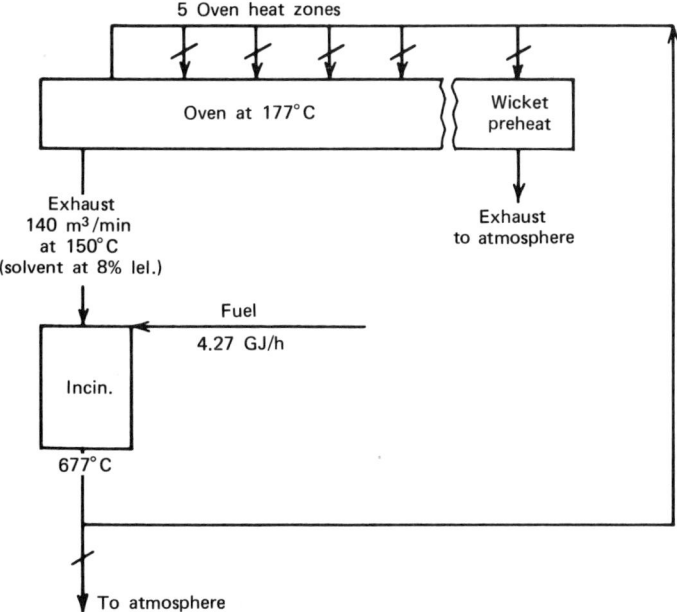

Figure 9. Practice before catalyst retrofit. To convert GJ to Btu, divide by 1.054×10^{-6}.

Catalyst performance at a number of can plant installations has been enhanced by proper maintenance. Annual analytical measurements show reduction of solvent hydrocarbons to be in excess of 90% for 3–6 yr (74), the equivalent of 12,000 to 30,000 operating hours. When propane was the only available fuel, the catalyst cost was recovered by fuel savings (vs thermal incineration prior to the catalyst retrofit) in two to three months. In numerous cases the fuel savings paid for the catalyst in 6 to 12 mo (Figures 9 and 10).

Can manufacturers who maintain air pollution control agency objectives practice an annual or biennial cleaning procedure during a weekend downtime. Both air lancing and an aqueous bath are utilized to remove noncombustible particulates that mask the active sites. Frequently, condensed organic material on the catalyst is removed via short-term (4–6 h) heating excursions to 370°C or 430°C; the organic matter is removed much like a self-cleaning oven. The gaseous and organic smoke, which is usually evolved from the first few cm of catalyst bed depth, is oxidized in the latter part of the bed. If allowed to operate too long at temperatures that promote condensation high boiling organic compounds, the subsequent carbon char that is formed may require temperatures of 480 to 540°C to convert the carbon to carbon monoxide for subsequent oxidation. The higher temperatures required for burn-offs should be approached in small (0–30°C) increments to bring about slow evolution and partial oxidation; this prevents autogenous combustion of local high concentrations of combustible material.

Day-to-day operating techniques that are employed by one large can manufacturer and are intended to prevent organic condensation are dictated by the use of a low-cost, well-established, sanitary coating for beer and beverage three-piece cans (75). Polybutadiene and oleoresinous sanitary coatings may have volatile resin monomers en-

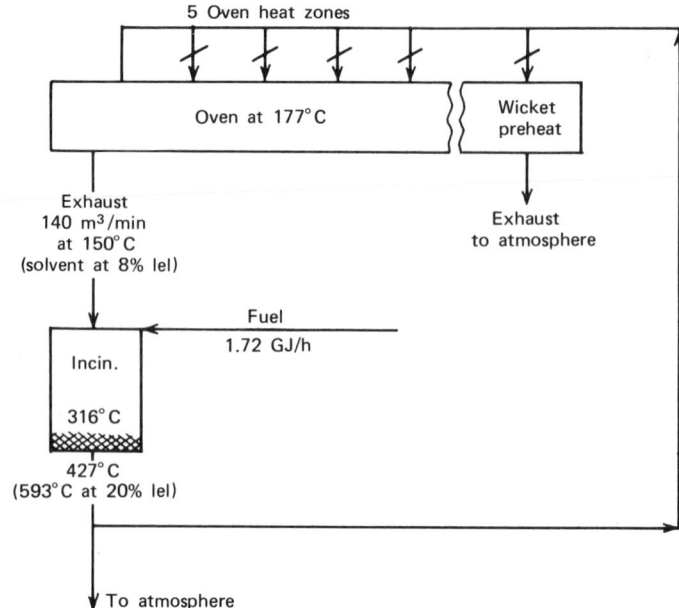

Figure 10. Practice after catalyst retrofit. To convert GJ to Btu, divide by 1.054×10^{-6}. Lel = lower explosive limit.

trained in the oven atmosphere as a result of rapid evaporation of solvent before polymerization takes place. A short (4 to 6 h) heating excursion up to a catalyst inlet temperature of 400°C after use of the coating usually burns off any condensed organic materials. It has become standard practice in some plants to turn the catalytic afterburner up to 370°C for these coatings versus the normal 315°C operating temperature for vinyls, acrylics, etc.

Coil Coating. Coil coating is the prefinishing of many sheet metal items with protective and decorative coatings that are applied by roll coating on one or both sides of a fast-moving metal strip (4). The metal strip (from 13 mm to 1.7 m in width) unwinding from a coil travels at rates of 30–150 m/min through the coating applicator rolls and bake ovens. It is rewound into a coil for transport to a forming operation for products that are to be used in cans, appliances, industrial and residential siding, shelving, cars, gutters, downspouts, etc.

The source of hydrocarbon emissions in coil coating is the coating application area and the cure-oven exhaust. The coatings include primers, finishes, and metal protective (5 μm) films or backers (4).

Catalytic oxidation of these oven exhaust fumes was readily done by industry 20 years ago, but the increasing use of siliconized coatings for weather durability caused severe masking problems for the all-metal, filter-mesh-like catalyst elements then available. Interest in catalytic afterburners waned until 1975 when increasing natural gas costs and new catalyst performance experience (dispersed-phase precious metal/alumina-on-ceramic honeycomb) offered economically attractive results. The newest catalytic afterburners, including thermal afterburner retrofits and new energy-efficient catalytic fume-control systems, have been accepted by aluminum siding manufacturers.

The hot oven exhaust (260–370°C) can be oxidized catalytically without preheat, but when the coater-area exhaust (at room temperature) is combined with the oven exhaust, a preheat burner becomes necessary. The greatest energy savings potential with the least capital investment is obtained by recycling of a portion of the hot, cleaned exhaust to the oven. This principle has been demonstrated at a number of can manufacturing plants and at least four coil coating facilities. One operator preheats oven make-up air, which has been taken from the oven cooler section by means of a heat exchanger; whereas others recycle directly to the oven (76). In one case, the heat energy for the dry-off oven is supplied by the catalytic incinerator exhaust (482°C) remaining after supplying most of the heat energy to operate the four zones in the oven (Fig. 11). The concentration of solvents in the exhaust is ca 12% lel (167°C temperature differential). The net fuel energy consumption is ca 20% of that required to fire the paint-bake and dry-off ovens without fume control.

Coil coaters operate their equipment continuously and, in most cases, operate their catalytic fume abaters 6000–7000 h/yr. Under these conditions the anticipated catalyst life is 3 years, with an annual aqueous solution cleaning. However, the catalyst may last no more than 2 years if frequent maintenance is needed, such as in-place air lancing every 30 to 90 d to remove noncombustible particulates. Frequent maintenance may be needed if coatings such as siliconized polyester (15 to 40% silicones) comprise 30% of the coatings put through the system.

Vehicle Assembly. Auto and light-duty truck assembly plants and their related subassembly parts plants apply large quantities of paint in spray booths. From 70–90% of the solvent vehicle in the paint is dissipated by the spray-booth vents; thus the bake oven exhausts emit relatively low concentrations of solvent hydrocarbons. Also, the concentration of organic substances (ppm) in the spray-booth exhaust is low because of the large air volume used, but the mass flow rate of solvent emissions is usually in excess of pollution regulation standards. The very large flow rates preclude any economical way of controlling the spray-booth exhausts with an add-on device. To date the use of Rule 66-exempt solvent systems (see below) with less restrictive emission criteria has allowed spray-booth operation with some minor production rate limits. The EPA has most recently proposed the use of water-borne paint and/or very high

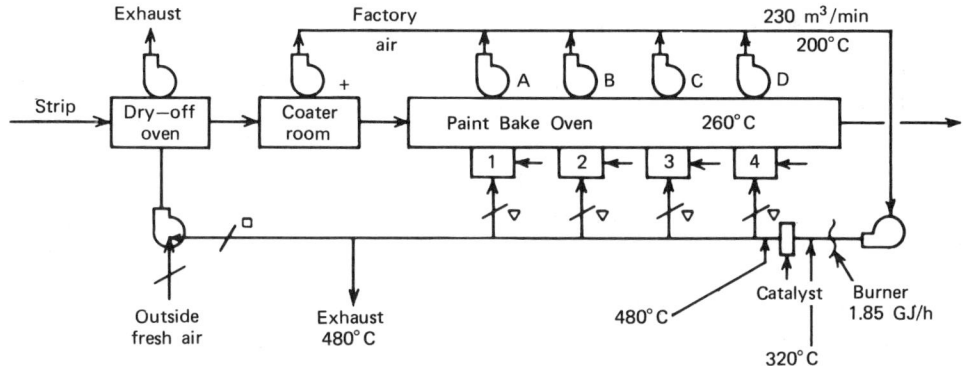

Figure 11. Cleaned exhaust recycle to heat paint bake oven. To convert GJ to Btu divide by 1.054×10^{-6}. ▽, dampers controlled by temperature demand in each of four zones; □, dampers controlled by temperature demand in dry-off oven; 1, 2, 3, 4, zone supply blowers and burners at minimum fire or off; A, B, C, D, zone exhaust blowers with manual dampers; +, 95 L/h solvents applied to strip. To convert m^3 to ft^3, multiply by 35.3.

solids-content solvent-borne paint for reducing the hydrocarbon emissions from spray booths (4).

Solvent hydrocarbons (exempt or not), when exposed to the temperatures required for curing in the bake oven, necessitate the use of a control device if the hydrocarbon emissions exceed 0.7 kg/h in those air quality control regions that enforce regulations similar to California's South Coast Air Quality Management District Rule 442, formerly Rule 66 (see Air pollution). Catalytic oxidation has been successfully applied in meeting these air quality regulations in California, Ohio, New Jersey, and New York.

If an exhaust stream is dilute, more catalyst per unit flow rate is needed to meet the usual requirment of 90% or more reduction of hydrocarbons. Higher mass transfer characteristics (ΔP) and higher operating temperatures (370–400°C), or both, may be needed to meet regulations. One reason is that a sizable (60–220°C) exotherm is not available to accelerate the rate of the oxidation reaction. Furthermore, a low oven exhaust concentration means the cleaned exhaust fume concentration must be very low to effect 90% or greater reduction.

A number of years ago catalyst beds were used widely by the auto industry in dip prime solvent-borne-paint bake ovens to reduce the solvent content in the oven atmosphere and limit condensation and resultant occasional fires in the oven insulation. This need has almost disappeared with the advent of water-borne dip prime electrophoretic deposition of paint. Generally, this paint method is superior and more costly, but it also easily meets typical hydrocarbon regulations. The small amount of organic material required in the water vehicle in the form of plasticizers, surfactants, and biocides can produce visible and odorous emissions that may be perceived as a nuisance, depending upon neighboring residents and the local air quality control agency. The higher cost of this paint method is partly owing to the need for more fuel energy to evaporate water and cure the coatings at slightly higher temperatures versus solvent-borne dip prime paints. Catalytic oxidation, based on the recently developed technology described earlier, is proving to be of value in control of the odorous fumes, but more importantly, it serves as part of the oven heater house which is designed to reduce fuel consumption. By passing 35–50% of the oven recirculation stream through the burner/catalyst system, a small amount of heat energy can be generated by oxidation of the organic materials present. Recycling of a large portion of the cleaned, hot stream supplies the heat energy demand of the oven zone but a small portion is exhausted through an air-to-air heat exchanger to preheat make-up air. A schematic diagram of a typical heat balance is shown in Figure 12. One operator of a system of this type has reported a 33% reduction in fuel energy compared to the previous system of heat exchangers which became clogged repeatedly with the condensed organic material in the stream (77). The application of catalysis to this use is practiced in two auto assembly plants (4 ovens with 16 heating zones). The same catalyst and type of system is designed into 14 ovens (139 heating zones) for assembly plants scheduled to begin production in early 1979 and 1980 (77).

Plywood Veneer Drying. In the drying of plywood veneers, large steam-heated or gas-fired ovens are employed to drive off the water absorbed during soaking of the logs in preparation for this high speed operation. The air pollution problem created by plywood veneer dryers arises from the small amount of pinenes, terpenes, wood resins, tars and fatty acids that are steam distilled out of the wet veneer at the 177–190°C operation temperature. Thus the dryer exhaust contains materials that condense

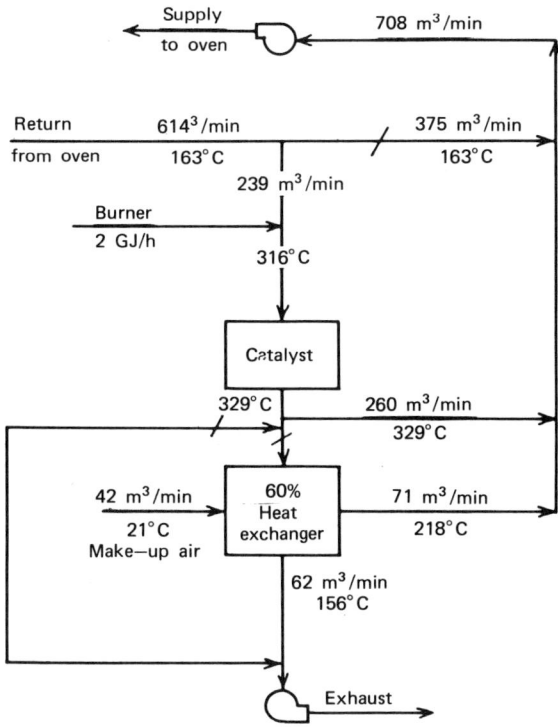

Figure 12. Schematic heat balance, bring-up zone prime bake oven. To convert GJ to Btu, divide by 1.054×10^{-6}. To convert m^3 to ft^3, multiply by 35.3.

at ambient temperatures forming a visible and odorous aerosol (blue haze) (see Laminated wood composites).

A plywood mill in Oregon has carried out catalytic oxidation of such fumes at 204–316°C for 4 yr. Elimination of the dryer exhaust opacity was actually attained at operating temperatures of 190°C, but small amounts of tiny wood splinters and fibers contained in the exhaust did not burn but eventually blocked the catalyst inlet face by forming a char. Raising the temperature to 204–316°C pyrolyzes the splinters without forming carbon char which allows the unit to function relatively maintenance-free. Preheating of the stream is accomplished with a natural gas or propane burner.

Filter Paper Processing. In the fabrication of fuel oil and air filters for vehicles such as motorcycles and diesel locomotives, heat processing of the filter paper is required to cure the resin (usually phenolic) with which the paper was impregnated (see Phenolic resins). The cure-oven exhaust, which contains water vapor, alcohols, and dimers and trimers of phenol, produces a typical blue haze aerosol with a pungent odor. The concentration of organic substances in the exhaust is usually rather low.

The paper-impregnation drying oven exhausts contain high concentrations (10–20% lel) of alcohols and some resin monomer. Vinyl resins and melamine resins, which sometimes also contain organic phosphate fire retardants, may be used for air filters (see Amino resins; Flame retardants). The organic phosphates could shorten catalyst life, depending upon the mechanism of reduction of catalyst activity. Mild

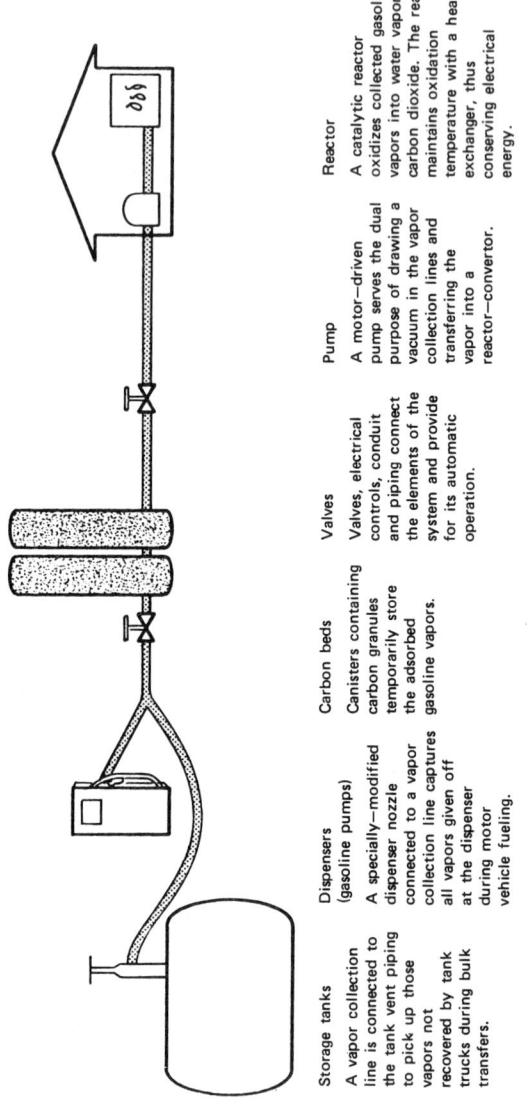

Figure 13. How the Vapox system works. Courtesy of Eneron, Inc.

Storage tanks
A vapor collection line is connected to the tank vent piping to pick up those vapors not recovered by tank trucks during bulk transfers.

Dispensers (gasoline pumps)
A specially-modified dispenser nozzle connected to a vapor collection line captures all vapors given off at the dispenser during motor vehicle fueling.

Carbon beds
Canisters containing carbon granules temporarily store the adsorbed gasoline vapors.

Valves
Valves, electrical controls, conduit and piping connect the elements of the system and provide for its automatic operation.

Pump
A motor-driven pump serves the dual purpose of drawing a vacuum in the vapor collection lines and transferring the vapor into a reactor-convertor.

Reactor
A catalytic reactor oxidizes collected gasoline vapors into water vapor and carbon dioxide. The reactor maintains oxidation temperature with a heat exchanger, thus conserving electrical energy.

acid leaching removes iron and phosphorus from partially deactivated catalyst and has restored activity in at least one known case (78).

Catalysis is utilized in the majority of new paper-filter cure ovens as part of the oven recirculation/burner system which is designed to keep the oven interior free of condensed resins and to provide an exhaust without opacity or odor.

The application of catalytic fume control to the exhaust of paper-impregnation dryers permits a net fuel saving by oxidation of easy-to-burn methyl or isopropyl alcohol, or both, at adequate concentrations to achieve a 110–220°C exotherm.

Gas-Fired Power Equipment. The selective catalytic reduction of NO_x with ammonia has been used on exhaust gases of gas-fired power equipment such as power boilers and peak-shaving turbines (79).

Inert Gas Generation. The inert gas that is generated by the catalytic reduction of the NO_x and O_2 in gas-fired engines is used in oil and gas fields for field pressure maintenance, so-called attic oil recovery, heat control, and H_2S stripping (80) (see Petroleum). This application of catalysis is very similar to that in nitric acid plants with the exception that effluent levels of 1 ppm NO_x are necessary to preclude piping corrosion. The same type of catalysts are used as in the nitric acid plants, and bed designs that achieve the desired NO_x levels are available from U.S. catalyst manufacturers.

Extremely low levels (<0.1 ppm) of NO_x and O_2 in inert gas have been produced by the reaction of the products of combustion from a fuel-rich natural gas burner over a platinum catalyst supported on a unitary ceramic honeycomb.

Fugitive Gasoline Vapors. A unique system for containing fugitive vapors from gasoline filling stations and from bulk loading stations utilizes a platinum catalyst on a honeycomb bed for oxidation of the excess vapors. Such a system is marketed under the trademark Vapox (Eneron, Inc.) (Fig. 13).

Nomenclature

C_p	= specific heat of gases, J/(kg·K)
D	= molecular diffusivity, m²/s
E	= substrate void fraction
f	= apparent friction factor
G	= gas mass flow rate, kg/(m²·s)
g_c	= gravitational constant, m/s²
h	= heat transfer coefficient, J/(m·s·K)
j_D	= mass transfer factor
j_H	= heat transfer factor
k	= thermal conductivity of gas, J/(m²·s·K)
kc	= mass transfer coefficient, m/s
K	= apparent reaction rate, mol/s
k_p	= physical transfer rate, mol/s
k_c	= chemical transformation rate, mol/s
L	= substrate bed depth, m
ΔP	= pressure drop across catalyst, kPa
Re	= Reynolds number
r_h	= hydraulic radius, m
S	= geometric surface area, m^{-1}
U	= linear gas velocity, m/s
μ	= gas viscosity, Pa·s
ρ	= gas density, kg/m³
X	= space velocity, s^{-1}

BIBLIOGRAPHY

1. U.S. Pat. 2,658,742 (Nov. 10, 1963), R. J. Ruff and R. Stuter (to Catalytic Combustion Corp.).
2. R. G. Lunche in B. R. Banerjee, ed., *Proceedings of the First National Symposium on Heterogeneous Catalysis for Control of Air Pollution,* National Air Pollution Control Administration, U.S. Dept. of Health, Education and Welfare, 1968, pp. 297–320.
3. *ICGI Booking Statistics,* Industrial Gas Cleaning Institute, Alexandria, Va., 1978.
4. *Control of Volatile Organic Emissions From Existing Stationary Sources,* EPA, Research Triangle Park, N.C.
5. *Ibid.,* Vol. I, EPA-450/2-76-028, Nov. 1976.
6. *Ibid.,* Vol. II, EPA-450/2-77-008, May 1978.
7. *Ibid.,* Vol. III, EPA-450/2-77-032, Dec. 1977.
8. *Ibid.,* Vol. IV, EPA-450/2-77-033, Dec. 1977.
9. *Ibid.,* Vol. V, EPA-450/2-77-034, Dec. 1977.
10. R. W. Rolke and co-workers, *Afterburners Systems Study,* EPA Contract EHS-D-71-3, Shell Development Company, Emeryville, Calif. 1972.
11. D. A. Dowden, "Applied Catalysis," *Surf. Sci. Lect. Int. Course 1974,* Vol. 2, IAEA, Vienna, pp. 215–271.
12. *Ibid.,* p. 227.
13. *Ibid.,* p. 225.
14. G. J. Barnes in R. F. Gould, ed., *Catalysts for the Control of Automotive Pollutants,* Vol. 143, *Advances in Chemistry Series,* American Chemical Society, Washington, D.C., 1975, pp. 72–84.
15. J. T. Kummer in Ref. 14, pp. 178–192.
16. Ref. 11, p. 245.
17. G. C. Bond, *Catalysis by Metals,* Academic Press, London, 1962, p. 66.
18. *Ibid.,* p. 447.
19. R. B. Anderson and co-workers, *Ind. Eng. Chem* **53,** 809 (1961).
20. G. R. Lester, J. F. Brennan, and J. Hoekstia in Ref. 14, pp. 24–31.
21. R. L. Klimisch, J. C. Summers, and J. C. Schlatter in Ref. 14, pp. 103–115.
22. L. C. Doelp, D. W. Koester, and M. M. Mitchell, Jr. in Ref. 14, pp. 133–146.
23. Ref. 10, p. 65.
24. J. C. Schlatter and K. C. Taylor, *J. Catal.* **49,** 42 (1977).
25. H. C. Yao, S. Japar, and M. Shelef, *J. Catal.* **50,** 407 (1977).
26. J. R. Anderson, *Structure of Metallic Catalysts,* Academic Press, New York, 1975.
27. K. Aika and co-workers, *J. Res. Inst. Catalysis Hokkaido Univ.* **24,** 54 (1976).
28. J. Wei and E. R. Becker in Ref. 9, pp. 116–132.
29. L. L. Hegedus and J. C. Summers, *J. Catal.* **48,** 345 (1977).
30. E. R. Becker and J. Wei, *J. Catal.* **46,** 365, 372 (1977).
31. T. Paryjczak and S. Karski, *Roczniki Chemii Ann. Soc. Chim. Polonorum.* **51,** 2215 (1977).
32. U.S. Pat. 3,470,019 (Sept. 30, 1969), R. Steele (to Matthey Bishop, Inc.).
33. W. H. Manogue and J. R. Katzer, *J. Catal.* **32,** 166 (1974).
34. J. J. Carberry and A. A. Kulkarni, *J. Catal.* **31,** 41 (1973).
35. R. Gauguin, M. Graulier, and D. Papee in Ref. 14, pp. 147–160.
36. U.S. Pat. 2,580,806 (Jan. 1, 1952), L. E. Maline (to E. J. Houdry).
37. U.S. Pat. 2,921,035 (Jan. 12, 1960), E. J. Houdry (to Oxycatalyst, Inc.).
38. U.S. Pat. 3,338,995 (Aug. 29, 1967), D. M. Sowards (to E. I. du Pont de Nemours & Co., Inc.).
39. U.S. Pat. 3,255,027 (June 7, 1966), H. Talsma (to E. I. du Pont de Nemours & Co., Inc.).
40. U.S. Pat. 3,977,090 (Aug. 31, 1976), D. M. Sowards (to E. I. du Pont de Nemours & Co., Inc.).
41. Private communication, G. J. K. Acres, Johnson Matthey & Company, Ltd., Wembley, England.
42. U.S. Pat. 3,231,520 (Jan. 25, 1966), R. J. Leak and H. J. LeBleu (to Texaco, Inc.).
43. U.S. Pat. 3,949,109 (April 6, 1976), J. J. McBride (to E. I. du Pont de Nemours & Co., Inc.).
44. U.S. Pat. 3,554,929 (Jan. 12, 1971), R. Aarons (to E. I. du Pont de Nemours & Co., Inc.).
45. L. C. Hardison and E. J. Doud, *Chem. Eng. Prog.* **73,** 31 (Aug. 1977).
46. T. K. Sherwood, *Ind. Eng. Chem.* **40,** 2077 (1950).
47. S. Ergun, *Chem. Eng. Prog.* **48,** 89 (1952).
48. W. M. Kays and A. L. London, *Compact Heat Exchangers,* 2nd ed., McGraw-Hill Book Company, New York, 1964.

49. Ref. 10, Chapt. 16.
50. M. R. Miller and H. J. Wilhoyte, *J. APCA* **17,** 791 (1967).
51. E. R. G. Eckert and R. M. Drake, Jr., *Analysis of Heat and Mass Transfer,* McGraw-Hill Book Company, New York, 1972.
52. J. R. Welty, C. E. Hicks, and R. E. Wilson, *Fundamentals of Momentum, Heat, and Mass Transfer,* 2nd ed., John Wiley & Sons, Inc., New York, 1976.
53. J. J. Carberry, *Chemical and Catalytic Reactor Engineering,* McGraw-Hill Book Company, New York, 1976.
54. J. M. Thomas and W. J. Thomas, *Introduction to the Principles of Heterogeneous Catalysis,* Academic Press, Inc., London, 1967.
55. C. N. Satterfield, *Mass Transfer in Heterogeneous Catalysis,* M.I.T. Press, Cambridge, 1970.
56. L. L. Hegedus, *AIChE J.* **21,** 849 (1975).
57. R. H. Heck, J. Wei, and J. R. Katzer, *AIChE J.* **22,** 477 (1976).
58. P. A. Nelson and T. R. Galloway, *Chem. Eng. Sci.* **30,** 1 (1975).
59. D. E. Mears, *Ind. Eng. Chem. Process Des. Dev.* **10,** 541 (1971).
60. J. Polidor and co-workers, *Int. Chem. Eng.* **10,** 124 (1970).
61. J. C. W. Kuo, C. R. Morgan, and H. G. Lassen, *Mathematical Modeling of CO and HC Catalytic Converter Systems,* SAE publ. SP-361, pub. 710289, 1971.
62. D. Lieberman, S. E. Voltz, and P. W. Snyder, *Ind. Eng. Chem. Prod. Res. Dev.* **13,** 166 (1974).
63. Ref. 10, p. 69.
64. R. M. J. Fierdorow, B. S. Chahar, and S. E. Wanke, *J. Catal.* **51,** 193 (1978).
65. F. M. Dautzenberg and H. B. M. Walters, *J. Catal.* **51,** 26 (1978).
66. Y. M. Lam and M. Boudart, *J. Catal.* **47,** 393 (1977).
67. S. E. Voltz and co-workers, *Ind. Eng. Chem. Prod. Res. Dev.* **12,** 294 (1973).
68. H. D. Cochran and co-workers, *Colloid and Interface Science,* Vol. III, Academic Press, Inc., New York, 1976, pp. 131–142.
69. M. B. Cutly and C. N. Kenney, *ACS Symp. Ser. 1978,* (Chem. React. Eng.—Houston), pp. 475–486.
70. L. Gonzalez-Tejuca and co-workers, *J. Phys. Chem.* **81,** 1399 (1977).
71. G. R. Gillepsie, A. A. Boyum, and M. F. Collins, *Chem. Eng. Prog.* **12,** 72 (1972).
72. Unpublished data, E. I. du Pont de Nemours & Co., Inc.
73. D. C. Tate, "Tall Oil," in A. Standen, ed., *ECT* 2nd ed., Vol. 19, Interscience Publishers, a division of John Wiley & Sons, Inc., New York, 1969, pp. 615–618.
74. Private communications, Continental Can Company and American Can Company.
75. Private communication, Crown Cork and Seal Co.
76. Private communications, Aluminum Co. of Canada and Clark Brothers Aluminum Siding Co.
77. Private communication, a major automobile manufacturer.
78. Unpublished data, E. I. du Pont de Nemours & Co., Inc.
79. J. M. Kline and Y. C. Lee, *Catalytic Reduction of Nitrogen Oxides with Ammonia: Utility Pilot Plant Operation,* EPA Contract No. 68-02-0292, Oct. 18, 1962.
80. W. F. Barstow and G. W. Watt, *Fifteen Years of Progress in Catalytic Treating of Exhaust Gas,* presented at Rocky Mountain Regional Meeting of the Society of Petroleum Engineers of AIME in Denver, Colorado, April 7–9, 1975.

JAMES H. CONKLIN
DONALD M. SOWARDS
JOHN H. KROEHLING
E. I. du Pont de Nemours & Co., Inc.

EXPECTORANTS, ANTITUSSIVES, AND RELATED AGENTS

Expectorants

Expectorants enhance the production of respiratory tract fluid and thus facilitate the mobilization and discharge of bronchial secretions. Historically, they have been divided into two classes based on their specific mechanisms of action. Expectorants that increase respiratory tract secretion by a direct effect on the bronchial secretory cells are called stimulant expectorants, whereas those that act by gastric reflex stimulation are called sedative expectorants. Unfortunately, many compounds classed as expectorants have been inadequately studied and their mechanisms of action are not known with certainty.

Expectorant preparations over the last 50 years have changed dramatically with respect to composition and characterization. For example, in 1941 numerous expectorant formulas were described that contained 20 or more ingredients (1). One such formula, which was recommended for the relief of cough due to the common cold, contained the following ingredients: thyme herb, horehound herb, grindellia, yerba santa, wild cherry bark, bloodroot, lobelia, squill, pleurisy root, life-everlasting, pipsisewa, mullein, comfrey, elecampane, ammonium chloride, menthol, eucalyptol, gaduol, cascarine, oil of white thyme, glycerol, honey, chloroform, alcohol, and sugar. Today only a few of the early natural expectorants are in widespread use, and even some of these have been chemically modified to improve their efficacy or physical characteristics.

An alphabetical list of all numbered compounds with their CAS Registry Numbers is given on pp. 557–558.

Guaiacols. Creosote, which is obtained from the pyrolysis of beechwood, and its active principles guaiacol (1) and creosol (2) have long been used in expectorant mixtures. The compounds are usually classed as direct-acting or stimulant expectorants, but their mechanisms of action have not been well studied. Creosol is obtained by the Clemmensen reduction of vanillin (qv) (2), whereas guaiacol can be prepared by a number of methods including the mercuric oxide oxidation of lignin (qv) (3), the zinc chloride reduction of acetovanillone (4), and the diazotization and hydrolysis of o-anisidine (5).

(1) guaiacol   (2) creosol

Because of its bitter taste and water insolubility, guaiacol has been chemically modified to improve its properties. Sulfonation provides a mixture of guaiacol-4- and 5-sulfonic acids which, as the potassium salt (3), is water soluble, comparatively tasteless, but less active than guaiacol. Treatment of the sodium salt of guaiacol with phosgene provides guaiacol carbonate (4) which has also lost the bitter taste of guaiacol, but is less water soluble.

The synthetic guaiacol derivative that has received by far the greatest accep-

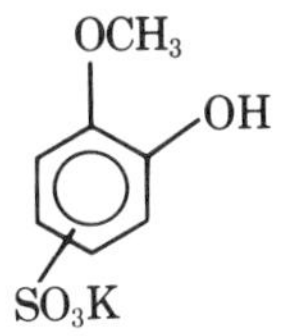

(3) potassium guaiacolsulfonate

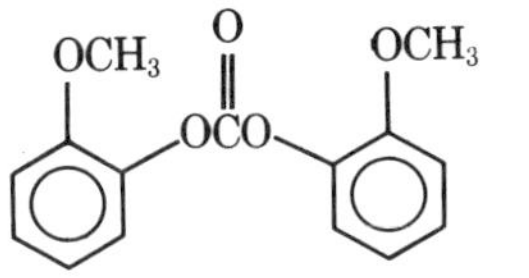

(4) guaiacol carbonate

tance as an expectorant is guaifenesin (5), formerly known as glyceryl guaiacolate. This compound is widely used today, both in single-entity cough preparations, and in combination with other active ingredients. Clinical studies carried out in the early 1940s indicated the usefulness of guaifenesin as an expectorant (6–7). More recently, guaifenesin has been shown to significantly increase the rate of clearance of inhaled radioactive particles in patients with chronic bronchitis (8). Guaifenesin may be prepared by coupling sodium guaiacolate with glyceryl monochlorohydrin.

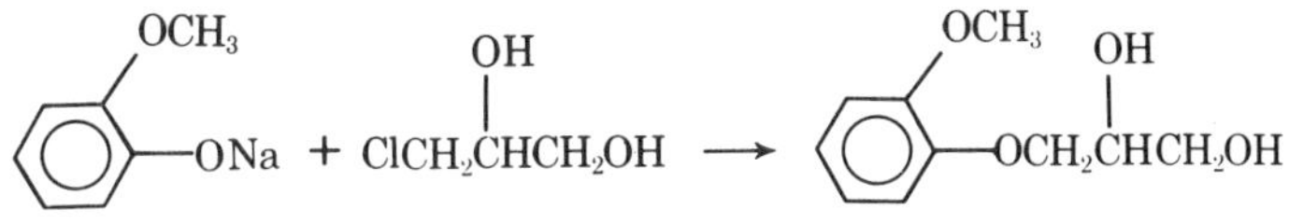

(5) guaifenesin

Volatile Oils. The use of volatile oils, sometimes called essential oils, as expectorants has been traced back 2000 yr to Pliny who used turpentine internally to relieve coughing (9). Yet in spite of their long use, very few objective studies have been carried out to determine their real effectiveness. Limited evidence suggests that the compounds act by direct stimulation of the bronchial secretory cells. Compounds in this category are administered by a number of different routes including oral; by aerosol; topical; as a cream; and sometimes in lozenges (see Oils, essential).

In general, the volatile oils are isolated from plant sources and are terpenoid (qv) in structure. They are purified by a combination of physical and chemical processes. Individual components of the oils are often isolated by crystallization or, in some cases, prepared synthetically.

Turpentine is one of the most familiar oils used in expectorant formulations. It is prepared by the process of rectification which consists of steam distillation of crude turpentine from sodium hydroxide. This removes the acidic and resinous components. The rectified oil contains primarily α-pinene (6) along with small amounts of β-pinene. Two common household oils used as expectorants are oil of lemon and oil of anise. Oil of lemon contains ca 90% limonene (7) and oil of anise, obtained by steam distilling the dried ripe fruit of *Pimpinella anisum* L., contains primarily anethole (8). Oil of

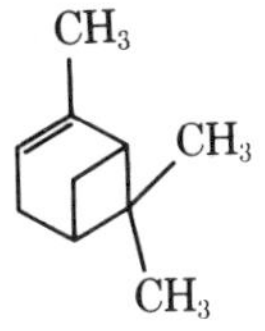

(6) α-pinene

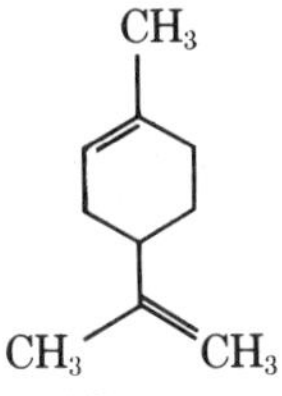

(7) limonene

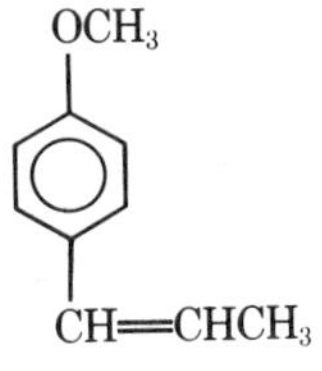

(8) anethole

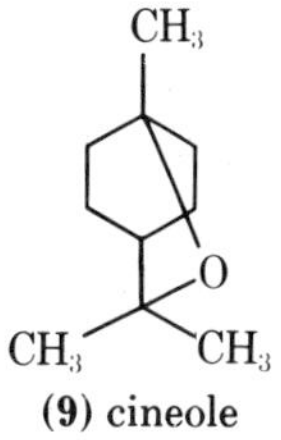

(9) cineole

eucalyptus was first introduced into European medicine during the 18th century, but received the greatest attention during the mid-19th century when it was used as a substitute for cinchona alkaloid antimalarials (see Alkaloids; Chemotherapeutics). During this period the expectorant properties were recognized. Oil of eucalyptus is prepared by steam distilling the leaves of *Eucalyptus globus* Labill and it generally contains not less than 70% cineole (eucalyptol) (9).

Terpin hydrate (10), one of the most well known expectorants, is isolated from crude pine rosin left after the distillation of volatile terpene hydrocarbons and alcohols. It is also manufactured from turpentine (α-pinene) by acid catalyzed hydration. Terpin may exist as cis and trans isomers, but only the cis isomer forms a stable, crystalline monohydrate.

(10) terpin hydrate

The three well-known oils camphor (11), menthol (12), and thymol (13) are often used in over-the-counter cough and cold preparations. Camphor is isolated from the camphor tree, *Cinnomomum camphora* T. Nees & Eberneier, or prepared synthetically from α-pinene or isoborneol. About 75% of the camphor sold in the United States is synthetic. Menthol, commercially the most important terpene alcohol, is obtained by crystallization from peppermint oil, or prepared synthetically in racemic form by the hydrogenation of thymol. Menthol's local anesthetic activity may contribute to its antitussive properties. Thymol, unlike the other terpenoids described, contains a fully aromatic ring. It is obtained from the essential oils of *Thymus vulgaris* L. or *Monarda punctata* L., or it can be prepared synthetically from p-cymene or m-cresol.

(11) camphor   (12) menthol   (13) thymol

Iodides and Other Inorganic Compounds. Inorganic compounds such as potassium iodide, hydriodic acid, antimony potassium tartrate, and ammonium chloride are thought to act by gastric reflex stimulation. Of these, only the iodides have been studied to any appreciable extent (10). A number of toxic reactions have been associated with both antimony potassium tartrate (11) and inorganic iodides (12–13). Reaction of iodine with glycerol (qv) produces a stable organic iodide mixture, iodinated glycerol (14), which has expectorant properties. Formulations of (14) contain no free or ionic iodine, but iodine is released metabolically.

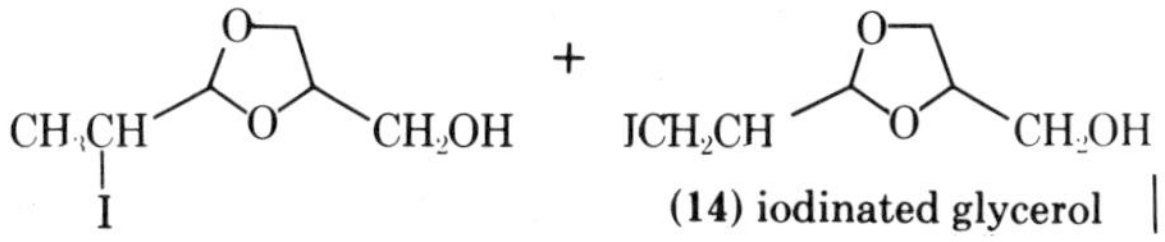

(14) iodinated glycerol

Miscellaneous Natural Products. The long list of natural products used in early cough preparations has diminished to the point where only a few are still mentioned. Four such natural products still occasionally used are squill, horehound, cocillana, and ipecac.

Squill is the dried, sliced bulb of *Urginea maritima* (L.) Baker and contains the glycosides scillaren A (**15**) and scillaren B [*1393-22-2*] (the mixture of glycosides remaining after separation of scillaren A). It has long been used in cough preparations and, oddly enough, red squill is also used as a rat poison (see Poisons, economic). As an expectorant, squill is usually administered in combination with other ingredients. It is an emetic when given in large doses and this property suggests that the expectorant effect is likely to be caused by reflex stimulation. Toxic effects include nausea, vomiting, and a digitalis-like action on the heart (14).

Horehound is the dried leaves and flowering tops of *Marrubium vulgare* L. Bickerman (15) reports that Serritus, in an early 16th century article on cough remedies, recommended the use of syrup of horehound to "purge through the sputa." Although it has frequently been used in cough lozenges, little has been reported concerning its efficacy. The active principle in horehound is the diterpene lactone, marrubiin (**16**).

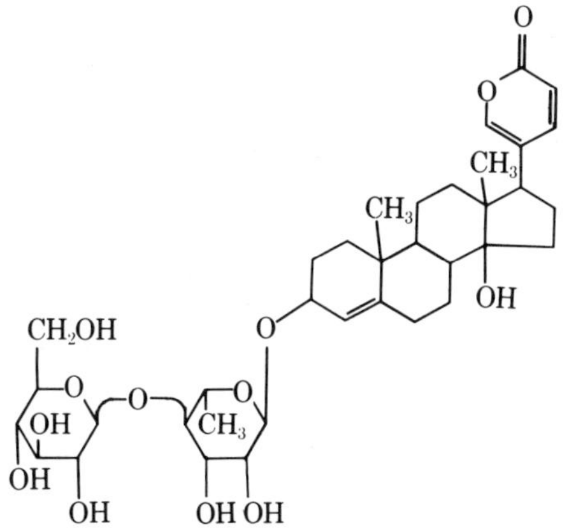

(15) scillaren A

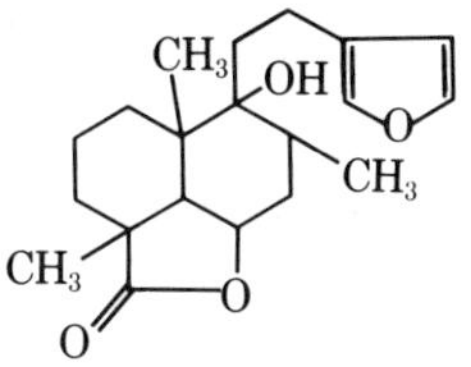

(16) marrubiin

546 EXPECTORANTS, ANTITUSSIVES, RELATED AGENTS

Ipecac is prepared from the dried roots and rhizomes of *Cephaelis ipecacuanha* (Brot:) A. Rich. and contains the alkaloids emetine (**17**) and cephaeline (**18**) in a ratio between 2:1 and 4:1. It has been used extensively in cough preparations and is believed to act by gastric reflex stimulation. Toxic effects include vomiting, irritation of the gastrointestinal tract, and cardiac arrhythmias (16).

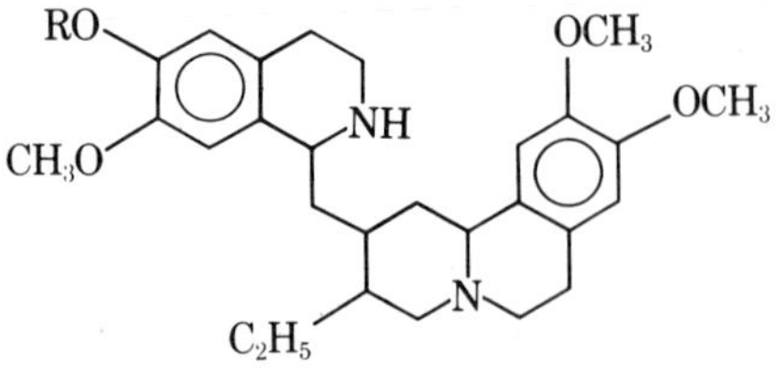

(**17**) emetine R = CH$_3$
(**18**) cephaeline R = H

Cocillana, the dried bark of *Guarea rusbyi* (Britt.) Rusby, was probably first used by the natives of the Bolivian Andes as an emetic-cathartic. It is often prescribed as an alternative to ipecac in the treatment of cough, and the emetic side effects at high doses suggest a mechanism of action similar to that of ipecac.

Mucolytics

Mucolytics are agents that reduce the viscosity of tenacious and purulent mucus, thus facilitating removal. Steam, sometimes in conjunction with surfactants or volatile oils, has long been used to decrease viscosity by physical hydration. Recently, however, agents that chemically depolymerize certain components of mucus have become available. Trypsin and other proteolytic enzymes have shown good clinical activity by virtue of their ability to cleave glycoproteins. Pancreatic dornase, which depolymerizes DNA found in purulent mucus, has also shown clinical utility (see Enzymes).

Several mucolytics reduce the viscosity of mucus by cleaving the disulfide bonds that maintain the gel structure. *N*-Acetyl-L-cysteine (**19**), introduced in 1963, and mesna (**20**), recently developed in Europe (17), are effective compounds in this class. Whereas most mucolytics must be administered by aerosol, carbocysteine (**21**), which contains a derivatized sulfhydryl group, has shown activity by the oral route (18–19).

Bromhexine (**22**) is a highly substituted aniline derivative which has been shown to decrease the viscosity and increase the volume of mucus. At least part of its activity has been attributed to its ability to fragment mucopolysaccharide fibers. Bromhexine is structurally related to the alkaloid vasicine (**23**), the active principle from the plant *Adhatoda vasica* Nees which has been used by the East Indians as an herbal medicine for the relief of cough. The pharmacological and clinical properties of bromhexine have

HSCH$_2$CHCOOH
|
NHCOCH$_3$

(**19**) *N*-acetyl-L-cysteine

HSCH$_2$CH$_2$SO$_3$Na

(**20**) mesna

HOOCCH$_2$SCH$_2$CHCOOH
|
NH$_2$

(**21**) carbocysteine

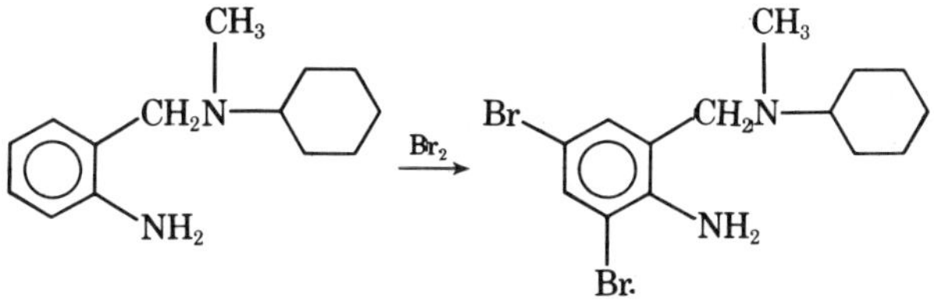

(22) bromhexine

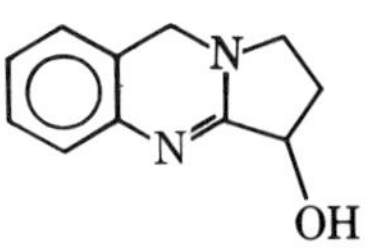

(23) vasicine

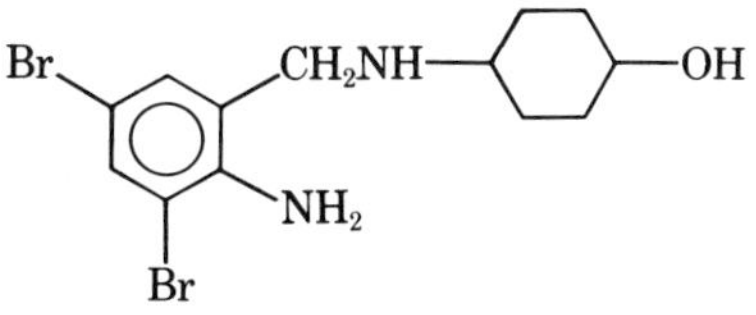

(24) ambroxol

been reviewed (20–21). Recently, a metabolite of bromhexine, ambroxol **(24)**, has also shown potent clinical activity (22). The preparation of bromhexine follows standard procedures to N-(2-aminobenzyl)-N-cyclohexyl-N-methylamine. This intermediate is then brominated to give **(22)** (23).

Antitussives

Records from antiquity describe the use of soothing syrups and herbal extracts to control excessive coughing. Gordon (24), in a review of ancient Hebrew medicine, reported the use of goat's milk fresh from the udder for acute and chronic cough. Galen, in the 2nd century AD, may have been the first to report a truly effective antitussive preparation, camphorated opium tincture (15). Through the centuries, cough remedies have remained a popular item and are found today in most medicine cabinets. Over 300 prescription and over-the-counter preparations are available (25), and it has been estimated that retail sales of nonprescription cough syrups, expectorants, and cough drops in 1977 exceeded 400 million dollars (26).

The Cough Reflex. Coughing is a protective reflex and is one of several important mechanisms for clearing the respiratory tract of excessive secretions and foreign debris, thus ensuring normal gas exchange and minimizing infection. It can be described as occurring in three phases: (*1*) a short deep inspiration of air into the lungs; (*2*) compression of the air by closure of the glottis, and contraction of the thoracic, abdominal, and diaphragmatic muscles; and (*3*) rapid expulsion of the air when the glottis opens. The intensity of the cough varies according to the force behind the expiration. In some cases the expired air moving through the trachea may achieve a linear velocity of >800 km/h (27). The cough reflex can be triggered by chemical, mechanical, or other stimuli to the sensory nerve endings of the respiratory tract. The mechanical receptors are generally confined to the mucosa of the large airway of the upper respiratory tract, especially the lower third of the trachea. The chemical receptors are widely distributed and respond to almost any irritant. Stimulation of the sensory receptors causes impulses to pass along afferent nerve pathways to the cough center in the medulla. Here they are coordinated and transmitted via efferent or motor pathways to abdominal and intercostal muscles, and the diaphragm. Recent evidence suggests that irritation of the bronchial mucosa may not lead directly to stimulation of the cough receptors. Instead, bronchoconstriction may result first, which then triggers the cough reflex (10,28). The intensity of the cough is regulated to some extent by the stretch receptors located in the alveolar walls of the lungs.

Coughing can be produced by a number of environmental factors and pathological disorders. The common cold is the most frequent cause of transient cough in children and adults, and cigarette smoking is the most common cause of chronic, persistent cough (29). The changing causative factors during the last 100 years are summarized in ref. 30.

Other factors that have an effect on cough and the expulsion of irritants from the respiratory tract include ciliary activity and the production of respiratory tract fluid and its physical characteristics. Under normal circumstances, the sweeping action of the cilia propels the demulcent fluid secreted by the goblet cells and the bronchial glands toward the glottis where it is swallowed or expectorated. Respiratory tract fluid is polymeric in nature, containing numerous polysaccharide units attached to a protein core. Large amounts of water and salt can be bound within the matrix, providing an ideal medium for transporting bacteria and debris away from the lungs. Pathological conditions or other factors which alter either ciliary activity, or the composition, amount, and physical properties of respiratory tract fluid may affect the frequency and productivity of coughing. Excellent reviews and monographs dealing with mucociliary clearance (31) and the role and physical characteristics of respiratory tract fluid (32–34) are available.

Theoretically, the act of coughing can be affected directly or indirectly by one or more of the following: (*1*) elimination of the ultimate cause of the cough by treating the responsible pathological condition, or by removing the anatomical or environmental irritant; (*2*) raising the threshold for stimulation of cough peripherally by anesthetizing sensory nerve endings in the respiratory tract; (*3*) interruption of the sensory impulses to the medulla; (*4*) specific depression of the cough center in the medulla; (*5*) interruption of conduction along the motor pathways; (*6*) nonspecific depression of the central nervous system; and (*7*) facilitating bronchial drainage and mucociliary clearance.

The first approach can be considered ideal if the necessary relief can be obtained quickly enough. Most therapeutic agents, however, act by one or more of mechanisms (*2*), (*4*), and (*7*).

The first pharmacological technique that permitted an objective evaluation of antitussive activity in animals was described by Ernst (35) in 1938. Since that time, a number of methods have been developed that produce experimental cough by stimulating the respiratory tract, the vagus nerve, or the medulla by chemical, mechanical, or electrical means. Unfortunately, the large number of response criteria, species variations, and other factors reduce the value of these methods for predicting clinical efficacy and potency. Meaningful comparisons of antitussive drugs against pathologic cough in humans are often difficult to obtain owing to varied etiologies and intrinsic variability associated with small patient populations. For this reason, artificially-induced cough is frequently used to measure antitussive activity clinically. A number of diverse chemical agents, such as citric acid, acetylcholine, acetic acid, sulfur dioxide, and ammonia have been used as stimulants. Several excellent reviews describe methods for the evaluation of antitussive drugs in animals and man (36–38).

Centrally Active Antitussives. Centrally active antitussives exert their effect by depressing the medullary cough center, thus raising the threshold for sensory cough impulses. The most well known compounds in this category are the narcotics. Unfortunately, many of them have the obvious disadvantage of addiction. In the last 40 years, molecular modifications of the morphine skeleton and the synthesis of totally new structures have produced more specific drugs without the disadvantage of addiction. Many of the synthetic compounds also possess other useful properties including local anesthetic and antispasmodic activity.

Narcotic Antitussives. Since its isolation in 1832, codeine (**25**) has been one of the most widely used and effective compounds for the treatment of cough. Though less potent than morphine (**26**), it has become the reference against which most antitussives are measured. Codeine, like morphine, is isolated from the opium poppy. However, the low yield of 0.7–2.5% does not provide sufficient material to meet commercial demands. The majority of the codeine sold today is prepared by methylating the phenolic hydroxy group of morphine. When prescribed for cough, the usual oral dose is 10–20 mg, three to four times daily. At these doses, adverse side effects are very few. Although the abuse potential for codeine is relatively low, the compound can substitute for morphine in addicts (39).

Molecular modifications of the morphine skeleton have produced numerous derivatives with antitussive properties, some of which have become commercially significant. Ethylmorphine (**27**), a simple homologue of codeine, is prepared by ethylating morphine. It is pharmacologically similar to codeine but is seldom used clinically. Pholcodine (**28**), the morpholinoethyl derivative of morphine, is used as an antitussive in a number of European countries. It is ca one and a half times as potent as codeine, has little or no analgesic activity, and produces minimal physical dependence. The compound is prepared by the aminoalkylation of morphine (40).

Hydromorphone (**29**) and hydrocodone (**30**) are, empirically, isomers of morphine

(**25**) R = CH_3, codeine
(**26**) R = H, morphine
(**27**) R = C_2H_5, ethylmorphine
(**28**) R = $CH_2CH_2N\diagup\diagdown O$, pholcodine

(**29**) R = H, hydromorphone
(**30**) R = CH_3, hydrocodone

and codeine, respectively. Hydromorphone can be prepared by catalytic rearrangement of morphine (41) or by oxidation of the aliphatic hydroxyl group of dihydromorphine (42). Hydrocodone can be similarly prepared. As antitussives, (29) is several times more active than morphine and (30) is slightly more active than codeine. Hydromorphone (29) has a much higher addiction potential than hydrocodone (30).

Dihydrocodeine (31), introduced in Germany over 50 years ago, and dihydrocodeinone enol acetate (37) both have clinical activity and addiction potential comparable to codeine.

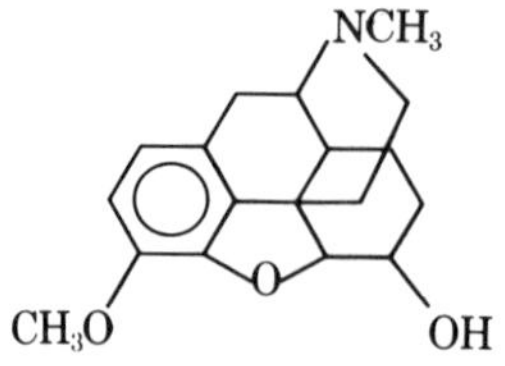

(31) dihydrocodeine

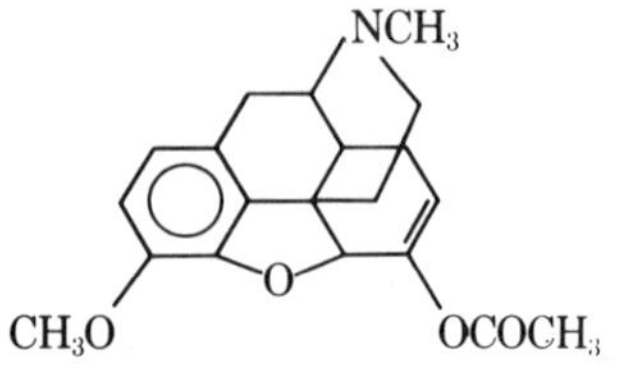

(32) dihydrocodeinone enol acetate

Recent modifications of the morphine skeleton have produced butorphanol (33) and drotebanol (34). In animal models, each compound has demonstrated antitussive activity much greater than that of codeine (43–44). Butorphanol is also a potent analgetic of the narcotic antagonist type (43). Both compounds possess a unique 14-hydroxy group.

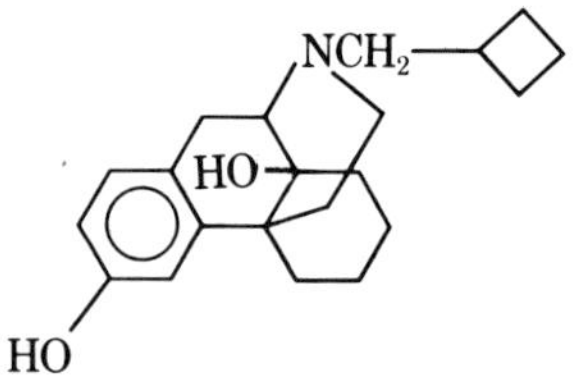

(33) butorphanol

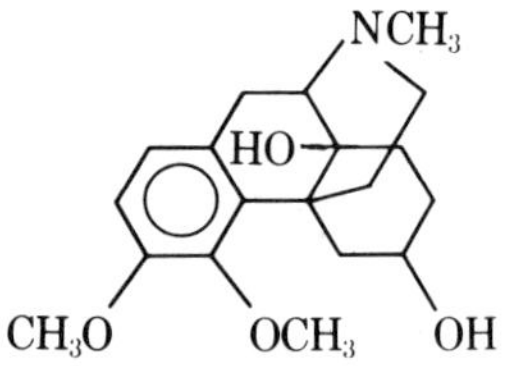

(34) drotebanol

Among the nonopiate narcotics, two compounds, methadone (35) and meperidine (36) have shown antitussive activity. Methadone is qualitatively morphine-like, and has demonstrated clinical activity against coughing induced by ammonia (45) and citric acid aerosol (46). Meperidine (36), although normally not thought of as an antitussive, has been shown to inhibit spasmodic coughing associated with bronchial asthma (47). The addiction potential of both of these agents limits their clinical use for treating cough. A convenient synthesis of methadone which avoids the production of undesirable isomeric intermediates has been described (48). Meperidine can be synthesized by a number of different routes, but the method starting with phenylacetonitrile is one of the most practical (49).

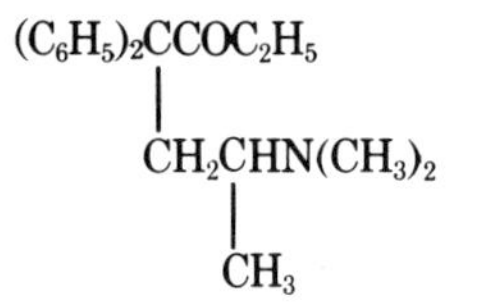

$(C_6H_5)_2CCOC_2H_5$
|
$CH_2CHN(CH_3)_2$
|
CH_3

(35) methadone

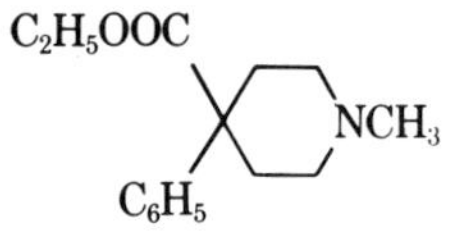

(36) meperidine

In addition to the usual side effects such as nausea, anorexia, and constipation associated with narcotics, most of them also diminish ciliary activity and produce a drying effect on the respiratory tract mucosa.

Nonnarcotic Antitussives. By far the most important centrally active, nonnarcotic antitussive is dextromethorphan (**37**). It is similar to codeine in terms of potency and mechanism of action; ie, it is a direct depressant of the cough center. It is unique in that even though it is structurally related to codeine, it is not addictive.

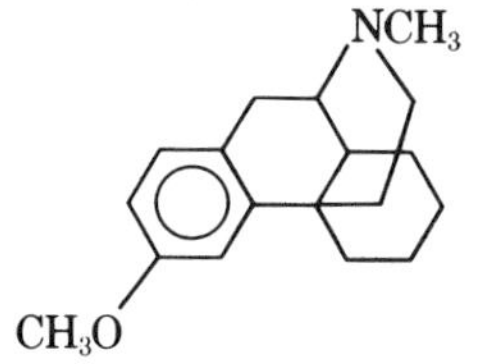

(**37**) dextromethorphan

The synthesis of dextromethorphan is an outgrowth of early efforts to synthesize the morphine skeleton. In 1946 Grewe (50–51) synthesized N-methylmorphinan (**38**), and soon thereafter the 3-hydroxy and the 3-methoxy analogues were prepared by the same method. Whereas the natural alkaloids of opium are optically active, ie, only one optical isomer can be isolated, synthetic routes to the morphine skeleton provide racemic mixtures (both optical isomers) which can be separated, tested, and compared pharmacologically. In the case of 3-methoxy-N-methylmorphinan, the levorotatory isomer was found to possess both analgetic and antitussive activity whereas the dextrorotatory isomer (dextromethorphan (**37**)) possessed only antitussive activity. Dextromethorphan, unlike most narcotics, does not depress ciliary activity, secretion of respiratory tract fluid, or respiration.

The Grewe synthesis of N-methylmorphinan (**38**), which paved the way for the preparation of dextromethorphan and numerous analogues, follows standard reactions to 2-methyl-1-benzyl-1,2,3,4,5,6,7,8-octahydroisoquinoline. Cyclization of this compound with phosphoric acid gave a mixture of isomers from which (**38**) was separated.

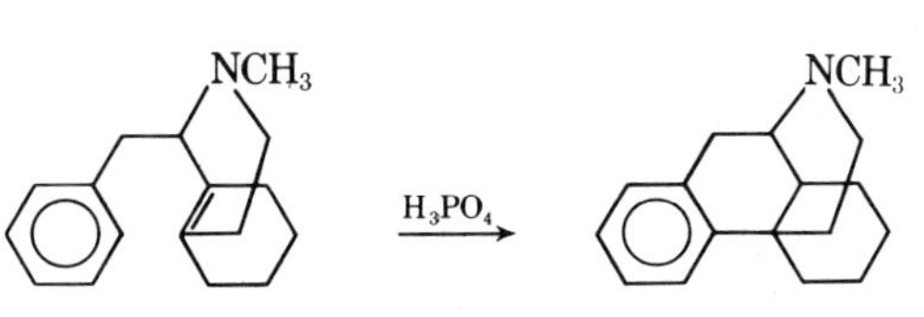

(**38**) N-methylmorphinan

The synthesis (52) and potent antitussive activity (53) of dextrorotatory 3-methyl-N-methylmorphinan, dimemorfan (**39**), have been reported. This compound was prepared by a modification of the Grewe process, and differs from dextromethorphan only by having a methyl group, rather than a methoxy group, in the 3 position.

Included in the nonnarcotic class of antitussives are many compounds that do not possess a morphine skeleton and which vary widely from each other with respect to structural features and pharmacological profiles.

Levopropoxyphene (**40**), the optical antipode of the dextrorotatory analgetic

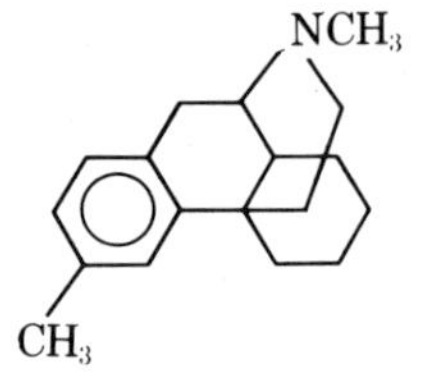

552 EXPECTORANTS, ANTITUSSIVES, RELATED AGENTS

(39) dimemorfan

propoxyphene, is an antitussive without analgetic activity (see Analgesics). It is widely used in the form of the 2-naphthalenesulfonate salt which has a less unpleasant taste than the hydrochloride salt. Clinical effectiveness has been demonstrated against pathological and artificially induced cough but the potency is somewhat less than codeine. The compound is reported not to cause addiction. Levopropoxyphene can be prepared (54) by first resolving β-dimethylamino-α-methylpropiophenone with dibenzoyl-(+)-tartaric acid. The resolved (+)-propiophenone (41) is then treated with benzylmagnesium chloride to give (42) which is converted by acylation to levopropoxyphene.

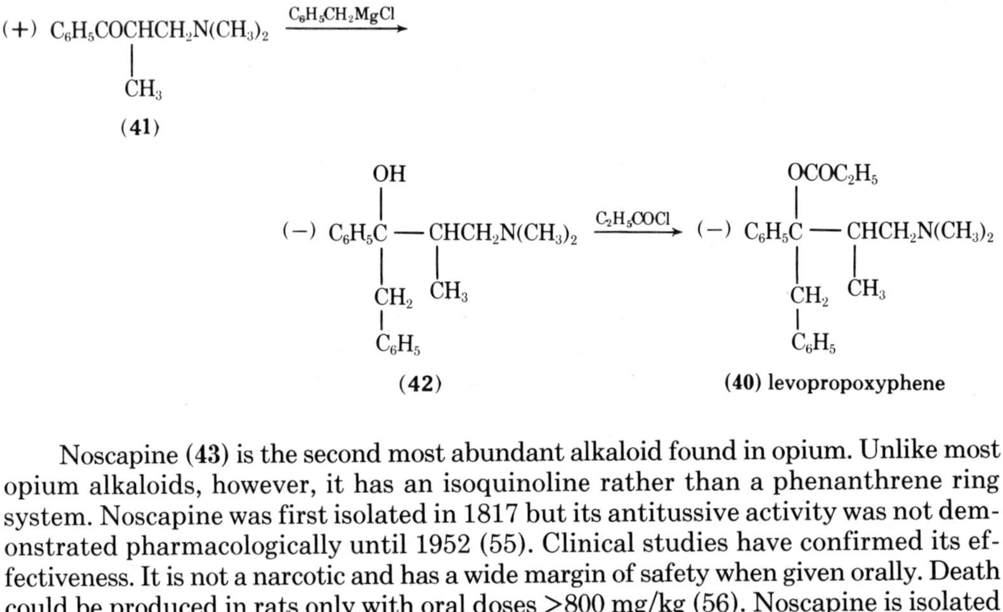

$$(+)\ C_6H_5COCHCH_2N(CH_3)_2 \xrightarrow{C_6H_5CH_2MgCl}$$
$$\quad\quad\quad\quad |$$
$$\quad\quad\quad\quad CH_3$$
$$\quad\quad\quad (41)$$

(42)   (40) levopropoxyphene

Noscapine (43) is the second most abundant alkaloid found in opium. Unlike most opium alkaloids, however, it has an isoquinoline rather than a phenanthrene ring system. Noscapine was first isolated in 1817 but its antitussive activity was not demonstrated pharmacologically until 1952 (55). Clinical studies have confirmed its effectiveness. It is not a narcotic and has a wide margin of safety when given orally. Death could be produced in rats only with oral doses >800 mg/kg (56). Noscapine is isolated from the water-insoluble residue remaining after processing opium for the manufacture of morphine.

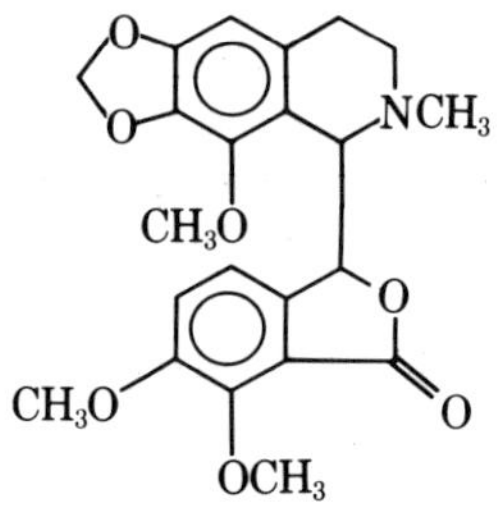

(43) noscapine

A unique compound which appears to have both central and peripheral antitussive effects is benzonatate (**44**). Structurally it is a derivative of *p*-aminobenzoic acid and contains a long poly(ethylene glycol) side chain. The peripheral effects are the result of local anesthetic action on the pulmonary stretch receptors. Clinical activity was first reported by Guiliano and Rossa (57).

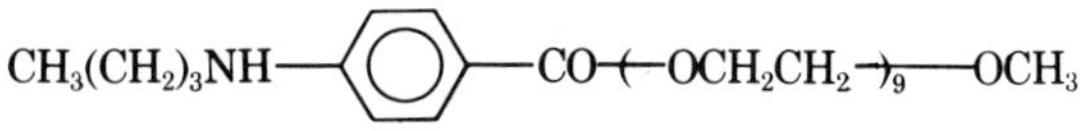

$$CH_3(CH_2)_3NH-\!\!\left\langle\bigcirc\right\rangle\!\!-CO\!\!-\!\!(OCH_2CH_2)_9\!\!-\!\!OCH_3$$

(**44**) benzonatate

Dimethoxanate (**45**) and pipazethate (**46**) are related phenothiazine derivatives that have shown antitussive activity. Unlike many phenothiazines, however, they do not produce central nervous system depression or analgesia at therapeutic doses. They are both somewhat less potent than codeine. It has been suggested that the unique side chain that is similar to, but shorter than, the one on benzonatate may be at least partly responsible for their antitussive effects. Both (**45**) and (**46**) are the result of molecular modifications of classical phenothiazines such as promethazine [60-87-7] which possess antitussive activity in addition to central nervous system depressant activity. Dimethoxanate can be prepared by the reaction of phenothiazine-10-carboxylic acid chloride with β-dimethylaminoethoxyethanol (58).

Chlophedianol (**47**) is the most potent antitussive in a series of compounds originally synthesized as potential antispasmodics. It is ca one third as active as codeine and has weak antispasmodic and local anesthetic activity. Although the onset of antitussive activity is slow, the duration is prolonged. Chlophedianol can be prepared by the method described in ref. 59.

Caramiphen (**48**) and carbetapentane (**49**) are structurally related antitussives which, like chlophedianol, also possess antispasmodic and local anesthetic activity.

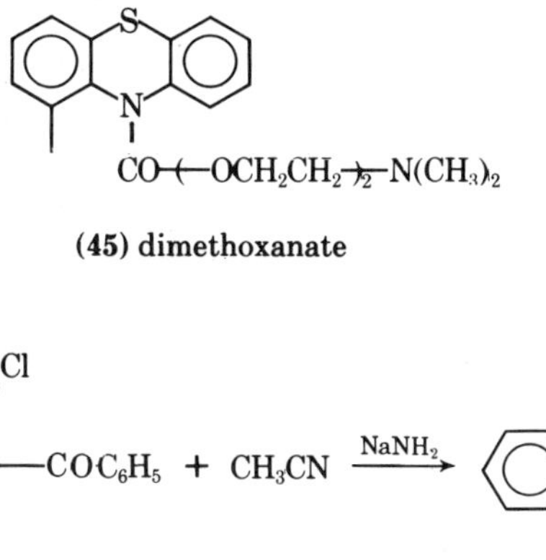

(**45**) dimethoxanate

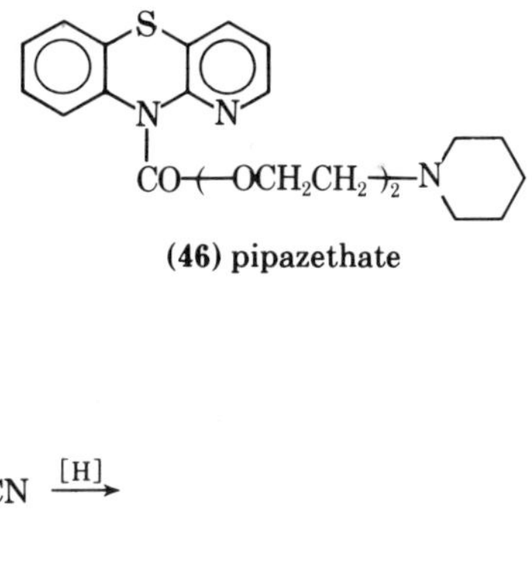

(**46**) pipazethate

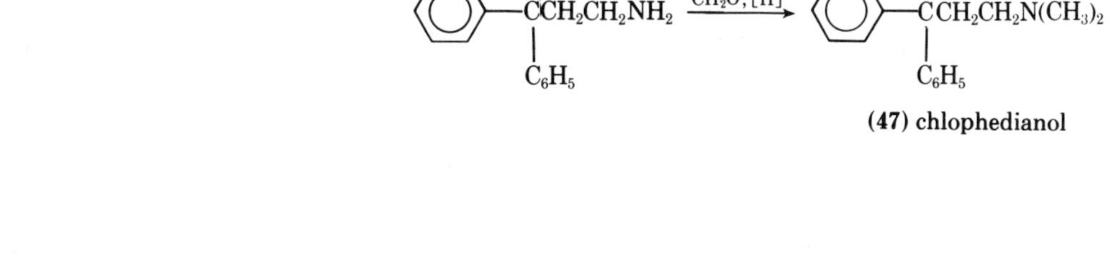

(**47**) chlophedianol

554 EXPECTORANTS, ANTITUSSIVES, RELATED AGENTS

They differ in that the ester side chain of carbetapentane is lengthened by an ethyleneoxy unit. Caramiphen is usually administered for antitussive activity as an ethanedisulfonate salt. It is comparable to codeine in potency but appears to have a longer duration of action. Carbetapentane is slightly less active than codeine. Both (48) and (49) produce typical atropine-like side effects including dryness of the mouth and visual disturbances. Caramiphen and carbetapentane can be synthesized (60–61) from a common intermediate, 1-phenylcyclopentanecarboxylic acid (50).

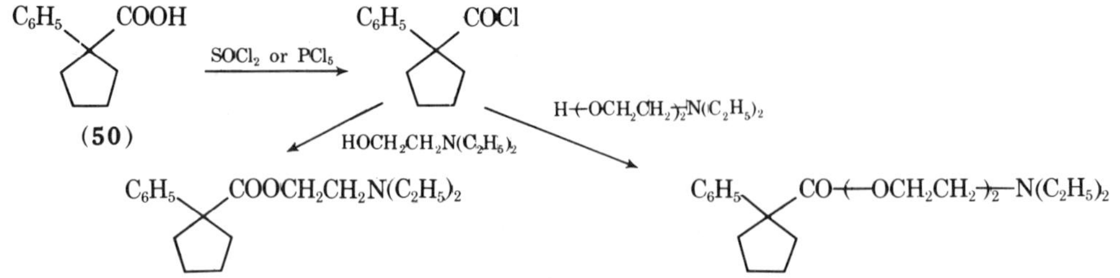

$$\text{(50)} \xrightarrow{SOCl_2 \text{ or } PCl_5} \underset{C_6H_5}{\overset{COCl}{\diagdown\diagup}}\text{cyclopentane}$$

with reagents $HOCH_2CH_2N(C_2H_5)_2$ giving (48) caramiphen: C_6H_5-cyclopentane-$COOCH_2CH_2N(C_2H_5)_2$

and with $H\text{-}(OCH_2CH_2)_2\text{-}N(C_2H_5)_2$ giving (49) carbetapentane: C_6H_5-cyclopentane-$CO\text{-}(OCH_2CH_2)_2\text{-}N(C_2H_5)_2$

Oxeladin (51), an antitussive developed in Great Britain, is structurally related to carbetapentane in that the cyclopentane ring has been broken at the 3,4-bond. It is also similar pharmacologically. Extensive clinical studies using cough drops and syrup containing oxeladin are described in ref. 62. The compound can be synthesized in the following way from phenylacetonitrile (63):

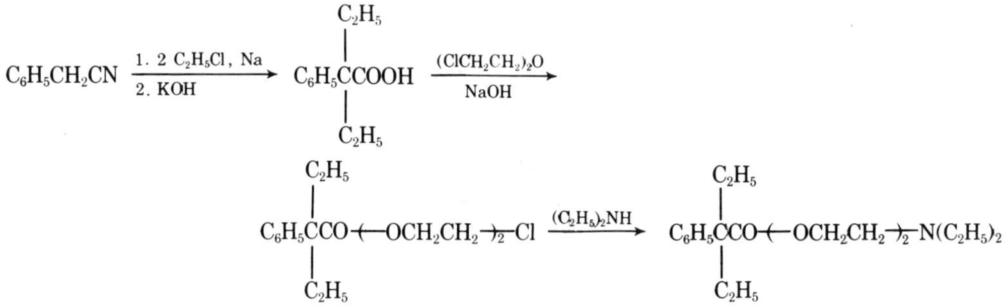

$$C_6H_5CH_2CN \xrightarrow[2.\ KOH]{1.\ 2\ C_2H_5Cl,\ Na} C_6H_5C(C_2H_5)_2COOH \xrightarrow[NaOH]{(ClCH_2CH_2)_2O}$$

$$C_6H_5C(C_2H_5)_2CO\text{-}(OCH_2CH_2)_2\text{-}Cl \xrightarrow{(C_2H_5)_2NH} C_6H_5C(C_2H_5)_2CO\text{-}(OCH_2CH_2)_2\text{-}N(C_2H_5)_2$$

(51) oxeladin

Diphenhydramine (52) was originally developed as an antihistamine and was first used clinically for this purpose in 1946 (see Histamine and antihistamine agents). In addition to this primary effect, however, central antitussive activity has also been demonstrated in animals (64–65) and in humans (66). Its antitussive activity is ca half that of codeine. Drowsiness is the most frequent side effect. Diphenhydramine can be prepared as follows (67):

$$(C_6H_5)_2CHBr + HOCH_2CH_2N(CH_3)_2 \xrightarrow{Na_2CO_3} (C_6H_5)_2CHOCH_2CH_2N(CH_3)_2$$

(52) diphenhydramine

Another antitussive with weak antihistaminic activity is the Japanese compound picoperine (53). This compound is a structural isomer of the well-known antihistamine tripelennamine and is more potent than codeine. The chemistry (68) and pharmacology (69) of picoperine have been reported.

Oxolamine (54) is sold in Europe. It is an oxadiazole, and its general pharmaco-

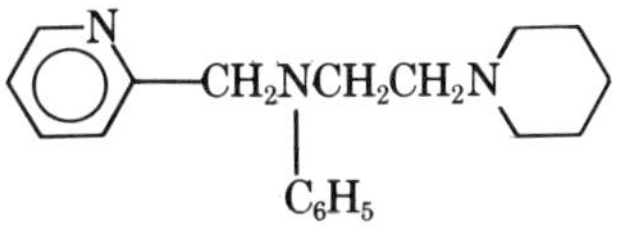

(53) picoperine

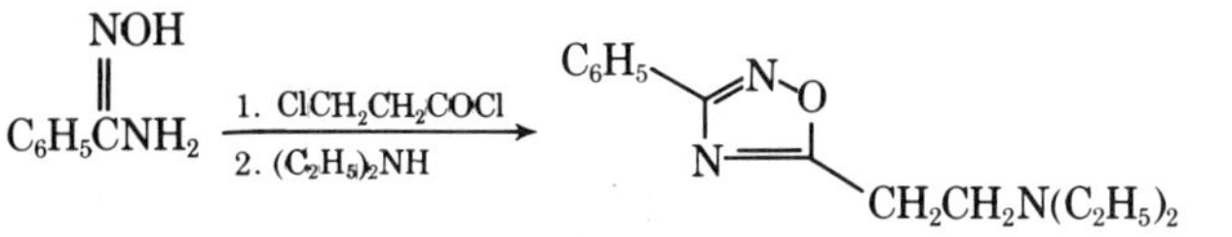

(54) oxolamine

logical profile is described in ref. 70. The compound posesses analgesic, antiinflammatory, local anesthetic, and antispasmodic properties, in addition to its antitussive activity. Although a central mechanism may account for some of the activity, peripheral inhibition of the cough reflex may be the predominant effect. The compound has been shown to be clinically effective, although it is less active than codeine (71–72). The synthesis of (54) is described in ref. 73.

Zipeprol (55) is another European antitussive with a wide range of pharmacological effects, including antispasmodic, antihistaminic, and local anesthetic activities (74–75). The compound can be prepared from 1-(2-methoxy-2-phenylethyl)piperazine and 3-methoxy-3-phenylpropylene oxide (74).

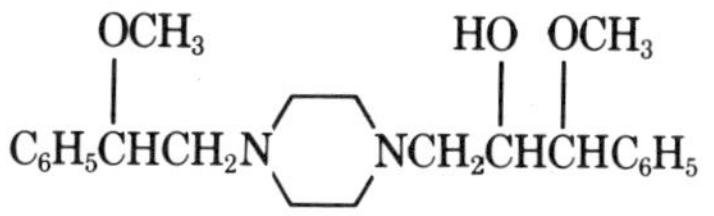

(55) zipeprol

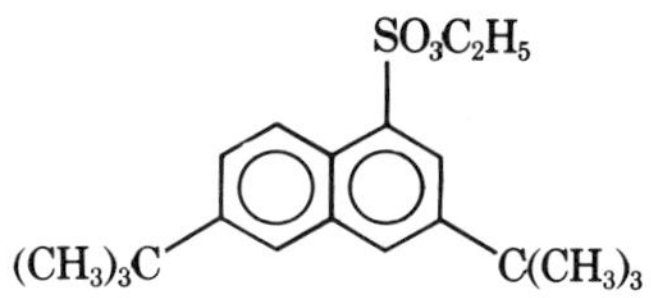

(56) ethyl dibunate

Ethyl dibunate (56), which is sold in Canada, is the ethyl ester of 3,6-(*tert*-butyl)-1-naphthalenesulfonic acid. It is structurally unrelated to most of the classical antitussives and is a selective central inhibitor of the cough reflex. Also significant is its low toxicity. The oral LD_{50} is greater than 5000 mg/kg in the rat. The clinical and pharmacological profile of this compound has been reviewed (77).

Diazepam (57) and clonazepam (58) suppress cough induced by electrical stimulation of the lower brainstem of cats (79). By intravenous administration, clonazepam and diazepam were found to be ca thirty-five times and six times more potent than codeine, respectively. Nevertheless, the compounds have not been widely used as antitussives in humans.

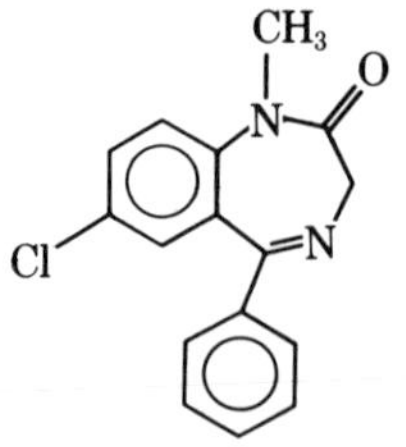

(57) diazepam

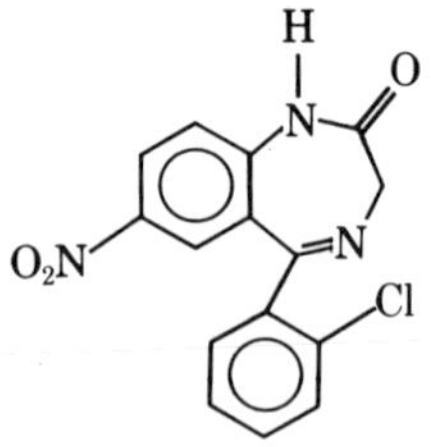

(58) clonazepam

Another recent addition to the long list of diverse structures reported to possess central antitussive activity is Δ^1-tetrahydrocannabinol (THC) (59), the major psychoactive component of marihuana (see Psychopharmacological agents). This compound was found to be comparable to codeine against electrically-induced cough in the anesthetized cat (79). Two other naturally occurring cannabinoids, cannabidiol, and cannabinol are inactive.

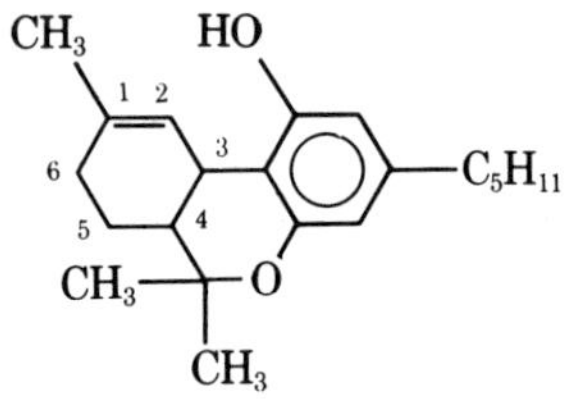

(59) Δ^1-tetrahydrocannabinol

Except for the addiction liability of some of the narcotic antitussives, side effects for most of the centrally acting compounds are relatively few and mild at therapeutic doses. Qualitative comparisons of both side effects and pharmacological profiles have been summarized for many of the compounds described above (80).

Peripherally Active Antitussives. In general, peripherally active antitussives act by raising the threshold for cough at the sensory nerve endings, or by facilitating bronchial drainage, mucociliary clearance, or both. Agents that act by the first

Table 1. Shipment Values for Cough or Cough and Cold Preparations

Product	Shipment values, millions of dollars[a]		
	1967	1972	1977
Cough preparations and expectorants (ethical)			
narcotic	27.4	32.5	56.7
nonnarcotic	15.9	28.8	40.2
Cough and cold combinations (ethical)	7.3	9.1	11.6
Cough and cold preparations (proprietary)			
cough syrups	44.5	65.5	84.0
capsules and tablets	38.3	71.0	98.6
lozenges	14.8	[a]	25.8
topical preparations	0.9	0.3	[a]
cough drops		[a]	[a]
other preparations	23.1	81.1	98.8

[a] Shipment value combined with value for other preparations.

Table 2. Alphabetical List of Compounds Referred to in the Text

Structure	Structure number	CAS Registry Number
N-acetyl-L-cysteine	(19)	[616-91-1]
ambroxol	(24)	[18683-91-5]
anethole	(8)	[104-46-1]
benzonatate	(44)	[104-31-4]
bromohexine	(22)	[611-75-6]
butorphanol	(33)	[42408-82-2]
camphor	(11)	[126-04-5]
caramiphen	(48)	[125-86-0]
carbetapentane	(49)	[77-23-6]
carbocysteine	(21)	[638-23-6]
cephaeline	(18)	[483-18-1]
chlophedianol	(47)	[791-35-5]
cineole	(9)	[470-82-6]
clonazepam	(58)	[1622-61-3]
codeine	(25)	[76-57-3]
creosol	(2)	[93-51-6]
dextromethorphan	(37)	[125-71-3]
diazepam	(57)	[439-14-5]
dihydrocodeine	(31)	[125-28-0]
dihydrocodeine enol acetate	(32)	[466-90-0]
dimemorfan	(39)	[36309-01-0]
dimethoxanate	(45)	[477-93-0]
α-[2-(dimethylamino)-1-methylethyl]-α-phenyl-benzeneethanol	(42)	[126-04-5]
3-(dimethylamino)-2-methyl-1-phenyl-1-propanone	(41)	[91-03-2]
diphenhydramine	(52)	[58-73-1]
drotebanol	(34)	[3176-03-2]
emetine	(17)	[483-18-1]
ethyl dibunate	(56)	[5560-69-0]
ethylmorphine	(27)	[76-58-4]
guaiacol	(1)	[90-05-1]
guaiacol carbonate	(4)	[553-17-1]
guaifenesin	(5)	[93-14-1]
hydrocodone	(30)	[125-29-1]
hydromorphone	(29)	[466-99-9]
iodinated glycerol	(14)	[5634-39-9]
limonene	(7)	[138-86-3]
levopropoxyphene	(40)	[2338-37-6]
marrubiin	(16)	[465-92-9]
menthol	(12)	[89-78-1]
meperidine	(36)	[57-42-2]
mesna	(20)	[19677-45-5]
methadone	(35)	[1095-90-5]
N-methylmorphinan	(38)	[3882-38-0]
morphine	(26)	[57-27-2]
noscapine	(43)	[128-62-1]
oxeladin	(51)	[468-61-1]
oxolamine	(54)	[959-14-8]
1-phenylcyclopentanecarboxylic acid	(50)	[77-55-4]
pholcodine	(28)	[509-67-1]
picoperine	(53)	[21755-66-8]
α-pinene	(6)	[80-56-8]
pipazetate	(46)	[2167-85-3]
potassium guaiacol sulfate	(3)	[1321-14-8]

Table 2 (*continued*)

Structure	Structure number	CAS Registry Number
scillaren A	(15)	[11003-70-6]
terpin hydrate	(10)	[2451-01-6]
Δ^1-tetrahydrocannibinol	(59)	[1972-08-3]
thymol	(13)	[89-83-8]
vasicine	(23)	[6159-55-3]
zipeprol	(55)	[34758-83-3]

mechanism include local anesthetics (qv) which desensitize the mucosa of the respiratory tract, and antispasmodics which relax the smooth muscles of the bronchi (see Neuroregulators). Most of these compounds also have a central antitussive component as a part of their pharmacological profile and have been previously discussed. Expectorants and mucolytics act primarily by the second mechanism.

Economic Aspects

Sales figures for expectorants and antitussives are usually combined under the general headings of cough preparations or cough and cold preparations. This is because antitussives and expectorants are frequently formulated together for the treatment of cough, or formulated with antihistamines and bronchodilators for the treatment of cold symptoms. Department of Commerce figures for the total value of manufacturers' shipments of cough or cough and cold preparations are shown in Table 1 (81–82).

At the retail level, sales of nonprescription cough syrups, elixirs, and expectorants at all outlets were $304.8 million in 1978 compared to $262.5 million in 1977 (83–84). Sales of cough drops, lozenges, troches, and gums were $159.5 million and $139.1 million for 1978 and 1977, respectively (83–84).

Among narcotic antitussives, codeine (**25**) is by far the leading product. However, sales of codeine-containing cough preparations have declined in recent years, probably due to the increased acceptance of nonnarcotic antitussives such as dextromethorphan (**37**). U.S. figures show drugstore and hospital purchases of codeine for antitussive use declining from 7.71 metric tons in 1972 to 6.38 t in 1977, and project a continued decline to 5.77 t in 1981 (85).

Health and Safety

Safety and efficacy data on a number of antitussives and expectorants have recently been reviewed by the FDA's Advisory Review Panel on Over-the-Counter (OTC) Cold, Cough, Allergy, Bronchodilator, and Antiasthmatic Products. Their conclusions and recommendations regarding the effectiveness, safety, labeling, and suitability for marketing of over-the-counter preparations have been reported (86). LD_{50} data for most of the compounds described in this article have been reported (87–88).

Table 2 is an alphabetical listing of all numbered compounds with their CAS Registry numbers.

BIBLIOGRAPHY

1. E. J. Belanger, *Drug and Specialty Formulas,* Chemical Publishing Co., Inc., New York, 1941, p. 166.
2. J. H. Fletcher and D. S. Tarbell, *J. Am. Chem. Soc.* **65,** 1431 (1943).
3. U.S. Pat 2,433,227 (Dec. 23, 1947), H. F. Lewis and I. A. Pearl (to Sulfite Products Corp.).
4. U.S. Pat. 3,057,927 (Oct. 9, 1962), D. W. Read (to Ontario Research Foundation).
5. Ger. Pat. 1,148,236 (May 9, 1963), G. H. Herbst (to Hoechst A.G.).
6. W. J. Connell, G. M. Johnston, and E. M. Boyd, *Can. Med. Assoc. J.* **42,** 220 (1940).
7. W. F. Perry and E. M. Boyd, *J. Pharmacol. Exp. Ther.* **73,** 65 (1941).
8. M. L. Thomson, D. Pavia, and M. W. McNicol, *Thorax* **28,** 742 (1973).
9. E. M. Boyd and G. L. Pearson, *Am. J. Med. Sci.* **211,** 602 (1946).
10. H. Salem and D. M. Aviado, eds., *International Encyclopedia of Pharmacology and Therapeutics,* Vol. I, Sect. 27, Permagon Press, Ltd., Oxford, Eng., 1970, p. 236.
11. S. C. Harvey in L. S. Goodman and A. Gilman, eds., *The Pharmacological Basis of Therapeutics,* MacMillan Publishing Co., Inc., New York, 1975, p. 928.
12. R. M. Harden, *Brit. Med. J.* **1,** 160 (1968).
13. G. E. Steffen, *J. Am. Med. Assoc.* **192,** 571 (1965).
14. A. Wade, ed., *Martindale The Extra Pharmacopoeia,* 27th ed., The Pharmaceutical Press, London, Eng., 1977, p. 604.
15. H. A. Bickerman, *Clin. Pharmacol. Ther.* **3,** 353 (1962).
16. Ref. 14, p. 601.
17. S. N. Steen and co-workers, *Clin. Pharmacol. Ther.* **16,** 58 (1974).
18. M. Aylward, *Current Medical Research and Opinion* **2,** 387 (1974).
19. A. Quevauvillier and co-workers, *Therapie* **22,** 485 (1967).
20. A. P. Launchbury in G. P. Ellis and G. B. West, eds., *Progress in Medicinal Chemistry,* Vol. 7, Appleton-Century-Crofts, New York, 1970, p. 44.
21. Ref. 10, Vol. III, p. 765.
22. *Arzneim. Forsch.* **28,** 887 (1978).
23. J. Keck, *Ann.* **662,** 171 (1963).
24. B. L. Gordon, *Ann. Med. Hist.* **4,** 219 (1942).
25. H. A. Bickerman in W. Modell, ed., *Drugs of Choice 1976–1977,* C. V. Mosby Co., St. Louis, Mo., 1976, p. 446.
26. J. Fishman, ed., *Product Marketing,* Vol. 8, Charleston Publishing Co., New York, 1978, p. F.
27. E. M. Greisheimer, *J. Am. Med. Women's Assoc.* **15,** 1165 (1960).
28. H. Salem and D. M. Aviado, *Am. J. Med. Sci.* **247,** 585 (1964).
29. R. S. Irwin, M. J. Rosen, and S. S. Braman, *Arch. Intern. Med.* **137,** 1186 (1977).
30. Ref. 10, p. 271.
31. A. Wanner, *Am. Rev. Respir. Dis.* **116,** 73 (1977).
32. M. J. Dulfano, ed., *Sputum, Fundamentals and Clinical Pathology,* Charles C Thomas, Springfield, Ill., 1973.
33. E. Puchelle, ed., *Rheology of Bronchial Secretions and Respiratory Functions,* Masson and Co., Paris, Fr., 1973.
34. *Br. Med. J.* **2,** 51 (1975).
35. A. M. Ernst, *Arch. Int. Pharmacodyn. Ther.* **58,** 363 (1938).
36. Ref. 10, p. 205.
37. N. B. Eddy and co-workers, *Codeine and Its Alternatives For Pain and Cough Relief,* World Health Organization, Geneva, Switz., 1970, p. 127.
38. K. Bucher, *Agents Actions* **4,** 377 (1974).
39. C. K. Himmelsbach, *J. Am. Med. Assoc.* **103,** 1420 (1934).
40. U.S. Pat. 2,619,485 (Nov. 25, 1952), P. Chabrier, P. R. L. Giudicelli, and C. H. Genot (to Laboratoires Dausse).
41. Ger. Pat. 623,821 (Jan. 6, 1936), H. Metzger (to Knoll A.G.).
42. U.S. Pat. 2,654,756 (Oct. 6, 1953), A. H. Homeyer and G. B. De la Mater (to Malinckrodt Chemical Works).
43. R. L. Cavanagh, J. A. Gylys, and M. E. Bierwagen, *Arch. Int. Pharmacodyn. Ther.* **220,** 258 (1976).
44. S. Kobayashi and co-workers, *Arzneim. Forsch.* **20,** 43 (1970).

45. U. Trendelenburg, *Acta. Physiol. Scand.* **21,** 174 (1950).
46. H. A. Bickerman and S. E. Itkin, *Clin. Pharmacol. Ther.* **1,** 180 (1960).
47. A. L. Barach and H. A. Bickerman, *Pulmonary Emphysema,* Williams and Wilkins Co., Baltimore, Md., 1956, p. 168.
48. N. R. Easton, J. H. Gardner, and J. R. Stevens, *J. Am. Chem. Soc.* **69,** 2941 (1947).
49. F. F. Blicke and co-workers, *J. Am. Chem. Soc.* **74,** 1844 (1952).
50. R. Grewe, *Naturwissenschaften* **33,** 333 (1946).
51. R. Grewe and A. Mondon, *Chem. Ber.* **81,** 279 (1948).
52. M. Murakamu and co-workers, *Chem. Pharm. Bull.* **20,** 1706 (1972).
53. Y. Kase and co-workers, *Arzneim. Forsch.* **26,** 353 (1976).
54. A. Poland, L. R. Peters, and H. R. Sullivan, *J. Org. Chem.* **28,** 2483 (1963).
55. C. A. Winter and L. Flataker, *Proc. Soc. Exp. Biol. Med.* **81,** 436 (1952).
56. C. A. Winter and L. Flataker, *Toxicol. Appl. Pharmacol.* **3,** 96 (1961).
57. V. Giuliano and G. Rossa, *Minerva Med.* **46,** 1502 (1955).
58. U.S. Pat. 2,778,824 (Jan. 22, 1957), C. Von Seemann (to American Home Products).
59. Brit. Pat. 815,217 (June 17, 1959), R. Lorenz, R. Grosswald, and H. Henecka.
60. Swiss Pat. 234,452 (Jan. 16, 1945), (to J. R. Geigy A.G.).
61. Brit. Pat. 753,779 (Aug. 1, 1956), H. G. Morren.
62. F. Kleibel, E. Steinhoff, and W. Wilde, *Therapiewoche* **15,** 1115 (1965).
63. U.S. Pat. 2,885,404 (May 5, 1959), V. Petrow, O. Stephenson, and A. M. Wild (to British Drug House Ltd.).
64. J. Wax, C. V. Winder, and G. Peters, *Proc. Soc. Exp. Biol. Med.* **110,** 600 (1962).
65. N. B. Eddy and co-workers, *Bull. WHO* **40,** 682 (1969).
66. L. S. Lilienfield, J. C. Rose, and J. V. Princiotto, *Clin. Pharmacol. Ther.* **19,** 421 (1976).
67. U.S. Pat. 2,421,714 (June 3, 1947), G. Rieveschl, Jr. (to Parke, Davis Co.).
68. Y. Kase and co-workers, *J. Med. Chem.* **13,** 704 (1970).
69. Y. Kase and co-workers, *Arzneim. Forsch.* **19,** 1916 (1969).
70. B. Silvesterni and C. Pozzatti, *Br. J. Pharmacol.* **16,** 209 (1961).
71. M. DeGregorio, *Panminerva Med.* **4,** 90 (1962).
72. K. Gary and W. T. Ulmer, *Arzneim. Forsch.* **17,** 240 (1967).
73. Ger. Pat. 1,097,998 (Jan. 26, 1961), (to Firma Angelini Francesco).
74. G. Rispat and co-workers, *Arzneim. Forsch.* **26,** 523 (1976).
75. D. Cosnier and co-workers, *Arzneim. Forsch.* **26,** 848 (1976).
76. Ger. Pat. 2,109,366 (Sept. 30, 1971), Y. Roland and co-workers (to Centre Europeen de Recherches Mauveray).
77. I. Shemano and J. M. Beiler in ref. 10, Vol. II, p. 615.
78. S. C. Wang, D. T. Chou, and M. C. Wallenstein, *Agents Actions* **7,** 337 (1977).
79. R. Gordon, R. J. Gordon, and R. D. Sofia, *Eur. J. Pharmacol.* **35,** 309 (1976).
80. Ref. 37, p. 239.
81. *Pharmaceutical Preparations, Except Biologicals, 1975,* Current Industrial Reports, Series: MA-28G(75)-1, U.S. Bureau of the Census, Department of Commerce, Washington, D.C., 1976.
82. *1972 Census of Manufacturers, Industry Statistics,* Vol. II, Part 2, U.S. Bureau of the Census, Department of Commerce, Washington, D.C., 1973, p. 14.
83. B. P. Johnson, ed., *Drug Topics,* Medical Economics Co., Oradell, N.J., June 1, 1979, p. 27.
84. B. P. Johnson, ed., *Drug Topics,* Medical Economics Co., Oradell, N.J., June 20, 1978, p. 36.
85. F. Streit and M. J. Nicolich, *Previous and Predicted Kilogram Purchases of Schedule II Drugs, 1972–1983,* Office of Medical and Professional Affairs, National Institute on Drug Abuse, Rockville, Md., 1979.
86. *Fed. Reg.* **41**(176), 38312 (1976).
87. M. Windholz, ed., *The Merck Index,* 9th ed., Merck & Co., Rahway, N.J., 1976.
88. E. J. Fairchild, ed., *Registry of Toxic Effects of Chemical Substances,* Vol. II, Dept. of HEW, Washington, D.C., 1977.

WILLIAM J. WELSTEAD, JR.
A. H. Robins Company

EXPLOSIVELY CLAD METALS. See Metallic coatings.

EXPLOSIVES AND PROPELLANTS

Explosives, 561
Propellants, 620

EXPLOSIVES

Propellants and explosives are chemical compounds or their mixtures that rapidly produce large volumes of hot gases when properly initiated. Propellants burn at relatively low rates measured in centimeters per second, whereas explosives detonate at rates of kilometers per second. Pyrotechnic materials evolve large amounts of heat but much less gas than propellants and explosives (see Pyrotechnics).

The energy liberated by explosives and propellants depends on the thermochemical properties of the reactants. As a rough rule of thumb, these materials yield about 1000 cm^3 of gas and 4.2 kJ (1000 cal)/g. A comparison of the characteristics associated with propellant burning, explosive detonation, and the performance of conventional fuels is shown in Table 1 (1–2). The most notable difference is the rate at which energy is evolved.

Although there are many compounds that explode when triggered by a suitable stimulus, most are either too sensitive or fail to meet cost and production-scale standards and requirements for safety in transportation and storage stability. Propellants and explosives in large-scale use are based mostly on a relatively small number of well-proven ingredients. Millions of kilograms are produced annually with an accident rate no greater than that encountered in many industrial operations. Propellants and

Table 1. Characteristics of Burning and Detonation[a]

Characteristics	Burning		Explosive detonation
	Fuel	Propellant	
typical material	coal–air	propellants	explosives
linear reaction rate, m/s	10^{-6}	10^{-2}	2–9×10^3
type of reactions	oxidation–reduction	oxidation–reduction	oxidation–reduction
time for reaction completion, s	10^{-1}	10^{-3}	10^{-6}
factor-controlling reaction rate	heat transfer	heat transfer	shock transfer
energy output, J/g[b]	10^4	10^3	10^3
power output, W/cm^2	10	10^3	10^9 [c]
most common initiation mode	heat	hot particles and gases	high temperature–high pressure shock waves
pressures developed, MPa[d]	0.07–0.7	0.7–7×10^2	7×10^3–7×10^4
uses	source of heat and electricity	controlled gas pressure, guns, and rockets	brisance, blast, munitions, civil engineering

[a] Ref. 1–2.
[b] To convert J to cal, divide by 4.184.
[c] This may be compared with the total United States electric generating capacity of about 30×10^6 kW.
[d] To convert MPa to psi, multiply by 145.

explosives for military systems are manufactured in the United States in government owned plants where they are also loaded into munitions. Composite propellants for large rockets are produced mainly by private industry as are small arms propellants for sporting weapons.

Explosives and propellants have relatively large amounts of available energy stored compactly and readily deliverable to perform many industrial and military operations. The power output depends on the rate at which energy is liberated and may vary by several orders of magnitude, depending on whether the reactions are those of propellant burning or explosive detonation. Propellants are used wherever a readily controllable source of energy is required for periods of time ranging from milliseconds in guns to seconds in rockets. The gases evolved are employed as a working fluid for propelling projectiles and rockets, driving turbines, moving pistons, shearing bolts and wires, operating pumps, and starting engines. Explosives are used wherever very rapid rates of energy application and high pressures are essential. They are employed to produce high-intensity shock waves in air, water, rock, and metal; for blasting, cratering, mining, and other civil engineering purposes; for metal forming, cutting, and fragmentation; in shaped charges and many specialty devices acquiring high rates of energy transmission; and for initiation of detonation phenomena. The terms burning and deflagration are used synonymously to describe propellant performance and to contrast it with the very much more rapid, violent, and destructive action associated with the explosion or detonation of explosives.

The development of explosives began in Europe with the formulation and use of black powder in about the middle of the 13th century (3–4), accelerated in the 19th century with the nitration of many compounds to produce high energy explosives, and greatly intensified during World War II with the development of many new compositions and their application for military and commercial use. The post World War II years have been marked by an enormous growth in scientific productivity in this field made possible by the availability of new electronic instrumentation, high speed photography, computers, the liberal monetary support of governments interested in military and space research, and the expanding opportunities in worldwide mining and civil engineering programs. The declassification and publication of many monographs, texts, symposia proceedings, and government publications contributed greatly to this progress (5–13).

General Characteristics

Exothermic oxidation–reduction reactions provide the energy released in both propellant burning and explosive detonation. The reactions are either internal oxidation–reductions, as in the decomposition of nitroglycerin and pentaerythritol tetranitrate, or reactions between discrete oxidizers and fuels in heterogeneous mixtures such as RDX (see below) and aluminum or ammonium perchlorate and a hydrocarbon fuel.

An activation energy of 125–250 kJ/mol (30–60 kcal/mol) is usually required to initiate the reaction. Once initiated, the heat evolved is sufficient to cause the reaction to continue and become self-sustaining. Most explosives and propellants are organic compounds or mixtures of compounds that contain carbon, hydrogen, oxygen, and nitrogen. Metallic fuels such as aluminum may be added to increase the heat of reaction. Industrial dynamites have traditionally used nitroglycerin, nitrocellulose, and inorganic salts as sources of oxygen but are now increasingly formulated with ammonium nitrate as the oxygen source. Composite propellants commonly use ammonium perchlorate to supply the oxygen required. The most common gaseous products of the oxidation–reduction reactions are hydrogen, water, carbon monoxide, carbon dioxide, and nitrogen. Other products depend on the reactants involved.

Propellants and explosives evolve about the same amount of energy which is considerably less than that produced in the combustion of common fuels. A high-performance rocket propellant produces 5–6.3 kJ/g (1.2–1.5 kcal/g). The detonation of nitroglycerin yields 6.3 kJ/g (1.5 kcal/g). By comparison burning a gram of carbon in air yields 33.5 kJ (8 kcal).

The specific stimulus that triggers an explosive depends on the material involved and the system environment. In most cases, the ultimate cause of the decomposition is heat. Gun and rocket propellants are initiated by igniter compositions that produce hot gases and solids at relatively low pressures. Explosives are detonated by high pressure shock waves. Mechanical impact, frictional forces, electric discharge, and other sources of heat may also act as initiating stimuli. The minimum quantity of energy required for initiation has a characteristic value that depends on the chemical characteristics of the material and its physical properties including mass, geometry, density, the degree of confinement, the rate at which energy is lost, the environment in which the energy release occurs, and the type of initiating stimulus provided. Explosives and propellants are often ranked in order of their sensitivity and response to a specific stimulus in a given environment (1–2,14).

Propellant Burning. Propellants generally operate at low pressures up to about 20 MPa (2900 psi) in rockets and up to about 689 MPa (100,000 psi) in high-performance guns (see Propellants). The process is characterized by a reaction front that moves in a direction normal to the exposed surface of the grain, proceeding from the outside to within in laminar layers. The rate of burning depends on the intrinsic rate of decomposition of the propellant formulation and the rate of heat transfer from the hot gases above the propellant surface. The rate of propellant burning is defined by:

$$dx/dt = bP^n$$

where x = a coordinate normal to the grain surface, t = time, b = a constant that depends on the propellant temperature and composition, p = pressure at which burning occurs, and n = pressure exponent that depends on the propellant composition and the pressure. Typical values of n range from 0.2 to 0.5 for low pressure rocket propellants and 0.7 to 1.0 for high pressure gun propellants.

Explosive Detonation. Detonations also proceed as a result of a reaction front moving in a direction normal to the surface of the explosive. However, detonation is a hydrodynamic phenomenon that differs in a fundamental sense from burning. Upon initiation, burning first occurs at an increasing rate for a period of time up to several microseconds. A high pressure shock wave is formed which passes through the explosive at high velocity. As it does so it causes exothermal decomposition of the explosive. The continued passage of the wave is supported by transfer of energy from the spent explosive to the unreacted explosive by shock compression. The rate of reaction depends on the rapid rate of transmission of a shock wave rather than on the relatively slow rate of heat transfer associated with propellant burning. The detonation rate (15) is stable and constant and is primarily governed by the physical and chemical properties of the explosive, its geometry, degree of confinement, and particularly its density ρ with which it varies in a linear fashion for most explosives:

$$D_i = D_o + M (\rho_i - \rho_o)$$

where D_i = linear detonation rate dx/dt at density ρ_i, D_o = linear detonation rate at density ρ_o, and M = a constant characteristic of the explosive composition. Typical values of D_o at ρ_o = 1.0 g/cm^3 are about 5000–6000 m/s. Values of M are about 3000–4000 m/s.

Deflagration to Detonation Transfer. The same compound or mixture may burn or detonate, depending on the type and intensity of initiation, the degree of confinement, and the physical and geometric characteristics of the material. Many explosives that ordinarily detonate, may burn under carefully controlled conditions involving gentle ignition to avoid shock-wave formation. Their burning characteristics are similar to those encountered in the burning of propellants of about the same energy level. Propellants containing colloid-treated nitrocellulose and nitroglycerin or ammonium perchlorate in a polymeric matrix normally burn quietly and controllably on initiation with conventional igniters. They are difficult to detonate under these conditions even when they contain relatively large quantities of highly energetic explosives such as RDX or HMX (see below). However, a detonation may occur if the propellant is initiated by a high-pressure, high-intensity shock wave, if the physical conditions are appropriate, eg, the presence of a large degree of porosity and heavy confinement or if the material is in the form of a large mass of finely divided material. Granular TNT initiated with black powder burns quietly if the TNT is spread in thin layers upon the ground. It is apt to detonate if piled up in a large mound. The disastrous explosion in 1947 of a cargo of fertilizer-grade ammonium nitrate packed in the hold of a burning freighter illustrates how conditions may cause an ordinarily inert material to detonate.

Since unwanted detonations of propellants that normally burn are likely to be catastrophic, the conditions at which deflagration to detonation transformations occur have been intensively studied. A simplified view of the process by which a deflagration is transformed to a detonation suggests that an initial burning occurs of the type associated with propellant combustion. This is followed by convective burning in which the hot gases penetrate the pores of the explosive. The combustion front may then build up into shock-wave pressures that produce a low-velocity detonation which is rapidly converted to a full-fledged detonation with its characteristic shock pressure and temperature. Conditions that minimize energy losses and increase the likelihood of build-up of a shock wave tend to enhance the likelihood of a deflagration converting to a detonation. The scaling up of explosive processes increases the possibility of detonation because of the mass effect, as does heavy confinement and the use of large concentrations of crystalline oxidizers (16–25).

Pyrotechnic Compositions. Pyrotechnic compositions engage in oxidation–reduction reactions that resemble those of propellants and explosives, but generally produce little or no gas (see Pyrotechnics). They are heterogeneous mixtures of a finely powdered metal, metal alloy, or organic fuel and inorganic oxidizers. Such compositions are commonly used for flares, signals, tracers, incendiaries, delays, igniters, heating mixtures, and in devices where the formation of much gas is unacceptable either because the gas pressure causes unwanted changes in the reaction rate or the system is not designed to withstand the pressure without rupturing. The properties of typical pyrotechnic and explosive compositions are compared in Table 2.

Table 2. Comparison of Properties of Pyrotechnic Compositions with Those of Some Explosives[a]

Composition	%	Heat of reaction, kJ/g[a]	Gas volume[b], cm³/g	Relative brisance, % TNT	Ignition temperature[c], °C	Impact test[d], % TNT
Pyrotechnic						
delay						
barium chromate	90					
boron	10	2.010	13	0	450	12
delay						
barium chromate	60					
zirconium–nickel alloy	26					
potassium perchlorate	14	2.081	12	0	485	23
flare						
sodium nitrate	38					
magnesium	50					
laminac	5	6.134	74	17	640	19
smoke						
zinc	69					
potassium perchlorate	19					
hexachlorobenzene	12	2.579	62	17	475	15
photoflash						
barium nitrate	30					
aluminum	40					
potassium perchlorate	30	8.989	15	15	700	26
High explosive						
TNT		4.560	710	100	310	100
RDX		5.694	908	140	260	35

[a] Ref. 26.
[a] To convert J to cal divide by 4.184.
[b] At STP.
[c] 5-second value.
[d] Pyrotechnic compositions produce only a mild ignition on impact.

Although pyrotechnic compositions are composed of inert ingredients (22,26–30), their accidental initiation during the manufacturing process may be accompanied by the same catastrophic consequences that attend explosive detonations. Mixtures of finely divided oxidizers and metals are sensitive to initiation by friction or by spark, particularly when in the powdered form. As little as 10 μJ (2.4 μcal) can initiate some pyrotechnic dust mixtures. Safety measures include mixing in liquid media, using nonsparking tools, maintaining the moisture content of the plant atmosphere at about 60% humidity, and using ionizing and grounding devices to reduce the possibility of electrostatic charge accumulation.

PRIMARY EXPLOSIVES

Explosives are commonly categorized as primary, secondary, or high explosives and propellants in order of decreasing sensitivity to energy input. Table 3 lists typical

Table 3. Typical Explosive Substances

Name	CAS Registry No.	Code	Formula	Use
ammonium nitrate	[6484-52-2]	AN	NH_4NO_3	solid oxidizer
ammonium perchlorate	[7790-98-9]		NH_4ClO_4	solid oxidizer
bis(N,N-trinitromethylurea)	[41407-46-9]	BTNEU	$[(NO_2)_3CCH_2NH]_2CO$	secondary high explosive
ammonium picrate	[131-74-8]	AP	(structure: picrate with ONH_4, O_2N, NO_2, NO_2)	secondary high explosive
2,4-diamino-1,3,5-trinitrobenzene	[1630-08-6]	DATB	(structure: benzene with NO_2, NH_2, O_2N, NO_2, NH_2)	secondary high explosive
diazodinitrophenol	[28655-69-8]	DDNP	(structure: O_2N, O—N=N, NO_2)	primary explosive
diethylnitramine dinitrate	[4185-47-1]	DINA	$(O_2NOCH_2CH_2)_2NNO_2$	secondary high explosive
ethylenedinitramine[a,b]	[505-71-5]	EDNA	$O_2NNHCH_2CH_2NHNO_2$	secondary high explosive
ethylene glycol dinitrate	[628-96-6]	EGDN	$O_2NOCH_2CH_2ONO_2$	liquid explosive
bis(2,2-dinitro-2-fluoroethyl) formal[a]	[17003-79-1]	FEFO	$[(NO_2)_2CFCH_2O]_2CH_2$	secondary high explosive
cyclotetramethylenetetranitramine	[2691-41-0]	HMX	(structure: ring with four NO_2 groups on N)	secondary high explosive
lead azide	[13424-46-9]		$Pb(N_3)_2$	primary explosive
lead styphnate	[15245-44-0]		(structure: styphnate) PbH_2O^{2+}	primary explosive
mannitol hexanitrate	[15825-70-4]	MN	H_2CONO_2–$HCONO_2$–$HCONO_2$–$HCONO_2$–$HCONO_2$–H_2CONO_2	primary explosive
mercury fulminate	[628-86-4]		$Hg(ONC)_2$	primary explosive

Table 3. (continued)

Name	CAS Registry No.	Code	Formula	Use
nitrocellulose	[9004-70-0, 9046-47-3]	NC	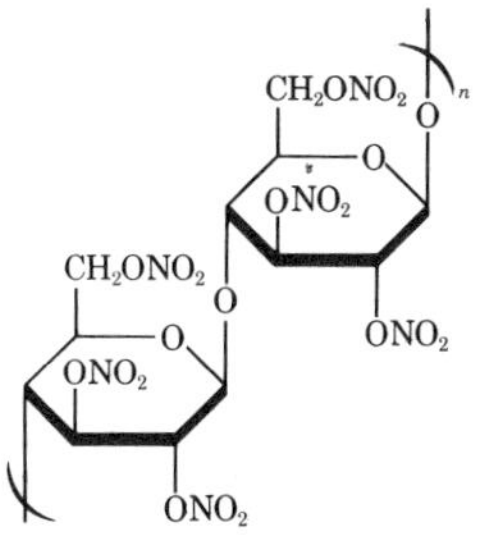 (structure shown)	secondary explosive used in propellants
nitroglycerin	[55-63-0]	NG	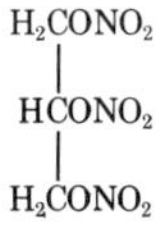 H_2CONO_2 $HCONO_2$ H_2CONO_2	liquid secondary explosive ingredient in commercial explosives and propellants
nitromethane	[75-52-5]	NM	CH_3NO_2	liquid secondary explosive
pentaerythritol tetranitrate	[78-11-5]	PETN	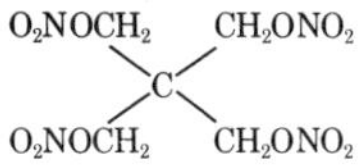 (structure shown)	secondary high explosive used as booster
picric acid	[88-89-1]	PA	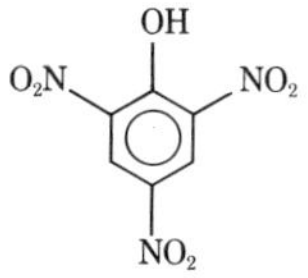 (structure shown)	secondary high explosive
cyclotrimethylenetrinitramine	[121-82-4]	RDX	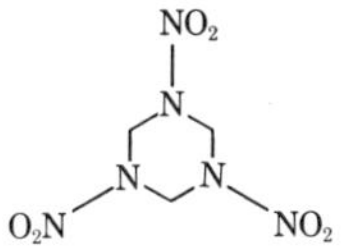 (structure shown)	secondary high explosive used as booster
trinitrophenylmethylnitramine	[479-45-8]	Tetryl	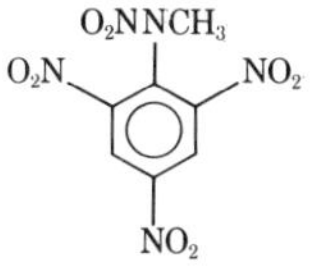 (structure shown)	secondary explosive used as booster
2,2,2-trinitroethyl 4,4,4-trinitrobutyrate[a]	[17543-76-9]	TNETB	$(NO_2)_3CCH_2$-$CH_2C(O)OCH_2C(NO_2)_3$	secondary high explosive
tetrazene	[31330-63-9]		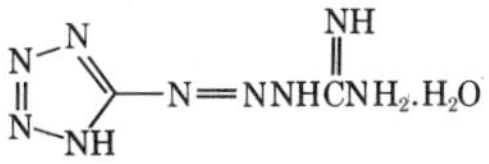 (structure shown)	primary explosive
tetranitromethane	[509-14-8]	TNM	$C(NO_2)_4$	liquid explosive
2,4,6-trinitrotoluene	[118-96-7]	TNT	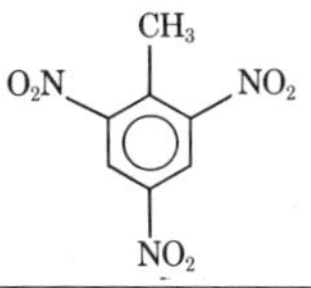 (structure shown)	secondary high explosive

[a] These explosives are not commonly used.
[b] Also called halcite.

explosives and their categories. Primary or initiator explosives are the most sensitive to heat, friction, impact, shock, and electrostatic energy. They have been studied in considerable detail because of their almost unique capability, even when present in very small quantities, to rapidly transform a low-energy stimulus into a high-intensity shock wave. Most recent evidence indicates that there is a minimum thickness or run-up distance before steady-state detonation occurs and a gradual transition from an unreacted shock to a stable detonation (31–40).

Primary explosives are used to initiate the next element in a series which consists of explosives of increasing mass and decreasing sensitivity. They are arranged in sequence to amplify the input stimulus to an output level of sufficient intensity to maximize the probability of initiating the main charge. Overall energy intensification is about ten million to one. Primary explosives are used in military detonators, commercial blasting caps, and in stab and percussion primers. They may be initiated electrically by penetration of the element with a firing pin (stab detonator) or by shock from an exploding wire. The explosives used in a detonator blasting cap are pressed into a metallic capsule and are designed so that the transformation from burning to a detonation shock wave can occur within very small diameters and lengths. A typical stab detonator consists of a primary charge which is readily ignited, an intermediate charge in which the transition from burning to detonation occurs, and a base charge which amplifies the energy output of the intermediate charge and assures initiation of the next element in the explosive series. The properties of commonly used primary explosives are shown in Table 4. The characteristics of other materials as initiators have also been studied (41–46).

Primary explosives have lower detonation rates and produce less energy than secondary explosives. They must be powders with good flow transfer and pressing characteristics to permit high-speed automatic loading of detonators, which are produced in very large quantities. Their high sensitivity to electrostatic energy requires special precautions in the detonator loading plants including atmospheric humidification, grounding of all equipment, and even the operators at times, and antistatic clothing and materials. Because of their sensitivity, initiator materials are generally loaded into detonators in the plant where they are made. When transportation is necessary, initiator explosives are shipped wet with water or water and alcohol, suspended in rubberized cloth bags, and well cushioned with sawdust or other inert media to prevent frictional initiation. The manufacture of detonators is accompanied by all necessary precautions to prevent the accumulation of substantial masses of explosive, dusting, and accidental initiation by static electricity or friction. However, detonators are safe to use with normal handling procedures once the initiating explosives have been pressed into the metal capsule (45–53).

Some primary explosives are also used in nondetonating stab and percussion primers to accomplish mechanical work or to initiate burning actions rather than detonation in pyrotechnic and propellant systems (36). The hot gases produced may be used to expand bellows, drive pins, and perform other types of controlled work or they may be used to initiate the igniter for propellant charges. These primers are also very small elements (eg, 4 mm dia by 2.5 mm long) which contain mixtures of an explosive ingredient and fuels and oxidizers to increase the amount of heat and gas evolved. Sometimes additional compounds and abrasives are incorporated to increase the sensitivity to mechanical action. A typical composition (NOL 130) consists of 15 wt % antimony sulfide, 20 wt % lead azide, 40 wt % basic lead styphnate, 20 wt % barium nitrate, and 5 wt % tetrazene.

Table 4. Properties of Primary Explosives

Property	Mercury fulminate	Lead azide	Silver azide	Normal lead styphnate	Diazo-dinitro-phenol	Tetrazene
molecular weight	285	291	150	468	210	188
color	grey	white	white	tan	yellow	light yellow
crystal density, g/cm^3	4.43	4.93	5.1	3.10	1.63	1.7
crystal form	orthorhombic	orthorhombic monoclinic		cubic	tabular	
melting point, °C	160 explodes		252	explodes	157	140–160 explodes
hygroscopicity						0.8
solubility in H$_2$O at 20°C, g/100 g	0.1					
heat of formation, kJ/g[a]	−0.925	−1.45	−2.07	17.9	4.00	1.13
heat of combustion, kJ/g[a]	3.93	2.64	4.34	5.24	13.58	
heat of detonation, kJ/g[a]	1.79	1.54	1.90	1.91	3.43	2.75
gas volume, cm^3/g at STP	316	308		368	876	
activation energy, kJ/mol[a]	29.8	172	146	259	230	
collision constant, log$_{10}$/s	10.8	14.0		22		
detonation rate, km/s	5.4	6.1	6.8	5.2	6.9	
at density, g/cm^3	4.2	4.8	5.1	2.9	1.60	
specific heat, J/(g·K)[a]	0.50	0.46	0.50	0.67		
compressive strength, MPa[b]		1.4–21				
thermal conductivity, W/(m·K) × 10^{-4}	1837	1256	837			
vacuum stability at 100°C, mL gas per g per 40 h at STP	explodes	<1	<1	<1	<1	>5
weight loss at 100°C, %	<1	<1	<1	<1	<5	<5
explosion temperature at 5 s, °C	190–260	345	290	265–280	195	160
effect of prolonged storage	detonates at 80°C	stable	stable	stable	stable	stable
relative impact test value, % TNT	5	11	18	8	15	5
friction pendulum	reacts	reacts	reacts			
static discharge max energy for nonignition, J[a]	0.07	0.01	0.007	0.001	0.25	0.036
relative energy output, % TNT						
lead block	50	40	45	40	110	50
ballistic mortar					95	
sand test	45	40		25	105	50
plate dent test		60				

[a] To convert J to cal, divide by 4.184.
[b] To convert MPa to psi, multiply by 145.

Only relatively few compounds can act as primary explosives and still meet the restrictive military and industrial requirements for reliability, ease of manufacture, low cost, compatibility, and long-term storage stability under adverse environmental conditions. Most initiator explosives are dense, metallo-organic compounds.

Mercury Fulminate. Mercury fulminate, mercuric cyanate, is a gray–white powder prepared in $\frac{1}{2}$–1 kg batches by what is essentially a large-scale laboratory process (36,38,41). In the Chandelen process a solution of one part mercury in 11 parts 57% nitric acid is poured into ten parts 95% ethyl alcohol where mercury fulminate is precipitated as fine crystals of about 99% purity. Although single crystals only flash when ignited, even a thin layer of crystals or slight confinement transforms ignition into detonation with considerable brisance. It is the most sensitive of initiating agents to impact and friction, although its sensitivity decreases as the density of the pressed mass increases. It is no longer used as a primary explosive in the United States.

Mercury fulminate decomposes when stored at elevated temperature. Although nonhygroscopic, it reacts with metals in the presence of water, liberating mercury, and forming metallic fulminates. After 30 months storage at 50°C, its purity is reduced from 99.75% to 90% and its initiating properties are greatly impaired. At a density of 3.6 g/cm^3, it has a calculated detonation temperature of 6900 K, a detonation pressure of 22 GPa (3.2×10^6 psi) and the following detonation products: 0.4 mol/kg CO, 3.3 mol/kg CO_2, 3.5 mol/kg N_2, 3.5 mol/kg Hg, and 3.3 mol/kg C.

Lead Azide. The azides are among the very few useful explosive compounds that do not contain oxygen. Lead azide is the primary explosive used in military detonators in the United States and has been intensively studied. It is very stable at ambient and at elevated temperatures, and has good flow characteristics. Lead azide has been stored for 25 mo at 50°C without change in purity or performance. It is compatible with most explosives and priming mixture ingredients. Since lead azide is less sensitive to ignition than mercury fulminate, a more readily ignitable material such as lead styphnate or NOL 130 is often used as a cover charge to ensure initiation. The crystalline forms of the azide are sensitive to impact and become even more sensitive if small amounts of foreign matter are present. Lead azide builds up to detonation velocity extremely rapidly on ignition. Even single crystals detonate when ignited. Although it has been reported that large crystals are supersensitive to shock, impact, and friction, little evidence supports this contention. Of the four polymorphic forms isolated (alpha to delta), the alpha is the most common. The detonation velocity of lead azide is relatively independent of its confinement in steel, aluminum, or brass, ranging from 4.52 km/s in steel to 4.85 km/s in brass sleeves at a density of 3.22 g/cm^3. Its calculated detonation temperature and pressure at a density of 4.0 g/cm^3 are 5600 K and 25 GPa (3.6×10^6 psi). Its major detonation products at the same density are 10.3 mol/kg N_2 and 3.4 mol/kg Pb (38–39,54–56).

Lead azide tends to hydrolyze at high humidities or in the presence of materials evolving moisture. The hydrazoic acid formed reacts with copper and its alloys to produce the very sensitive cupric azide [14215-30-6]. Appropriate protection must be provided by hermetic sealing and the use of noncopper or coated-copper metal (42,57–58).

Lead azide is not readily dead-pressed, ie, pressed to a point where it can no longer be initiated. However, this condition is somewhat dependent on the output of the mixture used to ignite the lead azide and the degree of confinement of the system. Since lead azide is a nonconductor, it may be mixed with flaked graphite to form a conductive mix for use in low-energy electric detonators. A number of different types of lead azide have been prepared to improve its handling characteristics and performance and to decrease sensitivity. In addition to the dextrinated lead azide commonly used in the United States, service lead azide, which contains a minimum of 97% lead azide and no protective colloid is used in the United Kingdom. Other varieties used include colloidal lead azide (3–4 μm), poly(vinyl alcohol)-coated lead azide, and British RD 1333 and RD 1343 lead azide which is precipitated in the presence of carboxymethyl cellulose (43,57,59–63).

Manufacture. Lead azide is made in small batches (eg, 5 kg) buffered by the reaction solutions of lead nitrate or lead acetate with sodium azide:

$$Pb(NO_3)_2 + 2\ NaN_3 \rightarrow 2\ NaNO_3 + PbN_6$$

Sodium azide is insensitive but highly toxic. Contact must be avoided with acid, with which it forms the dangerous hydrazoic acid, and with copper, lead, cadmium, silver, mercury, or their alloys, with which sensitive azides may be formed. Nucleating agents, such as poly(vinyl alcohol) (PVA), sodium carboxymethyl cellulose (CMC), or dextrin, may be added during precipitation to produce free-flowing crystals or rounded agglomerates required for the large-scale, automatic loading of detonators. The presence of hydrophilic polymeric substances also tends to eliminate the small possibility of spontaneous explosions occurring durring the precipitation process. Wetting agents may also be added (64).

All phases of the manufacturing process are conducted by remote control in stainless steel vessels using either distilled or demineralized water and filtered solutions. The overall precipitation time is about 60 min. In the manufacture of dextrinated lead azide, lead nitrate stock solution is prepared by dissolving lead nitrate, dextrin, and sodium hydroxide in water at pH 4.6–4.8. The solution is cooled, filtered, pumped to a storage tank, and allowed to settle for 8 h or longer. A sodium azide stock solution is similarly prepared. The precipitation vessel is a precisely made, open-topped, round-bottom, double-walled, polished stainless steel tilting pot equipped with an agitator, feed tubes, and a water-spray ring. The lead nitrate solution at 60°C is transferred to the precipitation vessel from a measuring tank and the sodium azide solution is added at a rate of about 2 L/min while maintaining a temperature of 60°C. Lead azide precipitates as free-flowing, fine white agglomerates. After settling, the mother liquor is decanted through a filter, collected, and neutralized with 30% sodium nitrite and then 30% nitric acid or with ceric ammonium nitrate to decompose the azide ion. Excess acid is neutralized with soda ash. Any soluble lead present is precipitated as the insoluble carbonate. The lead azide precipitate is washed repeatedly with water, vacuum filtered, and dried. Lead azide made without dextrin (RD 1333) usually contains more than 99% azide. When made with dextrin it contains about 92% lead azide, 4–5% lead hydroxide, 3% dextrin, and other impurities. Lead azide must be free of needle-shaped crystals longer than 0.1 mm. Dextrinated lead azide is somewhat more hygroscopic and less dense, sensitive, and efficient as an initiator than the 99% product.

The preparation of lead azide is hazard-free when conducted as described. A low incidence of explosions has been reported when precipitation is effected without a nucleating agent and about the same frequency of explosions during the screening process. Precautions must be taken during detonator loading to prevent dusting and to maintain a scrupulously clean operation. There is no evidence of toxicity of lead azide to workers engaged in its manufacture (41,43,61,64–67).

Silver Azide. Silver azide [13863-88-2], AgN_3, has received attention as a potential replacement for lead azide because it may be used in smaller quantities as an initiator and, therefore, offers the possibility of miniaturization of fuze components. Silver azide requires somewhat less energy for initiation than lead azide and fires with a shorter time delay. It is less apt to hydrolyze and more sensitive to heat. It is incompatible with sulfur compounds, with tetrazene, and with some metals, including copper. Silver azide is made in the same manner as lead azide, except that silver nitrate is used in the reaction with sodium azide (61,68–72).

Lead Styphnate. Lead styphnate, lead 2,4,6-trinitroresorcinate, is one of a number of compounds used in priming compositions to start the ignition-to-detonation process in the explosive sequence. Its sensitivity to stab, flame, heat, or impact ensures ignition of the true primary explosive. Lead styphnate is stable and noncorrosive. It is used as the top charge in stab primers, as a spot charge in electric detonators, or as a component of top charge mixtures. The addition of graphite enhances its electrical conductivity in systems designed for electrical initiation. Dry lead styphnate is the most sensitive of the primary explosives to electrostatic discharge. However, it is much less sensitive after being pressed into a detonator capsule.

Lead styphnate monohydrate is precipitated as the basic salt from a mixture of solutions of magnesium styphnate and lead acetate followed by conversion to the normal form by acidification with dilute nitric acid (73–77).

Diazodinitrophenol. Diazodinitrophenol is an orange–yellow compound made by diazotizing picramic acid, $NH_2(NO_2)_2C_6H_2OH$, with sodium nitrite and hydrochloric acid, washing the product with ice water, and recrystallizing it from hot acetone. It is almost nonhygroscopic, sensitive to impact and friction, and somewhat less stable than lead azide. Since it is about equivalent to TNT in brisance, it is more effective for some purposes than lead azide and is used as an initiator in commercial blasting caps (78–79).

Tetrazene. Tetrazene is a pale-yellow crystalline explosive made by adding sodium nitrite to a solution of 1-aminoguanidine hydrogen carbonate in dilute acetic acid at 30°C. The crystals are water-washed and dried at RT. The compound is stable up to about 75°C. It ignites readily and has a relatively high explosion energy. Tetrazene is used as a component in priming compositions.

SECONDARY EXPLOSIVES

Aliphatic Nitrate Esters

Aliphatic nitrate esters, such as glycerol trinitrate, ethylene glycol dinitrate, cellulose nitrate, and pentaerythritol tetranitrate, are among the most powerful explosives available. They are generally less stable than aromatic nitro compounds or nitramines because they tend to hydrolyze autocatalytically to form nitric and nitrous acids which further accelerate decomposition. Liquid nitrates, ie, nitroglycerin, are usually less stable than crystalline compounds because of the higher energy state of the liquid phase. The properties of the more commonly used aliphatic nitrate esters are listed in Table 5 (80–83).

Nitroglycerin. Nitroglycerin, NG, glyceryl trinitrate, is primarily used as an explosive in dynamites and a plasticizer for nitrocellulose in double- and multibase propellants. It is very sensitive to shock, impact, and friction and is employed only when desensitized with other liquids or absorbent solids or when compounded with nitrocellulose. When desensitized with other liquids, such as triacetin or dibutyl phthalate, it may be transported with care. It is readily soluble in many organic solvents

Table 5. Properties of Explosive Aliphatic Nitrate Esters

	NC[a]	NG	EGD	DEGN	BTN	TMETN	PETN
molecular weight		227	152	196	241	255	316
color	white	pale yellow	colorless	colorless	light yellow		white
density, g/cm	1.66	1.59	1.49	1.38	1.52	1.47	1.78
crystal form		rhombic					tetragonal
melting point, °C		13.2	−22.8	−11.3	−27	−3	141
hygroscopicity, % at 90% rh, 20°C	negligible	0.06	negligible		0.1	0.07	negligible
solubility in H_2O at 20°C, g/100 g		0.2	0.5	0.4	0.1	0.05	
oxygen balance, % to CO_2		3.5	0	−41	−17	−35	−10
viscosity at 20°C, mPa·s (= cP)		36	4.2	8	62	156	
heat of formation, kJ/g[b]		1.63	1.53	2.17	1.54	1.60	1.70
heat of combustion, kJ/g[b]		6.80	7.38	11.68	9.07	11.05	8.20
heat of detonation, kJ/g[b]		6.29	7.13	4.86	6.10	5.17	6.28
gas volume at STP, cm/g		715	740	880	840	855	784
activation energy, kJ/mol[b]	205	169	149				197
collision constant, $\log_{10}$/s	21.0	17.1	14.3				19.8
applicable temperature range, °C	90–135	75–105	140–170				160–225
detonation velocity, km/s	7.3	7.60	7.8–8.0	6.75		7.05	8.31
at density, g/cm³	1.20	1.59	1.49	1.38		1.47	1.77
detonation pressure, GPa[c]	25.30						32.00
at density, g/cm³	1.59						1.77
detonation temperature, K[d]	3470						3400
specific heat ratio[d]	2.82						2.55
vacuum stability at 100°C mL gas per g per 40 h at STP	>1 (90°C)	explode	explode	<1	1–3	1–3	<1
weight loss at 100°C, %	>1	3–5		1–3		1–3	>1
explosion temperature at 5 s, °C	170	220 (dec)	240	240	230 (dec)	235	225
effect of prolonged storage		dec 3–4 d at 75°					negligible after 6 d at 65°C
relative impact test value, % TNT	10	15		100		45	15
friction pendulum	explode	explode		explode			explode
static discharge max energy f/nonignition, J[b]	0.06						
relative energy output, % TNT							
lead block	130[e]	180	205	150		140	175
ballistic mortar	125[e]	140		125		135	145
sand test	105[e]	125	135	100	105	105	135

[a] Characteristics vary with nitrogen content:

% N	Mol wt	Oxygen balance	Hygroscopicity, %	Heat of formation	Heat of combustion	Explosion
12.5	$(272)_n$	−35	3	2.58	10.08	3.58
13.4	$(286)_n$	−29	2	2.35	9.68	4.04
14.14	$(297)_n$	−24	1	2.15	9.32	4.43

[b] To convert J to cal, divide by 4.184.
[c] To convert GPa to kilobars, multiply by 10.
[d] Calculated.
[e] For 13.4% nitrogen content.

and acts as a solvent for many explosive ingredients. It is completely miscible with homologous nitrate esters such as ethylene- and diethylene glycol dinitrates. It is sufficiently soluble in water (0.18 g/100 g H_2O at 20°C) to pose a contamination problem in the disposal of process water produced during its manufacture or used in formulations (84–85).

The sensitivity of nitroglycerin decreases with decreasing temperature. Solid nitroglycerin is less sensitive to shock than the liquid, although some tests indicate it to be more sensitive to friction because of intercrystalline contact. It has a relatively high freezing point (13.2°C), possibly accounting for the poorer physical properties of nitroglycerin–nitrocellulose propellants at very low temperatures unless freezing-point depressants are added. Ethylene glycol dinitrate (freezing point −22.8°C) is added to nitroglycerin-bearing dynamites for this purpose.

Pure nitroglycerin is a stable liquid at temperate conditions. It decomposes above 60°C to form nitric oxides which in turn catalyze further decomposition. Moisture increases the rate of decomposition under these conditions. Double- and multibase propellants containing nitroglycerin have substantially shorter usefulness at 65.5 and 80°C than do single-base propellants. The decomposition of nitroglycerin proceeds in accordance with the following equation:

$$2\ C_3H_5(NO_3)_3 \rightarrow 3\ CO + 2\ CO_2 + 6\ NO + 4\ H_2O + H_2CO$$

A rate equation applicable (86) to the liquid phase is $k = 10^{20.2}\ e^{-46,000/RT}$.

Unconfined nitroglycerin burns without exploding if present in thin layers and in small quantities but detonates if confined. Aerated nitroglycerin and other liquid explosives containing microbubbles are especially sensitive to shock. Nitroglycerin exhibits a low velocity propagation regime with a detonation rate of about 2000 m/s compared with its high energy detonation rate of about 7800 m/s. Low velocity detonation is uncommon in high explosives and is found primarily in nitroglycerin and its homologues and in some gelatin dynamites. The vapor pressure of nitroglycerin is high enough for temporary exposure to cause severe headaches, although day-to-day workers tend to become accustomed and immune to it. The accumulation of condensed pockets of nitroglycerin as may occur in some propellant air-drying systems should be prevented (84–85,87–88).

Manufacture. Nitroglycerin has traditionally been made by the batch process. However, the hazard of handling large quantities led to the development and wide use of continuous processes. In general, the methods used to make nitroglycerin are applicable to the manufacture of other liquid aliphatic polyhydroxy alcohol nitrates (82,89–93).

Nitroglycerin is made from very pure glycerol (qv) to ensure stability of the final product. Mixed acid (90% nitric acid and 25–30% oleum) is used in both the batch and continuous processes. The theoretical yield ratio of nitroglycerin from glycerol is 2.467:1. The actual yield using concentrated acid is about 2.36:1, slightly higher in the continuous process than in the batch process. The process factors affecting yield include solubility in spent acid and wash waters, acid concentration and composition, nitric acid–glycerol ratio, temperature, stirring efficiency, and the efficiency of separating the nitroglycerin from spent acid and wash fluids (**94**).

Batch Process. In the batch process, which has been widely replaced by continuous processes, the mixed acid charge is transferred to a steel reactor equipped with an impeller and brine-cooled coils. After cooling the acid to 20–22°C, the glycerol is added at a rate to permit the temperature of the mix to be maintained below 25°C.

The nitration is completed in ca 1 h. The mix is transferred to the separator tank, the lower layer of spent acid is drawn, and the nitroglycerin is freed of acid by washing with warm water. After separation of the supernatant water, the emulsion is transferred to the neutralizer, treated with a warm 2–3% solution of sodium carbonate followed by separation, and water-washed until the product is neutral to litmus (95).

The Biazzi Continuous Process. The equipment consists of a nitrator and acid separator with an associated drowning tank, soda washers with a separator, and water washers with an additional separator. Emulsification of the nitroglycerin increases safety since emulsions containing three or more parts of water to nitroglycerin are relatively insensitive to initiation. The stainless steel equipment is precision engineered and has highly polished inner surfaces. Operations are totally automatic, include numerous fail-safe features, controls, and signalling devices, and are monitored by a closed-TV circuit from a remote control bunker. The nitrator is a small cylindrical vessel equipped with a multiple bank of spiral, closely packed cooling coils and a high-speed turbostirrer. It has an outlet at the bottom for emergency drowning of the contents, an inlet for entry of displacement acid, and an emulsion overflow pipe attached ca 10 cm below the top. The nitrator cover has port holes for the impeller shaft, the glycerol nozzle, the mixed-acid feed pipe, and a fume pipe. A 250-L capacity nitrator produces ca 1700 kg of nitroglycerin per hour.

The mixed acid and glycerol are metered into the nitrator, quickly submerged, emulsified, and forced up the nitrator past the cooling coils. Part of the nitroglycerin overflows into the separator, and part returns to the vortex in the nitrator fluid. The temperature is kept at 10–20°C. The emulsified mixture of nitroglycerin and spent acid enters the acid separator where a slight rotating action imparted to the upper layer of the liquid breaks the emulsion and prevents overheating. The spent acid flows continuously through an overflow from the bottom of the acid separator to its storage tank. The nitroglycerin may also be separated in a specially designed centrifuge. The remaining acid is neutralized and the product washed with water until it passes the specification test for stability.

The Biazzi process is also used for the manufacture of other nitrate esters such as triethylene glycol dinitrate, butane triol trinitrate, and trimethylolethane trinitrate (metriol trinitrate) (96–98).

The Nitro Nobel Injector Process. In this process (99–100), an injector is used to mix the glycerol or other polyol with precooled nitration acid. The flow of the acid through the injector creates a vacuum so that a metered ratio of the polyhydroxy alcohol is sucked in through the throat, mixing the two quickly and thoroughly to form an emulsion. Increased safety is achieved since the quantity of alcohol being nitrated is automatically dependent on the quantity of acid entering the injector. The rate of acid flow is designed to create an injector vacuum of about 33 kPa (4.4 mm Hg). Glycerol at about 48°C is sucked into the injector where it reacts almost instantaneously with the acid. The emulsion produced is very rapidly cooled to 15°C and flows by gravity to a centrifuge where the nitroglycerin is separated from the spent acid on a continuous basis. The spent acid can be recycled or denitrated. The acidic-containing nitroglycerin is emulsified immediately with a water jet to form a nonexplosive mixture, neutralized with sodium carbonate, and washed. The stabilized nitroglycerin is emulsified by passage through an injector to form a nonexplosive water emulsion for safe temporary storage.

Environmental and Safety Considerations. Soda ash and water washes are the major sources of pollution from this process. These wastewaters may contain nitroglycerin, nitrates, and sulfates and vary in pH from acid for the first wash after nitration to alkaline for the washes with soda ash. The process water is first sent through catch basins to remove nonsoluble nitroglycerin. The overflow may be discharged without further treatment or sent to percolation-evaporation ponds or earthen sumps. Sediments can be decontaminated by detonation. Acidic waste water may be neutralized by passage through crushed lime beds. Caustic soda and sodium sulfide may also be used to decompose and dissolve nitroglycerin in the effluent, followed by use of an activated sludge system as the secondary treatment. Other methods investigated include biodegradation of the nitroglycerin, reverse osmosis, absorption by polymeric resins, treatment with lime and caustic, and oxidation with ozone and permanganate (101).

Safety has been greatly increased by the continuous nitration processes. Emulsification of nitroglycerin with water reduces the likelihood of detonation. Process sensors and automatic controls minimize the likelihood of runaway reactions. Detonation traps may be used to decrease the likelihood of propagation of an accidental initiation; eg, a tank of water into which the nitrated product flows and settles on the bottom.

Other Glycol Nitrates. Other liquid aliphatic nitrates have been used as explosive plasticizers for nitrocellulose (see Table 6). They are made like nitroglycerin by mixed-acid nitration.

The two most commonly used compounds are ethylene glycol dinitrate (nitroglycol) and diethylene glycol dinitrate. Their physical properties are listed in Table 5. Nitroglycol is an excellent solvent for low-grade nitrocellulose, is comparable to nitroglycerin in explosive energy, and is less sensitive and more stable. It can detonate at high and low velocities (8000 and 1000–3000 m/s). The relatively high volatility of nitroglycol precludes its use in military propellants, although it is an excellent substitute for nitroglycerin in dynamite. It also appears to be more toxic than nitroglycerin, probably because of its higher vapor pressure. The relative vapor pressures of various nitric esters at 15–55°C are listed in Table 7 (102).

Nitroglycol is made by nitration of ethylene glycol with mixed acid with a yield of ca 93%. The increasing use of ammonium nitrate–fuel oil (ANFO, see below) and slurry explosives to replace dynamites has greatly decreased the demand for nitroglycol (81–82,103).

Table 6. Explosive Plasticizers

Compound	CAS Registry No.	Code	Use
ethylene glycol dinitrate	[628-96-6]	EGD	as a freezing point depressant in low temperature dynamites
diethylene glycol dinitrate	[693-21-0]	DEGN	propellant, coolant
triethylene glycol dinitrate	[111-22-8]	TEGN	propellant, coolant
2-methyl-2-[(nitrooxy)methyl]-1,3-propanediol dinitrate ester	[3032-55-1]	TMETN	propellant, coolant
butanetriol trinitrate	[41407-09-4]	BTN	propellant, coolant
1,3-propanediol dinitrate	[3457-90-7]	PDN	propellant, coolant

Table 7. Vapor Pressures of Nitrate Esters[a], Pa[b]

Temperature, °C	Glycerol trinitrate	Ethylene glycol dinitrate	Diethylene glycol dinitrate	1,2-Propylene glycol dinitrate	1,3-Propanediol dinitrate
15	0.17	3.09	0.30	5.14	1.54
25	0.24	9.39	0.79	13.1	4.36
35	0.61	29.1	1.97	33.7	8.26
45	1.72	59.7	3.78	65.9	19.5
55	4.77	128	11.54	132.3	42.9

[a] Ref. 102.
[b] To convert Pa to µm Hg, multiply by 7.50.

Diethylene glycol dinitrate (DEGN) is the most widely used explosive plasticizer, other than nitroglycerin, in the formulation of military propellants for gun use. It is a better plasticizer for nitrocellulose than nitroglycerin. Because of its lower calorific value, it may be used to replace nitroglycerin to obtain cooler propellants, thereby decreasing erosion in gun tubes and increasing tube life. It also decreases muzzle flash and minimizes the need for flash reducers in propellant charges. It is employed in some double-base cast propellants because of its superior physical properties in case-bonded motors. Although DEGN is used in European propellants, it has not been generally used in United States gun propellants, because its greater volatility poses problems during prolonged high temperature storage. It is made like nitroglycerin with a yield of about 85% of theoretical. It is superior to nitroglycerin in terms of stability and handling safety, has a lower freezing point than nitroglycerin (-11.3 vs $13.2°C$), and has a high and low stable detonation rate (6760 m/s and 1800–2300 m/s, respectively). It does not appear to have harmful physiological effects (81–83,104–106).

Triethylene glycol dinitrate (TEGN) is an explosive plasticizer of low sensitivity that has been used in some nitrocellulose-based propellant compositions, often in combination with metriol trinitrate. *Butanetriol trinitrate* has been used occasionally as an explosive plasticizer coolant in propellants. Its physical properties are listed in Table 5. *Trimethylolethane trinitrate* (metriol trinitrate) is not very satisfactory as a plasticizer for nitrocellulose and must be used with other plasticizers such as metriol triacetate. Mixtures with nitroglycerin tend to improve the mechanical properties of double-base cast propellants at high and low temperatures. Metriol trinitrate has also been used in combination with triethylene glycol dinitrate as a plasticizer for nitrocellulose (107–108). Its physical properties are listed in Table 5.

Nitrocellulose. The exceptional properties of nitrocellulose, NC, cellulose nitrate, are derived from the polymeric and fibrous structure of the cellulose from which it is made. Nitrocellulose provides mechanical strength as well as readily available energy to gun and rocket propellants. It is manufactured by nitration of cellulose with mixed acid. The primary sources of cellulose for this purpose are wood pulp and short-fibered cotton hairs known as linters. Cellulose molecules may contain up to 3500 glucose anhydride units, whereas cotton linters and wood pulp, which have been chemically treated and partially degraded, have a degree of polymerization of 900–1300 units. Nitrocellulose may vary from 400–700 units because of the additional molecular breakdown that occurs during the nitration process (109–114). The properties of nitrocellulose are given in Table 5.

Nitration of cellulose may result in the addition of from one to three nitrate groups per glucose anhydride unit. The following compounds and their nitrogen content have been postulated:

Compound	Formula	Nitrogen, %
cellulose trinitrate	$C_6H_7O_2(ONO_2)_3$	14.15
cellulose dinitrate	$C_6H_8O_3(ONO_2)_2$	11.11
cellulose mononitrate	$C_6H_9O_4(ONO_2)$	6.76

Neither the mononitrate nor the dinitrate have been isolated as distinct compounds. The trinitrate cannot be prepared with mixed acid, but with nitric anhydride or a mixture of phosphoric and nitric acids. The nitrogen content of nitrocellulose prepared with mixed acid does not exceed 13.4–13.5%. As made conventionally, nitrocellulose retains the fibrous structure of cellulose although x-ray diffraction studies show a crystalline structure for the higher nitrogen grades. It may also be made in pelletized form by precipitation from solution in an organic solvent (eg, ethyl acetate). The nitrogen content of nitrocellulose is most important in defining the significant properties of nitrocellulose and the propellants made from it, including energy content and mechanical characteristics. Its effect on thermodynamic properties is shown in Table 8. Nitrocellulose used in propellants is most likely to have a nitrogen content of 12.6 or 13.15% although lower grades (12.0 and 12.2% N) have also been used. Blasting gelatin uses 11–12% nitrogen nitrocellulose, whereas the nitrocellulose in commercial products may vary from 8 to 11.5% nitrogen. Since it is impossible to manufacture nitrocellulose to an exact nitrogen content, the required nitrocelluloses are produced by careful blending (115–120).

Table 8. Explosive and Thermochemical Characteristics of Nitrocellulose as Functions of Its Nitrogen Content[a]

Characteristic	Nitrogen, %		
	12.60	13.15	14.0
brisance sand test, g	45	48	52
heat of combustion, kJ/g[b]	10.1	9.81	9.36
heat of formation, kJ/g[b]	2.58	2.41	2.14
heat of explosion, kJ/g[b]			
H_2O liquid	3.91	4.25	4.77
H_2O vapor	3.58	3.91	4.43
gas volume, cm^3/g[c]			
H_2O liquid	744	722	687
H_2O gas	919	893	853
approximate gas composition, H_2O liquid, %			
CO_2	21	24	
CO	46	44	
H_2	19	17	
N_2	14	15	
estimated explosion temperature, K	3060	3292	
impetus, J/g	998	1074	
ratio of specific heats	1.23	1.225	

[a] Ref. 115–117.
[b] To convert J to cal, divide by 4.184.
[c] At STP.

Nitrocellulose is among the least stable common explosives. At 125°C it decomposes to CO, CO_2, H_2O, N_2, and NO, primarily as a result of hydrolysis of the ester and intermolecular oxidation of the anhydroglucose rings. At 50°C the rate of decomposition of purified nitrocellulose is about 4.5×10^{-6} %/h, increasing by a factor of about 3.5 with each 10°C rise in temperature. Many values have been reported for the activation energy E and Arrhenius frequency factor Z of nitrocellulose. Typical values of E and Z are 205 kJ/mol (49 kcal/mol) and 10.21/s, respectively. The addition of small percentages of mildly alkaline compounds such as diphenylamine or diphenyldiethylurea greatly increases the stability. Compatibility with other substances must be well established before it is combined with new materials (121–128).

Dry nitrocellulose burns rapidly and furiously. It may detonate if present in large quantities or if confined. It is a dangerous material to handle in the dry state because of its sensitivity to friction, static electricity, impact, and heat. Nitrocellulose is always shipped wet with water or alcohol. The higher the nitrogen content the more sensitive it tends to be. Even nitrocellulose with 40% water detonates if confined and sufficiently activated (129).

Manufacture. The batch nitration processes have included the pot process, centrifugal process, the Thompson displacement process, and the mechanical dipper process. Semicontinuous nitration processes have also been developed for military and industrial grades. Cotton linters or wood pulp are nitrated with mixed acid followed by treatment with hot acidified water, pulping, neutralization, and washing. The finished product is blended for uniformity to a required nitrogen content. The controlling factors in the nitration process are the rates of diffusion of the acid into the fibers and of water from the fibers, the composition of the mixed acid, and the temperature (130–135).

Mechanical Dipper Batch Process. All raw materials are moved to the top floor of the plant by pump or conveyor and then proceed by gravity through the processing operations. A battery of mechanically stirred nitrators (dipping pots) are located on a floor above a bottom discharge centrifuge into which the nitrocellulose is dropped after completion of nitration. The acid–cellulose ratio may vary from 20:1 to 50:1, depending on the type of cellulose used and the nitrogen content of the product. The composition of the mixed acid is closely controlled and depends on the grade of nitrocellulose to be prepared and whether the starting material is cotton or wood pulp (136).

The linters or wood pulp or mixtures of the two are passed through picking rolls to form a fluffy mass of material and dried at 80–100°C to less than 1% moisture content to minimize dilution of acid and the possibility of fires in the nitrator. A measured volume of mixed acid corresponding to exactly one dipping charge is drawn into a measuring tank and into the stainless steel nitrator where it is stirred at high speed. The weighed cellulose is rapidly transferred by an operator to the pot where it is immediately drawn below the surface of the acid. The nitration of cellulose occurs very rapidly at first and then slows down as the maximum nitrogen content is approached in 15–20 min. The contents of the pot are then discharged by gravity into the centrifuge where the nitrocellulose is separated from the spent nitrating acid. The nitrators are staggered so that nitrations are carried out in sequence by a small team of workers moving from one to the other as nitrations are completed. Eight to twelve centrifuges may be used for a line of nitrators.

The filter cake from the wringer is washed to remove absorbed acid, transferred to a slurry tank of water and quickly submerged after which the nitrocellulose is pumped to the stabilization operation as a diluted water slurry. Exhaust systems are installed to protect personnel and equipment from acid fumes and water sprays and cyclone separators are used for acid fume recovery before venting to the air.

The procedures for purification are essentially the same for both the batch and continuous processes. The nitrocellulose is completely submerged in water and the contents brought to about 98°C with live steam; the acid boil is continued for 40–60 h for pyrocellulose (12.6 wt % N) to 60–96 h for guncotton (13.15% N). This is followed by several washings and neutral boils, pulping, and treatment with sodium carbonate. After poaching and further washings, the nitrocellulose slurry is screened and blended to the required uniform nitrogen content (137).

The large quantities of water used in this process may contain considerable amounts of acids and suspended solids and thus present a pollution problem.

Semicontinuous Nitration. This procedure introduces new techniques for nitration and wringing but purification procedures are essentially the same. The process increases safety, reduces the personnel involved, provides a substantial reduction in pollutants, and increases the uniformity of the product. It uses a multiple cascade system for nitration and a continuous wringing operation. The cellulose is automatically and continuously fed into the first of a series of pots at a controlled rate. It falls into the slurry of acid and nitrocellulose and is submerged immediately by a turbine-type agitator. The acid is delivered to the pots from tanks at a rate controlled by appropriate instrumentation based on the desired acid to cellulose ratio. The slurry flows successively by gravity from the first to the last of the nitration vessels through under- and overflow weirs to ensure adequate retention time during nitration. The overflow from the last pot is fully nitrated cellulose.

The centrifuge is a horizontal basket designed to operate so that the cake formed on the screen is pushed as an increment from the loading end of the basket to the discharge end by a pusher plate operating on a timed cycle. On completion, the nitrocellulose cake is discharged into water in a slurry tub on a lower floor, and purified by conventional procedures (138–140).

Although the purification techniques used with the semicontinuous system are similar to that for the batch system, much less polluted waste water is produced because much less acid is retained on the nitrocellulose that is discharged from the centrifuge (141–142).

Pentaerythritol Tetranitrate. The synthesis of pentaerythritol by Tollens and Wigand in 1891 was followed shortly thereafter by its nitration to pentaerythritol tetranitrate, PETN. Although the presence of as little as 0.01% occluded acid or alkali greatly accelerates decomposition, specification-grade PETN can be stored up to 18 mo at 65°C without significant deterioration. However, many materials have been found to be incompatible with PETN. The decomposition of PETN is autocatalytic with reported kinetic constants of $E = 196.6$ kJ/mol (47 kcal/mol) and $Z = 6.31 \times 10^{19}$/s. The decomposition products of PETN at 210°C are (wt %): 47.7 NO, 21.0 CO, 11.8 NO_2, 9.5 N_2O, 6.3 CO_2, 2.0 H_2, and 1.6 N_2. The vapor pressure of PETN as a function of temperature is:

$$\log_{10} P_{Pa} = 16.90 - 7380/K \quad (\log_{10} P_{mm\ Hg} = 14.78 - 7380/K)$$

The calorimetrically determined heat of detonation of the unconfined explosive is 6.1 kJ (1.47 kcal/g) and the calculated major products of detonation are (mol/mol PETN) 3.89 CO_2, 4.00 H_2O, 2.00 N_2, 0.22 mol CO, 0.90 $C_{(sol)}$. The detonation rate D

in centimeters per microsecond as a function of density d in grams per cubic centimeter between 0.4–1.65 may be expressed as $D = 3.19 + 3.7\,(d - 0.37)$. The thermal conductivity of PETN is 0.11 W/(m·K) (2.7×10^{-4} cal/(cm·s·°C)). It has a coefficient of linear expansion of $8.3 \times 10^{-5}/°C$, a compressive strength of about 10.3 MPa (1500 psi), a modulus of elasticity of 6895 MPa (1×10^6 psi) and a bulk modulus of 4826 MPa (7×10^5 psi). Additional physical properties are: specific heat 1.13 J/(g·°C) [0.27 cal/(g·°C)]; heat of fusion 318 J/g (76 cal/g); and heat of vaporization 305 J/g (73 cal/g). Maximum electrostatic energy for noninitiation of PETN powder is about 0.2 J (0.05 cal). Other characteristics are listed in Table 5 (143–157).

Pentaerythritol may be nitrated by a batch process at 15–25°C with concentrated nitric acid in a stainless steel vessel equipped with an agitator and cooling coils. The reaction temperature is kept at 15–25°C. The PETN is precipitated in a jacketed diluter by adding sufficient water to the solution at 15–25°C to reduce the acid concentration to about 30%. The crystals are vacuum filtered and washed with water followed by washes with water containing a small amount of sodium carbonate and then cold water. The water-wet PETN is dissolved in acetone at 50°C containing a small amount of sodium carbonate and reprecipitated with water; the yield is about 95%. Impurities include pentaerythritol trinitrate, dipentaerythritol hexanitrate and tripentaerythritol acetonitrate. Pentaerythritol tetranitrate is shipped wet in water–alcohol in packing similar to that used for primary explosives (144,156).

In the Biazzi continuous process, the reactants are continuously fed to a series of nitrators followed by separation of the PETN, water washing, solution in acetone at 50°C, neutralization with gaseous ammonia, and continuous precipitation by dilution with water. The overall yield is more than 95% of theory. The acetone and the spent acid, which is only slightly diluted, are readily recovered (144,157–158).

Pentaerythritol tetranitrate is used as a pressed base charge in blasting caps and detonators, as the core explosive in commercial detonating cord, and in sheet explosives. It is also mixed in various proportions with TNT to form the less sensitive pentolites; eg, PETN 50/TNT 50. It is easily initiated, its responses are reproducible, and it is readily available. However, it has been increasingly replaced by the more stable RDX for most military purposes (159–161).

Nitramines

The four most important nitramines are: cyclotrimethylenetrinitramine (RDX), cyclotetramethylenetetranitramine (HMX), nitroguanidine (NQ), and 2,4,6-trinitrophenylmethylnitramine (tetryl). Tetryl has been increasingly replaced by RDX. Both RDX and HMX are used as high energy explosives and are also incorporated in high performance rocket propellants. Nitroguanidine is employed almost exclusively in gun propellants. The characteristics of these and some less common nitramines are listed in Table 9 (162–163).

RDX and HMX. The properties of RDX and HMX are quite similar; HMX has a higher density and a higher detonation rate, yields more energy per unit volume and has a higher melting point and higher explosion and cook-off temperatures. The vapor pressure of RDX at various temperatures may be calculated from $\log_{10} P_{Pa} = 16.26 - 6785/K$ ($\log_{10} P_{mm\,Hg} = 14.14 - 6785/K$). Its specific heat is 1.26 J/(g·°C) [0.30 cal/(g·°C)], heat of vaporization is 489 J/g (11.7 cal/g) and thermal conductivity is 0.29 W/(m·K) [7×10^{-4} cal/(s·cm·°C)].

582 EXPLOSIVES AND PROPELLANTS

Table 9. Properties of Nitramines[a]

	NQ	EDNA	RDX	HMX	Tetryl
molecular weight	104	150	222	296	287
color	white	white	white	white	yellow
crystal density, g/cm^3	1.72	1.71	1.83	1.90 (beta)	1.73
crystal form	ortho-rhombic		ortho-rhombic	polymorphic	monoclinic
melting point, °C	245 (dec)	175 (dec)	204	286	129.5
hardness, Mohs			2.5	2.3	<1
solubility in water at 20°C, g/100 g	0.4		negligible	negligible	negligible
oxygen balance, % to CO_2	−31	−32	−22	−22	−47
heat of formation, kJ/g[b]	0.950	0.561	−0.277	−0.253	−0.067
heat of combustion, kJ/g[b]	8.35	10.36	9.46	9.43	12.24
heat of detonation, kJ/g[b]	3.01	5.33	5.54	5.67	4.63
gas volume at STP, cm^3/g	1,077	908	780 (calc)	755 (calc)	760
activation energy, kJ/mol[b]	87.5		197	221	161
collision constant, $\log_{10}$/s	7.5	12.8	18.3	19.7	15.4
applicable temperature range, °C		>185	215–300	270–295	210–260
detonation velocity, km/s	8.16	8.24	8.64	9.10	7.56
at density, g/cm^3	1.72	1.66	1.77	1.90	1.73
detonation pressure, GPa[c]	27.3	26.5	33.8	39.3	26.2
at density g/cm^3	1.78	1.51	1.77	1.90	1.73
detonation temperature, K (calc)			2590	2365	2915
explosion temperature at 5 s, °C	275 (dec)	190 (dec)	260	335	260
effect of prolonged storage	negligible	negligible at 50°C	negligible	negligible	negligible at 65°C
relative impact test value, % TNT	200	70	35	35 (beta)	50
friction pendulum	no explosion	no explosion	explodes	explodes	explodes
relative gap test value, % TNT	390	80	40	40	50
static discharge max energy for nonignition, J[b]			0.25	5	
performance, % TNT					
lead block	100	120	155	150	130
ballistic mortar	105	135	150	150	130
sand test	75	120	140	150	125
plate dent test	95	120	135	155	115
specific heat, J/(g·K)[b]			1.26	1.26	0.92
heat of vaporization, J/g[b]			490	368	

[a] For all five compounds: nonhygroscopic at 90% rh at 20°C; vacuum stability at 100°C, mL gas/(g·40 h), less than 1; % weight loss at 100°C, less than 1.

[b] To convert J to cal, divide by 4.184.

[c] To convert GPa to kilobars, multiply by 10.

Both RDX and HMX are white, stable, crystalline solids, somewhat less sensitive to impact than PETN. Both are much less toxic than TNT and may be handled with no physiological effect if appropriate precautions are taken to assure cleanliness of operations. Both RDX and HMX detonate to form mostly gaseous, low molecular weight products with little intermediate formation of solids. The calculated molar detonation products of RDX are: 3.00 H_2O, 3.00 N_2, 1.49 CO_2, and 0.02 CO. RDX has been stored for as long as ten months at 85°C without perceptible deterioration.

At present, HMX is the highest energy solid explosive produced on a large scale, primarily for military use. It exists in four polymorphic forms (164) of which the beta form is the least sensitive and most stable and the type required for military use. The mole fraction products of detonation of HMX in a calorimetric bomb are: 3.68 N_2, 3.18 H_2, 1.92 CO_2, 1.06 CO, 0.97 C, 0.395 NH_3, and 0.30 H_2.

Both RDX and HMX are substantially desensitized by mixing with TNT to form cyclotols (with RDX) and octols (with HMX) or by coating with waxes, synthetic polymers, and elastomeric binders. Most of the RDX made in the United States is converted to Composition B (60% RDX, 40% TNT, 1 pt wax). Composition A3 (RDX 91/wax 9) accounts for the next largest use, whereas HMX is used predominantly in maximum-performance explosives such as PBXN-5 and the octols (164–178).

Manufacture. The two most common processes for making RDX and HMX use hexamethylenetetramine (hexamine) as starting material. The Woolwich or direct nitrolysis process used in the United Kingdom proceeds according to:

hexamine + 6 HNO_3 → RDX + 3 CO_2 + 2 N_2 + 6 H_2O

The Bachmann process, now used exclusively in the United States, is a simplification of a series of complex reactions, which may be summarized as follows:

$$C_6H_{12}N_4 + 4\ HNO_3 + 2\ NH_4NO_3 + 6\ (CH_3CO)_2O \rightarrow 2\ RDX + 12\ CH_3COOH$$

In the Bachmann process an 80–84% yield is obtained, ca 10% of which is cyclotetramethylenetetranitramine (HMX). The Woolwich process gives a 70–75% yield containing only a trace of HMX (179–191).

In the Bachmann process, the reactants are mixed, and the slurry aged to complete the reaction and increase the yield. It is then diluted and simmered to destroy by-products.

The reaction vessels are stainless steel, jacketed, and temperature controlled. A solution of one part hexamine in 1.65 parts acetic acid, and a solution of 1.50 parts ammonium nitrate dissolved in 2.0 parts nitric acid and 5.20 parts acetic anhydride are added continuously to each of two reactors maintained at 65–72°C. The mixture is cycled in a loop in about 15 s from the reactor through a temperature-controlled jacketed reactor leg. The slurry containing RDX overflows continuously into the first of three aging tanks, kept at 65–72°C, and passes from one to the other over a period of about 24 min. The aged slurry is transferred to the first of a series of seven temperature-controlled simmer tanks. Dilution liquor is added to the first tank to produce an acetic acid concentration of about 63% in the simmer system. The hot slurry moves from one tank through the others in a gradual cooling process to hydrolyze undesirable by-products and precipitate the RDX from solution.

The RDX-acetic acid slurry is filtered and water-washed and the spent acetic acid is processed for recovery. The RDX is recrystallized from cyclohexanone. The particle size distribution must be carefully controlled to produce castable slurries of RDX and TNT with acceptable viscosity.

A continuous process for medium-scale production of RDX has also been developed by Biazzi (192) based on the Woolwich process. Hexamine and nitric acid are continuously fed to a series of jacketed, Freon-cooled, stainless steel nitrators. The mixture from the last nitrator is subjected to a stabilization process in a series of stainless steel decomposers. The unstable by-products are decomposed to form nitrous gases which are converted to 55% nitric acid in a gas absorption unit after separation of the RDX from the spent acid on a continuous vacuum filter. The RDX is water-washed and continuously stabilized to remove internal acidity.

A modification of the Bachmann process used to make RDX with the same starting materials and in similar equipment, is employed for the manufacture of HMX. The reaction temperature is lower (44 ± 1°C as compared to 68°C for RDX) and the raw materials are mixed in a two-step process. The yield of HMX per mole of hexamine is about 55–60%, as compared to 80–85% in the manufacture of RDX (193–196).

A typical HMX batch process starts with a reactor which contains a heel of acetic acid. The proper amounts of hexamine in glacial acetic acid, ammonium nitrate in nitric acid, and acetic anhydride are added in stages. The HMX slurry is transferred to the aging tanks, held for 30 min at 44°C, and pumped to a simmer tank where sufficient dilution liquor is added to reduce the concentration to 80% acetic acid. After dilution the temperature is raised to 110°C to decompose undesirable by-products and to improve the filtering characteristics of the HMX.

The slurry is filtered at about 60°C to retain RDX in solution. The filtered product is almost 99% HMX. The crystals are washed with cold water and crystallized from acetone, cyclohexanone, or both, depending on the particle distribution desired. Recrystallization also converts the more sensitive alpha crystals to the higher density beta form and reduces occluded acid to less than 0.02%. The RDX is recovered from the spent acid which is reclaimed (197–199). The pollutants resulting from these processes and from the manufacture of the acids used and the compositions employing RDX and RMX are primarily RDX, HMX, TNT, nitrate–nitrite nitrogen, and various compounds that increase the biochemical and chemical oxygen demand of the waste water (200–201).

Nitroguanidine. Nitroguanidine [556-88-7], NQ, has been used to some extent as an industrial explosive but not as a military explosive because of its relatively low energy content and difficulty of initiation. Although its detonation rate at comparable densities is almost as high as that of RDX, it has an anomalously low detonation energy and pressure, indicating a decomposition mechanism different from that of conventional explosives. It is approximately as effective as TNT. The high nitrogen content of the products of combustion and lower flame temperature per unit of energy accounts for its widespread use in multibase gun propellants to reduce barrel wear, muzzle blast, and flash. Many guanidine derivatives have been prepared as possible substitutes for nitroguanidine in propellants including triaminoguanidine [2203-24-9], nitroaminoguanidine [18264-75-0], and trinitroethylguanidine [8065-53-0] (202–206).

Nitroguanidine is stable and nonhygroscopic. It is produced in the alpha crystalline form with a bulk density of about 0.2 g/cm^3. The crystals are needlelike and often hollow and are about five μm in diameter and 15 μm long. Nitroguanidine appears to act as a stabilizer in nitrocellulose propellants by forming an addition compound with oxides of nitrogen. It is less sensitive than TNT to impact, friction, and shock. However, its brisance, power, and blast characteristics are similar to that of TNT. Nitroguanidine has been studied as a source of explosive energy in systems where low detonation pressures and low detonation rates are required. The rate of detonation at infinite diameter D_i in kilometers per second as a function of density d in g/cm^3 may be expressed by $D_i = 1.44 + 4.015\ d$. Nitroguanidine may be dead-pressed at densities greater than 1.63 g/cm^3. The estimated detonation products are (wt %): 38 N_2, 25 CO, 20 CO_2, 16 H_2O, and 1 H_2. Additional properties of nitroguanidine are presented in Table 10 (208–213).

Manufacture. Nitroguanidine may be made by several methods including the Welland process, the aqueous fusion process, the urea–ammonium nitrate process, and the Marquerol and Lorriette process. In all these processes guanidine nitrate is intermediate which is then dehydrated with sulfuric acid. When used in propellants, the average particle size of nitroguanidine has to be carefully controlled (214).

In the Welland process, calcium cyanamide is first made from calcium carbonate and converted to cyanamide by acidification of a water slurry. The dimer dicyandiamide which is then formed by filtration and evaporation of the filtrate is fused with ammonium nitrate to form guanidine nitrate. Dehydration with 96 percent sulfuric acid gives nitroguanidine. It is precipitated from the mixture by dilution and is filtered, washed, and dried. In the aqueous fusion process, calcium cyanamide is fused with ammonium nitrate in the presence of some water. The calcium nitrate produced is removed by precipitation with ammonium carbonate or carbon dioxide. The filtrate contains the guanidine nitrate which is recovered by vacuum evaporation and converted to nitroguanidine. Both operations can be run on a continuous basis. In a pilot process for producing guanidine nitrate molten urea is fused with ammonium nitrate in the presence of a silica gel catalyst. The guanidine nitrate is separated from the by-products by liquid–gas separation and its solubility in water and is converted to nitroguanidine by dehydration with sulfuric acid (215–219) (see Cyanamides).

In the Marquerol and Loriette process, nitroguanidine is obtained directly in about 90% yield from dicyandiamide by reaction with sulfuric acid to form guanidine sulfate followed by direct nitration with nitric acid (220).

Tetryl. 2,4,6-Trinitrophenylmethylnitramine (tetryl) has been used in pressed form mostly as a booster explosive and as a base charge in detonators and blasting caps because of its sensitivity to initiation by primary explosives and its relatively high energy content. It is highly stable losing virtually no weight on prolonged storage at 80°C. The calculated molar fraction products of detonation at a density of 1.70 g/cm^3 are: 4.16 $C_{(sol)}$, 2.66 CO_2, 2.50 H_2O, 2.50 N_2, and 0.17 CO. Its specific heat is 0.92 J/(g·K) [0.22 cal/(g·°C)] and thermal conductivity 0.25 W/(m·K) [6 × 10^{-4} cal/(cm·s·°C)]. Additional properties are presented in Table 9 (221–226).

In a batch process for making tetryl, dimethylaniline is dissolved in an excess of concentrated sulfuric acid (3 or 4 to 1) at 20–30°C to give dimethylaniline sulfate. The mixture is nitrated with mixed acid (67% nitric/16% sulfuric), first to 2,4-dinitrodimethylaniline at about 100°C, and finally to tetryl. The crude product is filtered, washed repeatedly with water, and then dissolved in acetone and recovered by evaporation of the acetone and filtration.

Continuous processes for the production of tetryl have also been developed (226–231).

Table 10. Properties of Nitroaromatic Explosives

Property	TNT	PA	AP	DATB	TATB	HNS	HNAB
molecular weight	227	229	246	243	258	450	452
color	beige	yellow	yellow	yellow	yellow	yellow	orange
density, g/cm^3	1.65	1.76	1.72	1.84	1.94	1.74	1.79
crystal form	ortho-rhombic	ortho-rhombic	rhombic	triclinic—ortho-rhombic	tri-clinic	ortho-rhombic	
melting point, °C	80.75	122.5	271 (dec)	286 (dec)	440 (dec)	415 (dec)	215
hygroscopicity at 20°C, % at 90% rh	0.03	0.05	0.1				
solubility in H$_2$O at 20°C, g/100 g	0.01	1.2	1.1				
oxygen balance, % to CO$_2$	−74	−45	−52				
heat of formation, kJ/g^a	0.293	0.938	1.56	0.502	0.599	−0.130	−0.536
heat of combustion, kJ/g^a	15.02	11.17	12.09				
heat of detonation, kJ/g^a	4.23	4.31	4.27	3.54	5.02	5.94	6.15
specific heat, J/(g·K)^a	1.38	1.09				1.05	1.68
heat of fusion, J/g^a	98.3	85.4					
heat of vaporization, J/g^a	339	385					
heat of sublimation, J/g^a	447			577	652	397	
thermal conductivity, W/(m·K)	0.54	0.26		0.25			
coefficient of linear expansion, °C % × 10^{-3}/°C	6.7			3.9		9.2	8.0
vapor pressure, kPa^b				45 (108°C)	220 (177°C)	50 (206°C)	
gas volume at STP, cm^3/g	710	740	680	625^c	651^c	590	
activation energy, kJ/mol^a	143	245		193	250	126	
collision constant, log$_{10}$/s	11.4	22.5		15.1	19.5	9.2	
detonation velocity, km/s	6.94	7.35	7.05	7.59	7.90	7.12	7.25
at density, g/cm^3	1.64	1.71	1.60	1.79	1.89	1.70	1.77
detonation pressure, GPa^d	18.9	26.5		25.9	29.1	26.2	
at density, g/cm^3	1.64	1.76		1.79	1.89	1.70	
vacuum stability at 100°C, cm^3 gas per g per 40 h at STP	<1	<1	<1	<1	<1	<1	<1
wt loss at 100°C in 48 h, %	<1	<1	<1	1% h	<1	<1	

Table 10. (continued)

Property	TNT	PA	AP	DATB	TATB	HNS	HNAB
explosion temperature, 5 s, °C	458 (dec)	320 (dec)	320 (dec)	(at 260°C)	315 (dec)		
relative impact test value, % TNT	100	100	120	>200	150	>200	>200
relative gap test value, % TNT	100		80		160		
relative energy output, % TNT							
lead block	100	100	100				
ballistic mortar	100	100	100				
sand test	100	110	85				
plate dent test	100	105	85	120	120	120	

[a] To convert J to cal, divide by 4.184.
[b] To convert kPa to mm Hg, multiply by 7.5.
[c] Calculated.
[d] To convert GPa to kilobars, multiply by 10.

Nitroaromatics

The commonly used nitroaromatic explosives contain three NO_2 groups, generally in the 1, 3, and 5 positions. Aromatics are most often nitrated to the trinitro stage with mixed acid. Further nitration is difficult, and aromatics with four or more nitro groups attached to the ring tend to be relatively unstable. The most extensively used explosive is trinitrotoluene (TNT); however, hexanitrostilbene (HNS), hexanitroazobenzene (HNAB), and di- and triaminotrinitrobenzene (DATB and TATB) have found increasing application because of their low sensitivity to impact, shock, and friction and their excellent stability at elevated temperatures. Ammonium picrate (AP) has been used in armor-piercing gun projectiles because of its insensitivity to impact and shock. The properties of the more commonly used nitroaromatics are in Table 10 (232–233).

Trinitrotoluene. α-2,4,6-Trinitrotoluene (TNT) is very stable and may be stored indefinitely at temperate conditions without deterioration. It does not show any evidence of significant decomposition even after having been cycled more than fifty times through the liquid–solid phases. The decomposition mechanism of TNT at elevated temperatures (200°C) is very complex, producing at least 25 different compounds as well as large amounts of undefined polymeric material. Trinitrotoluene is nonhygroscopic and relatively insensitive to impact, friction, shock, and electrostatic energy. It has been fired in high-acceleration gun projectiles with reported premature rates of less than one in a million. Its characteristics change only slightly at very low temperatures. Hot liquid TNT when confined is much more sensitive to impact, approaching the sensitivity of mercury fulminate. The vapor pressures of TNT and a number of related nitrotoluenes are shown in Table 11.

Like most other nitroaromatics, TNT is toxic. Its solubility in water (0.013 g/100 g at 20°C) is high enough to destroy aquatic life. The solidification point of military-grade TNT is about 80.2°C and that of pure TNT is 80.75°C. The viscosity of TNT is 0.008 Pa·s (8 cP) at 99°C. The viscosity of TNT slurries with RDX and other explosives has been studied intensively because of its importance in the cast loading of

Table 11. Vapor Pressure of Nitrotoluenes[a]

Compound	Constants[b]		
	A	B	C
TNT (solid)	17.5	8.42	13.76
(liquid)	14.3	13.3	51.62
p-nitrotoluene (liquid)	10.9	3.9	6.42
(solid)	10.2	2.6	
2,4-dinitrotoluene	9.1	1.5	−6.70

[a] Ref. 234.
[b] Constants are for the equation $\log_{10} P_{Pa} = A - 1000\, B/K - C \log_{10} K/1000$; to convert $\log_{10} P_{Pa}$ to $\log_{10} P_{mm\,Hg}$, subtract 2.122 from A.

projectiles. The enthalpy, ΔH, and density d of TNT as functions of temperature (t, °C) are represented by the following equations (235):

$$\Delta H\ (25°C\ to\ mp) = 0.045 + 0.24635\, t + 4.205 \times 10^{-4}\, t^2$$

$$d\ (g/cm^3)\ (83°–120°C) = 1.5446 - 1.016 \times 10^{-3}\, t$$

The following equations (236) may be used to calculate the maximum detonation rate D of pressed TNT as a function of density d (g/cm³):

Density range	Equation
0.9–1.53	$D_{inf} = 1872.7 + 3187\, d$
1.53–1.64	$D_{inf} = 6762.5 + 3187\, (d - 1.5342) - 25102\, (d - 1.5342)^2 + 115056\, (d - 1.5423)^3$

The major detonation products of heavily confined TNT as experimentally determined in a calorimetric bomb are: 3.65 $C_{(sol)}$, 1.98 CO, 1.60 H_2O, 1.32 N_2, 0.46 H_2, 0.16 NH_3, and 0.10 CH_4 (234–246).

Bombs and projectiles are filled with steam-melted TNT by the casting process. Melted TNT also serves as the liquid carrier for RDX, HMX, aluminum, ammonium nitrate, and other high-melting ingredients to form a wide range of castable slurries. Since TNT expands about 10–12% on liquefaction and contracts on solidification, cracks, and cavities tend to form in the cast TNT on cooling. Special techniques used to prevent this include slow-programmed cooling, hot probing, and temperature cycling following casting or compositional modifications to prevent crack and cavity formation (245,247–249).

Manufacture. Trinitrotoluene is made by batch or continuous process. Both processes are safe operations when conducted under proper conditions (250). They are relatively low cost, aided by the fact that the products are only slightly soluble in the mixed nitration acids and are, therefore, readily separable. The low solidification points of the intermediate and final products permit pumping them as oils from one stage to the next at readily attainable temperatures. Toluene is nitrated in a three-stage operation by using increasing temperatures and mixed-acid concentrations to successively introduce nitro groups to form mononitrotoluene (MNT), dinitrotoluene (DNT), and trinitrotoluene. Numerous other compounds are also formed including unsymmetrical isomers of TNT, and oxidation products such as tetranitromethane, nitrobenzoic acid, nitrocresol, and partially nitrated toluenes. Unless these impurities are removed, the TNT may be unstable at elevated temperatures and may also form low melting eutectics that separate over a period of time. The TNT is sellited or treated with 16% sodium sulfite which reacts preferentially with the unsymmetrical meta isomers to form the water-soluble sodium salts of the corresponding sulfonic acids. The formation of isomeric impurities and reaction by-products together with the losses in the selliting operation reduce the theoretical yield of alpha TNT to about 85–88% of theoretical (251–256).

Batch Process. The steps used in the three successive nitrations are similar and include acid mixing, addition of the oil, digesting (cooking) the reaction to completion, cooling and settling the mix, and separating the oil from the acid. The nitrators are made of stainless steel plate and are equipped with concentric cooling water coils, stainless steel propellers, thermometer wells, and fume ducts. Drowning tubs are provided near and below each nitrator. For a mononitration the nitrator is charged with mixed acid at 46–52°C prepared by fortifying the spent acid from the dinitration operation with 60% nitric acid. Toluene is transferred by gravity flow over a 20 min period during which the temperature is controlled at 57°C. The charge is digested and agitated for 15 min and allowed to settle and cool. The spent acid at the bottom is recovered. The entire operation takes about 65 min. The dinitration stage is similar. Spent acid from the trinitration operation is fortified with nitric acid in the dinitrator. The temperature is adjusted to 77°C and the mononitrotoluene is added over a 20 min period at 85°C. The charge is then allowed to settle and cool after which the dinitrotoluene oil is delivered under air pressure to the trinitrator as required. For trinitration the acid consists of oleum and a mix of concentrated sulfuric and nitric acids at 91°C. Dinitrotoluene is fed into the mixed acid over a period of 45 min at 91°C while the mix is agitated. The temperature is increased to 113°C at which the charge is digested and agitated for 50 min. The TNT is separated and moved by gravity to the purification operation.

Trinitrotoluene is purified by crystallization and water washing, neutralization, selliting, and filtering, a process which takes about 25 min. The spent sellite solution is deep red (red water) almost black, and contains the sulfonated derivatives of the unsymmetrical TNT isomers, and sodium sulfite, sulfate, nitrite, and nitrate. It is never discharged into streams because it is highly toxic. The TNT cake on the filter is washed, reslurried, pumped to the melter, where it is liquified with live steam. After decanting the excess water, the melted TNT is dried by bubbling hot air at about 100°C through the charge. After drying, molten TNT is transferred to a cast-iron, water-cooled, brass-faced flaker. After flaking, the product is ready for packaging and shipping (232,252–253,257–259).

Continuous Process. Continuous processes used in many countries are based on nitration of toluene with the organic phase flowing countercurrent to the acid phase. It has the advantages of lower costs, electronic controls, and fewer pollutants. A typical example is the Canadian Industries Limited (CIL) process. A series of eight nitrator–separator reactors are used to conduct the nitration in six stages. An agitator is fitted into a draft tube at the center of the nitrator to assure the required interphase mixing and circulation of the liquids over the cooling coils in the nitrator. The differential head created between the liquid on either side of the draft tube causes continuous fluid flow and internal recycling to occur between nitrator and separator. All the toluene is introduced in the first stage where the mononitration and a small amount of dinitration occur. The nitrobody stream moves from reactor to reactor through the separators, becoming increasingly nitrated in the process. Oleum (100% sulfuric acid plus SO_3) is added to the last stage and moves toward the first, decreasing in strength.

Nitric acid is added to each stage as needed to maintain the required mixed acid composition. Mononitration occurs in stage 1; dinitration occurs primarily in stages 2 and 3 and trinitration in the remaining stages. The temperature and the acid concentration of the mix in the reactors is progressively increased. Nominal temperatures for the six nitration stages (eight reactors) are 55, 60, 70, 80, 85, 90, 95, and 100°C, respectively.

The crude molten TNT is treated in a mixer-settler with countercurrent water washes, neutralized with sodium ash, sellited and separated from the sellite solution. It is then given a final wash, dewatered, dried, flaked, and packed as in the batch process (248,256,260–269).

Wastewater from TNT manufacture is contaminated with organic nitrobodies, nitrates, sulfates, and sodium sulfite and nitrite. Yellow water consisting of the washings formed by treating TNT with water to remove acid is recycled. Pink water is the washings produced in the manufacturing and loading plants. It contains dissolved TNT, dinitrotoluene, and isomers of TNT. The washings initially are colorless but turn pink if neutral or basic and exposed to sunlight. The dissolved products are removed by filtration through diatomaceous earth and activated carbon. Red water may be sold to paper mills for use in kraft paper products or concentrated by evaporation and incinerated to give crude sodium sulfate. Considerable progress has been made in the abatement of pollution resulting from the manufacture of TNT (246,259,270–272).

Picric Acid and Ammonium Picrate. Picric acid (2,4,6-trinitrophenol) (PA) is of historic interest as the first modern high explosive to be used extensively as a burster in gun projectiles. It was first obtained by nitration of indigo, and used primarily as a fast dye for silk and wool. It was not until Nobel's invention of the blasting cap and its use to detonate picric acid that its explosive properties were explored in detail. It offered many advantages: when compressed, it was used as a booster for other explosives, and when cast (melting point 122.5°C) it served as a burster in shell; it was stable, insensitive, nonhygroscopic, relatively nontoxic, and of high density when cast and could be made economically by simple nitration. Picric acid has an energy content somewhat greater than that of TNT and a higher detonation rate. Its calculated molar detonation products at a density of 1.76 g/cm^3 are: 3.16 $C_{(sol)}$, 2.66 CO_2, 1.50 N_2, 1.50 H_2O, and 0.18 CO. A major disadvantage is its tendency to form sensitive salts with calcium, lead, zinc, and other metals. Furthermore, its high melting point necessitated the use of superheated steam in shell-filling plants. Picric acid is no longer used as a military explosive.

It may be made by gradually adding a mixture of phenol and sulfuric acid at 90–100°C to a nitration acid containing a small excess of nitric acid. The picric acid crystals are separated by centrifuging, washed, and dried. The wash water is reused to decrease losses owing to the water solubility of the picric acid. A yield of about 225% of the weight of phenol is commonly obtained.

Ammonium picrate (explosive D) (AP) is at present used only where a high explosive is required that is particularly insensitive to shock. It has been employed in pressed form primarily as a burster in naval projectiles for armor penetration. The compound is very stable and does not form sensitive metallic salts. Its explosive characteristics are comparable to those of TNT. It is prepared by neutralizing a solution–suspension of picric acid in hot water with aqueous or gaseous ammonia. The salt crystallizes on cooling, is filtered, washed with cold water, and dried. Additional properties of picric acid and ammonium picrate are included in Table 10 (273–279).

Insensitive Nitroaromatic Explosives. The post World War II requirements for the use of temperature-resistant, insensitive explosives led to the synthesis of very stable compounds, exceptionally insensitive to impact and shock, which can withstand exposure at higher temperatures without initiation or exudation. They are too expensive to be widely used, although a number have been used for special applications (see Table 12). Physical properties are listed in Table 10 (280).

Hexanitrostilbene. Hexanitrostilbene, HNS, is used as a crystal-modifying additive in cast composition B. In conjunction with hexanitroazobenzene [*19159-68-3*] (HNAB), it has been used in an aluminum-sheathed mild detonation fuse designed to be stored up to 150–175°C. It may be prepared by adding a solution of TNT in tetrahydrofuran and methanol at 5°C to aqueous sodium hypochlorite. To this mixture a 20% solution of trimethylamine hydrochloride is added at 5–15% °C. Hexanitrostilbene precipitates, and is filtered and washed with methanol and acetone (yield ca 50%) (281–288).

Aminonitrobenzenes. Di- and triaminotrinitrobenzenes are highly insensitive to impact and are able to withstand prolonged exposure at elevated temperature. 1,3-Diamino-2,4,6-trinitrobenzene (DATB) is obtained by nitration of *m*-phenylenediamine at 120–140°C. 1,3,5-Triamino-2,4,6-trinitrobenzene (TATB) may be made by nitration of trichlorobenzene with mixed acid followed by treatment with ammonia in alcohol. It is used in molding powders employing techniques similar to those for making polymer-coated, plastic-bonded explosives (PBX) (289–293).

Safety and Environmental Considerations

In view of the catastrophic effects and the increasingly severe legal and economic implications of a disastrous explosion, a large amount of effort is devoted to the safety of operations, while keeping control costs at an acceptable level. This work has been given increased impetus in the United States by the military plant modernization now

Table 12. Low Sensitivity Explosives for High Temperature Use

Compound	Code	CAS Registry No.	Mp, °C	Vapor pressure, MPa
hexanitrostilbene	HNS	[*20062-22-0*]	415 (dec)	5.2 (206°C)
1,3-diamino-2,4,6-trinitrobenzene	DATB	[*1630-08-6*]	286 (dec)	4.5 (108°C)
1,3,5-triamino-2,4,6-trinitrobenzene	TATB	[*3058-38-6*]	440 (dec)	1.3 (150°C)
diaminohexanitrobiphenyl	DIPAM	[*17215-44-0*]	306 (dec)	
2,4,8,10-tetranitro-5*H*-benzotriazolo[2,1-a]benzotriazol-6-ium, hydroxide, inner salt	TACOT (Dupont)	[*25243-36-1*]	can withstand temperature to 316°C up to 8–9 h	
2,6-bis(picrylamino-3,5-dinitropyridine)		[*38082-89-2*]	360	

a To convert MPa to atm, divide by 0.101.

underway and the development of a variety of new or improved processes for making explosives (294–300).

Many governmental regulations control the classification, shipping, and handling of explosive materials. The publications in the field of safety have increased greatly and a number of important symposia have been held on the subject, including The Annual Explosive Safety Seminars conducted by the Explosives Safety Board of the United States Department of Defense (297–299,301–319).

Standard safety procedures include attention to such matters as intraline distances to minimize explosion propagation, cleanliness, nonsparking equipment and explosion-proof motors, and location in selected areas away from heavily populated areas. The development of continuous processes for explosive manufacture which utilize extensive automatic controls and minimize in-line quantities has contributed significantly to decreasing the hazards of operations (300,320–321).

Highly detailed and systematic methodologies have been developed to identify possible failure modes and to quantify their probability of occurrence and impact on operations, including hazards analyses, failure analyses, and risk or reliability assessment. They are characterized by the step-by-step detailed examination of a process, often using fault trees to specify and quantify the likelihood of undesirable events. Experiments are conducted simulating the actual application and environment of the explosives in the process to obtain valid quantitative data for calculation. The estimated probabilities of an accident occurring for each of the operations in a process are generally computerized to enable an assessment to be made of the overall hazard of the process and an attack launched on the most dangerous elements (322–328).

The concept of TNT equivalency has been increasingly used for evaluating the magnitude of an accidental explosion which may occur at some stage of an operation involving explosive materials. The TNT equivalency value, expressed as percent, is the relative weight of TNT, when tested as a hemispherical charge in the ground burst mode, which yields the same peak pressure or positive impulse as is produced at a given distance by the explosive or explosive system tested. The information obtained enables the designer to convert the data in barricade design manuals, which are usually presented in terms of surface bursts of TNT, to the explosive as it occurs in process operations. Some typical TNT equivalency values are presented in Table 13.

A great deal of experimental work has been done to quantify the hazards of an unwanted explosion. The vulnerability of structures and people to shock waves and fragment impact has been well established. This effort has also led to the design of protective structures superior to the conventional barricades and which permit considerable reduction in allowable safety distances. In the United States most of this work has been government sponsored and is reported in government manuals and the literature (296,318,335–344).

Environmental Effects. Federal Executive Orders 11507 and 11514 and the National Environmental Policy Act, PL 91-90 mandated rigorous federal standards for permissible quantities of contaminants added to the environment. These were further reinforced by strong state and municipal limitations on permissible pollutants (345–357).

Pollutants from explosives are primarily produced during manufacture of explosives and the acids used in nitration. They may also be produced during incorporation in munitions or industrial explosives and in clean-up and disposal operations.

Table 13. TNT Equivalency Values[a]

Explosives	Pressure, %	Impulse, %
primary		
lead azide (RD 1333)	30	35
lead styphnate	80	65
tetrazene	25	30
secondary		
nitroglycerin	135	120
guanidine nitrate, dry, as poured	100	65
nitroguanidine, dry, as poured	105	85
RDX slurry, settled	155	155
mixtures		
composition B	125	110
composition A5, in boxes	120	140
black powder	50	50
propellant		
composition N5, 10% moisture	75	70
composition N5, carpet rolls	100	80
M1 propellant, in shipping drum	110	90
M26 propellant, in shipping drum	120	100

[a] Refs. 329–334.

The wastewater effluent from a plant may be very acidic or very basic and high in oxygen demand, dissolved and particulate solids, soluble nitrates and sulfates, and oil and grease. Significant improvement may be made by good housekeeping practices, close process control, the separation of contaminated from clean process waters, and the application of proven techniques for pollution abatement (346,349–350,355,357–359).

Many procedures are being studied and used to reduce the level of air, water, and solid waste contaminants. Specific approaches and principles of attack for decreasing water contamination include: (1) minimizing the quantity of water leaving the plant by recycling process and cooling waters; (2) segregating and treating highly polluted waters before dilution; (3) using settling reservoirs for water treatment and removal of suspended particles by sedimentation; (4) centrifuging to remove suspended solids; (5) employing ion-exchange resins to concentrate pollutants; and (6) biologically denitrifying nitrates under anaerobic conditions.

Other procedures include the use of evaporators for handling small volumes of water and reverse osmosis (qv) to separate salt concentrations from waste waters. Both are usually uneconomical except in special situations. Table 14 lists the most common types of pollutants found in the manufacture of explosives, their effects, and various procedures for their reduction (346,350,359–364).

APPLICATIONS

Military

The single-component explosives most commonly used for military compositions are TNT, RDX or HMX, nitrocellulose, and nitroglycerin. The last two are used almost

Table 14. Pollution from Explosives Production[a]

Pollutant	Effect	Present and possible methods of reduction
Water soluble		
acids	toxic, corrosive	neutralization with limestone and landfill with solid products, biodenitrification of HNO_3, recycling and recovery
nitrates	toxic, increased solids content, eutrophication	biodenitrification, ion exchange
sulfates	increased solids content, odorific at anaerobic conditions	ion exchange, reverse osmosis, precipitation as the calcium or barium salts
phosphates	eutrophication	precipitation as calcium or rare earth phosphates
acetates and organic esters	toxic, increases dissolved oxygen demand, increased acidity	biodegradation, neutralization, incineration
nitrobodies (pink water)	toxic, discolored water	carbon absorption, polymer resin absorption, biodentitrification, electrolytic oxidation, catalyzed ozonolysis
red water	toxic, discolored water	concentration and sale, incineration, reduction and carbonation
Solid wastes		
propellants and explosives	dangerous, may be toxic, cannot be landfilled or open burned	fluidized-bed incineration, composting
inert contaminants	may be toxic, unsightly	dual-chamber or air curtain incineration, composting
process sludges	hazardous	incineration
contaminated activated carbon	pollutant if burned	thermal regeneration in indirectly heated rotary furnace, replacement with polymeric materials, solvent regeneration, use of powdered activated carbon followed by pyrolytic regeneration

[a] Ref. 361.

exclusively to make propellants (see under Propellants). Production of TNT far exceeds that of any other explosive. It is used as manufactured, as a base in binary slurries with other high melting explosives, or in ternary systems generally containing a binary mix and aluminum. The properties of formulations based on TNT are presented in Tables 15 and 16.

Munitions are filled with TNT and TNT-based slurry mixes by the melt-cast process. The compositions are liquified with plant steam to form a low viscosity fluid which is poured into munitions. Binary mixes are made by adding the high melting component to the liquified TNT. Composition B is the most widely used binary explosive and many studies have been made of its casting characteristics and its performance. It is made by adding water-wet RDX to TNT at 95–100°C, and then decanting excess water and evaporating the remainder. Other compounds may be added to decrease sensitivity and exudation and increase the mechanical strength of the cast.

Table 15. Explosive Mixtures Based on Ternary Systems Containing TNT and Aluminum[a]

Property	HBX 1	HBX 3	H 6	Tritonal 80/20	Minol 2	HTA 4
density, as cast, g/cm^3	1.70	1.86	1.75	1.77	1.65	2.00
heat of formation[b], kJ/g[c]	0.096	0.092	0.054	0.026		
heat of combustion[b], kJ/g[c]	16.24	18.80	16.61	18.74	13.22	
heat of detonation[b], kJ/g[c]	3.80	3.71	3.86			
gas volume[b], cm^3/g at STP	758	491	723			
detonation velocity, km/s	7.22	6.92	7.19	6.70	6.20	7.64
at density, g/cm^3	1.75	1.86	1.71	1.72	1.77	2.00
detonation pressure, GPa[d]			23.7	18.9		
at density, g/cm^3			1.73	177		
explosion temperature, 5 s, °C			275	470	260	300
relative impact test value, % TNT	75	70	75	75	95	105
relative gap test value, % TNT	80	90	75	80	120	
rifle bullet, uneffected, % TNT	75	80	80	100	55	40
compressive strength[b]						
stress at rupture, MPa[e]		28.9	24.8	14.5		23.4
compression at rupture, %		0.35	0.35	0.30		0.30
modulus of elasticity, kPa[f]		14.5	11.0	7.9		13.1
work to produce rupture, J/cm^3		0.07	0.06	0.02		0.04
tensile strength[b]						
stress at rupture, MPa[e]		2.8	2.6	1.7		2.1
elongation at rupture, %		0.02	0.20	0.03		0.02
modulus of elasticity, kPa[f]		14.5	12.4	10.7		12.4
work to produce rupture, J/cm^3		0.25	0.25	0.33		0.16
shear strength[g]						
stress at rupture, MPa[e]		7.58	5.86	5.00		7.92
charpy impact strength, J		0.16	0.17	0.16		0.18
coefficient of linear expansion, % × 10^{-3}°C	4.4	5.4	5.7	8.7		7.2
thermal conductivity, W/(m·K)	0.41	0.71	0.46	0.46		0.69
specific heat, J/(g·K)[c]	1.05	1.05	1.13	1.00		1.34
relative energy output, % TNT						
lead block					165	
ballistic mortar	135	110	135	130	145	
sand test	100		105	95	95	
fragmentation velocity		105	120	100		115

[a] For the compounds listed, vacuum stability at 100°C, cm^3 gas per g per 40 h, less than 1; friction pendulum test, no explosion.
[b] Molecular weight assumed to be 100 g.
[c] To convert J to cal, divide by 4.184.
[d] To convert GPa to kilobars, multiply by 10.
[e] To convert MPa to psi, multiply by 145.
[f] To convert kPa to atm, divide by 101.3.
[g] At load rate of 0.13 cm/min at 20°C.

Composition B or minor modifications are used in loading projectiles and warheads and as the starting material for making aluminized explosives. Other binary mixtures include the octols (HMX + TNT), cyclotols (RDX + TNT), pentolites (PETN + TNT), tetrytols (tetryl + TNT), amatols (ammonium nitrate + TNT), and picratols (ammonium picrate + TNT) (365–375).

Table 16. TNT Based Mixtures Without Aluminum

Property	Composition B	Cyclotol 75/25	Octol 75/25	Pentolite 50/50	Baratol 67/23	Amatol 80/20	Amatex
density, g/cm^3	1.74	1.77	1.81	1.71	2.52		1.61
heat of formation, kJ/g[a]	−0.042	−0.126	−0.109	1.01	2.97	2971	
heat of combustion, kJ/g[a]	11.67	10.98	11.20	6.48		4.19	
heat of detonation, kJ/g[a]	5.18	5.12	6.56	5.10	3.09	4.10	
gas volume at STP, cm^3/g		865		815		860	
activation energy, kJ/mol[a]	137					171	343
collision constant, log$_{10/s}$	13.3					15.5	
detonation velocity, km/s	7.90	8.17	8.38	7.52	5.00	5.20	6.83
at density, g/cm^3	1.69	1.71	1.80	1.64	2.53	1.60	1.68
detonation pressure, GPa[b]	25.9	31.3	34.3	28.0	14.0		24.0
at density, g/cm^3	1.72	1.74	1.84	1.68	2.53	1.60	1.68
detonation temperature,[c] K	2,750	2,710	2,580				
specific heat ratio	2.94	2.95	2.98				
vacuum stability at 100°C, cm^3 gas per g per 40 h	<1	<1		1–3	<1	<1	<1
relative impact test value, % TNT	60	40	50	45	120	85	100
relative gap test value, % TNT	60	70	60	45	255	150	105
explosion temperature, 5 s, °C	250	270	350	220	385	280	240
compressive strength[e]							
stress at rupture, MPa[d]	12.1	11.6	10.3	14.5	34.5	11.0	
compression at rupture, %	0.24	0.12	0.20				
modulus of elasticity, kPa[f]	8.1	20.9	9.3				
work to produce rupture, J/cm^3 [a]	0.02	0.01	0.01				
tensile strength[e]							
stress at rupture, MPa[d]	0.65	1.55	1.00				
elongation at rupture, %	0.01	0.11	0.01				
modulus of elasticity, MPa × 10^{-3} [d]	10.9	1.9	10.8				
work to produce rupture, J/cm^3 [a]	0.03	0.01	0.05				
shear strength[e]							
stress at rupture, MPa[d]	5.2	4.3	5.3			4.8	
charpy impact strength, J[a]	1.5	1.6	1.8				
coefficient of linear expansion, %/°C × 10^{-3}	7.8	6.8	7.0		4	6	
thermal conductivity, W/(m·K)	0.5					0.5	0.4
specific heat, J/(g·K)[a]	1.3					0.8	1.6
relative energy output, % TNT							
lead block	130			120		125	
ballistic mortar	135	140		135		130	110
sand test	120	120		125		80	

[a] To convert J to cal, divide by 4.184.
[b] To convert GPa to atm, divide by 10^{-4}.
[c] Calculated.
[d] To convert MPa to psi, multiply by 145.
[e] At load rate of 0.13 cm/min at 20 °C.
[f] To convert kPa to atm, divide by 101.3.

Ternary mixes with aluminum are made by first screening the finely divided aluminum, adding it to a melted RDX–TNT slurry, and stirring until the mix is uniform. A desensitizer and calcium chloride may be incorporated and the mixture cooled to ca 85°C after which it is poured. If composition B is the starting material, it is first melted and the necessary amount of TNT, aluminum, desensitizer, and calcium chloride are added to meet formulation requirements. The incorporation of aluminum increases the blast effect of explosives but decreases their rates of detonation, fragmentation, and shaped-charge performance. Typical TNT-based aluminized explosives are the tritonals (TNT + Al), ammonals (TNT, AN, Al), minols (TNT, AN, Al), and the torpexes and HBXs (TNT, RDX, Al) (376–377).

Casting is the most economical process for large-scale production of many munitions. It requires little capital equipment and lends itself to automation where the size of the operation makes it worthwhile. Gun projectiles, which have extremely large wartime requirements, may be filled automatically. However, although precise charges may be cast if considerable care is taken, the casting process for TNT-based explosives has a number of limitations in plant-loading operations, including the possibility of component segregation and nonuniformity of cast, the introduction of porosity and cavitation, and the difficulty of filling components with small openings. Casting cannot be used with high melting explosives except as a slurry nor can it be used to form explosive configurations of maximum mechanical strength and geometric stability (378).

High-speed automatic mechanical pressing is commonly used to volumetrically load small quantities of primary explosives into blasting caps and detonators and to make small explosive components. Primary explosives may be mixed with graphite to improve their flow and antistatic properties or they may be desensitized with waxes and stearates. Secondary explosives and explosive mixtures may be pressed to form booster pellets or to load components directly as in the case of armor-penetrating projectiles. Where the explosive is too sensitive in its pure crystalline state to permit press loading or lacks the required mechanical properties in its compressed state for subsequent use, it is coated with polymeric materials or waxes to form molding powders often referred to as plastic-bonded explosives (PBX). Desensitization is obtained when the explosive crystals are thoroughly and uniformly coated. A typical procedure for making PBX-type explosives involves making a lacquer of a solution of the organic polymer in a solvent (eg, ethyl acetate) and adding it to a water slurry of the explosive. The solvent is distilled under vacuum while the mix is agitated, precipitating the polymer on the explosive. The coated explosive forms small agglomerates as the solvent removal process continues. It is filtered, washed, and vacuum dried to form a free-flowing, dustless, high density powder. Bi- or trimodal particle size distributions of explosive are often used to improve the flow characteristics and packing density of the molding powder. In another coating technique, the required amount of low melting wax is added to a water slurry of the explosive at a temperature high enough to melt the wax. After agitation to distribute the wax on the crystals, the temperature is lowered, the water decanted, and the remaining mass filtered and dried (379–386).

The coated explosive powders may be mechanically pressed, using special procedures and tools including vacuum pressing and hot pressing, into many hitherto unattainable shapes. Hydrostatic pressing and isostatic pressing techniques are also used involving compression of the explosive by a fluid acting on the explosive through a flexible membrane. The explosives may be first deaerated to a pressure below 133

Pa (1 mm Hg) and then consolidated at pressures to 200 MPa (2000 atm) and temperatures to 120°C. Densities of 97–98% of the maximum theoretical are so produced. Final machining to shape is required. High-energy, polymer-bonded explosives prepared in this manner may be pressed into many configurations of high density, considerable mechanical strength, and good dimensional stability. They are generally prepared with 90–95% of RDX or HMX, and are much less sensitive to heat, friction, shock, and impact than pure RDX or HMX. Extrudable explosive mixtures consisting of a high-energy explosive in a plastic matrix have also been prepared which are either puttylike in consistency or may be cured to a rigid mass. The properties of a number of plastic-bonded explosives are presented in Table 17. Many inert formulations have also been developed for systems where it is desired to simulate the mechanical properties of live explosives (387–401).

Industrial

In the United States private corporations operate most of the government-owned plants that make explosives for military use. They also make explosives and explosive components in their own plants for industrial use. The quantity of industrial explosives used in the United States continues to increase but the type of explosives used has changed radically in the post World War II period. Between 1960 and 1975 explosive consumption increased from 531×10^3 metric tons to 1413×10^3 t. The use of black powder and liquid oxygen disappeared completely and dynamites declined significantly, whereas the dry and water-based ammonium nitrate consumption increased greatly (see Table 18) (402–403).

Dynamites. The first dynamite developed by Nobel, the guhr dynamite, contained about 75% nitroglycerin absorbed into and desensitized by 25% of diatomaceous earth (guhr). The inert guhr was replaced by a variety of fuel oxidizers to increase the energy of the dynamites. Nitroglycerin and sodium nitrate are the main oxidizers of many of these straight dynamites and may constitute up to 80% of the weight of the explosives. The strength of the dynamite is varied by modifying the percentage of nitroglycerin present and introducing other energetic oxidizers. A 20% straight dynamite contains 20% of nitroglycerin. Ethylene glycol dinitrate is usually mixed with the nitroglycerin (explosive oil) to decrease the freezing point of the composition. Other additives include carbonaceous fuels, antiacidic components and sulfur. Ammonia dynamites are similar to the straight dynamites with part of the solid oxidizer consisting of ammonium nitrate. The straight dynamites are of powdery consistency and subject to deterioration when exposed to water. Gelatin dynamites contain nitrocellulose gelatinized by the nitroglycerin present. Their consistency ranges from rubbery to plastic-powdery and their resistance to water is much greater than that of the straight dynamites. Blasting gelatin is the most powerful of the gelatin dynamites and contains 91% nitroglycerin, 8% low-nitrogen nitrocellulose, and an antiacidic compound. Straight gelatins and ammonia gelatins are similar to straight and ammonia dynamites. Semigelatins are low strength dynamites which are more plastic than rubbery. The dynamites are formulated for specific applications and have a wide range of densities (0.8–1.7 g/cm^3) and detonation rates (2000–7000 m/s). Permissible dynamites are granular, semigel, or gelled mixtures of low to medium strength that are designed to be used in underground coal mines with inflammable coal dust and methane–air atmospheres. They generally contain about 10% sodium chloride to reduce

Table 17. Plastic-Bonded Explosives

Property	Composition						
	A 3	C 4	HMX-KEL-F 95/5	LX-04-1	PBXN 3	PBX 9404	PBX 9010
density, g/cm³							
max theoretical							
as used	1.64	1.64	1.87	1.87	1.70	1.83	1.78
heat of formation[a], kJ/g[b]	−0.104	−0.138		0.899		−0.003	0.330
heat of detonation[a], kJ/g[b]	5.062	6.610[c]		5.481		5.481	6.150
detonation velocity, km/s	8.10	8.04	8.89	8.48	8.45	8.37	
at density, g/cm³	1.59	1.58	1.88	1.87	1.71	1.78	
detonation pressure, GPa[d]	27.7	25.7		35.4		32.8	
at density, g/cm³	1.59	1.58		1.84		1.78	
vacuum stability at 100°C, cm³/gas per g per 40 h	<1	<1	<1	<1	<1	<1	<1
explosion temperature, 5 s, °C	280	263		337	337	309	295
relative impact test value, % TNT	125	140	75	75	135	140	75
rifle bullet, unaffected, % TNT	100	60	0	0	0	100	0
compressive strength[f]							
stress at rupture, MPa[e]	3.8		28.2	7.2	68.9	16.9	9.3
compression at rupture, %	0.25		1.5	0.9	1.3	1.5	0.8
modulus of elasticity, MPa × 10⁻³ [e]	4.1		6.2	1.9	10.8	5.2	3.4
work to produce rupture, J/cm³ [b]	0.01		0.31	0.04	0.64	0.22	0.05
tensile strength[f]							
stress at rupture, MPa[e]	1.1		4.1	3.1	5.9	4.5	2.8
elongation at rupture, %	0.04		0.06	0.35	0.05	0.01	0.16
modulus of elasticity, MPa × 10⁻³ [e]	5.0		7.6	2.6	13.8	6.6	6.6
work to produce rupture, J/cm³ [b]	0.0003		0.002	0.01	0.002	0.002	0.003
shear strength[f]							
stress at rupture, MPa[e]	2.07		8.61	5.17	22.4	9.65	5.86
charpy impact strength, J[b]	0.15		0.18	0.37	0.17	0.18	0.18
coefficient of linear expansion, % × 10⁻³/°C	8.0	6.6	4.6	5.1	7.0	5.8	6.7
thermal conductivity, W/(m·K)		0.25		0.29		0.38	0.42
specific heat, J/(g·K)[b]						1.2	1.2
relative energy output, % TNT							
ballistic mortar	135	130					
sand test	115	115					
fragmentation velocity	125	125	140	130	125	140	130

[a] Molecular weight assumed to be 100 g.
[b] To convert J to cal, divide by 4.184.
[c] Calculated.
[d] To convert GPa to atm, divide by 10^{-4}.
[e] To convert MPa to psi, multiply by 145.
[f] At load rate of 0.13 cm/min at 20°C.

Table 18. Sale of Industrial Explosives and Blasting Agents in the United States, Metric Tons[a]

Year	Permissibles	Other high explosives	Water gels and slurries	Cylindrically packaged blasting agents	Other processed blasting agents and unprocessed ammonium nitrate	Black powder	Liquid oxygen	Total
1960	36	214	0	0	279	0.7	0.7	531
1965	34	246	0	0	571	0.4	2.5	854
1970	25	130	97	17	815	0.04	0	1084
1975	21	102	141	150	999	0	0	1413

[a] Ref. 102.

the flame temperature of the decomposition products. Dynamites are ordinarily sufficiently sensitive because of the nitroglycerin present that they may be initiated in small diameter charges by no. 6 or no. 3 blasting caps. The composition of typical dynamites is shown in Table 19. A list of typical substances used in dynamites is presented in Table 20 (404–413).

Ammonium Nitrate Explosives. Ammonium nitrate is the cheapest and safest source of readily deliverable oxygen for explosive applications. The extensive use of ammonium nitrate in ammonium nitrate–fuel oil (ANFO) and water-based commercial explosives has revolutionized the industry and displaced the nitroglycerin-based dynamites. Ammonium nitrate industrial explosives are low cost, safe, and versatile in performance and application and have better storage stability than dynamites. A large number of formulations is available for almost all purposes.

Pure ammonium nitrate decomposes in a complex manner in a series of progressive reactions with different thermochemical effects (Table 21). Oxygen is liberated from combination with combustibles only at temperatures above 300°C.

When a combustible material such as fuel oil is present in stoichiometric proportions (ca 5.6%) the energy evolved increases almost threefold:

$$3 \text{ NH}_4\text{NO}_3 + (\text{CH}_2)_n \rightarrow 3 \text{ N}_2 + 7 \text{ H}_2\text{O} + \text{CO}_2 + 428.8 \text{ kJ/g } (102.5 \text{ kcal/g})$$

Ammonium nitrate exists in five crystal forms which may transform from one to the other with accompanying volume, crystal structure and heat changes. It is very hygroscopic and deliquesces at relative humidities about 60%. Pure ammonium nitrate

Table 19. Composition of Typical Dynamites[a]

Component	Composition, wt %						
	NG	SN	AN	SC	NC	CCM	Other
straight dynamite (40%)	40	44				14	2
ammonia dynamite (40%)	14	33	36			12	1
blasting gelatin	92				7		1
straight gelatin (40%)	32	52			0.7	11	4
ammonia gelatin (40%)	21	49	14		0.4	9	7
granular permissible	9	5	65	10		10	1

[a] NG, nitroglycerin–explosive oil; SN, sodium nitrate; AN, ammonium nitrate; SC, sodium chloride; NC, nitrocellulose; CCM, combustible carbonaceous material; other antiacid components, sulfur, moisture % (404).

Table 20. Typical Components of Dynamites

Type	Primary purposes

oxidizers
  nitroglycerin
  nitrostarch
  ammonium nitrate
  sodium nitrate
  nitrocellulose
	energy modifiers

fuels
  sawdust
  wood metal
  flour
  wood pulp
  dextrin
  starch
  sulfur
  freezing point depressants
  ethylene glycol dinitrate
	energy modifiers and absorbents

sensitizers
  tetryl
  dinitrotoluene
  trinitrotoluene
  nitrostarch
  PETN
  smokeless powder
  pentolite
	increase initiability of small diameter charges (particularly ammonium nitrate explosives)

waterproofing compounds
  pregelatinized
  stearates
  silicon resins
  waxes
  swelling agents such as
    carboxymethyl cellulose
  liquid film forming compounds
	essential where ammonium nitrate or hygroscopic salts are present and in explosives for use in underwater demolition work

coolants
  ammonium nitrate
  sodium chloride
  ammonium chloride
  sodium bicarbonate
	reduce flame temperature to permit use in coal mines

antiacidic substances
  calcium carbonate
  magnesium carbonate
  zinc and magnesium oxides
	increase stability

is very stable and insensitive to impact and friction. It is impossible to initiate with conventional blasting caps. When mixed with organic materials such as hydrocarbons or cellulose, ammonium nitrate requires a powerful, high explosive booster for initiation. If the charge diameter is less than 7.5–13 cm when unconfined, or less than about 4 cm when heavily confined, the ammonium nitrate mix does not propagate. Although the total energy evolved by ANFOs is comparable to that of low-grade military explosives, the reaction rate is much slower and is more dependent on charge diameter and confinement. Ammonium nitrate has been used in military explosives such as amatols, ammonals, minols, and amatexes as a partial replacement for TNT or RDX.

Table 21. Thermochemical Data for the Decomposition of Ammonium Nitrate

Reaction	°C	Heat liberated[a], J/g[b]	Gas volume[c], cm^3/g	Temperature of reaction, K
$NH_4NO_3 \rightarrow NH_3 + HNO_3$	180	−2144	560	endothermic
$NH_4NO_3 \rightarrow N_2O + 2\,H_2O$	250	525	840	770
$NH_4NO_3 \rightarrow N_2 + 2\,H_2O + \tfrac{1}{2}\,O_2$	300	1465	981	1560

[a] At 25°C, constant volume, water vapor.
[b] To convert J to cal, divide by 4.184.
[c] At STP.

It is made from anhydrous ammonia and nitric acid and ranges from dense crystals to porous agglomerates (prills). Prills used in industrial explosives are made by spraying a 95% solution of the nitrate against a countercurrent stream of air in a prilling tower. The particles are dried in a series of rotary towers and coated to improve flow characteristics and moisture resistance. The properties of ammonium nitrate are listed in Table 22 (414–420).

ANFOs. Ammonium nitrate-fuel oil compositions (ANFOs) consist of 94% ammonium nitrate prills coated with an anticaking agent and 6% absorbed fuel oil. They are defined as dry blasting agents which are mixtures of nonexplosive ingredients used for blasting purposes that cannot be detonated by a no. 8 blasting cap. If they are sensitized by the addition of an explosive component so that they detonate when initiated by a no. 8 blasting cap, they are defined as blasting explosives. The ANFOs are usually initiated with a high explosive booster such as 50/50 pentolite or composition B.

The sensitivity of ANFOs to initiation is significantly affected by their composition, physical characteristics, and environment. Decreasing the particle size and density of the ammonium nitrate or increasing its porosity increases the sensitivity of the mix to initiation. Maximum sensitivity occurs at oil concentrations of ca 2–4%. Increasing the oil concentration beyond 4% results in substantial reductions in sensitivity to initiation. The presence of water decreases the sensitivity. The detonation velocity increases as the oil content increases to a maximum at ca 6% oil with the maximum velocity of about 4300 m/s being obtained with large diameter ANFO charges. The detonation velocity of ANFO charges increases with charge diameter up to ca 13 cm for compositions at conventional loading densities of 0.90–1.10 g/cm^3. Confinement also increases the detonation velocity. Maximum energy is obtained at an oil concentration of 5.5%; the addition of metallic fuels, such as aluminum or ferrosilicon, increases the energy content. Stabilizers and inhibitors may be added and the fuel oil may be dyed to identify specific compositions. The ANFOs may be mixed on site simply by adding oil to a bag of prills or they can be prepared in on-site trucks equipped for the purpose and then blown into boreholes (121–128).

Water-Based Blasting Agents and Explosives. Water gel and slurry explosives were developed to capitalize on the low cost of ammonium nitrate while increasing the available energy beyond that obtainable by ANFOs, improving initiability in small diameter charges, and eliminating the problem associated with the use of ANFOs under

Table 22. Properties of Ammonium Nitrate

Property	Value
molecular weight	80.0
color	white
density at 20°C, g/cm^3	1.725
specific volume, cm^3/g	0.580
melting point, °C	169.6
hygroscopicity at 90% rh/20°C	increase up to 156% after 8 d exposure
solubility at 20°C in H$_2$O g/100 g	66
oxygen content, %	60
oxygen balance to H$_2$O, %	20
hardness, Mohs	1.1
heat of formation, kJ/g^a	4.60
heat of combustion, kJ/g^a	2.62
heat of detonation (H$_2$O), kJ/g^a	2.63
heat of solution, kJ/g^a	−0.33
gas volume (H$_2$O vapor), cm^3 at STP	980
activation energy, kJ/mol^a	163–167
estimated flame temperature, °C	1,500
detonation velocity, km/s	2.70
at density, g/cm^3	0.98
detonation pressure, MPa^a	1,100
at density, g/cm^3	0.98
vacuum stability at 150°C, mL gas per g per 40 h at STP	less than 1
75°C international heat test, weight loss in 48 h	0
effect of long term storage	very stable below melting point in absence of organic matter
rifle bullet test, initiations, %	0
at density, g/cm^3	1.2
impact test	no action
relative energy output, % TNT	
ballistic pendulum	80
lead block	75
specific heat, J/mol^a	1.72 (0–31°C)
thermal conductivity, W/(m·K)	0.25
coefficient of thermal expansion at 20°C	9.82 × 10^{-4}
heat of fusion, kJ/g^a	0.075
latent heat of sublimation, kJ/g^a	2.18
maximum specific impulse in optimum formulations, NS/kg	1965

^{a} To convert J to cal, divide by 4.184.
^{b} To convert MPa to psi, multiply by 145.

wet conditions. The use of water-based formulations also increased field safety and economy and improved fume characteristics. These compositions, used largely in metal mining, are thickened suspensions of oxidizers, fuels, and a sensitizer dispersed in a saturated aqueous salt solution. Ammonium nitrate with or without other oxidizers, such as sodium nitrate, sodium perchlorate and ammonium perchlorate, are the

Table 23. Comparative Cratering Characteristics of ANFO, Slurry Explosives, and TNT[a]

Explosive	Detonation pressure, GPa[b]	Bulk specific gravity	Detonation velocity, km/s	Contains high explosives	Heat of detonation, kJ/g[c]	Excavated volume relative to equal wt of TNT
ANFO	6.0	0.93	4.56	no	3.76	1.0–1.1
AN slurry	10.4	1.40	6.05	yes	3.05	1.0–1.2
AN slurry (2% Al)	6.0	1.30	4.30	no	3.14	1.0–1.2
AN slurry (8% Al)	6.6	1.33	4.50	no	4.64	1.2–1.4
AN slurry (20% Al)	8.5	1.20	5.70	no	6.07	1.5–1.7
AN slurry (35% Al)	8.1	1.50	5.00	no	8.16	1.6–1.8
TNT	18.7	1.64	6.93		4.61	1.00

[a] Ref. 429.
[b] To convert GPa to atm, divide by 10^{-4}.
[c] To convert J to cal, divide by 4.184.

compounds dissolved and suspended in water. Explosive sensitizers, such as pentolite, methylamine nitrate, TNT, smokeless powder, nitrotoluene, and nitrostarch, may be incorporated. Inert sensitizers include finely divided aluminum, gas bubbles in suspension, gas enclosed in small glass spheres, and finely divided porous solids. Inert fuels, such as coal dust, urea, sulfur, and various types of hydrocarbons, are added. Guar gums, gelatin-forming compounds, such as carboxymethyl cellulose, resins, and synthetic thickeners, such as the polyacrylamides, are incorporated to thicken the mix. Cross-linking agents, such as sodium tetraborate and potassium dichromate, are used to control viscosity. Slurry explosives are made water resistant by the addition of hydrophilic colloids which bond the solid particles and prevent diffusion of water in and out of the system. Antifreezes, such as glycerol, methanol, and diethylene glycol, may be used. Typical slurry blasting agents or explosives may contain ammonium nitrate (30–70%), sodium nitrate (10–15%), calcium nitrate (15–20%), aliphatic amine nitrates (to 40%), aluminum (15–25%), TNT or other explosive sensitizer (5–15%), gellants (1–2%), stabilizers (0.1–2%), ethylene glycol (3–15%), and water (10–20%). Other ingredients present include oxidizers, fuels, explosives, chemical sensitizers, thickeners, cross-linking agents, and micropores (hollow glass beads). A wide range of viscosities from 1 Pa·s to 2 kPa·s (10–20,000 P) and high temperature stabilities are available.

The performance of several slurries in comparison with TNT and ANFO 94/6 is shown in Table 23. Aluminum-bearing slurries are among the highest energy producers of all industrial explosives. Slurries are extensively used in wet conditions and open-pit blasting. Their high loading densities and fluidity and a greater intrinsic energy output enable increased efficiencies to be attained in rock fragmentation. Slurry mixes have been studied for many military applications. They may be bulk mixed in a plant, transported to the site of operations, and pumped into boreholes after adding a thickening agent, or may be prepared on-site using a pump truck. Slurries may also be prepared hot in the plant and poured into lay-flat, sausage-shaped polyethylene bags for hand loading into bore holes (429–445).

The following lists government agencies cited in references.

Abbreviation	Name	Location
ADPA	American Defense Preparedness Association	Washington, D.C.
AMC (DARCOM)	U.S. Army Materiel Command (now Development and Readiness Command)	Alexandria, Va.
ARRADCOM	Armament Research and Development Command	Dover, N.J.
DODESB	Department of Defense Explosive Safety Board	Washington, D.C.
BRL	Ballistics Research Laboratory	Aberdeen, Md.
BuM	U.S. Bureau of Mines, Department of the Interior	Pittsburgh, Pa.
CIT	California Institute of Technology	Pasadena, Calif.
CPIA	Chemical Propulsion Information Agency	John Hopkins University, Laurel, Md.
DDC	Defense Documentation Center	Alexandria, Va.
EPA	Environmental Protection Agency	Washington, D.C.
ERDE (PERME)	Explosive Research and Development Establishment (now Propellants, Explosives and Rocket Motor Establishment)	Waltham Abbey, Essex, Eng.
ERL	Explosive Research Laboratory	no longer in existence
FA	Frankfort Arsenal	no longer in existence
LASL	Los Alamos Scientific Laboratory, University of Calif.	Los Alamos, N.M.
LRL (LLL)	Lawrence Radiation Laboratory (Lawrence Livermore Laboratory)	Livermore, Calif.
NASA	National Aeronautics and Space Administration, Scientific and Technical Information Facility	PO Box 8757, BWI Airport, Md.
NBS	U.S. National Bureau of Standards	Washington, D.C.
NOL (NSWC)	Naval Ordnance Laboratory (now Naval Surface Weapons Center)	White Oak, Md.
NOTS (NWC)	Naval Ordance Test Center (now Naval Weapons Center)	China Lake, Calif.
NOTS (NWC)	Naval Ordnance Test Center (now Naval Weapons Center)	Indian Head, Md.
NTIS	National Technical Information Center, U.S. Department of Commerce	Springfield, Va.
NWL	Naval Weapons Laboratory	Dahlgren, Va.
ONR	Office of Naval Research	Washington, D.C.
OSRD	Office of Scientific Research and Development	no longer in existence
PTA	Picatinny Arsenal	Dover, N.J.
SAN	Sandia Laboratories	Albuquerque, N.M.
USGPO	U.S. Government Printing Office	Washington, D.C.

BIBLIOGRAPHY

"Explosives" in *ECT* 1st ed., Vol. 6, by Wm. H. Rinkenbach, D. R. Cameron (Propellants), both of Picatinny Arsenal, U.S.A. Ordnance Department, and W. O. Snelling (Blasting Explosives), Trojan Powder Company; "Explosives" in *ECT* 2nd ed., Vol. 8, pp. 581–658, by Wm. H. Rinkenback, Consulting Chemist.

1. M. A. Cook, *Science of High Explosives,* Reinhold Publishing Corp., New York, 1958.
2. W. C. Davis, "Detonation Phenomena" in *12th Annual Symposium on Behavior and Utilization of Explosives in Engineering Design,* University of New Mexico, Albuquerque, N.M., 1972, p. 5.
3. *Black Powder Spec. MIL-P-223B,* USGPO, 1973.
4. R. A. Howard, *J. Soc. Ind. Archaeol.* **1,** 1 (1975).
5. D. B. Chedsey, *Goodbye to Gunpowder,* Crown Publishers, New York, 1963.
6. J. R. Partington, *A History of Greek Fire and Gunpowder,* Heffer, Cambridge, Eng., 1960.

7. A. Manucy, *Artillery Through the Ages,* U.S. Government Printing Office, Washington, D.C., 1949.
8. A. P. Van Gelder and H. Schlatter, *History of the Explosives Industry in America,* Oxford University Press, London, Eng., 1927.
9. A. Marshall, *Explosives,* Blakiston, Philadelphia, Pa., 1917.
10. R. Connor in W. A. Noyes, Jr., ed., *Chemistry in World War II,* Little, Brown, and Co., Boston, Mass., 1948.
11. W. S. Dutton, *One Thousand Years of Explosives from Wildfire to the H Bomb,* Winston, Philadelphia, Pa., 1960.
12. T. Urbanski, *Chemistry and Technology of Explosives,* Vols. 1–3, MacMillan, New York, 1964–1967.
13. C. W. Falterman, H. J. Gryting, and W. J. Griffith, *Development of an Explosive Capable of Withstanding at Least 600°F, NAVWEPS 9018,* 1966.
14. J. Taylor, *Detonations of Condensed Explosives,* Clarendon Press, Oxford, Eng., 1952.
15. W. F. McGarry and T. W. Stevens, *Detonation Rates of the More Important Military Explosives at Several Different Temperatures, TR 2383,* PTA, 1956.
16. M. W. Beckstead and co-workers, "Convective Combustion Modeling Applied to Deflagration-Detonation Transition" in *Proceedings of the 12th JANNAF Combustion Meeting Pub. No. 273,* CPIA, 1975.
17. A. Macek, *J. Chem. Phys.* **31,** 162 (1959).
18. W. H. Anderson and R. F. Chaiken, *ARS J.* **31,** 162 (1959).
19. J. E. Crump and E. W. Price, *ARS J.* **30,** 707 (1960).
20. S. Wachtel, "Prediction of Detonation Hazards in Solid Propellants," *paper presented at 145th National Meeting, Division of Fuel Chemistry, New York, 1963.*
21. C. M. Kintz, G. W. Jones, and C. B. Carpenter, *Explosions of Ammonium Nitrate Fertilizer on Board the S.S. Grand Camp and S.S. High Flyer at Texas City, Tex., R.I. 4245,* BuM, 1948.
22. *Ordnance Explosive Train Designers Handbook, NOLR 1111,* NOL, 1952.
23. R. R. Bernecker and A. Price, *J. Combust. Flame* **22,** 111, 161 (1974).
24. D. Price, J. F. Wehner, and G. E. Robertson, *Transition from Slow Burning to Detonation: Role of Confinement, Pressure Loading and Shock Sensitivity, TR 68-138,* NOL, 1968.
25. A. F. Belyaev and V. K. Bobolev, *Transition from Deflagration to Detonation in Condensed Phases, Rpt NO T-74-50028,* translated by US-Israel Binational Science Foundation, Washington, D.C., 1974.
26. *Military Pyrotechnics, AMCP 706-185 to 189, Engineering Design Handbook Series,* AMC, 1974.
27. K. O. Brauer, *Handbook of Pyrotechnics,* Chemical Publishing Co., New York, 1974.
28. H. Ellern, *Military and Civilian Pyrotechnics,* Chemical Publishing Co., New York, 1968.
29. *Proceedings of the International Pyrotechnics Seminars,* Denver Research Institute, University of Denver, Boulder, Colo., 1972–1976.
30. *Proceedings of the Conferences on Compatibility of Propellants, Explosives and Pyrotechnics with Plastics and Additives, Rpts. NC-OIP (1974) and NC-02P (1976),* ADPA, Washington, D.C.
31. F. P. Bowden and A. D. Yoffe, *Fast Reactions in Solids,* Academic Press, Inc., New York, 1958.
32. F. P. Bowden and A. D. Yoffe, *Initiation and Growth of Explosions in Liquids and Solids,* Cambridge University Press, New York, 1952.
33. A. S. Kompaneetz, *The Theory of Detonation,* Moscow, USSR, 1955.
34. H. Eyring and co-workers, *Chem. Rev.* **45,** 69 (1949).
35. T. J. Tucker, "Explosive Initiators" in *Behavior and Utilization of Explosives in Engineering Design,* 12th Annual Symposium, University of New Mexico, Albuquerque, N.M., 1972, pp. 175–188.
36. *Ordnance Explosive Train Designers' Handbook, NOLR 1111,* NOL, 1952.
37. *Explosive Trains, AMCP 706-179,* AMC, 1965.
38. M. A. Cook, *The Science of Industrial Explosives,* IRECO Chemicals, Salt Lake City, Utah, 1974.
39. *Principles of Explosive Behavior, AMCP 706-180,* AMC, 1972.
40. *Proceedings of the International Conference on Research in Primary Explosives, Explosives Research, and Development Establishment,* Waltham Abbey, Essex, Engl., 1975.
41. T. Urbanski, *Chemistry and Technology of Explosives,* Vol. 3, Pergamon Press, New York, 1967, p. 178.
42. *Proceedings of the Symposium on Lead and Copper Azides, Rpt. WAA/79/0216,* Explosives Research and Development Establishment, Waltham Abbey, Essex, Eng., 1966.
43. B. T. Federoff and O. Sheffield, *Encyclopedia of Explosives,* Vol. 1, A545 TR 2700, PTA, 1960.

44. C. N. E. Taylor and J. M. Jenkins, *Proceedings of Third Symposium on Chemical Problems Connected with the Stability of Explosives*, Vol. 3, Jonkoping, Sweden, 1973, p. 43.
45. Z. V. Harvalik, *Explosive Initiators, Electric, NAVORD 1487*, 1950.
46. A. M. Anzalone and A. Forsyth, *Characteristics of Explosive Substances for Application in Ammunition, TR 2179*, PTA, 1955.
47. W. G. Chace and H. K. Moore, eds., *Exploding Wires*, Vols. 1–3, Plenum Press, New York, 1959, 1962, 1964.
48. D. P. Donegan, "Trends in Electrostatic Precautions in Filling Factories" in *Minutes of 17th Explosives Safety Seminar, AD A036015*, DDC, 1976, p. 363.
49. A. F. Schlack, "Susceptibility of Electric Primers and Electrostatic Discharges" in *Minutes of the 15th Explosives Safety Seminar, AD-775 580*, NTIS, 1973, p. 637.
50. E. Demberg, "Spontaneous Detonation of Initiators" in *Minutes of the 15th Explosives Safety Seminar, AD-775 580*, NTIS, 1973, p. 681.
51. H. S. Leopold, *Effect of Wire Temperature Upon Initiation Times for Four Primary Explosives, TR 72-123*, NOL, 1972.
52. W. B. Leslie, R. W. Dietzel, and J. A. Searcy, "A New Inherently Safe Explosive for Low Voltage Detonator Applications" in *Proceedings of the 6th (International) Symposium on Detonation*, ONR, 1976, p. 144.
53. A. M. Anzalone, *Bibliography of the Electrical Initiation of High Explosives*, PTA, 1958.
54. F. D. Miles, *J. Chem. Soc.*, 2532 (1931).
55. C. Scott, *Proceedings of the 6th Symposium on Electro-Explosive Devices*, Franklin Institute, Philadelphia, Pa., 1969.
56. A. C. Forsyth and co-workers, *The Effects of Long Term Storage on Special Purpose Lead Azide, TR 4357*, PTA, 1972.
57. T. G. Blake, D. E. Seeger, and R. H. Stresau, "Lead Azide Precipitated with Polyvinyl Alcohol" in *2nd ONR Symposium on Detonation, Office of Naval Research*, Washington, D.C., 1955, p. 92.
58. I. Kabek and S. Urman, "Hazards of Copper Azide in Fuzes" in *Minutes of the 14th Annual Explosives Safety Seminar NTIS*, 1972, p. 533.
59. R. L. Wagner, *Lead Azide, Its Properties and Use in Detonators, TR 2662*, PTA, 1960.
60. R. L. Wagner, *Investigation of RD 1333 Lead Azide for Use in Detonators, Rpt. 18*, PTA, 1958.
61. H. Fair and R. Walker, eds., *The Inorganic Azides*, Vols. 1–2, Plenum Press, New York, 1977.
62. F. W. Davies and co-workers, "The Hugoniot and Shock Initiation Threshold of Lead Azide" in *6th Symposium (International) on Detonation*, ONR, 1976, p. 101.
63. *Spec. MIL-L-14758, Lead Azide, Special Purpose*, USGPO, 1968.
64. P. G. Fox, J. M. Jenkins, and G. W. C. Taylor, *Explosives* **17,** 8, 181 (1969).
65. W. H. Rinkenbach and J. D. Hopper, *A Study of the Explosive Characteristics of Commercially Prepared Lead Azide, TR 852*, PTA, 1937.
66. N. J. Blay, *Methods for the Assay of Lead Azide for Specification Purposes, Rpt. No. ERDE 11/M/67*, Explosives Research and Development Establishment, Waltham Abbey, Essex, Eng., 1967.
67. *Lead Azide U.S. Military Spec. L-3055A*, USGPO, 1962.
68. Brit. Pat. 781,440 (1957), G. W. C. Taylor.
69. Brit. Pat. 887,141 (Jan. 17, 1962), E. Williams, S. Peyton, and R. C. Harris.
70. T. Costain, *A New Method for Making Silver Azide, TR 4595*, PTA, 1974.
71. C. A. Taylor and W. H. Rinkenbach, *Army Ordnance* **5,** 824 (1925).
72. D. A. Young, *Recent Physico-Chemical Investigations on Silver Azide, NAVORD 5746*, 1958.
73. T. A. Bronner and H. J. Jackson, *Normal Lead Styphnate: Development of Standardized Preparatory Methods, TR 3079*, PTA, 1963.
74. *Lead Styphnate, Spec. JAN-L-757A*, USGPO, 1968.
75. *Lead Styphnate, Basic, Spec. MIL-L-16355A*, USGPO, 1967.
76. B. F. Husten, *Basic Lead Styphnate, Fine Milling, NAVORD OD 23996*, 1963.
77. T. Urbanski, *Chemistry and Technology of Explosives*, Vol. 3, Pergamon Press, New York, 1977, p. 213.
78. L. V. Clark, *J. Ind. Eng. Chem.* **25,** 663 (1933).
79. R. L. Grant and J. E. Tiffany, *J. Ind. Eng. Chem.* **37,** 661 (1945).
80. R. Barefoot, "Compatibility of Nitrate and Nitrate Esters" in *Conference on Compatibility of Propellants, Explosives, and Pyrotechnics with Plastics and Additives, Report NC-02P*, ADPA, 1976, p. 1-E-22.

81. P. Naoum, *Nitroglycerin and Nitroglycerin Explosives,* Williams and Wilkins Co., Baltimore, Md. 1928.
82. T. Urbanski, *Chemistry and Technology of Explosives,* Vol. 2, Pergamon Press, New York, 1965.
83. S. Kaye, ed., " Glycerol Trinitrate" in *Encyclopedia of Explosives and Related Items, TR 2700,* PTA, Vol. 3 p. 501, Vol. 6, p. 699, Vol. 8, p. N56.
84. R. W. Van Dolah, "Low Velocity Detonation in Nitroglycerin" in *5th Annual Explosives Safety Seminar on High Energy Propellants,* Armed Services Explosives Safety Board, Santa Monica, Calif., 1963.
85. C. Boyars, "Sensitivity and Desensitization of Nitroglycerin" in *Proceedings of 2nd Symposium on Chemical Problems Connected with the Stability of Explosives,* Jonkoping, Sweden, 1970, p. 197.
86. O. E. Waring and G. Krastins, *The Kinetics and Mechanism of Thermal Decomposition of Nitroglycerin, Report 5746,* NOL, 1958.
87. J. Taylor and C. R. L. Hall, *J. Phys. Coll. Chem.* **51,** 580, 593 (1947).
88. F. P. Bowden and A. D. Yoffe, *Initiation and Growth of Explosion in Liquids and Solids,* Cambridge University Press, Cambridge, New York, 1952.
89. Ger. Pat. 6208 (Dec. 24, 1878), R. Kurtz.
90. Brit. Pats. 284,700 (Nov. 15, 1928), 284,701 (Feb. 14, 1929), 299,384 (Mar. 14, 1929), (to Schmidt-Meissner).
91. U.S. Pat. 2,438,244 (Mar. 23, 1948).
92. Brit. Pat. 733,588 (July 13, 1955).
93. *Nitroglycerin, MIL STD Spec. N-246B,* USGPO, 1962.
94. C. S. Miner and N. N. Dalton, eds., *Glycerol, ACS Monograph No. 117,* Reinhold, New York, 1953.
95. H. J. Klassen and J. M. Humphreys, *Chem. Eng. Prog.* **49,** 641 (1953).
96. G. S. Biasutti, "Safe Manufacuture and Handling of Liquid Nitric Esters," *ACS Symposium Series No. 22,* ACS, Washington, D.C., 1975.
97. U.S. Pat. 3,987,381 (Oct. 19, 1976), (to International Standard Electric Corp.).
98. Personal communication, G. S. Biasutti, Vevey, Switz.
99. U.S. Pat. 2,237,522 (Mar. 6, 1956), A. Nilssen (to Nitroglycerin Aktiebolget, Sweden).
100. *Nitroglycerin Plants, S-71030,* Nitro Nobel A B ROC Division, Gyttorp, Sweden.
101. N. Shapira, ed., *Wastewater Treatment in the Military Explosives and Propellants Production Industry,* Vol. 3, ADPA, 1975, p. 103.
102. D. C. Crater, **21,** 7, 674 (1929).
103. W. H. Rinkenbach, *J. Ind. Eng Chem.* **18,** 1196 (1926).
104. *Ibid.,* **19,** 925 (1927).
105. R. A. Cooley, *Chem. Ind.* **59,** 645 (1946).
106. B. T. Federoff, *TR 2510,* PTA, 1958.
107. W. G. Clark, *Evaluation of 1,2,4-Butanetriol Tinitrate as the Liquid Explosive Plasticizer for Cast Double Base Propellant, Report 4,* PTA, 1960.
108. *Thermal Stability, Aging, and Exudation of METN Propellants,* Publication LL 77-29, CPIA, 1977.
109. F. D. Miles, *Cellulose Nitrate,* Interscience Publishers, New York, 1945.
110. G. A. Richter, *Wood Cellulose Nitrates for Munitions, Report No. 71,* OSRD, 1940.
111. *Cellulose, Cotton for Use in Explosives,* Spec. MIL-206A, USGPO, 1978.
112. *Cellulose, Wood Pulp (Sulfite),* Spec. MIL-216A, USGPO, 1976.
113. E. Ott and M. H. Spurlin, eds. *Cellulose and Cellulose Derivatives,* John Wiley & Sonc, Inc., New York, 1954.
114. J. C. Arthur, Jr., ed., *Cellulose Chemistry and Technology ACS Symposium Series No. 48,* ACS, Washington, D.C. 1977.
115. J. Jessup and E. J. Prosen, *J. Res.* **44,** *Nat. Bur. Stand.* **44,** 387 (1950).
116. J. Taylor and C. R. L. Hall, *J. Phys. Colloid Chem.* **51,** 593 (1947).
117. J. Taylor, C. R. L. Hall, and H. Thomas, *J. Phys. Colloid Chem.* **51,** 580 (1947); P. Baer and J. Frankle, *Simulation of Interior Ballistic Performance of Guns by Digital Computer Program, Report 1183,* BRL, 1962.
118. P. R. Miles, *Ind. Eng. Chem.* **29,** 492 (1937).
119. J. M. Goldman and E. H. Zeigler, Jr., "The Shock Nitration Process for Spherical Propellant Manufacture" in *Symposium on Processing Propellants, Explosives, and Ingredients,* ADPA 1977, p. 23.
120. H. M. Spurlin, *Trans. Electrochem. Soc.* **73,** 95 (1938).

121. *Nitrocellulose Degradation,* Classified Publication No. LS 76-39, CPIA, 1977.
122. J. L. Hoard, H. Taube, and O. N. Salmon, *Compatibility Relations of Nitrocellulose with Various Explosives, Semiexplosive and Nonexplosive Materials Pertinent to the Development of Special Propellant,* OSRD 5758, OSRD, 1945.
123. G. Gelertner and co-workers, *J. Phys. Chem.* **60,** 1260 (1956).
124. R. Klein and M. Menster, *J. Am. Chem. Soc.* **73,** 5888 (1951).
125. J. B. Levy, *J. Am. Chem. Soc.* **76,** 3254, 3790 (1954).
126. *Ibid.,* **77,** 2015 (1955).
127. R. W. Phillip, C. A. Orlick, and R. Steinberger, *J. Phys. Chem.* **59,** 1034 (1955).
128. E. K. Rideal and A. J. B. Robertson, "The Spontaneous Ignition of Nitrocellulose" in *Proceedings of 3th Symposium on Combustion, Flame, and Explosion Phenomena,* U.S. BuM, p. 536, 1948.
129. R. Van Dolah and S. Newman, *Nitrocellulose: A Review of Some of Its Fundamental Properties,* NOTS and Alleghany Ballistics Laboratory, Cumberland, Md., 1953.
130. T. Urbanski, *Chemistry and Technology of Explosives,* Vol. 2, Pergamon Press, New York, 1965, p. 362.
131. *Solid Propellants, AMCP 706-175,* U.S. Army Materiel Command, Washington, D.C., 1961.
132. *Military Explosives, TM9-1300-214,* USGPO, 1967.
133. B. T. Federoff and O. Sheffield, *Encyclopedia of Explosives and Related Items,* TR 2700 PTA, 1962, pp. 2 C 95 and 8N62.
134. J. R. DuPont, *Chem. Met. Eng.* **26,** 11 (1922).
135. *Industrial Nitrocellulose,* ICI Ltd, Nobel Division, Kynoch Press, Birmingham, Ala., 1961.
136. F. D. Miles and M. Milbourn, *J. Phys. Chem.* **34,** 2598 (1930).
137. *Nitrocellulose, U.S. Spec. MIL-N-244,* USGPO, 1975.
138. Private communication, J. C. Horvath, Radford Army Ammunition Plant, Radford, Va., 1978.
139. E. Dodgen, "Continuous Nitration of Cellulose: SNIA Viscosa Process" in *Symposium on Processing Propellants, Explosives, and Ingredients,* ADPA, 1977, p. 4.2-1.
140. *Hercules Chem.* 37, Wilmington, Del., (1959).
141. N. I. Shapira, *Explosives and Propellants Industry,* Vol. 3, ADPA, 1975, p. 69.
142. M. R. Olsen and R. K. Major, *Comparative Study of Dehydrating Processes in the Manufacture of Nitrocellulose,* United Technology Chemical Systems Division, Sunnyvale, Calif., 1975.
143. E. Berlow, R. H. Barth, and J. E. Snow, *The Pentaerythritols,* Reinhold Publishing Co., New York, 1926.
144. T. Urbanski, *Chemistry and Technology of Explosives,* Vol. 2, Pergamon Press, New York, 1965, p. 175.
145. A. B. Coates, E. Friedman, and L. P. Kuhn, *Characteristics of Certain Military Explosives, Rpt. 1507,* BRL, 1970.
146. S. D. Brewer, *Studies on the Stability of PETN and Pentolite, Rpt. 3983,* OSRD, 1944.
147. D. M. Coleman and R. N. Rogers, "Pentaerythritol Tetranitrate (PETN) Stability and Compatibility" in *Proceedings of Conference on Compatibility of Propellants, Explosives, and Pyrotechnics with Plastics and Additives,* ADPA, Washington, D.C. 1974, p. 11-B-1.
148. A. T. Blomquist, and J. F. Ryan, *Studies Related to the Thermal Stability of PETN, Rpt. 3566,* OSRD, 1944.
149. A. J. B. Robertson, *J. Soc. Chem. Ind. (London)* **67,** 221 (1948).
150. E. L. Lee and H. C. Hornig, Equation of State of Product Gases," p. 493 in *12th Symposium (Intl.) on Combustion,* The Combustion Institute, Pittsburgh, Pa., 1969, p. 493.
151. W. Fickett, *Detonation Properties of Condensed Explosives Calculated with an Equation of State Based on Intermolecular Potentials, Rpt. LA2712,* LRL, 1962.
152. E. A. Christian and H. E. Snay, *Analysis of Experimental Data on Detonation Velocities, NAVORD 1508,* Washington, D.C., 1956.
153. D. L. Ornellas, J. H. Carpenter, and S. L. Gunn, *Rev. Sci. Inst.* **37,** 907 (1966).
154. H. H. Cady, *Pentaerythritol Tetranitrate 2, Its Crystal Structure and Transformation to PETN 1,* American Crystallographic Association, Tulane University, New Orleans, La., 1970.
155. M. R. Kantz, *Pentaerythritol Tetranitrate: A Bibliography,* NTIS, Springfield, Va., 1965.
156. *Pentaerythritol Tetranitrate (PETN), U.S. Spec. Mil-P-00387B,* USGPO, 1967.
157. Private communication, G. S. Biasutti, Biazzi Co., Vevey, Switz.
158. J. W. Patterson and R. W. Minear, *State of the Art for the Inorganic Chemicals Industry: Commercial Explosives, PB-240 960,* Illinois Institute of Technology, Chicago, Ill., 1975.

159. J. Roth, "PETN" in *Encyclopedia of Explosives and Related Items, TR 2700,* Vol 8, ARRADCOM, 1978, p. 86.
160. R. E. Shear, *Detonation Properties of Pentolite, Rpt. 1159,* BRL, 1961.
161. W. Kegler and R. Schall, in *Proceedings of 4th Symposium (Intl.) on Detonation,* NOL, 1965, p. 496.
162. F. J. Hildebrandt and R. T. Schimmel, *Suitability of RDX Compositions for Replacing Tetryl in Booster Explosives, TR 4537,* PTA, 1973.
163. O. H. Johnson, *HMX as a Military Explosive, Rpt. 4371,* NAVORD, Washington, D.C., 1956.
164. H. H. Licht, "HMX (Octogen) and its Polymorphic Forms" in *Second Symposium on Chemical Problems Connected with the Stability of Explosives,* Jonkoping, Sweden, 1970, p. 169.
165. J. T. Rogers, *Physical and Chemical Properties of RDX and HMX, Control Rpt. 20-P-26A,* Holston Defense Corp., Kingsport, Tenn., 1962.
166. M. Rosen and C. Dickinson, *J. Chem. Eng. Data* **14,** 1, 120 (1969).
167. A. B. Coates, E. Friedman, and L. P. Kuhn, *Characteristics of Certain Military Explosives, TR 1507,* BRL, 1970.
168. D. Burrows, *Literature Review of the Toxicity of RDX and HMX,* U.S. Army Medical and Bioengineering Research and Development Lab., Washington, D.C., 1973.
169. F. W. Sunderman, *Hazards to the Health of Individuals Working with RDX, Rpt. 4174,* OSRD, 1944.
170. J. D. Cosgrove, *Chem. Commun.* **6,** 286 (1968).
171. A. Robertson, Jr., *Trans. Farday Soc.* **45,** 85 (1949).
172. G. C. Hale, Jr., *Am. Chem Soc.* **47,** 2745 (1925).
173. B. D. Paubron, *The Thermal Conductivity of RDX, MITS MP-76-50 Rev. 1,* Mason and Hangar-Silas Mason Co., Inc., 1976; C. L. Mader, *Detonation Properties of Condensed Explosives Computed Using the Becker-Kistiakowsky-Wilson Equation of State, LA 2900,* University of California, Los Alamos, N.M., 1963.
174. W. E. Bachman and co-workers, *The Impact Sensitivity of HMX and of RDX-HMX Mixtures, Rpt. 4099,* OSRD, 1944.
175. A. T. Blomquist, *The Polymorphism of HMX, Rpt. 1227,* OSRD, 1943.
176. D. L. Ornellas, *J. Phys. Chem.* **72,** 2390 (1968).
177. H. H. Cady and L. C. Smith, *Studies of the Polymorphs of HMX,* University of California, Los Alamos, N.M., 1962.
178. J. H. Bryden, *The Density of Crystalline Cyclotetramethylenetetranitramine (HMX), Rpt. 1562,* NOTS, 1957.
179. M. Carmack, I. Von, J. J. Leavitte, and F. A. Kuehl, Jr., *Studies of the Mechanism of Formation of Cyclonite, Rpt. 6628,* OSRD, 1946.
180. Private communication, L. Silberman and M. Baer, PTA, 1978.
181. *A Photographic Tour of the Holston Army Ammunition Plant,* Holston Defense Corp., Kingsport, Tenn.
182. W. E. Bachmann and co-workers, *J. Am. Chem. Soc.* **71,** 1842 (1945).
183. T. Urbanski, *Chemistry and Technology of Explosives,* Vol. 3, Pergamon Press, New York, 1967, p. 98.
184. W. H. Simmons, R. Forster, and R. C. Bowden, *Ind. Chem.* **24,** 530 (1948).
185. R. B. Herring, B. L. Beard, and R. Robbins, *Rpt. 20-1-14,* Holston Defense Corp., Kingsport, Tenn. 1952.
186. *RDX, U.S. Military Spec. R-398C,* USGPO, 1973.
187. G. F. Wright and co-workers, *Can. J. Res.* **27B,** 218, 469 (1949).
188. G. F. Wright in H. Feuer, ed., *The Chemistry of Nitro and Nitroso Groups,* Part 2, Wiley-Interscience, New York, 1969–1970.
189. J. Solomon, *A Study of the Nitrolysis of Hexamine to Increase HMX Yields,* Illinois Institute of Technology Chicago, Ill., 1973.
190. W. E. Bachmann, *The Preparation of HMX, OSRD Rpt. 1981,* OSRD, 1943.
191. A. H. Vroom and C. H. Winkler, *J. Can. Res.* **283,** 701 (1950).
192. Private communication, G. S. Biasutti, Biazzi Co., Vevey, Switz., 1979.
193. R. Robbins, *The Preparation, Properties, and Uses of HMX, Rpt. RR-GC-149,* Holston Defense Corp., Kingsport, Tenn., 1958.
194. M. Baer and co-workers, *Preparation of HMX on a Semi Plant Scale, TR 2183,* PTA, 1955.
195. W. E. Bachmann and co-workers, *J. Am. Chem. Soc.* **73,** 2769 (1951).

196. J. A. Hathaway and C. R. Buck, "Report of Absence of Health Hazards Associated with RDX Manufacture and Use in Shell Loading Plants" in *Minutes of 17th Explosives Safety Seminar ADA 036015,* DDC, 1976, p. 683.
197. *HMX, Military Spec. 45444B,* USGPO, 1974.
198. L. Silberman and S. M. Adelman, "Improved Yields in the Manufacture of HMX and RDX by the Bachmann Process" in *Symposium on Processing Propellants, Explosives, and Ingredients,* ADPA, 1977, p. 2.2-1.
199. G. H. Connor, "RDX/HMX Expansion Facility Process Technology," in *Symposium on Processing Propellants, Explosives, and Ingredients,* ADPA, 1977, p. 1.6-1.
200. *Safety, Pollution, and Conservation Energy Review (Spacer) for Munitions Plant Modernization, ARLCD-SP-77001,* ARRADCOM, 1977.
201. N. Shapira, ed., Wastewater Treatment in the Military Explosives and Propellants Production Industry, ADPA, 1975.
202. A. F. McKay, *Chem. Rev.* **51,** 301 (1952).
203. D. Price and A. R. Clairmont, Jr., "Explosive Behavior of Nitroguanidine" in *Proceedings of the 12th (Intl.) Symposium on Combustion,* Combustion Institute, Pa., 1969, p. 761.
204. L. D. Sadwin, *Science* **14,** 1164 (1964).
205. S. Levmore, *Air Blast Parameters and Others Characteristics of Nitroguanidine and Guanidine Nitrate, TR 4865,* PTA, 1975.
206. J. E. Flanagan, "Relationship of Nitramine Combustion Phenomena and Chemical Structure" in *Proceedings of the 13th JANNAF Combustion Meeting, Publication No. 281,* CPIA, 1976, p. 69.
207. J. P. Picard, D. Satriana, and F. J. Masuelli, *FRL TR 10,* PTA, 1960.
208. J. L. Block, *Thermal Decomposition of Nitroguanidine, NAVORD 2705,* NOL, 1953.
209. W. C. McCrone, *Anal. Chem.* **23,** 205 (1951).
210. A. F. McKay and co-workers, *Can. J. Chem.* **29,** 746 (1951).
211. C. L. Mader, *Detonation Properties of Condensed Explosives Computed Using the Becker-Kistiakowsky-Wilson Equation of State" LA-2900,* University of California, Los Alamos, N.M., 1963.
212. A. J. Tulis, "On Intermediate Explosive Compositions" in *Proceedings of the 5th (Intl.) Pyrotechnic Seminar,* Denver Research Institute, Colo., 1976, p. 522.
213. A. J. Tulis, "Sympathetic Detonation of Ammonium Perchlorate by Small Amounts of Nitroguanidine" in *Proceedings of the 6th Symposium (Intl.) on Detonation,* ONR, 1976.
214. *Nitroguanidine (Picrite), U.S. Military Spec. N-494A,* USGPO, 1963.
215. G. B. L. Smith, V. S. Sabetta, and O. F. Steinbach, Jr., *J. Ind. Eng. Chem.* **23,** 1124 (1934).
216. C. H. Nichols, *Evaluation of Technologies to Produce Nitroguanidine, TR 4566,* PTA, 1974.
217. T. Urbanski, *Chemistry and Technology of Explosives,* Vol. 3, Pergamon Press, New York 1967, p. 22.
218. V. Milan and co-workers, *The Preparation of High Bulk Density Nitroguanidine, Rpt. 3037,* NAVORD, 1957.
219. E. J. Pritchard and G. F. Wright, *J. Can. Res.* **25B,** 257 (1947).
220. P. Aubertein, *Mem. Poudres* **30,** 143 (1948).
221. F. Olsen, *Army Ordnance* **3,** 269 (1923).
222. *Military Explosives, TM9-1300-214,* USGPO, 1967.
223. A. B. Coates, E. Friedman, and L. P. Kuhn, *Characteristics of Certain Military Explosives, TR 1507,* BRL, 1970.
224. C. L. Mader, *Detonation Properties of Condensed Explosives Computed Using the Becker-Kistiakowsky-Wilson Equation of State, LA-2900,* University of California, Los Alamos, N.M., 1963.
225. D. Price and co-workers, *DDT Behavior of Tetryl and Picric Acids, NSWC DR 76-31,* NOL, 1976.
226. T. Urbanski, *Chemistry and Technology of Explosives,* Vol. 3, Pergamon Press, New York, 1967, p. 40.
227. C. J. Bain, *Army Ordnance* **6,** 435 (1926).
228. R. C. Elderfield, *Study of the British Continuous Tetryl Process, Rpt. 661,* OSRD, 1942.
229. T. B. Stanford, Jr., *The Determination of Tetryl and 2,3-, 2,4-, 2,5-, 2,6-, 3,4-, and 3,5-Dinitrotoluene Using High-Performance Liquid Chromatography,* Batelle Memorial Institute, 1977.
230. *Tetryl, U.S. Military Spec. T-338C,* USGPO, 1971.
231. N. Shapira, ed., *Wastewater Treatment in the Military Explosives and Propellants Industry,* Vol. 1, ADPA, 1975, p. 64.
232. T. Urbanski, *Chemistry and Technology of Explosives,* Vol. 1, Macmillan Co., New York, 1964.

233. S. M. Kaye, ed., *Encyclopedia of Explosives and Related Items*, Vols. 3 p. C501 and 8, p. N51, ARRADCOM, 1978.
234. A. B. Coates, E. Friedman, and L. P. Kuhn, *Characteristics of Certain Military Explosives, TR 1507*, BRL, 1970.
235. A. O. Long, *Viscosities of Some Castable High Explosives, NAVORD 2910*, 1953.
236. D. P. MacDougall, E. H. Eyster, and A. Layton, *The Visocity of Explosive Slurries*, Rpt. 3663, 1944, Rpt. 5625, OSRD, 1945.
237. J. C. Dacons, H. G. Adolph, and M. J. Kamlet, *J. Phys. Chem.* **74**, 3035 (1970).
238. *Military Explosives, TM 9-1300-214*, USGPO, 1967.
239. V. M. Titov and co-workers, "Investigation of Some Cast TNT Properties at Low Temperatures" in *6th Symposium (International) on Detonation*, ONR, 1976, p. 34.
240. A. J. B. Robertson, *Trans. Faraday Soc.* **44**, 977 (1948).
241. J. A. Hathaway, "A Review of Reported Dose-Related Effects Providng Documentation for a Work Place Standard" in *Proceedings of the 17th Explosive Safety Seminar*, ASESB, 1975, p. 693.
242. H. H. Cady and W. H. Rogers, *Enthalpy, Density, and Thermal Coefficient of Cubical Expansion of TNT, LA-2696*, LLL, 1962.
243. M. J. Urizar, E. James Jr., and L. C. Smith, "The Detonation Velocity of Pressed TNT in *3rd (International) Symposium on Detonation*, ONR, 1960, p. 337.
244. D. J. Ornellas, *J. Phys. Chem.* **72**, 2390 (1968).
245. H. J. Gryting and co-workers, *Additives for Controlling Cracking of Explosives Made With TNT, NAVORD 5595*, 1957.
246. O. Sandus and co-workers, *Mechanism of Formation of Pink Water, ARLCD-TR-78025*, ARRADCOM, 1978).
247. *Casting of TNT, Rpt. Proj. 90-1138C on Contract DA 11-173-ORD*, Armor Research Foundation, Chicago, Ill., 1952.
248. I. Dunstan, ed., *Joint US/UK Seminar on TNT Chemistry and Manufacture, Rpts. 26 (1971) and 106 (1972)*, ERDE.
249. W. R. Cox and V. G. Phillips *Investigations Relating to the Pourability of RDX Slurries (in TNT), Rpt. No. 2/R/55*, ERDE, 1955.
250. "TNT Explosion at Radford Army Ammunition Plant," news leaflet, Army Materiel Command, Sept. 1974.
251. E. De Beule, *Bull. Soc. Chem. Belge* **42**, 27 (1933).
252. *Trinitrotoluene, U.S. Military Spec. T-248C*, USGPO, 1974.
253. S. D. Stein, *The Problem of TNT Exudation, TR 2493*, PTA, 1958.
254. R. W. Heineman and S. J. Lowell, *Prevention of Exudation from Ammunition Items, TR 2675*, PTA, 1950.
255. M. Pollack, R. T. Schimmel, and S. J. Lowell, *Composition B4, A Non-Exuding Explosive Filler for Artillery Shell, TM-1149*, PTA, 1963.
256. *Composition B-4, U.S. Military Spec. C-46652*, USGPO, 1973.
257. E. Gilbert, "An Improved Procedure for Purifying TNT" in *Symposium on Processing Propellants, Explosives, and Ingredients*, p. 2.6-1.
258. C. D. Chandler and R. A. Mundy, "Purification of Crude TNT with Magnesium Sulfite" in *Abstracts of Papers Presented at the 7th Seminar on Nitroaromatic Chemistry*, ARRADCOM, 1977, p. 7
259. A. E. Tatyrek, *Treatment of TNT Munitions Wastewaters, The Current State of the Art, Rpt. 4909*, PTA, 1976.
260. *Standard Operating Manual for Manufacture of TNT*, U.S. Rubber Co., Joliet Arsenal, Joliet, Ill., 1954.
261. YOrlova, *The Chemistry and Technology of High Explosives*, Wright Patterson Air Force Base Translation, Dayton, Ohio, 1961.
262. Ger. Pats. 710,826 (1941), 732,742 (1943), J. Meissner.
263. A. B. Bofors, *TNT Manufacture by the Continuous Bofors-Norell Method*, brochure, Sweden, 1956.
264. S. Slemrod, *Ordnance*, 525 (1970).
265. W. T. Bolleter, "Recent Improvements in the Continuous TNT Manufacturing Process" in *Symposium on Processing Propellants, Explosives, and Ingredients*, ADPA, 1977, p. 2.4-1.
266. R. L. Goldstein, "Recent Developments in the Optimization and Control of Nitration in the Continuous Manufacture of TNT" in *Symposium on Processing Propellants, Explosives, and Ingredients*, ADPA, 1977, p. 2.5-1.

267. K. B. Maline and co-workers, "Dynamic Simulation of the Continuous TNT Process" in *Proceedings of the 1973 Summer Computer Simulation Conference,* Board of Simulation Conferences, 1973, p. 354.
268. M. Halek and C. McIntosh, "New Control Instrumentation for Manufacturing Energetic Materials" in *Symposium on Processing Propellants, Explosives, and Ingredients,* ADPA, 1977, p. 1.5-1.
269. H. C. Prime, *Chem Eng.* **71**(6), 126 (1964).
270. N. Shapira, ed., *Wastewater Treatment in the Military Explosives and Propellants Production Industry,* Vol. 1, ADPA, 1972, p. 55.
271. I. Forsten, *Environ. Sci. Technol.* **7,** 9, 806 (1976).
272. *Pollution Abatement and Conservation of Energy Review for Munitions Plant Modernization, Proc. in TR 2210,* PTA, 1976.
273. Y. Orlova, *The Chemistry and Technology of High Explosives,* Vol. 2, Wright Patterson Air Force Base, Dayton, Ohio, 1961, p. 337.
274. *Ammonium Picrate (Explosive D), U.S. Military Spec. JAN A-166C,* USGPO, 1975.
275. F. W. Brown, D. H. Kusler, and F. C. Gibson, *Sensitivity of Explosives to Initiation by Electrostatic Discharge, Rpt. RI 3852,* U.S. Bureau of Mines, Pittsburgh, Pa., 1946.
276. *Properties of Explosives of Military Interest, Engineering Design Handbook Series No. 707-177,* AMC, 1971.
277. J. M. Roth in S. M. Kaye, ed., *Encyclopedia of Explosives and Related Items, PATR 2700,* Vol. 8, ARRADCOM, 1978.
278. *Picric Acid, U.S. Military Spec. JAN-A-187,* USGPO, 1945.
279. R. L. Beauregard, *History of Navy Uses of Composition A-3 and Explosive D in Projectiles, TR 70-1,* NAVORD, 1970.
280. E. E. Kilmer, *J. Spacecr. Rockets* **5,** 10, 1216 (1968).
281. H. Heller and A. L. Bertram, *HNS-Teflon: A New Heat Resistant Explosive, TR 73-163,* NOL, 1973; J. C. Dacons and E. E. Kilmer, "HNS Specifications," *paper presented at Annual Meeting of the Pyrotechnics and Explosives Application Section,* ADPA, 1976.
282. D. G. Gould, "The Thermal Stability of Hexanitrostilbene as Determined by Precise Measurements of Detonation Velocity" in *Proceedings of Symposium on Compatibility of Plastics and Other Materials with Explosives, Propellants, and Pyrotechnics, 2-F-1,* ADPA, 1976.
283. S. A. Sheffield and D. E. Mitchell, "The Equation of State and Chemical Kinetics for Hexanitrostilbene (HNS) Explosive" in *6th Symposium (Intl.) on Detonation,* Office of Naval Research, Washington, D.C. 1976, p. 250.
284. E. E. Kilmer, *Detonating Cords Loaded with HNS Recrystallized from Acid and Organic Solvents, TR 75-142,* NSWC, 1975.
285. Brit. Pat., 1,513,221 (June 7, 1978), D. A. Salter, N. F. Scilly, and K. E. Watson.
286. A. C. Schwartz, *Application of Hexanitrostilbene HNS) in Explosive Components, SC-RR-710673,* SAN, 1972.
287. J. M. Holovka and others, "The Oxidation of TNT to Hexanitrostilbene (HNS)" in *Abstracts of Papers presented at the 7th Seminar on Nitroaromatic Chemistry,* ARRADCOM, 1977.
288. U.S. Pat. 3,669,176 (June 13, 1972), L. J. Syrop.
289. V. Evens, "Optimization of TATB Processing" in *Symposium on Processing Propellants, Explosives, and Ingredients,* ADPA, 1977.
290. R. K. Jackson and co-workers, "Initiation and Detonation Characteristics of TATB" in *Proceedings of Symposium on Compatibility of Plastics and Other Materials with Explosives, Propellants, and Pyrotechnics,* ADPA, 1976.
291. R. H. Pritchard, "Compatibility of TATB-PBM with Weapon Material" in *Symposium on Compatibility of Plastics and Other Materials with Explosives, Propellants, and Pyrotechnics, 2-F-1,* ADPA, 1976.
292. T. M. Massis and D. J. Gould, *Effects of Humidity on HNAB, 73-1120,* SAN, 1974.
293. A. G. Osborn, "TATB Formulation Processing" in *Symposium on Processing Propellant, Explosives, and Ingredients,* ADPA, 1977.
294. J. H. Rouse, "Liability for Torts Arising out of the Manufacture and Transportation of Explosives or Incendiary Ordnance by the U.S. Armed Forces" in *Minutes of the 15th Explosives Safety Seminar, NTIS AD 775 660,* 1973, p. 1205.
295. D. H. Chamberlain and R. H. Stresau, "Applications of Explosives to Manufacturing: Interaction of Safety, Legal and Economic Considerations" in *Behavior and Utilization of Explosives in Engi-*

neering Design, 12th Annual Symposium, University of New Mexico, Albuquerque, N.M., 1972, p. 155.
296. "Safety, Pollution Abatement and Conservation Energy Review (Spacer) for Munitions Plant Modernization" *paper presented to the Army Research Office, ARLCD-SP-7001,* ARRADCOM, 1977.
297. M. A. Cook, *Ind. Eng. Chem.* **56,** 2, 31 (1964).
298. D. S. Gaarder, ed., *Safety and Hazards of HE Propellants, Publication No. 284,* CPIA, 1977.
299. M. F. Smith, *Hazardous Materials Transportation: Part 1, General Studies, Tech. Report NTIN-PS-76-0331,* NTIS, 1976.
300. N. Dobbs, R. Parker, and R. Hendershot, *Approved Safety Concepts for use in Modernization of USA, MUCOM (now ARRCOM) Installations, TR 4429,* 1972.
301. T. C. George, *Regulations for Transportation of Explosives and Other Dangerous Articles by Land and Water in Rail Freight Service and by Motor Vehicle (Highway) and Water. Including Specifications for Shipping Containers, Interstate Commerce Commission Tariff No. 32,* ICC, 63 Vessey Street, New York.
302. F. G. Freund, *Regulations for Transportation of Explosives and Other Dangerous Articles by Motor, Rails, and Water, Including Specifications for Shipping Containers, ICC Tariff No. 11,* American Trucking Association, 1616 P Street, NW, Washington, D.C.
303. *Rules and Regulations for Military Explosives and Hazardous Munitions, CG108,* U.S. Coast Guard, 400 7th Street, Washington, D.C.
304. E. F. Johnson, *Official Air Transport Restricted Articles, Tariff No. 6-D,* National Airport, Washington, D.C.
305. *Safety Manual, AMC Regulation 385-100 NON,* AMC, 1970.
306. *DOD Contractors Safety Manual for Ammunition, Explosives and Related Dangerous Materials, DOD 4145.26M,* Washington, D.C., U.S. Department of Defense, 1968.
307. R. R. Watson, "The United Nations System of Classification of Explosives," in *Minutes of the 16th Explosives Safety Seminar, ADA 07557,* Vol. 1, NTIS, 1974, p. 73.
308. *The Annual Proceedings of the Explosives Safety Seminars conducted under the auspices of the Explosives Safety Board, U.S. Dept of Defense,* NTIS, 1959–1978.
309. *Ann. N.Y. Acad. Sci.* **152,** Art. 1 (1968).
310. *Explosives Hazard Classification Procedures, U.S. Army TB-700-2,* Washington, D.C. 1967.
311. W. E. Baker, "A Review of Hazard Classification Test Methods for Propellants" in *Minutes of the 17th Explosive Safety Seminar, ADA036016,* DDC, 1976, p. 1487.
312. S. Fleischnick, "Some Recent Approaches in Hazards Classification" in *Proceedings of the 15th Annual Explosive Safety Seminar, U.S. Dept of Defense,* NTIS, 1973, p. 1057.
313. R. L. Beauregard and H. J. Gryting, "Standard Procedures for Approving Explosives for Service Use" in *Proceedings of the 14th Annual Explosives Safety Seminar,* NTIS, 1972, p. 815.
314. *Methods for Evaluating Explosives and Hazardous Materials, BM-IC-8541,* BuM, Pittsburgh, Pa., 1972.
315. T. Small and P. V. King, *Pyrotechnics Hazard Classification and Evaluation Program, GE-MTSD-R-054,* General Electric Corp., Bay, St. Louis, Miss., 1971.
316. *Safety and Performance Tests for Qualification of Explosives, OD 44811,* Vol. 1, NAVORD, 1972.
317. *Symposium on Explosives and Hazards and Testing of Explosives, 145th National Meeting of the American Chemical Society, Division of Fuel Chemistry* **7,** 3 (1963).
318. B. Brown, "The Need for Revision of TB 700-2" in *Minutes of the 15th Explosive Safety Seminar,* NTIS, 1973, p. 1119.
319. F. L. McIntyre, "Classification and Its Pitfalls for Pyrotechnic Compositions" in *Minutes of the 16th Explosive Safety Seminar,* NTIS, 1974, p. 479.
320. R. S. Skaar, *Fundamentals of Safety for Processing, Handling and Storage of High Energy Materials, TP-2866,* NOTS, 1962.
321. M. Halik and C. McIntosh, "New Control Instrumentation for Manufacturing of Energetic Materials" in *Proceedings of Symposium on Processing Propellants, Explosives, and Ingredients,* ADPA, 1977, p. 1.5-1.
322. R. H. Richardson and co-workers, "Hazards Analysis Through Quantitative Interpretation of Sensitivity Testing" in *Proceedings of the International Conference on Sensitivity and Hazards of Explosives,* ERDE, 1963, p. 269.
323. C. A. Kot and R. T. Anderson, "Hazards Analysis of the DDC System for a Continuous TNT Manufacturing Process" in *Proceedings of the 14th Annual Explosives Safety Seminar, U.S. Dept. of Defense,* NTIS, 1972, p. 441.

324. F. T. Kristoff and J. Digiovanni, *A Hazards Analysis Study of the Continuous TNT Manufacturing Plant*, U.S. Army Radford Ammunition Plant, Radford, Va., 1971.
325. R. H. Richardson and J. M. Sutton, "Risk Analysis" in *Proceedings of the 14th Annual Explosives Safety Seminar U.S. Dept of Defense*, NTIS, 1972, p. 467.
326. *Hazards Analysis and Systems Safety*, brochure, Illinois Institute of Technology Research Institute, Chicago, Ill., 1971.
327. K. K. Arora and J. S. Chiappa, "Safety and Hazards Analysis of the Automated Production System for 155 mm and 8 Inch Propelling Charges" in *Minutes of the 16th Explosives Safety Seminar*, NTIS, 1974, p. 1313.
328. P. O. Chelsau, *Reliability Computation Using Fault Tree Analysis, TR 32-1542*, NASA, 1971.
329. F. L. McIntyre and P. Price, *TNT Equivalency of M10 Propellant, ARLCD-CR-78008*, ARRADCOM, 1978.
330. H. S. Napadensky and co-workers, *TNT Equivalency of Three Pyrotechnic Compositions, TR 4628*, PTA, 1974.
331. H. S. Napadensky, J. J. Swatosh, Jr., and D. R. Moreta, *TNT Equivalency Studies* in ref. 295.
332. J. Swatosh and H. Napadensky, *TNT Equivalency of Nitroglycerin, TRJ 6312*, Illinois Institute of Technology, Chicago Ill., 1973.
333. J. Swatosh, J. Cook, and P. Price, *Blast Parameters of M26E1 Propellant, TR 4901*, PTA, 1976.
334. G. Petino, *Detonation Propagation Tests on Aqueous Slurries of TNT, Composition B, M9, and M10 Propellants, TR 4584*, PTA, 1974.
335. I. G. Bowen, E. R. Fletcher, and D. R. Richmond, *Estimate of Man's Tolerance to the Direct Effects of Air Blast, DASA-2113*, Defense Atomic Support Agency, Washington, D.C., 1968.
336. J. Sperrazza and W. Wokinsakis, *Ann. N.Y. Acad. Sci.* **152**, 163 (1968).
337. J. C. Beyer, ed., *Wound Ballistics*, USGPO, 1962.
338. J. Healey and co-workers, *Primary Fragment Characteristics and Impact Effects on Protective Barriers, TR 4903*, PTA, 1975.
339. B. W. Jezek, "Suppressive Shielding for Hazardous Munitions Production Operations" in *Symposium on Processing Propellants, Explosives, and Ingredients*, ADPA, 1977, p. 5.6-1.
340. S. Glasstone, ed., *The Effects of Nuclear Weapons*, Supt. of Documents, USGPO, 1962.
341. *Structures to Resist the Effects of Accidental Explosions*, Dept. of the Army Technical Manual TM5-1300, Washington, D.C., 1969.
342. J. Healey and co-workers, *Design of Steel Structures to Resist the Effects of HE Explosions, TR 4837*, PTA, 1975.
343. *Chemical Rocket/Propellant Hazards,"* Vols. 1–2, CPIA, 1970–1971.
344. N. Dobbs and co-workers, "Design of Steel, Masonary and Precast Concrete Structures to Resist the Effects of HE Explosions" in *17th Annual Dept. of Defense Explosive Safety Seminar*, NTIS, 1976.
345. *Symposia on Environmental Pollution and Energy Research*, sponsored by the American Defense Preparedness Association, Washington, DC., ND-01P to ND-08P, 1970–1976; NC-02T, 3 Vols., 1975.
346. N. A. Shapira, ed., *Wastewater Treatment in the Military Explosives and Propellant Production Industry, Report NC-02T*, 3 Vols., ADPA, 1975.
347. "Safety, Pollution Abatement and Conservation Energy Review (Spacer) for Munitions Plant Modernization," paper presented to the Army Research Office, ARLCD-SP-7001, ARRADCOM, 1977.
348. *Pollution Abatement Engineering Programs for Munitions Plant Modernization: 5th Briefing for Senior Scientist Steering Group: TM 2170*, PTA, 1975.
349. B. Kozlorowski and J. Kucharski, *Industrial Waste Disposal*, Pergamon Press, New York, 1972.
350. J. W. Patterson and R. A. Minear, *State of the Art for the Inorganic Chemicals Industry: Commercial Explosives, 600/2-74-009b*, Environmental Protection Agency, Washington, D.C., 1975.
351. R. K. Andren and co-workers, "Removal of Explosives from Wastewater" *Proceedings of the 30th Purdue Industrial Waste Conference*, Purdue University, Lafayette, Ind., 1975.
352. A. M. Anzalone and co-workers, *Pollution Abatement—A Selected Bibliography with Abstracts and Indexes, ARLCD-SP-77003*, ARRADCOM, 1978.
353. *Annual Status Report on Environmental Programs and Activities*, Edgewood Arsenal, Md., 1977.
354. D. H. Rosenblatt, *Munitions Production Products of Potential Concern as Water Borne Pollutants, Phase I Technical Report USAMEERU-73-07*, Army Medical Environmental Engineering Research Unit, Edgewood Arsenal, Md., 1973.
355. H. E. Lund, *Industrial Pollution Control Handbook*, McGraw-Hill Book Co., New York, 1971.

356. L. L. Smith, *Propellant Plant Pollution Abatement, Status Report*, PTA, 1973.
357. L. L. Smith, *Propellant Plant Pollution Abatement Engineering Investigation to Develop Optimum Control Measures to Prevent Water Pollution, TR 4818*, PTA, 1975.
358. R. Zimmerman, "Industrial Wastewater Coefficient and Water Managment," paper presented at 28th Conference on Industrial Wastes, Purdue University, Ind., 1973.
359. L. R. Olsen and R. Major, *Comparative Study of Dehydrating Processes in the Manufacturing of Nitrocellulose*, 3 Vols., Chemical Systems Division, Sunnyvale, Calif., 1975.
360. F. D. Lonadier, W. H. Hedley, and L. D. Haws, "Biodegradability of Explosives" in *Minutes of the 15th Explosives Safety Seminar, AD 775660*, NTIS, 1973, p. 797.
361. Private communication, I. Forsten, PTA, Dover, N.J.
362. *Pollution Abatement and Conservation of Energy Review for Munitions Plant Modernization, TM 2210*, PTA, 1976.
363. L. R. Harris, *Abatement of High Nitrate Concentrations at Munitions Plants: A State of the Art Review, TR 4568* PTA, 1973.
364. E. Pregun, *Elimination of Sulfate Wastes, TM 2130*, PTA, 1973.
365. B. T. Federoff and O. E. Sheffield, "Loading and Fabrication of Explosives" in *Encyclopedia of Explosives and Related Items, TR2700*, Vol. 7, PTA, 1975, p. 146.
366. J. E. Ablard, *Composition B: A Very Useful Explosive, NAVSEA-03-TR-058*, Naval Sea Systems Command, Washington, D.C., 1977.
367. *Composition B, U.S. Military Spec. C-401D*, USGPO, 1974.
368. *Composition B-3, U.S. Military Spec. C-45113*, USGPO, 1958.
369. *Composition B-4, U.S. Military Spec. C-46652*, USGPO, 1962.
370. C. E. Jacobson and co-workers, *Evaluation of Materials for Use as Desensitizers in Composition B, TR 2425*, PTA, 1957.
371. R. Pellon and K. Russel, *Study of Melt Loading the 105mm M1 Projectile with Composition B Containing Grade B Wax, TR 4854*, PTA, 1975.
372. M. J. Margolin and E. A. Skettini, *Ammunition Loading Techniques, Explosives Dev. Sec. Rpt. 43*, PTA, 1958.
373. R. W. Heineman, *Control of Exudation, TR-2568*, PTA, 1958.
374. S. D. Stein, *The Problem of TNT Exudation, TR 2493*, PTA, 1958.
375. *Cyclotol, U.S. Military Spec. C-13477C*, USGPO, 1975.
376. J. E. Ablard, *HBX-1: Its History and Properties, NAVSEA-03-TR-021*, Washington, D.C., 1975.
377. *Explosive Compositions, HBX Type, U.S. Military Spec. E-22267A*, USGPO, 1963.
378. E. James, Jr., "Charge Preparation for Precise Detonation Studies" in *Proceedings of the 4th Symposium (Intl) of Detonation*, Office of Naval Research, Washington, D.C. 1965, p. 1.
379. *Ordnance Explosive Train Designer's Handbook, NOLR 1111*; NOL, 1952.
380. *Preparation and Properties of RDX Composition A, Rpt. 5626*, OSRD, 1945.
381. *Preparation and Properties of Plastic High Explosives, Rpt. 5633*, OSRD, 1946.
382. R. L. Beauregard, *History of Navy Usage of Composition A3 and Explosive D in Projectiles, TR 70-1*, NAVORD, 1970.
383. C. C. Misener, *Capabilities of the Explosive Loading Group of the Chemical Engineering Division, NAVORD 6873*, NOL, 1960.
384. E. James, *Development of Plastic Bonded Explosives, UCRL 12439-T*, University of California, 1965.
385. E. Y. McGann, *A Safety, Quality, and Cost Effectiveness Study of Composition A3 Press Loading Parameters, TR 76-1*, NWS, 1976.
386. S. M. Kaye, ed., *Encyclopedia of Explosives and Related Items, TR 2700*, Vol. 8, ARRADCOM, 1978, p. 60.
387. J. R. Polson, *Mechanical Pressing of Explosives*, Iowa Army Ammunition Plant, Iowa, 1973.
388. P. B. Archibald, *Ind. Eng.* **53**, 9, 737 (1961).
389. D. Kite, Jr., A. K. Behlert, and E. Jerzcerewski, *Plastic Bonded Explosives for use in Ammunition, PATM 2-2-62*, PTA, 1962.
390. J. McDevitt, "Processing Characteristics of Castable PBX Explosives" in *Symposium of Processing Propellants, Explosives, and Ingredients*, ADPA, 1977, p. 3.2-1.
391. C. D. Lind and co-workers, *Techniques for Injection Loading of PBXC-303(1) Explosive, TP 5615*, NAVWEPS, 1974.
392. E. James and L. C. Smith, *Plastic Bonded RDX: LA-1448*, University of California, 1952; *Composition C4, U.S. Military Spec. C45010A*, 1972.

393. H. L. Flaugh, *Properties of X-0280 and 0281, Rpt. 5981-MS*, LASL, 1975; Composition A5, U.S. Military Spec. E-14970B, 1976.
394. *General Specifications on Molding Powder, RM-252356*, Lawrence Livermore Laboratory, Calif.
395. *Specification LX-14, RM-253683*, Lawrence Livermore Laboratory, Calif.
396. *Plastic Bonded HMX (95/5) Powder*, U.S. Military Spec. P-50854, 1971.
397. *Powder, Molding Compound Explosive (PBX)*, U.S. Military Spec. P-14999, 1975.
398. *Powder, Molding, PBX 9404*, U.S. Military Spec. P-45446A, 1962.
399. *Powder, Molding, PBX 9010*, U.S. Military Spec. P-45447, 1960.
400. *RDX/Kel-F Resin Molding Powder*, U.S. Military Spec. R-48270, 1976.
401. B. M. Dobraetz, *Properties of Chemical Explosives and Explosive Simulants*, UCRL 51319 Rev. 1, Lawrence Livermore Laboratory, Calif., 1974.
402. *Apparent Consumption of Industrial Explosives and Blasting Agents in the United States, Annual Mineral Industry Surveys 1912–1975*, U.S. Dept. of Interior, Washington, D.C., 1976.
403. N. Shapira, ed., *Wastewater Treatment in the Military Explosives and Propellants Production Industry*, ADPA, 1975.
404. R. W. Watson, J. E. Hay, and R. W. Van Dolah, Commercial Explosives in the United States: Generalities and Some Details" p. 13 in *Symposium on Military Applications of Commercial Explosives, DREV M-2241/72*, Defense Research Establishment, Valcartier, Can., 1972, p. 13.
405. C. E. Gregory, *Explosives for North American Engineers*, Trans Tech Publications, Clausthal, Germany, 1973.
406. T. Urbanski, *Chemistry and Technology of Explosives*, Vol. 3, Pergamon Press, New York, 1964, p. 395.
407. J. Taylor and P. F. Gay, *British Coal Mining Explosives*, Newnes, London, Eng., 1958.
408. B. T. Federoff and O. Sheffield, *Encyclopedia of Explosives and Related Items, TR 2700 V. 5 D 1584 PTA*, 1972.
409. P. Naoum, *Nitroglycerin and Nitroglycerin Explosives*, Williams and Wilkens Co., Baltimore, Ind., 1928.
410. P. A. Richardson and C. M. Mason, *Active List of Permissible Explosives and Blasting Devices Approved Before July 1, 1970, Information Circular 8493*, BuM, 1960.
411. N. E. Hanna and co-workers, *Factors Influencing the Incendivity of Permissible Explosives*, I.C. 5867, BuM, 1961.
412. S. P. Howell, J. W. Paul, and L. Sherrick, *Progress of Investigations on Liquid Oxygen Explosives, Technical Paper 294*, BuM, 1923.
413. *Safety and Performance Characteristics of Liquid Oxygen Explosives, Bulletin No. 472*, BuM, 1941.
414. J. J. Yancik and G. B. Clark, "Some Detonation Properties of Ammonium Nitrate" in *Proceedings of 5th Annual Symposium on Mining Research*, University of Missouri, Rolla, Mo., 1960, p. 67.
415. G. W. Brown and E. J. Styskala, "Development of Stengel Process FGAN Used in Blasting Agents" in *Proceedings of 5th Annual Symposium on Mining Research*, University of Missouri, Rolla, Mo., 1960, p. 126.
416. B. Federoff and O. Sheffield, *Encyclopedia of Explosives and Related Items, p. A311, TR 2700*, PTA, 1960.
417. T. Urbanski, *Chemistry and Technology of Explosives*, Vol. 2, Pergamon Press, New York, 1965, p. 450.
418. J. W. Mellor, *Comprehensive Treatise on Inorganic and Theoretical Chemistry*, Suppl. 1, Longmans, Green and Company, Ltd., London, Eng., 1922–1937, p. 506.
419. J. G. Stites, M. D. Barnes, and R. F. McFarlin, "A Survey of the Physical and Chemical Characteristics of Fertilizer-Grade Ammonium Nitrate" in *Proceedings of 5th Annual Symposium on Mining Research*, University of Missouri, Rolla, Mo., 1960, p. 1.
420. R. F. Bruzewski and K. M. Kohler, "An Investigation of Some Basic Performance Parameters of Ammonium Nitrate Explosives" in *Proceedings of 4th Annual Symposium on Mining Research*, University of Missouri, Rolla, Mo., 1959, p. 175.
421. D. T. Baily and co-workers, "Ammonium Nitrate-Fuel Oil Systems: Their Density, Velocity, Strength and Sensitivity" in *Proceedings of 5th Annual Symposium on Mining Research*, University of Missouri, Rolla, Mo., 1960, p. 50.
422. S. R. Brinkley, W. G. Stoesm, and S. Mayers, "Some Effects of Particle Size Reduction on the Explosive Properties of ANFO Blasting Agent" in *Proceedings of 5th Annual Symposium on Mining Research*, University of Missouri, Rolla, Mo., 1960.

423. S. R. Brinkley and W. E. Gordon, *Explosive Properties of the Ammonium Nitrate Fuel–Oil System,* Hammer Coal Co., Cady, Ohio.
424. S. R. Brinkley and W. E. Gordon, "Explosive Properties of the Ammonium Nitrate–Fuel Oil System" in *Proceedings of 31st Intl. Congress of Industrial Chemistry,* Liege, Belg., 1958.
425. *Annual Proceedings of the International Symposia on Mining Engineering, Technical Bulletins 92, 94, 95, 97 and 98,* University of Missouri School of Mines, Rolla, Mo., 1956–1960.
426. *Proceedings of the Annual Mining Symposia,* University of Minnesota, Minneapolis, Minn.
427. J. J. Yancek, *ANFO Manual,* Monsanto Co., St. Louis, Mo., 1969.
428. W. E. Tournay and co-workers, "Some Studies in Ammonium Nitrate-Fuel Oil Compositions" *Proceedings of 4th Annual Symposium on Mining Research,* University of Missouri, Rolla, Mo., 1959, p. 164.
429. J. Briggs, "A Safer Blast for the Modern Army" in *Proceedings 14th Annual Explosives Safety Seminar,* U.S. Department of Defense Explosive Safety Board, NTIS, 1972, p. 313.
430. U.S. Pats. 2,930,685 (1960), 3,121,036 (1964), M. A. Cook and H. E. Farnham.
431. *Proceedings of the Symposium on Military Applications of Commercial Explosives, DREV M-2241/72,* Defense Research Establishment, Valcartier, Can., 1972.
432. R. A. Dick, *The Impact of Blasting Agents and Slurries on Explosive Technology, Rpt. I.C. 8560,* BuM, 1972.
433. M. A. Cook, *Science* **132,** 1105 (1960).
434. M. A. Cook, *J. Ind. Eng. Chem.* **60,** 7, 44 (1968).
435. J. F. Dixon, "Development of Physical Properties and Techniques Suitable for Commercial Exploitation of Slurry Explosives" in *Proceedings 4th Annual Symposium on Mining Research,* University of Missouri, Rolla, Mo., 1959, p. 124.
436. S. Levmore, *Principle Characteristics of the Gelled Slurry Explosive DBA-2214, TR 4237,* PTA, 1971.
437. R. A. Dick, *Factors in Selecting and Applying Commercial Explosives and Blasting Agents, I.C. Rpt. 9405,* BuM, 1968.
438. R. A. Dick, *Pit Quarry,* (July 1971).
439. M. A. Cook, "Water Compatible Ammonium Nitrate Explosives for Commercial Blasting" in *Proceeding of 4th Annual Symposium on Mining Research,* University of Missouri, Rolla, Mo., 1960, p. 101.
440. H. E. Farnham, "Large-Scale Use of Ammonium Nitrate Slurries by the Iron Ore Co. of Canada" in *Proceedings of 4th Annual Symposium on Mining Research,* University of Missouri, Rolla, Mo., 1960, p. 140.
441. A. M. Goldstein and E. N. Alter, in R. L. Whistles, ed., *Industrial Gums,* Academic Press, Inc., New York, 1959, Chapt. 14.
442. F. Smith and R. Montgomery, *The Chemistry of Plant Gums and Mucilages, Monograph Series 141,* Chapman and Hall, London, Eng., 1959, Ch. 16.
443. L. Penn and co-workers, *Determination of Equation-of-State Parameters for Four Types of Explosives, UCRL 5189,* LLL, 1975.
444. M. A. Cook, *The Science of Industrial Explosives,* Graphic Service and Supply, IRECO Chemicals, Salt Lake City, Utah, 1974.
445. M. A. Cook, *The Science of High Explosives,* Reinhold Publishing Co., New York, 1958.

General References

AMCP Engineering Design Handbooks, National Technical Information Service, NTIS, Department of Commerce, Springfield, Va.
Military Specifications; these are informative unclassified references describing characteristics of materials, method of sampling, testing, packaging, etc, eg, *MIL-STD-650, Explosives: Sampling, Inspection, and Testing,* U.S. Government Printing Office, Washington, D.C., 1962.
Chemical Propulsion Information Agency publications, (may be classified); they contain information relating to the explosive ingredients used in propellants as well as the entire field of propellants, rockets, and guns, Chemical Propulsion Information Agency, Applied Physics Laboratory/John Hopkins University, John Hopkins Road, Laurel, Mo.
Defense Documentation Center Reports, U.S. Department of Defense.
B. T. Federoff and O. E. Sheffield, eds., *The Encyclopedia of Explosives and Related Items, PATR 2700 CO. 1-8.*

A. H. Blatt, *Bibliography of OSRD Reports issued by Division 8 of the National Defense Research Committee, OSRD Report No. 6630,* 1946; Reports of the extensive work done in World War II, contain information still useful.

Symposia (Intl.) on Detonation, Office of Naval Research, Washington, D.C., Unclassified compilations of papers presented at the Symposia held every five years, starting in 1951, obtained from the clearing house for Federal Scientific and Technical Information, Springfield, Va., the official agency for providing unclassified research reports of various types to the public.

D. C. Ascani, bibliography on explosives in "The Literature of Chemical Technology" in *Advances in Chemistry Series No. 76,* American Chemical Society, Washington, D.C. 1968, p. 574.

Behavior and Utilization of Explosives in Engineering Design, 12th Annual Symposium, American Society Mechanical Engineers, University of New Mexico, Albuquerque, N.M., 1972.

Encyclopedia of Explosives, Ordnance Liaison Group, Durham, N.C., 1960.

Handbook of Foreign Explosives, FSTC 381-5042, AMC, 1965.

M. A. Cook, *The Science of High Explosives,* Reinhold Publishing Corp. New York, 1958.

C. J. Johansson and P. A. Persson, *Detonics of High Explosives,* Academic Press, Inc., New York, 1970.

J. Taylor, *Detonation in Condensed Explosives,* Clarendon Press, Oxford, Eng., 1952.

R. H. Cole, *Underwater Explosives,* Princeton University Press, Princeton, N.J., 1948.

E. W. Baker, *Explosives in Air,* University of Texas Press, Austin, Tex., 1973.

T. Urbanski, *Chemistry and Technology of Explosives,* Pergamon Press, New York, 1967.

M. A. Cook, *The Science of Industrial Explosives,* Ireco Chemicals, Utah, 1974.

S. Fordham, *High Explosives and Propellants,* Pergamon Press, London, Eng., 1966.

Military Explosives, TM-9-1300-214, USGPO, 1967.

Apparent Consumption of Industrial Explosives and Blasting Agents in the United States, Annual Mineral Industries Surveys 1912-1975, U.S. Dept. of Interior, Washington, D.C., 1976.

H. Schuck and R. Sohlmann, *The Life of Alfred Nobel,* Heinman, London, Eng. 1929.

E. Bergengren, *Alfred Nobel, the Man and His Work,* Thomas Nelson and Son, Ltd, London, Eng., 1960.

B. T. Federoff and O. Sheffield, *Encyclopedia of Explosives and Related Items, TR 2700,* Vol. 7, H114, PTA, 1975; A detailed chronology of the history of explosives.

J. C. Dacons, *New Heat Resistant Explosives, NOLTR 65-87,* 1965.

Technical Information on Military Specialties TACOT, E. I. du Pont de Nemours & Co., Inc., Wilmington, Del.

B. Wells, *High Energy Flexible Explosive, TR 4846,* PTA, 1976, TR 4713, PA, 1974.

Kegler and Schall, "Mechanical and Detonation Products of Rubber Bonded Sheet Explosives" in *4th Symposium (International) on Detonation,* ONR, 1965, p. 496.

New Explosive Specialties Brochure on Line Wave Generators, E. I. du Pont de Nemours & Co., Inc., Explosives Products Division, Wilmington, Del.

K. O. Brauer, *Handbook of Pyrotechnics,* Chemical Publishing Co., New York, 1974; provides a good description of special purpose components which use explosives, propellants or pyrotechnics as their source of energy.

S. A. Moses, "Explosive Components for Aerospace Systems," p. 235, G. Cohen, "Low Explosive Devices for Performing Mechanical Functions", p. 189, in *Behavior and Utilization of Explosives in Egnineering Design, Proceeding of 12th Annual Symposium,* University of New Mexico, Albuquerque, N.M., 1972.

O. A. Klamer, "Composition B Wax Study," *paper presented at 1974 Annual Meeting of the Loading Section—Ammunition Technology Div.,* American Defense Preparedness Association, Washington, D.C.

E. A. Timmons, "Pyrotechnic Systems and Devices Used in the NASA Space Shuttle" in *Proceedings of the Fifth International Pyrotechnics Seminar,* Denver Research Institute, University of Denver, Denver, Col., p. 503.

S. D. Stein, C. J. Horvat, and O. E. Sheffield, *Some Properties and Characteristics of HBX-3 and H6 Explosives, TR 2431,* PTA, 1957.

Report on Minol and Torpex, Report 4243, OSRD, 1944; early work on aluminized explosives.

E. H. Eyster and M. A. Paul, *Composition B (Cyclotol), Report 1167,* OSRD, 1943; early work on RDX-TNT compositions.

The Preparation and Properties of RDX-Composition A, Report 5626; OSRD, 1945; early work on plastic coated RDX.

T. W. Stevens, D. E. Seeger, and D. H. Stone, *Development of the M5 and M5A1 Demolition Blocks, TR 2332,* PTA, 1956.

Plastic Bonded HMX (95/5) Powder, U.S. Spec. MIL-P-50854, 1971.
Powder Molding Compound Explisive (PBX), U.S. Spec. MIL-P-14999, 1975.
Powder, Molding PBX 9404, U.S. Spec. MIL-P-45446A, 1962.
Powder, Molding PBX 9010, U.S. Spec. MIL-P-45447, 1960.
Composition B, U.S. Spec. MIL-401D, 1974.
Cyclotol, U.S. Spec. MIL-C-13477C, 1975.
H-6, HBX-1, HBX-3, U.S. Spec. MIL-E-22267A, 1963.
Octol, U.S. Spec. MIL-O-45445A, 1962.
RDX/KELF Resin Molding Powder, U.S. Spec. MIL-R-48270, 1976.
M. A. Cook, *The Science of Industrial Explosives, Graphic Services and Supply,* IRECO Chemicals, Salt Lake City, Utah 1974; a review of mechanisms of performance and in particular of slurry explosives.
S. M. Kaye, ed., "Nobel, Alfred Bernard" in *Encyclopedia of Explosives and Related Items* Vol. 8, N 165, TR 2700, ARRADCOM, 1978.
"Principles of Explosive Behavior" in *Engineering Design Handbook AMC 706-180,* AMC, 1972.

Victor Lindner
U.S. Army Armament Research and Development Command
(ARRADCOM)

PROPELLANTS

Propellants are mixtures of chemical compounds that produce large volumes of gas at controlled, predetermined rates. Their major applications are in launching projectiles from guns, rockets, and missile systems. Propellant-actuated devices are used to drive turbines, move pistons, operate rocket vanes, start aircraft engines, eject pilots, jettison stores from jet aircraft, pump fluids, shear bolts and wires, and act as sources of heat in special devices. They are applicable wherever a well-controlled force must be generated for a relatively short period of time. Solid propellants are compact, have a long storage life, and may be handled and used without exceptional precautions. Propellants are employed in guns in the form of dense grains or sheets of plasticized nitrocellulose which may contain other compounds such as nitroglycerin and diethylene glycol dinitrate to increase available energy and mechanical performance, inert liquid plasticizers such as dibutyl phthalate to improve physical and processing characteristics, explosives such as nitroguanidine or cyclotetramethylenetetranitramine (HMX) to improve performance, stabilizers such as diphenylamine and ethyl centralite to increase storage life and small amounts of inorganic additives to facilitate handling, improve ignitibility, and decrease muzzle flash. Propellants containing only plasticized nitrocellulose as energy source are referred to as single-base propellants. Double-base propellants also contain a liquid explosive plasticizer such as nitroglycerin. Multi- or triple-base propellants incorporate a crystalline explosive such as nitroguanidine in the double-base formulation. Double and triple-based nitrocellulose propellants are used in rockets as well as guns.

Rocket propellants are based either on nitrocellulose or on synthetic cross-linked polymeric binders such as polysulfides, polyurethanes, or polybutadienes. Nitrocellulose-based rocket propellants invariably contain an energetic liquid plasticizer and a stabilizer and may contain nonenergetic plasticizers, ballistic modifiers, inorganic oxidizing salts and organic explosives, metallic fuels, and other substances. Single-base propellants are not used in rockets because of their relatively low energy content, and unsatisfactory mechanical and combustion characteristics. Composite-modified double-base rocket propellants generally contain an inorganic crystalline oxidizer such as ammonium perchlorate and a metallic fuel, eg, aluminum in a double-base matrix.

Polymer-based rocket propellants, generally referred to as composite propellants, contain cross-linked polymers that act as a viscoelastic matrix for holding a crystalline inorganic oxidizer such as ammonium perchlorate and for providing mechanical strength. Many other substances may be added including metallic fuels, plasticizers, extenders, and catalysts. Polymer-based composite propellants are too erosive to be used in guns and because of the residues formed after repeated firings.

Nitrocellulose, the most important new explosive discovered during the 19th century, became useful as a propellant only after its tendency to decompose rapidly was overcome and its fibrous structure was eliminated to obtain controllable burning in a gun. Von Lenk (1862) found that surface acid and other degradation products of nitration could be removed by washing the nitrocellulose fibers. In 1865 Abel discovered that occluded acid could be neutralized by tearing and shredding the fibers with commercial beaters of the type used in the paper industry. Nobel (1889) suggested the addition of mildly alkaline organic compounds such as diphenylamine to combine with oxides of nitrogen formed during the exothermic decomposition of nitrocellulose. These compounds act as stabilizers, preventing autocatalysis of the decomposition reaction. The conversion of stabilized fibrous nitrocellulose to dense, nonporous gun propellant grains was finally accomplished between 1885 and 1889. Fibrosity was eliminated by a gelatinization or plasticization process after which the nitrocellulose could be converted to dense strands or sheets by extrusion or rolling. Vielle (1885) used ether–alcohol (1:1) for gelatinization. Nobel (1883) employed nitroglycerin to form a higher-energy solvent-free propellant. Abel and Dewar (1889) incorporated acetone as a solvent for nitroglycerin and nitrocellulose. These plasticization techniques are now widely used to make small nitrocellulose grains for guns and mortars and to some extent for small rockets. The newer ball-powder process developed by Olsen, Tibbitts and Kerone in 1936 is used to make nitrocellulose propellants chiefly for small arms.

Typical components of nitrocellulose propellants are listed in Table 1.

Following World War II, requirements for larger rocket propellant structures of complex shape could not be met by grain extrusion. The solution lay in the exploitation of cast propellants where the ingredients are mixed, cast into a mold, and cured into rigid configurations. Energy levels were increased by the addition of aluminum as a fuel and of high-energy explosive oxidizers such as HMX. Propellants have been cast in rigid bodies up to 5–6 m diameter and 35–40 m in height, weighing more than 1000 metric tons, and with total thrusts of the order of 1 GN·s (2.5×10^8 lbf·s). Propellant compositions are now made that withstand long-term storage and environmental stresses and meet the broad range of military, industrial, and research requirements (1–6).

Table 1. Typical Components of Nitrocellulose Propellants and Their Function

Component	Application
nitrocellulose	energetic polymeric binder
polyglycol diols	
nitroglycerin, metriol trinitrate, diethylene glycol dinitrate, triethylene glycol dinitrate, dinitrotoluene	plasticizers: energetic
dimethyl, diethyl or dibutyl phthalates, triacetin	plasticizers: fuels
diphenylamine, diethyl centralite, 2-nitrodiphenylamine	stabilizers
organic and inorganic salts of lead; eg, lead stannate, lead stearate, lead salicylate	ballistic modifiers
carbon black	opacifiers
lead stearate, graphite, wax	lubricants
potassium sulfate, potassium nitrate, cryolite (potassium aluminum fluoride)	flash reducers
ammonium perchlorate, ammonium nitrate	oxidizers inorganic
RDX, HMX, nitroguanidine and other nitramines	organic
aluminum	metallic fuels cross-linking catalysts
lead carbonate	defouling agents
tin	

The development of the castable composites started in 1942 at the Guggenheim Aeronautical Laboratory, California Institute of Technology, with the formulation of mixes containing ca 25% asphalt and 75% potassium perchlorate. These were rapidly superseded by a variety of cross-linked elastomer-based propellants with superior mechanical properties and increased available energy which contained as much as 90% crystalline filler in the polymeric matrix. The polymers that have found the greatest use as binders include polysulfides, polyurethanes, and polybutadienes. The polysulfides, developed in the late 1940s and early 1950s, start with low viscosity partially polymerized mercaptyl (-SH)-terminated compounds which are cured by oxidation by compounds such as *p*-quinone dioxime.

Other ingredients are sulfur as a cure catalyst, dibutyl phthalate as plasticizer, magnesium oxide to improve thermal stability, iron oxide and ferrocene as burning rate catalysts, and copper phthalocyanine as a burning rate suppressor. Polysulfide-based propellants have been replaced for most purposes by binders with superior characteristics. The polyurethanes were developed during the 1950s to take advantage of the long-chain polyalcohols which were becoming available in a wide molecular weight range. With diisocyanates these formed stable polymers which could be used in large, case-bonded rocket motors. Other components added include ferric acetylacetonate used as a cure catalyst, dioctyl sebacate and dioctyl adipate as plasticizers, sodium lauryl sulfate as an antifoaming agent, magnesium oxide and phenylnaphthylamine as stabilizers, ballistics modifiers such as metal oxides and copper chromite, and carbon black as a radiant heat diffuser. The presence of moisture tends to cause degradation of polyurethanes during high temperature storage. The polybutadienes were first made during the late 1950s by curing the liquid copolymer of butadiene and acrylic acid (PBAA) with epoxides and/or polyfunctional imines. The

PBAA binders were replaced during the 1960s by superior terpolymers of butadiene, acrylic acid, and acrylonitrile (PBAN). The PBANs have been extensively used in high performance rockets. During the early 1960s, polybutadiene prepolymers were developed with terminal carboxy groups. These were cured with epoxides or aziridines to form polymers that have better low temperature mechanical properties than the PBANs. Hydroxy-terminated polybutadienes cured with isocyanates were also developed during this period; they have superior processing properties and low temperature properties. Many other types of polymers have been proposed, but none have found widespread use. Typical components of composite propellants are listed in Table 2 (7–19).

A method for preparing large grains of cast nitrocellulose propellants was devised by Kincaid and Shuey during World War II. Casting comprises a two-stage solvation process in which small cylinders of a casting powder composed of single- or double-based nitrocellulose grains are joined by partial solution of the nitrocellulose in a casting solvent to form a monolithic grain. Inorganic oxidizers and fuels may be added. Ballistic modifiers such as lead stearate are used to control burning. Plasticizers such as triacetin (glyceryl triacetate) improve low temperature performance (1,20).

Slurry-cast propellants were developed to simplify the casting and reduce cost but have not found extensive use (21–24).

Selection

For the selection of gun and rocket propellants, the total amount of energy required has to be considered as well as the rate at which it must be delivered to meet system performance requirements and space available. The total energy delivered depends on the chemical energy of the propellant components, the characteristics of the products of combustion, the chemical equilibria which prevail among the reaction products, and the efficiency with which the system converts thermal to kinetic energy. The rate at which energy is produced depends on the intrinsic burning characteristics of the propellant, its burning surface area, and the operating pressure and temperature of the system in which the propellant is used. Control of the burning surface area is obtained by using appropriate grain geometries and the required number of grains. Uncontrolled burning can result in intolerably high pressures or, in the worst case, catastrophic detonation (25–26).

The thermochemical–thermodynamic factors affecting gun and rocket performance are essentially the same. The highest energy propellants produce the largest volume of gas per unit weight of propellant at the highest flame temperature. The selection of propellant compositions for maximum performance focuses on high density compositions that form highly exothermic low-molecular-weight combustion products which are stable with minimum dissociation at gun or rocket operating pressures. Many practical considerations limit the attainment of the theoretical maximum performance. High-flame-temperature propellants used in rockets may cause excessive nozzle erosion and dissociation of the gaseous products at the relatively low operating pressures in rocket chambers. Their use in guns causes excessive gun tube wear and muzzle flash. The incorporation of large percentages of nitroguanidine and the proposed use of nitramines such as RDX in gun propellants are intended to produce the maximum energy at the lowest possible flame temperature. The isochoric adiabatic flame temperatures of propellants in use ranges from ca 2000 to 3500 K. The impetus of gun propellants

Table 2. Typical Components of Composite Rocket Propellants

Binders	Characteristics
polysulfides	reactive group, mercaptyl(–SH), is cured by oxidation reactions; low solids loading capacity and relatively low performance; now mostly replaced by other binders
polyurethanes polyethers polyesters	reactive group, hydroxyl(–OH), is cured with isocyanates; intermediate solids loading capacity and performance
polybutadienes copolymer, butadiene and acrylic acid	reactive group, carboxy(–COOH) or hydroxyl(–OH), is cured with difunctional epoxides or aziridines
	intermediate solids loading capacity and better performance than polyurethanes; less than adequate cure stability and mechanical characteristics
terpolymers of butadiene, acrylic acid and acrylonitrile	superior physical properties and storage stability
carboxy-terminated polybutadiene	cured with difunctional epoxides or aziridines; have very good solids loading capacity, high performance and good physical properties
hydroxy-terminated polybutadiene	cured with diisocyanates; have very good solids loading and performance characteristics and good physical properties and storage stability
Oxidizers	
ammonium perchlorate	most commonly used oxidizer; it has a high density, permits a range of burning rates, but produces smoke in cold or humid atmosphere
ammonium nitrate	used in special cases only, it is hygroscopic and undergoes phase changes, has a low burning rate and forms smokeless combustion products
high energy explosives (RDX–HMX)	have high energy and density; produce smokeless products; have a limited range of low burning rates
Fuels	
aluminum	Al most commonly used; has a high density; produces an increase in specific impulse and smoky and erosive products of combustion
metal hydrides	provide very high impulse, but generally inadequate stability, give smoky products, and have a low density
Ballistic modifiers	
metal oxides	iron oxide most commonly used
ferrocene derivatives	permit a significant increase in burning rate
other	coolants for low burning rate and various special types of ballistic modifiers
Modifiers for physical characteristics	
plasticizers	improves physical properties at low temperatures, and processability; may vaporize or migrate; can increase energy if nitrated
bonding agents	improve adhesion of binder to solids

is ca 822 to 1196 J/g (275,000–400,000 (ft·lb)/lb). The specific impulse of high performance rocket propellants is ca 2455–2700 N·s/kg (250–275 lbf·s/lb) (27–28).

Performance Calculations. The energy evolved by a propellant may be estimated from its composition, its reaction products and the heats of formation of the reactants and the gases and solids produced. The composition and flame temperature of the products are determined from the applicable enthalpy-temperature and chemical equilibrium functions of the various molecular species and the operating conditions in the combustion chamber. The high temperature gas is expanded to propel a projectile in a gun or is passed through a nozzle to impart momentum to a rocket motor. The most important thermodynamic–thermochemical characteristics of propellant combustion products, in addition to gas volume and flame temperature, are their heat capacity and heat capacity ratio γ (c_p/c_v) and the covolume of the gases at high pressures. Rigorous calculations require the solution of numerous equations which describe the mass and enthalpy balance and the chemical equilibria of the reaction products at elevated temperatures and pressures. Many computer programs have been developed for predicting rocket or gun performance. The specific impulse of a rocket propellant may be predicted with an accuracy of 1 to 2% and thrust–time and pressure–time relationships to ca 3 to 5%. Projectile velocities in guns may be estimated to ca 1%. Simplified techniques are also available for obtaining first approximations of gun and rocket performance (27–30).

In addition to energy and burning rate considerations, a propellant must meet other criteria including mechanical characteristics, stability, sensitivity, cost of manufacture, and uniformity of performance. Typical factors affecting propellant selection are listed in Table 3.

Properties

Rocket propellants should have adequate mechanical properties to enable them to withstand the stresses imposed during handling and firing. They must be capable in many cases of performing satisfactorily after undergoing the thermal stresses produced during long-term exposure and cycling at temperature extremes. The development of many new high-energy rocket propellants emphasizes maximum toughness and low shock and impact sensitivity. Their mechanical properties depend on the characteristics of the binder, the percentage of solids present, and particle size distribution. Although the stresses on gun propellant grains are less severe because of their smaller size, they must withstand much higher weapon pressures and accelerations. The options for formulation of gun propellants are usually more limited than for rocket propellants because their products of composition must not foul or corrode a gun and should have a low flame temperature.

Because of the criticality of failure of a single grain in a rocket, rocket grains are subjected to a large number of tests and inspections to establish their mechanical and physical characteristics. Well-established laboratory methods determine the tensile and compressive strengths, the modulus in tension and compression, elongation under tension, and deformation under compression. Empirical techniques correlate the data obtained with field performance, particularly for rocket propellants. High rate of load application of tensile forces simulate those generated during ignition. Low-rate tests simulate the stresses produced by differential thermal expansion. Compression test data are related to the forces experienced by rocket grains supported at the aft end of the motor and the effects of high accelerational forces on gun propellant grains. Drop tests of loaded rocket motors, and vibrational, centrifugal, and sled tests that impose acceleration forces comparable to those expected in use are among the techniques employed for assessing the mechanical adequacy of propellants. The linear coefficient of thermal expansion is important in rocket systems in which the propellant is coated with an inhibitor or bonded to the motor. Typical values are 3.6–7.2×10^{-5} m/(m·K). Thermal diffusivities generally range between 7.7 and 15.5×10^8 m^2/s. Thermal con-

Table 3. Factors Affecting Propellant Selection for Guns and Rockets

Manufacturing characteristics
    availability and cost of raw materials and processing equipment
    simplicity and cost of manufacture and inspection
    manufacturing hazards
    propellant viscosity and flowability
Energy delivery requirements
    specific impulse or force
    loading density in terms of required burning characteristics
    metal parts requirements in terms of operating pressure over required temperature range
Temperature dependence
    ignition, pressure, burning rate, and thrust characteristics over temperature range
Mechanical characteristics over temperature range
Effect of high-low temperature cycling
Reliability of performance
    lot-to-lot variations in burning rate and pressure
    effect of small variations in metal parts on performance
    effect of small variations in composition and dimensions on performance
Long-term storage characteristics
    deformation changes
    performance changes
    moisture absorption
    exudation or migration of plasticizer
Mechanical characteristics, effects of
    long-term storage
    high-low temperature cycling
    acceleration forces
    rough handling
    case bonding
Compatibility
    with process equipment
    with personnel (toxicity)
    with metal and plastic parts and other components
    of reaction products with personnel, metal parts, and electronic equipment
    erosive effects of reaction products
System requirements
    smokeless exhaust
    combustion stability
    effect of exhaust plume on radar
    absence of ignition peaks or reinforcing pressure waves
    minimum gun smoke, flash and blast pressure
    detonation free in event of malfunction

ductivities are ca 0.22–0.33 W/(m·K)[0.13–0.19 (Btu·ft)/(h·ft^2·°F)]. Specific heat values are ca 1.26×10^3 J/(kg·K) (0.3 Btu/(lb·°F). Table 4 gives the important characteristics of nitrocellulose gun propellants. The relatively low thermal conductivity of propellants accounts for the fact that only the surface of the grain heats up during the burning process because the propellant burns away faster than heat is conducted to the interior. Low thermal conductivity may, however, cause severe thermal stresses that sometimes lead to cracking in large rocket grains when abrupt changes in storage temperatures occur.

Table 4. Thermochemical, Thermodynamic, and Performance Characteristics of Nitrocellulose Gun Propellants[a]

Characteristics	M1	M2	M5	M6	M8	M9	M10	M15	M17	M26	M30	M31	IMR
heat of explosion, J/g[b]	3140	4522	4354	3182	5192	5422	3936	3350	4019	4082	4082	3370	3601
heat of formation, $-\Delta H_f$, J/g[b]	2261	2366	2407	2261	1989	1989	2533	1256	1361	2114	1549	1465	2366
flame temperature, K, T_v	2435	3370	3290	2580	3760	3800	3040	2555	2975	3130	3090	2600	2835
impetus J/g[b]	911	1121	1091	956	1181	1142	1031	980	1088	1082	1090	1000	1007
heat capacity, C_v, J/(g·K)[b]	1.46	1.51	1.46	1.46	1.42	1.51	1.42	1.51	1.51	1.46	1.51	1.51	1.46
mean heat capacity products, J/(mol·K)[b]	1.84	1.76	1.76	1.80	1.76	1.72	1.80	1.88	1.88	1.80	1.80	1.88	1.80
mean mol wt of products, g/mol	22.0	25.1	25.4	22.6	26.8	26.4	24.6	21.5	23.1	24.1	23.2	21.6	23.9
specific heat ratio of gases	1.26	1.22	1.22	1.25	1.21	1.21	1.23	1.25	1.24	1.24	1.24	1.25	1.24
gas volume, mol/g	0.045	0.040	0.040	0.044	0.038	0.038	0.041	0.046	0.043	0.042	0.042	0.044	0.042
burning rate at 20°C, cm/s at 137.9 MPa[c]	7.6	12.7	14.0	8.4	17.8	23.0	11.4	10.2	14.0	11.4	12.2	7.9	
pressure exponent	0.66	0.73		0.66	0.81	0.85	0.67	0.66	0.60	0.85	0.70	0.65	
Compositions of combustion products, mol/g × 10^2													
CO	2.33	1.54	1.61	2.24	1.28	1.13	1.81	1.45	1.15	1.89			1.97
CO_2	0.19	0.51	0.48	0.22	0.66	0.74	0.40	0.14	0.25	0.33			0.32
H_2	0.88	0.31	0.34	0.78	0.19	0.15	0.44	0.92	0.57	0.52			0.55
H_2O	0.64	1.10	1.08	0.72	0.11	0.09	0.99	0.83	1.07	0.95			0.90
N_2	0.44	0.49	0.48	0.45	0.54	0.54	0.46	1.29	1.30	0.50			0.46

[a] At loading density of 0.2 g/cm^3.
[b] To convert J to cal, divide by 4.184.
[c] To convert MPa to psi, multiply by 145.

Rocket propellants must not contain sizable cracks, pores, or cavities. They are inspected with x-rays and ultrasonics (qv) and firings are conducted in interrupted burners and in reduced or full-scale rocket motors. Gun propellants are examined microscopically for porosity and tested in closed bombs for uniformity of burning (31–38).

Stability. The chemical safe life of all standard propellants is very satisfactory. The useful service life of gun propellants may be as long as 25 to 50 yr. The useful service life of rocket propellants may be significantly less than their chemical safe life if gassing occurs or significant physical changes take place. Generally such effects are produced by high temperature storage or high-low temperature cycling, particularly if moisture is present. Very little if any degradation occurs at ambient temperature conditions. Gas formation is associated mostly with nitrocellulose–nitroglycerin type propellants. Composite propellants are not as apt to form gas. Gassing can produce internal pressures which may crack a large rocket grain or cause propellant–inhibitor bond failure or propellant–motor bond failure unless the gas can diffuse through the grain as rapidly as it is produced. The likelihood of performance failure in standard rocket systems as a result of gassing is low because of the use of chemical stabilizers and the selection of compatible inhibitors, cements, and insulation materials.

The physical changes that may reduce the service life of composite propellants include binder depolymerization, oxidation, postcure hardening, or hydrolytic breakdown if exposed to high humidity environments. Storage at both high and low temperature and cycling may cause failure of the propellant-inhibitor bond or the propellant-insulator-motor bond and can cause propellant fatigue which would increase susceptibility to cracking on rocket acceleration at low temperatures. The procedures used for estimating the service life of rocket and gun propulsion systems include tests after storage at elevated temperatures under simulated field conditions, periodic surveillance tests of systems received after storage in the field, and extrapolation of the service life from the detailed data obtained (39–42).

Sensitivity. The sensitivity of propellants to external stimuli such as shock, impact, and friction is ordinarily very low. However, the formulation of high-energy rocket propellants which may contain large percentages of relatively sensitive high explosives, such as HMX together with ammonium perchlorate and aluminum, may significantly increase the possibility of a detonation in the event of a malfunction. This is most apt to occur if a high-velocity propellant fragment strikes the motor walls, detonates, and the detonation is propagated back to the main charge. The relative sensitivity of propellants to failure of this type is assessed by card gap tests and impact tests of propellant fired at increasing velocities at a steel plate (43–47).

Transition From Burning to Detonation. Since propellants consist largely of explosive components, the controllable burning process may change to an uncontrollable detonation under certain exceptional conditions. The transition from deflagration to detonation in explosives and propellants has been intensively studied (see Explosives), and the available evidence indicates that detonation occurs only if the conditions lead to the initiation and maintenance of a high-pressure shock wave. If mechanical breakup of a rocket propellant occurs during the burning process, the large burning surface produced results in a high rate of gas evolution with correspondingly high pressures. An increasingly steep pressure front is evolved, accompanied by a pressure wave which transforms to a shock wave. Steady-state detonation may occur shortly thereafter. Detonation may also occur if fragments produced during grain breakup rebound off the motor walls with sufficient kinetic energy to initiate an impact-sensitive propellant composition. Detonation in rocket propellants is more likely to occur with high-energy propellants that have a high crystalline filler content of energetic high explosives such as RDX or HMX. It may occur in gun propellants if high loading densities are used to attain maximum velocities and ignition is not rapid, uniform, and nearly simultaneous in the charge or if the grains can be substantially compacted or the entire propellant charge can be accelerated by localized ignition occurring at the charge base. A very high rate of pressure increase may then develop which produces a stress wave which is reflected from the base of the projectile back into the burning charge, reinforcing existing pressure waves and possibly leading to high-pressure shock waves. The potential for transition from burning to detonation in a gun may be minimized by selecting propellants that have high mechanical strength at all temperatures and using compact, well-supported charges. Design variations that can lead to a very rapid pressure buildup greater than about 6.9 GPa/s (10^6 psi/s) should be avoided (48–56).

The Burning Process

The mass rate of propellant burning at a given pressure and temperature depends on the amount of heat evolved during decomposition and the amount of heat transferred to the burning surfaces of the propellant from the hot gases above it. It is also influenced by the tangential velocity of the propellant gases and the radiation from the surroundings. Propellants burn in parallel layers so that the surface recedes in all directions normal to the original surface (Piobert's law). The geometry of the grain on completion of burning is similar to its geometry at the start. Propellant burning at high gun pressures proceeds more smoothly and is less subject to erratic behavior than burning at very low pressures because the conditions are appropriate for maximum energy transfer in minimum time. The burning rate at gun pressures usually varies somewhat less than the first power of the pressure. It changes more slowly at rocket pressures of 3.45–10.34 MPa (500–1500 psi), often to less than the square root of the pressure (57).

The composition of the propellant determines the rate of exothermic molecular breakdown at a given temperature and pressure. As the reaction rate increases, the rates of heat production and transfer increase with associated increases in the linear burning rate of the propellant. The heat evolved per gram of propellant is its heat of explosion, Q, or the calorific value. It may be readily calculated or experimentally determined in a calorimetric bomb (see Calorimetry). Values range from ca 2.09 kJ/g (500 cal/g) for "cool" propellants to ca 6.27 kJ/g (1.5 kcal/g) for maximum energy propellants. The flame temperatures and the burning rates of propellants of similar compositions are linearly related to their calorific values except at very low pressures. The presence of volatile solvents or water significantly reduces the propellant burning rate. The addition of crystalline oxidizers such as ammonium perchlorate or RDX also modifies the burning rate to a degree that depends on the physical and chemical characteristics of the compound and the percentage present. The burning rate increases ca 0.1 to 0.4%/°C as the temperature of the propellant increases (58–59).

The operating pressure of the system has the predominant effect on the burning rate of propellants. Photographic evidence shows that increasing the pressure decreases the distance between the flame zone in the gas phase and the propellant surface. The rate of heat transfer to the propellant surface increases accordingly. The reaction rate among the gaseous components of the zones above the propellant also increases in accordance with established relationships between pressure and the rate of gas reactions in equilibrium. The velocity of the gases passing over the propellant at the dynamic conditions prevailing in a rocket motor or in a gun tube may further increase the burning rate (erosive burning). When a turbulent flow of gas occurs behind the reaction zone, part of the turbulence may penetrate the zone and increase heat transfer to the propellant surface. Propellants with low burning rates are more readily susceptible to the erosion effects attributable to gas velocity than are high-burning-rate propellants (60–63).

Burning-Rate Equations. The design of propellants for gun or rocket performance requires a knowledge of the exact rate at which the products of combustion are produced under the prevailing conditions of pressure and temperature. Although burning rates may be estimated by various computational procedures, the required accuracy can only be obtained experimentally using strand burners or closed bombs. Burning-rate equations have been developed to describe the performance of solid propellants based on the assumption that all the exposed propellant surfaces are ignited simultaneously and burn at the same linear rate (57,64–68). For example:

$$r = a + bP \tag{1}$$

$$r = cP^n \tag{2}$$

where r is the linear burning rate, P is the pressure, n is the pressure exponent, and a, b, and c are constants that vary with temperature. Equation 1 is often used for propellants burning at high gun pressures, whereas equation 2 is associated with low-pressure rocket systems.

Burning Control. For metering propellant gas at a predetermined rate, a propellant composition is selected with the required burning rate at the operating pressures in the gun or rocket. Then the geometry of the propellant is so designed that the necessary burning surface is available to provide the required mass rate of gas evolution. The individual propellant grains may be very small and numerous as are the 0.01 cm diameter spherical grains used in some propellants for small arms, or they may be very large and of complex geometry as are the 3 to 5 m diameter structures used in some rocket boosters. Control of the total burning surface is achieved by establishing the number of grains to be used, their geometrical configuration, and in the case of rocket propellants, the cementing of noncombustible inhibitors on grain surfaces to prevent their burning and bonding the exterior surfaces to the motor wall. The effect of grain shape on performance of gun and rocket propellants is shown in Figure 1. Propellant grains that are designed to provide a relatively uniform rate of gas evolution during the burning process have "neutral" geometries and undergo "neutral" burning. End-burning grains are neutral burning, and single-perforated cylinders are almost neutral burning. Other geometries may be used to provide increasing or decreasing quantities of gas as burning occurs. Grains that increase in surface area during burning, eg, the seven perforated grains used in large-caliber gun propellant charges, are said to burn progressively. Grains that decrease in surface area, eg, spherical grains, burn regressively. Since regressive burning in a gun is undesirable, liquid or solid deterrents such as dibutyl phthalate, dinitrotoluene, and diphenylamine may be applied to the surface of a propellant such as used in small-caliber weapons to produce grains that burn more progressively than their geometry would otherwise permit. The geometry of rocket propellant grains is tailored to meet the specific performance required. It varies considerably and may be much more complex than that of gun propellants (69–72).

Experimental Determination of the Burning Rate. Although a number of procedures have been developed for calculating the burning rates of propellants, none is sufficiently accurate for final design purposes (73–74). Experimental determinations are made with the strand burner for rocket propellants and the closed bomb for gun propellants (73,75–76). The closed bomb is essentially a heavy-walled cylinder capable of withstanding pressures to 689 MPa (100,000 psi). It is equipped with a piezoelectric pressure gauge and the associated apparatus required to measure the total chamber pressure which is directly related to the force of the propellant. It also measures the rate of pressure rise, a function of pressure which is indirectly related to the linear burning rate of the propellant via its geometry. Tests are conducted on the propellant grains as manufactured, and the data are used to calculate the interior ballistics of closed-breech guns. Other devices, such as the Dynagun and the Hi-Low bomb have also been developed for the measurement of gun propellant performance (77–78).

The strand burner is a bomb pressurized with an inert gas to rocket pressures and equipped with auxiliary apparatus consisting primarily of electrical timers for determining the time to burn an accurately known distance on the propellant strand being tested. Tests are run on thin strands of propellant as extruded or machined from a grain. The data are directly converted to burning rates (79–80).

Mechanism. *Nitrocellulose-Based Propellants.* Much of the information available on the burning process of nitrocellulose propellants is based on the decomposition of nitrate esters and the reaction of oxides of nitrogen with the products of decomposition. A one-dimensional physicochemical description of a model of the burning of a double-base propellant at low (rocket) pressures is often used (see Fig. 2) and more complex models have been developed. The three reaction zones identified are: (*a*) the surface of the solid is a reaction zone or foam zone ca 0.01 cm in thickness and ranging up to ca 600 K. Molecular bond breakage occurs in this zone, primarily the O—N bond in cellulose nitrate–nitroglycerin type propellants. Large volatile molecules are produced such as aldehydes, alcohols, and low molecular weight oxygenated compounds. The rate of bond breakage depends on temperature in accordance with the applicable Arrhenius equation; (*b*) above the foam zone is a nonluminous gaseous reaction zone or fizz zone, several thousandths of a centimeter thick. This zone attains a temperature of ca 1500 K as a result of partial reaction among the materials ejected from the foam surface. Aldehydes and alcohols are converted to smaller molecules and nitrogen, water, carbon monoxide, carbon dioxide, and nitric oxide. About half the total heat evolved by the propellant is liberated in the fizz zone which is prominent at low pressures and disappears at high pressure; (*c*) the reaction continues to completion and thermodynamic equilibrium is established in the flame zone. This zone, ca 0.001 cm thick, defines the flame temperature of the propellant which may range from ca 1500 K for cool propellant to 3500 K for very hot ones. The nitric oxide reacts with the reaction products formed in the fizz zone to produce carbon monoxide, carbon dioxide, hydrogen, water, nitrogen, and a small percentage of other molecules (83–93).

Since the gaseous products are in thermodynamic equilibrium at the flame temperature, quite accurate calculations of the gas composition, maximum temper-

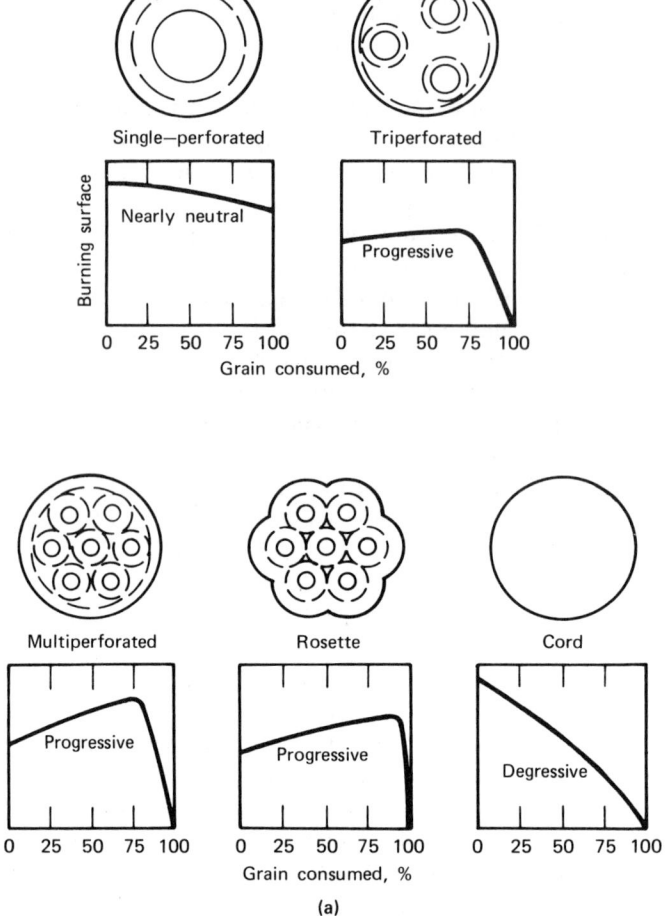

Figure 1. Effect of grain shape on (a) surface exposed during burning of gun propellants, and (b) pressure-time traces of rocket propellants. To convert kPa to psi, multiply by 0.145.

ature, and other thermodynamic properties may be readily derived from the propellant formulation and the thermodynamic characteristics of its components. Typical values of the thermochemical–thermodynamic characteristics of nitrocellulose propellants are given in Table 4. Calculations of the burning rates are not as rewarding. A number of models have been developed that consider the gas phase reactions and the rate of energy transfer from the gas phase to the propellant surface to be rate determining. Surface models and combination surface–gas-phase models have also been developed. The surface–gas-phase models have shown fairly good agreement between calculated and experimental burning rates, but not good enough for design purposes (94–99).

Additives. Although the burning rate of nitrocellulose propellants at high gun pressures is not significantly affected by the presence of additives, the addition of 1 to 2% of some metallic salts such as lead acetyl salicylate, lead stearate and lead stannate to propellants increases their burning rates at the much lower rocket pressures. The effect decreases to that of the unleaded propellant as the pressure increases

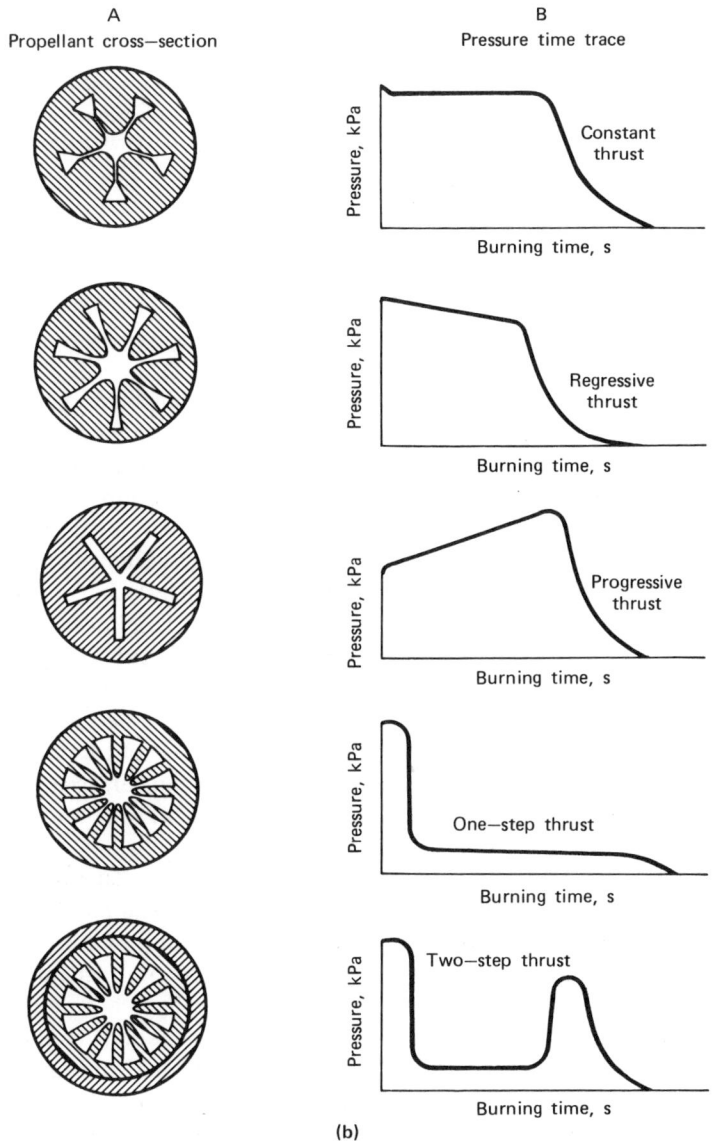

Figure 1. (*continued*)

so that burning rate-pressure curves are obtained with very low pressure exponents over limited pressure regions. The addition of solid oxidizers and metallic fuels tends to eliminate this catalytic effect. In plateau propellants, so named because of the shape of the log pressure-log burning rate curve, the catalytic effect disappears slowly, whereas in mesa propellants the catalytic effect disappears rapidly. The extent of the rate increase is affected markedly by the type and quantity of other components present (100–101).

Solid at ambient temperature	Solid-phase reaction zone (foam zone)	Nonluminous gas phase reaction zone (fizz zone)	Luminous flame reaction zone (flame zone)	Final flame equilibrium zone
Approximate dimension	10^{-2} cm	5×10^{-3} cm	10^{-3} cm	
Approximate temperature 300 K		600 K	1500 K	3000 K

Figure 2. One-dimensional model of propellant burning process (81).

Composite Propellants. A number of analytical models have been developed to permit estimations of the burning rates of composite propellants as a function of the factors that affect them most (102–110): calorific value, propellant temperature, chamber pressure, fuel–oxidant ratio, particle size distribution of the oxidizer, the presence of decomposition catalysts, and the incorporation of aluminum (in powder, wire, or staple form). The granular diffusion model postulates that the primary reaction zone of ammonium perchlorate propellants lies almost entirely in the gas phase. This zone is less than 0.01 cm thick at rocket pressures and its thickness decreases as the pressure increases. The oxidant and fuel are decomposed and converted into gases by pyrolysis or sublimation as a result of energy transferred primarily by thermal conduction from the gas phase to the propellant surface. The gases evolved leave pockets on the surface whose size is related to the particle size of the solid components. As burning occurs, the aluminum powder accumulates on the surface of the propellant and then agglomerates as clusters to form molten droplets up to 20 times the diameter of the individual particles in the propellant.

The Beckstead–Derr–Price model (Fig. 3) considers both the gas-phase and condensed-phase reactions. It assumes heat release from the condensed phase, an oxidizer flame, a primary diffusion flame between the fuel and oxidizer decomposition products, and a final diffusion flame between the fuel decomposition products and the products of the oxidizer flame. Examination of the physical phenomena reveals an irregular surface on top of the unheated bulk of the propellant that consists of the binder undergoing pyrolysis, decomposing oxidizer particles, and an agglomeration of metallic particles. The oxidizer and fuel decomposition products mix and react exothermically in the three-dimensional zone above the surface for a distance that depends on the propellant composition, its microstructure, and the ambient pressure and gas velocity. If aluminum is present, additional heat is subsequently produced at a comparatively large distance from the surface. Only small aluminum particles ignite and burn close enough to the surface to influence the propellant burn rate. The temperature of the surface is ca 500 to 1000°C compared with ca 300°C for double-base propellants.

Binders. Composite propellants are usually classified in terms of the binder used since it is the fuel that reacts with the oxidizer and has a fundamental effect on the ballistic, mechanical, and storage stability properties of the propellant. The effect of various percentages of binder on the specific impulse of aluminized ammonium perchlorate (AP) propellants is shown in Figure 4. The most commonly used binders are chemically cross-linked during the curing process. These polymers generally show

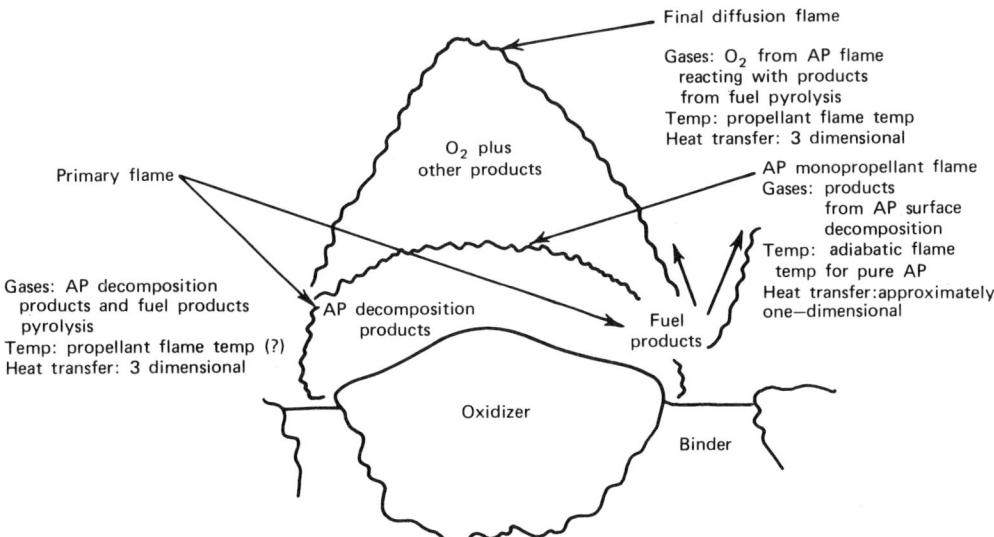

Figure 3. The postulated flame structure for an AP composite propellant, showing the primary flame followed by the final diffusion flame. AP = ammonium perchlorate.

better mechanical properties at temperature extremes than plasticized binders. The mechanical properties of a cross-linked polymeric propellant depends on the number of cross-links and dangling chains where only one end is attached to the cross-linked network. The degree of cross-linkage must be controlled to provide for polymer strength at elevated temperature while allowing for the required elasticity at low temperature. The addition of trifunctional components to the composition and control of the ratio of trifunctional to bifunctional units establishes the number of branch points in the polymer and prevents excessive cross-linking.

Binders must be fluid prepolymers even when filled with 85 to 90% of granular material. They must not react with the crystalline filler or other components and should polymerize or cross-link without the formation of gaseous reaction products. Binders must be chemically and physically stable over long periods of time under severe environmental and operational conditions of temperature change, vibration, and acceleration. They must form a durable and tough coating around the oxidizer and metallic ingredients and be capable of bonding to the interior wall of the motor after it has been suitably prepared by coating with an insulating liner and a bonding polymer. The rheological characteristics of the binder-filler and its pot life are critical in determining processability of the mix (111–115).

Oxidizers. The chemical and physical characteristics of the oxidizer affect the ballistic and mechanical properties of a composite propellant and its processability. Oxidizers are selected to provide the best combination of available oxygen, high density, low heat of formation, and maximum gas volume in reaction with binders. Increases in oxidizer content increase the density, the adiabatic flame temperature, and the specific impulse of a propellant up to a maximum. The most commonly used inorganic oxidizer in both composite and nitrocellulose-based propellants is ammonium perchlorate. Its decomposition mechanism has been thoroughly investigated. Ammonium nitrate has been used in slow burning propellants and where a smokeless

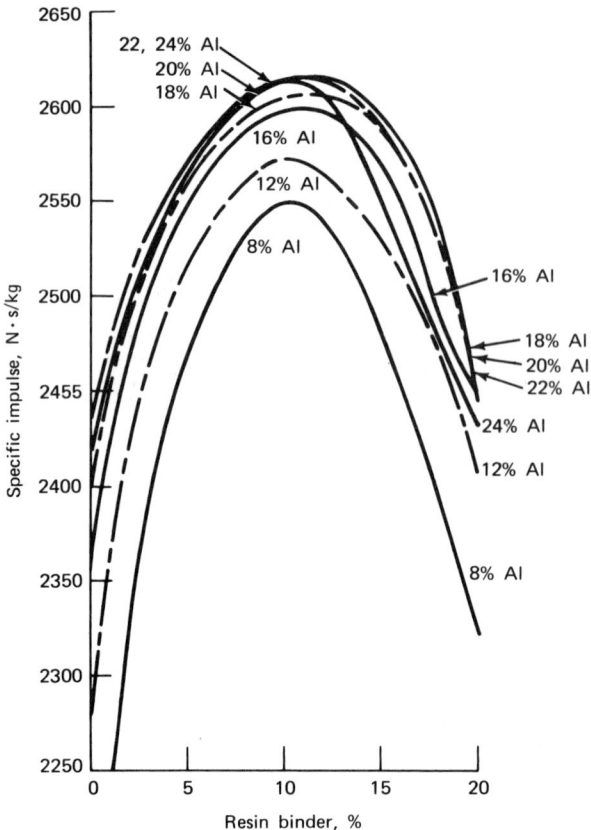

Figure 4. Impulse vs percent binder at various Al levels for a $\text{(CH}_2\text{)}_3$/Al/AP propellant (111). AP = ammonium perchlorate. To convert N·s/kg to lbf·s/lb, divide by 9.82.

exhaust is required (see Explosives). The characteristics of common inorganic oxidizers are listed in Table 5. In any homologous series, potassium perchlorate-containing propellants burn faster than those with ammonium perchlorate; ammonium nitrate propellants burn slowest.

Table 5. Properties of Some Common Inorganic Oxidizers

Oxidizer	Available oxygen	Melting point, °C	Density, g/cm³	Heat of formation, kJ/mol[a]	Heat capacity, J/(mol·K)[a]	Moles of gas per 100 g[b]
potassium perchlorate[c]	46.0		2.53	−433.4	112.5	0
ammonium perchlorate	34.0	dec	1.95	−290.3	128.0	2.55
ammonium nitrate[c]	20.0	169	1.72	−365.2	137.2	3.75
lithium perchlorate	60.6	236	2.43	−368.6	104.6[d]	0

[a] To convert J to cal, divide by 4.184.
[b] Gas produced by oxidizer other than that formed by reaction of oxygen with fuel components.
[c] Propellants with potassium perchlorate have relatively high burning rates [1.75 cm/s at 6.9 MPa/21°C (1000° psi)] and high burning rate exponents (0.6–0.7). Propellants with ammonium nitrate have very low burning rates (0.0125 cm/s).

Ammonium perchlorate is made by the reaction of sodium perchlorate with ammonia and hydrochloric acid. It is hygroscopic between ca 75 to 95% relative humidity and begins to deliquesce above 95% (116). It starts to decompose at 439°C and the decomposition may be catalyzed by metallic salts such as iron oxide and copper chromite at a lower temperature (117–119). Very finely divided ammonium perchlorate is more sensitive to impact and friction than the coarse material and the presence of hydrocarbons greatly increases the likelihood of a detonable reaction. The burning rate of ammonium perchlorate propellants is influenced by the particle size of the oxidizer (120–121); eg, it may increase by a factor of six by decreasing the average diameter from 400 to 1 μm. Bimodal distributions are generally used to load the binder with the maximum oxidizer content. Average particle size and particle size distribution affect the burning rate as well as the presence of other ingredients such as aluminum and catalysts. Particle size distribution of the perchlorate has a negligible effect on the pressure exponent and no effect on the specific impulse of the propellant (116,122–126).

Metallic fuels. Aluminum is most commonly used to increase the impulse of both composite and nitrocellulose-base propellants because of its highly exothermic reaction with the oxidizer. Its heat of reaction with oxygen is 10.25 kJ/g (2.450 kcal/g). Materials such as aluminum hydride, beryllium, beryllium hydride, and boron offer theoretical advantages in increased impulse, but are not used because of increased cost, toxicity, or long-term instability, or because their actual performance does not live up to their calculated performance. Increasing the aluminum content of a propellant increases its density. The flame temperature and specific impulse approach a maximum sharply near the stoichiometric ratio of metal–oxidizer–binder. The aluminum increases the hydrogen content of the reaction products and, by minimizing the formation of water vapor, reduces the energy losses caused by dissociation of water at elevated temperatures. Incorporation of aluminum staples to replace a small part of the aluminum powder may quadruple the burning rate while maintaining the specific impulse. The presence of aluminum in rocket propellants also reduces or eliminates combustion instability caused by the formation of pressure waves in the motor chamber which may resonate with one or more of the natural acoustic frequencies exhibited by the chamber. Aluminum-containing propellants deliver less than the calculated impulse because of two-phase flow losses in the nozzle caused by aluminum oxide particles. Combustion of the aluminum must occur in the residence time in the chamber to meet impulse expectations. As the residence time increases, the unburned metal decreases and the specific impulse increases. The solid reaction products also show a velocity lag during nozzle expansion and may fail to attain thermal equilibrium with the gas exhaust. An overall efficiency loss of 5 to 8% from theoretical may result from these phenomena. However, these losses are more than offset by the increase in energy produced by metal oxidation (127–135).

MANUFACTURE

Very large numbers of small perforated grains are used in gun propellant charges to provide the high mass rate of burning required to accelerate projectiles to maximum velocity in the relatively short distances of travel available in the gun tube. Gun propellant processes are designed to produce these grains in enormous quantities essentially by plasticizing nitrocellulose in simple mixers and extruding the soft propellant mix which can be readily cut to short grain lengths and dried to a hard, hornlike texture (136). Some gun propellants may be produced without a volatile solvent, eg, the NOSOL propellant used by the U.S. Navy. Although not commonly used in cannon propellants in the United States, stick propellant is mixed in the same way as granular propellant but cut to form long stands on extrusion. Uniformity of performance is obtained by control of the composition, the volatile material present, the grain dimensions, and by blending on a large scale. Some variation in burning characteristics is permissible, since the propellant charge can be modified to a limited extent to meet ballistic requirements by the addition or removal of propellant grains. Gun propellants are best evaluated by composition analysis and measurements of the heat of explosion, the grain dimensions, and the closed-bomb characteristics of relative force and quickness followed by confirmatory weapon firings.

Rocket systems pose problems that do not exist with gun propellants. Rocket grains are much larger and may have more complex shapes. Their burning characteristics must be more carefully controlled. The stresses imposed on large rocket grains as a result of mechanical and thermal shocks find no parallel in the small gun propellant grains. Rocket grains must be made free of flaws to avoid the possibility of internal burning and breakup. Composition and particle size of ingredients are carefully controlled; heats of explosion and strand burning rates are established; dimensional measurements made; physical properties determined; extensive radiographic examination for flaws and grain defects are conducted; ultrasonic inspection may be used to supplement x-ray examination; and small-scale and limited full-scale rocket motor firings made at high and low temperatures. In-process variations of a minor and not readily identifiable nature may produce significant changes in performance. Once the rocket grain is produced, it cannot be readily changed if it does not meet requirements. Differences in lot performance cannot be blended out as with gun propellants. As a result of the stringent requirements imposed on solid rocket propellants and the varied applications to which rockets are put, many different propellants have been developed.

Extruded Nitrocellulose Propellants. Nitrocellulose propellants are made with or without incorporation of a solvent as plasticizer by the five processes listed in Table 6. All require colloid treatment of the nitrocellulose to eliminate its fibrous structure and cause it to burn predictably in parallel layers. Mechanical working of the ingre-

Table 6. Nitrocellulose Propellant Manufacturing Processes

Process	Propellant use
solvent extrusion	cannon
	fast-burning rockets
	casting powder
	ignition powder
	rifles, small-caliber weapons, expulsion charges
solvent emulsion	rifles
	small-caliber weapons
solventless extrusion	small rockets
	cannon
solventless rolling	mortars
casting	small rockets
	large rockets

dients contributes to plasticization and uniformity of composition. Although the manufacture of extruded nitrocellulose propellants has been traditionally a batch process, continuous processes are being developed for single-, double-, and multibase propellants. The compositions of nitrocellulose-based gun propellants are shown in Table 7.

Solvent Extrusion Batch Process. Almost all standard gun propellants and small-webbed rocket propellant grains are made by the solvent extrusion process (137–139). Grains with webs greater than ca 1.30 cm are produced by solventless or casting processes. The removal of solvents from large-web grains would require long periods of time and could introduce stresses that would lead to grain cracking. A typical batch process (see Fig. 5) is outlined below for making triple-base propellants (M15, M17, M30, M31) with a high nitroguanidine content. It is similar to the method used to make double-base propellant. The manufacture of single-base propellants such as M1 and M6 compositions differs primarily in the mixing and drying operations.

Purified, blended, and centrifuged nitrocellulose of the required nitrogen content and wet with ca 30% water is received from the nitrocellulose plant. A charge is transferred to a double-acting hydraulic dehydration press and compressed at low pressure to remove some of the water. The remainder is removed by pumping 95% ethyl alcohol through the press. The final blocks, containing ca 18% alcohol, are broken up and screened to remove lumps or oversized particles. Mixing is usually conducted in a water-jacketed bladed mixer and consists essentially of solid–solid and solid–liquid incorporation, the solution of stabilizers and possibly ballistic modifiers, and the absorption of solvents and liquid plasticizers by the nitrocellulose. This operation is governed by the heat generated, the heat exchange characteristics, the method and sequence of incorporation and solvent addition, and the effects of specific equipment.

The premixing operation for double- and triple-base propellants is designed to incorporate safely the nitroglycerin in the nitrocellulose and to begin to distribute the remaining ingredients in a slow and uniform manner. A typical premixer has large clearances and applies relatively small mechanical forces. Half of the required amount of nitroguanidine and all of the alcohol-wet nitrocellulose are transferred to the mixer bowl, more alcohol is added if required, and, after preliminary mixing a nitroglycerin–acetone solution is distributed over the contents and mixed in for ca 10 min. The final mixing works the composition for an extended period of time until the ingredients are completely incorporated and plasticization occurs. The operation is generally conducted in a sigma-bladed, water-jacketed mixer to which several premix charges have been transferred. During the operation the remaining nitroguanidine and other ingredients are added including the necessary amounts of acetone and alcohol to meet the solvent requirements and obtain a satisfactory colloid. The temperature is maintained between 40 and 50°C, depending on the equipment and the colloid formation. The mix is cooled and discharged. All equipment has to be grounded and nonsparking tools used to avoid a solvent vapor–air explosion. The wet dough itself is nonflammable and almost nonexplosive except during the early stages of mixing.

Single-base propellants are mixed in a similar fashion by adding the ingredients to the nitrocellulose in the mixer together with the required amounts of ether and alcohol. The mixing time is about one-half hour and the temperature is kept below 25°C. The partly colloidal mixture looks like moist crude sugar. A maceration step may be included to increase homogeneity. After mixing, the dough-like composition (single-,

Table 7. Gun Propellant Compositions[a], wt %

Component	M1	M2	M5	M6	M8	M9	M10	M14	M15	M17	M26	M30	M31	NACO[a] type 1	IMR
nitrocellulose	85.0	77.5	82.0	87.0	52.2	57.8	98.0	90.0	20.0	22.0	67.5	28.0	20.0	93.6	100.0
(nitrogen, %)	(13.15)	(13.25)	(13.25)	(13.15)	(13.25)	(13.25)	(13.15)	(13.1)	(13.15)	(13.15)	(13.15)	(12.6)	(12.6)	(12.0)	(13.15)
nitroglycerin		19.5	15.0		43.0	40.0			19.0	21.5	25.0	22.5	19.0		
nitroguanidine									54.7	54.7		47.7	54.7		
ethyl centralite		0.6	0.6		0.6	0.70			6.0	1.5	6.0	1.5		1.2	
diphenylamine	1.0[b]			1.0[b]			1.0	1.0[b]							0.7[b]
2-nitrodiphenylamine													1.5		
dinitrotoluene	10.0			10.0	3.0			8.0							8.0[c]
dibutyl phthalate	5.0			3.0	1.2			2.0					4.5		
potassium nitrate		0.7	0.7			1.50					0.75				
barium nitrate		1.4	1.4								0.75				
potassium sulfate				1.0[b]			1.0								
lead carbonate	1.0[d]													1.2	1.0[b]
cryolite	1.0[d]													1.0[a]	
graphite		0.3	0.3			0.10[b]	0.10[b]		0.3	0.3		0.3	0.3		
n-butyl stearate										0.15[b]				3.0	

[a] All compositions are solvent extruded as grains except M8 which is solventless-rolled as sheet.
[b] On added basis.
[c] Added as a coating.
[d] If required, on added basis.
[e] As basic lead carbonate.

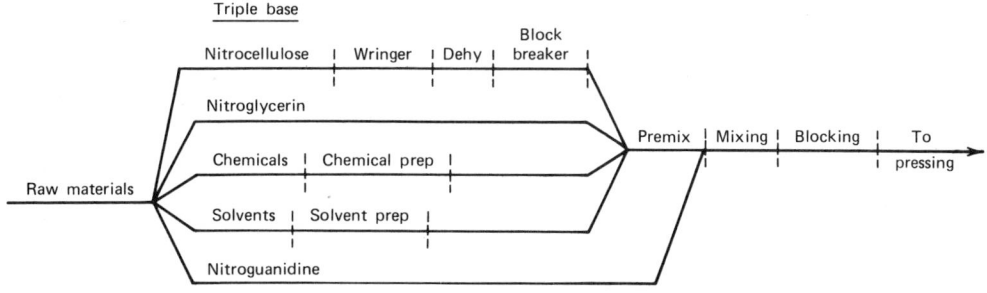

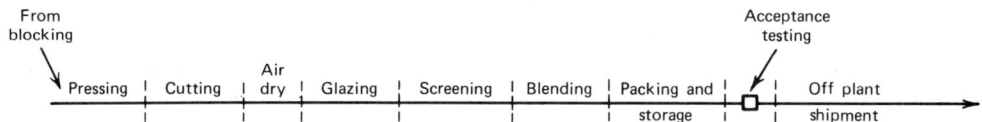

Figure 5. Solvent extrusion process for triple-base propellants. Courtesy of John Horvath, USA Radford Plant.

double-, or multi-based) is transferred to a vertical block press where it is consolidated after first purging the press chamber with an inert atmosphere of carbon dioxide or nitrogen. The block is ejected from the press and for single-base propellants is transferred to a screening press where the mass is forced through a series of coarse mesh screens. These operations remove air, lumps, and foreign particles from the mix, increase the uniformity of ingredient distribution, and improve the colloid-treating of the nitrocellulose and the density of the propellant (140–142). The strands from the screening press are again consolidated in a blocking press. The blocks of propellant, single-, double-, or triple-base, are transferred to a vertical or horizontal graining press and extruded at relatively low pressures of 10.3–17.2 MPa (1500 to 2500 psi) through dies designed to produce the required dimensions, with allowances made for shrinkage occurring during drying. To ensure safety during pressing, explosive mixtures of air and solvent have to be carefully excluded during the ramming operation.

The strands of perforated propellant are fed to a mechanical cutter and sliced to specified lengths. Grains of double and some single-base compositions are dried in trays with warm air at temperatures and for periods of time that depend on the composition and granulation of the propellant, the quantity of alcohol and acetone or ether present, and the characteristics of the drying system.

The drying process for single-base propellant is entirely different from that of double- and multibase propellant. First the alcohol–ether wet propellant is air-dried in transfer carts or large tanks to recover the solvents and reduce the volatile solvent content to ca 6%. The initial drying process is carefully controlled to prevent skin hardening or grain cracking. Temperatures are gradually raised to 50 to 65°C over a period of days, depending on grain size. The vapors are condensed and the solvent recovered. The propellant is then immersed in circulating water at 50 to 60°C. The solvent diffuses into the water which prevents case hardening of the propellant surface and permits the solvent to be removed more rapidly than exposure to air alone would permit. After a period of time up to 30 days for large single-base cannon grains, the temperature is slowly increased to reduce the solvent to a controlled minimum, depending on grain size. Final air drying at ca 55°C removes surface water.

The propellant may be tumbled in drums with a small amount of graphite to improve its flow characteristics and bulk density and to decrease the likelihood of formation of an electrostatic charge as well as to perform a degree of blending. It is then screened to remove foreign matter and blended into large lots which may vary in mass from 22 to 226 metric tons. The blending operation is essential to provide as homogeneous a lot as possible for ballistic uniformity. Although propellant grains are relatively insensitive to static electricity, propellant dust may be as sensitive as dry nitrocellulose. Fires in dry buildings and blending towers have been attributed in some cases to the electrostatic ignition of dust.

The wastewater produced in this process consists mostly of water used in cleanup and propellant conveyance and sorting operations, and is at present not subjected to treatment. Techniques such as the use of activated carbon and biological treatment are being investigated for the removal of solvents and dissolved organic compounds (143).

Continuous Solvent-Extrusion Process. *Single-Base Propellants.* Continuous processes for making solvent propellant for cannon offer substantial reductions in labor, hazard, and pollutants. A process for making ca 1100 metric tons of single-base propellant per month has been installed at the U.S. Army Radford Ammunition Plant. A schematic diagram for the continuous process is shown in Figure 6.

The main features in which the continuous process differs from the batch operation are as follows:

Thermal dehydration. Water-wet nitrocellulose on a continuous vacuum belt filter is vacuum dried followed by hot air transfusion (80°C) to reduce the moisture to less than 2%. After cooling, alcohol is sprayed on the nitrocellulose to a concentration of 15 to 20%.

Compounding. The alcohol-wet nitrocellulose is transferred from a surge feeder to a compounder by a continuous weigh-belt along with the other ingredients of the composition, which are also weighed and added automatically. Liquids are fed to the compounder by solvent-metering systems. The compounder is a water-jacketed horizontal rotary plow that blends the ingredients to produce a homogeneous premix paste of the dry, solid, and liquid components. The loosely mixed paste is fed by a conveyor to a heavy duty reciprocating screw mixer that is temperature controlled and specially designed to thoroughly mix and work the paste by forcing it past pins in the mixer barrel and out through a die plate. As the paste is extruded, it is cut into small pellets which are fed continuously to water-jacketed screw extruders and forced through multiple dies. The strands are cooled as they are extruded to facilitate cutting by an adjustable roll type cutter. After the cut grains are screened to remove clusters and odd sizes, the solvent is removed in the solvent recovery-water dry system where the propellant is treated first with hot inert gas and then hot water. Finally, in a series of air dryer units the moisture content is reduced from 12 to less than 0.8%.

The entire continuous automatic process is computer controlled by servo mechanisms that adjust flow rates, weights, temperatures, and other variables. It is heavily instrumented so that continuous performance information is available. Pressure relief is permitted wherever possible to minimize the likelihood of a detonation.

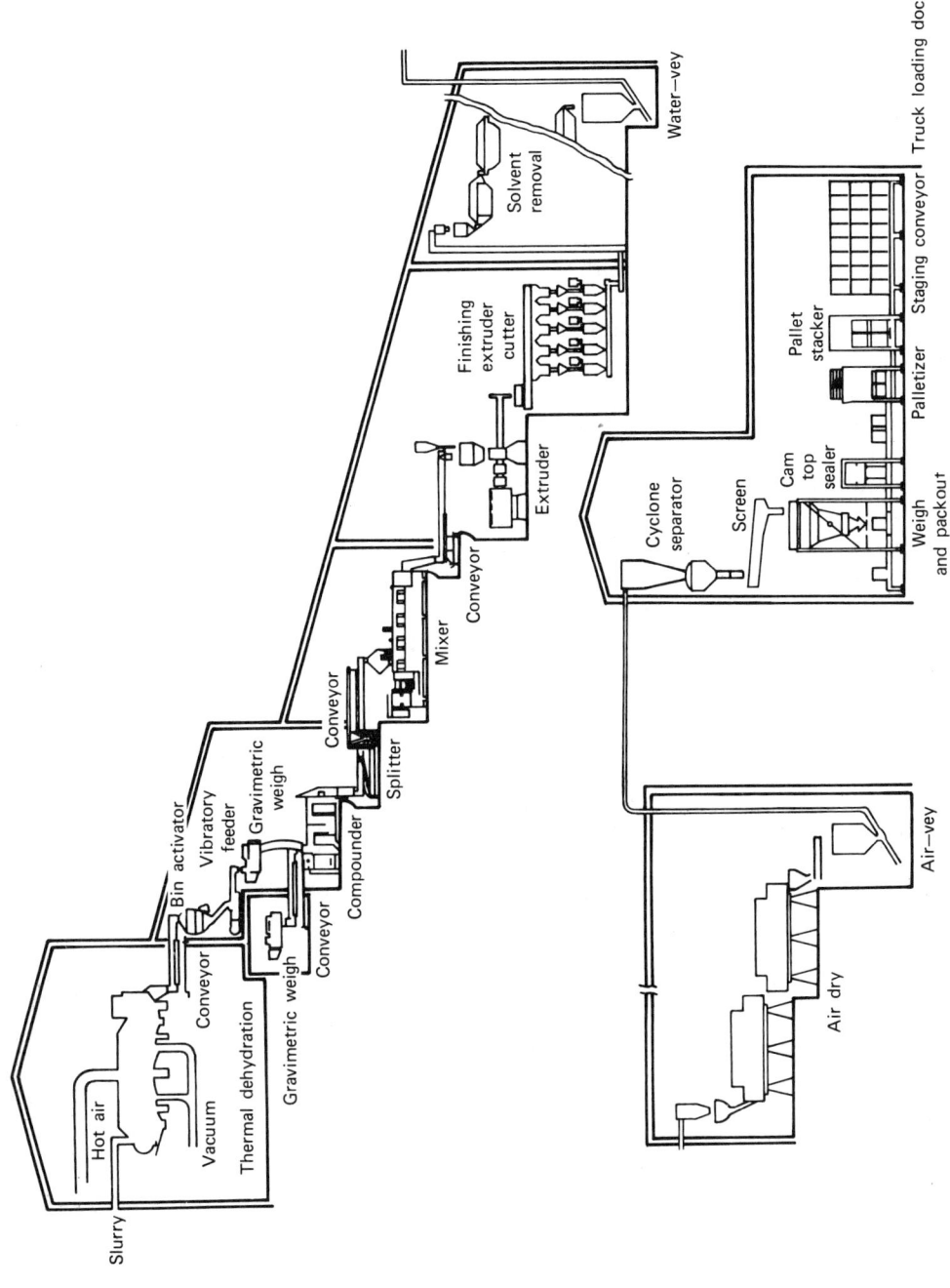

Figure 6. Continuous process for solvent-extruded single-base propellant (automated single base line). Vey = conveyor. Courtesy of John Horvath, USA Radford Plant.

Double- and Multibase Propellants. Continuous automatic processes are also under investigation for making double-base and nitroguanidine propellants. These are comparable to the single-base line with modifications to accommodate the use of nitroglycerin and a high crystalline salt content in multibase propellants. Smaller modules are used with varying numbers of individual units to form a balanced line (144–146).

After thermal dehydration (see above) of the nitrocellulose, the solvent mix consisting of nitroglycerin, acetone, and alcohol, and the dry chemical ingredients are continuously fed into a ribbon blender. A homogeneous mix is then introduced into the continuous mixer along with additional acetone and alcohol if required. The mixer has twin screws and paddles which provide an intensive shearing and forward movement action to the premix. The mixer feed is continuously moved by conveyor to a vertical screw-type extruder where it is extruded through multiple dies. A roll-type cutter consisting of two rotating cylinders in contact with each other is used to continuously cut the strands. The cut propellant is air dried on trays as in the batch process (144–146).

Detailed specifications describing the composition, performance, and methods of evaluation are sources of valuable information on extruded propellants (147–149).

Solventless Extrusion Process. The solventless process for making double-base propellant has been used in the United States primarily for the manufacture of rocket propellant grains with web thickness greater than ca 1.35–2.0 cm and for thin-sheet mortar (M8) propellant. For making small-web cannon propellant it offers such advantages as minimal dimensional changes after extrusion, the elimination of the drying process, and possibly better long-term ballistic uniformity since there is no loss of volatile solvent (137,150).

In the water-slurry process for making solventless propellants, developed by Lundholm and Sayers in England in 1889, explosive and nonexplosive liquid plasticizers and water-insoluble constituents are incorporated into nitrocellulose suspended as a slurry in a large volume of hot water. After removing the excess water, the resulting wet mass is partially dried and passed through heated rolls to thoroughly mix the colloidal composition into sheets of homogeneous propellant and remove any remaining water. These sheets may then be cut and rolled into scrolls for insertion in the press and extruded as propellant grains or cut up and used as a sheet or flake propellant. The details of a typical operation are shown in the flow chart in Figure 7.

A slurry tank is filled to ca one-third its capacity with water at 50°C. Nitrocellulose is added with agitation followed by the remaining amount of water required (the approximate ratio of water used to total propellant components is ca 10:1 and ca 20:1 on a nitrocellulose basis). After the nitrocellulose has been well dispersed, the required amount of plasticizer, such as nitroglycerin and other insoluble constituents, is added and the slurry is mixed for ca 30 min. The slurry is pumped in increments to a basket-type centrifugal wringer equipped with a wire-mesh screen. The time and rate of centrifuging are controlled to produce a cake containing ca 15% water. The centrifuged paste is transferred to a dry house for aging storage at ambient temperature. During the aging process the nitrocellulose absorbs and is partially gelatinized by the plasticizers. Water-soluble salts are then incorporated during a blending operation in a rotating drum.

The rolling operations which follow take place first on hot (95°C) differential-speed rolls which dry and colloid the paste and convert it into sheet form, and then on even-speed rolls which produce smoothly surfaced propellant sheets in which all ingredients have been uniformly incorporated. The roll gap in the differential roll is adjustable to produce sheets of various thicknesses, and rolling is continued until the moisture is reduced to a predetermined level, usually less than 0.5%. The sheet is then cut off the roll. Differential rolling is potentially hazardous and fires are not uncommon, although detonations are not apt to occur. Operations are generally conducted by remote control.

Typical even-speed rolls, about the same dimensions as the differential rolls, are highly polished, heated to ca 60°C, and revolve at ca 10 rpm. If rocket propellant grains are being made, the sheet is slit into strips and carpet-rolled to form a charge of large enough diameter to fit snugly into an extrusion press, which may be jacketed for temperature control and equipped with vacuum pumps for removal of air. The diameter of the press bore may be up to 60 cm and the press may be horizontal or vertical with pressures up to 103 MPa (15,000 psi) used for extrusion. The press is loaded with the propellant, evacuated, and extrusion is begun. On extrusion, the strand is cut into the required lengths. Single grains or multiple strands may be extruded at rates of 0.5 to 2.5 cm/min. The grains may be solid or have central perforations of various shapes, depending on the configuration of the die pin. The grains are visually inspected, annealed at elevated temperatures, and inspected by x-ray (151–152).

The composition and properties of typical solventless propellants are shown in Table 8.

The No-Roll Process. The No-Roll process has been proposed for making solventless double-base and multibase extrudable propellants. Reportedly it is safer and less costly than conventional procedures. A slurry is prepared of fibrous nitrocellulose, liquid nitrate ester plasticizers, and high explosive solids, if required, in a hydrocarbon medium such as heptane. The liquid medium is decanted and the remaining propellant mix dried, cured, and extruded in the desired grain shape. Typical explosive plasticizers used include metriol trinitrate, triethylene glycol dinitrate, and nitroglycerin. Solid explosives incorporated may be RDX, HMX, or nitroguanidine (153).

Ball Powder

Ball powder is typically used in small-caliber weapons such as 5.56-mm, 7.62-mm, and 20-mm projectiles. The product consists of spherically shaped or flattened ellipsoidal grains, ca 0.04–0.09 cm in diameter. The process permits the recovery and use of nitrocellulose from obsolete granular propellant (154–157). It eliminates the need for the conventional mixers, extruders, and cutters used to make granular and stick propellant and is relatively inexpensive to operate. It is safe because mixing and extrusion take place in the presence of water. The operations are flexible so that either single or double-base ball propellants of varying compositions and sizes may be produced. Typical compositional ranges and characteristics are shown in Table 9. The product has desirable flow characteristics because of particles shape so that small arms ammunition can be rapidly loaded by high-speed automatic equipment.

Batch Process. A flow chart for the batch operation is shown in Figure 8. Water-wet fresh or extracted nitrocellulose is transferred as a slurry to a graining still. Calcium carbonate is added to neutralize any free acid released by the dissolved nitrocellulose. The required amount of ethyl acetate is added as well as other soluble components

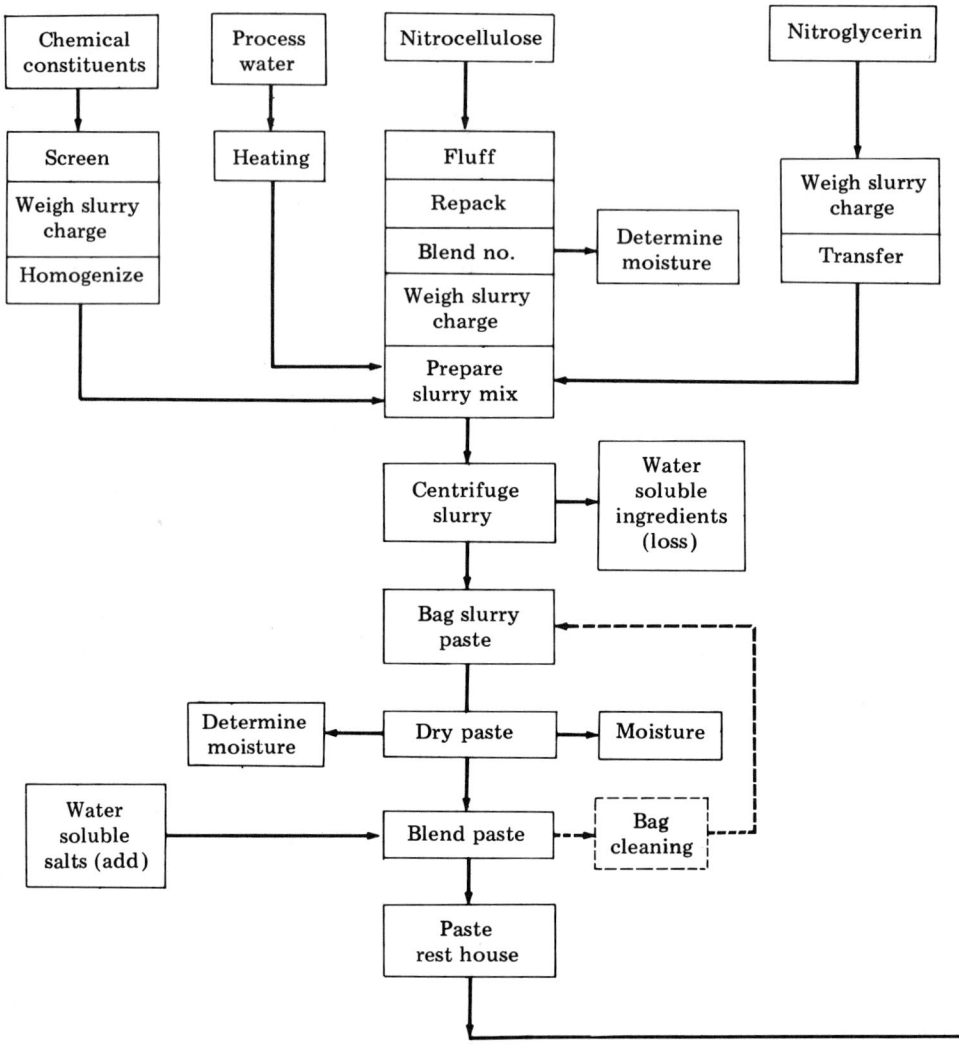

Figure 7. Slurry process for the manufacture of double-base, solventless rocket propellant.

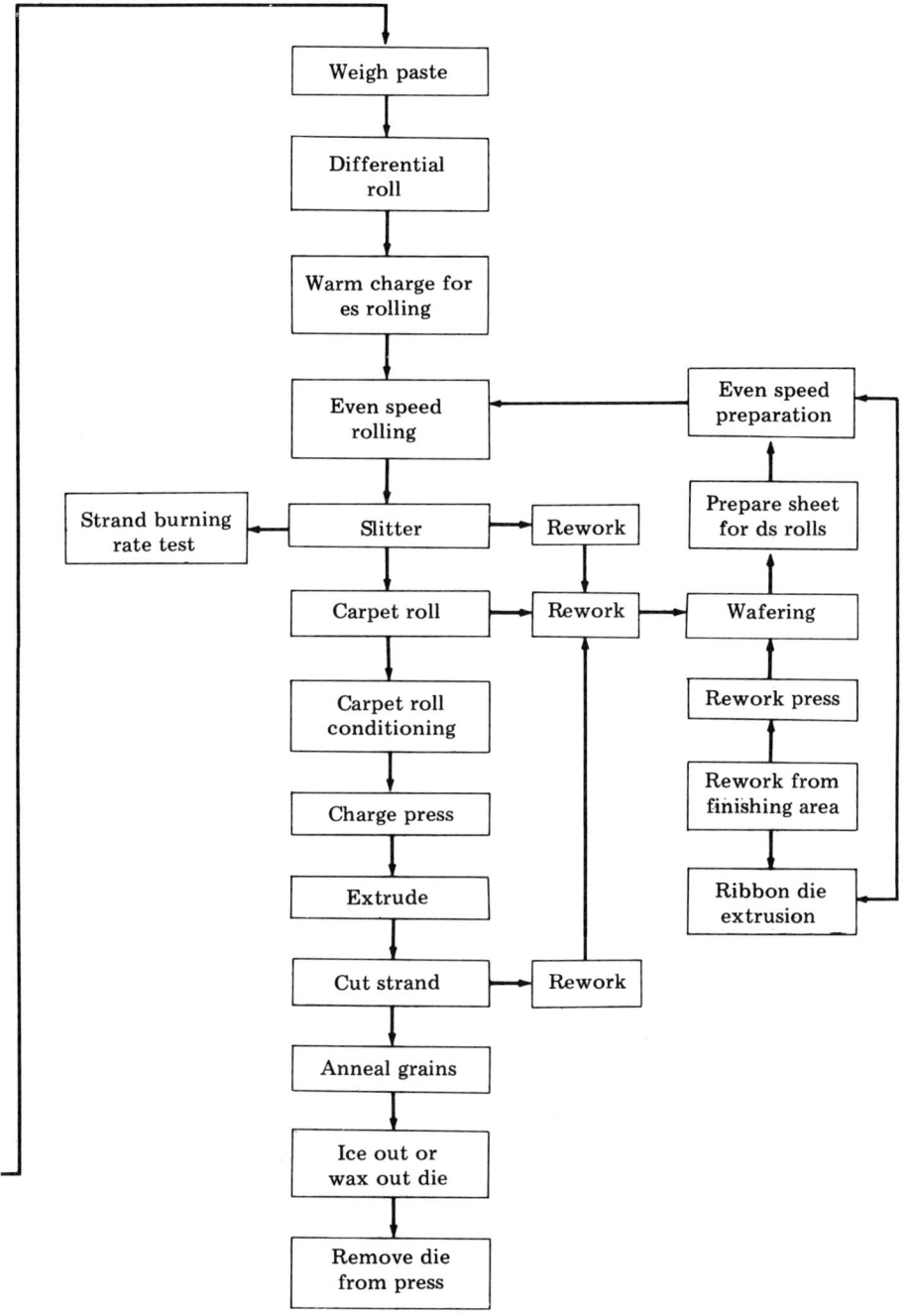

Figure 7. (*continued*)

Table 8. Composition and Properties of Typical Double-Base Solventless Propellant

	Extruded for	
	Rockets	Mortar sheets
Approximate composition wt %		
nitrocellulose	51.5	52.5
(nitrogen content)	(13.25)	(13.25)
nitroglycerin	43.0	43.0
potassium nitrate		1.0
diethyl phthalate	3.0	3.0
ethyl centralite	1.0	0.50
potassium sulfate	1.25	
carbon black	0.20	
wax	0.05	
Thermochemical properties		
flame, temperature, K, isochoric	3660	3695
isobaric	3010	
specific impulse, N·s/kg[c]	2317	
heat of explosion, J/g[a]	5108	5209
heat of combustion, J/g[a]	9295	
gas volume, mol/g	0.038	0.037
ratio of specific heats	1.22	1.21
Burning properties		
burning rate at 5.89 MPa[b] at 21°C, cm/s	1.52	
pressure exponent	0.68	
Products' composition, wt %		
hydrogen	6.5	
water	27.0	
carbon monoxide	33.0	
carbon dioxide	18.0	
nitrogen	14.0	
other	1.5	

[a] To convert J to cal, divide by 4.184.
[b] To convert MPa to psi, multiply by 145.
[c] To convert N·s/kg to lbf·s/lb, divide by 9.82.

such as diphenylamine. The contents are heated to ca 70°C and agitated to dissolve the nitrocellulose and form a lacquer. When the proper viscosity of the lacquer is attained, a protective colloid such as animal glue is added to form an emulsion of nitrocellulose globules and prevent their coalescence. Sodium sulfate is added ca one-half hour after the beginning of globule formation to extract water from the lacquer by establishing an osmotic pressure differential between the water-laden nitrocellulose globules and the concentrated salt solution in the still. Under these conditions, small spheres of dissolved nitrocellulose and the other soluble ingredients are formed. The ethyl acetate is distilled at 70 to 100°C, leaving spherical particles whose physical characteristics and solvent content are determined by the process conditions. The graining operation requires ca 1 to 1.5 h. Grain density and size are determined by the concentration of salt in solution, the temperature and time of the dehydration, agitation speed, and the rate of distillation of the ethyl acetate.

After graining, the slurry is water-washed to remove adherent salt solution and then wet-screened into the required particle sizes. Over- and under-size grains are

Table 9. Composition and Properties of Typical Ball Powder Propellants

	Value
Composition, wt %	
graphite, max	0.4
potassium nitrate	1.0–1.5
sodium sulfate, max	0.5
calcium carbonate, max	1.0
nitroglycerin	8.0–12.0
diphenylamine	0.75–1.50
dibutyl phthalate	3.5–7.5
total volatiles, max.	2.0
residual solvents, max.	1.2
dust and foreign matter, max	0.1
dinitrotoluene	as required
nitrocellulose	remainder
nitrogen, %	13.0–13.2
Physical properties	
hygroscopicity, max.	1.75
granulation	
through no. 20 (sieve, 840 μm) min, %	95
through no. 40 (sieve, 420 μm) max, %	50
through no. 45 (sieve, 350 μm) max, %	3
bulk density, g/cm^3	0.95–1.0
Thermochemical properties	
heat of explosion, J/g^a	3,350–3,768
flame temperature, K	2,700–3,000
volume of gaseous products at STP mol/g	0.04
impetus, J^a/g (ft·lb/lb)	1,000 (335,000)[b]

[a] To convert J to cal, divide by 4.184.
[b] Typical value.

returned to the graining operation for reworking. Nitroglycerin or an organic coating material are added following the transfer of the screened ball powder as a water slurry to a coating still. The water level is adjusted, the temperature increased to 60–65°C and a solution of nitroglycerin in ethyl acetate added with slow agitation to form an emulsion. The ethyl acetate is distilled under vacuum at 70 to 85°C, leaving the nitroglycerin present in a gradient of decreasing concentration from the surface toward the center of the spheres. If a deterrent is used such as dinitrotoluene in dibutyl phthalate, it is then transferred to the still, the slurry is slowly agitated at ca 75°C, cooled, transferred to a wash tub, and water-washed. The depth of penetration of the deterrent is determined by the temperature and time of the process. The coating cycle may take ca 24 to 30 h, depending on the propellant.

Rolling may be used for shape and size modification to increase the burning surface area of the propellant. A thick water slurry of the spheres is passed through a set of appropriately spaced even-speed polishing rolls, rotating toward each other. Centrifuging the slurry reduces excess moisture to ca 6%. Surface ballistic modifiers such as dinitrotoluene, tin oxide, potassium nitrate, or small amounts of water-soluble salts to meet requirements are added as an alcohol slurry by tumbling the propellant and additives in a jacketed "sweetie" barrel. The alcohol is removed with hot air and graphite is added to glaze the propellant. The moisture content is adjusted by the addition of water, if required, or drying.

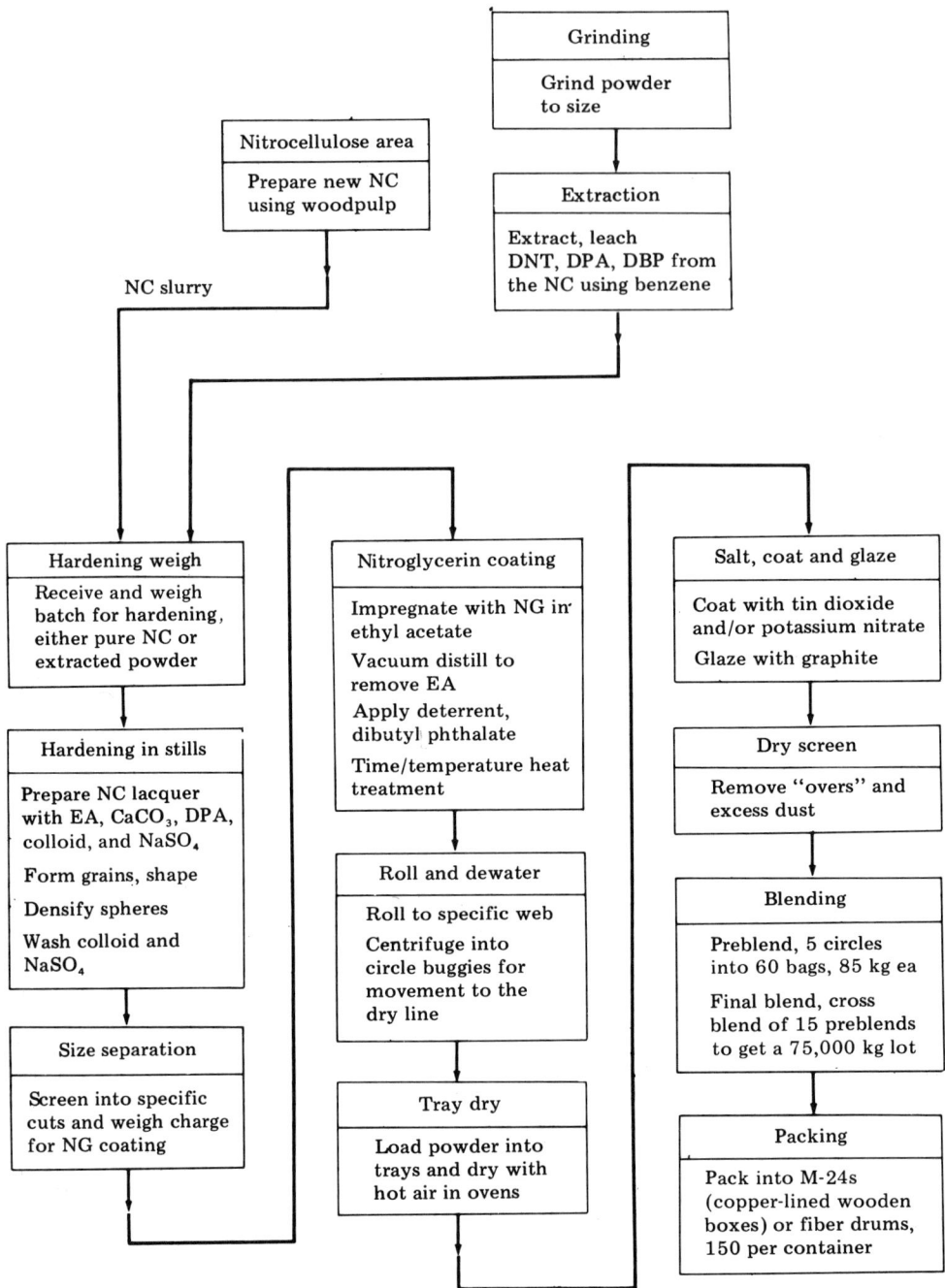

Figure 8. Ball powder batch process. NC = nitrocellulose, EA = ethyl acetate, DPA = diphenylamine, NG = nitroglycerine, DNT = dinitrotoluene, DBP = dibutylphthalate.

The propellant is dry-screened to remove dust and foreign material, and excess graphite. Unacceptable propellant from the dry-screen operation may be returned to the graining stage for reworking. The product is blended in large rotating barrels for ballistic uniformity.

Continuous Process. In the continuous process, developed by the Olin Corporation (158–159) and now used more extensively than the batch process, the nitrocellulose and stabilizing additives are dissolved in the solvent in a continuous screw mixer to form a dough-type lacquer. The lacquer is pumped continuously through filters to the graining operation where it is forced through an orifice plate with up to 10,000 openings. The lacquer is extruded as cylindrical strands that are cut with a rotary knife into grains with a length to diameter ratio of ca 1.5 to 1. The cut cylinders are flushed away from the exit side of the graining plate with a solution of water, colloid, and salt and are transferred to the shaping and dehydration lines. Automatic controls and monitors are used throughout.

The shaping and dehydration line is a very long bank of pipes connected by short-radius U-shaped bends that are hot-water jacketed to provide a gradually increasing temperature from ca 60 to 80°C. On passage through the pipe, the viscosity of the particles decreases and they become spherical and are dehydrated by the dissolved salt present to ca 10% entrained water. The large percentage of entrained solvent is removed by passing the lacquer/water through a series of evaporators. The propellant slurry is then passed over a screen and vacuum filter in series to recycle the salt and colloid solution and to wash residual colloid from the grains. After washing the propellant may be vacuum dried to remove excess water. It is size classified on a series of continuously rotating screens and may then be impregnated with nitroglycerin; ethyl acetate is removed by vacuum distillation. The product is coated with deterrent and rolled if necessary, by the same methods used in the batch process. Moisture is removed by a series of continuous vibrated semifluidized-bed dryers. Surface coatings are applied in a continuous drum or a batch barrel blender. Blending is carried out in a static internal tube-type blender or a large barrel blender. The propellant is packed in drums for shipment (160–167).

The wastewater in the ball powder processes arises primarily from the washing and wet screening process. Wash and screen waters are passed through clarifiers to remove suspended solids. The overflow from the clarifiers may be accumulated in lagoons. The wash waters contain a considerable amount of protective colloid, organic solvent, and sodium sulfate and have to be treated before they are discharged into local streams. The colloid foams in the effluent and increases the BOD by accumulating on the bottom of the collection ponds. Cooling water can be completely recycled and it is possible to design the washing and wet screening operations to decrease the contaminants in the plant effluent (163).

Cast Propellants

Nitrocellulose-Based. Cast nitrocellulose propellant is made by a two-step process (164–167). In the first stage casting powder is produced by techniques that are almost identical to those used for the manufacture of conventional solvent-extruded small-grain gun propellants. The second stage consolidates the casting powder by filling the interstices of the granules with a fluid plasticizer that diffuses into the powder and causes swelling and ultimate coalescence of the granules into a monolithic grain. The plasticizer generally consists of a mixture of an explosive energy-producing liquid such as nitroglycerin and an inert fluid such as triacetin. The process of consolidation is a physical one. No chemical reaction occurs and there is virtually no shrinkage during curing.

A typical high-performance cast propellant starts with a single- or a double-base casting powder consisting of 30% nitrocellulose, 10% plasticizer, 30% solid oxidizer, 28% metallic fuel, and 2% stabilizer. The final propellant composition contains ca 22% nitrocellulose, 32% plasticizer, 20% solid oxidizer, 20% fuel, and ca 2% stabilizer. The type and percentage of nitrocellulose significantly affects the mechanical characteristics of the propellant. Tensile strength and the modulus of elasticity increase and elongation decreases as the percentage of nitrocellulose increases. The mechanical properties improve as the nitrogen content of the nitrocellulose decreases from 13.15% to 12.6% and 12.0%; 12.6% N-containing nitrocellulose is most commonly used. Tougher propellants with favorable heats of formation and a satisfactory carbon-hydrogen-oxygen balance are obtained with low molecular weight nitrocellulose as a binder and the addition of compounds such as poly(ethylene glycol).

Since double-base propellants cannot be directly bonded to the walls of a rocket motor, an adhesive resin is sprayed into the interior while the motor is rotated. A small amount of casting powder may also be sprayed into the tacky resin. The liner is cured and becomes an integral part of the propellant charge after casting. The motor is fitted with the required casting attachments, placed in a casting pit, if necessary, and the casting powder dispenser and associated equipment are installed. The casting powder flows from a hopper through a distributor screen and a screen plate to disperse the powder uniformly into the motor. A high-velocity air stream may also be used to carry the powder into the motor through a series of tubes. Vibration of the motors during the filling process may be employed to achieve a high loading density. After the motor has been filled, it is evacuated to remove excess moisture and entrapped air.

The fluid plasticizer (solvent) consists generally of an explosive, an inert carrier, and a stabilizer. The system is evacuated to remove volatiles, moisture, and air and the plasticizer is then pressurized and passed slowly upward through the powder bed while the powder is held stationary by a pressure plate on the powder column. The solvent occupies the space between the granules. Casting solvent may also be added from the top of the mold.

The cast-loaded rocket motor is cured at 45 to 60°C for as long as two weeks, depending on grain size. The gelatinizing solvent and the casting grains mutually diffuse so that the final rocket grain is a tough, pore-free, sturdy structure (168–169). The compositions of several typical cast double-base and modified double-base propellants are given in Table 10. Cast propellant may also be made similarly in plastic inhibitor cases or uninhibited for use in cartridge loaded applications.

Slurry Casting. The two general types of slurry-cast propellants are based either on nitrocellulose as the propellant matrix (the nitrasols) or on PVC (the plastisols) (12,13). The nitrasols incorporate microspheres of densified, colloid-treated nitrocellulose 5 to 50 μm in average diameter made by a solution and graining process similar to that used to form ball powder. The dense granules resist plasticization until the propellant is cured at elevated temperatures. The nitrocellulose particles are dispersed under vacuum in a sigma-blade mixer in an energetic plasticizer such as nitroglycerin, metriol trinitrate, or triethylene glycol dinitrate to which a desensitizer–plasticizer

Table 10. Approximate Composition and Properties of Typical Nitrocellulose-Base Cast Propellants

	Type		
	Low energy	High energy	
	A	B	C
Composition, wt %			
nitrocellulose (12.6%N)	59.0	20.0	22.0
nitroglycerin	24.0	30.0	30.0
triacetin	9.0	6.0	5.0
dioctyl phthalate	3.0		
aluminum		20.0	21.0
HMX		11.0	
stabilizer	2.0	2.0	2.0
ammonium perchlorate		11.0	20.0
lead stearate	3.0		
Ballistic properties			
specific impulse, N·s/kg[a]	2062	2651	2602
burning rate at 6.9 MPa[b] at 20°C, cm/s	0.65	1.40	2.00
pressure exponent		0.45	0.40
pressure coefficient		0.025	0.04
Thermochemical-thermodynamic properties			
heat of explosion (J/g)[c]	2931	7718	7432
heat of formation, $-\Delta H$ J/g[c]		1570	1842
flame temperature, K	1925	3850	3900
mean heat capacity, (J/(g·K))[c]			
products	1.80	1.76	1.76
gases	1.80	1.26	1.21
mean molecular weight, g/mol			
products	21.8	27.9	28.9
gases	21.8	30.9	21.0
specifc heat ratio, gas	1.27	1.18	1.17
Combustion products composition, mol/100 g[d]			
C	2.12		
CO_2	0.31	0.05	0.07
CO	2.12	1.30	1.15
H_2	1.06	0.75	0.66
H_2O	0.66	0.27	0.33
N_2	0.43	0.49	0.38
Pb	0.004		
Al_2O_3		0.35	0.37
H		0.20	0.23
OH		0.05	
other		0.15	
HCl			0.10

[a] To convert from N·s/kg to lbf·/lb, divide by 9.82.
[b] To convert MPa to psi, multiply by 145.
[c] To convert J to cal, divide by 4.184.
[d] Major products only.

such as dibutyl phthalate and the other compounds required in the formulation have been added (170). Aluminum or other powdered metal may be added during mixing, followed by the addition of the solid oxidizer. After final mixing under vacuum at ca 40°C the slurry is poured into a mold and cured by solution of the nitrocellulose at elevated temperature to form a solid grain. Compositional ranges of typical nitrasols are given in Table 11.

In the plastisol process dense spherical particles of PVC are mixed with an equal weight of a long-chain aliphatic plasticizer such as dibutyl sebacate or 2-ethylhexyl adipate to form a uniform, creamy mix. The slurry is then poured into the mold together with the oxidizer, wetting agents, stabilizer, and rate accelerators, and heated to the relatively high temperatures of 150 to 180°C where it is maintained for a short period of time. PVC dissolves in the plasticizers to form a gel which becomes a viscoelastic solid at low temperature. High-viscosity plastisol mixes have also been formed directly into grains in a screw extruder in which curing may begin during the extrusion process. A typical composition is given in Table 11.

Polymer Based. The advantages of polymeric-based cast propellants are extensive range of performance characteristics, excellent thermal and mechanical stability, and relatively low cost. Furthermore, the facilities used in their manufacture do not compete with those required for making nitrocellulose propellants (171–176). Batch processes are generally employed; however, various operations may be automated (177–180).

The manufacturing operations for making different composite propellants are very similar, although a variety of polymeric binders may be used (181). Since the viscous propellant mix must flow uniformly and rapidly into all parts of the rocket motor assembly during the casting operation, the processibility of a formulation is fundamentally related to its rheological characteristics. These depend primarily on the cure characteristics of the polymeric binder, the volume of solids loaded into the binder, and the particle shape and size distribution of the solids. Relative humidity control at 40% or less is used in most of the process operations since degradation of the polymer, the liner insulation, and the bond between the liner and the propellant as well as shorter pot life and increased mix viscosity may occur in the presence of moisture. The perchlorate and the other components of the system may also be significantly affected.

A flow chart of a typical batch process is shown in Figure 9. The oxidizer most commonly used is ammonium perchlorate. It is rigidly controlled for total and surface moisture, impurities, and particle size and shape and may contain a flow additive such as tricalcium phosphate. Slow and high-speed grinding are accomplished by hammer mills which may be coupled to an air classifier to provide the range of particle size

Table 11. Typical Nitrasol and Plastisol Compositions

Component	Wt %
Nitrasol	
nitrocellulose (12.6 wt % N)	15–30
pentaerythritol trinitrate	0–50
ammonium perchlorate	20–30
aluminum	15–20
liquid explosives and inert plasticizers	10–30
stabilizer	1–2
Plastisol	
PVC	10
dibutyl sebacate	10
ammonium perchlorate	65
aluminum	15
stabilizers, accelerators, carbon black	0.5–2

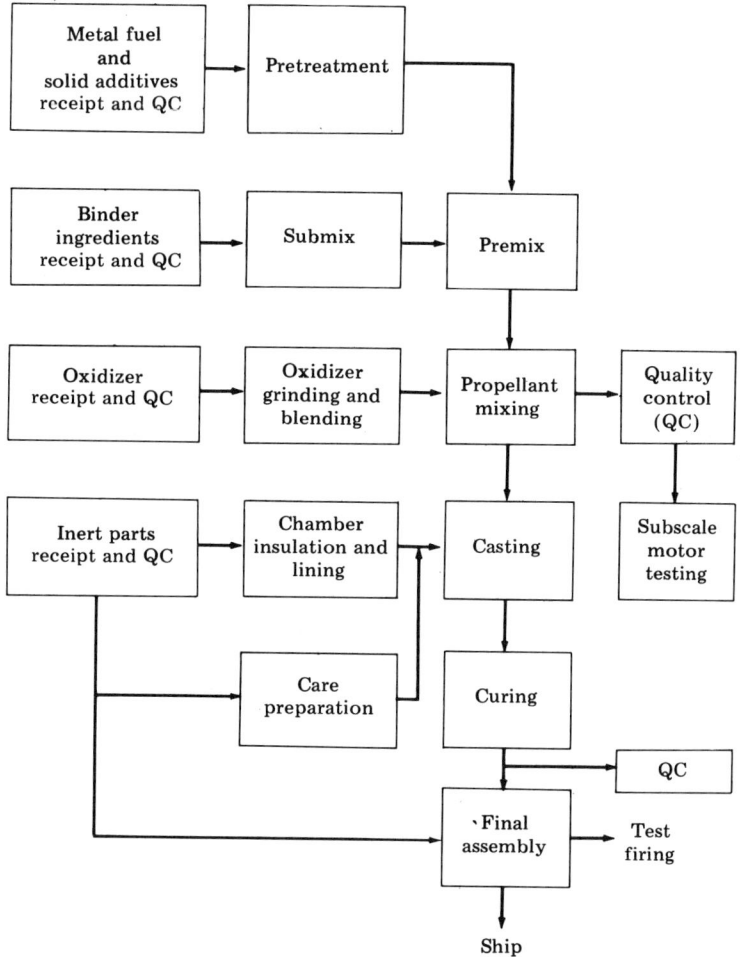

Figure 9. Batch process for cast-composite polymer-based propellants.

distributions required. Fluid energy pulverizers are also used (182). Typical particle size ranges from 3–9 μm for microatomizers to 20–160 μm for micropulverizers. The oxidizer is blended, screened, and transferred to a storage hopper for subsequent use.

In the premix operation a uniform slurry of all components, except the oxidizer, is prepared. The premixes may contain cross-linking, wetting, opacifying, and antifoaming agents, plasticizers, metallic fuels, catalysts, and curing compounds. Automated techniques ensure formulation uniformity and reproducibility. The polymer and other large-volume fluids required are pumped from the storage tanks to weigh tanks and then to the premix vessels. These may be up to 5000 L in capacity and equipped with turbine-driven agitators designed for the specific materials being handled. The secondary liquid components, including a portion of the curing agent are weighed, added, and mixed at a controlled temperature after first purging the premix vessel with nitrogen. The necessary solids other than the oxidizer are screened and added, followed by further mixing under nitrogen, and finally under vacuum to remove entrapped gases. Batch mixers are temperature controlled and designed to deaerate the viscous mass while imparting a shear action to ensure thorough and rapid incorporation of solids. They include relatively conventional horizontal mixers such as the sigma-blade dough mixer used in making nitrocellulose propellants, mixers with heavy duty bear claw blades and ribbon mixers. Vertical-change can planetary mixers are commonly used to meet requirements for increased mix capacity (182). Mixing times and temperatures are tightly controlled to maximize mix uniformity and minimize viscosity changes without accelerating the cure reactions to the stage where the pot life is excessively reduced. The mix temperature increases as a result of work input on the viscous mass and the exothermicity of the initial cure reaction. The required quantity of the premix is transferred to the mixer bowl which is moved into position and assembled to the mixer. Mixing is begun after purging with nitrogen. Where the process control conditions have been attained, the oxidizer is added followed by the curing agent. Mixing then proceeds under vacuum (183–186).

Propellants cast into rockets are commonly case-bonded to the motors to achieve maximum volumetric loading density. The interior of the motor is thoroughly cleaned, coated with an insulating material and then lined with a composition to which the propellant binder adheres under the environmental stresses of the system. The insulation material is generally a rubber-type composition, filled with silica, titanium dioxide, or potassium titanate. The liner generally consists of the same base polymer as is used in the propellant. It is usually applied in a thin layer and may be partially or fully cured before the propellant is poured into the rocket (187–191).

In the cast loading of large booster rockets, the motor is fixed in a vertical position and surrounded with necessary handling gear to facilitate subsequent operations. Very large motors are inserted in huge cylindrical pits. A shroud or similar enclosure may be used to surround the motor so that dry, warm air can be passed into it to preheat the motor and control the temperature of the casting and curing operations. The central mandrel required for grain geometry is inserted into the motor with controls for rigid alignment to close tolerances. The exact casting technique used depends on the rheological characteristics of the propellant and the quantity being processed. Several methods are commonly used for large grains including bayonet, bottom, and vacuum casting (171,192).

Upon completion of the casting operation, the motor is maintained at a closely controlled temperature–time regime to cure the propellant. Composite propellants are usually cured between 40 to 60°C. After curing, the core is withdrawn from the motor and the associated casting equipment removed. The final physical characteristics of the propellant are highly dependent on the cure conditions which in turn depend on the characteristics of the composition and the thermal conductivities of the metal and motor lining. Completeness of the cure is best determined by measuring the mechanical properties of the propellant. The reaction is finished when no change occurs on additional curing. The formulations and characteristics of a number of composite propellants are shown in Table 12 (171).

Safety. The facilities for the manufacture of composite cast nitrocellulose-based propellants incorporate the latest techniques for hazard detection and prevention and for damage control. The processes are monitored and controlled from central stations by a diversity of continuous-reading instrumentation. Closed circuit television is used for direct observation and deluge sprinkler systems with frangible seals on the nozzles are installed which can very rapidly respond to a fire. Battery power is available for emergencies. Pressure relief valves incorporating frangible disks, which are fragmented by small quantities of explosives, have been used in the lines to prevent pressure buildup and permit discharge if a fire occurs. Infrared detectors capable of sensing the light of burning propellant but indifferent to room light are mounted on the head of the batch mixers. Static pressure detector units to detect excess pressure may also be mounted in the mixer head. Both the light and pressure sensors can actuate a deluge system (174,193–194).

Black Powder

Black powder is mainly used as an igniter for nitrocellulose gun propellants and to some extent in safety blasting fuse, delay fuses and in firecrackers. Potassium nitrate black powder (74 wt % plus 15.6 wt % carbon, 10.4 wt % sulfur) is used for military applications. The slower-burning, less costly, and more hygroscopic sodium nitrate black powder (71.0 wt % plus 16.5 wt % carbon, 12.5 wt % sulfur) is used industrially.

The reaction products of black powder are complex (see Table 13) and vary with the conditions of initiation, confinement, and density.

The properties of black powder ignition systems for use in guns have received increasing attention as performance requirements have become more demanding. Black powder ignites and burns fast and has reproducibility and high heat flux. However, the solids produced in combustion are undesirable in some weapon systems. The reported thermochemical and performance characteristics vary greatly and depend on the source of material, its physical form, and the method of determination. Typical values are listed in Table 14.

The critical relative humidity of black powder is 60%. It gains ca 2% moisture in 48 h at 90% rh and 25°C. Ignitability decreases rapidly at ca the 3 to 4% moisture level. The structure of black powder granules also deteriorates during cycling through high humidity atmospheres. However, it can be stored satisfactorily for many years if dry. The hygroscopicity of black powder is caused by the carbon present and impurities in the potassium nitrate (195–205).

Manufacture. Since black powder is a mechanical mixture, its performance is critically dependent on the degree of intimacy of the components in the product. Thus the manufacture of black powder is essentially a procedure for bringing the ingredients into maximum mutual contact.

A detailed flow chart for the conventional process is presented in Figure 10. A typical procedure is as follows:

Dry potassium nitrate is pulverized in a ball mill. The sulfur is milled into cellular charcoal to form a uniform mix in a separate ball mill. The nitrate and the sulfur–charcoal mix are screened and then loosely mixed by hand or in a tumbling machine. Magnetic separators may be used to ensure the absence of ferrous metals. The preliminary mix is transferred to an edge-runner wheel mill with large and heavy cast iron wheels. A clearance between the pan and the wheels is needed for safety purposes. The magnitude of this gap also contributes to the density of the black powder granules obtained. Since the temperature increases during the milling operation and moisture evaporates, water is added to minimize dusting and improve incorporation of the nitrate into the charcoal. The milling operation requires ca 3 to 6 h.

Table 12. Approximate Composition and Properties of Typical Polymer-Based Cast Composite Propellants

	Propellant type[a]							
	Poly-sulfide	Poly-urethane	CTPB	CTPB	HTPB	PBAN	PBAA	Buta-diene
Composition, wt %								
ammonium perchlorate	63.0	70.0	73.0	63.0	70.0	69.0	68.0	
binder	36.0	21.0	12.0	10.0	12.0	11.0	15.0	14
MgO	1.0							
sulfur, added	0.2							
aluminum		8.0	15.0	17.0	18.0	15.0	16.0	
other		1.0						6
dioctyl adipate						4.0		
iron catalyst						1.0	1.0	
ammonium nitrate								80
HMX				10.0				
Ballistic properties								
specific impulse, N·s/kg[b]	2259	2406	2602	2602	2553	2602	2553	1866
burning rate at 6.9 MPa[c] at 20°C, cm/s	0.90	0.80	0.98	0.75	0.60	1.37	1.70	0.30
pressure exponent, n	0.45	0.20	0.30	0.30	0.30	0.33	0.20	0.50
pressure coefficient, c	0.012	0.07	0.06	0.03	0.03	0.06	0.14	0.004
Thermochemical-thermodynamic properties[d]								
heat of explosion, J/g[e]		4543	6280		6448	5966		2721
heat of formation, $-\Delta H_f$, H_f, J/g[e]	2156	2470	1842	1549		1999	1842	3768
flame temperature T_P, K	2375	2850	3500	3650	3450	3400	3300	1000
mean heat capacity, J/(g·K)[e]								
products	2.34		1.80	1.80	1.84			2.01
gases	2.34			1.97	1.97			2.01

mean molecular weight, g/mol							
products	25.2	24.8	28.1	26.9	28.1	27.2	19.5
gases	25.2			20.5			19.3
specific heat ratio, gas	1.20			1.19	1.19		
Combustion products, composition, mol/100 g[f]							
CO_2	0.28	0.15	0.08	0.06	0.07		0.37
CO	0.80	1.02	0.79	0.78	0.89	0.96	0.93
H_2	0.55	0.94	0.88	0.85	0.96	1.14	1.40
H_2O	1.17	1.00	0.73	0.57	0.64	0.38	1.37
N_2	0.23	0.31	0.30	0.40	0.30	0.28	1.00
Al_2O_3		0.14	0.26	0.29	0.30	0.25	
HCl	0.50	0.58	0.50	0.42	0.47	0.43	
H_2S	0.10						
H			0.12	0.17	0.13	0.09	
$AlCl + AlCl_2$			0.04		0.03		

[a] CTPB, carboxy-terminated polybutadiene; HTPB, hydroxy-terminated polybutadiene; PBAN, polybutadiene–acrylic acid–acrylonitrile; and PBAA, polybutadiene–acrylic acid.
[b] To convert N·s/kg to lbf·s/kg, divide by 9.82.
[c] To convert MPa to psi, multiply by 145.
[d] All gas volumes at standard temperature and pressure.
[e] To convert J to cal, divide by 4.184.
[f] Major products only.

Table 13. Approximate Composition of Reaction Products of Black Powder

Component	Wt %
Gases	
carbon dioxide	49
carbon monoxide	12
nitrogen	33
hydrogen sulfide	2.5
methane	0.5
water	1
hydrogen	2
Total	44
Solids	
potassium carbonate	61
potassium sulfate	15
potassium sulfide	14.3
potassium thiocyanate	0.2
potassium nitrate	0.3
ammonium carbonate	0.1
sulfur	9
carbon	0.1
Total	56

Table 14. Characteristics of Black Powder

Characteristic	Value
flame temperature, K (isochoric)	ca 2800
moles of gas per gram	0.0128–0.0159
heat of explosion (H_2O, liq), J/g[a]	3015–3140
impetus, J/g[a]	239–284
burning rate, cm/s at 6.9 MPa[b]	ca 1 to 1.5
temperature coefficient of pressure, %/°C	0.4
pressure exponent	0.25–0.5
ignition temperature, °C	450
activation energy, kJ/mol[a]	(87.9)

[a] To convert J to cal, divide by 4.184.
[b] To convert MPa to psi, multiply by 145.

The moist milled powder is transferred to a hydraulic press where it is consolidated in layers into cakes at pressures of ca 41.3 MPa (6000 psi) applied for ca 30 min. Each cake is ca 2.5 cm thick and 60 cm square. The density of the powder increases to 1.6 to 1.8 g/cm^3, depending on the pressure applied. The cakes are then transferred to a corning mill consisting of adjustable corrugated rollers which are cascaded so that a series of crushing actions occur with each crushing, followed by automatic screening to form a product which approximates the granulation requirements. The dust and fines which have been screened are recycled to the press feed or used in fuse powder or fireworks. Coarse material is recycled. The grains are polished, dried, graphite is added, and they are blended by tumbling in a large hardwood rotating drum. From 1360 to 2265 kg powder may be tumbled at 10–20 rpm for up to 8 h. Warm air may be forced through the barrel to assure drying and to decrease cycle time. The powder is screened before packing into air-tight metal drums (195,199–200).

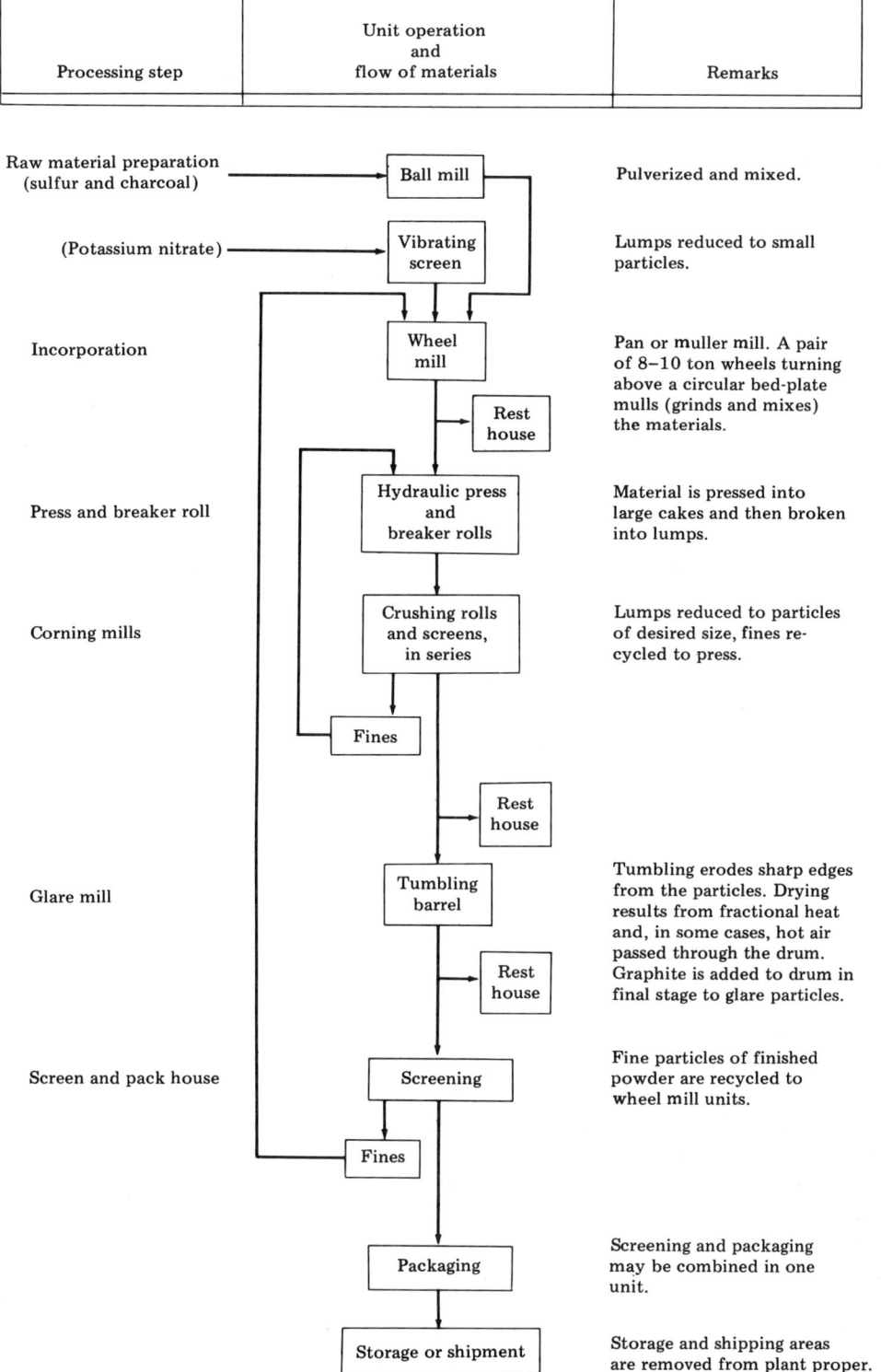

Figure 10. Process of manufacture of black powder.

A number of new techniques have been developed to eliminate the hazardous wheel milling process and reduce personnel exposure by increasing automation for the continuous transport of the product from operation to operation. The jet-mill air-attrition process which has no moving parts has replaced the wheel-mill operation of the conventional process. Potassium nitrate, sulfur, and charcoal are automatically weighed by transferring each ingredient to a weighing and mixing bin with a vibrating transporter. Air jets are applied to the bottom of the weigh–mix bin to blend the components. The air pressure is then increased to continuously transfer the mix pneumatically to a storage bin and then to the jet mill by air injection. A high velocity stream of air entering the mill forces the particles to collide and breaks them up by attrition. The product consists of a finely divided powder. The small particles exit through a cyclone separator where they are separated from the air, which is exhausted to the atmosphere. Coarse particles drop back to the attrition section where the milling action continues. The mill may be adjusted to produce powders of different granulations.

Pressing, corning, screening, and glazing are comparable to the conventional procedures except that automation is employed wherever possible. Deluge systems which are activated by ultraviolet light sensors and can respond in milliseconds are installed for additional safety. All operations are monitored and controlled from central process control areas. The presence of operators is restricted to the receipt of raw materials and packing of the final product. The pollution aspects of black powder manufacture are relatively insignificant in view of the small quantities made and since no wastewater or noxious fumes are produced (206–207).

Benite. Benite is an extrudable composition consisting of ca 60 parts of black powder in a matrix of ca 40 parts of plasticized nitrocellulose. It is used as a propellant igniter to reduce the residue formed compared to use of black powder alone. Since it can be extruded as strands, it permits a less obstructed flow of ignition gas and particles than granular black powder. Its approximate wt composition is nitrocellulose (13.15% N), 40%; potassium nitrate, 44%; sulfur, 6.5%; carbon, 9.5%; ethyl centralite, 0.5%. It is made by the single-base process, followed by air-drying to remove volatile solvents (208).

Felted Nitrocellulose Compositions. A combustible case containing the propellant charge offers tactical, logistic, and performance advantages in certain types of munitions such as those used for tank weapons, mortars, and howitzers. The case is rigid and completely combustible, replacing metallic cases or flexible cases with low mechanical strength. Since nitrocellulose itself cannot be molded into a structure with the desired mechanical characteristics, inert fibers and a resin are added to it. A typical composition by wt consists of nitrocellulose (N, 12.6%), 55%; kraft fiber, 9%; acrylic fiber, 25%; poly(vinyl acetate) resin, 10%; diphenylamine, 1%.

A common procedure for manufacture of a combustible case starts with the formation of a slurry of kraft fibers dispersed in water with a hydropulper-agitator. Nitrocellulose, an aqueous slurry of the acrylic fiber, an alcohol solution of the diphenylamine, and an aqueous emulsion of the resin are added. After mixing thoroughly, an aqueous solution of sodium aluminum sulfate is added to break the resin emulsion and precipitate the resin on the fibers. The mix is pumped to a storage tank, diluted with water to a solids content of ca 0.15% and aged for at least 24 h. The slurry is pumped from the storage tank to a felting die which is a hollow perforated form covered with wire mesh and contoured to the shape of the case to be formed. The die is lowered into the felting tank and vacuum applied to the die while the slurry is slowly stirred. The fibers and resin are deposited as a felt on the die screen. After a predetermined time to control the quantity of pulp deposited, the felt-covered die is raised from the tank while the vacuum continues to draw off water. A thin rubber sheet is placed over the felt to force out more water, the vacuum is cut off and a short pressure pulse is applied to the die through the line previously used for vacuum to expand the felt slightly so that it may be removed from the die. At this stage the felt form contains ca 50% water.

The wet, soft form is placed over the male member of a set of matched metal molds heated to about 110°C. The water is then pressed out and vaporized and the resin is set during the forming operation. A vacuum line draws off the water vapor from the heated part. After a brief dwell time, the mold is opened and the case removed and allowed to remain for at least 24 h at 20°C and 65% relative humidity to stabilize it before trimming to final dimensions. The case has a density of ca 0.85 g/cm^3 and a tensile strength of ca 24 MPa (3500 psi) (209–211) (see Felts).

BIBLIOGRAPHY

1. C. Boyars and K. Klager, eds., *Proceedings of Symposium on Propellants Manufacture, Hazards and Testing,* American Chemical Society, Washington D.C., 1969.
2. F. A. Williams, M. Barrere, and N. C. Huang, *Fundamental Aspects of Solid Propellant Rockets, Agardograph No. 116,* Tech Press, London, Eng., 1969.
3. R. L. Wilkins, *Theoretical Evaluation of Chemical Propellants,* Prentice-Hall, Inc., Englewood Cliffs, N.J., 1968.
4. *Static Firing of First Shuttle SRM, CPIA Bulletin 3,* Oct. 1977, p. 1.
5. F. Henden, "Review of Solid Propellants for Space Exploration," *Tech. Memo No. 33-254,* Jet Propulsion Laboratory, Pasadena, Calif., 1965.
6. F. A. Warren "Solid Propellant Technology" in R. A. Gess, ed., *Selected Reprint Series,* American Institute of Aeronautics and Astronautics, New York, 1970.
7. K. Klager and J. M. Wrightson, "Recent Advances in Solid Propellant Binder Chemistry" in *Proceedings of the Fourth Symposium on Naval Structural Mechanics,* Pergamon Press, Inc., New York, 1965, p. 47.
8. F. Arendale, "Chemistry of Propellants Based on Chemically Cross-linked Binders" in ref. 1, p. 67.
9. J. S. Jorczak and E. M. Fettes, *Ind. Eng. Chem.* **43,** 234 (1951).
10. R. T. Davis, "Ten Years of Polysulfide Propellant Aging" in *Bulletin of the Joint Meeting—JANNAF Panels on Physical Properties and Surveillance of Solid Propellants, Publication No. PP-13/SPSP-8,* CPIA, Johns Hopkins University, Md., 1960, p. 361.
11. P. J. Stedry and R. F. Landel, "Effects of Humidity on the Mechanical Properties of Selected Polysulfide and Polyurethane Propellants" in ref. 10, p. 393.
12. T. L. Smith and J. L. Coy, "The Storage Stability of Some Ammonium Perchlorate–Polyurethane Propellants" in *Bulletin of the Third Meeting—JANNAF Solid Propellant Surveillance Panel, Publication No. SPSP/5,* CPIA, 1958.
13. R. T. Davis, "Effects of Moisture on Storage Stability of Ammonium Perchlorate Propellants" in *Bulletin of the Fourth Meeting—JANNAF Solid Propellant Surveillance Panel, Publication No. SPSP/7,* SPIA, 1959.
14. A. E. Oberth and Bruenner, "Polyurethane Based Propellant" in ref. 1, p. 84.
15. E. J. Mastrolia and K. Klager, "Solid Propellants Based on Polybutadiene Polymers" in ref. 1, p. 122.
16. R. W. James, *Propellants and Explosives,* Noyes Data Corp., Park Ridge, N.J., 1974.
17. M. S. Cohen, "Advanced Binders for Solid Propellants—A Review" in *Proceedings of Symposium on Advanced Propellant Chemistry,* American Chemical Society, Washington, D.C., 1966.
18. "Hydrocarbon Binders—1974/1975," *CPIA Publication No. LS76-36,* CPIA, Johns Hopkins University, Md., 1976.

19. "Summary of Published Research on Binders and Propellants" in D. M. French, ed., *TR 392,* Naval Ordnance Station, 1973.
20. R. Steinberger and P. D. Drechsel, "Manufacture of Double-Base Propellant" in ref. 1, p. 1.
21. A. T. Camp, "Nitrocellulose Plastisol Propellants" in ref. 1, p. 29.
22. K. E. Rumbel, "Poly(vinyl chloride) Plastisol Propellants" in ref. 1, p. 36.
23. D. E. Boynton and J. W. Schowengerdt, *Chem. Eng. Prog.* **59**(3), 31 (1963).
24. *Solid Propellant Manual Publication M2,* CPIA, Johns Hopkins University, Md., 1969, (classified).
25. *Solid Propellant Selection and Characterization, Space Vehicle Design Criteria, Monograph No. SP 8064,* NASA, Md., 1971.
26. M. S. Zucrow, *Propulsion and Propellants, Engineering Design Handbook AMCP 706-282,* AMC, Alexandria, Va., 1963.
27. *Formulae and Calculated Thermochemical Values for Propellants,* Picatinny Arsenal, N.J., 1956.
28. B. Crawford and co-workers, *Rocket Fundamentals, Report No. 3992,* OSRD, 1944.
29. A. O. Decker, *Jet Propul.* **21**, (1956).
30. *Interior Ballistics, Report No. 6468,* OSRD, 1946.
31. *Methods for Determining the Tensile Properties of Solid Rocket Propellants,* SPIA, Johns Hopkins University, Md.
32. K. W. Bills, Jr. and J. H. Wiegand, *AIAA J.,* **1**, 2116 (1963).
33. G. Lianis, *ARS J.* **32**, 688 (1962).
34. J. W. Cole, "Non Destructive Testing of Large Rocket Motors; A State of of the Art Survey" in *Bulletin of the Joint Meeting—JANNAF Panel of Physical Properties and Surveillance of Solid Propellants, CPIA Publication No. PP-13/SPSP-8,* CPIA, Johns Hopkins University, Md., 1960, p. 345.
35. F. N. Kelley "Solid Propellant Mechanical Properties Testing, Failure Criteria and Aging" in C. Boyars and K. Klager, eds., *Propellants Manufacture, Hazards and Testing, Advances in Chemistry Series 88,* American Chemical Society, Washington, D.C., 1969, p. 188.
36. *Handbook For the Engineering Structural Analysis of Solid Propellants, CPIA Publication No. 214,* CPIA, Johns Hopkins University, Md., 1971.
37. *ICRPG Solid Propellant Mechanical Behavior Manual, CPIA Publication No. 21,* CPIA, Johns Hopkins University, Md., 1963.
38. *Nitrocellulose Propellants* and *Composite Propellants; Methods for Determining the Tensile Properties of Solid Rocket Propellants, SPIA/PP8,* Parts 1 and 2, CPIA, Johns Hopkins University, Md., 1956, 1957.
39. *Solid Propellant Aging, Mechanical Behavior and Grain Structural Integrity, CPIA Report No. LS 77-27,* CPIA, Johns Hopkins University, Md., 1977.
40. R. Stenson, *Factors Governing the Storage Life of Solid Propellant Rocket Motors, Report No. TN-47, Explosive Research and Development Establishment,* Waltham Abbey, England, 1972.
41. "Tools Required for a Meaningful Service Life Prediction," *Report—JANNAF Panel, CPIA Publication No. 259,* CPIA, Johns Hopkins University, Md., 1974.
42. C. Boyars, *ARS J.* **29**, 148 (1959).
43. R. B. Wetherell, J. M. Anderson, and J. M. Thacker, "Hazard Assessment for Solid Propulsion Systems" Vol. 2 in *Proceedings of the JANNAF Propulsion Meeting, CPIA Publication No. 293,* CPIA, Johns Hopkins University, Md., 1978, p. 47.
44. *Safety and Hazards of High Energy Propellants: A Bibliography, CPIA Publication No. 284,* CPIA, Johns Hopkins University, Md., March 1977.
45. "Solid Rocket Propellant Processing, Handling, Storage and Transportation" in A. Jensen, ed., *Chemical Rocket Propellant Hazards, CPIA Publication No. 194,* Vol. 2, CPIA, Johns Hopkins University, Md., 1970.
46. B. Brown, *Explosive Hazards of Composite Solid Propellants, Institute for Defense Analysis Report T-362,* Washington, D.C., 1968.
47. *Proceedings of Explosive Safety Seminar on High Energy Solid Propellants,* Armed Services Explosive Safety Board, Dept. of Defense, Washington, D.C., 1960.
48. W. H. Anderson and R. F. Chaiken, *ARS J.* **31**, 162 (1959).
49. J. E. Crump and E. W. Price, *ARS J.* **40**, 707 (1960).
50. S. Wachtel, "Prediction of Detonation Hazards in Solid Propellants," *paper presented at the 145th National Meeting of the Division of Fuel Chemistry, New York, 1963.*
51. F. L. Schuyler, *Analytical Investigations of Combustion Instability in Solid Propellant Rockets, Illinois Institute Technical Report No. IITRI-A 6002,* Chicago, Ill., 1963.

52. A. W. Horst, T. C. Smith, and S. E. Mitchell, "Key Design Parameters in Controlling Gun-Environment Pressure Wave Phenomena: Theory vs Experiment" in *Bulletin of the 13th JANNAF Meeting, CPIA Publication No. 281,* Vol. 1, CPIA, Johns Hopkins University, Md., 1976, p. 225.
53. J. J. Rocchio and co-workers, "Propellant Grain Tailoring to Reduce Pressure Wave Generation in Guns" in *Proceedings of the 12th JANNAF Combustion Meeting, CPIA Publication No. 273,* Vol. 1, CPIA, Johns Hopkins University, Md., 1975, p. 225.
54. E. V. Clarke, I. W. May, and J. R. Kelso, *Effects of Pressure Wave Dynamics on the Ballistic Performance of Guns, IMR 311,* BRL, Aberdeen, Md., 1974.
55. A. W. Horst, Jr., T. C. Smith, and S. E. Mitchell, *Experimental Evaluation of Three Concepts for Reducing Pressure Wave Phenomena in Navy 5 Inch 54-Caliber Guns: Summary of Firing Data, IHMR 76-258,* NOS, Indiana Head, Md., 1976.
56. A. W. Horst, Jr., and T. C. Smith, "The Influence of Propelling Charge Configuration on Gun Environment Pressure-Time Anomaly" in *Proceedings of the 12th JANNAF Combustion Meeting, CPIA Publication No. 273,* Vol. 1, CPIA, Johns Hopkins University, Md., 1975, p. 259.
57. J. Corner, *Theory of the Interior Ballistics of Guns,* John Wiley & Sons, Inc., New York, 1950.
58. R. Steinberger, "Preparation and Properties of Double Base Propellants," *paper presented at AGARD Panel: Chemistry of Propellants NATO, Paris, Fr., 1959.*
59. R. L. Glick, *AIAA J.* **5**(3), 585 (1967).
60. L. Green, Jr., *ARS J.* **24,** 9 (1954).
61. R. Schultz, L. Green, and S. S. Penner, "Studies of the Decomposition Mechanism, Erosive Burning, and Sonance and Resonance for Solid Composite Propellants," *paper presented at AGARD Panel: Combustion Panel Colloquium, NATO, Palermo, Italy 1958.*
62. R. C. Strittmater, E. M. Wineholt, and M. E. Holmes, *The Sensitivity of Double Base Propellant Burning Rate to Initial Temperature, MR-2593,* BRL, Aberdeen, Md., 1976.
63. M. M. Ibercu and F. A. Williams, *Combust. Flame* **24,** 185 (1975).
64. B. B. Grollman and C. W. Nelson, "Burning Rates of Standard Army Propellants in Strand Burner and Closed Chamber Tests" in *Bulletin of the 13th JANNAF Combustion Meeting, CPIA Publication No. 281,* Vol. 1, CPIA, Johns Hopkins University, Md., 1976, p. 21.
65. D. W. Reifler, *Linear Burning Rates of Ball Propellants Based on Closed Bomb Firings, CR 172,* BRL, Aberdeen, Md., 1974.
66. *Solid Propellant Manual M2,* CPIA, Johns Hopkins University, Md., 1969 (classified).
67. W. S. McEvan, *State of the Art of Combustion of Solid Propellants, AC/193,* NOTS, U.S. Navy, 1961.
68. B. T. Federoff and O. Sheffield, "Burning Characteristics of Propellants for Rockets" in *Encyclopedia of Explosives and Related Items, TR 2700,* Vol. 2, PTA, Dover, N.J., 1962.
69. J. A. Vanderkehove, *ARS J.* **29,** 483 (1959).
70. M. W. Stone, *Jet Propul.* **28,** 236 (1958).
71. F. R. W. Hunt, ed., *Internal Ballistics,* Philosophical Library, New York, 1951.
72. *Interior Ballistics of Guns, Engineering Design Handbook, AMCP 706-150,* DARCOM, U.S. Army, 1965.
73. A. O. Pallingston and M. Weinstein, *Method of Calculation of Interior Ballistic Properties of Propellants from Closed Bomb Data, Report No. 2005,* PTA, Dover, N.J., 1959.
74. B. L. Crawford, Jr., and co-workers, *Anal. Chem.* **19,** 630 (1947).
75. K. Atlas, *A Method of Computing Web for Gun Propellant Grains from Closed Bomb Burning Rates, Memo Report No. 73,* Naval Powder Factory, U.S. Navy, Indian Head, Md., 1954.
76. J. K. Domen, *Modernization of Closed Bomb Testing for Acceptance of Single Base Propellant, Report No. QA-X-016,* PTA, Dover, N.J., 1976.
77. E. B. Fisher and K. P. Tripper, "The Hi-Low Bomb: A New Device to Evaluate the Combustion Properties of Energetic Materials" in *Proceedings of the 12th JANNAF Combustion Meeting, CPIA Publication No. 273,* Vol. 1, CPIA, Johns Hopkins University, Md., 1975, p. 323.
78. M. J. Adams and H. Crier, "Determining the Dynamic Burning Rate of Gun Propellants Using the Dynagun Ballistic Simulator" in ref. 77, p. 303.
79. Ref. 52, p. 383.
80. "Method 801.1, Relative Quickness and Force of Propellants," "Method T803.1, Linear Burning Rate of Propellants, (Strand Burner Method)" in *U.S. Military Standard 286-B,* 1967.
81. B. L. Crawford, Jr., C. Huggett, and J. L. McBrady, *J. Phys. Colloid Chem.* **54,** 847 (1950).
82. C. Huggett, C. E. Bartley, and M. M. Mills, *Solid Propellant Rockets,* Princeton University Press, Princeton, N.J., 1960.

83. C. Fenimore, *The Final Reactions of the Burning of Nitrocellulose, Report No. R-464,* BRL, Aberdeen, Md., 1944.
84. G. K. Adams, *Selected Combustion Problems: Combustion Colloquium,* AGARD, NATO, Paris, 1953, p. 277.
85. R. E. Wilforce, W. S. Penner, and F. Daniels, *J. Phys. Colloid Chem.* **54,** 836 (1950).
86. O. K. Rice and R. Ginnel, *J. Phys. Colloid Chem.* **54,** 885 (1950).
87. R. G. Parr and B. L. Crawford, *J. Phys. Colloid Chem.* **54,** 929 (1950).
88. G. K. Adams, "The Chemistry of Solid Propellant Combustion: Nitrate Esters of Double Base Systems" in *Proceedings of the Fourth Symposium of Naval Structural Mechanics: Mechanics and Chemistry of Solid Propellants,* Pergamon Press, Inc., New York, 1967, p. 117.
89. J. G. Sotter, "Chemical Kinetics of the Cordite Explosion Zone," *paper presented at the 10th International Symposium on Combustion, The Combustion Institute,* The Williams & Wilkins Co., Baltimore, Md., 1964, p. 1405.
90. G. A. Heath and R. Hirst, "Some Characteristics of the High Pressure Burning of Some Double Base Propellants," *paper presented at the 8th International Symposium on Combustion,* The Williams & Wilkins, Co., Baltimore, Md., 1962, p. 711.
91. L. Dauerman and Y. A. Tajima, *AIAA J.* **6,** 688 (1968).
92. C. E. Kirby and N. P. Suh, *AIAA J.* **9**(2), 817 (1971).
93. M. M. Ibercu and F. A. Williams, "Mechanisms for the Study of Deflagration of Double Base Propellants" in *Proceedings of the 12th JANNAF Combustion Meeting, CPIA Publication No. 273,* Vol. 2, CPIA, Johns Hopkins University, Md., 1975, p. 283.
94. B. A. Burke, *A Procedures Manual for Computing Propellant Thermodynamic Properties, TR 63-202,* NOL, White Oak, Md, 1964.
95. H. N. Browne and M. M. Williams, *The Theoretical Computation of Equilibrium Compositions, Thermodynamic Properties and Performance Characteristics of Propellant Systems, TR 2434,* NOTS, China Lake, Calif., 1960.
96. *Thermodynamic Properties of Solid Propellants Systems, CPIA Publication No. M1,* CPIA, Johns Hopkins University, Md., 1957.
97. B. C. Crawford, ed., *Rocket Fundamentals, Report 3992,* OSRD, 1944.
98. A. O. Decker, *Jet Propul.* **21** (1956).
99. D. George and J. B. Levine, *Proceedings of the 13th JANNAF Combustion Meeting, CPIA Publication No. 281,* Vol. 2, CPIA, Johns Hopkins University, Md., 1976, p. 45.
100. R. F. Preckel, *AIAA J.* **3**, 346 (1965).
101. D. J. Hewken and co-workers, *Combust. Sci. Technol.* **2,** 307 (1971).
102. "The Burning Mechanism of Ammonium Perchlorate Propellants" M. Summerfield and co-workers in M. Summerfield, ed., *Solid Propellant Research,* Academic Press, New York, 1960.
103. C. K. Adams, B. H. Newman, and A. B. Robins, "The Combustion of Propellants Based Upon Ammonium Perchlorate," *Proceedings 8th Symposium (International) on Combustion,* The Williams & Wilkins Co., Baltimore, 1960, p. 693.
104. J. D. Hightower and E. W. Price, *Experimental Studies of the Combustion Zone of Composite Propellants,* paper presented at the 1st ICRPG/AIAA Propulsion Meeting, Washington, D.C., 1966.
105. J. A. Stein, P. L. Stang, and M. Summerfield, "The Burning Mechanism of Ammonium Perchlorate-Based Composite Propellants," *Aerospace and Mechanical Sciences Report No. 830,* Princeton Univ., 1969.
106. E. W. Price, "Recent Advances in Solid Propellant Combustion Instability," *Proceedings 12th Symposium (International) on Combustion,* The Williams & Wilkins Co., Baltimore, 1970, p. 101.
107. M. W. Beckstead, R. L. Derr, and C. F. Price, *AIAA J.* **8**(4), 2200 (1970).
108. M. W. Beckstead, R. L. Derr, and C. F. Price, "The Combustion of Solid Monopropellants and Composite Propellants," *Proceedings of the 13th Symposium (International) on Combustion,* The Combustion Institute, Pittsburgh, Pa., 1971.
109. N. S. Cohen, R. L. Derr, and C. E. Price, "Extended Model of Solid Propellant Combustion Based on Multiple Flames," *Proceedings of the 9th JANNAF Combustion Meeting,* CPIA 231, 1972, p. 25.
110. M. W. Beckstead, "A Model for Solid Propellant Combustion," *Proceedings of the 14th JANNAF Combustion Meeting,* Vol. 1, CPIA 292, 1977, p. 281.
111. M. S. Cohen, "Advanced Binders for Solid Propellants—A Review," *Proceedings of the Symposium on Advanced Propellant Chemistry,* American Chemical Society, Washington, D.C., 1966, p. 93.

112. M. Summerfield and T. D. Parker, "Interrelationships Between Combustion Phenomena and Mechanical Properties in Solid Propellant Rocket Motors," *Proceedings of the 4th Symposium on Naval Structural Mechanics,* Pergamon Press, New York, 1967 p. 75.
113. M. Shorr and A. J. Zaehringer, *Solid Rocket Technology,* John Wiley & Sons, Inc., New York 1967.
114. D. M. French, ed., *Summary of Published Research on Binders and Propellants,* TR 392, NOS, 1973.
115. *Hydrocarbon Binders 1974/1975,* CPIA, LS76-36, 1976.
116. R. Muracarf and L. L. Taylor, *The Hygroscopicity of Lithium and Ammonium Nitrates and Perchlorates, Progress Rpt. 20-347,* Jet Propulsion Lab. Institute of Technology, Pasadena, 1958.
117. P. W. M. Jacobs and H. M. Whitehead, *Chem. Rev.* **69,** 4, 551 (1969).
118. P. W. M. Jacobs and R. A. Jones, *AIAA J.* **5,** 829 (1976).
119. J. Wenograd, "Report on the Nature of Ammonium Perchlorate Decomposition," *Proceedings 6th ICRPG Combustion Conference,* Vol. 1, CPIA 192, 1969, p. 367.
120. J. P. Renie, J. A. Condon, and J. R. Osborn, "Oxidizer Size Distribution Effects," *Proceedings of the 14th JANNAF Combustion Meeting,* Vol. 1, CPIA 292, 1977, p. 325.
121. R. R. Miller, M. T. Donohue, and J. P. Peterson, "Ammonium Perchlorate Size Effects on Burn Rate—Possible Modification by Binder Type," *Proceedings of the 12th JANNAF Combustion Meeting,* Vol. 2, CPIA 273, 1975, p. 371.
122. E. K. Bastress, H. P. Hall, and M. Summerfield, *Modification of Burning Rates of Solid Propellants by Oxidizer Particle Size Control,* paper ARS 1597-61, ARS Solid Propellant Conference, Salt Lake City, Utah, 1961.
123. C. S. Brenner, *Oxidizer Particle Size Effects on the Ballistic Properties of a Composite Solid Propellant,* Paper ARS 1597-61, ARS Solid Propellant Conference, Salt Lake City, Utah, 1961.
124. J. P. Rome, J. A. Condon, and J. R. Osborn, "Oxidizer Size Distribution Effects," *Proceedings of the 14th JANNAF Combustion Meeting,* Vol. 1, CPIA 292, 1977, p. 325.
125. R. H. Waesche, "Workshop in the Relationship of Ammonium Perchlorate Decomposition to Deflagration," *Proceedings of the 7th JANNAF Combustion Meeting,* Vol. 1, CPIA 204, 1971, p. 15.
126. L. A. Povenille, "Summary Report on the Nature of Ammonium Perchlorate Deflagration," *Proceedings 6th ICRPG Combustion Conference,* Vol. I, CPIA 192, 1969 p 371.
127. H. Cheung and N. S. Cohen, *AIAA J.* **3**(2) 250 (1965).
128. A. Davis, *Combust. Flame* **7,** 359 (1963).
129. G. H. Markstein, *AIAA J.* **1,** 5 (1963).
130. D. J. Carlson and R. F. Hoglund, *AIAA J.* **2,** 80 (1964).
131. P. Kuentzmann, *Specific Impulse Losses of Metallized Solid Propellants,* FTD-NC-23-2717-74 Air Force Foreign Technology Division, Wright Patterson Air Force Base, Dayton, Ohio, 1974.
132. S. F. Sarner, *Propellant Chemistry,* Reinhold Publishing Co., 1966.
133. R. L. Derr, "Combustion Instability in Navy Rockets Revisited," *Proceedings of the 13th JANNAF Combustion Meeting,* CPIA 281, 1976, p. 185.
134. *Instability of Combustion of Solid Propellants,* Ad Hoc Group on Solid Propellant Instability of Combustion, Director of Defense Research and Engineering, Washington, D.C., 1959.
135. R. L. Derr and co-workers, "Combustion Instability Studies Using Metallized Solid Propellants: Part I, Experimental Verification of Particle Dumping Theory," *Proceedings JANNAF Combustion Meeting,* Vol. 1, CPIA 273, 1975.
136. A. M. Ball, "Solid Propellants," *Engineering Design Handbook,* Part 1, AMCP 706-175, USA Materiel Command, Washington, D.C., 1964.
137. T. Urbanski, *Chemistry and Technology of Explosives,* Vol. 3, Pergamon Press, New York, 1967, Chapts. 7–8.
138. R. Goldstein and K. Russel, *Process Engineering of Triple Base Propellants, DT-TR-8-16,* Picatinny Arsenal, Dover, N.J., 1960.
139. *The Preparation and Properties of Solvent Extruded Composite Propellant, OSRD 5576,* ERL, 1946.
140. R. P. Ayerst, "Some Observations on the Incorporation Stage in Colloidal Propellant Processing," *Proceedings Symposium on Processing Propellants, Explosives and Ingredients,* ADPA, 1977, p. 3.6-1.
141. L. T. Fan, S. J. Chem, and C. A. Watkins, *Ind. Eng. Chem.* **7,** 53 (1970).
142. R. J. Butters and co-workers, *Brit. Chem. Eng.* **15,** 41 (1970).
143. N. I. Shapiro and co-workers, *Wastewater Treatment,* Vol 3, ADPA, 1975, p. 120.

144. *A Continuous Automated Cannon Propellant Facility,* Radford Army Ammunition Plant, Radford, Va., 1970.
145. C. H. Johnson and F. T. Kristoff, "A Pressure Relief Venting Concept for Eliminating an Explosive Hazard in Automated Single Base Operations," *Proceedings Symposium on Processing Propellants, Explosives and Ingredients,* ADPA, 1977, p. 5.
146. W. J. Nolan, "Continuous Automated Single Base Propellant Process," *Proceedings Symposium on Processing Propellants, Explosives and Ingredients,* ADPA, 1977, p. 1.3-1.
147. *U.S. Specification Military Standard 652C,* Nov. 1974.
148. *Propellants, Solid: Sampling Examination and Testing,* MIL-STD-286B, 1971.
149. *Propellant, Artillery,* MIL-STD-P-207A, 1959
150. E. R. Csandady and E. Roberts, *Pilot Plant Development of Double Base Propellants,* paper presented at National Meeting of the American Institute of Chemical Engineers, Louisville, 1955.
151. S. Kaplowitz, "Ultrasonic Extrusion of Double Base Propellant" *Proceedings Symposium on Processing Propellants, Explosives and Ingredients,* ADPA, 1977, p. 4.3-1.
152. U.S. Pat. 3,960,993 (Feb. 27, 1975), C. E. Johnson and J. R. Luense (to USA).
153. C. E. Johnson, "No Roll Process for Making Solventless Double Base Propellants," *Proceedings Symposium on Processing Propellants, Explosives and Ingredients,* ADPA, 1977, p. 3.4-1.
154. *The Ball Powder Process,* Badger Army Ammunition Plant, Baraboo, Wis.
155. "Ball Powder" in B. T. Federoff and O. Sheffield, eds., *Encyclopedia of Explosives and Related Items,* Vol. 2, TR2700 PTA, 1962, p. B11.
156. T. Urbanski "Ball Grain Powder" in *Chemistry and Technology of Explosives,* Vol. 3, Pergamon Press, New York, 1967, p. 632.
157. T. R. Olive, *Chem. Eng.* **53,** 136 (1946); D. W. Reifler, "Optimum Performance of Deterred Ball Propellant as Determined By Computer Simulation," *Proceedings of the 10th JANNAF Combustion Meeting,* CPIA 243, 1973, p. 9; D. W. Reifler, *Linear Burning Rates of Ball Propellants Based on Closed Bomb Firings,* CR 172, Ballistic Research Laboratories, 1974.
158. J. J. O'Neil, Jr., *Ordnance* **41,** 365 (1956).
159. *The Continuous Ball Powder Production Process,* Olin Corp., Ill., 1974.
160. *Smokeless Powder Operation: St. Marks, Florida,* Winchester Group, Olin Corp., Ill.
161. R. H. Thied, "Design of Propellant Pilot Plants for Maximum Flexibility," *Proceedings of Symposium on Processing Propellants, Explosives and Ingredients,* ADPA, 1977.
162. A. E. Andrew, "Deflagration to Detonation of Smokeless Ball Powders and the Means Used to Reduce a Detonation Occurrence," *Proceedings of the 17th Explosive Safety Seminar,* 1975, p. 319.
163. *Wastewater Treatment in the Military Explosives and Propellants Production Industry,* Vol. 3, ADPA, 1975, p. 127.
164. R. Steinberger, "Preparation and Properties of Cast Double Base Propellants" in S. S. Penner and J. DuCarme, eds., *The Chemistry of Propellants,* Pergamon Press, Oxford, 1960, p. 246.
165. A. M. Ball, "Solid Propellants," *Engineering Design Handbook,* Army Materiel Command, AMCP 706-175, 1964.
166. R. Steinberger and P. D. Drechsel "Manufacture of Cast Double Base Propellant" in C. Boyars and K. Klager, eds., *Propellant Manufacture, Hazards and Testing,* American Chemical Society, Washington, D.C., 1969, p. 1.
167. D. E. Boynton and J. W. Schowengardt, *Chem. Eng. Prog.* **59,** 81 (1963).
168. R. Steinberger, "Advances in Double Based Propellant Launch Vehicles," *Proceedings of the 7th International Symposium on Space Technology and Science,* AGNE Publishing Inc., Tokyo, 1967, p. 63.
169. R. Steinberger, "Advances in Double Base Propellants for Launch Vehicles," in F. A. Warren, ed., *Solid Propellant Technology,* Vol. X, AIAA Selected Reprints, New York, 1970.
170. R. F. Preckel, *AIAA J.* **3,** 2346 (1965).
171. C. Boyars and K. Klager, eds., *Propellant Manufacture Hazards and Testing,* Advances in Chemistry Series 88, American Chemical Society, Washington, D.C., 1969.
172. *Solid Propellants and Motor Costs,* LS76-33, CPIA, 1976.
173. D. C. McGehee and P. K. Myers, *Chem. Eng. Prog. Symp., Am. Inst. Chem. Eng.* **62**(61), 19 (1966).
174. *Chem. Eng. News* **42,** 50 (Sept. 28, 1964).
175. Mishuck and Carleton, *IEC J.* **52,** 755 (1960).
176. W. P. Killian, *Chem. Eng. Prog.* **59,** 9 (1963).

177. G. A. Fluke "Composite Solid Propellant Processing Techniques" in Boyars and Klager, eds., *Propellant Manufacture, Hazards and Testing, Advances in Chemistry Series 88,* American Chemical Society, Washington, D.C., 1969, p. 165.
178. *Solid Propellant Selection and Characterization,* SP-8064, NASA, Cleveland, Ohio, 1971.
179. J. F. Tormey, "Processing and Manufacture of Composite Propellants," *Proceedings of the AGARD Colloquim, Advances in Tactical Rocket Propulsion,* CIRA Publications, Pelham, N.Y., 1968.
180. F. Skovgard and C. J. Barr, *Manufacture of Large Production Quantities of Precise Solid Propellant Mixes,* SAE Paper No. 650766, presented at SAE National Aeronautics and Space Engineering Meeting, 1965.
181. K. Klager and J. M. Wrightson "Recent Advances in Solid Propellant Binder Chemistry," *Proceedings of the 4th Symposium on Naval Structural Mechanics,* Pergamon Press, New York, 1965, p. 47.
182. L. W. Collins, *The Feasibility of Utilizing a Fluid Energy Mill in the Grinding of Potassium Perchlorate,* MLW 2444, Monsanto Research Corp, Miamisburg, Ohio, 1977.
183. H. E. Marsch, *IEC J.* **52,** 768 (1960).
184. W. F. Haite, *Chem. Eng. Prog. Symp., Am. Inst. Chem. Eng.* **62**(61), 1 (1966).
185. J. L. Anjier and co-workers, "Chamber Lining by Electrostatic Deposition," *Proceedings of AICHE, National Meeting,* New Orleans, La., p. 186.
186. "Solid Propellant Processing Factors in Rocket Motor Design," *Space Vehicle Design Criteria Monograph,* SP-8075, NASA, 1971.
187. C. Gustavson, T. W. Greenlee, and A. Ackley, *J. Spacecr. Rockets* **3**(3), 413 (1966).
188. G. W. Fust and A. O. Kays, "Adhesives for Bonding of Insulation in Solid Propellant Rocket Motors," *Applied Polymer Symposia,* Paper No. 3, John Wiley & Sons, Inc., New York, 1966, p. 219.
189. *260-SL Motor Propellant Tailoring and Liner Development,* CR-54473, NASA, 1965.
190. "Solid Rocket Motor Insulation," *Space Vehicle Design Criteria Monograph,* NASA, 1971.
191. E. A. Sienicki, and H. R. Schloss, eds., *Case Bonding and Development of Adhesives for Case Bonded Carboxyl Terminated Polybutadiene Propellants,* TR 4283, NOTS, 1967.
192. W. P. Killian, "Loading Composite Solid Propellant Rockets—Current Technology," *Proceedings Symposium on Selected Topics in Aerospace Chemistry, 64th National Meeting,* AICE, Fla., 1968.
193. *Explosives Hazard Classification Procedures,* TB 7002, NAVORD Inst 8020, 3 to 11A-1-47, DSAR8220.1, Depts of the Army, Air Force, and Defense Supply Agency, Washington, D.C., 1967.
194. D. S. Gaardner, ed., *Safety and Hazards of High Energy Propellants,* CPIA 284, 1977.
195. "Solid Propellants" *Engineering Design Handbook AMCP 706-175,* AMC, 1964.
196. *Black Powder, MIL Spec.-223-B,* 1973.
197. J. W. Burns, *Black Powder Manual,* NDRC, 1944.
198. *Ordnance Explosive Train Designers Handbook,* Rpt. 1111, NOL, 1952.
199. B. Federoff and O. Sheffield in *Encyclopedia of Explosives and Related Items,* Vol. 2, TR 2700, PTA, 1962, p. 165.
200. T. Urbanski in *Chemistry and Technology of Explosives,* Vol. 3, Pergamon Press, New York, 1967.
201. C. Campbell and G. Weingarten, *Trans. Faraday Soc.* **55,** 2221 (1951).
202. J. Isaksson and L. Rittfeldt, "Characterization of Black Powders," *Proceedings of 3rd Symposium on Chemical Problems Connected with the Stability of Explosives,* Jonkoping, Sweden, 1952, p. 242.
203. J. D. Blackwood and F. D. Bowden, *Proc. Royal Soc.* **213A,** 285 (1952).
204. F. A. Williams, "Black Powder: Combustion of Particles and Flame Spread Through Charges," *Proceedings of the 12th JANNAF Combustion Meeting,* vol. 1, CPIA 273, 1975.
205. J. V. White and co-workers, "Black Powder and Clean Burning Igniter Ignition Train Studies," *Proceedings 13th JANNAF Combustion Meeting,* CPIA 281, 1976, p. 405.
206. K. Lvold, "A New Production Process for Black Powder," *Proceedings 3rd Symposium on Chemical Problems Connected with the Stability of Explosives,* Jonkoping, Sweden, 1974, p. 266.
207. D. R. Mouta and co-workers, *Hazards Analysis of the Final Design of the Improved Black Powder Process,* Vols. 1–2, Rpt. J6329, Ill. Inst. of Tech.
208. E. Husselton and S. Kaplowitz, *Benite Manufacture,* DB-TR-5-60, Picatinny Arsenal, Dover, N.J., 1961.
209. S. Axelrod and G. Demitrack, *The Development of New Formulations for the Combustible Cartridge Case,* TR 2454 PTA, 1957.

210. E. Wurzel, *A Survey of Methods of Preparation of Combustible Cartridge Cases,* Report No. 102 PTA, 1960.
211. J. W. K. Gormley, *Ordnance* **47,** 231 (1962).

General References

The Annual Proceedings of the Joint Army-Navy-Air Force (JANNAF) Propulsion Meetings, the reports of the special committees, and the periodic literature surveys published by the Chemical Propulsion Information Agency are invaluable sources of information on all aspects of liquid and solid gun and rocket propellants. They may be classified.
The Proceedings of the International Symposia on Combustion, from 1928, Combustion Institute, Pittsburgh, Pa.
The International Symposia on Ballistics sponsored and published by the American Defense Preparedness Association (ADPA), Washington, D.C.
The Proceedings of the Symposium on Processing Propellants, Explosives, and Ingredients, ADPA, 1977, 1979.
The Proceedings of the Annual Seminars on Explosive Safety, sponsored by the U.S. Explosives Safety Board, Washington, D.C.
The Solid Propellant Manuals, issued by the Army Materiel Command, AMC706-175, 1961 and AMC 706-176, 1961.
CPIA/M2 Solid Propellant Manual, 1969, (classified, confidential).
F. X. Hartman, *Solid Propellants Safety Handbook,* Clearinghouse, Springfield, Va., 1965.
Military Specifications, US Military Standards 286-B, 1971, and *P-270A,* U.S. Government Printing Office, Washington, D.C., 1959.
Publications of the Chemical Propulsion Information Agency (CPIA) obtainable from Applied Physics Laboratory, Johns Hopkins Univ., Laurel, Md; (may be classified).
B. T. Federoff, and O. E. Sheffield, *"The Encyclopedia of Explosives and Related Items,"* TR 2700, PTA Vol. 1–8.
OSRD reports, Office of Scientific Research and Development, National Defense Research Council (NDRC).
A. H. Blatt, *Bibliography of OSRD Reports Issued by Division 8 of the National Defense Research Committee,* OSRD Rpt. No. 6630, 1946.
M. Shorr and A. J. Zaehringer, *Solid Rocket Technology,* John Wiley & Sons, Inc., New York, 1967.
B. L. Crawford, Jr., C. Huggett, and J. L. McBrady, "The Mechanism of Burning of Double Base Propellants," *J. Phys. Colloid Chem.* **54,** 847 (1950).
C. Huggett, C. E. Bartley, and M. M. Mills, *Solid Propellant Rockets,* Princeton University Press, N.J., 1960.

Texts and Monographs: Rocket Propellants

M. H. Smith "The Literature of Rocket Propulsion in *The Literature of Chemical Technology, Advances in Chemistry, Series No. 74,* American Chemical Society 1968, p. 581.
Monographs on rockets and rocket propellants by the National Aeronautics and Space Administration (NASA), Lewis Research Center, Cleveland. These include the following: *Solid Propellant Selection and Characterization,* Report SP-8064, 1971; *Solid Rocket Motor Performance,* Report SP-8039, 1971; *Solid Rocket Motor Igniters,* Report SP-8051, 1971; *Solid Rocket Motor Metal Cases,* Report SP-8025, 1970, and *Captive Fire Testing of Solid Rocket Motors,* Report SP-8041, 1971.
R. D. Geckler, AGARD, *Selected Combustion Problems,* Buttersworth, London, 1955.
M. Shorr and A. J. Zaehringer eds., *Solid Propellant Technology,* John Wiley & Sons, Inc., New York 1968.
F. Sarner, *Propellant Chemistry,* Reinhold Publishing Corp., New York 1966.
J. Taylor, *Solid Propellants and Exothermic Compositions,* Interscience Publishers, New York 1959.
G. P. Sutton, *Rocket Propulsion Elements,* 3rd ed, John Wiley & Sons, Inc., New York, 1963.
R. F. Gould, ed., *Propellant Manufacture, Hazards and Testing,* Advances in Chemistry Series 88, American Chemical Society, Washington, D.C., 1969.
Proceedings of the Symposium on Kinetics of Propellants, Division of Physical and Inorganic Chemistry, 112th Meeting of the American Chemical Society, New York 1957, published in *J. of Phys. Colloid Chem.* **54** (1950).

S. S. Penner, *Chemical Rocket Propulsion and Combustion Research,* Gordon and Breach Science Publishers, New York 1962.
F. A. Warren, "Solid Propellant Technology," in R. A. Gess, ed., Selected Reprint Series, *AIAA* New York (1970).
R. L. Wilkins, *Theoretical Evaluation of Chemical Propellants,* Prentice Hall, Inc., New York, 1963.
R. N. Wimpress, *Internal Ballistics of Solid Fuel Rockets,* McGraw Hill, New York, 1950.
C. Huggett, C. F. Bartley, and M. Mills, *Solid Propellant Rockets,* Princeton Univ. Press, N.J., 1960.
D. Altman, and co-workers, *Liquid Rocket Propellants,* Princeton Univ. Press, Princeton, N.J., 1960.
H. W. Ritchey, "Solid Propellants and the Conquest of Space," *Astronautics* **3,** 39, 75 (1958).
"Propellant Actuated Devices," *Engineering Design Handbook* AMCP 706-282, U.S. Army Materiel Command, Washington, D.C., 1963.
M. J. Zucrow, *Propulsion and Propellants, Engineering Design Handbook,* AMCP 706-282, U.S. Army Materiel Command, Washington, D.C., 1963.

Gun Propellants

J. Corner, *Theory of Interior Ballistics of Guns,* John Wiley & Sons, Inc. New York, 1950.
F. R. W. Hunt, ed., *Internal Ballistics,* Philosophical Library, New York, 1951.
"Interior Ballistics of Guns," *Engineering Design Handbook,* AMCP 706-150, DARCOM, 1965.
J. O. Hirschfelder, R. B. Kershner, and C. F. Curtis, *Interior Ballistics,* NDRC A-142, 1943.
C. F. Curtis, and J. W. Wrench, *Interior Ballistics: A Consolidation and Revision of Previous Reports. Interior Ballistics I and VII Inclusive,* NDRC Rpt A-397, OSRD Rpt. 6468, Geophysical Lab, Carnegie Inst. of Tech., Washington, D.C., 1945.
K. J. Laidler, *A Comparison of Interior Ballistic Systems,* NAVORD, Rpt. 750, 1947.
J. H. Grese "Ballistic Calculations" in *Encyclopedia Brittanica,* Vol. 3, 1976.
P. Baer, and J. Frankle, *Simulation of Interior Ballistic Performance of Guns by Digital Computer Program,"* Rpt. 1183, BRL, 1962.
C. Cranz, *Lehrbuck der Ballistic,* Springer, Berlin, 1926, Translated by NDRC, 1945.
W. H. Tschappat, *Text Book of Ordnance and Gunnery,* John Wiley & Sons, Inc., New York, 1945.
Journal of Ballistics, Memorials de Poudres, Annual Summaries of the Chemical Propulsion Information Agency (classified).

VICTOR LINDNER
U.S. Army Armament Research and Development Command
(ARRADCOM)

EXT. D&C DYES. See Colorants for food, drugs, and cosmetics.

EXTRACTION, LIQUID–LIQUID

Liquid–liquid extraction is sometimes also referred to as solvent extraction or merely as extraction. It is a separation process that depends on the transfer of the component to be separated (the consolute component) from one liquid phase to a second liquid phase that is immiscible with the first. In fractional extraction, there are two or more consolute components. Applications may be classified broadly as (1) purifications of the consolute component(s), eg, the preferential extraction of uranyl nitrate from an aqueous solution of mixed metal ions; and (2) purification of a liquid phase by extraction of a contaminant, eg, the extraction of phenols from aqueous industrial effluents.

Whether an extraction process is carried out in the laboratory or industry, it always involves contact of the liquid phases with an approach toward equilibrium, and separation of the contacted liquid phases. On the laboratory scale, this is done by shaking the phases in a separating funnel, leaving them to settle, and then separating the phases. On a larger scale, extraction is nearly always carried out continuously using a variety of equipment designs (see under Equipment). The process is shown schematically in Figure 1, which describes the standard nomenclature used in solvent extraction technology. The feed solution contains the consolute component C dissolved in a component A, and it is brought into contact with a solvent B. Mass transfer of C occurs between the two liquid phases; A and B are chosen to be immiscible or very sparingly miscible with each other (nonconsolute components). After contact and separation, the distribution of C between A and B is represented by a quantity ε known as the extraction factor of this component of C:

$$\varepsilon = m \frac{B}{A} = \frac{\text{quantity of C in phase B}}{\text{quantity of C in phase A}} \tag{1}$$

where m = distribution coefficient, B = quantity of solvent B, A = quantity of solvent A, and ε = extraction factor. Extraction factor ε plays a most important part in extraction calculations.

The separated extract from the single contact shown in Figure 1 can be subjected to a second liquid–liquid contact whereby the consolute component is extracted back into the raffinate phase under different conditions (eg, temperature). This process is known as stripping and the stripped solvent can then be used again for extraction. Alternatively, the solvent B can be recovered from the extract by a distillation or evaporation step.

The main advantage of liquid–liquid extraction lies in its extreme versatility because of the enormous range of choice of solvents. Whereas distillation and gas absorption processes impose rigid requirements as to the vapor pressures and solu-

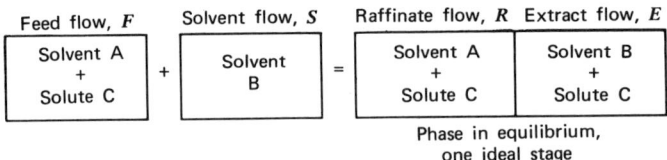

Figure 1. Single contact.

bilities of the various components, solvent extraction can be adapted to many more systems. It is increasingly common for solvent extraction processes to involve a chemical reaction of the extracted component with a component (the extractant) dissolved in the solvent; this has led to major new applications in hydrometallurgical processing (see Extractive metallurgy; Separation systems synthesis).

Solvent extraction is generally applied where direct separation methods, such as distillation and crystallization, have failed and where it provides a less expensive process than a competitive physical or chemical method (1).

Like many other chemical engineering unit operations, liquid–liquid extraction was developed in the 1920s and 1930s as part of the petrochemicals industry. Important applications were also developed for the recovery of vegetable oils and the purification of penicillin and other heat-sensitive pharmaceutical products that cannot be separated easily by distillation.

The growth of the nuclear industry has resulted in many applications of liquid–liquid extraction in the refining of uranium, plutonium, and other radioisotopes (see Nuclear reactors; Radioisotopes). Since about 1960, extractants for nonnuclear metals, in particular copper, cobalt, and nickel, have been developed.

A 1977 survey (2) listed 19 plants or proposed plants using liquid–liquid extraction for copper alone (see Copper).

The growth in the industrial application of liquid–liquid extraction has been paralleled by increased research aimed at an understanding of the fundamentals. In the 1920s and 1930s 1iquid–liquid extraction was usually treated as an equilibrium-stage process analogous to distillation. However, many types of extraction columns (eg, packed columns and spray columns) do not contain discrete stages, and this has led to increasing use of the transfer unit concept and more emphasis on mass transfer rate estimation using transport phenomena models (see Mass transfer). The importance of hydrodynamic effects such as droplet breakup and coalescence and axial mixing has been increasingly realized in the past decade. Finally, the chemistry of solvent extraction reactions has come under even closer scrutiny.

This article attempts to give a general outline of a complex and rapidly developing subject area. For further details, the reader is referred to books (1,3–12), review articles (13–22), and the *Proceedings of the Triennial International Solvent Extraction Conferences* (23–25). A useful listing of nomenclature recommended for liquid-liquid extraction technology is also available (26).

Principles

Physical Equilibria. Liquid–liquid equilibria are governed by the phase rule:

$$F = C - P + 2 \qquad (2)$$

where C is the number of components, P is the number of phases (two in this case), and F is the number of degrees of freedom, ie, independent variables permitted by the system.

A binary liquid–liquid system is, therefore, bivariant; eg, if the pressure and the temperature are fixed, the composition of each phase in equilibrium is fixed.

As the temperature of equilibration is increased, the equilibrium-phase compositions generally approach each other until the upper critical solution temperature, at which the two phases become identical, is reached. In some cases, there is also a lower critical solution temperature below which two liquid phases cannot coexist in equilibrium.

Ternary liquid–liquid systems are allowed three degrees of freedom according to the phase rule. At a given temperature and pressure, the composition of one phase determines the composition of the other phase at equilibrium. Figure 2 shows a typical equilateral triangular diagram for components A, B, and C. The compositions are read along axes inclined at an angle of 60° so that the sum of mass fractions x_A, x_B, and x_C must always be unity. The two phases in equilibrium are represented by points 1 and 2 on the envelope of the two-phase region and joined by a straight tie line as shown. A mixture M of a given composition separates into equilibrium phases whose composition is such that the appropriate tie line intersects point M on the diagram. The relative amounts of the separate phases can be estimated by the inverse lever rule, which is based upon the conservation of mass for the components:

$$\frac{\text{mass of phase 1}}{\text{mass of phase 2}} = \frac{\text{distance M} \rightarrow 2}{\text{distance M} \rightarrow 1} \tag{3}$$

As the amount of the consolute component C in the system is increased, the compositions of mixtures 1 and 2 increase and, for each new case, a new tie line must be drawn. In general, the tie lines on the triangular diagram are neither parallel nor horizontal, and the limit of the two-phase region at the plait point (shown as P in Fig. 2) is not necessarily at the highest point on the two-phase envelope.

In order to avoid the necessity of including a very large number of tie lines on a triangular diagram, a conjugate line PQR is usually provided. This allows a tie line to be drawn from any given point on either side of the phase envelope, as illustrated in the construction 1–Q–2 on Figure 2.

Another graphical method for representing equilibrium data is to plot the equilibrium conditions as the concentration of the solute in one phase against the concentration in the other phase. This is the most popular type of graph (see Fig. 9, below) and is discussed later as to its application to design calculation.

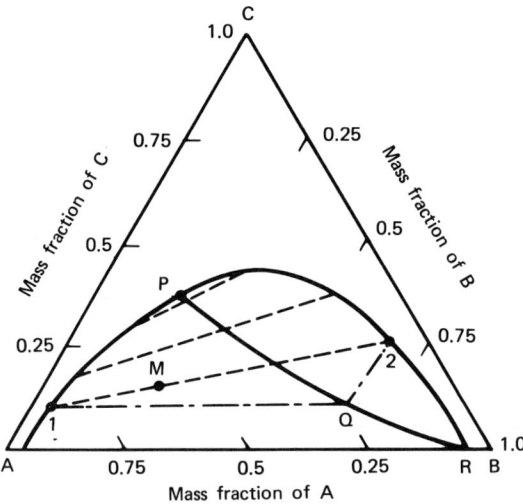

Figure 2. Typical equilibrium curve (ternary system) on triangular coordinates.

Other methods for plotting the equilibrium curve and tie-line data have been proposed and are described in ref. 1. All of these may be conveniently extrapolated to determine the plait point; however, slightly different points may be obtained, depending on the extent of the extrapolation required.

For those cases where the nonconsolute components A and B are immiscible or very nearly so, the triangular diagrams can be replaced by a simple plot of mass fractions of C in the two phases (x_{CB} vs x_{CA}) or mole fractions (x'_{CB} vs x'_{CA}). At large dilutions, the two compositions are proportional:

$$x_{CB} = m x_{CA} \tag{4}$$

$$x'_{CB} = m' x'_{CA} \tag{5}$$

and m and m' are the distribution ratios based on mass fractions and mole fractions, respectively. Distribution ratios can also be based on concentrations.

This linear relationship is referred to as the distribution law, or occasionally as the ideal distribution law. However, this does not indicate thermodynamic ideality. The activity coefficients γ_{CB} and γ_{CA} are not unity, but they do become independent of solution composition at high dilutions, and so it can be seen from equation 5 that:

$$m' = \gamma_{CB}/\gamma_{CA} \tag{6}$$

The prediction of liquid–liquid distribution data from activity considerations for nondilute systems with partial miscibility has been developed to a considerable extent (1,10). However, the number of systems tested is still rather limited, and these methods are recommended only as an approximate but inexpensive alternative to experimental measurement.

A large and ever-growing amount of experimental liquid–liquid equilibrium data are available (27–30). Data compilations are useful but are inevitably not up to date, and a thorough search for equilibrium data on a specific system should always involve access to the computer-indexed abstracts (eg, *Chemical Abstracts, Citation Index*) using the key words for the components of the system.

Equilibrium data can be obtained experimentally by manually shaking the solution and solvent and then separating the layers, followed by analysis. Recently, an automated device, the AKUFVE apparatus (31), has been developed for the collection of large quantities of equilibrium data.

Quaternary liquid–liquid equilibria are particularly important in fractional extraction operations involving two consolute components, B and C, and two nonconsolute components, A and D. A general graphical representation of equilibrium data for such systems is impossible, but if the concentrations of B and C are low and A and D are considered immiscible, the distribution law can be applied independently for each consolute component:

$$x_{BA} = m_B x_{BD} \tag{7}$$

$$x_{CA} = m_C x_{CD} \tag{8}$$

The selectivity or separation factor β_{BC} (analogous to relative volatility in distillation) is defined as the ratio of the distribution ratios:

$$\beta_{BC} = \frac{m_B}{m_C} = \frac{x_{BA} x_{CD}}{x_{CA} x_{CD}} \tag{9}$$

By convention, the components are chosen so that β_{BC} always exceeds unity. The greater the selectivity, the easier is the separation of B from C by fractional extraction with solvent A and D. As in the case of the distribution ratio, the selectivity can be defined in terms of mass fraction, mole fraction, or concentration units according to convenience.

Chemical Equilibria. In many cases, equilibria in liquid–liquid systems are affected by chemical changes; eg, (*1*) equilibrium with a chemical change occurring in the bulk of one or both of the phases but with components transferring between phases without chemical change; and (*2*) equilibrium with a chemical change occurring at the interface itself; no molecular species has significant solubility in both phases.

A classical example of type 1 is the distribution of benzoic acid between water and benzene (32). In the aqueous phase, the acid is partly ionized:

$$C_6H_5COOH \rightleftharpoons C_6H_5COO^- + H^+ \tag{10}$$

In the organic phase, partial association occurs:

$$2\,C_6H_5COOH \rightleftharpoons (C_6H_5COOH)_2 \tag{11}$$

Nernst in 1891 was the first to point out that the linear distribution law could still be applied to such a system on the basis of the concentrations of the consolute component which in this case is the unionized and unassociated species C_6H_5COOH. Hence it can be shown that:

$$c_a \propto \sqrt{c_b} \tag{12}$$

where c_a and c_b are the analytically determined concentrations of benzoic acid in the aqueous and benzene phases, respectively.

Another example of type 1 is the hydrolysis of a sparingly soluble ester:

$$RCOOR' + H_2O \rightleftharpoons RCOOH + R'OH \tag{13}$$

Equilibrium compositions must satisfy the law of mass action in each phase and also the mutual solubility relationships for all the components. This reaction is important in the fat-splitting process for the manufacture of glycerol (qv) and fatty acids by high temperature hydrolysis of fats or oils (see Fats and fatty oils).

The majority of hydrometallurgical extraction processes belong to type 2, in which an equilibrium is considered to exist at the interface between ionized species in the aqueous phase and unionized species in the organic phase (see also Catalysis, phase-transfer).

The extraction of uranyl nitrate by a kerosene solution of tributyl phosphate (TBP) is represented (33) by equation 14:

$$UO_2^{2+} + 2\,NO_3^- + 2\,TBP \rightleftharpoons UO_2(NO_3)_2 \cdot 2TBP \tag{14}$$

and hence:

$$\frac{C_{UO_2(NO_3)_2 \cdot 2TBP}}{C_{UO_2^{2+}}} \propto (C_{NO_3^-})^2 (C_{TBP})^2 \tag{15}$$

The TBP and nitric acid form an organic-soluble complex:

$$H^+ + NO_3^- + TBP \rightleftharpoons HNO_3 \cdot TBP \tag{16}$$

However, at high uranyl nitrate concentrations, the nitric acid is mainly displaced into the aqueous phase and the organic phase loading of uranium is limited only by the TBP concentration in accordance with equation 15 (see Uranium).

Numerous commercially available metal extractants act as liquid–liquid cation exchangers (7).

$$M^{n+} + n\,\overline{RH} \rightleftharpoons \overline{R_n M} + n\,H^+ \qquad (17)$$

Here, M^{n+} is the metal cation and RH is the extractant which is normally dissolved in an inert organic diluent such as kerosene. The overbar denotes species in the organic phase. $R_n M$ is an organic-soluble salt or complex of the metal and the extractant, which may be an acid or an acidic chelating agent (see Chelating agents).

The distribution ratio of the metal (organic to aqueous phases) decreases strongly as the hydrogen ion concentration in the aqueous phase increases. Thus the pH control is crucial; eg, a metal that is almost completely extracted into the organic phase at pH 4 can be reextracted by the aqueous phase at pH 1. Extraction equilibria for cation exchangers are often represented as plots of the percent of the metal extracted from the aqueous phase (for a given phase volume ratio and initial composition) as a function of pH (34).

As metal ion-exchange extraction processes of the type shown in equation 17 proceed, the liberation of hydrogen ions causes the distribution ratio of the metal (organic to aqueous phases) to decrease. This difficulty can be avoided by suitable buffering of the aqueous phase or, if the extractant is an acid, by the use of its sodium or ammonium salt. The latter procedure has been applied to the development of a cobalt–nickel separation process (35).

Interfacial Mass Transfer. Although the equilibrium relationships are important in determining the ultimate separation obtainable in a liquid–liquid system, the rate of equilibrium attainment is of equal importance in the design or operation of extraction processes. In a batch process, this rate determines the contact time required; in a continuous process, the required size of the equipment and the residence time (volume of equipment divided by volume flow rate of phases) depend on mass transfer rates as well as on equilibrium properties (see Mass transfer).

In the absence of any chemical reactions, the rate of transfer of a consolute component from one phase to another is governed by the diffusion laws. The diffusional mass transfer rate of solute per unit area, relative to a stationary medium, is given by Fick's first law:

$$N = -D\,\frac{\partial c}{\partial z} \qquad (18)$$

where N refers to transport in the z direction, c is the concentration of the consolute component, and D is its molecular diffusivity with respect to the solvent. It would also be proper to relate the flux to the activity gradient (36), particularly as activity coefficients in liquid–liquid systems are not usually unity. However, it is accepted practice to use concentration gradients in the definition of diffusivity. If the flux N is in $g/(cm^2 \cdot s)$, c is in g/cm^3, and z is in cm, then D has the units of cm^2/s. Values of D for many systems are available on the literature or can be predicted by various correlations which are accurate to within $\pm 10\%$ or better (37). In general, molecular diffusivities of solutes in liquids are within the range 10^{-6}–10^{-5} cm^2/s.

As might be expected from the low values of D, molecular diffusion is a very slow process. However, its effect is nearly always enhanced in practice by turbulent eddies and circulation currents. These provide almost perfect mixing in the bulk phases but their effect tends to be retarded in the vicinity of a liquid–liquid interface. Consequently, the important role of molecular diffusion is confined to very narrow regions, sometimes loosely referred to as films, on either side of the interface as shown in Figure 3.

In accordance with Fick's first law, the transfer across each film occurs in the direction of decreasing concentration. It is usually assumed that the solute concentrations c_{Ai} and c_{Bi} in the A-rich and B-rich phases at the interface are in equilibrium. If the system obeys the distribution law, then:

$$c_{Bi} = mc_{Ai} \tag{19}$$

where m is the appropriate distribution factor. This assumption of interfacial equilibrium is reasonable in practice, except for some cases where a physical barrier of a surface-active contaminant (38) or reaction product (39) accumulates at the interface. Then an allowance has to be made for interfacial resistance.

The thicknesses of the regions across which the concentration varies are on the order of 10^{-3} cm and depend upon hydrodynamic conditions. These thicknesses are extremely hard to measure directly and the flux of solute is, therefore, expressed in terms of mass transfer coefficients which combine the effects of molecular diffusivity and film thickness:

$$N = k_A(c_A - c_{Ai}) = k_B(c_{Bi} - c_B) \tag{20}$$

Here, N is the mass transfer rate per unit interfacial area.

The order of magnitude of k_A and k_B is usually within the range 10^{-3}–10^{-2} cm/s in liquid–liquid systems.

The flux may also be expressed in terms of bulk concentration driving forces. If c_A^* is defined as c_B/m, ie, the concentration that phase A would have if it were in equilibrium with phase B, then an overall mass transfer coefficient K_A based on phase A is defined by:

$$N = K_A(c_A - c_A^*) \tag{21}$$

Similarly:

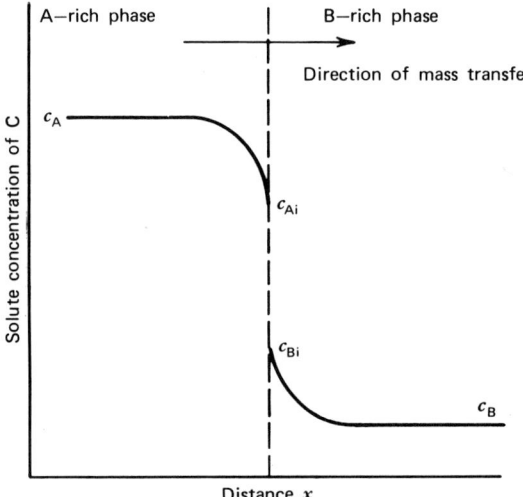

Figure 3. Concentration profiles near an interface. c_A = concentration of C in A-rich phase; c_B = concentration of C in B-rich phase.

$$N = K_B(c_B^* - c_B) \tag{22}$$

where

$$c_B^* = mc_A \tag{23}$$

A simultaneous solution of equations 20–23 gives the important relationships:

$$\frac{1}{K_A} = \frac{1}{k_A} + \frac{1}{mk_B} \tag{24}$$

$$\frac{1}{K_B} = \frac{m}{k_A} + \frac{1}{k_B} \tag{25}$$

There is a partial analogy between the above equations and the expressions for electrical resistances in series. However, the analogy is incomplete because the distribution factor m is often decisive in determining which phase contributes most to the overall mass transfer resistance. For example, in the transfer of acetic acid from dilute solutions in kerosene (phase A) to water (phase B), the distribution ratio m as defined in equation 23 is about 40. It can easily be seen from equations 24 and 25 that the kerosene-phase resistance must be the major contributor.

From equation 21 it can be seen that:

$$\left\{ \begin{array}{l} \text{mass transfer rate} \\ \text{per unit volume of} \\ \text{equipment} \end{array} \right\} = Na = K_A a (c_A - c_A^*) \tag{26}$$

where a is the interfacial area available per unit volume. If the transfer space contains a volume fraction h (the holdup) of the dispersed phase in the form of spherical droplets of diameter d_m, it can be shown that:

$$a = 6\,h/d_m \tag{27}$$

Thus a, and therefore the mass transfer effectiveness, can be maximized by increasing the holdup and reducing the drop size, subject to certain limits which are discussed under Hydrodynamic Factors in Extractor Design. Typically in a mixer–settler, a could have a value of 50 cm²/cm³ (or simply cm^{-1}) and, if K_A were 0.002 cm/s, the product $K_A a$ would be 0.1/s.

Considerable experimental and theoretical work has been done on mass transfer to and from an isolated drop moving through a continuous phase. Mass transfer coefficients are generally much greater for drops that circulate internally because of their motion through the surrounding liquid than for drops in which circulation is inhibited by the presence of surface-active contaminants. Table 1 summarizes (10) the main relationships that have been put forward for the dispersed-phase and continuous-phase mass transfer coefficients under circulating and noncirculating conditions.

Mass transfer during the formation of a drop and during its coalescence can be very important if the distance of free travel of a drop is relatively short. The average overall mass transfer coefficient K_{1f} during the formation (or coalescence) period θ_f is given by (1):

$$K_{1f} = \frac{0.805}{m}(D_2/\theta_f)^{0.5} \tag{28}$$

Table 1. Mass Transfer Coefficients in Freely Moving Single Droplets[a,b]

	Dispersed phase, k_1	Ref.	Continuous phase, k_2
circulating single drops	laminar circulation $k_1 \simeq 17.9\, D_1/d_m$	41	$k_2 = \sqrt{\dfrac{2\, D_2 u_t}{\pi d_m}}$
	turbulent circulation $k_1 = \dfrac{0.00375\, u_t}{1 + \mu_1/\mu_2}$	42	penetration theory, taking exposure time as d_m/u_t
rigid single drops[c]	$k_1 = 2\,\pi^2 D_1/3 d_m$	1	$\dfrac{k_2 d_m}{D_2} = 2.0 + 0.6 \left(\dfrac{u_t d_m \rho_1}{\mu_1}\right)^{0.5}\left(\dfrac{\mu_1}{\rho_1 D_1}\right)^{0.33}$

[a] Based on ref. 10.
[b] Symbols: k_1 and k_2 = mass transfer coefficients; D_1 and D_2 = molecular diffusivities; μ_1 and μ_2 = viscosities; ρ_1 and ρ_2 = densities; d_m = diameter of drop; u_t = terminal velocity of drop; subscript 1 = dispersed phase and subscript 2 = continuous phase.
[c] The value 0.6 varies between 0.56 and 0.70 (43–44).

where subscripts 1 and 2 refer to the dispersed and continuous phases, respectively, and m is the distribution ratio between the dispersed and continuous phases.

Although the equations given above and in Table 1 would suggest that the mass transfer coefficient is solely a function of external hydrodynamic factors, such as the Reynolds number, there are cases where the mass transfer process itself affects the mass transfer coefficient. If the interfacial tension is sensitive to the interfacial composition with respect to the consolute component, the so-called Marangoni effect can occur, giving pronounced interfacial instability and enhancement of the mass transfer rate. The formation of roll cells at a planar liquid–liquid interface by the Marangoni effect has been analyzed quantitatively (45). A system that is interfacially unstable for one direction of mass transfer may be stabilized for mass transfer in the reverse direction. An excellent review of these interfacial effects is available (46).

Mass Transfer with Chemical Reaction. Liquid–liquid extractions accompanied by a homogeneous chemical reaction in one phase are quite common, eg, nitrations and sulfonations (47).

The possible rate equations for extraction have been examined (48) with an irreversible second-order reaction in one phase, between a consolute component A and a nontransferring reactant B. In six cases the chemical rate constant increases with respect to the mass transfer coefficient. Concentration profiles for each case are shown on Figure 4a.

Case 1. Very slow reaction. Component A is almost at its equilibrium concentration in the reaction phase. The extraction rate is controlled by the rate of bulk reaction.

Case 2. Slow reaction. The concentration of A in the bulk phase is well below its equilibrium concentration; the rate of reaction between A and B is limited mainly by mass transfer of A to the bulk of the phase where reaction occurs.

Case 3. Fast reaction without depletion. Component B remains at uniform concentration but free A is no longer present in the bulk phase. Reaction is confined to a region near the interface.

Case 4. Very fast reaction without depletion. Similar to case 3 but the rate of extraction is now completely independent of the diffusional mass transfer coefficient but strongly dependent on the chemical rate constant.

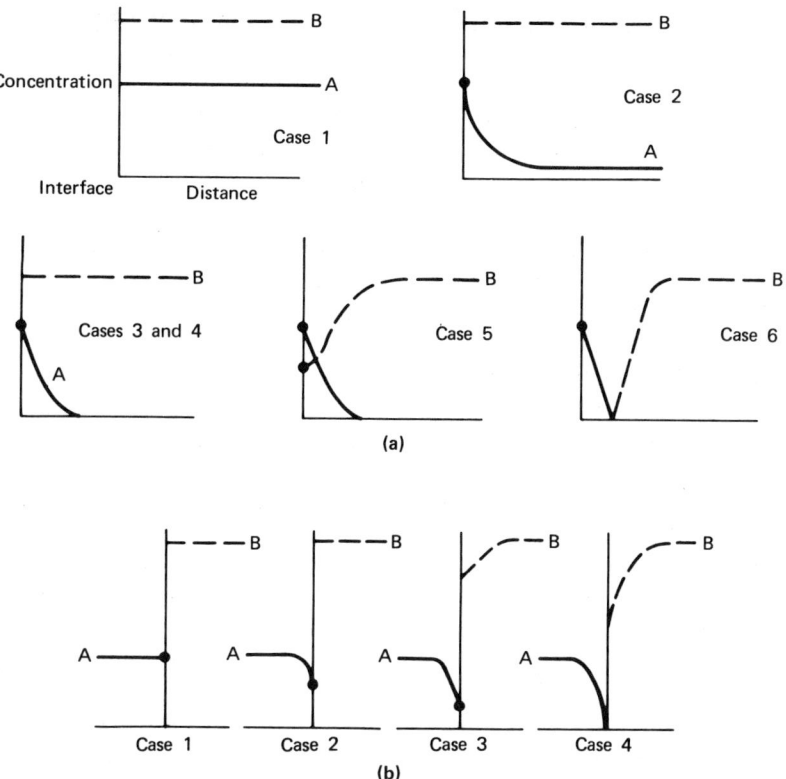

Figure 4. Concentration profiles for extraction with chemical reaction (case numbers increase with increasing reaction rate). (**a**) Homogeneous reaction A + B → products; (**b**) heterogeneous reaction A + B → products.

Case 5. Very fast reaction with depletion. The rate is so fast that component B is depleted near the interface.

Case 6. Instantaneous reaction. Reaction occurs in an infinitesimally thin reaction zone, near (but not at) the interface. The concentrations of both A and B tend to zero at this zone and the extraction rate is controlled by the diffusion of A and B to the zone. The rate is a large multiple of the diffusion-controlled rate for slow reaction (case 2).

In cases 3 to 6, the mass transfer rate exceeds that predicted from the theory of diffusion of A (48–50).

In the case of a surface reaction, the mass transfer rate can be considerably below the value estimated from diffusion theory, but no enhancement is possible. Figure 4b shows concentration profiles for the irreversible surface reaction between components A in one phase and B in the other (B in excess). The reaction product is not shown.

Case 1. Very slow surface reaction. The rate of transfer is completely controlled by the kinetics at the surface.

Case 2. Fast surface reaction. Significant depletion of A occurs at the interface and the rate is partly limited by diffusion of A.

682 EXTRACTION, LIQUID-LIQUID

Case 3. Very fast surface reaction. Depletion of both A and B occur at the interface. The overall rate equation includes diffusion in both phases as well as the interfacial rate constant.

Case 4. Instantaneous surface reaction. Component A is completely depleted at the interface. The transfer rate is limited solely by diffusional mass transfer of A and B to the interface.

Heterogeneous hydrometallurgical reactions of the type described in equation 17 usually cannot be treated as irreversible and it is necessary therefore to account for the transfer of reaction products as well as reactants (25). The complexity is further increased because, in many cases of hydrometallurgical extraction, the actual chemical steps are much more complex than the simple equation 17.

Calculation of Equilibrium Stages. In practice, multistage contacting of two immiscible phases can be arranged in a cocurrent, crosscurrent, or countercurrent manner, as illustrated in Figure 5. Multistage, cocurrent operation increases only the residence time and, therefore, does not increase the separation above that obtained in a single stage. Crosscurrent contact in which fresh solvent is added in each stage increases the separation beyond that obtainable in a single stage. However, it can easily be shown that the enhancement is not as great as can be obtained by countercurrent operations with a given amount of solvent, nor is it as economical. For purposes of calculation, it is important to distinguish between the two classes of the system: those in which the two phases are completely immiscible or in which the relative miscibility of the two phases is constant and independent of solute concentration, and those in which the relative miscibility of the two phases varies with the solute concentrations.

Crosscurrent Extraction. For the case of a partially miscible solvent system, the calculations for the equilibrium stages for a particular separation are carried out on a ternary diagram, and the method is well documented (1). For the case of solvents that are completely immiscible, the calculation can be simplified and carried out graphically (51) as shown in Figure 6. Assuming that a quantity A of solvent A' containing a solute at a mass ratio (C to A) of X_f is mixed with a quantity B of solvent B',

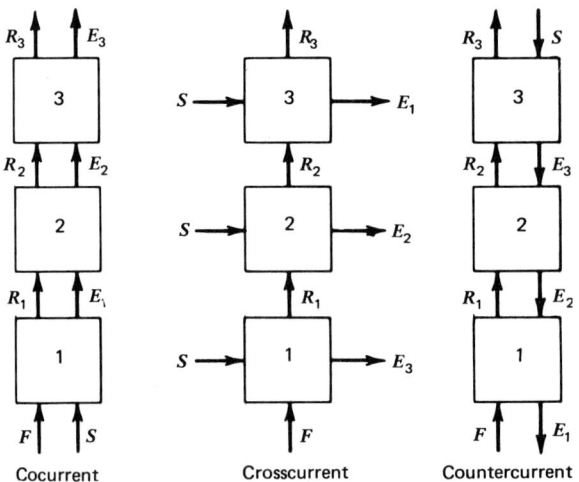

Figure 5. Arrangement of multistage contactors. F = feed flow (A-rich), R = raffinate flow, S = solvent flow (B-rich), E = extract flow.

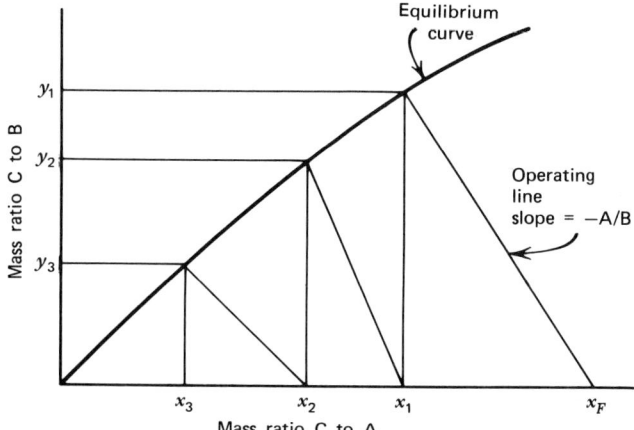

Figure 6. Crosscurrent extraction and multiple contact with immiscible solvents (51).

the mass ratios in the two exit phases may be given by the point (X_1, Y_1) on the equilibrium curve. From the material balance on the two phases, the point can be located by the following equation:

$$Y_1/(X_f - X_1) = A/B \tag{29}$$

The slope of the line between $(X_f, 0)$ and (X_1, Y_1) in Figure 6 is $-A/B$. The calculation procedure can be repeated as shown by constructing the line of slope $-A/B$ through the raffinate compositions to the equilibrium curve for successive treatments of the raffinate with a quantity of fresh solvent. This construction may also be done on a concentration plot if the systems are very dilute.

For the case where the solvents are completely immiscible and the ideal distribution law applies ($Y = mX$), the number of theoretical stages required for a particular crosscurrent separation can be calculated by the following equation (1,52):

$$N = \frac{\log (X_f/X_N)}{\log (1 + \varepsilon)} \tag{30}$$

where ε is the extraction factor mB/A.

Countercurrent Extraction. The best compromise between high extract concentration and high recovery is obtained by using multistage countercurrent extraction, as shown in Figure 7b. The feed entering stage 1 is brought into contact with a B-rich stream which has already been through the other stages, while the raffinate leaving the last stage has been in contact with fresh solvent. Because of the economic advantages of continuous countercurrent extraction, this type of operation is preferable for commercial-scale operation.

For the case of a partially miscible solvent system, design calculations can be carried out graphically on the triangular mass fraction diagram (53) as shown in Figure 7a. If the solvent–feed ratio, the feed composition (point F), and the extract composition (point E) are given, it is possible to determine points M and Δ on the diagram using the inverse lever rule. Stage-to-stage construction is then carried out from the feed end (stage 1) to the raffinate end (stage N) of the cascade making use of point Δ and the tie lines which are shown dotted in Figure 7a. The total number of tie lines

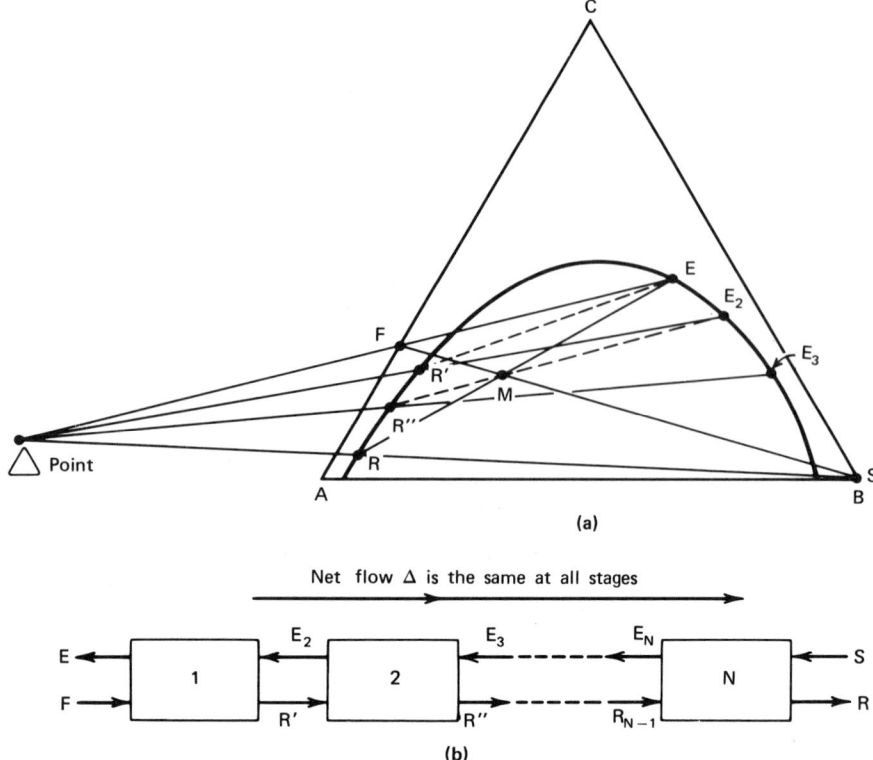

Figure 7. Countercurrent stagewise extraction. (a) Triangular diagram showing Hunter-Nash construction (53); (b) schematic representation.

drawn in proceeding from the feed to the raffinate composition (point R) is the required number of theoretical stages.

If the feed composition and solvent–feed ratio are held constant but the specified extract composition is increased, it can be shown that more stages are required. An upper limit to the extract composition is found when one of the lines to the Δ point (Figure 7a) coincides with a tie line. This is the so-called pinch effect which results in an infinite number of required stages. The pinch may occur at an intermediate composition in the cascade or it may be reached at the feed end (stage 1) or the raffinate end (stage N). If it occurs at the feed end, the maximum possible extract composition is that which is in equilibrium with the feed (see also Diffusion separation methods).

The pinch effect also determines the minimum solvent–feed ratio that can be used for given extract, feed, and solvent compositions. The economic optimum solvent–feed ratio is commonly one and one half to two times the minimum value and is determined for a given case by considering the capital costs (largely dependent on the number of stages) and the operating costs which are strongly influenced by the cost of solvent recovery (Fig. 8).

Further enrichment of the extract can be obtained by a reflux arrangement in which part of the extract is freed of component B by distillation and returned to stage 1 of the cascade as an A-rich phase. The feed enters the center of the cascade at the

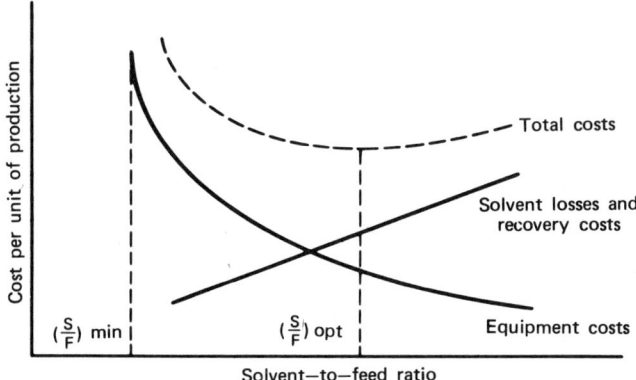

Figure 8. Costs for countercurrent extraction with given feed, solvent, and extract compositions.

appropriate stage and the design is to some extent analogous to that of a plate distillation column. Reflux can drastically increase the solvent recovery load and it may be economically preferable in a given case to increase the number of stages as an alternative to introducing reflux (54) (see also Distillation).

Stagewise design is considerably simplified when the nonconsolute components A and B are not significantly miscible. The mass flows of components A and B then remain constant from one stage to the next. If X represents the mass ratio of C to A in one phase and Y represents the mass ratio of C to B in the other, a material balance on the countercurrent system (Fig. 7) gives the relationship:

$$A(X_o - X) = B(Y_1 - Y) \tag{31}$$

where A and B are the constant mass flows of A and B in the respective streams, X_o is the mass ratio C to A in the feed and Y_1 is the mass ratio of C to B in the extract. The variation of Y with respect to X throughout the cascade is given by a straight line (the operating line) with a slope of A/B as shown in Figure 9. Also shown in this figure is the equilibrium curve for the system. At very low concentrations of C, the equilibrium mass ratios X and Y tend toward the values of mass fractions and, in accordance with equation 4, the equilibrium curves become straight.

The number of equilibrium stages is readily found by a stepwise procedure similar to that used in absorption and distillation problems. If we start at the point (X_o, Y_1), the first (horizontal) line represents the establishment of equilibrium in stage 1 to give $(X_1 Y_1)$. The second (vertical) line represents the solution of the material balance equation 31 with respect to X_1, giving the point $(X_1 Y_1)$ and so on. The construction in Figure 9 indicates between 3 and 4 ideal stages and in practice the designer would specify 4 ideal stages in this case.

When there is significant miscibility between A and B, the Hunter-Nash method is normally employed. However, a curved operating line on a mass fraction plot can be constructed from the triangular diagram (55) such as Figure 7. The operating curve is the locus of the compositions such as (R_1, E_2), (R_2, E_3), etc found from the Δ point plot in Figure 7. Once the operating and equilibrium curves are found, the stepwise procedure can be employed.

An advantage of the stepwise procedure over the Hunter-Nash method is that it can be modified for the case where equilibrium of the exit streams is not attained

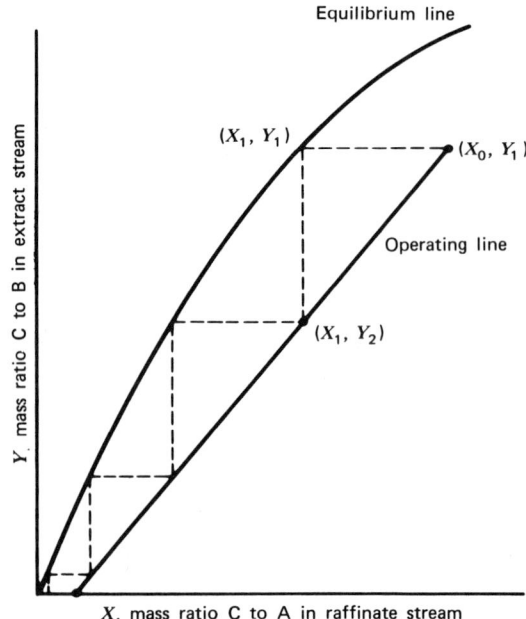

Figure 9. Countercurrent extraction with equilibrium stages where A and B are immiscible.

completely in each stage. By analogy with distillation, a Murphree stage efficiency can be defined for either phase. If a Murphree efficiency less than unity is specified, the stage-to-stage construction must be carried out using shorter steps than those shown in Figure 9, so that more stages are needed for a given separation than in the ideal case.

In cases where solvents are substantially immiscible and the distribution law holds ($m = Y/X$), the calculation can be further simplified and the following equations can be expressed (56–57) in terms of the extraction factor ε (= mB/A).

When $\varepsilon \neq 1$:

$$\frac{X_o - X_N}{X_o - Y_s/m} = \frac{\varepsilon^{N+1} - \varepsilon}{\varepsilon^{N+1} - 1} \tag{32}$$

or

$$N = \frac{\log\left[\left(\dfrac{X_o - Y_s/m}{X_N - Y_s/m}\right)\left(1 - \dfrac{1}{\varepsilon}\right) + \dfrac{1}{\varepsilon}\right]}{\log \varepsilon} \tag{33}$$

When $\varepsilon = 1$:

$$\frac{X_o - X_N}{X_o - Y_s/m} = \frac{N}{N + 1} \tag{34}$$

$$N = \frac{X_o - X_N}{X_N - Y_s/m} \tag{35}$$

where X_o is the solute mass ratio in the feed; X_N is the solute mass ratio in the raffinate after N equilibrium stages; and Y_s is the solute mass ratio in the entering solvent. These equations have been represented graphically (1,52). In the case of dilute systems, mass ratios may be replaced by concentrations in the above equations.

Fractional Extraction. Although solvent extraction is considered in its simplest form as the transfer of a consolute component C from a binary feed solution (A-rich) to a solution in another phase (B-rich), it is often used to separate one or more consolute components. This process is known as *fractional extraction*. In one type of application, a binary feed (C and A) is introduced into a countercurrent cascade through which a solvent B is circulated (see Fig. 10**a**). The solvent preferentially dissolves one component (eg, C) as it passes down the column. Reflux of the C product is provided. For the type of extraction to be practicable, the solvent should be only partially miscible with either A or C in order that two phases can exist in all parts of the column. This process is closely analogous to distillation of a binary mixture; a stagewise procedure has been designed on this basis (58).

If no solvent is available that satisfies the requirement of being sparingly miscible with each of the feed components, fractional extraction may be carried out using two solvents (Fig. 10**b**). The feed components C and D may be wholly miscible with either solvent as long as the two solvent streams are immiscible. Thus two-solvent fractional extraction involves two phases and four or more components.

Graphical stagewise design procedures based on three-dimensional quaternary equilibrium diagrams have been reviewed (1). If the two solvents (A and D) are immiscible and the two interfacial equilibria with respect to consolute components B and C (Fig. 10**b**) are unaffected by each other, a simplified technique (59) may be used. In this graphical method, stagewise calculations similar to those of Figure 9 are carried

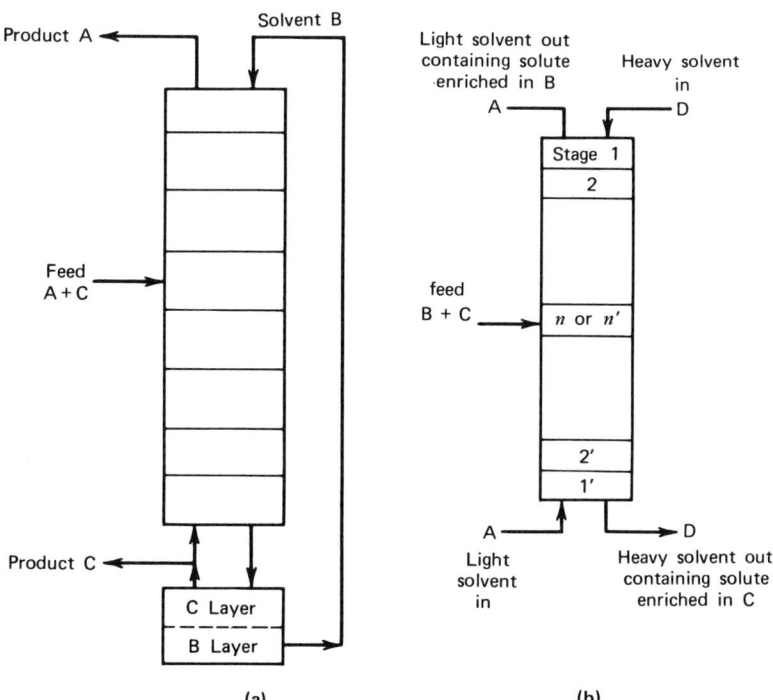

Figure 10. Fractional extraction: (**a**) one solvent; (**b**) two solvents.

out iteratively for each component until the compositions with respect to B and C at the feed plate are matched (60). Computerized stagewise design procedures also have been developed for multicomponent stagewise contact (61–63).

In case solvents are substantially immiscible, the distribution coefficients are independent of concentration for the solutes and for the central feed, the calculations can be greatly simplified (64–65):

$$R_B = \varepsilon_B^n \tag{36}$$

$$R_C = \varepsilon_C^n \tag{37}$$

$$N = (n + n' - 1) = \frac{2 \log \left(\frac{R_B}{R_C}\right)}{\log \beta_{B,C}} - 1 \tag{38}$$

where

R_B = rejection ratio of B

$\quad = \dfrac{\text{solute B leaving extractor in solvent A}}{\text{solute B leaving extractor in solvent D}}$

R_C = rejection ratio of C

$\quad = \dfrac{\text{solute C leaving extractor in solvent A}}{\text{solute C leaving extractor in solvent D}}$

$$\varepsilon_B = m_B \frac{A}{D} \qquad \varepsilon_C = m_C \frac{A}{D}$$

$\beta_{B,C} = m_B/m_C$ = relative distribution coefficient

N = total number of theoretical stages

For a symmetric separation, $R_B = 1/R_C$

$$\frac{D}{A} = \sqrt{m_B m_C} \tag{39}$$

$$\varepsilon_B \varepsilon_C = 1 \tag{40}$$

$$N = \frac{4 \log R_B}{\log \beta_{B,C}} - 1 \tag{41}$$

For multicomponent systems involving more than two solutes, the calculations can be handled by applying certain reasonable assumptions (60) for the separation of aromatic and paraffinic hydrocarbons.

Differential Contacting. Although the equilibrium stage concept has proved extremely useful in describing the performance of mixer-settlers and plate columns with discrete stages, it is not appropriate for spray towers, packed columns, etc, in which no discrete stages can be identified. In such differential types of contactor, equilibrium between phases is never reached and the mass transfer rate is, therefore, very important in the design procedure.

Consider a differential countercurrent contactor (Fig. 11) operating with a dilute solution of the consolute component C and with components A and B immiscible. Under these conditions, the volume flows of the A-rich and B-rich streams can be

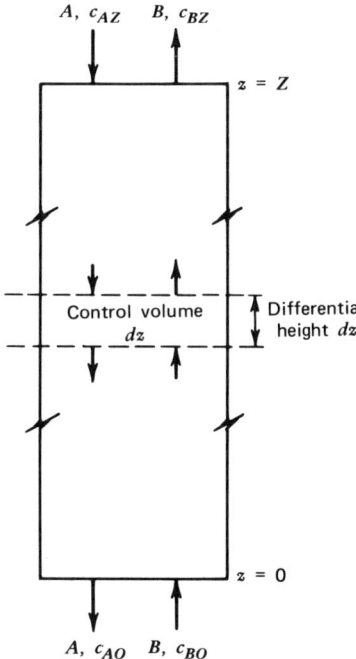

Figure 11. Mass transfer in a differential contactor.

assumed not to vary significantly with position in the contactor, and are taken to be A and B, respectively, expressed on the basis of unit cross-sectional area of contactor. The concentration of C in the A-rich stream is c_A and that in the B-rich stream is c_B.

A steady-state material balance can be carried out on a small section of length dz and volume dz (on the basis of unit cross-sectional area) in the contactor:

$$B dc_B = A dc_A = K_B a (c_B^* - c_B) dz \qquad (42)$$

Rearrangement and integration give a relationship for the contactor height in terms of the concentration change:

$$dz = \frac{B}{K_B a} \cdot \frac{dc_B}{c_B^* - c_B} \qquad (43)$$

$$Z = \left(\frac{B}{K_B a}\right) \cdot \int_{c_{BO}}^{c_{BZ}} \frac{dc_B}{c_B^* - c_B} \qquad (44)$$

The integral can be found graphically if the equilibrium line is curved. An analytical expression for the integral is available for the case where both the equilibrium and operating lines are straight (1).

$$\int_{c_{BO}}^{c_{BZ}} \frac{dc_B}{c_B^* - c_B} = \frac{1}{1-\varepsilon} \ln\left[\left(\frac{c_{BO} - c_{BZ}^*}{c_{BZ} - c_{BZ}^*}\right)(1-\varepsilon) + \varepsilon\right] \qquad (45)$$

where $\varepsilon = mB/A$. The integral is unitless and is known as the number of transfer units (NTU) based on the overall B-rich-phase driving force. Obviously the NTU, and hence the contactor length Z required, increase as the difference between c_{BZ} and c_{BO} is increased.

The factor $(B/K_B a)$ in equation 38 is known as the height of a transfer unit (HTU). It is a characteristic of the hydrodynamic conditions such as the flow rate B and the specific interfacial area a, but is independent of changes in c_B. It is important that the HTU be specified correctly in regard to the phase and driving force considered; in this case it relates to the overall mass transfer driving force in the B-rich phase. The HTU may vary with height because of changes in drop size etc; an average value is usually taken assuming no variation.

Mass transfer theory (eqs. 24–25) indicates that the overall mass transfer resistance $1/K_B$ consists of contributions from each phase, so that the overall HTU is also the sum of two contributions:

$$\left.\begin{array}{l} (HTU)_{OB} = \dfrac{B}{k_B a} + \dfrac{mB}{k_A a} \\[6pt] \quad = \dfrac{B}{k_B a} + \dfrac{mB}{A}\left(\dfrac{A}{k_A a}\right) \\[6pt] \quad = (HTU)_B + \varepsilon (HTU)_A \end{array}\right\} \qquad (46)$$

The heights of a transfer unit in each phase thus contribute to the overall heights of a transfer unit. Data on values of HTU for various types of countercurrent equipment have been reviewed (1,10).

In normal operating practice, the extraction factor mB/A is chosen to be not greatly different from unity, usually within the range of 0.5–2.

Although the stagewise model is not physically realistic for differential contactors, it is sometimes used. The number of equivalent theoretical stages N can be determined graphically using the stepwise construction as illustrated in Figure 9. For the case where both the equilibrium and operating lines are straight, it can be shown that:

$$\frac{N}{(NTU)_{OB}} = \frac{\varepsilon - 1}{\ln \varepsilon} \qquad (47)$$

If $\varepsilon = 1$, the number of theoretical stages is equal to $(NTU)_{OB}$.

It must be emphasized that equations 36–41 are applicable only to dilute, immiscible systems. If the amount of mass transfer is significant in comparison to the total flow rates, more complicated treatments of differential contactors are required (1).

In recent years, considerable attention has been paid to axial mixing effects in countercurrent contactors. Turbulent eddies or circulation currents have the effect of reducing the axial concentration gradients below the values predicted by equation 36 which is based only on convection (plug flow assumption). The net result of axial mixing is to reduce the real driving force $(c_B^* - c_B)$ for mass transfer and hence the NTU obtainable for a column of given height and true HTU (as defined by $B/k_B a$). The effect is illustrated in Figure 12, which also shows the sharp jumps in composition owing to mixing between the inlet streams and the phases already in the contactor.

Axial mixing is usually characterized by an axial dispersion coefficient E which has the same units as molecular diffusivity but can be a million times larger. The axial flux of dissolved solute owing to axial mixing is given by an equation analogous to equation 18:

$$N = -E\frac{\partial c}{\partial z} \qquad (48)$$

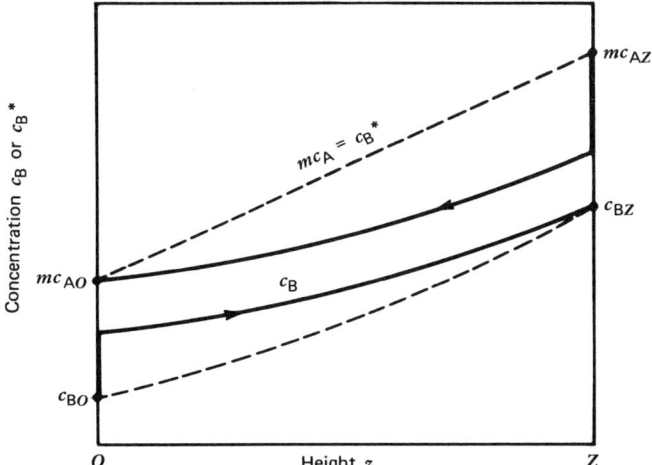

Figure 12. Effect of axial mixing on concentration profiles in a differential contactor. - - -, profiles with plug flow; —, profiles with axial mixing (arrows indicate directions of flow).

Values of E are commonly in the range 1–5 cm^2/s in packed and plate columns but much greater in spray columns because of circulation effects. The effect of axial dispersion on column performance depends on the axial Peclet number:

$$Pe = uZ/E \qquad (49)$$

where u is the superficial velocity of the phase in question and Z is the column height. Plug flow conditions can be assumed only if the axial Peclet number exceeds about 50, a condition that is not always fulfilled in extraction columns because of the relatively low flow velocities permitted. The effect of Pe on column performance, for different extraction factors and true HTU values has been theoretically predicted (66–67) (see also Adsorptive separation).

Experimental values of E have been obtained for most types of contactor (68–69). It is significant that E tends to increase with column diameter because of circulation effects (69). The Peclet number in a small diameter test column may be high enough for plug flow to be safely assumed, but the increase of E with diameter leads to a reduction in Pe and an increase in the apparent HTU (or height of an equivalent theoretical stage) as the column is scaled up. Hydrodynamic model testing of large columns is recommended (70) in order to remove uncertainties in scaling up.

Hydrodynamic Factors in Extractor Design. A successful extraction device should create droplets small enough to provide interfacial area for reasonably fast extraction, while allowing the two liquid phases to separate properly after extraction. In the case of countercurrent contactors there is a third hydrodynamic requirement, namely, that the flow of the two phases in opposite directions through the equipment should be stable.

The specific interfacial area in a droplet dispersion is determined by the dispersed-phase holdup h and the mean droplet diameter d_m in accordance with equation 27. In general, the droplet size is nonuniform. The mean drop diameter is defined for any size distribution as:

$$d_m = \frac{\sum_{i=0}^{N} d_i^3}{\sum_{i=0}^{N} d_i^2} \qquad (50)$$

so that the total area of the dispersion is equal to that of a similar volume of uniformly sized drops of diameter d_m.

Drops are initially formed when the dispersed phase flows into the continuous phase through a nozzle or an orifice. Detachment of the drop is favored by choice of an orifice material that is preferentially wetted by the continuous phase. At very low velocities of flow through the orifice, the drop size is determined by a balance between buoyancy forces and interfacial tension. At the higher velocities normally encountered in practice, inertial forces are also important. The data for formation of discrete drops at nozzles have been correlated (71) and the drop size distributions obtained studied when jets are initially formed at the nozzles and subsequently break up (72).

In turbulent agitation, the initially formed drops are rapidly broken down to smaller sizes by eddies. On the basis of turbulence theory (73):

$$d_m = K' \frac{\sigma^{0.6}}{\rho_c^{0.6} \psi^{0.4}} \quad (51)$$

where σ is the interfacial tension, ρ_c is the continuous-phase density, and ψ is the rate of energy dissipation per unit mass of fluid. This expression was originally proposed for very dilute dispersions ($h \to 0$) and for isotropic turbulence, but it has been found useful even when these restrictions do not apply. For stirred tanks at holdups up to 0.35, the dimensionless constant K' is $0.024(1 + 2.5\,h)$ (74). A larger value of K', 0.36 was obtained in reciprocating-plate extraction columns (75). In a pulsed perforated-plate column the value of K' is ca 0.18 (76–77). Despite the considerable variations of K' with respect to equipment type, equation 45 is useful in predicting the effects of changes in energy input and interfacial tension upon mean droplet size.

The dispersed-phase holdup h has just as much effect as the drop size d_m upon the specific interfacial area. In a batch extraction process, h is, of course, determined by the volumes of each phase added. However, in continuous operations the holdup is a complex function of hydrodynamic conditions. In a continuously operated, baffled stirred tank with upward, cocurrent flow, the holdup of the light dispersed phase is in general, less than its volume fraction in the feeds. As the impeller power input is increased, the holdup tends toward the value that would be expected from perfect mixing of the two feed streams (78).

Holdup in countercurrent equipment is governed by the superficial velocities U_D and U_C for each phase, ie, the volume flow of each phase divided by the total cross-sectional area. A slip velocity U_s is defined as follows (79):

$$U_s = \frac{U_D}{h} + \frac{U_C}{1-h} \quad (52)$$

In the absence of packing, U_s is a function of the terminal velocity of an isolated dispersed phase droplet and the holdup h. A chart developed for fluidized particles may be used for estimating U_s (80).

In a packed column, the slip velocity is enhanced by the reduced volume fraction (voidage ϵ) available for liquid flow:

$$U'_s = \frac{U_D}{\epsilon h} + \frac{U_C}{\epsilon(1-h)} \quad (53)$$

However, U'_s can be expressed as $u_o(1-h)$ where the characteristic velocity u_o is a function of system properties and the nature of the packing but independent of holdup (81).

As the flow rates are increased in countercurrent columns, the dispersed phase holdup h increases until it is so large that the two liquid phases are unable to pass each other. This condition is known as flooding, and can be observed either as an accumulation of the dispersed phase in a close-packed array of drops which eventually coalesces into a layer that fills the column, or as a severe entrainment of dispersed droplets in the exit continuous phase.

Since dU_D/dh and dU_C/dh are both zero at the flooding point (82), it can be shown from equation 47 that the holdup at flooding is:

$$h_f = \frac{(L^2 + 8L)^{0.5} - 3L}{4(1-L)} \tag{54}$$

where $L = U_D/U_C$. Thus for $L \simeq 1$, $h_f = 0.33$ and as $L \to \infty$, $h_f = 0.5$.

Flooding can be reached by exceeding the limits of flow rate as determined above by the slip velocity; alternatively, in externally agitated columns it can also be caused by increasing the agitation level (and hence decreasing U_s) at constant flow rates. A good review of flooding relationships for various types of equipment is given in ref. 10. The slip velocity approach does not satisfactorily account for flooding phenomena in a rotating-disk contactor (83).

The separation of two interdispersed phases by coalescence is an essential part of any solvent extraction operation. A drop of dispersed phase does not coalesce immediately on reaching a liquid–liquid interface because time is needed for drainage of the continuous phase from between the drop and the interface. In a multidrop system, a dispersion band is formed, with the drops becoming increasingly deformed as they approach the interface; drainage of the continuous phase must occur through interstices in the compacting assembly of drops. Coalescence occurs between drops as well as at the interface itself.

Coalescence is particularly slow from very fine drop dispersions such as those formed in stirred tanks. The settling tank in a mixer–settler unit is usually much larger than the mixing tank in order to accommodate the dispersion band. This takes the form of a wedge when the dispersion enters the settler from one side. Coalescence rates are enhanced by high interfacial tension and reduced by high continuous-phase viscosity. Coalescence is also very sensitive to the presence of surface-active impurities that can accumulate at the phase interface to form an objectionable deposit (crud).

The foregoing discussion shows that mass transfer rates per unit volume of equipment are improved by reducing the drop size, but at the expense of increased coalescence times and (in the case of countercurrent equipment) limitations on throughput because of flooding. A compromise has to be reached between mass transfer and hydrodynamic requirements. Many different types of solvent extraction equipment have been developed to meet the requirements of different processes and systems, but they all reflect this essential compromise.

Equipment and Processing

Industrial applications of solvent extraction have increased rapidly in the last 25 years. Simple mixer–settlers, packed columns, and spray columns were widely used in the process industries during the 1930s and 1940s. New and improved multistage and differential contactors employing mechanical energy input to achieve a high rate of mass transfer have been developed since the late 1940s and have found wide commercial application. In the past decade, several reviews (15,18,84–93) have appeared that describe various types of extractors.

Because of the great variety available, the choice of a commercial extractor for a new process can be bewildering. The following criteria should be taken into consideration when selecting a contactor (88,94) for a particular application: (1) stability and residence time, (2) settling characteristics of the solvent system, (3) number of stages required, (4) capital cost and maintenance, (5) available space and building height, and (6) throughput.

A qualitative chart (Fig. 13) of the economic operating range of various classes of extractors can be consulted when making a selection (95).

An extraction plant should operate at steady state in accordance with the flow sheet design for the process (see Fig. 14). However, this is not always the case. Fluctuation in the feed streams can cause changes in product quality unless a sophisticated system of feed-forward control is used (96). Upsets of operation caused by flooding in the column always force shutdowns. Therefore, interface control could be of utmost importance. The plant design should be based on (1) process control decisions made by trained technical personnel, (2) off-line analysis and/or limited on-line automatic analysis, and (3) control panels equipped with manual and automatic control for motor speed, flow, interface level, pressure, temperature, etc (see Instrumentation and control).

LABORATORY EXTRACTORS, PILOT-SCALE TESTING, AND SCALE-UP

Several laboratory units are useful in analysis, process control, and process studies.

The Craig contactor (97) is employed in qualitative and quantitative analysis and

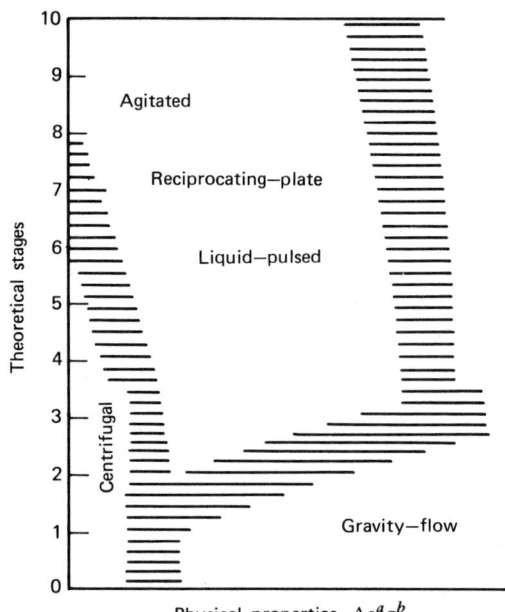

Figure 13. Economic operating range of extractors. Courtesy of Luwa A.G. (95). Superscripts a and b are constants.

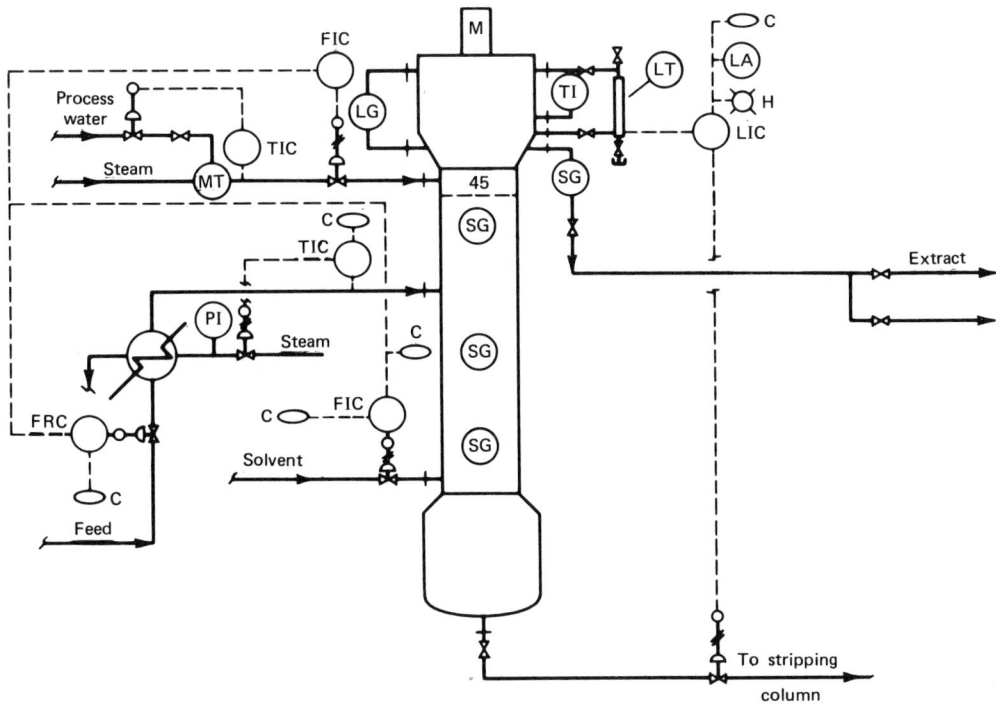

Figure 14. Flow diagram of process for an industrial extraction column (92). FRC = flow recording controller; FIC = flow indicating controller; TI = temperature indicator; TIC = temperature indicating controller; LT = level transmitter; LA = level alarm; LIC = level indicating controller; PI = pressure indicator; M = motor; MT = mixing tee; SG = sight glass; C = computer; H = annunciator at control panel, high limit.

for determining distribution coefficients of two or more components. A new automatic model with an operation capability of up to a thousand contacting stages is available. The AKUFVE contactor (31,98) incorporates a separate mixer and centrifugal separator. It is an efficient instrument for rapid and accurate measurement of partition coefficients, as well as for obtaining reaction kinetic data.

Miniature mixer–settlers which are continuous, bench-scale, multistage, countercurrent, liquid–liquid contactors (99) are particularly useful for the preliminary laboratory work associated with flow-sheet development and optimization since they give a known number of theoretical stages.

In a simple laboratory-scale, reciprocating-plate extraction column, a minimum height of an equivalent theoretical stage (HETS) of 7.1 cm and volumetric efficiencies of up to 532 per hour were achieved employing a methyl isobutyl ketone (MIBK)–acetic–water system (100). The column has been used successfully for countercurrent and fractional liquid extraction in the laboratory.

Because the factors relating to mass transfer and fluid dynamics of the system in an extractor are extremely complex, particularly for mixed solvents and feedstocks of commercial interest, pilot-scale testing remains an almost inevitable preliminary to a full-scale contactor design. These tests provide the following information: (*1*) total throughput and agitation speed; (*2*) HETS or HTU; (*3*) stage efficiency; (*4*) hydrodynamic conditions (droplet dispersion, phase separation, flooding, emulsion layer formation, etc); (5) selection of dispersed phase or direction of mass transfer; (6) solvent to feed ratio; (7) material of construction and its wetting characteristics; and (8) confirmation of desired separation (in cases where equilibrium data are not available).

For design of a large-scale commercial extractor, the pilot-scale extractor should be of the same type as that to be used on the large scale. Experimental data obtained from a mixer–settler or a pulsed column provide little information for design of mechanically agitated columns or centrifugal extractors.

In the present stage of knowledge, reliable scale-up for industrial-scale extractors still depends on correlations based on extensive performance data collected from both pilot-scale and large-scale extractors covering a wide range of liquid systems. Unfortunately, only limited data for a few types of large commercial extractors are available in the literature.

COMMERCIAL EXTRACTORS

Contactors can be classified according to the methods applied for interdispersing the phases and producing the countercurrent flow pattern. Figure 15 summarizes the classification of major types of commercial extractors; Table 2 summarizes their main characteristics.

Unagitated Columns. Spray columns (Fig. 16) are the simplest in construction but have very low efficiency because of poor phase contacting and excessive backmixing in the continuous phase. They generally provide one or, at the most, two equilibrium stages. Because of their simple construction, spray columns are still used in industry for operations, such as washing, treating, and neutralization, which often require no more than one or two stages.

Packed columns (Fig. 16) have better efficiency because of improved contacting and reduced backmixing. It is important that the packing material should be wetted by the continuous phase to avoid coalescence of the dispersed phase. To reduce the effects of channeling, redistribution of the liquids at fixed intervals is normally required in the taller columns.

Perforated-plate columns (Fig. 16) are operated semistagewise and are reasonably flexible and efficient. A perforated-plate tower, 2.13 m in diameter and 24.38 m in height, used for extraction of aromatics was reported to have the equivalent of ten theoretical stages (87).

Because of the simplicity and low cost of packed and perforated-plate columns they are still widely used in industry despite their low efficiency, particularly for processes requiring few theoretical stages and for corrosive systems where absence of mechanical moving parts is advantageous. A 12.2-m diameter, perforated-plate column with downcomers between each stage has been used for petroleum processing.

General features and applications of unagitated columns are summarized in Table 2.

Mixer–Settlers. Mixer–settlers are still widely used in the chemical process industry because of their reliability, flexibility, and high capacity. They are particularly economical for operations that require high capacity and few stages. Mixer–settlers with a capacity of up to 22.7 m^3/min (6000 gal/min) have been used in the mining in-

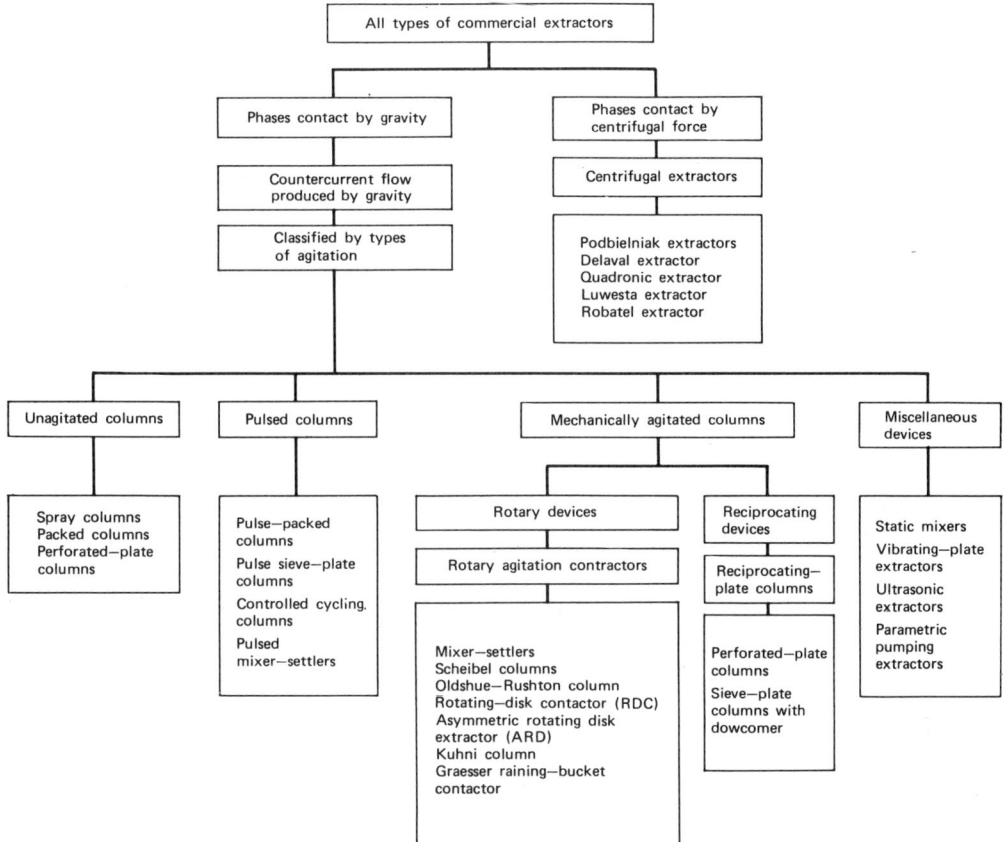

Figure 15. Classification of commercial extractors (92).

dustry. The main disadvantages of mixer-settlers are their size and the inventory of material held up in the equipment. In the past decade, considerable development work has been done to improve the contactors and many new devices have been reported.

The simple box-type mixer–settler (101) (Fig. 17) has been used extensively by the British Atomic Energy Authority for the separation and purification of uranium and plutonium (102). Interstage piping is avoided by use of a partitioned box construction and interstage pumping is not needed. The driving force for the flow is derived from the density difference between the stages.

In a pump-mix extractor interstage pumping (103) overcomes the flow limitations of the simple horizontal mixer–settler.

A new type of pump–mixer–settler has been developed by the Israeli Mining Industry (IMI) (104) and has been widely used in many process industries. A unit with a capacity of 8.33 m^3/min (2200 gal/min) has been used in phosphoric acid plants (105). The unique part of this design is that the pumping device is not required to act as the mixer, and the two phases are dispersed by a separate impeller mounted on a shaft running coaxially with the drive to the pump.

The General Mills mixer–settler (106) is a pump–mix unit designed for metal-

Table 2. Summary of Features and Fields of Industrial Application of Commercial Extractors[a]

Types of extractor	General features	Fields of industrial application
unagitated columns	low capital cost, low operating and maintenance cost, simplicity in construction, handles corrosive material	petrochemical, chemical
mixer–settlers	high-stage efficiency, handles wide solvent ratios, high capacity, good flexibility, reliable scale-up, handles liquids with high viscosity	petrochemical, nuclear, fertilizer, metallurgical
pulsed columns	low HETS, no internal moving parts, many stages possible	nuclear, petrochemical, metallurgical
rotary agitation columns	reasonable capacity, reasonable HETS, many stages possible, reasonable construction cost, low operating and maintenance cost	petrochemical, metallurgical, pharmaceutical, fertilizer
reciprocating-plate columns	high throughput, low HETS, great versatility and flexibility, simplicity in construction, handles liquids containing suspended solids, handles mixtures with emulsifying tendencies	pharmaceutical, petrochemical, metallurgical, chemical
centrifugal extractors	short contacting time for unstable material, limited space required, handles easily-emulsified material, handles systems with little liquid density difference	pharmaceutical, nuclear, petrochemical

[a] Ref. 92.

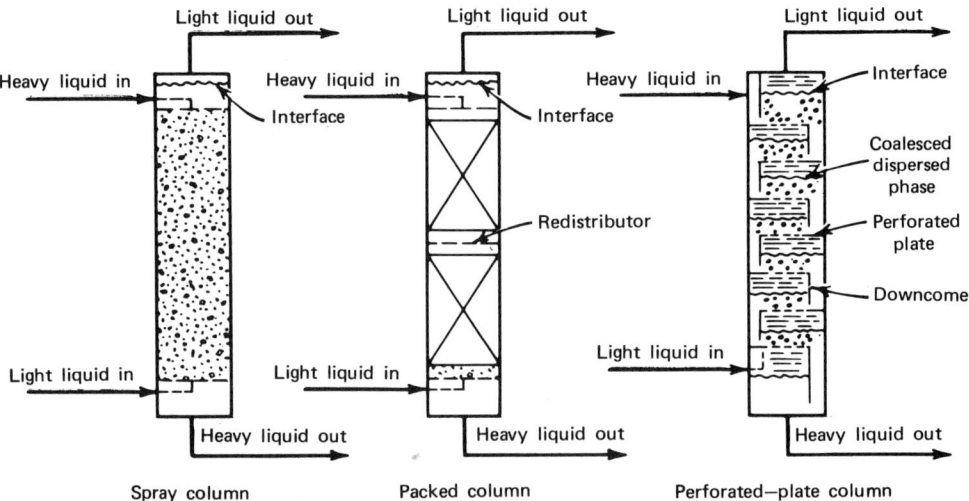

Figure 16. Unagitated column extractors.

lurgical extraction. It has a baffled cylindrical mixer fitted in the base, with a turbine that mixes and pumps the incoming liquids. The dispersion leaves from the top of the mixer and flows into a shallow rectangular settler designed for minimum holdup.

The Davy Powergas unit (107–109), as shown in Figure 18, also employs the pump–mix approach. The liquids run through a draft tube and are mixed and pumped

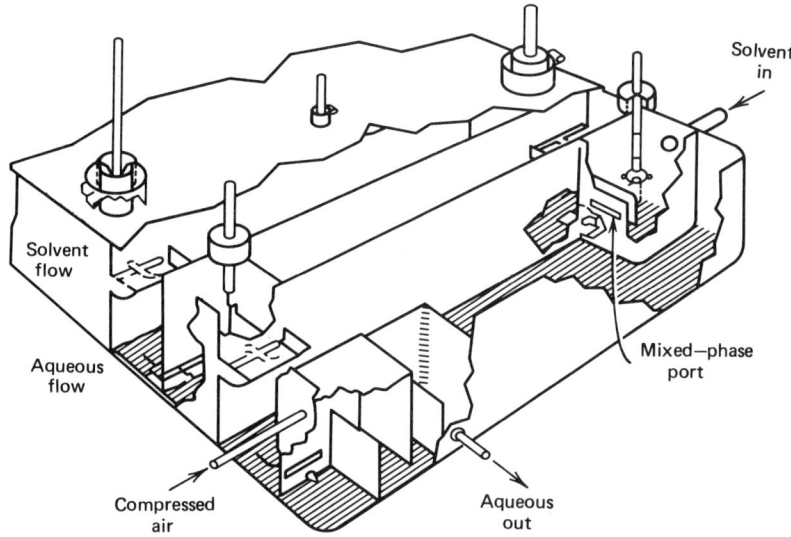

Figure 17. Box-type mixer–settler. Courtesy of the Institute of Chemical Engineers (101).

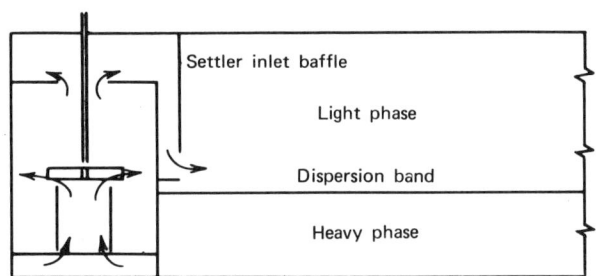

Figure 18. Davy Powergas mixer–settler. Courtesy of *Chemical Engineering* (*New York*) (15).

by an impeller running directly above the draft tube. The dispersion flows off the top of the mixer and down through a channel into a rectangular settler. The large units are used for copper extraction (11).

The Lurgi contactor (110), manufactured in the Federal Republic of Germany, consists of stacked mixer–settler units. Mixing and phase transfer take place in pumps attached to the sides of the settling column, and phases flow interstagewise by a complex arrangement of baffles within the settling zones. Since the contact time in each mixing stage is short, the contactor is more suitable for processes in which the rate of mass transfer is fast. It has a capacity of 1600 metric tons per hour and columns up to 3 m in diameter have been used for aromatic extractions.

Holmes and Narver, Inc. (111) are marketing a new type of mixer–settler. The contactor incorporates multicompartment mixers and also has many other features.

Motionless in-line mixers utilize the energy for mixing and dispersing from pressure drops resulting from flow. Performance data on static mixers (112) and on Sulzer mixers (113) have been reported.

The number and variety of the major mixer–settler designs described above show the extent of commercial interest in this type of contactor.

Scale-Up and Design. Mixer–settlers are relatively reliable when scaled up because they are practically free of interstage backmixing and stage efficiencies are high (ie, 80–90%). Various studies (36,75,114–115) have shown that (*1*) the rate of extraction is a function of power input, and (*2*) mixers can be reliably scaled up by geometric similitude at constant power input per unit mixer volume, eg, up to 200-fold of throughput (116–117).

The processes taking place in a settler are very complex. The flow capacity is characterized by a band of dispersion at the interface (see under Hydrodynamic Factors). The thickness of the band is a measure of the approach to flooding (15). The thickness of the band increases exponentially with increasing flow and settlers can be scaled up by factors of up to 1000 on this basis. The efficiency of a settler can be enhanced by minimizing turbulence and small drops, and maintaining linear velocities along the settler to a low value to avoid entrainment of small drops from the dispersion band.

In large, industrial mixer–settlers, the settlers usually represent at least 75% of the total volume of the units. Therefore, a practical means to increase the throughput is needed and thus the size of the settler could be reduced and the inventory of the solvent lowered.

Pulsed Column. The efficiency of sieve-plate or packed columns can be appreciably increased by the application of sinusoidal pulsation to the contents of the column. The application of the pulse increases both turbulence and interfacial areas and greatly improves the mass transfer efficiency compared with an unpulsed column. Axial mixing in a pulsed column is relatively small compared with mechanical rotary-agitated columns. This leads to a substantial reduction in HETS or HTU values.

Pulsed Packed Columns. A pulsed packed column consists of a vertical cylindrical vessel filled with packing. Both light and heavy liquids, dispersed in the form of drops, pass countercurrently through the column and are simultaneously moved up and down by means of a pulsating device connected to the bottom of the column through a side-entering pulse leg. Mechanical difficulties with the generation of the pulse formerly limited pulsed columns to comparatively small diameters. However, the installation of pulsed packed columns up to 2.7 m in diameter has recently been reported (118–119). Generation of pulsations by compressed air has received increasing attention (120). A detailed model of pulsed packed column behavior has been developed (119).

Pulsed Perforated-Plate Columns. The column is fitted with horizontal perforated plates or sieve plates which occupy the entire cross section of the column. The total free area of the plate is about 20–25%. The columns are generally operated at frequencies of 1.5–4 Hz with amplitudes of 0.635–2.54 cm.

Low axial mixing and high extraction efficiency owing to uniform distribution of energy over a cross section of the column and, hence, uniform distribution of drops in the column are the unique features of pulsed perforated-plate columns. They have been widely used in the nuclear industry up to 0.9 m in diameter. Even larger scale columns are being developed (121–122).

Several regions of operation can be distinguished, depending on the flow rate and intensity of pulsation (123) (Fig. 19). At lower pulsed volume velocities (amplitude

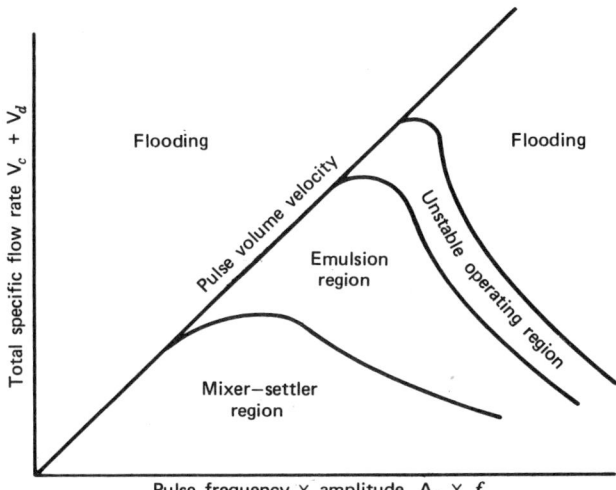

Figure 19. Regions of operation of a pulsed, perforated-plate column. Courtesy of *Chemical Engineering Progress* (123).

x, frequency x, column cross section area), a discrete layer of liquid appears between plates during each reversal of the pulse cycle. At higher pulsed volume velocities, there is little or no coalescence between the plates, and the column then behaves as a differential contactor. Extensive studies were made on flooding and mass transfer and development of empirical correlations for the column design (124–126) and on hydrodynamics and efficiency of various sizes in uranium extraction (121).

The controlled-cycling column (127) is an interesting development. It has a high throughput but no large-scale application has been reported.

Mechanically Agitated Columns. Mechanically agitated columns (see Fig. 20) are divided into rotary-agitated and reciprocating or vibrating-plate columns.

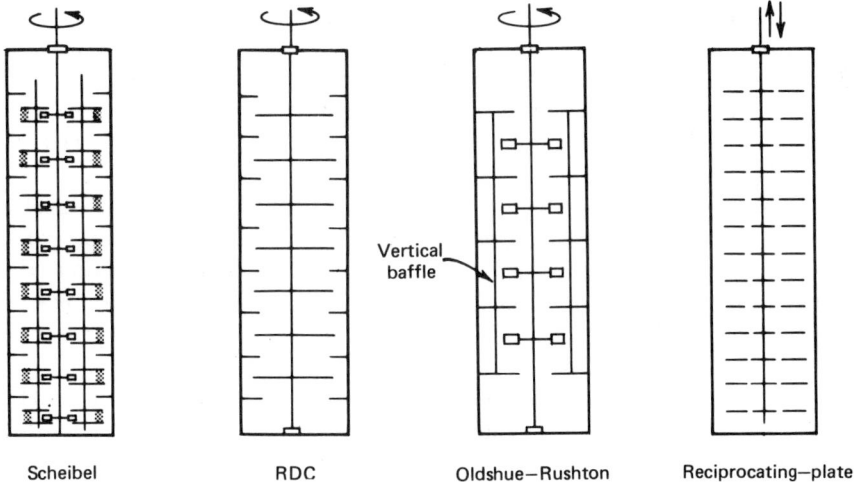

Figure 20. Typical mechanically agitated columns.

Rotary-Agitated Columns. Because of the mechanical advantages of rotary agitation, most modern differential contactors employ this method. Among the commercial rotary-agitated columns listed in Figure 15, the Scheibel column, the rotating disk contactor, and the Oldshue-Rushton multiple-mixer column are best known and have been proven in many industrial installations. Features and applications are given in Table 2.

An early model of the Scheibel column was developed in 1948 (128). Alternate compartments are agitated with impellers, whereas the others are packed with an open woven wire mesh. Capacity and mass transfer data on 2.5 cm, 7.6 cm, and 30.5 cm diameter columns using three different systems are given in refs. 36, 128, and 129.

A newer type of column, with or without wire mesh packing, using horizontal baffles (see Fig. 20) was developed in 1956 (130). It improves the HETS and permits a more efficient scale-up for large diameter columns.

Performance data for a 30.5 cm column, with and without wire mesh packing, have shown that the HETS of this type of column varies as the square root of the diameter. A third design (131) is basically similar but a pumping impeller instead of a turbine impeller is used in the mixing stage.

Although the Scheibel column has been widely used in the last 30 years, there are relatively few data available for the larger columns. However, the following general conclusions can be drawn from recent scale-up and performance data (92–93): the same stage efficiency can be maintained on scale-up, and total throughput can be increased by three and one half times at the expense of higher HETS. Scheibel columns up to 2.6 m in diameter are in service.

The rotating-disk contactor (RDC), developed in The Netherlands (132) in 1951, uses the shearing action of a rapidly rotating disk to interdisperse the phases (see Fig. 20). These contactors have been widely used throughout the world, particularly in the petrochemical industry for furfural and SO_2 extraction, propane deasphalting, sulfolane extraction for separation of aromatics from aliphatics, and carprolactam purification. Columns up to 4.3 m in diameter are in service.

An extensive study (133) provides an excellent theoretical framework for the scale-up of RDCs, and refs. 87 and 134 give design and performance information.

The Oldshue–Rushton column was developed (135) in the early 1950s and has been widely used in the chemical industry. It consists essentially of a number of compartments separated by horizontal stator-ring baffles, each fitted with vertical baffles and turbine-type impeller mounted in a central shaft (see Fig. 20). Columns up to 2.7 m in diameter have been reported in service (135–140).

The column is reported to have reliable predictability on scale-up (141); however, only few data are available (142).

The asymmetric rotating-disk (ARD) contactor (see Fig. 21) was developed in Czechoslovakia (91,95,134,143–145) and has been increasingly used in western Europe in recent years. Its design is aimed at retaining the efficient shearing action of the RDC by using the rotating disk to produce dispersion and to reduce backmixing by means of the coalescence–redispersion cycle produced in the separated transfer-settling zones.

The Kuhni contactor (146) has recently gained considerable commercial application in Europe (see Fig. 22). Its principal features are the use of a shrouded turbine impeller to promote radial discharge within the compartments, and a variable hole arrangement to allow flexibility of design for different process applications. Columns

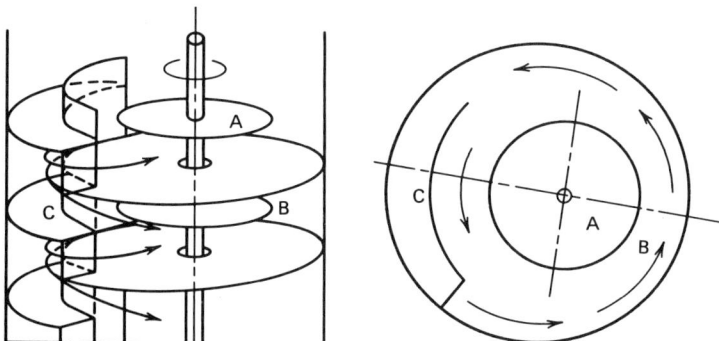

Figure 21. Asymmetric rotating-disk (ARD) extractor. A, rotating-disk rotor; B, mixing zone; C, settling zone. Courtesy of Luwa A.G. (143).

Figure 22. Kuhni extractor. Courtesy of Kuhni A.G.

up to 5.30 m in diameter have been constructed (15). Extensive studies on droplet size, holdup, and backmixing characteristics for a 15.2 cm diameter column have been reported (147–148).

The Graesser raining bucket contactor (149) is a horizontal design with the phases interdispersed by water wheel arrangements. The unit has the unusual feature of dispersing each phase into the other. It was developed for handling the difficult settling systems found in the coal tar industry, and it has proved attractive for other applications as well. It is also suitable for handling solid–liquid systems.

The Treybal contactor (150) was designed for semistagewise operation. The unit is claimed to have a low HETS and a high stage efficiency, 75–80% being maintained even with difficult extraction systems.

The Wirz column also involves a coalescence–redispersion mechanism. Experimental data have been reported (151) using columns of 10–51 cm in diameter.

Reciprocating-Plate Columns. These columns were developed over the last 40 years (152–159); general features and industrial applications are given in Table 2.

The open perforated reciprocating-plate columns (154) consist of a stack of perforated and baffle plates with a free area of about 58%. The central shaft supporting the plates is reciprocated by means of a drive mechanism located at the top of the column. Figure 23 shows the principal features of a pilot-scale model. A minimum HETS of 51 cm has been measured in a 0.91 m diameter column using a relatively difficult extraction system, o-xylene–acetic acid–water. Based on performance data on various column sizes (2.5, 7.6, 30.5, and 91 cm in dia), empirical equations for scale-up have been proposed (100,154,160–161). Hydrodynamics and axial mixing have been studied (75,162–163).

These columns have gained increasing industrial application in the pharmaceutical, petrochemical, and wastewater treatment industries (160–161), and columns up to 1 m in diameter are in service.

Another type (156) uses perforated plates with segmental passages for the continuous phase, with the dispersed phase passing through the relatively small perforations in the plates. Because of the segmental passages for the continuous phase, the throughput of the column is reported to be relatively higher than of pulsed or other types of extractors. Commercial applications of this type of column up to 51 cm in diameter have been reported in eastern Europe.

Mass transfer data are available for a column with reciprocated wire mesh packing

Figure 23. Five-cm diameter pilot-scale reciprocating-plate column. Courtesy of Julius Montz, G.m.b.H. (100).

on benzene–acetic acid–water, and on MIBK–acetic acid–water systems (158). Operating throughputs are significantly higher than those achieved with more conventional mechanically aided extractors, and high extraction rates are maintained. However, no commercial application has been reported.

The multistage vibrating-disk column consists of perforated plates which are reciprocated in the multistage compartments of the column (164). Axial mixing, mass transfer, and reaction data in gas–liquid contacting operations have been reported (165).

Centrifugal Extractors. In centrifugal extractors, residence time can be reduced and phase separation accelerated by application of centrifugal force. The units are compact, and relatively high throughput can be achieved in a small volume. They are used in chemically unstable (eg, extraction of antibiotics) or slow-settling systems. General features and fields of industrial application are given in Table 2.

Differential Centrifugal Contactors. The Podbielniak extractor was the first centrifugal unit introduced to industry in the early 1950s (166–167). Its design consists essentially of a perforated-plate extraction column that has been wrapped around a shaft which, in turn, is rotated to create a centrifugal force field that achieves a great reduction in the height and contacting time of a perforated-plate column (see Fig. 24). The extractors have been widely used in the pharmaceutical industry (eg, extraction of penicillin) and are increasingly used in other fields as well. Commercial units with throughput up to 98.4 m^3/h (26,000 gal/h) have been reported.

The Alfa-Laval extractor (168) can give up to 20 theoretical stages in one unit. The capacity of the standard unit depends on the systems being handled and ranges from 5.7 to 21 m^3/h (1500–5600 gal/h). Antibiotic extractions and petrochemical processing are typical applications.

The Quadronic extractor (169) is similar to the Podbielniak unit but has means for internally altering the perforated area of the cylindrical plates. It can achieve ten theoretical stages and a throughput of up to 83 m^3/h (22,000 gal/h) can be obtained in an extractor having a 1.5-m diameter rotor.

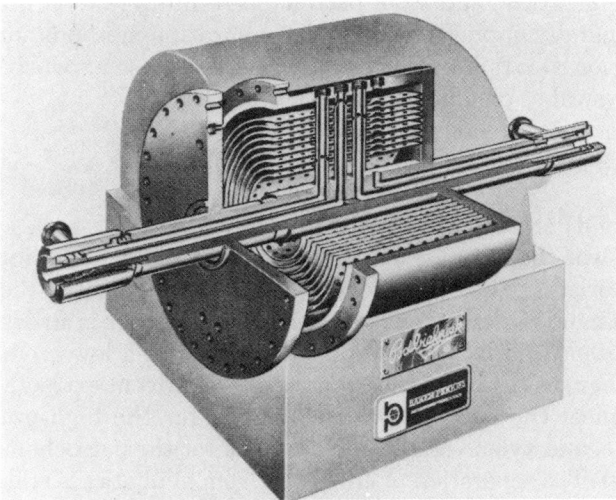

Figure 24. Podbielniak centrifugal extractor. Courtesy of Baker Perkins, Inc.

706 EXTRACTION, LIQUID–LIQUID

Discrete Stage Centrifugal Contactors. The Westfalia centrifugal extractor (170–171) is built on a vertical rotating principle, and is available with up to three contact stages. Its advantage is that the light phase does not have to be introduced under pressure. The capacity of the largest model has been reported as ranging from 7.6 m^3/h (2000 gal/h) with three stages to 49 m^3/h (13,000 gal/h) with a single stage.

The Robatel extractor (172) is a series of mixer–settlers stacked on their sides, with mixing in each stage being carried out by a stationary disk attached to the shaft while the mixing chamber revolves. The two phases pass into the separating chamber, which contains no coalescing plates, and then pass via a channel system into adjacent stages in countercurrent flow. The unit generally provides three to eight stages, and throughputs up to 6 m^3/h (1,600 gal/h) have been reported. These units have been extensively used in the nuclear industry.

Economics of Extraction Design

The economic considerations (93) for a solvent extraction include the capital costs, ie, equipment and inventory of material held within plant such as solvents, and the operating (prime) costs, ie, extractor operation, solvent recovery, and solvent losses.

Solvent recovery is usually the dominant factor because of large energy consumption (see Fig. 8). Many processes would be more economical with a larger number of extractor stages which would reduce the amount of solvent to be recovered and hence, the cost.

Industrial Applications

Industrial application of solvent extraction has increased rapidly over the last 25 years (173). The unique ability of solvent extraction to achieve separation according to chemical type rather than physical characteristics (eg, vapor pressure) has led to its application in processes that range from nuclear fuel enrichment and reprocessing to fertilizer manufacture, and from petroleum refining to food processing. Solvent extraction is generally applied where direct separation methods, such as distillation and crystallization (1), fail and where it provides a less expensive process than a competitive physical or chemical method.

ORGANIC PROCESSES

Petroleum and Petrochemical Processes. The first large-scale application of solvent extraction was the removal of aromatics from kerosene to improve its burning properties. The original Edeleanu process employed liquid sulfur dioxide as a solvent. Although the demand for lamp kerosene has dwindled, there is an ever-increasing need for jet fuel kerosene and lubricating oil, which require a low aromatic content (see Aviation and other gas turbine fuels). Furthermore, solvent extraction is used equally extensively to meet the ever-increasing demand for the high-purity aromatics of benzene, toluene, and xylene (BTX) as feedstock for the petrochemical industry (see BTX processing). The separation of aromatics from aliphatics is still one of the largest applications of solvent extraction today (see Petroleum, refinery processes).

Lubricating Oil Extraction. Aromatics are removed from lubricating oils to improve viscosity and chemical stability. The solvents used are furfural (174), phenol (175), and liquid sulfur dioxide. Owing to environmental concern over the toxicity and pollution potential of sulfur dioxide and phenol, most new plants are adopting furfural processes (see Furan derivatives).

Separation of Aromatic and Aliphatic Hydrocarbons. Aromatics extraction for aromatics production, jet fuel kerosene treatment, and enrichment of gasoline fractions is still one of the most important applications of solvent extraction. The features of various commercial processes are summarized in Table 3.

The Udex process (176) has been particularly popular in the United States until recently. The original process was used for making high-purity gasoline by removing aromatic hydrocarbons with aqueous diethylene glycol as solvent. However, in recent years the process has also been used for the manufacture of BTX. Aqueous tetraethylene glycol appears to be the best solvent (177).

The Sulfolane process (178–180), introduced by the Shell Company in 1962, has become one of the most important advances in large-scale aromatics production (see Fig. 25) and is used in many large units all over the world. Sulfolane, tetrahydrothiophene, 1,1-dioxide, $\overline{CH_2(CH_2)_3SO_2}$, is a strongly polar compound which is highly selective for aromatic hydrocarbons. It also has a much greater solvent capacity for hydrocarbons than glycol systems. Additional features are its high density, low heat capacity, and good stability. The sulfolane process uses the rotating-disk contactor (RDC), which is also a Shell development.

The Lurgi Arosolvan process (181) is becoming increasingly competitive with the well-established Udex and Sulfolane processes, and over a dozen commercial installations are in operation. Depending on the mixing component, two process arrangements are available. The first is based on the solvent N-methylpyrrolidinone (NMP) and water and the other on NMP and ethylene glycol. A polar mixing component such as water or glycol increases the selectivity of the solvent for aromatics. The Lurgi vertical multistage mixer–settler is used with towers up to 6 m in diameter and 35 m high.

A dimethyl sulfoxide (DMSO) process (182), which employs two separate extraction steps, has been developed by the Institut Français du Pétrole. The solvent is DMSO containing several percent water. Its selectivity and low viscosity allow the extraction to take place entirely at RT. In addition, DMSO is nontoxic and relatively inexpensive. The Kuhni column has been used for the process design and columns up to 2.7 m in diameter are in operation (84).

The Union Carbide process (183) is a recent development. A common feature of the processes mentioned above is the recovery of solvent from the extract by distillation. In the Union Carbide process, the solvent is recovered by means of a second extraction step. The process employs tetraethylene glycol as the solvent and features no water distillation, no extractor reflux distillation, and only a once-through distillation of aromatics. It is necessary, however, to distill the raffinate from the first extractor in order to recover the dissolved process solvents.

The Formex process (184), which employs N-formylmorpholine with a few percent water as solvent, has the flexibility to handle different feedstocks and product ranges. Either distillation or secondary extraction may be used, depending on the range of aromatics which are to be produced.

The Redex process (185) (recycle extract dual extraction) improves the octane

Table 3. Solvents for the Separation of Benzene–Toluene–Xylene (BTX) Mixtures from Light Feedstocks[a]

Solvent	Process	Solvent additives and reflux conditions	Operating temperature, °C	Contacting equipment	Comments
sulfolane	Shell process, Licensee: Universal Oil Products	sulfolane selectivity and capacity insensitive to water content caused by steam-stripping during solvent recovery; heavy paraffinic counter-solvent used	120	rotating-disk contactor, up to 4 m dia	the high selectivity and capacity of sulfolane leads to low solvent–feed ratios, and thus smaller equipment
glycol–water mixtures	Udex process, Universal Oil Products	solvent can be diethylene glycol and water, or a mixture of diethylene and dipropylene glycols and water, or tetraethylene glycol and water; light hydrocarbon reflux	150 for diethylene glycol and water	sieve-tray extractor	tetraethylene glycol and water mixtures are claimed to increase capacity by a factor of 4, and also require no antifoaming agent; the extract requires a two-step distillation to recover BTX
tetraethylene glycol	Union Carbide Corp.	the solvent is free of water; a dodecane reflux is used that is later recovered by distillation	100	reciprocating-plate extractor	the extract leaving the primary extractor is essentially free of feed aliphatics, and no further purification is necessary; two-stage extraction uses dodecane as a displacement solvent in the second stage
dimethyl sulfoxide (DMSO)	Institut Francais du Petrole	solvent contains up to 2% water to improve ambient selectivity; reflux consists of aromatics and paraffins	ambient	rotating-blade extractor, typically 10–12 stages	low corrosion allows use of carbon steel equipment; solvent has a low freezing point and is nontoxic; two-stage extraction has displacement solvent in the second stage
N-methyl-pyrrolidinone (NMP)	Arosolvan process, Lurgi	a polar mixing component, either water (12–20 wt %) or monoethylene glycol (40–50 wt %) must be added to the NMP to increase the selectivity and to decrease the boiling point of the solvent; the NMP–water processes use pentane countersolvent	NMP–glycol, 60; NMP–water, 35	vertical multistage mixer–settler, 24–30 stages, up to 8 m dia	the quantity of mixing component required depends on the aromatics content of the feed
N-formylmorpholine (FM)	Formex process, Snamprogetti	water is added to the FM to increase its selectivity, and also to avoid high reboiler temperatures during solvent recovery by distillation	40	perforated-tray extractor, FM density at 1.15 aids phase separation	low corrosion allows use of carbon steel equipment

[a] Courtesy of *Chemical Engineering* (New York) (16).

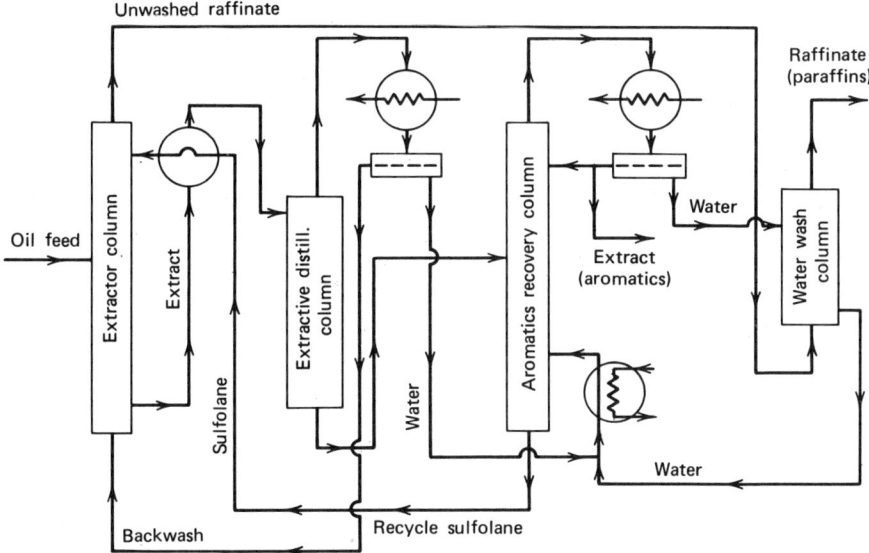

Figure 25. Aromatic separation, Sulfolane process (178–180).

number of diesel fuels by extracting an aromatic concentrate (see Gasoline and other motor fuels). The solvents include furfural–furfuryl alcohol–water mixtures, aqueous tetrahydrofurfuryl alcohol, and aqueous dimethylformamide.

Desulfurization. The sulfur compounds in petroleum oil include hydrogen sulfide, mercaptans, thiophenols, and thioethers in amounts ranging from a few tenths to several percent. Sulfur compounds have objectionable odors; they adversely affect the stability of the light distillate and have an unfavorable influence on antiknock and oxidation characteristics of gasoline. They are generally removed by multistage, countercurrent extraction with a relatively large volume of dilute alkali solution (see Petroleum, desulfurization).

Butadiene Separation. Solvent extraction is used in the separation of butadiene (qv) from other C_4 hydrocarbons in the manufacture of synthetic rubber. The butadiene is produced by catalytic dehydrogenation of butylene and the liquid butadiene is then extracted with an aqueous cuprammonium acetate solution with which the butadiene reacts to form a complex. Butadiene is then recovered by stripping from the extract. Extractive distillation is a competing process.

Caprolactam Extraction. Caprolactam is the monomer for nylon-6 (see Polyamides). Its purification is very important since fiber-grade caprolactam requires extremely high purity. Crude aqueous caprolactam (186) is purified by solvent extractions with aromatic hydrocarbons such as toluene as the solvent. The asymmetric rotating-disk contactors (130) as well as the Graesser contactors are used.

Extraction of C_8 Aromatics. The Japan Gas Chemical Company developed an extraction process (186–187) for the separation of p-xylene from its isomers using $HF-BF_3$ as an extraction solvent and isomerization catalyst. The highly reactive solvent imposes its own restrictions, but the new approach is claimed to be economically superior to conventional separation processes (see Xylenes and ethylbenzene).

Anhydrous Acetic Acid. In the manufacture of acetic acid by direct oxidation of a petroleum-based feedstock, solvent extraction has been used to separate acetic acid from the aqueous reaction liquor containing significant quantities of formic and propionic acids. Isoamyl acetate is used as a solvent to extract the aqueous feed to remove nearly all of the acetic acid that contained some water (188). The extract is then dehydrated by azeotropic distillation using isoamyl acetate as a water entrainer. It is claimed that the extraction step in this process affords substantial savings in plant capital investment and in operating cost (see Acetic acid).

Synthetic Fuel. Solvent extraction has many applications in synthetic fuel technology such as the extraction of Athabasca tar sands (qv) and Irish peat with n-pentane (189) and a process for treating coal under hydrogen with a solvent (190). In the latter, the coal is opened with a minimum hydrogen consumption so that solvent extracts valuable feedstock components before the coal is burned. Solvent extraction is used in coal liquification processes (175) and in synthesis-fuel refining (see Fuels, synthetic).

Pharmaceutical Processes. Solvent extraction has been used extensively in the pharmaceutical industry because many pharmaceutical intermediates and products are heat sensitive and cannot be processed by methods such as distillation, etc. However, few details of current commercial operations have been published.

Antibiotics. Solvent extraction is an important step in the recovery of many antibiotics such as penicillin, streptomycin, novobiocin, bacitracin, erythromycin, and the cephalosporins (see Antibiotics).

A good example is the manufacture of penicillin (191–192) by a batchwise fermentation. After filtration to remove mycelium, the aqueous fermentor broth is fed to a series of extractors for concentration and purification. Amyl acetate or n-butyl acetate is used as the extraction solvent. The penicillin is first extracted into the solvent from the broth at pH 2.0–2.5. The extract is then treated with a buffer solution at ca pH 6 to obtain a penicillin-rich aqueous solution. Finally, the pH is again lowered, and the penicillin is reextracted into the solvent to yield a pure concentrated solution. The extraction operation is complicated by the fact that the penicillin degrades rapidly as the pH is reduced, and it is necessary to perform the extraction at low pH as quickly as possible. Centrifugal extractors are generally used to ensure a short residence time.

Fractional extraction has been used in many processes for the purification and isolation of antibiotics from antibiotic complexes or isomers. An isopropanol–chloroform mixture and an aqueous Na_2HPO_4 buffer solution are the solvents (193), and a reciprocating-plate column is employed for the extraction process (92).

Vitamins. Solvent extraction is used extensively for preparation of heat-sensitive natural and synthetic vitamins. Natural vitamins A and D are extracted from fish liver oils and vitamin E from vegetable oils; liquid propane is the solvent. In the synthetic processes for vitamins A, B, C, and E, solvent extraction is generally used either in the separation steps for intermediates or in the final purification (see Vitamins).

Miscellaneous Processes. Solvent extraction is used for preparation of many products that are either isolated from naturally occurring materials or purified during synthesis. Among these are sulfa drugs, methaqualone, phenobarbital, antihistamines, cortisone, estrogens and hormones, and reserpine and alkaloids. Common solvents for these applications are chloroform, isoamyl alcohol, diethyl ether, and methylene chloride.

Distribution coefficient data for drug species are important for the design of solvent extraction procedure. They can be determined with a continuous solvent extraction system (AKUFVE) (194).

Food Processing. Solvent extraction is used in food processing (qv). Industrial refining of fats and oils with propane is known as the Solexol process (195). Vegetable oils (qv) are refined by extraction using furfural as solvent (196). Solvent extraction is used in many protein manufacturing processes, such as manufacture of fish protein by extracting the ground fish with isopropyl alcohol (197).

Miscellaneous Organic Processes. Solvent extraction has found application for many years in the coal-tar industry; eg, extraction of phenols from coal-tar distillates by washing with caustic soda solution. Dissociation extraction is applied to the separation of m- from p-cresol (198) and of 2,4- from 2,5-xylenol in tar-acid fraction (199). Although the coal-tar industry is declining, work is in progress in several locations to use solvent extraction for the direct manufacture of chemicals from coal (200) (see Coal, carbonization).

Treatment of Industrial Effluents. Solvent extraction appears to have great potential in the field of effluent treatment, both for the economic recovery of valuable materials and for their removal to comply with environmental requirements.

The Phenox process (201–202) removes phenol from the effluent of catalytic cracking in petroleum refinery. Extraction processes may show a small profit from the value of the extracted phenols from ammoniacal coke-oven liquor (203). Acetic acid is recovered by extraction from dilute waste streams (204). Oils are recovered by extraction from oily wastewater from petroleum and petrochemical operations. Solvent extraction is employed commercially for the recovery of valuable by-products from the effluents produced in the wool industry (205) and is applied in the same way in the pharmaceutical industry (206). A combination of solvent extraction and wet air oxidation is used for treatment of a toxic pharmaceutical effluent prior to discharge for biological treatment (206). Several solvent extraction schemes have been reported (207–208) for organic industrial waste-water treatment. A comprehensive review of wastewater treatment by solvent extraction has been reported (209) (see Wastes, industrial and municipal).

Extractive Reactions. Aromatic nitration in the manufacture of intermediates for dyestuffs, pharmaceutical products, plastics, and explosives is an example of an important industrial liquid–liquid organic reaction (210–211).

Difficult Separations. Difficult separations are frequently expensive because they involve high operating costs. Solvent extraction processes for separating substances having separation factors as low as 0.95 or 1.05 can be economically feasible by reducing the solvent recovery load (212); eg, the separations of m- and p-cresol, linoleic and abietic components of tall oil (qv), and the production of heavy water (see Deuterium).

Parametric pumping, using a temperature gradient established in a bed by a heat source, which gives an extremely high separation, has been applied to liquid–liquid extraction (213).

INORGANIC PROCESSES

The first major application of liquid–liquid extraction in inorganic chemical technology was the separation of uranium and plutonium from nuclear reactor fission products in the late 1940s (214). A few years later, it was successfully applied to the extraction of uranium from ore leach liquors as an alternative to liquid–solid ion exchange. Since then, many other hydrometallurgical uses of liquid–liquid extraction have been developed (2,17), as well as a number of applications involving nonmetallic inorganic products (see Extractive metallurgy).

Inorganic compounds are mostly insoluble or very sparingly soluble in organic liquids, and therefore, inorganic liquid–liquid extraction usually requires an extractant that reacts with the inorganic compound (or ion) to give an organic-soluble compound, as illustrated in equations 14–17.

The extractant is used in solution in an organic diluent (also referred to as the carrier solvent) which is usually regarded as being chemically inert and has physical properties that facilitate the handling of the liquid phases. A low viscosity, high flash point, and low entrainment and evaporation losses are desirable. Losses of diluent and extractant are an important part of the operating cost of any solvent extraction process, whereas the in-process inventory of diluent and extractant can contribute significantly to the capital cost. For reasons of cost, diluents tend to be cuts from the distillation of petroleum with flash points in the 40–80°C range, rather than pure compounds. The relative proportion of aliphatic and aromatic molecules in the diluent has been found to affect the rates and equilibria in the case of oxime extraction of copper (215) which suggests that the diluent cannot always be regarded as chemically inert.

An ideal diluent would be capable of retaining in solution both the loaded (ie, complexed) and unloaded extractant but many diluents which are otherwise satisfactory allow a third phase to form as extraction proceeds. This can be prevented by the presence of a few percent of a third component, the modifier, in addition to the extractant and the diluent. For example, tributyl phosphate (TBP) is often used as a modifier in association with di(2-ethylhexyl) phosphoric acid.

Nuclear-Fuel Reprocessing. Spent fuel from a nuclear reactor contains ^{238}U, ^{235}U, ^{239}Pu, ^{232}Th, and a large number of other radioactive isotopes (fission products) (see Nuclear reactors). Spent fuel is reprocessed in order to separate the isotopes from the fission products for reuse as fuel (214,216). In the Purex process, the spent fuel is dissolved in nitric acid and extracted with tributyl phosphate in 30 vol % solution in an aliphatic diluent such as a kerosene. Uranium and plutonium are initially present in the aqueous nitric acid phase in the hexavalent state as $UO_2(NO_3)_2$ and $PuO_2(NO_3)_2$ and are extracted according to equation 15. The fission products remain in the aqueous raffinate. The loaded organic phase containing decontaminated uranium and plutonium is treated with an aqueous strip solution containing ferrous sulfamate that reduces Pu to its trivalent state which is preferentially extracted by the aqueous phase. Finally, the uranium is stripped from the organic phase by a dilute acid solution. Tributyl phosphate is used under different conditions in the Thorex process which provides decontamination followed by a uranium–thorium separation. The extraction of uranium and plutonium by methyl isobutyl ketone (hexone) formed the basis of the Redox process which was the first solvent extraction process used for nuclear materials (see Nuclear reactors).

Extraction equipment for nuclear-fuel reprocessing should operate for long periods without maintenance and be capable of remote control, and the amount of retained radioactive material should not be great enough for critical conditions to be approached. The air-pulsed column has been popular as a differential contactor providing a very large number of stages, whereas compact box-type mixer settlers (see Fig. 15) have been developed to minimize shielded space requirements when relatively few equilibrium stages are needed.

Uranium Extraction. Solvent extraction is used as an alternative or a sequel to ion exchange (qv) in the selective removal of uranium from ore leach liquors (see Uranium and uranium compounds). These liquors differ from the feed to a nuclear fuel reprocessing plant in that they are relatively dilute in uranium (0.5–2 g/L), contain sulfuric rather than nitric acid, and are only slightly radioactive. The Amex process, developed in the United States in the 1950s, uses a tertiary amine extractant such as Alamine 336 in a kerosene diluent with a few percent of a higher alcohol as modifier (3). Extraction and stripping both require only a few stages and conventional mixer–settlers are used.

Several processes have been developed (217) in which solid–liquid ion-exchange treatment is initially applied to the leach liquors, with the ion-exchange eluate being concentrated and purified by liquid–liquid extraction.

Uranium is present in phosphate rock in small amounts (50–200 ppm) which are nevertheless interesting in view of the large deposits of phosphate rock throughout the world. Several processes have been proposed for recovery of uranium as a by-product of the fertilizer industry by solvent extraction of the black acid (30% H_3PO_4) obtainable by the wet process (218) (see Fertilizers).

Copper. The recovery of copper from ore leach liquors as a stage in the hydrometallurgical route to the pure metal is one of the largest applications of liquid–liquid extraction (2,219) (see Copper).

The most common type of copper leach liquor fed to a liquid–liquid extraction unit is produced by dilute sulfuric acid leaching and contains 1–5 g copper per liter. This concentration is too low for electrowinning to be economical and the purpose of liquid–liquid extraction is to raise the concentration of copper as well as to purify it.

A typical extraction circuit for acid leach liquor is shown in Figure 26 (215). The extraction is carried out at pH 2–4, whereas stripping of copper back to the aqueous phase occurs at a pH on the order of 0, in accordance with cation-exchange equation 17. Chelating extractants in aliphatic diluents are mainly used in this type of application.

Mixer–settlers are used because only a few equilibrium stages are needed for almost complete extraction of copper. The degree of agitation in the mixers must be

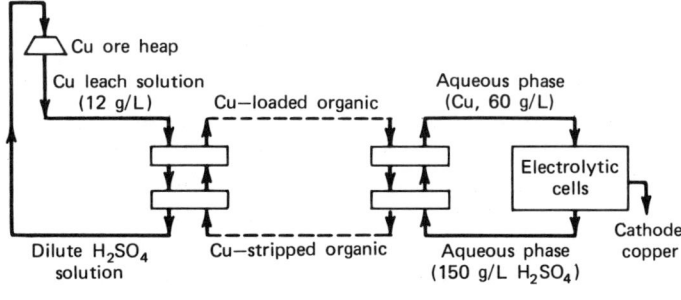

Figure 26. Liquid–liquid extraction of copper. Courtesy of *Engineering Mining Journal* (215).

carefully controlled (220) in order to minimize entrainment losses (carry-over) of the organic phase. Such losses are generally on the order of 0.05 m^3/t (13 gal/t) of cathode copper produced.

Nickel and Cobalt. Nickel and cobalt can both be separated from copper in sulfate leach liquors by extraction with acid extractants at the appropriate pH; however, the separation of nickel from cobalt is rather difficult. A separation process has been developed using di(2-ethylhexyl) phosphoric acid (D2EHPA) in two pulsed columns. The feed liquor is extracted in the first column at 60°C to give an organic phase containing Co and Ni. This phase is then contacted in the second column at room temperature with a cobalt-rich aqueous phase that removes the nickel from the organic phase, leaving only cobalt (35).

Nickel and cobalt are recovered from chloride leach liquors by liquid–liquid extraction using amine extractants; this forms the basis of the Falconbridge process which is operated in Norway (221).

Other Metal Extraction Applications. Numerous process applications of liquid–liquid extraction of other metals have been reported. Recent reviews (2,17,219) indicate that the situation is changing rapidly, and at the time of writing there are processes using liquid–liquid extraction for zinc, tungsten, vanadium, tantalum–niobium, platinum group metals, zirconium–hafnium, rare earths, yttrium, chromium, and gold.

Ocean-bed nodules containing manganese, copper, nickel, and other metals are becoming increasingly important. Several possible hydrometallurgical treatment processes for nodules have been examined (222) and it appears that liquid–liquid extraction is likely to play a very important part, irrespective of which process alternative is eventually adopted (see Ocean raw materials).

Phosphoric Acid. Wet-process phosphoric acid contains a number of impurities (eg, fluorine, metals, organic matter) that reduce its usefulness for many applications. A liquid–liquid extraction process for purification of the crude acid has been developed by IMI and successfully implemented in Mexico (223) and elsewhere. Several other processes employing liquid–liquid extraction are also in use (224).

In the IMI process (225), the solvent is an alcohol or an ether in which phosphoric acid is moderately soluble. The organic solvent is used in a closed extraction-stripping circuit with the stripping carried out at an elevated temperature. The resulting aqueous phosphoric acid solution can be concentrated by evaporation to 95% H_3PO_4 without scaling (see Phosphoric acid).

Salt–Acid Processes. Liquid–liquid extraction can be used to influence an aqueous-phase equilibrium of the type:

$$MX + HY \rightleftharpoons MY + HX \qquad (55)$$

where M is a metallic cation and X and Y are anions. If, eg, the acid HX can be removed continuously from the aqueous phase by liquid–liquid extraction, the reaction is encouraged to proceed to the right. This principle has been applied to phosphoric acid manufacture (226). Recently, a process has been developed by IMI for manufacturing sodium bicarbonate and hydrochloric acid from brine and carbon dioxide (227). A tertiary amine extractant is used to remove hydrochloric acid as it is formed, then the amine hydrochloride is stripped with a magnesium hydroxide solution. Finally, hydrochloric acid is recovered from the magnesium chloride by thermal decomposition, with the magnesium oxide being hydrolyzed and returned to the process.

Acid Recovery. Liquid–liquid extraction is of considerable use in effluent treatment (see above) and has been successfully applied in Sweden to the recovery of nitric and hydrofluoric acids from stainless steel pickling solutions (228). Extraction is carried out in a pulsed column using 75% TBP in kerosene.

Nomenclature

a	= specific interfacial area, m^{-1} or cm^{-1}
A	= quantity or flow of component A′, kg or kg/s
B	= quantity or flow of component B′, kg or kg/s
c	= concentration, kg/m^3 or g/cm^3
d_m	= Sauter mean drop diameter, m or cm
D	= molecular diffusivity, m^2/s or cm^2/s
E	= axial dispersion coefficient, m^2/s or cm^2/s
h	= holdup of dispersed phase
h_f	= holdup at flooding condition
HETS	= height of an equivalent theoretical stage, m or cm
HTU	= height of a transfer unit, m or cm
NTU	= number of transfer units
k	= film mass transfer coefficient, m/s or cm/s
K	= overall mass transfer coefficient, m/s or cm/s
K'	= dimensionless constant in equation 51
L	= ratio of dispersed to continuous phase superficial velocities
m	= distribution ratio based on mass fraction
m'	= distribution ratio based on mole fraction
N	= flux, kg/(m^2·s) or g/(cm^2·s) in equations 18–26
N	= number of theoretical stages elsewhere
N	= mass transfer rate per unit interfacial area
Pe	= Peclet number
u_t	= terminal drop velocity, m/s or cm/s
u	= velocity, m/s or cm/s
U_C	= superficial velocity, continuous phase, m/s or cm/s
U_D	= superficial velocity, dispersed phase, m/s or cm/s
U_S	= slip velocity, m/s or cm/s
ϵ	= voidage of packing
x	= mass fraction
x'	= mole fraction
X	= mass ratio C/A in A-rich phase
Y	= mass ratio C/B in B-rich phase
z	= distance along contactor, m or cm
Z	= total column height of contactor, m or cm
	= contactor length
β_{BC}	= selectivity of B relative to C, see equation 9
γ	= activity coefficient
ε	= extraction factor
μ	= viscosity, Pa·s (10 P)
ρ	= density, kg/m^3 or g/cm^3
σ	= interfacial tension, N/m or dyn/cm
ψ	= power dissipation per unit mass, W/kg
θ_f	= formation time, s

Subscripts

f	= feed
N	= stage N

1	= dispersed phase or stage 1
2	= continuous phase or stage 2
i	= interface
s	= solvent
o	= A-rich stream entering stage 1
Δ	= at Δ point
A	= component A
B	= component B
D	= component D
C	= component C
DB	= solute D in solvent B
DA	= solute D in solvent A
CA	= solute C in solvent A
CB	= solute C in solvent B

BIBLIOGRAPHY

"Liquid–Liquid Extraction" under "Extraction" in *ECT* 1st ed., Vol. 6, pp. 122–140, by E. G. Scheibel and A. J. Frey, Hoffmann-La Roche Inc.; "Extraction, Liquid–Liquid" in *ECT* 1st ed., Suppl. 1, pp. 330–365, by Marcel J. P. Bogart, the Lummus Company; "Liquid–Liquid Extraction" under "Extraction" in *ECT* 2nd ed., Vol. 8, pp. 719–761, by E. G. Scheibel, Cooper Union School of Engineering and Science.

1. R. E. Treybal, *Liquid Extraction,* 2nd ed., McGraw-Hill Book Co., New York, 1963.
2. D. S. Flett, *Chem. Ind. (London)* (17), 706 (1977).
3. W. J. Jamrack, *Rare Metal Extraction by Chemical Engineering Techniques,* Pergamon Press, Oxford, Eng., 1963.
4. H. A. C. Mackay and co-eds., *Solvent Extraction Chemistry of Metals,* Macmillan, London, Eng., 1965.
5. A. S. Kertes and Y. Marcus, eds., *Solvent Extraction Research,* Wiley-Interscience, New York, 1969.
6. S. Hartland, *Countercurrent Extraction,* Pergamon Press, Oxford, Eng., 1970.
7. Y. Marcus, ed., *Solvent Extraction Reviews,* Vol. 1, Marcel Dekker, Inc., New York, 1971.
8. C. Hanson, ed., *Recent Advances in Liquid–Liquid Extraction,* Pergamon Press, New York, 1971.
9. A. Marinsky and Y. Marcus, eds., *Ion Exchange and Solvent Extraction,* Vol. 3, Marcel Dekker, Inc., New York, 1973.
10. G. S. Laddha and T. E. Degaleesan, *Transport Phenomena in Liquid Extraction,* Tata-McGraw Hill, New Delhi, India, 1976.
11. G. M. Ritcey and A. W. Ashbrook, *Solvent Extraction: Principles and Application to Process Metallurgy,* Elsevier, Amsterdam, The Netherlands, 1978.
12. T. Sekine and Y. Hasegawa, *Solvent Extraction Chemistry: Fundamentals and Applications,* Marcel Dekker, New York, 1977.
13. R. G. Bautista, "Hydrometallurgy" in *Advances in Chemical Engineering,* Vol. 9, Academic Press, Inc., New York, 1974.
14. D. S. Flett and D. R. Spink, *Hydrometallurgy* **1,** 207 (1976).
15. P. J. Bailes, C. Hanson, and M. A. Hughes, *Chem. Eng. (N.Y.)* **83**(2), 86 (1976).
16. *Ibid.,* (10), 115 (1976).
17. *Ibid.,* (18), 86 (1976).
18. J. C. Godfrey, C. Hanson, and M. J. Slater, *Chem. Ind. (London)* (17), 713 (1977).
19. P. J. Bailes, *Chem. Ind. (London)* (17), 724 (1977).
20. H. A. C. McKay and R. K. Webster, *Chem. Ind. (London)* (17), 737 (1977).
21. C. Hanson, *Chem. Eng. (N.Y.)* **75**(18), 76 (1968).
22. A. Baniel, *Isr. J. Chem.* **14,** 252 (1975).
23. J. G. Gregory, B. Evans, and P. C. Weston, eds., *Proc. Intl. Solvent Extraction Conf., 1971,* Vols. 1–2, Society of Chemical Industry, London, Eng., 1971.
24. G. V. Jeffreys, ed., *Proc. Intl. Solvent Extraction Conf. 1974,* Vols. 1–3, Society of Chemical Industry, London, Eng., 1974.
25. B. H. Lucas, ed., *Proc. Intl. Solvent Extraction Conf. 1977,* Vols. 1–2, Canadian Institute of Mining and Metallurgy, Montreal, Can., 1979.

26. N. M. Rice, *Chem. Ind.* (London) (17), 718 (1977).
27. A. W. Francis in A. Siedell and W. F. Linke, eds., *Solubilities c   organic and Organic Compounds*, 3rd ed., Suppl., D. Van Nostrand Co., Princeton, N.J., 1952.
28. R. H. Perry, C. H. Chilton, and S. D. Kirkpatrick, *Chemical Engineers' Handbook*, 4th ed., McGraw-Hill Book Co., New York, 1963, pp. 14–50.
29. R. E. Treybal in R. H. Perry and C. H. Chilton, eds., *Chemical Engineers' Handbook*, 5th ed., McGraw-Hill Book Co., New York, 1973, pp. 15–71.
30. A. W. Francis, *Liquid–Liquid Equilibria*, Wiley-Interscience, New York, 1963.
31. J. Rydberg, H. Reinhardt, and J. O. Liljenzin, "Experience with the AKUFVE Solvent Extraction Equipment" in ref. 9, p. 111.
32. S. Glasstone, *Textbook of Physical Chemistry*, 2nd ed., Van Nostrand, New York, 1946, p. 737.
33. J. W. Codding, W. O. Haas, and F. K. Hermann, *Ind. Eng. Chem.* **50,** 145 (1958).
34. L. F. Cook and W. W. Szmokaluk in ref. 23, p. 451.
35. G. M. Ritcey, A. W. Ashbrook, and B. H. Lucas, *CIM Bulletin*, **68**(1), 111 (1975).
36. A. E. Karr and E. G. Scheibel, *Chem. Eng. Prog. Symp. Ser.* **50**(10), 73 (1954).
37. R. C. Reid and T. K. Sherwood, *The Properties of Gases and Liquids*, McGraw-Hill Book Co., New York, 1958.
38. K. P. Lindland and S. G. Terjesen, *Chem. Eng. Sci.* **5,** 1 (1956).
39. G. A. Yagodin and co-workers in ref. 25, paper 25e.
40. G. Thorsen and S. G. Terjesen, *Chem. Eng. Sci.* **17,** 137 (1962).
41. A. E. Handlos and T. Baron, *AIChE J.* **3,** 127 (1957).
42. R. Higbie, *Trans. Am. Inst. Chem. Eng.* **31,** 365 (1935).
43. R. M. Griffith, *Chem. Eng. Sci.* **12,** 198 (1960).
44. M. Linton and K. L. Sutherland, *Chem. Eng. Sci.* **12,** 214 (1960).
45. C. V. Sternling and L. E. Scriven, *AIChE J.* **5,** 514 (1959).
46. H. Sawistowski in ref. 8, p. 292.
47. C. Hanson in ref. 8, p. 429.
48. M. M. Sharma and A. K. Nanda, *Trans. Inst. Chem. Eng.* **46,** 44 (1968).
49. P. V. Danckwerts, *Gas–Liquid Reactions*, McGraw-Hill Book Co., New York, 1970.
50. G. Astarita, *Mass Transfer with Chemical Reaction*, Elsevier, Amsterdam, The Netherlands, 1967.
51. T. W. Evans, *Ind. Eng. Chem.* **26,** 860 (1934).
52. A. J. V. Underwood, *Ind. Chemist* **10,** 128 (1934).
53. T. G. Hunter and A. W. Nash, *J. Soc. Chem. Ind.* (*London*) **53,** 95T (1932).
54. S. Smith, *Br. Chem. Eng.* **12,** 447 (1967).
55. K. A. Varteressian and M. R. Fenske, *Ind. Eng. Chem.* **28,** 928 (1936).
56. A. Kremser, *Natl. Pet. News* **22**(21), 42 (1930).
57. M. Souders and G. G. Brown, *Ind. Eng. Chem.* **24,** 519 (1932).
58. J. O. Maloney and A. E. Schubert, *Trans. Am. Inst. Chem. Eng.* **36,** 741 (1940).
59. A. J. P. Martin and R. L. M. Synge, *Biochem. J.* **35,** 91 (1941).
60. E. G. Scheibel, *Chem. Eng. Prog.* **44,** 681, 777 (1948); *Pet. Refiner* **38**(9), 227 (1959).
61. D. N. Hanson, J. H. Duffin, and G. F. Somerville, *Computation of Multistage Separation Processes*, Reinhold, New York, 1962.
62. J. W. Tierney and co-workers in ref. 23, p. 1051.
63. L. Boyadzhiev and G. Angelov in ref. 24, p. 2651.
64. A. Klinkenberg, *Chem. Eng. Sci.* **1,** 86 (1951); *Ind. Eng. Chem.* **45,** 653 (1953).
65. A. Klinkenberg, H. A. Hauwerien, and G. H. Reman, *Chem. Eng. Sci.* **1,** 93 (1951).
66. T. Miyauchi and T. Vermeulen, *Ind. Eng. Chem. Fundam.* **2,** 113 (1963).
67. T. Misek and V. Rod in ref. 8, Chapt. 7.
68. J. Ingham in ref. 8, Chapt. 8.
69. A. M. Rosen and V. S. Krylov, *Chem. Eng. J.* **7,** 85 (1974).
70. A. M. Rosen and V. S. Krylov, *Chem. Eng. Sci.* **22,** 407 (1967).
71. C. B. Hayworth and R. E. Treybal, *Ind. Eng. Chem.* **42,** 1174 (1950).
72. R. M. Christiansen and A. N. Hixson, *Ind. Eng. Chem.* **49,** 1017 (1957).
73. J. O. Hinze, *AIChE J.* **1,** 289 (1955).
74. J. W. Van Heuven and W. J. Beek in ref. 23, p. 70.
75. M. H. I. Baird and S. J. Lane, *Chem. Eng. Sci.* **28,** 947 (1973).
76. T. Miyauchi and H. Oya, *AIChE J.* **11,** 395 (1965).

77. V. Khemangkorn, G. Muratet, and H. Angelino in ref. 25, paper 29B.
78. M. Seewald, *Ind. Eng. Chem.* **53,** 597 (1961).
79. L. Lapidus and J. C. Elgin, *AIChE J.* **3,** (63) (1957).
80. F. A. Zenz, *Pet. Refiner* **36**(8), 147 (1957).
81. R. Gayler, N. W. Roberts, and H. R. C. Pratt, *Trans. Inst. Chem. Eng.* **31,** 57 (1953).
82. J. D. Thornton, *Chem. Eng. Sci.* **5,** 201 (1956).
83. J. Landau and R. Houlihan, *Can. J. Chem. Eng.* **52,** 338 (1974).
84. D. H. Logsdail and L. Lowes in ref. 8, Chapt. 5.
85. J. H. Perry and C. H. Chilton, *Chemical Engineers' Handbook,* 5th ed., McGraw-Hill Book Co., New York, 1973, pp. 12–1.
86. R. B. Akell, *Chem. Eng. Prog.* **62**(9), 50 (1966).
87. G. H. Reman, *Chem. Eng. Prog.* **62**(9), 56 (1966).
88. C. Hanson, *Chem. Eng. (N.Y.)* **75**(18), 76; **75**(19), 135 (1968).
89. C. J. Mumford, *Br. Chem. Eng.* **13,** 981 (1968).
90. G. C. Warwick, *Chem. Ind.* (London), 403 (May 1973).
91. J. Oldshue in ref. 7.
92. T. C. Lo, "Recent Developments on Commercial Extractors" *paper presented at Engineering Foundation Conference on Mixing Research, Rindge, N.H.,* Aug. 1975, The Engineering Foundation, New York, 1975.
93. T. C. Lo in P. Schweitzer, ed., *Handbook of Separation Techniques for Chemical Engineers,* McGraw-Hill Book Co., New York, 1979, Section 1.10.
94. K. H. Reissinger and J. Schroter, *Chem. Eng.* **85**(25), 109 (1978).
95. J. Marek, technical report, Luwa A.G., Zurich, Switz., Mar. 1970.
96. S. Ochia, *Automatica* **13,** 435 (1977).
97. L. C. Craig, D. Craig, and E. G. Scheibel in A. Weissberger, ed., *Techniques of Organic Chemistry,* Vol. 3, Part 1, 2nd ed., Interscience, New York, 1956, pp. 149–393.
98. H. Reinhardt and J. Rydberg, *Chem. Ind.* (London) **11,** 488 (1970).
99. M. M. Anwar, C. Hanson, and M. W. T. Pratt, *Chem. Ind.* (London) **9,** 1090 (1969).
100. T. C. Lo and A. E. Karr, *paper presented at Engineering Foundation Conference on Mixing Research, Andover, N.H.,* Aug. 9–13, 1971; *Ind. Eng. Chem. Process Des. Dev.* **11**(4), 495 (1972).
101. L. Lowes and M. J. Larkin, *IChemE Symposium Series No. 26,* Institute of Chemical Engineers, London, Eng., 1967, p. 111.
102. B. F. Warner, *Proc. 3rd U.N. Conf. on Peaceful Uses of Atomic Energy, Geneva,* **10,** 224, 1964.
103. B. V. Coplan, J. K. Davidson, and E. L. Zebroski, *Chem. Eng. Prog.* **50**(8), 403 (1954).
104. J. Mizrahi, E. Barnea, and D. Meyer in ref. 24, Vol. 1, p. 141.
105. IMI staff in ref. 23, Vol. 2, pp. 1, 386.
106. D. W. Ager and E. R. Dement, *Proceedings of the International Symposium on Solvent Extraction in Metallurgical Processes,* p. 27, Technologisch Instituut K. VIV, Antwerp, Belgium, 1972, p. 27.
107. G. C. I. Warwick, J. B. Scuffham, and J. B. Lott in ref. 23, Vol. 2, p. 1373.
108. G. C. I. Warwick and J. B. Scuffham in ref. 106, p. 36.
109. I. D. Jackson and co-workers, *IChemE Symposium Series No. 42,* Paper No. 15, Institute of Chemical Engineers, London, Eng., 1975.
110. W. Mehner, G. Hoehfeld, and E. Mueller in ref. 23, Vol. 2, p. 1265.
111. Exhibition during International Solvent Extraction Conference ISEC 1974, Lyon, Fr., 1974.
112. *Kenics Static Mixers, Bulletin KTEK-5,* Kenics Corp., Danvers, Massachusetts, 1972.
113. *Chem. Eng. (N.Y.)* **80**(7), 111 (1973).
114. S. A. Miller and C. A. Mann, *Trans. Am. Inst. Chem. Eng.* **40,** 709 (1944).
115. A. W. Flynn and R. E. Treybal, *AIChE J.* **1,** 324 (1955).
116. A. D. Ryon, F. L. Daley, and R. S. Lowry, *Chem. Eng. Prog.* **55**(10), 71 (1959).
117. B. F. Warner, joint symposium, *The Scaling-up of Chemical Plant and Processes,* London, Eng., 1957, p. 44.
118. *The Bronswerk Technical Bulletin on Pulsed Packed Column,* Bronswerk, P. C. E. S., Amersfoort, The Netherlands.
119. N. U. Spaay, A. J. F. Simons, and G. P. ten Brink in ref. 23, Vol. 1, p. 281.
120. M. H. I. Baird and G. M. Ritcey in ref. 24, Vol. 2, p. 1571.
121. H. Rouyer and co-workers in ref. 24, Vol. 3, p. 2339.
122. *Liquid–Liquid Extraction in C.E.A. Establishments, Bulletine 25/74,* Commissariat a l'Energie Atomique, Genas, Fr., 1974.

123. G. Sege and F. W. Woodfield, *Chem. Eng. Prog.* **50**(8), 396 (1954).
124. J. D. Thornton, *Br. Chem. Eng.* **3**, 247 (1958).
125. J. D. Thornton, *Trans. Inst. Chem. Eng.* **35**, 316 (1957).
126. D. H. Logsdail and J. D. Thornton, *Trans. Inst. Chem. Eng.* **35**, 331 (1957).
127. M. E. Weech and B. E. Knight, *Ind. Eng. Chem., Process Des. Dev.* **6**, 480 (1967); **7**, 156 (1968); **2**, 1199 (1963).
128. U.S. Pat. 2,493,265 (Jan. 3, 1950), E. G. Scheibel (to Hoffmann-La Roche Inc.); *Chem. Eng. Prog.* **44**(9), 681 (1948).
129. E. G. Scheibel and A. E. Karr, *Ind. Eng. Chem.* **42**(6), 1048 (1950).
130. U.S. Pat. 2,856,362 (Sept. 2, 1958), E. G. Scheibel (to Hoffmann-La Roche Inc.); *AIChE J.* **2**, 74 (1956).
131. U.S. Pat. 3,389,970 (June 25, 1968), E. G. Scheibel.
132. G. H. Reman, *Proceedings of the 3rd World Petroleum Congress, The Hague, The Netherlands,* Sect. III, p. 121 (1951).
133. C. P. Strand, R. Olney, and G. H. Ackerman, *AIChE J.* **8**, 252 (1962).
134. T. Misek, *Rotating Disc Extractors and Their Calculation,* State Publishing House of Technical Literature, Prague, GDR, 1964.
135. J. Y. Oldshue and J. H. Rushton, *Chem. Eng. Prog.* **48**(6), 297 (1952).
136. J. Y. Oldshue, *Biotech. Bioeng.* **8**(1), 3 (1966).
137. R. Bibaud and R. Treybal, *AIChE J.* **12**, 472 (1966).
138. H. F. Haug, *AIChE J.* **17**, 585 (1971).
139. J. Ingham, *Trans. Inst. Chem. Eng.* **50**, 372 (1972).
140. T. Miyauchi, H. Mitsutake, and I. Harase, *AIChE J.* **12**, 508 (1966).
141. J. Y. Oldshue, private communication, Mixing Equipment Co., Inc., P.O. Box 1370, Rochester, N.Y., 1970.
142. J. Y. Oldshue, F. Hodgkinson, and J. C. Pharamond in ref. 24, Vol. 2, p. 1651.
143. T. Misek and J. Marek, *Br. Chem. Eng.* **15**, 202 (1970).
144. J. Marek and co-workers, *paper presented at the Society of Chemical Industry Symposium, Bradford, U.K., 1967.*
145. B. Seidlova and T. Misek in ref. 24, Vol. 3, p. 2365.
146. A. Fischer, *Verfahrenstechnik* **5**, 360 (1971).
147. J. Ingham and co-workers in ref. 24, Vol. 2, p. 1299.
148. D. Hody, *Chemische Rundschau* **28**, 9 (1975).
149. Brit. Pats. 860,880 (Feb. 15, 1961); 972,035 (Oct. 7, 1964); 1,037,573 (July 27, 1966), J. Coleby.
150. U.S. Pat. 3,325,255 (June 13, 1967), R. E. Treybal.
151. J. Leisibach, *Chem. Ing. Technol.* **37**, 205 (1965).
152. U.S. Pat. 2,011,186 (Aug. 13, 1935), W. J. D. Van Dijck.
153. A. E. Karr, *AIChE J.* **5**, 446 (1959).
154. A. E. Karr and T. C. Lo in ref. 23, Vol. 1, p. 299.
155. N. Issac and R. L. DeWitte, *AIChE J.* **4**, 498 (1958); *Dechema Monogr.* **32**, 218 (1959).
156. J. Prochazka and co-workers, *Br. Chem. Eng.* **16**, 42 (1971).
157. D. Elenkov and co-workers, *Khim. Inst. Sof.* **4**, 181 (1966).
158. R. Wellek and co-workers, *Ind. Eng. Chem. Process Des. Dev.* **8**, 515 (1969).
159. K. Tojo, T. Miyanami, and T. Yano, *J. Chem. Eng. Jpn.* **7**, 123 (1974).
160. A. E. Karr and T. C. Lo in ref. 25, paper 8a.
161. A. E. Karr and T. C. Lo, *Chem. Eng. Prog.* **72**(11), 68 (1976).
162. M. H. I. Baird, R. G. McGinnis, and G. C. Tan in ref. 23, Vol. 1, p. 251.
163. S. D. Kim and M. H. I. Baird, *Can. J. Chem. Eng.* **54**, 81 (1976).
164. K. Takeba, *Preprint of the 10th General Symposium of the Society of Chemical Engineers, Japan,* p. 124, 1971.
165. K. Miyanami, K. Tojo, and T. Yano, *J. Chem. Eng. Jpn.* **6**, 518 (1973).
166. W. J. Podbielniak, *Chem. Eng. Prog.* **49**(5), 252 (1953).
167. D. B. Todd and G. R. Davis in ref. 24, Vol. 3, p. 2379.
168. E. Broadwell, *paper presented at the Society of Chemical Industry Symposium, Bradford, U.K. 1967.*
169. C. M. Doyle and co-workers, *Chem. Eng. Prog.* **64**(12), 68 (1968).
170. H. Eisenlohr, *Dechema Monogr.* **19**, 222 (1951).
171. *Paper presented at the Society of Chemical Industry Symposium, Bradford, U.K., 1967.*

172. C. Bernard, P. Michel, and M. Tarnero in ref. 23, Vol. 2, p. 1282.
173. J. Coleby in ref. 8, Ch. 4.
174. *Hydrocarbon Process. Pet. Refiner,* 191 (Sept. 1972).
175. J. M. Fox, *Hydrocarbon Process. Pet. Refiner,* 2 (Sept. 1963).
176. D. Read, "Production of High-Purity Aromatics for Chemicals," *paper presented at American Petroleum Institute Meeting, San Francisco, Calif., May 1952.*
177. T. S. Hoover, *Hydrocarbon Process.* **12,** 69 (1969).
178. H. Voetter and W. C. G. Koster, *Proc. World Pet. Cong.* **III,** 131 (1963).
179. F. S. Beadmore and W. C. G. Kosters, *J. Inst. Pet.* **49,** 469 (1963).
180. W. C. G. Koster, *Inst. Chem. Eng. Symp. Liquid Extraction,* Newcastle-upon-Tyne, U.K., 1967.
181. E. Müller and G. Hoehfeld, *Proc. 7th World Pet. Cong.* **IV,** 13, 1967.
182. *Hydrocarbon Process. Pet. Refiner,* 185 (Sept. 1972).
183. G. S. Somekh in ref. 23, Vol. 1, p. 323.
184. E. Cinelli, S. Noe, and G. Paret, *Hydrocarbon Process. Pet. Refiner,* 141 (Apr. 1972).
185. A. L. Benham and co-workers, *Hydrocarbon Process. Pet. Refiner* **46**(9), 134 (1967).
186. J. Coleby in C. Hanson, ed., *Recent Advances in Liquid–Liquid Extraction,* Pergamon Press, Oxford, Eng., 1971.
187. G. R. Herrin and E. H. Martel, *Chem. Eng. (N.Y.)* included in *Trans. Inst. Chem. Eng. (London)* (253), 319 (1971).
188. E. Lloyd-Jones, *Chem. Ind. (London),* 1590 (1967).
189. F. Panzner, S. R. M. Ellis, and T. R. Bott in ref. 25, paper 15g.
190. G. H. Bryer, *Proceedings of the International Solvent Extraction Conference ISEC 1977,* paper no. 31a, The Canadian Institute of Mining and Metallurgy Ottawa, Can., 1977.
191. A. L. Edler, ed., *Chem. Eng. Prog. Symp. Ser.* **66,** (1970).
192. J. R. E. Hoover and C. H. Nash, "Antibiotics–β-Lactams" in M. Grayson and D. Eckroth, eds., *Encyclopedia of Chemical Technology,* 3rd ed., Vol. 2, John Wiley & Sons, Inc., New York, 1978, p. 893.
193. U.S. Pat. 3,572,750 (Sept. 8, 1970), A. E. Karr (to Hoffmann-La Roche Inc.); A. E. Karr and co-workers, Hoffmann-La Roche Inc., Nutley, N.J., unpublished reports, 1970.
194. S. S. Davis and co-workers, *Chem. Ind. (London),* 677 (Aug. 1976).
195. H. J. Passino, *Ind. Eng. Chem.* **41,** 280 (1949).
196. S. W. Gloyer, *Ind. Eng. Chem.* **40,** 228 (1948).
197. *Chem. Eng. Prog.* **67**(5), (1971).
198. M. W. T. Pratt and J. Spokes in ref. 25, paper 31C.
199. J. Coleby, *Symposium on Solvent Extraction,* Institute of Chemical Engineers, Newcastle-upon-Tyne, U.K., 1967.
200. L. Crainger and W. S. Wise, *Chem. Br.* **4,** 12 (1968).
201. W. L. Lewis and W. L. Hartin, *Hydrocarbon Process.* **46**(2), 131 (1967).
202. *Manual on Disposal of Refinery Wastes,* American Petroleum Institute, Washington, D.C., 1969, Chapt. 10.
203. W. E. Carbone and co-workers, *Blast Furnace Steel Plant,* May 1958.
204. *Chem. Eng. (N.Y.),* 58 (Mar. 15, 1976).
205. P. Ramsden, *paper presented at Institute of Chemical Engineering Research Meeting on Solvent Extraction, Bradford, U.K., 1965.*
206. T. C. Lo, *paper presented at Engineering Foundation Conference on Mixing Research, Rindge, N.H., Aug. 17–22, 1975,* The Engineering Foundation, New York, 1975.
207. K. W. Wong and J. M. Prausnitz, *AIChE J.* **20**(6), 1187 (1974).
208. C. C. Hewes, W. H. Smith, and R. R. Devison, *AIChE Symp. Ser.* **70**(144), 54 (1974).
209. P. R. Kiezyk and D. Mackay, *Can. J. Chem. Eng.* **49,** 747 (1971).
210. L. F. Albright, *Chem. Eng. (N.Y.)* **73**(9), 169 (1966).
211. C. Hanson in ref. 8, Chapt. 12.
212. E. G. Scheibel, *Chem. Eng. Prog.* **62**(9), 66 (1966).
213. P. Wankat, *Ind. Eng. Chem. Fund.* **12,** 372 (1973).
214. J. T. Long, *Engineering for Nuclear Fuel Reprocessing,* Gordon and Breach, Inc., New York, 1967.
215. K. J. Murray and C. J. Bouboulis, *Eng. Min. J.* **174**(7), 74 (1973).
216. A. Naylor and M. J. Larkin in ref. 23, p. 1356.
217. J. Dasher in ref. 24, p. 2817.

218. R. C. Ross, *Eng. Min. J.* **176**(12), 80 (1975).
219. D. S. Flett, *Solvent Extraction of Non-Ferrous Metals: A Review 1975–1976, Report No. LR265(ME)*, Warren Spring Laboratory, Department of Industry, U.K., 1977.
220. A. K. Biswas and W. G. Davenport, *International Series in Materials Science and Technology,* Vol. 20, Pergamon Press, Oxford, Eng., 1976.
221. P. G. Thornhill, E. Wigstol, and G. Van Veert, *J. Metals* **23**(7), 13 (1971).
222. J. C. Agarwal and co-workers, *J. Metals* **28**(4), 24 (1976).
223. I. Raz, *Chem. Eng. (N.Y.)* **81**(13), 52 (1974).
224. J. F. McCullough, *Chem. Eng. (N.Y.)* **83**(26), 101 (1976).
225. R. Blumberg, D. Gonen, and D. Meyer in ref. 8, Chapt. 3.
226. Isr. Pat. 21,071 (Oct. 1, 1965), (to Israel Mining Industries).
227. R. Blumberg, J. E. Gai, and K. Hajdu in ref. 24, p. 2787.
228. U. Kuylenstierna and H. Ottertun in ref. 24, p. 2803.

TEH C. LO
Hoffmann-La Roche Inc.

MALCOLM H. I. BAIRD
McMaster University

EXTRACTION, LIQUID–SOLID

The extraction from a solid material of a soluble component by means of a solvent, liquid–solid extraction or leaching, is an industrial operation which predates large-scale chemical technology, having been applied in earlier centuries to extract alkaline salts from wood ashes. In that application the operation was referred to as lixiviation, but the term leaching, according to *The Oxford English Dictionary,* was already used in 1796 to describe the process of percolating a liquid through solid material. The history of leaching goes back to Roman times (1).

The solid can be contacted with the solvent in a number of different ways but traditionally that part of the solvent retained by the solid is always referred to as underflow or holdup, whereas the solid-free, solute-laden solvent separated from the solid after extraction is called the overflow. Figure 1 represents a typical multistage countercurrent liquid–solid extraction process, in this case based on immersion and decantation. Fresh solid enters the first stage and fresh solvent enters the final stage; the latter is gradually enriched in solute until it leaves the extraction battery as overflow from the first stage.

This operation can be subdivided into two types: (*1*) extractions that occur because of the solubility of the solute in or its miscibility with the solvent, eg, oilseed extraction, and (*2*) extractions where the solvent must react with a constituent of the solid material in order to produce a compound soluble in the solvent, eg, the extraction of metals from metalliferous ores (2). The two categories are subject to different constraints, the first being rate-controlled by diffusional phenomena whereas the

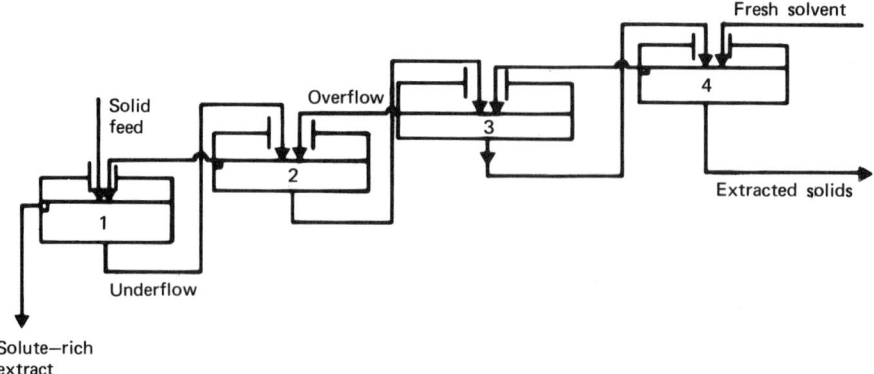

Figure 1. Liquid–solid extraction process based on immersion and decantation.

second is more frequently governed by the kinetics of the chemical reaction producing the solute. These differences are reflected in the different techniques used to analyze and carry out the operations; our principal concern here is with the extraction of solutes without the aid of chemical (or bacterial) reaction (see Applications).

Diffusion

Where extraction occurs by virtue of a solid diffusing from the interior of a solid particle, the diffusional mechanism involved provides important information for process design (3–4). The concept of simultaneous slow and rapid diffusion (5–6) is particularly relevant to anisotropic materials such as sugarcane as well as those materials where the rate of extraction depends considerably on the extent to which internal barriers to diffusion have been broken down by mechanical or thermal pretreatment.

Other investigations have proceeded from the assumption that a single diffusion coefficient (7) or a function thereof (8) can satisfactorily represent the extraction process. It is important to consider liquor convection in any study of diffusion rates (9). Permitting solvent to diffuse into the interior of the material being subjected to extraction enhances the extraction rate (10).

The nature of the diffusional process is illustrated by applying the integrated form of Fick's second law of diffusion:

$$\frac{\delta c}{\delta t} = D \frac{\delta^2 c}{\delta x^2} \tag{1}$$

Integration gives:

$$E = \frac{c_t}{c_o} = \frac{8}{\pi^2} \sum_{m=0}^{\infty} \frac{1}{(2m+1)^2} \exp\left[\frac{-D(2m+1)^2 \pi^2 t}{l^2}\right] \tag{2}$$

where the constraints are as defined in ref. 4. This equation reduces to

$$E = \frac{c_t}{c_o} = \frac{8}{\pi^2} \exp\left[\frac{-D\pi^2 t}{l^2}\right] \tag{3}$$

when the Fourier number Dt/l^2 is greater than 0.1. Equation 3 produces a linear semilogarithmic plot of the degree of extraction E against the Fourier number for $E < 0.7$, provided that diffusivity D and thickness l of the section being extracted are constant.

Although equations 2 and 3 apply to a uniform isotropic slab rather than to the irregularly shaped particles frequently encountered in extraction processes, the basic relationship between the degree of extraction and extraction time as expressed in equation 3 have been successfully used (4) in studies of the effect of time and particle thickness on the progress of extraction. The thickness l, in this case, is the particle thickness during extraction, not always identical with the thickness of the prepared material, eg, if the solvent has a swelling effect on the material.

With modern computational techniques this analysis (4) can be used to develop a process model (6) but the simplification of the model to one using a single diffusivity is still favored (11). The use of an extraction coefficient offers a more direct approach (12).

Rigorous analysis of an extraction process requires a somewhat more complex approach. In the first place, application of the equation for diffusion from a solid depends on whether the principal resistance to mass transfer is in the material being extracted or in the bound liquor. This is assessed by comparing the mass transfer coefficients for the solid and for the bed. Equation 1 must then be modified to take into account the fact that a finite amount of solvent is used (13) as well as the effect of the percolation velocity and of the particle size distribution on extraction. Application of diffusion theory to the analysis of a countercurrent extractor is discussed in ref. 14.

Process Design

Solvent. Solvent choice is in the first place determined by the chemical structure of the material to be extracted, the rule that "like dissolves like" providing useful guidance. Thus vegetable oils (qv), consisting essentially of triglycerides of fatty acids, are normally extracted with hexane whereas for the free fatty acids, which are more polar than the triglycerides, more polar alcoholic solvents are used. Where a choice of solvent (other than water) exists on the grounds of comparable solubility of the solute in question, the following criteria are likely to be considered:

Selectivity. Considering the purity of the recovered extract, with consequences for possible further upgrading steps, maximum selectivity consistent with solvent capacity is desirable. This concept has been extensively applied in liquid–liquid extraction (15) (see Extraction, liquid–liquid extraction).

Physical Properties. Low surface tension facilitates wetting of the solids in the first extraction stage, whereas low viscosity improves diffusion in the solvent phase. Low solvent density is desirable as it results in a reduced mass of solvent held up in the solid being extracted. A high boiling solvent with a high latent heat of evaporation requires recovery conditions that may be adverse for thermally sensitive extracts and will increase the cost of solvent recovery.

Thermal Stability. At the processing temperatures (including those in the solvent recovery section) the solvent should be completely stable in order to avoid expensive solvent purification and losses. Contamination of the solvent-free solute with any solvent breakdown products must be avoided, especially in food products.

Hazards. The solvent should be nontoxic and nonhazardous (ie, nonflammable and nonexplosive).

Cost. The cost of fresh solvent is reflected in the working capital and in the operating cost in the form of solvent make-up charges.

In the case of aqueous extraction, pH gives some control over selectivity but greater control over selectivity can be exercised with an organic solvent. This can be achieved by a choice between solvents having different selectivities for specific solutes or by the use of mixed solvents, eg, changing the water content of alcoholic solvents. When using a combination of solvents, the solution properties are determined by the various interactive forces involved in the mixing of the liquids. Regular solution theory may be used to estimate the resultant solvent properties (16).

The short-chain alcohols are an important class of solvents. They are generally used in admixture with water and a large range of compositions is therefore available. Another large class of extraction solvents are the halogenated hydrocarbons. Of the aliphatic hydrocarbons, only hexane is widely used. Benzene was formerly important but is no longer so. A solvent of increasing interest is liquid carbon dioxide which appears to be suitable for extracting flavor components from plants (17). It requires operation under pressure of up to 7 MPa (ca 70 atm) (critical pressure, p_c = 7.3 MPa) which is reflected in the cost of the extraction plant, though solvent recovery seems likely to be more economical than for conventional solvents.

In chemical leaching, the thermodynamics of the leaching reaction must be considered in terms of the redox potential (E)–pH diagram, the so-called Pourbaix diagram, which can be constructed from standard free energy data (18). This approach provides the basis for choosing the leaching conditions (acidic vs basic, oxidizing vs reducing) but deals only with the equilibrium situation. A kinetic study is required to provide information on the leaching rate.

Mass Balance. The mass balance is considered under the assumption that sufficient time has been allowed for the system to reach equilibrium.

In liquid–solid extraction the mass balance differs in one crucial respect from that normally used as the basis for calculating gas–liquid and liquid–liquid separations: the two phases are not immiscible or partially miscible, but are rather two streams, overflow and underflow, based on the same solvent. The freely flowing stream, the overflow, progressively depletes the bound liquor, or underflow, of solute in a multistage operation; however, the normal concept of equilibrium between two phases cannot be applied. Nevertheless, it is convenient to consider the system as comprising two liquid pseudo-phases and an inert solid phase consisting of the solute-free solid.

The holdup of bound liquor plays a vital role in the estimation of separation performance. In practice both static and dynamic holdup are measured in a process study, other parameters of importance being the relationship of holdup to drainage time and the percolation rate. The results of such studies permit conclusions to be drawn about the feasibility of extraction by percolation, the holdup of different bed heights of material prepared for extraction, and the relationship between solute content of the liquor and holdup.

If the structure of the solid is such that tne percolation rate is very low (in the case of oilseeds a minimum percolation rate of 3×10^{-3} m/s is normally required), extraction by immersion may be more effective. Percolation rate measurements and the means of using the information produced have been reported (19–20), the former also showing the effect of solute concentration on holdup which plays an important part in determining the solute concentration in the liquor leaving the extractor (19). In simplified cases holdup may be taken as constant throughout the extractor and the analogous McCabe-Thiele analysis for constant molar overflow (as in binary distillation) may then be applied (see Distillation).

Mathematical analysis designed to arrive at an expression for the number of stages required for countercurrent extraction is based on mass balances for the total solution and for the solute. The derivations of the equations used for graphical or analytical solution of problems in separations such as distillation, liquid–liquid and liquid–solid extraction are given in numerous texts (21–22).

For a set of N contactors (Fig. 2) in which the overflow is represented by V and the underflow (holdup) by L, it is possible to write overall and solute mass balances, provided that solute concentrations of the various streams are known (22). Since $y_i^* = x_i$ by virtue of the nature of the relationship between the two liquid phases, both operating and equilibrium lines are rectilinear when holdup is constant. A graphical construction therefore becomes superfluous and equation 4 is used to calculate the number of stages required for a given separation:

$$N = \frac{\log (y_b - y_b^*)(y_a - y_a^*)}{\log (y_b - y_a)(y_n^* - y_a^*)} \quad (4)$$

Since the McCabe-Smith analysis (22) excludes the first stage for the reason that the solid enters that stage solvent-free, the number of theoretical stages required becomes $N + 1$.

Alternatives to the McCabe-Smith analysis are to be found in the literature. An analytical method (23), more complex than that represented by equation 4, enables the first stage to be included in the calculation. A rapid method (24), based upon an overall mass balance and the use of finite difference equations, leads to a simple expression for the number of stages.

Variable Underflow. In practise most cases of interest exhibit variable underflow. This is normally greatest at that point in the process where the concentration of solute in the solvent is highest. Consequently, the operating line derived from the mass balance equations for the McCabe-Thiele diagram has a slope varying from stage to stage and equation 4 is then no longer applicable. Manipulation of the mass balance equations (25) permits use of a modified McCabe-Thiele diagram to arrive at the number of theoretical stages required for extraction. Stage-by-stage calculations also can be used for this purpose (21).

A coordinate system devised by Jaenecke has been extensively used in liquid–liquid extraction (15) and is also appropriate to the analysis of liquid–solid extraction involving variable holdup; in this case the diagram plotted is sometimes referred to as a modified Ponchon-Savarit diagram (22). In liquid–solid extraction the coordinates to be plotted are the ratio: solute-free solid to solute plus solvent as ordinate Y, and the ratio: solute to solute plus solvent as abscissa X. This type of plot gives the user a choice of scales for the axes to suit the range of data. The data of liquor holdup against extract concentration are then plotted on these coordinates to give the underflow line; the overflow, being solid-free, is given by the line $Y = 0$, ie, the abscissa (see Fig. 3). Since solute concentration in the underflow on a solid-free basis equals solute concentration in the overflow, the tie lines must be vertical. Other diagrams (26) can also be constructed for the solution of extraction problems and for deriving the number of equilibrium stages required for a given extraction.

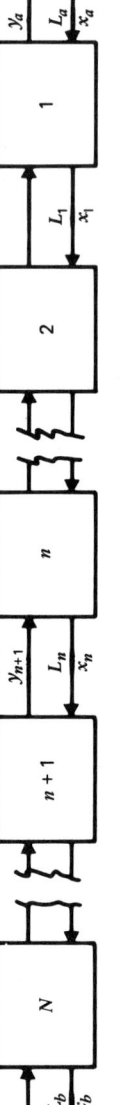

Figure 2. Countercurrent flow in a liquid–solid extraction battery. Courtesy McGraw-Hill Book Co. (22).

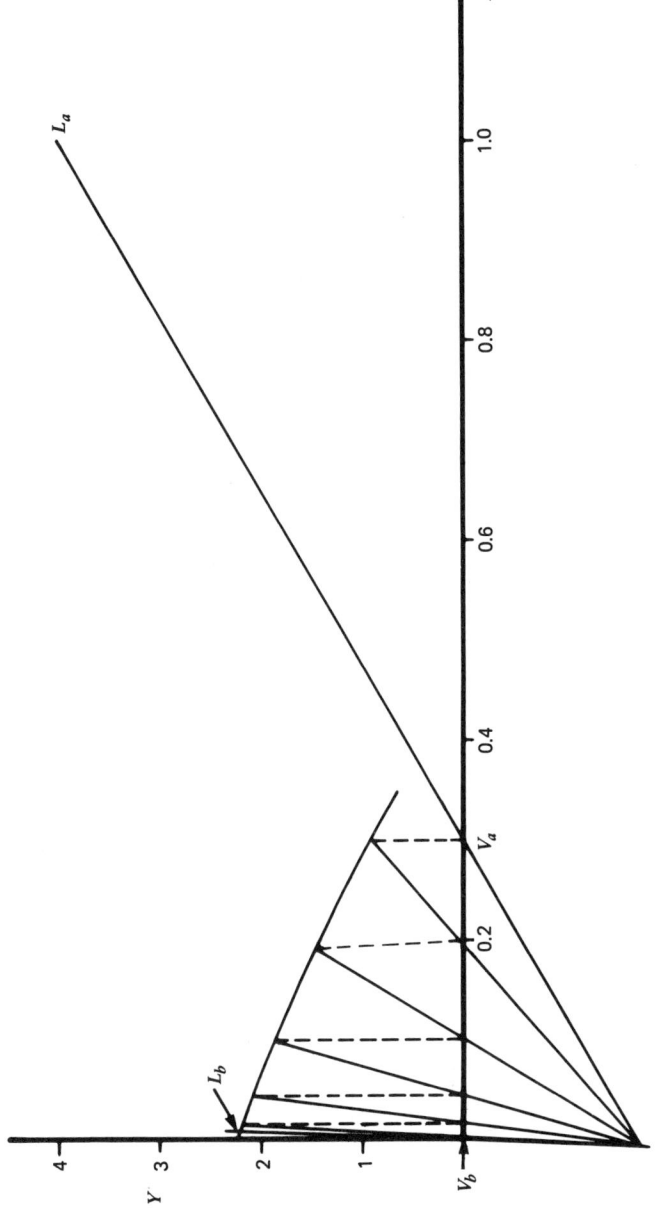

Figure 3. Liquid–solid extraction with variable underflow: use of Jaenecke coordinates. L_a, L_b = solids entering and exhausted solids leaving extractor, respectively. V_a, V_b = solvent phase leaving and entering extractor, respectively. See also Figure 2.

$$X = \frac{\text{solute}}{\text{solute + solvent}}, \quad Y = \frac{\text{solute-free solid}}{\text{solute + solvent}}.$$

The transfer unit concept can be applied to solid–liquid extraction (27). However, it is difficult to obtain values of the physical characteristics required to calculate the height of a transfer unit, and thus to use the concept for extractor design. This concept is best used for evaluating the performance of a given differential extractor, particularly when such an extractor is to be used for processing a range of raw materials.

Method of Extraction. Percolation and extraction rate data provide guidance on whether extraction should be by percolation or immersion and how much extraction time is needed to give an acceptable approach to equilibrium. For a percolation system to be viable both rates need to be high. Even in that case, however, an immersion process may be preferred. Where percolation rates are so high that extraction becomes inefficient, upward flow through a bed can be applied (28). A percolation process can be carried out either stagewise or in a differential contactor. For an immersion process a stagewise process is often more practical, particularly where a low extraction rate requires an extended residence time best effected in a series of contacting stages or where the nature of the solid material does not facilitate continuous transport. Combination of percolation with immersion has been recommended (29) for extraction of solids containing high levels of solute. When measuring the various rates, the conditions in the extractor to be used should be matched closely if reliable estimates of extraction and drainage times are to be obtained. The importance of allowing for carry-over of solvent from one stage to the next when using a moving extractor must be borne in mind (30).

With a differential (stepless) extractor, backmixing makes reliable scale-up prediction difficult. Residence time distribution measurements can be used in extractors of this type as a means of judging their effectiveness (31). This is at a maximum when mean linear velocity is high relative to the dispersion coefficient, both values being obtained from the residence time distribution curve.

If an extraction is accompanied by chemical reaction, a stagewise operation is preferred, eg, a semibatch operation suitably controlled to give continuity of output. The residence time is determined by the reaction rate.

Process Analysis. The calculations discussed above generally serve to guide the process design of an extraction plant. However, for an analysis of an existing plant (frequently associated with process optimization studies), a mathematical process model based on either empirical correlations (20) or on a complete process analysis (32) can be used.

The steps required to complete the operation of producing solvent-free solute are as important as the preparation. The recovery method depends largely on the solute and to some extent on the solvent. Solvent recovery is often energy intensive and a complete process energy analysis is recommended to reduce costs. With an organic solvent, recovery from the exhausted solids is also important and is generally more difficult than recovery from a liquid. The use of superheated solvent vapor to remove the solvent from the exhausted solids (33) is an interesting development stimulated by the need to economize on energy.

Extractors

The range of devices used in liquid–solid extraction reflects the diverse nature of the development from batch to truly continuous operation. Dump or heap leaching is at one end of the scale, whereas at the other end the equipment used to extract materials such as oilseeds and sugar beets shows the extent to which the problems of transporting solids have dominated equipment development.

The generally slow reaction rates encountered in chemical leaching of ores favored the use of batch extraction at an earlier stage and subsequently of stepwise extraction. At its simplest, this type of extraction takes the form of dump or heap leaching which is still used for the extraction of low-grade ores. The material to be processed is deposited in the milled state on a false bottom inside retaining walls and solvent is sprayed over it. The solute-rich liquor is collected after it has percolated through the dumped material. In the absence of any form of agitation liquid–solid contact efficiency is not high. Agitation, mechanical or by air, is provided in tanks known as pachucas or in autoclaves (2). Autoclaves are used where high temperature leaching significantly improves process performance; eg, for bauxite leaching temperatures of up to 180°C are used. Autoclave capacities may be as high as 300 m^3 (80,000 gal).

Tank and autoclave leaching can both be carried out in a stepwise continuous manner. The Dorr leaching system is used for large-scale operations in a series of circular flat-bottomed tanks. Autoclaves used in a multistage operation are generally mounted at a slight angle to the horizontal in order to facilitate drainage.

The extractor contributes substantially to the capital and operating cost of the whole plant. In view of the fact that these costs are influenced strongly by plant capacity and solvent requirements, efforts are frequently made to reduce the extraction load in order to increase extractor capacity and also reduce specific solvent requirements. In the main these efforts have concentrated on removing a part of the solute by prepressing prior to extraction, a technique that is feasible in the processing of material of plant origin where the solute may be contained in cells that can be ruptured by heat or pressure, or both.

The pot extractor is a batch extraction plant that offers the small-scale processor the advantage of carrying out extraction and solvent recovery from the exhausted solids in one vessel. These extractors are normally provided with mechanical agitation; however, in the case of coarser solids from which the solute is extracted relatively easily, the agitation may only be required during solvent recovery. Pot extractors range from 2 to 10 m^3 (500–2500 gal), beyond which the battery system discussed below is more attractive.

The diffuser battery, which had its origins in beet sugar extraction technology, is a semibatch extractor operating on a cyclical basis (see Sugar). The individual units in the battery (Fig. 4) (34) are charged sequentially with the solids to be extracted and the extracting liquor flow is designed to provide apparent countercurrent flow. The number of diffusers in the battery depends on extraction conditions and equilibria, an additional diffuser above the estimated number being required to permit cyclical operation. When the battery consists of more than four units, a close approximation to countercurrent flow is achieved. The size of individual units depends on factors related to the mass balance, the contact time required, and the hydrodynamic behavior of the bed of solids. Since each unit in the battery changes its position in the extraction sequence, at the end of each cycle the battery piping must be designed to facilitate this re-routing. Consideration must then be given to optimal vessel lay-out if process labor is to be used efficiently.

Diffuser batteries have in many cases been superseded by fully continuous devices but are still applied where a temperature gradient across the extraction operation as a whole, or extraction under pressure, is required.

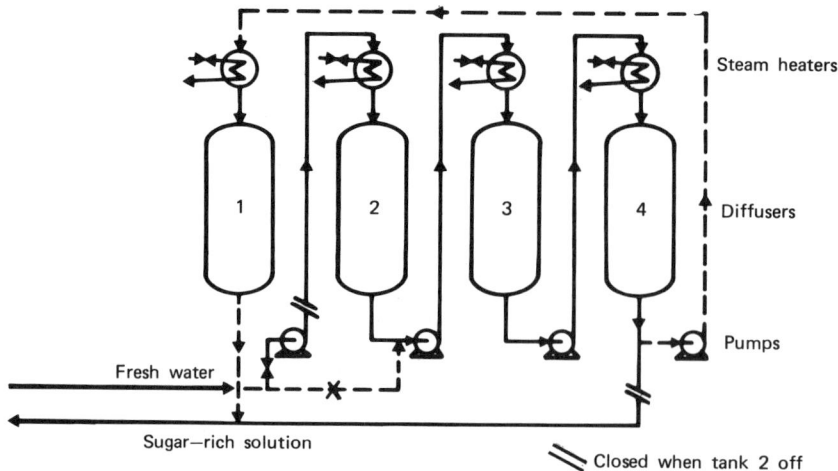

Figure 4. Shanks system for leaching of sugar from sugar beets. Courtesy of McGraw-Hill Book Co. (34).

Continuous Extractors. The development of fully continuous extractors was the logical sequel to the widespread use of the diffuser battery, once the problem of transport of the solid in a form suitable for extraction had been solved. Although continuous extraction may be operated cocurrent, crosscurrent, or countercurrent, considerations of extraction depth and of solvent consumption generally preclude the first two modes. The countercurrent extraction devices considered here may operate on either the percolation or the immersion principle, with the percolation rate of the comminuted solid playing an important part in determining the choice.

Percolation. Percolation extraction evolved in the form of basket extraction. This led to the continuous rotary extractor and various forms of the band extractor, with discrete stepwise liquid–solid contact forming the underlying principle of design. In the *Rotocel extractor* (Fig. 5), the material to be extracted is fed continuously as a slurry in the extraction solvent or as a dry feed to sector-shaped cells arranged around a horizontal rotor (35). The cells have a perforated base to permit drainage of the solvent into stage basins from which it is pumped to the next cell on the countercurrent principle. In the last cell, where fresh solvent is supplied, an extended drainage period is provided (by allowing a proportionately larger arc of the rotary motion for this cell) and thereafter the extracted solids are dumped. The miscella, in addition to being filtered by the bed of material being extracted, is filtered over a tent screen before complete solvent removal. A schematic view of the Rotocel flow arrangements is given in Figure 6. Rotary extractors similar in principle to the Rotocel are also offered by other equipment manufacturers and it has been claimed (36) that filtration of the miscella by a bed of flakes (in the case of oilseed extraction) results in miscella containing less than 5 ppm suspended solids. This method of clarification is widely applied. A simplified version of such an extractor for use in smaller extraction plants (35) has recently become available.

The endless-belt extractor is in principle of operation closely related to the Rotocel. Extraction time and percolation rate, established experimentally for the material as comminuted, determine the belt speed and the amount of drainage area required.

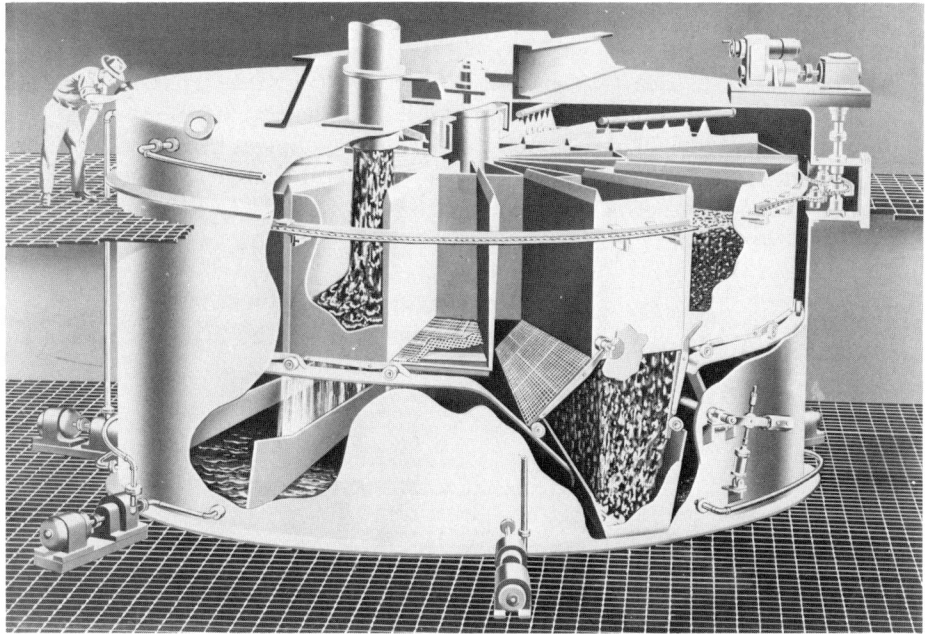

Figure 5. Rotocel percolation extractor. Courtesy of Dravo Corporation.

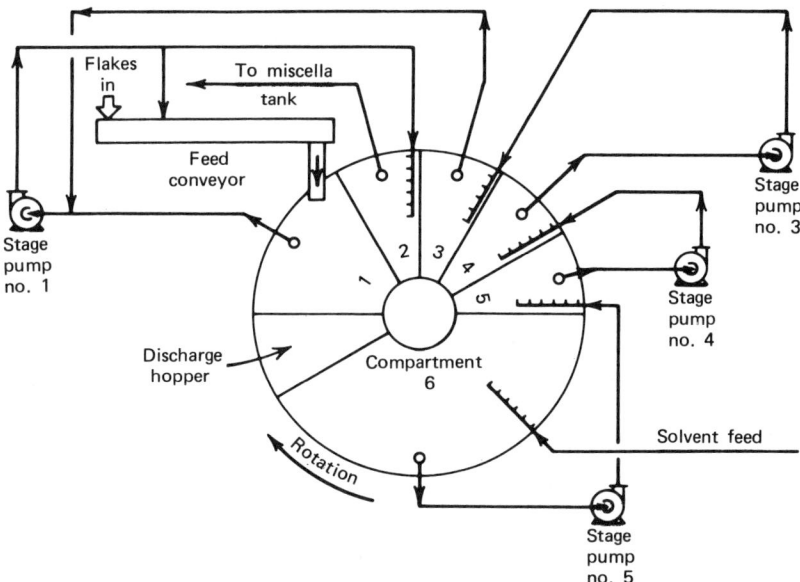

Figure 6. Downflow Rotocel flow diagram. Courtesy of Dravo Corporation.

Since bed height is virtually fixed by the mechanical design of the extractor, these parameters control the plant capacity. A low percolation rate could make the required drainage area prohibitively large. Solvent, which can be fed by spraying or simply from overflow weirs, may be used in simple countercurrent manner or, where the percolation rate is high, may also be recycled internally in order to improve the approach to equilibrium.

The Lurgi frame-belt extractor (Fig. 7) has a two-tier system in which the solid material travels the length of the extractor while being extracted in an upper series of compartments (frame buckets) having the perforated endless belt serving as a false bottom. The bed is then partially drained of solvent and thereafter discharged into a lower series of compartments where extraction continues with increasingly leaner solvent until a final drainage zone is reached before discharge of the exhausted solids.

The De Smet belt extractor (Fig. 8) uses a single endless belt to hold the material being extracted. The risk of solvent migration is minimized by using rakes that pen-

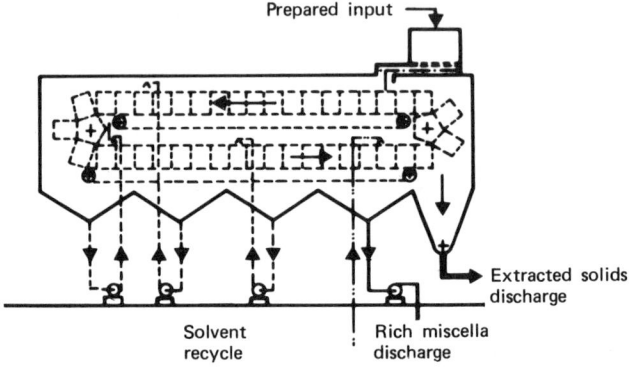

Figure 7. Lurgi frame belt extractor. Courtesy of Lurgi Umwelt und Chemotechnik G.m.b.H.

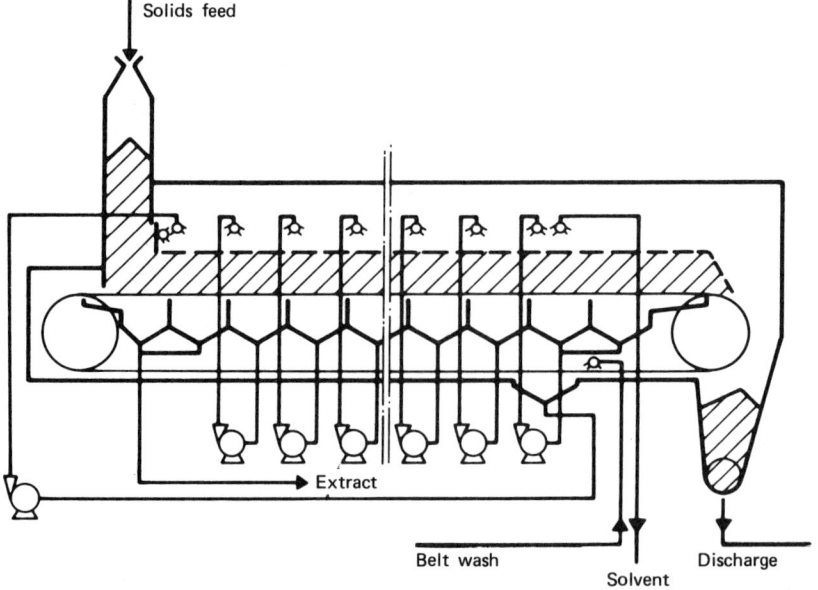

Figure 8. De Smet continuous-belt extractor. Courtesy of Extraction de Smet, S.A.

etrate the upper 150 mm of the bed (overall bed height 1.3–1.8 m) to form ridges of solid material at intervals. The rakes also serve the purpose of breaking up the upper layer of the bed in order to maintain steady percolation conditions. The belt moves discontinuously, thus providing a clearly defined extraction period followed by a drainage period which continues until the following spray is reached.

Other extractors operating on the percolation principle are discussed in ref. 37. Percolation extraction is used extensively in oilseed extraction and in the extraction of alkaloids.

Immersion. The main advantage of immersion extraction is the ability to handle finely ground material. Since only the drainage rate must be considered, the operation is simpler. Immersion extraction may also be used where the percolation rate of the material to be extracted is too great for effective diffusion within the bed. Its most extensive application is to be found in the sugar industry but it is also used in the extraction of oilseeds having a high oil content, in the extraction of small quantities (eg, pigments and pharmaceuticals from plant materials), and sometimes in combination with percolation.

The stepwise extraction by immersion can be carried out by a number of techniques analogous to the mixer–settler widely used in liquid–liquid extraction. Extraction followed by decantation is effective where the holdup of liquid by the material being extracted is relatively low after settling; however, where this is not the case the mixing stage can be followed by centrifugal separation (qv). More fully continuous methods of immersion extraction are used in the sugar industry.

Continuous immersion extractors were at first constructed in tower form and such designs have maintained an important place in the sugar industry. As indicated earlier, the key to this development was the continuous transport of solids which was achieved in a number of ways.

The BMA diffusion tower (3) has a central shaft fitted with a series of inclined plates that serve to direct movement of the solid material. The tower shell is also fitted with a series of staggered guide plates that serve the same purpose. Another tower extractor, in this case designed and built by Wolf, also employs the principle of wings attached to the central shaft to transport the solid material to be extracted up the tower. In both cases sugar beet cossettes are fed to the base of the tower. Heights of 10–15 m are commonly employed, and different capacities are achieved by variations in tower diameter. In either form of construction, power consumption for a tower of 5.5 m diameter (capacity: 3000 metric tons of beets per day) is of the order of 40 kW (3).

The De Danske Sukkerfabriker (DDS diffuser extractor) (Fig. 9) may be regarded as a tower extractor having its axis turned about 80°. The extractor is normally installed at a slope of one-seventh and a double screw in the housing is used to transport solids. The operating temperature is reached by employing jacket heating, thus avoiding the requirement for preheating. The dimensions of the DDS diffuser and its power consumption are broadly speaking similar to those of the tower extractors discussed above.

The use of these immersion extractors is contingent upon the ability to transport solids without excessive backmixing. They need less space than percolation extractors (3) and less power for the band drive and liquor circulation.

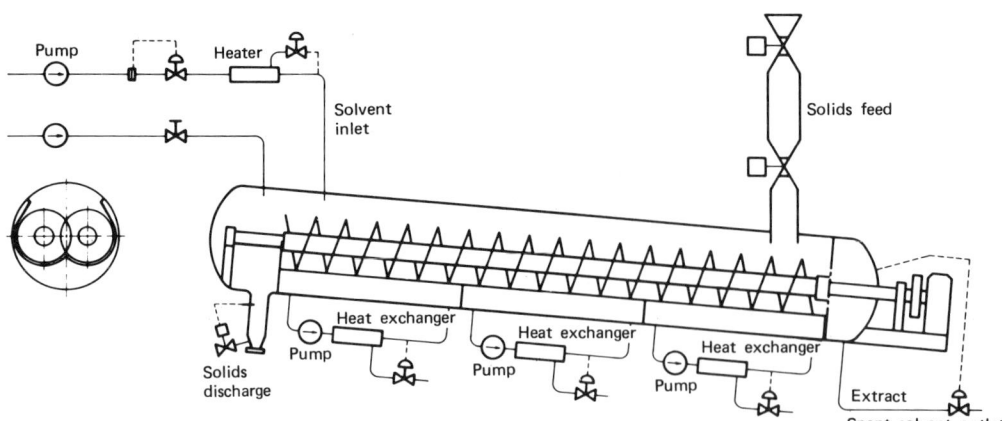

Figure 9. De Danske Sukkerfabriker (DDS) diffuser. Courtesy of A/S Niro Atomizer.

Applications

The extraction of sugar from sugar beets (3) and of vegetable oils from oilseeds are by far the largest applications in terms of scale of liquid–solid extraction. In recent years, the use of liquid–solid extraction to recover sugar from sugarcane has replaced crushing. A major industrial application in the last thirty years has been the extraction of coffee beans to satisfy the demand for soluble coffee (qv). Furthermore, on a smaller scale, flavors and essences are extracted from natural sources, adsorbed vegetable oil is recovered from spent bleaching earth by means of solvents such as hexane, and turpentine and wood resins are recovered from wood. Alcoholic solvents are used to remove residual oil from fish meal or fish waste as a step in the production of fish protein concentrate. Solvents of similar polarity remove soluble saccharides and related compounds from deoiled oilseed meals where such meals are to be upgraded to protein concentrates, an application of increasing importance (see Soybeans). Alcoholic solvents are used for the extraction of oleoresins from various spice sources, though it is claimed (38) that ethylene dichloride is a more suitable solvent in some cases because of greater specificity.

The extraction of sugar from sugar beet first assumed industrial importance when diffusion batteries were successfully introduced in the 1860s. The preparation of the beet for extraction attracted much attention (3) because of the need to improve sugar recovery and reduce contact time, and further development work on the process led to the evolution of a number of systems operating in continuous manner and giving a rich liquor containing 12–15% sugar. A variety of contactors has been developed to serve the industry. This technology is equally applicable to the extraction of fruit juices (qv) where the fruit can be sliced (27). For the extraction of sugarcane raw material preparation is important because of the anisotropic nature of the sugarcane (see Sugar).

Only with the change to hexane as solvent has extraction become the predominant process for edible oil production. The principal incentives to abandon the screw-press process were higher oil recovery, particularly in the case of cottonseed and soybeans, and the lower labor requirement of the solvent extraction process. The process does, however, call for efficient solvent removal from oil and deoiled meal on grounds of safety and cost (39). A more recent development has been the removal of carbohydrates from deoiled soybean meal, generally by further extraction with a polar solvent.

Solvent extraction of materials of plant origin is frequently preceded by a thermal treatment which serves to rupture cell walls and thus expedites extraction. Such treatment may affect the products adversely and operating conditions are therefore subject to appropriate constraints. This applies equally to the solvent removal stage.

The extraction of the constituents of roasted ground coffee beans is an example of a process involving extraction of soluble compounds as well as those solubilized by chemical reaction, in this case hydrolysis above 100°C. Temperature profile and residence time then play an important role in determining product yield and the flexibility of the diffuser battery makes it an appropriate extractor for this type of process. Fineness of grind of the roasted coffee beans not only affects extraction efficiency but also alters the nature of the solubles and colloids entering the extracting liquor (40).

The extraction of the soluble components of the tea leaf at first sight appears to be closely analogous to the extraction of coffee beans but in tea extraction a significant part of the fraction dissolved is found on the surface of the leaf. Tea extraction is, therefore, predominantly a washing process and diffusion plays a correspondingly smaller role. The extraction of hops, using methylene dichloride, methanol, or hexane is carried out widely and extensive development work has been done on the use of liquid (17) as well as supercritical (41) carbon dioxide to replace the conventional solvents.

The removal of carbohydrates from extracted soybean meal, generally by further extraction with a polar solvent, is another recent development.

The solvent extraction of bitumen from North American tar sands (qv) (42) may develop into a major application of liquid–solid extraction. The raw material consists of lumps or aggregates of sand suspended in bitumen, the content of which in the case of Athabascan material is on average 12%. Recovery of the bitumen by liquid–solid extraction is, therefore, primarily a washing operation and the high viscosity of the bitumen can be expected to put special demands on effective liquid agitation in an extraction device if concentration gradients in the solvent phase are not to be allowed to become rate controlling. In view of the high solids content of the oil sands and the nature of these solids, ecologically satisfactory disposal of tailings will be an important aspect of process development.

Supercritical gas extraction is an operation falling very clearly into the class of purely diffusional operations. The solvent power of supercritical carbon dioxide, as of several other gases above their critical point, has been studied for a number of systems and the extraction of hops, and of caffeine from coffee beans, is now being developed on industrial scale. Solute is recovered from the extract phase by temperature or pressure cycling. A comprehensive review (43) has recently been published.

Chemical Leaching. Chemical leaching, in which the transfer of a valuable component from its solid substrate to the extracting solvent takes place as a result of a chemical reaction between solvent and solute, is applied widely (44) but in volume and growth the application in hydrometallurgy is the most significant (45–46). A wide range of metals occurring naturally in the native, oxide, or sulfide state can be recovered by chemical leaching, which can be applied to the ore bodies as mined and crushed or to concentrates (see Extractive metallurgy). This process avoids the air pollution problems presented by the roasting of ore bodies (47).

Chemical leaching differs fundamentally from liquid–solid extraction by virtue of the importance of the thermodynamics and kinetics of the chemical reactions required to effect solution (see Separation systems synthesis). In chemical leaching the reaction rate may change as leaching proceeds (47); such changes obviously affect both equilibrium and rate.

Bacterial leaching is a special form of chemical leaching, the bacteria primarily serving to assist in oxidation of primary leaching products (45–46). Most widely studied of the bacteria capable of enhancing leaching in this manner are *Thiobacillus ferrooxidans* and *T. thiooxidans*, and these have been used commercially in recovering copper and uranium from low-grade ores or in scavenging operations.

Although bacterial leaching can thus be used to recover valuable metals from waste otherwise uneconomic to process, it can also cause environmental problems, particularly in dump leaching, where discoloration owing to iron precipitation and lack of growth of vegetation are objectionable features. Underground, *in-situ* leaching can cause water pollution.

Safety Factors and Environmental Considerations

Factors that constitute a hazard in extraction plants are solvent flammability, solvent content of the work atmosphere and of the products in the case of edible material, and dust content of the work atmosphere. The latter is both a health and a safety hazard (48–51). General safety and environmental standards must therefore be applied (52) (see Plant safety; Solvents, industrial) and accumulation of static electricity must be particularly guarded against (53).

Recommendations published annually by ACGIH should be consulted for currently acceptable limits of the threshold limit values (TLV) for solvents and dusts, although outside the United States the user may find that national regulations are not identical with ACGIH recommendations.

The permissible levels of solvent residues and emissions have repeatedly been lowered in recent years. The toxicology of benzene has been reviewed in depth (54–55) and an emergency exposure limit of 1 ppm has been ordered in the United States (56). A rigorous specification has been prepared for industrial hexane, widely used in oilseed extraction, by FAO–WHO (57) and, in the light of evidence of a relationship between atmospheric contamination by hexane and the incidence of polyneuropathy (58), the TLV–TWA (time-weighted average) recommended by ACGIH for hexane emissions has been reduced from 500 to 100 ppm (59–60).

Halogenated solvents have also come under pressure from regulatory authorities. The use of trichlorethylene for the extraction of edible products has recently (61) been severely restricted in the United States because of the suspected risk of cancer, although this view is controversial (62) (see Chlorocarbons; Industrial hygiene and toxicology). TLV–TWA recommendations for various halogenated solvents are also being reviewed (60). In the case of alcoholic solvents the standards set are less stringent. A FAO–WHO report (63) discusses interactions of solvents with food and also deals with impurities in solvents. In the United Kingdom a new classification of solvents commonly used in processing materials to be used in the food industry has recently been issued (64). The effects of toxicological considerations on process design have been assessed (65).

Liquid effluents and gaseous emissions are subject to the usual environmental considerations (see Wastes, industrial). Disposal as landfill of extracted solids of no intrinsic value is often resorted to; incineration of organic material that has been subjected to aqueous extraction may be practicable if the extracted material has been extensively dewatered (66) (see Incinerators).

Nomenclature

C_o, C_t = concentration at time o, t
D = diffusivity
E = concentration ratio = C_t/C_o
l = thickness of particle being extracted
L_a, L_n = underflow to 1st stage and from nth stage
m = positive integer
N = number of stages in extractor
n = stage in extractor
t = time
$V_{a,b}$ = overflow from and to 1st and nth stage resp.
x = distance in direction of flow
x_a, x_b = concentrations in underflow
X = ratio solute to solute plus solvent
y_a, y_b = concentrations in overflow
y_i^* = equilibrium concentration corresponding to actual concentration y_i
Y = ratio solute-free solid to solute plus solvent

BIBLIOGRAPHY

"Liquid–Solid Extraction" under "Extraction" in *ECT* 1st ed., Vol. 6, pp. 91–122, by Frank Lerman, The Vulcan Copper & Supply Co.; "Liquid–Solid Extraction" under "Extraction" in *ECT* 2nd ed., Vol. 8, pp. 761–775, by E. G. Scheibel, The Cooper Union for the Advancement of Science and Art.

1. M. B. Donald, *Trans. Inst. Chem. Eng.* **15,** 77 (1937).
2. F. Habashi, *Principles of Extractive Metallurgy*, Vol. II, Gordon & Breach, London, Eng., 1970.
3. F. Schneider, ed., *Technologie des Zuckers*, 2nd ed., M & H Schaper, Hannover, FRG, 1968.
4. J. O. Osburn and D. L. Katz, *Trans. Am. Inst. Chem. Eng.* **40,** 511 (1944).
5. G. Karnofsky, *J. Am. Oil Chem. Soc.* **26,** 564, 570 (1949).
6. P. W. Rein and E. T. Woodburn, *Chem. Eng. J.* **7**(1), 41 (1974).
7. F. P. Plachco and M. E. Lago, *Chem. Eng. Sci.* **31,** 1085 (1976).
8. J. H. Krasuk, J. L. Lombardi, and C. D. Ostrovsky, *Ind. Eng. Chem. Process Des. Dev.* **6**(2), 187 (1967).
9. G. W. Genie, *Int. Sugar J.* **75,** 67 (1973).
10. H. B. Coats and G. Karnofsky, *J. Am. Oil Chem. Soc.* **27,** 51 (1950).
11. F. P. Plachco and M. E. Lago, *Ind. Eng. Chem. Process Des. Dev.* **15**(3), 361 (1976).
12. H. Bruniche-Olsen, *Sugar Technol. Rev.* **1**(1), 3 (1969).
13. J. Crank, *The Mathematics of Diffusion*, Oxford University Press, Oxford, Eng., 1957.
14. F. P. Plachco and J. H. Krasuk, *Ind. Eng. Chem. Process Des. Dev.* **9,** 419 (1970).
15. R. E. Treybal, *Liquid Extraction*, McGraw-Hill Book Co., London, Eng., 1963.
16. J. H. Hildebrand, J. M. Prausnitz, and R. L. Scott, *Regular and Related Solutions*, van Nostrand, New York, 1970.
17. D. R. J. Laws and co-workers, *J. Inst. Brew. London* **83**(1), 39 (1977).
18. A. R. Burkin, *The Chemistry of Hydrometallurgical Processes*, E. & F. N. Spon, London, Eng., 1966.
19. J. D. Keane and C. T. Smith, *J. Am. Oil Chem. Soc.* **35,** 199 (1958).
20. H. Tomschke, M. Meiners, and E. Frohnert, *Tech. Mitt. Krupp Werksber.* **35**(1), 9 (1977).

21. J. M. Coulson and J. F. Richardson, *Chemical Engineering,* 2nd ed., Vol. I, Pergamon Press, Oxford, Eng., 1970.
22. W. L. McCabe and J. C. Smith, *Unit Operations in Chemical Engineering,* 3rd ed., McGraw-Hill Book Co., London, Eng., 1976.
23. J. A. Grosberg, *Ind. Eng. Chem.* **42**(1), 154 (1950).
24. N. H. Chen, *Chem. Eng. (N.Y.)* **71**(24), 125 (Nov. 23, 1964).
25. K. Schoenemann and Th. Voeste, *Fette Seifen* **54**(7), 385 (1952).
26. H. Sawistowski and W. Smith, *Mass Transfer Process Calculations,* Interscience Publishers, New York, 1963.
27. R. Dousse and E. Ugstad, *Lebensm. Wiss. Technol.* **8**(6), 255 (1975).
28. *Food Process.* **36**(3), 71 (1975).
29. E. Bernardini and M. Bernardini, *Riv. Ital. Sostanze. Grasse* **52**(8), 271 (1975).
30. W. Kehse, *Fette Seifen Anstrichmittel* **77**(1), 15 (1975).
31. D. Schliephake, B. Mechias, and E. Rheinefeld, *Zucker* **29**(7), 378 (1976).
32. P. W. Rein, *Sugar J.* **39**(7), 15 (1976).
33. K. Weber, *Fette Seifen Anstrichmittel* **76,** 495 (1974).
34. C. J. King, *Separation Processes,* McGraw-Hill Book Co., New York, 1971, p. 195.
35. K. W. Becker, *AIChE Symp. Ser.* **64**(86), 60 (1968).
36. W. Kehse, *Chem. Ztg. Chem. Appar.* **94**(2), 56 (1970).
37. E. D. Milligan, *J. Am. Oil Chem. Soc.* **53,** 286 (1976).
38. A. G. Mathew and co-workers, *Flavour Ind.* **2**(1), 23 (1971).
39. H. P. J. Jongeneelen, *J. Am. Oil Chem. Soc.* **53,** 291 (1976).
40. M. Sivetz and H. E. Foote, *Coffee Processing Technology,* Vol. I, AVI Publishing Co., Westport, Conn., 1963.
41. K. Zosel, *Angew. Chem.* **90,** 748 (1978).
42. D. E. Cormack and co-workers, *Can. J. Chem. Eng.* **55**(5), 572 (1977).
43. P. Hubert and O. Vitzthum, *Angew. Chem.* **90,** 756 (1978).
44. R. N. Rickles, *Chem. Eng. (N.Y.)* **72**(6), 157 (Mar. 15, 1965).
45. R. G. Bautista, in Thomas B. Drew, ed., *Advances in Chemical Engineering,* Vol. 9, Academic Press, Inc., New York, 1974.
46. A. R. Burkin, ed., *Leaching and Reduction in Hydrometallurgy,* The Institution of Mining and Metallurgy, London, Eng., 1975.
47. A. R. Burkin, *Proc. Roy. Soc. London Ser. A* **338,** 419 (1974).
48. W. Kelm, *Zucker* **30**(3), 120 (1977).
49. R. Wasmund, *Z. Zuckerind. Boehm.* **27,** 581 (1977).
50. K. N. Palmer, *Dust Explosions and Fires,* Chapman and Hall, London, Eng., 1973.
51. J. Pepys, "Hypersensitive Diseases in the Lungs Due to Fungi and Organic Dusts" monograph in *Allergy,* Vol. 4, Karger, New York, 1969.
52. J. E. Heilman, *J. Am. Oil Chem. Soc.* **53,** 293 (1976).
53. D. H. Napier, *Inst. Chem. Eng. Symp. Ser.* **34,** 170 (1971).
54. R. Snyder and J. J. Kocsis, *Crit. Rev. Toxicol.* **3,** 265 (1975).
55. S. Laskin and B. D. Goldstein, eds., *J. Toxicol. Environ. Health* (Suppl. 2), (1977).
56. *Fed. Reg.* **43,** 5918 (1978).
57. *Specifications for the Identity and Purity of Some Extraction Solvents and Certain Other Substances,* FAO Nutr. Meet. Rep. Series 48B, WHO/FOOD ADD/70, 40, FAO, New York, 1971.
58. G. Abbritti and co-workers, *Br. J. Ind. Med.* **33**(2), 92 (1976).
59. *Threshold Limit Values for Chemical Substances and Physical Agents in the Working Environment with Intended Changes for 1973,* American Conference of Governmental Industrial Hygienists, Washington, D.C., 1973.
60. *Threshold Limit Values for Chemical Substances and Physical Agents in the Working Environment with Intended Changes for 1978,* American Conference of Governmental Industrial Hygienists, Washington, D.C., 1978.
61. *Fed. Reg.* **42,** 49,464 (1977).
62. *U.S. Dept. of Health NIOSH Current Intelligence Bulletin 20,* U.S. Government Printing Office, Washington, D.C., Jan. 1978.
63. Evaluation of Food Additives, 14th Report of Joint FAO/WHO Expert Committee on Food Additives, *WHO Tech. Rep. Ser.* 462, Geneva, Switz., 1971.

64. *Food Additives and Contaminants Committee on the Review of Solvents in Food SAC/REP/25,* H.M.S.O. London, Eng., 1978.
65. W. R. Payne, *Chem. Eng. (N.Y.)* **85**(10), 83 (1978).
66. M. T. Bond and L. W. Canter in *Proc. 24th Ind. Waste Conf. Purdue Univ.,* (1969).

WOLF HAMM
Unilever Limited

EXTRACTIVE METALLURGY

Extractive metallurgy deals with the extraction of metals from naturally occurring compounds and their refinement to a purity suitable for commercial use. These operations, known as the winning and refining of metals, follow the mining and beneficiation of the ore, and they precede the fabricating processes. The selection and design of the extractive processes depend on the raw materials available, and the conditions for the refining steps are related to the ultimate use of the metal so that mining, extraction, and fabrication are closely interrelated.

The production of a metal is usually achieved by a sequence of chemical processes that is represented as a flow sheet. Limited numbers of unit processes are commonly used in extractive metallurgy, but the combination of these steps and the precise conditions of operations vary significantly from metal to metal. Even for the same metal they vary with the type of ore or raw material. The technology of extraction processes was developed in an empirical way, and in many instances, technical innovations remain more advanced than scientific understanding of the processes.

The scientific basis of extractive metallurgy is inorganic physical chemistry, mainly chemical thermodynamics (qv) and kinetics. Metallurgical engineering relies on basic chemical engineering science, material and energy balances, and heat and mass transport. Metallurgical systems, however, are often very complex and scale-up from the bench to the commercial plant is more difficult than for other chemical processes.

Extractive metallurgy is usually divided into three principal areas: (*1*) pyrometallurgy consists of processes that use high temperatures to carry out smelting and refining reactions; (*2*) hydrometallurgy is characterized by the use of aqueous solutions and inorganic solvents to achieve the desired reactions; and (*3*) electrometallurgy applies electrical energy to extract and refine metals by electrolytic processes carried out either at high temperature or in aqueous solution. This classification is followed here to present a general description of the main types of metallurgical operations. An integrated metallurgical flow sheet may include pyrometallurgical, hydrometallurgical, and electrometallurgical steps. Specific examples are presented, but the reader is referred to articles in the Encyclopedia concerning each metal for a complete description of its extraction and refining technology.

The Origin of Extractive Metallurgy. A brief review of the development of extractive metallurgy provides some insight into the characteristics of the present technology.

Seven metals were known and widely used in antiquity: copper, gold, silver, tin, lead, iron, and mercury. The first three, which occur naturally in the metallic state, were discovered ca 10,000 years ago.

The invention of the first extractive metallurgy process is difficult to locate geographically or in time, but it probably started in the mountains of western Asia, in the areas that are now Iraq and Iran. There is tangible evidence of the smelting of copper by the Summerians and Egyptians as early as ca 4000 BC. Discovery of the reduction of copper ore by fire was certainly accidental, but the observation of the relationship between the blue azurite [$Cu(OH)_2.2CuCO_3$] or the green malachite [$Cu(OH)_2.CuCO_3$] and the red metal was a truly remarkable scientific discovery. The ancient peoples also knew how to treat copper sulfide ores.

Gold was used extensively in antiquity and the operation of gold mines and mills by the Egyptians is well documented. Mercury also was known to the ancients; its preparation from cinnabar (HgS) and its use in the amalgamation process for the recovery of gold were discovered early. Silver was found associated with gold, and the gold–silver alloy called electrum was commonly used by the ancients. Silver also was obtained from argentiferrous lead ores, and the cupellation process for the separation of precious metals from base metals was another early metallurgical achievement.

The production of metallic lead from galena (PbS) is a simple metallurgical operation, and both the metal and its compounds were used in antiquity. Tin was used mainly in bronze, probably obtained by mixing the tin ore, cassiterite (SnO_2) with the copper ore. Some iron of meteoric origin had been found by the ancients, but the use of wrought iron was also widespread. It was obtained easily by reduction of iron ore by carbon followed by hammering of the sponge iron and forging.

During the Middle Ages an active mining industry developed in Central Europe, but it produced few technological breakthroughs. Several books on mining and metallurgy were written. The extensive treatise *De Re Metallica,* by Agricola (1), illustrates the state of the art of the industry in the mid-sixteenth century.

Further progress was made in the eighteenth and early nineteenth centuries with the discovery of most metals and the development of experimental chemistry. The modern metallurgical industry was born with the invention of steelmaking in 1856 (see Steel). Industrial processes for making zinc, aluminum, and copper followed before the end of the century. They made possible the industrial revolution and the development of an industrial society relying heavily on the use of metals.

The Sources of Metals. All metals come originally from the Earth's crust where they are present in natural deposits. These ore deposits, which result from a geological concentration process, consist mainly of metallic oxides and sulfides from which the metals can be extracted. Sea water and brines are another natural source of metals (eg, magnesium) (see Chemicals from brine; Ocean raw materials). Metal extracted from a natural source is called primary metal. Scrap is also an important raw material for the metallurgical industry; the metal produced by a recycling process is called secondary metal (see Recycling).

The concentration of most metals in the Earth's crust is very low. Even for abundant elements, such as aluminum and iron, extraction from common rock is not economically feasible. An ore is a metallic deposit from which the metal can be economically extracted. The amount of valuable metal in the ore is the tenor, or ore grade, usually given as the wt % of metal or oxide. For precious metals, it is given in grams per metric ton or troy ounces per avoirdupois ton. The tenor and the type of metallic compounds are the main characteristics of an ore. The economic feasibility of its processing, however, depends also on the nature, location, and size of the deposit, the availability and cost of a suitable extraction process, and the market price of the metal.

The valuable metal is present in the ore in a form called the ore mineral, an inorganic substance of given composition. The ore mineral may be native metal or a chemical compound of the metal. Most ores contain several ore minerals. The minerals without economic value are called the gangue minerals. Since the designation as ore minerals and gangue minerals depends on the economics of processing the ore, the same mineral may be classified in either group. In particular, silica, alumina, and iron oxides are usually gangue minerals except when their concentration is high enough to warrant their extraction.

The direct treatment of an ore by a chemical extraction process is usually not economically feasible. Instead the first step after mining is the ore preparation, also called ore dressing, milling, mineral dressing, or beneficiation. The purpose of ore dressing is to separate the ore minerals from the gangue minerals by physical methods. Milling processes yield a concentrate containing the valuable metals to be recovered and a tailing containing the gangue minerals to be discarded.

The first step in ore preparation is the liberation of finely disseminated minerals. This is accomplished by comminution, or crushing and grinding the ore (see Size reduction). It is followed by screening and classification (see Size classification). The main step of mineral beneficiation is concentration, achieved by making use of different physical properties of the mineral particles: magnetic and electrostatic separation (see Magnetic separation; Electromigration), gravity concentration (qv) and flotation (qv).

Table 1 illustrates the average metal content of the Earth's crust, ore deposits and concentrates. With the exception of the recovery of magnesium from sea water, alkali metals from brines, and the solution mining of some copper and uranium, ores are processed through mills, and concentrates are the raw materials for the extraction of primary metals.

Pyrometallurgy

The essential operations of an extractive metallurgy flow sheet are the decomposition of a metallic compound to yield the metal and the physical separation of the reduced metal from the residue. This is usually achieved by a simple reduction or by controlled oxidation of the nonmetal with simultaneous reduction of the metal (as illustrated below by the matte smelting and converting processes).

In a simple pyrometallurgical reduction, the reducing agent, R, combines with the nonmetal, X, in the metallic compound, MX, according to a substitution reaction of the type:

$$MX + R \rightarrow M + RX \tag{1}$$

The product RX and the gangue material constitute the residue. In order to achieve a spontaneous process, the free-energy change for equation 1 must be negative. This condition is obtained not merely by the choice of the appropriate reducing agent, but

Table 1. Metal Content of Igneous Rocks, Ore Deposits and Concentrates, wt %

Metal	Igneous rocks	Ore deposits	Concentrates
aluminum	8.13	27–29	
iron	5.01	30–60	55
magnesium	2.09		
titanium	0.63	2.5–25	
manganese	0.10	45–55	
nickel	0.020	1.5–3	10
vanadium	0.017	1.6–4.5	
copper	0.010	0.5–5	30
uranium	0.008	0.1–0.9	
tungsten	0.005		50
zinc	0.004	10–30	50
lead	0.002	5–10	70
cobalt	0.001	1–11	
beryllium	0.001		3–4
molybdenum	2×10^{-4}	0.6–1.8	50
tin	4×10^{-4}	1.5	40–75
mercury	5×10^{-5}	0.1–0.5	
silver	2×10^{-6}	0.04–0.08	
platinum	5×10^{-7}	0.001	
gold	10^{-7}	0.001	

also by adjusting the chemical activities of the reactants and products and by selecting the optimum temperature and pressure for a suitable thermodynamic driving force and acceptable reaction rates.

The selection of a particular type of reduction depends on the technical feasibility and the economics of the process as well as on physico-chemical considerations. In particular: (a) the reducing agent should be inexpensive relative to the value of the metal to be reduced; (b) the product of the reaction should be easily separated from the metal, and it should be easily contained and safely recycled or disposed of; and (c) the physical conditions for the reaction should be such that a suitable reactor can be designed and operated economically.

Reduction processes are characterized either by the reducing agent selected or by the physical form of the metallic product, namely liquid, solid, or gaseous. Since the separation of reaction products determines the choice and design of the furnace, the processes are classified here according to the physical state of the reduced metal.

Usually, the ore or concentrate cannot be reduced to the metal in a single operation and a preparation process is needed to modify the physical or chemical properties of the raw material. Furthermore, most pyrometallurgical reductions do not yield a pure metal and an additional step, refining, is needed to achieve the chemical purity that is specified for the commercial use of the metal.

The preparation, reduction, and refining operations are very much interdependent, and for a given metal they must be considered as parts of a single process. To illustrate the principles of extractive metallurgy, however, it is convenient to classify the various operations in these categories.

Preparatory Processes. Several pyrometallurgical operations are used to change the chemical and physical properties of the ore or concentrate in order to make it more suitable for the main extraction process. These preparatory processes are characterized by the fact that their chemistry, which involves mainly gas–solid reactions, is relatively simple.

Drying and Calcination. The simplest pyrometallurgical operation is the evaporation of free water and the decomposition of hydrates and carbonates. A typical reaction is the decomposition of pure limestone ($CaCO_3$) to calcium oxide and carbon dioxide:

$$CaCO_3(s) \rightarrow CaO(s) + CO_2(g) \qquad (2)$$

where the symbols (s) and (g) represent the solid and gaseous state, respectively (see Lime and limestone). This reaction is strongly endothermic, and the equilibrium pressure of CO_2, which is equal to the equilibrium constant of the reaction, increases exponentially with temperature.

Lime is an important raw material for the metallurgical industry. It is used primarily as a flux in smelting and converting, but it is also a neutralizing agent for hydrometallurgical processes. The calcination of magnesite ($MgCO_3$) yields magnesia (MgO) which is an essential raw material for furnace refractories. The calcination of dolomite ($CaCO_3 \cdot MgCO_3$) also yields a calcine used as a feed for the preparation of magnesium metal.

Other examples of similar processes are the decomposition of precipitated aluminum trihydroxide to alumina, which is the feed for the electrolytic production of aluminum metal, and the drying of wet sulfide concentrates in preparation for flash roasting.

Roasting of Sulfides. Most nonferrous metals occur in nature mainly as sulfides, which cannot be easily reduced directly to the metal. Burning metallic sulfides in air transforms them into oxides or sulfates which are more easily reduced. The sulfur is released as sulfur dioxide, as shown by the following typical reaction for a divalent metal, M:

$$MS(s) + \tfrac{3}{2} O_2(g) \rightarrow MO(s) + SO_2(g) \qquad (3)$$

This reaction is strongly exothermic and proceeds spontaneously from left to right for most common metallic sulfides under normal roasting conditions, namely in air [$p_{SO_2} + p_{O_2}$ = ca 20 kPa (0.2 atm)] at temperatures ranging from 650 to 1000°C.

The physical chemistry of the roasting process is more complex than indicated by equation 3 alone. Indeed, sulfur trioxide is also formed,

$$SO_2(g) + \tfrac{1}{2} O_2(g) \rightarrow SO_3(g) \qquad (4)$$

and the oxide-sulfate equilibrium must be considered.

$$MO(s) + SO_3(g) \rightarrow MSO_4(s) \qquad (5)$$

Very little SO_3 is produced at 1000°C, but the equilibrium of equation 4 shifts to the right at lower temperatures favoring the formation of sulfates according to equation 5. This behavior can be illustrated by the schematic diagram in Figure 1, which represents the equilibria and predominance areas of the different species as a function of the reciprocal temperature and the oxygen pressure for a given sulfur dioxide pressure.

Most roasting is carried out to obtain an oxide for reduction by carbon or carbon

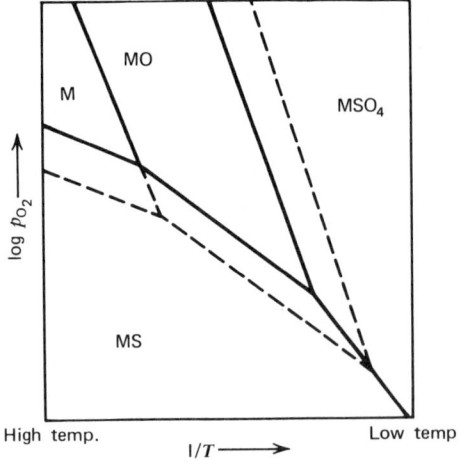

Figure 1. Equilibria and predominance areas at constant SO_2 pressure as function of temperature for the system M-S-O (2). —, $p_{SO_2} = 101$ kPa (1 atm); - - - -, $p_{SO_2} = 10$ kPa (0.1 atm).

monoxide, or for leaching in sulfuric acid solution followed by electrowinning. This roasting to completion, which eliminates all the sulfur and produces the metal oxide, is called dead roast or sweet roast. Incomplete roasting is used to remove excess sulfur in preparation of copper and nickel sulfides for the matte smelting process described later. A sulfating roast yields water-soluble sulfates in preparation for leaching.

The sulfur dioxide produced by the process is usually converted to sulfuric acid (or sometimes liquefied), and the design of modern roasting facilities takes into account the need for an efficient and environmentally clean operation of the acid plant.

Chlorination. In some instances, the extraction of a pure metal is more easily achieved from the chloride than from the oxide. Oxide ores and concentrates react at high temperature with chlorine gas to produce volatile chlorides of the metal. This reaction can be used for the common nonferrous metals, but it is particularly useful for refractory metals like titanium and zirconium and for reactive metals like aluminum.

The chlorination of titanium, for instance, requires a reducing agent, such as carbon,

$$TiO_2(s) + C(s) + 2\,Cl_2(g) \rightarrow TiCl_4(g) + CO_2(g) \qquad (6)$$

This reaction is slightly exothermic, but additional heat is required to maintain the operating temperature at 800 to 900°C.

Sintering and Pelletizing. The beneficiation of ores by physical techniques requires the liberation of metal-bearing minerals, and this is usually achieved by reducing the raw material to a very fine size. The high-throughput reduction processes, such as the iron blast furnace, however, can not handle a finely divided feed. Sintering and pelletizing are techniques for agglomerating finely divided solid particles into a coarse material suitable for charging a blast furnace (see Pelleting and briquetting).

Sintering consists of heating a mixture of fine materials to an elevated temperature without complete fusion. Surface diffusion and some incipient fusion cause the solid particles in contact with one another to adhere and form larger aggregates. In the processing of hematite or magnetite fines to prepare a suitable feed for the iron blast furnace, the oxide ore is mixed with 5 to 8% of coal or coke to supply the fuel requirement. Often limestone is added to produce the self-fluxing sinter, and it is calcined during the sintering process. In the processing of sulfide ores, such as the treatment of lead sulfide concentrates, the contained sulfur acts as the fuel after ignition of the charge, and a single process achieves both roasting and sintering of the feed.

Pelletizing was developed more recently for low-grade iron ores. In particular, taconite ores consist of finely divided magnetite or hematite mixed with the gangue. The agglomeration of this material after fine comminution and separation is accomplished by balling in a rotating device such as a disk, drum, or cone followed by heating to a final temperature of ca 1300°C, a process called induration. The induration achieves not only the agglomeration of the fines, but also the vaporization of free or combined water, the calcination of limestone and the oxidation of magnetite (Fe_3O_4) to hematite (Fe_2O_3). The pelletizing process has many advantages over sintering, and it also is being applied now to high-grade ores. The introduction of digital computers has made possible the completely automatic control of operations of modern pelletizing plants. The induration process can also be carried out under reducing conditions to yield partially reduced pellets for the blast furnace or fully reduced pellets for direct charging to steelmaking furnaces.

Equipment. Drying and calcination are usually carried out in various types of kilns such as rotary kilns, shaft furnaces, and rotary hearths. The induration step in a pelletizing plant requires similar equipment, but sometimes the charge passes through successive stages: travelling grates for drying, a rotary hearth for preheating, and a shaft furnace for the chemical reactions and sintering (see also Drying; Furnaces).

The Dwight-Lloyd continuous sintering machine is commonly used. It consists of a series of grates mounted as an endless travelling belt. The feed is spread as a bed (10–20 cm deep), and the top is ignited as it passes through a fuel-fired ignition box. Air passes through the bed into a suction box and fan below the grate, and the combustion zone progresses through the bed as material moves to the discharge end of the belt. Downdraft machines are used for sintering iron ores, and updraft machines are preferred for sinter-roasting of lead and zinc concentrates.

The older but still widely used roasting furnace is the multiple-hearth roaster consisting of 8–12 hearths enclosed in a cylindrical shell. The ore or concentrate fed at the top drops from hearth to hearth by the moving action of a rotating rabble attached to a central shaft. The sulfide particles are roasted by contact with the rising gases and are discharged at the bottom as calcine. In the suspension roaster, which is derived from the multiple-hearth furnace, the pulverized concentrate is dried on the top hearths, drops through a combustion chamber, and is collected at the bottom. If the roasting process occurs while a very fine concentrate is mixed with the gas stream in the hot chamber, it is called flash roasting.

Fluid-bed roasters are well-suited for gas–solid reactions, and their use in metallurgical operations developed appreciably since about 1950. Fluid-bed roasters have no moving parts in the hot zone, and the temperature and gas composition can be easily controlled. Air is pumped upward through a perforated hearth at a rate sufficient to expand the bed of sulfide particles, and the gas–solid reaction proceeds very fast (Fig. 2). The fluid-bed reactor is unsuitable for treating extremely fine materials or for particles that agglomerate upon heating.

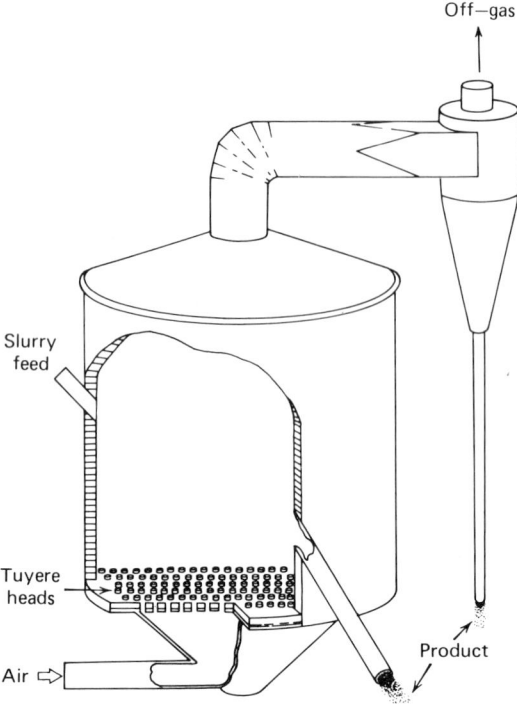

Figure 2. Cutaway view of a fluid-bed roaster.

Reduction to Liquid Metal. Reduction to liquid metal is the most common of metal reduction processes. It is the preferred method and is used for metals of moderate melting point and low vapor pressure. Because most metallic compounds are fairly insoluble in molten metals, the separation of the liquified metal from a solid residue or from another liquid phase of different density is usually complete and relatively simple. Because the product is in condensed form, the throughput per unit volume of reactor is high and the number and size of the units is minimized. The common furnaces for production of liquid metals are the blast furnace, the reverberatory furnace, the converter, the flash smelting furnace, and the electric-arc furnace (see Furnaces, electric). Their characteristics are briefly outlined in relation to the process for which they are designed.

The Iron Blast Furnace. The reduction of iron oxides by carbon in the iron blast furnace is not only the most important of all extractive processes, but it is also the cornerstone of all industrial economies (see also Iron). It is remarkable that the modern blast furnace is the result of a gradual evolution through many centuries. Progress, of course, has been accelerated during the past thirty years. A better understanding of the reactions that take place in the furnace has made possible a more efficient operation through better preparation of the burden, higher blast temperature, and sometimes, increased pressure. This has resulted in doubling the furnace capacity while reducing the coke consumption from ca one ton to ca half a ton per ton of iron produced.

The iron blast furnace is a circular shaft furnace ca 30 m high with a maximum internal diameter of ca 9 m. It is a steel shell lined with refractory bricks. The furnace is charged at the top with ore, coke, and fluxes. Preheated air is introduced through the tuyères above the hearth at the bottom of the furnace. At the tuyère level, the coke burns in the air according to the reactions:

$$C(s) + O_2(g) \rightarrow CO_2(g) \tag{7}$$

$$CO_2(g) + C(s) \rightarrow 2\,CO(g) \tag{8}$$

Equation 7, which is strongly exothermic, is the main source of heat for the process. Equation 8, called the Boudouard reaction, supplies the reducing gas, carbon monoxide; it is endothermic, and the value of its equilibrium constant determines the composition of the gas phase in the furnace.

The iron blast furnace is a countercurrent reactor in which the iron oxides descending through the upper part of the stack are heated and reduced in stages:

$$3\,Fe_2O_3(s) + CO(g) \rightarrow 2\,Fe_3O_4(s) + CO_2(g) \tag{9}$$

$$Fe_3O_4(s) + CO(g) \rightarrow 3\,FeO(s) + CO_2(g) \tag{10}$$

$$FeO(s) + CO(g) \rightarrow Fe(s) + CO_2(g) \tag{11}$$

In the lower part of the stack where the temperature is >1000°C, the equilibrium of equation 8 is shifted to the right and it proceeds simultaneously with equation 11. The combination of both gives

$$FeO(s) + C(s) \rightarrow Fe(s) + CO(g) \tag{12}$$

a reaction referred to as the direct reduction of wustite (FeO) as compared to equation 11 which describes the indirect reduction.

The fluxes consisting mainly of limestone or sometimes dolomite are heated and calcined on their way down the shaft. In the lower part of the furnace, called the bosh, they melt with the gangue material to form the slag. The iron blast furnace slag consists mainly of lime, alumina, and silica with smaller amounts of magnesia, manganese oxide, iron oxide, and calcium sulfide. The lime is added as a flux to produce a slag with a melting point of 1300–1400°C and a relatively low viscosity. The reduced iron is also molten at the tuyère level; it drops as globules through the slag and settles at the bottom of the furnace. Figure 3 summarizes the reactions in an iron blast furnace.

Other reactions proceed in the lower part of the furnace, namely the dissolution of carbon in molten iron, and the reduction of manganese oxide and some silica with dissolution of manganese and silicon in the metal.

$$C(s) \rightarrow C(l\,Fe) \tag{13}$$

$$MnO(l\,slag) + C(l\,Fe) \rightarrow Mn(l\,Fe) + CO(g) \tag{14}$$

$$SiO\,(l\,slag) + 2\,C(l\,Fe) \rightarrow Si(l\,Fe) + 2\,CO(g) \tag{15}$$

where the symbol l represents a liquid phase.

The molten slag and the molten iron, which is called hot metal or pig iron, are tapped from the hearth of the blast furnace. A modern blast furnace yields 5,000–9,000 t iron in 24 h. The compositions of the pig iron and the slag are determined by the furnace temperature, the composition of the ore and the flux added. Pig iron always contains 3.5–4.5 wt % carbon, variable amounts of silicon, manganese, sulfur, and phosphorus.

The main technique for regulating the operating temperature of the blast furnace

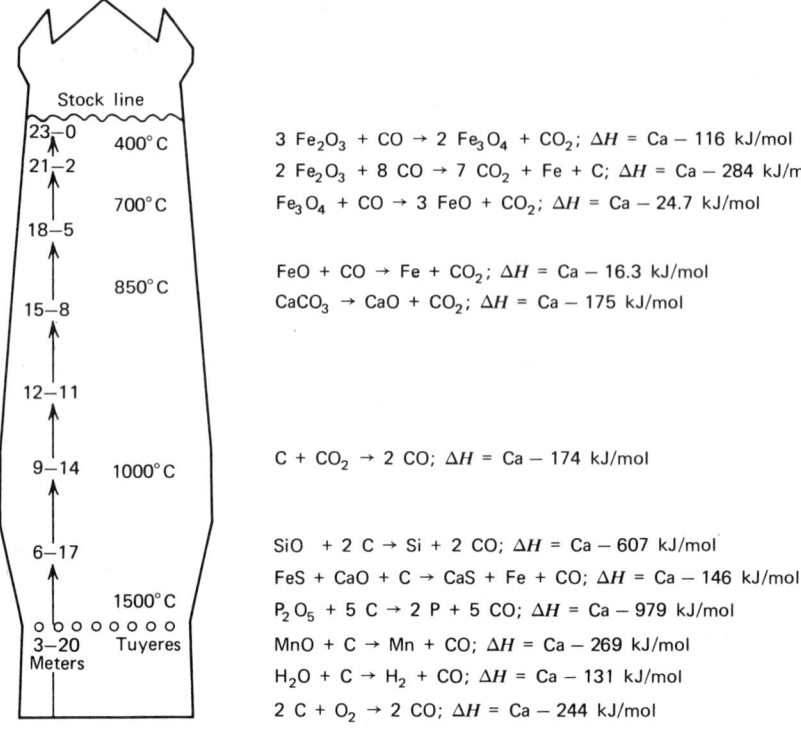

Figure 3. Schematic diagram indicating some of the chemical reactions in the iron blast furnace (3). To convert J to cal, divide by 4.184.

is the preheating of the blast. In recent years, progress has been made with the addition of natural gas or oil to the blast, oxygen enrichment and addition of steam. Preheating and prereduction of the ore is also considered.

Reduction Smelting of Nonferrous Metals. The standard method of lead smelting is reduction of an oxide in a blast furnace that is similar to the iron blast furnace in principle, but very different in design and operation (see Lead). It consists of a shaft of rectangular cross-section, 6–9 m high, 5–7 m long, and tapering in width from ca 2.5 m at the top to ca 1.5 m at the bottom. The feed, added at the top, consists mainly of roasted and sintered lead concentrates, coke and fluxes to form a slag with the gangue material. Air is blown through tuyères near the bottom, and coke burns according to equations 7–8 given earlier. The reduction proceeds by contact between the hot reducing gas and the lump charge.

$$PbO(s) + CO(g) \rightarrow Pb(l) + CO_2(g) \tag{16}$$

Because of its low mp (327°C), the lead is molten and trickles down to the crucible below the tuyères. The slag melts when it reaches the high temperature zone of the tuyères and floats on top of the lead. The slag consists mainly of SiO_2, CaO, FeO, with lesser amounts of Al_2O_3, MgO, ZnO, and other oxides. Because most lead ores contain zinc, and because zinc-plant residues are often charged into lead blast furnaces, the amount of zinc in the slag is sufficient to justify its recovery by treatment in a separate fuming furnace. The molten lead, called lead bullion, is alloyed with other easily reducible metals, mainly copper, tin, antimony, arsenic, gold, and silver present in the charge. The capacity of typical furnaces ranges from 100–400 t/d.

Many nonferrous metals can be extracted by reduction smelting: copper, tin, nickel, cobalt, silver, antimony, bismuth, and others. Blast furnaces are sometimes used for the smelting of copper or tin, but reverberatory furnaces are more common for metals other than lead.

The reverberatory furnace consists of a shallow hearth (usually rectangular), side-walls, end-walls, and a roof or arch. The charge rests on the hearth and the furnace is fired with gas, oil, or pulverized coal which burns in the space between the charge and the roof. Heat transfer is achieved mainly by radiation directly from the flame and by reflection from the arched refractory roof. There is little or no reaction between the charge and the waste gases which leave the furnace at very high temperature. The feed often consists of fine-particle concentrates, and the reducing agent is usually coke. Both are added through ports on the side or the top of the furnace. Slag and metals are removed through tap holes, and the smelting can be carried out either continuously or as a batch process. Reverberatory furnaces are made in many sizes from small furnaces holding only one metric ton of metal to the large ones employed in copper smelting (see below) which hold several hundred tons.

Electric heat provided by a resistance or by an electric arc can be substituted for the burning of a fuel. Electric furnaces can be designed in a variety of shapes and are more versatile than fuel-heated furnaces since the furnace atmosphere can be controlled independently of the chemistry of the combustion (see Furnaces, electric).

Matte Smelting and Converting. Rich oxide ores of copper or roasted sulfide concentrates can be treated by reduction smelting in a blast or reverberatory furnace to yield an impure copper known as black copper (see Copper). Except for a few plants treating mainly copper scrap, this process is now obsolete. Most primary copper is produced by matte smelting, an operation yielding a molten sulfide of copper and iron, called matte, which is further oxidized in the converting step to yield metallic copper. A very similar process is used for the extraction of nickel (see Nickel).

Copper is a fairly noble metal that is easily reduced from its compounds. The most important copper mineral is chalcopyrite, $CuFeS_2$, and the separation of copper from iron is one of the objectives of the smelting operation. The sulfur in the ore has value as a fuel as well as a reducing agent. The traditional matte smelting, which evolved from earlier empirical practices, is carried out in a reverberatory furnace in which a solid charge consisting of concentrates, calcines, converter slags, and fluxes is heated to 1150–1250°C. This melting produces two separate immiscible liquid phases, a slag floating on top of a copper-rich matte. The matte is a sulfide phase consisting mainly of Cu_2S and FeS that ranges from 30–60 wt % Cu. The smelting slag is a mixture of iron oxides and silica with smaller amounts of alumina and lime. This slag is usually discarded, and it should contain as little copper as possible.

The electric furnace is an alternative to the reverberatory furnace in environmentally sensitive areas where the electricity costs are not too high. The electric furnace is versatile, it produces small volumes of effluent gases, and the SO_2 concentration can be easily controlled. Its operating costs, however, are high.

Since the mid-nineteen-sixties most new matte smelting capacity has been provided by flash smelting furnaces. In flash smelting, dry concentrates are blown with oxygen, hot air, or a mixture of both, into a hearth-type furnace. Inside the furnace, the sulfide particles react rapidly with the gases, causing a large evolution of heat and controlled partial oxidation of the concentrates. The main reactions, which are exothermic, can be summarized by:

$$CuFeS_2(s) + 5/4\ O_2(g) \rightarrow 1/2\ (Cu_2S.FeS)(l) + 1/2\ FeO(l) + SO_2(g) \tag{17}$$

$$FeS(s) + 3/2\ O_2(g) \rightarrow FeO(l) + SO_2(g) \tag{18}$$

$$2\ FeO(l) + SiO_2(s) \rightarrow 2\ FeO.SiO_2(l\ slag) \tag{19}$$

The reacting particles melt rapidly, and the droplets fall to the slag layer. The sulfide drops settle through it to form the matte phase. Any oxidized copper is reduced to the matte by the reaction:

$$Cu_2O(l\ slag) + FeS(l) \rightarrow FeO(l\ slag) + Cu_2S(l) \tag{20}$$

Flash smelting is efficient in that the fuel value of the sulfur and iron in the charge is fully used, and the productivity (8 to 12 metric tons of charge per day and per square meter of hearth) is higher than that of the reverberatory or electric furnace.

In addition to copper, iron and sulfur, the matte produced by smelting contains oxygen and various impurities (As, Bi, Ni, Pb, Sb, Zn, Au, Ag, etc) depending on the composition of the concentrate. The converting operation removes the iron and the sulfur along with some of the other impurities. The molten matte is charged to a converter where it is oxidized by blowing air. This is universally done in a Pierce-Smith converter, which consists of a cylindrical steel shell lined with magnesite refractory bricks. The shell is placed horizontally and rotates to achieve the different positions for charging, blowing through submerged tuyères, and pouring. Typical inside dimensions are 4 m dia and 9 m in length for an output of 100–200 t/d.

The converting operation takes place in two stages. In the first, the slag-forming stage, the main reaction is the oxidation of FeS to FeO and Fe_2O_3 which combine with a silica flux to form a silicate slag by reactions similar to the ones described by equations 18–19. The matte is added in several steps, each addition followed by oxidation and discharge of the slag. In the second copper-making stage, the remaining sulfur is oxidized to sulfur dioxide.

$$Cu_2S(l) + O_2(g) \rightarrow 2\ Cu(l) + SO_2(g) \tag{21}$$

Some copper oxide is formed which then reacts with the remaining sulfide to form metallic copper:

$$2\ Cu_2O(l) + Cu_2S(l) \rightarrow 6\ Cu(l) + SO_2(g) \tag{22}$$

The equilibrium constants for these reactions are such that copper is not appreciably oxidized by oxygen until most sulfur has been removed. This makes possible the production of blister copper which is 98.5 to 99.5% Cu and is low in both sulfur (0.02 to 0.1%) and oxygen (0.5 to 0.8%). The converter slag, however, contains a significant amount of copper and must be recycled to the smelting stage.

All the operations in the winning of copper from sulfide ores are controlled oxidations with air or oxygen. An important effort has been made to carry out the operation as a continuous single-step process. Two of these new processes have now reached the commercial stage: the Noranda process, which is carried out in a long cylindrical vessel, and the Mitsubishi process, which uses three furnaces interconnected by a continuous flow of matte and slag.

Reduction to Gaseous Metal. The volatile metals can be reduced to metallic vapors. These are easily and completely separated from the residue before being condensed to a liquid or a solid product in a container physically separated from the reduction reactor. Reduction to gaseous metal is possible for zinc, mercury, cadmium, and the alkali and alkaline earth metals, but industrial practice is significant only for zinc, mercury, magnesium, and calcium.

Zinc is produced by reduction of zinc oxide (usually a calcine obtained by roasting zinc sulfide concentrates) with carbon in the absence of air at 1200–1300°C, well above the boiling point of the metal (906°C).

$$ZnO(s) + C(s) \rightarrow Zn(g) + CO(g) \tag{23}$$

The original process, patented in 1810 and known as the Belgian retort process, was the main process for winning zinc for more than a century, and a few plants are still in operation. The reaction is carried out in small retorts producing daily batches of 25–50 kg of zinc per retort. The gases produced by equation 23 exit through a condenser where the liquid zinc is collected and carbon monoxide is vented and burns in air. The retorts and condensers are made of fireclay and are placed horizontally in a fuel-fired furnace. This process is labor intensive, and it has a relatively poor metallurgical efficiency.

Vertical retorts that can be operated continuously have been developed. They are either fuel-fired (New Jersey process) or heated by an electric arc (St. Joe process). Their capacity ranges from 5 to 25 t zinc per day per retort. They are superior to the horizontal retort process from the viewpoints of labor requirements and metallurgical efficiency, but the capital costs and in some cases the energy costs are higher. The development in the 1950s of a zinc–lead blast furnace (Imperial Smelting Process) is the most recent breakthrough in zinc smelting. Both lead bullion and zinc metal are recovered by roasting and reducing mixed lead and zinc concentrates. The main feature of the process is the rapid cooling of zinc vapor from 1100 to 550°C in a spray of molten lead in a specially designed condenser. This prevents reoxidation of the zinc vapor by carbon dioxide in the furnace gases. This process was adopted quickly all over the world, but its success did not slow the replacement of zinc smelting facilities by electrolytic plants described later (see Zinc and zinc alloys).

A vacuum-retort process (Pidgeon process) was used during World War II for the production of magnesium and calcium. To avoid the problems of cooling magnesium vapor in the presence of carbon dioxide, silicon, in the form of ferrosilicon, is used as the reducing agent instead of carbon:

$$2\,CaO.MgO(s) + Fe_xSi(s) \rightarrow 2\,Mg(g) + (CaO)_2.SiO_2(s) + x\,Fe(s) \tag{24}$$

At 1200°C, the equilibrium pressure of magnesium vapor for this reaction is well below atmospheric pressure, and a high vacuum must be maintained in the condenser. The reduction was carried out in small retorts made from Cr–Ni–Fe alloys and placed horizontally in a furnace. A new process using a vertical electric-arc furnace (Magnetherm process) has been developed recently, but most magnesium is produced by electrolysis as described later (see Magnesium and magnesium alloys).

Reduction to Solid Metal. Metals with very high melting points cannot be reduced in the liquid state. Because the separation of a solid metallic product from a residue is usually difficult, the raw material is purified before reduction. Tungsten and molybdenum, for instance, are prepared by reduction of a purified oxide (WO_3, MoO_3) or a salt ($[NH_4]_2MoO_4$) by hydrogen following a reaction such as:

$$WO_3(s) + 3 H_2(g) \rightarrow 3 H_2O(g) + W(s) \tag{25}$$

The metallic product consists of fine particles which are fabricated into various shapes by the techniques of powder metallurgy (see Tungsten and tungsten alloys; Molybdenum and molybdenum alloys; Powder metallurgy).

Very reactive metals (eg, titanium or zirconium) which, in the liquid state, react with all the refractory materials available to contain them, also require reduction to solid metal. Titanium is produced by metallothermic reduction of its chloride (obtained by the chlorination process described earlier) with liquid magnesium at 750°C (Kroll process).

$$TiCl_4(g) + 2 Mg(l) \rightarrow Ti(s) + 2 MgCl_2(l) \tag{26}$$

A steel reaction vessel is partly filled with solid magnesium, sealed, and flushed with helium or argon. The reactor is placed in a furnace and heated to melt the magnesium. Pure liquid titanium tetrachloride is fed slowly to the vessel, where it vaporizes and is reduced by molten magnesium. Equation 26, which is exothermic, proceeds to completion. Some of the liquid magnesium chloride is drained, and the vessel is cooled to room temperature. The reaction mass, interlocking crystals of metallic titanium, magnesium chloride and some magnesium metal, is removed by boring. The magnesium chloride and the magnesium are separated from the titanium by vacuum distillation or by leaching with dilute acid. About 700–1400 kg of titanium is produced in one batch. This process is summarized in the flow sheet of Figure 4 (see Titanium and titanium alloys).

Less reactive metals can also be produced by reduction below their melting point. Several processes have been developed for the reduction of iron oxides to solid iron; they are referred to as direct reduction processes. The reducing agents used commercially are hydrogen, carbon monoxide, natural gas, carbon, and carbonaceous fuels.

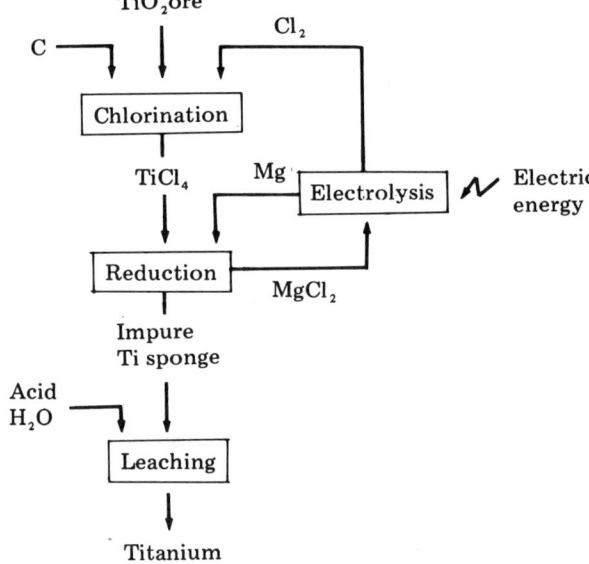

Figure 4. Flow sheet for the process of separating magnesium chloride and magnesium from titanium by vacuum distillation or by leaching with dilute acid.

The reaction can be carried out in a rotary kiln, a shaft furnace or, in the case of gaseous reduction, in a fluidized-bed reactor. These processes require a relatively pure iron ore; the metallic product is impure metal (90–95 wt % Fe) which is usually charged to a steelmaking furnace. The advantage of the direct reduction process is that it can be operated on a small scale with a capital investment much lower than that of a blast furnace plant (see Iron by direct reduction).

Refining Processes. All the reduction processes discussed above yield an impure metal that still contains some of the minor elements present in the concentrate (eg, cadmium in zinc) or some elements introduced during the smelting process (eg, carbon in pig iron). These impurities must be removed from the crude metal in order to meet the specifications for its use. Refining operations may be classified according to the kind of phases involved in the process: separation of a vapor from a liquid or solid, separation of a solid from a liquid, or transfer between two liquid phases. In addition, they may be characterized by whether or not they involve oxidation–reduction reactions.

Volatilization. In this simplest separation process, the impurity or the base metal is removed as a gas. Lead containing small amounts of zinc is refined by batch vacuum distillation of the zinc. Most of the zinc produced by the smelting processes described earlier contains lead and cadmium and it must be purified for commercial use. Crude zinc is refined by a two-step fractional distillation. In the first column, zinc and cadmium are volatilized from the lead residue, and in the second column cadmium is removed from the zinc.

Impurities can be removed by formation of a gaseous compound. This is illustrated by the fire refining of copper. Sulfur is removed from the molten metal by oxidation with air and evolution of sulfur dioxide. Oxygen is then removed by reduction with C, CO, H_2, or CH_4 in the form of natural gas, reformed natural gas, or wood. Final impurity contents of ca 0.001 wt % S and 0.05–0.2 wt % O are usually achieved.

The iodide process (van Arkel-de Boer process) is a volatilization process involving transfer of an involatile metal as its volatile compound. It is the process adopted for the purification of titanium. The reaction of iodine gas with the impure metal at 175°C yields gaseous titanium iodide and leaves the impurities in the solid residue.

$$Ti(s) + 2\ I_2(g) \rightarrow TiI_4(g) \qquad (27)$$

The equilibrium is reversed at high temperature. The iodide is decomposed by passing the vapor over an electrically heated wire (1300–1400°C), yielding purified solid titanium and iodine gas which is recycled. A similar reaction for the iodides of zirconium, hafnium, and silicon, for instance, allows the purification of these metals.

Precipitation. In the simplest case, the solubility of an impurity in the liquid metal changes with temperature, and it precipitates as a solid phase upon cooling. For instance, the removal of iron from tin and the removal of copper from lead are achieved by precipitation of a solid phase. When the solid is lighter than the liquid, it floats as a dross on the surface of the melt where it is easily removed by scraping. The process is called drossing.

The precipitation of a solid phase can also be achieved by chemical reaction between the impurity and a precipitating agent to form a solid compound insoluble in the molten metal. The refining of crude lead is an example of this process. Most copper is removed as a copper dross upon cooling of the molten metal, but the removal of the residual copper is achieved by adding sulfur to precipitate copper sulfide. The precious metals are separated by adding zinc to liquid lead to form solid intermetallic compounds of zinc with gold and silver (Parkes process), from which they can be recovered by further treatment (see Lead).

Slag Refining. The principle of removing unwanted constituents by transfer into a slag phase has been illustrated above in the description of smelting processes. It is also used for refining operations in which the liquid metal is maintained in contact with a slag or a molten salt. This second immiscible liquid is usually more oxidizing than the metallic phase and selective oxidation of the impurities renders them soluble in the slag or molten salt. Impurities that are less easily oxidized remain in the liquid metal.

Slag refining is effective for the purification of the less reactive nonferrous metals: gold, silver, copper, lead, and others. For instance, in the recovery of precious metals from anode slimes (see the discussion of copper electrorefining below), one of the steps is a fusion, usually in a small reverberatory furnace, where impurities are removed by oxidation with air and dissolution into successive slags: an oxide slag, a sodium carbonate slag, and a sodium nitrate slag. The product of this operation is a silver–gold alloy (ca 90% Ag and 9% Au) called doré metal, which is further refined by electrolysis. In the refining of lead, one of the steps is the removal of arsenic, antimony and tin in a molten mixture of sodium hydroxide and sodium nitrate (Harris process). Slag refining of copper is effective in removing impurities less noble than copper, but it has been replaced by electrorefining for the production of primary copper.

Steelmaking. Steelmaking is the most important slag refining process by the economic value of its product (see Steel). Pig iron contains up to 4% carbon, ca 1% manganese, 1% silicon, and variable amounts of phosphorus and sulfur. The removal of these impurities is based on preferential oxidation and control of the slag-metal equilibrium.

Modern steelmaking started with the invention of the Bessemer process in 1856. Blowing air through the bottom of a converter containing pig iron oxidizes silicon first, then manganese and finally carbon. The reactions are exothermic, and the heat released is sufficient to maintain the final product, a low carbon steel, molten at ca 1600°C. The original Bessemer process uses an acid slag (high in silica) in which silicon oxide and manganese oxide are soluble. Carbon escapes as carbon monoxide. Phosphorus is also oxidized, but it does not transfer into an acid slag. The Thomas or Basic Bessemer process uses a basic slag (high in lime) which removes phosphorus and sulfur by the following reactions:

$$2 \text{ P(l Fe)} + 3 \text{ CaO(l slag)} + 5 \text{ FeO(l slag)} = \text{P}_2\text{O}_5 \cdot 3\text{CaO(l slag)} + 5 \text{ Fe(l)} \tag{28}$$

$$\text{S(l Fe)} + \text{CaO(l slag)} \rightarrow \text{CaS(l slag)} + \text{O(l Fe)} \tag{29}$$

Steel produced by the converter process contains dissolved nitrogen and oxygen. It is usually deoxidized by small additions of elements like aluminum, manganese, and silicon. Vacuum degassing is used to remove residual oxygen, nitrogen and hydrogen.

The Siemens-Martin or open-hearth process was developed ca 1880. It uses a reverberatory furnace which is charged with solid steel scrap, limestone, iron ore, and molten pig iron. Electric-arc furnaces are also used; they provide the higher temperatures and flexibility of operation required for the production of alloy steels.

Shortly after World War II, the top-blown converter was invented in Austria (L-D process), and oxygen lancing was applied to the open-hearth process. Top-blown oxygen steelmaking is common practice today. The process is autogeneous. Larger converters (250–300 t capacity) can handle 25–35% scrap. Rotating converters like the Kaldo and Rotor processes are also recent developments (4) (see Steel).

Hydrometallurgy

The treatment of ores by dissolution in aqueous solutions is a fairly simple operation which imitates natural leaching processes. Our understanding of aqueous physical chemistry dates from the late nineteenth and early twentieth centuries, and the development of hydrometallurgy started at the same time.

Hydrometallurgical processes are preferred when the pyrometallurgical route is impossible or impractical—for instance, when the metal to be extracted is more reactive than the impurities to be removed (eg, aluminum) or when the grade of the ore is very low and can not be upgraded by physical beneficiation (eg, gold and uranium). In some respects, hydrometallurgy can be described as wet analytical chemistry carried out on a large scale. An infinite number of flow sheets can be designed with many unit operations so that all but a few reactive metals can be extracted from a complex ore and recovered at any desired level of purity. A viable hydrometallurgical process, however, must achieve that goal at an economically acceptable cost.

In recent years, the decrease in grade and the increasingly complex nature of available ores and the need for high-purity metallic products has favored the development of hydrometallurgy. Concern for air pollution caused by pyrometallurgical plants and the cost of preventing it as compared to relative ease of controlling water pollution by hydrometallurgical plants have furthered this trend.

The two main steps of hydrometallurgical flow sheets are leaching, or dissolution of the metal in a suitable aqueous solvent, and recovery or precipitation of the metal or its compound from solution. In exceptional cases like the treatment of sea water or brines for their metal values, the first step has already been achieved by nature. Most hydrometallurgical processes, however, are much more complex. They involve one or several steps for the purification of the solution before the recovery of a pure product becomes feasible. The chemistry of the removal of impurities is in principle similar to that of the recovery of the valuable metals. In addition, several steps are usually required to achieve the physical separation of the solid phase from the liquid: washing, clarification, thickening, filtering, drying, evaporation, etc. Last but not least, the solvent is usually too valuable to be discarded, and it must be regenerated and recycled. Most hydrometallurgical plants operate most efficiently as a closed circuit which has the advantage of limiting water pollution.

Leaching Chemistry. The purpose of the leaching operation is to dissolve the desired mineral and to separate it from the gangue material. The reaction should be selective and fast, and the solvent should be inexpensive or easily regenerated. Several leaching agents are commonly used.

Water. Because most metallic chlorides and sulfates are fairly soluble in water, water can be used to leach calcines from chloridizing or sulfating roasts.

Acid Solutions. Dilute sulfuric acid is the most important solvent for oxide ore and for dead-roasted sulfide concentrates. For instance, the leaching of zinc oxide is described by the following equation, written in the ionic form:

$$ZnO(s) + 2\,H^+(aq) \rightarrow Zn^{2+}(aq) + H_2O(l) \qquad (30)$$

Other acids (eg, hydrochloric or nitric acid) can also be used, but seldom are employed because of higher costs and corrosion problems.

Alkaline Solutions. The most important example of alkaline leach is the digestion of aluminum hydroxide from bauxite by a sodium hydroxide solution at 160–170°C (Bayer process).

$$Al(OH)_3(s) + 2\,OH^-(aq) \rightarrow 2\,AlO_2^-(aq) + 4\,H_2O(l) \tag{31}$$

Complex-Forming Solutions. The solubility of a metal can be enhanced by the formation of a complex with a suitable ligand as illustrated by the dissolution of copper oxide in solutions of ammonium hydroxide and ammonium carbonate.

$$CuO(s) + 4\,NH_4^+(aq) + 2\,OH^- \rightarrow Cu(NH_3)_4^{2+}(aq) + 3\,H_2O(l) \tag{32}$$

The alkalinity of the solution prevents the attack of a carbonate gangue which would be soluble in an acid medium.

Oxidizing Solutions. In many leaching processes the mineral must be oxidized, as for instance, in the leaching of copper sulfides by ferric sulfate or ferric chloride solutions.

$$Cu_2S(s) + 2\,Fe^{3+}(aq) \rightarrow CuS(s) + Cu^{2+}(aq) + 2\,Fe^{2+}(aq) \tag{33}$$

$$CuS(s) + 2\,Fe^{3+}(aq) \rightarrow S(s) + Cu^{2+}(aq) + 2\,Fe^{2+}(aq) \tag{34}$$

Oxidizing and Complex-Forming Solutions. The leaching of gold (or silver) can be achieved only by oxidation of the metal by air and formation of a stable cyanide complex.

$$Au(s) + 2\,CN^-(aq) + \tfrac{1}{2}\,O_2(g) + H_2O(l) \rightarrow Au(CN)_2^-(aq) + 2\,OH^-(aq) \tag{35}$$

Another important example is the leaching of nickel sulfide under ammonia and oxygen pressure to form the nickel hexammine complex ion (Sherritt-Gordon Process).

$$NiS(s) + 2\,O_2(g) + 6\,NH_3(g) \rightarrow Ni(NH_3)_6^{2+}(aq) + SO_4^{2-}(aq) \tag{36}$$

Leaching Techniques. The leaching techniques vary depending on the type of reaction and the characteristics of the ore. Dissolution occurs at the interface between the solid and the solution; its rate depends on the rate of transport of the leaching agent from the solution to the solid and on the rate of chemical reaction. These rates are enhanced by increasing the surface area of the solid, usually by fine grinding of the ore or concentrate. For some low grade material, little preparation is justified beyond fracture and rubblization of the rock.

In Situ Leaching. Copper and uranium ores are sometimes leached *in situ* (in place) by circulating acidified mine water through the underground deposit. This process is known as solution mining.

Heap and Dump Leaching. Heap leaching is practiced mainly for oxide copper ores. The crushed ore is placed in large heaps and sprayed with the leaching solution. Old tailings containing copper values as sulfides are also leached in a similar way known as dump leaching.

In situ and dump leaching rely on the presence of iron in natural waters. The iron is oxidized by contact with the atmosphere, and then the ferric ions act as oxidant for the dissolution of sulfides by reactions similar to equations 33–34. Bacterial activity in the bed has been found to enhance reaction rates. This type of leaching, however, is a slow process carried out over periods of several years.

Percolation Leaching. Ground material coarse enough to permit circulation of a solution through a bed of particles can be leached by percolation of the solvent through the material placed in a tank, or vat. The process usually takes several days.

Agitation Leaching. Very fine ore products, slimes, are treated by agitation of a pulp, which is a suspension of the ore particles in the solution. The process is carried out in a tank with mechanical or air agitation. In a leaching process for which oxygen is needed, eg, gold leaching, air agitation is preferred. Residence time is several hours.

Pressure Leaching. Leaching at elevated temperature (90–250°C) is carried out in autoclaves under pressure. In some instances the purpose is to modify equilibrium conditions and to increase the solubility of the ore (eg, digestion of bauxite). In cases requiring a gaseous reagent, eg, oxygen or ammonia, the higher pressure increases the solubility of the gas and enhances the reaction rates. The increased temperature also increases reaction rates, and residence times are one or two hours.

Liquid-Solid Separation. The separation of the solution containing the dissolved metal (pregnant solution) from the leach residue is the final step of the leaching. Most leaching operations are carried out in several tanks in series to permit countercurrent flow: the pregnant solution leaves at one end after contacting fresh ore, and the depleted solids exit at the other end after contacting the fresh solvent. The additional steps of washing of the residue, clarification of the solution, and filtering are required.

Recovery of Metal from Solution. The same chemical principles govern the recovery of an element or compound from solution, whether it is the main metal to be processed or an impurity to be removed. The various methods are discussed briefly with examples of both types.

Electrowinning. When it is possible, electrolytic deposition is the most efficient way of recovering a valuable metal from solution. It is quite selective and usually yields a pure product which can be marketed directly as cathodes, or after casting into commercial shapes. It is, however, the most expensive method. See Electrometallurgy below.

Precipitation. The precipitation of aluminum trihydroxide in the recovery step of the Bayer process is achieved either by lowering the temperature or by diluting the pregnant liquor and reducing its pH. Both methods reverse the direction of equation 31, but seeding with previously precipitated crystals is required in order to initiate nucleation.

The removal of copper from the pregnant nickel solution in the Sherritt-Gordon process is an example of purification by precipitation of a fairly insoluble compound. First, in the "copper boil" step, ammonia is driven off by heating the solution. This causes precipitation of some copper sulfide. The residual copper is removed by adding hydrogen sulfide for the chemical precipitation of more copper sulfide.

$$Cu^{2+}(aq) + H_2S(g) \rightarrow CuS(s) + 2\ H^+(aq) \tag{37}$$

Cementation. A metal can be removed from solution by displacing it with a more active metal. This simple and cheap method has been commonly used to recover copper from dilute solution (1–3 g/L) using scrap iron or tin cans as a cheap reducing agent.

$$Cu^{2+}(aq) + Fe(s) \rightarrow Cu(s) + Fe^{2+}(aq) \tag{38}$$

A similar reaction achieves the precipitation of gold (or silver) from cyanide solution by using zinc powder.

$$2\,Au(CN)_2^-(aq) + Zn(s) \rightarrow 2\,Au(s) + Zn(CN)_4^{2-}(aq) \tag{39}$$

Cementation is also an efficient way of purifying a pregnant solution by removing impurities that are less active than the metal being processed. This is the case in removing copper, cadmium, cobalt, and nickel from pregnant zinc solutions prior to electrowinning.

The cementation of gold and the purification of the zinc electrolyte are usually carried out in cylindrical vessels with mechanical agitation. The cementation of copper is carried out in long narrow tanks called launders, in rotating drums, or in an inverted cone precipitator (see Copper).

Gas Reduction. The use of a gaseous reducing agent is attractive because it produces the metal as a powder that can easily be separated from the solution. Carbon dioxide, sulfur dioxide, and hydrogen can be used to precipitate copper, nickel, and cobalt, but only hydrogen reduction is applied on an industrial scale. In the Sherritt-Gordon process, the excess ammonia is removed during the purification to achieve a 2/1 ratio of NH_3/Ni in solution. Nickel powder is then precipitated by

$$Ni(NH_3)_2^{2+}(aq) + H_2(g) \rightarrow Ni(s) + 2\,NH_4^+(aq) \tag{40}$$

The reaction proceeds in an autoclave at 200°C under ca 3 MPa (30 atm) of hydrogen pressure.

Ion Exchange. Metallic ions can be removed from an aqueous solution by exchange with ions at the surface of an organic resin. Both anionic and cationic exchangers are available. Anionic ion-exchange resins are used for concentration and purification of the dilute pregnant solutions obtained by leaching uranium ores with sulfuric acid. The ion-exchange reaction can be represented by:

$$UO_2(SO_4)_3^{4-}(aq) + 4\,RX(s) \rightarrow R_4UO_2(SO_4)_3(s) + 4\,X^-(aq) \tag{41}$$

where X^- is an anion, usually NO_3^- or Cl^- and R represents the organic resin. After the resin is loaded, it is eluted with a strong solution of the X^- ion. This reverses the equilibrium conditions for equation 41, transfers the complex uranyl ion back to an aqueous phase and regenerates the resin. The technology for this process is borrowed from the water-treatment industry using mainly packed columns (see Ion exchange; Water). The resin-in-pulp process is typical of uranium hydrometallurgy; the leaching pulp and the ion-exchange resin are agitated together and both reactions proceed in the same reactor. Then the residue and the loaded resin are separated.

Solvent Extraction. This liquid–liquid extraction process, well known in the chemical industry, was first used in extractive metallurgy for the processing of uranium. When a dilute solution of uranium is contacted with an extractant like di(2-ethylhexyl) phosphoric acid (EHPA) or R_2HPO_4, dissolved in kerosene, the uranyl ion is transferred to the organic phase.

$$2\,R_2HPO_4(org) + UO_2^{2+}(aq) \rightarrow UO_2(R_2PO_4)_2(org) + 2\,H^+(aq) \tag{42}$$

The pregnant organic solvent is stripped by agitation with a strong carbonate solution which removes uranium as the stable $UO_2(CO_3)_3^{4-}$ aqueous complex.

More recently, the development of selective extractants for copper (LIX 64N is the most commonly used) has made economically feasible the extraction of copper from dilute solutions (3 g Cu/L and its transfer to a more concentrated sulfuric acid solution (30 g Cu/L, 150 g H$_2$SO$_4$/L), from which it is recovered by electrowinning. The extraction reaction is:

$$\text{Cu}^{2+}(\text{aq}) + 2\,\text{HR}(\text{org}) \rightarrow \text{CuR}_2(\text{org}) + 2\,\text{H}^+(\text{aq}) \tag{43}$$

The process is often called liquid ion exchange (LIX).

In addition to its use for processing of nuclear fuels, solvent extraction is applied to cobalt–nickel separation among others. Both extraction columns and mixer-settler apparatus are in use (see Extraction, liquid–liquid).

Hydrometallurgical Flow Sheets. The various operations described above can be combined in many ways to design hydrometallurgical processes. This is illustrated by a few examples discussed below.

Aluminum. All primary aluminum is presently produced by molten salt electrolysis (see Electrometallurgy below), which requires a feed to the reduction cell of high purity alumina. The Bayer process is a chemical purification of the bauxite ore by selective leaching of aluminum according to equation 31. Other oxide constituents of the ore, namely silica, iron oxide, and titanium oxide remain in the residue, known as red mud. No solution purification is required and pure aluminum hydroxide is obtained by precipitation; it is calcined to yield pure alumina (see Aluminum).

Uranium. The hydrometallurgical treatment of uranium ores is a concentration and purification process. Typical ore grade is 0.1 to 0.5% U$_3$O$_8$, and pregnant solutions contain ca 1 g U$_3$O$_8$/L. The dissolution requires the presence of an oxidant, either oxygen or a ferric salt.

$$\text{U}_3\text{O}_8(\text{s}) + 6\,\text{H}^+(\text{aq}) + \tfrac{1}{2}\,\text{O}_2(\text{g}) \rightarrow 3\,\text{UO}_2^{2+}(\text{aq}) + 3\,\text{H}_2\text{O}(\text{l}) \tag{44}$$

The solvent is a solution of either sulfuric acid or sodium carbonate which form stable complex uranyl ions: $\text{UO}_2(\text{SO}_4)_2^{2-}$, $\text{UO}_2(\text{SO}_4)_3^{4-}$, $\text{UO}_2(\text{CO}_3)_3^{4-}$. The pregnant solution is concentrated and purified by ion exchange or by solvent extraction, yielding a stripping solution of ca 50 g U$_3$O$_8$/L. Uranium is then precipitated chemically, and pure U$_3$O$_8$ is obtained by calcination (see Uranium).

Nickel. Most nickel is produced by smelting sulfide ores. Several hydrometallurgical processes have been developed. Among them the Sherritt-Gordon process stands out as the first successful commercial application of pressure hydrometallurgy to a complex feed. The raw material is a pentlandite (NiS.FeS) concentrate. Nickel is leached selectively in an ammoniacal solution according to equation 36, iron remains in the residue as Fe$_2$O$_3$. Copper is removed from solution by the purification steps mentioned earlier with equation 37. Cobalt remains in solution and nickel is recovered by hydrogen reduction following equation 40. The flow sheet in Figure 5 is an excellent illustration of the combination of several hydrometallurgical operations into a complex process.

Electrometallurgy

The use of electricity for winning and refining metals could not have been developed without an understanding of the basic principles of electrochemistry and the availability of cheap industrial electric power. This explains why electrometallurgy is only ca one hundred years old.

Electrometallurgy is a more powerful tool than the other extractive processes. It provides a way of supplying energy to the system in a way that enables a reaction

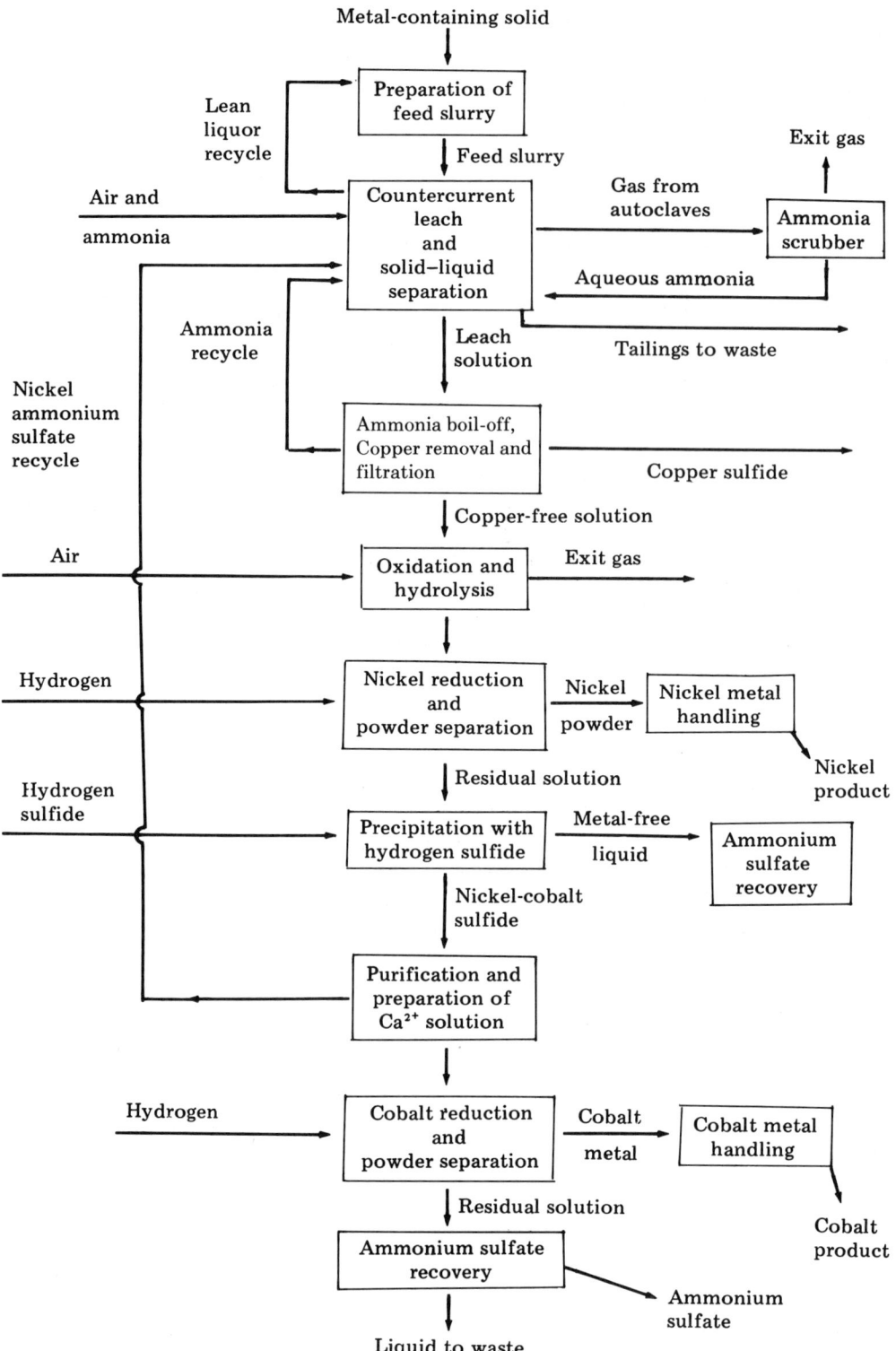

Figure 5. Flow sheet for Sherritt-Gordon process for production of nickel and cobalt metals from sulfide ore.

to proceed against its chemical affinity and in the direction of its electrochemical affinity. Chemically reactive metals are more easily recovered by electrometallurgy than by chemical reduction, and they can be obtained very pure. Electrometallurgy, however, has practical limitations: First, electricity is an expensive form of energy. Second, the electrochemical process occurs at an electrolyte–metal interface, so that the output of an electrochemical reactor is directly proportional to the area of that interface. Improving the energy efficiency and increasing the electrode area per unit volume of reactor remain a challenge for the electrometallurgical engineer.

The theoretical energy requirement for an electrochemical process is given by

$$nFE = -\Delta G \tag{45}$$

where E is the reversible electromotive force (emf), ΔG is the Gibbs free energy change, n is the number of electrons involved in the process and F is Faraday's constant (96,483 coulombs per equivalent). For an electrolytic process, ΔG is positive, and E is negative, which means that energy must be supplied to the system. The actual voltage to be applied to the cell, V, is larger than the reversible emf because of the irreversibility of the process, namely electrochemical overpotentials, η, and ohmic losses due to the resistance of the circuit, R

$$V = -E + \sum_i \eta_i + RI \tag{46}$$

where I is the current. The voltage efficiency, ϵ_V, is defined as the ratio between the arithmetic value of the emf and the applied voltage.

$$\epsilon_V = \frac{-E}{V} \tag{47}$$

The theoretical amount of metal produced by electrolysis is directly proportional to the amount of electricity according to Faraday's law. Because of losses by chemical or electrochemical processes, the actual amount is less. It is characterized by the current efficiency ϵ_I, defined by

$$\epsilon_I = \frac{nF}{It} \tag{48}$$

where t is the time required to produce one mole of metal. Combining equations 47–48 leads to the relationship between ΔG and the energy actually used, Q, and the definition of the energy efficiency, ϵ_Q:

$$Q = VIt = \frac{\Delta G}{\epsilon_I \epsilon_V} = \frac{\Delta G}{\epsilon W} \tag{49}$$

Usually the specific energy consumption is reported in kW·h/kg of product, q:

$$q = \frac{V}{\epsilon_I} \frac{nF}{M} \tag{50}$$

where M is the atomic weight of the metal in kg.

Electrowinning from Aqueous Solutions. Electrowinning is the recovery of a metal by electrochemical reduction of one of its compounds dissolved in a suitable electrolyte. Various types of solutions can be used, but sulfuric acid and sulfate solutions are preferred because they are less corrosive than others and the reagents are fairly cheap. From an electrochemical viewpoint, the high mobility of the hydrogen ion leads to high conductivity and low ohmic losses, and the sulfate ion is electrochemically inert under usual conditions.

The generalized flow sheet of an aqueous electrowinning process consists of at least three main steps: (1) The metal is put into solution by leaching of a calcine (roasted concentrate of sulfide ore) or by direct leaching of low grade ores containing oxidized minerals or weathered sulfides. (2) The pregnant solution is purified to remove any metallic impurities more noble than the metal to be electrowon, or any impurities that could reduce the current efficiency. Sometimes, the pregnant solution is treated by solvent extraction to produce a more concentrated electrolyte. (3) The purified solution is fed to the electrolysis tanks where the metal is plated on a cathode and oxygen is evolved at an inert anode, usually made of lead. The sulfuric acid is regenerated by the anodic process and is recycled to leaching.

Zinc. The electrowinning of zinc on a commerical scale started in 1915, and its share of the zinc production grew steadily. At present, most new facilities are electrolytic plants. The success of the process is due to its ability to handle complex ores and to produce, after purification of the electrolyte, high-purity zinc cathodes at an acceptable cost. Over the years, there have been only minor changes in the chemistry of the process to improve zinc recovery and solution purification, but major improvements have been made in the areas of process instrumentation and control, automation and prevention of water pollution (see Zinc).

Zinc sulfide ores are concentrated by flotation, roasted, and then leached in sulfuric acid. The leaching involves at least two stages. In a neutral leach, an excess of calcine maintains a pH of ca 5; with aeration and sometimes addition of manganese dioxide, iron is removed by precipitation of ferric hydroxide. The solid residue undergoes further dissolution of the zinc in one or several stages of cold or hot acid leach, and the final residue goes to a lead smelter. Recent improvements have resulted in the precipitation of iron as goethite (FeOOH) or jarosite ($NaFe_3(SO_4)_2(OH)_6$). Impurities such as arsenic, antimony, and germanium coprecipitate with iron. Copper, cadmium, cobalt, and nickel are removed by cementation on zinc powder (see Fig. 6).

The reversible potential for zinc reduction (−0.763 V) is much more cathodic than the potential for hydrogen evolution, and the two reactions proceed simultaneously. The efficient cathodic deposition of zinc depends upon maintaining a high hydrogen overvoltage. Metals more noble than zinc are plated on the cathode and promote hydrogen evolution and zinc corrosion by setting up local galvanic cells, thereby reducing the electrochemical yield of zinc. Current efficiencies slightly above 90% are achieved in modern plants by careful purification of the electrolyte to bring the concentration of the most harmful impurities, germanium, arsenic and antimony, down to ca 0.01 mg/L. Addition of organic surfactants like glue, improves the quality of the deposit and the current efficiency.

The zinc electrolyte contains 60–160 g zinc/L and ca 100 g free sulfuric acid/L. It is electrolyzed between electrodes suspended vertically in lead or plastic-lined (eg, poly(vinyl chloride)) tanks of concrete or wood. The insoluble anodes are of lead with small amounts of silver. The anodic reaction produces oxygen and regenerates sulfuric acid which is recycled to the acid leach. The cathodes are aluminum sheets, from which the zinc deposits are stripped every 24–48 h. The effective cathode area varies from about 1.5 m^2 in traditional plants to 2.6 m^2 for the new "jumbo" cathodes. The current density is usually 300–600 A/m^2. The electrodes are in parallel in each cell and the cells

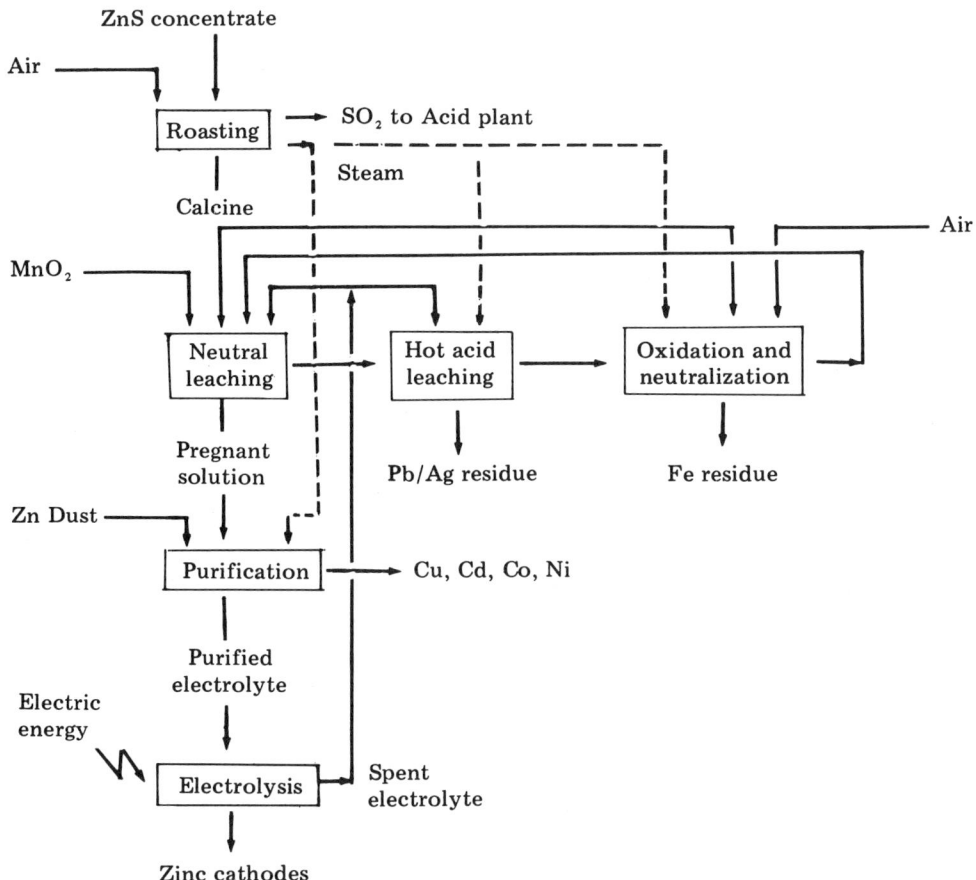

Figure 6. Removal of copper, cadmium, cobalt, and nickel by cementation on zinc powder.

are connected in series. The average voltage across a cell is ca 3.5 V. The reversible emf calculated for the cell reaction:

$$Zn^{2+}(aq) + H_2O(l) \rightarrow Zn(s) + \tfrac{1}{2} O_2(g) + 2 H^+(aq) \qquad (51)$$

is ca 2 V; the remaining 1.5 V is accounted for by the anodic overvoltage (about 0.9 V), the cathodic overvoltage, and the ohmic drop in the electrolyte and at the electrode contacts. The voltage efficiency is ca 57% and the energy efficiency is usually slightly above 50%. The energy consumption is 3.3 kW·h/kg zinc. The relatively high energy cost of the process is acceptable in view of the special high grade zinc produced (99.995% zinc).

Copper. A copper electrowinning process was developed commercially in 1912 for the treatment of lean ores, but only ca one tenth of the world copper output is presently electrowon. Since copper is easily extracted by smelting, some copper is produced by dissolution of roasted copper oxide or sulfate followed by electrodeposition. Most electrowon copper, however, comes from the direct leaching of low-grade oxidized copper ores. In recent years, the application of solvent extraction technology and the development of liquid ion-exchange reagents specifically for copper have made possible the treatment of low-grade raw material through dump leaching or solution mining, followed by concentration of the copper solution by solvent extraction to a level suitable for electrolysis. The electrowinning of copper is carried out in sulfate solutions containing 25–35 g/L copper and ca 100 g/L free acid at current densities equal to ca 200 A/m^2. The anodes are made of lead alloyed with antimony, and copper is deposited on copper starting sheets (see Copper).

Other Nonferrous Metals. Cadmium, and in much smaller quantities, thallium and indium, are by-products of zinc and are recovered by electrolysis in sulfate solutions. When associated with copper, cobalt is also electrowon by a similar process. About 10% of the primary nickel output is produced by electrolysis in chloride and sulfate solutions. The relatively small outputs of metallic chromium and manganese are mainly through electrometallurgy. These metals are more difficult to electroplate because of their more complex chemistry and higher electrochemical activity. The close control of the process variables required is achieved by separating the anolyte and catholyte by a diaphragm. Gallium is produced in the liquid state or as an amalgam by electrolysis of a caustic solution.

Electrolysis of Molten Salts. Metals more active than zinc and manganese can not be recovered by electrodeposition from aqueous solutions. Most of them, however, can be electrowon from a molten electrolyte. Usually, a compound of the metal to be electrodeposited is dissolved in a mixture of salts of more active metals in order to achieve a low melting point, a suitable viscosity and density, and a high conductivity. Solid electrodeposits obtained from molten salts are mostly dendritic or powdery, and they are difficult to separate from the melt. Reduction of the metal as a separate liquid phase with a density different from that of the electrolyte is preferable. The recombination of reaction products is the most common cause of faradaic inefficiency. Because recombination is often related to the solubility of the reduced metal in the molten electrolyte, the cells must be designed to protect the reduced metal from contact with the anodic reaction products and with air. The cell electrolyte can not be recycled through leaching and purification as in aqueous electrowinning, and the feed must be carefully purified and dehydrated.

Aluminum. The electrowinning of aluminum, developed in 1885 by Hall in the United States and by Héroult in France, is the single most important electrometallurgical process and remains today the only commercial process for the production of aluminum. Aluminum is extracted from bauxite ores by digestion in a caustic soda solution to dissolve the alumina as sodium aluminate and leave behind a red mud containing iron oxide and other major impurities. After crystallization and calcining, the purified alumina is fed to the cells where it is dissolved in an electrolyte consisting mainly of cryolite ($AlF_3 \cdot 3NaF$) and other fluorides in smaller amounts (eg, CaF_2 and LiF). The cells, called pots, are made of steel lined with insulating material and carbon bricks that make electrical contact to the molten aluminum cathode. The fused salt electrolyte floats on top of the molten metal and carbon anodes dip into the bath from above, leaving a spacing of ca 5 cm to the cathode. Two types of anodes, prebaked blocks and self-baking Söderberg anodes, are used in modern plants.

The electrolytic decomposition of alumina yields oxygen which reacts with the carbon anode for an overall cell reaction:

$$Al_2O_3(\text{melt}) + 3/2\ C(s) \rightarrow 2\ Al(l) + 3/2\ CO_2(g) \tag{52}$$

The theoretical value of the reversible electromotive force for this reaction is 1.5 V, but the cell operates under a total voltage from ca 4 to 4.5 V. The difference is accounted for by the overvoltage at the electrodes (including the anode effect) and the ohmic drop through the electrolyte and the electrode contacts. The heat generated by irreversible electrode processes and by ohmic losses keeps the operating temperature of the cell at the desired level of ca 960°C. Current densities range from 0.5 to 1.2 A/cm^2, and the total current of modern pots is ca 100,000 A. Alumina is periodically added to the melt to maintain its concentration between 2 and 5%. The current efficiency of modern plants is ca 90%; the main losses are due to the reoxidation of aluminum by carbon dioxide, yielding alumina, and carbon monoxide. The energy consumption is ca 14 kW·h/kg of aluminum. Improvements in the energy efficiency of modern plants have been achieved mainly by a better control of the electrolysis through the use of computers to monitor the pot conditions.

The chemistry of the process has not changed since its invention, but the electrolysis of an all-chloride mixture in a cell of a new design is presently being tested on a small commercial scale (see Aluminum).

Magnesium. The electrowinning of magnesium, another well-established process, accounts for ca 80% of the metal output. The electrolyte is a mixture of chlorides of potassium, sodium, calcium, and magnesium. Because the liquid metal has a lower density than the electrolyte, it floats to the surface. The cells are designed so that the reduced metal does not come into contact with air or with the chlorine gas produced at the anode. The temperature of the cells is kept between 700 and 750°C by external heating. The cell voltage ranges from 5 to 8 V, and the energy consumption is ca 22 kW·h/kg of magnesium (see Magnesium).

Other Metals. All the sodium metal produced today comes from electrolysis of sodium chloride melts in Downs cells. The cell consists of a cylindrical steel cathode separated from the graphite anode by a perforated steel diaphragm. Lithium is also produced by electrolysis of the chloride in a process similar to that used for sodium. The other alkali and alkaline earth metals can be electrowon from molten chlorides, but thermochemical reduction is preferred commercially. The rare earths can also be electrowon but only the mixture known as misch metal is prepared in tonnage quantity by electrochemical means. In addition, beryllium and boron are produced by electrolysis on a commercial scale in the order of a few hundred tons per year. Processes have been developed for electrowinning titanium, tantalum, and niobium from molten salts; these metals, however, are obtained as a powdery deposit which is not easily separated from the electrolyte so that further purification is required.

Electrorefining. Electrolytic refining is a purification process in which an impure metal anode is dissolved electrochemically in a solution of a salt of the metal to be refined, and then recovered as a pure cathodic deposit. Electrorefining is a more efficient purification process than other chemical methods because of its selectivity. In particular, for metals such as copper, silver, gold, and lead, which exhibit little irreversibility, the operating electrode potential is close to the reversible potential, and a sharp separation can be accomplished, both at the anode where more noble metals do not dissolve and at the cathode where more active metals do not deposit (see also Electrochemical processing).

Since the same electrochemical reaction proceeds in opposite directions at the anode and the cathode, the overall chemical change is a small change in the activity of the metal at the two electrodes. Therefore, the reversible emf is practically zero. There is some polarization of the electrodes during electrolysis, but the main component of the voltage across the operating cell is the ohmic drop through the electrolyte and at the electrode contacts. Normal cell voltages are ca 0.2 V. The power consumption is correspondingly very small, and electrorefining is much less sensitive to the cost of electric power than other electrometallurgical processes. When a diaphragm is used to separate the anodic and cathodic solutions, the cell voltage increases up to ca 1.2 V, and the power consumption rises accordingly.

Copper. The first electrolytic copper refinery was started in 1871, making it the oldest commercial electrometallurgical process (see Copper). Today, most copper is electrorefined. The refining process insures that the metal meets the specifications of the electrical industry which are stringent because minor quantities of some impurities lower the electrical conductivity of copper markedly; furthermore, silver and gold, common constituents of copper ores, follow along through all the pyrometallurgical steps. Their removal by the electrolytic refining is an important economic asset of the process.

Impure copper is cast in the shape of anodes ca 0.9 m by 1.0 m and 3.5–4.5 cm thick, weighing 300–400 kg. These anodes are cast with lugs that support them from the walls of the cell or from bus bars and make electrical contact. The starting cathodes are pure copper sheets, made by electrodeposition on smooth starting blanks of copper or titanium. The electrode spacing from anode center to anode center is ca 10 cm. The current density is usually 200 to 250 A/m^2. The anodes are consumed and replaced at regular intervals of 20 to 28 d, and two successive cathodes are produced during the same period. The solution contains ca 45 g/L of copper as copper sulfate and ca 200 g/L of sulfuric acid. The temperature is maintained at 55–60°C to lower the resistance of the electrolyte which is circulated through the cell. The lead or plastic-lined tanks of wood or concrete used hold ca 30 to 35 anodes and one more cathode than anode.

Metals less noble than copper, such as iron, nickel, and lead, dissolve from the anode. The lead precipitates as lead sulfate in the slimes. Other impurities such as arsenic, antimony, and bismuth remain partly as insoluble compounds in the slimes and partly as soluble complexes in the electrolyte. Precious metals, such as gold and silver, remain as metals in the anode slimes. The bulk of the slimes consist of particles of copper falling from the anode, and also contain insoluble sulfides, selenides or tellurides. These slimes are processed further for the recovery of the various constituents. Because the cathode potential is only slightly more cathodic than the equilibrium potential for the deposition of copper, metals less noble than copper do not deposit but accumulate in solution. This requires periodic purification of the electrolyte to remove nickel sulfate, arsenic, and other impurities.

Nickel. Most nickel is also refined by electrolysis. Nickel, however, does not behave reversibly at an electrode, and therefore both copper and nickel dissolve at the potential required for anodic dissolution. To prevent plating of the dissolved copper at the cathode, a diaphragm cell is used, and the anolyte is circulated through a purification circuit before entering the cathode compartment (see Nickel).

Other Metals. Although most cobalt is refined by chemical methods, some is electrorefined. Lead and tin are fire-refined, but a better removal of impurities is achieved by electrorefining. Very high purity lead is produced by an electrochemical process using a fluosilicate electrolyte. A sulfate bath is used for purifying tin. Silver is produced mainly by electrorefining in a copper nitrate electrolyte, and gold is refined by chemical methods or by electrolysis in a chloride bath.

The electrorefining of many metals can be carried out with molten-salt electrolytes, but these processes are usually expensive and have found little commercial use in spite of possible technical advantages. The only application on an industrial scale is the electrorefining of aluminum by the three-layer process. The density of the molten-salt electrolyte is adjusted so that a pure molten aluminum cathode floats on the electrolyte, which in turn floats on the impure anode consisting of a molten copper-aluminum alloy. The process is used to manufacture high purity aluminum.

BIBLIOGRAPHY

1. G. Agricola, *De Re Metallica*, Basil, 1556, translated by H. C. Hoover and L. L. Hoover, Dover Publications, Inc., New York, 1950.
2. T. Rosenqvist, *Principles of Extractive Metallurgy*, McGraw-Hill Book Co., New York, 1974.
3. J. Newton, *Extractive Metallurgy*, John Wiley & Sons, Inc., New York, 1959.
4. J. R. Boldt, Jr., and P. Queneau, *The Winning of Nickel*, Van Nostrand Reinhold Co., Princeton, 1967.

General References

Refs. 2–4.
C. B. Alcock, *Principles of Pyrometallurgy*, Academic Press, London, 1976.
A. K. Biswas and W. G. Davenport, *Extractive Metallurgy of Copper*, Pergamon Press Ltd., Oxford, 1976.
J. R. Boldt, Jr., and P. Queneau, *The Winning of Nickel*, Van Nostrand Reinhold Co., Princeton, 1967.
A. R. Burkin, *The Chemistry of Hydrometallurgical Processes*, D. Van Nostrand, Princeton, N.J., 1966.
W. H. Dennis, *Extractive Metallurgy*, Philosophical Library, New York, 1965.
J. D. Gilchrist, *Extractive Metallurgy*, Pergamon Press Ltd., Oxford, 1967.
"General Principles" in F. Habashi, *Principles of Extractive Metallurgy*, Vol. 1, Gordon and Breach, New York, 1969.
"Hydrometallurgy" in F. Habashi, *Principles of Extractive Metallurgy*, Vol. 2, Gordon and Breach, New York, 1969.
C. R. Hayward, *An Outline of Metallurgical Practice*, Van Nostrand Reinhold Co., New York, 1952.
A. Kuhn, ed., *Industrial Electrochemical Processes*, Elsevier Scientific Publishing Co., Inc., Amsterdam, 1971.
H. H. Kellogg "Pyrometallurgy" in D. N. Lapedes, ed., *McGraw-Hill Encyclopedia of Science and Technology*, Vol. 10, McGraw-Hill Book Company, New York, pp. 125–132.
P. Duby "Electrometallurgy" in D. N. Lapedes, ed., *McGraw-Hill Encyclopedia of Science and Technology*, Vol. 4, McGraw-Hill Book Company, New York, pp. 564–566.
C. L. Mantell, *Electrochemical Engineering*, McGraw-Hill Book Co., New York, 1960.
R. H. Parker, *An Introduction to Chemical Metallurgy*, 2nd ed., Pergamon Press Ltd., Oxford, 1978.
R. D. Pehlke, *Unit Processes of Extractive Metallurgy*, Elsevier Scientific Publishing Co., Inc., New York, 1973.
N. N. Sevryukov, B. A. Kuzmin, and E. V. Chelischev, *General Metallurgy*, Mir Publishers, Moscow, 1969.
N. N. Sevryukov, *Nonferrous Metallurgy*, Mir Publishers, Moscow, 1975.
G. D. Van Arsdale, *Hydrometallurgy of Base Metals*, McGraw-Hill Book Co., New York, 1953.
"Metallurgie" in K. Winnacker and L. Küchler, eds., *Chemische Technologie*, Vol. 6, Hanser Verlag, Munich, 1973.

PAUL DUBY
Columbia University

F

FACE POWDER. See Cosmetics.

FANS AND BLOWERS

Fan is the generic term for low pressure air- and gas-moving devices using rotary motion. They are subdivided into centrifugal and axial-flow types, depending on the direction of air flow through the impeller. In centrifugal fans, the air is introduced into the center of a revolving wheel or rotor with peripheral blades. Air is drawn through the blades and forced out in centrifugal flow into a scroll or volute housing where a portion of the kinetic energy is converted to pressure or static head. In axial-flow fans the air continues to move directly forward through the fan along the axis of the shaft. Kinetic energy is imparted to the air by the shape and arrangement of the blades. After discharge through the blades, although the general flow direction is still forward, a spiral component of velocity generally has been added to the air. A propeller-type fan is the most common axial-flow fan but more complicated types are in use where the blades resemble vanes in a turbine.

Blower is a term applied to a centrifugal fan generally when it is used to force air through a system under positive pressure. It generally implies a fan developing a reasonably high static pressure of at least 500 Pa (several in. of water). High speed centrifugal blowers (≥ 3600 rpm) are also available in one or more stages to compress air to pressures of 108–150 kPa (1–7 psig). The term blower is also applied to relatively low pressure positive displacement compressors of the rotary lobe, screw, or sliding vane types where the discharge pressure is usually less than 205 kPa (15 psig). Positive displacement blowers are outside the scope of this article.

When a fan is placed at the end of a system so that most of the system pressure drop is on the suction side of the fan, it is commonly called an exhaust fan or an exhauster. This term may also be applied to a ventilating fan whose primary function is to exhaust air from a room or an open hood.

Centrifugal compressors or turbocompressors are high volume centrifugal devices capable of gas compression varying from 105 to >1500 kPa (0.5 to several hundred psig). They generally consist of a number of stages of alternating rotating and stationary turbine blades and turn at very high speeds (see High pressure technology).

The total pressure produced by a fan can be measured with an impact probe pointed directly upstream. The pressure so measured is a combination of both the static pressure and the kinetic energy pressure equivalent. Static pressure can be measured with a properly designed static wall tap or with the static pressure parts of a pitot tube. It represents the true pressure head exclusive of velocity effects. The difference between the total (impact) pressure and the static pressure is the velocity pressure or velocity head. Pressure readings are normally expressed in millimeters or inches of water (1 mm water = 9.807 Pa; 1 in. water = 249.1 Pa) and are referred to atmospheric pressure (101.3 kPa) as the reference base. Thus barometric pressure must be added to obtain absolute pressure. The total pressure rise produced by a fan is the difference in total pressure between the fan outlet and inlet. The fan static pressure is the total pressure rise for the fan reduced by the discharge velocity pressure. Inlet velocity head is assumed to be zero for fan rating purposes.

CENTRIFUGAL FANS

Figure 1 shows parts and names commonly associated with centrifugal fan components. The rotation of the wheel causes air between the blades to be rotated. The

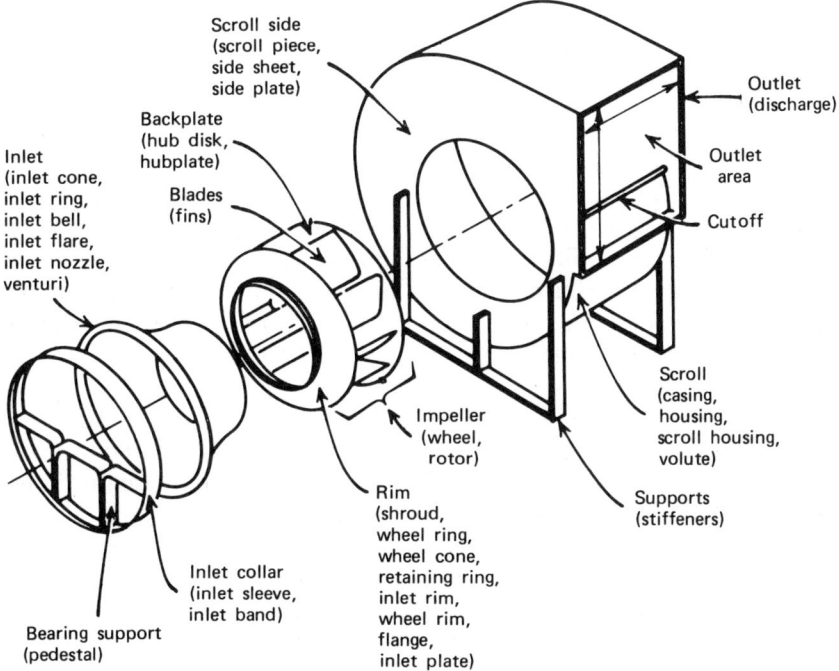

Figure 1. Component parts of a centrifugal fan (1).

resulting centrifugal force causes this air to be compressed and ejected radially from the wheel. The compression results in an increase in static pressure in the fan scroll. The static pressure produced at the blade tips depends on the ratio of the velocity of air leaving the tips to the velocity of air entering at the heel of the blades. Thus the longer the blades, the greater the static pressure developed by the fan at a constant speed.

As the air leaves the blade tips, it contains kinetic energy by virtue of its velocity. The directional component of this velocity is both rotative and radial. When the fan blades are inclined forward, these components are cumulative. With backward-inclined blades, the components are in opposition. The purpose of the fan volute or scroll-shaped casing is to convert a portion of the kinetic energy of the air leaving the blades into static pressure.

Design operating efficiencies of fans under test conditions are in the range of 40–80%. Actual efficiency can be affected appreciably by the arrangement of inlet and outlet duct connections.

The air power in W (hp) of a fan is given by equation 1:

$$\text{air power} = Q\Delta p \quad (\text{in units of hp, } 144\, Q'\Delta p'/33{,}000) \tag{1}$$

where Q is the volume of gas handled in m^3/s (for hp, Q' in ft^3/min), and Δp is the pressure rise across the fan, Pa (for hp, $\Delta p'$ in psi). In many fan installations, the velocity head of the fan discharge is wasted. In such cases, the fan static pressure may be used in equation 1 instead of the total pressure. Fan efficiency is expressed by equation 2:

$$\text{efficiency} = \frac{\text{air power}}{\text{shaft power}} \tag{2}$$

Performance Testing

Although fan performance characteristics can be roughly estimated during the early stages of design, fan efficiency losses and slip cannot be estimated accurately from theory alone. Therefore, the exact performance characteristics of a new fan design must be determined by testing. Test conditions must be carefully controlled, such as provision for steady and uniform flow of air approaching the fan inlet since any inlet disturbances can affect performance. For this reason, fan field tests are seldom reliable and most testing is performed in a laboratory on a test block following procedures set forth in AMCA standard 210 (2) for performance. AMCA standard 300 (3) covers sound testing.

Standard 210 specifies several test methods. Figure 2 illustrates one of the available methods and a typical performance curve. Fans designed for a duct as illustrated have a section of straight discharge duct attached. Straightening vanes are provided to eliminate swirl, reduce turbulence, and aid flow equilization across the duct. Air flow is determined with a pitot traverse while the fan is operated at a constant speed. The measured pressures are corrected for duct losses back to fan outlet conditions. (It should be understood that a fan performs in accordance with the performance curve only if there is an equivalent duct actually present to convert velocity head efficiently to static head.) At the end of the test, the duct is blanked to measure discharge pressure and shaft power at a shutoff (no flow) condition. The opposite

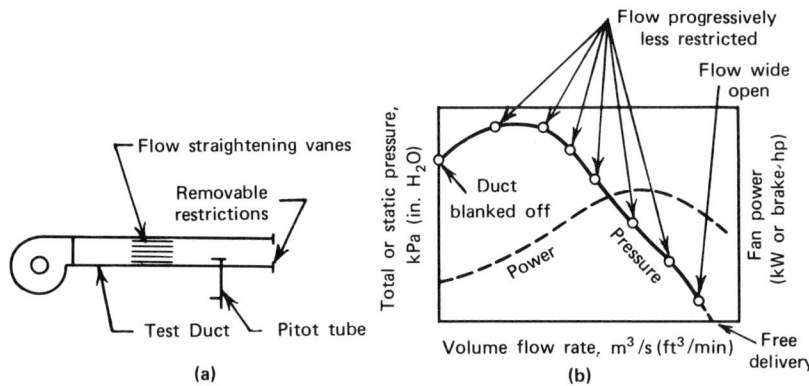

Figure 2. Fan performance: (**a**) a typical test arrangement; (**b**) performance curve.

extreme of the curve, free delivery (equivalent to duct removal), is extrapolated from nearly wide-open conditions. Intermediate points at sufficiently close intervals to define the curve would be measured by replacing the blank at the end of the duct with restricting orifices of varying cross section.

Types and Characteristics

Figure 3 illustrates the four basic fan wheel and blade designs and their performance curves.

Forward-Curved Blades. In the forward-curved design, both the heel and tip of the blade are curved forward in the direction of rotation. Air leaves the tip of the wheel at a velocity greater than the wheel-tip speed. Blades are generally quite shallow and spaced much close together than in other blade designs; 24–64 blades are typical. For a given fan duty, the wheel would have the smallest diameter and operate at the lowest speed of the various blade types. Such fans are commonly used for low pressure, high volume ventilating applications. As such, the wheel is often constructed of lightweight, low cost materials. Its mechanical efficiency is generally somewhat lower than that of the backward-curved blade fan. The pressure curve has a dip to the left of the peak which can cause operating problems (fan instability). Flow control in this region is difficult. Highest efficiency is reached to the right of the pressure peak, usually at 40–50% of wide-open flow. The fan is usually operated and rated to the right of the pressure peak. Power rises continually toward free delivery, which must be considered in motor selection.

Backward-Curved Blades. In the backward-curved design, the blades incline backward (opposite to rotation direction) from the point of heel attachment on the wheel. The single-thickness blades may be either straight or curved, usually 12–16 blades to a wheel. Air leaves the blade at a velocity less than wheel-tip speed since the increasing flow passage through the blade provides for expansion of the air. This feature improves the mechanical efficiency over that of the forward-curved blade. The deep blades lend themselves to developing a high static pressure. Wheel diameter and speed are generally higher for a given performance than the forward-curved blade.

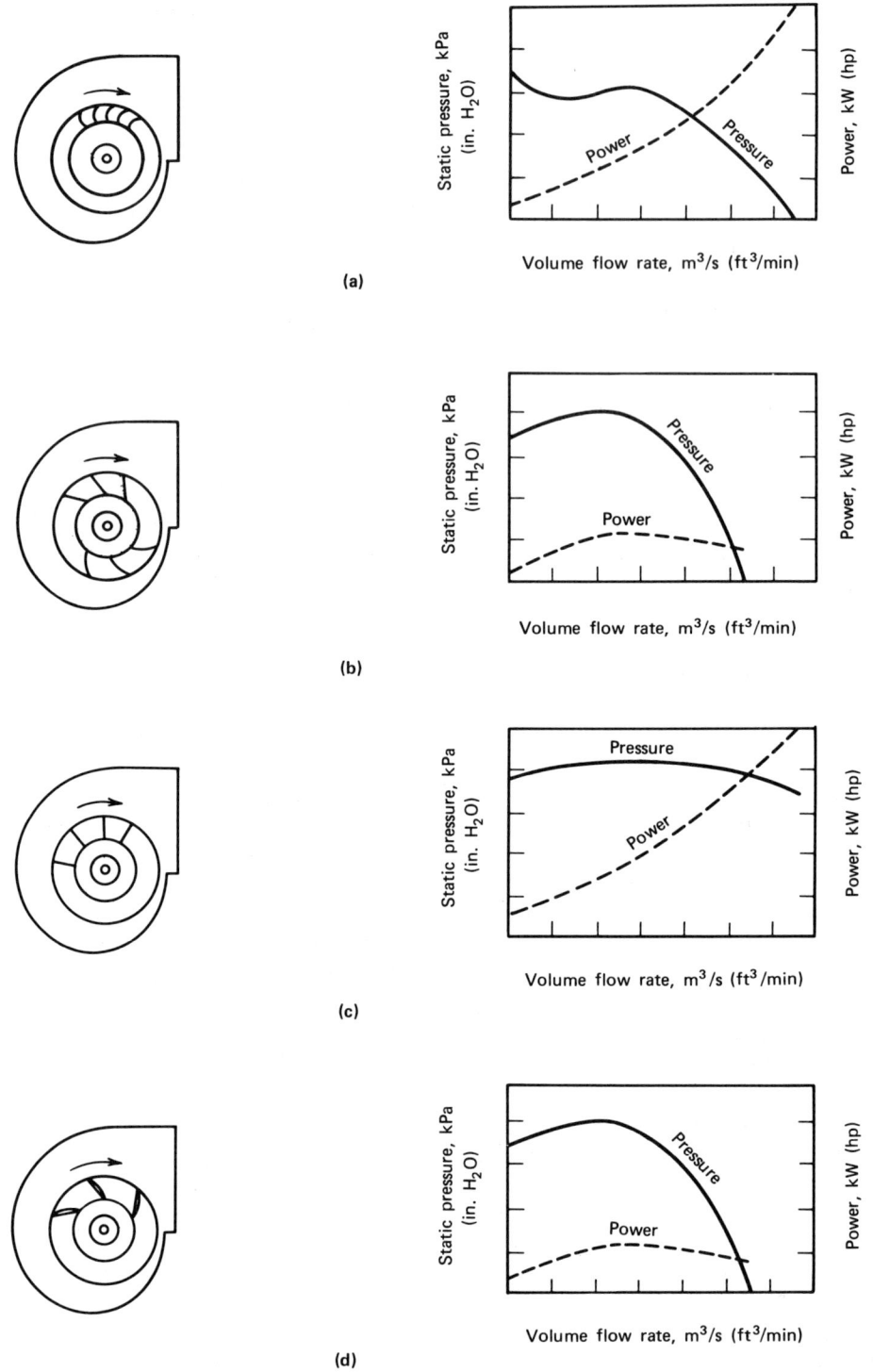

Figure 3. Shape of fan blades and typical performance curves: (**a**) forward-curved blades; (**b**) backward-inclined blades, at left, straight backward blades at top, curbed backward blades at bottom; (**c**) straight radial blades; (**d**) airfoil blades.

Close clearance and alignment of the wheel with the inlet bell are important aspects of this design to obtain maximum efficiency, especially at high static pressures. (Inaccurate clearances allow leakage of compressed air back to the suction side of the wheel.) The pressure curve rises somewhat from shutoff with increase in flow until a maximum pressure is reached. The maximum efficiency is reached at 50–65% of wide-open volume. The power curve reaches a maximum near the point of peak efficiency and then tends to drop off slightly with increased flow resulting in a nonoverloading design from the standpoint of motor sizing.

Straight Radial Blades. The straight radial blade design is the simplest of all centrifugal fans and also the least efficient. However, the wheel can be designed with great mechanical strength and is easily repaired. It is useful for two different applications. For a given speed, it tends to develop a higher static pressure than other wheel designs and thus is attractive for high speed, high pressure fans compressing air to 108–120 kPa (1–3 psig). Such designs often have pressure performance curves that are fairly flat to 70% of the wide-open flow. This is desirable for applications where a constant output pressure is needed, such as for primary combustion air. Another use is in material handling. The blades can be coated with abrasion-resistant coatings or equipped with replaceable liners. When large objects must occasionally be handled (such as a loose bag from a bag-filter house), the rims of the wheel may be omitted entirely and the blades supported with stiff struts from the hub. The shape of the power curve leads to overloading characteristics.

Airfoil Design. The airfoil design is similar to the backward-curved blade, except that it is designed for maximum mechanical efficiency. Each blade is composed of two pieces, with the upper surface contoured to reduce air friction and provide for most efficient compression of the air. For a given performance, this fan has the highest rotational speed of any of the wheel designs. The scroll is usually designed for the most efficient conversion of velocity head to static pressure. Performance characteristics are generally similar to a backward-curved blade but power requirements are somewhat less. Such fans are generally more expensive to construct and are used only in larger sizes with higher pressures or large flow volumes where reduced operating cost justifies the increased initial expense.

Fan Laws and Their Applications

Manufacturers' performance ratings are generally based on atmospheric pressure at sea level, 20°C, and 50% rh. Changes in temperature, gas density, and fan speed affect the performance. Fan laws predict these effects. Some authorities (4) list as many as 10 fan laws relating variables such as size, speed, capacity, gas density, discharge pressure, power, efficiency, and sound level. For most fan users, the following four laws take care of the necessary computations. When fan speed is changed: (*1*) the capacity (flow rate) varies directly with the speed ratio; (*2*) discharge pressure varies directly with the square of the speed; (*3*) power varies directly with the cube of the speed (at constant inlet density with no change in temperature, absolute pressure, or composition); and (*4*) discharge pressure and power requirements (at a constant capacity and fan speed) vary directly with the gas density.

The fan laws can be used to construct performance curves. However, one common failing is the attempt to apply the fan laws individually. Since each point on an existing fan curve is associated with a specific (but unnoted) mechanical efficiency, the efficiency of the point remains unchanged only when all of the applicable fan laws have been used in calculating an equivalent point for the new curve. In Figure 4, the solid lines represent performance curves for a fan operating at 1800 rpm. In order to determine the characteristic curve for the fan operating at 2100 rpm, the pressure corresponding to point A is multiplied by the speed ratios squared, $(2100:1800)^2$, to give a new pressure value corresponding to point B'. However, this point does not lie on the new fan curve because the rule applying to fan capacity has not been applied. The fan capacity at point A must be multiplied by the ratio of fan speeds (2100:1800) to obtain the new fan capacity, moving the rating point B' to the right to point B. In the same way, a new point on the power curve is obtained by multiplying original point C by the ratio of the speeds to the third power for the new power coordinate and the flow by the first power of the speed ratio for the new fan capacity at which this power applies.

Fan Selection

A fan is selected according to its location in the air-flow system, system performance and control characteristics, cost, efficiency, control stability, flexibility, and noise level (see Insulation, acoustic; Noise pollution). Location in the flow system is very important. A fan operating on the highest-density inlet gas available is smaller and less expensive, has lower operating costs and requires less maintenance. This is partly due to the fourth fan law which states that the fan pressure varies directly with the density. Since a system has a required pressure drop at a given flow, a fan operating on a low density gas has to be operated at a higher speed than if it were operating on a more dense gas. Because of the higher speed, the low density fan requires more power, and bearing life is shorter. Frequently, a fixed mass of air must be moved through the system rather than a fixed volume. Under this condition, the capacity of a low density

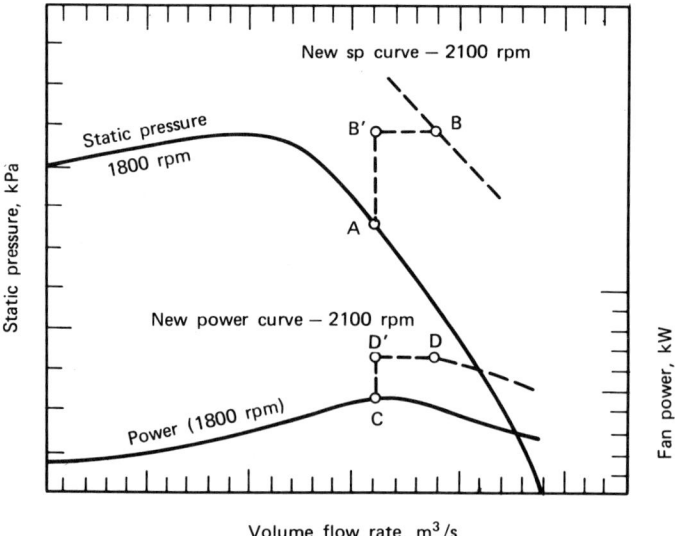

Figure 4. Application of fan laws to construct a performance curve for a fan operating at a speed of 2100 rpm from a test performance curve at 1800 rpm; sp = static pressure.

fan has to be greater than that of a high density fan. These facts point out the benefits of using a forced-draft fan located near the air introduction point of the system. Inlet density is reasonably high and most of the system pressure drop occurs on the discharge side of the fan. Placement of an exhaust fan at the end of the system with most of the system pressure drop on the suction side ensures that the fan will handle a lower density gas. Similarly, if throttling flow control is used, or if the air is to be heated, it is desirable to place the throttling damper or the air heater on the fan discharge.

System flow resistance as a function of flow rate is needed to select the proper fan size. For calculation of system pressure drop see refs. 5–8. The resistance pressure curve for a typical system (see Fig. 5a) shows that the pressure required to force air through the system increases with the flow rate. The pressure–volume curve of a proposed centrifugal fan is also shown. (This fan curve must be drawn for the anticipated fan inlet density expected at its location in the system.) The point of intersection of these two curves locates the flow rate and pressure rise at which the fan and system operate. The intersection shown represents a desirable operating combination for fan and system. The system curve intersects the fan curve in the middle of its maximum efficiency range and also at a point where the fan pressure produced varies smoothly but distinctly in a constant trend with flow rate which is desirable for flow control.

For air-flow control, the system may contain a control valve or damper that automatically or manually modulates system pressure drop. The dotted curves in Figure 5a on each side of the system resistance curve might represent operating extremes of the system resistance as the control valve is varied from maximum to minimum opening. These curves also intersect the fan curve at desirable operating portions of its range both for efficiency and flow control.

If a much larger fan as in Figure 5b had been considered so that the system resistance curve intercepted the fan curve close to its pressure peak, flow control would

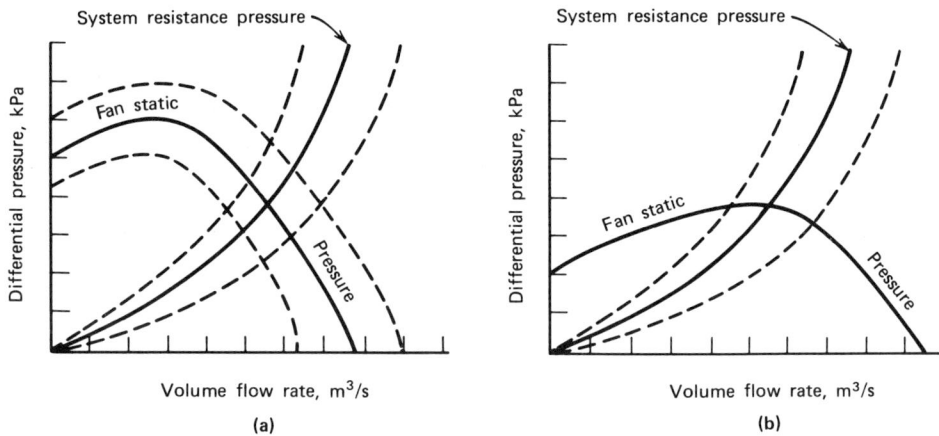

Figure 5. Selection of fan size. (**a**) Desirable sizing. The system resistance curve intersects the fan curve near its maximum efficiency. Dotted lines on each side of system resistance represent changes in system resistance from a flow-control element and also intersect the fan curve at desirable points for good flow control. Dotted fan curves represent performance at lower and higher speeds which also intersect system resistance curves at desirable locations. (**b**) Selection of a fan essentially too large for the system. The intersection of the system curve near the peak of the fan curve results in poor system flow control and perhaps surging.

be much poorer. Fan pressure rise changes very little over the anticipated flow control range so that larger changes in flow volume accompany small changes in system pressure drop. In addition, fan pressure decreases on both a flow rise and decrease. This is a situation likely to cause surging and out-of-phase hunting between the fan and an automatic control system. Higher flow rates may be required for future expansion. Lower flow rates may also be desirable seasonally. These flow changes might best be achieved through changes in speed. The dotted lines in Figure 5a on each side of the fan curve represent higher and lower wheel speeds for which this fan is suitable.

The wisest fan choice is frequently not the cheapest fan. A small fan operates well on its curve but may not have adequate capacity for maximum flow control, future needs, or process upset conditions. It may be so lightly constructed that it is operating near its peak speed with no provision for speed increases in the future, if needed. As fan size is increased, efficiency generally improves and wheel speed is lower. These factors decrease operating cost and provide reserve capacity for the future. However, as pointed out, it is possible to oversize a fan and impair its performance.

Noise level has to be considered in fan selection. Most manufacturers provide tables of operating ranges of quietest operation. There is no set fan discharge velocity that is applicable to all fans to ensure quiet operation. Fans do not operate as quietly when throttled back as when they are allowed to handle substantial quantities of air. Figure 6 illustrates the range of quiet operation of a specific airfoil fan as a function of outlet velocity and discharge pressure. Outlet velocity (and hence fan capacity) must be allowed to increase with static pressure to stay in the quiet region. Table 1 lists typical fan outlet velocities for quiet operation. Industrial process fans with backward-inclined blades should usually be selected with discharge velocities somewhat higher than those for quiet operation to achieve best all around performance and to provide pressure reserve.

Duct Connections. Performance curves are measured under ideal laboratory conditions. However, to obtain the same performance curve from a fan in a field installation, the system must approach the characteristics of the test conditions at least in that part of the system close to the fan. Both inlet and outlet duct connections can influence fan performance significantly. These connections can actually change the

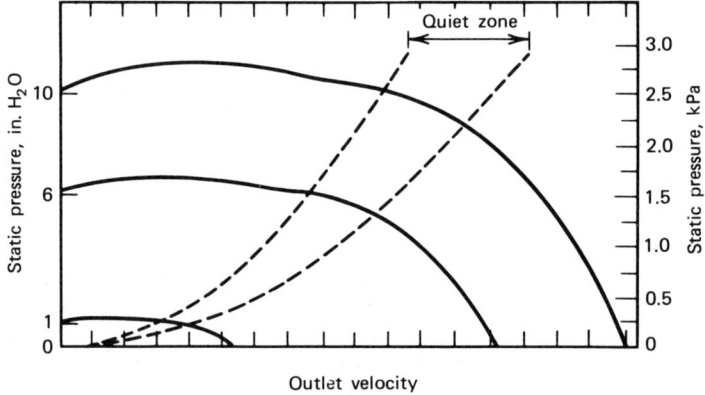

Figure 6. Typical quiet operating range for an airfoil fan.

Table 1. Typical Fan Outlet Velocities for Quiet Operation[a]

Static pressure		Forward-curved fan		Flow-nozzle airfoil fans[b]	
kPa	in. of water	m/s	ft/min	m/s	ft/min
0.25	1	8.1–10.4	1600–2050	4.3–7.4	850–1450
0.50	2	11.2–14.4	2200–2840	6.4–10.2	1250–2000
0.75	3			7.6–12.7	1500–2500
1.0	4			8.6–14.5	1700–2850
1.2	5			9.4–16.3	1850–3200
1.5	6			10.7–17.8	2100–3500
1.7	7			11.7–19.3	2300–3800
2.0	8			12.7–20.3	2500–4000
2.2	9			13.5–21.8	2650–4300
2.5	10			14.2–22.9	2800–4500
2.7	11			14.7–24.4	2900–4800
3.0	12			15.2–25.4	3000–5000

[a] Recomputed from data of New York Blower Co.
[b] Somewhat higher outlet velocities should normally be used for industrial processes with backward-inclined blade fans.

shape of the fan curve. Therefore, no single correction factor can account for the performance change over its entire range. Although poor outlet connections affect performance, improper inlet connections generally hurt performance more, reducing it the most near free-delivery conditions and the least at peak pressure.

Poor performance can result from fan inlet eccentric or spinning flow, and discharge ductwork that does not permit development of full fan pressure. Sometimes inlet restrictions starve a fan and limit performance. To obtain rated performance, the air must enter the fan uniformly over the inlet area without rotation or unusual turbulence. This allows all portions of the fan wheel to do equal work. If more air is distributed to one side of the wheel, such as with an elbow on the inlet (Fig. 7a), the work performed by the lightly loaded portions of the wheel is reduced and capacity is decreased by 5–10%. The use of an inlet box duct on a fan as in Figure 7b can reduce capacity by as much as 25%.

Spinning or vortex flow of air entering a fan can have as much effect on performance characteristics as the installation of inlet vanes to provide for reduced flow. If the air spins in the direction of wheel rotation, the bite of the blades on the air is reduced and both air flow and pressure are reduced. If the spin opposes wheel rotation, the wheel must overcome the momentum of the air: power requirements increase and efficiency is reduced. Spiral flow in a duct can be set up by a series of bends and elbows forming a corkscrew path, cyclones, and tangential inlets. Figure 7c illustrates a restriction too close to an inlet which also reduces performance. A full diameter inlet duct straight for 10 diameters is desirable. Where such inlet connections are not possible, corrective devices should be provided in the ductwork. Spiral flow can be eliminated with the use of eggcrate straightening vanes. Turning vanes in the ductwork can largely eliminate problems of eccentric flow. The turning vanes of Figure 7d reduce the 25% capacity loss of Figure 7b to around 5%.

The velocity of air discharging from a fan is not uniform across the discharge outlet but tends to be higher toward the outside of the scroll as shown in Figure 8. The discharge duct evens the velocity distribution into the standard turbulent-flow distri-

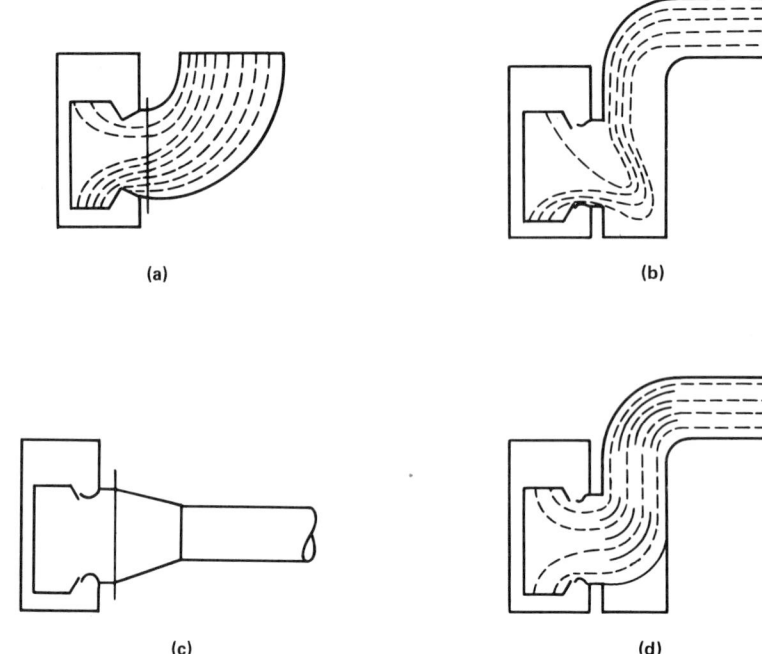

Figure 7. Effect of poor fan-inlet duct connections on fan performance. (**a**) Elbow on fan inlet giving uneven distribution of air to fan wheel. (**b**) Fan inlet box producing uneven distribution of air to fan wheel causing up to 25% fan capacity loss. (**c**) Free flow of air to fan wheel restricted by high inlet duct resistance reducing fan capacity. (**d**) Addition of turning vanes to fan inlet box design of (**b**) providing improved air distribution to the wheel and reducing capacity loss to about 5%.

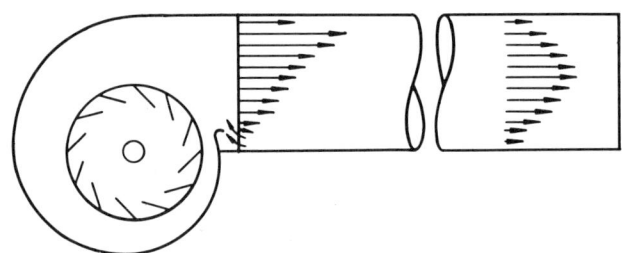

Figure 8. Illustration of variation of velocity of air at the outlet of a centrifugal fan and the function filled by several diameters of straight discharge duct in converting velocity head to static head and establishing normal turbulent flow distribution. Bends or obstructions at the discharge outlet cause turbulence and prevent conversion of velocity head to static pressure.

bution some distance downstream and converts part of the discharge velocity to static pressure. If a fan is operated without an outlet duct (discharging into a large plenum or the atmosphere), it loses 1–1.5 velocity heads. Such a loss must be added to the calculated system resistance. The addition of a straight discharge duct for several diameters and of the same size as the fan outlet can obviate this loss. An expanding outlet can increase static pressure beyond the curve performance if an efficient expander (included angle no more than 17°) is used. An elbow placed directly on a fan discharge destroys most of the velocity pressure leaving the fan.

Where correction of either inlet or outlet deficiencies is not feasible, allowances have to be made for these problems in performance selection criteria which are frequently suggested by the manufacturer. Suggestions for improving fan duct connections and effect factors to be applied to fan performance curves are given in AMCA publication 201 (9) and others (10–14).

Flow Control. In many applications, it is desirable to be able to change the quantity of air being handled through the system. The need to change the flow may be frequent, such as every few minutes or every hour, or less frequent such as daily or weekly. Infrequent changes might correspond to changes in a manufacturing process or to seasonal factors. The choice of control method can be influenced by the frequency with which the flow must be changed. In order to control flow, either the system characteristics or the fan characteristics must be changed. Generally, flow control affects the energy input to the fan. Low-cost control devices generally result in reduced fan efficiency and increased power consumption. Thus, if flow reduction is to occur for a long time with powerful fans, more energy-efficient control devices should be considered.

The simplest and cheapest control device is a damper, butterfly valve, or an orifice placed in the duct to throttle the flow and change the system resistance characteristics. As the flow is throttled more, system resistance is increased as illustrated in Figure 5a. To produce a higher discharge pressure, flow through the fan has to decrease. The power input is reduced but the energy expended in pressure drop across the throttling element is wasted. As mentioned under Fan Selection, placement of a throttling device on the discharge side of a fan is often preferable since the density of the air entering the fan is not reduced.

Changing centrifugal fan characteristics usually results in greater energy savings than changing system characteristics. If fan pressure can be reduced together with flow, the most desirable method of energy conservation is to change the speed, since that leaves the efficiency unchanged. If fan capacity is to be changed only infrequently, speeds of belt-driven fans can be adjusted easily with sheave changes. Where frequent speed changes are required, variable-speed motors and drives (electric or hydraulic) are the best but the most expensive. Multispeed motors and motors with step speed control can be used when infinitely variable control is not needed. The effects of speed control on a fan can be predicted from the fan laws. An alternative to speed change with axial-flow fans is blade-pitch control.

Inlet-vane control can be used to change the shape of the fan performance curve through imparting spin to the air entering the fan. As more spin is imparted, less energy can be transferred to the air from the blades and static pressure output is reduced. Figure 9 illustrates how the performance is reduced as more and more spin is imparted to the inlet air. Each setting of the inlet vanes has a separate power curve. The intersection of the system curve with the various fan pressure curves is shown and their equivalent power. The power required with inlet-vane control is usually intermediate between that required with throttling control and speed control.

Motor and Drive. The preferred prime mover for a fan is usually an electric motor. For fans of low to moderate power, V-belt drives are frequently employed to transmit the power. This permits selection of fans that can be operated over a wide range of speeds rather than being limited to motor synchronous speeds. Furthermore, change of speed is less expensive with V-belt drives. However, fans requiring powerful motors, 37–75 kW (50–100 hp) and higher, are generally directly connected to the motor and driven at synchronous speed.

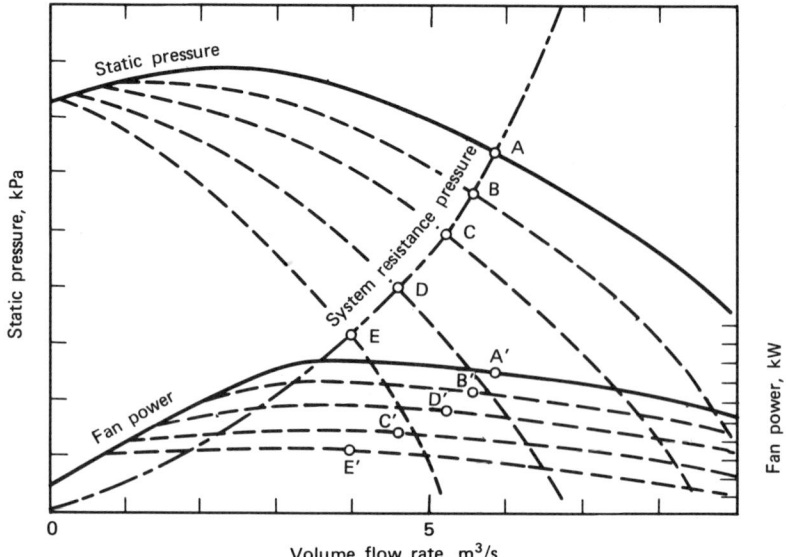

Figure 9. Control of fan performance with inlet vane control. Solid lines marked A and A′ show normal performance without vanes (vanes wide open). As vanes are progressively closed, static and power curves are modified as indicated by dotted lines. Intersection of the system resistance curve with these reduced pressure curves at points B, C, D, and E show how imparting more spin to the inlet air reduces flow. Projecting points A to E vertically downward to the corresponding power curve locates fan power points A′ through E′. Power savings achieved over throttling control can be estimated by projecting points B through E vertically downward to the A′ power curve and comparing the value with that from the proper reduced power curve. To convert kPa to in. H_2O, divide by 0.249; to convert m^3/s to ft^3/min, multiply by 2119.

When selecting the motor, power requirements, effect of temperature changes on load, and motor starting current and torque have to be considered. Calculation of system air-flow resistance is subject to some error and cannot always be predicted precisely. Therefore, the fan power predicted by the intersection of the fan and system curves may not be precise. If the system resistance is higher, it may be necessary to speed up the fan which will make it draw more power. If the resistance is less than anticipated, the flow increases (unless dampered) also resulting in higher power consumption. A general rule is to size the motor for the power required for a system pressure drop both 25% greater and less than that predicted. Air temperature can also affect power requirements. A fan normally operated on a hot gas may have to be started when the system is cold. Under such conditions, the inlet gas density is much higher. The fan develops more head and a greater mass of air is delivered. Unless the system flow can be throttled back until normal operating temperatures are reached, the motor has to be sized for the cold-starting conditions based on density ratios, often two or three times normal running power.

In starting a fan, the air power increases gradually with speed which is a desirable starting load. However, in large heavy fans, considerable torque is required to overcome the fan wheel inertia (referred to as WR^2, where W is the mass of the wheel and shaft and R is the radius of gyration). Figure 10 illustrates typical fan wheel and motor torques as a function of system speed during the starting process. Fan torque is that required for overcoming wheel inertia and for running power for the speed attained.

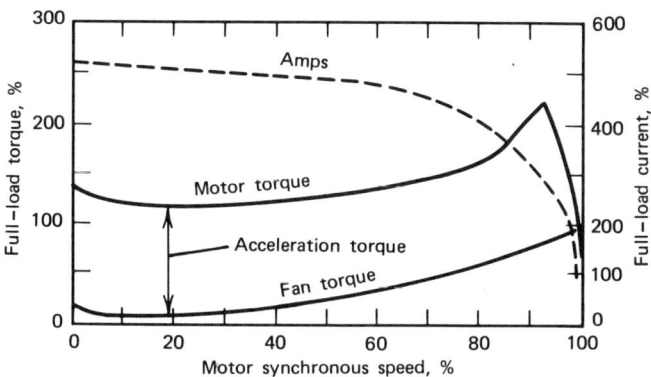

Figure 10. Typical plot of fan and motor starting-torque curves.

Motor torque at every point on the starting curve must be greater than fan torque and their vertical difference in Figure 10 is the torque available for acceleration. If the motor is started across the line and the length of time required to reach full load is too long (usually 10 s is desirable) the motor may become overheated and overload controls shut off the power. Thus the fan cannot be started. On long power lines, the inrush of starting current may also drop line voltage sufficiently so that fan and motor are too slow in coming up to speed. Alternatively, reduced-voltage starters can be used, which permit extended starting periods without motor overheating, or special motors with winding that can be bypassed during starting can be used. In calculating the starting time of a fan (15), in addition to the WR^2 of the fan wheel, the flywheel effect of large drive sheaves and the motor rotor itself must also be included.

Other Selection Problems. Additional considerations can arise when fans must handle solids or gases of low density, or must be operated in parallel or series. A complicated flow system involving several fans in parallel, all of which are in series with a common exhaust fan, can lead to surging and vibration unless selected carefully. Maximum tip speed, bearing types, single- and double-inlet fans, and wheel and shaft natural frequency and rigidity must also be considered.

Low Density Gases. A fan may have to operate on low density gas because of temperature, altitude, gas composition (high water vapor content of the gas can be a cause of low density), reduced process pressure, or a combination of such causes. To develop a required pressure, the fan has to operate at a considerably higher speed than it would at atmospheric pressure and hence it must operate much closer to top wheel speed. Bearing life is shorter, and the fan tends to vibrate more or can be overstressed more easily by a slight wheel unbalance. Abrasion of the blades from dust particles is more severe. Therefore, a sturdier fan is needed for low density gas service.

Near its top speed, a fan may operate at a speed that is near or above the natural frequency of the wheel and shaft. Under such conditions, the fan can vibrate badly even when the wheel is clean and properly balanced. Manufacturers often do not check the natural frequency of the wheel and shaft in standard designs; however, many have suitable computer programs for such calculations. Frequency calculations should be made on large high speed fans. The first critical wheel and shaft speed of a fan that is subject to wheel deposits or out-of-balance wear should be about 25–50% above the normal operating speed. The problem of starting a fan designed to run on low density gas is discussed under Motor Selection.

Mechanical Considerations. The mechanical design of a fan and the various forces that fan parts must be designed to withstand are discussed in ref. 16. The forces result from a combination of fluid, inertial, and vibrational effects.

Tangential forces from air compression act on the blades and are transmitted through the fan hub to the shaft in the form of a resisting torque. Axial thrust may be developed on the fan wheel and shaft because of pressure differences about the wheel and the directional change in momentum of the air at the wheel inlet. The net unbalanced axial thrust must be taken by a thrust bearing which transmits it through the bearing supports to the fan foundation. At maximum fan efficiency, the radial fluid forces acting on the wheel are nearly balanced, but the volute can be correctly designed for only one rating condition. Therefore, as fan operation departs from maximum efficiency, unbalanced radial thrust increases which must be carried by the bearings to the foundation. Centrifugal forces also act on the wheel. If the center of the wheel and shaft rotation does not coincide exactly with the center of the rotating mass, a flexural force produces bending of the shaft which is apparent as vibration. In a rotating elastic system, dangerous vibrations are likely to occur at critical speeds. The application of repeated external forces such as flow surging or wheel unbalance excites the elastic structure and causes it to vibrate. If the excitation frequency is close to the natural frequency, resonance can occur with large amplitude vibrations. All of these forces must be carried by the bearings. Thus it is common to use heavier components as fans are called on to operate at higher speeds or higher pressure differentials. Many fan designs, are available in different construction strengths designated Class I, II, III, and IV. The higher numbers denote a fan capable of operation at a higher speed and higher pressure rise. The required class of the fan needed must be considered in its selection.

Bearings used on fans may be either sleeve or antifriction type and must be designed to withstand loads due to dead weight, unbalance, and rotor thrust and be able to operate at the intended maximum speed without excessive heating (see Bearing materials). When natural convection from the bearings is inadequate, some other cooling method must be provided. Lubricating oil may be circulated through an external cooler or the pillow blocks may be cored with passages for forced circulation of air or water. Fans operated at high temperatures increase the bearing cooling problem caused by heat conduction along the shaft. A small external fan wheel on the shaft, called a heat slinger, is frequently provided, or forced-circulation water cooling is used. In addition to the bearings of fans operating on hot, low density gas at high pressure rise, special attention is needed to ensure high rigidity of the wheel and shaft. Fan wheels should be balanced both statically and dynamically, eg, in the field with chalk and weights (17). Elaborate electronic test instruments are also available. An unbalanced condition causes a vibrational displacement of the bearings which is frequently checked. Table 2 lists typical displacements of fans operating at various speeds and various degrees of unbalance.

Small volume fans, usually designed with an air inlet on only one side of the wheel and casing, are known as single-inlet fans. With an enclosed wheel, the fan hub is fastened to a solid backplate which supports the blades. Larger capacity fans can be either single- or double-inlet fans. A double-inlet fan has an inlet on both sides of the wheel and casing. The hub is usually fastened to a common backplate midway between the two inlets. A double-inlet fan is generally more efficient or runs at a slower speed than a large single-inlet fan for the same capacity since the air is better distributed

Table 2. Typical Fan-Bearing Vibrational Displacement as a Function of Wheel Speed and Degree of Unbalance[a]

Wheel speed, rpm	Bearing displacement, μm, Qualitative degree of unbalance			
	Smooth	Fair	Rough	Very rough
600	50	100	200	380–500
900	38	70	150	200–250
1200	25	50	115	150–200
1800	19	38	90	125–180
3600	10	18	65	100–125

[a] Ref. 17.

over the width of the wheel. Finally, air volumes are reached with large fans such that only double-inlet designs are feasible.

Vibration in a fan may be caused by mechanical problems or by the flowing air (surging, poor fan-curve operating position, poor design of fan-duct connections resulting in poor air distribution, etc) as discussed under Fan and Ductwork System Operating Problems. A double-inlet fan is expected to have little axial unbalance since the symetrical design of the air flow between the two halves of the wheel tends to result in a balancing of opposing forces. Such fans are frequently supplied with bearings suitable for only small thrust loads. Poor inlet ductwork arrangements can result in excessive thrust if unequal air flows are provided to opposite sides of the wheel. An unsteady changing of air-flow unbalance alternately between inlets can set up an alternating thrust pattern which can be very damaging to bearings designed for low thrust load. Mechanical vibration and elastic deformation problems and diagnostic techniques for structural inadequacies in fan design are discussed in ref. 18.

Fans in Parallel or Series. Occasionally two or more fans are operated in parallel or in series. The composite operating curve can be drawn by adding the two fan curves. In the case of parallel fans, points of equal pressure rise on the individual fan curves are noted and the combined flow for the two fans is plotted against this pressure. For fans in series, the pressure and gas density produced at the inlet to the second fan by the first fan must be calculated over the flow range of interest. The characteristic curve for the second fan must be drawn for small incremental changes in inlet density and the air volume handled by the first fan must be corrected to the inlet density of the second fan. At this corrected flow and inlet density, the pressure rise of the second fan is noted. The sum of the pressure rise of the two fans minus the friction losses between the fans is plotted against the inlet flow to the first fan. The process is repeated for other flow rates until a combined curve is developed.

Figure 11 shows combined parallel and series curves developed from two individual fan curves. Problems of unsteady operation can result from using two or more fans in parallel or in series. Figure 11 illustrates the use of two fans of different size but with similar performance curves. In the combined curves, the intersection of the system curve with the combined pressure curve for the parallel fans occurs to the right of the combined pressure peak which results in stable fan operation. Had the intersection been to the left of the pressure peak, such as point A, this could have corresponded to at least two different flow rates on the performance curve of each of the fans (illustrated by points B). When operating in this range, each fan is hunting in flow

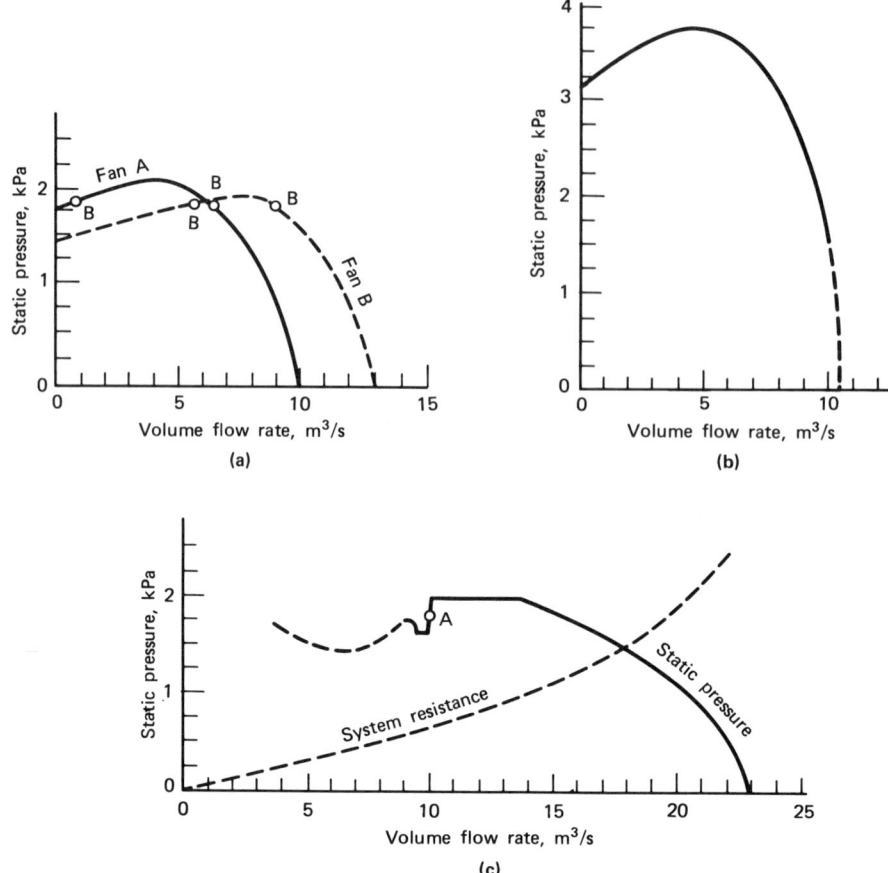

Figure 11. Operation of fans in parallel or series: (**a**) performance curves of fans A and B of different size to be considered for parallel or series operation; (**b**) combined curve of fans A and B in series; (**c**) combined curve of fans A and B in parallel. The performance curves of the combined fans in **b** and **c** are obtained by adding the curves in **a**. Parallel operation of these fans at a system resistance intersection in the flat, level portion of the fan curve and especially to the left of point A will result in unstable operation. This occurs because point A is satisfied by two different operating combinations of fan A and B denoted by points B in **a**. Thus the fans hunt between these combinations. To convert kPa to in. H_2O, divide by 0.249; to convert m^3/s to ft^3/min, multiply by 2119.

between these two points and operation is unstable. If the curve of one fan could produce a higher pressure than the other, operating back toward the peak of the higher pressure fan would produce a point where the higher pressure fan would blow air backward through the lower pressure fan, resulting in large momentary pulsations in flow rate. Similar problems can arise with fans in series when operating in undesirable portions of the combined pressure curve.

Fans may be used in parallel in a single air system because meeting the total air requirements would require an unusually large single fan and motor. In addition, multiple fans give flexibility; ie, when less air is required, one or more fans can be shutdown. More often, several fans are used because they have different duties to perform. A combustion operation might be equipped with a primary-air fan, a secondary-air fan, and a third fan to lower the temperature of the effluent gas following completion of a high-temperature reaction. These three fans in parallel might all operate in series with a single exhaust fan. The flow quantity of each fan and its static pressure is different and it discharges to a slightly different point in a common duct system. Ensuring the compatibility of such a multiplicity of fans is a complicated task. To ignore the possibility of a problem can result at times in a badly pulsing system.

Fan and Ductwork System Operating Problems

Important problems of fan and ductwork systems are vibration, effect of ductwork on performance, and fan and system pulsation. Vibration results from out-of-balance rotating parts as previously discussed, flow pulsation, or an inadequate foundation. As indicated under mechanical considerations, many fan forces are transmitted to the foundation even by a smoothly operating fan. Therefore, it is important that the fan foundation be designed to withstand the forces with little movement. With concrete foundations, a ratio of foundation mass to fan mass of 7–10 to 1 has often been recommended but is not a satisfactory design. With both concrete and structural steel supports, it is important that the supporting structure natural frequency should be at least one half or preferably one and one-half times the frequency of the disturbing forces (19). Such separation of fan oscillations and support frequency prevents resonance of fan and foundation or support structure. A frequently used alternative is to mount the fan and motor carried on a common frame on vibration isolators (springs or rubber or cork pads, depending on wheel speed) and join the fan to the ductwork with flexible connections. The effect of the flow system on the performance curve (inlet and outlet connections and spiral flow) has been discussed. Pulsation problems resulting from multiple fans have also been mentioned, but the interaction of fan and system can lead to serious flow pulsation problems resulting from many causes.

Fans may surge or give pulsing flow because they are operated in an unstable portion of their performance curve, eg, to the left of the peak discharge pressure. Poor inlet or outlet connections can cause a modification in the shape of the fan curve from nonuniform wheel loading and result in unstable operation when not expected. A resonating duct system can amplify such pulsation. A flow system has frequency properties that can result in acoustic waves (not necessarily audible). Such waves can excite fan performance if it is operating close to unstable portions of the curve or is not well isolated from system surges. In addition, the flow system can amplify or reinforce acoustic waves generated by compression in the fan. By its nature, a fan does not generate a smooth steady pressure on the flowing gas. Rather, the gas is subjected to a number of high-frequency pressure pulses which is generally some multiple of the number of blades and the wheel rotational speed. Design features of the system can be excited by these pulses. The dimensions of a chamber or duct affect the frequency response of acoustic phenomena. Pressure pulses flowing through a duct of constant cross section, open at both ends, or partially closed at one or more ends, can produce standing waves in the chamber, referred to as organ pipe resonance. Large, flat duct panels can vibrate, amplifying pressure pulses or producing frequencies of their own. Such pulsation is frequently referred to as flutter-panel vibration. Inadequately secured large dampers and gates in gas breeching may produce similar effects.

A series of small ducts acting as flow resistance interconnecting enlarged chambers (referred to as acoustic resistance and capacitance) can act as a Helmholtz resonator (20), particularly if the fan is operating near a portion of its curve where surging can result. Branch duct lines, either flowing or stagnant, can also act as resonators. Combustion burners and atmospheric openings can also generate or amplify pulses. Burners frequently produce pulsation or can amplify it if the combustion air-flow surges. Atmospheric openings permit pulsing flow in and out from the atmosphere, tending to enlarge the amplitude of the pulses. Much work on frequency response of systems, analogue simulation, and damping has been performed at the Southwest Research Institute for the Southern Gas Association. References 21–22 are recent discussions of fan system pulsation problems.

Aside from structural and mechanical vibrational problems of the fan itself, a number of alternatives are available to improve the operating stability of a pulsing fan and resonating system. To the extent that duct connections to the fan are causing pulsation, these can generally be eliminated by providing several diameters of straight duct before and after the fan, using turning vanes where necessary, and flow straighteners to eliminate spiral flow. Isolation of the fan acoustically from the flow system often eliminates the interaction between the two. A pressure resistance which is equal to or greater in magnitude than the pulsation effectively dampens pulsations. Flow-control dampers for the system can often serve this purpose. Dampers on a fan outlet can isolate the fan from the effect of pulsations in the downstream ductwork, but the fan and upstream system can still be in resonance. Upstream dampers can isolate the fan from upstream pulsations. However, the dampers should be placed a short distance upstream or straightening vanes should be used with the damper so that the fan performance is not influenced by damper turbulence. Inlet-vane control is not effective for isolating a fan from its system since little pressure drop occurs across the inlet vanes. Furthermore, vane control reduces stability when it is operated in the nearly closed position (less than 20–30% open).

Acoustic damping for a resonating system is also achieved by changing the lengths of straight ducts and chambers (to change standing-wave frequency), inserting baffles, providing inlet gas diffusers, and inserting limiting flow orifices at atmospheric openings or on low pressure fans. Perforated-plate flow baffles, duct bends, and small duct flow resistances can all serve as acoustic resistances. Burners frequently pulsate when supplied with too much air, or a liquid fuel is inadequately atomized or too little energy is expended in mixing fuel and air.

AXIAL-FLOW FANS

In axial-flow fans, the air-flow direction parallels the rotational axis of the wheel, in contrast to the centrifugal fan in which flow through the wheel is radial to the shaft. The axial-flow fan is often the first choice when large quantities of air or gas are to be exhausted at relatively low pressures. They are frequently used without connecting ductwork for exhausting air through a wall or roof or with only a limited amount of duct connections for forcing air through a device such as an air-cooled heat exchanger or an induced- (or forced-) draft cooling tower.

The Air Moving and Conditioning Association classifies axial-flow fans into propeller, tube-axial, and vane-axial (see Fig. 12). Some authorities recognize a fourth type, called a disk fan, of which a common household electric fan is typical. Although air flow through an axial fan is parallel to the shaft, it frequently has a rotative or helical

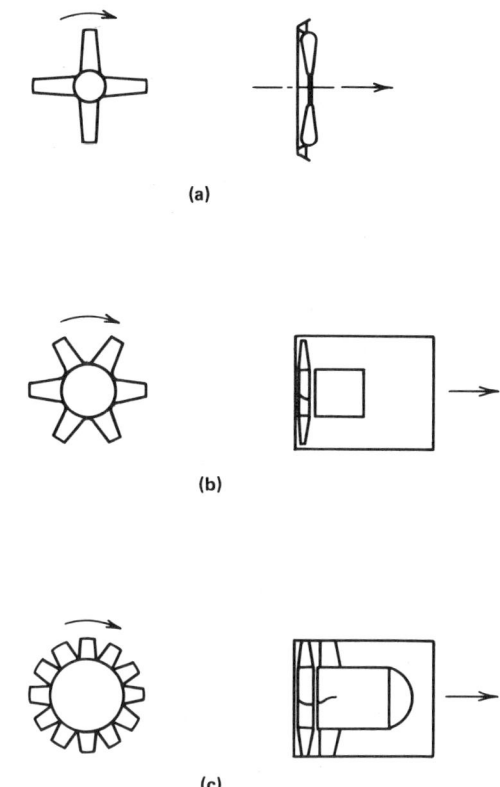

Figure 12. Types of axial-flow fans (23): (**a**) propeller fans; (**b**) tube-axial fans; (**c**) vane-axial fans.

pattern. This rotative energy is usually wasted unless stationary vanes as in vane-axial fans are provided to convert this kinetic energy to static pressure. Typical pressure operating ranges are 0–250 Pa (0–1 in. of water) for propeller fans, 60–620 Pa (¼–2½ in. of water) for tube-axial fans, and 125–2000 Pa (½–8 in. of water) for vane-axial fans. Vane-axial fans can be designed with nonoverloading power requirements, but propeller and tube-axial fans are normally not operated close to shutoff because of increasing power needs with reduced flow.

Design Elements

Ideal conditions are obtained in the design of an axial-flow fan when energy transfer from the blade to the gas is uniform along the length of the blade, resulting in uniform pressure generation, minimum losses, and maximum efficiency and stability. Since the blade linear velocity varies with position from tip to hub, attainment of a uniform pressure rise along the blade at different radii requires variation of the blade angle from hub to tip. The choice of blade section is dictated by the required aerodynamic characteristics and varies in practice from cast or molded precise airfoil profiles to formed materials to single-thickness plate materials. Hub size is increased for higher pressure designs where it is impractical to generate equal pressures nearer the center of the wheel. Low pressure designs have hubs ranging from 1/3 to 1/2 wheel diameter whereas hubs in higher pressure designs may occupy 75–85% of the tip diameter. The number of blades must also be increased as pressure rise is increased; 3 to 5 may be used with lower pressure designs, as many as 24 with higher pressures. Close clearance between blade tips and fan housing is a stringent requirement to prevent backflow losses at the housing wall. High pressure designs require clearances of less than 0.79 mm. The cylindrical housing of a vaneaxial fan may be cast or rolled. To attain the close clearance at the blade tips, either very careful forming or machining is required. Inlet and outlet connections are carefully designed to minimize turbulence and connecting inlet and outlet ducts should be straight for at least 2–3 diameters to avoid undue effect on fan performance.

Capacity Control

Any of the methods discussed previously under Fan Flow Control can be used to control the axial-flow fan capacity such as speed control, variable inlet vanes, and dampering. Throttling dampers should generally be placed sufficiently far upstream or downstream so that they do not destroy fan performance. A frequently used alternative is variable blade-pitch control. The capacity or pressure produced can be varied by rotating the blades to change the bite they take on the air. Axial fans with variable pitch control can be obtained with manual, remote, and automatic adjustments. With manual adjustment, the fan must be shut off and access is required to the blades. With the other methods, fan pitch can be changed in operation.

Propeller Fans

These fans may have typically from 2 to 6 blades mounted on a central shaft and revolving within a narrow mounting ring, either driven by belt drive or directly connected. The form of the blade in commercial units varies from a basic airfoil to simple flat or curved plates of many shapes. The wheel hub is small in diameter compared to the wheel. The blades may even be mounted to a spider frame or tube without any hub. The housing surrounding the blades can range from a simple plate or flat ring to a streamlined or curved bell–mouth orifice.

Some type of close-clearance shroud at the blade tips is desirable to prevent air recirculation from the discharge side of the blades back to the suction side. A propeller fan with no shroud has fairly low efficiency because of air recirculation. A curved orifice-like ring greatly reduces air recirculation and improves efficiency. An angle-shaped ring essentially eliminates recirculation, and optimum efficiency is achieved with an angle-like ring with streamlined edges.

Figure 13 shows a typical performance curve for a propeller fan. Such fans generally operate from free delivery up to a static head of 125–250 Pa ($\frac{1}{2}$–1 in. of water). If driven at tip speeds of 61–81 m/s (12,000–16,000 ft/min), they are capable of discharge pressures of up to 375 Pa ($1\frac{1}{2}$ in. of water) but the noise level may be objectionable or exceed OSHA standards for employee exposure (see Noise pollution). Power requirements for most propeller fans increase as flow decreases and static pressure increases. They are generally applied under operating conditions to the right of the point where the pressure–volume curve tends to flatten with decreased flow. The performance curve in Figure 13 is for a propeller fan with a small pitch angle of around

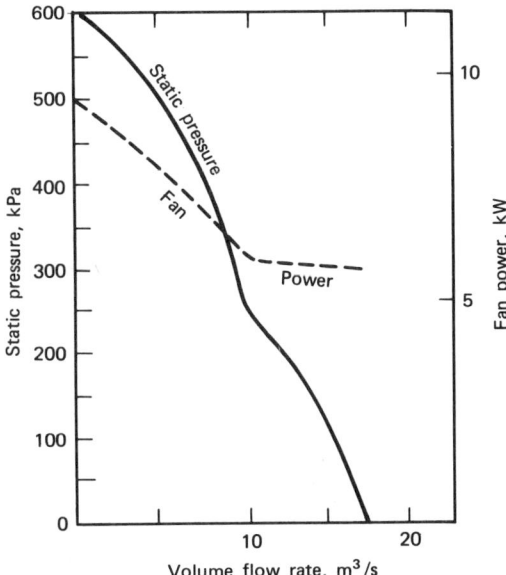

Figure 13. Typical performance curve for propeller fans. To convert kPa to in. H₂O, divide by 0.249; to convert m³/s to ft³/min, multiply by 2119; to convert kW to hp, multiply by 1.341.

15° at the tip. Performance of such a fan is perfectly stable over the entire range. This situation holds for pitch angles up to 17° and with careful design up to 20°. With steeper pitches, instability (a dip in the curve) occurs at the inflection point of the pressure curve. This corresponds to the stall condition of an airplane wing at which the wing angle ceases to develop lift. With no air delivery from a propeller fan, the air flow consists of two eddies entering the impeller near the hub on opposite sides and discharging in both directions near the tip. This action increases both static pressure and power requirements resulting in the steep climb in pressure with rapid decrease in flow shown to the left of the inflection point.

The small-pitch-angle fan has the advantage of high efficiency and stable operation but its capacity is low, its speed is high, and it is apt to be noisy. Steep-pitch fans deliver large volumes, can operate at lower speeds, produce less noise, and can be structurally designed for higher pressures. Noise level generally varies with the tip speed to the sixth power. Many large propeller fans are of the medium pitch (17–20°) type with wide blades occupying more than 50% of the projected area of the circle. When such fans are used in applications where air volume requirements can be expected to vary seasonally, they are often designed with variable-pitch propellers, which can be changed either manually with the fan off or automatically and remotely with the fan running. The primary application of conventional propeller fans is in movement of relatively large air volumes at low static pressure or free delivery conditions.

For general air circulation and exhaust without ducts, disk fans are frequently used that have plain or curved blades similar to an ordinary household fan. Such fans can move large volumes at slower speeds with low noise levels but produce very little pressure boost.

Tube-axial Fans

The tube-axial fan is a refinement of the propeller fan in both wheel design and mechanical strength, with improved capacity, pressure level, and efficiency. Designs are often capable of operating over a greater range of speeds. The cheapest fans may have an open-type propeller wheel with the motor enclosed in a tube if directly connected. (Belt-drive models are also available.) In more refined types, the blades are shorter and of airfoil cross section mounted on a large diameter hub which may approach 50% of the wheel diameter. The hub and motor tube are normally of the same diameter and reduce the back flow of higher pressure air which might recycle through less effective central portions of the wheel if a smaller hub were utilized. The performance curve (Fig. 14) may have a dip to the left of the pressure peak which would constitute an unstable region for fan operation and which should be avoided. Commercial models are available with static pressures up to 750 Pa (3 in. of water). The general range of application is for pressures of 125–375 Pa (½–1½ in. of water) permitting use of appreciable ductwork. Maximum efficiencies are in the range of 65–75%. Principal applications are in industrial processes and ventilation requiring moderate static pressures and the need for simplicity of fan installation in a straight duct.

Vane-axial Fans

The vane-axial fan is the result of developing the propeller fan into a precision machine using refined aerodynamic principles and precise manufacturing procedures and control. Where such principles and techniques are applied, excellent capacity, pressure, efficiency, and sound emission levels are attained. Some units with mechanical efficiencies above 90% have been developed but many commercial units have efficiencies no higher than 80%. High efficiency vane-axial fans are more efficient than comparable centrifugal fans. They have been used for energy conservation in Europe for a number of years, but it is only recently that U.S. industry has shown interest in large vane-axial fans for energy conservation in applications such as electric power boiler service (24–25).

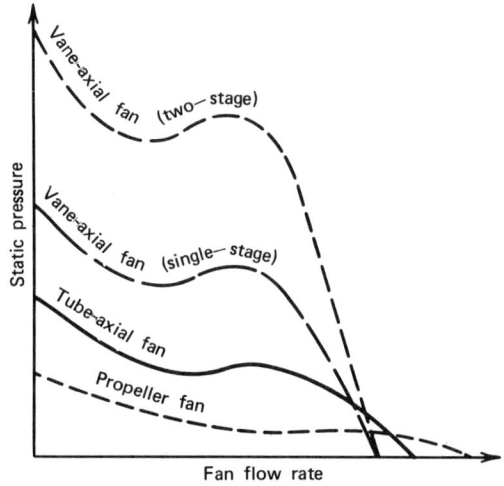

Figure 14. Performance characteristic comparison of axial-flow fan types.

The vane-axial fan wheel has short, stubby airfoil blades mounted on a hub which may be as large as 75% of the wheel diameter (see Fig. 15). The air leaving the axial-flow wheel has an appreciable rotational component. This rotational energy can be converted to static pressure in a suitably designed set of stationary straightening vanes. The straightening vanes are shaped to pick up the air leaving the wheel blades without shock. Although straightening vanes of airfoil cross section are theoretically desirable, vanes formed of pressed heavy sheet metal are less expensive. The motor is enclosed in a housing with the same diameter as the hub and has either a rounded cap or a bullet-shaped tail to reduce eddy losses. The straightening vanes surround the motor housing and can serve as structural supports for the housing. Generally, the number of guide vanes exceeds the number of propeller vanes by one, with the numbers selected so that there is no common divisor for the number of hub vanes and guide vanes. This minimizes flow pulsation and noise. Single-stage fans can develop pressures to 1.5 kPa (6 in. of water) with some designs going as high as 2.25 kPa (9 in. of water). Standard designs are available either belt-driven or directly connected to motors with speeds as high as 3450 rpm. In addition, two-stage units have been developed that produce considerably higher pressures but have received little industrial use. Typical performance curves for propeller, tube-axial, and vane-axial fans are compared in Figure 14. These curves show a dip to the left of the pressure peak. Axial-flow fans can be designed that do not have such dips, but those that do have such dips should be operated to the right of the pressure peak. The major advantage of the vaneaxial fan is its compactness and convenience of use in inline ducts plus its better efficiency when carefully designed. The higher manufacturing precision required generally eliminates any cost savings that might result from its smaller size.

APPLICATIONS

Fans and blowers are the most widely used mechanical device for moving air and gases in both large and small volumes in a tremendous range of applications (26), including ventilation, mechanical draft for combustion (including forced- and induced-draft fans and primary- and secondary-air fans), local exhaust for fume and dust containment at hoods and equipment enclosures; forced- and induced-draft

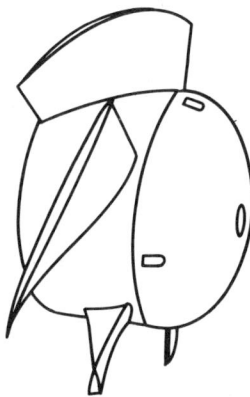

Figure 15. Hub and axial-flow wheel for a vane-axial fan.

cooling for spray towers, cooling towers and ponds, and air-cooled heat exchangers, and conveying of solids (see also Heat exchange technology). Other applications include air or gas movement in dryers (qv), gas-recirculation fans, air supply for air curtains and air-blast operations and a great many miscellaneous process industry uses often involving hot and corrosive gases. The range of performance required by fans for these various applications is enormous. Most ventilating applications require pressures ranging from 25 to 1500 Pa (0.1–6 in. of water). Frequently, noise level is an important consideration. Induced-draft fans must often handle gases of 150–425°C containing various levels of suspended erosive particles. Such fans are frequently equipped with replaceable wear pads of abrasion-resistant materials or are coated with wear-resistant surfaces.

Forced-draft fans generally operate on clean air and at pressures from a few hundred Pa to as high as 20 kPa (80 in. of water) for pressurized furnaces. Backward-inclined blading is used almost exclusively for high efficiency. Blades with airfoil contours give improved structural strength, higher efficiency, and lower sound levels in large fans. Conveying systems in which the solids pass through the fan almost always use low-speed wheels of the radial-blade paddle–wheel type of construction. In the area of hot and corrosive process gas handling, fan designs are adapted to the specific need of the process. Frequently stainless steel or other alloy construction is required. Where dilution of the gas with atmospheric air is objectionable, the fan shaft is equipped with a stuffing box, a rubber labyrinth seal, or even a purged rotary seal depending on the degree of contamination control required. Occasionally fans must handle gases with sticky or tarry particulates where solids buildup can occur. Continuous or intermittent flushing of the fan with a liquid spray in the fan inlet is helpful. In addition to deliberate flushing, fans may be called upon to handle gases containing mist or entrained liquid droplets. A large percentage of such mist may be collected and agglomerated in the fan, particularly if operated at a high tip speed. Liquid-handling fans must be equipped with oversize motors as the acceleration of the liquid within the fan can utilize considerable power. Particular attention must be paid to the corrosiveness of the wet–dry environment within the fan. The presence of chloride ions and high wheel stress can lead to stress corrosion cracking in stainless-steel wheels. Although elimination of the chlorides is the best solution, the use of much lower wheel-tip speeds and wheels that can be stress-relieved to remove residual fabrication stresses is often helpful. Fans with rubber and polymeric coatings are often useful in moist environments, but special considerations in fan design are necessary to assure thorough bonding of such coatings to the wheel. Buildup of liquid within the fan casing can also be a problem with liquid-handling fans. The use of bottom–horizontal discharge designs with a large discharge duct drain is generally more satisfactory than a small fan-housing drain.

The choice between axial-flow and centrifugal fans in certain applications is by no means clear-cut. Axial fans have an advantage when compactness is important and when straight-through flow benefits the installation. They also frequently have higher efficiency which is important in energy conservation. Advantages of the centrifugal fan are better ability to cope with fluctuating operating conditions, conditions that could result in unstable fan operation; greater ability to vary fan performance through speed changes; better access to the fan motor and greater facility to provide sturdy structural support for the fan and motor; generally lower noise level; and its natural adaptability to design situations where a 90° turn in gas direction is desirable.

BIBLIOGRAPHY

1. *ASHRAE Handbook and Product Directory—1975 Equipment,* American Society of Heating, Refrigeration, and Air Conditioning Engineers, Inc., New York, 1975, Chapt. 3, p. 3.1.
2. *Laboratory Methods of Testing Fans for Rating Purposes, AMCA Standard 210,* Air Moving and Conditioning Association, Arlington Heights, Ill., 1974.
3. *Test Code for Sound Rating Air Moving Devices, AMCA Standard 300,* Air Moving and Conditioning Association, Arlington Heights, Ill., 1967.
4. R. Jorgensen, *Fan Engineering,* 7th ed., Buffalo Forge Co., Buffalo, N.Y., 1970, pp. 231–233.
5. *Fundamentals Governing the Design and Operation of Local Exhaust Systems, ANSI Standard A9.2—1971,* American National Standards Institute, Inc., New York, 1972.
6. Industrial Ventilation, 15th ed., American Conference of Governmental and Industrial Hygienists, Lansing, Mich., 1978.
7. J. L. Alden and J. M. Kane, *Design of Industrial Exhaust Systems for Dust and Fume Removal,* 4th ed., Industrial Press, New York, 1970.
8. Ref. 4, Chapt. 3–4.
9. *AMCA Fan Application Manual, AMCA Publication 201,* Air Moving and Conditioning Association, Arlington Heights, Ill., 1973, Sect. 1.
10. J. W. Markert, *Heat. Piping Air Cond.* **42,** 100 (Oct. 1970).
11. D. H. Cristie, *ASHRAE Trans.* **77,** 84 (1971).
12. L. S. Marks and E. A. Winzenburger, *Trans. ASME* **54**(21), 213 (1932).
13. D. G. Traver, "System Effects on Centrifugal Fan Performance" in *Fan Application—Testing and Selection Symposium Papers, San Francisco, Calif., Jan. 19–22, 1970,* ASHRAE, New York, 1972, pp. 1–6.
14. H. F. Farquhar, "Outlet Ducts—Effect on Fan Performance" in ref. 13, pp. 7–10.
15. Ref. 4, pp. 295–313.
16. Ref. 4, pp. 271–283.
17. Ref. 4, pp. 285–288.
18. J. W. Martz and R. R. Pfahler, *Hydrocarbon Process.* **54,** 57 (June 1975).
19. A. H. Karabinis and T. J. Fowler, "Design Considerations for Dynamically Loaded Equipment Foundations," *paper presented at meeting of the American Concrete Institute, Houston, Tex., Oct. 29–Nov. 3, 1978.*
20. W. R. Heath and W. R. Elliot, *Trans. ASME J. Appl. Mech* **13,** 291 (Dec. 1946).
21. A. N. Bolton, "Pressure Pulsations and Rotating Stall in Centrifugal Fans" in *Vibration and Noise in Pump, Fan, and Compressor Installations—Conference Proceeding of the University of Southampton, England, Sept. 16–18, 1975,* published for Institute of Mechanical Engineers, London, Eng., by Mechanical Engineers Publishing, Ltd., New York, 1975, pp. 175–181.
22. J. K. Hay and J. W. Martz, *Mech. Des.* **47,** 113 (Feb. 20, 1975).
23. Ref. 1, p. 3.2.
24. C. E. Wagner, *Combustion* **47,** 20 (Már. 1976).
25. C. C. Curley and P. Olesen, *Combustion* **48,** 23 (Sept. 1976).
26. Ref. 4, Chapt. 14–22.

General References

T. Baumeister, Jr., *Fans,* McGraw-Hill Book Co., New York, 1935; a good treatise on fans as of the date of publication including mechanical and structural design.
W. C. Osborne, *Fans,* 2nd ed., Pergamon Press, New York, 1977.
ASHRAE Handbook and Product Directory—1975 Equipment, American Society of Heating, Refrigeration, and Air-Conditioning Engineers, Inc., New York, 1975, Ch. 3.
J. K. Salisbury, ed., *Kent's Mechanical Engineers Handbook,* 12th ed., Power Volume, John Wiley & Sons, Inc., New York, 1950, pp. 1-57–1-96.
T. Baumeister, ed., *Standard Handbook for Mechanical Engineers,* 8th ed., McGraw-Hill Book Co., New York, 1967, pp. 14-69–14-82.
R. Pollak, "Selecting Fans and Blowers," *Chem. Eng.* **80,** 86 (Jan. 22, 1973).
J. B. Graham, "Methods of Selecting and Rating Fans" in *Fan Application—Testing and Selection Symposium Papers, San Francisco, Calif., Jan. 19–22, 1970,* ASHRAE, New York, 1972, pp. 15–24.

R. J. Aberbach, "Fans—A Special Report," *Power* **112,** S2-S24 (Mar. 1968).

AMCA Fan Application Manual, Section 2—*Troubleshooting, AMCA Pub. 202,* 1972; Section 3—*Field Performance Measurements, AMCA Pub. 203,* 1976, Air Moving and Conditioning Association, Arlington Heights, Ill.

R. C. Myers "Industrial Fans—Guidelines for a Successful Installation," *Iron Steel Eng.* **53,** 38 (Oct. 1976).

J. Thompson "Understanding Fan Performance Curves," *Plant Eng.* **31,** 91 (May 26, 1977).

R. E. Perry, "Operation, Maintenance and Repair of Industrial Centrifugal Fans," *Combustion* **47,** 7 (Feb. 1976).

Noise control

S. G. Tetorka, "Calculating Employee Noise Exposure Levels Generated by Large Industrial Fans" in *Industrial Pollution Control Measures and Instrumentation—Proceedings of a Specialty Conference, New Jersey Institute of Technology, Newark, N.J., Mar. 22–23, 1976,* Technomic Publishing Co., Westport, Conn., 1976, pp. 295–303.

W. Neise, "Noise Reduction in Centrifugal Fans—A Literature Survey," *J. Sound Vib.* **45,** 375 (Apr. 8, 1976).

J. B. Graham, "Industrial Fan Selection and Installation," *Can. Min. J.* **96,** 69 (Oct. 1975).

R. C. Mellin, "Selection of Minimum Noise Fans for a given Pumping Requirement," *Noise Control Eng.* **4,** 35 (Jan.–Feb. 1975).

M. I. Schiff, "Noise Control for Draft Fans," *Actual Specif. Eng.* **34,** 62 (July 1975).

M. J. Huggan, "Control of Higher Pressure Fan Noise," *Noise Control Vib. Reduct.* **6,** 178 (June 1975).

T. W. Rimmer and R. J. Anderson, "In-duct Measurement of Centrifugal Fan Sound Power," *ASHRAE J.* **18,** 52 (Sept. 1976).

M. J. Crocker, "In-duct Sound Power Measurement Systems," *ASHRAE Trans.* **80**(Part 2), 82 (1974).

W. M. Deeprose, "Fan Noise Generation and its Control," *Chart. Mech. Eng.* **21,** 64 (Nov. 1974).

W. A. Smith, J. K. O'Malley, and A. H. Phelps, "Reducing Blade Passage Noise in Centrifugal Fans," *ASHRAE Trans.* **80**(Part 2), 45 (1974).

M. Bartenwefer and co-workers, "Noise Reduction in Centrifugal Fans by Means of an Acoustically-Lined Casing," *Noise Control Eng.* **8,** 100 (May–June 1977).

BURTON B. CROCKER
Monsanto Company

FAST COLOR SALTS. See Azo dyes.

FATS AND FATTY OILS

Fats and fatty oils are made up predominantly of triesters of glycerol [56-81-5] (qv) with fatty acids and commonly are called triglycerides. The designations fats or fatty oils are used merely for convenience in that customarily fats are solids at ambient temperatures and oils are liquids. Both classes of compounds are triglycerides differing only in melting point. To avoid confusion the general term fat is used in this article irrespective of the physical state of the material under discussion. The term lipid also finds considerable use in the literature where it is employed interchangeably with fat.

Fats are distributed widely in nature. They are derived from vegetable, animal, and marine sources and often are by-products in the production of vegetable proteins or fibers and animal and marine proteins. Fats of all types have been used throughout the ages as foods, fuels, lubricants, and starting materials for other compounds. This wide utility results from the unique chemical structures and physical properties of fats. The chemical structures of fats are very complex owing to the combinations and permutations of fatty acids that can be esterified at the three (enzymatically nonequivalent) hydroxyl groups of glycerol. A generalized triglyceride has structure (**1**), without regard to optical activity.

When R = R′ = R″, the trivial name of the triglyceride is derived from the parent acid by means of a termination -in, eg, for stearic acid where R = R′ = R″ = $C_{17}H_{35}$, the triglyceride is called tristearin [555-43-1] (**2**). If, on the other hand, R and R″ are different, the center carbon is asymmetric and the chiral glyceride molecule can exist in two enantiomeric forms. An excellent discussion of this subject is available (1).

Many naturally occurring fats are made up of fatty acids with chain length greater than 12 carbon atoms; the vast majority of vegetable and animal fats are made up of fatty acid molecules of more than 16 carbon atoms. Marine fats (and some *Cruciferae* fats) are characterized by their content of longer-chain (up to C_{24}) fatty acids. Thus, because the fatty acid portions of the triglycerides make up the larger proportion (ca 90% fatty acids to 10% glycerol) of the fat molecules, most of the chemical and physical properties result from the effects of the various fatty acids esterified with glycerol.

Naturally occurring fats contain small amounts of soluble, minor constituents: pigments (carotenoids, chlorophyll, etc), sterols (phytosterols in plant fats, cholesterol in animal fats), phospholipids, lipoproteins, glycolipids, hydrocarbons, vitamin E (tocopherols), vitamin A (from carotenes), vitamin D (calciferol), waxes (qv) (esters of long-chain alcohols and fatty acids), ethers and degradation products of fatty acids, proteins, and carbohydrates. Most of these minor compounds are removed in processing and some are valuable by-products.

$$\begin{array}{cc}
\text{H}_2\text{COCR} & \text{H}_2\text{COC(CH}_2)_{16}\text{CH}_3 \\
| & | \\
\text{HCOCR}' & \text{HCOC(CH}_2)_{16}\text{CH}_3 \\
| & | \\
\text{H}_2\text{COCR}'' & \text{H}_2\text{COC(CH}_2)_{16}\text{CH}_3 \\
(\mathbf{1}) & (\mathbf{2})
\end{array}$$

796 FATS AND FATTY OILS

Roughly two-thirds of the world's production of fats is utilized for human food. Fats are concentrated sources of energy (38.9 kJ/g or 9.3 kcal/g), carriers for vitamins (qv) and other fat-soluble compounds, as well as sources of essential fatty acids that are required by most organisms. Nonedible uses depend mainly on the properties of the various fatty acids in fats although some triglycerides are directly used as specialized lubricants. The many derivatives of fatty acids (see Carboxylic acids) manufactured are used in surface coatings, plastics, detergents, lubricants, etc, where the long hydrocarbon chains confer needed plasticity, surface activity, or lubricity. In some instances, eg, drying oils (qv), the reactive unsaturation of certain fatty acids is exploited, and by means of catalyzed, oxidative polymerization, tough, flexible surface coatings are developed (see Coatings; Paint).

Composition

Glycerides. The number of triglycerides in a given natural fat is a function of the number of fatty acids present and the specificity of the enzyme systems involved in the particular fat synthesis reactions. In his excellent treatise on triglyceride analysis, Litchfield points out that many plant seed fats have the potential to provide 125–1,000 different triglycerides, animal fats contain potentially 1,000–64,000 triglyceride species, and butterfat could generate 2,863,288 triglycerides from 142 different fatty acids (2). In other work, Litchfield developed information on the taxonomic patterns arising from the fatty acid distributions in triglycerides of natural fats (3). Some practical applications of these studies relate mainly to the use of pancreatic lipase to identify a fat type, or to estimates of triglyceride composition from fatty acid composition and the application of some of the distribution hypotheses.

Fatty Acids. Most of the fatty acids in fats are esterified with glycerol to form glycerides. However, in some fats, particularly where abuse of the raw material has occurred leading to enzymatic activity, considerable (>5%) free fatty acid (FFA) is found. Hydrolysis occurs in the presence of moisture. This reaction is catalyzed by some enzymes, acids, bases, and heat. Most producers of fats attempt to prevent the formation of free fatty acid because certain penalties are assessed if they are present in the trading of crude and refined fats (see Specifications). As in the case of the triglycerides, the number of known fatty acids is very large. In 1965 it was noted that ca 900 vegetable and 500 animal fats had been analyzed (4). Seed lipids research in the USDA laboratories (5), as well as the unusual fatty acids in plants (6) have been reviewed. Table 1 lists fatty acids prevalent in fats of economic consequence, and their principal sources.

Phospholipids. Phospholipids (3) occur in most natural fats with differing amounts and compositions depending on the source of the fat. Owing to their complexity, these fat-soluble, biologically important compounds also have presented some intriguing analytical problems to chemists and biochemists (7). From a technical standpoint, phospholipids, eg, from soybean (qv), are composed mainly of lecithin (qv), cephalin (phosphatidylethanolamine—similar to lecithin but with 2-aminoethanol substituted for choline (qv)) or phosphatidylinositol (where inositol (see Vitamins) is linked through its 1-hydroxyl). These complex mixtures (2–3% in soybean oil) are hydrated during the degumming step, removed and dried (see Processing). These products are sold as commercial lecithin used in margarines, confections, and shortenings where a fat-soluble emulsifier is required.

$$\begin{array}{c} \text{O} \\ \| \\ H_2COC(CH_2)_mCH_3 \\ | \quad \text{O} \\ | \quad \| \\ HCOC(CH_2)_nCH_3 \\ | \quad \text{O} \\ | \quad \| \\ H_2COPOR \\ | \\ O^- \end{array}$$

(3) phospholipid $m, n = 10\text{–}16$

lecithin [8002-43-5], soybean lecithin [8030-76-0] $\overset{+}{C}H_2CH_2\overset{R}{N}(CH_3)_3$

cephalin [4537-76-2] where $m = n = 16$ $CH_2CH_2NH_2$

phosphatidylinositol [28154-40-7]

where $m = n = 14$,
ammonium salt.

Antioxidants. The most commonly occurring natural antioxidants (qv) in vegetable fats are the tocopherols (vitamin E active). These derivatives of 6-chromanol (2H-1-benzopyran-6-ol,3,4-dihydro-) are not synthesized by mammals and occur in their fats only through ingestion of plant materials and vegetable fats. Antioxidants tend to protect fats by inhibiting autoxidation and subsequent rancidity. A certain percentage of the tocopherols are removed in some refining steps and are recovered as by-products of vegetable-oil processing. An excellent review of tocopherols in foodstuffs is ref. 8. Another compound of consequence is sesamol, a sterol found in sesame oil. The lecithins and gossypol (a pigment of cottonseed oil) also have been reported to have antioxidant characteristics, and this may reflect an ability to interact with heavy metals.

Pigments. The major pigments of fats are the carotenoids. Palm oil, usually bright reddish-orange, contains as much as 0.2% β-carotene. Many seed oils, particularly if processed from immature seeds, also contain significant levels of chlorophyll pigments that lend a greenish tinge to the fats. Cottonseed oil is heavily colored by gossypol-type (phenolic) pigments. Most of these pigments are removed in the alkali refining and bleaching steps. A few pigments (fixed) are difficult to remove in processing and may result from heat or oxidative abuse of the fat-containing raw materials or the crude fats themselves. The carotenoid pigments are mainly decolorized by heat, light or oxidative treatment but the quinones generated by oxidation of the tocopherols generally cause darkening of fats (see Pigments).

Vitamins. The principal components in vegetable fats with vitamin activity are the tocopherols. Vitamin A is found in butterfat and in fish oils. The carotenes (provitamin A) are found at significant levels in palm oil, in butterfat and as traces in other fats. Vitamin D is found primarily in some fish oils (see Vitamins).

Table 1. Fatty Acids of Economically Significant Fats

Fatty acid	CAS Registry No.	Common name	Designation[a]	Principal sources
butanoic	[107-92-6]	butyric	4:0	butter
hexanoic	[142-62-1]	caproic	6:0	butter
octanoic	[124-07-2]	caprylic	8:0	coconut
decanoic	[334-48-5]	capric	10:0	coconut
dodecanoic	[143-07-7]	lauric	12:0	coconut, palm kernel
tetradecanoic	[544-63-8]	myristic	14:0	coconut, palm kernel, butter
hexadecanoic	[57-10-3]	palmitic[b]	16:0	palm, cottonseed, butter, animal fat, marine fats
cis-9-hexadecenoic	[373-49-9]	palmitoleic	16:1(9c)	butter, animal fats
octadecanoic	[57-11-4]	stearic[b]	18:0	butter, animal fats
cis-9-octadecenoic	[112-80-1]	oleic[b]	18:1 (9c)	olive, tall oil, peanut, canbra[c], animal fats, butter, marine fats
cis,cis-9,12-octadecadienoic	[60-33-3]	linoleic[b]	18:2 (9c, 12c)	safflower, sunflower, corn, soy, cottonseed
cis,cis,cis-9,12,15-octadecatrienoic	[463-40-1]	linolenic	18:3 (9c, 12c, 15c)	linseed
cis,cis,cis,cis-6,9,12,15-octadecatetraenoic	[20290-75-9]		18:4 (6c, 9c, 12c, 15c)	marine fat
cis,trans,trans-9,11,13-octadecatrienoic	[506-23-0]	α-eleostearic	18:3 (9c, 11t, 13t)	tung
12-hydroxy-cis-9-octadecenoic	[141-22-0]	ricinoleic	18:1 (9c) 12-OH	castor
cis-9-eicosenoic	[29204-02-2]	gadoleic	20:1 (9c)	marine fat
cis-11-eicosenoic	[5561-99-9]		20:1 (11c)	rapeseed
all cis-5,8,11,14-eicosatetraenoic	[506-32-1]	arachidonic	20:4 (5c, 8c, 11c, 14c)	animal, marine fats
all cis-8,11,14,17-eicosatetraenoic	[24880-40-8]		20:4 (8c, 11c, 14c, 17c)	marine fats
all cis-5,8,11,14,17-eicosapentaenoic	[10417-94-4]		20:5 (5c, 8c, 11c, 14c, 17c)	marine fats
docosanoic	[112-85-6]	behenic	22:0	
cis-11-docosenoic	[506-36-5]	cetoleic	22:1 (11c)	marine fats
cis-13-docosenoic	[112-86-7]	erucic	22:1 (13c)	rapeseed
all cis-7,10,13,16,19-docosapentaenoic	[24880-45-3]		22:5 (7c, 10c, 13c, 16c, 19c)	marine fats

Table 1 (*continued*)

Fatty acid	CAS Registry No.	Common name	Designation[a]	Principal sources
all *cis*-4,7,10,13,16,19-docosahexaenoic	[6217-54-5]		22:6 (4c, 7c, 10c, 13c, 16c, 19c)	marine fats

[a] Number of carbon atoms: number of double bonds (geometric isomerism).
[b] Constituent of most fats.
[c] Low erucic rapeseed.

Sterols. Most of the unsaponifiables (see Analytical and Test Methods) in vegetable and animal fats are sterols. The animal fats predominantly contain cholesterol and most vegetable fats contain only traces of this sterol. Plant sterols—collectively called phytosterols—are made up mainly of sitosterols and stigmasterol but some individual vegetable fats contain additional phytosterols. The pattern of typical sterols has been suggested as useful in detecting adulteration of one oil with another (9). Sterols are of minor importance in the technology of fats. Normally they are removed in the refining and deodorization steps. The gross mixtures recovered from these processes have been utilized as a source of certain phytosterols that are used as raw materials in the pharmaceutical industry. A good summary of current knowledge on sterols in fats is ref. 10.

Minor Constituents. In addition to the materials listed above, waxes, hydrocarbons, ketones, aldehydes, and mono- and diglycerides are found in fats and oils at varying but low levels. The waxes (qv) in some seed oils (eg, corn, sunflower and safflower) are troublesome and are removed in processing to prevent haze formation in the finished products. The ketones and aldehydes probably arise from oxidative damage and can cause flavors and odors in fats. The mono- and diglycerides result from hydrolytic reactions either in the raw materials or during processing but do not pose particular problems in end products. The hydrocarbons are mainly analytical curiosities and are of no technological consequence (11).

Characteristics and Occurrence

Milk fats are distinguished by low unsaturation and the presence of a wide variety of saturated fatty acids of short chain length. Butterfat is the only important member; it is relatively expensive and is used only for edible purposes. Recently, cows have been fed encapsulated polyunsaturated fat (safflower oil) which prevents bacterial reduction of the fat in the rumen and leads to polyunsaturated milk and meat (12) (see Pet and other livestock feeds).

Lauric acid oils are distinguished by very low unsaturation and a high content of lauric acid as well as other short-chain acids. They melt sharply at relatively low temperatures and are relatively light colored and low in nonglyceride constituents. Lauric acid oils are derived from seeds of cultivated or noncultivated palms. The important members are coconut and palm kernel oils; they are used for edible products and also for soap-making.

Vegetable butters are low in unsaturation and contain principally C_{14}, C_{16}, and

C_{18} acids. The relatively low and very sharp melting points of these butters are the result of even distribution of saturated and unsaturated fatty acids, rather than the presence of low molecular weight acids as in lauric acid oils. Vegetable butters are derived from seeds of tropical trees (see Vegetable oils). Cocoa butter is the most important member; it is expensive and is used principally in confectionary (see Chocolate and cocoa).

Land-animal fats are relatively low in unsaturation and contain principally C_{16} and C_{18} acids. The unsaturated acids consist almost entirely of oleic and linoleic acids. With saturated and unsaturated acids evenly distributed, they exhibit gradual melting and relatively high melting points. If derived from undamaged materials, they are light and low in nonglyceride constituents; inedible grades are often dark. Lard (from hogs) and tallow (from cattle and sheep) are derived principally from the body fat as the by-product of meat packing (see Meat products). They are relatively cheap and are used as edible fats, as a source of commercial fatty acids (see Carboxylic acids) and for soap-making. The term grease refers to the softer inedible fats used principally by soap makers (see Soap).

Oleic–linoleic acid oils are of medium, but rather variable unsaturation (iodine value (IV) varies from ca 50 for palm oil to ca 120 for corn oil and up to 145 for safflower oil) with no fatty acids more unsaturated than linoleic (2 double bonds); the fatty acids are predominantly C_{18} acids. These oils are normally liquid in the raw form but are frequently hydrogenated to produce plastic fats. The crude seed oils are relatively high in nonglyceride substances. These fats are derived principally from the fruit pulp of perennial plants (palm, olive) and the seeds of cultivated annual plants (eg, cotton, peanut, corn, safflower, sesame, and sunflower seeds). The oils are of medium price and are used primarily for edible purposes. Plant breeders have been particularly successful in developing special varieties of some plants, eg, high-oleic safflower oil (IV ca 90) is now commercially available. Further, growers have taken advantage of the temperature—IV relationship in sunflower and now grow seed with relatively low IV oil in warmer climatic zones (see Soybeans and other seed proteins).

Erucic acid oils resemble the oleic–linoleic acid oils in composition except that the predominant unsaturated fatty acid is a C_{22} acid, erucic acid, and there is a minor proportion (6–10%) of linolenic acid. The commercially important oilseeds are: rapeseed (colza), ravison and mustard seed. Until a few years ago considerable regular rapeseed was grown in Canada and Europe as an oilseed. Concern about possible physiological effects of erucic acid in rapeseed oil led to the development of so-called zero erucic rapeseed varieties. This oil (called canbra or canola oil in Canada) and the limiting of erucic acid levels in edible products in Canada and the European Economic Community (EEC) has limited production of the high erucic acid rapeseed varieties. Some high erucic acid rapeseed is grown as a source of oil for industrial purposes (see Uses and Derivatives). It is used in the United States industrially and has not, to date, been allowed in edible products except as a fully hydrogenated product.

Linolenic acid oils, derived almost wholly from the seeds of cultivated annual plants, are generally similar to the oleic–linoleic acid oils but are distinguished from the latter by containing the more highly unsaturated acid, linolenic acid (18:3). The most important members are soybean and linseed oils, which are medium-priced oils. Soybean oil is used both in edible products and as a drying oil, and linseed oil is used exclusively as a drying oil (except in the Baltic region, where it is also used as an edible oil) (see Driers and metallic soaps). Linolenic acid oils or other oils containing a substantial proportion of unsaturated acids with more than 2 double bonds have been considered less desirable than other fats for the manufacture of edible products because of their flavor instability or tendency toward flavor reversion after deodorization.

Conjugated acid oils are used only as drying oils for which they are particularly suited because of their high content of unsaturated fatty acids with conjugated double bonds. The commercially important members, tung and oiticica oils, are derived from the seeds of subtropical trees. They are relatively high priced, and with the increased use of synthetic surface coatings these oils are becoming increasingly rare in commerce.

Marine oils are distinguished by their considerable content of fatty acids that vary considerably in chain length both above and below 18 carbons and by the presence of highly unsaturated acids (4 or more double bonds), together with a considerable content of saturated acids (as much as 25% of the total acids). Although used for edible purposes, as drying oils, and after hydrogenation, for soapmaking (including the manufacture of metallic soap), the diversity of their component fatty acids prevents them from being the most highly desirable products for any particular purpose. They are generally the cheapest of all fats and oils. Large quantities are used in the fat-liquoring of leather (qv). Unlike the land-animal fats, marine oils are not a by-product of food processing. The commercially important members are derived from small oily fishes, such as California sardine (pilchard), menhaden and herring, which are taken principally for their oil. Whale oil is becoming increasingly rare as overhunting has severely decimated the numbers of these large mammals. Many international and local restrictions have been placed on the killing of all types of whales in an attempt to allow this resource to replenish itself. Fish-liver oils, derived from an entirely different species of fish (since oily fish generally have livers with a low oil content and vice versa), were formerly important as a source of natural vitamin A.

Hydroxy-acid oils are represented by castor oil which consists principally of glycerides of ricinoleic (12-hydroxyoleic) acid. Castor oil (qv) is used (after dehydration) as a drying oil of the conjugated type, and for the manufacture of a number of specialty products.

Miscellaneous fats include many unusual fatty acid compositions that have been the subjects of extensive reviews (5–6). In these extensive studies, many olefinic, acetylenic, and oxygenated fatty acids have been found and characterized; to date, none of these oils have been commercially exploited but potential exists for some if the crops from which they are derived become agronomically promising.

The classification, sources, and production of many commercially significant fats are listed in Table 2. Pertinent properties and compositional data are shown in Table 3.

Physical Properties

Viscosity. Although log viscosity and log temperature yields nearly a straight line function, linearity is obtained by use of the equation $\log n = a + 10^6 b \cdot t^{-3}$ ($t = °C$) for the fats with various iodine values listed in Table 4 (19). The data in Table 4 are obtained with a Couette-type viscometer at differing shear rates above the samples'

Table 2. Classification, Sources, and Production of Commercially Significant Fats

Fat	Type	Source	Principal production areas	Estimated 1978 production,[a] 1000 metric tons
Vegetable				
babassu	lauric acid	*Orbignya (Attalea) speciosa*	Brazil	95
castor	hydroxy acid	*Ricinus communis*	Brazil, India	375
coconut	lauric acid	*Cocos nucifera*	Philippines, Indonesia, India	3,188
corn	oleic–linoleic acid	*Zea mays*	U.S., Europe, Argentina	440
cottonseed	oleic–linoleic acid	*Gossypium hirsutum, Gossypium barbadense*	USSR, U.S., China, India, Pakistan, others	3,298
linseed	linolenic acid	*Linum usitatissimum*	Argentina, Canada, India, U.S., USSR	916
oiticica	conjugated acid	*Licania rigida*	Brazil	14
olive[b]	oleic–linoleic acid	*Olea europaea*	Italy, Spain, Greece, North Africa	1,796
palm	oleic–linoleic acid	*Elaeis guineensis*	Malaysia, Africa, Indonesia	3,740
palm kernel	lauric acid	*Elaeis guineensis*	Malaysia, Africa, Indonesia	602
peanut (groundnut)	oleic–linoleic acid	*Arachis hypogaea*	India, Africa, China, U.S., others	3,336
rapeseed	erucic acid (oleic–linoleic acid)[c]	*Brassica campestris, Brassica napus*	Canada, Europe, India, China	2,908
safflower	oleic–linoleic acid	*Carthamus tinctorius*	U.S., India, others	275
sesame	oleic–linoleic acid	*Sesamum indicum*	India, China, Africa	694
soybean	linoleic–linolenic acid	*Soya max*	U.S., China, Brazil, others	11,250
sunflower	oleic–linoleic acid	*Helianthus annus*	USSR, U.S., Argentina, Europe, others	4,572
tung	conjugated acid	*Aleurites fordii, A. montana*	China, Argentina, others	110
Animal				
butter	milk	*Bos gaurus*	worldwide	4,830
lard	land animal	*Sus domesticus*	worldwide	4,700
tallow and grease	land animal	*Bos taurus, Ovis aries*	worldwide	5,175
Marine				
whale	marine	*Cetacea*	worldwide	40
sperm whale	marine	*Physeter catodon*	worldwide	110
fish (liver)	marine	*Clupea harenqus, C. pilchardus, Sardinops corrulea, Brevoortia tyrannus*	Norway, Peru, Japan, U.S.	930

[a] Ref. 13.
[b] Includes olive residue oil at 10% of total.
[c] Low erucic varieties.

melting points (see Rheological measurements). All samples show Newtonian behavior. These data point out the relationship of viscosity to both molecular weight and unsaturation. Viscosity is useful for assessing the changes occurring in fats as they are oxidatively or thermally polymerized. Both types of polymerization (viscosity increase) are key to the use of fat in the surface-coating industry but are to be avoided in food processes (mainly frying) where fats are used as heat transfer media.

Surface and Interfacial Tension. Surface tension data for fats appear to be in the range of other organic compounds (20–40 mN/m (= dyn/cm)). One report (20) lists data for cottonseed oil (35.4 mN/m at 20°C) and coconut oil (33.4 mN/m at 20°C). Also noted is an increase of surface tension with increasing chain length and a decrease with increasing temperature. Interfacial tension data (21) were given for peanut oil (29.92 mN/m), cottonseed oil (29.76 mN/m), and soybean oil (30.92 mN/m). For an interesting discussion of monolayer behavior of fats see ref. 22. Surface activity is not a particularly important feature of fats but is particularly important in many derivatives (eg, soaps, detergents, and derivatives).

Density (Specific Gravity). Specific gravities of fats in the liquid state do not differ much for most of the common fats, and are 0.914–0.964 at 15°C. A compilation is available (23). An equation developed for specific gravity of liquid oils is: specific gravity = 0.8475 + 0.00030 (saponification number) + 0.00014 (iodine value) (20). Over the normal processing temperature range of ca 65–260°C, the density decreases linearly by ca 0.64 g/(L·°C). Densities of fats in the solid state are much higher (1.00–1.06 kg/L) than are those of liquid fats.

Melting and Freezing Points. Melting and freezing points of most commercial fats and oils are, at best, crude indicators of the product under examination. Pure triglycerides of single fatty acids and some well-characterized mixed triglycerides have well-defined melting and freezing points but these data are difficult to interpret because of the problem of polymorphism. The data up to 1950 are interpreted in ref. 24 and reviewed in ref. 25. Melting points of various polymorphs of simple triglycerides range from −44.6 to 73.1°C depending on the fatty acids and polymorphs present. More complex mixed triglycerides cover a greater mp range (24).

Because most fats are made up of complex mixtures of hundreds of triglycerides, it is easy to understand why melting and freezing points are so inexact. In commercial practice the Wiley melting point method (26) is widely used for many fats. For some special products (highly hydrogenated or fractionated), the capillary method (27) is employed. Freezing points, as such, are not often used. The congeal point (set point) method provides some control in the hydrogenation of fats where it is principally used (28). Some general information on final melting points of average samples of commercial fats is shown in Table 5.

Smoke, Fire, and Flash Points. These values depend mainly on the content of impurities (solvents, fatty acids, and mono- and diglycerides) in fats. The important thermal reactions that often enter into the buying and selling specifications of fats are: smoke point, the temperature, under standardized test conditions (29), at which a thin continuous smoke stream is observed; flash point, the temperature at which a flash appears when a test flame is applied under standard conditions (29); and fire point, the temperature where the flame continues to burn for 5 s after application of the test flame (29). A useful graph is shown in Figure 1 (20).

Crystal Structure. In many instances, triglycerides exist in polymorphic forms. The simplified depiction of the various polymorphs from x-ray diffraction studies is

Table 3. Properties and Composition of Commercially Significant Fats[a]

Fat	Iodine value	Saponification value	12:0	14:0	16:0	18:0	18:1	18:2	18:3	Other acids
Vegetable										
babassu	13–16	247–253	44–45	15–16.5	5.8–8.5	2.5–5.5	12–16	1.4–2.8		8:0, 4.1–4.8 wt %; 10:0, 6.6–7.6 wt %
castor[b]	84–88	178–180			0.8–1.1	0.7–1.0	2.0–3.3	4.1–4.7	0.5–0.7	ricinoleic, 87.7–90.4 wt %; dihydroxystearic, 0.6–1.1 wt %; 20:1, 0.3–0.8 wt %
coconut	7.5–12	250–264	44–51	13–18.5	7.5–10.5	1–3	5–8.2	1.0–2.6		8:0, 7.8–9.5 wt %; 10:0, 4.5–9.7 wt %
corn	116–140	188–198			7	3	43	39		
cottonseed[c]	90–112	189–198		1.5	22	5	19	50		
linseed	168–204	188–196			6	4	13–37	5–23	26–58	
oiticica	205–220	188–193		10–12	10–12	10–12				licanic, 73–83 wt %; unsaturated, 5–16 wt %[d]
olive	76–90	186–196		1.3	7–16	1.4–3.3	64.5–84.5	4–15		
palm	35–61	195–205		0.6–2.4	32–45	4–6.3	38–53	6–12		
palm kernel	14–24	245–255	47–52	14–17.5	6.5–8.8	1–2.5	10.5–18.5	0.7–1.3		8:0, 2.7–4.3 wt %; 10:0, 3.0–7.0 wt %
peanut	84–102	188–195		0.5	6–11.4	3–6	42.3–61	13–33.5		20:0, 1.5 wt %; 20:1+2, 1–1.5 wt %; 22:0, 3–3.5 wt %
rapeseed										
regular	94–106	168–179			1–4.7	1–3.5	13–38	9.5–22	1–10	22:1 (erucic), 40–64 wt %
low erucic (canbra or canola)				1.5	4–5	1–2	55–63	20–31	9–10	22:1, 1–2 wt %
safflower										
regular	126–152	175–195			6.4–7.0	2.4–2.8	9.7–13.1	76.9–80.5		
high oleic	90–100	175–195			4–8	4–8	74–79	11–19		20:1, 0.5 wt %[e]
sesame	104–116	187–193			7.2–7.7	7.2–7.7	35–46	35–48		
soybean	117–140	189–195			2.3–10.6	2.4–6	23.5–30.8	49–51.5	2–10.5	[e]
sunflower	113–143	186–194			3.5–6.5	1.3–3	14–43	44–68		
tung[f]	160–175 (248–252)[g]	189–195			4	1	8	4	3	α-eleostearic (conjugated 18:3), 80 wt %

Animal								
butter[c]	25–38	218–235	3	10	25	11	28.5	2.5	4:0, 4 wt %; 6:0, 2 wt %; 8:0, 1 wt %; 10:0, 2.5 wt %
lard	53–77	190–202		0.9–2.1	22.4–31	16.5–23.7	38.3–44.4	4.5–8.8	16:1, 1.3–3.6 wt %; 18:3 + 20:1, 1–2.3 wt %
tallow (beef)	35–48	193–202		3–6	25–37	14–29	26–50	1–2.5	
Marine									
whale	110–130	183–198		4–8	7–12	1–3	28–32	1–2	16:1, 7–18 wt %; 20:1, 12–20 wt %; 20:5, 1–4 wt %; 22:1, 4–18 wt %; 22:5, 0.5–3 wt %; 22:6, 1–5 wt %
herring	120–160	192		3–8	8–13	1–3	17–22	1–4	16:1, 6–9 wt %; 20:1, 9–15 wt %; 20:5, 6–9 wt %; 22:1, 11–16 wt %; 22:5, 1–4 wt %; 22:6, 6–8 wt %
sardine (pilchard)	160–190	191		8	16	2–4	10–15	1	16:1, 8–10 wt %; 20:1, 3–6 wt %; 20:4, 1–2 wt %; 20:5, 12–17 wt %; 22:1, 3–6 wt %; 22:5, 2–4 wt %; 22:6, 10–14 wt %
sardine (Peruvian)	170–190	191		8	19	3	10–15	1	16:1, 8–12 wt %; 20:1, 2 wt %; 20:4, 1 wt %; 20:5, 18–24 wt %; 22:1, 1 wt %; 22:5, 2 wt %; 22:6, 4 wt %
menhaden	150–195	190		7–8	17–29	3–4	13–16	1	16:1, 7–10 wt %; 18:4, 2–4 wt %; 20:1, 1–2 wt %; 20:4, 1–2 wt %; 20:5, 10–13 wt %; 22:1, 2 wt %; 22:5, 2–3 wt %; 22:6, 9–14 wt %

[a] Ref. 14.
[b] Ref. 15.
[c] Ref. 16.
[d] Ref. 17.
[e] Ref. 18.
[f] Constants and composition of natural fats and oils, Ashland Chemical Co., Columbus, Ohio, 1969.
[g] By special method (17).

Table 4. Viscosity of Various Neutral and Deodorized Oils and Fats in the Liquid State Measured at Different Temperatures[a]

Oil or fat	IV[d]	Viscosity[b], mPa·s(= cP)			Constants[c]	
		20°C	40°C	60°C	a	b
soybean oil	134	60	28	15	−0.073	46.6
medium chain triglycerides	0		21	11	−0.306	50.1
sunflower seed oil	132	63	29	16	−0.038	44.8
corn oil	122	70	30	16	−0.142	49.9
coconut oil	9		27	14	−0.242	51.0
hydrogenated soybean oil, mp 28°C	101		33	18	−0.148	51.1
butterfat	38		34	17	−0.151	51.2
groundnut oil	89	81	36	19	−0.080	50.5
olive oil	83	82	35	17	−0.102	50.1
hydrogenated cottonseed oil, mp 32°C	76		45	23	−0.166	55.9
rapeseed oil	104	93	41	21	−0.023	50.1
lard olein	73		36	18	−0.151	51.9
palm olein	64		37	19	−0.145	52.2
palm oil	51		37	19	−0.192	53.8
lard	63		36	19	−0.068	48.2
hydrogenated rapeseed oil, mp 32°C	81		49	24	−0.140	56.0

[a] Courtesy of *J. Am. Oil Chem. Soc.* (19).
[b] Standard deviation of replicates 1%.
[c] Constants in the equation $\log \eta = a + 10^6 bt^{-3}$ ($t = °C$).
[d] IV = Iodine Value.

Table 5. Melting Points of Average Samples of Fats

Fat or oil	mp, °C	Fat or oil	mp, °C
babassu oil	26	palm oil (refined)	40
beef tallow	50	palm kernel oil	29
Borneo tallow	38	peanut oil	13
butterfat	37	castor oil[a]	87
cocoa butter	36	cottonseed oil[b]	60.0
coconut oil	26	sardine oil[b]	57.5
cottonseed oil	11	soybean oil[b]	66.5
lard, prime steam, U.S.	45		

[a] Hydrogenated oil; IV, 0.5.
[b] Hydrogenated oil; IV, 10.

reproduced in Figure 2 (25). Crystal structure is very important to the properties of margarines, shortenings and specialty fats. The very unstable α form is readily transformed to the more stable β' form which in some triglycerides is higher melting (more stable) than the β form. In single-acid triglycerides, the order is α → β' → β but some mixed triglycerides show a lower melting (less stable) β form. This is further complicated by the existence of multiple β' and β forms depending upon the detailed triglyceride structures at hand. An additional view of these complications can be found in ref. 30 which is a study of polymorphism in a series of single-acid triglycerides where

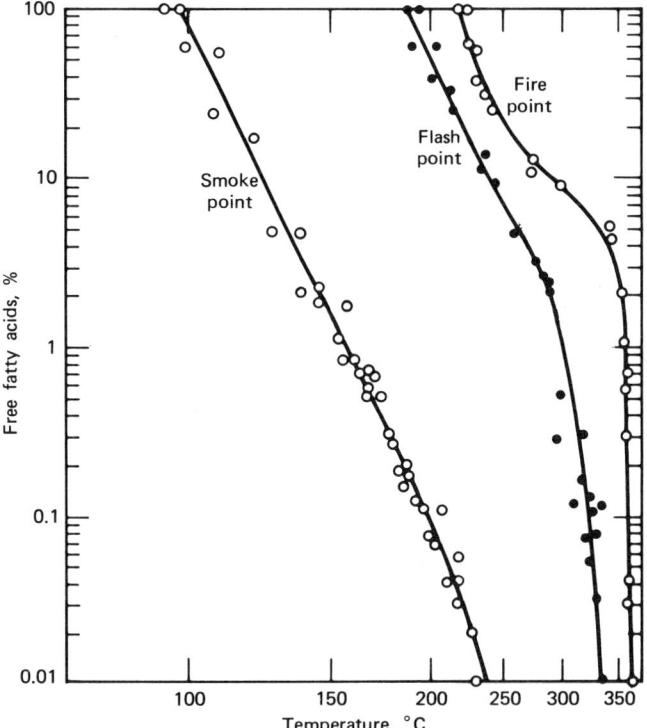

Figure 1. Smoke, fire, and flash points of miscellaneous crude and refined fats and oils, as functions of free fatty acids (20).

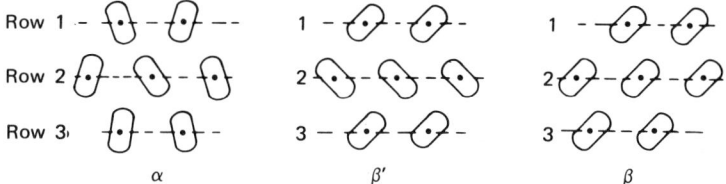

Figure 2. Cross-sectional structures of long-chain compounds (25).

the data were developed using differential scanning calorimetry (dsc) (see Analytical methods). A more extensive compilation of melting points of polymorphic forms of many triglycerides is ref. 14.

Specific Heat and Heat of Fusion. Specific heat and heat of fusion are related to the size and shape of the molecules in question. The specific heat of liquid fats increases with chain length but decreases with unsaturation, and there is a progressive increase of specific heat with increasing temperature (20). Also, heat of fusion increases with chain length and decreases with unsaturation. Some representative values of heat of fusion are shown in Table 6 (20). Heats of fusion of the pure triglycerides in the β form have been determined with dsc (31–32). Ref. 32 suggests the values in Table 6 may be 1–10% high.

It has been proposed (14) that the specific heat Cp can be calculated readily and with satisfactory precision:

Table 6. Heats of Fusion of Representative Samples of Fats[a]

Fat	CAS Registry No.	Heat of fusion, kJ/kg[b]
trilaurin[c]	[538-24-9]	193.3
trimyristin[c]	[555-45-3]	210.5
tripalmitin[c]	[555-44-2]	222.2
tristearin[c]	[555-43-1]	228.0
cottonseed oil[d]	[8001-29-4]	86.2
cottonseed oil, partially hydrogenated[e]		114.6
cottonseed oil, hydrogenated[f]		185.4

[a] Ref. 20.
[b] To convert kJ to kcal, divide by 4.184.
[c] β form.
[d] 108.3 IV.
[e] 59.5 IV.
[f] 0.85 IV.

For liquid fats: $C_p = 1.93 + 0.0025\ t$ ($t = 15–60°C$).

For tallow, palm oil, and partially hydrogenated fats: $C_p = 1.99 + 0.0023\ t$ ($t = 40–70°C$).

And for highly saturated fats (IV < 10): $C_p = 1.92 + 0.003\ t$ ($t = 60–80°C$) (to convert J to cal, divide by 4.184).

Vapor Pressure. Vapor pressures of fats are exceedingly low, eg, in mPa (1 Pa = 7.5 μm Hg): tristearin, 0.133 at 253°C and 6.66 at 313°C; trilaurin 0.133 at 188°C and 6.66 at 244°C; soybean and olive oils, 0.133 at 254°C and 6.66 at 308°C.

Thermal Conductivity. Fats are relatively poor conductors of heat. The thermal conductivities vary little for different oils, and range from ca 167 mW/(m·K) at 20°C to 163 mW/(m·K) at 100°C.

Heat of Combustion. The following formula can be used for calculation of the approximate heat of combustion of fatty oils, in kJ (1 kJ = 0.239 kcal)/g, in terms of constant volume at 15°C: heat of combustion = 47.6 − (0.0042 × iodine value) − (0.038 × saponification value) (33). A value of 39.4 kJ/g (9500 cal/g) is ordinarily taken for common edible fats such as lard and cottonseed oil. Additional compilations are available (20,34).

Solubility and Miscibility. At temperatures above their melting points, fats and oils are freely miscible with most organic solvents except alcohols. Castor oil exhibits the peculiarity of free miscibility with alcohols and limited miscibility with hydrocarbons at usual ambient temperatures. At temperatures far below their melting points, fats are only very slightly soluble. Near their critical temperatures and pressures, organic solvents exhibit anomalous behavior with respect to their miscibility with fats; eg, liquid propane becomes incompletely miscible and may be used as an agent for liquid–liquid extraction.

Ordinary refined liquid oils dissolve about 0.07% of their own weight of water at −1°C and ca 0.14% at 32°C. The solubility of liquid oils in water is extremely low. Liquid oils dissolve ca 92% of their own volume of carbon dioxide at 64°C and ca 62% at 140°C. The solubility of other gases, such as nitrogen, oxygen, hydrogen, and carbon monoxide, increases with increasing temperature of the oil (20). Air dissolves in liquid oils to the extent of ca 8 vol % at 30°C and 13 vol % at 150°C. In all cases there is a linear relationship between solubility and temperature.

Refractive Index. The refractive index n is easily determined on small samples; it increases with molecular weight and it has an approximately linear increasing relationship with the degree of unsaturation of neutral fats. Impurities (fatty acids, mono- and diglycerides, oxidation products, and conjugated olefinic bonds) cause positive deviations in n. However, n is widely used in quality control to check purity of materials and to follow and control hydrogenation and isomerization procedures. Roughly, n_D^{60} ranges from 1.4468 (0 IV) to 1.4568 (100 IV) to 1.4687 (200 IV) for many neutral oils. Added information on the use of n in process control is available (20,35–36).

Absorption Spectra. Most triglycerides do not absorb in the visible or significantly in near uv regions but nonconjugated, unsaturated fats absorb in the far uv (37). All fats have reasonably similar ir spectra (38). The carbonyl stretching band can be used to estimate total triglyceride if no other ester bands are present. A principal use of ir is the estimation of isolated trans unsaturation using the 965–975 cm^{-1} band (39). Although saturated fats and those with isolated olefinic bonds do not absorb in the near uv, this region is used for the detection of conjugated polyunsaturation occurring naturally in some oils or induced by catalytic isomerization. Principal absorption regions of interest are those for conjugated dienes (232 nm), conjugated trienes (260–280 nm, triplets) and tetraenes (290–320 nm, triplets). The location of these peaks varies depending upon the geometrical isomers present. The usual methylene-interrupted polyolefins in fats (—CH=CHCH$_2$CH=CH—) can be conjugated using heat and strong bases (eg, KOH in glycerol) and then are analyzed in the uv (40). Principal absorption bands for conjugated dienes and trienes are found in the 233 nm and 260–280 nm regions. Some care must be exercised, however, in quantitatively applying this method as differences in absorption do occur depending on the geometrical isomers present (20,40).

X-ray diffraction of crystalline fats leads to sharply defined patterns that are useful in the study of polymorphism (2,20,24–25,38). The powder method is simple and yields patterns of lines corresponding to long and short spacings. The α, β', and β polymorphs show strong, short spacing lines at ca 0.415, 0.42, 0.38, and 0.46 nm (38). The method also has had limited use in detecting positional isomers or in distinguishing enantiomorphic triglycerides from their racemates (2).

High resolution nmr spectra are useful in detecting the presence of functional groups. Examples of typical spectra and applications are given (2). Broad line nmr (particularly pulsed) is finding utility for the rapid determination of solid fat content. An excellent review is available (41). This method shows promise as a replacement for the tedious, but necessary, dilatation procedure of determining solid fat index (SFI) of margarine, shortening, and confectionery fats.

Optical rotatory dispersion (ord) spectroscopy allows the study of optical activity of compounds at various wavelengths. There are no practical consequences of chirality but the topic is of considerable interest to those concerned with the biosynthesis and metabolism of triglycerides (see Analytical methods).

Chemical Properties

The chemical reactions of fats are principally those of esters, olefins and hydrocarbons. Triglyceride esters are usually hydrolyzed with acid or base catalysts to yield soaps (fatty acid salts) (see Soaps; Surfactants and detersive systems) or fatty acids (see Carboxylic acids), and glycerol (qv) is recovered. Triglycerides also undergo interesterification or alcoholysis with added alcohols (usually with basic catalysts, eg, sodium methoxide). Amines and other nucleophilic reagents react readily with triglycerides to yield amides and other derivatives. The principal reactions of the long-chain olefins are oxidation (called autoxidation when it occurs spontaneously in air), reduction (usually accomplished by treatment with hydrogen in the presence of metal catalyst), and cis–trans isomerization (which also can occur during catalytic hydrogenation). Unsaturated fats react with all common reagents for olefins. Autoxidation is avoided in most fat processing, storage, and distribution systems as the complicated breakdown products resulting from peroxide decomposition lead to undesirable, odorous ketones, aldehydes, and shorter-chained acids. These characteristic autoxidation reactions of fats probably have been studied more widely than any other aspect of fat chemistry. Many detailed reviews are available (42–44).

Manufacture and Processing

Commercially important fat sources include oilseeds, fruit pulps, animals, and fish. In oilseeds the fat is concentrated in the kernel and varies widely as to amount, eg, soybeans (18–22%), sunflower (42–63%), peanut (46–50%), cottonseed (30–38%), coconut (63–70%), rapeseed (22–49% in whole seed), linseed (33–43%), palm kernel (40–52%), corn (50% in germ), castor seed (65–70%), and safflower (46–54%). The two principal fruit pulps (fat contents) are palm (30–55%) and olive (38–58%). The fat content of animal and fish tissue depends upon the part of the organism rendered. Land-animal fatty tissue ranges from 60–90% fat. Whole fish contain 10–20% fat.

Oilseed Processing. As most oilseeds are dried to a narrow moisture range (7–13% H_2O depending on the seed type) before storage, further drying normally is not required. The oilseeds are carefully cleaned of tramp metal, sticks, stones, and weed seeds (dockage), using magnets, screens, and aspirator systems. Depending upon the seed type, oil content, hull content, etc, one of a variety of processing systems is employed. High hull, high fat–content oilseeds often are dehulled (decorticated) using various mechanical schemes (cf cottonseed). This reduces the fiber content of the meal produced which often has a significant value in the marketplace as food for both humans and animals.

The decorticated kernels are cooked (conditioned) with steam and prepressed in screw presses (expellers) where the oil content is reduced to 10–15%. There are many manufacturers of such presses. Details on their construction and performance are available (45). The press cake is flaked through large steel rollers, often with moisture adjustment, and is conveyed to a solvent-extraction plant. Many varieties of such plants exist but the principles employed are essentially common to all types (see Extraction). The flakes are countercurrently contacted with solvent; hexane is almost universally employed except for castor seed where the higher boiling heptane is preferred. Many mechanical extraction systems (eg, conveyors, buckets, rotating baskets, and moving solvent sprays) are employed to reduce the oil content of the flakes below

1%. The flakes go to a solvent-stripper that removes all traces of solvent. This meal is processed further (eg, meal from soybeans and cottonseed). The oil–solvent mixture (called miscella) can be processed *per se* (see Miscella Refining) after filtration, but the vast majority of the plants remove the solvent in a vacuum stripper and store the dried, crude oil. Some high hull, low fat-content oilseeds (eg, soybeans) are decorticated, cooked, flaked and solvent-extracted directly as outlined above. Prepressing is not economical for lower fat-content materials. On the other hand, high fat-content materials normally cannot be handled directly in many solvent plants although some systems have been developed for this purpose. The equipment available and solvent extraction techniques are discussed in a series of recent reviews covering most of the oilseeds (46).

Recovery of Oil from Fruit Pulps. *Palm Fruit.* One of the outstanding accomplishments of the last decade in the fat industry is the improvement in palm oil quality. For many years the harvesting, handling and processing of palm fruit led to fat of poor color and odor with high free fatty acid (FFA). Now, systematic approaches to the elimination of the factors causing lower oil quality have allowed the palm oil producers to offer high-quality fat with less than 2% FFA. Thus this fat, which at one point was considered only fit for soap-making and other industrial uses, now commands a price similar to other edible fats in world commerce. Surveys are available that discuss the processing details and precautions (47). In summary, good harvesting, expeditious fruit handling, and prompt sterilization (with steam) minimize the enzymatic (lipase) attack leading to high FFA. Fat flavor, color, and oxidative stability are improved by using good manufacturing practices common to most edible fat operations.

Rendering of Animal Fats. For a description of the wet-rendering and dry-rendering methods packing houses use in the United States for the recovery of lard, tallow, and greases, see the article entitled Meat products. Rendering methods for other animal fats, including marine oils, are essentially similar; however, whale and fish oils are invariably wet-rendered, often continuously, with centrifugal separation of the fatty and aqueous phases. Open hydraulic presses, cage presses, and continuous screw presses are all used for the final recovery of oil from rendering residues, and the latter are often solvent-extracted after pressing.

Degumming. Many fats are not degummed prior to refining but soybean oil, with 1–3% phosphatides, often is treated with water to recover commercial lecithin which is widely used as an edible emulsifier. The resulting, degummed oil is then processed further. Degumming for lecithin recovery is accomplished by treating the filtered, crude oil with ca 2% water with agitation at 70°C after which a high speed centrifuge is utilized to separate the oil and lecithin sludge. Both product streams are dried in vacuum driers; the oil goes to storage; the lecithin is treated in various manners depending on the end product desired.

Refining. Crude fats or crude fats combined with degummed oils are refined to remove free fatty acids, gums (phosphatides), and some pigments.

Alkali Refining. Although originally carried out in batch kettles, almost all edible oils are now processed in continuous systems. With crude soybean oil the steps are: (*1*) treatment with ca 0.05% concentrated phosphoric acid to precoagulate (condition) the gums; (*2*) addition of ca 0.1% (1.13 sp gr or 17° Bé) NaOH solution in excess of the amount required to neutralize the free fatty acids; (*3*) adequate mixing and holding to hydrate the gums; (*4*) heating to ca 75°C to cause the emulsion to break; and (*5*) centrifugation through a suitable high-speed refining centrifuge. The soapstock is transferred for further processing. The oil containing traces of soap and moisture is heated, treated with 10–20% water, mixed, and passed through a water-wash centrifuge. This oil is dried *in vacuo*, cooled, and taken to storage or to the bleaching step (48). Recent, detailed reviews are available (48–49). There are some variations required in the alkali refining of other fats. Conditions and precautions are described (48–49). The major difference occurs in the handling of lauric fats (coconut and palm kernel) and palm oil where the fats and caustic are preheated before caustic addition and then quickly centrifuged with minimum dwell time to minimize hydrolysis.

In Europe, the Zenith process (49) is utilized in which degummed oil is allowed to percolate in droplets through a column of dilute sodium hydroxide solution. No washing is required, saponification is minimized, and the fat is taken directly to the bleaching step.

Steam Refining. In this technique, fats are freed of phosphatides (degummed), bleached if necessary, and stripped with sparging steam under high vacuum so that the free fatty acids are removed by distillation. Several variations of the technique have been described (50) particularly for the refining of palm oil. Also, the scheme is more economical for high FFA fats and does reduce pollution problems connected with alkali refining.

Miscella Refining. This process is very similar to alkali refining except that the fat is dissolved in a solvent (51). Usually the process is carried out at the solvent extraction plant where the oil extracted from the seed by hexane is not stripped of solvent but is taken directly to the refinery where it is treated with gum conditioners and alkali, heated, mixed, and centrifuged before bleaching and removal of the solvent. If desired, other unit operations such as hydrogenation and winterization can be conducted before the solvent is evaporated.

Other Refining Techniques. Some industrial oils are refined with strong acid or with solvent partition methods.

Bleaching and Decolorization. Most edible and some industrial applications require fats with very little color which can be derived from oxidation products, carotenoids or chlorophyll. Thus, in addition to the decolorizing occurring in the refining step, adsorptive bleaching is widely used. Bleaching can also be achieved oxidatively or thermally but these methods have disadvantages, eg, formation of odorous compounds and polymers.

Adsorption. Adsorptive bleaching follows the Freundlich equation $X/m = Kc^2$ where X = amount of adsorbate, m = amount of adsorbent, c = amount of unadsorbed pigments, and K is a constant unique to each system (52). Thus as m increases, c decreases (see Adsorptive separation).

Bleaching practices in Europe (53) and the United States (54) have been described in detail. Fats are mixed with 0.2–2.5% neutral- or acid-activated earths (bentonite clays) heated to 80–110°C with agitation in open bleachers or preferably *in vacuo*, held for as much as 15 min and filtered. Easily bleached materials are treated with neutral (least costly) earths; more difficult oxidized pigments are removed with acid-activated earths; green chlorophyll colors often require the addition of up to 10% activated charcoal. These adsorbents retain some fat: clays about 30 wt %; and carbon up to 150 wt %. Thus for economic reasons, minimum amounts are utilized. Filters used include open, plate-and-frame types, and completely automatic, closed, self-cleaning, leaf filters. The former are labor intensive in that they must routinely be cleaned and redressed with filter cloths. The leaf filters can be made essentially automatic but some mechanical and fouling problems can be encountered which require special cleaning and backflushing techniques. Occluded fat is removed from the filter presses by blowing the filter cake with steam, hot air or nitrogen depending on the fat type being processed. This scheme only removes 50–75% of the fat. Other approaches employ solvents such as hexane or detergent solutions to strip the fat from the clay, but this can become very expensive and require added capital equipment.

Disposal of spent earth can pose problems if it contains considerable unsaturated fat. Owing to autoxidation, the spent earth can overheat and smolder unless it is used as landfill and quickly buried (see Wastes, industrial).

Chemical Bleaching. Methods using reagents are not employed in processing edible fats with the exception of lecithin where hydrogen peroxide or benzoyl peroxide is used. The beneficial oxidative bleaching of the pigments in fats is accompanied by undesirable oxidation of unsaturated fatty acids and leads to poorly flavored products (see Bleaching agents). Some heat-bleaching or destruction of carotenoids and adsorption on catalyst occurs in the hydrogenation process. Also, some heat-bleaching occurs at the higher temperatures encountered in the deodorization step.

Hydrogenation. This is one of the most important processes in the fat industry because it allows the production of many different functional edible fats (eg, margarines, shortenings, and confectionary fats) from a wide variety of liquid or partially solid fats. Depending on the particular end use, hydrogenation, actually partial hydrogenation, yields products that are interchangeable for many applications. Thus many raw materials can be substituted for one another, depending on supply and economic demand, without large variations in product quality or performance. In addition to the conversion of liquid fats to plastic fats for edible purposes, hydrogenation decreases the polyunsaturation of fats, particularly by the elimination of linolenic (18:3) acid, and thereby reduces susceptibility to autoxidation. In the production of industrial fats (particularly soaps and other derivatives) the presence of unsaturation is undesirable because of odors.

Equipment and Procedures. Commercial hydrogenation is mainly carried out in batch reactors (called convertors in the U.S.) equipped with gas-dispersing agitators, heating and cooling coils, gas handling systems, catalyst introduction and removal equipment, and attendant safety and operational controls. These vessels normally are tall cylinders (height: diameter, ca 2:1) with a top-entering stirrer shaft holding at least one turbine agitator near the bottom and another near the surface of the liquid. Capacities are ca 5–30 t.

The well-refined and bleached fat is introduced, vacuum-deaerated as it is heated (some convertors have external preheating systems and deaerators to reduce residence time) and the catalyst is introduced. Normally, wet or dry reduced nickel on a support is added at a predetermined concentration dependent upon the type of product to be produced. For most edible applications 0.02–0.15 wt% Ni is employed. High-purity hydrogen gas is introduced through a sparge ring at the bottom of the convertor and the agitator is started at high speed. As hydrogenation commences, the temperature rises. This is an exothermic reaction and convertors often are equipped with automatic cooling systems for temperature control. Good temperature control is required for many products. Convertors can be operated in a number of modes. Some have recirculating gas systems whereby headspace hydrogen is drawn off, scrubbed and recompressed to be reintroduced through the sparge ring. Make-up hydrogen can be added. Others are operated in a dead-end fashion where the gas is introduced via the sparge ring and then recirculated down from the headspace by the turbine. A third approach uses a hydrogen bleed system wherein a small amount of hydrogen is allowed to escape from the reactor and fresh gas is continuously introduced through the sparge ring. Normal operating pressures are 200–700 kPa (1–6 atm) gauge and most pressure vessels are rated to only 1 MPa (10 atm) (36). Other operating parameters include process times of 1–4 h at 120–190°C, depending on the degree of hydrogenation desired. End points are normally assessed by refractive index although other tests may be applied (see Analytical and Test Methods). A few continuous hydrogenation systems have been described (36). Although other laboratory and pilot-scale continuous systems have been described (55–56), very few commercial installations are known (36). Inherently, continuous systems have lower selectivity than batch systems. In addition, stock changes are cumbersome. Thus, unless a manufacturer is producing large volumes of a single hydrogenated product, continuous processes are not utilized.

Principles. Hydrogenation is based on the metal-catalyzed addition of hydrogen to the olefinic double bonds of a variety of fatty acid chains in a large number of triglycerides. If only oleic acid triglycerides were present, then hydrogen addition would yield stearic acid triglycerides; however, in the presence of hydrogen and some metal catalysts (notably nickel) cis–trans isomerization also can occur. The latter reaction increases the melting point of the fats (oleic acid 18:1 (9c) mp 16.3°C, elaidic acid 18:1 (9t) mp 43.7°C, stearic acid 18:0 mp 69.6°C), although not to the degree that saturation does. If a mixture of linoleic, oleic, and saturated acid triglycerides are hydrogenated (eg, cottonseed oil) then competitive reactions occur. In this case, the extent of the reaction of linoleic to yield oleic (and elaidic) acid compared to the reaction of oleic to yield stearic acid is called selectivity. Normally, for certain applications high selectivity (rate of linoleic reaction: rate of oleic reaction, >50:1) is desired. Similar situations apply in the case of linolenic-containing fats (eg, soybean oil) but the complexity introduced with an additional more reactive, triunsaturated fatty acid is obvious. Also, the definition of selectivity must be qualified in that there are multiple competitive reactions possible (36,57). With nickel catalysts linolenic and linoleic acids are reduced at somewhat similar rates and considerable trans isomers are formed. Reaction conditions that promote high selectivity also promote higher trans-isomer formation. An example of the effects of selectivity is shown in Figure 3 (58) and Figure 4 (36). The steeper solid-content (SCI) curve (S_I = 50) describes the more desirable product if it is to be used in margarines or filled confectionary fats. For bakery applications where harder fats have some utility, the less selectively hardened products (S_I = 4) are desirable. For salad and table oils soybean oil is often selectively hydrogenated with nickel catalyst to about 110 IV and winterized (see below) to remove the solid fractions (mostly mixed glycerides containing saturated and trans isomers). This practice reduces the linolenic acid content below 2–3%, provides a clear oil at refrigerator temperatures and improves the oxidative stability.

The use of copper catalysts has been suggested (36) to avoid some of the processing problems noted above but the disadvantages, such as low hydrogenation rates and the prooxidant effect of residual traces of copper, make copper catalysts unattractive for most applications.

Catalysts and Hydrogen. Catalysts can be prepared by wet or dry reduction of nickel salts. Most catalysts are supplied on a carrier, eg, diatomaceous earth (see Diatomite). The nickel salt is precipitated on the carrier before reduction. Catalysts for edible fat processing are readily poisoned by traces of impurities in the oil feedstocks

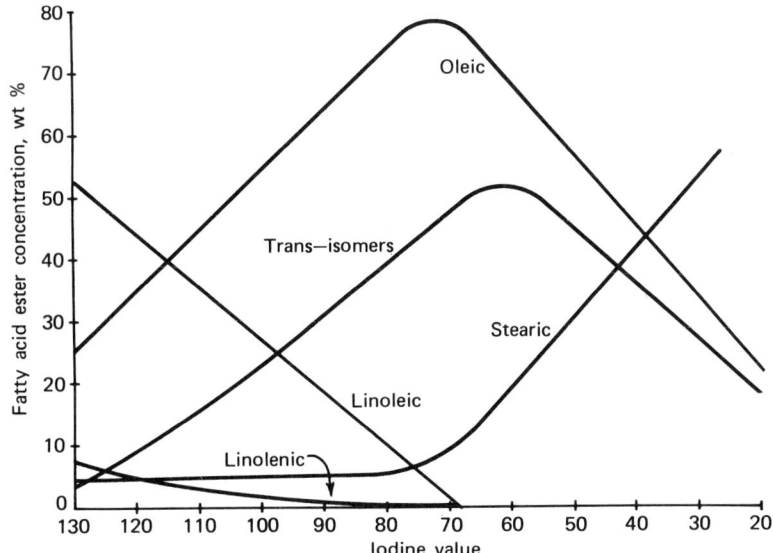

Figure 3. Hydrogenation of soybean oil under trans-isomer promoting conditions of 142 kPa (1.4 atm), 216°C, 0.025% Ni. Courtesy of Harshaw Chemical Company (58).

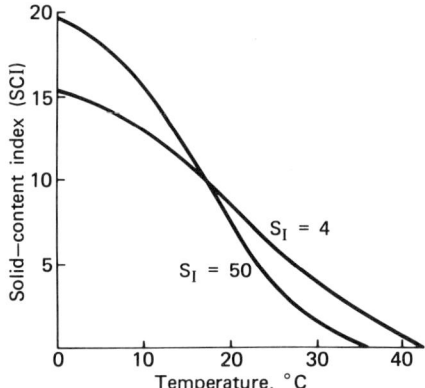

Figure 4. SCI-temperature curves for soybean oil hydrogenated to iodine value 95 under conditions of high selectivity ($S_I = 50$) and of low selectivity ($S_I = 4$) (36).

or by gaseous impurities from the hydrogen stream (see Catalysis). The former include phosphatides and soaps from improper alkali refining and oxidation products from poor handling, bleaching, or both. The latter are mainly hydrogen sulfide and carbon monoxide from the production of hydrogen (59). Hydrogen (qv) of high purity (99.5%) is required and can be obtained from gas plants based on the catalytic steam reforming of natural gas. Special scrubber systems are provided to remove water, carbon monoxide and hydrogen sulfide. Smaller installations can utilize electrolytic hydrogen plants and one installation has been described where liquid hydrogen, which is a by-product of the production of liquid oxygen, is utilized (60). As energy costs increase and natural gas becomes more scarce, other previously used hydrogen manufacturing schemes may become attractive, especially those based on coal.

Catalyst Recovery. When hydrogenation is completed in batch operations, the oil is cooled and recirculated through a high capacity filter, or black press, until a sample is clear of nickel. This oil is mixed with a small quantity of citric acid solution, often automatically in a surge tank, treated with a low level of clay and heated or post-bleached before it is filtered. The cake from the black press is recovered and reused a number of times, often in combination with smaller amounts of fresh catalyst. Exhausted catalyst can be recovered and reprocessed for its nickel content.

Winterization. This term is applied to the process where fats are cooled to remove solid fractions that otherwise would cause clouding when the liquid fat is held at refrigerator temperatures. Historically, cottonseed oil that had been stored in tanks over the winter periods deposited solids (stearine) and the liquid portion called winter oil was then decanted and sold for mayonnaise manufacture or bottling. Currently, the process is carried out in large, stirred vessels with cooling coils. Warm, solid-free oils are slowly cooled under carefully controlled conditions with slight agitation to promote slow growth of large crystals. When crystallization is complete, filtration or centrifugation, or both, are employed to separate the liquid and solid fractions. Liquid fats so prepared are normally rated on the basis of a cold test, ie, the length of time for the fat to show a cloud at 0°C. As noted, this winterization process originally was developed for cottonseed oil but now it is widely applied. Much of the hydrogenated, winterized soybean oil of approximately 110 IV is sold as bottled salad oil in the United States. A discussion of the application of this process and other winterization techniques to many fat types is found in ref. 61.

Interesterification. Natural fats are not random with respect to the types of fatty acids esterified at the different glycerol hydroxyl groups. However, if such fats are heated to ca 80°C in the presence of strong bases such as sodium methoxide or a sodium–potassium alloy, base-catalyzed interesterification occurs and the fatty acids are randomly esterified at the three glycerol hydroxyls. This procedure changes the melting point, the solid fat index and often the crystalline habit of the fat. An additional technique is the use of lower reaction temperatures which leads to directed interesterification, ie, as higher melting glycerides are formed they crystallize and do not react further. This displaces the equilibrium of the ensuing reactions and yields different triglyceride mixtures with varying properties. Many applications of this technique have been described (62). The method is widely utilized in Europe for the production of tailor-made fats and margarine oils (63) but has not been adopted to a large degree in the United States except for the treatment of lard.

Deodorization. Almost all edible fats are deodorized before they are consumed. Olive oil of the cold-pressed type is the principal exception because it is valued for its natural color and flavor. A few prime animal fats also are utilized in an undeodorized state as their native flavors and odors are considered desirable in some applications. Deodorization is a process of steam distillation at relatively high temperatures (210–275°C) conducted at low pressures (133–800 Pa or 1–6 mm Hg). The triglycerides with very low vapor pressures are relatively unaffected and the odoriferous impurities, such as lower molecular weight hydrocarbons, aldehydes, ketones, and fatty acids resulting from oxidative deterioration, are removed by steam distillation. Also, free fatty acids are reduced to a level of 0.02–0.05% and sterols and some tocopherols are also removed. Many pigments, particularly those of the carotenoid group, are decolorized at the high temperatures employed.

Three types of deodorization equipment are employed: batch, continuous, and semicontinuous. In the batch deodorizer, fat is added, vacuum is applied, the system is heated, usually by steam coils, and sparge steam is introduced at the bottom of the vessel. Retention times are 3–8 h and stripping steam volume is 5–15% based on the weight of fat. When the deodorization period is completed, the oil is cooled and discharged to storage through a polishing filter. Because of the long time periods involved and high utility requirements, batch units are quite inefficient.

Continuous deodorizers (Fig. 5) operate so that the fat flows through deaerating, heating, stripping, and cooling sections in a uniform manner. The retention time is controlled by the physical size of the units and flow rates. Stirring steam is applied in all sections to aid heat transfer and prevent overheating at the heat transfer surfaces. A much larger volume of stripping steam is introduced through sparging systems in the highest temperature, stripping segment of the continuous deodorizer. Sparge steam requirements (1–5%) and retention times (15–120 min) are much lower and heating

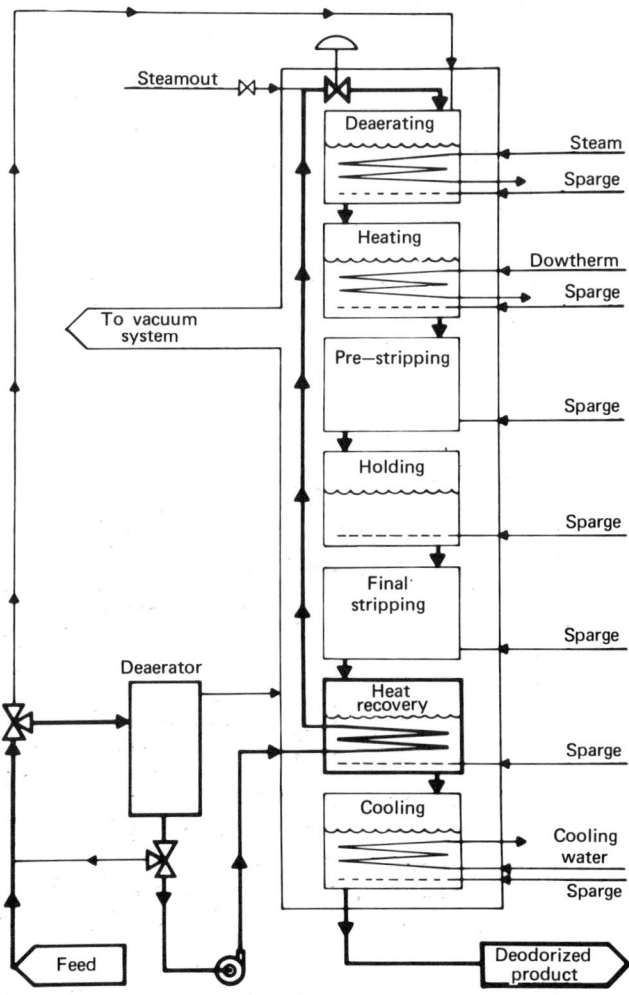

Figure 5. Schematic diagram of double-shell deodorizer (EMI Corporation) (64).

costs per kilogram of product are less than in batch deodorizers. The continuous unit is very practical when only a few products are produced in a plant but they are less flexible when numerous product changes are made. Note particularly the oil-to-oil heat recovery system which provides for heating of incoming feedstock through coils in the first cooling section (Fig. 5).

The semicontinuous deodorizer (Fig. 6) operates by treating each tray of oil as a separate entity. By sequencing the draining and filling of each segment of the deodorizer, a finite batch of product is moved down through the various segments shown. The fat is deaerated in the measuring tank and preheated. It then flows into the top tray where it is raised to the deodorization temperature (210–275°C) by heat exchange media (many U.S. units employ Dowtherm A whereas European and Japanese units use high pressure steam) (see Heat exchange technology). The heated oil drops to the stripping unit where specially designed mixing systems intimately contact the oil with a large percentage (ca 75%) of the stripping steam. The oil then drops to the first cooling tray where a closed-loop heat exchange system recovers heat to be used in heating feedstock. Final cooling is done with water from an external source, normally a cooling tower. The deodorized oil is removed by a discharge pump and filtered before it is sent to storage. Current United States continuous and semicontinuous units are designed to process as much as 14 t/h which reduces the cost of deodorization greatly. Despite this, with the large volumes of steam required for 3–4 stage vacuum-ejector systems (up to 4.5 t/h), depending on pressures and sparge rates) and the high temperature requirements of up to 275°C, deodorization is the most energy-intensive step in the processing of fats.

By-Products, Waste Materials and Pollution Control. The origin of by-products, the handling of waste materials and pollution control are important factors in the manufacture of fats. Several detailed overviews of these areas are the sources of much of the material summarized below (65–68).

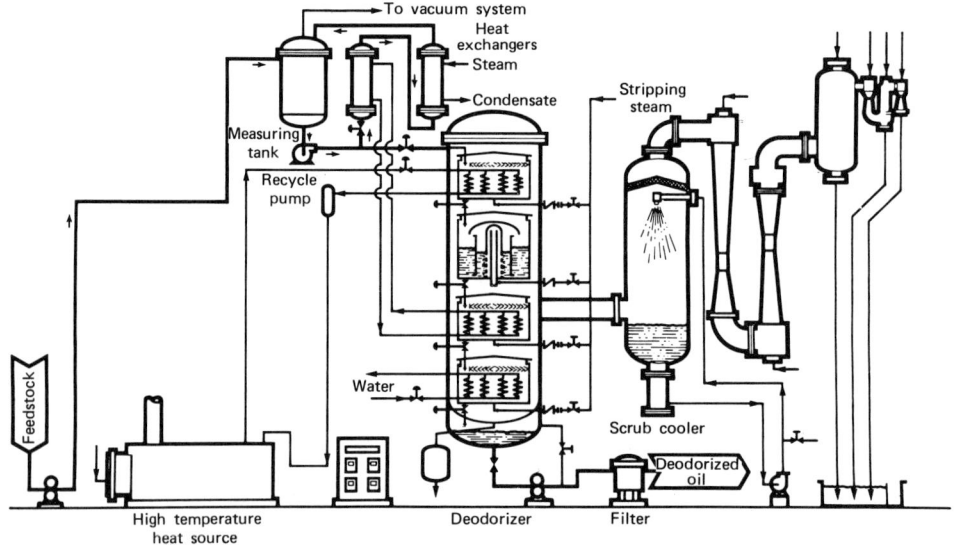

Figure 6. Votator semicontinuous deodorizing plant (with heat recovery). Courtesy of Process Equipment Division, Chemetron Corp., subsidiary of Allegheny Ludlum Industries.

The major by-products in the manufacture of fats are: residual gums and meals that can separate from crude oil; soapstock from alkali refining; used bleaching clays and oil recovered from bleaching clays; oils and catalyst metals recovered from the hydrogenation steps and distillate from the deodorization process. Additionally, there are certain quantities of washout, ie, fats utilized to clean out pipes and vessels between batches to prevent cross-contamination of products as well as materials recovered from spills, packaging or filling operations. The latter group of materials, if of good quality, can be reworked into fully hydrogenated or relatively low IV basestocks for blending with other fats, eg, in the production of compound shortenings.

Degumming of filtered oils with water produces gum phosphatides, referred to as lecithins, that are subsequently dried in a swept-surface vacuum dryer, taking care to control the temperatures (80–95°C), pressures (6.7–40 kPa or 50–300 mm Hg) and residence times (1–2 min) to yield a low moisture (<1% H_2O) light-colored, good-flavored product (69). Depending on the efficiency of the initial degumming, fat may be readded to the dried lecithin to provide the viscosity required (soy lecithin is normally 30–40% soybean fat).

Soapstock and washwater from alkali refining is acidulated usually with sulfuric acid, either in batch vessels or continuously and the acid oil (fatty acids, gums, and neutral oil) are separated and dried. Some acid oils, such as animal fats, coconut, palm kernel and palm oil, are sold for industrial use in soap, metal working, and fatty acids. Most vegetable acid oils are sold as high energy components for animal feeds. The aqueous phase is neutralized and subjected to secondary treatment to remove glycerol and other water-soluble contributors to BOD before discharge into sewers or waterways (see Wastes, industrial).

Spent clay from the bleaching operations contains 20–35 wt % fat which can be recovered by steaming the clay in water suspension or solvent extraction. As neither of these methods is economically attractive, most spent clay is used directly as landfill. In refineries contiguous with solvent extraction operations, clays can be combined with material going to the solvent extractor; the clay then becomes part of the meal. This practice has limitations, in that it reintroduces poorer quality fat into the system and may increase mineral content of the feedstuff to unacceptable levels.

Hydrogenation catalysts are reused for certain products; however, there are practical limits as the catalyst becomes poisoned by impurities and loses its activity and selectivity. Such spent catalyst can be reworked to recover nickel. Some refiners have their own catalyst plants; others sell spent catalyst back to the catalyst manufacturers. No attempt is made to recover residual fat.

By-products from deodorization are neutral oil, fatty acids, sterols, tocopherols and the minor odoriferous components that are removed during the high vacuum steam distillation. To aid recovery of these materials, most modern deodorizers are equipped with oil entrainment scrubbers (Fig. 6) that reduce the amount of material by 80–90 wt % going to the barometric condensers. The 10–20 wt % that does go through the scrubber ends up in the hot well where it is skimmed off. Some plants recirculate water from the hot well directly through the cooling towers. This carries some fatty materials into the towers causing fouling problems and creating a source of odors. Various corrective approaches are utilized including: oxidizing agents and carbon added to the cooling water (65); use of heat exchangers between the condenser feedwater and the cooling tower water to prevent fatty buildup in the tower (66); and the use of so-called biological cooling towers (70).

Depending on the type of fat that is processed, the deodorizer distillates are sold for whatever value they contain. For example, soybean distillates are good sources of tocopherols and plant sterols but they are not particularly good for fatty acid production. Distillates, if they are low in pesticide residues, can be used as a high energy source for animal feeds (see Pet and other livestock feeds). They also are used in formulating oil-well drilling aids (see Petroleum). Some distillates, particularly those low in sterols and tocopherols, are used in fatty acid production. However, when economics are unfavorable for any of the above disposal routes, it has been suggested that distillates can be burned for their energy value by adding as much as 10% distillate to fuel oil (66).

The main waste material from fat manufacture is water and much attention has been devoted to the problems of water treatment (67–68). Techniques for fat removal include gravity separators, dissolved air flotation, and surface adsorption. Biological treatment in aerated lagoons or activated sludge systems, addition of flocculating agents (qv) and filtration are used before the effluent is discharged to public waters. In many instances, joint treatment is employed whereby the plant discharges raw or partially treated water into municipal systems. General requirements include removal of floatable fat and adjustment of pH to 5.0–10.5. Many municipalities also suggest equalization of wastewater discharge to prevent overloads of organic materials into sewers. Details of BOD reduction at various stages of treatment in United States plants are available (68) along with specific information on the types of treatments employed (see Water, industrial water treatment).

Additional United States experiences with various waste treatment systems have been discussed in considerable detail (71) as have EPA guidelines (72) for effluents from the United States edible fat industries. In addition, discussions are available on odor control in Canada (73) and on water treatment in Europe (74).

Production and Shipment

World production (1977–1978) of fats was centered mainly in vegetable sources (67%) with 31% derived from animals and only 2% from fish. Estimated 1978 tonnage of the major fats is shown in Table 2 along with the principal sources. Data on the world's major oilseed production (75) is used for the following estimates. The estimated world soybean crop (1977–1978) of 77.5×10^6 metric tons was produced mainly in the U.S. (68%), China (16%) and Brazil (13%). The major cottonseed (24.7×10^6 t) producers were: the USSR (20%), the U.S. (20%), China (15%), and India (10%). Sunflower seed (12.7×10^6 t) was produced in the USSR (47%), Argentina (12%), and the U.S. (10%). The major shelled peanut (11.2×10^6 t) producers are: India (38%), China (14%), and the U.S. (11%). Rapeseed (8×10^6 t) was produced in Canada (25%), India (20%), China (16%), and Europe (31%). Copra (from coconuts) is produced mainly in the Philippines (54%) and Indonesia (19%).

The oil seeds are major items of export. Thus many of the fats derived from them are produced in importing countries. Soybean oil in 1977/78 (11.6×10^6 t) was produced in the U.S. (40%), western Europe (20%), Brazil (16%), and China (9%). Cottonseed oil (3.0×10^6 t) was produced (1977–1978) in the USSR (25%), the U.S. (22%), China (10%), and India (7%). Peanut oil (2.6×10^6 t) was produced (1977–1978) mainly in India (46%), China (12%), and Senegal (8%).

Sunflower oil (4.2×10^6 t) was produced (1977–1978) largely in the USSR (46%), western Europe (13%), and Argentina (10%). Additional data on other fats are available (75). Actual net exports of major oilseeds, oils, and fats on a fat basis are shown in Table 7 (76). An excellent perspective of trends in world fat production is shown in Figure 7 (77) and another view is offered by the details in Table 8 (78).

Shipment of fats varies depending on the type and quality of the fat, the quantity and the economics involved. Bulk shipments of crude, refined and edible grade fats

Table 7. World Net Exports Oilseeds, Oils, and Fats, 1000 Metric Tons, Fat Basis[a]

	77/78[b]	76/77[b]	75/76[b]	74/75[b]	73/74[b]
Edible					
soybeans and oil	5,164	4,569	4,467	3,380	3,864
cottonseed and oil	485	405	343	436	375
peanuts and oil	560	685	691	515	533
sunflower seed and oil	1,200	702	679	754	862
rapeseed and oil	775	917	751	650	708
sesame oil	90	83	101	96	117
olive oil	195	249	217	200	223
copra and coconut oil	1,500	1,369	1,839	1,282	935
palm kernels and oil	360	364	409	389	359
palm oil	2,150	1,967	1,915	1,736	1,309
butterfat	640	662	622	603	664
lard	450	455	374	415	413
fish oils	470	534	537	506	548
Nonfood					
linseed and oil	410	377	271	254	293
castorseed and oil	200	186	244	155	243
tung oil	40	47	58	47	46
tallow and grease	1,790	1,853	1,556	1,526	1,492
Total	*16,479*	*15,424*	*15,074*	*12,944*	*12,984*

[a] Ref. 76.
[b] For October to September crop year.

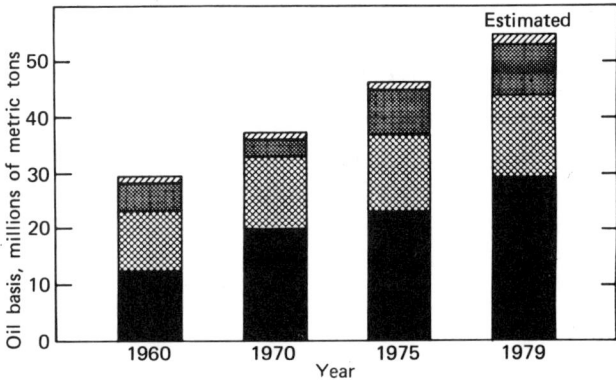

Figure 7. World production of oils and fats 1960–1975 with 1979 estimated (77). ▨, Marine oils; ▩, tree crops (includes palm, palm kernel, coconut, babassu, oiticica, and tung oils); ▩, animal fats (includes butter, lard and tallow, and greases); ■, annual field crops (mainly soybean, peanut, cottonseed, rapeseed, and sunflower oils).

Table 8. World Fat Production 1975–1978 with 1979 Estimated[a], Million Metric Tons

Fat	1975–1976	1976–1977	1977–1978	Estimated 1978–1979
cottonseed	2.8	2.9	3.3	3.2
peanut	3.6	3.2	3.0	3.5
soybean	10.2	9.1	11.0	11.9
sunflower	3.6	3.7	4.6	4.8
rapeseed	2.9	2.3	3.0	3.4
olive	1.8	1.3	1.4	1.4
coconut	3.3	3.1	3.1	2.8
palm	3.1	3.3	3.4	3.9
other edible vegetable	2.0	2.0	1.9	2.2
industrial	1.4	1.3	1.6	1.5
animal	14.0	14.5	14.7	14.6
marine	1.1	1.0	1.0	1.0
Total	49.8	47.7	52.0	54.2

[a] Ref. 78.

are often transported in clean, seagoing vessels. Water and rust must be avoided. Where feasible, eg, within a given landmass, bulk fats can be barged if waterways are available but most often they are moved by rail cars (27–68 t) or tank truck (9–18 t). Many fat processors own their equipment and maintain needed cleanliness to avoid contamination of the fats. Careful loading and unloading is necessary to avoid aeration and overheating. Solid or partially solidified fats are heated with low pressure steam or hot-water coils to assure that the fat is homogenous before loading or unloading. Another precaution is the technique of bottom-loading. When liquid fat is being introduced into any empty vessel, the best practice to avoid aeration is the use of a bottom-loading pipe. Storage tanks should be constructed with integral pipes of this type. Trucks, ships and tank cars should be bottom-loaded using a special loading pipe at the terminal. Techniques for handling and storage have been discussed for both crude (79) and edible fats (80). A further consideration for the protection of polyunsaturated fats is the use of a nitrogen atmosphere both in storage and shipping. Nitrogen is in-

Table 9. 1977 United States Fat Utilization, Edible Products[a], Quantity of Fat Utilized, 1000 Metric Tons

Fat	Salad and cooking oils	Margarines	Shortening
soybean	1508	719	1035
cottonseed	182	20	73
corn	130	110	
peanut	90		
safflower	11	3.6	
olive	24		
palm			168
coconut			35
lard		33	84
beef fats		3.2	339
all other	126		
Average retail price, ¢/kg	164	112	121

[a] Ref. 81.

troduced into the suction side of the loading pumps feeding the storage or transport vessels. In some cases, a nitrogen headspace is maintained in tanks by appropriate pressure and relief valve systems. Most transporting vessels depend mainly on the excess nitrogen sparged through the bottom-loading pipe. For other than bulk shipment, drums (208 L or 55 gal), lined cartons (23 kg) and tins (3.8–14 L or 1–5 gal) are used. Retail packages of liquid fats range from 0.47 to 3.8 L (0.12–1 gal). Most United States plastic (solid) household shortenings are sold in 1.36-kg tins although some 0.45 kg parchment-wrapped blocks are offered.

Economic Aspects

Total fat consumption (all products) in the United States was nearly 36.3 kg per capita in the 1976–1977 October–September season. In Table 9 are listed amounts and average prices of the principal edible products derived from the major fats utilized in the United States (81) for the calendar year 1977.

Use of butter for this period was 431,000 t at an average price of $2184/t (99.1 ¢/lb) and lard usage was 360,000 t at $469/t (21.3 ¢/lb). All United States edible usage for the crop year 1976–1977 is estimated to be 24.2 kg per person and nonedible use is 11.8 kg per person. Major nonedible categories were: soap 423,000 t, drying oils 250,000 t, animal feeds 656,000 t, fatty acids 813,000 t, and other 412,000 t. Principal fat sources for this category were inedible tallow, coconut, soybean, and linseed (82). Prices for crude fats are highly variable and the markets are often very volatile. Climatic changes, weather conditions particularly in the critical planting and harvesting seasons, demand for oilseed proteins, governmental actions, farmer planting plans, and general worldwide economic situations all combine to affect the price structure of fats. At the same time, there has been a growing demand for fats from the developing nations. Some impression of the year to year price volatility can be obtained from the data in Table 10.

Specifications, Standards and Quality Control

Fats are marketed and traded based on quality. In the United States primary processors sell crude fats based on standards established in trading rules developed by various trade associations, eg, the National Soybean Processors Association (NSPA), the National Cottonseed Products Association (NCPA), the National Renderers Association (NRA), and the National Institute of Oilseed Processors (NIOP), among others. These rules also can be used for international trade but often the rules are those of international trade associations such as the U.K. Federation of Oils, Seeds, and Fats Associations, Ltd. (FOSFA) and the Netherlands Oils, Fats, and Oilseed Trade Association (NOFOTA). Major quality considerations for crude oils are: flashpoint, free fatty acid (FFA), color, moisture and impurities (M&I), unsaponifiables, refining loss, bleachability, and odor. Often the rules establish premiums or discounts based on the deviation of a given shipment from an established quality standard. Degummed,

Table 10. Fat Prices of the United States and Europe[a]

Fat	Grade	Container	Location	1969–1972	1973[b]	1974[b]	Average price, $/t 1975[b]	1976[b]	1977[b]
soybean	crude	tanks	Decatur, Ill.	285	695	676	402	525	539
soybean	crude	tanks	Neth. mills	257	727	694	422	573	578
soybean	refined	tanks	New York	320	728	728	518	644	598
cottonseed	crude	tanks	Mississippi valley	312	734	699	599[c]	514[c]	540[c]
cottonseed	U.S. PBSY	tanks	Rotterdam, Neth.	340	835	823	604	629	630
butter	A-92 score	bulk	Chicago, Ill.	1511	1493	1578	1748[c]	2010[c]	2182[c]
coconut	Phil. crude	tanks	Rotterdam, Neth.	288	1008	499	374	579	600
corn	crude	tanks	Decatur, Ill.	397	804	798	714[c]	569[c]	677[c]
fish	refined	drums	New York	296	429	771		456[c]	507[c]
fish	any origin		NW Europe	200	517	405	349	468	452
lard	ref. qual.	tank	U.K.	250	547	537	457	604	613
linseed	any origin	tank	Rotterdam, Neth.	205	1023	846	561	508	400
palm	crude	tankwagon	New York	274	489	511	386	516	585
palm	crude	bulk	NW Europe	228	628	508	389	527	564
peanut	crude	tankcar	SE U.S.	386	912	930	745	707	699
peanut	any origin	bulk	Rotterdam, Neth.	374	948	880	678	823	1009
sunflower	any origin	tank	Rotterdam, Neth.	311	845	864	593	657	618
rapeseed	crude	tank	Neth. mills	242	645	674	406	562	583
olive	edible	drums	Spain	752		2380	2228	2201	2349
tallow	U.S. bleachable fancy	bulk	Rotterdam, Neth.	186	459	340	369	416	449
tallow	U.S. bleachable fancy	bulk	Chicago, Ill.	195	413	308	313[c]	331[c]	377[c]

[a] Ref. 75, 82–83.
[b] October of previous year to September.
[c] Calendar year.

partially refined, refined, and edible grades all have additional requirements that are spelled out in the rules. Often a buyer adds unique specifications for a particular fat that become the basis for negotiations with the sellers. This is particularly true of specialty edible fats, eg, salad and cooking oils, margarine fats, shortenings, confectionary fats, and fats for filled dairy products. Fats for industrial use are covered under specifications listed by the NSPA and NIOP as well as those of the American Society for Testing Materials (ASTM).

Although the American Oil Chemists' Society (AOCS) does not have rules for fat trading, it was first established by a group of analysts concerned with developing uniform methods for the analysis of fats and related materials. The methods used by the referee chemists and their referee laboratories are the AOCS *Official and Tentative Methods* (26) wherever applicable. Additional information on this subject is given in ref. 84.

Another current approach to the development of international fat standards is that of the FAO–WHO Codex Alimentarius Commission Committee on Fats and Oils. This group with representatives from all cooperating countries is working towards the adoption of standards for many fats and fat-based products. In addition, it has recommended analytical methods and as many as eight alternatives for some 48 different tests (85).

The United States standards and rules for fat trading include the following categories: identity and type; free fatty acids (FFA); refining loss (86); color (87); moisture, insolubles, and unsaponifiable material (MIU); and iodine value (IV).

Quality control in the processing of fats involves many of the checks on crude fats listed above. In addition, samples are taken from the work-tank as representative of a large lot of crude fat in order to set up optimum refining conditions. Analysts determine FFA to aid in setting up the caustic treatment level; they assess the refining loss to determine if excess caustic or other special treatment is required; and they check the laboratory bleached oil to be sure that the correct type and amount of bleaching earth is employed.

If hydrogenated stocks are produced, the quality control chemists must check the products for the correct IV, mp, and the solid fat index (SFI). Finally, all finished deodorized products, either pure fats or blends, must be checked to assure that composition, IV, FFA, peroxide value (PV), color, flavor, odor, and appearance are within specification. Failure to meet any standard is grounds for rejection and leads to costly rework, usually into lesser-valued stocks.

Analytical and Test Methods

In the United States most of the analytical and test methods for fats are described (26). In addition, there are specific tests and methods published by the ASTM and the Association of Official Analytical Chemists (AOAC). Analytical methods are published by various groups in many countries. A partial list of such groups includes: International Union for Pure and Applied Chemistry (IUPAC), International Organization for Standardization (ISO), Codex Alimentarius Commission (CAC), International Olive Oil Council (IOOC), European Economic Community (EEC), British Standards Institution (BSI), and Deutschen Gesellschaft für Fettwissenschaft (DGF). Ref. 88 is an excellent text covering classical methodology. Key analytical methods for fats are those listed below.

Gas–Liquid Chromatography. The determination of fatty acid composition by AOCS method Ce 1-62 (26) allows the analyst to quickly determine the type of fat as well as a calculated IV and an estimate of the saponification value. Only a few mg of sample is required. The fatty acids of the triglycerides are converted into methyl esters by AOCS method Ce 2-66 (26) and rapidly and quantitatively analyzed.

Iodine Value (IV). According to the definition in AOCS method Cd 1-25 (26), IV is expressed as the number of centigrams of iodine absorbed per gram of fat (% iodine absorbed). This is an empirical measure of the amount of iodine reacting under carefully controlled conditions with the nonconjugated bonds of the fat. The Wijs method employed uses iodine monochloride reagent with a thiosulfate back titration and is rather involved. It does, however, provide laboratories without glc a convenient method for checking the unsaturation of many fats, and is particularly useful in identifying fats as they typically fall into well-defined IV ranges.

Saponification Value. This is the weight (mg) of potassium hydroxide required to saponify (hydrolyze) 1 g of fat as outlined in AOCS method Cd 3-25 (26).

Flash Point. This is the temperature at which a sample will flash when a test flame is applied under the conditions of AOCS method Cc 9b-55 (26). It is particularly used to assess the presence of traces of residual solvent in crude oils. Such volatile materials are very hazardous if introduced into a processing plant not equipped to handle solvents.

Moisture and Impurities. Moisture is detected by codistillation with toluene as in AOCS method Ca 2a-45, by gravimetric techniques (Ca 2b-38, Ca 2c-25, and Ca 2d-25) or by Karl-Fischer titration as in AOCS Method Ca 2e-55 (26). Insoluble impurities are determined gravimetrically by AOCS method Ca 3-46 (26).

Unsaponifiable Matter. AOCS methods employed are Ca 6a-40 and Ca 6b-53 (26), depending on the fat type. The first method covers fats with normal levels of unsaponifiables; the latter covers those with high levels, eg, marine fats. The technique involves careful hydrolysis with ethanolic alkali and extraction of unsaponifiables into petroleum or ethyl ether. The extracts are assayed gravimetrically after evaporation of solvents.

Free Fatty Acid (FFA). AOCS method Ca 5a-40 (26) involves the direct titration of fats with standard NaOH solution. For most fats the % FFA is expressed as oleic acid. Coconut and palm kernel fat FFA are calculated as lauric acid, and palm oil FFA as palmitic acid.

Refining Loss. Crude fats are checked for loss of free fatty acid, fat and impurities when treated with alkali solutions under the conditions of AOCS method Ca 9a-52 (26) which is a carefully controlled laboratory, batch refining technique.

Bleaching Test. Refined fat from the above test can be directly used in AOCS methods Cc 8a-52 and 8b-52 (26) for cottonseed and soybean fats, respectively. The fats are bleached with AOCS official bleaching earths as described in the method and the filtered fat is checked for color by AOCS methods Cc 13b-45 or Cc 13c-50 (26).

Peroxide Value (PV). This measure of fat quality is covered by AOCS method Cd 8-53 (26) and is defined as the milliequivalents of peroxide per kilogram of sample that oxidizes KI under the stated conditions.

Cold Test. As utilized to measure the resistance to crystallization of a sample of liquid fat, AOCS method Cc 11-53 provides an index of the time a liquid fat remains clear under the conditions of the test (quiescent at 0°C).

Melting Point. Melting point methods are the capillary tube (AOCS Cc 1-25) and Wiley (AOCS Cc 2-38) (26). The first method is the straightforward technique employed in organic chemistry; the second is a more empirical test employing a disk of solidified fat suspended at an alcohol–water interface in which reproducibility depends on the technique of the analyst in running the test and observing the end point.

Congeal Point. This AOCS method (Cc 14-59) (26) is a measure of the solidification temperature of the fat. The sample, heated until clear, is cooled in a constant temperature bath (15 or 20°C depending on the congeal point range) until crystallization is started and then it is transferred to a constant temperature (20°C) air bath and observed with a thermometer at 1 min intervals. The maximum temperature observed (rise owing to the heat of crystallization) is the congeal point. This value is unique for most fats and yields a reproducible number useful in judging end points of hydrogenation.

Solid Fat Index (SFI). This is an empirical measure of the solid fat content as carried out by AOCS method Cd 10-57 (26). The technique depends on the changes in specific volume measured in precision dilatometers at various standard temperatures and it requires skilled technicians, accurate temperature control and meticulous experimental technique to yield reproducible results. It is a very important method as many specialty fats are traded on the basis of multiple SFI points. A good brief discussion on the method is available (89) as is a more complete text (24).

Fat Stability, Active Oxygen Method (AOM). AOCS method Cd 12-57 measures the time in hours for a sample to reach a predetermined PV under the conditions of the test (26). Samples are held at $97.8 \pm 0.2°C$ and clean air is bubbled through them. The value determined is the number of hours necessary to reach the end point. This is often 100 meq of peroxide/kg but other values can be employed. This value is assumed to be related to the resistance of a fat to oxidative deterioration but it has not correlated well with other quality indicators.

Many other specific tests are outlined in great detail (26) and the index should be consulted if a particular subject is of interest (see also Analytical methods). Some newer instrumental approaches now under active consideration for adoption as tentative methods by the Technical Committees of the AOCS include: improvements in atomic absorption spectroscopy for heavy metals; gas–liquid chromatographic techniques for various minor components in fats; Mettler dropping-point determination of melting points; high performance liquid chromatography for the detection of various constituents in fats; mass spectroscopy applied to many problems of fat chemistry; nmr spectroscopy applied to structural problems (high-resolution) and to the determination of SFI (broad-line). These and other newer methods now have wide application in fat chemistry and only have to undergo the usual collaborative studies before adoption by the AOCS and others as official methods.

Health and Safety Factors

Fats have no particular technical problems with respect to health and safety. If extremely overheated, traces of glycerol in fats can decompose to form acrolein (qv) which is a toxic and lachrymatory substance. Polymeric substances developed by heating unsaturated fats either in the presence or absence of air are toxic to animals when included in their diets at significant levels. However, such substances normally are found only in fats that have been abused oxidatively or thermally to a degree that would make them completely unpalatable to man. A few minor fatty acid constituents in some fats (eg, sterculic acid) in cottonseed oil have been shown to have deleterious effects on some organisms (see Carboxylic acids, survey). In recent history, the dietary studies with animals on rapeseed oil and more particularly on erucic acid (90) have shown some effects of fatty infiltration into heart tissue and long-term lesions there. These rather definitive studies have led to the limitation of erucic acid levels in edible products in Canada and the EEC. Furthermore, these results were also responsible for the conversion of much of the Canadian rapeseed crop from regular erucic level seed (40–50%) to the so-called low-erucic (or zero erucic) canbra (canola) varieties (0–5% erucic acid).

Many other studies aimed at clearly defining the role of dietary fats in several endemic diseases of man have been conducted but, to date, there is no clear agreement as to the effect of any fat type with the exception, perhaps, of the requirement for sufficient dietary linoleic acid to prevent essential-fatty-acid deficiency diseases. For every proponent of a particular hypothesis relative to fats or fat types and various diseases, there appears to be equally vocal opponents. Some recent reviews (91) summarize many points of view in this area and, additionally, show the intensity and diversity of opinions in this field. A great deal of credance also has been given by some to documents and testimony leading up to United States dietary goals (92). However, significant scientific reasons have been cited for disregarding the generality of these goals particularly with respect to dietary fats. For a good overview see the references (93).

Uses and Derivatives

Most of the principal uses of fats are described earlier in this article. The principal edible uses are in salad and cooking oils, shortenings, margarines, filled dairy products (see Milk products) and prepared foods. Major industrial use areas are: soaps and detergents (see Surfactants and detersive systems), paints (qv), plastic additives (see Alkyd resins), and lubricants (qv) (see also Carboxylic acids).

Edible emulsifiers derived from fats include: mono- and diglycerides, lactylated mono- and diglycerides, propylene glycol monoesters, sorbitan stearate, poly(oxyethylene) sorbitan stearate, succinylated monoglycerides, acetylated monoglycerides, and polyglycerol esters of fatty acids (see Emulsions).

BIBLIOGRAPHY

"Fats and Fatty Oils" in *ECT* 1st ed., Vol. 6, pp. 140–172, by A. E. Bailey, The Humko Co.; "Fats and Fatty Oils" in *ECT* 2nd ed., Vol. 8, pp. 776–811, F. A. Norris, Swift & Company.

1. C. R. Smith, Jr. in F. D. Gunstone, ed., *Topics in Lipid Chemistry,* Vol. 3, John Wiley & Sons, Inc., New York, 1972, pp. 89–124.
2. C. Litchfield, *Analysis of Triglycerides,* Academic Press, Inc., New York, 1972, 355 pp.
3. C. Litchfield, *Fette Seifen Anstrichm.* **75,** 223 (1973).
4. T. P. Hilditch, *J. Am. Oil Chem. Soc.* **42,** 745 (1965).
5. I. A. Wolff, *Science* **154,** 1140 (1966).
6. C. R. Smith, Jr., *Prog. Chem. Fat Other Lipids* **11,** 139 (1970).
7. G. B. Ansell, J. N. Hawthorne, and R. M. C. Dawson, eds., B. B. A. Library, Vol. 3: *Form and Function of Phospholipids,* 2nd ed., Elsevier Scientific Publishing Co., Amsterdam, The Netherlands, 1973, pp. 1–7.
8. J. C. Bauernfeind, *CRC Crit. Rev. Food Sci. Nutr.* **8,** 337 (Mar. 1977).

9. J. Eisner and D. Firestone, *J. Assoc. Off. Agric. Chem.* **46,** 542 (1963).
10. H. Seher in R. Paoletti, G. Jacini, and G. Porcellati, eds., *Lipids—Technology,* Vol. 2, Raven Press, New York, 1976, pp. 293–313.
11. A. Kuksis, *Biochem.* **3,** 1086 (1964).
12. L. J. Cook and co-workers, *Nature (London)* **228,** 178 (1970).
13. *USDA Foreign Agriculture Circular FOP 25-77,* USDA, Washington, D.C., Dec. 1977.
14. J. Baltes, *Gewinnung und Verarbeitung von Nahrungsfetten,* P. Parey, Berlin 1975, 255 pp.
15. R. G. Binder and co-workers, *J. Am. Oil Chem. Soc.* **39,** 513 (1962).
16. H. Pardon, *Analysis of Edible Fats,* Paul Parey, Hamburg, FRG, 1976, pp. 330–331.
17. E. S. Perry, W. H. Weber, and B. F. Daubert, *J. Am. Chem. Soc.* **71,** 3720 (1949).
18. T. H. Applewhite, *J. Am. Oil Chem. Soc.* **43,** 406 (1966).
19. A. J. Haighton, K. Van Putte, and L. F. Vermaas, *J. Am. Oil Chem. Soc.* **49,** 153 (1972).
20. D. Swern, ed., *Bailey's Industrial Oil and Fat Products,* 3rd ed., John Wiley & Sons, Inc., New York, 1964, 1103 pp.
21. R. O. Feuge, *J. Am. Oil Chem. Soc.* **24,** 49 (1947).
22. G. L. Gaines, Jr., in I. Prigogine, ed., *Insoluble Monolayers at Liquid-Gas Interfaces,* Vol. 1, Wiley-Interscience, New York, 1966, pp. 238–249.
23. G. D. Fasman, ed., *Handbook of Biochemistry and Molecular Biology,* 3rd ed., CRC Press, Cleveland, Ohio, 1975, p. 502.
24. A. E. Bailey, *Melting and Soldification of Fats,* Interscience Publishers, New York, 1950, 357 pp.
25. E. S. Lutton, *J. Am. Oil Chem. Soc.* **49,** 1 (1972).
26. W. E. Link, ed., *Official and Tentative Methods of the American Oil Chemists' Society,* 3rd ed., American Oil Chemists' Society, Champaign, Ill., 1977, method Cc 2-38.
27. *Ibid.,* method Cc 1-25.
28. *Ibid.,* method Cc 14-59.
29. *Ibid.,* method Cc 9a-48.
30. J. W. Hagemann and co-workers, *J. Am. Oil Chem. Soc.* **52,** 204 (1975).
31. R. A. Yoncoskie, *J. Am. Oil Chem. Soc.* **44,** 446 (1967).
32. J. W. Hampson and H. L. Rothbart, *J. Am. Oil Chem. Soc.* **46,** 143 (1969).
33. S. H. Bertram, *Chem. Tech. (Dordrecht)* **1,** 101 (1946).
34. H. Adriaanse, *Heats of Combustion and Physical Constants of Normal Saturated Fatty Acids and Their Methyl Esters,* G. Van Soest, Amsterdam, The Netherlands, 1960, 102 pp.
35. E. G. Lantondress, *J. Am. Oil Chem. Soc.* **44,** 154A, 156A, 192A (1967).
36. J. W. E. Coenen, *J. Am. Oil Chem. Soc.* **53,** 382 (1976).
37. R. G. Binder, L. A. Goldblatt, and T. H. Applewhite, *J. Org. Chem.* **30,** 2371 (1965).
38. D. Chapman, *The Structure of Lipids,* John Wiley & Sons, Inc., New York, 1965, 323 pp.
39. Ref. 26, method Cd 14-61.
40. Ref. 26, method Cd 7-58.
41. J. C. van den Enden and co-workers, *Fette Seifen Anstrichm.* **80,** 180 (1978).
42. W. O. Lundberg, ed., *Autoxidation and Antioxidants I and II,* Wiley-Interscience, New York, 1962, 1156 pp.
43. K. U. Ingold, *Chem. Rev.* **61,** 563 (1961).
44. H. Wexler, *Chem. Rev.* **64,** 591 (1964).
45. J. A. Ward, *J. Am. Oil Chem. Soc.* **53,** 261 (1976); L. H. Tindale and S. R. Hill-Haas, *J. Am. Oil Chem. Soc.* **53,** 265 (1976).
46. E. Bernardini, *J. Am. Oil Chem. Soc.* **53,** 275 (1976); R. P. Hutchins, *J. Am. Oil Chem. Soc.* **53,** 279 (1976); W. Stein and F. W. Glaser, *J. Am. Oil Chem. Soc.* **53,** 283 (1976); E. D. Milligan, *J. Am. Oil Chem. Soc.* **53,** 286 (1976).
47. B. DeRamecourt, *J. Am. Oil Chem. Soc.* **53,** 256 (1976); E. Bernardini, *The New Oil and Fat Technology,* 2nd ed., Technologie, Rome, Italy, 1973, pp. 219–223.
48. R. A. Carr, *J. Am. Oil Chem. Soc.* **53,** 347 (1976).
49. B. Braae, *J. Am. Oil Chem. Soc.* **53,** 353 (1976).
50. F. E. Sullivan, *J. Am. Oil Chem. Soc.* **53,** 358 (1976); A. M. Gavin, K. T. Toeh, and G. Carlin, *J. Am. Oil Chem. Soc.* **54,** 312A (1977); A. M. Gavin, *J. Am. Oil Chem. Soc.* **54,** 528 (1977).
51. G. C. Cavanagh, *J. Am. Oil Chem. Soc.* **53,** 361 (1976).
52. J. C. Cowan, *J. Am. Oil Chem. Soc.* **53,** 382 (1976).
53. H. B. W. Patterson, *J. Am. Oil Chem. Soc.* **53,** 339 (1976).
54. E. H. Goebel, *J. Am. Oil Chem. Soc.* **53,** 342 (1976).

55. K. D. Mukerjee, I. Kiewitt, and M. Kiewitt, *J. Am. Oil Chem. Soc.* **52,** 282 (1975).
56. H. J. Schmidt, *J. Am. Oil Chem. Soc.* **47,** 134 (1970).
57. E. N. Frankel and H. J. Dutton in F. D. Gunstone, ed., *Topics in Lipid Chemistry,* Vol. 1, John Wiley & Sons, Inc., New York, 1970, pp. 161–276.
58. Harshaw Chemical Co., *J. Am. Oil Chem. Soc.* **52,** 3A (1975).
59. B. Drozdowski and M. Zajac, *J. Am. Oil Chem. Soc.* **54,** 595 (1977).
60. Frank E. Sullivan Co., *J. Am. Oil Chem. Soc.* **50,** 393A (1973).
61. H. P. Kreulen, *J. Am. Oil Chem. Soc.* **53,** 393 (1976).
62. H. H. Hustedt, *J. Am. Oil Chem. Soc.* **53,** 390 (1976).
63. K. F. Gander, *J. Am. Oil Chem. Soc.* **53,** 417 (1976).
64. C. T. Zehnder, *J. Am. Oil Chem. Soc.* **53,** 364 (1976).
65. K. S. Watson and C. H. Meierhoefer, *J. Am. Oil Chem. Soc.* **53,** 437 (1976).
66. C. Svensson, *J. Am. Oil Chem. Soc.* **53,** 443 (1976).
67. G. Choffel, *J. Am. Oil Chem. Soc.* **53,** 446 (1976).
68. G. N. McDermott, *J. Am. Oil Chem. Soc.* **53,** 449 (1976).
69. W. Van Nieuwenhuyzen, *J. Am. Oil Chem. Soc.* **53,** 425 (1976).
70. W. M. Neuner and E. K. Holt, *J. Am. Oil Chem. Soc.* **52,** 17A (1975).
71. E. F. Harp, *J. Am. Oil Chem. Soc.* **52,** 4A (1975); W. C. Seng and G. M. Kreutzer, *J. Am. Oil Chem. Soc.* **52,** 9A (1975); M. K. Cantrell and H. F. Keller, Jr., *J. Am. Oil Chem. Soc.* **52,** 13A (1975).
72. W. J. Lacy, G. Keeler and G. R. Webster, *J. Am. Oil Chem. Soc.* **52,** 20A (1975).
73. J. O'Keefe, *Food Can.* **35,** 29 (1975).
74. H. Jennewein and E. Trunzer, *Fette Seifen Anstrichm.* **80,** 127 (1978).
75. *Oil World* **XXI**(45), 1023 (1978).
76. *Ibid.,* (17), 412 (1978).
77. *Fats and Oils Situation (FOS) Bulletin 293,* Economics, Statistics and Cooperatives Service, U.S. Department of Agriculture, Washington, D.C., 1978, p. 30.
78. *Oilseeds and Products (FOP) Bulletin 9-78,* Foreign Agriculture Service, U.S. Department of Agriculture, Washington, D.C., 1978, p. 3.
79. J. P. Burkhalter, *J. Am. Oil Chem. Soc.* **53,** 332 (1976).
80. L. M. Wright, *J. Am. Oil Chem. Soc.* **53,** 408 (1976); G. M. R. Johansson, *J. Am. Oil Chem. Soc.* **53,** 410 (1976).
81. Ref. 77, *Bulletin 291.*
82. *U.S. Fats and Oils Statistics 1961–76, Statistical Bulletin No. 574,* Economics Research Service, U.S. Department of Agriculture, Washington, D.C., 1977.
83. Ref. 77, *Bulletin 290.*
84. W. E. Link, *J. Am. Oil Chem. Soc.* **53,** 232 (1976).
85. *Amendments to Methods of Analysis Appearing in Recommended International Standards,* FAS/WHO Cx/FO 78/10, New York, 1978.
86. Ref. 26, method Ca 9a-52.
87. F. A. Norris in ref. 20, pp. 601–635.
88. V. C. Mehlenbacher, *The Analysis of Fats and Oils,* The Garrard Press, Champaign, Ill., 1960, 616 pp.
89. T. J. Weiss, *Food Oils and Their Uses,* The Avi Publishing Co., Inc., Westport, Conn., 1970, pp. 9–17.
90. A. A. M. Abdellatif and R. O. Vles, *Nutr. Metab.* **12,** 285 (1970) cited in U. M. T. Houtsmuller, *Fette Seifen Anstrichm.* **80,** 162 (1978).
91. R. Reiser, *Am. J. Clin. Nutr.* **26,** 524 (1973); A. Keys, F. Grande, and J. T. Anderson, *Am. J. Clin. Nutr.* **27,** 188 (1974); R. Reiser, *Am. J. Clin. Nutr.* **27,** 228 (1974); G. V. Mann, *N. Engl. J. Med.* **297,** 644 (1977); J. Rivers, *Nature (London)* **270,** 2 (Nov. 3, 1977); A. S. Truswell, *Am. J. Clin. Nutr.* **31,** 977 (1978); J. McMichael, Abstract No. 32, 69th Annual Meeting, American Oil Chemists' Society, St. Louis, Mo., May 14–18, 1978.
92. *Dietary Goals for the United States,* U.S. Senate Committee on Nutrition and Human Needs, Washington, D.C., 1977.
93. *Nutr. Today,* 10 (Nov.–Dec. 1977); G. A. Leveille, *Food Nutr. News* **49,** 1 (1977) reproduced wholly from a statement to the USDA 1978 Food and Agricultural Conference; A. E. Harper, *Am. J. Clin. Nutr.* **31,** 310 (1978); K. Oster, *Assoc. Food Drug Off. U.S. Q. Bull.* **42,** 250 (1978).

General References

D. Swern, ed., *Bailey's Industrial Oil and Fat Products,* 4th ed., John Wiley & Sons, Inc., New York, 1979; an important reference on fats and oils used worldwide.

E. W. Eckey, *Vegetable Fats and Oils,* ACS Monograph Series, Reinhold Publishing Corp., New York, 1954, 836 pp.; an older but classic text on vegetable fats, their sources, properties, and processing.

R. T. Holman, ed., *Progress in the Chemistry of Fats and Other Lipids,* Pergamon Press, London, Eng., Vol. 1, 1952, through Vol. 16, 1978; a continuing series.

Fats and Fatty Oils, in the series, C. A. Price, ed., *Reports on the Progress of Applied Chemistry,* Academic Press, London, Eng.; a series of excellent reviews up to **60** (1975).

M. E. Stansby, ed., *Fish Oils: Their Chemistry, Technology, Stability, Nutritional Properties, and Uses,* Avi Publishing Co., Inc., Westport, Conn., 1967, 440 pp.

A. Kuksis, ed., "Fatty Acids and Glycerides" in D. J. Hanahan, ed., *Handbook of Lipid Research,* Vol. 1, Plenum Press, New York, 1978, 469 pp.

A. J. Vergroesen, ed., *The Role of Fats in Human Nutrition,* Academic Press, London, Eng., 1975, 494 pp.

M. L. Meara, *Physical Properties of Oils and Fats,* British Food Manufacturing Industries Research Association, 1978, No. 110, 34 pp.; scientific and technical surveys; an excellent tabulation of published information; other volumes of this series are also worthy of attention.

THOMAS H. APPLEWHITE
Kraft, Inc.

FATTY ACIDS. See Carboxylic acids.

FATTY ACIDS FROM TALL OIL. See Carboxylic acids.

FEEDSTOCKS

Petrochemical feedstocks are derived almost exclusively from natural gas or petroleum. Therefore, the use of these materials as feedstocks is competitive with their use in the so-called energy sector, ie, with the raw materials for space heating, transportation fuels, industrial fuels, or electricity generation. In the ideal free-market economy, the choice of alternatives to the disposition or use of a material is objectively settled by economics. A material flows (in a commercial sense) to the applications offering adequate economic return.

For the petrochemical producer, the selection of the proper feedstock is largely an economic matter, involving as well a knowledge of supply logistics. Consider, for example, the planning for a new manufacturing plant to make ethylene (qv), the petrochemical building block produced in largest tonnage worldwide. The process could be based primarily on one of the following feedstocks: ethane, propane, *n*-butane, light straight-run gasoline (qv), refinery-manufactured raffinates, full range naphtha, light gas oil, or any of a variety of heavy gas oils. Each of these materials has been selected recently as the feed to one or more commercial ethylene pyrolysis units in the

832 FEEDSTOCKS

United States. Added to this list are a number of potential feedstocks, undergoing evaluation by R & D or process design groups, that include crude oil, petroleum residual fractions, and other unusual feedstocks such as shale oil, coal, biomass products, Brazilian rubber tree latex, etc. The competitive ethylene business requires large, efficiently designed (and expensive) producing plants. The need for efficiency, ie, good control of energy flows, pyrolysis yield, and recovery, tends to severely limit the flexibility of the producing unit with respect to changes in the feedstock. There are some feed interchanges that can be made: ethane and propane can be interchanged fairly broadly, and the light straight-run gasoline/raffinate interchange would not incur serious penalties. But within narrow limits, the feedstock design choice, once made, sets the actual feed requirement unless major plant equipment revisions are made. Therefore, the prospective ethylene producer must select the feedstock with care and understanding of supply logistics and economics not only from a chemical framework, but also from the framework of the alternative energy sector.

In the United States in recent years, forces other than those based on free-market economics are being applied to direct the flow of natural gas and petroleum in the market place. In the case of natural gas, regulatory control has existed since 1954, when the U.S. Supreme Court ruled that the 1938 Natural Gas Act established price control for gas in interstate commerce (see Gas, natural). However, producer sales and pipeline sales within the producing state, ie, intrastate gas, were not federally regulated until the Natural Gas Policy Act (NGPA) of 1978. Therefore, the acquisition cost of one important petrochemical feedstock, ethane, most of which is recovered from natural gas, was for years dependent upon whether the ethane was in inter- or intrastate gas. The difference in cost is significant. Typically in 1977–1978, interstate natural gas might have been valued to the producer at \$0.33–0.43/GJ (\$0.35–0.45/10^6 Btu) as opposed to ca \$2.13/GJ (\$2.25/10^6 Btu) for intrastate gas. World energy economics is based on price schedules set by the Organization of Petroleum Exporting Countries (OPEC) who quote in U.S. dollars per barrel (159 L or 42 gal) for a so-called marker crude. A barrel of average crude oil has a heating value of ca 6.1 GJ (5.8 × 10^6 Btu). Thus crude oil priced at the U.S. Gulf Coast at \$12 per barrel (or \$75.5/m^3) equates to alternative fuels at \$1.96/GJ (\$2.07/10^6 Btu). Relative to its fuel value, the raw material cost of ethane (exclusive of the recovery cost, delivery, and profit) was either 2.0 or 11.0 ¢/kg, depending on the classification of the natural gas from which it was recovered. This is a considerable swing in raw material cost for a feedstock that is used almost exclusively for making ethylene, which was priced at 26.5–28.6 ¢/kg at the time.

With price controls extended to intrastate gas by the NGPA of 1978, one might think that some simplification of feedstock supply economics had been effected. However, the NGPA established ≥20 classifications of natural gas, with detailed and often complex price-setting rules, and with escalation formulas and deregulation plans that differ for many of the classifications (1).

Other regulatory forces that can direct the flow of feedstock raw material counter to economic incentives are the set of allocation regulations that can be invoked when potential shortages are considered to constitute an emergency. Priorities for the use of natural gas were published in the Energy Supply and Environmental Coordination Act (ESECA) of 1974. The use of natural gas in the manufacture of chemicals had high priority status during the 1973–74 shortages. However, in a new emergency standby allocation system recently developed by the DOE, petrochemical feedstocks have lost their priority position. If the allocation system is activated they would be treated the same as fuels (2).

The application of price setting and allocating regulations to crude petroleum and several related products also tends to complicate, if not distort, the economics at the interface of the energy/feedstock alternative. It seems likely that controls of some nature will continue as long as the United States is required to import large volumes of petroleum from a world with perceived political or economic instabilities. Currently, ca 45% of the U.S. petroleum supply is based on foreign sources. Put another way, 22% of the energy budget of the United States is imported. Based on concerns of national security and the U.S. trade balance, and fears by other western nations that the United States is consuming too large a share of world oil production, there have been various proposals for reducing oil imports.

In summary, the selection of a chemical feedstock almost invariably involves its diversion from the alternative of a fuel use. Political and other forces frequently tend to modify idealized economic forces. However, many of the politically-derived constraints or incentives are transitory. For example, in the mid-1960s, when foreign crude oil was priced considerably lower than U.S. domestic crude petroleum, knowledge of the detailed regulations of the Mandatory Oil Import Program (MOIP), administered by the Oil Import Administration (OIA) of the U.S. Department of the Interior, was important to the economic selection of petrochemical feedstocks. Now the importance of the MOIP has all but vanished (although OIA regulations have not). Economic realities today are based on the fact that foreign crude oil is several dollars per barrel (1 \$/barrel = 6.25 \$/m^3) more costly than some classifications of price-controlled domestic crude oil and, on a GJ-equivalent basis (1 GJ = 0.949×10^6 Btu), much more costly than the controlled price of some classifications of domestic natural gas. The administration of the U.S. government has embarked on a program to deregulate domestic crude oil by 1981, but refiners and petrochemical producers face uncertain future feed costs under declining domestic production, import quotas, entitlements, and allocations.

Definitions

Confusion can arise over the definition of feedstock because the petrochemical industry is characterized by interunit (and often intercompany) transfers in which a product from one processing stage is often the feedstock for the next. OIA regulations, for example, define ethylene as a feedstock and the next generation of products, eg, ethylene oxide, as petrochemicals, but this definition was adopted to affect the distribution of economic benefits under the MOIP. A more generally accepted set of definitions follows.

Petrochemical feedstocks are mainly hydrocarbons that are transferred from the energy sector to the chemical sector. A more explicit statement is that they are transferred from operations commonly performed in preparation of usable energy to operations in which they are first used as raw materials in processes involving chemical change or physical separation before eventual end use. An example involving a chemical change is the steam pyrolysis of naphtha to make ethylene; an example involving physical separations is the extraction of an aromatic concentrate from a refinery-produced reformate with subsequent distillation of the extract to high purity benzene. Recent trends in raw materials supply make the definition of petrochemical feedstocks more complex. For the present, however, the definition covers essentially all situations of commercial importance. A tabular representation of the definition is given in Table 1.

The classifications in Table 1 presently serve well as a framework for most technical or commercial concerns. However, developments that lead to a need for definitional flexibility appear to be underway. For example, in the gasification of coal or shale, the product gas containing CO and H_2 can naturally be burned as a fuel or it could be converted catalytically to primary petrochemicals (eg, ammonia (qv)) or to petrochemical products (eg, methanol (qv)). The gasification product containing CO and H_2 would be taken as the feedstock, whereas in practice today a similar mixture of CO and H_2 is an intermediate stream from a natural gas or petroleum-derived feedstock.

The Contribution of Feedstocks to the U.S. Economy

The primary petrochemicals in Table 1 and others produced in lesser quantities are often called primary building blocks. They are processed through intermediate chemicals to an enormous number of products. Although no formal government definition of a petrochemical industry occurs as a census industry classification, several aspects of the flow of raw material to finished products have been quantified (3). Estimated 1978 values of interstage flows and product consumption are given in Table 2.

Table 1. Petrochemical Feedstocks and Principal Primary Petrochemicals

	1978 Volume, 10^6 t coe^a	CAS Registry No	1978 Production, 10^6 t
Feedstock			
natural gas	17.7		
natural gas liquids	19.4		
ethane			
propane			
butane			
petroleum products	43.1		
refinery gases			
naphtha			
catalytic reformate			
coal tar	1.6		
Important primary petrochemicals			
ammonia		[7664-41-7]	15.4
ethylene		[74-85-1]	12.8
propylene		[115-07-1]	6.5
benzene		[71-43-2]	5.0
toluene		[108-88-3]	2.9
butadiene		[106-99-0]	1.6
p-xylene		[106-42-3]	1.6
o-xylene		[95-47-6]	0.5

^{a} coe = crude oil equivalent. The heating (calorific) value of "average" crude oil here follows the definition of the 24-nation Organization for Economic Cooperation and Development (OECD), ie, 1.0 t coe = 41.9 GJ = 10^{10} cal (1.0 bbl coe = 6.1 GJ = 5.8×10^6 Btu).

Table 2. Flow of Industrial Organic Chemicals from Feedstock to Products, Estimated for United States, 1978

Stage in manufacturing process, and interstage shipment value, $ billion	Value of product consumed, $ billion
feedstocks	
|	
5	
↓	
chemical building blocks	
|	
10	
↓	
industrial organic chemicals →	4[a]
|	
24	
↓	
chemical sector industries with organic chemical inputs →	27
|	
40	
↓	
industries that are major users of products from organic chemicals sector: agriculture, automotive, construction, footwear, household appliances, paper and allied products, printing and publishing, textiles →	530 (total shipments by the selected industries)

[a] Including exports.

In 1978 feedstocks valued at $5 billion were used to produce $10 billion worth of chemical building blocks. These building blocks are the major intermediates for the manufacture of industrial organic chemicals. Of the $28 billion of industrial organic chemicals shipped, $24 billion, or ca 86% were used as input to other chemical sectors or to other industries. These included plastics and resins, synthetic rubber, chemical fibers (qv), rubber and plastic products, household cleaning products, and toilet preparations, all having a value of an estimated $67 billion of total output including $40 billion input to other industries. The remaining $4 billion of organic chemical shipments were largely exports.

The petrochemical product grouping expected to grow most rapidly in the next few years is plastics materials and resins. This conclusion was reached in a study sponsored by the Petrochemical Energy Group, an industry trade organization, to estimate future potential of businesses that are dependent on the petrochemical industry (4). Business sector SIC (standard industrial classification) 2821 is expected to grow at a rate of 10.3% per year during the period 1976–1983 compared to an expansion of the gross national product (GNP) by 4.3%/yr. The predominant importance of six building blocks to the plastics and resins sector is demonstrated in Table 3.

Although there are ca 40 distinct plastics, some very expensive and highly specialized, the six performance materials listed in Table 3 account for the bulk of plastics production in the United States. When all chemicals and allied products are considered, the list of large tonnage primary petrochemicals becomes eight, including ammonia and p-xylene (see Table 1). The primary petrochemicals are produced from a narrow field of economically acceptable feedstocks. Feedstocks must be reliably available and must make the building blocks at competitive cost, or they are not used in today's large plants.

836 FEEDSTOCKS

Table 3. Petrochemicals for Commercial Plastics and Resins; Requirements for the Originating Petrochemicals; U.S. Production Data for 1978 in 10^6 Metric Tons

Plastic or resin, 10^6 t/yr	Important uses	Intermediate petrochemicals	Primary petrochemicals consumed, 10^6 t/yr
polyethylene 5.13	packaging film, garbage bags, bottles, pipe		ethylene 5.44
poly(vinyl chloride) 2.60	siding, pipe, conduit, flooring	vinyl chloride, dioctyl phthalate (plasticizer)	ethylene 1.30 propylene 0.09 o-xylene[a] 0.09
polystyrene 1.77	appliances, toys, packaging, furniture, foam cups	styrene, polybutadiene	ethylene 0.55 benzene 1.49 butadiene 0.07
polypropylene 1.39	packaging film, auto parts, appliances, luggage, battery cases		propylene 1.53
polyurethanes 0.91	insulation, bedding, car seats, furniture	aniline, toluene diisocyanate, propylene oxide, methylenediphenyl diisocyanate	propylene 0.45 benzene 0.08 toluene 0.15
phenolics 0.74	plywood, insulation, foundry, appliances, electrical parts	phenol, formaldehyde	benzene 0.54
unsaturated polyesters 0.55	fiberglass boats, artificial marble, transportation parts	maleic anhydride, phthalic anhydride, propylene oxide, styrene	ethylene 0.08 propylene 0.08 benzene 0.28 o-xylene 0.11
acrylonitrile–butadiene—styrene (ABS) 0.51		acrylonitrile, styrene	ethylene 0.09 propylene 0.13 benzene 0.25 butadiene 0.11

[a] See Xylenes and ethylbenzene.

Demand Patterns for Feedstocks

Two series of questions are important in assessing the impact of the diversion of petrochemical feedstocks from the nation's energy supply: (1) what are the present requirements, including the associated consumption of various hydrocarbon fuels and other forms of energy by the petrochemical industry; and (2) what are the trends in feedstock consumption and what would be the consequences of the extrapolation of these trends? Since the 1973–1974 energy crisis there has been widespread concern about the continued availability of feedstocks. Many inquiries have been made into feedstock alternatives, eg, the 1977 conference on chemical feedstock alternatives sponsored by the AIChE and the NSF (5).

The feedstock and energy demands of the petrochemical industry are only a small percentage of the total consumption of natural gas and petroleum. In Table 4 the feedstock consumption of the U.S. chemical industry is given for 1965–1977, with an estimate for 1978. In addition to the net feedstock consumption of 67 × 10^6 t crude oil equivalent (coe) in 1978 there is 65×10^6 t/yr coe total energy consumed by the industry. Total energy is all fuels plus electricity consumed. The combined feedstock plus energy use of 132×10^6 t/yr coe is equal to 5.6×10^9 GJ/yr (5.3×10^{18} Btu/yr). This is about 7% of the estimated total U.S. energy consumption in 1978 of 84 GJ (80 $\times 10^{18}$ Btu). It is the basis for the industrial production shown in Table 2.

The growth in feedstock consumption during the period 1965–1978 can be seen in Table 4. Both the gross and net demands for hydrocarbons are listed. For feedstocks only, usage increased from 3.7% to 4.3% of the total consumption of petroleum and natural gas. The absolute growth in petrochemical feedstocks during the period was 4.6%/yr. In a report filed with the Department of Energy by the Chemical Manufacturers' Association on behalf of 104 manufacturers in category SIC 28, it is stated that aggregate energy consumption per unit of product produced had declined in 1978 by 18% compared to 1972 (6). For comparison the total usage of petroleum and natural gas in the U.S. grew by 3.2%/yr.

A comparison of the petrochemical feedstocks demands for 1978 for Western Europe and the United States is shown in Table 5. Western Europe uses a feedstock mix that is weighted much more toward the heavy liquid side and, as a consequence, returns a higher proportion of the gross feed to the energy sector.

The extrapolation of historical growth figures has doubtful validity. However, it may be useful to do so to examine broad perspectives of a potential problem. On this basis, it can be readily calculated that if the petrochemical industry continues its 4.6%/yr growth until the year 2000, the combined feedstock and energy needs will be 15.1×10^9 GJ/yr (14.3×10^{18} Btu/yr) or 1.0×10^6 t/d coe (6.8×10^6 b/d coe). A demand this high would be just barely credible on the basis of the historical use pattern of total solid materials per capita since 1900 (7). However, the extrapolation demonstrates that growth of petrochemicals demand could continue at present rates (no matter how unlikely when viewed historically), and the feedstocks could be obtained from traditional natural gas and petroleum sources.

Note that no implication is being made here that some feedstocks in the future will not be obtained from sources called unconventional today. Indeed, energy experts almost unanimously predict that new source material such as coal liquids, shale oil,

Table 4. Demand for Petrochemical Feedstocks, 10^6 t coe[a]

Feedstock	1965	1970	1975	1977	1978 (estd)
natural gas	9.5	14.9	17.8	18.3	17.7
ethane	3.1	6.8	8.8	9.3	9.6
propane, butanes	7.2	10.1	8.9	8.9	9.8
refinery offgases	3.8	4.7	5.3	6.0	5.7
naphthas, including coal tar	12.8	16.2	19.2	21.9	27.0
gas oils and heavier	3.4	5.1	8.6	11.7	12.0
Total gross	39.8	57.8	68.6	76.1	81.8
less by-products returned to energy sector	3.4	8.1	10.9	13.7	14.8
Total net	36.4	49.7	57.7	62.4	67.0
Total net					
as 10^6 m^3/d	0.11	0.15	0.18	0.19	0.21
(10^6 b/d[b] coe)	(0.71)	(0.92)	(1.12)	(1.21)	(1.30)

[a] Crude oil equivalent. See Table 1.
[b] b/d = barrels per day.

Table 5. United States and Western Europe Demand for Petrochemical Feedstock in 1978

	Estimated feedstock demand in 1978			
	United States		Western Europe	
Feedstock	10^6 t coe	%	10^6 t coe	%
natural gas	17.7	22	6.4	9
natural gas liquids	19.4	24	1.1	1
refinery offgases	5.7	7	1.4	2
naphtha including coal tar	27.0	32	51.6	70
gas oils and heavier	12.0	15	13.6	18
Total gross	81.8	100	74.1	100
by-products returned to energy sector	14.8	18	21.1	28
Total net	67.0	82	53.0	72
Total net as 10^6 m^3/d	0.21		0.16	
(10^6 b/d^a coe)	(1.3)		(1.0)	

^{a} b/d = barrels per day.

biomass, etc, will be commercially developed in the United States before the year 2000. Any need based on traditional natural gas and petroleum can probably be satisfied by alternative feedstocks. However, the economics of feedstock choice logically should determine whether the new source material is consumed in the energy sector or is used as a petrochemical feedstock.

In the growth projection, demands for natural gas and natural gas liquids have been combined with that for heavier petroleum feedstocks. Availability analyses have not been made for each type of feedstock, although this detail is certainly checked carefully by a petrochemical manufacturer before committing capital funds to a new olefins plant. If the lighter feeds are not available, the principal primary petrochemicals can be made commercially from naphtha or heavier feedstocks, eg, petrochemical industries existed in Puerto Rico and Europe before natural gas was widely available.

Selection of Feedstocks

The choice of a feedstock is an important decision in planning for petrochemicals because large capital investments and the kind and volume of by-products generated in the plant depend on it. From the end of World War II until the late 1960s the U.S. petrochemicals industry enjoyed the benefits of low-cost natural gas liquids (NGL) (butanes, propane, and ethane). Feedstocks such as propane or ethane/propane mixes were abundantly available and were purchased as required to produce the desired ratio of ethylene, propylene, and other primary petrochemicals. This era has ended. The supply of gas liquids has not kept up with the need for feedstocks for light olefin manufacture. Petrochemical manufacturers have built more plants with so-called heavier feedstocks (ie, higher molecular weight feeds), with boiling ranges that characterize them as naphthas, light gas oil (ie, the kerosene boiling range), or heavy gas oils having boiling points as high as 450°C at their 90% point. In 1976 nearly 75% of U.S. ethylene production was based on NGL, but only one of 13 light olefins plants announced or under construction was designed for NGL feedstock. The changing pattern of feedstock sources is shown in Table 6 (8).

Table 6. Light Olefin Feedstocks, Projection of U.S. Production Patterns to 1990

Year	Total ethylene production, 10^6 t/yr	Percent of ethylene from		
		NGL[a]	Naphtha	Gas oils
1976	10	74	14	12
1980	14.5	60	20	20
1990	23.6	30	30	40

[a] NGL = natural gas liquid.

Form Value. There are a number of commonly used empirical relations that aid in the selection of preferred feedstocks for desired products. Many of these relate to a property of the feed called form value, a useful concept that is usually intuitively applied by experts in the petrochemical industry. The form value of a feedstock is not an exact property, but in a qualitative sense it refers to its molecular structure and whether the feed is a good one or a poor one for the desired petrochemical product.

Feedstock naphthas can be ranked according to their suitability in making certain primary petrochemicals, using the paraffin, isoparaffin, and cycloparaffin content of the feed as a parameter (see Table 7).

Characterization Factor. A relationship which has been correlated with form value for several processes is characterization factor (CF) (9).

$$CF = \frac{1.216 \, (bp, K)^{1/3}}{\text{sp gr}_{15.6}^{15.6}}$$

Feedstocks are often compared at roughly similar boiling points because they are normally produced in gas plant or refinery distilling units where cuts between fractions are made under boiling point control. Under these conditions, high CF will in general be associated with low specific gravity and high paraffin/isoparaffin content. Feedstocks with relatively high CF tend to be good for ammonia and light olefin manufacture and poor for BTX manufacture as noted in Table 7 (see BTX processing).

To illustrate the detail of feedstock comparisons, suppose one were to develop a plan for the best use of an unprocessed natural gas stream between the alternatives of a fuel and a feedstock for ethylene. In addition, suppose that it is practical to separate and recover as many of the natural gas liquids as desired, ie, it is economically practical to recover the contained pentanes-plus, butanes, propane, and ethane before selling the residual natural gas (methane). In the fuel markets there are assumed to be ready demands at equitable prices for each condensate and liquefied petroleum gas (qv) (LPG) component, and the natural gas is purchased on the basis of its energy

Table 7. Feedstock Naphthas[a], Comparison of Values by Form

Product	Basis for comparison	Typical mid-boiling point, °C	Hydrocarbon-type, characterization		
			Paraffin	Isoparaffin	Cycloparaffins
ethylene	yield	120	best	fair	poor
ammonia, hydrogen	yield	120	best	good	fair
benzene	yield	93	poor	poor	good

[a] Ref. 9.

content so that ethane left in the natural gas stream also is sold at an equitable price. A petrochemical manufacturer needs some of the liquids as feedstock for ethylene, and is willing to pay for their fuel value equivalent plus a reasonable return for the cost of recovery, handling, and shipping. If there are no restraints or economic distortions in the market, the manufacturer's acquisition costs of condensate, n-butane, propane, or ethane are set by their respective values in the fuel markets. (The term fuel value is used here in a broad concept.) Normally the preferred market in the United States for condensate and n-butane is motor gasoline (qv). As motor gasoline components the materials have values higher than their direct-burning fuel value. In this supposition, sometimes there is surplus n-butane (especially in local situations) compared to the ability to blend it into gasoline; the remaining market alternatives are fuel (eg, in a refinery or large chemical plant) or petrochemical feedstock. In practice there also would be other fuel-sector alternatives for the NGL producer, for example, LPG sold to farmers, to industrial consumers, to motor-fuel consumers, to the residential market, etc, and possibly as feedstock for substitute natural gas (SNG) manufacture (see Fuels, synthetic). The additional outlets make optimum marketing in the fuel sector more complex, but the detail need not be considered in this example of the fuel sector/petrochemical sector interface.

The above premises set the stage for a discussion of feedstock selection in which form value enters. The best yield of ethylene in commercial steam-cracking (pyrolysis) plants from the three potential feedstocks is roughly 37 wt % from n-butane, 46 wt % from propane, and 79 wt % from ethane (see Ethylene; Olefins). Since, in this example, ethylene is the only desired product, it is clear that ethane is the preferred feedstock as long as it is not inordinately more difficult to steam-crack the ethane than the other feedstocks. It does require a somewhat higher temperature in the cracking coils (ie, the reaction zone) of the pyrolysis furnaces to achieve the optimum yield per pass with ethane, but the cost associated with this operational difference is not significant relative to the large yield differences. It is therefore apparent that ethane is the preferred feedstock for making ethylene without associated coproducts, ie, in a comparison of the three feedstocks, ethane has by far the best form value for ethylene.

In this supposition it is simple to make an economic comparison of the three feedstocks for ethylene manufacture because the feed costs are almost the same per unit weight, and there are large yield differences for the single desired petrochemical product. To compare feedstocks in the general case, a detailed and accurate economic balance must be made. The large olefins plants built today in the United States are designed to produce as much as 600,000 to 700,000 t/yr of ethylene. If the feed is gas oil they would consume up to five times that weight in feedstock annually (3.5×10^6 t/yr). In very rough terms, gas oil feed is valued at ca 12 ¢/kg, so that an error of 0.1% in feedstock value amounts to ca $400,000/yr. In the routine of the engineer-economist, feedstocks and processes are rank-ordered by comparing unit feed costs, yields, operating costs, by-product values, and the capital costs of the conversion units. Pyrolysis yields for many different feedstocks under a variety of operating conditions are shown in ref. 10, and many useful economic comparisons of both feedstocks and processes for ammonia, light olefins and BTX aromatics are given in ref. 11.

Feed Properties. Accurate feedstock evaluations are based on feed properties data and accurate predictions of the yields of primary products and all by-products in the commercial plants. The feed properties needed to evaluate naphthas as olefin plant feedstocks are: specific gravity; initial boiling point; distillation temperatures for 10 vol %, 50 vol %, 90 vol %; final boiling point; compositional analysis, vol %, n-paraffins, isoparaffins, cycloparaffins, and aromatics; and sulfur, wt %. A detailed component analysis by chromatography would be useful if available. With heavier gas-oil feeds it is not practical to obtain the compositional analysis. Instead, it is useful to obtain average molecular weight and C:H ratio. If laboratory test results are available for the properties listed above, in almost all cases it is possible for cracking experts to give a first estimate of the quality of the feed. This is done from in-house correlations, from yield data published for numerous feed types (10), and by comparing known commercial yields from feeds with similar properties.

When accurate yield data are needed for steam pyrolysis at specified cracking temperatures, cracking coil residence time, feed partial pressure, etc, most operating companies can obtain the data from established empirical correlations between feed properties and yields. These correlations predict yield data with more accuracy than can be obtained from a single plant test run. In the absence of this type of correlation, the usual manner in which yield data are obtained is by experimental cracking of relatively small quantities of the feedstock in pilot plants.

The inaccuracies inherent to the scale-up between pilot plant and commercial plants are reduced by establishing correlations between pilot and large plants with known feeds at standard operating conditions. Some companies have made correlations from hundreds of sets of data and expect high accuracy from their feedstock evaluations. However, correlations that work well within the span of feedstock types and operating conditions covered, may err when the correlation is extrapolated for an unusual feedstock.

Other Evaluations. Other evaluations of feedstock quality are available to those who do not have a pilot plan nor a file of empirical correlations. Major engineering contractors who design and build olefins plants, and several consultant firms specializing in light olefins technology, have pilot-sized pyrolysis testing equipment and computer-based programs correlating feedstock properties with yields. Many feedstock evaluations can be made quickly and accurately from their existing experience base. If the feedstock is uncommon, however, the calculated predictions will be of uncertain accuracy and actual testing in a pilot plant is suggested. To evaluate the data there must be a careful check of the types of feedstocks that have been tested earlier, of the claims that are being made about the accuracy of the material balances and yield determinations, and over what ranges of feed composition or operating conditions there is assumed to be acceptable correlations with full-scale commercial units.

Reformer Naphtha

Desirable feedstocks for reformers are high in cycloparaffins (naphthenes, N) and aromatics (A) content specifically in the C_6–C_8 range. If the reformer is used to make gasoline, the C_6 fraction is normally not reformed. This is because the benzene produced is not a desirable gasoline component even though there would be a net gain in unleaded octane barrels across the reformer. If the reformer is used to make chemicals, lighter feeds (C_6–C_7) are preferred since benzene is usually the preferred product.

A good reforming naphtha contains at least 45 vol % of N + A. This naphtha yields 30–40 vol % C_6–C_8 aromatics. A typical BTX reformer using a naphtha with a 150°C end point produces 4–8 vol % benzene, 14–18 vol % toluene, and 12–14 vol % xylenes (see BTX processing).

A reformer converts cyclohexane and methylcyclopentane to benzene. Minor amounts of benzene are produced by the dehydrocyclization of C_6 paraffins (see Hydrocarbons, C_1–C_6). If the component analysis of a C_6 naphtha is available, a good estimate of the benzene potential can be made by using the following rules of thumb.

Feed	Product benzene, vol %
benzene	100
cyclohexane	80
methylcyclopentane	40

Crudes are changing toward those with less N + A in the naphtha cuts. Middle East crudes have N + A contents in the range of 24–32 vol %. African and American crudes run 32–50 vol % N + A. Reformers that operate at lower pressures and higher temperatures promote more dehydrocyclization of paraffins. The trend to this type of reformer is an attempt to make up for the lack of N + A in the newer, more plentiful crudes.

Very generally, the value of a higher BTX potential is 50–60 ¢/t (6–7 ¢/bbl) per volume percent N + A in an unleaded gasoline pool where octanes are needed. It falls to 15–20 ¢/t (2–3 ¢/bbl) when octanes or BTX are not needed. The lower value represents savings from feedstock volume and operating costs.

Alternative Feedstocks

The present seems to be the transition from an era of abundant natural gas and petroleum to one in which raw material prices are increasing significantly and long term supply arrangements are uncertain. Thus it is natural for planners and researchers to be actively looking for alternative feedstocks that put an economic ceiling on costs or promise long term security in supply. Much inquiry is focusing on raw material from tar sands, coal, shale, biomass, and waste as substitute feedstocks to the chemical industry (see Fuels, synthetic; Tar sands; Oil shale; Fuels from biomass; Fuels from waste).

In the reports of the President's Materials Policy Commission (the Paley Commission) of 1952, President Truman noted his determination not to "allow shortages of materials to jeopardize our national security nor to become a bottleneck to our economic expansion" (12). Several sections of the report bear directly on petrochemical feedstocks; two are the following:

"Chemical industry also, above all other industries, has a great capacity for adapting itself to variations in raw materials, because to a large extent it can work out methods for using raw materials interchangeably (13).... Synthetic oil, probably first from shale and later from coal, will come into commercial production within a decade or so—perhaps sooner" (14).

The Paley Commission misjudged the timing of the essential changes to the feedstock patterns of the petrochemical industry, although proper recognition should be given to the production of synthetic crude oil from the bitumen of the Athabasca tar sands: first commercial production began in 1967 by Great Canadian Oil Sands, Ltd. However, the product is mixed with other refining crudes and has not made a significant change in the petrochemical industry.

Because of the large project sizes and the enormous capital requirements associated with proposals to develop synthetic fuels from coal or shale, it is expected that the first commercial ventures will be connected with the energy sector (see Fuels, synthetic). Then, chemical companies probably will make arrangements to buy some of the synthetic material as feedstock. This sequence was noted as a likely development by the experts at the AIChE/NSF conference (5). Several groups at the Conference agreed that alternative raw materials under study could be considered as potential feedstocks in a technical sense, but that their use could not be supported economically. It was expected that in general it would be better to burn the new source material for its energy content and shift a somewhat higher percentage of the available natural gas, NGL or petroleum (qv) to feedstock use. This conclusion is a consequence of the superior form value of today's feedstocks as precursors for primary petrochemicals.

Coal. The first commercial, clean fuel made in the United States from coal (qv) probably will be a substitute for natural gas; coal is gasified more easily than liquefied. Commercial gasification plants, eg, Lurgi, have been operating worldwide since the 1930s (15). In the United States, the economics of utility power systems that link gasifiers with generating units is becoming satisfactory compared to environmentally acceptable coal-fired power plants (16).

The crude gas from a steam- and oxygen-fed Lurgi gasifier typically has the following composition (15): CO_2, 28 vol %; CO, 19 vol %; H_2, 39 vol %; CH_4, 10–11 vol %; and N_2, 1 vol %. Also included are H_2S, naphtha, phenols, etc. The crude gas can be processed further in a shift-conversion unit in which the CO content is changed to whatever is desired in a downstream synthesis. Comparisons of ammonia manufacturing costs based on natural gas reforming, naphtha reforming, partial oxidation of fuel oil, and coal gasification are given in ref. 17. Many possibilities of chemical synthesis starting with synthesis gas, including Fischer-Tropsch synthesis and direct synthesis to methanol are described in ref. 11. By 1990 the petrochemical industry will be faced with shortages of natural gas and petroleum feedstocks and will have to turn to coal, according to one study. The production costs of aromatics, methanol, and possibly even ammonia from coal may be competitive with the costs from, eg, naphtha, but plant investment costs will be much higher (18).

Oil Shale. Oil shale (qv) is a sedimentary rock containing a solid organic material called kerogen. Kerogen, largely insoluble in petroleum solvents, thermally decomposes to oil, gas, water, and residual carbon. The composition of kerogen (19) from the Green River formation in Colorado, Utah, and Wyoming is: carbon, 80.5 wt %; hydrogen, 10.3 wt %; nitrogen 2.4 wt %; sulfur 1.0 wt %; and oxygen 5.8 wt %.

In a hypothetical chemical refinery, the basis is a mine producing 91,000 t/d of shale containing 125 L/t (30 gal/short ton) of oil (20). The raw material is retorted by an indirect heating method; naphthas are cracked to produce petrochemicals; heavy oil is coked to make a fuel oil; by-product gas from the retorting and processing supplies plant heat, hydrogen, and power. The hypothetical refinery could produce the following products: ammonia 110×10^3 t/yr; ethylene 215×10^3 t/yr; propylene 100×10^3 t/yr; butadiene 35×10^3 t/yr; and BTX aromatics 85×10^3 t/yr. In addition, the refinery would produce 3×10^6 t/yr of fuel oil and significant quantities of alumina, soda ash, coke, and sulfur.

Shale deposits in the Green River formation have been assessed at 90×10^9 t coe. This could theoretically supply the chemical feedstock needs of the United States for centuries (see Table 4). Utilization of this resource awaits the development of the need, and the refinement of the economics in competition with other alternatives.

Peat. There are widespread reserves of peat in the United States, the equivalent of 35×10^9 t coe (21). Peat characteristically has a high air-dried moisture content (35–50 wt %) and a high oxygen content (32–38 wt %) which gives it a low heating value. It also has the characteristic of being particularly easy to gasify and studies are being directed along these lines with the possibility of using the gas as the starting material for chemical syntheses as well as its potential use in gas power plants (22) (see Lignite).

Biomass. Proponents of biomass utilization point out that, as fossil hydrocarbon supplies approach depletion, the petrochemical industry will have to base its raw material demands on renewable biomass. This material is stored solar energy (qv) available on a renewable basis. A potential in the order of 500×10^6 t/yr coe in the United States is considered to be optimistic (23). It is not clear to what extent biomass, eg, wood, would be used as a chemical feedstock, although the production of phenol (qv) from lignin (qv) might even today be competitive with phenol from coal. A detailed energy balance and process economics on the conversion of cornstalk residues to ethanol via hydrolysis, fermentation, and distillation is given in ref. 24. The economics are acceptable for the capital investment required at the ethanol market price in 1979. To meet the annual U.S. requirement of ethanol (9×10^5 t/yr) would require <5% of the available corn residues (see Fermentation). It is suggested that other chemical intermediates (acids, aldehydes, and alcohols), produced by use of different microorganisms in the fermenters, also could be economically feasible today (24).

Ethanol can be converted to ethylene by dehydration or to butadiene by dehydrogenation/dehydration. Both processes are used commercially today although the scale is relatively small. They require a low-cost source of ethanol or special circumstances to be attractive.

It has been estimated that the Brazilian rubber tree *Euphorbia* could be grown in the U.S. as a source of liquid hydrocarbons at a cost similar to that of naphtha from petroleum (25) (see Fuels from biomass; Chemurgy).

BIBLIOGRAPHY

1. A. Stuart, *Fortune,* 86 (Feb. 12, 1979).
2. D. P. Burke, *Chem. Week* **125,** 5 (July 4, 1979).
3. F. S. Magnusson, *Petrochemical Feedstocks,* Chemicals and Rubber Program, Office of Basic Industries, Materials Division, Industry and Trade Administration, U.S. Department of Commerce, Washington, D.C., Nov. 1978.
4. "The Petrochemical Industry and the U.S. Economy," *Report to the Petrochemical Energy Group,* C74903, A. D. Little, Inc., Boston, Dec. 1978.
5. *Proceedings Conference on Chemical Feedstock Alternatives, Houston, Oct 2–5, 1977,* Amer. Inst. Chem. Engrs., New York, 1978.
6. *Report to DOE pursuant to Energy Policy and Conservation Act,* MCA, Washington, D.C., April 18, 1979.
7. M. G. Marbach, *Constraint to Olefin Growth?,* paper at Chemical Marketing Research Association, Montreal, Nov. 2–4, 1977.
8. D. F. Fridley, Symposium on the U.S. Polymer Industry, paper at American Chemical Society meeting, Chicago, Aug. 1977; *Chem. Eng. News* **55,** 18 (Sept. 12, 1977).
9. W. L. Nelson, *Oil Gas J.* **77,** 112 (Jan. 8, 1979); **77,** 108 (Jan. 15, 1979).
10. S. B. Zdonik, E. J. Green, and L. P. Hallee, *Manufacturing Ethylene,* The Petroleum Publishing Co., Tulsa, 1971.
11. A. M. Brownstein, *Trends in Petrochemical Technology,* The Petroleum Publishing Co., Tulsa, 1976.

12. W. S. Paley, ed., *Resources for Freedom,* Report to U.S. President Truman by the President's Materials Policy Commission, U.S. Government Printing Office, 5 Vols., June 1952.
13. *Ibid.,* Vol. 2, p. 103.
14. *Ibid.,* Vol. 1, p. 107.
15. T. J. Pollaert, *Synthetic Fuels from Coal,* Vol. 1, presented at Coal Technology '78, Conference Papers, Industrial Presentation, Houston, 1978, p. 766.
16. R. Whitaker, *EPRI Journal* **4**(3), 6 (April 1979).
17. L. J. Buividas, J. A. Finneran, and Q. J. Quartulli, *Chem. Eng. Prog.* **70,** 21 (Oct. 1974).
18. J. R. Dosher, Ref. 5, p. 29.
19. J. W. Hand, Ref. 5, p. 4.
20. J. W. Hand, Ref. 5, p. 5.
21. U.S. Department of Interior, *U.S. Recoverable Energy Resources,* Energy Perspectives, Vol. 2, June 1976.
22. R. M. McGhee, private communication, Science Applications, Inc., Houston, May 1979.
23. K. V. Sarkanan, ref. 5, p. 113.
24. J. L. Gaddy and co-workers, *Ethanol from Agriculture Residues,* presented at 86th National AIChE Meeting, Houston, April 1–5, 1979.
25. *Chem. Week* 44, (Sept. 19, 1979).

General References

Evaluation of World's Important Crudes, Petroleum Publishing Co., Tulsa, 1973.
G. H. Cummings and W. B. Franklin eds., *Declining Domestic Reserves—Effect on Petroleum and Petrochemical Industry,* AIChE Symposium Series no. 127, Vol. 69, New York, 1973.
O. H. Hammond and R. E. Baron, "Synthetic Fuels: Prices, Prospects, and Prior Art," *Am. Sci.* **64,** 407 (1976).
P. H. Spitz and G. N. Ross, "What is Feedstock Worth," *Hydrocarbon Process.* **55,** 143 (April 1976).
"Feedstock Options of the Future," *Eur. Chem. News,* **32**(846), 30 (July 21, 1978).
"U. S. Oil Demand: Details," *Oil Gas J.* **77,** 107 (Jan. 29, 1979).
"W. European Petrochemical Industry and M. East Feedstocks," *Oil Gas J.* **77,** 80 (Jan. 27, 1979).
J. B. O'Hara and co-workers, "Use of Coal for Petrochemical Feedstock," *Hydrocarbon Process.* **57,** 117 (Nov. 1978).

F. A. M. Buck
King, Buck & Associates

Merritt G. Marbach
Shell Chemical Company

FELTS

Felt is a homogeneous fibrous structure created by interlocking fibers using heat, moisture, and pressure. Wool or wool combined with other fibers may be meshed using mechanical work and chemical action while the fibrous mass is kept warm and moist; no weaving is used to make this type of felt, which is often called a *pressed felt*. These felts may be combined with resins or chemicals or laminated with other materials, and many industrial applications result from the cutting, molding, or shaping of such felts.

Needled felts are strong, mechanically interlocked structures created by barbed needles penetrating and compressing synthetic fibers. Such felts may incorporate a woven base (foundation). Needled felts are used for products such as ink rollers or filters, or where chemical inertness or thermal stability may be required. Needled felts, with or without bases, are also used on paper machines for pressing or drying operations.

Woven felts include those felts made with a fabric of a special weave. Wool is usually one of the fibers of these felts, enabling the use of heat, moisture, and mechanical action to create a compact, interlaced structure that may have a deep nap or pile of surface fibers. Applications for these felts are similar to needled felts, although the conventional woven felt is now infrequently used on the paper machine.

Some useful properties of felts include: thermal and chemical stability, depending upon fiber content; high permeability and porosity; vibration and shock absorption; and wear resistance. These and other properties are manipulated by the engineer in fabricating mechanical parts and by the felt designer and paper manufacturer in providing the best felt for a particular paper and machine position. The following describes the manufacture and application of papermaking felts. In addition, a brief description of some other industrial uses of felt is included.

Function of Felts on the Paper Machine

In papermaking, felts of matted hair or wool were used as early as the eleventh century: a mold of woven cloth was dipped into a suspension of paper fibers so that a sheet could be formed; after most of the water had drained away, the mold was pressed against a felt and the sheet transferred or "couched" from the mold to the felt. The paper sheet was then allowed to air dry. By the eighteenth century, woven wool cloths were fulled or felted for the hand manufacture of paper. In 1799, the first paper machine was invented, and soon continuous felts were manufactured for pressing and drying operations (see Paper).

The two paper machine configurations most commonly used today, the Fourdrinier and multicylinder are shown in Figures 1–2. Papermaking generally follows the same principles for both machine types and most paper and board: (*1*) a web of paper is formed from an aqueous suspension of fibers (furnish) by gravity and suction through the screen or fabric; (*2*) the web is transferred to the pressing section where more water is removed by pressure and vacuum; (*3*) the sheet enters the dryer section, where steam-heated dryers and hot air complete the drying process; and (*4*) if desired, the sheet may be finished by coating or calendering before it is wound onto the reel.

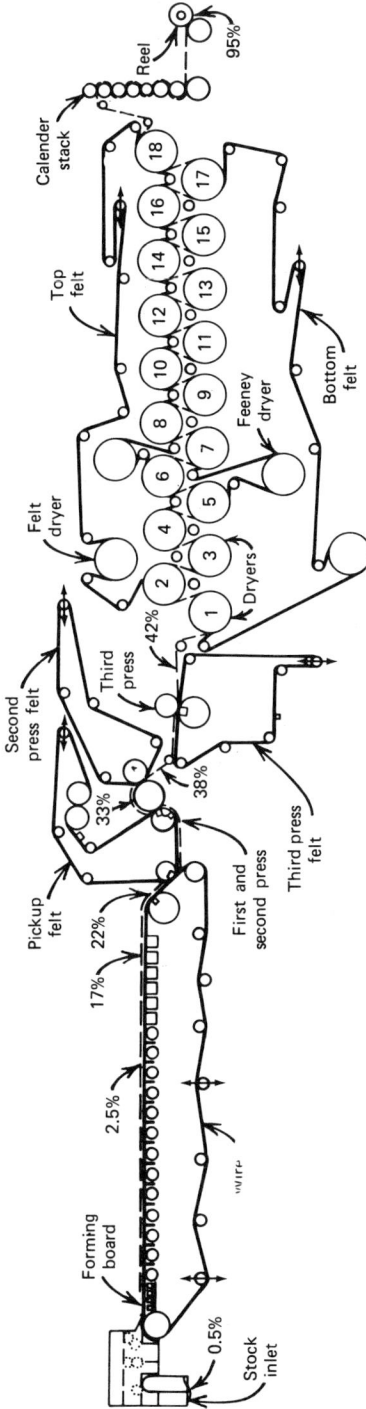

Figure 1. Fourdrinier paper machine. Consistency of paper web represented as percent solids.

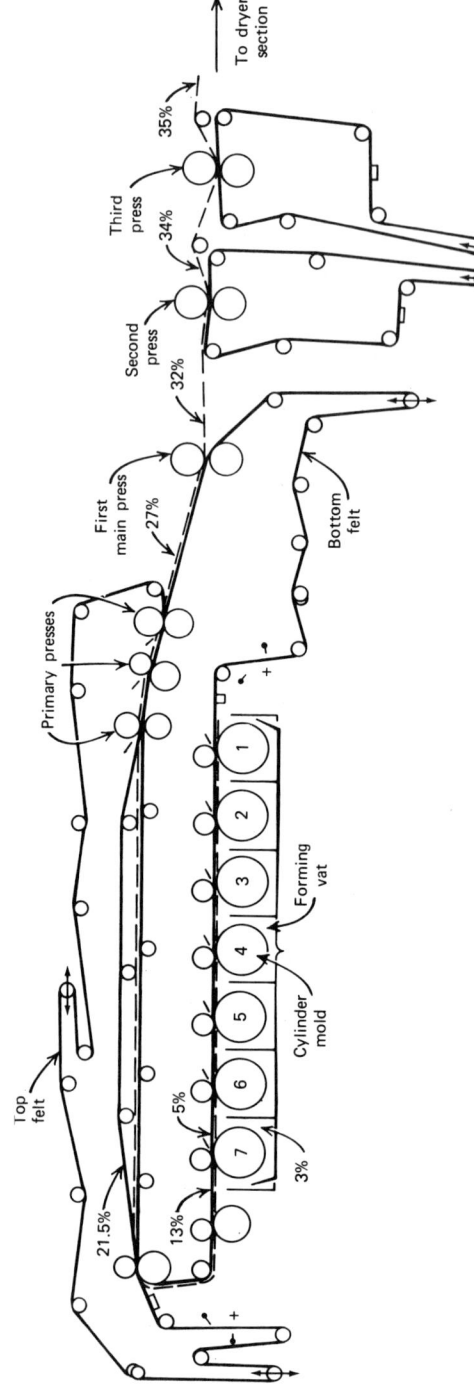

Figure 2. Multicylinder paper machine. Consistency of paper web represented as percent solids.

Manufacture of Felts for Papermaking

Design. Each felt is custom-made and carefully monitored by quality control throughout manufacture.

The following factors are considered in design:

Finish depends upon the competitive demands of the paper grade produced. The fineness of the felt, commensurate with machine type and production rates, is mainly determined by the market in which the paper is sold.

Drainage (for pressing) is the ability of the felt to handle water on the machine. The type of sheet, its filler, machine speed, press roll covers, and conditioning equipment are some variables that must be considered when analyzing drainage requirements. Drainage can be related to permeability, a standard measure of openness. For press felts, permeability gives an indication of expected drainage. For dryer felts and fabrics, it gives an indication of air passage and air pumping characteristics, both of which affect drying. Drainage is also considered with respect to void volume, which is defined in terms of fabric and fiber density; a press felt must have adequate void volume to absorb the water extracted from the sheet in the press nip. Moreover, felt compressibility should be low enough to maintain this adequate void volume.

Durability refers to the life of felts and relates to factors such as: strength, the ability of the felt to withstand operating tension and local strains such as wads going through the press; dimensional stability; and resistance to chemical degradation, heat, abrasion, and compaction.

Other factors such as the ability to run and guide without wrinkling and ease of cleaning must be considered during design.

Felt Classifications. In Tables 1–2, press felts and dryer felts are divided into groups based on constructional characteristics; each construction is divided into classes based upon the degree of fineness or coarseness; applications are also designated. Paper machine felt classifications vary internationally. The press felt categories are those currently recognized in the United States.

Materials. Depending upon the type of felt, the designer may vary the following: type of material; yarn form, weight, and count; weave; amount and fineness of batt (fibrous web); seam, if any; and chemical treatment, if any.

Both natural and synthetic fibers can be used. Table 3 summarizes some important fiber properties. In drying operations, for example, the high temperatures that could degrade the yarns must be accounted for. No one fiber is best for all applications. Basic fiber types and forms can be combined by blending or twisting operations during manufacturing. For example, a fine synthetic multifilament yarn can be twisted with a blended spun yarn for improved tensile strength.

One fiber type is frequently used for a machine direction yarn, and another is used in the cross-machine direction. In press felts, for example, use is made of polyester's excellent resistance to stretching and nylon's superior resistance to friction. Nylon could thus be selected for cross-machine yarns that are more subject to abrasive wear. Since machine direction yarns must withstand the tension on the paper machine, polyester is sometimes used to prevent excess stretching under high load conditions.

Manufacturing Processes. The five general manufacturing categories of yarn making, weaving, burling and joining, needling, and finishing are summarized in Figure 3. Not all felts go through all processes.

Table 1. Classification of Press Felts

Generic name	Constructional characteristics	Classes	Application
conventional press felt	Traditional felt woven of spun yarn in the machine and cross-machine directions and then mechanically felted into final form. Fiber content is predominantly wool with small percentages of synthetics.	coarse	For dissolving pulps, 9 point, roofing, flooring, sulfite, sulfate, or groundwood pulps.
		medium	For kraft, news, wrapping, groundwood specialties, Yankee tissue bottoms, cylinder tissue, Harper machines,[a] binders board, multicylinder tops, bottoms, chip and test liner.
		fine	For lightweight kraft, fine papers, and Fourdrinier board.
		superfine	For Yankee tissue and toweling pickup and various grades of fine paper requiring good sheet finish.
		extra superfine	For fine papers requiring extreme finish.
batt-on-base	Base fabric woven of spun yarns in the machine and cross-machine directions into which is needled a web of fiber batt. Fiber content up to 100% synthetic in batt and base.	coarse	Same as for conventional felts.
		asbestos cement	For top and bottom positions of special cylinder machines making asbestos cement products such as siding, shingles, interior and exterior wallboard and high and low pressure pipe.
		medium	Same as for conventional felts.
		fine	Same as for conventional felts.
		superfine	Same as for conventional felts.
knuckle-free or fillingless	Base fabric of spun machine direction yarns but without cross-machine direction yarns into which is needled a fiber batt. Fiber content up to 100% synthetic in batt and base.	medium	For a wide range of paper and board grades on all positions of both Fourdrinier and multi-ply board machines.
		fine	Same as above except where better sheet finish required.
		superfine	Same as above except for even better finish. Also for Yankee pickup.
		condenser	For condenser papers, this very fine fabric is specially finished to minimize conductivity.
batt-on-mesh	Three basic types: (1) Base fabric woven of fine synthetic multifilament yarns in the machine and cross-machine directions into which is needled a fiber batt. Fiber content is always 100% synthetic.	medium	Same as for conventional felts.
		fine	Same as for conventional felts.
		superfine	Same as for conventional felts.

Table 1 (continued)

Generic name	Constructional characteristics	Classes	Application
	(2) A very permeable, single-layer weave base fabric woven of synthetic multifilament machine direction yarns and synthetic monofilament cross-machine direction yarns into which is needled a fiber batt. Fiber content is always 100% synthetic.		
	(3) Base fabric woven of treated synthetic multifilament yarns in the machine and cross-machine directions into which is needled a fiber batt. The yarn resins when cured produce a relatively incompressible yarn structure. Fiber content is always 100% synthetic.		
combination	A two-layer weave, rigid base fabric of large void volume. Base composed mainly of synthetic monofilament with a portion of multifilaments into which is needled a fiber batt. Fiber content is usually 100% synthetic.		For all types of presses including plain presses for wide range of paper grades; best applied to latter presses or last press before dryer; for suction pickup as well as high speed Yankee pickup or first suction presses to eliminate shadow marking.
nonwoven	Construction without machine or cross-machine direction yarns. It is an all-needled fiber batt. Fiber content is usually 100% synthetic.		For suction, plain, grooved roll, and shrink sleeve presses primarily for fine paper and board machines where good finish is desired.
batt-on-base (with treated base yarns)	Base fabric of resin-treated yarns in the machine and cross-machine directions into which is needled a fiber batt. Fiber content is always 100% synthetic.	medium fine	Same as for conventional felts. Same as for conventional felts.

^a Paper machines.

Yarn Making. Synthetic multifilaments and monofilaments are usually prepared by melting the polymer and forcing it through fine extrusion dies. After extruding, the filaments are stretched to ensure orientation of the molecular chains. This builds high strengths at low elongation into the material. Spun yarns of cotton, asbestos, synthetics, or wool are prepared by standard textile techniques applied in the following order: blending to evenly mix the various components; carding to align the fibers somewhat parallel; and spinning to draft or stretch, twist, and wind the yarn onto bobbins.

Table 2. Classification of Dryer Clothing

Generic name	Constructional characteristics	Application
woolen dryer felt	Fabric woven of spun yarn in the machine and cross-machine directions and then heavily fulled (compacted) to produce a dense, smooth structure. Fiber content can be 100% wool[a] or wool and synthetics.[b]	for cigarette and condenser tissue and photographic paper
conventional dryer felt	Two- or three-layer weave of spun yarns in the machine and cross-machine directions. Fiber content can be cotton[a] and/or synthetics, or a combination of cotton, synthetics, and asbestos.[a]	for all grades of paper
needled batt-on-base dryer felt	Base fabric woven of spun yarns in the machine and cross-machine direction, into which is needled a fiber batt. Fiber content of batt is usually 100% synthetic; base may be cotton and/or synthetic.	for fine papers
needled batt-on-mesh dryer felt	Base fabric woven of synthetic multifilament yarns in the machine direction and monofilament yarns in the cross-machine direction, into which is needled a fiber batt. Fiber content is always 100% synthetic.	for fine papers
open mesh dryer felt	A two- or three-layer weave. Multifilament or spun yarns are woven in the machine direction; spun yarns or asbestos content yarns are woven in the cross-machine direction. Fiber content varies from high cotton-low synthetic to spun Nomex polyamide[c] machine direction yarns; cross-machine direction yarns are often combination asbestos wire.	for most grades of paper
dryer fabric	A two- or three-layer weave. Multifilaments and/or monofilaments are woven in the machine and cross-machine directions. Fiber content is always 100% synthetic.	for all grades of paper

[a] qv.
[b] See Fibers, chemical.
[c] See Aramid fibers; Polyamides.

Weaving. Yarns are woven into different patterns, chosen for dimensional stability, strength, finish characteristics, drainage ability, wear, and bulk. Felts can be woven into two forms, either as a piece of flat material that is later joined together, or in endless form without a seam. Before actual weaving, the warp yarns must be wound onto the warp beam (dressing). During drawing-in, warp ends are individually drawn through the eyes of the heddles in the loom harnesses that control the weave pattern.

Burling and Joining. Most felts are next inspected and burled, during which minor weaving and yarn imperfections are corrected. Felts that are not woven endless are joined by hand. The woven fabric then goes to needling (if it is to be a needled base) or to finishing and treating.

Needling. Mechanical bonding created by needling replaces fulling in conventional felts by creating controlled contraction and stabilization of the felt and by establishing final weight and size characteristics. A needled felt is composed of two parts: the base fabric and batt. The construction of the base fabric (weave, fiber content, etc) helps determine strength and stability. The batt, formed by carding fibers, is laid on the base just before passing through the needling machine. The number of batt

Table 3. General Fiber Properties

Fiber	Form available			Tensile strength	Abrasion resistance	Chemical resistance		Comments	Relevant article in ECT
	Staple	Multi-filament	Mono-filament			Acids	Alkalies		
dacron	X	X	X	excellent next to nylon	good to excellent next to nylon	good	fair to good	Good all-round fiber. Subject to moist heat hydrolysis. Do not use with strong alkalies.	Polyesters
orlon	X	X		good	above average	good	fair	Good combination of abrasion and heat resistance. Use where moist heat conditions prevail. Special heat finishing increases stability.	Acrylonitrile polymers
nylon	X	X	X	excellent	excellent	poor	good to excellent	Excellent abrasion resistance. Best of all fibers. Should be heat-set for high temperature uses. Do not use with strong acids.	Polyamides
nomex	X	X		good to excellent	good to excellent	fair	good	Good stability. Heat resistant.	Polyamides
polypropylene	X	X	X	excellent	excellent	excellent	excellent	Stretches under load even at low temperatures.	Olefin polymers
wool	X			fair	average	poor	poor	Can be felted.	Wool
cotton	X			average	average	poor	excellent		Cotton

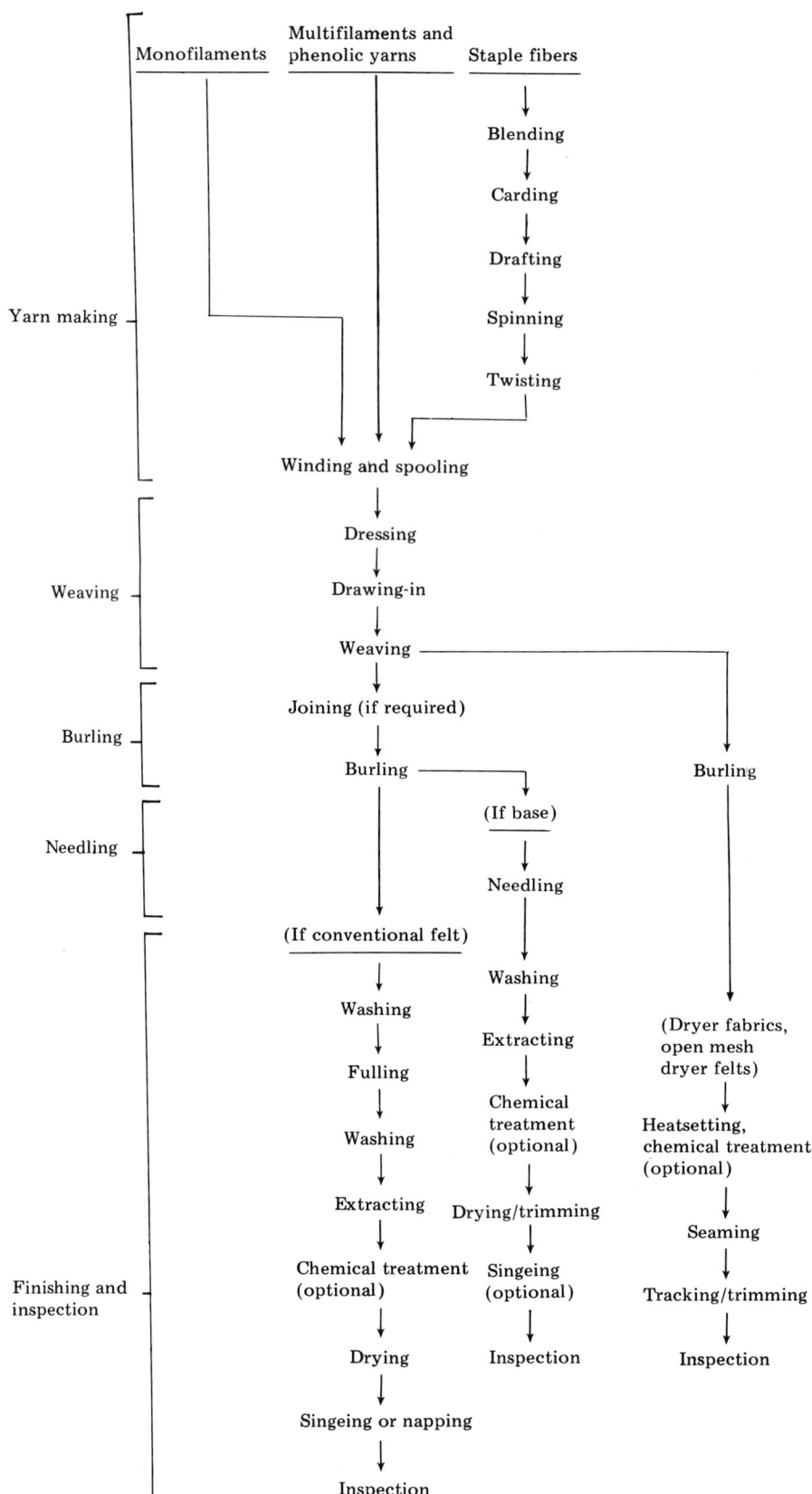

Figure 3. Common manufacturing processes.

layers may be varied and applied to either or both sides. Batt characteristics as well as the amount of penetration of the needles and the number of passes through the needle loom determine the finish, openness, and other characteristics of the felt.

The needles are equipped with tiny barbs facing towards the needle point and are fastened on a vertically reciprocating plate. As they descend, the barbs grasp a few batt fibers and force them into the base fabric (Fig. 4). The fibers are thus locked in the body of the felt and are partially oriented perpendicularly to the plane of the fabric. Each square centimeter of felt may be subjected to more than 400 needle penetrations.

Finishing and Treating. Conventional press felts are not needled but wet with a soap solution and then *fulled* or felted on a rotary fulling mill. This rotary consists of a set of driven rolls through which the felt runs continuously in a roped form. Just before the felt enters the nip of the rolls, it passes through a throat or vertical slot, which can be adjusted in width. The crowding and squeezing that occurs from this combined action causes the felt to shrink in width. As the felt emerges from the nip, it enters an open-ended box called a trap, which has a hinged vane at the top. This vane can be swung down to decrease the size of the opening at the back of the trap box. This results in a partial lengthwise restriction on the felt and is used to control the rate of length shrinkage. Shrinkage may be as great as 20% of the original length and 50%–60% of the starting width (Fig. 5).

This process of controlled felting determines final size and tension of the finished felt by mechanically bonding the individual wool fibers. A felted cloth of wool will change size for two main reasons: (1) wool fibers have scales pointing in one direction, thereby creating a directional friction effect that allows the fibers to move in one direction more easily than the other; (2) wool fibers possess the ability to stretch and recover from stretch when wet. As fulling progresses, the fibers become increasingly intertwined, changing the felt in length and width.

For both conventional and needled felts, the trademark line is applied next, and the felt is washed, extracted, and possibly treated with proprietary chemicals. Some treatments chemically modify the fibers and others coat the fibers and adhesively bond the yarns or fabric structure. Such treatments may be used to retard bacterial or chemical degradation, fiber shedding, abrasive wear, or compaction, or to enhance drainage, wet-up after installation, and stability.

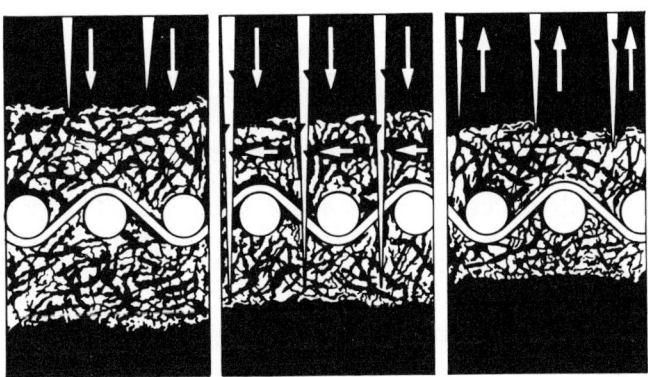

Figure 4. Needling.

856 FELTS

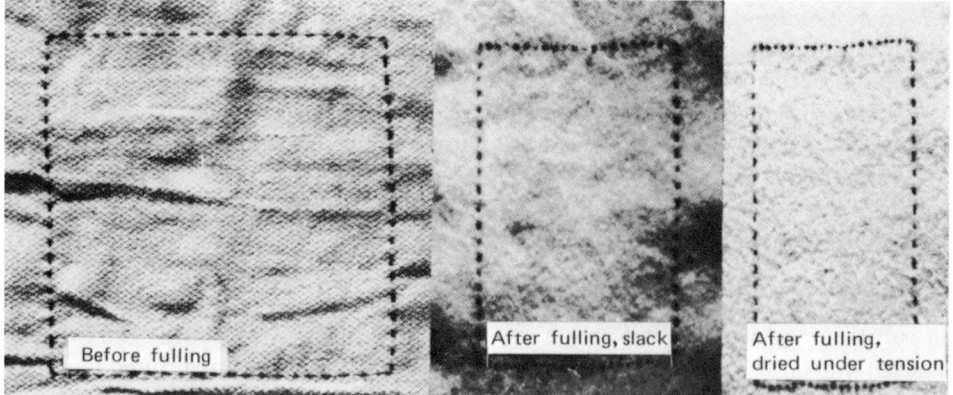

Figure 5. Conventional felt before and after fulling.

Felts are dried around a large steam-heated cylinder and trimmed if required; conventional felts may undergo napping, for extra cushioning or to increase water removal capacity. Singeing may be required to eliminate long surface fibers. Careful inspection follows to check manufacturing specifications, including proper length, width, and tension.

Other Finishing Processes. All-synthetic felts may be heat-set to size and tension. Dryer felts usually require a seam, since few are manufactured in an endless configuration; most are sold with a seam to join mechanically to the paper machine at installation.

Application of Press Felts

Evolution of Press Felt Constructions. In the past twenty years, press felt development has changed dramatically. Before 1960, most press felts were conventionally woven felts, and the plain press and the suction press were commonly found on paper machines. The economic value of the felts was measured by how many days the felt ran on the machine, and most felts, when removed, were worn out. Modern felts of highly synthetic constructions have grown from new manufacturing processes such as needling and from increased knowledge of press efficiency. These highly synthetic felts are seldom removed from the machine because they are worn out; they are generally removed because they are filled or compacted to the point where they no longer handle water uniformly or lack the drainage capabilities to maintain a high level of pressing efficiency.

As the concept of pressing has changed from the plain press to a suction press to the fabric press or shrink sleeve and to the grooved plain press or grooved suction press, changes in felt design have been made simultaneously. The conventional woven felt described in Table 1 is for the most part obsolete, although it remains used on some older machines that have not been updated. Its yarns and fibers are horizontally oriented.

The first significant departure from the conventional woven felt was the batt-on-base felt, a needled felt. The base of this felt resembles a conventional felt, with both machine and cross-machine direction spun yarns. Resistance to water flow within the base therefore remains high and pressing efficiency is limited by the buildup of hydraulic pressure. However, this needled felt is a substantial advance over conventional felts because the needling creates batt fibers that are more or less vertically oriented. Thus flow resistance in the vertical direction is much lower than in the conventional woven felt. Needled (batt-on-base) felts are widely used today; many are completely synthetic. If they can be kept open, their running time can be very long.

The knuckle-free or fillingless felt described in Table 1 has no cross-machine direction yarns, thereby lowering resistance to water flow in the machine direction and removing a major trap for paper fines and dirt. This was followed in its development by the various batt-on-mesh designs, which use an open mesh screen-type base fabric of multifilament and sometimes monofilament yarns. Such yarns reduce water flow resistance in the base and create a soft, pliable structure, the latter characteristic being desirable for ease of installation in certain machines. The batt-on-mesh felt with multifilaments in the machine direction and monofilaments in the cross-machine direction provides a very stable structure, a highly permeable base, and an excellent finish for fine paper applications on certain presses.

A further design development, the combination felt, has low flow resistance in the base and high water storage capability. These are the most effective press felts today to provide significant sheet dryness gains and improved moisture profiles on all types of presses including plain presses.

A recent development is the nonwoven felt, which eliminates the base weave to give the best possible openness and permeability; the absence of any type of organized yarn structure eliminates sheet marking and filling with dirt and paper fines, while providing the most uniform pressure distribution from felt to sheet in the press nip. The nonwoven felts are applicable to suction, plain, grooved roll, and shrink sleeve presses primarily for fine paper and board machines where good finish is desired (see Nonwoven textiles).

Press Felts for Papers and Board. Table 4 lists some felts used on typical machines; the example given are represented with large ranges in basis weights. Nomenclature for presses and felts can generally be seen from Figures 1–2. The following discussion mentions machines important for producing some selected paper grades, requirements for specific felt positions, and general characteristics of such felts.

Fine Paper. This category generally includes coated and uncoated publication, tablet, bond, litho, copying, business, cover stock, photographic, machine-glazed and modified-finish tissue, one-time carbon, and numerous other fine specialty grades.

Many two- and three-press open-draw Fourdriniers are still used for the production of fine paper. In the early 1950s, when the demand for fine paper grades exceeded the capability of open draw machines, the suction transfer came into use and provided a better transfer with fewer sheet breaks and, therefore, higher speeds. Next to be developed were the inverted first press and three-roll inclined press, the latter currently the most widely used configuration.

The suction-pickup felt for fine paper requires good openness, resistance to compaction, and enough surface fineness to promote a water-film bond between the sheet and felt; subsequent surface tension improves pickup at the wire and retards sheet loss. The pickup position of any fine paper machine configuration uses a relatively open base structure with a uniform layer of relatively permeable face-side batt. The batt is introduced to the inside surface to control an existing or potentially dangerous

Table 4. Felt Application Chart, Typical Felts on Their Representative Machines

Paper grade	Machine	Position	Machine speed, m/s (ft/min)	Press load, kN/m (pli[a])	Weight per unit area, kg/m² (oz/ft²)
fine paper	three-roll inclined press	suction pickup	6.1–14 (1200–2800)	35–61 (200–350)	0.92–1.1 (3.0–3.7)
		second press	6.1–14 (1200–2800)	39.4–65.7 (225–375)	0.98–1.2 (3.2–3.8)
		third press	6.1–14 (1200–2800)	44–79 (250–450)	0.98–1.2 (3.2–4.0)
tissue	Yankee Fourdrinier (two-felt)	Yankee pickup	7.6–23 (1500–4500)	58–96 (330–550)	1.0–1.2 (3.4–3.9)
		Yankee bottom	7.6–23 (1500–4500)	30.6–44 (175–250)	0.79–.88 (2.6–2.9)
board	multicylinder	top	0.76–3.3 (150–650)	26–44 (150–250)	0.61–0.82 (2.0–2.7)
		bottom	0.76–3.3 (150–650)	26–44 (150–250)	0.76–0.92 (2.5–3.0)
		press	0.76–3.3 (150–650)	35–149 (200–850)	0.76–1.4 (2.5–4.5)
pulp	three-press Fourdrinier	third press (grooved)	1.3–2.5 (250–500)	175–263 (1000–1500)	1.2–1.3 (3.9–4.3)
asbestos cement	two-cylinder	bottom	0.56–0.81 (110–160)	18–44 (100–250)	1.4–1.5 (4.5–5.0)

[a] pli = pounds-force per lineal inch.

wear condition. The first press felt of open draw machines have characteristics similar to the pickup designs.

At the second and third presses, less water and fines are available in the sheet and the potential filling problem is reduced. The base fabrics can be closed up to present a smoother surface for uniform pressing. Because of the higher press loadings associated with the latter presses, the thickness of the batt layer is increased to eliminate the possibility of base or roll strike through. All constructions can be used on all machines, with emphasis on all-synthetic felts with batt-on-mesh, combination, or nonwoven constructions.

Newsprint and Offset. Newsprint is usually a lighter weight sheet than the average fine paper grade, and generally runs at higher speeds. Stock for newsprint is primarily composed of groundwood and is characterized by short fibers and a percentage of long fibered chemical pulp for added strength. Machines used for these grades are similar to those producing fine paper, with the three-roll inclined press and three-roll direct pickup press predominating.

Pickup felts for newsprint and offset should be permeable and not susceptible to filling: hence, batt-on-mesh designs are preferable. The three-roll direct pickup press does not require pickup quality to the same extent as pickup felts for other configurations. For other positions on all machine types, synthetic batt-on-base, batt-on-mesh, and combination felts are applicable. Compared to fine paper, these felts can be slightly heavier on the latter presses because of increased loading.

Tissue. Machines producing tissue are often referred to as Yankee Fourdriniers and produce grades of toilet, facial, napkin, and toweling tissues. The Yankee Fourdrinier with two felts has traditionally predominated, but there is a trend toward the single felt because it eliminates the bottom felt with its accompanying equipment and maintenance requirements. Pickup is generally not vacuum-assisted; thus characteristics are different from the suction pickup for fine paper. The Yankee pickup must be hard and dense with a superfine or very fine surface so that good pickup can be achieved with a uniform surface water film on the pickup. Batt-on-base Yankee pickup for a two-felt configuration has a synthetic content of 40–100%. Combination felts provide good pickup without requiring as much shower water to condition the felts, compared to other constructions.

The conventional woven Yankee bottom felt has been replaced mostly by the needled felt. Since openness, free drainage, and cushion are important, this felt is bulky and coarse compared to the Yankee pickup. Treated yarns in a batt-on-mesh or batt-on-base construction help maintain openness and drainage. Synthetic content is 50–100%.

Multi-ply Board. Grades included in this category are container board, folding boxboard, nonfolding boxboard, solid bleached board, and other board such as gypsum board. Many paperboard grades have a top liner, underliner, filler, and backliner, and are produced on machines having more than three cylinder vats. Most machines commonly have seven or eight cylinders.

The bottom felt on multicylinder machines, which picks up the web plies, must be open to allow for drainage, smooth enough to impart a desirable finish, strong and stable because of the higher press loadings and higher speeds, and resistant to the action of high pressure showers, which can wear the inside of the felt.

The top felt is very much like the bottom felt in construction, but is usually much lighter. Finish is more important, thus yarns are smaller. The top felt must also be open to prevent sheet-crushing. Tension requirements are not as severe as on the bottom felt. It is important that a balance in design of the top and bottom felts be achieved, so that both can handle the water adequately.

For the press felts, higher loaded presses, mainly grooved, are increasing in use. Therefore, heavier weight felts are needed to absorb more water from the sheet without crushing, and to impart good finish. New constructions such as batt-on-mesh, combination, an nonwoven felts are largely replacing batt-on-base.

Pulp. The pulp (qv) sheet is rarely used as a finished product and is generally converted to other grades. Pulp sheets used outside of the paper industry result in varied finished products including rayon (qv) yarn, cigarette filters (see Cellulose acetate), cellophane (see Film and sheeting materials), photographic and x-ray films (see Photography), plastic articles (see Plastics), lacquers, explosives (qv), and nonwovens (see also Cellulose). Most pulp machines are three-press Fourdriniers. To aid in drying, a short section of four to eight dryers sometimes precedes the third or hot press. High press loadings are also typical, particularly on the third press, which might have a loading as high as 263 kN/m (1500 pli). In some cases, up to 50% dryness out of the press section has been achieved.

It is particularly important that felts used on pulp machines provide good drainage; hence, such felts are coarse and open. As with most felts, long life is important. Pulp felts must also be bulky to withstand heavy press loadings. Finish is generally not a requirement. Needled felts have replaced conventional woven felts in most cases. Synthetic content is increasing, and all-synthetic batt-on-mesh and combination constructions are being used extensively.

Other Industrial Applications of Felt

Felt can be fabricated to produce parts to precise specifications for many diverse functions. Parts may be molded under heat and pressure, often resin-impregnated for stability and to provide special properties. For special shapes, felts may easily be cut without fraying and denser grades may be machined. Fiber content and felt construction can be varied for the application; impregnations and laminations are used to increase adaptability. Some varied functions of these fabricated felts are listed below.

The vibration and shock absorbing properties of wool felt suggest its application as a mounting under precision instruments; denser felts are used to accommodate heavier usage (see Wool). Felt can also be an effective sound absorber when applied as a surfacing, such as to line an air conditioner or a section of a typewriter or computer. Sound absorption depends on fiber content, felt construction, density, and air permeability (see Noise pollution; Insulation, acoustic).

The natural capillarity of felt enables it to be a liquid reservoir and medium for the transfer of fluids to lubricate bearings and other mechanical parts, particularly in closed systems. Felt seals may be used to retain grease or oil, while eliminating dust, moisture, and contaminants; other sealing applications include use in vacuum cleaners and weather stripping. From the wide variety of fibers and constructions, felts can be adapted to many types of wet and dry filtration (qv), such as respirators as well as filters for products ranging from vegetable oils (qv) to adhesives (qv). Another application is polishing felts; felts used in precision polishing of optical lenses provide a delicate abrasive surface that is long lasting and uniform.

Acknowledgement

Much of the text and all tables and figures are derived from *Paper Machine Felts and Fabrics,* Albany International Corp., Albany, New York, 1976; this is a fine reference for further reading.

BIBLIOGRAPHY

"Felt" in *ECT* 1st ed., Vol. 6, pp. 313–316, by Ivor Griffith, Philadelphia College of Pharmacy and Science.

General References

J. B. Casey, *Pulp and Paper; Chemistry and Chemical Technology,* 2nd ed., 3 Vols. Interscience Publishers, Inc., New York, 1961.
G. H. Ireland, *Paperboard on the Multi-Vat Cylinder Machine.* Chemical Publishing Company, Inc., New York, 1968.
Nuttall, G. H. *Theory and Operation of the Fourdrinier Paper Machine.* S. C. Phillips & Co., Ltd., London, 1968.
W. E. Becker, "Designing with Felt," *Mach. Des.,* 3 (June 26, 1969).
H. W. LoFaro, "Engineers Guide to Felt," *Mater. Eng.,* 34 (Nov. 1971).
F. M. Bolam, ed., *Design of the Wet End of the Papermachine,* Monographs in Paper and Board Making, Vol 3, Ernest Benn Limited for the Technical Section, The British Paper and Board Makers' Association, London, 1971.

J. D. Parker, *The Sheet-Forming Process,* Special Technical Association Publication, no. 9. Technical Association of the Pulp and Paper Industry, Atlanta, 1972.

J. F. Atkins, ed., *Paper Machine Wet Press Manual,* TAPPI Monograph Series, no. 34. Technical Association of the Pulp and Paper Industry, Atlanta, 1972.

Papermachine Clothing, Monographs in Paper and Board Making, Vol 5. Ernest Benn Limited for the Technical Section, The British Paper and Board Industry Federation, London, 1974.

J. Heller, *Papermaking.* Watson-Guptill Publications, New York, 1978.

MARY K. PORTER
Consultant, Troy, N. Y.
and the Albany International Corp.

FERMENTATION

In the last fifteen years, the fermentation industries have expanded both in number of products prepared on a commercial scale and in fermentation capacity: the number of products has more than doubled, and the volume increased about fivefold. Although more than 250 companies are using fermentation processes in the manufacturing operations, and the products of interest are quite dissimilar entities (ranging from amino acids to enzymes), the technology used for their production is somewhat standardized.

Although the word fermentation referred originally to the anaerobic metabolism of organic compounds by microorganisms or their enzymes to produce products simpler than the starting material, the modern definition is that of any microbial action controlled by man to make useful products. Some of the substances produced from carbohydrates on a commercial scale by anaerobic microbial metabolism include ethanol (qv) and lactic acid (see Hydroxy carboxylic acids) and products containing these materials such as beer (qv), wine (qv), vinegar, pickles, and sauerkraut (see also Food processing). Products from aerobic metabolism include antibiotics (qv), organic acids (see Carboxylic acids), enzymes (qv), and vitamins (qv). These have made the fermentation industry a multibillion dollar operation and a strong competitor to the chemical industries (see also Chemurgy).

Molds, yeasts (qv), bacteria, and streptomycetes are among the microorganisms used in fermentation processes today (see Microbial transformations). In comparison with chemical processes, most of the successful fermentation operations have some degree of simplicity, a rather uniform series of techniques, and the possibility of using the same manufacturing facilities to make many chemically unrelated substances. Other advantages may also include less expensive raw materials, an energy conservation aspect in that biological reactions are carried out at lower temperatures and have lower energy barriers and a higher degree of efficiency than may be achieved in a multistep chemical synthesis; and a higher degree of specificity and quality of final product than obtainable by chemical synthesis (which is often more costly if at all feasible, eg, vitamin B_{12}).

Table 1. A Century of Growth of Industrial Fermentations[a]

Year of commercialization	Fine chemicals	Enzymes	Therapeutic substances	Microbial transformations	Other
1880–1900	lactic acid				bakers' yeast
1900–1910	ethanol glycerol	amylases			
1910–1920	acetone n-butanol	invertase			
1920–1930	citric acid				
1930–1940	gluconic acid	proteinases	riboflavin	sorbose acyloins	
1940–1950	itaconic acid 2-keto-D-gluconic acid	cellulases pectinases	vitamin B_{12} penicillin G bacitracin streptomycin 7-chlortetracycline tyrothricin gramicidin polymyxin neomycin		
1950–1960	kojic acid glutamic acid lysine	glucose oxidase catalase	amphotericin B colistin cycloheximide cycloserine erythromycin griseofulvin kanamycin novobiocin nystatin oleandomycin penicillin V semisynthetic penicillins oxytetracycline polymyxins tetracycline demeclocycline variotin viomycin	steroid oxidations	gibberellins dextran single-cell protein
1960–1970	valine	glucose isomerase glucamylase lipases lactase	amphomycin blasticidin S cactinomycin candicidin cephalosporins dactinomycin fusidic acid gentamicins hygromycin B kitasatomycin lincomycin mikamycins monensin pimaricin polyoxins pristinamycins ribostamycin rifamycins siomycin	dihydroxy-acetone	xanthan 5'-nucleosides bioinsecticide

Table 1 (*continued*)

Year of commercialization	Fine chemicals	Enzymes	Therapeutic substances	Microbial transformations	Other
1970–1977	comenic acid	microbial enzyme as rennet replacement	spectinomycin thiostrepton tylosin vancomycin adriamycin	sterol cleavage	ribose
		melibiase dextranase	avoparcin bambermycin bicyclomycin bleomycin candidin capreomycin daunorubicin enduracidin fortimicin josamycin lasalocid lividamycin macarbomycin mepartracin medicamycin mocimycin myxin quebemycin sagamicin salinomycin sisomicin tetranactin thiopeptin tobramycin validamycin A	xylitol aspartic acid malic acid	zearalenol

[a] Courtesy of *Chemtech* (1).

The fermentation industries in the United States (and most of the industrialized world) date back to about 1880 when a small-scale lactic acid production was started in Avery, Massachusetts. As indicated in Table 1, during the initial phase of development (1880–1910) fermentation products had immediate application in foods (lactic acid and bakers' yeast (see Bakery processes)), as a raw material for chemical processing (ethanol) and in the textile industry (amylases for desizing of textiles). Certain practices were based on experience rather than science; although the industries prospered, the element of "practical witchcraft" limited development of the science.

Wartime needs for acetone (qv) in 1914–1918 stimulated some practical research and led to the establishment of a substantial scientific base for the fermentation industries' development from 1920–1940. The requirements of the Weizmann process (acetone–butanol fermentation) for sterile media, pure culture technique, and an understanding of the bacterial physiology required the combined efforts of biochemists, engineers, bacteriologists, and chemists to bring this rather complicated fermentation from laboratory scale to production. This experience demonstrated the value of a team approach in solving fermentation problems.

Development of chemical industries in the United States in the post-World War I era also had a stimulating effect on the fermentation industries, as efforts were initiated to develop processes that were new and different from those already in operation in Europe. Early success of the mycological (mold) processes for production of citric acid, gluconic acid, and continued success of the acetone–butanol fermentation process convinced some chemical producers that microbial processes could be dependable and economically successful.

The interest in vitamins to supplement human and animal diets stimulated the search for fermentation processes for riboflavin, thiamine, vitamin C, and other therapeutically useful compounds. This enthusiasm increased when the first reports on penicillin suggested that this microbial product would have tremendous impact on the therapy of infectious diseases.

The 1945–1960 period saw further exploitation of fermentation technology for production of antibiotics, enzymes, and additional organic acids with commercial potential. Some of the more remarkable practical developments included: microbial processes for glutamic acid and lysine; commercial production of the anti-inflammatory agents including cortisone, hydrocortisone, prednisolone, and triamcinolone, based on selective microbial oxidation of the 1,2-cyclopentenophenanthrene nucleus; and large-scale preparation of yeast and other organisms as protein sources in human and animal diets.

The list of new fermentations activated during the 1960–1980 period include about 50 antibiotics, and several enzymes, polysaccharides, flavor-enhancing purine nucleosides, microbial insecticides, and fragrances. Of special interest are widespread application of glucose isomerase to produce high fructose syrups which have displaced the more expensive sucrose in many industrial food applications, and the use of immobilized microbial cells to produce amino acids and malic acid on an industrial scale (see Enzymes, immobilized). More than half of the protein-coagulating proteases used in manufacturing various cheeses in the United States in 1980 are mold enzymes which have replaced and supplemented the limited amount of calf-stomach-derived rennet for this purpose.

There are currently more than 150 companies worldwide using fermentation processes for production of fine chemicals (qv) and therapeutic substances, and nearly 100 companies manufacturing yeast for food or feed purposes (see Enzymes, therapeutic). Some of these companies are listed in Tables 2 and 3. Although many are multinational in character, they are mentioned only once in the listings in order to reduce confusion in identification, eg, Pfizer Inc. has fermentation operations in at least 7 countries, and in several countries the subsidiary company is a joint effort between Pfizer and a local partner (see Citric acid).

A number of the products mentioned in Table 2 are intended as starting materials for chemical processing. Some of these are listed in Table 4. In some instances, fermentation is the only route to the starting material, whereas in others, it has proved to be the most economical.

Table 2. Fermentation Products, 1977[a,b]

Product	Some producers
Antibiotics	
adriamycin	42
amphomycin	79
amphotericin B	124
avoparcin	7
azalomycin F	115
bacitracin	11; 13; 39; 62; 79; 90; 94; 108; 123
bambermycins	56
bicyclomycin	49
blasticidin S	66
bleomycin	88
cactinomycin	41
candicidin B	39: 79; 94
candidin	79
capreomycin	37
cephalosporins	9; 20; 23; 30; 42; 46; 48; 51; 65; 74; 76; 90; 98; 101; 107; 124; 132; 136
chromomycin A$_3$	132
colistin	13; 14; 39; 70; 73; 107
cycloheximide	66; 140
cycloserine	5; 62
dactinomycin	83; 105
daunorubicin	42; 107
destomycin	80
enduracidin	132
erythromycin	1; 6; 8; 12; 22; 23; 30; 31; 37; 44; 65; 76; 95; 98; 101; 111; 134; 140
fortimicins[c]	73
fumagillin	1; 25
fungimycin	79
fusidic acid	75
gentamicins	23; 25; 118
gramicidin A	79; 94; 141
gramicidin J(S)	80; 89
griseofulvin	23; 51; 63; 75; 88; 132
hygromycin B	18; 76; 132
josamycin	116; 146
kanamycins	6; 20; 23; 38; 80; 88; 98; 107
kasugamycin	15; 59; 80; 116
kitasatamycin	136
lasalocid[d]	57
lincomycin	140
lividomycin	69; 107
macarbomycin	80
mepartricin[c]	122
midecamycin	80
mikamycins	14; 67
mithramycin	95
mitomycin C	20; 73
mocimycin[e]	50
monensin	76
myxin	57

Table 2 (*continued*)

Product	Some producers
neomycins	10; 11; 18; 23; 42; 79; 88; 94; 95; 107; 111; 124; 132; 140
novobiocin	80; 107; 140
nystatin	7; 23; 25; 98; 107; 124
oleandomycin	95
oligomycin B	105
paromomycins	93
penicillin G	3; 4; 5; 10; 13; 14; 15; 17; 20; 23; 46; 50; 51; 52; 55; 56; 64; 76; 80; 83; 90; 95; 98; 107; 121; 124; 128; 136; 145
penicillin V	1; 3; 4; 5; 14; 17; 18; 20; 37; 46; 50; 51; 56; 76; 80; 90; 95; 107; 124; 136; 145
penicillins (semisynthetic)	3; 10; 14; 15; 17; 20; 23; 30; 41; 46; 49; 50; 64; 65; 75; 80; 83; 95; 98; 103; 124; 132; 136; 145
pentamycin	89
pimaricin	50
polymyxins	39; 90; 95
polyoxins	66; 116
pristinamycins	107
quebemycin[e]	73; 107
ribostamycin	80
rifamycins	12; 47; 53
sagamicin[c]	73
salinomycin	66
siccanin	115
siomycin	120
sisomicin[c]	41; 118
spectinomycin	1; 66; 140
spiramycin	73; 107
streptomycins	6; 10; 23; 50; 51; 80; 83; 90; 95; 107; 128; 145
Tetracyclines	
chlortetracycline	7; 9; 23; 36; 42; 55; 64; 78; 101; 104; 107; 120; 122; 132
demeclocycline	7; 30; 98; 101; 107; 132
oxytetracycline	9; 18; 23; 25; 30; 45; 50; 63; 65; 71; 72; 74; 95; 99; 101; 104; 123; 134
tetracycline	3; 7; 9; 10; 12; 14; 20; 21; 22; 23; 26; 30; 31; 36; 42; 44; 45; 53; 55; 56; 59; 61; 64; 65; 77; 90; 95; 96; 101; 103; 104; 107; 111; 122; 123; 124; 132; 134; 140
Miscellaneous antibiotics	
tetranactin[f,g]	27
thiopeptin[e]	49; 66
thiostrepton	124
tobramycin[c]	18; 76
trichomycin	49
tylosin	37; 76
tyrothricin	17; 79; 94; 141
tyrocidine	79
uromycin	105
validamycin A	132
vancomycin	76
variotin	74; 88
viomycin	136
virginiamycin	106
Enzymes	
amylases	18; 19; 29; 30; 48; 50; 54; 80; 85; 87; 90; 100; 102; 109; 110; 115; 125; 141
amyloglucosidase	29; 32; 48; 50; 51; 85; 90; 141
anticyanase	133

Table 2 (*continued*)

Product	Some producers
L-asparaginase	41; 73
catalase	19; 35; 43; 73; 136
cellulase	3; 19; 50; 80; 82; 85; 90; 109; 110; 141
dextranase	90; 115
diagnostic enzymes	73; 85; 136
esterase–lipase	141
glucanase	90; 141
glucose dehydrogenase	82
glucose isomerase	8; 29; 32; 50; 63; 85; 87; 90; 125
glucose oxidase	19; 35; 43; 50; 85; 87
glutamic decarboxylase	73
gumase	100
hemicellulase	85; 90; 110; 141
hespiriginase	133
invertase	50; 78; 82; 115; 125; 139; 141
lactase	50; 85; 90; 141
lipase	19; 110; 133; 136; 141
microbial enzyme as rennet replacement	34; 81; 85; 90; 95; 141
naringinase	133; 141
pectinase	19; 85; 90; 109; 110; 115; 133; 136; 141
pentosanase	110; 141
proteases	18; 19; 25; 50; 66; 73; 80; 82; 85; 90; 95; 100; 109; 110; 115; 132; 136; 141
streptokinase–streptodornase	7
uricase	90
Organic acids	
citric acid	16; 19; 28; 85; 95; 107; 114; 127; 129
comenic acid	95
erythorbic acid	49
gluconic acid	16; 20; 49; 73; 82; 95; 100; 144
itaconic acid	95; 107
2-keto-D-gluconic acid	73; 82; 95
α-ketoglutaric acid	2; 73
lactic acid	19; 29; 33; 107
malic acid	2; 73; 133
urocanic acid	2; 133
Solvents	
ethanol	48; 50; 52; 73; 102; 116; 130
2,3-butanediol	50
Vitamins and growth factors	
gibberellins	1; 63; 73; 76; 83; 98; 132
riboflavin	83
vitamin B_{12}	6; 25; 42; 51; 83; 107; 108; 110
zearalanol	62
Nucleosides and nucleotides	
5′-ribonucleotides and nucleosides	2; 13; 73; 132; 147
orotic acid	73
ara-A-(9-(β-D-arabinofuranosyl)adenine	93
6-azauridine	73
Amino acids	
L-alanine	2; 73; 133
L-arginine	2; 73
L-aspartic acid	2; 21; 73; 91; 107; 133
L-citrulline	2; 73; 133
L-glutamic acid	2; 13; 73; 84; 90; 116; 126; 132; 137; 138; 142; 143; 144
L-glutamine	2; 73

Table 2 (*continued*)

Product	Some producers
L-glutathione	73; 136
L-histidine	2; 73
L-homoserine	73
L-isoleucine	2; 73; 133
L-leucine	2; 73
L-lysine	2; 40; 73; 107
L-methionine	73; 133
L-ornithine	2; 73
L-phenylalanine	73; 133
L-proline	2; 73; 133
L-serine	2; 73
L-threonine	2; 73
L-tryptophan	73; 133
L-tyrosine	73
L-valine	2; 67; 133
Miscellaneous products and processes	
acetoin	50
acyloin	68; 83
anka-pigment (red)	131
blue-cheese flavor	34
desferrioxamine	47
dextran	3; 18; 80; 97
diacetyl (from acetoin)	50
dihydroxyacetone	82; 95; 111; 118; 133; 141
ergocornine	17; 42; 108; 112
ergocristine	17; 42; 112
ergocryptine	17; 42; 108; 112
ergometrine	108
ergotamine	42; 60; 108
Bacillus thuringiensis, insecticide	1; 73; 107; 113
lysergic acid	42; 60
paspalic acid[f]	17; 113
picibanil[f,h]	27
ribose[f]	2
scleroglucan[f]	91
sorbose (from sorbitol)	57; 58; 82; 92; 95; 132
starter cultures	34; 83; 85
sterol oxidations	50; 86; 117; 119; 140
steroid oxidations	47; 50; 51; 82; 107; 108; 117; 118; 124; 140; 145
xanthan	12; 18; 78; 83; 91; 95; 107; 135

[a] Courtesy of *Chemtech* (1).
[b] Index to fermentation companies:
  1. Abbott Laboratories, North Chicago, Ill.
  2. Ajinomoto Company, Tokyo, Japan
  3. Aktiebolaget Astra, Södertalje, Sweden
  4. Aktiebolaget Fermenta, Strangas, Sweden
  5. Aktiebolaget KABI, Stockholm, Sweden
  6. Alembic Chemical Works Co., Ltd., Baroda, India
  7. American Cyanamid, Wayne, N.J.
  8. Anheuser-Busch, Inc., St. Louis, Missouri
  9. Ankerfarm S.p.A., Milano, Italy
  10. Antibioticos S.A., Madrid, Spain
  11. Apothekernes Laboratorium für Specialpraeparater A/S, Oslo, Norway
  12. Archifar S.p.A., Milano, Italy
  13. Asahi Chemical Industry, Tokyo, Japan
  14. Banyu Pharmaceutical Company, Tokyo, Japan

Table 2 (*continued*)

15. Beecham Pharmaceutical Company Ltd., Surrey, Eng.
16. Joh. A. Benckhiser G.m.b.H., Ludwigshafen/Rhein, FRG
17. Biochemie G.m.b.H., Kundl/Tirol, Austria
18. Biogal, Debrecen, Hungary
19. C. H. Boehringer Sohn, Ingelheim/Rhein, FRG
20. Bristol-Myers Company, Syracuse, N.Y.
21. Carlo Erba S.p.A., Milano, Italy
22. Chemibiotic Ltd., Inishannon, Ireland
23. China National Chemicals Import and Export Corporation, Peking, People's Republic of China
24. Chinese Petroleum Corporation, Taipei, Republic of China
25. Chinoin, Budapest, Hungary
26. Chong-Kun-Dong Corporation, Seoul, South Korea
27. Chugai Pharmaceutical Company, Tokyo, Japan
28. Citrique Belge, Tienen, Belgium
29. Clinton Corn Processing Company, Clinton, Iowa
30. Companhia Industrial Produtora de Antibioticos S.A.R.L. (CIPAN) Lisboa, Portugal
31. Compania Espanola de la Penicillina y Antibioticos S.A., Aranjuez, Spain
32. CPC International Inc., Argo, Ill.
33. Croda Bowmans Chemicals Ltd., Cheshire, Eng.
34. Dairyland Food Laboratories Inc., Waukesha, Wisc.
35. Dawe's Laboratories Inc., Chicago Heights, Ill.
36. Diaspa S.p.A., Coranna, Italy
37. Dista Products Ltd., Liverpool, Eng.
38. Dong-Myung Industrial Company, Ltd., Seoul, South Korea
39. Dumex Ltd., Copenhagen, Denmark
40. Eurolysine Company, Paris, Fr.
41. Farbenfabriken Bayer A.G., Wuppertal, FRG
42. Farmitalia S.p.A., Milano, Italy
43. Fermco Biochemics Inc., Elk Grove Village, Ill.
44. Fermentfarma S.p.A., Milano, Italy
45. Fermic S.A. de S.V., Ixtapalapa, Mexico
46. Fermion Oy, Tapiola, Finland
47. Fervet S.p.A. (division of CIBA-Geigy Ltd.) Torre Annunziata, Italy
48. Finnish State Alcohol Monopoly, Helsinki, Finland
49. Fujisawa Pharmaceutical Company, Osaka, Japan
50. Gist-Brocades, Delft, Holland
51. Glaxo Laboratories Ltd., Greenford, Eng.
52. Grain Processing Corporation, Muscatine, Iowa
53. Gruppo Lepetit S.p.A., Milano, Italy
54. Hailsun Chemical Company, Ltd., Taipei, Republic of China
55. Hindustan Antibiotics Ltd., Pimpri, India
56. Hoechst A.G., Frankfurt Hoechst, FRG
57. Hoffmann-La Roche Inc., Nutley, N.J.
58. F. Hoffman-La Roche and Company, Basle, Switz.
59. Hokko Kagaku Koygo Company, Tokyo, Japan
60. Huhtamäki Oy, Turku, Finland
61. ICN-Chimica S.p.A., Milano, Italy
62. IMC Chemicals Group Inc., Terre Haute, Ind.
63. Imperial Chemical Industries Ltd., Manchester, Eng.
64. I.S.F. S.p.A., Rome, Italy
65. Istituto Biochemico Italiano, Milano, Italy
66. Kaken Chemical Company, Tokyo, Japan
67. Kanegafuchi Chemical Industries, Osaka, Japan
68. Knoll G.m.b.H., Ludwigshafen, FRG
69. Kowa Company, Nagoya, Japan
70. Kayaku Antibiotics Research Company Ltd., Tokyo, Japan

Table 2 (*continued*)

71. Krakow Pharmaceutical Works ("Polfa"), Krakow, Poland
72. KRKA Pharmaceutical and Chemical Works, Novo Mesto, Yugoslavia
73. Kyowa Kakko Kogyo Company, Tokyo, Japan
74. Lark S.p.A., Milano, Italy
75. Leo Pharmaceutical Products, Ballerup, Denmark
76. Eli Lilly and Company, Indianapolis, Ind.
77. Linson Ltd., Dublin, Ireland
78. Lohmann and Company A.G., Cuxhaven, FRG
79. H. Lundbeck and Company, Valby, Denmark
80. Meiji Seika Kaisha Ltd., Tokyo, Japan
81. Meito Sangyo Company Ltd., Tokyo, Japan
82. E. Merck, Darmstad, FRG
83. Merck and Company, Inc., Rahway, N.J.
84. Mi-Won, Seoul, South Korea
85. Miles Laboratories, Inc., Elkhart, Ind.
86. Mitsubishi Chemical Industries, Tokyo, Japan
87. Nagase and Company Ltd., Tokyo, Japan
88. Nihon Kayaku Company, Tokyo, Japan
89. Nikken Chemicals Company Ltd., Tokyo, Japan
90. Novo Industri A/S, Bagsvaerd, Denmark
91. Orsan S.A., Paris, Fr.
92. Pao Yeh Chemical Company, Ltd., Taipei, Republic of China
93. Parke, Davis and Company, Detroit, Mich.
94. S. B. Penick and Company, Lyndhurst, N.J.
95. Pfizer, Inc., New York
96. Pharmachim Antibiotic Works, Razgrad, Bulgaria
97. Pharmacosmos, Valby, Denmark
98. Pierrel S.p.A., Milano, Italy
99. Pliva Pharmaceutical and Chemical Works, Zagreb, Yugoslavia
100. Premier Malt Products, Inc., Milwaukee, Wisc.
101. Proter S.p.A., Milano, Italy
102. Publicker Industries, Inc., Philadelphia, Pa.
103. Quimasa S.A., Sao Paulo, Brazil
104. Rachelle Laboratories Inc., Long Beach, Calif.
105. Reanal, Budapest, Hungary
106. Recherche et Industrie Thérapeutique, Genval, Belg.
107. Rhone-Poulenc S.A., Paris, Fr.
108. G. Richter, Budapest, Hungary
109. Rohm G.m.b.H., Darmstadt, FRG
110. Rohm and Haas, Philadelphia, Pa.
111. Roussel-UCLAF, Romainville, Fr.
112. Sandoz-Wander A.G., Basle, Switz.
113. Sandoz, Inc., Hanover, N.J.
114. San Fu Chemical Company Ltd., Taipei, Republic of China
115. Sankyo Company Ltd., Tokyo, Japan
116. Sanraku Ocean Company Ltd., Tokyo, Japan
117. Schering A.G., West Berlin, FRG
118. Schering Corporation, Bloomfield, N.J.
119. G. D. Searle and Company, Skokie, Ill.
120. Shionogi and Company Ltd., Osaka, Japan
121. Sociedade Produtora de Leveduras Seleccionadas, Matozinhos, Portugal
122. Societa Prodotti Antibiotici (SPA), Milano, Italy
123. Societé Chimique Pointet Girard, Villeneuvela-Garenne, France
124. E. R. Squibb and Sons, Inc., Princeton, N.J.
125. Standard Brands, Inc., New York
126. Stauffer Chemical Company, Westport, Conn.

Table 2 (continued)

127. John and E. Sturge Ltd., Birmingham, Eng.
128. Surabhai Chemicals Ltd., Baroda, India
129. Tai Nan Fermentation Industrial Company Ltd., Taipei, Republic of China
130. Taiwan Sugar Corporation, Taipei, Republic of China
131. Taiwan Tobacco and Wine Monopoly Bureau, Taipei, Republic of China
132. Takeda Chemical Industries Ltd., Osaka, Japan
133. Tanabe Seiyaku Company Ltd., Osaka, Japan
134. Tarchomin Pharmaceutical Works ('Polfa'), Warsaw, Poland
135. Tate and Lyle Ltd., Yorkshire, Eng.
136. Toyo Jozo Company Ltd., Tokyo, Japan
137. Tsin Tsin Foods Company, Taipei, Republic of China
138. Tung Hai Industrial Fermentation Company Ltd., Taipei, Republic of China
139. Universal Foods Corporation, Milwaukee, Wisc.
140. The Upjohn Company, Kalamazoo, Mich.
141. Wallerstein Laboratories, Inc., Morton Grove, Ill.
142. Wei Chuan Foods Corporation, Taipei, Republic of China
143. Wei Wang Industrial Fermentation Company Ltd., Taipei, Republic of China
144. The World Champion Company Ltd., Taipei, Republic of China
145. Wyeth Laboratories, Philadelphia, Pa.
146. Yamanouchi Pharmaceutical Company, Tokyo, Japan
147. Yamasa Shoyu Company Ltd., Choshi, Japan

[c] Used in human infections.
[d] Used in coccida control in poultry.
[e] Growth promotants in feeds.
[f] New fermentation product.
[g] Insecticide.
[h] Stimulates immune system.

Microbiological Aspects

The most important consideration in industrial fermentations is the selection of the proper microorganism. This choice must provide suitable stability of the process, whereupon the engineering and development aspects of process design may be initiated. Stock cultures of microorganisms useful in industrial fermentations are usually maintained by the manufacturer, and less frequently in commercial collections or collections at academic or government laboratories.

Since the microorganism is so important to the success of the process, many methods have been evaluated for preservation of some of these cultures (1). Most of the successful methods are based on storage under conditions where viability is maintained but multiplication is suppressed. These methods include the older technique of lyophilization of cell suspensions (introduced by Wickerham in 1940 (2), now used in more than 3000 laboratories) and the more modern procedure of storing the cell suspension at the temperature of liquid N_2 ($-196°C$). The latter procedure is favored at present since a much higher survival rate of the cells is obtained (3). Alternative methods including storage of cell suspensions at $-16°C$ (commercial freezer temperature) or storage on agar slants in a refrigerator ($+4°C$) are satisfactory for only certain microorganisms.

Since the strains of the microorganisms used in commercially important fermentations are usually carefully selected, the owner has a financial investment that must be recognized. Thus most manufacturers are reluctant to deposit their useful

Table 3. Some Companies Growing Yeast for Food and Feed[a]

Bakers' yeast

Argentina:	Compania Argentina de Levaduras S.A. Jc, Buenos Aires; Levaduras de Pan S.A. I.C.A., Buenos Aires
Australia:	Fermentation Industries PTY, Ltd., Granville, N.S.W.; Mauri Brothers and Thomson, Ltd., Sydney
Austria:	Vereinte Maunter Markhofsch Presshefe Fabriken, Vienna
Belgium:	Gist & Spiritusfabrieken Bruggeman, Ghent
Brazil:	Usina Itaiquara de Acucar e Alcool S.A., Sao Paulo
Canada:	The Fleischmann Company, Montreal; Lallemand Inc., Montreal
Chile:	Hoffman Prado y Co., Ltd., Santiago; Legrand Cia, Ltd., Santiago; Levaduras y Fermentos S.A., Santiago; Levaduras Collico Limited, Valdivia
Columbia:	Compania Nacional de Levaduras Levapan Ltd., Bogota
Denmark:	Danish Fermentation Industry, Ltd., Copenhagen
Egypt:	Egypt Starch Products and Yeast Company, Alexandria
Finland:	The Finnish State Alcohol Monopoly (ALKA), Helsinki; Lahden Polthimo Oy, Lahti
France:	Establishments Fould-Springer, Maisons-Alfort (Seine); Societe Industrielle Lasaffre, Marcq-en-Baroeul; Societe Lesaffre, Lille; Syndicat des Producteurs de Levure Aliment de France, Paris
FRG	Asbeck, Hamm; Asmussen, Elshorn; Giegold, Schwartzenbeck; Lindenmeyer & Co., Heilbronn; Norddeutsche Hefe Industrie G.m.b.H., Hamburg-Wandsbeck; Ostfriesische Hefefabriken, Leer; Pliser G.m.b.H., Darmstadt; Uniferm G.m.b.H., Werne; Weineger, Rettsteig
Guatemala:	Central Productos Alimenticios Universal & Cia, Ltd., Guatemala City
Holland:	Gist-Brocades, N.V., Delft
India:	Dhampur Sugar Mills; Indian Yeast Company; Mohan Mekins
Iran:	Iran Meyeh Company, Teheran
Japan:	Chuetsu Yeast Industries, Ltd., Nagoaka; Kanegafuchi Chemical Industry Co., Ltd., Osaka; Kyowa Hakko Kogyo Co., Ltd., Tokyo; Nihon Tensaitoh, Hokkaido; Oriental Yeast Company, Ltd., Tokyo; Sankyo Company, Ltd., Tokyo; Toyo Jozo Fermentation Company, Shizuoka
Mexico:	Acidos Organicos S.A., Mexico D.F.; Industria Mexicana de Alimentos S.A. de C.V., Mexico D.F.; Levadura Aztica S.A., Mexico D.F.; Leviatan y Flor S.A., Mexico D.F.
Peru:	Fleischmann Peruana Inc., Lima; Red Star del Peru S.A., Lima
Philippines:	Basic Foods Corporation, Manila; Standard Brands of the Philippines Inc., Manila
Republic of Korea:	Chun Ho Yeast Company, Seoul; Cho Hung Chemical Works, Seoul; Jeil-Universal Ltd., Seoul
South Africa:	Anchor Yeast (Pty) Ltd., Johannesburg; Union Yeast Products Ltd., Johannesburg
Spain:	Compania de Azucares y Alcoholes S.A., Barcelona; Compania de Industrias Agricolas S.A., Valladolid; Union Alcoholera Espanola S.A., Madrid
Sweden:	Aktiebolaget S.J.A., Stockholm
Tunisia:	Societa Tunisianne de L'Industrie Laitiere, Tunis
Turkey:	Mayadag, Istanbul; Pak Gida Sinayii ve Ticaret S.A., Istanbul; Uludag Fenni Maya Ltd., Sti., Istanbul
United Kingdom:	The Distillers Co. (Yeast) Ltd., Surrey
United States:	American Yeast Company, Baltimore, Md.; Anheuser-Busch, Inc., St. Louis, Missouri; Federal Yeast Company, Baltimore, Md.; Fleischmann Laboratories division of Standard Brands, Inc., New York; Universal Foods Corporation, Milwaukee, Wisc.
Uruguay:	Levadura Uruguaya S.C.I., Montevideo
Zaire:	Leveudrie du Zaire (Leza), Kinhasa
Zambia:	Lee Yeast Company, Lusaka

Table 3. (continued)

Food and feed yeast	
Belgium:	Raffinerie Tirelemontoise, Tienen
Finland:	Metaslüton Teollisuus Oy, Äänekosi; Ynthneet Paperitehtaat OY, Jämsänkosko
France:	Societe Franciase de Petroles B.P., Lavera
United States:	Amber Laboratories, Inc., Juneau, Wisc.; Amoco Foods Company, Chicago, Ill.; Boise-Cascade, Inc., Portland, Oregon; Diamond V Mills, Inc., Cedar Rapids, Iowa; Lake States Yeast Company, Rhinelander, Wisc.; Stauffer Chemical Company, Westport, Conn.; Standard Brands, Inc., New York

[a] Courtesy of *Chemtech* (1).

Table 4. Fermentation Substrates and Products Used as Starting Materials for Chemical Syntheses

Starting material	Products
cephalosporin C	7-aminocephalosporanic acid → many semisynthetic cephalosporins[b]
demeclocycline	minocycline
kanamycin	amikacin
oxytetracycline	doxycycline
penicillins G and V	6-aminopenicillanic acid → semisynthetic penicillins[a]
rifamycin B	rifampin
sisomicin	netilmicin
β-sitosterol and cholesterol	androstane-3,17-dione and androst-1-ene-3,17-dione → steroid analogues

[a] Currently in clinical use: amoxicillin, ampicillin, azidocillin, bacampicillin, carbenicillin, indanylcarbenicillin, carfecillin, cloxacillin, cyclacillin, dicloxacillin, epicillin, flucloxacillin, hetacillin, mecillinam, methicillin, nafcillin, oxacillin, phenethicillin, piperacillin, pivampicillin, pivmecillinam, propicillin, sulbenicillin, talampicillin, and tricarcillin.
[b] Currently in clinical use: cefamandole, cefachlor, cefaparole, cefatrizine, cefazolin, cefoxitin, ceftezole, cefuroxime, cephacetrile, cephalexin, cephaloridine, cephalothin, and cephapirin, cephradine.

microorganisms in commercial collections (24). Currently, there is some interest in obtaining patents that cover certain strains of microorganisms that have unusual attributes for industrial fermentations. The interpretation of the *Patent Office Regulations* by the courts is still in progress and the outcome is far from certain. In the meantime some consulting laboratories are in the business of developing new strains of microorganisms of particular value to manufacturers, eg, Panlabs, Inc. (specializing in penicillin- and cephalosporin-producing microorganisms), and Cetus Corporation (specializing in organisms useful for antibiotic production) (see Genetic engineering). For example, in one case, improved cultures capable of producing 50 g/L of penicillin G (when grown under selected fermentation conditions) have been developed. This is a fivefold increase in production level over those reported as recently as 1970 (5) (see Antibiotics).

For antibiotic-producing fermentations, the careful selection of the microorganism appears to be the most important method for increasing productivity of the process. Other factors which are important include formulation of the medium used for growth of the microorganism, and changes in engineering technology. New developments in microbial genetics (6) may lead to technology that can be generally applied to all microorganisms of interest for industrial processes.

Fermentation Operations

The microorganisms used in industrial fermentations require various nutrients for growth, including carbohydrates, nitrogen-containing compounds (or ammonium salts), growth factors, vitamins, and minerals. In most of the fermentations mentioned in Table 2, these nutritional requirements are met by including in the medium peptones, yeast products, agricultural materials, eg, cornsteep liquor (steepwater), soybean meal, cottonseed flour, and buffer salts. The costs of these substances frequently influence the selection of feedstock and the cost of the finished product, and sometimes an effort must be made to find cheaper nutrients in order to keep costs within reasonable limits. This has been true in recent years as the cost of sucrose rose tenfold and that of soybean meal fivefold (7). Substitutes were found to replace the more expensive grades. In a few instances strains of the microorganism have been developed that use less costly nutrients, eg, acetate or ethanol, instead of sucrose. Lipids including soybean oil, cottonseed oil, and related substances have been used widely for both foam control in aerated fermentation processes and as a nutrient for these aerobic organisms (7) (see Defoamers).

Specific precursors are sometimes added to fermentations to increase the rate of formation and amount of product formed. These include substances such as cobalt salts for the vitamin B_{12} fermentation, sulfate salts for penicillin and cephalosporin fermentations not very widely used 1979, and phenylacetic acid and phenoxyacetic acid for fermentations producing penicillin G (benzylpenicillin) and penicillin V (phenoxymethylpenicillin), respectively. In a few fermentations, unusual substances are added to direct the fermentation; apparently they are not incorporated into the product, and probably act as selective inhibitors. Among the more interesting are barbiturates for the rifamycin B fermentation and methionine for some of the cephalosporin fermentation processes.

In most fermentation processes, the nutrient medium is first sterilized to inactivate or kill extraneous microorganisms which may be present. This sterilization is usually accomplished by heating at high temperatures (eg, 150°C) for a few minutes, or at lower temperatures (eg, 126°C) for longer periods. Heat is often supplied by introducing live steam into the nutrient medium contained in the fermentor, or in a continuous process by means of heat exchangers through which the nutrient medium is pumped at high velocity. There are a few fermentations where the combination of large inoculum, high concentration of nutrients, and relatively short fermentation cycle exist, and only pasteurization or partial sterilization may be needed (see Sterile techniques). An example of this is the conversion of D-sorbitol to L-sorbose by *Gluconobacter suboxydans* where the fermentation cycle may be 24 h and the concentration of substrate more than 20 wt % which is also likely to inhibit the growth of many microorganisms. Another example in the production of lactic acid by *Lactobacillus delbruekii* where the 50°C incubation temperature and relatively high substrate concentration seem sufficient to discourage any contaminating microorganisms.

In most fermentation operations, especially those for the production of antibiotics, any material that enters the fermentor during the process must be sterilized. Air may be sterilized by passage through filters packed with glass wool or through bacteriological filters (Millipore), and nutrients are sterilized by autoclaving, eg, heating at 126°C. If the supplements are substances that have been previously sterilized by crystallization from solvents, eg, steroids, further treatment may not be needed, especially if the contact time between the microbial cells and the added substrate is less than 24 h and a heavy growth of the microorganism is present.

Once the fermentor has been charged with medium and the sterilization process completed, the unit is inoculated with a rapidly growing culture of the particular microorganism. Inocula for the commercial fermentations are usually produced in several stages, and the final volume may be 1–10% of the fermentation volume. The procedures used in preparation of this inoculum determine in large part the success of the process, and care must be taken to evaluate the inoculum before use. Usually organisms growing in their logarithmic phase of multiplication are used, and those that have passed into a stationary growth phase are not favored since they may result in process variation.

Experience accumulated in the past 25 years suggests that the desired metabolic changes taking place in the fermentation frequently have a rather narrow temperature optimum, sometimes as limited as ±0.3°C. The temperature of the fermenting mass is usually controlled by cooling coils submerged in the fermenting medium, and this usually permits control of ±0.5°C. In some processes, the optimal incubation temperature for growth of the microorganism may be several degrees higher than the optimal temperature for product formation, and the temperature controller may be programmed to take these situations into account.

The physiological state of the microorganism during the fermentation can be influenced by changes in the pH of the fermenting mass. Product yields can often be increased through control of the pH of the growing culture, and this too can be programmed based on the pH of the cell suspension and product yield. It is now common practice to control the pH to within 0.1 pH unit in large-scale fermentation operations.

In aerobic fermentations, an ample supply of sterile air to the fermentor is needed. This may be introduced into the liquid phase through a perforated pipe (sparger) or by other means. Many aerobic fermentation vessels are equipped with baffles and agitators so that the introduced air can quickly reach the solubility maximum in the liquid phase. In some fermentations the dissolved oxygen demand is a factor limiting the growth of the organism (and the formation of product). This situation can be remedied by either increasing the aeration–agitation levels or by using an air stream containing additional oxygen (8). In some fermentation units (380 m^3 or 100,000 gal capacity) the aeration rate may be 8.5 m^3/min (300 cfm) of sterilized air, and this can be a major expense in the operations.

Vessels for current fermentation operations range in size from ≤380 L (≤100 gal) for production of specialized products, including enzymes and vaccines, to 570 m^3 (150,000 gal). These tanks can be mounted vertically or horizontally depending upon the technology available, the capital investment considered, and the versatility desired with the equipment (9–11). There seem to be many different designs that can be successfully used for the same type of fermentation process (see also Vaccine technology).

A flow sheet of a typical fermentation process for production of penicillin G is shown in Figure 1. Spores of *Penicillium chrysogenum* are used to inoculate 100 mL of medium in a 500-ml Erlenmeyer flask, and the inoculated flasks are placed on a

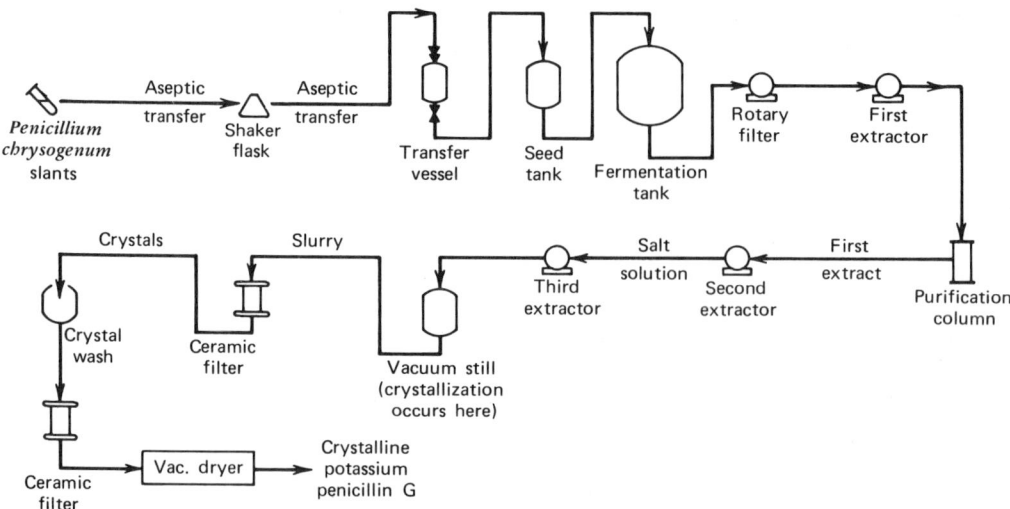

Figure 1. Flow sheet for penicillin G process.

rotary shaker (250 rpm, 5-cm displacement) located in a 25°C incubator. After 4 d of incubation, the contents of the flask are transferred to 2 L of medium (in a 4-L flask). This second flask fermentation is incubated on the shaker for 2 d and the contents are transferred to 500 L of medium in an 800-L stainless steel tank. This tank is equipped with air spargers, agitators, cooling coils for temperature control, and antifoam agent addition devices. After 3 d of incubation, the contents of the tank are used to inoculate 180 m^3 (48,000 gal) of fermentation medium in a 250-m^3 (66,000-gal) fermentor. This larger fermentor is equipped with devices for continuous addition of sterile glucose syrups, pH control (automatic addition of NaOH and H$_2$SO$_4$ to keep the pH at a preset value), foam-sensing devices to activate automatic addition of antifoam agents (eg, animal or vegetable oils) (see Defoamers), and metering pumps for continuous addition of sterile phenylacetic acid. After 7 d of incubation, the contents of the fermentor are filtered (all the penicillin is in the cell-free liquid), and the filtrate is passed through a series of Podbelniak extractors where the antibiotic is extracted into amyl or butyl acetate. The penicillin G in the amyl acetate is transferred back into aqueous solvent by extraction with phosphate buffer, and the potassium penicillin G is crystallized from a n-butanol–water mixture. The potassium penicillin G may be further purified and used as such, may be converted into procaine penicillin G, or may be converted enzymatically or chemically to 6-aminopenicillanic acid (a starting material for synthesis of semisynthetic penicillins).

The complexity of the equipment involved in a penicillin or other antibiotic-producing process is exemplified by the equipment shown in Figures 2 and 3. A model of a 5-fermentor unit with 230-m^3 (60,000-gal) fermentors is shown in Figure 2, and a method of recording data in Figure 3.

The final cell suspension from many commercial fermentations may contain as much as 10% of the desired product, though usually it is less than 5%. The purification process includes recovering the 5% of product in a commercially desirable form, and disposing of the 95% extraneous materials. This is usually accomplished by a series of physical and chemical treatments including chilling, filtering, evaporating, pre-

Figure 2. Model of fermentation plant. Courtesy of Hoffmann-La Roche Inc.

cipitating, extracting, adsorption on macroporous and ion-exchange resins, and crystallization (12). The purification process may cost more than the fermentation operation, particularly in the case of pharmaceutical materials such as antibiotics and vitamins.

Economic Aspects

The volume of materials produced varies considerably with the market. Among those products where the volume is valued at more than $50,000,000 or the production is more than 23,000 metric tons per year are: antibiotics (bacitracin, cephalosporins, doxorubicin, erythromycin, gentamicins, kanamycins, kasugamycin, lincomycins, monensin, neomycins, penicillins, rifamycins, tetracyclines, and tylosin); enzymes (amylases, diagnostic enzymes, glucose isomerase, microbial rennet, and proteases); organic acids (citric acid and gluconic acid); solvents (ethanol); vitamins (vitamin B_{12}); amino acids (monosodium glutamate, L-lysine, and L-aspartic acid); and miscellaneous (xanthan, ergot alkaloids, steroids by transformations, and L-sorbose for synthesis of vitamin C).

The production of 270,000 t of monosodium glutamate (in 16 countries with ca 50 manufacturing plants), and 200,000 t of citric acid (qv) (in 7 countries with 20 plants) represent perhaps the largest volume items. On the other hand, the production of ca

Figure 3. Control area for fermentation unit. Courtesy of Novo Industri A/S.

25,000 t of penicillins (in 18 countries with 35 producers) is also significant (see also Enzymes; Antibiotics; Amino acids; Gums; Alkaloids; Steroids; Vitamins).

Many of the fermentations are operated on a very large scale with fermentation volumes approaching 250,000 L (66,000 gal) per batch. Since the concentration of microbial metabolite in some of these fermentations reaches 100 g/L, a single batch will supply 25 t of useful material. Some of the data on volume of fermentation products and their values are summarized in Table 5.

Future Prospects

Although the continued expansion of the fermentation industry activities during the past decade has not brought it to a crossroad in long-term planning, and there are still periodic shortages of fermentation-produced substances, eg, penicillin, tetracycline, bacitracin, apparently some experts consider the prospects for continued expansion are limited because:

(1) Presently nearly 100 antibiotics (qv) are sold commercially. Many of them would ordinarily be considered as obsolete but with the increase in requirements to

Table 5. Information on Volume of Fermentation Products, 1979

Product	Estimated production, metric tons	Price, $/kg	Estd. no. of world producers (U.S. producers)
citric acid	200,000	1.57	5 (2)
erythromycin	1,800	110.00	10 (4)
gluconic acid	27,000	1.46	5 (2)
L-lysine	27,000	12.20	2 (0)
L-monosodium glutamate	250,000	2.16	8 (1)
L-sorbose (for ascorbic acid)	27,000	ca 0.88	5 (2)
potassium benzylpenicillin and potassium phenoxymethylpenicillin	27,000	22.00	18 (4)
tetracycline hydrochloride	9,100	33.00	23 (5)
vitamin B_{12}	11	8000.00	6 (1)

be met for new compounds for approval by various government agencies, a number of the older ones continue to survive. At least 25 new antibiotics will be approved for clinical and animal health use in the next 5 years, and probably half of these will be new chemical entities (the other half will be chemical derivatives of currently available antibiotics). On the other hand, production of these new antibiotics together with the existing group will require improvement in production processes (or expansion of current facilities by perhaps as much as 40%).

(2) Many of the fermentation industry's manufacturing facilities are approaching the replacement period. (The original facilities built in 1943–1948 were replaced in 1960–1965, and these are now again becoming ready for replacement.) The new equipment will be more sophisticated, containing labor-saving devices, and computer control (13).

(3) The potential use of microbial cells for use as protein in human and animal diets has been studied for nearly 20 years (see Foods, nonconventional). The conflicting reports (14–15) leave the impression that both political and scientific objectives must be reconciled before the product will be widely accepted for many purposes. Nevertheless, the success of the Amoco product (in the U.S.), Tortuin (a yeast product), and the potential of the production of *Methylophilus methylotrophus* in the world's largest fementor (ca 5700 m^3 or 1,500,000 gal) at the ICI plant in Billingham, U.K., should eventually reassure those who are uncertain about the commercial potential of this fermentation operation.

(4) The potential for improving the efficacy of chemical syntheses by including enzymatic steps has been realized with the steroids (qv) and is now being generally accepted for other difficult chemical syntheses (16). This application should increase in the next few years as more organic chemists and chemical engineers become aware of the advantages of these combined operations. Some new processes that are under study include the formation of L-lysine from aminocaprolactam (17–18) or 1-chlorocyclohexene (19), and production of D-tartaric acid by enzymic hydrolysis of *cis*-epoxysuccinic acid (prepared from maleic acid by peroxide oxidation) (20).

Because of current government controls frequently more than 20% of the cost of a new manufacturing facility has to be invested in waste disposal systems. In some instances, the effluent gases are incinerated or treated with ozone to eliminate odors, and effluent water is of higher quality than that entering the process. These needs will probably be multiplied in the future, raising additional problems for the manufacturer. Some of the problems can be resolved by using new raw materials and others by choice of different strains of the microorganisms. It is indeed a challenge for both microbial geneticist and bioengineer.

As the fermentation industry embarks on its second century of operation, there are opportunities for diversifications that were not considered as recently as 25 years ago. At that time the newest developments were the production of antibiotics and the use of microorganisms as replacements for chemical operations in complicated chemical syntheses. Now, uses of microorganisms are considered for preparation of peptide hormones, eg, somatostatin, insulin (qv), and growth hormone, substances that are quite foreign to the microbiologist's domain. This potential, application of genetics to fermentations, is so large that there may be a complete revolution in the fermentation industries in the next 20 years (6) (see also Genetic engineering).

BIBLIOGRAPHY

"Fermentation" in *ECT* 1st ed., Vol. 6, pp. 317–375, by George I. de Becze, Schenley Distillers, Inc.; "Fermentation" in *ECT* 2nd ed., Vol. 8, pp. 871–880, by Ralph F. Anderson, International Minerals & Chemical Corporation, Bioferm Division.

1. D. Perlman, *Chemtech* **7**, 434 (1977).
2. L. J. Wickerham, *Wallerstein Lab. Commun.* **5**, 165 (1942).
3. R. J. Heckly, *Adv. Appl. Microbiol.* **24**, 1 (1978).
4. M. Pearson, *The Million Dollar Bugs,* G. P. Putnam's Sons, New York, 1969.
5. R. P. Elander, L. T. Chang, and R. W. Vaughn, *Ann. Rep. Ferm. Proc.* **1**, 1 (1977).
6. O. K. Sebek and A. I. Laskin, eds., *Genetics of Industrial Microorganisms,* American Society for Microbiology, Washington, D.C., 1979.
7. C. Ratledge, *Ann. Rep. Ferm. Proc.* **1**, 49 (1977).
8. M. C. Flickinger and D. Perlman, *Appl. Environ. Microbiol.* **39**, (1979).
9. L. P. Tannen and L. K. Nyiri in H. J. Peppler and D. Perlman, eds., *Microbial Technology,* Vol. II, Academic Press, Inc., New York, 1979, pp. 331–374.
10. L. K. Nyiri and M. Charles, *Ann. Rep. Ferm. Proc.* **1**, 365 (1977).
11. W. H. Bartholomew and H. Reisman in ref. 9, pp. 463–496.
12. P. Belter in ref. 9, pp. 403–432.
13. W. A. Weigand, *Ann. Rep. Ferm. Proc.* **2**, 43 (1978).
14. A. I. Laskin, *Ann. Rep. Ferm. Proc.* **1**, 151 (1977).
15. J. Litchfield in H. J. Peppler and D. Perlman, eds., *Microbial Technology,* Vol. I, Academic Press, Inc., New York, 1979, pp. 93–155.
16. J. B. Jones, C. J. Sih, and D. Perlman, eds., *Applications of Biochemical Systems in Organic Chemistry,* Wiley-Interscience, New York, 1976.
17. T. Fukumura, *Agric. Biol. Chem.* **40**, 1695 (1976).
18. U.S. Pat. 3,770,585 (Nov. 6, 1973), T. Fukumura (to Toray Industries, Inc.).
19. U.S. Pat. 3,796,632 (Mar. 12, 1974), T. Fukumura (to Toray Industries, Inc.).
20. U.S. Pat. 3,957,579 (May 18, 1976), E. Sato and A. Yanai (to Toray Industries, Inc.).

General References

H. J. Peppler and D. Perlman, eds., *Microbial Technology,* 2nd ed., Volumes I and II, Academic Press, Inc., New York, 1979.

D. I. C. Wang and co-authors, *Fermentation and Enzyme Technology,* John Wiley & Sons, Inc., New York, 1979.

G. L. Solomons, *Materials and Methods in Fermentation,* Academic Press, Ltd., London, Eng., 1969.

A. H. Rose, ed., *Economic Microbiology,* Vols. 1–5, Academic Press, Ltd., London, Eng., 1977–1981.

DAVID PERLMAN
University of Wisconsin

FERRICYANIDES. See Iron compounds.

FERRITES

The term ferrite has been misused but is commonly used as a generic term describing a class of magnetic oxide compounds that contain iron oxide as a major component as distinct from the metallurgical term which applies to the compound Fe_3C (see Iron compounds; Magnetic materials). More specifically, there are several crystal structure classes of compounds loosely defined as ferrites, such as spinel, magnetoplumbite, garnet, and perovskite structures.

Although there are many characterizations of specific technical interest depending on the application, one property is shared by all materials designated as ferrites, namely the existence of a spontaneous magnetization (a magnetic induction in the absence of an external magnetic field).

Figure 1 shows diagrammatically what happens when a ferrite experiences an external field. Starting at point 1, as the magnetic field H (A/m) increases, the magnetic induction B (tesla), increases as magnetic domains in the material align themselves with the field. This ratio of induction vs field taken at the low field is known as the relative initial permeability (μ) of the material. In accordance with the definitions adopted by the IEC (International Electrotechnical Commission), the relative initial permeability is defined as:

$$\mu_i = \frac{1}{\mu} L_{H \to o}^{im} \frac{B}{H}$$

where μ_o is the permeability constant *in vacuo* of $4\pi \times 10^{-7}$ Wb/(A·m) (or 15.79×10^{-13} G/Oe). Consequently, μ_i is a dimensionless quantity. This relative permeability makes a wide range of applications possible. As the field increases, more and more

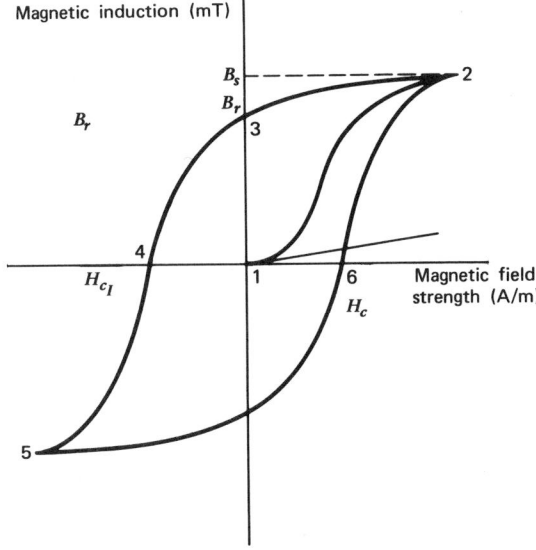

Figure 1. Typical magnetic hysteresis loop for ferrite materials. To convert A/m to Oe, divide by 79.58; to convert mT to G, multiply by 10.

magnetic domains in the material align themselves with that field until no further alignment or domain wall growth is possible. At this point, the material is said to be saturated and has a saturation magnetization equal to B_s. This parameter is most important where large magnetic induction (flux) must be handled (eg, ferrite transformers).

Reduction of the magnetic field causes the magnetic induction to follow the path shown from 2 to 3 in Figure 1. At point 3 the external field is zero and the magnetic induction (B_r) at this point is known as the remanent magnetization. Applications such as ferrite memory cores (stable magnetic state) and permanent magnets (supplying magnetic flux) benefit from high values of B_r.

If the magnetic field is increased in the negative sense, the magnetic induction follows the line from 3 to 4; at point 4 the field is sufficient to reverse the flux in the material. Point 4 is known as the demagnetization field (H_c) of the material. It is a critical parameter for permanent magnet applications (must be very high) because it is a measure of the resistance of the ferrite to demagnetizing forces.

Continuing to increase the magnetic field in the negative direction causes saturation of the material (point 5). Further increase in the positive sense causes the hysteresis loop to be closed along path 5-6-2 until it is saturated again at the positive value.

Structure and Chemistry

The magnetic properties of ferrites derive directly from the electron configuration of the ions and their interactions with each other. Thus investigation of their crystal structure is fundamental to the understanding of these materials. Although the specific structures differ, they can all be considered to be composed of two sublattices: a rigid anion lattice composed of the relatively large ($r = 0.132$ nm) oxygen anions and the cation sublattice formed by the filling of holes (interstitial sites) with the smaller cations.

Spinel Ferrites. This important class of ferrites has the general composition AB_2X_4 and is isostructural with the mineral spinel $MgAl_2O_4$. The structure (see Fig. 2) (1) is a cubic close packing of the anions (X), with a variety of A and B cations capable of filling the interstitial sites. The smallest crystallographic unit cell which has the required cubic symmetry contains eight formula units of AB_2X_4. Each unit cell has two types of interstitial sites that can be occupied by the A and B cations; thus there are 64 tetrahedral sites (8 of which are occupied) and 32 octahedral sites (16 of which are occupied). Calculations based on the assumption that the oxygen anions are rigid spheres (2) show that the interstitial tetrahedral and octahedral sites have radii in the following ranges:

$$0.055 < r_{tet} < 0.067 \text{ nm}$$

$$0.070 < r_{oct} < 0.075 \text{ nm}$$

As can be seen from Table 1, a wide variety of transition metal cations could fit into these interstitial sites and the d electron configuration ranges from d^0 to d^{10}. Thus it becomes possible to make a large number of spinel ferrite compounds, each having specific magnetic interactions. The difference in the electron environment at the

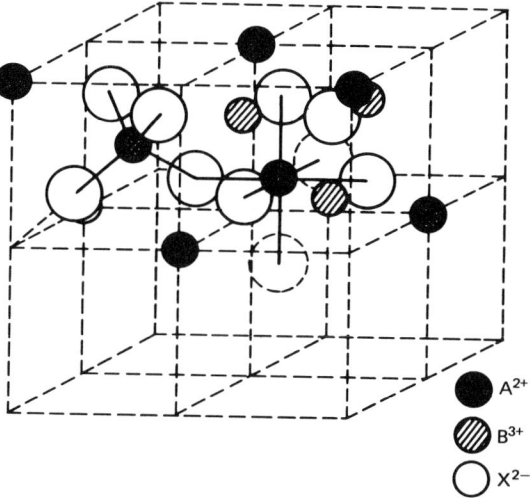

Figure 2. Spinel structure (adapted from ref. 1).

fourfold (tetrahedral) and eightfold (octahedral sites) and the interaction of that environment with the d electrons of the various transition metal cations results in the magnetic properties of the spinel ferrites.

The spinel structure may be formed from various anions including sulfur (thiospinels), chlorine (halospinels), and oxygen. The cations in the structure must satisfy both size and neutrality charge considerations. For a simple spinel $A^{m+}B_2^{n+}O_4$ this would mean that $m + 2n = 8$. Although there are innumerable spinel compounds of scientific and fundamental interest, the materials of greatest technical and commercial interest are the oxide spinels based on oxides of iron (ferrospinels). Within this group, materials such as lithium ferrite $Li_{0.5}^{1+}Fe_{2.5}^{3+}O_4$ and the whole family of $M^{2+}Fe_2^{3+}O_4$ spinels are found. The latter can be formed from any of the divalent transition metal cations that have ionic radii small enough to allow them to fit in the interstitial sites (eg, M^{2+} = Mg, Mn, Fe, Co, Ni, Zn, Cu, etc).

Because the electron environment surrounding the A and B sites is different, the $M^{2+}Fe_2^{3+}O_4$ designation for the ferrospinels must be further clarified in order to define the location of the M^{2+} and Fe^{3+} cations. Thus the formula is written with the cations that are located on the octahedral sites (B sites) contained in brackets. There are three possible distributors for the cations: (1) the normal spinel $M^{2+}[Fe_2^{3+}]O_4$ in which all of the M^{2+} cations reside on the tetrahedral sites (eg, zinc ferrite [12063-19-3], $Zn^{2+}[Fe_2^{3+}]O_4$); (2) the inverse spinel $Fe^{3+}[M^{2+}Fe^{3+}]O_4$ in which the M^{2+} cations are on octahedral sites (eg, nickel ferrite [12168-54-6], $Fe^{3+}[Ni^{2+}Fe^{3+}]O_4$); and (3) the mixed spinel in which the divalent cations are distributed between both sites. The site preference of the divalent cations is discussed in refs. 1, 3–5.

The spontaneous magnetization (M_s) arising in the spinel ferrites is a result of the interactions of transition metal cations and the oxygen anions. Consequently, the magnetic properties can be considered to be the sum of the magnetic interactions on the A-site sublattice and the B-site sublattice. In the spinel ferrites these two sublattices have magnetic moments that are oppositely aligned and the magnetic moment

Table 1. Ionic Radii of Transition Metal Ions of Various d-Electron Configurations, nm

Element	d^0	d^1	d^2	d^3	d^4	d^5	d^6	d^7	d^8	d^9	d^{10}
titanium	Ti^{6+} VI[a] 0.0605	Ti^{3+} VI 0.067	Ti^{2+} VI 0.086								
vanadium		V^{4+} VI 0.059	V^{3+} VI 0.0640	V^{2+} VI 0.079							
tantalum		Ta^{4+} VI 0.066	Ta^{3+} VI 0.067								
niobium		Nb^{4+} VI 0.069	Nb^{5+} VI 0.070	Nb^{2+} VI 0.071							
chromium		Cr^{5+} IV 0.0350	Cr^{4+} IV 0.044 VI 0.055	Cr^{3+} VI 0.0615							
molybdenum		Mo^{5+} VI 0.063	Mo^{4+} VI 0.0650	Mo^{3+} VI 0.067							
manganese				Mn^{4+} VI 0.0540	Mn^{3+} VI LS[b] 0.058 HS 0.065	Mn^{2+} VI LS 0.067 HS 0.0820					

Element						
tungsten	W^{4+} VI 0.0650					
iron		Fe^{3+} IV HS 0.049 VI LS 0.055 HS 0.0645	Fe^{2+} IV HS 0.063 VI LS 0.061 LS 0.0770			
cobalt			Co^{3+} VI LS 0.0525 HS 0.061	Co^{2+} VI LS 0.065 VI HS 0.0735		
nickel				Ni^{3+} VI LS 0.056 VI HS 0.060	Ni^{2+} VI 0.0700	
copper						Cu^{2+} VI 0.073
cobalt						Co^{2+} IV 0.084
zinc						VI 0.095 Zn^{2+} IV 0.060 VI 0.0745
gallium						Ga^{3+} VI 0.0620

^{a} Roman numeral indicates coordination number.
^{b} LS = low spin; HS = high spin.

per molecule can be calculated from the distribution of the cations between the A and B sites (6). For example, the mixed spinel Fe_3O_4, $Fe^{3+}[Fe^{2+}Fe^{3+}]O_4$, has a magnetic moment of 4.6×10^{-23} J/T (5 Bohr magnetons, μ_B) on the tetrahedral (A-site sublattice) caused by the five d-electrons of Fe^{3+} and a magnetic moment of $4+5$ on the octahedral (B-site) sublattice caused by the Fe^{2+} (four d-electrons) and the Fe^{3+} (five d-electrons). These two sublattices are oppositely aligned and thus the net magnetic moment is 3.7×10^{-23} J/T (4 Bohr magnetons).

Because the magnetic saturation is dependent on the site location and the d-electron structure of the transition metal cations, it is possible to systematically alter the net magnetic moment by chemical substitutions. The classic example is the increase of the magnetic moment in the mixed zinc ferrites. In this case, the addition of nonmagnetic zinc ions, which prefer to occupy the tetrahedral sites, reduces the magnetic moment on the A-site sublattice and thus increases the net magnetic moment with increasing zinc substitution (see Fig. 3) (7). As the zinc content increases beyond about 0.5, however, a number of the Fe^{3+} ions do not have magnetic neighbors, and this leads to a reduction of the spontaneous magnetization.

As already mentioned, a great variety of oxide materials form the spinel structure. However, nickel–zinc–ferrite, $Ni_{1-x}Zn_xFe_{2-\delta}O_4$, and manganese–zinc–ferrite, $Mn_{1-x}Zn_xFe_{2+\delta}O_4$, account for most applications.

Many of the nickel–zinc–ferrites are formulated with an iron deficiency (δ) in order to keep the magnetic losses low and the resistivity high ($>10^6$ $\Omega\cdot$cm). The manganese–zinc–ferrites, on the other hand, have a slight excess of iron in order to optimize permeability and magnetic saturation.

Crystal structure and specific composition also play an important role in determining the magnetocrystalline anisotropy in magnetic oxides. This anisotropy field represents the internal field in the crystal structure which tends to line up the magnetization vector in a preferred (easy) direction. Any external field which would tend to move the magnetization vector to a different crystalline direction must be great enough to overcome the anisotropy field.

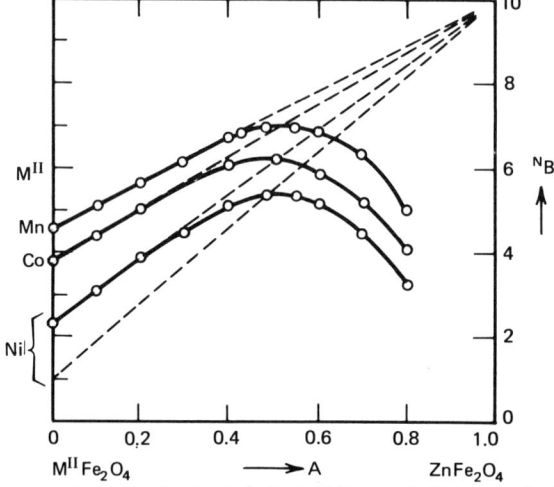

Figure 3. Saturation moment of mixed spinels (adapted from ref. 7). A = mole fraction; N_B = magnetic moment in Bohr magnetons. To convert μ_B to J/T, multiply by 9.274×10^{-24}.

It is desirable to make cubic spinel ferrite materials which have the highest inductance (high relative permeability) and are relatively easy to magnetize and demagnetize at high frequencies. These materials are used as inductors and high frequency transformers. Materials with the highest permeability are those for which the anisotropy constant K_1 is approximately zero (8–10) and the compositional regions where K_1 is very low have been delineated.

In addition to the major crystal chemical interactions of the d-electrons, a number of dopants have specific effects on the magnetic properties of spinel ferrites. For example, it has been well documented that the addition of small amounts of CaO (0.1 mol %) and SiO_2 (0.02 mol %) greatly reduce the eddy current losses in ferrites (11). The effects of silica on the density, power losses, and microstructure of manganese–zinc–ferrites have been demonstrated (12) and the location of the silica has been definitely determined by dissolving away the ferrite matrix. Other dopants such as B_2O_3, ZnO_2, and TiO_2 have been investigated and their effects on the temperature coefficient of permeability and permeability disaccommodation have been noted (13). The latter quantity, which is a measure of the percent change of permeability with time, is of particular interest when precise stable inductors must be designed for long service life.

Hexagonal Ferrites. The hexagonal ferrites (14–16) are a group of ferromagnetic oxides in which the principal component is Fe_2O_3 in combination with a divalent oxide (BaO, SrO, or PbO) and a divalent transition metal oxide.

As can be seen from Figure 4 (17), phases are designated as W, Y, Z, and M hexagonal ferrites (18–19). S has the cubic spinel structure. By convention, those that contain a second divalent cation are identified by the structure type and the cation, eg, $BaZn_2Fe_{16}O_{27}$ [12009-04-0] = Zn_2–W hexagonal ferrite.

Most hexagonal ferrite materials in use today as permanent magnet materials are of the M hexagonal type which is isostructural with the mineral magnetoplumbite [12173-91-0], $PbFe_{7.5}Al_{0.5}Ti_{0.5}O_{19}$. This structure can be considered as being built of oxygen (anion) layers which have hexagonal and cubic packing (see Fig. 5). The fifth layer contains a barium or other divalent cation in place of one fourth of the oxygen

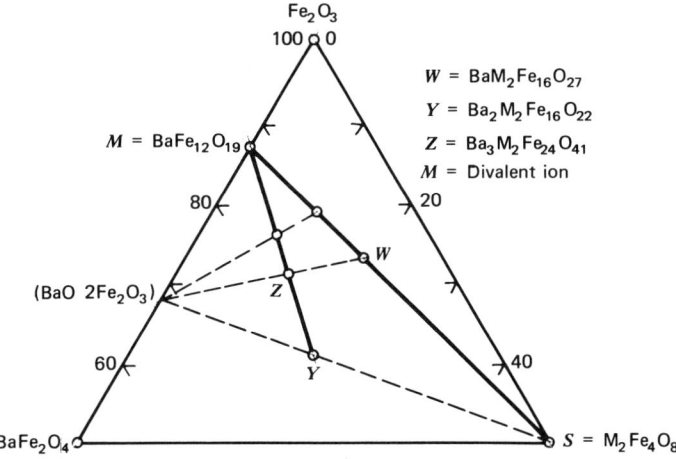

Figure 4. Composition diagram for hexagonal ferrites (adapted from ref. 17).

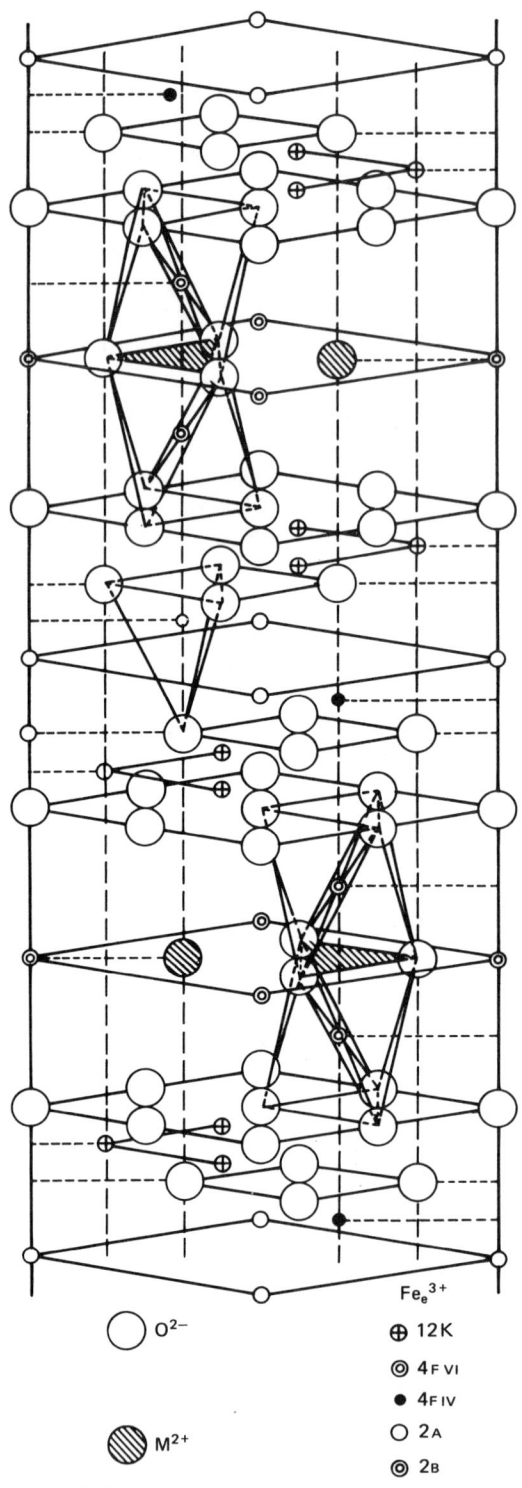

Figure 5. Crystal structure of M-type ferrites $MFe_{12}O_{19}$ (M = Ba, Sr, etc) (adapted from ref. 20).

ions. The resultant structure contains interstitial sites having a fourfold (tetragonal), fivefold (trigonal bipyramid), and sixfold (octahedral) coordination. The octahedral sites are of three different types. Consequently, the cations have five different types of environments (17).

Based upon principles of magnetic superexchange (21), it is possible to assign directions to the magnetic moments of the various ions in the crystal structure. Since the important parameter is the net magnetic moment, it has been assumed in Figure 5 that the magnetic moment of the ferric (Fe^{3+}) ion in the trigonal bipyramid site has its moment pointing up. Summing the individual moments of the magnetic cations over the cell results in a net spontaneous magnetization of eight magnetic moments or 37.1×10^{23} J/T (40 μ_B) ($Fe^{3+} = 4.6 \times 10^{-23}$ J/T or 5 μ_B), and a calculated 12.5×10^{-5} (Wb·m)/kg in good agreement with the 13.75×10^{-5} (Wb·m)/kg obtained from the extrapolation of measured values (14).

The competing, magnetically opposed moments of the ferric ions result in the net low saturation value of the hexagonal M-type ferrite (ie, the high magnetic moment of one sublattice is diluted by the oppositely aligned moment of the other sublattice). This is quite similar to the case encountered in the spinel ferrites where the opposing moments on the two magnetic sublattices also lead to a similar value of the saturation magnetization.

In contrast to the case of the spinel ferrites, where the object is to produce a material with the lowest possible value of the magnetocrystalline anisotropy (typically $0-10^{-11}$ J/cm^3 at room temperature) in order to maximize permeability and reduce hysteresis losses, the M-type hexagonal ferrites are useful because of their high anisotropic value (typically 3×10^{-1} J/cm^3) (22). This large anisotropy, caused by the crystal structure, makes it difficult to turn the magnetization away from the easy direction of magnetization, parallel to the c-axis. Furthermore, because of the high anisotropic value, if the size of the magnetic particles is kept below a certain value (1.6 μm for barium ferrite) the particles are single domain and very difficult to demagnetize.

Garnets. The garnets represent a series of compounds having the general structure $M_3Fe_5O_{12}$ and are isostructural with the ideal mineral $M_3Al_2Si_3O_{12}$. Although cubic, the garnet structure is considerably more complicated than the spinel. The unit cell contains eight formula units or 160 ions. Within the structure there are 24 tetrahedral and 16 octahedral sites. These sites can accommodate the small Fe cation ($r = 0.055$ nm) and other cations of similar size. Additionally, there are 24 dodecahedral sites (eightfold coordination) that can accomodate Y, La, Ca, the rare earths, and other large cations (23–24).

Again, as was the case with both the hexagonal and spinel ferrites, there are two magnetic sublattices opposed to each other. The wide variety of cations that can be substituted into the lattice allow specific material properties to be engineered. The most widely known magnetic compounds having this structure are yttrium–iron–garnet [*12063-56-8*], $Y_3Fe_5O_{12}$ (25), and gadolinium–iron–garnet [*12063-50-2*], $Gd_3Fe_5O_{12}$ (26).

Preparation and Processing

Most ferrites are prepared as ceramic materials by what is commonly known as standard ceramic processing. In this technology (see Fig. 6) the constituent raw ma-

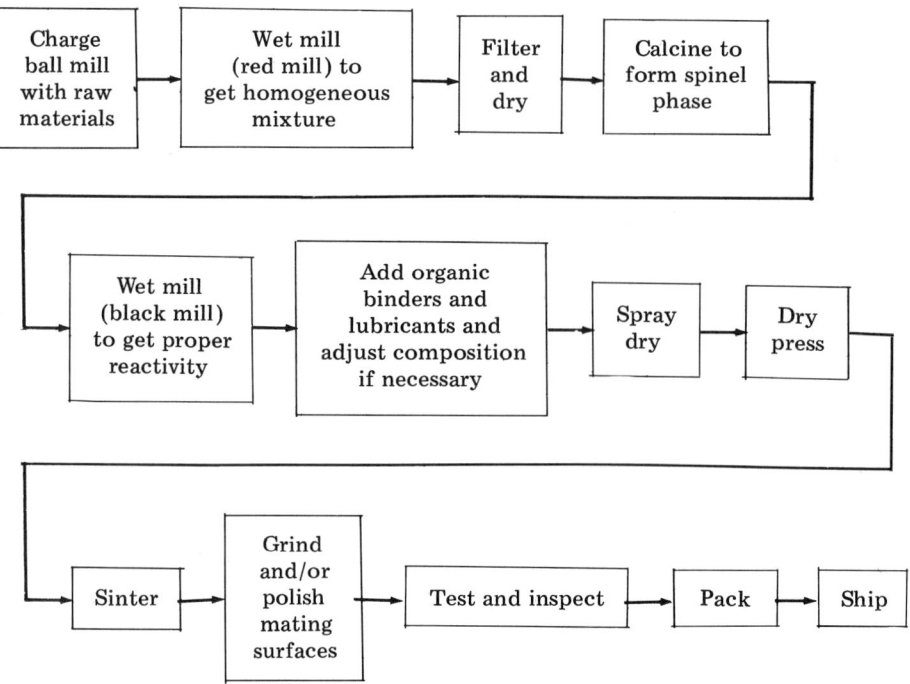

Figure 6. Standard ceramic processing.

terials, oxides, hydroxides, or carbonates, are weighed and first milled in a steel mill using steel balls as the milling media and water as the carrier. This milling is often referred to as red milling because of the characteristic red color imparted by the iron oxide. In red milling, the raw materials are mixed to yield a homogeneous mixture on a macroscopic scale. Other mixing methods may also be employed such as dry mixing of raw materials. However, red milling gives the most uniform mixing and may also result in some size reduction leading to better reactivity in the calcining step. In the calcining (sometimes called presintering) reaction, the raw materials are heated to 1073–1573 K and form the ferrite compound. The carbonates decompose and react by solid-state diffusion to form the final compound.

In the case of the nickel–zinc–spinel ferrites, the powder is calcined at a temperature of ca 1300 K to yield an agglomerated, friable powder that is essentially 100% converted to the spinel phase. However, in the case of the manganese–zinc–ferrites, the calcining conditions are such that the material is 50–85% converted to spinel. Time and temperature are the most important control parameters in the calcining step and the process itself is monitored by x-ray diffraction, surface area of the calcined powder, and sometimes by measuring the inductance of a toroid pressed from the calcined powder. The calcining step must be closely controlled because the properties of the calcined powder determine the reactivity, sinterability, and shrinkage during the subsequent sintering cycle.

Following the calcining reaction, the material is milled again, ie, black-milled. This term is also generic to ferrite processing and is applied to the second milling operation. It has come into use because many ferrites are black after calcination. However, it is equally applied to nickel–zinc–ferrites which retain a reddish-brown color.

The purpose of this milling is to further homogenize the material and to reduce the particle size to permit subsequent pressing and sintering. The milling itself can be carried out in a variety of ways. Many manufacturers wet-ball mill with steel balls in a manner analogous to the first milling (red milling). Others may mill the powder dry in an attritor-type mill which gives a higher throughput and a dry powder. The main objective, however, is to get a finely divided powder that can be slurried and spray dried.

Following the second milling, the material must be granulated so that it will be free flowing and can be dry pressed into the desired shape. For small quantities, this can be accomplished by spraying in a small amount of an organic binder while agitating the powder in a pan or in a V-type blender. However, the best method for producing ferrite powder in volume is to add a binder such as poly(ethylene glycol) or poly(vinyl alcohol) at 1–4 wt % and sufficient water to form a slurry that is ca 65–70 wt % ferrite. The slurry is spray dried to yield a dry powder consisting of small spherical particles having a narrow size distribution (27).

Very thin parts, such as used in memory cores, may be formed by tape casting or doctor blading (28) followed by punching the desired shape. Parts that have a high length-to-diameter ratio may be formed by either extrusion (antenna rods) or by isostatic pressing (29).

In the sintering process, the ceramic material is densified and the final magnetic properties are developed. Some materials such as the iron-deficient nickel–zinc–ferrites and the M-type hexagonal ferrites may be fired in air because all the cations exist at their highest valence state. However, this is not the case with the manganese–zinc ferrites. As mentioned previously, in order to get the highest permeability and lowest losses, it is necessary to have the lowest possible magnetocrystalline anisotropy. This reduction is accomplished by anisotropy compensation; the best method of compensation is by control of the amount of ferrous iron (Fe^{2+}) in the crystal lattice.

The phase field of the spinel structure with respect to temperature and atmosphere has been identified in addition to the equilibrium lines that correspond to the lines of constant Fe^{2+} content (see Fig. 7) (30–32). The implication of this plot is quite clear, namely, the oxygen content of the kiln must be closely controlled to maintain the desired ferrous iron content. This control presents quite a significant technological challenge at the lower temperatures during the cooling down portion of the sintering cycle.

For example, if it were desired to produce a material that contained 1.8 wt % Fe^{2+}, this material could be sintered in an equilibrium atmosphere containing ca 2% O_2 at 1673 K (1400°C). However, as the ferrite parts are cooled down, it becomes necessary to continuously lower the oxygen content of the atmosphere in order to maintain the 1.8 wt % Fe^{2+} value. For example, at 1473 K the equilibrium atmosphere would have to contain no more than 0.1% O_2, and further along the cooling curve at 1223 K the oxygen content of the atmosphere would have to be reduced to 0.001% O_2, which presents a considerable technological problem.

Attempts to continuously vary the oxygen partial pressure in static kilns in such a fashion that it follows an equilibrium atmosphere line throughout the sintering cycle have had only limited success. Most spinel ferrites are fired in continuous (tunnel) kilns and the firing cycle is divided into distinct zones (see Fig. 8). The highest temperature zone is called the sintering zone where the material densifies and the microstructure is fully developed. Typical temperatures for the sintering zone are in the range of 1550–1725 K; sintering times may range from 20 min to 12 h.

892 FERRITES

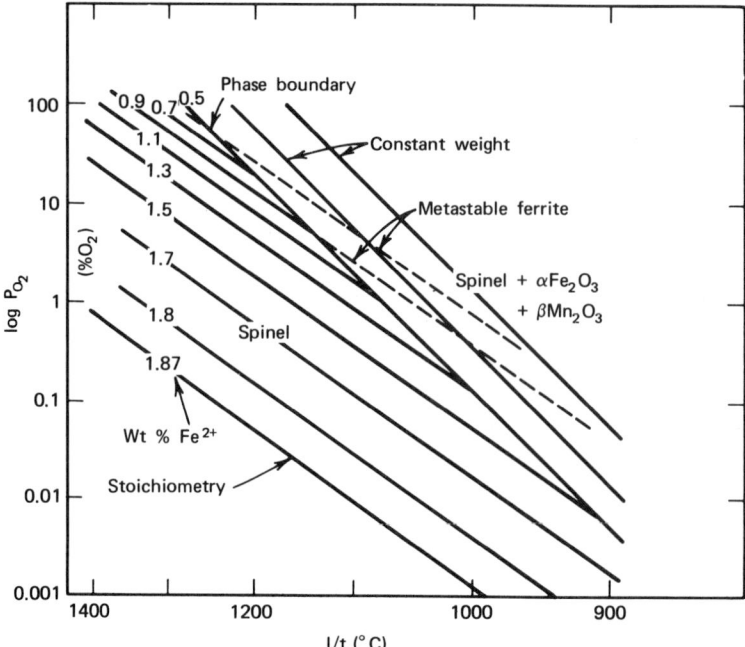

Figure 7. P_{O_2} vs $1/t$ for $Mn_{0.60}$–$Zn_{0.34}$–$Fe_{2.12}$–O_4 (adapted from ref. 30).

The next zone in the kiln is called the anneal or equilibration zone, where the temperature is dropped to 1375–1575 K and the oxygen content of the atmosphere is lowered by the introduction of nitrogen gas. At this elevated temperature the reaction kinetics are still rapid, the ferrite equilibrates quickly with the atmosphere, and the desired ferrous iron level is established. Following the annealing step, the parts are cooled as rapidly as possible and the oxygen content of the atmosphere is reduced still further. Although it is technologically very difficult to obtain a low enough oxygen content to stay on the equilibrium line, at the lower temperatures the reaction kinetics are slower and the tendency to reoxidize is minimized.

In an attempt to improve chemical homogeneity, a wet-chemical process was designed (33) in which an aqueous solution was prepared containing the metal cations. Addition of a strong base (eg, NaOH) precipitated an intermediate hydroxide which was subsequently oxidized by bubbling air through the suspension. The result was a homogeneous fine-particle ferrite which was examined in some detail (34). A similar type of process (35) used an ammonium bicarbonate–ammonium hydroxide mixture as the precipitating agent followed by conventional calcining.

The preparation of ferrite compounds by the cryochemical method has also been investigated (36). In this technique, an aqueous solution is sprayed into a chilled liquid (eg, hexane) where the droplets freeze into beads ca 0.4 mm dia (37). These pellets are removed from the liquid and placed in a freeze dryer where the moisture is removed by sublimation. The resultant pellets are converted to the spinel by calcining.

The preparation of the hexagonal ferrites by wet-chemical precipitation (38), topotactic reaction (39), and fluidized-bed reaction has been investigated. However, the most common method of technological importance is still standard ceramic pro-

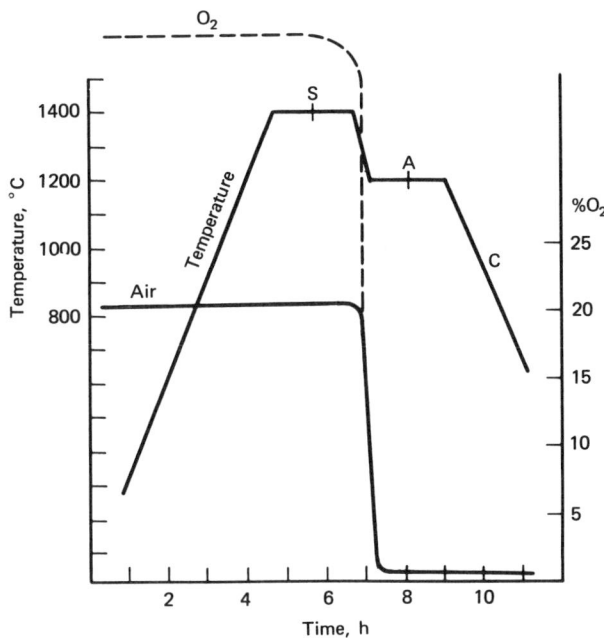

Figure 8. Schematic firing cycle for manganese–zinc–ferrites. S = sintering; A = annealing; C = cooling.

cessing. As was outlined above, the raw materials are mixed, presintered, and the reacted material milled. Initially, the iron oxide used in the preparation of hexagonal ferrites was the same high-purity material used in the preparation of spinel ferrites. However, because of the high iron oxide content (86 wt %) of the hexagonal materials and their relatively lower selling price per kilogram compared with linear ferrite materials, much effort has been devoted to the identification and utilization of lower-grade raw materials. At present, hexagonal materials are made with iron oxides produced by the reclamation of the acidic pickling liquors used in the steel industry (Ruthner process), synthetic oxides produced by the Lurgi process, and mineral iron oxides. These materials are less expensive but also much less pure and require close process control (40).

Important areas of process control are the composition and the presintering conditions. The mixed or wet-milled material is presintered in a rotating or oscillating tube calciner. Typically these units are 2–3 m dia and 20–30 m long with throughput capacities of 500–1500 kg/h. The calcining step is especially critical because it determines to a large extent the properties of the magnet after sintering. At a typical calcining temperature of 1500–1650 K the material reacts completely to form the hexagonal phase. If calcining takes place at a lower temperature, the magnetic properties are not affected adversely but the calcined material is too soft and the subsequent milling step gives a very fine particle size. This leads to difficulty in pressing and a very high shrinkage during sintering. If, on the other hand, the sintering temperature is too high, the particles are too hard and the particle size after milling is rather coarse. Although this does not cause a pressing problem, after sintering the particles are too large and the shrinkage and coercive force are both too low.

After calcining the material must be milled to reduce the particle size to the range of 1 μm in order to obtain single-domain properties. Large ball mills are employed having capacities of 10 m³ (ca 2600 gal) or more using steel balls and a water medium. During this step reclaimed scrap material and modifiers or dopants are added.

Fabrication of the milled powder into parts can take place by a number of methods depending on the degree of magnetic alignment desired. For the lowest-grade material, the milled powder is spray dried and then dry pressed into the required shape. In these materials, the individual particles are randomly aligned with respect to each other, resulting in an isotropic magnet in which the magnetic properties are the same in all directions.

Anisotropic magnets have the best magnetic properties. They are prepared by dry or wet pressing the material in the presence of an external magnetic field which causes the individual magnetic particles to align themselves with that field. The dry-pressing technique is quite similar to that used for preparing isotropic magnets, except that pressing takes place in the presence of a magnetic field. However, because of friction between adjacent powder particles, the alignment is still not optimum; however, dry pressing is rapid and hence relatively cheap.

Wet pressing, on the other hand, gives the highest degree of alignment with the field because the individual particles are much freer to rotate under its influence. The slurry coming from the ball mill is pumped into a die cavity where the individual particles are aligned by an external magnetic field. When alignment is essentially complete, the water is removed by applying a vacuum to the die cavity, and a very fine filter paper prevents the powder from being pulled out with the water. The design and construction of wet-pressing dies is more expensive and the pressing time is longer than for dry pressing.

Sintering of dry-pressed parts can take place immediately after forming. However, wet-pressed parts must be carefully dried to remove most of the residual moisture before being placed in the kiln. Drying under controlled conditions may take from 10 to 300 h, depending on size and shape.

The pressed parts are sintered in air at 1400–1650 K to yield a dense ceramic material. In order to minimize the grain growth that occurs during sintering, the firing temperature is kept as low as possible. The effect of processing parameters on magnetic properties is being studied (41–43). Following the sintering operation, the parts must be ground to meet the tight geometrical tolerances necessary for loudspeaker rings and arc-shaped segments used in d-c motors.

Applications

Spinel Ferrites. The applications of the spinel ferrites are myriad but can be grouped into several main categories: memory cores, and linear, power, and recording-head applications.

Magnetic-Core Memories. Magnetic-core memories are based on switching small toroidal cores of spinel ferrite between two stable magnetic states. These two magnetic states represent the equivalent of 0 and 1 in binary logic. Because energy is only necessary when switching from one state to the other, the magnetic core memory is used where nonvolatility of the stored data is important, eg, loss of power does not result in loss of information. Furthermore, because the memory consists of toroidal cores strung on wires, there are no moving parts and hence core memories are also used in applications where ruggedness and reliability are necessary, eg, military applications.

Initially, these toroidal cores were 2.03 mm in diameter when the technology was developed in the early 1950s but the necessity to reduce switching time and switching current has led to a decrease in core size. Currently memory cores of 0.40 mm and 0.36 mm in diameter are available (44–45).

Such low drive (400 mA) cores are operable over a range of 218 to 398 K and, because of the low temperature coefficient of the core, material can be used in normal commercial operation at 273–323 K without special compensation circuits.

Linear Applications. The linear or low signal applications are those in which the magnetic field in the ferrite is well below the saturation level and the relative magnetic permeability can be considered constant over the operating conditions (see Fig. 9). Consequently, there is a linear relationship between the excitation current and the magnetic induction in the core. At these low signal levels, the magnetic core losses are low and the operating temperature of the core changes only slightly.

The manganese–zinc–ferrite materials characteristically have higher relative permeabilities, higher saturation magnetization, lower losses, and lower resistivities (ca 10 Ω·cm). Since the ferrimagnetic resonance frequency is directly related to the permeability (46), the usual area of application is below 2 MHz. Furthermore, because of the lower resistivity, which is a direct result of the presence of Fe^{2+} and Fe^{3+} in the spinel lattice, eddy current losses can become significant.

At low signal levels, ferrite cores are used either as transformers, low frequency and pulse transformers, or low energy inductors. As inductors, the manganese–zinc–ferrites find numerous applications in the design of telecommunications equipment where they must provide a specific inductance over specific frequency and temperature ranges. Nickel–zinc–ferrites with lower saturation magnetization, generally lower relative magnetic permeabilities, and lower resistivities (ca 10^6 Ω·cm), experience ferrimagnetic resonance effects at much higher frequencies than the manganese–zinc–ferrites. Consequently, they find application at frequencies from 2 to 70 MHz (46).

By adjustment of the nickel–zinc ratio it is possible to prepare a series of materials covering the relative permeability range of 10–2000. These materials are used as high

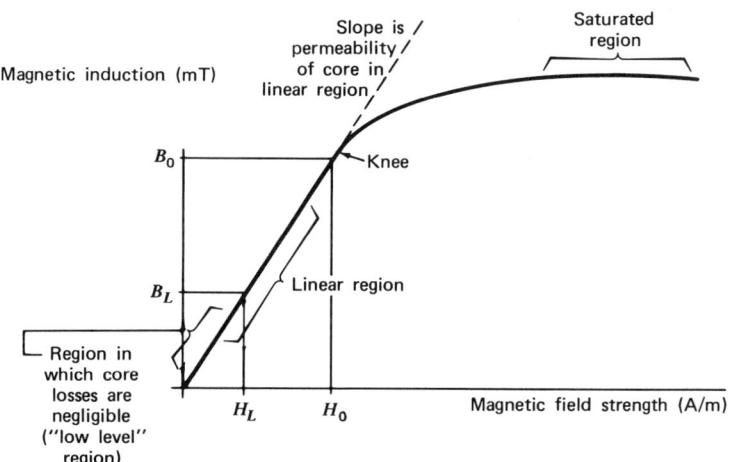

Figure 9. Region for linear operation. The point H_0, B_0 is the upper limit of the linear region. The region between the origin and point $H_L B_L$ is schematically the region where magnetic losses are low.

frequency inductors, antenna rods, high frequency power transformers, and pulse transformers. A variety of materials have been developed to serve these applications (see Table 2).

The IEC standard series of pot cores and other specially shaped cores have been developed to allow the best possible tradeoff between technological considerations (cost, ease of assembly, ease of adjustment, temperature, and time stability of the assembly) and the intrinsic properties of the ferrite. Careful selection of the proper material and the proper shape permits the design of stable, precise inductors for use in high frequency inductor capacitor filters and tuned circuits (46–47).

Materials such as 3C8 (Ferroxcube) or the equivalent grades available from Indiana General (8000 Series), Siemens (N27, T26), and TDK (H7Cl) have been developed to be used in power-transformer applications. Here, the core excitation is high enough to drive the material beyond the linear range causing the effective permeability to vary significantly from the small signal value used in the design of low-level inductors and transformers. Furthermore, losses in the core itself and in the windings cause self heating and result in the transformer operating at a temperature well above ambient. This limits core operation in power transformer applications.

Switched-Mode Power Supplies. The lower magnetic losses of ferrite materials and its higher resistance (10 $\Omega \cdot$cm) compared with laminated transformer steel (50 $\mu\Omega \cdot$cm) permits ferrite cores to be used as the transformer element in high frequency power supplies. Commonly known as switched-mode power supplies, they operate at a frequency of 15–30 kHz and offer higher efficiencies and smaller size than comparable laminated steel transformers.

Operating transformers at these high frequencies overcomes the lower saturation flux density of the ferrite (330 mT or 3300 G) compared with that of laminated transformer steels (1800 mT or 18,000 G). This can be seen clearly from the equation for flux density (48):

$$B = \frac{E \times 10^8}{4.44 \, f N_p A_e}$$

where E = volts rms (root mean square), f = frequency in Hz, A_e = the effective area in cm^2, and N_p = turns on the primary side of the transformer.

As the frequency of operation increases, the magnetic induction (B) in the transformer core decreases, and thus the lower saturation magnetization of the ferrite is no longer a problem. In addition, at the higher frequencies the size of the transformer and the number of turns can be reduced.

Operation of a switched-mode power supply requires that the primary voltage be rectified and smoothed before being switched at high frequency. Detailed design information on the use of ferrite cores in this application is available from a number of sources (49).

In addition to the ferrite transformer, a number of chokes and filters with ferrite components are necessary and thus the switched-mode power supply represents a growing market for spinel ferrites.

Audiovisual Applications. The television and audio markets use large quantities of ferrite material that does not have to meet very high technological requirements. These materials are used in yoke rings for the deflection coils for television picture tubes, flyback transformers, and various convergence and pincushion distortion corrections as well as antenna rods, etc. With 1978 color television production at 9.6 million

Table 2. Properties and Applications of Manganese Zinc Ferrites[a]

Characteristic	Unit	Ferroxcube grade						
		4C4	3D3	3B9	3B7	3C8	3E2A
initial permeability at 298 K	μ_i	125	750	1800	2300	2700	5000
saturation flux density at 298 K, mT[b]	B_s	300	380	320	380	440	360
coercive force, mA/m	H_c	37.7	12.6	3.8	2.5	2.5	1.0
residual flux density, mT[b]	B_r					100	
loss factor at 4 kHz	$\tan \delta / \mu_i$						
100 kHz	$\tan \delta / \mu_i$	35×10^{-6}		$\leq 1 \times 10^{-6}$	$\leq 1 \times 10^{-6}$		
500 kHz	$\tan \delta / \mu_i$	35×10^{-6}	$\leq 14 \times 10^{-6}$	$\leq 5 \times 10^{-6}$	$\leq 5 \times 10^{-6}$	$\leq 10 \times 10^{-6}$	$\leq 10 \times 10^{-6}$
1 MHz	$\tan \delta / \mu_i$	$\leq 40 \times 10^{-6}$	$\leq 30 \times 10^{-6}$	25×10^{-6}	25×10^{-6}		
5 MHz	$\tan \delta / \mu_i$	60×10^{-6}		120×10^{-6}	120×10^{-6}		
temperature factor	$\dfrac{\Delta \mu_i}{\mu_i^2 \Delta T}$	-6.0×10^{-6} min $+6.0 \times 10^{-6}$ max 278–328 K	$+1.0 \times 10^{-6}$ min $+3.0 \times 10^{-6}$ max 243–343 K	$+0.9 \times 10^{-6}$ min $+1.9 \times 10^{-6}$ max 243–343 K	-0.6×10^{-6} min $+0.6 \times 10^{-6}$ max 253–343 K		-2×10^{-6} min $+2 \times 10^{-6}$ max 293–343 K
power losses at 16 KHz, 200 mT[b] at 25°C	mW/cm³					≤ 110	
100°C						≤ 100	
disaccommodation factor 10–100 min	$\dfrac{\Delta \mu_i}{\Delta \mu_i^2 t_1 \log t_2/t_1}$	$<15 \times 10^{-6}$	12×10^{-6}	$\leq 2.5 \times 10^{-6}$	$\leq 3.5 \times 10^{-6}$		
hysteresis losses at 4 kHz, t⁻¹	η_B		$\leq 1.8 \times 10^{-3}$	$\leq 1.1 \times 10^{-3}$	$\leq 1.1 \times 10^{-3}$		
Curie temperature	K (°C)	>573 (>300)	>423 (>150)	>418 (>145)	>443 (>170)	>483 (>210)	>443 (>170)
uses		1–20 MHz, filter coil applications, high frequency wide band and pulse transformers	200 kHz–2.5 MHz, matches polystyrene capacitors, inductors, pulse transformers	audio to 300 kHz, inductors and tuned transformers	audio to 300 kHz, inductors used with low temperature coefficient capacitors, pulse and power transformers	high flux density applications, switched-mode power supplies	wide band and pulse transformers, inductors

[a] Courtesy of Ferroxcube.
[b] To convert millitesla to gauss, multiply by 10.

898　FERRITES

sets and an estimated material content of ca 1 kg/set, audiovisual applications represent a large segment of the spinel ferrite market (50).

Recording Heads. Manganese–zinc and nickel–zinc–spinel ferrites are used in magnetic recording heads for duplicating magnetic tapes and the recording of digital information (see Magnetic tape). At present, the largest market for ferrite recording heads is in the field of digital recording. For this application, the recording-head components are machined out of a solid block of ferrite and then bonded together with a precisely defined separation between the two halves. As the recording head is excited by a current passing through the winding, the magnetic flux lines fringe out in the vicinity of the gap and saturate the recording media. To retrieve this information the recording head reads it inductively, ie, as the saturated region on the media passes under the gap it induces a circulating flux in the recording head which passes through the coil and produces an output signal.

Recording-head material must be strong enough to withstand the machining, bonding, and lapping operations during fabrication. It must also be dense, otherwise porosity in the recording head may damage the media with a subsequent loss of stored information. For these reasons, most recording heads are fabricated from polycrystalline nickel–zinc–ferrite; operating frequencies of 100 kHz to 2.5 GHz preclude the use of metallic alloy heads because of eddy-current losses. These nickel–zinc–ferrites are prepared by standard ceramic processing and fired in air to yield a dense (99.6%),

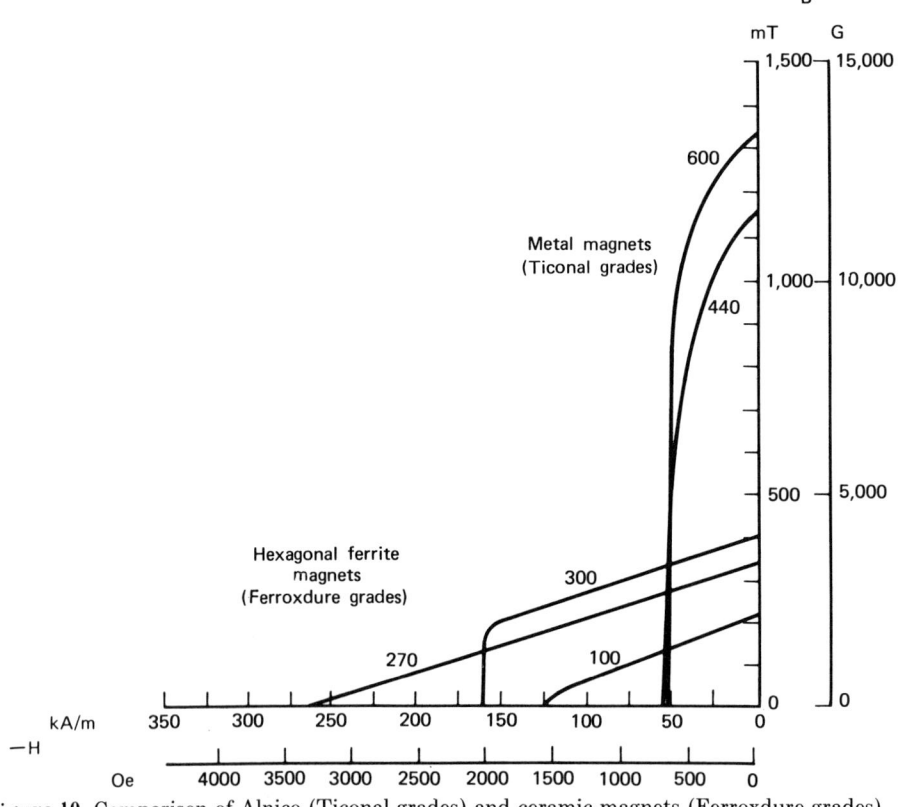

Figure 10. Comparison of Alnico (Ticonal grades) and ceramic magnets (Ferroxdure grades). —H = magnetic field; B = magnetic induction.

Table 3. Properties of Ceramic Magnets[a]

Approximate chemical composition	Max BH product $(BH)_{max}$, kJ/m³[b]		Remanence, B_r, mT[c]		Coercivity, H_{cB}, kA/m[d]		Polarization, H_{cJ}, kA/m[d]		B and H at $(BH)_{max}$		Saturation field strength[e], H_{sat}, kA/m[d]
	Typical	Min	Typical	Min	Typical	Min	Typical	Min	B_d, mT[c]	H_d, kA/m[d]	
isotropic											
FXD 100-22/22[f], BaFe$_{12}$O$_{19}$	7,6	7,2	220	210	135	130	220				800[g]
anisotropic											
FXD 270-34/33[f], SrFe$_{12}$O$_{19}$	21,5	19,9	340	330	255	247	334	318	165	131	1114
FXD 330-37/25[f], SrFe$_{12}$O$_{19}$	25,5	23,9	370	360	239	223	247	231	180	143	876
FXD 360-39/21[f], SrFe$_{12}$O$_{19}$	28,7	27,1	390	380	199	183	207	191	200	143	716
FXD 370-39/25[f], SrFe$_{12}$O$_{19}$	27,9	27,1	390	380	235	223	247	231	190	151	876
FXD 380-39/28[f], SrFe$_{12}$O$_{19}$	28,7	27,1	390	380	263	247	279	263	200	143	955
FXD 300-40/16[f], BaFe$_{12}$O$_{19}$	29,5	27,8	400	390	159	143	163	147	240	123	557

[a] Courtesy of Ferroxcube.
[b] To convert kJ/m³ to MGOe, multiply by 0.126.
[c] To convert mT to G, multiply by 10.
[d] To convert kA/m to oersted, multiply by 12.57.
[e] Minimum.
[f] Ferroxcube designation.
[g] Typical.

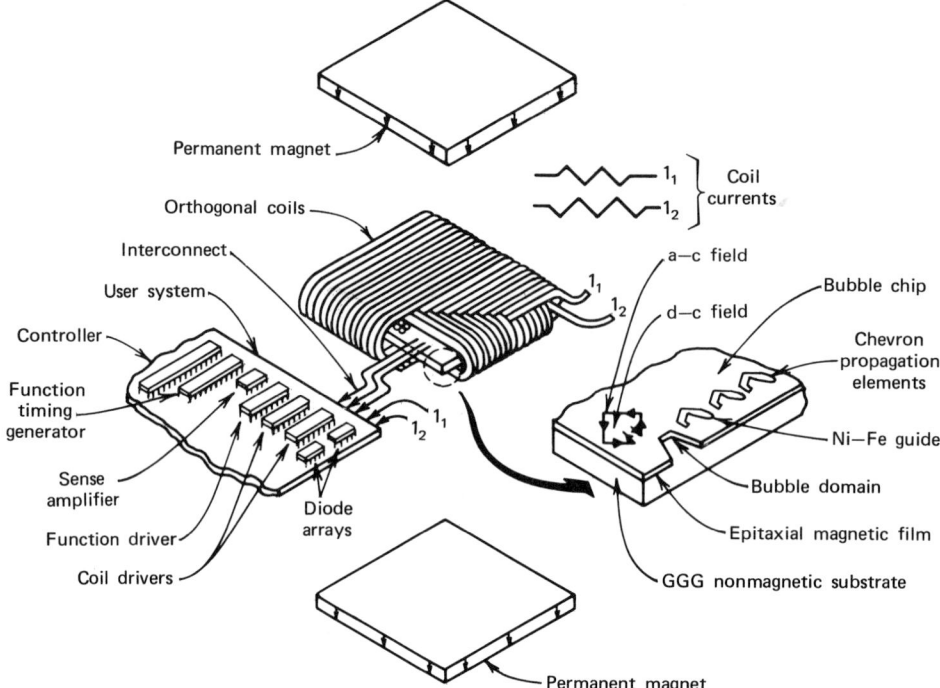

Figure 11. Schematic of bubble memory.

machinable, fine-grained (10–20 μm) ceramic. Typical relative permeabilities at 100 kHz are 1500–2100 and the saturation magnetization is approximately 350 mT (3500 G).

Manganese–zinc material would be much more attractive from a performance standpoint because of the higher relative permeability (5000 at 100 kHz) and the higher saturation magnetization (470 mT). However, these materials require a complicated sintering cycle (see under Processing) and are not as strong as nickel–zinc. Furthermore, the sintered density is too low (93–98%). A number of investigators have examined methods for preparing suitable materials, including vacuum sintering (51), hot pressing (52–53), and the growth of single-crystal material. It is now possible to prepare hot-pressed manganese–zinc–ferrites by uniaxial or isostatic hot pressing with the magnetic values noted above and densities in excess of 99%. These ferrites are finding applications in high performance digital and video recording applications.

Hexagonal Ferrites. The unique properties of hexagonal ferrites are low cost, low density, and high coercive force. Comparison with the conventional Alnico-type metal magnets (see Fig. 10) reveals that they would be best applied where a lower flux density can be tolerated and a higher coercive force is needed.

Consequently, the ceramic magnet is finding increasing application in d-c permanent magnet motors, especially in automotive applications. Such applications as window lift, blower, and windshield-wiper motors are using hexagonal ferrite segments to provide the flux.

A number of standard grades of barium and strontium ferrite material have been developed to satisfy market needs; some of the important properties are summarized in Table 3.

Other applications of hexagonal ferrites are in loudspeakers where, because of lower weight and lower cost, they replace (with redesign) the more expensive Alnico magnets which can contain as much as 38 wt % cobalt.

The increased utilization of millimeter wavelengths has led to continued interest in the hexagonal ferrites for use in self-resonant isolators where the strong magnetocrystalline anisotropy permits a resonator without large d-c magnetic biasing fields (54–55).

Hexagonal ferrites are also used as magnetic biasing components in magnetic bubble memories. In this application, two thin (1 mm) plates of ferrite are bonded together to form a subassembly; one plate is made of isotropic hexagonal ferrite material and the other of manganese–zinc–spinel ferrite. The two subassemblies are placed inside the bubble memory, one above and one below the memory chip (see Fig. 11). The hexagonal material provides the field necessary to maintain stable bubble domains in the memory and the spinel ferrite makes the field uniform. In addition, the temperature coefficient of the hexagonal material closely matches the temperature coefficient of the garnet material in the memory, assuring operation over a wide temperature range.

This technology, in rapid development since the early 1960s, utilizes each of the ferrite materials discussed here to provide a large capacity, solid-state, nonvolatile memory (see Magnetic materials).

BIBLIOGRAPHY

"Ferrites" in *ECT* 2nd ed., Vol. 8, pp. 881–901, by George Economos, Allen-Bradley Company.

1. A. Broese van Groenov, P. F. Bongers, and A. L. Stuijts, *Mater. Sci. Eng.* **3,** 317 (1968–1969).
2. J. Smit and H. P. J. Wijn, *Ferrites,* John Wiley & Sons, Inc., New York, 1959, p. 143.
3. J. D. Dunitz and L. E. Orgel, *J. Phys. Chem. Solids* **3,** 31B (1957).
4. D. S. McClure, *J. Phys. Chem. Solids* **3,** 311 (1957).
5. G. Blasse, *Philips Res. Rep. Suppl.,* 3 (1964).
6. Ref. 1, p. 149.
7. Ref. 1, p. 151.
8. U. Enz, *Inst. Elec. Eng. Pap* **3700,** 246 (Jan. 1962).
9. K. O'Hata, *J. Phys. Soc. Jpn.* **18,** 695 (1968).
10. B. Hoekstra, *J. Appl. Phys.* **49,** 4902 (1979).
11. T. Akashi, *Trans. Jpn. Inst. Met.* **2,** 171 (1961).
12. A. D. Giles and F. F. Westerndorp, *J. Phys. Colloq. Suppl. Cl* **38**(4), Cl-317 (1977).
13. T. G. W. Stijntjes and co-workers in Y. Hoshino, S. Iida, and M. Sugimoto, eds., *Ferrites: Proceedings of the International Conference,* University Park Press, 1971, p. 195.
14. J. J. West and co-workers, *Philips Tech. Rev.* **13**(7), 194 (Jan. 1952).
15. A. L. Stuijts, G. W. Rathenall, and G. W. Weber, *Philips Tech. Rev.* **16**(4–5), 141 (1954).
16. U.S. Pat. 2,762,777 (Sept. 11, 1956), J. J. Went, G. W. van Oosterhout, and E. W. Gorter (to Hartford National Bank and Trust Co., trustee for North American Philips).
17. Ref. 1, p. 177.
18. P. B. Braun, *Philips Res. Rep.* **12,** 491 (1957).
19. Ref. 1, Chapt. 9; W. D. Townes and co-workers, *Z. Krist* **125,** 437 (1967).
20. G. Albanese, *J. Phy. Colloq. Suppl. Cl* **38**(4), Cl-85 (1977).
21. P. W. Anderson, *Phys. Rev.* **79,** 705 (1950).
22. A. L. Stuijts, G. W. Rathenall, and G. H. Weber, *Philips Tech. Rev.* **16**(4–5), 141 (1954).
23. K. J. Standley, *Oxide Magnetic Materials,* Oxford University Press, New York 1962, p. 107.
24. H. Chang, ed., *Magnetic Bubble Technology,* IEEE Press, New York, 1975, p. 176.
25. J. Geller and M. A. Gilleo, *Acta Cryst.* **10,** 239 (1957).
26. F. Bertaut and F. Forrat, *C. R. Paris* **242,** 382 (1956).

27. J. S. Reed and R. B. Runk in F. F. Y. Wang, ed., *Treatise on Materials Science and Technology*, Vol. 9, Academic Press, Inc., New York, 1976, p. 71.
28. J. C. Williams in ref. 27, p. 173.
29. G. F. Austin and G. D. McTaggart in ref. 27, p. 35.
30. J. M. Blank, *J. Appl. Phys. Suppl.* **32,** 3785 (Mar. 1961).
31. P. I. Slick, *Proceeding of the International Conference on Ferrites, Japan, July 1970*, p. 81.
32. R. Morineall and M. Paulus, *IEEE Trans. Mag.* **11,** 1352 (Sept. 1975).
33. T. Takadas and M. Kiyama in ref. 31, p. 96.
34. T. Akashi and co-workers in ref. 31, p. 96.
35. A. Goldman and A. M. Laing, *J. Phys. Colloq. Suppl. Cl.* **38**(4), 297 (1977).
36. P. K. Gallagher and co-workers, *Bull. Am. Ceram. Soc.* **52,** 842 (1973).
37. R. Schnettler and D. W. Johnson in ref. 31, p. 121; A. L. Micheli, *IEEE Trans. Magnetics* **MAG 6**(3), 1970.
38. T. Takada and M. Kiyana, ref. 31, p. 69; C. D. Mee and Jeschke, *J. Appl. Phys.* **34,** 127 (1963); K. Haneda and H. Kosima, *J. Appl. Phys.* **44,** 3760 (1973); K. Haneda, C. Miyakawa, and H. Kojima, *J. Am. Ceram. Soc.* **57,** 354 (1974).
39. U.S. Pat. 3,822,210 (July 2, 1974), K. Iwase and co-workers.
40. C. A. M. van den Broek, *Ceramurgi* **3**(3), 115 (1977).
41. H. G. Richter and H. E. Dietrich, *IEEE Trans. Mag.* **MAG-4**(3), 263 (1968).
42. H. Stablein and J. Willibrand, *IEEE Trans. Mag.* **MAG-2**(3), 459 (1966).
43. F. J. Esper in ref. 35 p. 69.
44. *Digital Design,* 14 (Feb. 1976).
45. J. A. Rajachman in ref. 43, p. 409.
46. E. C. Snelling, *IEEE Spectrum* 42, (Jan. 1972); 26, (Feb. 1972).
47. J. M. van der Poel, *Electron. Des.* 58 (Feb. 1976).
48. *Ferroxcube Linear Ferrite Magnetic Design Manual,* Ferroxcube Corp., Saugerties, N.Y., 1971.
49. *SMPS Transformer Design Manual,* Ferroxcube, Corp., Saugerties, N.Y., 1978.
50. *Telev. Digest,* 10 (July 2, 1979).
51. Y. Shichijo and E. Takama in ref. 31, p. 210.
52. A. Ikeda and co-workers in ref. 31, p. 337.
53. C. Buthker and T. Berben, *J. Phys. Suppl.* **38,** 341 (Apr. 1977).
54. S. Hayashi in ref. 31, p. 542.
55. D. M. Bolle and L. R. Wicker, *IEEE Trans. Magnetics* **11,** 907 (May 1975).

THOMAS G. REYNOLDS, III
Ferroxcube Corporation

FERROCYANIDES. See Iron compounds.

ERRATA

VOLUME 9

Page	Line	For:	Read:
218	Table 6		
	4	DL	many vendors
	5	CH	many vendors
	6	porcine rennin/pepsin	calf rennet/porcine pepsin
	14	CH	P
	14	bovine	*Endothia parasitica*
273	structure (6)	bottom of structure